# Manual of Grasses for North America

# Manual of Grasses for North America

*Edited by*  Mary E. Barkworth, Laurel K. Anderton, Kathleen M. Capels, Sandy Long, and Michael B. Piep

*Illustrated by*  Cindy Talbot Roché, Linda Ann Vorobik, Sandy Long, Annaliese Miller, Bee F. Gunn, Christine Roberts, and Mary S. Blankenship

Intermountain Herbarium and Utah State University Press
Utah State University
Logan, Utah 84322

Utah State University Press
Logan, Utah 84322-7800

Manufactured in United States of America

Cover design by Barbara Yale-Read

Library of Congress Cataloging-in-Publication Data
Manual of grasses for North America
edited by Mary E. Barkworth, Laurel Anderton, Kathleen Capels, Sandy Long, Michael Piep.
p. cm.

ISBN 978-0-87421-686-8 (pbk. : alk. paper)
1. Grasses–North America–Identification.
2. Grasses–United States–Identification.
3. Grasses–Canada–Identification.
1. Barkworth, Mary E., 1941-
QK495.G74M23 2007 584'.9097–dc22
2007025375

9 8 7 6 5 4 3 2 1
Printed in
on acid-free paper

# Contents

# Preface

The original goal of the Grass Manual Project (GMP) was to develop a single-volume work modeled on Hitchcock's *Grasses of the United States*. When the GMP became part of the Flora of North America Project, the contributors were asked to develop more detailed descriptions than originally envisioned. The resulting two volumes, *Flora of North America* volumes 24 and 25 (Barkworth et al. 2003, 2007), have been well received, but several individuals commented that there was still a need for a more compact and less expensive presentation of the information. This volume seeks to meet that need.

**Design:** The design of this manual was dictated by the need to save space while keeping all the illustrations. It quickly became evident that, even after reducing each illustration to a quarter of its original size and using abbreviations in the descriptions, ruthless cutting was required. We eliminated all the citations, reduced the comments to habitat information that would aid in identification, and modified the layout. This still led to a volume of over 1,000 pages, so we eliminated the subfamily and species descriptions. We have not abbreviated words in the keys because of the critical importance of the keys in the absence of species descriptions.

**Organization:** The order of the treatments, illustrations, and distribution maps reflects, to the extent that is feasible, current thinking on phylogenetic relationships. These are, admittedly, best known at the subfamily level and least well known (and least amenable to a linear arrangement) at the species level. Nevertheless, this arrangement makes it easier to obtain an overall picture of the morphology and distribution of a tribe or group of related genera than would an alphabetical arrangement.

A bipartite number is used to indicate the location of each genus, the part before the "decimal" indicating its tribal membership, the part after the decimal, its placement within the tribe. These numbers form part of the header on each page. In the treatments, the name of each species is followed by two page numbers: an italicized number for the illustration page, and an underlined number for the distribution map page. At the end of the text material there is a brief "Literature Cited" section. It contains only references that were used for the first time in preparing this volume. The "Literature Cited" section is followed first by the illustrations, then by the distribution maps. Tripartite numbers are associated with each illustration and map. The first two parts of these numbers correspond to the tribal and generic numbers that are used as page headers on the treatment pages. The third part of each number indicates the position of the species within the text material for its genus. The index lists the page numbers for the written treatments (in bold type), illustrations (in italics), and distribution maps (underlined). It also shows how this volume treats names likely to be encountered in other publications.

**Changes:** There are some differences between the taxonomic treatment presented in this *Manual* and that in the two *Flora of North America* volumes on which it is based. In addition, some illustrations have been added or enhanced and many maps modified to reflect new distributional information. New taxa recognized include *Festuca roemeri* var. *klamathensis* (new taxon), *Hopia* (*Panicum*, in part), *Phanopyrum* (*Panicum*, in part), *Zuloagaea* (*Panicum*, in part), *Phragmites* subspp. *americanus* and *berlandieri* (new taxa), *Trisetum montanum* (*T. spicatum*, in part), and *T. projectum* (*T. spicatum*, in part). The parenthetical names indicate how these taxa were treated in FNA 24 or 25.

Each of the new taxa has been illustrated by Dr. C.T. Roché. She has also prepared additions to some of the previous illustrations in order to help clarify the revised key leads. The primary illustrator of *Neyraudia*, both in this volume and in FNA 25, was Mary S. Blankenship. Barkworth apologizes for failing to acknowledge her work in FNA 25.

Many of the maps have been modified. Most of the changes are minor and reflect additional records. The sources of information for significant changes have been checked. In a few instances, leaders of regional projects, having failed to locate a voucher for a particular record, have suggested that the record be deleted. We accepted such recommendations and urge leaders of other projects to ask us about records that seem questionable. The maps on the Web (http://herbarium.usu.edu/webmanual/ are updated when changes are made.

**Citations:** The citations in the general bibliographies of FNA 24 and 25 are available at http://utc.usu.edu/grassbib.htm. This site also provides citations for some additional articles on grasses, including some of those compiled by Dr. J.F. Veldkamp. The sources of geographic information used in preparing the distribution maps are available at http://utc.usu.edu/atlas/geogbib.htm. Works cited in this volume but not the two precursor volumes are presented at the end of the written material.

**Preparation:** The treatments in this volume, unless otherwise indicated, are essentially identical to those in *Flora of North America* volumes 24 and 25 and should be credited to the author(s) cited in the footnotes. This manual reflects the editorial efforts of all those involved with producing the two grass volumes in the *Flora of North America* series. Barkworth is responsible for all the changes from those volumes in the taxonomic treatments, keys, and descriptions. Anderton edited the comments for most taxa and designed and prepared the book for publication. The citation for this volume is:

Barkworth, M.E., L.K. Anderton, K.M. Capels, S. Long, and M.B. Piep (Eds.). 2007. *Manual of grasses for North America north of Mexico*. Utah State University Press. 627 pp.

# Abbreviations

abx . . . . . . . . . . . . . abaxial
adx . . . . . . . . . . . . . adaxial
adxly . . . . . . . . . . adaxially
ann . . . . . . . . . . . . . annual
anth . . . . . . . . . . . . anthers
apc . . . . . . . . . . . . . . apices
aur . . . . . . . . . . . . . auricles
ax . . . . . . . . . . . . . . axillary
bas . . . . . . . . basal, basally
bisx . . . . . . . . . . . . bisexual
bld . . . . . . . . . . . . . . blades
bldless . . . . . . . . . bladeless
br . . . . . . . . . . . . . branches
brchd . . . . . . . . . branched
brchg . . . . . . . . . branching
brlets . . . . . . . . . branchlets
bu sc . . . . . . . . . . bud scales
cal . . . . . . . . . . . . . calluses
car . . . . . . . . . . . caryopses
cent . . . . . . . . . . . . . central
ces . . . . . . . . . . . . cespitose
clm . . . . . . . . . . . . . . culms
clstgn . . . . . . . . . cleistogenes
col . . . . . . . . . . . . . . collars
cmp . . . . . . . . . complements
crwn . . . . . . . . . . . . crowns
dis . . . . . . . . disarticulation
emb . . . . . . . . . . . embryos
emgt . . . . . . . . . . emarginate
epdm . . . . . . . . . epidermes
exvag . . . . . . . extravaginal
flt . . . . . . . . . . . . . . .florets
fmb . . . . . . . . . . . .fimbriae
fnctl . . . . . . . . . . functional
fol . . . . . . . . . . . . . foliage
ftl . . . . . . . . . . . . . . . fertile

glab . . . . . . . . . . . glabrous
glm . . . . . . . . . . . . glumes
hrb . . . . . . . . . . herbaceous
infl . . . . . . . . inflorescences
infvag . . . . . . . infravaginal
intnd . . . . . . . . . internodes
invag . . . . . . . . intravaginal
jnct . . . . . . . . . . . junction
lat . . . . . . . lateral, laterally
lf . . . . . . . . . . . . . . . . leaf
lig . . . . . . . . . . . . . ligules
lm . . . . . . . . . . . . . lemmas
lo . . . . . . . . . . . . . . lower
lod . . . . . . . . . . . lodicules
lvs . . . . . . . . . . . . leaves
memb . . . . . . membranous
mid . . . . . . . . . . . . middle
mrg . . . . . . . . . . . margins
mrgl . . . . . . . . . . marginal
nd . . . . . . . . . . . . . nodes
occ . . . . . . . . . occasionally
ov . . . . . . . . . . . . ovaries
pal . . . . . . . . . . . . . paleas
pan . . . . panicles, paniculate
ped . . . . . . . . . . . pedicels
pedlt . . . . . . . . . pedicellate
per . . . . . . . . . . . perennial
pist . . . . . . . . . . . pistillate
pl . . . . . . . . . . . . . . plants
plcsp . . . . . . . pluricespitose
pri . . . . . . . . . . . . . primary
prcp . . . . . . . . . . pericarp
prphl . . . . . . . . . prophylls
psdlig . . . . . . .pseudoligules
psdpet . . . . . . psuedopetioles
psdspk . . . . . pseudospikelets

psinva . . pseudointravaginal
rchl . . . . . . . . . . . rachillas
rchs . . . . . . . . . . . rachises
rcm . . . . . racemes, racemose
rcmly . . . . . . . . . racemosely
rdcd . . . . . . . . . . . reduced
rdg . . . . . . . . . . . . ridges
rdgd . . . . . . . . . . ridged
rdmt . . . . . . . . . rudiments,
rudimentary
rhz . . rhizomes, rhizomatous
sd . . . . . . . . . . . . . seeds
sdlg . . . . . . . . . . seedling
sec . . . . . . . . . . . secondary
sht . . . . . . . . . . . . shoots
shth . . . . . . . . . . sheaths
smt . . . . . . . . . . sometimes
spklt . . . . . . . . . . spikelets
spnd . . . . . . supranodal
sta . . . . . . . . . . . . stamens
st g . . . . . . . . . starch grains
stln . . . stolons, stoloniferous
stmd . . . . . . . . staminodes
stmt . . . . . . . . . . staminate
strl . . . . . . . . . . . . . sterile
sty . . . . . . . . . . . . . styles
sx . . . . . . . . . . . . . sexual
tml . . . . terminal, terminally
ult . . . . . . . . . . . . ultimate
unilat . . . . . . . . unilateral
unisx . . . . . . . . . unisexual
up . . . . . . . . . . . . upper
usu . . . . . . . . . . . usually
wd . . . . . . . . . . . . woody

*Taxonomic Treatments*

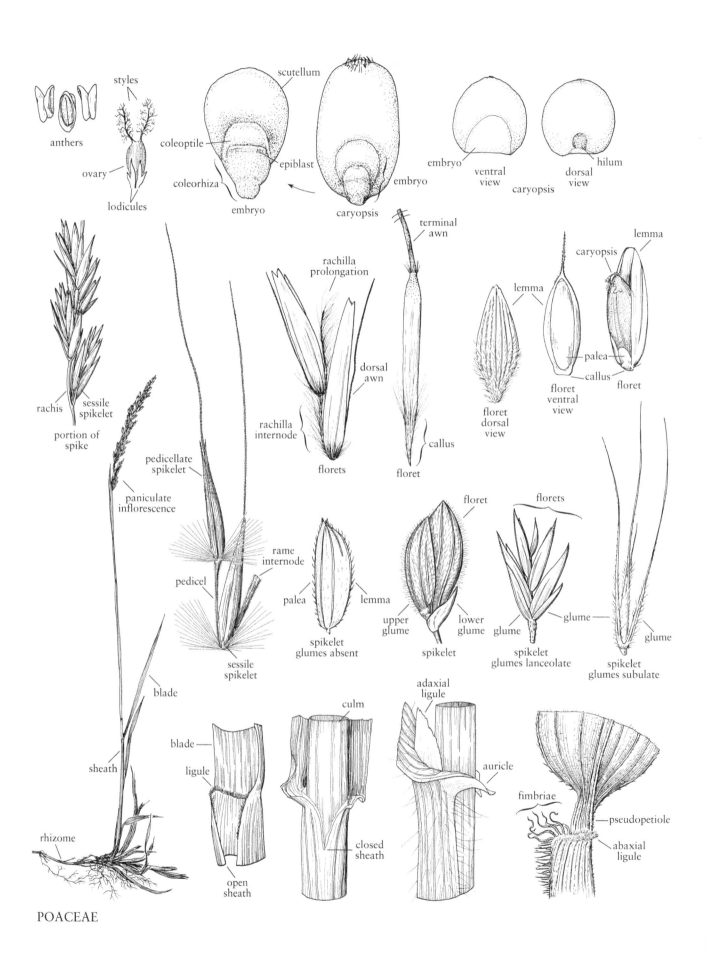

anthers

styles

ovary

lodicules

scutellum

coleoptile

coleorhiza

epiblast

embryo

embryo

caryopsis

embryo

ventral view

hilum

dorsal view

caryopsis

rachis

sessile spikelet

portion of spike

pedicellate spikelet

paniculate inflorescence

pedicel

rame internode

sessile spikelet

blade

sheath

rhizome

rachilla prolongation

dorsal awn

rachilla internode

florets

terminal awn

callus

floret

lemma

floret dorsal view

caryopsis

lemma

palea

callus

floret

floret ventral view

palea

lemma

spikelet glumes absent

upper glume

lower glume

floret

spikelet

glume

floret

florets

glume

spikelet glumes lanceolate

glume

spikelet glumes subulate

adaxial ligule

culm

blade

ligule

closed sheath

open sheath

auricle

fimbriae

pseudopetiole

abaxial ligule

POACEAE

# POACEAE Barnhart[1]

# GRAMINEAE Adans., alternate name

- Grass Family

Pl ann or per; usu terrestrial, smt aquatic; tufted, mat-forming, csp, plcsp, or with solitary *culms* (flowering stems), rhz and stln often well developed. **Clm** ann or per, hrb or wd, usu erect or ascending, smt prostrate or decumbent for much of their length, occ climbing, rarely floating; **nd** prominent, smt concealed by lf shth; **intnd** hollow or solid, bases meristematic; **brchg** from the bas nd only or from bas, mid, and up nd; **bas brchg** exvag or invag; **up brchg** invag, exvag, or infvag. **Lvs** alternate, 2-ranked, each composed of a shth and bld encircling the clm or br; **shth** usu open, smt closed with mrg fused for all or part of their length; **aur** (lobes of tissue extending beyond the margins of the sheath on either side) smt present; **lig** usu present at the shth-bld jnct, particularly on the adx surface, abx lig common in the *Bambusoideae*, memb, smt ciliate, adx lig usu present, of memb to hyaline tissue, a line of hairs, or a ciliate membrane; **bld** usu linear to lanceolate, occ ovate to triangular, bases smt *pseudopetiolate* (having a petiole-like constriction), venation usu parallel, smt with evident cross veins, occ divergent. **Infl** (*synflorescences*) usu compound, composed of simple or complex aggregations of pri infl, aggregations pan, spicate, or rcm or of spikelike br, often with an evident *rachis* (central axis), pri infl *spikelet, pseudospikelet*, or spklt equivalents; **infl br** usu without obvious bracts. **Spklt** with (0–1)2(3–6) *glumes* (empty bracts) subtending 1–60 flt, glm and flt distichously attached to a *rachilla* (central axis); **psdspklt** with bud-subtending bracts below the glm. **Glm** usu with an odd number of veins, smt awned. **Flt** bisx, stmt, or pist, usu composed of a *lemma* (lower bract) and *palea* (upper bract), lod, and reproductive organs, often lat or dorsally compressed, smt round in cross section; **lm** usu with an odd number of veins, often awned, bases frequently thick and hard, forming a cal, backs rounded or keeled over the midvein, awns usu 1(–3), arising bas to tml; **pal usu** with 2 major veins, with 0 to many additional veins between the major veins, smt also in the mrg, often keeled over the major veins; **lod** (0)2–3, inconspicuous, usu without veins, bases swelling at anthesis; **sta** usu 3, smt 1(2) or 6+, filaments capillary, anth versatile, usu all alike within a flt, smt 1 or 2 evidently longer than the others; **ov** 1-loculed, with (1)2–3(4) sty or sty br, stigmatic region usu plumose. **Fr** car, prcp usu dry and adhering to the sd, smt fleshy or dry and separating from the sd at maturity or when moistened; **emb** ⅓ as long as to almost equaling the car, highly differentiated with a *scutellum* (absorptive organ), a sht with lf primordium covered by the *coleoptile* (shoot sheath), and a root covered by the *coleorhiza* (root sheath); **hila** punctate to linear.

The *Poaceae* or grass family includes approximately 700 genera and 11,000 species. This volume treats 10 subfamilies, 25 tribes, 240 genera, and 1377 species. Of these, all the subfamilies, 22 tribes, 139 genera, and 906 species are native to the *Manual* region; 2 tribes, 78 genera, and 290 species have become established in the region. The remaining taxa include ornamental species; species grown for research; species that, if introduced to the region, would pose a threat to important agricultural species; and a few species that have been found in the region but have not become established. Most species in the last category were found on ballast dumps near ports around the turn of the last century.

Grasses constitute the fourth largest plant family in terms of number of species. Nevertheless, the family is clearly more significant than any other plant family in terms of geographic, ecological, and economic importance. Grasses grow in almost all terrestrial environments, including dense forests, open deserts, and freshwater streams and lakes. There are no truly marine grasses, but some species grow within reach of the highest tides.

In addition to being widely distributed, grasses are often dominant or co-dominant over large areas. This is reflected in the many words that exist for grasslands, words such as meadow, palouse, pampas, prairie, savanna(h), steppe, and veldt. Not surprisingly, grasses are of great ecological importance as soil stabilizers and as providers of shelter and food for many different animals.

The economic importance of grasses to humans is almost impossible to overestimate. The wealth of individuals and countries is dependent on the availability of such sources of grain as *Triticum* (wheat), *Oryza* (rice), *Zea* (corn or maize), *Hordeum* (barley), *Avena* (oats), *Secale* (rye), *Eragrostis* (tef), and *Zizania* (wild rice). Most countries invest heavily in research programs designed to develop better strains of these grasses and the many other grasses that are used for livestock, soil stabilization, and revegetation. Developing improved grasses for recreation areas, such as playing fields, golf courses, and parks, is also a major industry, and increasing recognition of the aesthetic value of grasses is reflected in their prominence in horticultural catalogs.

There are, of course, grasses that are considered undesirable, but even the most obnoxious grasses may be

[1]Lynn G. Clark and Elizabeth A. Kellogg

well-regarded over a portion of their range. For instance, *Bromus tectorum* (cheatgrass) is a noxious, fire-prone invader of western North American ecosystems; it is also welcomed as a source of early spring feed in some parts of the *Manual* region. *Cynodon dactylon* (bermudagrass) is listed as a noxious weed in some jurisdictions; in others it is valued as a lawn grass.

Although grasses are widespread and often dominant in open areas, all evidence points to an origin of the family in forests, most likely in the Southern Hemisphere, at least 55–70 mya. Recent evidence from phytoliths (isolated silica bodies commonly produced inside the epidermal cells of grasses and some other plants) embedded in fossil coprolites strongly suggests that grasses evolved earlier in the Cretaceous than previously thought. Living representatives of the three earliest lineages of the grass family, together comprising about 30 species, are perennial, broad-leaved plants of relatively small stature, native to tropical or subtropical forests in South America, Africa, southeast Asia, some Pacific Islands, and northern Australia. The major diversification of the family probably occurred in the mid-Cenozoic, and was associated with climatic changes that produced more open habitats. All major lineages of the grass family were present by the middle of the Miocene; $C_4$ photosynthesis in grasses had also evolved by then.

# Key to Tribes[1]

1. Leaf blades with divergent veins; spikelets unisexual and dimorphic, the pistillate lemmas with uncinate hairs (*Pharoideae*; p. 7) . . . . . . . . . . . . . . . . . . . . . . . . . . . . . . . . . . . . . . . . . . . . . . . . . . . . . . 1. *Phareae*
1. Leaf blades with parallel veins; spikelets bisexual, unisexual, or modified into plantlets, the pistillate lemmas never with uncinate hairs.
    2. Culms perennial, woody or herbaceous, often developing complex branching systems from the upper nodes; leaves on the upper portion of the culms, or distal on the branches, usually pseudopetiolate (*Bambusoideae*).
        3. Culms woody, to 30 m tall; leaves strongly dimorphic, those of the main culms (*culm leaves*) with expanded sheaths and often with reduced, non-photosynthetic blades, those of the branches (*foliage leaves*) with abaxial ligules; blades of the distal leaves not folding at night or under stress; florets bisexual; plants native or introduced, often cultivated (p. 8) . . . . . . . . . . . . . . . . . . . . . . . . . 2. *Bambuseae*
        3. Culms herbaceous, to 3.5 m tall or climbing; leaves not strongly dimorphic; blades of the distal leaves often folding at night or under stress; florets unisexual; plants known only in cultivation in the *Manual* region (p. 11) . . . . . . . . . . . . . . . . . . . . . . . . . . . . . . . . . . . . . . . . . . . . . . . . 3. *Olyreae*
    2. Culms usually annual, sometimes facultatively perennial, rarely woody, sometimes branching from the upper nodes but the branching system not complex; leaves usually not pseudopetiolate.
        4. Spikelets almost always with 2 florets, the lower florets in the spikelets always sterile or staminate, frequently reduced to lemmas, occasionally missing, the upper florets bisexual, staminate, or sterile, unawned or awned from the lemma apices or, if the lemmas bilobed, from the sinuses; glumes membranous and the upper lemma stiffer than the lower lemma, or both florets reduced and concealed by the stiff to coriaceous glumes; rachilla not prolonged beyond the second floret (*Panicoideae*, in part).
            5. Glumes flexible, membranous, the lower glumes usually shorter than the upper glumes, sometimes missing, the upper glumes usually subequal to or exceeded by the upper floret; lower lemmas membranous; upper lemmas usually coriaceous to indurate, sometimes membranous; upper paleas similar in texture; spikelets usually single or in pairs, occasionally in triplets and all pedicellate, often shortly so) (p. 267) . . . . . . . . . . . . . . . . . . . . . . . . . 25. *Paniceae*
            5. Glumes stiff, coriaceous to indurate, often subequal, at least 1 and usually both exceeding the upper floret (excluding the awn); both lemmas hyaline; paleas hyaline or absent; most spikelets in pairs or triplets, at least 1 spikelet in each group usually sessile; pedicels shorter or only a little longer than the sessile spikelets (p. 320) . . . . . . . . . . . . . . . . . . . . 26. *Andropogoneae*
        4. Spikelets either with other than 2 florets or, if with 2, the lower floret bisexual or the upper floret awned from the back or base of the lemma, or the spikelets bulbiferous; glumes usually membranous; lemmas scarious to indurate; rachilla sometimes prolonged beyond the distal floret.
            6. Spikelets with 1 floret; lemmas terminating in a 3-branched awn (the lateral branches sometimes greatly reduced); callus well developed; ligules usually of hairs, sometimes ciliate membranes, the cilia longer than the membranous base (*Aristidoideae*; p. 257) . . . . . . . . . . . . . . . . 21. *Aristideae*
            6. Spikelets with more than 1 floret or, if only 1, the lemma not terminating in a 3-branched awn; callus development various; ligules various.

[1]Mary E. Barkworth

7. Spikelets with 1 sexual floret or the spikelets bulbiferous; glumes absent or less than ¹/₄ as long as the adjacent floret; lower glumes, if present, without veins, upper glumes, if present, veinless or 1-veined.

   8. Upper glumes present, 1-veined; lower glumes absent or much shorter than the upper glumes and lacking veins (*Poöideae*, in part; p. 18) . . . . . . . . . . . . . . . . . . . . . . 6. *Brachyelytreae*

   8. Both glumes absent or lacking veins.

      9. Inflorescences 1-sided spikes; triangular in cross section, (*Poöideae*, in part; p. 18) . . . . . . . . . . 7. *Nardeae*

      9. Inflorescences panicles; spikelets laterally compressed or terete.

         10. Culms aerenchymatous, 20–500 cm long; plants of wet places, often emergent, sometimes floating; lemmas of the bisexual or pistillate florets 3–14-veined; paleas 3–10-veined (*Ehrhartoideae*, in part; p. 12) . . . . . . . . . . . . . . . . . . . . . . . . . . 5. *Oryzeae*

         10. Culms not aerenchymatous, 2–300 cm tall; plants of wet or dry habitats but not emergent or floating; lemmas of the bisexual or pistillate florets 1–3-veined; paleas 2-veined.

            11. Culms 2–19 cm tall; plants of cold or damp habitats, not rhizomatous; sheaths of the flag leaves closed for at least ¹/₂ their length; caryopses exposed at maturity (*Poöideae*, in part; p. 81) . . . . . . . . . . . . . . . . . . . . . . . 14. *Poeae* (in part)

            11. Culms 5–300 cm tall; plants usually of warm or dry habitats, often rhizomatous; sheaths of the flag leaves open to the base; caryopses not exposed at maturity (*Chloridoideae*, in part; p. 186) . . . . . . . . . . . . . . 17. *Cynodonteae* (in part)

7. Spikelets usually with more than 1 sexual floret; usually with 2 glumes, 1 or both glumes often longer than ¹/₄ the length of the adjacent floret and/or with more than 1 vein, always longer in taxa with 1 sexual floret.

   12. Lemmas unawned, flabellate or with (5)7–15 awnlike teeth (*Chloridoideae*, in part).

      13. Plants not viscid, usually perennial; ligules present, composed of hairs (p. 249) . . . . 18. *Pappophoreae*

      13. Plants viscid annuals; ligules absent (p. 251) . . . . . . . . . . . . . . . . . . . . . . . . . . . . . 19. *Orcuttieae*

   12. Lemmas awned or unawned, lanceolate, rectangular, or ovate, apices entire, mucronate, bilobed, or bifid, occasionally 4-lobed or 4–5-toothed, sometimes erose.

      14. Cauline leaf sheaths closed for ¹/₂ their length or more; glumes usually exceeded by the distal florets, sometimes greatly so (*Poöideae*, in part).

         15. Spikelets 5–80 mm long, not bulbiferous; lemmas usually awned, often bilobed or bifid, veins convergent distally; ovary apices hairy (p. 45) . . . . . . . . . . . . . 12. *Bromeae*

         15. Spikelets 0.7–60 mm long, sometimes bulbiferous; lemmas often unawned, not both bilobed/bifid and with convergent veins; ovary apices usually glabrous.

            16. Lemma veins (4)5–15, usually prominent, parallel distally; spikelets 2.5–60 mm long, not bulbiferous (p. 20) . . . . . . . . . . . . . . . . . . . . . . . . . . . . . 9. *Meliceae*

            16. Lemmas veins 1–9, often inconspicuous, usually convergent distally; spikelets 0.7–18(20) mm long, sometimes bulbiferous (p. 81) . . . . . . . . . . 14. *Poeae* (in part)

      14. Cauline leaf sheaths open for at least ¹/₂ their length; glumes exceeding or exceeded by the distal florets.

         17. Spikelets with 1 floret; lemmas terminally or subterminally awned, the junction of the awn and lemma conspicuous; rachillas notprolonged beyond the base of the floret (*Poöideae*, in part; p. 28) . . . . . . . . . . . . . . . . . . . . . . . 10. *Stipeae* (in part)

         17. Spikelets with 1–60 florets; lemmas unawned or awned, awns basal to terminal, if terminal or subterminal, the lemma-awn junction not conspicuous; rachillas often prolonged beyond the base of the distal floret.

            18. Ligules, at least of the flag leaves, of hairs, a ciliate ridge or membrane bearing cilia longer than the basal ridge or membrane; leaves usually hairy on either side of the ligule; auricles absent.

               19. Lemmas of the fertile florets with 3–11 inconspicuous veins, never glabrous, if with 3 veins, pilose throughout or with transverse rows of tufts of hair, if with 5–11 veins, the margins pilose proximally, the hairs not papillose-based; lemma apices usually bilobed or bifid and awned or mucronate from the sinus, if acute to acuminate, the lemmas pilose; awns twisted proximally (*Danthonioideae*; p. 253) . . . . 20. *Danthonieae*

               19. Lemmas of the fertile florets usually with 1–3 conspicuous veins, sometimes with 3 inconspicuous veins or 5–11 veins, often glabrous, if with 3 veins, usually glabrous throughout or hairy over the veins, sometimes the margins with papillose-based hairs; lemma

apices acute to obtuse, bilobed, or 4-lobed, often mucronate or awned from the sinuses; awns usually not twisted.

20. Lemmas 1–11-veined, veins glabrous or hairy, margins without papillose-based hairs; rachillas and calluses not pilose, sometimes strigose or strigulose; basal internodes of the culms not persistent, not swollen and clavate (*Chloridoideae*, in part; p. 186) ................................ 17. *Cynodonteae* (in part)

20. Lemmas 3(5)-veined, veins glabrous, margins sometimes with papillose-based hairs; rachillas or calluses pilose or the basal internodes of the culms persistent, often swollen and clavate (*Arundinoideae*, in part; p. 183) .................... 16. *Arundineae* (in part)

18. Ligules membranous, if ciliate, the cilia shorter than the membranous base; leaves usually glabrous on either side of the ligule; auricles present or absent.

21. Inflorescences panicles or unilateral racemes, not spikelike, without spike-like branches; spikelets solitary, the lowest 0–4 florets in a spikelet sterile or staminate, the distal florets sexual.

22. Spikelets with (1)2–25 bisexual florets; all lemmas similar in size and shape; glumes and lemmas membranous (*Centothecoideae*).

23. Culms 35–150 cm tall; spikelets with (2)3–26 florets, including the lowest (0)1–4 sterile or staminate florets; lower glumes(1)2–9-veined (p. 264) ................... 22. *Centotheceae*

23. Culms 150–400 cm tall; spikelets with 2–4 florets, including the lowest sterile floret; glumes 0–1-veined (p. 265) ........................................ 23. *Thysanolaeneae*

22. Spikelets with 1 bisexual or unisexual floret; lemmas of the sterile florets usually differing in size and shape from those of the sexual floret; glumes membranous, lemmas of the sexual florets firmer.

24. Lemmas of the lower florets coriaceous, at least the upper exceeding the sexual floret (*Ehrhartoideae*, in part; p. 11) ...... 4. *Ehrharteae*

24. Lemmas of the lower florets membranous, often both much shorter than the sexual floret, sometimes subequal to it, sometimes only 1 sterile floret present (*Poöideae*, in part; p. 81) ................................... 14. *Poeae* (in part)

21. Inflorescences panicles, racemes, or spikes; spikelets sometimes in pairs or triplets, sterile florets, if any, distal to the bisexual or pistillate florets.

[⇐revert to left, Ed.]

25. Lemmas with 1–3 or 9–11 conspicuous veins; sheaths open; blade cross sections with Kranz leaf anatomy (*Chloridoideae*, in part; p. 186) ....................................... 17. *Cynodonteae* (in part)

25. Lemmas with (1)3–15 often inconspicuous veins, if with 3 conspicuous veins, the sheaths closed; sheaths open or closed; blade cross sections without Kranz leaf anatomy.

26. Inflorescences spikes or spikelike; spikelets 1–5+ per node, at least 1 spikelet sessile or subsessile (*Poöideae*, in part).

27. Upper glumes 5–9-veined; spikelets subsessile and solitary at the nodes; auricles absent (p. 44) .................................................................... 11. *Brachypodieae*

27. Upper glumes 1–5-veined; spikelets 1–5+ per node, usually at least 1 sessile at each node, sometimes highly reduced branches present; auricles present or absent.

28. Inflorescences with 1–5 spikelets at a node, if 3, usually with 1 sessile and 2 pedicellate spikelets, if 1, the spikelet tangential to or embedded in the rachis, with 2 glumes, the glumes facing each other; ovaries with hairy apices; auricles often present (p. 53) ..... 13. *Triticeae* (in part)

28. Inflorescences spikelike panicles with highly reduced branches, or spikes with spikelets radial to the rachises and all but the terminal spikelet with only 1 glume, or spikes with spikelets tangential to the rachises and having 2 glumes adjacent to each other; ovaries with glabrous apices; auricles usually absent (p. 81) ........................... 14. *Poeae* (in part)

26. Inflorescences panicles, with no sessile spikelets.

29. Caryopses with a thick pericarp forming a distinct apical knob or beak at maturity; lemmas 3(5)-veined (*Poöideae*, in part; p. 19) ..................................... 8. *Diarrheneae*

# 1. PHAROIDEAE L.G. Clark & Judz.[1]

The *Pharoideae* has one tribe, the *Phareae,* three genera, and twelve species. It is pantropical. In the Americas, it is represented by one genus, *Pharus,* that extends from Florida to Uruguay and Argentina. The *Pharoideae* is a basal lineage of the *Poaceae,* and the first subfamily in which an adaxial ligule and true spikelets are found.

## 1. PHAREAE Stapf[2]

Pl per; rhz, smt csp or stln; monoecious. **Clm** ann, 10–300 cm, erect to decumbent; **intnd** usu solid. **Lig** scarious, smt ciliolate; **psdpet** present, twisted, placing the abx surface of the bld upmost; **bld** linear to oblong, not folding or drooping at night, lat veins diverging obliquely from the midveins, cross venation evident. **Infl** pan, usu espatheate; **ult br** with 1–2 pist spklt and 1 tml, stmt spklt; **dis** beneath the pist flt and in the pan br. **Spklt** unisx, heteromorphic, usu in stmt-pist pairs on brlets, with 1 flt; **rchl** not prolonged beyond the flt. **Stam spklt** pedlt, smaller than the pist spklt, lanceolate to ovate, caducous; **glm** unequal; **lo glm** absent or much shorter than the up glm; **up glm** somewhat shorter than the flt; **lod** minute or absent; **anth** 6. **Pist spklt** sessile or shortly pedlt, terete, smt inflated; **glm** unequal to subequal, shorter than the flt, scarious, entire, smt persistent; **lm** chartaceous, becoming coriaceous, veins 5 or more, mrg involute or utriculate, partly or wholly covered with uncinate hairs, not terminating in a brchd awn; **pal** 2-veined; **lod** absent; **stl** 1, 3-brchd. **Car** oblong to linear; **hila** as long as the car.

The *Phareae* include three genera, all of which grow in tropical and subtropical forests. The tribe is represented by one genus in the Western Hemisphere, *Pharus.*

## 1.01 PHARUS P. Browne[3]

Pl per, some apparently monocarpic; rhz, smt csp or stln; monoecious. **Clm** 10–130 cm, erect to decumbent; **intnd** solid, frequently with prominent prop roots at the lo nd. **Shth** open, glab; **lig** usu scarious, smt memb and ciliate; **psdpet** conspicuous, twisted 180° distally, inverting the bld; **bld** linear to ovate, usu broad, usu tessellate, lat veins diverging obliquely from the midvein. **Infl** tml pan, ovate, open; **rchs** terminating in a stmt spklt or naked; **br** with uncinate hairs, spklt appressed. **Spklt** unisx, dimorphic, sexes paired or pist spklt solitary, with 1 flt; **rchl** not prolonged beyond the flt; **dis** above the glm and in the pan br. **Stam spklt** smaller than the pist spklt, attached below the pist spklt on appressed ped; **lo glm** shorter than the up glm or absent; **lm** longer than the glm, ovate, 3-veined; **lod** 3, minute; **anth** 6. **Pist spklt** larger than the stmt spklt, subsessile, elongate; **glm** subequal, lanceolate, (3)5–9(11)-veined, purple or green; **lm** cylindrical, longer than the glm, indurate, involute, with uncinate hairs over at least a portion of the surface, 7-veined, mrg inrolled, concealing the pal; **lod** absent; **stmd** 6, minute; **sty** 1, 3-brchd, stigmas hispid.

*Pharus* includes eight species. It extends from central Florida through Mexico to Argentina and Uruguay, and grows in moist to wet lowland forests. One species, *Pharus glaber,* is native to the *Manual* region.

The uncinate hairs and disarticulating panicle branches of *Pharus* promote dispersion by attaching to the coats of passing animals. The inverted, pseudopetiolate leaf blades and oblique venation make the genus easy to distinguish, even in its vegetative state.

[1]Grass Phylogeny Working Group　[2]Mary E. Barkworth　[3]Emmet J. Judziewicz and Gerald F. Guala

### 1. **Pharus glaber** Kunth UPSIDEDOWN GRASS [p. *349*, <u>507</u>]

*Pharus glaber* grows on limestone-influenced sand in the hammocks of central Florida. Only two remaining populations are known in the United States, but the species is still widely present elsewhere in the Neotropics.

## 2. BAMBUSOIDEAE Luerss.[1]

The *Bambusoideae* includes two tribes, the woody *Bambuseae* and the herbaceous *Olyreae*. Their range includes tropical and temperate regions of Asia, Australia, and the Americas, primarily Central and South America. Three species of *Bambuseae* are native to the *Manual* region; there are no native species of *Olyreae*.

Members of the *Bambusoideae* grow in temperate and tropical forests, high montane grasslands, along riverbanks, and sometimes in savannahs. They are mainly forest understory or margin plants with a limited ability to reproduce, disperse, or survive outside their forest environment. Many have relatively small geographic ranges, and there is a high degree of endemism. The conservation status of most bamboos is not known; all are intrinsically vulnerable because of their breeding behavior and reliance upon a benign forest habitat. Only the $C_3$ photosynthetic pathway is found in the subfamily.

1. Culms woody, usually taller than 1 m, developing complex vegetative branching from the upper nodes; abaxial ligules present on the foliage leaves, rarely present on the culm leaves . . . . . . . . . . . . . . . . . . . . . . .2. *Bambuseae*
1. Culms herbaceous, usually shorter than 1 m; complex vegetative branching not developed; abaxial ligules not present . . . . . . . . . . . . . . . . . . . . . . . . . . . . . . . . . . . . . . . . . . . . .3. *Olyreae*

## 2. BAMBUSEAE Nees[2]

Pl per; rhz, shrubby to arborescent, self-supporting to climbing; **rhz** well developed, pachymorphic or leptomorphic, rarely both. **Clm** per, wd, to 30 m tall, intnd usu hollow, initially unbrchd and bearing thickened overlapping clm lvs, subsequently developing fol lvs on the complex br systems from buds at the intnd bases. **Clm lvs** thickened, usu early deciduous; **aur** and/or **fim** often present; abx **lig** usu lacking; adx **lig** present; **bld** poorly to well developed, erect or reflexed, the base as wide as or narrower than the shth apex, smt pseudopetiolate. **Fol lvs: aur** and **fim** present or absent; abx and adx **lig** present; **psdpet** nearly always present; **bld** deciduous, venation parallel, cross venation often evident, particularly at the base. **Infl** determinate or indeterminate, bracteate or ebracteate, rcm to pan, composed of psdspk or spklt. **Spklt** or **psdspklt** with 1 to many flt, the lo flt(s) often strl, the others bisx; **glm** often subtending the buds; **lm** often unawned, usu multiveined; **lod** usu 3, with vascular tissue; **anth** usu 3 or 6, smt fewer or up to 7, very occ many; **ov** glab or pubescent; **sty** or **sty br** 1–4; **Car** with or without a thickened fleshy prcp.

The *Bambuseae* has about 80–90 genera and around 1400 species. They are most abundant in Asia and South America, but are also found in Africa, Australia, Central America and North America. One genus, *Arundinaria*, is native to the *Manual* region, where it is represented by 3 species. Many other genera and species are cultivated in the region for their ornamental value. These often persist for decades; some have become established beyond the original planting. Identification of these introduced bamboos is hampered by the lack of taxonomic studies in their countries of origin, particularly studies of vegetative features, and the large number of taxa that have not yet been described. Identification is further hindered by their infrequent flowering. Only a few of the cultivated genera are treated here.

1. Rhizomes pachymorphic, short, thicker than the culms; culms forming separate, well-defined clusters . . . . . . . . . . . . . . . . . . . . . . . . . . . . . . . . . . . . . . . . . . . . . . . . . . . . . . . .2.02 *Bambusa*
1. Rhizomes leptomorphic, long, thinner than the culms; culms solitary, loosely clumped, or both.
    2. Culm internodes grooved their whole length, often doubly sulcate above the branches; branches usually without compressed internodes at the base; spikelets sessile . . . . . . . . . . . . . . . . . . . .2.03 *Phyllostachys*
    2. Culm internodes mostly terete, flattened or shallowly sulcate above the branches; branches usually with 1–5 compressed internodes at the base, sometimes without any compressed internodes; spikelets pedicellate.
        3. Culm leaves persistent or deciduous, usually with auricles; culms to 3 cm thick; culm buds open, margins not fused; plants native in the *Manual* region . . . . . . . . . . . . . . . . . . . . . . . . .2.01 *Arundinaria*
        3. Culm leaves persistent, usually without auricles; culms to 1.5 cm thick; culm buds closed, margins fused; plants cultivated in the *Manual* region, occasionally escaped . . . . . . . . . . . . .2.04 *Pseudosasa*

[1]Grass Phylogeny Working Group  [2]Christopher M.A. Stapleton

## 2.01 ARUNDINARIA Michx.[1]

Pl arborescent or subarborescent, spreading or loosely clumped; **rhz** leptomorphic. **Clm** 0.5–8 m tall, to 3 cm thick, erect; **nd** not swollen; **spnd rdg** not prominent; **intnd** terete to slightly flattened or shallowly sulcate above the br. **Clm lvs: shth** persistent or deciduous, mostly glab, abx surfaces sparsely pilose towards the mrg and apc, mrg ciliate; **aur** usu present; **bld** erect or becoming reflexed, narrowly triangular to strap-shaped, abx surfaces sparsely pilose; **lvs at tips of new sht** crowded into distinctive fan-shaped clusters or *topknots*, bld expanded as on the fol lvs. **Brch cmp** of 1 pri br and 0–2 subequal sec br on young clm, rebrchg to produce to 40+ sec br on older clm. **Fol lvs: shth** persistent on the lo br nd; **aur** usu present; **fim** to 10 mm; **bld** finely cross veined abxly, acuminate, bld of the ult brlets often smaller, crowded into flabellate clusters of 3–7 lvs. **Infl** open rcm or pan; **dis** below and between the flt. **Spklt** 3–7 cm, with 6–12 flt, bas flt occ strl, lat compressed. **Glm** 1–2, shorter than the lowest lm; **lm** to 2 cm, smt awned, awns about 4 mm; **anth** 3; **sty** 3; **pal** 2-keeled, not exceeding the lm.

*Arundinaria* is a north-temperate genus with three native North American species.

1. Primary branches with 0–1 compressed basal internodes; culm internodes usually sulcate; culm leaves deciduous . . . . . . . . . . . . . . . . . . . . . . . . . . . . . . . . . . . . . . . . . . . . . . . . . . . . . . . . . . . . 1. *A. gigantea*
1. Primary branches with 2–5 compressed basal internodes; culm internodes usually terete; culm leaves persistent to tardily deciduous.
  2. Foliage blades coriaceous, persistent, abaxial surfaces densely pubescent or glabrous, strongly cross veined; primary branches usually longer than 50 cm, basal nodes developing secondary branches; topknot blades 20–30 cm long . . . . . . . . . . . . . . . . . . . . . . . . . . . . . . . . . . . . . . . . . . . . . 2. *A. tecta*
  2. Foliage blades chartaceous, deciduous, abaxial surfaces pilose or glabrous, weakly cross veined; primary branches usually shorter than 35 cm, basal nodes not developing secondary branches; topknot blades 9–22.5 cm long . . . . . . . . . . . . . . . . . . . . . . . . . . . . . . . . . . . . . 3. *A. appalachiana*

### 1. Arundinaria gigantea (Walter) Muhl. RIVER CANE, GIANT CANE [p. *349*, <u>507</u>]

*Arundinaria gigantea* forms extensive colonies in low woods, moist ground, and along river banks. It was once widespread in the southeastern United States, but cultivation, burning, and over-grazing have destroyed many stands.

### 2. Arundinaria tecta (Walter) Muhl. SWITCH CANE [p. *349*, <u>507</u>]

*Arundinaria tecta* grows in swampy woods, moist pine barrens, live oak woods, and along the sandy margins of streams, preferring moister sites than *A. gigantea*. It grows only on the coastal plain of the southeastern United States.

### 3. Arundinaria appalachiana Triplett, Weakley & L.G. Clark HILL CANE [p. *349*, <u>507</u>]

*Arundinaria appalachiana* grows on moist to dry slopes and in seeps. It is restricted to the southern Appalachians and upper piedmont.

## 2.02 BAMBUSA Schreb.[2]

Pl usu arborescent, in well defined or rather loose clumps; **rhz** pachymorphic, with short necks. **Clm** 0.5–30(35) m tall, 0.5–18(20) cm thick, wd, per, usu self-supporting; **nd** not swollen; **spnd rdg** obscure; **intnd** terete, usu thinly covered initially with light-colored wax. **Brch cmp** usu with a dominant pri cent br and 2 smaller co-dominant lat br, usu similar at all nd; **bu sc** 2-keeled, thickened, initially closed at the back and front; **br** all subtended by bracts, higher order brlets at the lo nd smt thornlike. **Clm lvs** usu promptly deciduous, initially lightly waxy, smt with short, stiff hairs, subsequently losing the wax and becoming glab; **aur** usu well developed; **fim** usu present; **bld** triangular to broadly triangular, usu erect. **Fol lvs: shth** usu deciduous from the lo nd of the br, persistent at the distal nd; **bld** to 30 cm long, to 6 cm wide, not distinctly cross veined. **Infl** usu spicate, rarely capitate, bracteate; **prphyl** 2-keeled, narrow. **Psdspklt** 1–5 cm, with 3–12 flt; **dis** above the glm and below the flt, rapid; **rchl intnd** usu long. **Glm** several, subtending the buds; **lm** narrowly ovate, acute, unawned; **pal** not exceeding the lm, 2-keeled, not winged; **anth** 6; **ov** usu suboblong; **sty** short, with (2)3–4 plumose br.

*Bambusa* is a tropical and subtropical genus of 75–100+ species. It is native to southern and southeastern Asia, but is widely cultivated and naturalized throughout the tropics. *Bambusa vulgaris* and *B. multiplex* grow widely in Florida and Texas, having spread after being planted as ornamentals. Other species are known only in cultivation. This treatment includes a few of the more commonly cultivated species.

1. Branchlets of the lower branches recurved, hardened, thornlike . . . . . . . . . . . . . . . . . . . . . . . . . . . . . . . . . . 1. *B. bambos*
1. Branchlets of the lower branches not thornlike.
  2. Culm sheath auricles well developed, to 5 cm long . . . . . . . . . . . . . . . . . . . . . . . . . . . . . . . . . . . . . . . . 2. *B. vulgaris*

[1]Lynn G. Clark and J.K. Triplett  [2]Christoper M.A. Stapleton

2. Culm sheath auricles absent or poorly developed.
   3. Culm internodes antrorsely hispid; culms 0.5–7 m tall, broadly arched above . . . . . . . . . . . . . . . . . 3. *B. multiplex*
   3. Culm internodes glabrous; culms 6–15 m tall, erect . . . . . . . . . . . . . . . . . . . . . . . . . . . . . . . . . . . . 4. *B. oldhamii*

### 1. Bambusa bambos (L.) Voss GIANT THORNY BAMBOO [not illustrated]

*Bambusa bambos* is native to India and Indochina, but is cultivated throughout the tropics. It was the first bamboo species to be given a scientific name, being described as treelike, thorny, and a source of tabashir, lumps of pure silica that form in the internodal cavities.

### 2. Bambusa vulgaris Schrad. *ex* J.C. Wendl. COMMON BAMBOO [not illustrated]

*Bambusa vulgaris* probably originated in tropical Asia. It is now the most widely cultivated tropical bamboo, largely because of the ease with which the branches and culm sections take root. Many different cultivars exist.

### 3. Bambusa multiplex (Lour.) Raeusch. ex Schult. & Schult. f. HEDGE BAMBOO [p. 349]

*Bambusa multiplex* is native to southeast Asia. It is now widely planted around the world. The dense foliage with many leaves on each branchlet makes it well suited to hedging. A large number of cultivars are available.

### 4. Bambusa oldhamii Munro OLDHAM'S BAMBOO [p. 350]

*Bambusa oldhamii* is native to low-lying areas of eastern China and Taiwan. It is the most commonly grown large, clump-forming bamboo in the United States, where it is grown mostly in Florida and California.

## 2.03 PHYLLOSTACHYS Siebold & Zucc.[1]

**Pl** shrublike to arborescent, in open or dense, spreading clumps or thickets; **rhz** leptomorphic. **Clm** 3–10(20) m tall, 3–10(15) cm thick, self-supporting, erect or nodding, diffuse or plcsp, rarely solitary; **nd** slightly swollen; **rdg** above prominent; **intnd** strongly flattened for their whole length, doubly sulcate above the br, glab, smooth. **Br** 2(3) per midclm nd, unequal, initially erect, becoming deflexed, bas intnd not compressed. **Clm lvs** coriaceous, very quickly deciduous; **bld** usu strap-shaped and narrow, usu reflexed. **Fol lvs: shth** deciduous; **bld** small to medium-sized, usu glossy and thickened, indistinctly cross veined. **Infl** open or congested, smt spicate to subcapitate, fully bracteate, bracts usu bearing a small bld at the apex. **Spklt** or **psdspklt** with 2 to several flt, the upmost rdmt. **Lm** lanceolate; **pal** not exceeding the lm, strongly to very weakly 2-keeled, often bifid; **anth** 3; **sty** or **sty br** 3.

*Phyllostachys* is a hardy, temperate, Asiatic genus of at least 50 species, native mainly to China. It is the most distinct genus of hardy temperate bamboos, of enormous economic importance in eastern Asia, and increasingly valued in North America and Europe. Many species and a large number of cultivars have been introduced.

All species are ornamental, especially those having cultivars with colored culms. Almost all species are likely to be invasive.

1. Neither auricles nor fimbriae present on any culm leaves . . . . . . . . . . . . . . . . . . . . . . . . . . . . . . . . . . . . 1. *P. aurea*
1. Auricles present on the upper culm leaves; fimbriae present on all culm leaves . . . . . . . . . . . . . . . . . . . 2. *P. bambusoides*

### 1. Phyllostachys aurea Carrière *ex* Rivière & C. Rivière FISHPOLE BAMBOO, GOLDEN BAMBOO [not illustrated]

*Phyllostachys aurea* is native to China, but it is widely cultivated in temperate and subtropical regions. In North America, it grows as far north as Vancouver, British Columbia, in the west and Buffalo, New York, in the east. The young shoots are very palatable, even when raw, but the mature culms are very hard when dried. They are sometimes used for fishpoles. This species differs from other species of *Phyllostachys*, including those with brighter yellow culms, in having a raised collar below the nodes and irregularly compressed basal culm nodes.

### 2. Phyllostachys bambusoides Siebold & Zucc. GIANT TIMBER BAMBOO, MADAKE [p. 350]

*Phyllostachys bambusoides*, a widely cultivated species, is hardy to −17°C. Several cultivars are available, differing in the color of their culms and leaves.

## 2.04 PSEUDOSASA Makino *ex* Nakai[2]

**Pl** shrublike, spreading or loosely to densely clumped; **rhz** leptomorphic. **Clm** 0.5–13 m tall, to 4 cm thick, self-supporting, erect or nodding, plcsp; **nd** not or slightly swollen; **spnd rdg** not evident; **intnd** mainly terete, only slightly flattened immediately above the br, glab, with light wax below the nd. **Br** initially 1–3, erect to arcuate, often short, cent br dominant, with compressed bas nd, br fully shthed, lat br arising either from the bas nd or from more distal nd, shth and prphl more or less glab, persistent, tough. **Clm lvs** coriaceous and very persistent; **bld** erect or reflexed, narrowly triangular to strap-shaped. **Fol lvs: shth** persistent; **bld** cross veined, medium to large for the size of the clm, without mrgl necrosis in winter, their arrangement random. **Infl** rcm or pan; **br** subtended by much rdcd or quite substantial bracts. **Spklt** 2–20 cm, with 3–30 flt; **rchl** sinuous; **dis** below the flt. **Glm** 2, shorter than the first lm; **lm** to 1 cm; **anth** 3; **sty** 3; **pal** 2-keeled.

*Pseudosasa* includes about 36 species, all of which are native to China, Japan, and Korea.

---

[1]Christopher M.A. Stapleton and Mary E. Barkworth    [2]Christopher M.A. Stapleton

**1. Pseudosasa japonica** (Siebold & Zucc. *ex* Steud.) Makino *ex* Nakai JAPANESE ARROW BAMBOO, METAKE, YADAKE [p. *350*]

*Pseudosasa japonica* is a widely cultivated ornamental species that used to be grown for arrows in Japan. There are no known wild populations. It forms a tough and effective screen, and has become naturalized in British Columbia and the eastern United States.

## 3. OLYREAE Kunth[1]

Pl usu per; csp, stln. Clm per, 3–350 cm or climbing, not wd; nd often with 2 thick circumferential rdg, with more elastic tissue between; br usu not developed at the mid and up nd. Lvs often crowded towards the clm tips; abx **lig** absent; adx **lig** memb; psdpet 1–2 mm, not twisted; bld usu persistent, usu folding at night or when stressed, venation parallel, cross venation smt evident, particularly at the base, the bases and apc often asymmetric. Infl spicate or pan, usu produced at the mid and up nd of the lfy clm; dis above or below the glm. Spklt unisx and dimorphic, usu mixed within an infl, with 1 flt. Pist spklt on clavate ped; glm usu exceeding the flt, many-veined; lm usu coriaceous to indurate, pale when immature, mottled with dark spots or uniformly dark when mature, glab or with non-uncinate hairs, unawned; pal usu shorter than and enclosed by the lm; lod 3; sty br 2(3). Car dry; emb small relative to the car; hila usu linear. Stam spklt deciduous; glm usu lacking; lm memb or hyaline; lod 3 or absent; anth 3, 2, or multiples of 6.

The *Olyreae* is primarily a New World tribe that extends from Mexico to Argentina and southern Brazil. It includes about 20 genera and 110 species.

### 3.01 LITHACHNE P. Beauv.[2]

Pl per; loosely csp. Clm 6–75 cm, not woody; nd numerous, swollen. Shth open; adx **lig** memb; psdpet 1–2 mm; bld lanceolate to ovate, folding downwards at night or when stressed, bases and apc asymmetric. Infl rcm, partly enclosed by the leaf shth; ax rcm unisx or bisx, if bisx, the stmt spklt below 1 to several pist spklt, ped clavate; tml rcm stmt, ped filiform. Spklt with 1 flt. Pist spklt much larger than the stmt spklt, flt borne on a persistent peglike rchl intnd; gl exceeding the flt, subequal, several-veined; lem helmet-shaped, indurate, lat compressed, the mrg from overlapping the pal mrg to nearly concealing the pal; sty br 2, plumose. Stam spklt lanceolate, early deciduous; gl absent; lem translucent, 3-veined; anth 3.

*Lithachne* has four species. Its primary range extends south from Mexico and the Caribbean islands to Ecuador.

**1. Lithachne humilis** Soderstr. SMALL LITHACHNE [p. *350*]

*Lithachne humilis* is endemic to Honduras. It is sold as an ornamental in the United States.

## 3. EHRHARTOIDEAE Link[3]

The *Ehrhartoideae* encompasses three tribes, one of which, the *Oryzeae*, is native to the *Manual* region; the *Ehrharteae* is represented by introduced species. The third tribe, *Phyllorachideae* C.E. Hubb., is native to Africa and Madagascar. There are approximately 120 species in the *Ehrhartoideae*. They grow in forests, open hillsides, and aquatic habitats.

Molecular data provide strong support for the close relationship of the *Oryzeae* and *Ehrharteae*. Morphologically, they are characterized by spikelets that have a distal unisexual or bisexual floret with up to two proximal sterile florets and, frequently, six stamens in the staminate or bisexual florets.

1. Spikelets with 2 sterile florets below the functional floret, both well-developed, at least the upper sterile floret as long as or longer than the functional floret; glumes from $^1/_2$ as long as the spikelets to exceeding the florets; culms not aerenchymatous; plants of dry to damp habitats . . . . . . . . . . . . . . . . . . . . . . 4. *Ehrharteae*
1. Spikelets with 0–2 sterile florets below the functional floret, when present, sterile florets $^1/_8$–$^9/_{10}$ as long as the functional floret; glumes absent or highly reduced; culms aerenchymatous; plants of wet habitats . . . . . . . . . . . . . . . . . . . . . . . . . . . . . . . . . . . . . . . . . . . . . . . . . . . . . . . . . . . . . . . . . . 5. *Oryzeae*

### 4. EHRHARTEAE Nevski[4]

Pl ann or per. Clm ann, (1)6–200 cm, smt wd, not aerenchymatous, smt brchd above the base. Shth open, usu rounded on the back, glab or not, smt scabrous; col frequently with tuberculate hairs; aur usu present, often ciliate; **lig** usu memb, smt a memb rim or of hairs; psdpet not present; bld linear, venation parallel, cross venation not evident, abx surfaces with microhairs and variously shaped silica bodies, cross sections non-

[1,2]Mary E. Barkworth  [3]Grass Phylogeny Working Group  [4]Mary E. Barkworth

Kranz; **1st sdlg lvs** with well-developed, erect bld. **Infl** tml, pan or unilat rcm; **dis** above the glm, flts falling as a cluster. **Spklt** solitary, terete or lat compressed, with 3 flt, lo 2 flt strl, tml flt bisx, at least the up strl flt as long as or longer than the bisx flt; **rchl** smt shortly prolonged beyond the base of the bisx flt. **Glm** 2, from $^1/_2$ as long as to exceeding the flt, (3)5–7-veined. **Stl flt: lm** coriaceous, 5–7-veined, awned or unawned; **pal** lacking. **Bsx flt: lm** lanceolate or rectangular, firmly cartilaginous to coriaceous, 5–7-veined, veins inconspicuous, apc entire, unawned; **pal** 0–2(5)-veined; **lod** 2, free; **anth** 1–6; **sty** 2, fused or free to the base, stigmas linear, plumose. **Car** ellipsoid; **hila** linear, at least $^1/_2$ as long as the car; **emb** up to $^1/_3$ the length of the car, waisted, without an epiblast, with a scutellar tail and a minute mesocotyl intnd.

The number of genera recognized in the *Ehrharteae* varies from one to four. The largest genus, *Ehrharta*, is native to Africa, the other three being Australasian. Only *Ehrharta* has been found in the *Manual* region.

## 4.01 EHRHARTA Thunb.[1]

**Pl** ann or per; synoecious. **Clm** 6–200 cm, smt wd, erect to decumbent, smt brchd above the base, usu pubescent; **bas brchg** invag. **Lvs** bas or bas and cauline; **shth** terete, open; **aur** present, often ciliate; **lig** 0.5–3 mm, truncate, memb or of hairs; **bld** linear to lanceolate, smt disarticulating from the shth. **Infl** rcm or pan; **pri br** spreading to ascending; **dis** above the glm, not between the flt. **Spklt** 2–17 mm, solitary, pedlt, terete or lat compressed, with 3 flt, lo 2 flt strl, at least the up equaling or exceeding the distal flt, distal flt bisx. **Glm** from about $^1/_2$ as long as to exceeding the flt, (3)5–7-veined. **Stl flt** consisting only of lm; **stl lm** firmer than the glm, glab or pubescent, stipitate or non-stipitate, smooth to rugose, unawned or awned, lowest lm often with lat earlike appendages at the base, up lm subequal to or longer than the distal flt. **Bsx flt: lm** often indurate at maturity, glab, 5–7-veined, keeled, unawned, smt mucronate; **pal** thinner than the lm, 1–2(5)-veined; **anth** (1–5)6, yellow; **sty** 2, fused or free to the base, white or brown. **Car** lat compressed.

*Ehrharta* is a genus of approximately 25 species, most of which are native to southern Africa. Three species, all from southern Africa, are established in California.

1. Upper glumes $^3/_4$–$^9/_{10}$ the length of the spikelets; sterile lemmas hairy, not transversely rugose . . . . . . . . . . . . . . . . . . . . . . . . . . . . . . . . . . . . . . . . . . . . . . . . . . . . . . . . . . . . . . . . . . . 1. *E. calycina*
1. Upper glumes less than $^3/_4$ the length of the spikelets; sterile lemmas glabrous or sparsely hispidulous, often transversely rugose distally.
   2. Lower sterile lemmas unawned . . . . . . . . . . . . . . . . . . . . . . . . . . . . . . . . . . . . . . . . . . . . . . . . . 2. *E. erecta*
   2. Lower sterile lemmas awned, the awns 2–20 mm long . . . . . . . . . . . . . . . . . . . . . . . . . . . . . . 3. *E. longiflora*

### 1. Ehrharta calycina Sm. PERENNIAL VELDTGRASS [p. 351, 507]

*Ehrharta calycina* is native to southern Africa. It was introduced to Davis, California, as a drought-resistant grass for rangelands, but it is unable to withstand heavy grazing. It is now common on the coastal sand dunes at San Luis Obispo and San Diego, California, and has been reported from Nevada and Texas.

### 2. Ehrharta erecta Lam. PANIC VELDTGRASS [p. 351, 507]

*Ehrharta erecta* was introduced to California from South Africa. It prefers shady, somewhat moist locations, and is best known from the eastern San Francisco Bay area, San Diego, and the campus of the University of California at Riverside.

### 3. Ehrharta longiflora Sm. LONGFLOWERED VELDTGRASS, ANNUAL VELDTGRASS [p. 351, 507]

*Ehrharta longiflora* is a southern African species, well-established in Australia that, in the *Manual* region, is established near Torrey Pines State Park in southern California. It is said to prefer shaded areas on hillsides and disturbed areas, such as gardens and roadsides. It usually grows in sandy to loamy soils.

## 5. ORYZEAE Dumort.[2]

**Pl** ann or per; synoecious or monoecious. **Clm** ann, 20–500 cm tall, aerenchymatous, smt floating. **Lvs** aerenchymatous; **aur** present or absent; **lig** memb or scarious, smt absent; **psdpet** smt present; **bld** with parallel veins, cross venation not evident; abx **bld epdm** with microhairs and transversely dumbbell-shaped silica bodies; **1st sdlg lf** without a bld. **Infl** usu pan, smt rcm or spikes; **dis** below the spklt, not occurring in cultivated taxa. **Spklt** lat compressed or terete, with 1 bisx or unisx flt, if unisx, pist and stmt spklt in the same or different pan, smt with 2 strl flt below the sx flt, these no more than $^1/_2$($^9/_{10}$) the length of the ftl flt; **rchl** not prolonged. **Glm** absent or highly red, forming an annular ring or lobes at the ped apc; **stl flt** $^1/_8$–$^1/_2$($^9/_{10}$) as long as the spklt; **ftl lm** 3–14-veined, memb or coriaceous, apc entire, unawned or with a tml awn; **pal** similar to the lm, 3–10-veined, 1-keeled; **lod** 2; **anth** usu 6(1–16); **sty** 2, bases fused or free, stigmas linear, plumose. **Fr** usu car, smt achenes, ovoid, oblong, or cylindrical; **emb** of the F+FP or F+PP type, small or elongate, with or without a scutellar tail; **hila** usu linear.

[1]Mary E. Barkworth  [2]Edward E. Terrell

The *Oryzeae* include about 10–12 genera and 70–100 species. Its members are native to temperate, subtropical, and tropical regions. *Oryza sativa* is one of the world's most important crop species. Four genera are native to the *Manual* region; two are introduced. Molecular data (Guo and Ge 2005) suggest including *Hygroryza* and *Zizaniopsis* in *Luziola*.

1. Lemma margins free; fruits achenes, ellipsoid, obovoid, ovoid or subglobose, beaked with a shell-like pericarp.
    2. Lemmas of the pistillate spikelets awned; plants emergent, more than 1 m tall . . . . . . . . . . . . . . . . . . 5.05 *Zizaniopsis*
    2. Lemmas of the pistillate spikelets unawned; plants emergent and less than 1 m tall or submerged aquatics . . . . . . . . . . . . . . . . . . . . . . . . . . . . . . . . . . . . . . . . . . . . . . . . . . . . . . . . . . . . . . . . 5.06 *Luziola*
1. Lemmas and paleas clasping along their margins; fruits caryopses, cylindrical or laterally compressed, not beaked.
    3. Spikelets unisexual; caryopses terete . . . . . . . . . . . . . . . . . . . . . . . . . . . . . . . . . . . . . . . . . . . . . 5.03 *Zizania*
    3. Spikelets bisexual; caryopses laterally compressed or terete.
        4. Sterile florets present below the fertile floret, $^1/_8$–$^1/_2$ ($^9/_{10}$) as long as the spikelets . . . . . . . . . . . . . . . . . . 5.01 *Oryza*
        4. Glumes absent.
            5. Leaf blades aerial, not pseudopetiolate, linear to broadly lanceolate; spikelets pedicellate, without stipelike calluses; lemmas unawned; widespread native species . . . . . . . . . . . . . . . . . . . . . . 5.02 *Leersia*
            5. Leaf blades floating, pseudopetiolate, elliptic to ovate or ovate-lanceolate; spikelets on stipelike calluses (1)2–10 mm long; lemmas awned; aquatic ornamental species, not known to be established in the *Manual* region . . . . . . . . . . . . . . . . . . . . . . . . . . . . . . . . . . . . . . . . . . . 5.04 *Hygroryza*

## 5.01 ORYZA L.[1]

**Pl** ann or per; usu aquatic, rooted and emergent or floating, smt terrestrial; rhz and/or csp; synoecious. **Clm** to 3.3(5) m, erect, decumbent, or prostrate, smt rooting at the lo nd, aerenchymatous, emergent or immersed, brchd or unbrchd. **Lvs** cauline and bas; **shth** open, lo shth often slightly inflated, up shth not inflated; **aur** usu present; **lig** memb, often veined; **psdpet** absent; **bld** linear to narrowly lanceolate, flat, mrg smooth or scabridulous. **Infl** tml pan; **dis** above the glm, beneath the strl flt in wild taxa, spklt of cultivated taxa not disarticulating. **Spklt** bisx, lat compressed, with 3 flt, lo 2 flt strl, tml flt fnctl. **Glm** absent or rdcd to lobes at the ped apc; **stl flt** glmlike, 1.2–10 mm, $^1/_8$–$^1/_2$ ($^9/_{10}$) as long as the spklt, linear or subulate to narrowly ovate, coriaceous, 1-veined, acute to acuminate; **fct flt: cal** usu inconspicuous and flat to rounded, smt conspicuous and stipelike, glab; **lm** coriaceous or indurate, with vertical rows of tubercles separated by longitudinal furrows, 5-veined, keeled, mrg clasping the mrg of the pal, apc obtuse or acute to acuminate, awned or unawned; **pal** with surfaces similar to the lm, 3-veined, unawned; **lod** 2; **anth** 6; **sty** 2, bases fused or not, stigmas lat exserted, plumose. **Car** lat compressed; **emb** usu $^1/_4$–$^1/_3$ as long as the car; **hila** linear.

*Oryza* is a tropical and subtropical genus of about 20 species that grow in shallow water, swamps, and marshes in seasonally inundated areas, or along streams, rivers, or lake edges. *Oryza sativa* (rice) is one of the three most economically valuable cereals, and constitutes a major portion of the diet for half of the world's population. In the *Manual* region, *O. sativa* is cultivated and several weedy forms have become established. These are thought to be derived from introgression between *O. sativa* and *O. rufipogon* and *O. punctata*.

1. Ligules truncate to rounded, 1.5–10 mm long; sterile florets 1.2–2 mm long; disarticulation scar centric or slightly eccentric . . . . . . . . . . . . . . . . . . . . . . . . . . . . . . . . . . . . . . . . . . . . . . . . . . . . . . . . . . 3. *O. punctata*
1. Ligules acute, 4–45 mm long; sterile florets 1.3–10 mm long; disarticulation scar lateral.
    2. Anthers 1–2.5 mm long; spikelets persistent; lemmas usually unawned, plants not rhizomatous; auricles absent or to 5 mm long; blades 5–20 mm wide . . . . . . . . . . . . . . . . . . . . . . . . . . . . . . . . . . . . 4. *O. sativa*
    2. Anthers 3.5–7.4 mm long; spikelets deciduous; lemmas awned; plants usually rhizomatous; auricles absent or to 15 mm long; blades 7–50 mm wide.
        3. Caryopses 5–7 mm long; lemma-awn junctions purplish, pubescent; lemma awns 4–16 cm long; plants cespitose or rhizomatous; auricles absent or to 7 mm long . . . . . . . . . . . . . . . . . . . . 2. *O. rufipogon*
        3. Caryopses 7.5–8.5 mm long; lemma-awn junctions similar in color to the lemmas, glabrous; lemma awns 2.6–8 cm long; plants strongly rhizomatous; auricles present, to 15 mm long . . . . . . 1. *O. longistaminata*

### 1. Oryza longistaminata A. Chev. & Roehr.
LONGSTAMEN RICE [p. *351*]

*Oryza longistaminata* is native to Africa; it has not yet been found in North America. It is included here because its establishment in North America would seriously impact North American agriculture. The U.S. Department of Agriculture considers it a noxious weed; plants found growing in the United States should be reported to that agency.

### 2. Oryza rufipogon Griff. RED RICE, BROWNBEARD RICE [p. *351*, <u>*507*</u>]

*Oryza rufipogon* is native to southeast Asia and Australia, where it grows in shallow, standing or slow-moving water, along irrigation canals, and as a weed in rice fields. It hybridizes readily with *O. sativa*, forming partially fertile hybrids. This makes it a serious threat to rice growers.

[1]Mary E. Barkworth and Edward E. Terrell

**3. Oryza punctata** Kotschy *ex* Steud. RED RICE [p. *351*]

*Oryza punctata* is an African grass that is not established in North America. It is considered a noxious weed by the U.S. Department of Agriculture; if any populations are found in the United States, that agency should be notified.

**4. Oryza sativa** L. RICE [p. *351*, *507*]

*Oryza sativa* is cultivated in California, Arkansas, Texas, Louisiana, Mississippi, and Florida and is sometimes found as an adventive in moist or wet places, particularly in the southeastern United States, but it is not established in the *Manual* region. It is derived from, and crosses with, *O. rufipogon* (Londo et al. 2006).

## 5.02 LEERSIA Sw.[1]

Pl usu per, rarely ann; terrestrial or aquatic; rhz or csp; synoecious. Clm 20–150 cm (occ longer in floating mats), erect or decumbent, often rooting at the nd, brchd or unbrchd. Lvs equitably distributed along the clm; shth open; aur absent; lig memb; psdpet absent; bld aerial, linear to broadly lanceolate, flat or folded, smt involute when dry. Infl tml pan, usu exserted, ax pan smt present; dis beneath the spklt. Spklt bisx, with 1 flt; flt lat compressed, linear to suborbicular in sideview. Glm absent; cal not stipelike, glab; lm and pal subequal, chartaceous to coriaceous, ciliate-hispid or glab, tightly clasping along the mrg; lm 5-veined, obtuse or acute to acuminate, smt mucronate, usu unawned; pal 3-veined, unawned; lod 2; anth 1, 2, 3, or 6; sty 2, bases fused, stigmas lat exserted, plumose. Car lat compressed; emb about ⅓ as long as the car; hila linear.

*Leersia* is a genus of about 17 aquatic to mesophytic species, growing primarily in tropical and warm-temperate regions. Five species are native to the *Manual* region. *Leersia* is closely allied to *Oryza*.

1. Spikelets 1.5–2 mm long, glabrous; plants not rhizomatous . . . . . . . . . . . . . . . . . . . . . . . . . . . . . . . . . . . . . . . 1. *L. monandra*
1. Spikelets 2.5–6.5 mm long, usually ciliate on the margins and keel, and glabrous or pubescent elsewhere; plants rhizomatous.
  2. Spikelets nearly as wide as long . . . . . . . . . . . . . . . . . . . . . . . . . . . . . . . . . . . . . . . . . . . . . . . . . . . . . 2. *L. lenticularis*
  2. Spikelets not more than ½ as wide as long.
    3. Anthers 2; spikelets 2.5–3.6 mm long; panicle branches single at all nodes . . . . . . . . . . . . . . . . . . . . . 3. *L. virginica*
    3. Anthers 3 or 6; spikelets 3.2–6.5 mm long; panicle branches 1–2 or more at the lower nodes, single at the upper nodes.
      4. Panicles exserted, 5–15 cm long; branches appressed to ascending, spikelet-bearing to near the base; anthers 6; spikelets 3.2–4.7(5) mm long . . . . . . . . . . . . . . . . . . . . . . . . . . . . . . . . . . . 4. *L. hexandra*
      4. Panicles exserted or enclosed, 10–30 cm long; branches spreading on exserted panicles, naked on the lower ⅓; anthers 3; spikelets (4)4.2–6.5 mm long . . . . . . . . . . . . . . . . . . . . . 5. *L. oryzoides*

**1. Leersia monandra** Sw. BUNCH CUTGRASS, CANYONGRASS, CEDAR WHITEGRASS [p. *351*, *507*]

*Leersia monandra* grows in rather dry, rocky, limestone soils in open woods, grasslands, and bluffs, from Texas and Florida south to the Yucatan Peninsula, Mexico, and the Antilles. It is also sold as an ornamental. With heavy grazing, *L. monandra* tends to disappear, surviving only in areas where shrubs provide protection.

**2. Leersia lenticularis** Michx. CATCHFLY GRASS, OATMEAL GRASS [p. *351*, *507*]

*Leersia lenticularis* grows in river bottoms and moist woods of the midwestern and southeastern United States. It flowers from July to November. Ohio and Maryland list it as an endangered species.

**3. Leersia virginica** Willd. WHITE CUTGRASS, WHITEGRASS, LÉERSIE DE VIRGINIE [p. *351*, *507*]

*Leersia virginica* grows in moist places in woods and along stream courses east of the Rocky Mountains. The western Wyoming record may represent an introduction. *Leersia virginica* flowers from July to October.

**4. Leersia hexandra** Sw. SOUTHERN CUTGRASS [p. *352*, *507*]

*Leersia hexandra* is found in wet areas, usually in fresh water along streams and ponds, where it sometimes forms floating mats. It grows in the southeastern United States and throughout much of the neotropics; the California record probably represents a recent introduction.

**5. Leersia oryzoides** (L.) Sw. RICE CUTGRASS, LÉERSIE FAUX-RIZ [p. *352*, *507*]

*Leersia oryzoides* grows in wet, heavy, clay or sandy soils, and is often aquatic. It is found across most of southern Canada, extending south throughout the contiguous United States into northern Mexico, and flowers from July to October. It has also become established in Europe and Asia.

## 5.03 ZIZANIA L.[2]

Pl ann or per; aquatic, usu rooted in the substrate; smt rhz or stln; monoecious. Clm to 5 m, erect and emergent or floating. Lvs concentrated on the lo portion of the stem or evenly distributed; shth open, not inflated; lig memb or scarious, glab; psdpet absent; bld flat, aerial or floating, scabrous or smooth. Infl tml pan; br usu unisx, lo br stmt, up br pist, mid br smt with stmt and pist spklt intermixed; ped apc cupulate; dis beneath the spklt, in cultivated strains dis delayed, the spklt tending not to shatter until harvested. Spklt unisx, with 1 flt. Glm absent; cal inconspicuous; lm 5-veined; pal 3-veined; lod 2, memb. Stam spklt pendant, terete or appearing so; lm memb; pal memb, loosely enclosing the st; anth 6. Pist spklt terete; lm chartaceous or coriaceous, mrg involute and clasping the mrg of the pal, apc acute to acuminate, smt awned, awns tml,

[1]Grant L. Pyrah   [2]Edward E. Terrell

slender, scabridulous; **sty** 2, bases not fused, stigmas lat exserted, plumose. **Car** cylindrical; **emb** linear, often as long as the car; **hila** linear. $x = 15$. Name from the Greek *zizanion*, a weed growing in grain.

*Zizania* includes three North American and one eastern Asian species. *Zizania aquatica* and *Z. palustris* are important constituents of aquatic plant communities in North America, providing food and shelter for numerous animal species. *Zizania palustris* is also an important food source for humans. *Zizania texana* is federally listed as an endangered species in the United States.

1. Culms decumbent, completely immersed or the upper parts of the culm emergent; known only from the San Marcos River in Hays County, Texas . . . . . . . . . . . . . . . . . . . . . . . . . . . . . . . . . . . . . . . . . . . . . . . . . . . . 3. *Z. texana*
1. Culms usually erect at maturity, rarely completely immersed; plants not known from Texas.
  2. Plants rhizomatous, perennial; middle branches of the panicles with both staminate and pistillate spikelets, other branches with either staminate or pistillate spikelets; plants cultivated as ornamentals . . . . . . . . . . . . . . . . . . . . . . . . . . . . . . . . . . . . . . . . . . . . . . . . . . . . . . . . . 4. *Z. latifolia*
  2. Plants without rhizomes, annual; all panicle branches unisexual, with either staminate or pistillate spikelets; plants native and widespread, also cultivated for grain.
    3. Lemmas of the pistillate spikelets flexible and chartaceous, dull or sublustrous, bearing short, scattered hairs, these not or only slightly more dense towards the apices; aborted pistillate spikelets 0.4–1 mm wide; pistillate inflorescence branches usually divaricate at maturity . . . . . . . . . . . . 1. *Z. aquatica*
    3. Lemmas of the pistillate spikelets stiff and coriaceous or indurate, lustrous, glabrous or with lines of short hairs, the apices more densely hairy; aborted pistillate spikelets 0.6–2.6 mm wide; pistillate inflorescence branches usually appressed at maturity, or with 1 to few, somewhat spreading branches . . . . . . . . . . . . . . . . . . . . . . . . . . . . . . . . . . . . . . . . . . . . . . . . . . . . 2. *Z. palustris*

## 1. Zizania aquatica L. [p. *352*, <u>*507*</u>]

*Zizania aquatica* is native from the central plains to the eastern seaboard. It is sometimes planted for wildfowl food. The records from western North America reflect such plantings. Most, possibly all, have since died out.

1. Plants to 5 m tall; blades (5)10–75 mm wide; pistillate spikelets 7–24 mm long; awns to 10 cm long. . . . . . . . . . . . . . . . . . . . . . . . . . . . . . . . . . . . . . . var. *aquatica*
1. Plants 0.2–1 m tall; blades 3–12(20) mm wide; pistillate spikelets 5–11 mm long; awns 1–8 mm long . . . . . . . . . . . . . . . . . . . . . . . . . . . . var. *brevis*

### Zizania aquatica L. var. aquatica SOUTHERN WILDRICE, ZIZANIE AQUATIQUE [p. *352*]

*Zizania aquatica* var. *aquatica* grows in fresh or somewhat brackish marshes, swamps, streams, and lakes. Its native range extends from southeastern Minnesota to southern Maine, and south to central Florida and southern Louisiana.

### Zizania aquatica L. var. brevis Fassett ESTUARINE WILDRICE [p. *352*]

*Zizania aquatica* var. *brevis* is known from tidal mud flats along the St. Lawrence River, about 80 km up- and downstream from Quebec City, and from a small delta along the northern shore of the northwest Miramichi River estuary in New Brunswick.

## 2. Zizania palustris L. [p. *352*, <u>*507*</u>]

*Zizania palustris* grows mostly to the north of *Z. aquatica*, but the two species overlap in the Great Lakes region, eastern Canada, and New England. It is cultivated as a crop in some provinces and states. All records from the western part of the *Manual* region reflect deliberate plantings; none are known to have persisted.

1. Lower pistillate branches with 9–30 spikelets; pistillate part of the inflorescence 10–40 cm or more wide, the branches ascending to widely divergent; plants 1–3 m tall; blades 10–40+ mm wide . . . . . . . . var. *interior*
1. Lower pistillate branches with 2–8 spikelets; pistillate part of the inflorescence 1–8(15) cm wide, the branches appressed or ascending, or a few branches somewhat divergent; plants to 2 m tall; blades 3–21 mm wide . . . . . . . . . . . . . . . . . . . . . var. *palustris*

### Zizania palustris L. var. interior (Fassett) Dore INTERIOR WILDRICE [p. *352*]

*Zizania palustris* var. *interior* grows on muddy shores and in shallow water, mainly in the north central United States and adjacent Canada. It resembles *Z. aquatica* in its vegetative characters, and *Z. palustris* var. *palustris* in its pistillate spikelets.

### Zizania palustris L. var. palustris NORTHERN WILDRICE, ZIZANIE DES MARAIS, FOLLE AVOINE, RIZ SAUVAGE [p. *352*]

*Zizania palustris* var. *palustris* grows in the shallow water of lakes and streams, often forming extensive stands in northern lakes. It has been introduced to British Columbia, Nova Scotia, Idaho, Arizona, and West Virginia for waterfowl food.

## 3. Zizania texana Hitchc. TEXAS WILDRICE [p. *352*, <u>*507*</u>]

*Zizania texana* grows only in the headwaters of the San Marcos River, in San Marcos, Texas. It is officially listed as an endangered species in the United States.

## 4. Zizania latifolia (Griseb.) Turcz. *ex* Stapf ASIAN WILDRICE [p. *352*]

*Zizania latifolia* is native to Asia, extending from northeast India and Russia through China and Myanmar to Korea and Japan. In its native range, it grows in the shallow waters of lakes and swamps, forming large patches. The rhizomes and basal parts of the culms of *Z. latifolia* are edible, and become swollen when infected with the fungus *Ustilago esculenta* Henn. The infection also prevents the plants from flowering and fruiting. If infected plants were introduced into North America, the fungus might also infect the native species of *Zizania* and likewise prevent their flowering, a possibility that should be strenuously resisted. Plants of *Z. latifolia* should not be brought into North America.

## 5.04 HYGRORYZA Nees[1]

Pl per; aquatic, producing long, floating clm; synoecious. Clm 50–150 cm, spongy, developing adventitious roots at the nd, brchd; br erect, lfy. Lvs cauline, glab, veins tessellate; shth open, inflated, serving as floats; lig absent or hyaline; psdpet present; bld elliptic, ovate, ovate-lanceolate, or oblong. Infl tml pan, aerial, lowermost br whorled; dis beneath the spklt cal. Spklt bisx, lat compressed, with 1 flt. Glm absent or an annular rim; cal (1)2–10 mm, stipelike, glab, jnct with the ped marked by a tan constriction; lm 5-veined, mrg clasping the pal, apc acuminate, awned, awns tml, antrorsely scabridulous; pal similar to the lm, 3-veined, 1-keeled, acute-acuminate, unawned; lod 2, glab; anth 6; sty 2, bases not fused, stigmas lat exserted, plumose. Car terete, fusiform; emb small; hila linear, almost as long as the embryo.

*Hygroryza* is a unispecific Asian genus that grows in India, Ceylon, and throughout southeast Asia. It forms floating masses, often of considerable extent, in lakes and slow-moving streams, and is sometimes a weed in rice.

1. **Hygroryza aristata** (Retz.) Nees ASIAN WATERGRASS, WATER STARGRASS [p. *352*]

*Hygroryza aristata* is native to tropical Asia, where it has occasionally been used as forage for cattle. It is sold for ponds and aquaria, where its long, feathery, adventitious roots have a decorative effect, but it has the potential to become a significant weed problem in the southern United States.

## 5.05 ZIZANIOPSIS Döll & Asch.[2]

Pl per or ann; aquatic, rooted and emergent; rhz; monoecious. Clm 1–4 m, erect or decumbent, smt rooting at the nd. Lvs bas and cauline; shth open, somewhat lat compressed; lig scarious; psdpet absent; bld flat or folded at the base, lanceolate. Infl tml pan, stmt and pist spklt on the same br, stmt spklt proximal, pist spklt distal; dis beneath the spklt; ped apc cupulate. Spklt unisx, lat compressed to subterete, lm mrg not clasping the pal, with 1 flt. Glm absent; cal glab. Stam lm memb, 5–7-veined, acuminate or tml awned; pal similar to the lm, 3-veined; lod 2; anth 6. Pist lm memb, 7-veined, tml awned; pal similar to the lm, 3-veined, awned or unawned; sty 2, bases fused, stigmas tml exserted, plumose. Car ellipsoid or obovoid, beaked by the persistent sty base; prcp shell-like, partially free from the sd, smooth, coriaceous or crustaceous; sd oblong, subterete, or 2-angled; emb bas; hila linear.

*Zizaniopsis* grows from the southern United States to Argentina. Its five species grow in wet habitats. Only *Zizaniopsis miliacea* is native to and found in the *Manual* region.

1. **Zizaniopsis miliacea** (Michx.) Döll & Asch. GIANT CUTGRASS, WATER MILLET [p. *353*, *507*]

*Zizaniopsis miliacea* grows in shallow, fresh- or brackish-water marshes, swamps, streams, lakes, and ditches. It is most common on the eastern coastal plain of the United States, extending south to Florida and west to Illinois, Oklahoma and Texas.

## 5.06 LUZIOLA Juss.[3]

Pl per; aquatic, usu rooted, smt floating; stln, smt mat-forming; monoecious. Clm 10–100+ cm, erect or prostrate, smt rooting at the nd, brchd, emergent or immersed. Lvs cauline; shth open, not inflated or somewhat inflated; lig hyaline; psdpet present or absent; bld flat, linear to lanceolate or narrowly elliptic, glab, pubescent, or scabrous. Infl pan, rcm, or spikes, exserted or enclosed, stmt and pist spklt usu in separate infl, pist infl at the lo or mid nd, stmt infl usu tml; dis below the spklt. Spklt unisx, lat compressed to subterete, with 1 flt. Glm absent; cal glab; lm and pal subequal, ovate or lanceolate, memb or hyaline, unawned; lod 2. Stam lm and pal obscurely few- to several-veined; anth 6–16. Pist lm 5–14-veined, mrg not clasping the mrg of the pal, unawned; pal 3–10-veined; sty 2, bases fused, stigmas lat or tml exserted, plumose. Car ovoid, ellipsoid, or subglobose, beaked by the persistent sty bases; prcp shell-like, partially free from the sd, smooth or striate, crustaceous; sd ovoid to subglobose; emb bas; hila linear.

*Luziola* has 12 species. They range from the southeastern United States to Argentina. Only *L. fluitans* is native to the *Manual* region. All three species in the region are emergent or immersed in shallow, fresh to brackish water.

1. Culms prostrate, usually immersed; leaves floating or streaming in currents, 1–5(8) cm long, usually more numerous towards the ends of the culms; pistillate inflorescences mostly included in the sheaths, only the stigmas visible . . . . . . . . . . . . . . . . . . . . . . . . . . . . . . . . . . . . . . . . . . . . . . .1. *L. fluitans*
1. Culms suberect to erect, from fully emergent to immersed; leaves not conspicuously floating or streaming, longer than 6 cm, basal or scattered along the culms; pistillate inflorescences all or mostly exserted, their branches and spikelets evident.
   2. Pistillate florets 3–5 mm long; achenes striate . . . . . . . . . . . . . . . . . . . . . . . . . . . . . . . . . . . . . . .2. *L. bahiensis*
   2. Pistillate florets 2–2.5 mm long; achenes smooth . . . . . . . . . . . . . . . . . . . . . . . . . . . . . . . . . . . . . .3. *L. peruviana*

[1]J.K. Wipff    [2,3]Edward E. Terrell

**1. Luziola fluitans** (Michx.) Terrell & H. Rob. [p. 56]
SILVERLEAF GRASS, WATERGRASS [p. 353, 507]

*Luziola fluitans* grows in fresh to slightly saline lakes and streams in the southeastern United States and eastern Mexico. It is most common in the coastal plain, and also occurs in the Piedmont.

**2. Luziola bahiensis** (Steud.) Hitchc. BRAZILIAN WATERGRASS [p. 353, 507]

*Luziola bahiensis* is native from the Caribbean south to Argentina. It has been found at scattered locations in southern Louisiana, Mississippi, Alabama, and northwestern Florida. It grows in wet places or shallow water along streams and lakes.

**3. Luziola peruviana** J.F. Gmel. PERUVIAN WATERGRASS [p. 353, 507]

*Luziola peruviana* has been found at scattered locations from Texas to Florida. It is native to the Caribbean, Central America, and South America, and grows in wet places and shallow water along streams and lakes.

# 3. POÖIDEAE Benth.[1]

The subfamily *Poöideae* includes approximately 3300 species, making it the largest subfamily in the *Poaceae*. It reaches its greatest diversity in cool temperate and boreal regions, extending across the tropics only in high mountains.

1. Inflorescences 1-sided spikes, the spikelets radial to and partially embedded in the rachises; spikelets with 1 floret each . . . . . . . . . . . . . . . . . . . . . . . . . . . . . . . . . . . . . . . . . . . . . . . . . . . . . . . . 7. *Nardeae*
1. Inflorescences panicles, racemes, or 2-sided spikes with spikelets radial or tangential to the rachises, sometimes embedded in the axes, never both radial and embedded; spikelets with 1–30 florets.
  2. Cauline leaf sheaths closed for at least ³/₄ their length; lemmas longer than (4.5)6.5 mm or awned or with prominent, parallel veins.
    3. Ovary apices glabrous; styles fused at the base, divergent, naked on the lower portion, plumose distally; lemmas often with a purplish band in the distal ¹/₂, usually unawned; distal 1–3 florets often reduced to lemmas, the lower 1–2 lemmas often enclosing the terminal lemmas; lodicules about 0.2–0.5 mm long, truncate, fleshy, without a distal membranous portion . . . . . . . . . . . . . . . . . . . . 9. *Meliceae*
    3. Ovary apices hairy; styles separate and plumose to the base; lemmas usually without a purplish band, sometimes with purplish bases, usually awned; distal 1–2 florets sometimes reduced, each separate with lemma and palea; lodicules usually more than 1 mmlong, fleshy at the base, with a distal membranous portion . . . . . . . . . . . . . . . . . . . . . . . . . . . . . . . . . . . . . . . . . . 12. *Bromeae*
  2. Cauline leaf sheaths usually open for most or all of their length; if the sheaths closed, the lemmas shorter than 7 mm, unawned and with lemma veins inconspicuous and converging distally.
    4. Inflorescences usually spikes or spikelike racemes, sometimes panicles, lateral spikelets on pedicels less than 3 mm long; if inflorescences with 1 spikelet per node, the spikelets tangential to the rachises or pedicellate and the lemmas unawned or terminally awned; ovary apices hairy.
      5. Glumes unequal, exceeded by the lowest lemmas, lanceolate, apices obtuse to acuminate or mucronate, rarely awned; inflorescences spikelike racemes, all spikelets pedicellate; pedicels 0.5–2.5 mm long . . . . . . . . . . . . . . . . . . . . . . . . . . . . . . . . . . . . . . . . . . . . . . . . . . 11. *Brachypodieae*
      5. Glumes equal to unequal, sometimes absent, frequently exceeding the lowest lemmas, subulate to lanceolate, ovate, or obovate, apices truncate to acuminate, frequently awned; inflorescences usually spikes or spikelike, with 1 or more sessile spikelets per node, sometimes a panicle; pedicels absent or up to 4 mm long . . . . . . . . . . . . . . . . . . . . . . . . . . . . . . . . . . . 13. *Triticeae*
    4. Inflorescences usually panicles, sometimes racemes with pedicels more than 2.5 mm long, or spikes with 1 spikelet per node and the spikelets radial or tangential to the rachises; if spikelets 1 per node and tangential, the lemmas awned from midlength to subapically, never terminally, if spikes with radial spikelets, the lemmas unawned or awned, awns basal to terminal; ovary apices usually glabrous, sometimes hairy.
      6. Lower glumes absent or highly reduced; inflorescences panicles . . . . . . . . . . . . . . . . . . . . . . . 6. *Brachyelytreae*
      6. Lower glumes usually well-developed, sometimes present only on the terminal spikelets; inflorescences panicles, racemes, or spikes.
        7. Caryopses beaked; blades tapering both basally and apically, midveins usually eccentric . . . . . . . . . . . . . . . . . . . . . . . . . . . . . . . . . . . . . . . . . . . . . . . . . . . . . . . . . . . 8. *Diarrheneae*
        7. Caryopses not beaked; blades usually tapering only apically, midveins usually centric.
          8. Spikelets with 1 floret; lemmas terminally awned, the junction of the lemma and awn abrupt, evident; glumes equal to or longer than the florets . . . . . . . . . . . . . . . 10. *Stipeae* (in part)
          8. Spikelets with 1–22 florets; lemmas unawned or dorsally to terminally awned, if terminally awned, the transition from lemma to awn gradual, not evident; glumes absent or shorter than to longer than the adjacent florets.

[1]Grass Phylogeny Working Group

9. Lemmas membranous, bidentate or bifid; both surfaces of the leaf blades deeply ribbed; ovary apices hairy; culms with solid internodes; plants cultivated or established at a few locations . . . . . . . . . . . . . . . . . . . . . . . . . . . . . . . . . . 10. *Stipeae* (in part)
9. Lemmas hyaline to membranous, entire or minutely bidentate; leaf blades rarely deeply ribbed on both sides; ovary apices usually glabrous; culms usually with hollow internodes; plants mostly native or established throughout the *Manual* region, sometimes cultivated . . . . . . . . . . . . . . . . . . . . . . . . . . . . . . . . . . 14. *Poeae*

## 6. BRACHYELYTREAE Ohwi[1]

**Pl** per; with knotty rhz. **Clm** ann, not brchg above the base; **intnd** solid. **Shth** open, mrg not fused; **col** glab, without tufts of hair at the sides; **aur** absent; **lig** scarious, not ciliate, those of the up and lo cauline lvs usu similar; **psdpet** absent; **bld** tapering bas and distally, venation parallel, cross venation not evident, sec veins parallel to the midvein; **x-sec** non-Kranz, with arm and fusoid cells; **epdm** without microhairs, cells not papillate. **Infl** tml pan. **Spklt** scarcely compressed, with 1 flt, flt bisx; **rchl** prolonged beyond flt base; **dis** above the glm, beneath the flt. **Glm** unequal, lanceolate; **lo glm** absent or highly red; **up glm** less than ¼ as long as the

flt, 1-veined; **flt** 8–12 mm, dorsally compressed; **cal** rounded, antrorsely hairy, hairs 0.2–0.5 mm; **lm** coriaceous, unawned, rounded dorsally, 5-veined, veins converging distally, apc entire; **pal** subequal to the lm, 2-veined, rdgd over the veins; **lod** 2, glab, veined; **anth** 3; **ov** glab; **sty** 2, elongate, hairy, bases free. **Car** grooved, sty persistent; **hila** linear; **emb** less than ½ as long as the car.

There is one genus, *Brachyelytrum*, in the *Brachyelytreae*. It is anomalous within the *Poöideae* in having arm and fusoid cells and broad seedling leaves.

## 6.01 BRACHYELYTRUM P. Beauv.[2]

**Pl** per; rhz, rhz knotty. **Clm** 28–102 cm, erect, not brchd above the bases; **intnd** solid; **nd** glab or retrorsely pubescent. **Lvs** mostly cauline; **shth** open; **aur** absent; **lig** memb; **lo lf bld** absent or red; **up lf bld** flat, tapering both bas and apically. **Infl** tml pan, contracted; **br** appressed, with 1–3(5) spklt. **Spklt** pedlt, terete to dorsally compressed, with 1 flt; **rchl** prolonged beyond the flt base, glab; **dis** above the glm, beneath the flt. **Glm** 1 or 2; **lo glm** 0.1–1.1 mm, smt absent; **up glm** 0.2–7 mm, clearly exceeded by the flt; **flt** 8–12 mm; **cal** about 0.8 mm, blunt, with hairs; **lm** memb to coriaceous,

scabrous, enclosing the pal, 5-veined, tapering, awned, awns tml, lm-awn transition gradual; **awn** 9.5–32.5 mm, longer than the lm bodies, straight, scabrous; **pal** subequal to the lm, 2-veined; **lod** 2, veined; **anth** 3, yellow; **sty** 2, bases free, white. **Car** linear, longitudinally grooved, apc beaked, pubescent; **hila** linear.

*Brachyelytrum* includes three species, two native to eastern North America and one to eastern Asia. The ranges of the two North American species overlap but, although they often grow closely together, neither mixed populations nor apparent hybrids have been found.

1. Lemmas hispid, hairs 0.2–0.9 mm long, visible at 10× magnification; anthers 3.5–6 mm long; awns 13–17(20) mm long . . . . . . . . . . . . . . . . . . . . . . . . . . . . . . . . . . . . . . . . . . . 1. *B. erectum*
1. Lemmas scabrous, scabrules 0.08–0.14(0.2) mm long; anthers 2–3.5 mm long; awns (14)17–24(26) mm long . . . . . . . . . . . . . . . . . . . . . . . . . . . . . . . . . . . . . . . . . . . . . . 2. *B. aristosum*

### 1. Brachyelytrum erectum (Schreb.) P. Beauv.
SOUTHERN SHORTHUSK, BRACHYELYTRUM DRESSÉ [p. *353*, 507]

*Brachyelytrum erectum* grows in woodlands, occasionally over limestone bedrock, and in moist woods and forests. It extends from Ontario east to Newfoundland, and in the United States from Minnesota to New England and south to the Gulf Coast and Florida.

### 2. Brachyelytrum aristosum (Michx.) P. Beauv. *ex* Branner & Coville [p. 61] NORTHERN SHORTHUSK [p. 353, 507]

*Brachyelytrum aristosum*, like *B. erectum*, grows in moist woods and forests, but its primary distribution is more northern, extending from Newfoundland west to Minnesota and south through the Appalachian Mountains to the junction of Tennessee, North Carolina, and Georgia.

## 7. NARDEAE W.D.J. Koch[3]

**Pl** per; csp. **Clm** ann, to 60 cm; **intnd** hollow. **Shth** open, mrg not fused; **col** glab, without tufts of hair at the sides; **aur** absent; **lig** scarious, not ciliate, those of the up and lo cauline lvs usu similar; **psdpet** not present; **bld** filiform, venation parallel, cross venation not evident,

sec veins parallel to the midvein; **x-sec** non-Kranz, without arm or fusoid cells, adx **epdm** with bicellular microhairs, not papillate. **Infl** tml spikes, 1-sided, spklt solitary, radial to the rchs; **rchs** with the spklt partially embedded. **Spklt** not compressed, triangular in cross

[1]Stephen N. Stephenson  [2]Stephen N. Stephenson and Jeffery M. Saarela  [3]Mary E. Barkworth

section, with 1 flt, flt bisx; **rchl** not prolonged beyond the flt base; **dis** above the glm, beneath the flt. **Glm** absent or vestigial; **lo glm** a cupular rim; **up glm** absent or vestigial; **flt** 5–10 mm, not compressed; **cal** poorly developed, glab; **lm** chartaceous, 3-veined, angled over the veins, most strongly so over the lat veins, apc entire, awned, awns tml, not brchd, lm-awn transition gradual, not evident; **pal** subequal to the lm, hyaline, 2-keeled; **lod** absent; **anth** 3; **ov** glab; **sty** 1. **Car** fusiform, sty bases not persistent; **hila** linear, more than $^1/_2$ as long as the car; **emb** about $^1/_6$ the length of the car.

There is only one genus, *Nardus*, in the *Nardeae*.

## 7.01 NARDUS L.[1]

**Pl** per; csp. **Clm** 3–60 cm, erect; **bas brchg** invag. **Lvs** mostly bas; **shth** open; **aur** absent; **lig** memb, entire, rounded; **bld** filiform, tightly convolute, **epdm** with bicellular microhairs. **Infl** tml spikes, 1-sided, spklt in 2 rows, loosely to closely imbricate; **rchs** terminating in a bristle; **dis** below the flt. **Spklt** triangular in cross section, with 1 flt, flt bisx. **Lo glm** a highly red, cupular rim; **up glm** absent or vestigial; **flt** 5–10 mm; **lm** linear-lanceolate to lanceolate-oblong, chartaceous, enveloping the pal, 3-veined, awned; **pal** hyaline, 2-veined, 2-keeled; **lod** absent; **anth** 3; **sty** 1.

*Nardus* is a unispecific European genus. Its relationships to other genera are unclear.

### 1. Nardus stricta L. MATGRASS, NARDE RAIDE [p. *353*, *507*]

*Nardus stricta* is a widespread xerophytic and glycophytic species in Europe, usually growing in open areas on sandy or peaty soils. In the *Manual* region, it is found in scattered locations from upper Michigan to Newfoundland and Greenland, and in Oregon and Idaho, where it is listed as a state noxious weed. The stiff, sharp leaves make it unpalatable; hence it tends to survive in areas of heavy grazing. This, combined with its broad ecological range, makes its potential for spreading in western rangelands a matter of concern.

## 8. DIARRHENEAE C.S. Campb.[2]

**Pl** per; rhz. **Clm** ann, not brchg above the base. **Shth** open, mrg not fused; **col** glab, without tufts of hair at the sides; **aur** smt present; **lig** stiff, scarious, ciliolate, those of the up and lo cauline lvs usu similar; **psdpet** absent; **bld** tapering both bas and apically, midveins usu eccentric, venation parallel, cross venation not evident; **x-sec** non-Kranz, without arm or fusoid cells; **epdm** without microhairs or with unicellular microhairs, cells not papillate. **Infl** tml pan. **Spklt** lat compressed, pedlt, with (2)3–5(7) flt, distal flt(s) rdcd and strl, smt concealed by the subtml flt; **rchl** not prolonged beyond the tml, strl flt; **dis** above the glm and beneath the flt. **Glm** 2, 1–5-veined, at least the up glm longer than $^1/_4$ the length of the adjacent flt; **flt** lat compressed; **cal** glab or with a few hairs, rounded; **lm** lanceolate, cartilaginous to thinly coriaceous, 3(5)-veined, veins inconspicuous, apc unawned, smt mucronate; **pal** from $^1/_2$ as long as to subequal to the lm, 2-veined; **lod** 2, memb, ciliate; **anth** (1)2(3); **sty** 2, bases free. **Car** obliquely ellipsoid, prcp thick, easily peeled away at maturity, forming a conspicuous knob or beak, sty not persistent; **hila** linear; **emb** $^1/_4$–$^1/_3$ as long as the fruits.

There are 1–2 genera in the *Diarrheneae*. The tribe is sometimes placed in the *Bambusoideae*, sometimes in the *Poöideae*.

## 8.01 DIARRHENA P. Beauv.[3]

**Pl** per; rhz, rhz 1.5–5 mm thick, scaly. **Clm** 48–131 cm tall, 1–3 mm thick, slender and arching, unbrchd, usu clumped, rarely solitary. **Lvs** bas concentrated or proximal; **shth** open, longer than the intnd, mrg narrowly hyaline, entire, smt ciliate; **col** cartilaginous, thickened, light green or yellowish, somewhat flared marginally; **aur** smt present; **lig** stiffly memb, rounded, ciliolate; **bld** flat, tapering bas, long-tapering apically, midveins usu eccentric. **Infl** pan, contracted, exserted, arching, rcm distally; **br** 1 or 2 per nd, ascending or appressed, terminating in a spklt. **Spklt** cylindrical when young, lat compressed at maturity, with (2)3–5(7) flt, distal flt rdcd and strl, smt including an additional rdmt flt; **dis** above the glm and beneath the flt. **Glm** unequal, chartaceous, lanceolate, glab, keeled, smt scabridulous near the keels distally, mrg entire or ciliate, apc acute; **lo glm** $^1/_3$–$^2/_3$ shorter than the up glm, less than $^1/_3$ as long as the adjacent lm, 1–3(5)-veined; **up glm** (3)5-veined; **cal** glab or with a few hairs, hairs about 0.5 mm; **lm** mostly chartaceous, veins 3, prominent, convergent, mrg hyaline, entire, smt ciliate, apc sharply cuspidate, cusps 1–2 mm; **pal** from $^1/_2$ as long as to slightly shorter than the lm, chartaceous, keeled, sides narrowly hyaline; **lod** about 1.5 mm, lanceolate to elliptic, apc ciliate; **anth** 2, yellow. **Car** prominently beaked, sty bases usu persistent, prcp loose, at least partially.

*Diarrhena* is an odd and distinctive genus whose relationships are not clear. Two of its approximately six species grow in the woodlands of eastern North America; the remainder occupy similar habitats in eastern Asia. Although *D. americana* and *D. obovata* grow in similar habitats and overlap in their ranges, no intermediates have been found.

[1,2]Mary E. Barkworth  [3]David M. Brandenburg

1. Calluses pubescent on all but the lowest mature lemma; lemma of the lowest floret in each spikelet
   (6)7.1–10.8 mm long, widest below the middle, tapering gradually to the apex; mature fruits 1.3–1.8
   mm wide, gradually tapering to a blunt beak . . . . . . . . . . . . . . . . . . . . . . . . . . . . . . . . . . . . . . .1. *D. americana*
1. Calluses glabrous on all mature lemmas; lemma of the lowest floret in each spikelet 4.6–7.5 mm long,
   widest near or above the middle, abruptly contracted to the apex; mature fruits 1.8–2.5 mm wide,
   abruptly contracted to a bottlenose-shaped beak . . . . . . . . . . . . . . . . . . . . . . . . . . . . . . . . . . . . .2. *D. obovata*

### 1. Diarrhena americana P. Beauv. AMERICAN BEAKGRAIN [p. 354, <u>507</u>]

*Diarrhena americana* is restricted to the United States, where it grows in rich, moist woods from Missouri to Maryland and south to Oklahoma and Alabama. Its range is primarily to the east of the range of *D. obovata*.

### 2. Diarrhena obovata (Gleason) Brandenburg OBOVATE BEAKGRAIN [p. 353, <u>508</u>]

*Diarrhena obovata* is restricted to the *Manual* region, growing in rich woodlands from South Dakota to Ontario and New York and south to Texas, Tennessee, and Virginia. It is most common in the prairie states.

## 9. MELICEAE Endl.[1]

Pl usu per, smt ann; csp, smt rhz. Clm ann, not wd, not brchg above the base; intnd hollow. Shth closed at least ¾ their length; col without tufts of hair on the sides; aur smt present; lig hyaline, glab, often lacerate, occ ciliate, those of the lo and up cauline lvs usu similar; psdpet absent; bld linear to narrowly lanceolate, venation parallel, cross venation smt evident; x-sec non-Kranz, without arm or fusoid cells; epdm without microhairs, smt papillate. Infl tml pan or rcm; dis above the glm and beneath the flt or below the glm. Spklt 2.5–60 mm, not viviparous, slightly to strongly lat compressed, with 1–30 flt, proximal flt bisx, distal 1–3 flt usu strl, smt pist, smt rdcd and amalgamated into a knob- or club-shaped *rudiment*; rchl prolonged beyond the base of the distal flt. Glm exceeded by the distal flt, shorter than to longer than the adjacent lm, mostly memb, scarious distally, 1–11-veined, apc usu rounded to acute; flt lat or dorsally compressed; cal blunt, glab or with hairs; lm of sx flt rectangular or ovate, mostly memb, scarious distally, often with a purplish band adjacent to the scarious apc, (4)5–15-veined, veins not converging distally, often prominent, unawned or awned, awns not brchd, apc entire to bilobed or bifid, awns straight, subtml or from the sinuses; pal from shorter than to longer than the lm, similar in texture, 2-veined, veins keeled, smt winged; lod 2, fleshy, usu connate into a single structure, without a memb wing, truncate, not ciliate, not or scarcely veined; anth 1, 2, or 3; ov glab; sty 2-brchd, bases persistent, br plumose distally. Car ovoid to ellipsoid, longitudinally grooved or not; hila usu linear; emb less than ⅓ as long as the car.

There are approximately 130 species and 8 or 9 genera in the *Meliceae*. *Melica* and *Glyceria*, the two largest genera, are well represented in North America. *Pleuropogon* and *Schizachne* are primarily North American, but extend into eastern Asia. Members of the tribe are most easily recognized by the combination of closed leaf sheaths, scarious lemma apices, and non-converging lemma veins.

1. Calluses hairy; lemmas awned, awns 8–15 mm long, twisted, divergent to slightly geniculate . . . . . . . . . . . 9.03 *Schizachne*
1. Calluses glabrous; lemmas unawned or awned, awns to 12 mm long, straight.
   2. Inflorescences racemes; palea keels winged, the wings notched and awned . . . . . . . . . . . . . . . . . . . . 9.04 *Pleuropogon*
   2. Inflorescences usually panicles, racemes in depauperate specimens; palea keels not winged or the
      wings entire and unawned.
      3. Lower glumes 1-veined, 0.3–4.5 mm long; disarticulation always above the glumes; lemmas
         unawned, never with hairs more than 1 mm long; culms never with cormous bases; distal florets
         in the spikelets sometimes reduced, not forming a morphologically distinct rudiment; plants of
         wet meadows and streamsides . . . . . . . . . . . . . . . . . . . . . . . . . . . . . . . . . . . . . . . . . . . . . . 9.01 *Glyceria*
      3. Lower glumes 1–9-veined, 2–16 mm long; disarticulation above or below the glumes; lemmas
         sometimes awned, sometimes with hairs longer than 1 mm; culms sometimes with cormous
         bases; distal florets in the spikelets often forming a morphologically distinct rudiment; plants of
         drier or well drained habitats . . . . . . . . . . . . . . . . . . . . . . . . . . . . . . . . . . . . . . . . . . . . . . . 9.02 *Melica*

## 9.01 GLYCERIA R. Br.[2]

Pl usu per, rarely ann; rhz. Clm (10)20–250 cm, erect or decumbent, freely rooting at the lo nd, not cormous based. Shth closed for at least ¾ their length, often almost entirely closed; lig scarious, erose to lacerate; bld flat or folded. Infl tml, usu pan, smt rcm in depauperate specimens, br appressed to divergent or reflexed. Spklt cylindrical and terete or oval and lat compressed, with 2–16 flt, tml flt in each spklt strl, red; dis above the glm, below the flt. Glm much smaller than to equaling the adjacent lm, 1-veined, obtuse or acute, often erose; lo glm 0.3–4.5 mm; up glm 0.6–7 mm; cal glab; lm memb to thinly coriaceous, rounded over the back, smooth or

[1]Mary E. Barkworth  [2]Mary E. Barkworth and Laurel K. Anderton

scabrous, glab or hairy, hairs to about 0.1 mm, 5–11-veined, veins usu evident, often prominent and rdgd, not or scarcely converging distally, apical mrg hyaline, smt with a purplish band below the hyaline portion, apc acute to rounded or truncate, entire, erose, or irregularly lobed, unawned; **pal** shorter than to longer than the lm, keeled, keels smt winged; **lod** thick, smt connate, not winged; **anth** (1)2–3; **ov** glab; **sty** 2-brchd, br divergent to recurved, plumose distally.

*Glyceria* includes approximately 35 species, all of which grow in wet areas. All but five species are native to the Northern Hemisphere. The genus is represented in the *Manual* region by 13 native and 4 introduced species, plus 3 named hybrids.

Culm thickness is measured near midlength of the basal internode; it does not include leaf sheaths. Ligule measure-ments reflect both the basal and upper leaves. Ligules of the basal leaves are usually shorter than, but similar in shape and texture to, those of the upper leaves. The number of spikelets on a branch is counted on the longest primary branches, and includes all the spikelets on the secondary (and higher order) branches of the primary branch. Pedicel lengths are measured for lateral spikelets on a branch. Lemma characteristics are based on the lowest lemmas of most spikelets in a panicle. There is often considerable variation within a panicle.

Tsvelev (2006) recognized four additional species in the region, *Glyceria davyi* Tzvelev, a segregate of *G. leptostachya*, *G. mexicana* (Kelso) Beetle and *G. neogaea* Steud, both segregates of *G. striata*; and *G. texana* Tzvelev, which resembles *G. septentrionalis* var. *arkansana*.

1. Spikelets laterally compressed, lengths 1–4 times widths, oval in side view; paleal keels not winged (sects. *Hydropoa* and *Striatae*).
    2. Upper glumes 2.5–5 mm long, longer than wide.
        3. Blades 3–7 mm wide; culms 2.5–4 mm thick, 60–90 cm tall; anthers 0.7–1.2 mm long . . . . . . . . . 2. *G. alnasteretum*
        3. Blades 6–20 mm wide; culms 6–12 mm thick, 60–250 cm tall; anthers (1)1.2–2 mm long . . . . . . . . . . . 3. *G. maxima*
    2. Upper glumes 0.6–3.7 mm long, if longer than 3 mm, then shorter than wide.
        4. Panicles ovoid to linear; panicle branches appressed to strongly ascending; ligules of the upper leaves 0.5–0.9 mm long.
            5. Panicles 5–15 cm long, 2.5–6 cm wide, ovoid, erect . . . . . . . . . . . . . . . . . . . . . . . . . . . . . . . . . 4. *G. obtusa*
            5. Panicles 15–25 cm long, 0.8–1.5 cm wide, linear, nodding . . . . . . . . . . . . . . . . . . . . . . . . . . . 5. *G. melicaria*
        4. Panicles pyramidal; panicle branches strongly divergent or drooping; ligules of the upper leaves 1–7 mm long.
            6. Lemma apices almost flat; anthers 3; veins of 1 or both glumes in each spikelet usually extending to the apices . . . . . . . . . . . . . . . . . . . . . . . . . . . . . . . . . . . . . . . . . . . . . 1. *G. grandis*
            6. Lemma apices prow-shaped; anthers 2; veins of both glumes terminating below the apices.
                7. Glumes tapering from below midlength to the narrowly acute (< 45°) apices; lemma lengths more than twice widths . . . . . . . . . . . . . . . . . . . . . . . . . . . . . . . . . . . . . . . . 6. *G. nubigena*
                7. Glumes narrowing from midlength or above to the acute (≥ 45°) or rounded apices; lemma lengths less than twice widths.
                    8. Spikelets (2.5)3–5 mm wide; lemma veins evident but not raised distally; palea lengths 1.5–1.8 times widths . . . . . . . . . . . . . . . . . . . . . . . . . . . . . . . . . . . 10. *G. canadensis*
                    8. Spikelets 1.2–2.9 mm wide; lemma veins distinctly raised throughout; palea lengths 1.5–3.5 times widths.
                        9. Lemmas 2.5–3.5 mm long; glume lengths about 3 times widths, glume apices broadly acute; lower glumes 1.5–2 mm long; upper glumes 2–2.6 mm long . . . . . . . . . . . 7. *G. pulchella*
                        9. Lemmas 1.2–2.2 mm long; glume lengths up to twice widths, glume apices rounded or acute; lower glumes 0.5–1.5 mm long; upper glumes 0.6–1.5 mm long.
                            10. Blades 2–6 mm wide; anthers 0.2–0.6 mm long; culms 1.5–3.5 mm thick . . . . . . . . . . . 8. *G. striata*
                            10. Blades 6–15 mm wide; anthers 0.5–0.8 mm long; culms 2.5–8 mm thick . . . . . . . . . . . . . 9. *G. elata*
1. Spikelets cylindrical and terete, except at anthesis when slightly laterally compressed, lengths more than 5 times widths, rectangular in side view; paleal keels usually winged distally (sect. *Glyceria*).
    11. Lemmas tapering from near midlength to the acuminate or narrowly acute apices; paleas exceeding the lemmas by 0.7–3 mm; palea apices often appearing bifid, the teeth 0.4–1 mm long . . . . . 13. *G. acutiflora*
    11. Lemmas not tapered or tapering only in the distal ¼, apices truncate, rounded, or acute; paleas shorter or to 1(1.5) mm longer than the lemmas; palea apices not or shortly bifid, the teeth to 0.5 mm long.
        12. Lemma apices with 1 strongly developed lobe on 1 or both sides, entire to crenulate between the lobes; blades 3–12 cm long; primary panicle branches 1.5–9.5 cm long . . . . . . . . . . . . . . . . . 17. *G. declinata*
        12. Lemma apices not or more or less evenly lobed; blades 5–30 cm long; primary panicle branches 3–18 cm long.
            13. Lemmas 5–8 mm long.
                14. Anthers 0.6–1.6 mm long; lemma apices usually slightly lobed or irregularly crenate . . . . . . . . . . . . . . . . . . . . . . . . . . . . . . . . . . . . . . . . . . . . . . 15. *G. ×occidentalis* (in part)

14. Anthers 1.5–3 mm long; lemma apices usually entire . . . . . . . . . . . . . . . . . . . . . . . . . . . 16. *G. fluitans*
13. Lemmas 2.4–5 mm long.
   15. Lemmas usually smooth between the veins, if scabridulous the prickles between the
       veins smaller than those over the veins.
      16. Lemmas usually acute, sometimes obtuse, entire or almost so; adaxial
         surfaces of the midcauline blades usually densely papillose, glabrous . . . . . . . . . . . . . 11. *G. borealis*
      16. Lemmas truncate to obtuse, crenate; adaxial surfaces of the midcauline blades
         rarely densely papillose, sometimes sparsely hairy.
         17. Culms 73–182 cm tall; pedicels 0.7–1.7 mm . . . . . . . . . . . . . . . 12. *G. septentrionalis* (in part)
         17. Culms 25–80 cm tall; pedicels 1–6 mm . . . . . . . . . . . . . . . . . . . . . . . . . . 18. *G. notata*
   15. Lemmas scabridulous or hispidulous between the veins, the prickles between the
       veins similar in size to those over the veins.
      18. Lemma apices acute.
         19. Lemmas 2.4–4.8 mm long; pedicels 0.7–1.7 mm long; plants from
            east of the Rocky Mountains . . . . . . . . . . . . . . . . . . . . . . 12. *G. septentrionalis* (in part)
         19. Lemmas 4.5–5.9 mm long; pedicels 1.5–8 mm long; plants from
            west of the Rocky Mountains . . . . . . . . . . . . . . . . . . . . . . 15. *G.* ×*occidentalis* (in part)
      18. Lemma apices truncate to obtuse.
         20. Pedicels 0.7–1.7 mm long; anthers 0.5–1.8 mm long; plants from
            east of the Rocky Mountains . . . . . . . . . . . . . . . . . . . . . . 12. *G. septentrionalis* (in part)
         20. Pedicels 2–5 mm long; anthers 0.3–0.9 mm long; plants from
            British Columbia and the Pacific states . . . . . . . . . . . . . . . . . . . . . . . . . 14. *G. leptostachya*

## *Glyceria* sect. **Hydropoa** (Dumort.) Dumort.

*Glyceria* sect. *Hydropoa* includes approximately five species. Three species grow in the *Manual* region; one is introduced. They grow along streams and at the edges of lakes and ponds.

### 1. Glyceria grandis S. Watson AMERICAN GLYCERIA, AMERICAN MANNAGRASS [p. *354*, 508]

*Glyceria grandis* grows throughout most of the *Manual* region on banks and in the water of streams, ditches, ponds, and wet meadows. It is similar to *G. maxima*, differing primarily in its shorter, flatter lemmas and shorter anthers. It is also confused with *G. elata* and *Torreyochloa pallida*. It differs from the former in having acute glumes with long veins, more evenly dark florets, flatter lemma apices, and paleal keel tips that do not point towards each other, and from the latter in its closed leaf sheaths and 1-veined glumes.

1. Spikelets 3.2–6.4 mm long, with 4–8 florets . . . . . . . var. *grandis*
1. Spikelets 6–10 mm long, with 5–10 florets . . . . . . var. *komarovii*

### Glyceria grandis S. Watson var. grandis GIANT GLYCERIA, GIANT MANNAGRASS, GLYCÉRIE GÉANTE [p. *354*]

*Glyceria grandis* var. *grandis* is the more widespread of the two varieties, growing throughout the range of the species.

### Glyceria grandis var. komarovii Kelso [p. *354*]

*Glyceria grandis* var. *komarovii* is restricted to Alaska and Yukon Territory.

### 2. Glyceria alnasteretum Kom. ALEUTIAN GLYCERIA [p. *354*, 508]

*Glyceria alnasteretum* is included in this treatment with some hesitation, based on two specimens collected at Signal Point, Attu Island, Alaska in 1945. Further investigation is called for; the habitat of the Attu Island plants was unusual for the species.

### 3. Glyceria maxima (Hartm.) Holmb. TALL GLYCERIA, ENGLISH WATERGRASS, GLYCÉRIE AQUATIQUE [p. *354*, 508]

*Glyceria maxima* is native to Eurasia. It grows in wet areas, including shallow water, at scattered locations in the *Manual* region. It is an

excellent fodder grass, and may have been planted deliberately at one time. At some sites, the species appears to be spreading, largely vegetatively. It differs from *G. grandis* in its firmer, more prow-tipped lemmas as well as its larger lemmas and usually larger anthers.

## *Glyceria* sect. **Striatae** G.L. Church

Members of *Glyceria* sect. *Striatae* grow along streams, in swamps, and in shallow, fresh water. The section includes seven species, all of which are native to the *Manual* region.

### 4. Glyceria obtusa (Muhl.) Trin. ATLANTIC MANNAGRASS [p. *354*, 508]

*Glyceria obtusa* is a distinctive species that grows in wet woods, swamps, and shallow waters, primarily on the eastern seaboard of North America, from Nova Scotia and New Brunswick to South Carolina.

### 5. Glyceria melicaria (Michx.) F.T. Hubb. MELIC MANNAGRASS, GLYCÉRIE MÉLICAIRE [p. *354*, 508]

*Glyceria melicaria* grows in swamps and wet soils. Its range extends from southeastern Ontario east to Nova Scotia, south to Illinois and the northeastern United States and, in the Appalachian Mountains, to northern Georgia. **Glyceria** ×**gatineauensis** Bowden is a sterile hybrid between *G. melicaria* and *G. striata*.

### 6. Glyceria nubigena W.A. Anderson GREAT SMOKY MOUNTAIN MANNAGRASS, GREAT SMOKY MOUNTAIN GLYCERIA [p. *354*, 508]

*Glyceria nubigena* is known only from moist areas of balds and high ridges in the Great Smoky Mountains of North Carolina and Tennessee.

### 7. Glyceria pulchella (Nash) K. Schum. BEAUTIFUL GLYCERIA, MACKENZIE VALLEY MANNAGRASS [p. *354*, 508]

*Glyceria pulchella* grows in marshes, muskegs, ponds, and ditches, from central Alaska and the Northwest Territories to southern British Columbia and central Manitoba. It resembles *G. striata* and *G. elata*, differing in having somewhat stiffer and straighter panicle branches, plus larger spikelets and florets.

**8. Glyceria striata** (Lam.) Hitchc. RIDGED GLYCERIA, GLYCÉRIE STRIÉE [p. 355, <u>508</u>]

*Glyceria striata* grows in bogs, along lakes and streams, and in other wet places. Its range extends from Alaska to Newfoundland and south into Mexico. Larger specimens are easy to confuse with *G. elata*. The differences between the two are evident in the field; they are not always evident on herbarium specimens. *Glyceria striata* also resembles *G. pulchella*, but it has somewhat more lax panicle branches in addition to smaller spikelets and florets.

*Glyceria* ×*gatineauensis* Bowden is a sterile hybrid between *G. striata* and *G. melicaria*. It resembles *G. melicaria* but has longer, less appressed panicle branches, and is a triploid. It was described from a population near Eardley, Quebec.

*Glyceria* ×*ottawensis* Bowden is a sterile hybrid between *G. striata* and *G. canadensis*. Known only from the original populations near Ottawa, it is intermediate between the two parents.

**9. Glyceria elata** (Nash) M.E. Jones TALL MANNAGRASS [p. 355, <u>508</u>]

*Glyceria elata* grows in wet meadows and shady moist woods, from British Columbia east to Alberta and south to California and New Mexico. It is probably introduced in Georgia, and is not known from Mexico. Although similar to and sometimes found with *G. striata*, the two species show no evidence of hybridization. Their differences in growth habit and stature are evident in the field.

*Glyceria elata* differs from *G. grandis* in having rounded glumes with veins that terminate below the apices, more readily disarticulating florets, and greener lemmas with more prow-shaped apices, as well as in having paleal tips that point towards each other. It also resembles *G. pulchella*, but has somewhat more lax panicle branches than that species, in addition to smaller spikelets and florets.

**10. Glyceria canadensis** (Michx.) Trin. [p. 355, <u>508</u>]

*Glyceria canadensis* is an attractive native species that grows in swamps, bogs, lakeshore marshes, and wet woods throughout much of eastern North America. It is now established in western North America, having been introduced as a weed in cranberry farms. Sterile hybrids with *G. striata* are called *G.* ×*ottawensis* Bowden.

1. Lemmas 2.4–4 mm long; spikelets 5–8 mm long, with 4–10 florets; lower glumes 1.6–2.4mm long; upper glumes acute . . . . . . . . . . . . . . . . . . . . var. *canadensis*
1. Lemmas 1.8–2.5 mm long; spikelets 3–5 mm long, with 2–5 florets; lower glumes 0.6–1.3 mm long; upper glumes usually rounded, sometimes acute . . . . . . var. *laxa*

**Glyceria canadensis** (Michx.) Trin. var. **canadensis**
CANADIAN GLYCERIA, CANADIAN MANNAGRASS, RATTLESNAKE MANNAGRASS, GLYCÉRIE DU CANADA [p. 355]

*Glyceria canadensis* var. *canadensis* grows throughout the range of the species. The spikelets bear some resemblance to those of *Bromus briziformis*, otherwise known as rattlesnake brome, hence the vernacular name "rattlesnake mannagrass".

**Glyceria canadensis** var. **laxa** (Scribn.) Hitchc. LIMP MANNAGRASS [p. 355]

*Glyceria canadensis* var. *laxa* grows in swamps, bogs, and wet woods, primarily along the eastern seaboard of North America from Nova Scotia to northeastern Tennessee.

## Glyceria R. Br. sect. Glyceria

*Glyceria* sect. *Glyceria* includes about 15 species. Seven species grow in the *Manual* region, three of which are introduced. In addition, there is one named hybrid. They grow in and beside shallow, still or slowly moving fresh water, such as along the edges of lakes and ponds and in low areas in wet meadows.

**11. Glyceria borealis** (Nash) Batch. BOREAL GLYCERIA, BOREAL MANNAGRASS, GLYCÉRIE BOREALE [p. 355, <u>508</u>]

*Glyceria borealis* is a widespread native species that grows in the northern portion of the *Manual* region, extending through the western mountains into northern Mexico. It grows along the edges and muddy shores of freshwater streams, lakes, and ponds. In the southern portion of its range, it is restricted to subalpine and alpine areas. It differs from *G. notata* in having acute lemmas and, usually, densely papillose midcauline leaves.

**12. Glyceria septentrionalis** Hitchc. NORTHERN GLYCERIA, NORTHERN MANNAGRASS, GLYCÉRIE SEPTENTRIONALE [p. 355, <u>508</u>]

*Glyceria septentrionalis* is native and restricted to eastern North America. It grows in shallow water or very wet soils. It resembles *G. notata* in its rather short, truncate to rounded lemmas, but tends to have fewer spikelets on its branches. In addition, the veins of its leaf sheaths appear completely smooth, even under high magnification. That said, many specimens will be hard to identify if their provenance is not known.

1. Lemmas hispidulous over the veins, hairs about 0.1 mm long . . . . . . . . . . . . . . . . . . . . . . . . . . . . var. *arkansana*
1. Lemmas scabrous over the veins, prickles about 0.05 mm long . . . . . . . . . . . . . . . . . . . . . . . var. *septentrionalis*

**Glyceria septentrionalis var. arkansana** (Fernald) Steyerm. & Kučera [p. 355]

*Glyceria septentrionalis* var. *arkansana* grows in roadside ditches and on the edges of swamps, lakes, and ponds in the flood plain of the Mississippi River, from southern Illinois and Indiana to the Gulf coast. There is also one record from central Tennessee. *Glyceria texana* Tsvelev resembles this taxon (Tsvelev 2006).

**Glyceria septentrionalis** Hitchc. var. **septentrionalis** [p. 355]

*Glyceria septentrionalis* var. *septentrionalis* grows throughout the range of the species, but is less common in the lower floodplain of the Mississippi River and Kentucky than var. *arkansana*. It is found in shallow water or wet soils.

**13. Glyceria acutiflora** Torr. CREEPING MANNAGRASS [p. 355, <u>508</u>]

*Glyceria acutiflora* grows in wet soils and shallow water of the northeastern United States. Its long paleas make *G. acutiflora* the most distinctive North American species of sect. *Glyceria*.

**14. Glyceria leptostachya** Buckley NARROW MANNAGRASS [p. 355, <u>508</u>]

*Glyceria leptostachya* grows in swamps and along the margins of streams and lakes, on the western side of the coastal mountains from southern Alaska to San Francisco Bay. It is similar to the European *Glyceria notata*, differing primarily in its tendency to have fewer spikelets on its branches. *Glyceria davyi* Tsvelev resembles this taxon (Tsvelev 2006).

**15. Glyceria ×occidentalis** (Piper) J.C. Nelson WESTERN MANNAGRASS [p. 355, <u>508</u>]

*Glyceria* ×*occidentalis* has hitherto been considered an uncommon native species that grows along lakes and streams and in marshy areas of western North America. It differs from other native species in the region except for *G. fluitans* in its longer lemmas and anthers. *G. fluitans* is one of its parents.

**16. Glyceria fluitans** (L.) R. Br. WATER MANNAGRASS, GLYCÉRIE FLOTTANTE [p. 355, <u>508</u>]

*Glyceria fluitans* is a Eurasian species. It has been collected from British Columbia to California on the west coast, in South Dakota, and from Newfoundland to Pennsylvania on the east coast. In Europe, it grows in rich, organic, wet soils, often near *G. notata*, with which it hybridizes. In western North America, it has been confused with *G. ×occidentalis*, from which it differs in its longer lemmas and anthers.

### 17. Glyceria declinata Bréb. LOW GLYCERIA [p. 356, 508]

*Glyceria declinata* is a European species that is established in many parts of the *Manual* region. It is invading vernal pools in California. *Glyceria declinata* has been confused with *G. ×occidentalis*. The most reliable distinguishing characteristics are the lateral lemma lobes of *G. declinata* and its rather short, straight panicle branches.

### 18. Glyceria notata Chevall. MARKED GLYCERIA [p. 356, 508]

*Glyceria notata* is a Eurasian species that has been reported from scattered locations in the *Manual* region; the reports have not been verified. In Europe, *G. notata* grows in rich, organic, wet soils, often near *G. fluitans*, with which it hybridizes. It is more tolerant of trampling than *G. fluitans*.

---

## 9.02 MELICA L.[1]

**Pl** per; csp or soboliferous, not or only shortly rhz. **Clm** (4)9–250 cm, smt forming a bas corm; **nd** and **intnd** usu glab. **Shth** closed almost to the top; **aur** smt present; **lig** thinly memb, erose to lacerate, usu glab, those of the lo lvs shorter than those of the up lvs; **bld** flat or folded, glab or hairy, particularly on the adx surfaces, smt scabrous. **Infl** tml pan; **pri br** often appressed; **sec br** appressed or divergent; **ped** either more or less straight or sharply bent below the spklt, scabrous to strigose distally; **dis** below the glm in species with sharply bent ped, above the glm in other species. **Spklt** with 1–7 bisx flt, terminating in a strl structure, the *rdmt*, composed of 1–4 strl flt; **rdmt** smt morphologically distinct from the bisx flt, smt similar but smaller. **Glm** memb or chartaceous, distal mrg wide, translucent; **lo glm** 1–9-veined; **up glm** 1–11-veined; **cal** glab; **lm** memb bas, smt becoming coriaceous at maturity, glab or with hairs, (4)5–15-veined, usu unawned, smt awned, awns to 12 mm, straight; **pal** from ½ as long as to almost equaling the lm, keels usu ciliate; **lod** fused into a single, col-like structure extending ½–⅔ around the base of the ov; **anth** (2)3. **Car** usu 2–3 mm, smooth, glab, longitudinally furrowed, falling from the flt when mature.

*Melica* includes approximately 80 species, which grow in all temperate regions of the world except Australia, usually in shady woodlands on dry stony slopes. The species are relatively nutritious, but are rarely sufficiently abundant to be important as forage.

Nineteen species grow in the *Manual* region and two European species are grown as ornamentals. Many of the seventeen native species merit such use.

In the following key and descriptions, unless otherwise stated, comments on the panicle branches apply to the longest branches within the panicle; glume widths are measured from side to side, at the widest portion; lemma descriptions are for the lowest floret in the spikelets; and rachilla internode comments apply to the lowest internode in the spikelets.

1. Spikelets disarticulating below the glumes; pedicels sharply bent just below the spikelets.
    2. Lemmas with hairs.
        3. Lemmas with hairs on the lower portion of the lemmas, the hairs twisted . . . . . . . . . . . . . . . . . . 15. *M. montezumae*
        3. Lemmas with hairs on the marginal veins, the hairs not twisted . . . . . . . . . . . . . . . . . . . . . . . . 18. *M. ciliata*
    2. Lemmas glabrous, sometimes scabridulous to scabrous.
        4. Rudiments acute to acuminate, similar to but smaller than the bisexual florets.
            5. Spikelets broadly V-shaped when mature, 5–13 mm wide; upper glumes 6–18 mm long . . . . . . . . . . . 13. *M. stricta*
            5. Spikelets parallel-sided when mature, 1.5–5 mm wide; upper glumes 5–8 mm long . . . . . . . . . . . . . 14. *M. porteri*
        4. Rudiments clublike, not resembling the bisexual florets.
            6. Rudiments at an angle to the rachilla; panicle branches with 2–5 spikelets . . . . . . . . . . . . . . . . . . . 16. *M. mutica*
            6. Rudiments in a straight line with the rachilla; panicle branches with 5–20 spikelets.
                7. Panicle branches often divergent to reflexed; glumes unequal, lower glumes shorter and more ovate than the upper glumes . . . . . . . . . . . . . . . . . . . . . . . . . . . . . . . . . . . . . . . . . . . . . . . . . 17. *M. nitens*
                7. Panicle branches strongly ascending to appressed; glumes subequal in length and similar in shape . . . . . . . . . . . . . . . . . . . . . . . . . . . . . . . . . . . . . . . . . . . . . . . . . . . . . . . . . . . . . . . 19. *M. altissima*
1. Spikelets disarticulating above the glumes; pedicels more or less straight.
    8. Rudiments truncate to acute, not resembling the lowest florets.
        9. Bisexual florets 1(2); paleas almost as long as the lemmas.
            10. Rudiments shorter than the terminal rachilla internode; bisexual lemmas scabridulous, sometimes hairy . . . . . . . . . . . . . . . . . . . . . . . . . . . . . . . . . . . . . . . . . . . . . . . . . . . . . . . . 1. *M. torreyana*
            10. Rudiments longer than the terminal rachilla internode; bisexual lemmas glabrous, sometimes scabrous . . . . . . . . . . . . . . . . . . . . . . . . . . . . . . . . . . . . . . . . . . . . . . . . . . . . . . . . . 2. *M. imperfecta*
        9. Bisexual florets 2–7; paleas ½–¾ the length of the lemmas.
            11. Culm bases not forming distinct corms . . . . . . . . . . . . . . . . . . . . . . . . . . . . . . . . . . . . . . . . 6. *M. californica*
            11. Culm bases forming distinct corms.
                12. Glumes usually less than ½ as long as the spikelets; ligules 0.1–2 mm long; corms connected to the rhizomes by a rootlike structure . . . . . . . . . . . . . . . . . . . . 3. *M. spectabilis* (in part)
                12. Glumes from (½)⅔ as long as to equaling the spikelets; ligules 2–6 mm long; corms almost sessile on the rhizomes . . . . . . . . . . . . . . . . . . . . . . . . . . . . . . . . . . . 4. *M. bulbosa* (in part)

[1]Mary E. Barkworth

8. Rudiments tapering, smaller than but otherwise similar to the lowest florets in shape.
　13. Lemmas awned.
　　14. Awns shorter than 3 mm.
　　　15. Panicle branches appressed; lemmas usually with 0.7–1.3 mm hairs on the margins . . . . . . . . . . . . . . . . . . . . . . . . . . . . . . . . . . . . . . . . . . . . . 7. *M. harfordii* (in part)
　　　15. Panicle branches widespread to reflexed; lemmas glabrous . . . . . . . . . . . . . . . . . . . . . 8. *M. geyeri* (in part)
　　14. Awns 3–12 mm long.
　　　16. Panicle branches 4–6 cm long, appressed or ascending; blades 2–6 mm wide . . . . . . . . . . . . . . 9. *M. aristata*
　　　16. Panicle branches 7–11 cm long, spreading to reflexed; blades 5–12 mm wide . . . . . . . . . . . . 10. *M. smithii*
　13. Lemmas unawned.
　　17. Lemmas strongly tapering and acuminate, the veins usually hairy . . . . . . . . . . . . . . . . . . . . . . . 11. *M. subulata*
　　17. Lemmas acute to obtuse, the veins hairy or not.
　　　18. Lemmas pubescent, the hairs on the marginal veins clearly longer than the hairs elsewhere . . . . . . . . . . . . . . . . . . . . . . . . . . . . . . . . . . . . . . . . . . . . . . 7. *M. harfordii* (in part)
　　　18. Lemmas glabrous, scabrous, or pubescent, never with clearly longer hairs on the marginal veins.
　　　　19. Rachilla internodes swollen when fresh, wrinkled when dry . . . . . . . . . . . . . . . . . . . . . 12. *M. fugax*
　　　　19. Rachilla internodes not swollen when fresh, not wrinkled when dry.
　　　　　20. Panicle branches with 5–15 spikelets; paleas about ¹/₂ as long as the lemmas; culms not forming corms . . . . . . . . . . . . . . . . . . . . . . . . . . . . . . . . 5. *M. frutescens*
　　　　　20. Panicle branches with 1–6 spikelets; paleas from ²/₃ as long as to equaling the lemmas; culms forming corms.
　　　　　　21. Panicle branches 3–11 cm long, divergent to reflexed, flexuous; lowest rachilla internodes 2–3 mm long . . . . . . . . . . . . . . . . . . . . . . . 8. *M. geyeri* (in part)
　　　　　　21. Panicle branches 2–6.5 cm long, usually appressed to ascending, straight, sometimes strongly divergent and flexuous; lowest rachilla internodes 1–2 mm long.
　　　　　　　22. Ligules 0.1–2 mm long; glumes usually less than ¹/₂ the length of the spikelets; corms not attached directly to the rhizomes . . . . . 3. *M. spectabilis* (in part)
　　　　　　　22. Ligules 2–6 mm long; glumes from (¹/₂) ²/₃ as long as to equaling the spikelets; corms almost sessile, directly attached to the rhizomes . . 4. *M. bulbosa* (in part)

## 1. Melica torreyana Scribn. Torrey's Melic [p. 356, 508]

*Melica torreyana* grows from sea level to 1200 m, in thickets and woods in California. It is common throughout chaparral areas and coniferous forests but, on serpentine soils, grows only in shady locations. The shape and size of the rudiments make *M. torreyana* unique among the species found in North America.

## 2. Melica imperfecta Trin. Little California Melic [p. 356, 508]

*Melica imperfecta* grows from sea level to 1500 m, on stable coastal dunes, dry, rocky slopes, and in open woods, from California and southern Nevada south to Baja California, Mexico. Plants vary with respect to size, panicle shape, and pubescence, but no infraspecific taxa merit recognition.

## 3. Melica spectabilis Scribn. Purple Oniongrass [p. 356, 508]

*Melica spectabilis* grows in moist meadows, flats, and open woods, from 1200–2600 m, primarily in the Pacific Northwest and the Rocky Mountains. It is often confused with *M. bulbosa*, differing in its shorter glumes, "tailed" corm, and the more marked and evenly spaced purplish bands of its spikelets.

## 4. Melica bulbosa Geyer *ex* Porter & J.M. Coult. Oniongrass [p. 356, 508]

*Melica bulbosa* grows from 1370–3400 m, mostly in open woods on dry, well-drained slopes and along streams. It is restricted to the western half of the *Manual* region. It differs from *M. spectabilis* in its sessile corm and longer glumes. In addition, in *M. bulbosa* the

spikelets have purplish bands which appear to be concentrated towards the apices; in *M. spectabilis* the bands appear more regularly spaced. It differs from *M. californica* in its more narrowly acute spikelets, more strongly colored lemmas, and lack of corms, and from *M. fugax* in not having swollen rachilla internodes.

## 5. Melica frutescens Scribn. Woody Melic [p. 356, 508]

*Melica frutescens* grows from 300–1500 m in the dry hills and canyons of southern California, Arizona, and adjacent Mexico.

## 6. Melica californica Scribn. California Melic [p. 356, 508]

*Melica californica* grows from sea level to 2100 m, in a wide range of habitats, from dry, rocky, exposed hillsides to moist woods. It differs from *M. bulbosa* in its more obtuse spikelets and less strongly colored lemmas, as well as in not having corms.

## 7. Melica harfordii Bol. Harford Melic [p. 356, 509]

*Melica harfordii* grows primarily in the Pacific coast ranges from Washington to California, as well as in the Sierra Nevada and a few other inland locations, usually on dry slopes or in dry, open woods. The awns often escape attention because they do not always extend beyond the lemma.

## 8. Melica geyeri Munro Geyer's Oniongrass [p. 356, 509]

*Melica geyeri* grows to 2000 m, primarily in dry, open woods, in Oregon and California. Its large size and open panicle distinguish *M. geyeri* from most other North American species of *Melica*.

1. Lemma apices awned, awns 0.5–2 mm long . . . . . . var. *aristulata*
1. Lemma apices unawned . . . . . . . . . . . . . . . . . . . . . . . var. *geyeri*

**Melica geyeri** var. **aristulata** J.T. Howell [p. 356]

*Melica geyeri* var. *aristulata* grows in Marin, and possibly Shasta, counties in California.

**Melica geyeri** Munro var. **geyeri** [p. 356]

*Melica geyeri* var. *geyeri* grows throughout the range of the species.

**9. Melica aristata** Thurb. *ex* Bol. AWNED MELIC [p. 356, 509]

*Melica aristata* grows from 1000–3000 m in open fir and pine woods. It is restricted to the *Manual* region, being native on the west coast. It has also been found as an introduction in Kentucky. *Melica aristata* is easily distinguished from most species of *Melica* by its conspicuous awns.

**10. Melica smithii** (Porter *ex* A. Gray) Vasey SMITH'S MELIC [p. 356, 509]

*Melica smithii* grows in cool, moist woods from British Columbia and Alberta south to Oregon and Wyoming and, as a disjunct, from the Great Lakes region to western Quebec. It often forms colonies in the eastern portion of its range.

**11. Melica subulata** (Griseb.) Scribn. ALASKAN ONIONGRASS, TAPERED ONIONGRASS [p. 356, 509]

*Melica subulata* grows from sea level to 2300 m in mesic, shady woods. Its range extends from the Aleutian Islands of Alaska through British Columbia to California, east to South Dakota and Colorado.

**12. Melica fugax** Bol. LITTLE MELIC [p. 357, 509]

*Melica fugax* grows at elevations to 2200 m on dry, open flats, hillsides, and woods, from British Columbia to California and east to Idaho and Nevada. It is usually found on soils of volcanic origin. *Melica fugax* is often confused with *M. bulbosa*, but its rachilla internodes are unmistakable, being swollen when fresh and wrinkled when dry.

**13. Melica stricta** Bol. ROCK MELIC [p. 357, 509]

*Melica stricta* grows from 1200–3350 m on rocky, often dry slopes, sometimes in alpine habitats. Its range extends from Oregon and California to Utah. Two varieties are recognized, more on their marked geographical separation than on their morphological divergence.

1. Paleas about $^3/_4$ the length of the lemmas; anthers 2–3 mm long . . . . . . . . . . . . . . . . . . . . . . . . . . . var. *albicaulis*
1. Paleas about $^1/_2$ the length of the lemmas; anthers 1–2 mm long . . . . . . . . . . . . . . . . . . . . . . . . . . . . . var. *stricta*

**Melica stricta** var. **albicaulis** Boyle [p. 357]

*Melica stricta* var. *albicaulis* is restricted to the mountains of southern California.

**Melica stricta** Bol. var. **stricta** [p. 357]

*Melica stricta* var. *stricta* is more widespread than var. *albicaulis*, growing throughout the range of the species except in the mountains of southern California.

**14. Melica porteri** Scribn. PORTER'S MELIC [p. 357, 509]

*Melica porteri* grows on rocky slopes and in open woods, often near streams. It grows from Colorado and Arizona to central Texas and northern Mexico. Living plants are sometimes confused with *Bouteloua curtipendula*; the similarity is superficial.

1. Panicle branches flexible, ascending to strongly divergent; glumes purplish-tinged . . . . . . . . . . . . . . . . var. *laxa*
1. Panicle branches straight, appressed; glumes green or pale . . . . . . . . . . . . . . . . . . . . . . . . . . . . . var. *porteri*

**Melica porteri** var. **laxa** Boyle [p. 357]

*Melica porteri* var. *laxa* grows from southern Arizona east to the Chisos Mountains, Texas, and south to northern Mexico.

**Melica porteri** Scribn. var. **porteri** [p. 357]

*Melica porteri* var. *porteri* grows from northern Colorado to Arizona and central Texas, and south to the Sierra Madre Occidental, Mexico.

**15. Melica montezumae** Piper MONTEZUMA MELIC [p. 357, 509]

*Melica montezumae* grows primarily in shady locations in the mountains of western Texas and adjacent Mexico.

**16. Melica mutica** Walter TWO-FLOWER MELIC [p. 357, 509]

*Melica mutica* grows in moist or dry areas in open woods and thickets, from Iowa and Texas east to Maryland and Florida. It is unique among the North American species in having a clublike rudiment at a sharp angle to the rachilla.

**17. Melica nitens** (Scribn.) Nutt. *ex* Piper THREE-FLOWER MELIC [p. 357, 509]

*Melica nitens* grows in dry to moist woodlands, often in rocky areas with rich soil. It grows primarily from Minnesota to Pennsylvania and southwest to Texas.

**18. Melica ciliata** L. CILIATE MELIC, SILKY-SPIKE MELIC, HAIRY MELIC [p. 357]

*Melica ciliata* is grown as an ornamental in North America and is not known to have escaped. It is native to Europe, northern Africa, and southwestern Asia, where it grows on damp to somewhat dry soils.

**19. Melica altissima** L. TALL MELIC, SIBERIAN MELIC [p. 357, 509]

*Melica altissima* is native to Eurasia. It is grown as an ornamental in North America and is reported to have escaped and become established in Oklahoma and Ontario. In its native region, it grows in the moist soils of shrubby thickets and forest edges, and on rocky slopes.

---

# 9.03 SCHIZACHNE Hack.[1]

**Pl** per; loosely csp. **Clm** 30–110 cm, glab, often decumbent at the base; **nd** glab, becoming dark. **Shth** closed almost to the top; **lig** memb, mrg often united; **bld** folded or loosely involute, glab or pilose. **Infl** pan or rcm, with 4–20 spklt; **br** straight and appressed to lax and drooping. **Spklt** slightly lat compressed, with 3–6 flt; **dis** above the glm and beneath the flt. **Glm** exceeded by the lowest lm in each spklt, chartaceous, often anthocyanic below, the up $^1/_3$ hyaline; **cal** rounded, with hairs; **lm** chartaceous, slightly scabrous, 7–9-veined, veins parallel, conspicuous, apc scarious, bifid, awned from below the teeth, awns 8–15 mm, divergent or

[1]Jacques Cayouette and Stephen J. Darbyshire

slightly geniculate; **pal** shorter than the lm, 2-veined, veins ciliate, keeled; **lod** truncate; **anth** 3; **ov** glab. **Car** 3.2–3.8 mm, smooth, shiny, falling free of the lm and pal.

### 1. Schizachne purpurascens (Torr.) Swallen FALSE MELIC, SCHIZACHNÉ POURPRÉ [p. *358*, *509*]

In North America, *Schizachne purpurascens* grows in moist to mesic woods, from south of the tree line in Alaska and northern Canada

*Schizachne* is a unispecific genus that extends across North America in boreal regions and southwards in the montane areas. It also grows from the Ural Mountains of Russia to Kamchatka and Japan.

through the Rocky Mountains to New Mexico in the west, and to Kentucky and Maryland in the east.

## 9.04 PLEUROPOGON R. Br.[1]

**Pl** ann or per; csp or rhz. **Clm** 5–160 cm, erect or geniculate at the base, glab; **bas brchg** exvag. **Shth** closed almost to the top; **lig** memb; **bld** flat to folded, adx surfaces with prominent midribs. **Infl** tml, rcm, rarely pan. **Spklt** lat compressed, with 5–20(30) flt, up flt red; **dis** above the glm and beneath the flt. **Glm** unequal to subequal, shorter than the adjacent lm, memb to subhyaline, mrg scarious; **lo glm** 1-veined; **up glm** 1–3-veined; **rchl intnd** in some species swollen and glandular bas, the glandular portion turning whitish when dry; **cal** rounded, glab; **lm** thick, hrb to memb, 7(9)-veined, veins parallel, mrg scarious, apc scarious,

entire or emgt, midvein smt extended into an awn, awns straight; **pal** subequal to the lm, 2-veined, keeled over each vein, keels winged, with 1 or 2 awns or a flat triangular appendage; **lod** 2, completely fused; **anth** 3, opening by pores; **ov** glab.

*Pleuropogon* is a genus of five hydrophilous species, one circumboreal in the arctic, the other four restricted to the Pacific coast of North America.

The flat, triangular paleal appendages differ from bristly or flattened awns in being wider at the base, and smooth rather than scabrous.

1. Paleal keels each with 2 awns, the lower awn 1–3 mm long, the upper awn 0.3–1 mm long; lemmas
   3.5–5 mm long; plants of the arctic (subg. *Pleuropogon*) . . . . . . . . . . . . . . . . . . . . . . . . . . . . . .5. *P. sabinei*
1. Paleal keels each with 1 awn 3–9 mm long, or a triangular appendage; lemmas 4.5–10 mm long; plants
   of the Pacific Northwest and California (subg. *Lophochlaena*).
   2. Lowest lemma in each spikelet 4.5–7.5 mm long; culms 15–95 cm tall; caryopses 2.5–3.1 mm long.
      3. Paleal keels unawned, with a triangular appendage; rhizomes absent or poorly developed;
         rachilla internodes with a glandular swelling at the base . . . . . . . . . . . . . . . . . . . . . . . . . . . . .1. *P. californicus*
      3. Paleal keels with an awn 3–9 mm long, without a triangular appendage; rhizomes strongly
         developed; rachilla internodes without a glandular swelling at the base . . . . . . . . . . . . . . . . . . . . . . .3. *P. oregonus*
   2. Lowest lemma in each spikelet 8–10 mm long; culms mostly 100–160 cm tall; caryopses 3.5–6 mm
     long.
      4. Lemma awns 0.2–4 mm long; pedicels usually erect, rarely reflexed, the spikelets erect or
         ascending at maturity . . . . . . . . . . . . . . . . . . . . . . . . . . . . . . . . . . . . . . . . . . . . .2. *P. hooverianus*
      4. Lemma awns (5)9–20 mm long; pedicels reflexed, the spikelets pendent at maturity . . . . . . . . . . . . . . .4. *P. refractus*

### 1. Pleuropogon californicus (Nees) Benth. *ex* Vasey CALIFORNIA SEMAPHOREGRASS [p. *358*, *509*]

*Pleuropogon californicus* is a Californian endemic with two varieties.

1. Plants annual or facultative perennials; lemmas
   usually with awns 5–11 mm long, rarely unawned;
   paleal appendages 0.5–2.5 mm long; spikelets 15–30
   mm long . . . . . . . . . . . . . . . . . . . . . . . . . . . var. *californicus*
1. Plants perennial; lemmas unawned, sometimes
   mucronate, mucros to 1.5 mm long; paleal
   appendages 0.5–1 mm long; spikelets 25–60 mm
   long . . . . . . . . . . . . . . . . . . . . . . . . . . . . . . . . . var. *davy*

### Pleuropogon californicus (Nees) Benth. *ex* Vasey var. californicus ANNUAL SEMAPHOREGRASS [p. *358*]

*Pleuropogon californicus* var. *californicus* grows in vernal pools, marshy grasslands, orchards, and roadside ditches in California, from southern Humboldt County south to San Luis Obispo County, and east to Amador County.

### Pleuropogon californicus var. davyi (L.D. Benson) But DAVY'S SEMAPHOREGRASS [p. *358*]

*Pleuropogon californicus* var. *davyi* is the more restricted of the two varieties, being known only from vernal pools, sloughs, and marshy grasslands in Mendocino and Lake counties, California.

### 2. Pleuropogon hooverianus (L.D. Benson) J.T. Howell HOOVER'S SEMAPHOREGRASS [p. *358*, *509*]

*Pleuropogon hooverianus* grows in wet and marshy areas, usually in shady locations. Several of the populations are around redwood groves. It is known only from Mendocino, Sonoma, and Marin counties in California. It is listed as rare by the state of California.

### 3. Pleuropogon oregonus Chase OREGON SEMAPHORE-GRASS [p. *358*, *509*]

*Pleuropogon oregonus* grows in swampy ground, wet meadows, and stream banks. It is known from only a few locations in Union and Lake counties, Oregon. In 1975 it was thought to be extinct, but a population has since been discovered in Lake County. The species is listed as threatened by the state of Oregon.

[1]Paul P.H. But

**4. Pleuropogon refractus** (A. Gray) Benth. *ex* Vasey
NODDING SEMAPHOREGRASS [p. *358*, <u>*509*</u>]

*Pleuropogon refractus* grows in wet meadows, riverbanks, and shady places, from sea level to about 1000 m. Its range extends from British Columbia south to California.

*5. Pleuropogon sabinei* R. Br. FALSE SEMAPHOREGRASS, PLEUROPOGON DE SABINE [p. *358*, <u>*509*</u>]

*Pleuropogon sabinei* grows in open, wet places, frequently partially submerged, around lakes, ponds, marshy areas, and riverbanks. Its range extends from eastern Siberia and the Altai Mountains to northern Alaska, Canada, and Greenland.

## 10. STIPEAE Dumort.[1]

Pl usu per; usu tightly to loosely csp, smt rhz. Clm ann or per, not wd, smt brchg at up nd. Lvs bas concentrated to evenly distributed; shth open, mrg not fused, smt ciliate distally, bas shth smt concealing ax pan (*cleistogene*), smt wider than the bld; col smt with tufts of hair at the sides extending to the top of the shth; aur absent; lig scarious, often ciliate, cilia usu shorter than the base, lig of the lo and up cauline lvs smt differing in size and vestiture; psdpet absent; bld linear to narrowly lanceolate, venation parallel, cross venation not evident, cross sections non-Kranz, without arm or fusoid cells; epdm of adx surfaces smt with unicellular microhairs, cells not papillate. Infl usu tml pan, occ rdcd to rcm in depauperate pl, smt 2–3 pan developing from the highest cauline nd. Spklt usu with 1 flt, smt with 2–6 flt, lat compressed to terete; rchl not prolonged beyond the base of the flt in spklt with 1 flt, prolonged beyond the base of the distal flt in spklt with 2–6 flt, prolongation hairy, hairs 2–3 mm; dis above the glm and beneath the flt. Glm usu exceeding the flt(s), always longer than ¼ the length of the adjacent flt, 1–10-veined, narrowly lanceolate to ovate, hyaline or memb, flexible; flt usu terete, smt lat or dorsally compressed; cal usu well-developed, rounded or blunt to sharply pointed, often antrorsely strigose; lm lanceolate, rectangular, or ovate, memb to coriaceous or indurate, 3–5-veined, veins inconspicuous, apc entire, bilobed, or bifid, awned, lm-awn jnct usu conspicuous, awns 0.3–30 cm, not brchd, usu tml and centric or eccentric, smt subtml, caducous to persistent, not or once- to twice-geniculate, if geniculate, proximal segment(s) twisted, distal segment straight, flexuous, or curled, not or scarcely twisted; lod 2 or 3; anth 1 or 3, smt differing in length within a flt; ov glab throughout or pubescent distally; sty 2(3–4)-brchd. Car ovoid to fusiform, not beaked, prcp thin; hila linear; emb less than ⅓ the length of the car.

The tribe *Stipeae* includes about 15 genera and approximately 500 species. It grows in northern Africa, Australia, South and North America, and Eurasia. In Australia, South America, and Asia, it is often the dominant grass tribe over substantial areas. It is not present in southern India, and is represented by only one native species in southern Africa. Most species grow in arid or seasonally arid, temperate regions.

The hybrid genus ×*Achnella* is not included in the key; it is treated on p. 40.

1. Spikelets with 2–6 florets . . . . . . . . . . . . . . . . . . . . . . . . . . . . . . . . . . . . . . . . . . . . . . . . . . . 10.01 *Ampelodesmos*
1. Spikelets with 1 floret.
  2. Paleas sulcate, longer than the lemmas; lemma margins involute, fitting into the paleal groove; lemma apices not lobed . . . . . . . . . . . . . . . . . . . . . . . . . . . . . . . . . . . . . . . . . . . . . . 10.09 *Piptochaetium*
  2. Paleas flat, from shorter than to longer than the lemmas; lemma margins convolute or not overlapping; lemma apices often lobed or bifid.
    3. Prophylls exceeding the leaf sheaths; plants cultivated as ornamentals.
      4. Panicles contracted; lemma awns once-geniculate . . . . . . . . . . . . . . . . . . . . . . . . . . . . . . . . . . 10.05 *Macrochloa*
      4. Panicles open; lemma awns twice-geniculate . . . . . . . . . . . . . . . . . . . . . . . . . . . . . . . . . . . . . . . 10.06 *Celtica*
    3. Prophylls concealed by the leaf sheaths; plants native, introduced, sometimes cultivated as ornamentals.
      5. Flag leaf blades up to 12 mm long; basal leaves overwintering . . . . . . . . . . . . . . . . . . . . . . . . . . 10.10 *Oryzopsis*
      5. Flag leaf blades more than 10 mm long; basal leaves not overwintering.
        6. Plants with multiple stiff branches from the upper nodes; pedicels sometimes plumose; species cultivated as ornamentals in the *Manual* region . . . . . . . . . . . . . . . . . . . . . . . . 10.15 *Austrostipa*
        6. Plants not branching at the upper nodes, or with a few, flexible branches; pedicels never plumose; species native, established introductions, or cultivated as ornamentals.
          7. Apices of the leaf blades sharp and stiff; caryopses obovoid, often with 3 smooth ribs at maturity; cleistogenes usually present . . . . . . . . . . . . . . . . . . . . . . . . . . . . . 10.14 *Amelichloa*
          7. Apices of the leaf blades acute to acuminate, never both sharp and stiff; caryopses fusiform, ovoid or obovoid, without ribs; cleistogenes sometimes present.
            8. Lemma margins strongly overlapping their whole length at maturity, lemma bodies usually rough throughout, apices not lobed; paleas ¼–½ the length of the lemmas, without veins, glabrous . . . . . . . . . . . . . . . . . . . . . . . . . . . . . . . . . . . . . . . . . . . . . 10.12 *Nassella*

[1]Mary E. Barkworth

8. Lemma margins usually not or only slightly overlapping for some or all of their length at maturity, strongly overlapping in some species with smooth lemmas, lemma bodies usually smooth on the lower portion, apices often 1–2-lobed; paleas from ⅓ as long as to equaling or slightly exceeding the lemmas, 2-veined at least on the lower portion, usually with hairs or both lemmas and paleas glabrous.
　　9. Calluses 1.5–6 mm long, sharply pointed; plants perennial or annual, if perennial, awns 65–500 mm long, if annual, awns 50–100 mm long; panicle branches straight.
　　　　10. Lower ligules densely hairy, upper ligules less densely hairy or glabrous; plants perennial . . . . . . . . . . . . . . . . . . . . . . . . . . . . . . . . . . . . . . . . . . . 10.13 *Jarava* (in part)
　　　　10. Ligules glabrous or inconspicuously pubescent, lower and upper ligules alike in vestiture; plants perennial or annual.
　　　　　　11. Plants perennial; florets 7–25 mm long; awns scabrous or pilose on the first 2 segments, the terminal segment scabrous, or if pilose, the hairs 1–3 mm long . . . . . . . . . . . . . . . . . . . . . . . . . . . . . . . . . . . . . . . . 10.08 *Hesperostipa*
　　　　　　11. Plants annual or perennial, if annual, the florets 4–7 mm long and the awns not plumose, if perennial, the florets 18–27 mm long and the awns plumose on the terminal segment, the hairs 5–6 mm long . . . . . . . . . . . . . . . . . . 10.07 *Stipa*
　　9. Calluses 0.1–2 mm long, blunt to sharply pointed; plants perennial; awns 1–70 mm; panicle branches straight or flexuous.
　　　　12. Florets usually dorsally compressed at maturity, sometimes terete; paleas as long as or longer than the lemmas and similar in texture and pubescence; lemma margins separate for their whole length at maturity . . . . . . . . . . . . . . 10.04 *Piptatherum*
　　　　12. Florets terete or laterally compressed at maturity; paleas often shorter than the lemmas, sometimes less pubescent, sometimes as long as the lemmas and similar in texture and pubescence; lemma margins often overlapping for part or all of their length at maturity.
　　　　　　13. Glumes without evident venation, glume apices rounded to acute; plants subalpine to alpine, sometimes growing in bogs . . . . . . . . . . . . . . . . . . . . 10.03 *Ptilagrostis*
　　　　　　13. Glumes with 1–3(5) evident veins or the glume apices attenuate; plants growing from near sea level to subalpine or alpine habitats, not growing in bogs.
　　　　　　　　14. Lemma bodies with evenly distributed hairs of similar length or completely glabrous, sometimes with longer hairs around the base of the awn; basal segment of the awns sometimes with hairs up to 2 mm long . . . . . . . . . . . . . . . . . . . . . . . . . . . . . . . . . . . . . . . 10.02 *Achnatherum*
　　　　　　　　14. Lemma bodies with hairs to 1 mm long over most of their length, with strongly divergent hairs 3–8 mm long on the distal ¼, or the basal segment of the awns with hairs 3–8 mm long . . . . . . . . . . . 10.13 *Jarava* (in part)

## 10.01 AMPELODESMOS Link[1]

Pl per; csp, rhz. Clm 60–350 cm, ann, intnd solid. Lvs mostly bas; clstgn not developed; prphl shorter than the shth; shth open; lig memb, ciliate; bld initially flat, becoming involute, bases becoming indurate and curved. Infl pan, loosely contracted, somewhat 1-sided. Spklt pedlt, lat compressed, with 2–6 flt; rchl hairy, hairs 2–3 mm, prolonged beyond the distal flt; dis above the glm and beneath the flt. Glm subequal, more than ½ as long as the adjacent lm, scarious or chartaceous, 3–5-veined, awn-tipped; flt 10–12 mm; cal 0.2–0.5 mm, rounded, strigose; lm coriaceous, smooth, 5–7-veined, mostly glab, hairy over and adjacent to the bas ½ of the midvein, hairs 1–2 mm, apc bidentate or bilobed, mucronate or awned from the sinuses, lm-awn jnct not conspicuous; pal subequal to the lm, 2-keeled, keels extending as teeth, flat between the keels; lod 3, lanceolate, memb, ciliate; anth 3, 6–8 mm; ov pubescent distally; sty 2, white. Car fusiform, subterete, grooved adxly, not ribbed; hila linear; stch gr simple.

*Ampelodesmos* is a unispecific, xerophytic genus that is native to the Mediterranean. It is now established in California. It is somewhat similar in overall shape to *Cortaderia*, but differs in its membranous ligules, drooping and somewhat one-sided panicles, and deeply ribbed leaves.

[1]James P. Smith, Jr.

## 1. Ampelodesmos mauritanicus (Poir.) T. Durand & Schinz MAURITANIAN GRASS [p. *358*, <u>509</u>]

*Ampelodesmos mauritanicus* is sparingly established in California: in dry oak woodlands in Napa County, and beneath a mixed evergreen canopy on Mount St. Helena in Sonoma County. It is cultivated in other parts of the United States. The plants dry out rapidly in the summer, making them fire-prone. The amount of seed set varies substantially between years. In its native range, which lies along the drier portions of the Mediterranean coast, the leaves and culms are used for mats, vine ties, brooms, baskets, and thatching.

## 10.02 ACHNATHERUM P. Beauv.[1]

**Pl** per; tightly to loosely csp. **Clm** 10–250 cm, erect, not brchg at the up nd; **bas brchg** exvag or invag. **Shth** open, mrg often ciliate distally; **clstgn** not present in bas lf shth; **col** smt with hairs on sides; **aur** absent; **lig** hyaline to memb, glab or pubescent, smt ciliate; **bld** flat, convolute, or involute, apc acute, flexible, bas bld not overwintering, flag lf bld more than 10 mm long. **Infl** pan, usu contracted, smt 2+ at tml nd; **br** usu straight, smt flexuous. **Spklt** usu appressed, with 1 flt; **rchl** not prolonged; **dis** beneath flt. **Glm** exceeding flt, usu lanceolate, 1–7-veined, acute, acuminate, or obtuse; **flt** usu terete, fusiform or globose, smt somewhat lat compressed; **cal** 0.1–4 mm, blunt to sharp, usu strigose; **lm** stiffly memb to coriaceous, smooth, usu hairy, smt glab, hairs on the lm body to 6 mm, evenly distributed, hairs on the up ¼ smt somewhat longer than those below, not both markedly longer and more divergent, apical hairs to 7 mm, lm mrg usu not or only weakly overlapping, firmly overlapping in some species with glab lm, usu with 0.05–3 mm lobes, smt unlobed, lobes usu memb and flexible, smt thick, apc awned, lm-awn jnct evident; **awn** 3–80 mm, centric, readily deciduous to persistent, usu scabrous to scabridulous, smt hairy in whole or in part, if shorter than 12 mm, usu deciduous, not or once-geniculate and scarcely twisted, if longer than 12 mm, usu persistent, once- or twice-geniculate and twisted below, tml segment usu straight, smt flexuous; **pal** from ⅓ as long as to slightly longer than the lm, usu pubescent, 2-veined, not keeled over veins, flat between veins, veins usu terminating below apc, smt prolonged 1–3 mm, apc usu rounded; **lod** 2 or 3, memb, not lobed; **anth** 3, 1.5–6 mm, smt penicillate; **ov** with 2 sty br, br fused at base. **Car** fusiform, not ribbed, sty bases persistent; **hila** linear, almost as long as car; **emb** ⅕–⅓ the length of the car.

*Achnatherum* is one of the larger and more widely distributed genera in the *Stipeae*. Its size is difficult to estimate because its boundaries are still unclear. Of the 28 species in the *Manual* region, only *A. splendens* is introduced.

Glume widths are the distance between the midvein and the margin. Floret lengths include the callus, but not the apical lobes. Floret thickness refers to the thickest part of the floret.

1. Awns persistent, basal segments pilose, at least some hairs 0.5–8 mm long.
    2. Flag leaves with ligules 3–8 mm long; lemmas with 1 apical lobe, the lobe to 0.1 mm long, thick, coriaceous . . . . . . . . . . . . . . . . . . . . . . . . . . . . . . . . . . . . . . . . . . . . . . . . . . . . . . . . . . . . .9. *A. thurberianum*
    2. Flag leaves with ligules 0.3–3 mm long; lemmas usually with 2 apical lobes, sometimes not lobed, lobes to 1 mm long, thin, membranous.
        3. Basal awn segments with hairs of mixed lengths, the longer hairs scattered among the shorter hairs; apical lemma hairs longer than most basal awn hairs.
            4. Florets 8–9 mm long; glumes 1.3–1.9 mm wide from midvein to margin . . . . . . . . . . . . . 7. *A. latiglume* (in part)
            4. Florets 5–7.5 mm long; glumes 0.6–1 mm wide from midvein to margin.
                5. Calluses 0.5–0.7 mm long; paleas ½–¾ as long as the lemmas; palea apices with hairs usually about 1 mm long . . . . . . . . . . . . . . . . . . . . . . . . . . . . . . . . . . . . . . . . . .4. *A. nevadense*
                5. Calluses 0.8–1.2 mm long; paleas ⅖–⅗ as long as the lemmas; palea apices with hairs usually less than 1 mm long . . . . . . . . . . . . . . . . . . . . . . . . . . . . . . . . . . . . . . . . .5. *A. occidentale* (in part)
        3. Basal awn segments with hairs that gradually and regularly decrease in length distally; apical lemma hairs usually similar in length to the longest basal awn hairs, sometimes longer on the adaxial side.
            6. Basal blades curling with age, forming circular arcs; paleas ¼–⅓ as long as the lemmas; panicles 7–11 cm long . . . . . . . . . . . . . . . . . . . . . . . . . . . . . . . . . . . . . . . . . . . . . 18. *A. curvifolium*
            6. Basal blades straight to lax, not forming circular arcs; paleas ⅖–⅘ as long as the lemmas; panicles 5–30 cm long.
                7. Florets 5.5–7.5 mm long; paleas ⅖–⅗ as long as the lemmas; glumes less than 1 mm wide from midvein to margin . . . . . . . . . . . . . . . . . . . . . . . . . . . . . . . . . . . . . . . . .5. *A. occidentale* (in part)
                7. Florets 8–9 mm long; paleas ⅗–⅘ as long as the lemmas; glumes 1.3–1.9 mm wide from midvein to margin . . . . . . . . . . . . . . . . . . . . . . . . . . . . . . . . . . . . . . . . . . . . . .7. *A. latiglume* (in part)
1. Awns deciduous or persistent, basal segments scabrous or with hairs shorter than 0.5 mm.
    8. Lemmas evenly hairy, hairs 1.2–6 mm long, hairs on the lemma body usually not evidently shorter than those at the apices.

[1]Mary E. Barkworth

9. Awns persistent.
   10. Plants sterile, the anthers indehiscent, with few pollen grains (see discussion of hybrids on
      p. 35) . . . . . . . . . . . . . . . . . . . . . . . . . . . . . . . . . . . . . . . . . hybrids of 26. *Achnatherum hymenoides* (in part)
   10. Plants fertile, the anthers dehiscent, with many pollen grains.
      11. Sheaths not becoming flat and ribbonlike with age; blades usually involute and 0.2–0.4
         mm in diameter, 0.5–1 mm wide when flat; awns twice-geniculate . . . . . . . . . . . . . . . . . . 21. *A. pinetorum*
      11. Sheaths becoming flat and ribbonlike with age; blades 0.5–1.5 mm in diameter when
         convolute, to 7 mm wide when flat; awns once- or twice-geniculate.
         12. Awns twice-geniculate, culms 3–6 mm thick . . . . . . . . . . . . . . . . . . . . . . 10. *A. coronatum* (in part)
         12. Awns once-geniculate, culms 0.8–2 mm thick . . . . . . . . . . . . . . . . . . . . . . 11. *A. parishii* (in part)
9. Awns rapidly deciduous.
   13. Florets at least 4.5 mm long, fusiform, anthers sometimes indehiscent.
      14. Anthers dehiscent, the pollen grains well formed . . . . . . . . . . . . . . . . . . . . . . . . . . . . . 22. *A. webberi*
      14. Anthers indehiscent, the pollen grains poorly formed.
         15. Anthers dimorphic, 1 longer than the other 2; lemmas with 7 veins . . . . . . . . . . . . see 10.11 ×*Achnella*
         15. Anthers all alike; lemmas with 5 veins (see discussion of hybrids on p. 35)
         . . . . . . . . . . . . . . . . . . . . . . . . . . . . . . . . . . . . . . hybrids of 26. *Achnatherum hymenoides* (in part)
   13. Florets 2.5–4.5 mm long, usually ovoid to obovoid, sometimes fusiform, anthers dehiscent.
      16. Panicle branches terminating in a pair of spikelets on conspicuously divaricate, unequal
         to subequal pedicels, most shorter pedicels at least ¹/₂ as long as the longer pedicels . . . . . 26. *A. hymenoides*
      16. Panicle branches terminating in a pair of spikelets on loosely appressed, unequal
         pedicels, most shorter pedicels less than ¹/₂ as long as the longer pedicels.
         17. Panicles 0.5–2.8 cm wide, branches 0.5–5 cm long, strongly ascending; spikelets
            evenly distributed over the branches . . . . . . . . . . . . . . . . . . . . . . . . . . . . . . . . 27. *A. arnowiae*
         17. Panicles 7–15 cm wide, branches 5–8 cm long, ascending to strongly divergent;
            spikelets confined to the distal ¹/₂ of the branches . . . . . . . . . . . . . . . . . . . . . . . . 28. *A. contractum*
8. Lemmas glabrous or with hairs 0.2–1.5(2) mm long at midlength, glabrous or with hairs distally,
the hairs at midlength often evidently shorter than those at the lemma apices.
   18. Apical lemma hairs 2–7 mm long, usually 1+ mm longer than those at midlength.
      19. Calluses sharp; paleas ¹/₃–¹/₂ as long as the lemmas . . . . . . . . . . . . . . . . . . . . . . . . . . 19. *A. scribneri*
      19. Calluses blunt to acute; paleas ¹/₂–⁹/₁₀ as long as the lemmas.
         20. Awns twice-geniculate; culms 3–6 mm thick . . . . . . . . . . . . . . . . . . . . . . 10. *A. coronatum* (in part)
         20. Awns once-geniculate; culms 0.8–2 mm thick . . . . . . . . . . . . . . . . . . . . . . 11. *A. parishii* (in part)
   18. Apical lemma hairs absent or to 2.2 mm long, usually less than 1 mm longer than those at
   midlength.
      21. Awns 5–12 mm long, readily deciduous, not or only once-geniculate.
         22. Lemmas glabrous.
            23. Panicles lax, the branches flexuous, diverging . . . . . . . . . . . . . . . . . . . . . . . . 24. *A. wallowaense*
            23. Panicles erect, the branches straight, ascending to appressed . . . . . . . . . . . . . . . . . 25. *A. hendersonii*
         22. Lemmas pubescent.
            24. Culms 30–250 cm long; plants cultivated ornamentals . . . . . . . . . . . . . . . . . . . . . 1. *A. splendens*
            24. Culms 15–25 cm long; plants native in the *Manual* region . . . . . . . . . . . . . . . . . . . . 23. *A. swallenii*
      21. Awns 10–80 mm long, persistent, once- or twice-geniculate.
         25. Terminal awn segment flexuous.
            26. Panicles contracted, all branches straight, appressed or strongly ascending; ligules
               on the flag leaves to 1.5 mm long . . . . . . . . . . . . . . . . . . . . . . . . . . . . . . . . 15. *A. aridum*
            26. Panicles open, the lower branches flexuous, ascending to widely divergent; ligules
               on the flag leaves to 4.5 mm long . . . . . . . . . . . . . . . . . . . . . . . . . . . . . . . . 16. *A. eminens*
         25. Terminal awn segment straight or slightly arcuate.
            27. Panicle branches flexuous, ascending to strongly divergent; spikelets pendulous
            . . . . . . . . . . . . . . . . . . . . . . . . . . . . . . . . . . . . . . . . . . . . . . . . . . . . . . . .17. *A. richardsonii*
            27. Panicle branches straight, usually appressed to ascending, sometimes divergent;
               spikelets appressed to the branches.
               28. Flag leaves with a densely pubescent collar, the hairs 0.5–2 mm long; paleas
                  ²/₃–³/₄ as long as the lemmas . . . . . . . . . . . . . . . . . . . . . . . . . . . . . . . . . . . . 12. *A. robustum*
               28. Flag leaves glabrous or sparsely pubescent on the collar, the hairs shorter
                 than 0.5 mm; paleas from ¹/₃ as long as to longer than the lemmas.
                 29. Lemma apices 2-lobed, lobes 1–3 mm long; palea veins extending
                    beyond the palea body, reaching to the tips of the lemma lobes . . . . . . . . . . . 2. *A. stillmanii*

29. Lemma apices unlobed or with lobes to 1.2 mm long; palea veins terminating before or at the palea apices.
   30. Apical lemma lobes thick, stiff, about 0.1 mm long; florets somewhat laterally compressed .................................. 8. *A. lemmonii*
   30. Apical lemma lobes membranous, 0.1–1.2 mm long; florets terete.
      31. Lower cauline internodes densely pubescent for 3–9 mm below the nodes, more shortly and less densely pubescent elsewhere ..... 13. *A. diegoense*
      31. Lower cauline internodes glabrous or slightly pubescent to 5 mm below the nodes, usually glabrous elsewhere.
         32. Glumes subequal, the lower glumes exceeding the upper glumes by less than 1 mm.
            33. Paleas $^3/_5$–$^9/_{10}$ as long as the lemmas, the apical hairs exceeding the apices; blades 0.5–2 mm wide; awns 12–25 mm long ............................. 3. *A. lettermanii*
            33. Paleas $^1/_3$–$^2/_3$ as long as the lemmas, the apical hairs usually not exceeding the apices; blades (0.5)1.2–5 mm wide; awns 19–45 mm long ..................... 6. *A. nelsonii*
         32. Glumes unequal, the lower glumes exceeding the upper glumes by 1–4 mm.
            34. Apical lemma hairs erect; lemma lobes 0.5–1.2 mm long ......................................... 14. *A. lobatum*
            34. Apical lemma hairs divergent to ascending; lemma lobes 0.2–0.5 mm long ........................ 20. *A. perplexum*

## 1. Achnatherum splendens (Trin.) Nevski JIJI GRASS [p. 359]

*Achnatherum splendens* is native from the Caspian Sea to eastern Siberia and south through central Asia to the inner ranges of the Himalayas. It was reported to be "sparingly cultivated" in the United States in 1951.

## 2. Achnatherum stillmanii (Bol.) Barkworth
STILLMAN'S NEEDLEGRASS [p. 359, 509]

*Achnatherum stillmanii* grows at scattered locations in coniferous forests in northern California, at 900–1500 m, possibly being edaphically restricted. Its combination of large size, long, narrow lemma lobes, and paleal morphology distinguish it from all other North American species of *Achnatherum*.

## 3. Achnatherum lettermanii (Vasey) Barkworth
LETTERMAN'S NEEDLEGRASS [p. 359, 509]

*Achnatherum lettermanii* grows in meadows and on dry slopes, from sagebrush to subalpine habitats, at 1700–3400 m in the western United States; it is not known from Mexico. It resembles and often grows with *A. nelsonii*. It tends to differ from that species in being more tightly cespitose, having finer leaves, blunter calluses, longer paleas, and occupying shallower or more disturbed soils. Its long paleas also distinguish it from *A. perplexum*.

## 4. Achnatherum nevadense (B.L. Johnson) Barkworth
NEVADA NEEDLEGRASS [p. 359, 509]

*Achnatherum nevadense* grows in sagebrush and open woodlands, from Washington to south-central Wyoming and south to California and Utah. Most of the apical lemma hairs of *A. nevadense* appear longer than the lowermost awn hairs. This is the best character for distinguishing *A. nevadense* and *A. occidentale* subsp. *californicum* from *A. occidentale* subsp. *pubescens*. *Achnatherum nevadense* differs from *A. occidentale* subsp. *californicum* in the shape of the boundary between the glabrous and strigose portions of the callus. In addition, *A. nevadense* is usually pubescent below the lower cauline nodes, and has paleas that are longer in relation to the lemmas. *Achnatherum nevadense* also resembles *A. latiglume*, but the latter species has blunter calluses and paleas that tend to be thicker and somewhat longer in comparison to the lemmas than those of *A. nevadense*.

## 5. Achnatherum occidentale (Thurb.) Barkworth [p. 359, 509]

*Achnatherum occidentale*, which extends from British Columbia to California, Utah, and Colorado, varies considerably in pubescence and size. The three subspecies recognized here occasionally occur together.

1. Terminal awn segment usually pilose; culms 0.3–1 mm thick, glabrous even on the basal internodes; glumes often purplish ................... subsp. *occidentale*
1. Terminal awn segment usually scabrous or glabrous, occasionally pilose at the base; culms 0.5–2 mm thick; glumes usually green.
   2. First 2 awn segments scabrous or pilose with hairs of mixed lengths; apical lemma hairs longer than the basal awn hairs ............. subsp. *californicum*
   2. First 2 awn segments pilose, the hairs gradually and evenly becoming shorter towards the first geniculation; apical lemma hairs similar in length to the basal awn hairs ................. subsp. *pubescens*

## Achnatherum occidentale subsp. californicum (Merr. & Burtt Davy) Barkworth CALIFORNIA NEEDLEGRASS [p. 359]

*Achnatherum occidentale* subsp. *californicum* grows from Washington through Idaho to southwestern Montana and south to California and Nevada, with disjunct records from south-central Wyoming and southwestern Utah. Its elevation range is 2000–4000 m. It intergrades with *A. nelsonii* and *A. occidentale*. The scattering of longer hairs among shorter hairs on the basal awn segments, combined with the long apical lemma hairs, give florets of subsp. *californicum* a more untidy appearance than those of the other two subspecies. In this it resembles *A. nevadense*, but differs from that species in the shape of the boundary between the glabrous and strigose portions of the callus, in usually being glabrous below the lower cauline nodes, and in having paleas that are shorter in relation to the lemmas. Plants with scabrous awns are often confused with *A. nelsonii* subsp. *nelsonii*; they differ in having sharper calluses, a more elongated extension of the glabrous callus area into the strigose portion of the callus, and, usually, longer awns.

## Achnatherum occidentale (Thurb.) Barkworth subsp. occidentale WESTERN NEEDLEGRASS [p. *359*]

*Achnatherum occidentale* subsp. *occidentale* grows above 2400 m, primarily in California. It differs from *A. occidentale* subsp. *pubescens* in having culms that are glabrous throughout and awns with terminal segments that are usually pilose.

## Achnatherum occidentale subsp. pubescens (Vasey) Barkworth COMMON WESTERN NEEDLEGRASS [p. *359*]

*Achnatherum occidentale* subsp. *pubescens* grows from Washington to California and eastward to Wyoming, at 1300–4700 m. It is the most widespread and variable subspecies of *A. occidentale*, intergrading with subsp. *californicum*, *A. nelsonii*, and *A. lettermanii*. It differs from the latter two in its shorter paleas and its pilose awns.

## 6. Achnatherum nelsonii (Scribn.) Barkworth [p. *359*, 509]

*Achnatherum nelsonii* grows in meadows and openings, from sagebrush steppe and pinyon-juniper woodlands to subalpine forests, at 500–3500 m. It flowers in late spring to early summer, differing in this respect from *A. perplexum*. It differs from *A. lemmonii* in having wider leaf blades, shorter paleas, and membranous lemma lobes, and from *A. nevadense* and *A. occidentale* in its scabrous awns and the truncate to acute boundary of the glabrous tip of the callus with the callus hairs. See also comments under *A. lettermanii*, p. 32.

The two subspecies intergrade to some extent. There is also intergradation with *Achnatherum occidentale*, possibly as a result of hybridization and introgression.

1. Calluses blunt, dorsal boundary of the glabrous tip and the callus hairs almost straight to rounded; awns 19–31 mm long . . . . . . . . . . . . . . . . . . . . . . . subsp. *dorei*
1. Calluses sharp, dorsal boundary of the glabrous tip and the callus hairs acute; awns 19–45 mm long
. . . . . . . . . . . . . . . . . . . . . . . . . . . . . . . subsp. *nelsonii*

## Achnatherum nelsonii subsp. dorei (Barkworth & J.R. Maze) Barkworth DORE'S NEEDLEGRASS [p. *359*]

*Achnatherum nelsonii* subsp. *dorei* grows from the southern Yukon Territory to California and Wyoming. In regions where both subspecies grow, subsp. *dorei* is at higher elevations than subsp. *nelsonii*. It differs from *A. robustum* in the sparsely hairy collars of its flag leaves. Many reports from New Mexico and Arizona are based on *A. perplexum*, which differs in having sparse, narrow inflorescences and slightly recurved glumes. The two also differ in flowering time, *A. nelsonii* subsp. *dorei* flowering in late spring to early summer and *A. perplexum* in the fall.

## Achnatherum nelsonii (Scribn.) Barkworth subsp. nelsonii NELSON'S NEEDLEGRASS [p. *359*]

*Achnatherum nelsonii* subsp. *nelsonii* intergrades with subsp. *dorei* in Montana and Wyoming, and with *A. occidentale* subsp. *pubescens* in California. Its range extends from Idaho and Montana south to Nevada. It tends to grow at lower elevations than subsp. *dorei*.

## 7. Achnatherum latiglmume (Swallen) Barkworth WIDE-GLUMED NEEDLEGRASS [p. *360*, 509]

*Achnatherum latiglume* usually grows on dry slopes in yellow pine forests of southern California. It resembles *A. nevadense* and *A. occidentale*, but the latter two species have sharper calluses, and their paleas tend to be thinner and somewhat shorter relative to the lemmas than those of *A. latiglume*.

## 8. Achnatherum lemmonii (Vasey) Barkworth LEMMON'S NEEDLEGRASS [p. *360*, 510]

*Achnatherum lemmonii* grows in sagebrush and yellow pine associations, from southern British Columbia to California and east to Utah. It has been confused in the past with *A. nelsonii*; it differs in

having narrower leaves, laterally compressed florets with a thick apical lobe, and longer paleas.

1. Lower sheaths and culms glabrous or pubescent, not tomentose, the hairs to 0.2 mm long . . . . . . . . . subsp. *lemmonii*
1. Lower sheaths and culms tomentose, the hairs 0.4–0.6 mm long . . . . . . . . . . . . . . . . . . . . . . subsp. *pubescens*

## Achnatherum lemmonii (Vasey) Barkworth subsp. lemmonii [p. *360*]

*Achnatherum lemmonii* subsp. *lemmonii* grows throughout the range shown on the map, on both serpentine and non-serpentine soils.

## Achnatherum lemmonii subsp. pubescens (Crampton) Barkworth [p. *360*]

*Achnatherum lemmonii* subsp. *pubescens* is restricted to serpentine soils in southern California.

## 9. Achnatherum thurberianum (Piper) Barkworth THURBER'S NEEDLEGRASS [p. *360*, 510]

*Achnatherum thurberianum* grows in canyons and foothills, primarily in sagebrush desert and juniper woodland associations in the western United States, at 900–3000 m. Its long ligules and pilose awns make it one of the easier North American species of *Achnatherum* to identify.

## 10. Achnatherum coronatum (Thurb.) Barkworth CRESTED NEEDLEGRASS [p. *360*, 510]

*Achnatherum coronatum* grows on gravel and on rocky slopes, mostly in chaparral associations of the Coast Range from Monterey County, California, to Baja California, Mexico. It differs from *A. diegoense* in its mostly glabrous internodes and longer paleas, and from *A. parishii*, an inland species, in its twice-geniculate awns, more robust habit, and more sparsely pubescent paleas. Occasional plants combine the characteristics of both species.

## 11. Achnatherum parishii (Vasey) Barkworth [p. *360*, 510]

*Achnatherum parishii* grows from the coastal ranges of California to Nevada and Utah, south to Baja California, Mexico, and to the Grand Canyon in Arizona. It differs from *A. coronatum* in its once-geniculate awns, more densely pubescent paleas, and smaller stature; from *A. scribneri* in its shorter, blunter calluses and more abundant lemma hairs; and from *A. perplexum* in its longer lemma hairs.

1. Basal sheath margins glabrous or hairy distally, hairs to 0.5 mm long; culms 14–35 cm tall . . . . . subsp. *depauperatum*
1. Basal sheath margins hairy distally, hairs 1–3.2 mm long; culms 20–80 cm tall . . . . . . . . . . . . . . . . . subsp. *parishii*

## Achnatherum parishii subsp. depauperatum (M.E. Jones) Barkworth LOW NEEDLEGRASS [p. *360*]

*Achnatherum parishii* subsp. *depauperatum* grows in gravel and on rocky slopes, in juniper and mixed desert shrub associations, from central Nevada to western Utah. It differs from *A. webberi* in its persistent awns and thicker leaves that tend to curl when dry, and from *A. parishii* subsp. *parishii* in its smaller stature, glabrous or shortly hairy sheath margins, and densely hairy paleas.

## Achnatherum parishii (Vasey) Barkworth subsp. parishii PARISH'S NEEDLEGRASS [p. *360*]

*Achnatherum parishii* subsp. *parishii* grows on dry, rocky slopes, in desert shrub and pinyon-juniper associations, from the coastal ranges of California to northeastern Nevada, eastern Utah, and the Grand Canyon in Arizona. Its range extends into Baja California, Mexico. It differs from *A. coronatum* in its shorter culms and once-geniculate awns, and from subsp. *depauperatum* in its longer culms, hairy sheath margins, and sparsely hairy paleas.

### 12. Achnatherum robustum (Vasey) Barkworth
SLEEPYGRASS [p. *360*, *510*]

*Achnatherum robustum* grows on dry plains and hills, in open woods and forest clearings, and along roadsides, from Wyoming through Colorado to Arizona, New Mexico, and northern Mexico. Records from Kansas represent recent introductions. *Achnatherum robustum* is sometimes confused with *A. nelsonii* subsp. *dorei* and *Nassella viridula*, but it differs from both in the densely hairy collars of its flag leaves. It has potential as an ornamental grass. The English name refers to the effect some samples have on livestock, especially horses and cattle.

### 13. Achnatherum diegoense (Swallen) Barkworth SAN DIEGO NEEDLEGRASS [p. *360*, *510*]

*Achnatherum diegoense* grows in chaparral and coastal sage scrub, on rocky soil near streams or the coast, at 0–350 m, on the Channel Islands of Santa Barbara County, California, and, on the mainland, in Ventura and San Diego counties south into Baja California, Mexico.

### 14. Achnatherum lobatum (Swallen) Barkworth
LOBED NEEDLEGRASS [p. *360*, *510*]

*Achnatherum lobatum* grows on rocky, open slopes in pinyon-pine and white fir associations at 2100–2800 m. It differs from *A. scribneri* in its shorter apical lemma hairs and blunt calluses, and from *A. perplexum* in having longer lemma lobes and erect apical hairs.

### 15. Achnatherum aridum (M.E. Jones) Barkworth
MORMON NEEDLEGRASS [p. *360*, *510*]

*Achnatherum aridum* grows on rocky outcrops, in shrub-steppe and pinyon-juniper associations, from southeastern California to Colorado and New Mexico, at 1200–2000 m. It has not been found in Mexico.

### 16. Achnatherum eminens (Cav.) Barkworth
SOUTHWESTERN NEEDLEGRASS [p. *360*, *510*]

*Achnatherum eminens* grows on dry, rocky slopes and valleys in the mountains of the southwestern United States and northern Mexico, primarily in desert scrub, at 600–2600 m. It is superficially similar to *Nassella cernua*, but differs in its longer, glabrous ligules, not or weakly overlapping lemma magins, pubescent paleas, and distribution.

### 17. Achnatherum richardsonii (Link) Barkworth
RICHARDSON'S NEEDLEGRASS [p. *360*, *510*]

*Achnatherum richardsonii* grows in open woodlands and grasslands, often on sand or gravel, from Yukon Territory to northern Colorado. Its elevation range is 1000–3100 m.

### 18. Achnatherum curvifolium (Swallen) Barkworth
CURLYLEAF NEEDLEGRASS [p. *361*, *510*]

*Achnatherum curvifolium* grows on cliffs and in disturbed, rocky, limestone habitats. It is known from few locations in the *Manual* region, but is more common in northern Mexico. It differs from other species of *Achnatherum* in the *Manual* region in its combination of curly leaves and hairy awns.

### 19. Achnatherum scribneri (Vasey) Barkworth
SCRIBNER'S NEEDLEGRASS [p. *361*, *510*]

*Achnatherum scribneri* grows on rocky slopes, in pinyon-juniper and ponderosa pine associations at 1500–2700 m. It appears to be disjunct in Utah; this may reflect a lack of collecting. *Achnatherum scribneri* differs from *A. parishii*, *A. robustum*, *A. perplexum*, and *A. lobatum* in having sharp calluses.

### 20. Achnatherum perplexum Hoge & Barkworth
PERPLEXING NEEDLEGRASS [p. *361*, *510*]

*Achnatherum perplexum* grows on slopes in pinyon-pine associations of the southwestern United States and adjacent Mexico,

at 1500–1700 m. It flowers in late summer to early fall. It differs from *A. scribneri* in the glabrous collar margins of its basal leaves and its blunt calluses; from *A. nelsonii* and *A. lettermanii* in its unequal glumes; from *A. lettermanii* in its relatively short paleas; and from *A. lobatum* in its shorter lemma lobes and ascending to divergent apical lemma hairs.

### 21. Achnatherum pinetorum (M.E. Jones) Barkworth
PINEWOODS NEEDLEGRASS [p. *361*, *510*]

*Achnatherum pinetorum* usually grows on rocky soil, in pinyon-juniper to subalpine associations, at 2100–3300 m in the western United States. It differs from *A. webberi* in its longer, persistent awns, and from *A. lettermanii* in its sharp calluses and longer lemma hairs.

### 22. Achnatherum webberi (Thurb.) Barkworth
WEBBER'S NEEDLEGRASS [p. *361*, *510*]

*Achnatherum webberi* grows in dry, open flats and on rocky slopes, often with sagebrush, at 1500–2500 m. It differs from *A. hymenoides* in its cylindrical floret and non-saccate glumes, from *A. pinetorum* and *A. parishii* subsp. *parishii* in its shorter, deciduous awns, and from *A. parishii* subsp. *depauperatum* in its narrower blades.

### 23. Achnatherum swallenii (C.L. Hitchc. & Spellenb.) Barkworth SWALLEN'S NEEDLEGRASS [p. *361*, *510*]

*Achnatherum swallenii* grows on open, rocky sites, frequently with low sagebrush, in Idaho and western Wyoming, at 1500–2200 m. It is a dominant species in parts of eastern Idaho, although it is poorly represented in collections.

### 24. Achnatherum wallowaense J.R. Maze & K.A. Robson WALLOWA NEEDLEGRASS [p. *361*, *510*]

*Achnatherum wallowaense* grows in shallow, rocky soil at scattered localities, from 1000–1600 m, in the Wallowa and Ochoco mountains, Oregon.

### 25. Achnatherum hendersonii (Vasey) Barkworth
HENDERSON'S NEEDLEGRASS [p. *361*, *510*]

*Achnatherum hendersonii* grows in dry, rocky, shallow soil, in sagebrush or ponderosa pine associations. It is known from only three counties: Yakima and Kittitas counties, Washington, and Crook County, Oregon. It often grows with *Poa secunda*, occupying more disturbed sites possibly because of its different root structure.

### 26. Achnatherum hymenoides (Roem. & Schult.) Barkworth INDIAN RICEGRASS [p. *361*, *510*]

*Achnatherum hymenoides* grows in dry, well-drained soils, primarily in the western part of the *Manual* region and northern Mexico. Specimens from further east may be introduced; it is unknown whether they have persisted. Native Americans used the seeds of *Achnatherum hymenoides* for food. It is also one of the most palatable native grasses for livestock. Several cultivars have been developed for use in restoration work and as an ornamental. It forms natural hybrids with other members of the *Stipeae*. See discussion on p. 35.

### 27. Achnatherum arnowiae (S.L. Welsh & N.D Atwood) Barkworth ARNOW'S RICEGRASS [p. *361*, *510*]

*Achnatherum arnowiae* grows in pinyon-juniper, sagebrush, and mixed desert shrub communities in Utah, at 1400–2000 m.

### 28. Achnatherum contractum (B.L. Johnson) Barkworth CONTRACTED RICEGRASS [p. *361*, *510*]

*Achnatherum contractum* grows in rocky grasslands. It is a fertile derivative of a *Piptatherum micranthum* × *Achnatherum hymenoides* hybrid. Immature specimens of *A. hymenoides* differ from *A. contractum* in having pedicel pairs in which the shorter pedicel is more than half as long as the longer pedicel.

**29. Achnatherum ×bloomeri** (Bol.) Barkworth and other hybrids involving *A. hymenoides* [p. 362, <u>510</u>]

Numerous natural hybrids exist between *Achnatherum hymenoides* and other members of the *Stipeae*. Using the treatment adopted here, these hybrids may have as the second parent any of *A. occidentale* (all subspecies), *A. thurberianum*, *A. scribneri*, *A. robustum*, *Jarava speciosa*, *Nassella viridula* and *Piptatherum micranthum*. Evidence from herbarium specimens suggests that *A. hymenoides* also forms sterile hybrids with other species of *Achnatherum*. The name **Achnatherum ×bloomeri** applies only to hybrids between *A. hymenoides* and *A. occidentale* subsp. *occidentale*, but plants keying here may include any of the other interspecific hybrids. They all differ from *A. hymenoides* in having more elongated florets and 10–20 mm long awns, and from their other parent, in most instances, in having longer lemma hairs and more saccate glumes. Identification of the second parent is best made in the field.

The intergeneric hybrid with *Nassella viridula* is treated as ×*Achnella caduca* (see p. 40). It differs from *Achnatherum hymenoides* in its longer glumes and florets, and from other *A. hymenoides* hybrids in having a readily deciduous awn. No binomial has been proposed for the hybrid with *Jarava speciosa*. The fertile intergeneric hybrid involving *Piptatherum micranthum* is treated above as *A. contractum*.

## 10.03 PTILAGROSTIS Griseb.[1]

**Pl** per; tightly csp, not rhz. **Clm** 10–60 cm tall, 0.4–1.2 mm thick, erect, glab, not brchg at the up nd; **nd** 1, exposed; **bas brchg** invag; **prphl** shorter than the shth. **Lvs** mostly bas, not overwintering; **clstgn** not developed; **shth** open for most of their length, glab, smooth to somewhat scabrous; **aur** absent; **lig** 0.2–3 mm, memb, rounded to acute, glab, not ciliate; **bld** convolute, 0.2–0.6 mm in diameter, apc stiff, flag lf bld longer than 1 cm. **Infl** tml pan; **br** capillary, often flexuous, smt straight, glab or sparsely hirtellous; **dis** above the glm, beneath the flt. **Spklt** 3.4–7 mm, with 1 flt; **rchl** not prolonged beyond the flt. **Glm** subequal, hyaline, mostly purplish, venation not evident, apc rounded to acute; **flt** slightly shorter than the glm, terete; **cal** 0.1–0.8 mm, blunt, hairy; **lm** thickly memb, smooth, hairy over the bas portion or throughout, hairs 0.2–0.4 mm, mrg flat, not overlapping at maturity, apc lobed, lobes 0.1–1 mm, memb, apc with a single tml awn, lm-awn jnct evident; **awn** 5–30 mm, centric, persistent, once- or twice-geniculate, smt weakly so, scabrous or hairy, hairs to 2 mm; **pal** slightly shorter to slightly longer than the lm, hairy, hairs to 0.5 mm, 2-veined, not keeled over the veins, flat between the veins, veins ending before the apc, apc rounded; **lod** 3, free, memb; **anth** 3, 0.4–3.3 mm, smt penicillate; **ov** glab; **sty** 2, white, free to the base. **Car** 2–5 mm, fusiform, not ribbed.

*Ptilagrostis* is an alpine and subalpine genus of about 9 species. It grows in central Asia and the high mountains of western North America, sometimes in bogs. Two species are native to the *Manual* region.

1. Awns not hairy; panicles loosely contracted . . . . . . . . . . . . . . . . . . . . . . . . . . . . . . . . . . . . . . . . . . . . . . 1. *P. kingii*
1. Awns hairy, hairs on the lowest segment 1–2 mm long; panicles open or loosely contracted . . . . . . . . . . . . . . . 2. *P. porteri*

**1. Ptilagrostis kingii** (Bol.) Barkworth KING'S PTILAGROSTIS, SIERRA PTILAGROSTIS [p. 362, <u>510</u>]

*Ptilagrostis kingii* grows along damp streambanks and in wet meadows of the Sierra Nevada, at elevations from 2700–3500 m. It differs from most species in the genus in its scabridulous, rather than plumose, awns and short lemma lobes.

**2. Ptilagrostis porteri** (Rydb.) W.A. Weber PORTER'S PTILAGROSTIS, ROCKY-MOUNTAIN PTILAGROSTIS [p. 362, <u>510</u>]

*Ptilagrostis porteri* grows on hummocks of poorly drained wetlands, at 2700–3600 m, in central Colorado. It is often associated with *Salix* spp. and *Deschampsia cespitosa*. There are 29 known populations.

## 10.04 PIPTATHERUM P. Beauv.[2]

**Pl** per; csp or soboliferous, smt rhz. **Clm** 10–140(150) cm, erect, usu glab, usu smooth; **nd** 1–6; **brchg** invag or exvag at the base, not brchg above the base; **prphl** concealed by the lf shth. **Lvs** smt bas concentrated; **clstgn** not present; **shth** open, glab, smooth to scabrous; **aur** absent; **lig** 0.2–15 mm, memb to hyaline; **bld** 0.5–16 mm wide, flat, involute, valvate, or folded, often tapering in the distal ⅓, apc acute to acuminate, not stiff, bas bld not overwintering, smt not developed, flag lf bld well developed, longer than 1 cm. **Infl** 3–40 cm, tml pan, open or contracted; **br** straight or flexuous, usu scabrous, rarely smooth; **ped** often appressed to the br. **Spklt** 1.5–7.5 mm, with 1 flt; **rchl** not prolonged beyond the flt; **dis** above the glm, beneath the flt. **Glm** from 1 mm shorter than to exceeding the flt, subequal or the lo glm longer than the up glm, memb, 1–9-veined, veins evident, apc obtuse to acute or acuminate; **flt** 1.5–10 mm, usu dorsally compressed, smt terete; **cal** 0.1–0.6 mm, glab or with hairs, blunt; **lm** 1.2–9 mm, smooth, coriaceous or stiffly memb, tawny or light brown to black at maturity, 3–7-veined, mrg flat, separated and parallel for their whole length at maturity, apc not lobed or lobed, glab or hairy, hairs about 0.5 mm, not spreading, awned, lm-awn jnct evident; **awns** 1–18(20) mm, centric, often caducous, almost straight to once- or twice-geniculate, scabrous; **pal** as long as or slightly longer than the lm, similar in texture and pubescence, 2(3)-veined, not keeled over the veins, flat between the veins, veins terminating near the apc, apc often pinched; **anth** 3, 0.6–5 mm, smt penicillate; **sty** 2 and free to their bases, or 1 with 2–3 br. **Car** glab, ovoid to obovoid; **hila** ½ as long as to equaling the car.

[1,2]Mary E. Barkworth

*Piptatherum* has approximately 30 species, most of which are Eurasian. They extend from lowland to alpine regions, and grow in habitats ranging from mesic forests to semideserts.

1. Basal leaf blades 0–2 cm long; cauline leaf blades 8–16 mm wide; florets 4.5–7.5 mm long . . . . . . . . . . . . 6. *P. racemosum*
1. Basal leaf blades 4–45 cm long; cauline leaf blades 0.5–10 mm wide; florets 1.5–6 mm long.
   2. Lemmas and calluses usually glabrous, occasionally sparsely pubescent; florets 1.5–2.5 mm long.
      3. Blades 0.5–2.5 mm wide, often involute; panicles 5–20 cm long, the lower nodes with 1–3 branches . . . . . . . . . . . . . . . . . . . . . . . . . . . . . . . . . . . . . . . . . . 4. *P. micranthum* (in part)
      3. Blades 2–10 mm wide, flat; panicles 10–40 cm long, the lower nodes usually with 3–7 branches, sometimes with 15–30+ branches . . . . . . . . . . . . . . . . . . . . . . . . . . . . . . . . . . . . 7. *P. miliaceum*
   2. Lemmas evenly pubescent; calluses hairy; florets 1.5–6 mm long.
      4. Awns 3.9–15 mm long, persistent, once- or twice-geniculate.
         5. Primary panicle branches straight, appressed; awns 3.9–7 mm long; florets 3–6 mm long . . . . . . . . . . 1. *P. exiguum*
         5. Primary panicle branches somewhat flexuous, often divergent; awns 5–15 mm long; florets 2.2–4.5 mm long . . . . . . . . . . . . . . . . . . . . . . . . . . . . . . . . . . . . . . 2. *P. canadense*
      4. Awns 1–8 mm long, caducous, often absent from herbarium specimens, straight or arcuate.
         6. Awns 4–8 mm long; florets 1.5–2.5 mm long . . . . . . . . . . . . . . . . . . . . . . . . 4. *P. micranthum* (in part)
         6. Awns 1–2.5 mm long; florets 2.2–4.5 mm long.
            7. Lower panicle branches straight; ligules 0.5–2.5 mm long . . . . . . . . . . . . . . . . . . . . . . . 3. *P. pungens*
            7. Lower panicle branches flexuous; ligules 1.8–5.5 mm long . . . . . . . . . . . . . . . . . . . . . . 5. *P. shoshoneanum*

### 1. Piptatherum exiguum (Thurb.) Dorn LITTLE PIPTATHERUM [p. 362, <u>510</u>]

*Piptatherum exiguum* grows on rocky slopes and outcrops in upper montane habitats, from central British Columbia to southwestern Alberta and south to northern California, Nevada, Utah, and northern Colorado.

### 2. Piptatherum canadense (Poir.) Dorn CANADIAN PIPTATHERUM, ORYZOPSIS DU CANADA [p. 362, <u>511</u>]

*Piptatherum canadense* grows in grasslands and open woods, from the British Columbia–Alberta border east to Newfoundland, extending south into the Great Lakes region and the northeastern United States. Its persistent, longer awns distinguish *P. canadense* from *P. pungens*.

### 3. Piptatherum pungens (Torr.) Dorn SHARP PIPTATHERUM [p. 362, <u>511</u>]

*Piptatherum pungens* grows in sandy to rocky soils and open habitats, mainly in Canada and the northeastern United States. Its deciduous, shorter awns distinguish it from *P. canadense*. The awns fall off so rapidly that it is sometimes mistaken for *Milium* or *Agrostis*, but the only perennial species of *Milium* in the *Manual* region has much wider leaf blades, and no species of *Agrostis* has such stiff lemmas and well-developed paleas.

### 4. Piptatherum micranthum (Trin. & Rupr.) Barkworth SMALL-FLOWERED PIPTATHERUM [p. 362, <u>511</u>]

*Piptatherum micranthum* grows on gravel benches, rocky slopes, and creek banks in the western *Manual* region. The combination of small, dorsally compressed florets and appressed pedicels distinguishes this species from all other native North American *Stipeae*.

### 5. Piptatherum shoshoneanum (Curto & Douglass M. Hend.) P.M. Peterson & Soreng SHOSHONE PIPTATHERUM [p. 362, <u>511</u>]

*Piptatherum shoshoneanum* is known best from eastern Idaho, where it grows in the canyons of the Middle Fork of the Salmon River and its tributaries. It has also been found in the Belted Range of southwestern Nevada. It usually grows in moist crevices of igneous, metamorphic, or sedimentary cliffs and rock walls.

### 6. Piptatherum racemosum (Sm.) Eaton MOUNTAIN RICEGRASS [p. 363, <u>511</u>]

*Piptatherum racemosum* usually grows in deciduous woods, and less often in open pine woods, in rocky, mountainous areas. The absence of basal blades and the dark, shiny lemmas distinguish it from all other North American *Stipeae*. It is highly palatable to livestock, but is never sufficiently abundant to be important as forage.

### 7. Piptatherum miliaceum (L.) Coss. SMILO GRASS [p. 363, <u>511</u>]

*Piptatherum miliaceum* is a Eurasian introduction that is now established in several parts of the world. Within the *Manual* region, *P. miliaceum* is known from Arizona and California, growing in disturbed sites. It has also been found on a ballast dump in Maryland.

1. Panicle branches loosely whorled, lower nodes with 3–7 branches, all spikelet-bearing . . . . . . . . . . subsp. *miliaceum*
1. Panicle branches densely whorled, lower nodes with 15–30+ branches, some with highly reduced or no spikelets . . . . . . . . . . . . . . . . . . . . . . . . . . . . . subsp. *thomasii*

### Piptatherum miliaceum (L.) Coss. subsp. miliaceum [p. 363]

*Piptatherum mileaceum* subsp. *miliaceum* is the most common of the two subspecies, and the only one known to be established in the *Manual* region.

### Piptatherum miliaceum subsp. thomasii (Duby) Soják [p. 363]

*Piptatherum miliaceum* subsp. *thomasii* has a native range similar to that of subsp. *miliaceum*, except that it does not grow in semidesert regions. In the *Manual* region, it is known only from cultivated specimens.

## 10.05 MACROCHLOA Kunth[1]

**Pl** per; **csp**, not **rhz**. **Clm** 60–170 cm, mostly smooth, scabrous beneath the pan; **bas brchg** invag; **prphl** longer than the shth, 2-awned, awns 2–3 cm, velutinous; **nd** concealed by the shth. **Lvs** bas concentrated; **clstgn** not developed; **shth** open to the base, smooth, glab or hairy, hairs smt curly, **mrg** extending into 2 awnlike

[1]Francisco M. Vázquez

extensions; **aur** absent; **lig** truncate, velutinous; **bld** conduplicate or convolute, about 1 mm in diameter. **Infl** pan, contracted, erect. **Spklt** 26–30 mm, with 1 flt; **rchl** not prolonged beyond the base of the flt; **dis** above the glm, beneath the flt. **Glm** exceeding the flt, linear to lanceolate, gradually attenuate; **flt** 9–12 mm, terete; **cal** well developed, sharp; **lm** thickly memb to somewhat indurate, tan to brown, pubescent, mrg flat, not overlapping at maturity, apc bifid, awned from between the teeth, teeth 1–6 mm, linear, scarious, awns once-geniculate, first segment twisted, tml segment straight,

scabrous; **pal** equaling or exceeding the lm, pubescent, 2-veined, not keeled over the veins, flat between the veins, veins terminating before the apc, apc scarious, thinner than the pal body; **lod** 3, glab, ovate, acute, posterior lod smaller than the lat lod; **anth** 3, 10–15 mm, penicillate; **ov** glab; **sty** 1, with 2 pilose br. **Car** fusiform; **hila** linear, about as long as the car.

*Macrochloa* includes one to two species. It is native to the Mediterranean region, where it grows in basic and argillaceous soils. One species is cultivated in the *Manual* region.

### 1. Macrochloa tenacissima (L.) Kunth ESPARTO GRASS [p. 363]

*Macrochloa tenacissima* is native to the western Mediterranean regions of Europe and Africa. It is rarely cultivated in the United

States. It was used until the 1800s as a source of fiber for paper in Spain; currently it is used for erosion control in the Mediterranean basin, and for making hats and baskets.

## 10.06 CELTICA F.M. Vázquez & Barkworth[1]

**Pl** per; csp, not rhz. **Clm** 100–250 cm, erect, smooth, glab, upmost nd often exposed; **bas brchg** invag; **prphl** exceeding the subtending lf shth, awned, ciliate or glab. **Lvs** bas concentrated; **clstgn** not developed; **shth** open to the base, smooth, glab except at the throat; **aur** absent; **lig** memb, rounded, abx surfaces densely pubescent, mrg ciliate; **bld** flat to involute, abx surfaces smooth, adx surfaces scabrous or hirtellous. **Infl** pan, nodding, open. **Spklt** 25–32 mm, with 1 flt; **rchl** not prolonged beyond the base of the flt; **dis** above the glm, beneath the flt. **Glm** lanceolate, exceeding the flt, 3-veined; **flt** 14–16 mm, terete to lat compressed; **cal**

sharp, strigose; **lm** coriaceous, evenly pubescent, hairs 1–2 mm, mrg flat, overlapping at maturity, apc bifid, awned from between the teeth, teeth scarious; **awn** persistent, twice-geniculate, first segment twisted, tml segment straight; **pal** subequal to or longer than the lm, memb, dorsally pubescent, veins forming 2 awnlike extensions; **lod** 3, glab, lanceolate, posterior lod larger than the lat lod; **anth** 3, penicillate; **ov** glab; **sty** 2. **Car** fusiform; **hila** linear, about as long as the car.

*Celtica* is a unispecific genus that was formerly included in *Stipa* or *Macrochloa*.

### 1. Celtica gigantea (Link) F.M. Vázquez & Barkworth GIANT FEATHERGRASS [p. 363]

*Celtica gigantea* is native to the western and southern portions of the Iberian Peninsula and northern Africa. It is grown as an ornamental

in the *Manual* region; no attempt has been made to determine which of the infraspecific taxa is grown.

## 10.07 STIPA L.[2]

**Pl** ann or per; tufted or csp, not rhz. **Clm** 10–200 cm, hrb, not brchg at the up nd; **bas brchg** usu invag; **prphl** shorter than the shth. **Lvs** mostly bas; **clstgn** usu not developed; **shth** open; **aur** absent; **lig** memb, smt stiffly so, up and lo lig similar or up lig longer than those below; **bld** prominently ribbed, usu tightly convolute when dry. **Infl** tml pan, usu contracted. **Spklt** 12–90 mm, with 1 flt; **rchl** not prolonged beyond the base of the flt; **dis** above the glm, beneath the flt. **Glm** much longer than the flt, hyaline to memb, usu acuminate, 1–3-veined; **flt** 3–27 mm, terete to slightly lat compressed; **cal** (1)1.5–6 mm, sharp or blunt, antrorsely hairy; **lm** coriaceous to indurate, tan to brown, smooth, glab or hairy, hairs smt uniformly distributed, smt in lines, mrg flat, slightly overlapping at maturity, apc

awned, lm-awn jnct evident; **awn** 50–500 mm, persistent, usu once- or twice-geniculate, smt plumose in whole or in part, bas segment often strongly twisted; **pal** from shorter than to subequal to the lm, glab, 2-veined, not keeled, flat between the veins, apc smt scarious, smt similar in texture to the body; **lod** 2 or 3, glab or pilose; **anth** 3; **sty** 2(3,4), free at the base, if 3 or 4, then 1 or 2 distinctly shorter. **Car** fusiform, not ribbed.

As treated here, *Stipa* is a genus of 150–200 species, all of which are native to Eurasia or northern Africa. Until recently, the genus was interpreted as including almost all species of *Stipeae* with cylindrical florets. In some parts of the world, this broader interpretation still prevails. Two species of *Stipa* grow in the *Manual* region.

1. Plants perennial; glumes 60–90 mm long; awns plumose on the distal segment, hairs 5–6 mm long . . . . . . 1. *S. pulcherrima*
1. Plants annual; glumes 12–20 mm long; awns glabrous on the distal segment . . . . . . . . . . . . . . . . . . . . . . . . . 2. *S. capensis*

[1]Francisco M. Vázquez  [2]Mary E. Barkworth

### 1. Stipa pulcherrima K. Koch  BEAUTIFUL FEATHERGRASS [p. *363*]

*Stipa pulcherrima* is native from France and Germany to Armenia, Azerbaijan, Turkey, and Iran. Its long, plumose awns make it a striking ornamental in the *Manual* region.

### 2. Stipa capensis Thunb. [p. *363*, <u>*511*</u>]

*Stipa capensis* is known from two locations in Riverside County, California: one in Palm Springs, and the other near the mouth of Chino Canyon.

## 10.08 HESPEROSTIPA (M.K. Elias) Barkworth[1]

Pl per; csp, not rhz. Clm 12–110 cm, erect, not brchg at the up nd; prphl shorter than the shth. Lvs not overwintering, not bas concentrated; clstgn not developed; shth smooth; aur absent; lig memb, frequently ciliate; bld 4–40 cm long, 0.5–4.5 mm wide, usu tightly involute, adx surfaces conspicuously rdgd, apc narrowly acute, not sharp. Infl tml pan, contracted or open. Spklt 15–60 mm, with 1 flt; rchl not prolonged beyond the base of the flt; dis above the glm and beneath the flt. Glm 15–60 mm long, 2–4 mm wide, tapering from near the base to a hairlike tip; flt 7–25 mm, narrowly cylindrical; cal 2–6 mm, sharp, densely strigose distally; lm indurate, smooth, mrg flat, slightly overlapping at maturity, the up portion fused into a papillose, ciliate crwn, awned, lm-awn jnct distinct; awn 50–225 mm, persistent, twice-geniculate, often weakly so, lo segments twisted and scabrous to pilose, tml segment not twisted, usu scabridulous or pilose; pal equal to the lm, flat, pubescent, coriaceous, 2-veined, veins terminating at the apc, apc indurate, prow-tipped; anth 3, 1.2–9 mm. Car fusiform, not ribbed.

*Hesperostipa* is a North American endemic genus with five species, four of which are found in the *Manual* region. The fifth species, from southern Mexico, is known only from the type specimen.

1. Awns pilose on all segments, the terminal segment with hairs 1–3 mm long . . . . . . . . . . . . . . . . . . . . 2. *H. neomexicana*
1. Awns scabrous to strigose on the first 2 segments, the terminal segment scabridulous.
   2. Lemmas usually evenly white-pubescent, sometimes glabrous immediately above the callus; lower
     ligules often lacerate . . . . . . . . . . . . . . . . . . . . . . . . . . . . . . . . . . . . . . . . . . . . . . . 1. *H. comata*
   2. Lemmas unevenly pubescent with brown to beige hairs; lower ligules not lacerate.
     3. Florets 8.5–14(17) mm long; awns 50–105 mm long; lower nodes usually glabrous, occasionally
       evenly pubescent . . . . . . . . . . . . . . . . . . . . . . . . . . . . . . . . . . . . . . . . . . . . . 3. *H. curtiseta*
     3. Florets 15–25 mm long; awns 90–190 mm long; lower nodes usually with lines of pubescence . . . . . . . . 4. *H. spartea*

### 1. Hesperostipa comata (Trin. & Rupr.) Barkworth  NEEDLE-AND-THREAD [p. *364*, <u>*511*</u>]

*Hesperostipa comata* is found primarily in the cool deserts, grasslands, and pinyon-juniper forests of western North America. The two subspecies overlap geographically, but are only occasionally sympatric. Both are primarily cleistogamous.

1. Terminal awn segment 40–120 mm long, sinuous to curled at maturity; lower cauline nodes usually concealed by the sheaths; panicles often partially enclosed in the uppermost sheath at maturity . . . . . . . . . . . . . . . . . . . . . . . . . . subsp. *comata*
1. Terminal awn segment 30–80 mm long, straight; lower cauline nodes usually exposed; panicles usually completely exserted at maturity . . . subsp. *intermedia*

#### Hesperostipa comata (Trin. & Rupr.) Barkworth subsp. comata [p. *364*]

*Hesperostipa comata* subsp. *comata* grows on well-drained soils of cool deserts, grasslands, and sagebrush associations, at elevations of 200–2500 m. It is widespread and often abundant in western and central North America, particularly in disturbed areas. It is similar to *H. neomexicana*, differing primarily in having awns that are either not hairy or have hairs that are no more than 0.5 mm long, and in having thinner, longer ligules.

#### Hesperostipa comata subsp. intermedia (Scribn. & Tweedy) Barkworth [p. *364*]

*Hesperostipa comata* subsp. *intermedia* is found in pinyon-juniper woodlands, at elevations of 2175–3075 m, in the Sierra Nevada and Rocky Mountains, from southern Canada to New Mexico. It resembles *H. curtiseta*, but differs in its evenly pubescent lemmas and its often lacerate ligules.

### 2. Hesperostipa neomexicana (Thurb.) Barkworth  NEW MEXICAN NEEDLEGRASS [p. *364*, <u>*511*</u>]

*Hesperostipa neomexicana* grows in grassland, oak, and pinyon pine associations, from 800–2400 m, usually in well-drained, rocky areas in the southwestern United States and adjacent Mexico. It differs from *H. comata* subsp. *comata* in its longer awn hairs and shorter ligules.

### 3. Hesperostipa curtiseta (Hitchc.) Barkworth  SMALL PORCUPINEGRASS [p. *364*, <u>*511*</u>]

*Hesperostipa curtiseta* grows on light to clay loams in the prairies and northern portion of the central plains and northern intermontane grasslands, at elevations from 750–2050 m. It differs from *H. comata* subsp. *intermedia* in having unevenly pubescent lemmas and non-lacerate ligules, and from *H. spartea* in its smaller size, usually glabrous or evenly pubescent culm nodes, usually glabrous sheaths, and shorter florets.

### 4. Hesperostipa spartea (Trin.) Barkworth  PORCUPINEGRASS [p. *364*, <u>*511*</u>]

*Hesperostipa spartea* grows at elevations of 200–2600 m, primarily in the grasslands of the central plains and southern prairies of the *Manual* region. In its more northern locations, it tends to grow on sandy soils. It was once a common species, but its habitat is now intensively cultivated. It differs from *H. curtiseta* in its larger size, unevenly pubescent culm nodes, usually ciliate lower sheaths, and longer florets.

[1]Mary E. Barkworth

## 10.09 PIPTOCHAETIUM J. Presl[1]

Pl per; csp, not rhz. **Clm** 4–150 cm, usu erect, smt decumbent, glab, not brchd above the base; **bas brchg** invag; **prphl** shorter than the shth, mostly glab, keels usu with hairs, apc bifid, teeth 1–3 mm; **clstgn** not developed. **Shth** open to the base, mrg glab; **lig** memb, decurrent, truncate to acute, smt highest at the sides, smt ciliate; **bld** convolute to flat, translucent between the veins, often sinuous distally. **Infl** tml pan, open or contracted, spklt usu confined to the distal ¹/₂ of each br. **Spklt** 4–22 mm, with 1 flt; **rchl** not prolonged beyond the base of the flt; **dis** above the glm, beneath the flt. **Glm** subequal, longer than the flt, lanceolate, 3–7(8)-veined; **flt** globose to fusiform, terete to lat compressed; **cal** well developed, sharp or blunt, glab or antrorsely strigose, hairs yellow to golden brown; **lm** coriaceous to indurate, glab or pubescent, striate, particularly near the base, smooth, papillose, or tuberculate, often smooth on the lo portion and papillate to tuberculate distally, mrg involute, fitting into the grooved pal, apc fused into a crwn, awned, lm often narrowed below the crwn, crwn usu ciliate; **awn** caducous to persistent, usu twice-geniculate, first 2 segments usu twisted and hispid, tml segment straight and scabridulous; **pal** longer than the lm, similar in texture, glab, sulcate between the veins, apc prow-tipped; **lod** 2 or 3, memb, glab, blunt or acute; **anth** 3; **ov** glab; **sty** 2. **Car** terete to globose or lens-shaped.

*Piptochaetium* is primarily South American, being particularly abundant in Argentina. It has 27 species. Four species are native in the *Manual* region; two South American species are established at a single location in Marin County, California.

1. Florets 6.5–22 mm long; culms 40–130 cm tall.
    2. Lemmas hairy; awns 19–35 mm long . . . . . . . . . . . . . . . . . . . . . . . . . . . . . . . . . . . . . . . . . . . . . . . . 1. *P. pringlei*
    2. Lemmas glabrous; awns 40–120 mm long.
        3. Florets 7–13 mm long; awns 40–75 mm long . . . . . . . . . . . . . . . . . . . . . . . . . . . . . . . . . . . 2. *P. avenaceum*
        3. Florets 13.5–22 mm long; awns 62–120 mm long . . . . . . . . . . . . . . . . . . . . . . . . . . . . . . 3. *P. avenacioides*
1. Florets 2.3–5.5 mm long; culms 20–95 cm tall.
    4. Lemmas golden brown, hairy, the hairs easily rubbed off . . . . . . . . . . . . . . . . . . . . . . . . . . . 4. *P. fimbriatum*
    4. Lemmas dark brown, glabrous.
        5. Awns 10–16 mm long; blades 0.8–1.5 mm wide; distal margin of the lemma crowns straight . . . . . . . . . . 5. *P. setosum*
        5. Awns 15–25 mm long; blades 0.2–0.4 mm wide; distal margin of the lemma crowns sometimes slightly to strongly revolute . . . . . . . . . . . . . . . . . . . . . . . . . . . . . . . . . . . . . . . 6. *P. stipoides*

### 1. Piptochaetium pringlei (Beal) Parodi PRINGLE'S SPEARGRASS [p. *364*, *511*]

*Piptochaetium pringlei* grows in oak woodlands, often on rocky soils, in the southwestern United States and northwestern Mexico. It differs from *P. fimbriatum* in having longer florets and sharper calluses.

### 2. Piptochaetium avenaceum (L.) Parodi BLACKSEED SPEARGRASS [p. *364*, *511*]

*Piptochaetium avenaceum* grows in open oak and pine woods, often on sandy soils, primarily in the southeastern United States. It differs from *P. avenacioides* in its smaller size and more widespread distribution, and from *P. leianthum* (Hitchc.) Beetle, a species of northeastern Mexico, in it larger size.

### 3. Piptochaetium avenacioides (Nash) Valencia & Costas FLORIDA SPEARGRASS [p. *364*, *511*]

*Piptochaetium avenacioides* grows in dry woods, generally on sandy ridges. It is endemic to Florida, growing primarily in the central peninsula. It differs from *P. avenaceum* in its larger size and more restricted distribution.

### 4. Piptochaetium fimbriatum (Kunth) Hitchc. PINYON RICEGRASS [p. *364*, *511*]

*Piptochaetium fimbriatum* is an attractive species that grows in oak and pinyon woods of the southwestern United States and adjacent Mexico, and merits consideration as an ornamental. It differs from *P. pringlei* in having shorter florets and blunt calluses.

### 5. Piptochaetium setosum (Trin.) Arechav. BRISTLY RICEGRASS [p. *364*, *511*]

*Piptochaetium setosum* is native to central Chile. There is an established population in Marin County, California, that grows intermingled with *P. stipoides*, another South American species. The seeds might have been brought in by birds, as the area was a bird refuge at one time.

### 6. Piptochaetium stipoides (Trin. & Rupr.) Hack. STIPOID RICEGRASS [p. *364*, *511*]

*Piptochaetium stipoides* is native to South America. There is one known population in the *Manual* region, in Marin County, California, which grows with *P. setosum* in a meadow adjacent to an old dirt road. The seeds might have been brought in by birds, as the area was a bird refuge at one time.

## 10.10 ORYZOPSIS Michx.[2]

Pl per; csp, not rhz. **Clm** 25–65 cm, erect or spreading, **bas brchg** exvag; **prphl** not visible; **nd** glab. **Lvs** mostly bas; **clstgn** not developed; **shth** open, glab; **aur** absent; **lig** memb, longest at the sides or rounded, ciliate; **bld** of bas lvs 30–90 cm, remaining green over winter, erect when young, recumbent in the fall, bases twisted, placing the abx surfaces upmost, cauline lf bld red, flag lf bld 2–12 mm, conspicuously narrower than the top of the shth. **Infl** pan, contracted. **Spklt** 5–7.5 mm, with 1 flt; **rchl** not prolonged beyond the base of the flt; **dis**

[1,2]Mary E. Barkworth

above the glm, beneath the flt. **Glm** subequal, 6–10-veined, apc mucronate; **flt** terete to lat compressed; **cal** usu less than $^1/_5$ the length of the flt, blunt, distal portions pilose; **lm** coriaceous, pubescent at least bas, 3–5(9)-veined, mrg strongly overlapping at maturity, awned, lm-awn jnct conspicuous, lobed, lobes 0.1–0.2 mm; **awn** more or less straight, deciduous; **pal** similar to the lm in length, texture, and pubescence, concealed by

### 1. Oryzopsis asperifolia Michx. ROUGHLEAF RICEGRASS, WINTER GRASS, ORYZOPSIS À FEUILLES RUDES [p. *365*, <u>511</u>]

*Oryzopsis asperifolia* grows in both deciduous and coniferous woods, generally on open, rocky ground in areas with well-developed duff. It is listed as endangered or threatened in Indiana, Ohio, New Jersey, Maryland, and Virginia. Leaf development is unusual in that the

the lm, 2-veined, flat between the veins; **lod** 2, free, memb, 2-veined; **anth** 3; **sty** 1, with 2 br; **ov** glab. **Car** falling with the lm and pal.

*Oryzopsis* is treated here as a unispecific genus that is restricted to North America. The North American species previously included in *Oryzopsis* have been transferred to *Achnatherum* and *Piptatherum*.

leaves start to develop in midsummer, the blades growing upright. As the year progresses, they bend over, but stay alive and green through winter and the following spring.

## 10.11 ×ACHNELLA Barkworth[1]

**Pl** per; csp, not rhz. **Clm** to 90 cm, erect, glab, not brchg at the up nd; **prphl** shorter than the shth. **Shth** mostly glab, mrg sparsely ciliate, hairs longer distally; **clstgn** not present in the bas shth; **col** without tufts of hair at the sides; **aur** absent; **lig** scarious, glab; **bld** of bas lvs to 40 cm long, 1–1.5 mm wide, convolute when dry, tapering to the narrowly acute apc, flag lf bld longer than 10 mm, bases about as wide as the top of the shth. **Infl** pan. Spklt 6–8.5 mm, with 1 flt; **rchl** not prolonged beyond the base of the flt; **dis** above the glm, beneath the flt. **Glm** saccate-lanceolate, tapering from above midlength into elongate apc, midveins extending to or

### 1. ×Achnella caduca (Beal) Barkworth DROPAWN [p. *365*, <u>511</u>]

*×Achnella caduca* is a sterile hybrid between *Achnatherum hymenoides* and *Nassella viridula* that occurs infrequently in Montana and Wyoming. It differs from *Achnatherum hymenoides* in its longer glumes and florets; from *Nassella viridula* in its more

nearly to the apc, apc narrowly acute to acuminate; **lo glm** 6–8.5 mm; **up glm** 6.5–7.5 mm, slightly narrower than the lo glm; **flt** 4–5 mm, lengths more than 3 times the widths; **cal** about 0.7 mm, blunt; **lm** coriaceous, evenly hairy throughout, hairs 1–2 mm, apical hairs not longer than those below; **awn** twisted, not or once-geniculate, readily deciduous; **pal** more than $^2/_3$ as long as the lm, veins ending at the apc; **anth** poorly developed, indehiscent, smt with 1 or 2 penicillate hairs.

*×Achnella* comprises sterile hybrids between *Achnatherum* and *Nassella*. Only one such hybrid is known; it is restricted to the *Manual* region.

saccate glumes, longer lemma hairs, and well-developed palea; and from other hybrids involving *Achnatherum hymenoides* in its readily deciduous awn.

## 10.12 NASSELLA (Trin.) E. Desv.[2]

**Pl** usu per, rarely ann; usu csp, occ rhz. **Clm** 10–175(210) cm, smt brchd at the up nd, br flexible; **prphl** not evident, shorter than the shth. **Lvs** mostly bas, not overwintering; **shth** open; **clstgn** smt present; **aur** absent; **lig** memb, smt pubescent or ciliate; **bld** of bas lvs 3–60 cm long, 0.2–8 mm wide, apc narrowly acute to acute, not sharp, flag lf bld 1–80 mm, bases about as wide as the top of the shth. **Infl** tml pan, smt partially included at maturity. Spklt 3–22 mm, with 1 flt; **rchl** not prolonged beyond the base of the flt; **dis** above the glm, beneath the flt. **Glm** longer than the flt, narrowly lanceolate or ovate, bas portion usu purplish at anthesis, color fading with age, (1)3–5-veined, smt awned; **flt** usu terete, smt slightly lat compressed; **cal** blunt or sharp, glab or antrorsely strigose; **lm** usu papillose or tuberculate, at least distally, smt smooth throughout, glab or variously hairy, strongly convolute, wrapping 1.2–1.5 times around the car, apc not lobed,

fused distally into *crowns*, these often evident by their pale color and constricted bases; **crwn** mostly glab, rims often bearing hairs with bulbous bases; **awn** tml, centric or eccentric, deciduous or persistent, usu twice-geniculate, second geniculation often obscure; **pal** up to $^1/_2$ as long as the lm, glab, without veins, flat; **lod** 2 or 3, if 3, the third somewhat shorter than the other 2; **anth** 1 or 3, if 3, often of 2 lengths, penicillate; **ov** glab; **sty** 2, bases free. **Car** glab, not ribbed; **hila** elongate; **emb** to $^2/_5$ as long as the car.

*Nassella* is now interpreted as including at least 116 species, the majority of which are South American. There are eight species in the *Manual* region, one of which is introduced; two additional species treated here were found in the region at one time, but have not become established. The strongly convolute lemmas distinguish *Nassella* from all other genera of *Stipeae* in the Americas and, in combination with the reduced, ecostate, glabrous paleas, from all other genera in the tribe.

[1,2]Mary E. Barkworth

1. Florets 1.5–3 mm long; blades 0.2–1.5 mm wide, usually tightly convolute.
   2. Florets widest about midlength; awns 45–100 mm long, almost centric . . . . . . . . . . . . . . . . . . . . . . . . 7. *N. tenuissima*
   2. Florets widest near the top; awns 7–35 mm long, eccentric.
      3. Awns 15–35 mm long; lemmas strongly tuberculate, particularly distally . . . . . . . . . . . . . . . . . . . . . 8. *N. trichotoma*
      3. Awns 7–10 mm long; lemmas smooth . . . . . . . . . . . . . . . . . . . . . . . . . . . . . . . . . . . . . . . . . . . 9. *N. chilensis*
1. Florets 3.4–13 mm long; blades 0.4–8 mm wide, flat to convolute.
   4. Terminal segment of the awns cernuous.
      5. Awns 12–55 mm long, 0.1–0.2 mm thick at the base . . . . . . . . . . . . . . . . . . . . . . . . . . . . . . . . 4. *N. lepida*
      5. Awns 50–110 mm long, 0.2–0.3 mm thick at the base . . . . . . . . . . . . . . . . . . . . . . . . . . . . . . . 6. *N. cernua*
   4. Terminal segment of the awns straight.
      6. Florets 3.4–5.5 mm long; lemmas not constricted below the crown; awns 19–32 mm long . . . . . . . . . 10. *N. viridula*
      6. Florets 6–13 mm long; lemmas constricted below the crown; awns 30–120 mm long.
         7. Lemmas hairy between the veins at maturity . . . . . . . . . . . . . . . . . . . . . . . . . . . . . . . . . . . . . . . . 5. *N. pulchra*
         7. Lemmas glabrous between the veins at maturity.
            8. Crowns usually wider than long, the rims with hairs to 0.5 mm long; florets widest just
               below the crowns . . . . . . . . . . . . . . . . . . . . . . . . . . . . . . . . . . . . . . . . . . . . . . . . . . . . . . . . 1. *N. neesiana*
            8. Crowns usually longer than wide, the rims with hairs 1–2 mm long; florets widest near or
               slightly above midlength.
               9. Florets 6.5–13 mm long; crowns often flaring distally; plants native to Texas and
                  adjacent states . . . . . . . . . . . . . . . . . . . . . . . . . . . . . . . . . . . . . . . . . . . . . . . . . . 2. *N. leucotricha*
               9. Florets 6–8 mm long; crowns more or less straight-sided; plants introduced, established
                  in California . . . . . . . . . . . . . . . . . . . . . . . . . . . . . . . . . . . . . . . . . . . . . . . . . . . . . 3. *N. manicata*

## 1. Nassella neesiana (Trin. & Rupr.) Barkworth
URUGUAYAN TUSSOCKGRASS [p. *365*, <u>511</u>]

*Nassella neesiana* is native to South America, growing from Ecuador to Argentina, primarily in steppe habitats. It was found on ballast dumps in Mobile, Alabama but has not persisted in the *Manual* region. It has become established in Australia, where it is considered a noxious weed.

## 2. Nassella leucotricha (Trin. & Rupr.) R.W. Pohl
TEXAN NASSELLA, TEXAN NEEDLEGRASS [p. *365*, <u>511</u>]

*Nassella leucotricha* grows from the southern United States into northern Mexico, mostly in open grasslands, but it is also found in woodlands. It provides good spring forage and increases in abundance with moderate grazing, primarily because of its cleistogenes. It differs from *N. manicata* in its longer florets and less strongly developed crowns. The sharp callus easily sticks to skin and clothing, and can cause wounds, especially in the mouths of grazing animals.

## 3. Nassella manicata (E. Desv.) Barkworth ANDEAN
TUSSOCKGRASS [p. *365*, <u>511</u>]

*Nassella manicata* is native to South America, growing in the foothills of the Andes Mountains. It is established in California, growing in disturbed sites, including grazed meadows and old gold tailings. *Nassella manicata* differs from both *N. leucotricha* and *N. pulchra* in its shorter florets and more strongly developed crowns.

## 4. Nassella lepida (Hitchc.) Barkworth FOOTHILLS
NASSELLA, FOOTHILLS NEEDLEGRASS [p. *365*, <u>511</u>]

*Nassella lepida* usually grows on dry hillsides in chaparral habitats, from California into northern Mexico. It differs from *N. cernua* in its shorter, thinner awns and more numerous spikelets. It occasionally hybridizes with *N. pulchra*.

## 5. Nassella pulchra (Hitchc.) Barkworth PURPLE
NASSELLA, PURPLE NEEDLEGRASS [p. *365*, <u>511</u>]

*Nassella pulchra* grows in oak chaparral and grassland communities of the coast ranges and Sierra foothills of California, extending south into Mexico. It resembles *N. manicata*, but has longer florets and less strongly developed crowns. *Nassella pulchra* and *N. cernua* sometimes hybridize.

## 6. Nassella cernua (Stebbins & Love) Barkworth
CERNUOUS NASSELLA, NODDING NEEDLEGRASS [p. *365*, <u>511</u>]

*Nassella cernua* grows in grasslands, chaparral, and juniper associations of the inner coast ranges of California and Baja California, Mexico. Small specimens resemble *N. lepida*, but have longer and thicker awns and fewer florets. Large specimens resemble *N. pulchra*, but have thinner awns with cernuous, rather than straight, terminal segments. *Nassella cernua* is superficially similar to *Achnatherum eminens*, but differs in its shorter ligules, strongly overlapping lemma margins, glabrous paleas, and geographic distribution.

## 7. Nassella tenuissima (Trin.) Barkworth FINELEAVED
NASSELLA [p. *365*, <u>512</u>]

*Nassella tenuissima* grows on rocky slopes in oak or pine associations and in open, exposed grasslands. Its native range extends from the southwestern United States into northern Mexico. It is now also established in the San Francisco Bay area. It is an attractive species, but it readily escapes from cultivation into nearby disturbed sites.

## 8. Nassella trichotoma (Nees) Hack. *ex* Arechav.
SERRATED TUSSOCKGRASS, YASS TUSSOCKGRASS [p. *365*, <u>512</u>]

*Nassella trichotoma* is a native of South America, and has been accidentally introduced into the United States. Because it is on the U.S. Department of Agriculture's noxious weed list, all known populations have been eliminated. New populations should be reported to the Department.

## 9. Nassella chilensis (Trin.) E. Desv. CHILEAN
TUSSOCKGRASS [p. *365*, <u>512</u>]

*Nassella chilensis* is an Andean species that was once collected from a ballast dump in Portland, Oregon. It is not established in the *Manual* region.

## 10. Nassella viridula (Trin.) Barkworth GREEN
NASSELLA, GREEN NEEDLEGRASS [p. *365*, <u>512</u>]

*Nassella viridula* grows in grasslands and open woods, frequently on sandy soils. It is the most widespread species of *Nassella* in North America. It differs from *Achnatherum robustum* in its tightly convolute lemmas and in having glabrous to sparsely pubescent collars on its flag leaves. It differs from the hybrid with *Achnatherum hymenoides*, ×*Achnella caduca*, in its less saccate glumes, shorter lemma hairs, and shorter paleas.

## 10.13 JARAVA Ruiz & Pav.[1]

Pl per; csp, smt rhz, rhz forming knotted bases. **Clm** 15–200 cm, not brchg at the up nd; **bas brchg** invag or exvag; **prphl** not evident, shorter than the lf shth. **Lvs** mostly bas, not overwintering; **shth** open to the base; **clstgn** not present; **col** with tufts of hair on either side; **aur** absent; **lig** memb, truncate or shortest in the center and rounded, edges usu ciliate, hairs at the outer edges often longer than the cent memb portion, lig of the lo lvs glab or hairy, smt densely hairy, those of the up lvs glab or sparsely hairy; **bld** usu convolute, apc narrowly pointed, flag lvs longer than 10 mm. **Infl** pan, often partially included in the up lf shth; **br** straight. **Spklt** 5.5–24 mm, with 1 flt; **rchl** not prolonged beyond the flt; **dis** above the glm, beneath the flt. **Glm** unequal, usu longer than the flt, smt shorter, hyaline, 0–5-veined; **flt** narrowly lanceoloid, terete; **cal** 0.2–1.6(3) mm, acute, less than or equaling the flt diameter, antrorsely strigose distally, hairs white; **lm** thickly memb, bas $^2/_3$ scabrous or shortly pubescent, distal $^1/_3$ often bearing a pappus of ascending to strongly divergent 3–8 mm hairs, smt glab or with appressed hairs shorter than 1 mm, mrg not or only slightly overlapping at maturity, apc not fused into a crwn, lobes to 0.2 mm, with a single, tml awn, lm-awn jnct conspicuous; **awn** 9–45(80) mm, persistent or deciduous, scabrous, weakly once- or twice-geniculate, first segment scabrous or pilose, tml segment glab or pilose, smooth or scabrous; **pal** $^1/_3$–$^1/_2$ as long as the lm, flat between the veins, memb to hyaline, glab or sparsely pubescent, 2-veined, veins poorly developed, apc rounded to irregular; **lod** 2–3, the third, if present, red; **anth** 3. **Car** fusiform, not ribbed; **hila** linear.

As treated here, *Jarava* is a genus of approximately 50 species, all of which are native to South America. *Jarava speciosa* grows as a disjunct in the southwestern United States, its broadest distribution being in South America. *Jarava ichu* and *J. plumosa* have been found as escapes from cultivation in California.

1. Awns 35–80 cm long, the basal segment pilose; ligules of the basal leaves softly and densely hairy . . . . . . . . . . 3. *J. speciosa*
1. Awns 9–30 mm long, the basal segment scabridulous or smooth; ligules of the basal leaves glabrous or almost so.
    2. Glumes clearly exceeding the florets; pappus hairs 3–4 mm long . . . . . . . . . . . . . . . . . . . . . . . . . . . . . . 1. *J. ichu*
    2. Glumes from shorter than to subequal to the florets; pappus hairs 5–8 mm long . . . . . . . . . . . . . . . . . . 2. *J. plumosa*

### 1. Jarava ichu Ruiz & Pav. PERUVIAN NEEDLEGRASS [p. 366]

*Jarava ichu* is native to Mexico, Costa Rica, Venezuela, Colombia, Ecuador, Peru, Bolivia, and Argentina. It is abundant in much of this range. In the *Manual* region, it is sold as an attractive ornamental. The species could become a problem, because it is self-compatible and produces a large quantity of wind-dispersed seeds.

### 2. Jarava plumosa (Spreng.) S.W.L. Jacobs & J. Everett PLUMOSE NEEDLEGRASS [p. 366, 512]

A native of Argentina, Chile, and Uruguay, *Jarava plumosa* was collected in Berkeley, California in 1983. It is not known to be established in the *Manual* region. In its native range, it often grows on poor, unstable soils.

### 3. Jarava speciosa (Trin. & Rupr.) Peñail. DESERT NEEDLEGRASS [p. 366, 512]

*Jarava speciosa* grows on rocky slopes in canyons of arid and semiarid regions of the southwestern United States and northern Mexico, and in Chile and northern to central Argentina. The reddish brown leaf bases, differing lower and upper ligules, and the pilose, once-geniculate awns make *Jarava speciosa* an easy species to recognize in North America. It is also an attractive species, well worth cultivating. The growth of young shoots and flowering is stimulated by fire.

## 10.14 AMELICHLOA Arriaga & Barkworth[2]

Pl per; csp. **Clm** erect, with 2–3 nd, not brchg at the up nd; **bas brchg** invag; **prphl** concealed by the lf shth, winged over the keels, apc bifid, teeth 0.5–3.5 mm. **Lvs** mostly bas; **shth** open, smooth, glab; **clstgn** often present, spklt of clstgn 0.5–1 mm long, with thin glm shorter than the flt, flt unawned or with rdcd awns; **aur** absent; **lig** scarious, rounded to acute, ciliate; **bld** stiff, involute, apc stiff, brown, sharply pointed, bld of the flag lvs 5–13 cm long, bases similar in width to the top of the shth. **Infl** pan, the main pan tml, apparently wholly chasmogamous. **Spklt** with 1 flt; **dis** above the glm, beneath the flt. **Glm** exceeding the flt, acute to acuminate, 1–5-veined; **flt** fusiform, terete; **cal** antrorsely strigose, blunt; **lm** pubescent, often more densely and/or more persistently so over the midvein and lat veins, hairs on the proximal portion about 0.7–2 mm, hairs on the distal portion often longer; **crwn** not developed; **awn** once- or twice-geniculate, scabrous, persistent; **pal** $^3/_4$ as long as to almost equaling the lm, flat, hairy, hairs 0.2–1 mm, veins terminating at or near the apc, apc similar in texture to the body; **lod** 3; **anth** 3, anth smt all of equal size and more than 2 mm, smt 1 longer than 2 mm and 2 much shorter, smt all shorter than 2 mm; **ov** glab; **sty** with 2 br, united at the base, stigmas plumose. **Car** obovoid, with 3 smooth, longitudinal ribs at maturity, stylar bases 1–2 mm, persistent, smt eccentric; **hila** linear, about as long as the car.

*Amelichloa* includes five species, four of which are South American. The fifth species, *A. clandestina*, grows in northern Mexico. Two species are established in the *Manual* region. A

[1,2]Mirta O. Arriaga

third species, *A. caudata*, was found on ballast dumps near Portland, Oregon, at the turn of the twentieth century; it is not established in the region. All species are avoided by cattle because of their sharply pointed leaves. The cleistogamous panicles, which may be at or below ground level, remain a source of seeds unless the plants are completely uprooted. *Amelichloa caudata* and *A. clandestina* are a greater problem than *A. brachychaeta* in this regard, because they appear to produce such panicles more frequently.

1. Mature caryopses with inclined, eccentric stylar bases; lemmas glabrous or hairy between the lateral and marginal veins, glabrous between the midvein and the lateral vein, even at the base . . . . . . . . . . . . . . . . . 2. *A. caudata*
1. Mature caryopses with erect, usually centric stylar bases; proximal ¹/₂ of the lemmas pubescent between the lateral and marginal veins, at least initially, usually also between the midvein and lateral veins.
    2. Florets 4–5.5 mm long; ligules 0.2–0.6 mm long; anthers 2–3 mm long . . . . . . . . . . . . . . . . . . . . 1. *A. brachychaeta*
    2. Florets 5.5–8 mm long; ligules 0.5–1.5 mm long; anthers 3–4 mm long . . . . . . . . . . . . . . . . . . . . . . 3. *A. clandestina*

### 1. Amelichloa brachychaeta (Godr.) Arriaga & Barkworth PUNA NEEDLEGRASS [p. 366, 512]

*Amelichloa brachychaeta* has been found at a few locations in California, where it is listed as a noxious weed. It is native to Uruguay and Argentina.

### 2. Amelichloa caudata (Trin.) Arriaga & Barkworth SOUTH AMERICAN NEEDLEGRASS [p. 366, 512]

*Amelichloa caudata* is native to South America. It was collected, as *Stipa litoralis* Phil., on ballast dumps near Portland, Oregon, early in the twentieth century. Although it has not become established in the *Manual* region, it has done so in Australia. It is a potentially invasive weed.

### 3. Amelichloa clandestina (Hack.) Arriaga & Barkworth MEXICAN NEEDLEGRASS [p. 366, 512]

*Amelichloa clandestina* is native from northern Mexico to Colombia. It has been accidentally introduced to pastures and roadsides in Texas, and is now established there. Reports from California may reflect misidentifications of *A. brachychaeta*, which differs from *A. clandestina* in its shorter ligules, florets, and anthers.

## 10.15 AUSTROSTIPA S.W.L. Jacobs & J. Everett[1]

Pl facultative per; to 2.5 m, smt shrublike, often with knotty bases. **Clm** often persistent, smt geniculate at the base, smt with stiff br at the up nd; **bas brchg** usu exvag; **prphl** not evident. **Lvs** not overwintering; **shth** open to the base, mrg smt extending beyond the base of the lig, the extensions often conspicuously hairy, smt grading into the lig; **aur** not present; **lig** memb; **bld** flat, convolute, or terete, usu scabrous, smt pubescent, those of the flag lvs longer than 10 mm, bases usu as wide as the top of the shth, smt narrower. **Infl** pan, smt contracted, rchs persistent or disarticulating whole at maturity; **ped** smooth or scabrous, smt hairy, hairs to 3 mm; **dis** above the glm, smt also at the base of the pan. **Spklt** with 1 flt; **rchl** not prolonged. **Glm** usu longer than the flt, narrow, more or less keeled, hyaline to chartaceous, 1–5(7)-veined, apc usu acute or acuminate, rarely muticous or mucronate, remaining open after the flt falls; **flt** at least 5 times longer than wide; **cal** strigose, usu sharp; **lm** usu coriaceous or indurate, 3–5(7)-veined, mrg thin, usu convolute, rarely involute, apc glab or pilose, with 0–2 minute to small, memb lobes, awned, lm-awn jnct evident; **awn** once- or twice-geniculate; **pal** from shorter than to subequal to the lm, flat between the veins, 0–2-veined; **lod** 3 or 2; **anth** 3, frequently penicillate; **ov** glab. **Car** fusiform, terete; **hila** linear, nearly as long as the car.

*Austrostipa* is a genus of 63 species, all of which are native to Australasia. Two species are cultivated in the *Manual* region.

1. Plants shrubby; panicle branches and pedicels plumose, hairs 1.5 mm or longer . . . . . . . . . . . . . . . . . . 1. *A. elegantissima*
1. Plants bamboolike; panicle branches and pedicels not plumose, if hairy then the hairs to 0.3 mm long . . . 2. *A. ramosissima*

### 1. Austrostipa elegantissima (Labill.) S.W.L. Jacobs & J. Everett AUSTRALIAN FEATHERGRASS [p. 366]

*Austrostipa elegantissima* is native to southern Australia. It is cultivated as an ornamental in the United States.

### 2. Austrostipa ramosissima (Trin.) S.W.L. Jacobs & J. Everett PILLAR-OF-SMOKE, AUSTRALIAN PLUMEGRASS, STOUT BAMBOOGRASS [p. 366]

*Austrostipa ramosissima* is native to eastern Australia. It is cultivated in the United States and southern British Columbia. In its native range *A. ramosissima* is drought tolerant, but prefers moist soils and well-drained gullies near forest or woodland margins.

[1]Surrey W.L. Jacobs

# 11. BRACHYPODIEAE Harz[1]

Pl ann or per; rhz or csp. **Clm** ann, not wd, ascending to erect or decumbent, smt brchg above the base; **intnd** hollow. **Shth** open, mrg overlapping for most of their length; **col** without tufts of hair on the sides; **aur** absent; **lig** memb, entire or toothed, smt shortly ciliate, those of the lo and up cauline lvs usu similar; **psdpet** absent; **bld** linear to narrowly lanceolate, venation parallel, cross venation not evident, without arm or fusoid cells, cross sections non-Kranz, epdm without microhairs, not papillate. **Infl** tml, spikelike rcm, spklt subsessile, solitary at all or most nd; **ped** to 2.5 mm. **Spklt** terete to slightly lat compressed, with (3)5–24 flt, distal flt smt red, strl; **dis** above the glm, beneath the flt; **rchl** prolonged beyond the base of the distal flt. **Glm** unequal, ¹/₂ as long as to equaling the adjacent lm, lanceolate, lo glm 3–7-veined, up glm 5–9-veined; **flt** subterete to slightly lat compressed; **cal** glab, not well developed; **lm** lanceolate, usu memb, rounded dorsally, (5)7–9-veined, veins not converging distally, inconspicuous, apc entire, obtuse or acute, unawned or tml awned; **pal** shorter than to slightly longer than the lm; **lod** 2, not veined, distal mrg ciliate or the apc puberulent; **anth** 3; **ov** with hairy apc; **sty** 2, bases free. **Car** with hairy apc, longitudinally grooved; **hila** linear; **emb** about ¹/₆ the length of the car.

The only genus of this tribe, *Brachypodium*, has sometimes been included in the *Bromeae* or *Triticeae*. It now appears that *Brachypodium* is an isolated genus within the *Poöideae*, hence its treatment here as the sole genus in a distinct tribe.

## 11.01 BRACHYPODIUM P. Beauv.[2]

Pl per or ann; rhz or csp, rhz often extensively brchd. **Clm** 5–200 cm, erect or decumbent, often rooting at the lo nd, smt brchd above the base; **nd** often pubescent. **Lvs** not bas concentrated; **shth** open, mrg overlapping, not fused; **aur** absent; **lig** memb, entire, toothed, or ciliate; **bld** flat or convolute, often attenuate. **Infl** spikelike rcm, most or all nd with 1 spklt, smt some with 2–3, most or all spklt appressed to strongly ascending; **dis** above the glm, beneath the flt. **Spklt** 14–80 mm, terete to lat compressed, with (3)5–24 flt. **Glm** unequal, ¹/₂ as long as to equaling the adjacent lm, lanceolate, memb, apc obtuse to acuminate, lo glm 3–7-veined, up glm 5–9-veined; **lem** usu memb, smt coriaceous at maturity, rounded on the back, (5)7–9-veined, apc obtuse or acute, unawned or tml awned; **pal** shorter than to slightly longer than the lm, with 2 well-developed veins, smt with minor veins in between, keeled over the well-developed veins, keels strongly ciliate; **lod** 2, oblong, attenuate distally, mrg ciliate or apc puberulent; **anth** 3; **sty** 2, free to the base, white. **Car** oblong, flattened, apc pubescent; **hila** linear.

*Brachypodium* is a genus of about 18 species, with about 15 species in Eurasia, centered on the Mediterranean, and three in the Western Hemisphere, centered in Mexico. All five species in the *Manual* region are Eurasian.

1. Plants annual; spikelets laterally compressed; anthers 0.5–1.1 mm long . . . . . . . . . . . . . . . . . . . . . . . . . . . . . 1. *B. distachyon*
1. Plants perennial; spikelets terete or subterete; anthers 2.8–6 mm long.
    2. Lemma awns 7–15 mm long, as long as or longer than the lemmas . . . . . . . . . . . . . . . . . . . . . . . . . . . . 2. *B. sylvaticum*
    2. Lemma awns absent or to 7 mm long, shorter than the lemmas.
        3. Blades with all veins more or less equally prominent on the adaxial surfaces . . . . . . . . . . . . . . . . . 3. *B. phoenicoides*
        3. Blades with the primary veins separated by finer secondary veins on the adaxial surfaces.
            4. Leaf blades flat, dark green, abaxial surfaces scabrous, not shiny; lemmas usually hairy . . . . . . . . . . 4. *B. pinnatum*
            4. Leaf blades involute or flat, light green, abaxial surfaces smooth or almost so, conspicuously
                shiny; lemmas usually glabrous . . . . . . . . . . . . . . . . . . . . . . . . . . . . . . . . . . . . . . . . . . . . . . 5. *B. rupestre*

### 1. Brachypodium distachyon (L.) P. Beauv. PURPLE FALSEBROME [p. *366*, <u>512</u>]

*Brachypodium distachyon* is native to dry, open habitats in southern Europe. It is now established in California and is known from scattered locations elsewhere in the *Manual* region. It is also established in Australia, where it grows in dry, disturbed areas on sandy or rocky soils.

*Brachypodium distachyon* differs from other species of *Brachypodium* in being a cleistogamous annual with shorter pedicels and anthers, laterally compressed spikelets, and fewer spikelets per raceme.

### 2. Brachypodium sylvaticum (Huds.) P. Beauv. SLENDER FALSEBROME [p. *366*, <u>512</u>]

*Brachypodium sylvaticum* is native to Eurasia and northern Africa, where it grows in woods and other shady places. In the *Manual* region, it is established in Oregon, where it is an aggressive weed that has spread to several western counties since its first discovery near Eugene, Lane County, in 1939. In December 2003 it was discovered in California. An additional report from Virginia has been confirmed. *Brachypodium sylvaticum* has sometimes been cultivated as an ornamental, but because of its ability to become an aggressive weed, such use is discouraged in the *Manual* region.

[1]Mary E. Barkworth   [2]Michael B. Piep

**3. Brachypodium phoenicoides (L.) Roem. & Schult.**
THINLEAF FALSEBROME [p. *367*, 512]

*Brachypodium phoenicoides* is native to dry, usually open, and often sandy habitats of the northern Mediterranean region. In the *Manual* region, it is currently known only from Sonoma County, California, where it is established on coastal sand dunes.

**4. Brachypodium pinnatum (L.) P. Beauv.** HEATH
FALSEBROME, TOR GRASS [p. *367*, 512]

*Brachypodium pinnatum* is native to Eurasia. It is reportedly established in Sonoma County, California, and in Massachusetts. In its native range it prefers open woodlands, forest edges, and grassland habitats, and is fairly tolerant of hot, dry conditions.

**5. Brachypodium rupestre (Host) Roem. & Schult.**
TUFTED FALSEBROME [p. *367*]

*Brachypodium rupestre* is native to Europe and northern Turkey; it is not known to be established in the *Manual* region. Specimens of this species were grown by the Natural Resources Conservation Service in Pima County, Arizona and Bernalillo County, New Mexico; they were distributed either as *Brachypodium* sp. or as *B. cespitosum* (Host) Roem. & Schult.

## 12. BROMEAE Dumort.[1]

Pl ann or per; usu csp, smt rhz. Clm ann, not wd, not brchg above the base; intnd usu hollow, rarely solid. Shth closed, mrg united for most of their length; col without tufts of hair on the sides; aur smt present; lig memb, smt shortly ciliate, those of the up and lo cauline lvs usu similar; psdpet absent; bld linear to narrowly lanceolate, venation parallel, cross venation not evident, without arm or fusoid cells, cross sections non-Kranz, epdm without microhairs, not papillate. Infl usu tml pan, smt rdcd to rcm in depauperate pl; dis above the glm and beneath each flt. Spklt 5–80 mm, not viviparous, terete to lat compressed, with 3–30 bisx flt, distal flt smt rdcd; rchl prolonged beyond the bases of the distal flt. Glm usu unequal, rarely more or less equal, exceeded by the distal flt, usu longer than ¹/₄ the length of the adjacent flt, lanceolate, 1–9(11)-veined; flt terete to lat compressed; cal glab, not well developed; lm lanceolate to ovate, rounded or keeled over the midvein, hrb to coriaceous, 5–13-veined, veins converging somewhat distally, apc usu minutely bilobed to bifid, rarely entire, usu awned, smt unawned, awns unbrchd, tml or subtml, usu straight, smt geniculate; pal usu shorter than the lm; lod 2, glab, not veined; anth 3; ov with hairy apc; sty 2, bases free. Car narrowly ellipsoid to linear, longi-tudinally grooved; hila linear; emb about ¹/₆ the length of the car.

There are three genera in the *Bromeae*. One genus, *Bromus*, grows in the *Manual* region.

## 12.01 BROMUS L.[2]

Pl per, ann, or bien; usu csp, smt rhz. Clm 5–190 cm. Shth closed to near the top, usu pubescent; aur smt present; lig memb, to 6 mm, usu erose or lacerate; bld usu flat, rarely involute. Infl pan, smt rcm in depauperate specimens, erect to nodding, open to dense, occ 1-sided; br usu ascending to spreading, smt reflexed or drooping. Spklt 5–70 mm, terete to lat compressed, with 3–30 flt; dis above the glm, beneath the flt. Glm unequal, usu shorter than the adjacent lm, always shorter than the spklt, glab or pubescent, usu acute, rarely mucronate; lo glm 1–7(9)-veined; up glm 3–9(11)-veined; lm 5–13-veined, rounded to keeled, glab or pubescent, apc entire, emgt, or toothed, usu tml or subtml awned, smt with 3 distinct awns or unawned; pal usu shorter than the lm, ciliate on the keels, adnate to the car; anth (2)3.

*Bromus* grows in temperate and cool regions. It includes 100–400 species. Of the 52 species in the *Manual* region, 28 are native and 24 are introduced. The native perennial species provide considerable forage for grazing animals, with some species being cultivated for this purpose. The introduced species, all but three of which are annuals, range from sporadic introductions to well-established members of the region's flora. Many are weedy and occupy disturbed sites. Some are used for hay; others have a floret structure that can injure grazing animals.

In the keys and descriptions, the distances from the bases of the subterminal lemma awns to the lemma apices are measured on the most distal florets in a spikelet.

1. Lemmas strongly keeled, at least distally; spikelets strongly laterally compressed; lower glumes 3–7(9)-veined . . . . . . . . . . . . . . . . . . . . . . . . . . . . . . . . . . . . . . . . . . . . . . . . . . . . . . . . . . . sect. *Ceratochloa*
1. Lemmas rounded over the midvein; spikelets terete to moderately laterally compressed; lower glumes 1–5-veined.
  2. Awns, if present, arising less than 1.5 mm below the lemma apices; lemma apices entire, emarginate, or with teeth less than 1 mm long.
    3. Lower glumes 1–3-veined; upper glumes 3–5-veined; plants perennial or annual, if annual, the lower glumes 1-veined and the upper glumes 3-veined . . . . . . . . . . . . . . . . . . . . . . . . . . . . . . . . . . sect. *Bromopsis*
    3. Lower glumes 3–5-veined; upper glumes 5–9-veined; plants annual or biennial, if biennial, the upper glumes 7-veined and/or the lateral veins of the lemmas prominently ribbed . . . . . . . . . . . sect. *Bromus* (in part)
  2. Awns arising 1.5 mm or more below the lemma apices, lemma apices entire, emarginate, or with teeth to 5 mm long.

[1]Mary E. Barkworth  [2]Leon E. Pavlick† and Laurel K. Anderton

4. Awns usually geniculate, sometimes only divaricate, lemma teeth 2–3 mm long, usually aristate, sometimes only acuminate . . . . . . . . . . . . . . . . . . . . . . . . . . . . . . . . . . . . . . . . . . . . . . . . . . . . . . sect. *Neobromus*

4. Awns straight, arcuate, or divaricate, not geniculate, sometimes absent; lemma teeth absent or to 5 mm long, acuminate.

    5. Lower glumes 1–3-veined; upper glumes 3–5-veined; spikelets with parallel or diverging sides in outline, often widening distally; lemma apices bifid, teeth (0.8)1–5 mm long . . . . . . . . . . . . . . . . . . sect. *Genea*

    5. Lower glumes 3–5-veined; upper glumes 5–9-veined; spikelets with parallel or converging sides in outline; lemma apices entire to bifid, teeth less than 1 mm long, apices sometimes split and teeth appearing longer . . . . . . . . . . . . . . . . . . . . . . . . . . . . . . . . . . . . . . . . . . . . . . . . . sect. *Bromus* (in part)

## Bromus sect. Ceratochloa

1. Lemmas unawned or with awns to 3.5 mm long; lemmas usually glabrous, sometimes pubescent distally, veins prominent for most of their length . . . . . . . . . . . . . . . . . . . . . . . . . . . . . . . . . 1. *B. catharticus* (in part)

1. Lemmas awned, awns (2)4–17 mm long; lemmas pubescent or glabrous, veins obscure or prominent.

    2. Lower panicle branches shorter than 20 cm, with 1–3 spikelets on the distal ¹⁄₂, sometimes confined to the tips; culms 3–7 mm thick.

        3. Lower panicle branches shorter than 20 cm, spreading to drooping . . . . . . . . . . . . . . . . . . . . . . . . 2. *B. sitchensis*

        3. Lower panicle branches shorter than 10 cm, stiffly ascending . . . . . . . . . . . . . . . . . . . . . . . . . . . 4. *B. aleutensis*

    2. Lower panicle branches usually shorter than 10 cm, with 1–5 spikelets variously distributed; culms less than 4 mm thick.

        4. Upper glume about as long as the lowest lemma in each spikelet; lemmas glabrous or pubescent distally or throughout, the marginal hairs, if present, longer than those elsewhere . . . . . . . . . . . . . . . 3. *B. arizonicus*

        4. Upper glume shorter than the lowest lemma in each spikelet; lemmas glabrous or pubescent only on the margins or throughout, if throughout, the marginal hairs similar in length to those elsewhere.

            5. Panicles dense; spikelets crowded, overlapping, usually longer than the pedicels and branches; culms 20–70 cm tall, sometimes geniculate at the base; blades glabrous; ligules 1–6 mm long . . . . . 5. *B. maritimus*

            5. Panicles loose to compact; spikelets not crowded or overlapping, shorter than at least some pedicels and branches; culms 30–120(180) cm tall, erect or decumbent; blades glabrous or hairy; ligules 1–4 mm long.

                6. Lemmas and sheath throats glabrous . . . . . . . . . . . . . . . . . . . . . . . . . . . . . . . . . . . . . . 7. *B. polyanthus*

                6. Lemmas and/or sheath throats with hairs.

                    7. Lemmas 9–13-veined, veins often raised and riblike distally or throughout . . . . . . . 1. *B. catharticus* (in part)

                    7. Lemmas 7–9-veined, veins usually not raised or riblike . . . . . . . . . . . . . . . . . . . . . . . . . . . 6. *B. carinatus*

## Bromus sect. Bromopsis

1. Plants rhizomatous.

    2. Culms 30–90 cm long, forming distinct clumps; rhizomes short . . . . . . . . . . . . . . . . . . . . . . . . . . . . . . 9. *B. riparius*

    2. Culms 50–135 cm long, single or few together; rhizomes short to long-creeping.

        3. Lemma backs sparsely to densely hairy throughout, or on the lower portion and margins, or along the marginal veins and keel; cauline nodes and leaf blades pubescent or glabrous; awns usually present, to 7.5 mm long, sometimes absent . . . . . . . . . . . . . . . . . . . . . . . . 10. *B. pumpellianus* (in part)

        3. Lemma backs usually glabrous, occasionally sparsely puberulent at the base and sometimes on the margins; cauline nodes and leaf blades usually glabrous, rarely hairy; awns absent or to 3 mm long . . . . . . . . . . . . . . . . . . . . . . . . . . . . . . . . . . . . . . . . . . . . . . . . . . . . . . . . . . . . . . . . . . . 8. *B. inermis*

1. Plants not rhizomatous.

    4. Anthers (3.5)4–6(6.8) mm long; awns 2.5–7.5 mm long; plants of the Yukon River drainage of Alaska . . . . . . . . . . . . . . . . . . . . . . . . . . . . . . . . . . . . . . . . . . . . . . . . . . . . . . . . . 10. *B. pumpellianus* (in part)

    4. Anthers 1–7 mm long; awns 1–12 mm long; plants of various locations in the *Manual* region, if in the Yukon River drainage of Alaska, anthers 1–1.4 mm long.

        5. Culms with 9–20 nodes; collars and throats densely pilose; auricles 1–2.5 mm long on most lower leaves . . . . . . . . . . . . . . . . . . . . . . . . . . . . . . . . . . . . . . . . . . . . . . . . . . . . . . . . . . . . . 11. *B. latiglumis*

        5. Culms with (1)2–9 nodes; collars and throats pubescent or glabrous; auricles, if present, of various lengths.

            6. Most lower glumes within a panicle 3-veined, sometimes some 1-veined.

                7. Most upper glumes within a panicle 5-veined, sometimes some 3-veined.

                    8. Awns 1.5–3 mm long; anthers 1.5–2.5 mm long; ligules 0.5–1 mm long . . . . . . . . . . . . . . . . . . 13. *B. kalmii*

                    8. Awns 3–7 mm long; anthers 3–6 mm long; ligules to 4.2 mm long.

9. Glumes glabrous; ligules glabrous ............................................ 12. *B. laevipes*
9. Glumes usually pubescent, rarely glabrous; ligules usually pubescent or pilose, sometimes glabrous.
    10. Margins of the glumes and lemmas often bronze-tinged; ligules to 1.5 mm long; auricles usually present on the lower leaves, rarely absent ................. 14. *B. pseudolaevipes*
    10. Margins of the glumes and lemmas not bronze-tinged; ligules 1–3 mm long; auricles sometimes present ....................................... 17. *B. grandis* (in part)
7. Most upper glumes within a panicle 3-veined, sometimes some 5-veined.
    11. Culms 70–180 cm tall; awns 3–8 mm long; anthers 3–6 mm long.
        12. Lower leaf sheaths pilose, hairs 2–4 mm long; blades glabrous or with pilose margins ................................................. 16. *B. orcuttianus* (in part)
        12. Lower leaf sheaths densely pubescent, hairs to 1 mm long; blades densely pubescent.
            13. Blades 7.5–16.5 cm long; culm nodes 1–2(3) ................... 15. *B. hallii* (in part)
            13. Blades (13)18–38 cm long; culm nodes 3–7 ...................... 17. *B. grandis* (in part)
    11. Culms 30–100 cm tall; awns 1–4 mm long; anthers (1)1.5–4 mm long.
        14. Leaf blades often glaucous; glumes usually glabrous, rarely slightly pubescent ...... 18. *B. frondosus*
        14. Leaf blades not glaucous; glumes usually pubescent, rarely glabrous.
            15. Midrib of the culm leaves abruptly narrowed just below the collar; auricles frequently present on the lower leaves; plants of western Texas ....... 19. *B. anomalus* (in part)
            15. Midrib of the culm leaves not abruptly narrowed just below the collar; auricles absent; plants of western North America, including Texas ....... 20. *B. porteri* (in part)
6. Most lower glumes within a panicle 1-veined, sometimes some 3-veined.
    16. Upper glumes within a panicle consistently 5-veined; collars with a dense line of hairs; lower sheaths often sericeous; ligules 0.4–1 mm long .......................... 21. *B. nottowayanus*
    16. All or most upper glumes within a panicle 3-veined, sometimes some with 2 additional faint lateral veins; collars glabrous or hairy, hairs evenly distributed over the surface, not in a dense line; lower sheaths glabrous or hairy, not sericeous; ligules to 6 mm long.
        17. Plants annual; lemmas glabrous; ligules pubescent ............................ 22. *B. texensis*
        17. Plants perennial; lemmas usually pubescent on the backs and/or margins, sometimes glabrous; ligules usually glabrous, sometimes pubescent or pilose.
            18. Awns (4)6–12 mm long; ligules 2–6 mm long ............................. 23. *B. vulgaris*
            18. Awns 1–8 mm long; ligules to 4 mm long.
                19. Blades densely pubescent on both surfaces, 7.5–16.5 cm long; anthers 3–6 mm long; awns 3.5–7 mm long .................................... 15. *B. hallii* (in part)
                19. Blades glabrous or hairy on 1 or both surfaces, (3)5–60 cm long, if 7.5–16.5 cm long and densely pubescent on both surfaces, then anthers 1–4 mm long and/or awns 1–4 mm long.
                    20. Panicle branches appressed to slightly spreading; culm nodes 1–4.
                        21. Awns 2–5 mm long; anthers 2–3.5 mm long; blades flat .......... 26. *B. suksdorfii*
                        21. Awns (4)5–8 mm long; anthers 3–6.5 mm long; blades sometimes involute.
                            22. Culms 90–150 cm tall; ligules 1–3 mm long ........ 16. *B. orcuttianus* (in part)
                            22. Culms 50–100 cm tall; ligules to 1.5 mm long ................. 25. *B. erectus*
                    20. Panicle branches ascending to drooping; culm nodes (1)2–8.
                        23. Midrib of the culm leaves abruptly narrowed just below the collar; auricles frequently present on the lower leaves; plants of western Texas ....................................... 19. *B. anomalus* (in part)
                        23. Midrib of the culm leaves not abruptly narrowed just below the collar; auricles sometimes present; plants of various distribution, including Texas.
                            24. Glumes usually pubescent, rarely glabrous.
                                25. Upper glume mucronate ....................... 27. *B. mucroglumis*
                                25. Upper glume not mucronate.
                                    26. Awns (1)2–3(3.5) mm long; blades 2–6 mm wide .. 20. *B. porteri* (in part)
                                    26. Awns 3–7(8) mm long; blades 3–19 mm wide.
                                        27. Anthers 3–6 mm long; ligules densely pubescent to pilose .................. 17. *B. grandis* (in part)
                                        27. Anthers 2–4(5) mm long; ligules glabrous.
                                          28. Ligules 2–4 mm long ................... 24. *B. pacificus*

               28. Ligules 0.5–2 mm long . . . . . . . . . . . . . . . . . 28. *B. pubescens*
       24. Glumes usually glabrous, sometimes pubescent.
           29. Ligules 2–3.5 mm long; auricles present . . . . . . . . . . . . . . . . 30. *B. ramosus*
           29. Ligules 0.4–2 mm long; auricles sometimes present.
               30. Lemma margins and backs usually pubescent,
                   sometimes nearly glabrous; awns 2–4 mm long;
                   anthers 1.8–4 mm long . . . . . . . . . . . . . . . . . . . . . . . . . 29. *B. lanatipes*
               30. Lemma margins conspicuously hirsute or densely
                   pilose, at least along the lower ¹/₂, the backs
                   glabrous at least on the lower lemmas in a spikelet;
                   awns 3–5 mm long; anthers 1–2.7 mm long.
                   31. Backs of all lemmas glabrous; anthers 1–1.4
                       mm long; upper glumes 7.1–8.5 mm long . . . . . . . . . 31. *B. ciliatus*
                   31. Backs of the upper lemmas in a spikelet hairy;
                       anthers 1.6–2.7 mm long; upper glumes
                       8.9–11.3 mm long . . . . . . . . . . . . . . . . . . . . . . 32. *B. richardsonii*

## Bromus sect. Neobromus

This section includes one species, 33. *Bromus berteroanus.*

## Bromus sect. Genea

1. Lemmas 20–35 mm long . . . . . . . . . . . . . . . . . . . . . . . . . . . . . . . . . . . . . . . . . . . . . . . . . . . . . . . . . . . . . . . . . . 34. *B. diandrus*
1. Lemmas 9–20 mm long.
   2. Spikelets usually shorter than the panicle branches; panicle branches ascending to spreading or
      drooping.
      3. Lemmas 14–20 mm long; panicles with spreading, ascending, or drooping branches, rarely with
         any branches with more than 3 spikelets . . . . . . . . . . . . . . . . . . . . . . . . . . . . . . . . . . . . . . 35. *B. sterilis*
      3. Lemmas 9–12 mm long; panicles with drooping branches, often with 1 or more branches with
         4–8 spikelets . . . . . . . . . . . . . . . . . . . . . . . . . . . . . . . . . . . . . . . . . . . . . . . . . . . . . . . . . 36. *B. tectorum*
   2. Spikelets longer than the panicle branches; panicle branches ascending to spreading, never drooping.
      4. Some panicle branches 1–3+ cm long, most branches visible . . . . . . . . . . . . . . . . . . . . . . . . 37. *B. madritensis*
      4. Panicle branches 0.1–1 cm long, usually not readily visible . . . . . . . . . . . . . . . . . . . . . . . . . . 38. *B. rubens*

## Bromus sect. Bromus

1. Lemmas inflated, 6–8 mm wide; unawned or with awns up to 1 mm long; spikelets ovate . . . . . . . . . . . . 39. *B. briziformis*
1. Lemmas not inflated, 1–7 mm wide; awns 2–25 mm long, rarely absent; spikelet shape various.
   2. Lemma margins inrolled at maturity; floret bases visible at maturity; rachilla internodes visible at
      maturity; caryopses sometimes thick, strongly inrolled.
      3. Anthers 2.5–5 mm long; awns straight; spikelets often purple-tinged; lower leaf sheaths with soft,
         appressed hairs . . . . . . . . . . . . . . . . . . . . . . . . . . . . . . . . . . . . . . . . . . . . . . . . . . 40. *B. arvensis* (in part)
      3. Anthers 0.7–2 mm long; awns straight or flexuous; spikelets not purple-tinged; lower leaf sheaths
         glabrous, loosely pubescent and glabrate, or evenly covered with stiff hairs.
         4. Lower leaf sheaths glabrous or loosely pubescent and glabrate; lemmas 6.5–8.5(10) mm long,
            margins evenly rounded; awns straight or flexuous . . . . . . . . . . . . . . . . . . . . . . . . . . . . . 41. *B. secalinus*
         4. Lower leaf sheaths evenly covered with stiff hairs; lemmas 8–11.5 mm long, margins bluntly
            angled; awns straight . . . . . . . . . . . . . . . . . . . . . . . . . . . . . . . . . . . 42. *B. commutatus* (in part)
   2. Lemma margins not inrolled at maturity; floret bases concealed at maturity; rachilla internodes
      concealed at maturity; caryopses thin, weakly inrolled or flat.
      5. Lemmas 4.5–6.5 mm long, margins sharply angled; caryopses longer than the paleas . . . . . . . . . . . . . . 43. *B. lepidus*
      5. Lemmas 6.5–20 mm long, margins rounded or slightly to strongly angled; caryopses equaling or
         shorter than the paleas.
         6. Awns arising less than 1.5 mm below the lemma apices, erect or weakly divaricate, not twisted
            at the base.
            7. Panicle branches shorter than the spikelets; lemmas chartaceous, with prominent ribs over
               the veins, often concave between the veins; anthers 0.6–1.5 mm long . . . . . . . . . . . 44. *B. hordeaceus* (in part)
            7. At least some panicle branches longer than the spikelets; lemmas coriaceous, veins obscure
               or distinct, not ribbed; anthers 0.7–5 mm long.

8. Lower leaf sheaths with soft, appressed hairs; anthers 2.5–5 mm long; panicles 11–30 cm long . . . . . . . . . . . . . . . . . . . . . . . . . . . . . . . . . . . . . . . . . . . . . . . . . . . . . . 40. *B. arvensis* (in part)

8. Lower leaf sheaths with stiff hairs; anthers 0.7–3 mm long; panicles 4–16 cm long.

   9. Anthers 0.7–1.7 mm long; rachilla internodes 1.5–2 mm long; lemmas 8–11.5 mm long, margins bluntly angled . . . . . . . . . . . . . . . . . . . . . . . . . . . . . . . . . . . . . . . 42. *B. commutatus* (in part)

   9. Anthers 1.5–3 mm long; rachilla internodes 1–1.5 mm long; lemmas 6.5–8 mm long, margins rounded . . . . . . . . . . . . . . . . . . . . . . . . . . . . . . . . . . . . . . . . . . . . . . . . 45. *B. racemosus*

6. Awns arising 1.5 mm or more below the lemma apices, erect to strongly divaricate, often twisted at the base.

  10. Panicle branches shorter than the spikelets, slightly curved or straight, panicles erect.

    11. At least the upper lemmas in each spikelet with 3 awns . . . . . . . . . . . . . . . . . . . . . 46. *B. danthoniae*

    11. All lemmas 1-awned.

      12. Lemmas 11–20 mm long; spikelets 20–50 mm long.

        13. Spikelets usually single at the nodes; glumes glabrous or puberulent; panicles strongly contracted, even at maturity . . . . . . . . . . . . . . . . . . . . . . 47. *B. caroli-henrici*

        13. Spikelets often 2 or more at each node; glumes pilose; panicles contracted when immature, more open with age . . . . . . . . . . . . . . . . . . . . . . . . . . . . 48. *B. lanceolatus*

      12. Lemmas 6.5–11 mm long; spikelets 11–25 mm long.

        14. Lemmas 1.5–2 mm wide; panicles obovoid, branches sometimes verticillate . . . . 49. *B. scoparius*

        14. Lemmas 3–5 mm wide; panicles usually ovoid . . . . . . . . . . . . . . . . . 44. *B. hordeaceus* (in part)

  10. At least some panicle branches as long as or longer than the spikelets, sometimes sinuous; panicles nodding.

    15. Lower glumes 7–10 mm long; upper glumes 8–12 mm long; panicle branches conspicuously sinuous; awns erect to weakly spreading; lemma margins rounded . . . . . . . 50. *B. arenarius*

    15. Lower glumes 4–7 mm long; upper glumes 5–8 mm long; panicle branches sometimes sinuous; awns erect to strongly divergent; lemma margins slightly to strongly angled above the middle.

      16. Anthers 2.5–5 mm long; spikelets often purple-tinged; culms 80–110 cm tall . . . . . . . . . . . . . . . . . . . . . . . . . . . . . . . . . . . . . . . . . . . . . . . . . . . . . . . . . 40. *B. arvensis* (in part)

      16. Anthers 1–1.5 mm long; spikelets not purple-tinged; culms 20–70 cm tall.

        17. Lemmas with hyaline margins 0.3–0.6 mm wide, slightly angled above the middle; branches somewhat drooping, sometimes sinuous, often with more than 1 spikelet . . . . . . . . . . . . . . . . . . . . . . . . . . . . . . . . . . . . . . . . . . . . . 51. *B. japonicus*

        17. Lemmas with hyaline margins 0.6–0.9 mm wide, strongly angled above the middle; branches not drooping or sinuous, usually with 1 spikelet . . . . . . . . . . 52. *B. squarrosus*

## Bromus sect. Ceratochloa (P. Beauv.) Griseb.

*Bromus* sect. *Ceratochloa* is native to North and South America, and contains about 25 species. It is marked by polyploid complexes; the major one in North America is the *Bromus carinatus* complex.

### 1. Bromus catharticus Vahl RESCUE GRASS [p. 367, 512]

1. Awns absent or to 3.5 mm long . . . . . . . . . . . . . var. *catharticus*
1. Awns (5)6–10 mm long . . . . . . . . . . . . . . . . . . . . . . var. *elatus*

### Bromus catharticus Vahl var. catharticus [p. 367]

*Bromus catharticus* var. *catharticus* is native to South America. It has been widely introduced in the *Manual* region as a forage crop and is now established, particularly in the southern half of the United States. It usually grows on disturbed soils.

### Bromus catharticus var. elatus (E. Desv.) Planchuelo [p. 367]

*Bromus catharticus* var. *elatus*, a native of South America, now grows in disturbed soils in central California. It has also been reported from ballast dumps in Oregon. Although originally published as var. *elata*, the correct name is var. *elatus*.

### 2. Bromus sitchensis Trin. SITKA BROME, ALASKA BROME [p. 367, 512]

*Bromus sitchensis* grows on exposed rock bluffs and cliffs, and in meadows, often in the partial shade of forests along the ocean edge, and on road verges and other disturbed sites. Its range extends from the Aleutian Islands and Alaska panhandle through British Columbia to southern California. It resembles *B. aleutensis*, the two sometimes being treated as conspecific varieties.

### 3. Bromus arizonicus (Shear) Stebbins ARIZONA BROME [p. 367, 512]

*Bromus arizonicus* grows in dry, open areas and disturbed ground of the southwest, usually below 2000 m. Its range extends from California and southern Nevada into Arizona, New Mexico, and northern Mexico.

### 4. Bromus aleutensis Trin. ex Griseb. ALEUT BROME [p. 367, 512]

*Bromus aleutensis* grows in sand, gravel, and disturbed soil along the Pacific coast and on some lake shores of central British Columbia. It has also been found further east in disturbed sites such as road edges. *Bromus aleutensis* may represent a modified version of *B. sitchensis*, in which reproduction occurs at a relatively early developmental state in response to the climatic conditions of the Aleutian Islands. *Bromus aleutensis* intergrades with *B. carinatus* var. *marginatus* to the south.

**5. Bromus maritimus** (Piper) Hitchc. MARITIME BROME [p. 367, 512]

*Bromus maritimus* grows in coastal sands from Lane County, Oregon, to Los Angeles County, California.

**6. Bromus carinatus** Hook. & Arn. [p. 367, 512]

*Bromus carinatus* is native from British Columbia to Saskatchewan and south to Mexico. It has been introduced to various more eastern locations and to the southern Yukon Territory. The two varieties recognized here are sometimes recognized as species.

1. Most awns 8–17 mm long . . . . . . . . . . . . . . . . . . var. *carinatus*
1. Most awns 4–7 mm long . . . . . . . . . . . . . . . . var. *marginatus*

**Bromus carinatus** Hook. & Arn. var. **carinatus**

CALIFORNIA BROME [p. 367]

*Bromus carinatus* var. *carinatus* is primarily coastal and grows in shrublands, grasslands, meadows, and openings in chaparral and oak and yellow pine woodlands. It ranges from southern British Columbia through Washington, Oregon, and California to Baja California, Mexico, and extends eastward through Arizona to New Mexico. It intergrades with var. *marginatus*, which tends to grow at higher elevations and extends further inland.

**Bromus carinatus** var. **marginatus** (Nees) Barkworth & Anderton MOUNTAIN BROME [p. 367]

*Bromus carinatus* var. *marginatus* is primarily an inland species and grows on open slopes, grass balds, shrublands, meadows, and open forests, in montane and subalpine zones. It grows from British Columbia to Saskatchewan, south throughout the western United States, and also extends into northern Mexico. Its elevational range is 350–2200 m in the northern part of its distribution, and 1500–3300 m in the south. A variable species, it intergrades with *B. carinatus* var. *carinatus* to the west, *B. aleutensis* to the north, and *B. polyanthus* to the southeast.

**7. Bromus polyanthus** Scribn. COLORADO BROME, GREAT-BASIN BROME [p. 367, 512]

*Bromus polyanthus* grows on open slopes and in meadows. It is found primarily in the central Rocky Mountains. It is not known from Mexico. It intergrades with *B. carinatus* var. *marginatus*.

## Bromus sect. Bromopsis Dumort.

*Bromus* sect. *Bromopsis* is sometimes incorrectly called sect. *Pnigma* Dumort. It is native to Eurasia as well as to North and South America, and has about 90 species.

**8. Bromus inermis** Leyss. SMOOTH BROME, HUNGARIAN BROME, BROME INERME [p. 368, 512]

*Bromus inermis* is native to Eurasia, and is now found in disturbed sites in Alaska, Greenland, and most of Canada as well as south through most of the contiguous United States. It has been used for rehabilitation, and is planted extensively for forage in pastures and rangelands from Alaska and the Yukon Territory to Texas. It is similar to *B. pumpellianus*, differing mainly in having glabrous lemmas, nodes, and leaf blades, but lack of pubescence is not a consistently reliable distinguishing character. *Bromus inermis* differs from *B. riparius*, a recently introduced species, primarily in its shorter or nonexistent awns.

**9. Bromus riparius** Rehmann MEADOW BROME [p. 368]

*Bromus riparius* is an Asian species that was introduced to the United States in the late 1950s for cultivation as a pasture grass. Various cultivars are now grown, mainly in Canada and the northwestern United States. It appears to differ from *B. inermis* and *B. pumpellianus* in having shorter culms on average, longer awns than *B. inermis*, and shorter rhizomes than *B. pumpellianus* subsp. *pumpellianus*. Its distribution in the *Manual* region is not known.

**10. Bromus pumpellianus** Scribn. ARCTIC BROME [p. 368, 512]

The range of *Bromus pumpellianus* extends from Asia to North America, where it includes Alaska, the western half of Canada, the western United States as far south as New Mexico, and a few other locations eastward. It is sometimes treated as a subspecies of *B. inermis*. It differs from that species primarily in its tendency to have pubescent lemmas, nodes, and leaf blades.

Two subspecies that differ in morphology and distribution are described below. Both strongly resemble the recently introduced *Bromus riparius*, differing in the case of *B. pumpellianus* subsp. *pumpellianus* in having longer rhizomes, or, in the case of *B. pumpellianus* subsp. *dicksonii*, in having a more restricted distribution.

1. Panicles usually open; plants cespitose, sometimes shortly rhizomatous; culms ascending, often geniculate; nodes glabrous or pubescent; plants of the Yukon River drainage . . . . . . . . . . . . . . . . subsp. *dicksonii*
1. Panicles contracted to open; plants rhizomatous; culms erect; nodes usually pubescent; plants of the range of the species . . . . . . . . . . . . . . . . subsp. *pumpellianus*

**Bromus pumpellianus** subsp. **dicksonii** W.W. Mitch. & Wilton [p. 368]

*Bromus pumpellianus* subsp. *dicksonii* grows in shallow, rocky soils of river banks and bluffs in the Yukon River drainage of Alaska. Apart from the more restricted distribution, it is not clear how this subspecies differs from the introduced *B. riparius*.

**Bromus pumpellianus** Scribn. subsp. **pumpellianus** [p. 368]

*Bromus pumpellianus* subsp. *pumpellianus* grows on sandy and gravelly stream banks and lake shores, sand dunes, meadows, dry grassy slopes, and road verges.

**11. Bromus latiglumis** (Scribn. *ex* Shear) Hitchc. HAIRY WOODBROME, FLANGED BROME, BROME À LARGES GLUMES [p. 368, 512]

*Bromus latiglumis* grows in shaded or open woods, along stream banks, and on alluvial plains and slopes. Its range is mainly in the north-central and northeastern United States and adjacent Canadian provinces.

**12. Bromus laevipes** Shear CHINOOK BROME [p. 368, 512]

*Bromus laevipes* grows from northern Oregon to southern California. It grows in shaded woodlands and on exposed brushy slopes, at 300–1500 m.

**13. Bromus kalmii** A. Gray KALM'S BROME, BROME DE KALM [p. 368, 512]

*Bromus kalmii* grows in sandy, gravelly, or limestone soils in open woods and calcareous fens. Its range centers in the north-central and northeastern United States and adjacent Canadian provinces.

**14. Bromus pseudolaevipes** Wagnon WOODLAND BROME [p. 368, 513]

*Bromus pseudolaevipes* grows in dry, shaded or semishaded sites in chaparral, coastal sage scrub, and woodland-savannah zones, from near sea level to about 900 m, in central and southern California. It is not known from Mexico.

**15. Bromus hallii** (Hitchc.) Saarela & P.M. Peterson HALL'S BROME [p. 368, 513]

*Bromus hallii* grows in southern California on dry, open or shaded hillsides, rocky slopes, and in montane pine woods, from 1500–2700 m.

**16. Bromus orcuttianus** Vasey ORCUTT'S BROME [p. 368, <u>513</u>]

*Bromus orcuttianus* grows on dry hillsides and rocky slopes, and in open pine woods and meadows in the mountains, from 500–3500 m. It is found in the western United States, including Washington, Oregon, California, Nevada, and Arizona. It is not known from Mexico.

**17. Bromus grandis** (Shear) Hitchc. TALL BROME [p. 368, <u>513</u>]

*Bromus grandis* grows on dry, wooded or open slopes, at elevations of 350–2500 m. Its range extends from central California into Baja California, Mexico.

**18. Bromus frondosus** (Shear) Wooton & Standl. WEEPING BROME [p. 368, <u>513</u>]

*Bromus frondosus* grows in open woods and on rocky slopes, at 1500–2500 m. Its range extends from Colorado, Arizona, and New Mexico into Mexico.

**19. Bromus anomalus** Rupr. *ex* E. Fourn. MEXICAN BROME [p. 369, <u>513</u>]

*Bromus anomalus* grows on rocky slopes in western Texas and adjacent Mexico. Many records of this species in the *Manual* region are here treated as *B. porteri*, a closely related species that has sometimes been included in *B. anomalus*. The main difference is that *B. anomalus* has auricles, and culm leaves with midribs that are narrowed just below the collar.

**20. Bromus porteri** (J.M. Coult.) Nash NODDING BROME [p. 369, <u>513</u>]

*Bromus porteri* grows in montane meadows, grassy slopes, mesic steppes, forest edges, and open forest habitats, at 500–3500 m. It is found from British Columbia to Manitoba, and south to California, western Texas, and Mexico. It is closely related to *B. anomalus*, and has often been included in that species. It differs chiefly in its lack of auricles, and in having culm leaves with midribs that are not narrowed just below the collar.

**21. Bromus nottowayanus** Fernald VIRGINIA BROME [p. 369, <u>513</u>]

*Bromus nottowayanus* is native to the east-central and eastern United States from Iowa to New York, south to Oklahoma, northern Alabama, and Virginia. It grows in damp, shaded woods, often in ravines and along streams.

**22. Bromus texensis** (Shear) Hitchc. TEXAS BROME [p. 369, <u>513</u>]

*Bromus texensis* grows in openings in brushy areas on rocky ground. It is rare, found only in southern Texas and northern Mexico.

**23. Bromus vulgaris** (Hook.) Shear COMMON BROME [p. 369, <u>513</u>]

*Bromus vulgaris* grows in shaded or partially shaded, often damp, coniferous forests along the coast, and inland in montane pine, spruce, fir, and aspen forests, from sea level to about 2000 m. Its range extends from coastal British Columbia eastward to southwestern Alberta and southward to central California, northern Utah, and western Wyoming.

**24. Bromus pacificus** Shear PACIFIC BROME [p. 369, <u>513</u>]

*Bromus pacificus* grows in moist thickets, openings, and ravines along the Pacific coast from southeastern Alaska to northern California, with a few occurrences further inland.

**25. Bromus erectus** Huds. MEADOW BROME, UPRIGHT BROME, BROME DRESSÉ [p. 369, <u>513</u>]

*Bromus erectus* is native to Europe. In the *Manual* region, it grows on disturbed soils, often over limestone. It is established in the eastern United States and Canada, and has been reported from other locations where it has not persisted.

**26. Bromus suksdorfii** Vasey SUKSDORF'S BROME [p. 369, <u>513</u>]

*Bromus suksdorfii* grows on open slopes and in open subalpine forests, at about 1300–3300 m, from southern Washington to southern California.

**27. Bromus mucroglumis** Wagnon SHARPGLUME BROME [p. 369, <u>513</u>]

*Bromus mucroglumis* grows at 1500–3000 m in the southwestern United States and northern Mexico.

**28. Bromus pubescens** Muhl. *ex* Willd. CANADA BROME [p. 369, <u>513</u>]

*Bromus pubescens* grows in shaded, moist, often upland deciduous woods. Its range is centered in the eastern half of the United States, and extends northward to southern Manitoba, Ontario, and Quebec, westward in scattered locations to Arizona, and southward to eastern Texas and western Florida.

**29. Bromus lanatipes** (Shear) Rydb. WOOLY BROME [p. 369, <u>513</u>]

*Bromus lanatipes* grows in a wide range of habitats at 800–2500 m, from Wyoming through the southwestern United States to northern Mexico.

**30. Bromus ramosus** Huds. HAIRY BROME [p. 369]

*Bromus ramosus* is native to Asia, Europe, and northern Africa. It is included here based on a report that it is found sporadically in the southern and eastern United States; specimens to substantiate this report have not been located.

**31. Bromus ciliatus** L. FRINGED BROME, BROME CILIÉ [p. 370, <u>513</u>]

*Bromus ciliatus* grows in damp meadows, thickets, woods, and stream banks across almost all of northern North America except the high arctic, extending further south mainly through the western United States to Mexico.

**32. Bromus richardsonii** Link RICHARDSON'S BROME [p. 370, <u>513</u>]

*Bromus richardsonii* grows in meadows and open woods in the upper montane and subalpine zones, at 2000–4000 m in the southern Rocky Mountains, and at lower elevations northwards. Its range extends from southern Alaska to southern California and northern Baja California, Mexico; it is found as far east as Saskatchewan, South Dakota, and western Texas.

## Bromus sect. Neobromus (Shear) Hitchc.

*Bromus* sect. *Neobromus* has two species, both of which are native to South America. *Bromus berteroanus* has become established in the *Manual* region.

**33. Bromus berteroanus** Colla CHILEAN CHESS [p. 370, <u>513</u>]

*Bromus berteroanus* is from Chile, and can now be found in dry areas in western North America, including British Columbia, Montana, California, Nevada, Arizona, southwestern Utah, and Baja California, Mexico.

## Bromus sect. Genea Dumort.

*Bromus* sect. *Genea* is native to Europe and northern Africa; five of its six species are established in the *Manual* region.

### 34. Bromus diandrus Roth GREAT BROME, RIPGUT GRASS [p. 370, 513]

*Bromus diandrus* is native to southern and western Europe. It is now established in North America, where it grows in disturbed ground, waste places, fields, sand dunes, and limestone areas. It occurs from southwestern British Columbia to Baja California, Mexico, and eastward to Montana, Colorado, Texas, and scattered locations in the eastern United States. The common name 'ripgut grass' indicates the effect it has on animals if they consume the sharp, long-awned florets of this species. *Bromus diandrus*, as treated here, includes *B. rigidus*.

### 35. Bromus sterilis L. BARREN BROME [p. 370, 513]

*Bromus sterilis* is native to Europe, growing from Sweden southward. In the *Manual* region, it grows in road verges, waste places, fields, and overgrazed rangeland. It is widespread in western and eastern North America, but is mostly absent from the Great Plains and the southeastern states.

### 36. Bromus tectorum L. CHEATGRASS, DOWNY CHESS [p. 370, 513]

*Bromus tectorum* is a European species that is well established in the *Manual* region and other parts of the world. It grows in disturbed sites, such as overgrazed rangelands, fields, sand dunes, road verges, and waste places. In the southwestern United States, *Bromus tectorum* is considered a good source of spring feed for cattle, at least until the awns mature. It is highly competitive and dominates rapidly after fire, especially in sagebrush areas. It now dominates large areas of the sagebrush ecosystem of the western *Manual* region.

### 37. Bromus madritensis L. COMPACT BROME [p. 370, 513]

*Bromus madritensis* is native to southern and western Europe. It is now established in North America, and grows in disturbed soil, waste places, banks, and road verges in southern Oregon, California, and Arizona.

### 38. Bromus rubens L. FOXTAIL CHESS, RED BROME [p. 370, 513]

*Bromus rubens* is native to southern and southwestern Europe. It now grows in North America in disturbed ground, waste places, fields, and rocky slopes, from southern Washington to southern California, eastward to Idaho, New Mexico, and western Texas. It was found in Massachusetts before 1900 in wool waste used on a crop field; it is not established there. The record from New York represents a rare introduction; it is not known whether it is established.

## Bromus L. sect. Bromus

*Bromus* sect. *Bromus* has about 40 species that are native to Eurasia, northern Africa, and Australia; 14 species have been introduced to the *Manual* region.

### 39. Bromus briziformis Fisch. & C.A. Mey. RATTLESNAKE BROME [p. 370, 513]

*Bromus briziformis* grows in waste places, road verges, and overgrazed areas. It is native to southwest Asia and Europe, and is adventive in the *Manual* region, occurring from southern British Columbia to as far south as New Mexico, and in scattered locations eastward. The unique shape of its spikelets has led to its use in dried flower arrangements and as a garden ornamental.

### 40. Bromus arvensis L. FIELD BROME [p. 370, 514]

*Bromus arvensis* grows along roadsides and in fields and waste places at scattered locations in the *Manual* region. It is native to southern and south-central Europe.

### 41. Bromus secalinus L. RYEBROME, BROME DES SEIGLES [p. 370, 514]

*Bromus secalinus* is native to Europe. It is widespread in the *Manual* region, growing in fields, on waste ground, and along roadsides.

### 42. Bromus commutatus Schrad. MEADOW BROME, HAIRY CHESS [p. 370, 514]

*Bromus commutatus* grows in fields, waste places, and road verges. It is native to Europe and the Baltic region. In the *Manual* region, it is found mainly in the United States and southern Canada.

### 43. Bromus lepidus Holmb. SCALY BROME [p. 371, 514]

*Bromus lepidus* grows in fields and waste places. A native of Europe, it is reported from New York and Massachusetts. It probably also occurs elsewhere in the *Manual* region. *Bromus lepidus* differs from *B. hordeaceus* subsp. *pseudothominei* in the wide apical notch of its lemmas, and the length of its caryopses relative to the paleas.

### 44. Bromus hordeaceus L. LOPGRASS, BROME MOU [p. 371, 514]

*Bromus hordeaceus* is native to southern Europe and northern Africa. It is weedy, growing in disturbed areas such as roadsides, fields, sandy beaches, and waste places, and can be found in many locations in the *Manual* region.

1. Lemmas (7)8–11 mm long, usually pubescent or pilose.
   2. Awns more than 0.1 mm wide at the base, straight, erect; culms (3)10–70 cm long . . . subsp. *hordeaceus*
   2. Awns less than 0.1 mm wide at the base, often divaricate or recurved at maturity; culms 15–25(60) cm long . . . . . . . . . . . . . . . . subsp. *molliformis*
1. Lemmas 6.5–8(9) mm long, glabrous or pubescent.
   3. Culms (3)10–70 cm long; panicles up to 10 cm long, usually with more than 1 spikelet; lemmas usually glabrous; caryopses usually as long as the paleas; habitat various . . . . . . . . . . subsp. *pseudothomineii*
   3. Culms 2–16 cm long; panicles 1–3 cm long, often reduced to 1 spikelet; lemmas pubescent or glabrous; caryopses shorter than the paleas; plants of maritime or lacustrine sands . . . . . subsp. *thominei*

### Bromus hordeaceus L. subsp. hordeaceus [p. 371]

*Bromus hordeaceus* subsp. *hordeaceus* grows throughout the range of the species, being most prevalent in southwestern British Columbia, the western United States, and the northeastern coast.

### Bromus hordeaceus subsp. molliformis (J. Lloyd *ex* Billot) Maire & Weiller [p. 371]

*Bromus hordeaceus* subsp. *molliformis* grows in California and other scattered locations, including Idaho, New Mexico, and southern Michigan.

### Bromus hordeaceus subsp. pseudothominei (P. M. Sm.) H. Scholz [p. 371]

*Bromus hordeaceus* subsp. *pseudothominei* grows sporadically throughout the range of the species in the *Manual* region. It often resembles *B. lepidus* in lemma characteristics (e.g., length, smoothness, and margin angle), so that either may be misinterpreted.

## Bromus hordeaceus subsp. thominei (Hardouin) Braun-Blanq. [p. *371*]

*Bromus hordeaceus* subsp. *thominei* grows along the Pacific coast of Canada, from the Queen Charlotte Islands to Vancouver Island, as well as at inland locations in British Columbia; it has also been recorded from California, Massachusetts, and Rhode Island.

## 45. Bromus racemosus L. SMOOTH BROME, BROME À GRAPPES [p. *371*, 514]

*Bromus racemosus* grows in fields, waste places, and road verges. It is native to western Europe and the Baltic region, and occurs throughout much of southern Canada and the United States.

## 46. Bromus danthoniae Trin. *ex* C.A. Mey. THREE-AWNED BROME [p. *371*, 514]

*Bromus danthoniae* is native from the western Asia to southern Russia and Tibet. It was collected in 1904 in Ontario; no other North American collections are known.

## 47. Bromus caroli-henrici Greuter [p. *371*, 514]

*Bromus caroli-henrici* is native to Mediterranean Europe. In the *Manual* region, it grows in open, disturbed areas in Butte and Yolo counties, California. It has been misidentified as *B. alopecuros* Poir. It differs from that species in having 1, not 2–3, spikelets at the rachis nodes, and acuminate, not broadly triangular, lemma teeth.

## 48. Bromus lanceolatus Roth LANCEOLATE BROME [p. *371*, 514]

*Bromus lanceolatus* grows in waste places, and is also cultivated as an ornamental. It has been introduced to the *Manual* region from southern Europe, and is reported from scattered sites such as Yonkers, New York (wool waste); College Station, Texas; and Pima County, Arizona.

## 49. Bromus scoparius L. BROOM BROME [p. *371*, 514]

*Bromus scoparius* is native to southern Europe. It grows in waste places. In the *Manual* region, it has been recorded from California and New York.

## 50. Bromus arenarius Labill. AUSTRALIAN BROME [p. *371*, 514]

*Bromus arenarius* grows in dry, often sandy slopes, fields, and waste places. Native to Australia, it is now widely scattered throughout California, and is also recorded from Oregon, eastern Nevada, Arizona, New Mexico, Texas, and Pennsylvania.

## 51. Bromus japonicus Thunb. JAPANESE BROME [p. *371*, 514]

*Bromus japonicus* grows in fields, waste places, and road verges. It is native to central and southeastern Europe and Asia, and is distributed throughout much of the United States and southern Canada, with one record from the Yukon Territory.

## 52. Bromus squarrosus L. SQUARROSE BROME [p. *371*, 514]

*Bromus squarrosus* grows in overgrazed pastures, fields, waste places, and road verges. Native to central Russia and southern Europe, it can be found mainly in southern Canada and the northern half of the United States.

---

# 13. TRITICEAE Dumort.[1]

**Pl** ann or per; smt csp, smt rhz. **Clm** ann, not wd, usu erect, not brchg above the base; **intnd** hollow or solid. **Shth** usu open, those of the bas lvs smt closed; **col** without tufts of hair on the sides; **aur** usu present; **lig** memb or scarious, smt ciliolate, those of the up and lo cauline lvs usu similar; **psdpet** absent; **bld** linear to narrowly lanceolate, venation parallel, cross venation not evident, without arm or fusoid cells, surfaces without microhairs, not papillate, cross sections non-Kranz. **Infl** usu spikes or spikelike rcm, with 1–5 sessile or subsessile spklt per nd, occ pan, smt with morphologically distinct strl and bisx spklt within an infl; **ped** absent or to 4 mm; **dis** usu above the glm and beneath the flt, smt in the rchs, smt at the infl bases. **Spklt** usu lat compressed, smt terete, with 1–16 bisx flt, the distal (or only) flt smt strl; **rchl** smt prolonged beyond the base of the distal flt. **Glm** unequal to equal, shorter than to longer than the adjacent flt, subulate, lanceolate, rectangular, ovate, or obovate, 1–5-veined, absent or vestigial in some species; **flt** lat compressed to terete; **cal** glab or hairy; **lm** lanceolate to rectangular, stiffly memb to coriaceous, smt keeled, 5(7)-veined, veins not converging distally, inconspicuous, apc entire, lobed, or toothed, unawned or awned, awns tml, unbrchd, lm-awn junction not evident; **pal** usu subequal to the lm, smt considerably shorter or slightly longer than the lm; **lod** 2, without venation, usu ciliate; **anth** 3; **ov** with hairy apc; **sty** 2, bases free. **Car** ovoid to fusiform, longitudinally grooved, not beaked, prcp thin; **hila** linear; **emb** about $^1/_3$ as long as the car.

The *Triticeae* are primarily north-temperate in distribution. The tribe includes 400–500 species, among which are several important cereal, forage, and range species. Its generic treatment is contentious due to the prevalence of natural hybridization, introgression, and polyploidy.

The following key does not include intergeneric hybrids; they are treated in the text on the following pages: ×*Triticosecale* (p. 59), ×*Pseudelymus* (p. 63), ×*Elyhordeum* (p. 63), ×*Elyleymus* (p. 75), ×*Pascoleymus* (p. 76), and ×*Leydeum* (p. 79). In the field, they can usually be detected by their intermediate morphology and sterility. In sterile plants, the anthers are indehiscent, somewhat pointed, and tend to remain on the plants.

Measurements of rachis internodes and spikelets should be made at midspike.

[1]Mary E. Barkworth

1. Spikelets 2–7 at all or most nodes.
    2. Spikelets 3 at each node, the central spikelets sessile, the lateral spikelets usually pedicellate, sometimes all 3 spikelets sessile in cultivated plants; spikelets with 1 floret, usually only the central spikelet with a functional floret, the florets of the lateral spikelets usually sterile and reduced, in cultivated plants all florets functional or those of the lateral spikelets functional and those of the central spikelet reduced . . . . . . . . . . . . . . . . . . . . . . . . . . . . . . . . . . . . . . . . . . . . . . . . . . . . . . . . . . 13.01 *Hordeum*
    2. Spikelets usually other than 3 at each node, if 3, all 3 sessile; spikelets with 1–11 florets, if 1 floret, additional reduced or sterile florets present distal to the functional floret in at least 1 spikelet per node.
        3. Lemmas strongly keeled, keels conspicuously scabrous distally, scabrules 0.5–0.8 mm long; lemma awns straight . . . . . . . . . . . . . . . . . . . . . . . . . . . . . . . . . . . . . . . . . . . . . . . . . . . . . . . . . . . . . . 13.05 *Secale*
        3. Lemmas rounded proximally, sometimes keeled distally, keels not or inconspicuously scabrous distally; lemma awns straight, flexuous, or variously curved.
            4. Plants annual, weedy; spikelets with only 1 bisexual floret . . . . . . . . . . . . . . . . . . . . . . . 13.04 *Taeniatherum*
            4. Plants perennial, usually not weedy; spikelets usually with more than 1 bisexual floret.
                5. Lemma awns (0)1–120 mm long; anthers 0.9–6 mm long; blades with well-spaced, unequally prominent veins on the adaxial surfaces . . . . . . . . . . . . . . . . . . . . . . . . . . . . 13.13 *Elymus* (in part)
                5. Lemmas usually unawned or with awns up to 7 mm long, if awns 16–35 mm long, anthers 6–8 mm; blades usually with closely spaced, equally prominent veins on the adaxial surfaces.
                    6. Disarticulation in the spikelets, beneath the florets; plants sometimes cespitose, often rhizomatous . . . . . . . . . . . . . . . . . . . . . . . . . . . . . . . . . . . . . . . . . . . . . . . . 13.17 *Leymus* (in part)
                    6. Disarticulation tardy, in the rachises; plants cespitose, not rhizomatous . . . . . . . . . . . 13.19 *Psathyrostachys*
1. Spikelets 1 at all or most nodes.
    7. Spikelets usually more than 3 times the length of the middle rachis internodes, usually divergent, sometimes ascending; rachis internodes 0.2–5.5 mm long.
        8. Glumes with 2 prominent keels, keels with tufts of hair . . . . . . . . . . . . . . . . . . . . . . . . . . . . . . . . 13.03 *Dasypyrum*
        8. Glumes initially with 1 keel, sometimes 2-keeled at maturity, keels glabrous or hairy, hairs never in tufts.
            9. Plants annual; anthers 0.4–1.4 mm long; spikes 0.8–4.5 cm long . . . . . . . . . . . . . . . . . . . . . . 13.02 *Eremopyrum*
            9. Plants perennial; anthers 3–5 mm long; spikes 1.3–15 cm long . . . . . . . . . . . . . . . . . . . . . . . . 13.09 *Agropyron*
    7. Spikelets ½–3 times the length of the middle rachis internodes, appressed or ascending; rachis internodes 3–28 mm long.
        10. Glumes subulate to narrowly lanceolate, tapering from below midlength, 1(3)-veined at midlength.
            11. Glumes lanceolate, tapering to acuminate apices from near midlength or below, keels curving to the side distally; plants always rhizomatous . . . . . . . . . . . . . . . . . . . . . . . . . . 13.15 *Pascopyrum*
            11. Glumes subulate to lanceolate, tapering from below midlength, keels straight or almost so; plants often rhizomatous . . . . . . . . . . . . . . . . . . . . . . . . . . . . . . . . . . . . . . . . . . . 13.17 *Leymus* (in part)
        10. Glumes lanceolate, rectangular, ovate, or obovate, narrowing beyond midlength, often in the distal ¼, (1)3–5(7)-veined at midlength.
            12. Plants annual; glumes often with lateral teeth or awns, midveins smooth throughout.
                13. Glumes rounded over the midveins; plants weedy . . . . . . . . . . . . . . . . . . . . . . . . . . . . 13.07 *Aegilops*
                13. Glumes keeled over the midveins; plants cultivated, sometimes escaping . . . . . . . . . . . . . . 13.08 *Triticum*
            12. Plants perennial; glumes without lateral teeth or awns, midveins sometimes scabrous.
                14. Glumes stiff, truncate, obtuse, or acute, unawned; glume keels smooth proximally, usually scabrous distally . . . . . . . . . . . . . . . . . . . . . . . . . . . . . . . . . . . . . . . . . . . . . . . 13.20 *Thinopyrum*
                14. Glumes flexible, acute to acuminate, sometimes awn-tipped; glume keels usually uniformly smooth or scabrous their whole length, sometimes smooth proximally and scabrous distally.
                    15. Spikelets distant, not or scarcely reaching the base of the spikelet above on the same side of the rachis; anthers 4–8 mm long . . . . . . . . . . . . . . . . . . . . . . . . 13.10 *Pseudoroegneria*
                    15. Spikelets usually more closely spaced, reaching midlength of the spikelet above on the same side of the rachis; anthers 0.7–7 mm long . . . . . . . . . . . . . . . . . . . 13.13 *Elymus* (in part)

# 13.01 HORDEUM L.[1]

Pl summer or winter ann or per; csp, smt shortly rhz. Clm to 135(150) cm, erect, geniculate, or decumbent; nd glab or pubescent. **Shth** open, pubescent or glab; **aur** present or absent; **lig** hyaline, truncate, erose; **bld** flat to more or less involute, more or less pubescent on both sides. **Infl** usu spikelike rcm, smt spikes, all customarily called *spikes*, with 3 spklt at each nd, cent spklt usu sessile, smt pedlt, ped to 2 mm, lat spklt usu pedlt, ped curved or straight, smt all 3 spklt sessile in cultivated pl; **dis** usu in the rchs, the spklt falling in triplets, cultivated forms generally not disarticulating. **Spklt** with 1 flt. **Glm** awnlike, usu exceeding the flt. **Lat spklt** usu strl or stmt, often bisx in cultivated forms; **flt** pedlt, usu rdcd; **lm** awned or unawned. **Cen spklt** bisx; **flt** sessile; **rchl** prolonged beyond the flt; **lem** ovate, glab to pubescent, 5-veined, usu awned, rarely unawned; **pal** almost equal to the lm, narrowly ovate, keeled; **lod** 2, broadly lanceolate, mrg ciliate; **anth** 3, usu yellowish. **Car** usu tightly enclosed in the lm and pal at maturity.

*Hordeum* is a genus of 32 species that grow in temperate and adjacent subtropical areas, at elevations from 0–4500 m. The genus is native to Eurasia, the Americas, and Africa, and has been introduced to Australasia. The species are confined to rather moist habitats, even on saline soils. The annual species occupy seasonally moist habitats that cannot sustain a continuous grass cover. Eleven species of *Hordeum* grow in the *Manual* region: six are native, three are established weeds, and two are cultivated and occasionally persist as weeds.

Spike measurements and lemma lengths, unless stated otherwise, do not include the awns.

1. Plants perennial.
    2. Culms usually with a bulbous swelling at the base; auricles to 5.5 mm long, well developed. . . . . . . . . 10. *H. bulbosum*
    2. Culms not bulbous-based; auricles absent or no more than 1 mm long.
        3. Glumes of the central spikelet flattened near the base . . . . . . . . . . . . . . . . . . . . . . . . . . . 6. *H. arizonicum* (in part)
        3. Glumes of the central spikelet usually setaceous throughout, rarely flattened near the base.
            4. Glumes 15–85 mm long, divergent to strongly divergent at maturity . . . . . . . . . . . . . . . . . 5. *H. jubatum* (in part)
            4. Glumes 7–19 mm long, divergent or not at maturity.
                5. Anthers of the central spikelet 0.8–4 mm long; auricles absent . . . . . . . . . . . . . . . . . . . . 4. *H. brachyantherum*
                5. Anthers of the central spikelet 3.5–5 mm long; auricles present on the basal leaves . . . . . . . . . . 8. *H. secalinum*
1. Plants annual.
    6. Auricles to 8 mm long, well developed even on the upper leaves; lemmas of the lateral florets 6–15 mm long.
        7. Rachises disarticulating at maturity; glumes of the central spikelets ciliate; lemmas of the central florets to 2 mm wide, with awns 20–40 mm long; lateral spikelets staminate . . . . . . . . . . . . . . . . . . . . . 9. *H. murinum*
        7. Rachises usually not disarticulating at maturity; glumes of the central spikelets pubescent; lemmas of the central florets at least 3 mm wide, unawned or with awns 30–180 mm long; usually 1 or both lateral spikelets at a node seed-forming . . . . . . . . . . . . . . . . . . . . . . . . . . . . . . . . . . 11. *H. vulgare*
    6. Auricles usually absent or to 0.3 mm long; lemmas of the lateral florets 1.7–8.5 mm long.
        8. Glumes bent, strongly divergent at maturity.
            9. Glumes of the central spikelets not flattened, (15)35–85 mm long . . . . . . . . . . . . . . . . . . . . 5. *H. jubatum* (in part)
            9. Glumes of the central spikelets slightly flattened towards the base, 11–28 mm long . . . . 6. *H. arizonicum* (in part)
        8. Glumes straight, ascending to slightly divergent at maturity.
            10. Lemmas of the lateral spikelets with awns 3–8 mm long . . . . . . . . . . . . . . . . . . . . . . . . . . . . . 7. *H. marinum*
            10. Lemmas of the lateral spikelets unawned or with awns no more than 3 mm long.
                11. Glumes of the central spikelets setaceous to slightly flattened near the base.
                    12. Spikes 4–8 mm wide; lemmas of the central spikelets with awns 3–12 mm long; ligules 0.3–0.8 mm long . . . . . . . . . . . . . . . . . . . . . . . . . . . . . . . . . . . . . . . . . 3. *H. depressum*
                    12. Spikes 6–20 mm wide; lemmas of the central spikelets with awns 10–22 mm long; ligules 0.6–1.8 mm long . . . . . . . . . . . . . . . . . . . . . . . . . . . . . . . . 6. *H. arizonicum* (in part)
                11. Glumes of the central spikelets distinctly flattened near the base.
                    13. Lemmas of the lateral spikelets 1.7–4.4 mm long, usually unawned, rarely with awns to 1.2 mm long; sheaths with stripes of hairs . . . . . . . . . . . . . . . . . . . . . . . . . 1. *H. intercedens*
                    13. Lemmas of the lateral spikelets 2.5–5.7 mm long, usually awned, with awns to 1.8 mm long; sheaths glabrous . . . . . . . . . . . . . . . . . . . . . . . . . . . . . . . . . . . . . . . 2. *H. pusillum*

## 1. Hordeum intercedens Nevski BOBTAIL BARLEY [p. 372, <u>514</u>]

*Hordeum intercedens* grows in vernal pools and flooded, often saline river beds and alkaline flats. It is restricted to southwestern California, including some of the coastal islands, and northwestern Baja California, Mexico.

## 2. Hordeum pusillum Nutt. LITTLE BARLEY [p. 372, <u>514</u>]

*Hordeum pusillum* grows in open grasslands, pastures, and the borders of marshes, and in disturbed places such as roadsides and waste places, often in alkaline soil. It is native, widespread, and often common in much of the *Manual* region. Its range extends into northern Mexico, but it is not common there.

[1]Roland von Bothmer, Claus Baden†, and Niels H. Jacobsen

### 3. **Hordeum depressum** (Scribn. & J.G. Sm.) Rydb. LOW BARLEY [p. 372, 514]

*Hordeum depressum* grows in vernal pools and ephemeral habitats, often in alkaline soil. It is restricted to the western United States.

### 4. **Hordeum brachyantherum** Nevski [p. 372, 514]

*Hordeum brachyantherum* is native to the Kamchatka Peninsula and western North America, and has been introduced to a few locations in the eastern United States. It grows in salt marshes, pastures, woodlands, subarctic woodland meadows, and subalpine meadows. Two subspecies are recognized here, but there is so much overlap in their morphological variation that unambiguous determination of many specimens is impossible in the absence of a chromosome count. They are sometimes treated as two species.

1. Basal sheaths usually glabrous, sometimes sparsely pubescent; anthers 0.8–3.5 mm long; culms often robust, sometimes slender . . . . . . . . . . . subsp. *brachyantherum*
1. Basal sheaths usually densely pubescent; anthers 1.1–4 mm long; culms usually slender . . . . . subsp. *californicum*

### **Hordeum brachyantherum** Nevski subsp. **brachyantherum** MEADOW BARLEY, NORTHERN BARLEY, ORGE À ANTHÈRES COURTES, ORGE DES PRÉS [p. 372]

*Hordeum brachyantherum* subsp. *brachyantherum* grows in pastures and along streams and lake shores, from sea level to 4000 m. Its range extends from Kamchatka through western North America to Baja California, Mexico. It is also known from disjunct locations in Newfoundland and Labrador and the eastern United Sates.

### **Hordeum brachyantherum** subsp. **californicum** (Covas & Stebbins) Bothmer, N. Jacobsen & Seberg CALIFORNIA BARLEY [p. 372]

*Hordeum brachyantherum* subsp. *californicum* is restricted to California. It grows on dry and moist grass slopes, in meadows and rocky stream beds, along stream margins, and around vernal pools, in oak woodlands and disturbed ground, and in serpentine, alkaline, and granitic soils, up to 2300 m. Records from outside California, and many from inside California, are based on misidentified specimens, usually of *H. brachyantherum* subsp. *brachyantherum*.

### 5. **Hordeum jubatum** L. [p. 372, 514]

*Hordeum jubatum* grows in meadows and prairies around riverbeds and seasonal lakes, often in saline habitats, and along roadsides and in other disturbed sites. It is native from eastern Siberia through most of North America to Mexico, growing at elevations of 0–3000 m. It has been introduced to South America, Europe, and central Asia. It is grown in Russia and other areas outside its native range as an ornamental. In its native range, it is a weedy species.

*Hordeum jubatum* shows a wide range of variation in almost all characters; most such variation is not taxonomically significant.

1. Glumes of the central spikelet 15–35 mm long; lemma awns of the central spikelets 11–35 mm long . . . . . . . . . . . . . . . . . . . . . . . . . . . . . . . . subsp. *intermedium*
1. Glumes of the central spikelet 35–85 mm long; lemma awns of the central spikelets 35–90 mm long . . . . . . . . . . . . . . . . . . . . . . . . . . . . . . . . subsp. *jubatum*

### **Hordeum jubatum** subsp. **intermedium** Bowden INTERMEDIATE BARLEY [p. 372]

*Hordeum jubatum* subsp. *intermedium* is most abundant in the dry prairies of the northern Rocky Mountains and northern plains, growing at 0–3000 m. It also grows, as a disjunct, in southern Mexico.

### **Hordeum jubatum** L. subsp. **jubatum** FOXTAIL BARLEY, SQUIRRELTAIL BARLEY, SQUIRRELTAIL GRASS, ORGE AGRÉABLE, QUEUE D'ÉCUREUIL, ORGE QUEUE D'ÉCUREUIL [p. 372]

*Hordeum jubatum* subsp. *jubatum* is the more widespread of the two subspecies, extending from eastern Siberia through most of North America to northern Mexico. Native in western and northern portions of the *Manual* region, it is considered to be adventive in the eastern and southeastern portion of its range. It grows in moist soil along roadsides and other disturbed areas, as well as in meadows, the edges of sloughs and salt marshes, and on grassy slopes.

### 6. **Hordeum arizonicum** Covas ARIZONA BARLEY [p. 372, 514]

*Hordeum arizonicum* grows in saline habitats, along irrigation ditches, canals, and ponds in the southwestern United States and northern Mexico.

### 7. **Hordeum marinum** Huds. [p. 372, 514]

*Hordeum marinum* is native to Eurasia, where it grows in disturbed habitats. It has become established in similar habitats in western North America, and in scattered locations elsewhere. Two subspecies are recognized.

1. Lower glumes of the lateral spikelets usually setaceous, not winged . . . . . . . . . . . . . . . . subsp. *gussoneanum*
1. Lower glumes of the lateral spikelets with a flattened wing, the wings 0.5–2.3 mm wide . . . . . . . . . . subsp. *marinum*

### **Hordeum marinum** subsp. **gussoneanum** (Parl.) Thell. MEDITERRANEAN BARLEY, GENICULATE BARLEY [p. 372]

*Hordeum marinum* subsp. *gussoneanum* grows in grassy fields, waste places, and open ground. It was introduced to North America from the Mediterranean area, and it is now an established weed, especially in western North America.

### **Hordeum marinum** Huds. subsp. **marinum** SEA BARLEY [p. 372]

*Hordeum marinum* subsp. *marinum* is native to Eurasia, where it grows in disturbed habitats. Although it has been reported occasionally from the *Manual* region, it has not become established.

### 8. **Hordeum secalinum** Schreb. FALSE-RYE BARLEY, MEADOW BARLEY [p. 373]

*Hordeum secalinum* is native to Europe, where it grows in moist, saline areas, often in coastal meadows. It does not grow in the *Manual* region; reports from North America were based on specimens of *Hordeum brachyantherum*, from which *H. secalinum* differs in having auricles on the basal leaves and longer anthers.

### 9. **Hordeum murinum** L. [p. 373, 514]

*Hordeum murinum* is native to Eurasia, where it is a common weed in areas of human disturbance. It is now an established weed, primarily in the southwestern part of the *Manual* region. Prostrate plants are associated with grazing. Three subspecies are recognized.

1. Central spikelets sessile to subsessile; lemmas of the central florets subequal to those of the lateral florets, the awns longer than those of the lateral florets; paleas of the lateral florets almost glabrous . . . . subsp. *murinum*
1. Central spikelets pedicellate; lemmas of the central florets from subequal to shorter than those of the lateral florets, the awns from shorter to longer than those of the lateral florets; paleas of the lateral florets scabrous to hairy.
    2. Lemmas of the central florets much shorter than those of the lateral florets; paleas of the lateral florets scabrous on the lower $^1/_2$; anthers of the central and lateral florets similar in size . . . subsp. *leporinum*

2. Lemmas of the central florets about equal to those of the lateral florets; paleas of the lateral florets distinctly pilose on the lower ¹/₂; anthers of the central florets 0.2–0.6 mm long, those of the lateral florets 1.2–1.8 mm long . . . . . . . subsp. *glaucum*

## Hordeum murinum subsp. glaucum (Steudel) Tzvelev
SMOOTH BARLEY [p. *373*]

*Hordeum murinum* subsp. *glaucum* grows in grasslands, fields, and waste places. It is native to the eastern Mediterranean area. It is now common in arid areas of the western United States, and is also known from scattered locations elsewhere in the *Manual* region.

## Hordeum murinum subsp. leporinum (Link) Arcang.
MOUSE BARLEY [p. *373*]

*Hordeum murinum* subsp. *leporinum* grows in waste places, roadsides, and disturbed areas in arid regions. It is native to the Mediterranean region. It is now established in the *Manual* region, being most common in the western United States.

## Hordeum murinum L. subsp. murinum WALL BARLEY, FARMER'S FOXTAIL, WAY BARLEY [p. *373*]

*Hordeum murinum* subsp. *murinum* grows in waste places that are somewhat moist. It is native to Europe. Within the *Manual* region, it

has the most restricted distribution of the three subspecies, being found from Washington to Arizona, and in scattered locations from Maine to Virginia.

## 10. Hordeum bulbosum L. BULBOUS BARLEY [p. *373*]

*Hordeum bulbosum* is native to the eastern Mediterranean and western Asia. In the *Manual* region, it is known as an occasional escape from breeding programs. In its native range it is found in a wide range of habitats, from wet meadows to dry hillsides, roadsides, and abandoned fields.

## 11. Hordeum vulgare L. BARLEY, ORGE, ORGE VULGAIRE [p. *373*, 514]

*Hordeum vulgare* is native to Eurasia. Plants in the *Manual* region belong to the cultivated subspecies, **H. vulgare** L. subsp. **vulgare**. *Hordeum vulgare* subsp. *vulgare* was first domesticated in western Asia. It is now grown in most temperate parts of the world. In the *Manual* region, it occurs as a cultivated species that is often found as an adventive in fields, roadsides, and waste places throughout the region, not just at the locations shown on the map.

## 13.02 EREMOPYRUM (Ledeb.) Jaub. & Spach[1]

**Pl** ann. **Clm** 3–40 cm, geniculate. **Shth** open for most of their length; **aur** present, often inconspicuous; **lig** 0.4–2 mm, memb, truncate; **bld** 1–6 mm wide, flat, linear. **Infl** distichous spikes, 0.8–4.5 cm, with 1 spklt per nd, usu erect when mature; **rchs intnd** flat, mrg glab or with hairs, hairs white; **mid intnd** 0.5–3 mm; **dis** in the rchs, at the nd beneath each spklt, or at the base of each flt. **Spklt** 6–25 mm, including the awns, more than 3 times the length of the intnd, divergent, lat compressed, with 2–5 bisx flt, strl flt distal or absent. **Glm** equal, 4–19 mm, including the awns, coriaceous, becoming indurate, 1-keeled initially, smt 2-keeled at maturity,

keels glab or hairy, never with tufts of hair, bases slightly connate, apc tapering to a sharp point or straight awn; **lm** 5–24 mm, coriaceous, rounded bas, keeled distally, 5-veined, unawned or shortly awned; **pal** usu shorter and thinner than the lm, 2-keeled, ciliate or scabrous distally, keels smt prolonged into 2 toothlike appendages; **anth** 3, 0.4–1.3 mm, yellow; **ov** pubescent; **sty** 2, free to the base.

*Eremopyrum* includes 5–10 species that grow in steppes and semidesert regions from Turkey to central Asia and Pakistan. Three species have been found in North America; only *E. triticeum* is widepsread.

1. Glumes 1-veined and 2-keeled at maturity; lemmas of the first floret in each spikelet pubescent on the lower ¹/₂, glabrous distally, the other lemmas glabrous; disarticulation beneath the florets, sometimes at the base of the spikes . . . . . . . . . . . . . . . . . . . . . . . . . . . . . . . . . . . . . . . . . . . . . . . . . . . . . . . 1. *E. triticeum*
1. Glumes usually 3–5-veined and 1-keeled; lemmas of all florets alike in their pubescence or lack thereof; disarticulation at the rachis nodes.
    2. Glume bases straight; spikes 1.3–2.8 cm wide . . . . . . . . . . . . . . . . . . . . . . . . . . . . . . . 2. *E. bonaepartis*
    2. Glume bases arcuately curved; spikes 0.9–1.8 cm wide . . . . . . . . . . . . . . . . . . . . . . . . . . . 3. *E. orientale*

## 1. Eremopyrum triticeum (Gaertn.) Nevski ANNUAL WHEATGRASS [p. *373*, 514]

*Eremopyrum triticeum* is known primarily from scattered disturbed sites in western North America, from southern Canada to Arizona and New Mexico. Like most weeds, it is probably more widely distributed than herbarium records indicate. It is tolerant of alkaline soils, and is summer-dormant.

## 2. Eremopyrum bonaepartis (Spreng.) Nevski [p. *373*, 514]

In the *Manual* region, *Eremopyrum bonaepartis* is known only from a few collections in Arizona.

## 3. Eremopyrum orientale (L.) Jaub. & Spach [p. *373*, 514]

*Eremopyrum orientale* has been collected from southern Manitoba, growing with *E. triticeum*, and has been reported from southeastern British Columbia and New York. It is not known to be established in the *Manual* region.

[1]Signe Frederiksen

## 13.03 DASYPYRUM (Coss. & Durieu) T. Durand[1]

Pl ann or per; shortly rhz if per. **Clm** 20–100 cm. **Shth** open; **aur** present, often inconspicuous; **lig** 0.3–1 mm, memb, truncate; **bld** 1–5 mm wide, flat, linear, cent vein distinct on the abx side. **Infl** tml spikes, 4–12 cm long including the awns, 0.6–2 cm wide excluding the awns, compressed, dense, with 1 spklt per nd; **rchs intnd** flat, mrg ciliate, hairs white; **mid intnd** 1–3 mm; **dis** in the rchs, at the nd beneath each spklt. **Spklt** 25–75 mm including the awns, 7–22 mm excluding the awns, more than 3 times the length of the rchs intnd, usu divergent, smt ascending, lat compressed, with 2–4 flt, the lo 2 flt usu bisx, the tml flt strl; **rchl intnd** below the lo flt shorter than those below the tml flt. **Glm** equal, to 40 mm including the awns, to 8 mm excluding the awns, coriaceous, usu 5-veined, strongly 2-keeled, keels with 1–3 mm hairs, mrg unequal, stiff, translucent, apc tapering into scabrous awns; **bsx lm** 9–13 mm excluding the awns, lanceolate, keeled, usu 5-veined, apc acuminate, awned, awns to 60 mm; **stl lm** smaller, awns to 10 mm; **pal** narrowly lanceolate, memb, 2-veined, 2-keeled; **lod** 2, free, memb, ciliate or glab; **anth** 3, 4–7 mm, yellow; **ov** pubescent; **sty** 2, free to the base.

*Dasypyrum* is a Mediterranean genus of two species; only one has been collected in the *Manual* region. The hairy, 2-keeled glumes make the genus easily distinguishable from other genera in the *Triticeae*.

### 1. Dasypyrum villosum (L.) P. Candargy
MOSQUITOGRASS [p. *373*, <u>515</u>]

*Dasypyrum villosum* is native from southern Europe to Turkey, the Crimea, and the Caucasus. The only known North American record is a collection made in Philadelphia County, Pennsylvania, in 1877.

## 13.04 TAENIATHERUM Nevski[2]

Pl ann. **Clm** (5)10–55(70) cm, erect, glab; **nd** 3–6. **Lvs** evenly distributed; **shth** open, usu glab; **aur** 0.1–0.5 mm, rarely absent; **lig** memb, truncate; **bld** flat to involute. **Infl** spikes, erect; **nd** 4–24(28), each with 2(3, 4) spklt; **intnd** 0.5–3.5 mm. **Spklt** with 2(3) flt, the lowest flt in each spklt bisx, the distal flt(s) highly rdcd, strl; **dis** above the glm. **Glm** 5–80 mm, equal, awnlike, erect to spreading or reflexed, bases connate. **Bsx flt: lm** 5-veined, glab or scabrous, mrg flat, scabrous, apc tml awned, awns 20–110 mm, longer than the lm, divergent, often cernuous; **pal** as long as the lm, keels antrorsely ciliate, apc truncate; **lod** 2, lobed, ciliate. **Rdcd flt: lm** 3-veined, awned; **pal** absent; **anth** 3, yellow to purple. **Car** narrowly elliptic, with an adx groove, apc pubescent.

*Taeniatherum* has one species, which is native to Eurasia.

### 1. Taeniatherum caput-medusae (L.) Nevski
MEDUSAHEAD [p. *374*, <u>515</u>]

*Taeniatherum caput-medusae* is native from Portugal and Morocco east to Kyrgyzstan. It usually grows on stony soils, and flowers from May–June (July). It is an aggressive invader of disturbed sites in the western United States, where it has become a serious problem on rangelands. It has been found as a rare introduction at several sites in the eastern United States, but may not persist there. It is listed as a noxious weed by the U.S. Department of Agriculture.

## 13.05 SECALE L.[3]

Pl ann, bien, or short-lived per; csp when per. **Clm** 25–120(300) cm. **Shth** open; **aur** usu present, 0.5–1 mm; **lig** memb, truncate, often lacerate; **bld** flat or involute. **Infl** lat compressed, distichous spikes; **mid intnd** 2–4 mm, with 1 spklt per nd, spklt strongly ascending; **dis** in the rchs, below the spklt, rchs not or tardily disarticulating in cultivated strains. **Spklt** 10–18 mm, with 2(3) flt; **flt** bisx. **Glm** 8–20 mm, shorter than the adjacent lm, linear to subulate, scabrous, mrg hyaline, 1-veined, keeled, keels terminating in an awn, awns to 35 mm; **lm** 8–19 mm, strongly lat compressed, strongly keeled, keels conspicuously scabrous distally, scabrules 0.6–1.3 mm, apc tapering to a scabrous awn, awns 2–50 mm; **anth** 3, 2.3–12 mm, yellow.

*Secale* has three species. All are native to the Mediterranean region and western Asia; two species have been collected in the *Manual* region. *Secale cereale* is cultivated as a crop and used along roadsides to prevent soil erosion, and is established in the *Manual* region. *Secale strictum* has been cultivated experimentally and is not established.

×*Triticosecale* is an artificially derived hybrid between *Triticum* and *Secale* that is now widely cultivated (see p. 59).

1. Plants annual or biennial; rachises not or tardily disarticulating; lemmas 14–18 mm long . . . . . . . . . . . . . . . . . 1. *S. cereale*
1. Plants perennial; rachises readily disarticulating; lemmas 8–16 mm long . . . . . . . . . . . . . . . . . . . . . . . . . . . . 2. *S. strictum*

[1]Signe Frederiksen  [2]J.K. Wipff  [3]Mary E. Barkworth

**1. Secale cereale** L. Rye, Seigle, Seigle Cultivé [p. *374, 515*]

*Secale cereale* is one of the world's most important cereal grasses; it is also widely used in North America for soil stabilization and, particularly in Canada, for whisky. When dry, the spike is often distinctly nodding.

**2. Secale strictum** (C. Presl) C. Presl [p. *374*]

In 1951 it was reported that *Secale strictum* had become established around the Agricultural Experiment Station in Pullman, Washington, but it is no longer present there. Prior to 1931, the station worked on development of a *S. cereale* × *S. strictum* strain that would combine the perennial habit with good seed production. The attempt had been abandoned by 1931, but hybrid seed had been distributed as 'Michael's Grass'.

## 13.06 ×TRITICOSECALE Wittm. *ex* A. Camus [p. *374*][1]
### Triticale

**Pl** ann. **Clm** to 130 cm, erect, straight or geniculate at the lowest nd. **Lvs** mainly cauline; **lig** 2–4 mm, memb, truncate to rounded. **Infl** tml, distichous spikes, with solitary spklt; **intnd** 3–5 mm, densely pilose, at least on the edges. **Spklt** 10–17 mm, with 2–4 flt, distal flt usu rdcd. **Glm** 9–12 mm, asymmetrically keeled, keels stronger and smt conspicuously ciliate distally, apc retuse to acute, awned, awns 3–4 mm; **lm** 10–15 mm, lat compressed, keeled, keels smt ciliate distally, tml awned, awns 3–50 mm; **anth** 3, yellow.

*×Triticosecale* comprises hybrids between *Secale* and *Triticum*. Natural hybrids between the two genera are rare, but Triticale, which consists of cultivars derived from artificial hybrids between *S. cereale* and *T. aestivum*, is becoming an increasingly important cereal crop. The existing species names in *×Triticosecale* do not apply to Triticale because they involve different species of *Triticum*.

## 13.07 AEGILOPS L.[2]

**Pl** ann. **Clm** 14–80 cm, usu glab, erect or geniculate at the base, with (1)2–4(5) nd. **Shth** open; **aur** ciliate; **lig** 0.2–0.8 mm, memb, truncate; **bld** 1.5–10 mm wide, linear to linear-lanceolate, flat, spreading. **Infl** tml spikes, with 2–13 spklt, usu with 1–3 additional rdmt spklt at the base; **intnd** 6–12 mm; **dis** either at the base of the spikes or in the rchs, the spklt falling attached to the intnd above or below. **Spklt** solitary at each nd, ¹/₂–2(3) times the length of the intnd, tangential to the rchs, appressed or ascending, the up spklt(s) smt strl; **ftl spklt** 5–15 mm, with 2–7 flt, the distal flt often strl. **Glm** ovate to rectangular, rounded on the back, scabrous or pubescent, with several prominent veins, midveins smooth throughout, apc truncate, toothed, or awned, smt indurate at maturity; **lm** rounded on the back, apc toothed, frequently awned; **pal** chartaceous, 2-keeled, keels ciliate; **anth** 3, 1.5–4 mm, not penicillate; **ov** with pubescent apc. **Car** lanceolate to lanceolate-ovate.

*Aegilops* has about 23 species, and is native to the Canary Islands, as well as from the Mediterranean region to central Asia. Four species are established in the *Manual* region; only *A. cylindrica* is widespread. The introductions occurred at the end of the nineteenth or beginning of the twentieth century. Three other species have been collected in the region; they are not known to have persisted.

In the key, spike and spikelet lengths exclude the rudimentary spikelets and awns.

1. Glumes unawned, or with a single awn to 2 cm long; spikes narrowly cylindrical to moniliform, not ovoid; disarticulation in the rachises, the spikelets falling attached to the internodes above.
  2. Spikes cylindrical to slightly moniliform . . . . . . . . . . . . . . . . . . . . . . . . . . . . . . . . . . . . . 1. *A. tauschii*
  2. Spikes distinctly moniliform.
    3. Glumes mostly glabrous, the veins setulose; lemmas of the apical spikelets with awns to 4 cm long; spikelets with 2–5 florets, the distal 1 or 2 sterile . . . . . . . . . . . . . . . . . . . . . . . . 2. *A. ventricosa*
    3. Glumes appressed-velutinous; lemmas of the apical spikelets with awns 3–8.5 cm long; spikelets with 4–7 florets, the distal 2 sterile . . . . . . . . . . . . . . . . . . . . . . . . . . 3. *A. crassa*
1. Some glumes with awns 2–8 cm long; spikes narrowly cylindrical to ovoid, not moniliform; disarticulation near the base of the spikes, at least initially.
  4. Spikes narrowly cylindrical, about 0.3 cm wide . . . . . . . . . . . . . . . . . . . . . . . . . . . . . 4. *A. cylindrica*
  4. Spikes subcylindrical to ovoid, widest at the base, 0.4–1.3 cm wide.
    5. Upper spikelets 7–9 mm long; lemmas of the lower fertile florets with 2–3 teeth, 1 tooth sometimes extending into an awn up to 10 mm long . . . . . . . . . . . . . . . . . . . . . . 5. *A. triuncialis*
    5. Upper spikelets 4–5 mm long; lemmas of the fertile florets 2–3-awned, awns 5–40 mm long.
      6. Rudimentary spikelet(s) usually 1, occasionally 2; spikes gradually tapering distally . . . . . . . . . . . 6. *A. geniculata*
      6. Rudimentary spikelets 3, occasionally 2; spikes abruptly contracted distally to a narrow cylinder . . . . . . . . . . . . . . . . . . . . . . . . . . . . . . . . . . . . . . . . . . . . . . . . . . 7. *A. neglecta*

[1]Mary E. Barkworth  [2]Sandra M. Saufferer

**1. Aegilops tauschii** Coss. Tausch's Goatgrass, Rough-Spiked Hardgrass [p. *374*, <u>515</u>]

*Aegilops tauschii* is a weed of disturbed areas. In the *Manual* region, it is known only from Riverside County, California; Cochise County, Arizona; and an old collection from Westchester County, New York. It is native from the Caucasus and southern shores of the Caspian Sea, eastward to Kazakhstan and western China, and southward to Iraq and northwestern India.

**2. Aegilops ventricosa** Tausch Swollen Goatgrass, Belly-Shaped Hardgrass [p. *374*, <u>515</u>]

In the *Manual* region, *Aegilops ventricosa* was collected once in New Castle County, Delaware. It is native to the Mediterranean area. It occasionally forms hybrids with *Triticum durum*.

**3. Aegilops crassa** Boiss. Persian Goatgrass [p. *374*, <u>515</u>]

The single record of *Aegilops crassa* for North America is a specimen collected "from about the Yonkers Wool Mill [in Yonkers, New York]" in 1898. The species is native from Egypt to central Asia.

**4. Aegilops cylindrica** Host Jointed Goatgrass [p. *374*, <u>515</u>]

*Aegilops cylindrica* is a widespread weed in North America, being particularly troublesome in winter wheat. It usually grows in disturbed sites, such as roadsides, fields, and along railroad tracks. It is native to the Mediterranean region and central Asia, and is adventive in other temperate countries. Its apparent absence from Canada is somewhat remarkable.

Hybrids between *Aegilops cylindrica* and *Triticum aestivum*, called ×**Aegilotriticum sancti-andreae** (Degen) Soó [p. *374*], have been found in various parts of North America. They often have a few functional seeds which can backcross to either parent. For this reason, *A. cylindrica* is considered a serious weed in many wheat-growing areas within the *Manual* region.

**5. Aegilops triuncialis** L. Barbed Goatgrass [p. *374*, <u>515</u>]

North American collections of *Aegilops triuncialis* are from disturbed sites, mostly roadsides and railroads. The native range of the species extends from the Mediterranean area east to central Asia and south to Saudi Arabia. Specimens from the *Manual* region belong to **Aegilops triuncialis** L. var. **triuncialis**.

**6. Aegilops geniculata** Roth Ovate Goatgrass [p. *375*, <u>515</u>]

In the *Manual* region, *Aegilops geniculata* is known only from Mendocino County, California, where it usually grows along roadsides. It is native from the Mediterranean area to central Asia. In California, it grows in silty clay.

**7. Aegilops neglecta** Req. *ex* Bertol. Three-Awned Goatgrass [p. *375*, <u>515</u>]

*Aegilops neglecta* is native around the Mediterranean and in western Asia. It has been collected in Arlington County, Virginia, and near Corvallis, Oregon; the Oregon record indicates it is persisting from previous cultivation and becoming weedy.

# 13.08 **TRITICUM** L.[1]

**Pl** ann. **Clm** 14–180 cm, solitary or brchd at the base; **intnd** usu hollow throughout in hexaploids, usu solid for about 1 cm below the spike in diploids and tetraploids, even if hollow below. **Shth** open; **aur** present, often deciduous at maturity; **lig** memb; **bld** flat, glab or pubescent. **Infl** usu tml spikes, distichous, with 1 spklt per nd, occ brchd; **intnd** (0.5)1.4–8 mm; **dis** in the rchs, the spklt usu falling with the internd below to form a wedge-shaped diaspore, smt falling with the adjacent intnd to form a barrel-shaped diaspore, domesticated taxa usu non-disarticulating, or disarticulating only under pressure. **Spklt** 10–25(40) mm, usu 1–3 times the length of the intnd, appressed to ascending, with 2–9 flt, the distal flt often strl. **Glm** subequal, ovate, rectangular, or lanceolate, chartaceous to coriaceous, usu stiff, tightly to loosely appressed to the lo flt, with 1 prominent keel, at least distally, keels often winged and ending in a tooth or awn, a second keel or prominent lat vein present in some taxa; **lm** keeled, chartaceous to coriaceous, 2 lowest lm usu awned, awns 3–23 cm, scabrous, distal lm unawned or awned, awns to 2 cm; **pal** hyaline-memb, splitting at maturity in diploid taxa; **anth** 3. **Car** tightly (hulled

wheats) or loosely (naked wheats) enclosed by the glm and lm, lm and pal not adherent; **endosperm** flinty or mealy.

*Triticum* is a genus of approximately 25 wild and domesticated species that is native to western and central Asia. It was first cultivated in western Asia at least 9000 years ago and is now the world's most important crop, being planted more widely than any other genus.

Only *Triticum aestivum*, *T. durum*, and *T. spelta* are grown commercially in North America, *T. aestivum* being by far the most important. The remaining species in this treatment are those most frequently grown by North American plant breeders and wheat researchers. None of the species has become an established part of the North American flora, but they may be encountered as escapes near agricultural fields and research stations, or along transportation routes.

The width of a spike is the distance from one spikelet edge to the other across the two-rowed side of the spike; its thickness is the distance across the frontal face of spikelet, from one edge to the other. The spike and spikelet measurements do not include the awns. The glumes are measured from the base to the shoulder, and do not include any toothed tip.

1. Culms usually hollow to the base of the spikes; glumes with only 1 keel, this often developed only in the upper $\frac{1}{2}$ of the glumes.
  2. Glumes loosely appressed to the lower florets; rachises not disarticulating, even under pressure . . . . . . . . 11. *T. aestivum*
  2. Glumes tightly appressed to the lower florets; rachises disarticulating under pressure.
    3. Spikes strongly flattened; glumes acute . . . . . . . . . . . . . . . . . . . . . . . . . . . . . . . . . 5. *T. timopheevii* (in part)
    3. Spikes almost cylindrical; glumes truncate . . . . . . . . . . . . . . . . . . . . . . . . . . . . . . . . . . . . 12. *T. spelta*

[1]Laura A. Morrison

1. Culms partially to completely solid 1 cm below the spikes; glumes with 1 fully developed keel, sometimes with a second keel.
    4. Rachises not disarticulating, even under pressure; glumes loosely appressed to the lower florets.
        5. Glumes chartaceous.
            6. Glumes 6–13 mm long; rachises not enlarged at the base of the glumes; spikelets usually producing 1 caryopsis . . . . . . . . . . . . . . . . . . . . . . . . . . . . . . . . . . . . . . . . . . 4. *T. monococcum* (in part)
            6. Glumes 20–40 mm long; rachises enlarged at the base of the glumes; spikelets producing 2–3 caryopses . . . . . . . . . . . . . . . . . . . . . . . . . . . . . . . . . . . . . . . . . . . . . . . . . . . 9. *T. polonicum*
        5. Glumes coriaceous.
            7. Glumes awned, awns 1–6 cm long; spikes thicker than wide, never branched at the base . . . . . . . 10. *T. carthlicum*
            7. Glumes toothed, teeth to 0.3 cm long; spikes about as wide as thick, sometimes branched at the base.
                8. Spikes 4–11 cm long, never branched at the base; plants 60–160 cm tall; blades usually glabrous; endosperm usually flinty . . . . . . . . . . . . . . . . . . . . . . . . . . . . . . . . . . . . . 7. *T. durum*
                8. Spikes 7–14 cm long, sometimes branched at the base; plants 120–180 cm tall; blades hairy; endosperm mealy . . . . . . . . . . . . . . . . . . . . . . . . . . . . . . . . . . . . . . . . . . . . 8. *T. turgidum*
    4. Rachises disarticulating spontaneously or with pressure; glumes usually tightly appressed to the lower florets.
        9. Paleas splitting at maturity; spikelets 10–17 mm long.
            10. Spikelets elliptical to ovate; rachis internodes 1.4–2.5 mm long; rachises disarticulating with pressure . . . . . . . . . . . . . . . . . . . . . . . . . . . . . . . . . . . . . . . . . . . 4. *T. monococcum* (in part)
            10. Spikelets rectangular; rachis internodes 3–5 mm long; rachises disarticulating spontaneously.
                11. Caryopses blue or amber or red mottled with blue; third lemma in each spikelet, if present, usually unawned; blades blue-green, hairs stiff, those on the veins longer than those between; anthers 3–6 mm long . . . . . . . . . . . . . . . . . . . . . . . . . . . . . . . 1. *T. boeoticum*
                11. Caryopses red; third lemma in each spikelet, if present, awned, the awns up to 10 mm long; blades yellow-green, hairs soft, of uniform length; anthers 2–4 mm long . . . . . . . . . . . . . . . 2. *T. urartu*
        9. Paleas not splitting at maturity; spikelets 10–25 mm long.
            12. Spikelets oblong to rectangular; rachises disarticulating spontaneously; glumes unequally 2-keeled, the more prominent keel winged to the base . . . . . . . . . . . . . . . . . . . . . . . . . . . . 3. *T. dicoccoides*
            12. Spikelets elliptical to ovate; rachises disarticulating only with pressure; glumes with 1 prominent keel, the keel not winged to the base.
                13. Spikelets 16–18 mm long; rachis internodes 1.5–2.5 mm long; spikes always wider than thick; culms partially solid to hollow for 1 cm below the spikes . . . . . . . . . . . . . . 5. *T. timopheevii* (in part)
                13. Spikelets 10–16 mm long; rachis internodes (0.5)2–5 mm long; spikes variously shaped, from cylindrical to wider than thick; culms usually solid for 1 cm below the spike . . . . . . . . 6. *T. dicoccum*

### 1. Triticum boeoticum Boiss. WILD EINKORN [p. *375*]

*Triticum boeoticum* is a wild diploid wheat that is native from the Balkans through the Caucasus to Iran and Afghanistan and south to Iraq. It is morphologically similar to and, in its native range, sometimes sympatric with *T. urartu*, another wild diploid wheat.

### 2. Triticum urartu Thumanjan *ex* Gandilyan RED WILD EINKORN [p. *375*]

*Triticum urartu* is the wild diploid wheat that contributed the **A** haplome to the durum and bread wheat evolutionary lines. It does not have a diploid domesticated form. It is morphologically similar to *T. boeoticum*, but has a more limited distribution, being known from disjunct regions in Turkey, Lebanon, Armenia, western Iran, and eastern Iraq.

### 3. Triticum dicoccoides (Körn.) Körn. *ex* Schweinf. WILD EMMER [p. *375*]

*Triticum dicoccoides* is the wild counterpart of *T. dicoccum*, and is an ancestor of both *T. durum* and *T. aestivum*. *Triticum dicoccoides* is native to the Fertile Crescent.

### 4. Triticum monococcum L. EINKORN, SMALL SPELT, PETIT ÉPEAUTRE [p. *375*]

*Triticum monococcum* is the domesticated derivative of *T. boeoticum*. Its primary range extends from the Balkans and Romania through the Crimea and Caucasus to northern Iraq and western Iran, and south to

northern Africa. It was originally introduced to the *Manual* region as a food crop, but is now used primarily for plant breeding.

### 5. Triticum timopheevii (Zhuk.) Zhuk. TIMOPHEEV'S WHEAT [p. *375*]

*Triticum timopheevii* is the domesticated derivative of *T. araraticum* Jakubz. It is established in Georgia, Armenia, and northeastern Turkey. It differs from other species of *Triticum* in its long leaf hairs and their relatively higher density.

### 6. Triticum dicoccum Schrank *ex* Schübl. EMMER, FARRO, FAR [p. *375*]

*Triticum dicoccum* is the domesticated derivative of *T. dicoccoides*. It was once grown fairly extensively in central and southern Europe, southern Russia, northern Africa, and Arabia, because it can withstand poor, waterlogged soils. It is rarely grown now. It was introduced to the *Manual* region as a feed grain and forage for livestock. Currently, its primary use in the region is for plant breeding; it is also sold for human consumption as farro in specialty food markets.

### 7. Triticum durum Desf. DURUM WHEAT, MACARONI WHEAT, HARD WHEAT, BLÉ DUR [p. *375*]

*Triticum durum* is a domesticated spring wheat that is grown in temperate climates throughout the world. In the *Manual* region, it is grown in the Canadian prairies and northern Great Plains as a spring wheat, and in the southwestern United States and Mexico as a winter wheat. *Triticum durum* is typically used for macaroni-type pastas, semolina, and bulghur.

### 8. Triticum turgidum L. Rivet Wheat, Cone Wheat, Blé Poulard [p. *375*]

*Triticum turgidum* is the tallest of the wheats, and differs from other species of domesticated wheat in having branched-spike forms. It is grown primarily in southern Europe, northern Iraq, southern Iran, and western Pakistan. As treated here, *T. turgidum* is a narrowly distributed taxon of minor importance in plant breeding.

### 9. Triticum polonicum L. Polish Wheat [p. *375*]

*Triticum polonicum* is a minor, durum-like, spring wheat species. It is grown in the Mediterranean basin and central Asia on a small scale. In the *Manual* region, it is grown principally for plant breeding. It differs from other domesticated wheats in its unusually long, chartaceous glumes and lemmas.

### 10. Triticum carthlicum Nevski Persian Wheat [p. *376*]

*Triticum carthlicum* is of evolutionary interest because, morphologically, its spikes resemble those of *T. aestivum* rather than

those of free-threshing tetraploid wheats such as *T. durum*, *T. turgidum*, and *T. polonicum*. It is still occasionally cultivated in Georgia, Armenia, Azerbaijan, northern Iraq, and Iran because of its resistance to drought, frost, and ergot infection.

### 11. Triticum aestivum L. Wheat, Bread Wheat, Common Wheat, Soft Wheat, Blé Cultivé, Blé Commun [p. *376*]

*Triticum aestivum* is the most widely cultivated wheat. Both winter and spring types are grown in the *Manual* region. In addition to being grown for bread flour, *T. aestivum* cultivars are used for pastry-grade flour, Oriental-style soft noodles, and cereals.

### 12. Triticum spelta L. Spelt, Dinkel, Épeautre, Grand Épeautre [p. *376*]

In the *Manual* region, *Triticum spelta* is grown for the specialty food and feed grain markets. It is known for yielding a pastry-grade flour not suitable for bread making unless mixed with *T. aestivum*, the bread-quality flour. The ability of *Triticum spelta* to break under pressure into barrel-shaped units similar to those found in *Aegilops cylindrica* distinguishes it from all other members of *Triticum*.

## 13.09 AGROPYRON Gaertn.[1]

Pl per; densely to loosely csp, smt rhz. **Clm** 25–110 cm, geniculate or erect. **Shth** open; **aur** usu present; **lig** memb, often erose. **Infl** spikes, usu pectinate; **mid intnd** 0.2–3(5.5) mm, bas intnd often somewhat longer. **Spklt** solitary, usu more than 3 times as long as the intnd, usu divergent or spreading from the rchs, with 3–16 flt; **dis** above the glm and beneath the flt. **Glm** shorter than the adjacent lm, lance-ovate to lanceolate, 1–5-veined, asymmetrically keeled, a secondary keel smt present on the wider side, keels glab or with hairs, hairs not tufted, apc acute and entire, smt awned, awns to 6 mm; **lm** 5–7-veined, asymmetrically keeled, acute to awned, awns to 4.5 mm; **pal** from slightly shorter than to

exceeding the lm, bifid; **anth** 3, 3–5 mm, yellow. **Car** usu falling with the lm and pal attached.

*Agropyron*, it is now agreed, should be restricted to perennial species of *Triticeae* with keeled glumes, i.e., *A. cristatum* and its allies, or the "crested wheatgrasses". The excluded species are distributed among *Pseudoroegneria*, *Thinopyrum*, *Elymus*, *Eremopyrum*, and *Pascopyrum*. The genus is now widespread in western North America, frequently being used for soil stabilization on degraded rangeland and abandoned cropland, because it is highly tolerant of grazing and provides good spring forage.

This treatment recognizes two species within the *Manual* region, a very broadly interpreted *Agropyron cristatum* and a traditionally interpreted *A. fragile*.

1. Lemmas usually awned, awns 1–6 mm long; spikelets diverging from the rachises at angles of 30–95°;
     spikes narrowly to broadly lanceolate, rectangular, or ovate in outline . . . . . . . . . . . . . . . . . . . . . . . . . . . . 1. *A. cristatum*
1. Lemmas unawned, sometimes mucronate; spikelets diverging from the rachises at an angle of less than
     30(35)°; spikes linear to narrowly lanceolate in outline . . . . . . . . . . . . . . . . . . . . . . . . . . . . . . . . 2. *A. fragile*

### 1. Agropyron cristatum (L.) Gaertn. Crested Wheatgrass, Agropyron Accrêté, Agropyron à Crête [p. *376*, <u>515</u>]

*Agropyron cristatum* is native from central Europe and the eastern Mediterranean to Mongolia and China. Because it is easy to establish, *Agropyron cristatum* has often been used to restore productivity to areas that have been overgrazed, burned, or otherwise disturbed. This ability, combined with its high seed production, tends to prevent establishment of most other species, both native and introduced.

### 2. Agropyron fragile (Roth) P. Candargy Siberian Wheatgrass [p. *376*]

*Agropyron fragile* is native from the southern Volga basin through the Caucasus to Turkmenistan and Mongolia. It is more drought-tolerant than *A. cristatum*. Within the *Manual* region, *A. fragile* appears to be uncommon outside of experimental plantings. This may change as more cultivars become available.

## 13.10 PSEUDOROEGNERIA (Nevski) Á. Löve[2]

Pl per; usu csp, smt rhz. **Clm** 30–100 cm, usu erect, smt decumbent or geniculate. **Lvs** evenly distributed; **shth** open; **aur** well developed; **lig** memb; **bld** flat to loosely involute. **Infl** tml spikes, erect, with 1 spklt per nd; **intnd** (7)10–20(28) mm at midlength, lo intnd often longer

than those at midlength. **Spklt** (8)12–25 mm, 1.1–1.5(2) times the length of the intnd, usu appressed, smt slightly divergent, with 4–9 flt; **dis** above the glm and beneath the flt. **Glm** unequal, from shorter than to slightly longer than the lowest lm in the spklt, lanceolate to

[1]Mary E. Barkworth   [2]Jack R. Carlson

oblanceolate, (3)4–5(7)-veined, usu acute to obtuse, occ truncate, narrowing beyond midlength, veins prominent; **lm** inconspicuously 5-veined, unawned or tml awned, awns straight to strongly bent and divergent; **anth** 4–8 mm.

### 1. Pseudoroegneria spicata (Pursh) Á. Löve BLUEBUNCH WHEATGRASS [p. *376*, 515]

*Pseudoroegneria spicata* is primarily a western North American species, extending from the east side of the coastal mountains to the western edge of the Great Plains, and from the Yukon Territory to northern Mexico. It grows on medium-textured soils in arid and semiarid steppe, shrub-steppe, and open woodland communities, and was one of the dominant species in grassland communities of the Columbia and Snake river plains. It is still an important forage plant in the northern portion of the Intermountain region. Several cultivars have been developed.

The above observations make it clear that the awned and unawned phases of *Pseudoroegneria spicata* are of little taxonomic significance, despite their evident morphological difference. If it is considered necessary to distinguish between them, the awned phase can be called **Pseudoroegneria spicata** (Pursh) Á. Löve f. **spicata** and the unawned phase **P. spicata** f. **inermis** (Scribn. & J.G. Sm.) Barkworth.

Plants with densely pubescent leaves are known from the east slope of the Cascade Mountains in Washington. Plants with nearly as densely pubescent leaves are found elsewhere in southern Washington and

*Pseudoroegneria* includes 15–20 species, one of which is North American and the remainder either Eurasian or Asian.

northeastern Oregon. Such pubescent plants may be called **Pseudoroegneria spicata** f. **pubescens** (Elmer) Barkworth.

*Pseudoroegneria spicata* used to be confused with *Elymus wawawaiensis*, from which it differs in its more widely spaced spikelets and wider, less stiff glumes. The two species are geographically sympatric, but *P. spicata* grows in medium- to fine-textured loess soils, and *E. wawawaiensis* in shallow, rocky soils. *Pseudoroegneria spicata* may also be confused with *E. arizonicus*, particularly with immature specimens of that species or specimens mounted so that they appear to have erect, rather than drooping, spikes. It differs in having shorter, truncate ligules and generally thicker culms than *E. arizonicus*, and in having a distribution that extends much further north.

*Pseudoroegneria spicata* has been suggested as one of the parents in numerous natural hybrids with species of *Elymus* in the *Manual* region. These hybrids are usually mostly sterile, but development of even a few viable seeds permits introgression to occur, as well as the formation of distinctive populations. It is often difficult to detect such hybrids, particularly if they involve the unawned form of *Pseudoroegneria*. The named hybrids are treated under ×*Pseudelymus* (see below). Others are discussed under the *Elymus* parent.

## 13.11 ×PSEUDELYMUS Barkworth & D.R. Dewey[1]

**Pl** per; smt rhz. **Clm** 50–80 cm, erect, glab. **Lvs** not bas concentrated; **shth** glab or puberulent; **aur** present; **lig** truncate. **Infl** distichous spikes, with 1(2) spklt(s) per nd. **Spklt** appressed, with 3–5 flt. **Glm** unequal, linear-lanceolate to lanceolate; **lm** awned or unawned; **pal** slightly shorter than to slightly longer than the lm; **anth** indehiscent.

### 1. ×Pseudelymus saxicola (Scribn. & J.G. Sm.) Barkworth & D.R. Dewey [p. *376*]

×*Pseudelymus saxicola* consists of hybrids between *Pseudoroegneria spicata* and *Elymus elymoides*. It is a rather common hybrid in western North America. It differs from *E. albicans*, which is thought

×*Pseudelymus* comprises hybrids between *Pseudoroegneria* and *Elymus*. Only one species is treated here. Another species, *E. albicans*, is thought to be a similar hybrid, but it is treated as a species because it is frequently fertile.

to be derived from hybrids between *P. spicata* and *E. lanceolatus*, in lacking rhizomes, having longer awns on its glumes and lemmas, and having disarticulating rachises. It differs from *E.* ×*saundersii* in its longer glume and lemma awns.

## 13.12 ×ELYHORDEUM Mansf. *ex* Tsitsin & K.A. Petrova[2]

**Pl** per; usu csp, occ shortly rhz. **Infl** tml, spikes or spikelike, with 1–3(7) spklt per nd, lat spklt usu shortly pedlt, cent spklt sessile or nearly so; **dis** tardy, at the rchs nd and beneath the flt. **Spklt** with 1–4 flt. **Glm** subulate to narrowly lanceolate, usu awned; **lm** usu awned; **anth** strl. **Car** rarely formed.

×*Elyhordeum* is the name given to hybrids between *Elymus* and *Hordeum*. These hybrids are fairly common. All appear to be sterile, i.e., they do not produce good pollen or set seed.

### 1. ×Elyhordeum dakotense (Bowden) Bowden [p. *377*]

×*Elyhordeum dakotense* refers to hybrids between *Elymus canadensis* and *Hordeum jubatum*. They are known only from Brookings, South Dakota.

Interspecific hybrids between *Elymus elymoides* or *E. multisetus* and other species of *Elymus* resemble the ×*Elyhordeum* hybrids in having tardily disarticulating, spikelike inflorescences and awned glumes and lemmas, but are more likely to have solitary spikelets, even at the lowest node. Distinguishing between them and ×*Elyhordeum* hybrids, without knowledge of other species of *Triticeae* at a site, is challenging.

[1,2]Mary E. Barkworth

## 2. ×**Elyhordeum macounii** (Vasey) Barkworth & D.R. Dewey [p. *377*]

×*Elyhordeum macounii* consists of hybrids between *Elymus trachycaulus* and *Hordeum jubatum*. It is quite common in western and central North America. Backcrosses to *E. trachycaulus* may have non-disarticulating rachises; they are likely to be identified as *E. trachycaulus*, falling between subsp. *trachycaulus* and subsp. *subsecundus*.

## 3. ×**Elyhordeum pilosilemma** (W.W. Mitch. & H.J. Hodgs.) Barkworth [p. *377*]

×*Elyhordeum pilosilemma* is a hybrid between *Elymus macrourus* and *Hordeum jubatum* that occurs in many locations where the two parental species co-occur. It is very similar to ×*E. jordalii*, a hybrid between *E. macrourus* and *H. brachyantherum*.

## 4. ×**Elyhordeum jordalii** (Melderis) Tzvelev [p. *377*]

×*Elyhordeum jordalii* consists of hybrids between *Elymus macrourus* and *Hordeum brachyantherum*. It grows in the Brooks Range, Alaska, near settlements on the lowlands south of the range where it has a weedy habit. It resembles ×*E. pilosilemma*, which differs in having *H. jubatum* as the *Hordeum* parent.

## 5. ×**Elyhordeum schaackianum** (Bowden) Bowden [p. *377*]

×*Elyhordeum schaackianum* consists of hybrids between *Elymus hirsutus* and *Hordeum brachyantherum*. It is known only from Attu Island, Alaska, and the Queen Charlotte Islands, British Columbia.

## 6. ×**Elyhordeum stebbinsianum** (Bowden) [p. *377*]

×*Elyhordeum stebbinsianum* consists of hybrids between *Elymus glaucus* and *Hordeum brachyantherum*. They have been found at scattered locations in western North America.

## 7. ×**Elyhordeum iowense** R.W. Pohl [p. *377*]

×*Elyhordeum iowense* is a hybrid between *Elymus villosus* and *Hordeum jubatum* that has been found at scattered locations in the central plains. It probably occurs elsewhere, but is unlikely to be common, because *E. villosus* usually grows in more shady locations than *H. jubatum*.

## 8. ×**Elyhordeum arcuatum** W.W. Mitch. & H.J. Hodgs. [not illustrated]

×*Elyhordeum arcuatum* is probably a hybrid between *Elymus sibiricus* and *Hordeum jubatum*. It was described from disturbed sites around Palmer, Alaska, from which it has since been eliminated. No additional reports are known. There is no illustration because the type specimens could not be located.

## 9. ×**Elyhordeum montanense** (Scribn. *ex* Beal) Bowden [p. *377*]

×*Elyhordeum montanense* applies to hybrids between *Elymus virginicus* and *Hordeum jubatum*. It is often found in disturbed areas where both parental taxa grow. Short-awned specimens may reflect the involvement of *E. submuticus* rather than *E. virginicus*, two taxa that have sometimes been treated as conspecific.

## 10. ×**Elyhordeum californicum** (Bowden) Barkworth [p. *377*]

×*Elyhordeum californicum* consists of hybrids between *Elymus elymoides* or *E. multisetus* and *Hordeum brachyantherum* subsp. *brachyantherum*. It was described on the basis of specimens collected in California. It seems probable that it will be found at many locations where the two parents grow together.

# 13.13 ELYMUS L.[1]

Pl per; smt csp, smt rhz, smt stln. Clm 8–180(220) cm, usu erect to ascending, smt strongly decumbent to prostrate, usu glab. Lvs usu evenly distributed, smt somewhat bas concentrated; shth open for most of their length; aur often present; lig memb, usu truncate or rounded, smt acute, entire or erose, often ciliolate; bld 1–24(25) mm wide, abx surfaces usu smooth or scabrous, smt with hairs, adx surfaces scabrous or with hairs, particularly over the veins, usu with unequal, not strongly ribbed, widely spaced veins, smt with equal, strongly ribbed, closely spaced veins. Infl spikes, usu exserted, with 1–3(5) spklt per nd, intnd (1.5)2–26 mm; rchs with scabridulous, scabrous, or ciliate edges. Spklt usu appressed to ascending, smt strongly divergent or patent, with 1–11 flt, the lowest flt usu fnctl, strl and glmlike in some species, the distal flt often rdcd; dis usu above the glm and beneath each flt, smt also below the glm or in the rchs. Glm usu 2, absent or highly rdcd in some species, usu equal to subequal, smt unequal, usu linear-lanceolate to linear, setaceous, or subulate, smt oblanceolate to obovate, (0)1–7-veined, smt keeled over 1 vein, not necessarily the cent vein, keel vein smt extending into an awn; lm linear-lanceolate, obscurely 5(7)-veined, apc acute, often awned, smt bidentate, teeth to 0.2 mm, smt with bristles, bristles to 10 mm, awns tml or from the sinus, straight or arcuately divergent, not geniculate; pal from shorter than to slightly longer than the lm, keels scabrous or ciliate, at least in part; anth 3, 0.7–7 mm. Car with hairy apc.

As treated here, *Elymus* is a widespread, north-temperate genus of about 150 species. It includes *Sitanion* Raf. and *Roegneria* K. Koch, but moves some taxa that others include in *Elymus* to *Leymus*, *Pascopyrum*, *Pseudoroegneria*, and *Thinopyrum*. Thirty-two species of *Elymus* are native to the *Manual* region. Of the seven non-native species treated, one is established (*E. repens*), two are distributed as forage (*E. dahuricus* and *E. hoffmannii*), two are known from ballast dumps and are not established (*E. tsukushiensis* and *E. ciliaris*), and two (*E. caninus* and *E. semicostatus*) have been attributed to the *Manual* region but specimens documenting the reports have not been located. Eight named, naturally occurring, intrageneric hybrids are described at the end of the treatment. They are not included in the key. Other interspecific hybrids undoubtedly exist. Because many of the hybrids are partially fertile, backcrossing and introgression occurs. Intergeneric hybrids are treated under ×*Elyhordeum* (p. 63), ×*Elyleymus* (p. 75), and ×*Pseudelymus* (p. 63); most are sterile.

All species of *Elymus* are alloploids that combine one copy of the St haplome present in *Pseudoroegneria* with at least one other haplome. So far as is known, all species that are native to North America, as well as many species native to northern Eurasia, are tetraploids with one additional haplome, the H genome from *Hordeum* sect. *Critesion* (Raf.) Nevski. Many

[1]Mary E. Barkworth, Julian J.N. Campbell, and Björn Salomon

Asian species combine the **St** haplome with the **Y** haplome, which has no known diploids; such species are sometimes placed in *Roegneria*. This treatment includes two such species, *E. ciliaris* and *E. semicostatus*. In addition, the treatment includes two hexaploid species, *E. tsukushiensis* and *E. dahuricus*, that combine all three haplomes. *Elymus repens* and *E. hoffmannii*, the other two hexaploid species in this treatment, basically combine two copies of the **St** haplome with one of the **H** haplome, but the molecular data for *E. repens* point to a more complex situation.

In the key, unless otherwise stated, the following conventions are observed: the number of culm nodes refers to the number of nodes above the base; measurements of spikes include the awns, while measurements of spikelets, glumes, and lemmas do not; rachis internodes are measured in the middle of the spike; glume widths of lanceolate to linear glumes are measured at the widest point, and those of linear to setaceous glumes about 5 mm above the base of the glumes; the number of florets in a spikelet includes the distal reduced, sterile florets; dates of anthesis, when provided, are for the central range of each species.

The curvature of the lemma awns increases with maturity, and may vary within a spike. If a plant appears to have at least some strongly curved lemma awns, it should be taken through the "strongly curved" side of the key.

1. Spikelets 1 at all or most nodes; glumes with flat, non-indurate bases, glume bodies linear-lanceolate to obovate, margins hyaline, scarious, or chartaceous; lemmas awned or unawned [for opposite lead, see p. 66].
  2. Anthers 3–7 mm long; plants often strongly rhizomatous, sometimes not or only weakly rhizomatous.
      3. At least some lemmas with strongly divergent, outcurving, or recurved awns.
          4. Culms prostrate to decumbent and geniculate, 20–50 cm tall; plants of subalpine and alpine habitats . . . . . . . . . . . . . . . . . . . . . . . . . . . . . . . . . . . . . . . . . . . . . . . . . . . . . . 32. *E. sierrae* (in part)
          4. Culms erect or decumbent only at the base, (15)40–130 cm tall; plants of valley and montane, but not subalpine or alpine, habitats.
              5. Plants strongly rhizomatous; blades 1–3 mm wide . . . . . . . . . . . . . . . . . . . . . . . . . . . 34. *E. albicans*
              5. Plants cespitose or weakly rhizomatous; blades 1.5–6 mm wide.
                  6. Spikes often drooping to pendent at maturity; rachis internodes 11–17 mm long; plants of the southwestern United States . . . . . . . . . . . . . . . . . . . . . . . . . . . . . . . . . . . . . . . 29. *E. arizonicus*
                  6. Spikes erect to slightly nodding at maturity; rachis internodes 5–12 mm long; plants of the northwestern contiguous United States . . . . . . . . . . . . . . . . . . . . . . . . . 33. *E. wawawaiensis*
      3. Lemmas unawned or with straight to flexuous awns.
          7. Lemmas 12–14 mm long; plants not or weakly rhizomatous.
              8. Palea keels straight or slightly outwardly curved below the apices, apices about 0.2 mm wide between the vein ends . . . . . . . . . . . . . . . . . . . . . . . . . . . . . . . . . . . . . . . 10. *E. glaucus* (in part)
              8. Palea keels distinctly outwardly curved below the apices; apices 0.3–0.7 mm wide between the vein ends . . . . . . . . . . . . . . . . . . . . . . . . . . . . . . . . . . . . . . . . . . . . . . 39. *E. semicostatus*
          7. Lemmas 7–12 mm long; plants not, weakly, or strongly rhizomatous.
              9. Glumes keeled distally, keels smooth and inconspicuous proximally, scabrous and conspicuous distally; lemmas glabrous.
                  10. Adaxial surfaces of the blades usually sparsely pilose, sometimes glabrous, veins smooth, the primary veins separated by secondary veins; plants strongly rhizomatous . . . . . . . 35. *E. repens*
                  10. Adaxial surfaces of the blades glabrous, veins smooth or scabrous, all veins more or less equally prominent; plants slightly to moderately rhizomatous . . . . . . . . . . . . . . . . 36. *E. hoffmannii*
              9. Glumes not keeled or keeled throughout their length, keels smooth or scabrous throughout, sometimes hairy, conspicuous or not; lemmas glabrous or hairy.
                  11. Plants strongly rhizomatous; glumes 5–9 mm long; lemmas densely to sparsely hairy or glabrous . . . . . . . . . . . . . . . . . . . . . . . . . . . . . . . . . . . . . . . . . . . 27. *E. lanceolatus* (in part)
                  11. Plants cespitose or weakly rhizomatous; glumes 6–19 mm long; lemmas glabrous or pubescent, never densely hairy.
                      12. Spikelets usually at least twice as long as the internodes; internodes 4–12 mm long; glumes often awned, sometimes unawned; blades usually lax . . . . . . . . . 10. *E. glaucus* (in part)
                      12. Spikelets from shorter than to almost twice as long as the internodes; internodes 9–27 mm long; glumes unawned; blades usually straight . . . . . . . . . . . . . . . . . . . . . 28. *E. stebbinsii*
  2. Anthers 0.7–3 mm long; plants usually not or weakly rhizomatous, sometimes strongly rhizomatous.
      13. Culms prostrate or strongly decumbent at the base; disarticulation in the rachises or beneath the florets; plants of subalpine, alpine, and arctic habitats.
          14. Glumes unawned or with awns to 1 mm long; plants of arctic habitats . . . . . . . . . . . 26. *E. alaskanus* (in part)
          14. Glumes awned, awns 3–30 mm long; plants of subalpine and alpine habitats.

15. Anthers 1–1.6 mm long; internodes 2.5–5(7) mm long; disarticulation initially in the rachises; spikelets appressed to ascending . . . . . . . . . . . . . . . . . . . . . . . . . . . . . . . . . 31. *E. scribneri*

15. Anthers 2–3.5 mm long; internodes 5–15 mm long; rachises not disarticulating; spikelets ascending to divergent . . . . . . . . . . . . . . . . . . . . . . . . . . . . . . . . . . . . . . 32. *E. sierrae* (in part)

13. Culms usually ascending to erect, sometimes geniculate or weakly decumbent at the base; disarticulation beneath the florets; plants of sea level to subalpine habitats.

16. Lemmas with coarse, stiff, marginal hairs up to 1 mm long; paleas ²/₃–⁴/₅ as long as the lemmas, with wide, rounded apices . . . . . . . . . . . . . . . . . . . . . . . . . . . . . . . . . . . . . . . . . . . . 38. *E. ciliaris*

16. Lemmas with the marginal hairs, if present, similar to those elsewhere on the lemma; paleas ³/₄ as long as to slightly longer than the lemmas, tapering to the apices.

17. Lemmas awned, awns 7–40 mm long.

18. Lemma awns strongly arcuate to outcurving or recurved.

19. Spikes 8–12 cm long, straight, erect or inclined; blades 2–4 mm wide . . . . . . . . . . . . 30. *E. bakeri*

19. Spikes 7–30 cm long, flexuous, nodding to pendent; blades 5–14 mm wide

. . . . . . . . . . . . . . . . . . . . . . . . . . . . . . . . . . . . . . . . . . . 13. *E. sibiricus* (in part)

18. Lemma awns usually straight or flexuous, or, if shorter than 10 mm, sometimes weakly curving.

20. Glumes with hairs on the adaxial (inner) surface, these often inconspicuous . . . . . . 23. *E. caninus*

20. Glumes glabrous on the adaxial (inner) surface.

21. Palea keels distinctly outwardly curved below the apices, winged, not or scarcely extending beyond the intercostal region; apices 0.3–0.5 mm wide . . . . . . . . . . . . . . . . . . . . . . . . . . . . . . . . . . . . . . . . . . . . 37. *E. tsukushiensis*

21. Palea keels straight or slightly outwardly curved below the apices, not winged, often extending beyond the intercostal region, sometimes forming teeth; apices 0.1–0.3 mm wide.

22. Glumes 1.8–2.3 mm wide, margins 0.2–0.3 mm wide . . . . . 22. *E. trachycaulus* (in part)

22. Glumes 0.4–1.5(2) mm wide, margins 0.1–0.2 mm wide.

23. Spikes erect or almost so, 0.5–2 cm wide . . . . . . . . . . . . . . 10. *E. glaucus* (in part)

23. Spikes nodding to pendent, 2–5 cm wide . . . . . . . . . . . . . 13. *E. sibiricus* (in part)

17. Lemmas unawned or with awns up to 7 mm long.

24. Plants strongly rhizomatous . . . . . . . . . . . . . . . . . . . . . . . . . . . . . . 27. *E. lanceolatus* (in part)

24. Plants not or only shortly rhizomatous.

25. Glumes ¹/₃–²/₃ as long as the adjacent lemmas.

26. Glumes 0.8–1.8 mm wide, lanceolate, margins subequal; lemmas evenly hairy or glabrous distally . . . . . . . . . . . . . . . . . . . . . . . . . . . . . . . . . . 25. *E. macrourus*

26. Glumes 1.5–2 mm wide, oblanceolate to obovate, margins unequal; lemmas glabrous, evenly hairy, or more densely hairy distally . . . . . 26. *E. alaskanus* (in part)

25. Glumes ³/₄ as long as to slightly longer than the adjacent lemmas.

27. Glumes 3(5)-veined; glume margins unequal, the wider margins 0.3–1 mm wide, usually widest in the distal ¹/₃; lemma awns 0.5–3 mm long . . . . . 24. *E. violaceus*

27. Glumes 3–7-veined, glume margins equal, 0.1–0.5 mm wide, widest at or slightly beyond midlength; lemmas unawned or with awns to 40 mm long.

28. Glumes 1.8–2.3 mm wide, margins 0.2–0.3 mm wide . . . . . 22. *E. trachycaulus* (in part)

28. Glumes 0.4–1.5(2) mm wide, margins 0.1–0.2 mm wide . . . . . . . 10. *E. glaucus* (in part)

1. Spikelets 2–3(5) at all or most nodes; glumes often with subterete to terete, indurate bases, sometimes with flat, non-indurate bases, glume bodies linear-lanceolate to setaceous or subulate, margins usually firm, sometimes hyaline or scarious; lemmas usually awned, awns up to 120 mm long [for opposite lead, see p. 65].

29. Rachises disarticulating at maturity; glumes 10–135 mm long including the awns, sometimes split longitudinally, flexuous to outcurving from near the base; lowest floret in each spikelet sometimes sterile; blades 1–6 mm wide.

30. Glume awns split into 3–9 divisions; lemma awns about 0.2 mm wide at the base; rachis internodes 3–5 mm long . . . . . . . . . . . . . . . . . . . . . . . . . . . . . . . . . . . . . . . . . . 20. *E. multisetus*

30. Glume awns entire or split into 2–3 divisions; lemma awns about 0.4 mm wide at the base; rachis internodes 3–10(15) mm long . . . . . . . . . . . . . . . . . . . . . . . . . . . . . . . . . . . . 21. *E. elymoides*

29. Rachises not disarticulating at maturity; glumes 0–43 mm long including the awns, entire, straight or outcurving from well above the base; lowest floret in each spikelet functional; blades 2–25 mm wide.

31. Glume bodies with 0–1(2) veins, linear or tapering from the base, 0.1–0.6 mm wide, 0–24 mm long including the awns, often differing in length by more than 5 mm, persistent after the florets disarticulate; rachis internodes 0.1–0.3(0.4) mm thick at the thinnest sections, often with green lateral bands.

 32. Spikelets widely divergent to patent at maturity; lemma awns usually straight, rarely slightly curving; glumes vestigial or 1–3 mm long, occasionally some unequal glumes up to 10(20) mm long and 0.1–0.2 mm wide but with no distinct vein; spikes more or less erect . . . . . . . . . . . . . . . . . . . . . . . . . . . . . . . . . . . . . . . . . . . . . . . . . . . . . . . . 19. *E. hystrix*

 32. Spikelets usually appressed, never widely divergent; lemma awns straight or curving; glumes sometimes vestigial, usually 1–24 mm long, 0.1–0.6 mm wide, often with 1(2) distinct veins; spikes erect, nodding, or pendent.

  33. Glumes 12–30 mm long including the awns, subequal; lemma awns straight to moderately curving; spikes erect to slightly nodding.

   34. Spikelets (6)9–15(22) mm long excluding the awns, each with 2–5 florets; lemma awns moderately outcurving at maturity; glumes (0.2)0.3–0.5(0.7) mm wide . . . . . . . . . . . . . . . . . . . . . . . . . . . . . . . . . . . . . . . . 9. *E. interruptus* (in part)

   34. Spikelets 18–40 mm long excluding the awns, each with 3–8 florets; lemma awns straight to slightly curving at maturity; glumes 0.1–0.3(0.6) mm wide.

    35. Anthers 2.5–4 mm long; lemmas scabrous-hispid to thinly strigose, at least distally; spikes 4–12 cm long; internodes 3–6 mm long, without green lateral bands, with hispid dorsal angles . . . . . . . . . . . . . . . . . . . . . 14. *E. pringlei*

    35. Anthers 4.5–6 mm long; lemmas smooth, glabrous; spikes 9–20 cm long; internodes (5)7–15(22) mm long, with green lateral bands, glabrous except for the ciliolate margins . . . . . . . . . . . . . . . . . . . . . . . . . . . . . . . . . . . 15. *E. texensis*

  33. Glumes 0–15(30) mm long including the awns, usually differing in length by at least 4 mm, 1 or both shorter than 12 mm, sometimes both essentially absent; lemma awns outcurving at maturity; spikes more or less nodding.

   36. Rachis internodes 4–6(9) mm long; lemmas hirsute to strigose, at least near the margins, awns 20–35 mm long; sheaths glabrous; plants not glaucous or moderately glaucous . . . . . . . . . . . . . . . . . . . . . . . . . . . . . . . . . . . . . . 18. *E. diversiglumis*

   36. Rachis internodes (4)6–13(18) mm long; lemmas glabrous or pubescent, awns (8)10–30(35) mm long; sheaths glabrous or villous; plants usually glaucous, sometimes strongly so.

    37. Lemmas usually glabrous, veins occasionally hispidulous near the apices, awns (8)10–20(25) mm long; spikelets with (3)4–5 florets; rachis internodes (4)6–10(12) mm long, without green lateral bands, glabrous; adaxial surfaces of the blades usually villous; plants strongly glaucous . . . . . . . . 16. *E. svensonii*

    37. Lemmas usually hairy, awns (10)20–30(35) mm long; spikelets with 3(5) florets; rachis internodes (5)7–13(18) mm long, with green lateral bands and hispid dorsal angles; adaxial surfaces of the blades glabrous or short-pilose; plants somewhat glaucous . . . . . . . . . . . . . . . . . . . . . . . . . . . . . . 17. *E. churchii*

31. Glume bodies with 2–5(8) veins, widening or parallel-sided above the base, (0.2)0.3–2.3 mm wide, 4–43 mm long including the awns, equal or subequal, persistent or disarticulating; rachis internodes 0.1–0.8 mm thick at the thinnest sections, usually lacking green lateral bands.

 38. Glumes bases more or less terete, indurate, and without veins for 0.5–4 mm; glume bodies exceeding the adjacent lemmas by 1–5 mm or indistinguishable from the glume awns; lemma awns usually straight, occasionally contorted on the lower spikelets; rachis internodes (1.5)2–5(8) mm long.

  39. Glumes persistent, glume bodies (0.2)0.3–0.8(1) mm wide, with 2–4 veins, the basal 0.5–2 mm straight or slightly curving; lemmas with hairs or scabrous; spikelets with 1–3(4) florets; spikes nodding, exserted from the sheath.

   40. Adaxial surfaces of the blades densely villous with fine whitish hairs, rarely just pilose on the veins, dark glossy green; spikes 4–12 cm long; internodes (1.5)2–3(4) mm long; spikelets with 1–2(3) florets; lemmas usually villous, sometimes glabrous, sometimes scabrous, 5.5–9 mm long, 0.5–1.5 mm longer than the paleas; anthesis usually in early June to early July . . . . . . . . . . . . . . . . . . . . . 5. *E. villosus*

   40. Adaxial surfaces of the blades glabrous or scabrous, dull green; spikes 7–25 cm long; internodes 3–5(8) mm long; spikelets with 2–3(4) florets; lemmas

hispidulous or scabrous, 7–14 mm long, 1–5 mm longer than the paleas; anthesis usually in late June to late July . . . . . . . . . . . . . . . . . . . . . . . . . . . . . . 6. *E. riparius*

39. Glumes disarticulating, glume bodies (0.5)0.7–2.3 mm wide, with (2)3–5(8) veins, the basal 1–4 mm clearly bowed out; lemmas often glabrous, sometimes scabrous; spikelets with 2–5(6) florets; spikes erect, exserted or sheathed.

41. Spikes (0.5)0.7–2.2(2.5) cm wide including the awns, exserted or sheathed; glume awns 0–10(15) mm long; spikelets appressed to slightly spreading; blades usually glabrous or scabridulous.

42. Lemma awns 5–15(20) mm long at midspike; blades of all leaves usually spreading or lax and flat, those of the lower leaves not markedly larger or more persistent than those of the upper leaves; anthesis in mid-June to mid-August, usually 1–2 weeks earlier than sympatric *E. curvatus* . . . . . . . . . . . . . . 3. *E. virginicus*

42. Lemma awns 0.5–3(4) mm long at midspike; upper blades usually ascending and somewhat involute, blades of the lower leaves relatively short, narrow, and senescing earlier than those of the upper leaves; anthesis usually in late June to early August, 1–2 weeks later than sympatric *E. virginicus* . . . . . . . . . . . . . . . . . . . . . . . . . . . . . . . . . . 4. *E. curvatus*

41. Spikes (1.7)2.2–4.5(5.5) cm wide including the awns, exserted; glume awns (10)15–30 mm long; spikelets spreading; blades glabrous or villous.

43. Spikes with (6)9–16(20) nodes; internodes 4–7 mm long, about 0.3 mm thick at the thinnest portion; blades lax, dark glossy green under the glaucous bloom; auricles 2–3 mm long, often purplish black, at least in the central range of the species; anthesis usually in mid-May to mid-June . . . . . . . 1. *E. macgregorii*

43. Spikes with (10)18–30(36) nodes; internodes 3–5 mm long, 0.3–0.8 mm thick at the thinnest portion; blades lax, or ascending and involute, usually dull green, with or without a glaucous bloom; auricles 0–2 mm long, usually purplish brown; anthesis usually in mid-June to late July . . . . . . . . . . 2. *E. glabriflorus*

38. Glume bases flat and veined or, if subterete to terete, indurate and without veins for less than 1 mm; glume bodies shorter than or subequal to the lowest lemmas; lemma awns usually flexuous to curving, sometimes straight; rachis internodes (2)3–14 mm long.

44. Glumes with more or less terete bases, without hyaline or scarious margins, always awned, awns (5)8–25(27) mm long or the glume bodies indistinguishable from the awns; spikelets (1)2–3(5) per node, spreading, not or rarely purplish; cauline nodes usually concealed by the sheaths.

45. Spikes erect to slightly nodding, internodes (5)8–14 mm long; glumes 0.2–0.5(0.7) mm wide; lemmas 7–10 mm long, usually smooth or scabrous, occasionally hirtellous, especially near the margins, awns 15–22 mm long, straight to moderately outcurving; blades 3–9 mm wide; culms (40)60–100(120) cm tall, nodes usually exposed . . . . . . . . . . . . . . . . . . . 9. *E. interruptus* (in part)

45. Spikes usually nodding to pendent, sometimes erect, internodes (2)3–8(12) mm long; glume bodies (0.2)0.4–1.6 mm wide; lemmas 8–15 mm long, glabrous or uniformly hairy, awns (10)15–40(50) mm long, moderately to strongly outcurving; blades 3–24 mm wide; culms (40)60–180(220) cm tall, nodes usually concealed by the leaf sheaths.

46. Rachis internodes (2)3–5(7) mm long; spikelets 2(3) at most nodes, occasionally 1 or up to 5 at some nodes; paleas acute; blades (3)4–15(20) mm wide, usually firm and somewhat involute, dull green, drying grayish . . . . . 7. *E. canadensis*

46. Rachis internodes 5–12 mm long; spikelets 2 per node; paleas narrowly truncate; blades (8)10–20(24) mm wide, flat, lax, dark green . . . . . . . . . . . . . . . 8. *E. wiegandii*

44. Glumes with flat bases and hyaline or scarious margins, usually awned, awns 1–10 mm long, sometimes unawned; spikelets (1)2(3) per node, appressed to divergent, sometimes purplish; cauline nodes mostly exposed.

47. Anthers 0.9–1.7 mm long; glumes 3–8 mm long; lowest lemmas 3–6 mm longer than the glumes, densely scabridulous to scabrous, awns usually outcurving; spikelets with (3)4–5(7) florets; spikes 2–5 cm wide, nodding to pendent; cauline nodes glabrous . . . . . . . . . . . . . . . . . . . . . . . . . . . . . . . . . . . 13. *E. sibiricus* (in part)

47. Anthers 1.5–4.5 mm long; glumes (4.5)6–14(19) mm long; lowest lemmas from shorter than to 2.5 mm longer than the glumes, smooth, sometimes hairy, awns straight, flexuous, or outcurving; spikelets with 2–4(7) florets; spikes

(0.2)0.5–2.5 cm wide, erect, nodding, or pendent; cauline nodes occasionally with short hairs.

48. Glume bodies (6)9–14(19) mm long; lemmas 8–16 mm long, awns usually straight to flexuous; auricles usually present, to 2.5 mm long . . . . . . . . . 10. *E. glaucus* (in part)
48. Glume bodies (4.5)6–10(11) mm long; lemmas 5–14 mm long, awns flexuous to moderately outcurving; auricles often absent, or to 1.5 mm long.
   49. Lemmas with hairs, the marginal hairs markedly longer than those elsewhere; paleas acute; spikes nodding to pendent; rachis internodes 3–12 mm long; leaves usually deep green; plants native to the Pacific coastal mountains . . . . . . . . . . . . . . . . . . . . . . . . . . . . . . . . . . . 11. *E. hirsutus*
   49. Lemmas smooth, scabrous, or hispid, the marginal hairs, if present, not markedly longer than those elsewhere; paleas obtuse or truncate; spikes erect to slightly nodding; rachis internodes 3–6 mm long; leaves usually pale green, sometimes glaucous; plants introduced . . . . . . . . . . . . 12. *E. dahuricus*

## 1. Elymus macgregorii R. Brooks & J.J.N. Campb. EARLY WILDRYE [p. 377, 515]

*Elymus macgregorii* grows in moist, deep, alluvial or residual, calcareous or other base-rich soils in woods and thickets, mostly east of the 100th Meridian in the contiguous United States. It used to be confused with *E. glabriflorus* or *E. virginicus*, but reaches anthesis about a month earlier than sympatric populations of those species. It hybridizes with several species, particularly *E. virginicus* and *E. hystrix*.

## 2. Elymus glabriflorus (Vasey *ex* L.H. Dewey) Scribn. & C.R. Ball SOUTHEASTERN WILDRYE [p. 377, 515]

*Elymus glabriflorus* grows in open woods, thickets, and tall grasslands, sometimes spreading into old fields and roadsides. It is found in most of the southeastern United States, but is rare north of Maryland. Anthesis is usually 2–4 weeks later than in *E. virginicus* and other sympatric taxa. It differs from *E. villosus* in having erect spikes and glumes that are bowed out and disarticulate at maturity.

## 3. Elymus virginicus L. VIRGINIA WILDRYE, ÉLYME DE VIRGINIE [p. 377, 515]

*Elymus virginicus* is widespread in temperate North America, growing as far west as British Columbia and Arizona. It is divided here into four intergrading varieties.

1. Spikelets hispidulous to villous-hirsute, usually glaucous; anthesis usually in early July to mid-August . . . . . . . . . . . . . . . . . . . . . . . . . var. *intermedius*
1. Spikelets usually glabrous or scabrous, glaucous or not; anthesis usually in mid-June to late July.
   2. Spikes partly sheathed; glumes 1–2.3 mm wide, strongly indurate and bowed out in the basal 2–4 mm; plants not glaucous, becoming yellowish brown or occasionally somewhat purplish at maturity . . . . . . . . . . . . . . . . . . . . . . . . . . . .var. *virginicus*
   2. Spikes exserted; glumes (0.5)0.7–1.5(1.8) mm wide, moderately indurate and bowed out in the basal 1–2 mm; plants usually glaucous, becoming yellowish or reddish brown at maturity.
      3. Culms usually 70–100 cm tall, with 6–8 nodes; blades 3–15 mm wide, flat; spikes 4–20 cm long, not strongly glaucous; glumes indurate only in the basal 1 mm . . . . . . . . . . var. *jejunus*
      3. Culms usually 30–80 cm tall, with 4–6 nodes; blades 2–9 mm wide, often becoming involute; spikes 3.5–11 cm long, often strongly glaucous; glumes usually indurate in the basal 1–2 mm . . . . . . . . . . . . . . . . . var. *halophilus*

## Elymus virginicus var. halophilus (E.P. Bicknell) Wiegand [p. 377]

*Elymus virginicus* var. *halophilus* grows in the moist to damp soil of dunes and brackish marsh edges along the northern Atlantic coast, from Nova Scotia to North Carolina.

## Elymus virginicus var. intermedius (Vasey *ex* A. Gray) Bush [p. 377]

*Elymus virginicus* var. *intermedius* grows in moist, base-rich soil in open forests and thickets, especially on rocky, gravelly, or sandy banks of larger streams. It grows primarily from the central and southern Great Plains, through the central Mississippi and Ohio valleys, to the northeastern United States and adjacent Canada.

## Elymus virginicus var. jejunus (Ramaley) Bush [p. 377]

*Elymus virginicus* var. *jejunus* grows in moist to dry, sometimes alkaline or saline soil, in open, rocky, or alluvial woods, grasslands, glades, and disturbed places. It occupies the western range of the species, except for the Intermountain region.

## Elymus virginicus L. var. virginicus [p. 377]

*Elymus virginicus* var. *virginicus* grows in moist to damp or rather dry soil, mostly on bottomland or fertile uplands, in open woods, thickets, tall forbs, or weedy sites. It is widespread and abundant in the eastern range of the species, but also overlaps with var. *jejunus* in the Great Plains, east to Texas and Manitoba. It occasionally hybridizes with sympatric *Elymus* species, including *E. riparius*, and even with *Hordeum*.

## 4. Elymus curvatus Piper AWNLESS WILDRYE [p. 378, 515]

*Elymus curvatus* grows in moist or damp soils of open forests, thickets, grasslands, ditches, and disturbed ground, especially on bottomland. It is widespread from British Columbia and Washington, through the Intermountain region and northern Rockies, to the northern Great Plains.

## 5. Elymus villosus Muhl. *ex* Willd. DOWNY WILDRYE [p. 378, 515]

*Elymus villosus* grows in moist to moderately dry, often rocky soils in woods and thickets, especially in calcareous or other base-rich soils, but it is also frequent on drier, sandy soils or damper, alluvial soils in glaciated regions. It extends from the Great Plains east to southern Quebec, northern New York, and Vermont south to Texas, Georgia, and South Carolina.

*Elymus villosus* has sometimes been confused with hairy plants of *E. canadensis* and *E. glabriflorus*. The hairs of *E. villosus* are fine, whitish, and consistently dense on the leaf blades, typically spreading in the spikelets; the hairs of *E. canadensis* and *E. glabriflorus* are typically stouter and more appressed in the spikelets. The only proven natural hybrid is with *Hordeum jubatum* (see ×*Elyhordeum*, p. 63).

**6. Elymus riparius** Wiegand  EASTERN RIVERBANK
WILDRYE, ÉLYME DES RIVAGES [p. 378, 515]

*Elymus riparius* grows in moist, usually alluvial and often sandy soils
in woods and thickets, usually along larger streams and occasionally
along upland ditches. It is widespread in most of temperate east-
central North America.

*Elymus riparius* is sometimes confused with *E. canadensis*, but that
species has curving awns. It hybridizes occasionally with several other
taxa, especially *E. virginicus* var. *virginicus* and *E. hystrix*, but the
hybrids produce only late, depauperate spikes or none at all.

**7. Elymus canadensis** L.  GREAT PLAINS WILDRYE, ÉLYME
DU CANADA [p. 378, 515]

*Elymus canadensis* grows on dry to moist or damp, often sandy or
gravelly soil on prairies, dunes, stream banks, ditches, roadsides, and
disturbed ground, or, especially to the south, in thickets and open
woods near streams. It is widespread in most of temperate North
America, extending from the southwestern Northwest Territories to
Coahuila, Mexico, being especially common in the Great Plains.

*Elymus canadensis* differs from *E. riparius* in having curved rather
than straight awns, and from *E. wiegandii* in its less robust habit and
narrower leaves. It can hybridize with *E. glabriflorus*, *E. virginicus*, *E.
hystrix* and allies, *E. glaucus*, *E. trachycaulus*, *Pseudoroegneria
spicata*, and other species. The three varieties recognized here show
clear differences in their typical expression and evidence some
geographic separation, but they may prove to be artificial reference
points within a more or less continuous variation.

1. Lemmas usually villous or hispid; spikes nodding to
   almost pendent; internodes 4–7 mm long, often
   strongly glaucous . . . . . . . . . . . . . . . . . . . . . . var. *canadensis*
1. Lemmas usually smooth or scabridulous,
   occasionally hirsute; spikes usually nodding,
   occasionally almost erect; internodes 3–4 mm long,
   not strongly glaucous.
   2. Glumes not clearly indurate or bowed out at the
      base, awns 10–20 mm long; lemmas smooth or
      scabridulous, awns usually 20–30 mm long,
      moderately outcurving; spikes 6–20 cm long
      . . . . . . . . . . . . . . . . . . . . . . . . . . . . var. *brachystachys*
   2. Glumes often slightly indurate and bowed out at
      the base, awns 15–25 mm long; lemmas
      occasionally hirsute, awns 30–40 mm long, often
      strongly outcurving; spikes 15–25(30) cm long
      . . . . . . . . . . . . . . . . . . . . . . . . . . . . . . . . . var. *robustus*

### Elymus canadensis var. brachystachys (Scribn. & C.R. Ball) Farw. [p. 378]

*Elymus canadensis* var. *brachystachys* is widespread in the southern
Great Plains from Nebraska to Mexico, where anthesis is from March
to early June. It also occurs sporadically as far north as southern
Canada, from British Columbia to Quebec.

### Elymus canadensis L. var. canadensis [p. 378]

*Elymus canadensis* var. *canadensis* is widespread across the northern
range of the species, where anthesis is from June to August, but it
is also frequent as far south as Arizona, New Mexico, and Oklahoma.
Tentatively included here are *E. canadensis* var. *glaucifolius* (Muhl.)
Torr., which is strongly glaucous, with scabrous blades and hirsute or
scabrous lemmas, and *E. canadensis* var. *villosus* Bates, which has
villous leaves and occurs rarely in the northern Great Plains.

### Elymus canadensis var. robustus (Scribn. & J.G. Sm.) Mack. & Bush [p. 378]

*Elymus canadensis* var. *robustus* grows mostly in the east-central
range of the species, from Illinois and Ohio to Oklahoma and
Nebraska, locally becoming the most common variety. Anthesis can
be earlier than in other sympatric *E. canadensis* varieties.

**8. Elymus wiegandii** Fernald  NORTHERN RIVERBANK
WILDRYE, ÉLYME DE WIEGAND [p. 378, 515]

*Elymus wiegandii* grows in moist or damp, rich, alluvial soil,
especially on sandy river terraces and in woods and thickets, primarily
from Saskatchewan through much of the Great Lakes region to Nova
Scotia and Connecticut. It is similar to *E. riparius* and overlaps with
it in range and habitat within the Great Lakes region, where there are
a few plants that appear to be hybrids between the two. It is often
confused with sympatric *E. canadensis* and *E. diversiglumis*, but has
a distinctive robust, broad-leaved habit. It is intermediate between the
two in spike density and glume development.

**9. Elymus interruptus** Buckley  SOUTHWESTERN WILDRYE
[p. 378, 515]

*Elymus interruptus* grows in dry to moist, rocky soil, often in canyons,
open woods, and thickets, in the southwestern United States and
northern Mexico. Apparent intermediates between *E. interruptus* and
*E. canadensis* have been collected north of the documented range of
typical *E. interruptus* in Arizona, New Mexico, and Iowa.

**10. Elymus glaucus** Buckley  COMMON WESTERN
WILDRYE, BLUE WILDRYE [p. 378, 515]

*Elymus glaucus* grows in moist to dry soil in meadows, thickets, and
open woods. It is widespread in western North America, from Alaska
to Saskatchewan, and south to Baja California and New Mexico. It is
also sporadic, sometimes appearing transitional to *E. trachycaulus*,
from the northern Great Plains to southern Ontario and New York
and, as a disjunct, on rocky sites in the Ozark and Ouachita
mountains.

Populations can differ greatly in morphology, especially in rhizome
development, leaf width, pubescence, and the prevalence of solitary
spikelets. Rhizomatous plants are more common on unstable slopes
or sandy soils. Plants with solitary spikelets are more common on
poor soil or in shade. They are often confused, particularly in the
herbarium, with *E. stebbinsii* or *E. trachycaulus*. They differ from *E.
stebbinsii* in their shorter anthers and awned glumes. Distinction from
*E. trachycaulus* can be difficult with herbarium specimens, but is
generally easy in the field, *E. glaucus* having more evenly leafy culms,
laxer and wider blades, more tapered glumes that are almost always
awned, and shorter anthers than the sympatric *E. trachycaulus*.

There are reports of natural hybrids with several other species of
*Elymus*, including *E. elymoides*, *E. multisetus* (see *E.* ×*hansenii*, p.
75), *E. trachycaulus*, and *E. stebbinsii*. These hybrids often appear at
least partially fertile. *Elymus glaucus* can also form intergeneric
hybrids with *Leymus* and *Hordeum* (see ×*Elyleymus*, p. 75, and
×*Leydeum*, p. 79).

The following three subspecies appear to be morphologically,
ecologically, and geographically distinct.

1. Lemma awns (0)1–5(7) mm long; glume awns 0–2
   mm long . . . . . . . . . . . . . . . . . . . . . . . . . . . . subsp. *virescens*
1. Lemma awns (5)10–30(35) mm long; glume awns
   (0.5)1–9 mm long.
   2. Blades 4–17 mm wide, adaxial surfaces glabrous
      or strigose, occasionally pilose to hirsute with
      hairs of fairly uniform length; glume awns
      (0.5)1–5(9) mm long . . . . . . . . . . . . . . . . . . . subsp. *glaucus*
   2. Blades 3–8 mm wide, densely short-pilose with
      scattered longer hairs; glume awns 3–8 mm long
      . . . . . . . . . . . . . . . . . . . . . . . . . . . . . . . . . subsp. *mackenziei*

### Elymus glaucus Buckley subsp. glaucus [p. 378]

*Elymus glaucus* subsp. *glaucus* grows throughout the range of the
species, from sea level to 2500 m. It is absent from the area where *E.
glaucus* subsp. *mackenziei* grows. It resembles *E. hirsutus*, differing in
its erect spikes and in the pattern of its lemma pubescence. It also
resembles the introduced *E. dahuricus*, from which it differs in its
palea shape.

## Elymus glaucus subsp. mackenziei (Bush) J.J.N. Campb. [p. *378*]

*Elymus glaucus* subsp. *mackenziei* grows on limestone clifftops, rocky ledges, and glades, in open woods and thickets. It is known only from Arkansas, Missouri, and Oklahoma, at scattered sites in the Ozark Mountains and at Rich Mountain in the Ouachita Mountains. This subspecies is remarkably disjunct, at least 500 miles from the nearest known *E. glaucus* to the west and north.

## Elymus glaucus subsp. virescens (Piper) Gould [p. *378*]

*Elymus glaucus* subsp. *virescens* usually grows in relatively dry or rocky soils along cliffs, bluffs, slopes, shores, and river banks, and in coniferous forests, chaparral, and other woodlands along the coast from Alaska to central California, from sea level to 1200 m.

## 11. Elymus hirsutus J. Presl NORTHWESTERN WILDRYE [p. *379*, *515*]

*Elymus hirsutus* grows in moist to damp or dry soils in woods, thickets, and grasslands. Its range extends along the coastal mountains from the Aleutian Islands to northern Oregon, and inland to eastern British Columbia. Plants in the southern part of the range tend to have villous leaves and more erect spikes with shorter, straighter awns.

Intermediates exist between *Elymus hirsutus* and *E. glaucus*, but the more pendent spikes, lemma pubescence pattern, and shorter glumes of *E. hirsutus* enable most specimens to be readily identified. *Elymus hirsutus* occasionally hybridizes with *Leymus mollis* and *Hordeum brachyantherum*.

## 12. Elymus dahuricus Turcz. ex Griseb. [p. *379*]

*Elymus dahuricus* is widespread in temperate central and eastern Asia. It has been introduced for reclamation in some parts of western North America. It differs from *E. glaucus* in its palea shape. Its distribution in the *Manual* region is not known. Several varieties have been described in Asia; only **Elymus dahuricus** Turcz. *ex* Griseb. var. **dahuricus** has been introduced to North America.

## 13. Elymus sibiricus L. SIBERIAN WILDRYE [p. *379*, *515*]

*Elymus sibiricus* grows in dry to damp grasslands and thickets, on slopes, eroding river banks, mud flats, coastal benches, dunes, clearings, and other disturbed areas, in southern Alaska, the southern Yukon Territory, the southwestern MacKenzie District in the Northwest Territories, and central British Columbia.

## 14. Elymus pringlei Scribn. & Merr. MEXICAN WILDRYE [p. *379*]

*Elymus pringlei* grows on moist slopes and canyons, in pine and deciduous tree woods, at 1500–2300 m in the Sierra Madre Orientale of eastern Mexico, and is similar to *E. texensis* and *E. interruptus*. It is included here because it seems likely that it also grows in southern Texas, having been collected in Coahuila, Mexico, 54 miles from the border, near Big Bend National Park.

## 15. Elymus texensis J.J.N. Campb. TEXAS WILDRYE [p. *379*, *515*]

*Elymus texensis* is known only from calcareous bluffs and hills in juniper woods and grassy areas on the Edwards Plateau of southwest Texas. It is known from only three collections. It is similar to the Mexican species *E. pringlei*, but differs in its larger anthers, larger, less pubescent spikelets, and in its longer, glabrous rachis internodes with green lateral bands.

## 16. Elymus svensonii G.L. Church SVENSON'S WILDRYE [p. *379*, *516*]

*Elymus svensonii* grows in dry, rocky soils in open woods of the interior low plateaus, mostly along bluffs of the Kentucky River and its tributaries in the bluegrass region of Kentucky, and along bluffs of the Cumberland River and its Caney Fork in the central basin of Tennessee. Most sites are on Ordovician limestone, but its discovery by Natural Heritage programs in Kentucky along the Green River on Mississippian limestone, and in Tennessee along the Piney River on Silurian limestone, suggest that it may be more widespread. It has been a candidate for federal protection in the United States.

*Elymus svensonii* hybridizes naturally with *E. hystrix*, *E. virginicus* and other species of *Elymus*. It differs from *E. churchii* in having less open spikes, shorter awns, more florets per spikelet, and more pubescent, glaucous foliage.

## 17. Elymus churchii J.J.N. Campb. CHURCH'S WILDRYE [p. *379*, *516*]

*Elymus churchii* grows in dry, rocky, often relatively base-rich soils, in open woods on ridges, and on bluffs and river banks. Its range includes the central Ouachita Mountains and the western Ozark Mountains in Arkansas, Oklahoma, and Missouri.

*Elymus churchii* used to be included in *E. interruptus*. It differs from *E. svensonii* in its more open spikes, longer awns, fewer florets per spikelet, and less pubescent, less glaucous foliage. Occasional intermediates with *E. canadensis* and *E. hystrix* exist.

## 18. Elymus diversiglumis Scribn. & C.R. Ball UNEQUAL-GLUMED WILDRYE [p. *379*, *516*]

*Elymus diversiglumis* grows in moist to dry, often base-rich and alluvial soils, in open woods, woodland margins, and thickets in the northern Great Plains, from Saskatchewan and Manitoba to Wyoming, Wisconsin, and Iowa.

*Elymus diversiglumis* usually reaches anthesis 2–4 weeks earlier than sympatric populations of *E. canadensis*. Introgressant populations involving *E. diversiglumis*, *E. canadensis*, and *E. hystrix* are known.

## 19. Elymus hystrix L. BOTTLEBRUSH GRASS, GLUMELESS WILDRYE [p. *379*, *516*]

*Elymus hystrix* grows in dry to moist soils in open woods and thickets, especially on base-rich slopes and small stream terraces. It grows throughout most of temperate eastern North America, extending west to Manitoba and Oklahoma.

*Elymus hystrix* hybridizes with most eastern species of *Elymus*. Introgression may account for the considerable variation in glume development and spikelet appression among these species. Within the ranges of *E. diversiglumis*, *E. svensonii*, and *E. churchii*, there appear to be frequent introgressants between these species and *E. hystrix*.

## 20. Elymus multisetus (J.G. Sm.) Burtt Davy BIG SQUIRRELTAIL [p. *379*, *516*]

*Elymus multisetus* grows in dry, often rocky, open woods and thickets on slopes and plains, from central Washington and Idaho to southern California, Colorado, and northwestern Arizona, and from sea level to 2000 m. It has also been reported from Baja California, Mexico. It usually grows in less arid habitats than *E. elymoides* subsp. *elymoides*, but the two taxa are sometimes sympatric.

A wide belt of introgression between *Elymus multisetus* and *E. elymoides* subsp. *elymoides* has been reported from southeastern California to southern Nevada, but not in other areas where they are sympatric. There are also probable hybrids with *E. glaucus* and *Pseudoroegneria spicata*.

## 21. Elymus elymoides (Raf.) Swezey [p. *380*, *516*]

*Elymus elymoides* grows in dry, often rocky, open woods, thickets, grasslands, and disturbed areas, from sagebrush deserts to alpine tundra. It is widespread in western North America, from British Columbia to northern Mexico and the western Great Plains, and introduced in western Missouri, Illinois, and Kentucky. It is often dominant in overgrazed pinyon-juniper woodlands. Although palatable early in the season, the disarticulating, long-awned spikes irritate grazing animals later in the year.

*Elymus elymoides* intergrades with *E. multisetus* in parts of its southern range. It differs from *E. scribneri* in having more than one spikelet per node, narrower glumes, and less tardily disarticulating rachises. Hybrids with several other species in the *Triticeae* are known; they can often be recognized by their tardily disarticulating rachises. Named interspecific hybrids (pp. 74–75) (and the other parent) are *E. ×saundersii* (*E. trachycaulus*), *E. ×pinalenoensis* (*E. arizonicus*), and possibly *E. ×hansenii* (*E. elymoides* or *E. multisetus* × *E. glaucus*). Hybrids with *E. sierrae* have not been named; they are common where the two species are sympatric. They have broader glume bases, shorter glume awns, and longer anthers than *E. elymoides*.

1. Rachis nodes with 3 spikelets, the central spikelet usually with 2 fertile florets, the florets of the lateral spikelets rudimentary to awnlike; lemma awns 15–30 mm long . . . . . . . . . . . . . . . . . . . . . . subsp. *hordeoides*
1. Rachis nodes usually with 2 spikelets, each spikelet usually with (1)2–4(5) fertile florets; lemma awns 15–120 mm long.
   2. No spikelets appearing to have 3 glumes, the lowermost floret in each spikelet well developed; paleas rarely with the veins extended as bristles . . . . . . . . . . . . . . . . . . . . . . . . . . . . . subsp. *brevifolius*
   2. One or more of the spikelets at most nodes appearing to have 3 glumes, the lowest 1–2 florets sterile and glumelike; paleas usually with the veins extended as bristles.
      3. Glumes with awns 15–70 mm long, all glumes entire . . . . . . . . . . . . . . . . . subsp. *californicus*
      3. Glumes with awns 35–85 mm long, one of the glumes at most nodes with the awn split into 2 or 3 divisions . . . . . . . . . . . . . . subsp. *elymoides*

### Elymus elymoides subsp. brevifolius (J.G. Sm.) Barkworth LONGLEAF SQUIRRELTAIL [p. 380]

*Elymus elymoides* subsp. *brevifolius* has a wide ecological and elevation range, extending from the arid Sonoran Desert to subalpine habitats, from 600–3500 m. It extends further south than the other subspecies, into northern Mexico; it is rare in Canada.

### Elymus elymoides subsp. californicus (J.G. Sm.) Barkworth CALIFORNIA SQUIRRELTAIL [p. 380]

*Elymus elymoides* subsp. *californicus* grows in mid-montane to arctic-alpine habitats in western North America, at elevations of 1500–4200 m. Plants transitional to subsp. *elymoides* occur where the two are sympatric.

### Elymus elymoides (Raf.) Swezey subsp. elymoides COMMON SQUIRRELTAIL [p. 380]

*Elymus elymoides* subsp. *elymoides* grows in desert and shrub-steppe areas of western North America, extending to the western edge of the Great Plains and, as an adventive, occasionally further east. It is frequently associated with disturbed sites.

### Elymus elymoides subsp. hordeoides (Suksd.) Barkworth [p. 380]

*Elymus elymoides* subsp. *hordeoides* grows in dry, rocky, often shallow soils, particularly in *Artemisia rigida–Poa secunda* communities, from eastern Washington and Idaho to northern California and Nevada. It resembles some *Elymus–Hordeum* hybrids.

### 22. Elymus trachycaulus (Link) Gould [p. 380, 516]

*Elymus trachycaulus* grows from sea level to 3300 m, usually in open or moderately open areas, but sometimes in forests. It extends from the boreal forests of North America east through Canada to Greenland and south into Mexico. It also grows, as an introduction, in Asia and Europe. It exhibits considerable variability in the presence or absence of rhizomes, the length and density of the spike, awn development on the glumes and lemmas, and glume venation. The variability in these features has often been used to circumscribe infraspecific taxa, but most such taxa, even though locally distinctive, appear to intergrade. Some of the features appear to be strongly influenced by environmental factors. For instance, plants growing in forested areas of northwestern North America tend to be slightly rhizomatous, more gracile, and later-flowering that those in adjacent, more exposed areas. Plants growing at higher elevations tend to have glumes with more widely spaced veins and broader, often unequal margins, resembling *E. violaceus* in these respects.

*Elymus trachycaulus* differs from *E. stebbinsii* in having shorter anthers, shorter internodes, and glumes that are sometimes awned. It may be confused, particularly in the herbarium, with specimens of *E. glaucus* having solitary spikelets at all the spike nodes; it usually differs in having shorter anthers and less acuminate glumes. When, as is sometimes the case, the two species grow together, *E. trachycaulus* can be distinguished by its stiffer leaves. *Elymus trachycaulus* also resembles *E. macrourus* and *E. alaskanus*, but its glumes are longer relative to the lemmas. It also has less hairy rachillas than most plants of those species. It differs consistently from *E. caninus* in having glumes that are glabrous on the adaxial (inner) surface, in a chromosome interchange, and in its molecular characteristics. It also tends to have a more erect spike.

*Elymus trachycaulus* has been implicated in several interspecific and intergeneric hybrids. Named interspecific hybrids (pp. 74–75) (and the other parent) are *E. ×cayouetteorum* (*E. canadensis*), *E. ×palmerensis* (*E. sibiricus*), *E. ×pseudorepens* (*E. lanceolatus*), and *E. ×saundersii* (*E. elymoides*). Hybrids with *E. hystrix* have been named ×*Agroelymus dorei* Bowden; the appropriate combination has not been made in *Elymus*. Named intergeneric hybrids are ×*Elyhordeum macounii* (*Hordeum jubatum*), ×*Elyleymus jamesensis* (*Leymus mollis*), and ×*Elyleymus ontariensis* (*Leymus innovatus*). Hybrids with *Elymus elymoides*, *E. multisetus*, and *Hordeum jubatum* have brittle rachises and tend to be awned. Others are harder to recognize.

1. Lemma awns 17–40 mm long, longer than the lemma body, straight; spikes somewhat 1-sided . . . . . . . . . . . . . . . . . . . . . . . . . . . . . . subsp. *subsecundus*
1. Lemmas unawned or with awns to 24 mm long, shorter or longer than the lemma body, straight or curved; spikes 2-sided.
   2. Lemma awns 9–24 mm long . . . . . . . *E. trachycaulus* hybrids
   2. Lemmas unawned or with awns to 9 mm long, the awns sometimes curved.
      3. Spike internodes 8–15 mm long; spikes 8–25 cm long; glumes unawned or with straight awns to 2 mm long; spikelet bases usually visible; lemmas unawned or with straight awns to 40 mm long . . . . . . . . . . . subsp. *trachycaulus*
      3. Spike internodes 4–5 mm long; spikes 5–10 cm long; glumes awned, awns 1.8–4 mm long; spikelet bases usually concealed; lemmas awned, awns 2–3 mm long, slightly curved . . . . . . . . . . . . . . . . . . . . . . subsp. *virescens*

### Elymus trachycaulus subsp. subsecundus (Link) Á. Löve & D. Löve ONE-SIDED WHEATGRASS [p. 380]

*Elymus trachycaulus* subsp. *subsecundus* grows primarily in the Great Plains. It differs from plants of *E. glaucus* with solitary spikelets, in its 1-sided spike and stiffer, more basally concentrated leaves.

### Elymus trachycaulus (Link) Gould subsp. trachycaulus SLENDER WHEATGRASS, ÉLYME À CHAUMES RUDES, AGROPYRE À CHAUMES RUDES [p. 380]

*Elymus trachycaulus* subsp. *trachycaulus* grows throughout the habitat and range of the species, and exhibits considerably more variation than subsp. *subsecundus*, particularly in glume venation and the spacing of spikelets in the spikes. Plants with glumes having 5–7 well-developed, narrowly spaced veins are restricted to lower elevations and the southern portion of the subspecies range; northern plants and plants at higher elevations generally have 3–5 weakly developed and widely spaced veins. In at least some instances, plants with widely spaced spikelets appear to be associated with more shady habitats.

**Elymus trachycaulus** subsp. **virescens** (Lange) Á. Löve & D. Löve [p. *380*]

*Elymus trachycaulus* subsp. *virescens* is restricted to Greenland. It is very consistent in its morphology.

## 23. Elymus caninus (L.) L. BEARDED WHEATGRASS [p. *380*]

*Elymus caninus* is native to Eurasia; it is not known to be established in the *Manual* region. It differs from *E. ciliaris* and *E. tsukushiensis* in having flatter glumes that are longer in relation to the lemmas, and palea keels that are straight or almost straight below the apices. The hairs on the inside of the glumes of *E. caninus* are difficult to see. Nevertheless, this is the single most reliable morphological character for distinguishing *E. caninus* from all other species of *Elymus* in this treatment.

## 24. Elymus violaceus (Hornem.) Feilberg ARCTIC WHEATGRASS, ÉLYME LATIGLUME [p. *380*, 516]

*Elymus violaceus* grows in arctic, subalpine, and alpine habitats, on calcareous or dolomitic rocks, from Alaska through arctic Canada to Greenland, and south in the Rocky Mountains to southern New Mexico. In western North America, it forms intermediates with *E. scribneri*, *E. trachycaulus*, and *E. alaskanus*. It is treated here as including *E. alaskanus* subsp. *latiglumis* [≡ *Agropyron latiglume*], *E. alaskanus* being restricted to plants with relatively short glumes that are often found in valleys and at lower elevations than *E. violaceus*.

## 25. Elymus macrourus (Turcz. *ex* Steud.) Tzvelev NORTHERN WHEATGRASS [p. *380*, 516]

*Elymus macrourus* grows on river banks and bars, lake shores, and hillsides in northwestern North America. Outside of North America, it grows across the Russian arctic, and extends south into the boreal forest. Plants growing on shifting river banks and bars often appear rhizomatous, as the lower internodes elongate in response to the disturbed substrate. Plants of *E. macrourus* differ from *E. alaskanus* in the shape of their glumes and their narrower glume margins, and from *E. trachycaulus* in their relatively short glumes and evidently hairy rachilla segments.

*Elymus macrourus* is one of the parents in both *E.* ×*palmerensis* and ×*Elyhordeum pilosilemma*.

## 26. Elymus alaskanus (Scribn. & Merr.) Á. Löve [p. *380*, 516]

*Elymus alaskanus* extends across the high arctic of North America to extreme eastern Russia. This treatment interprets *E. alaskanus* as having relatively short glumes. Large specimens resemble *E. macrourus*, but differ in the shape of their glumes and in their wider glume margins. *Elymus alaskanus* differs from *E. trachycaulus* in its greater cold tolerance and the distal widening of its glume margins. There is some intergradation, particularly with *E. violaceus* and *E. trachycaulus*, but these species have longer glumes. Moreover, in western North America, *E. violaceus* is restricted to rocky habitats at or above treeline, whereas *E. alaskanus* is often associated with valleys and flat areas.

1. Glumes glabrous, scabrous or sparsely hairy, hairs
   to about 0.2 mm long; lemmas glabrous or with
   hairs to about 0.2 mm long . . . . . . . . . . . . . . subsp. *alaskanus*
1. Glumes and lemmas densely hairy, hairs 0.2–0.5 mm
   long . . . . . . . . . . . . . . . . . . . . . . . . . subsp. *hyperarcticus*

**Elymus alaskanus** (Scribn. & Merr.) Á. Löve subsp. **alaskanus** ALASKAN WHEATGRASS [p. *380*]

*Elymus alaskanus* subsp. *alaskanus* grows on river banks and hillsides, primarily north of 50° N latitude.

**Elymus alaskanus** subsp. **hyperarcticus** (Polunin) Á. Löve & D. Löve HIGH-ARCTIC WHEATGRASS [p. *380*]

*Elymus alaskanus* subsp. *hyperarcticus* grows on river banks and hillsides. It extends from the Lake Taymyr basin in arctic Russia across northern North America to Greenland.

## 27. Elymus lanceolatus (Scribn. & J.G. Sm.) Gould [p. *380*, 516]

*Elymus lanceolatus* grows in sand and clay soils and dry to mesic habitats. It is found primarily in the western half of the *Manual* region, between the coastal mountains and 95° W longitude, with the exception of *E. lanceolatus* subsp. *psammophilus*, which extends around the Great Lakes. Three subspecies are recognized, primarily on the basis of their lemma and palea pubescence.

*Elymus lanceolatus* hybridizes with several species of *Triticeae*. *Elymus albicans* is thought to be derived from hybridization with the awned phase of *Pseudoroegneria spicata*. Judging from specimens of controlled hybrids, hybridization with *E. trachycaulus* and unawned plants of *P. spicata* probably occur, but would be almost impossible to detect without careful observation in the field.

1. Lemmas densely hairy, hairs flexible, some 1 mm
   long or longer . . . . . . . . . . . . . . . . . . . subsp. *psammophilus*
1. Lemmas glabrous or with stiff hairs shorter than 1
   mm.
   2. Lemmas with hairs, not scabrous . . . . . . . . subsp. *lanceolatus*
   2. Lemmas smooth, sometimes scabrous distally,
      mostly glabrous, sometimes the lemma margins
      hairy proximally . . . . . . . . . . . . . . . . . . . . subsp. *riparius*

**Elymus lanceolatus** (Scribn. & J.G. Sm.) Gould subsp. **lanceolatus** THICKSPIKE WHEATGRASS [p. *380*]

*Elymus lanceolatus* subsp. *lanceolatus* grows in clay, sand, loam, and rocky soils, and is widely distributed in the western *Manual* region. It differs from the octoploid *Pascopyrum smithii* in having more evenly distributed leaves and acute glumes that tend to taper from midlength or higher, rather than acuminate glumes that tend to taper from below midlength. In addition, the midvein of the glumes of *E. lanceolatus* is straight, whereas that of *P. smithii* "leans" to the side distally.

**Elymus lanceolatus** subsp. **psammophilus** (J.M. Gillett & H. Senn) Á. Löve SAND-DUNE WHEATGRASS [p. *380*]

*Elymus lanceolatus* subsp. *psammophilus* tends to grow in sandy soils. It was described from around the Great Lakes, but similar plants have been found scattered throughout the western range of the species, almost always in association with sandy soils. Those from the Yukon and northern British Columbia tend to be shorter and have smaller spikelets and spikelet parts than those from Washington and Saskatchewan, but there is considerable overlap in these characters.

**Elymus lanceolatus** subsp. **riparius** (Scribn. & J.G. Sm.) Barkworth STREAMBANK WHEATGRASS [p. *380*]

*Elymus lanceolatus* subsp. *riparius* grows throughout most of the western part of the range of *E. lanceolatus*, being more common in mesic habitats and clay soils than the other two subspecies.

## 28. Elymus stebbinsii Gould [p. *381*, 516]

*Elymus stebbinsii* is restricted to California, where it grows on dry slopes, chaparral, and wooded areas, at elevations below 1600 m. It differs from other *Elymus* species primarily in its combination of long anthers and solitary spikelets. It is often confused with *E. glaucus* and *E. trachycaulus* with solitary spikelets. It differs from both in its longer anthers, and from most representatives of *E. glaucus* in its acute, but unawned, glumes.

1. Lemmas awned, awns 8–28 mm long; lower leaf
   sheaths rarely pubescent; spikelets 13–22 mm long
   . . . . . . . . . . . . . . . . . . . . . . . . . . . . . subsp. *septentrionalis*
1. Lemmas unawned or with awns to 8(12) mm long;
   lower leaf sheaths pubescent or glabrous; spikelets
   17–29 mm long . . . . . . . . . . . . . . . . . . . . . . subsp. *stebbinsii*

**Elymus stebbinsii** subsp. **septentrionalis** Barkworth
NORTHERN STEBBINS' WHEATGRASS [p. *381*]

*Elymus stebbinsii* subsp. *septentrionalis* grows primarily in the Sierra Nevada. Its range extends from near the Oregon border to Tulare County, California, and includes the coastal mountains north of San Francisco Bay.

**Elymus stebbinsii** Gould subsp. **stebbinsii** STEBBINS' WHEATGRASS [p. *381*]

*Elymus stebbinsii* subsp. *stebbinsii* is best known from the coastal mountains south of San José. It also grows at scattered locations from the central Sierra Nevada south to the Transverse Mountains.

**29. Elymus arizonicus** (Scribn. & J.G. Sm.) Gould ARIZONA WHEATGRASS [p. *381*, *516*]

*Elymus arizonicus* grows in moist, rocky soil in mountain canyons of the southwestern United States and northern Mexico. When mature, the drooping spike and solitary spikelets make *E. arizonicus* easy to identify. Immature specimens, or those mounted so that the spike appears erect, are easily mistaken for *Pseudoroegneria spicata*, but they have thicker culms and longer ligules, more basal leaves, and wider leaf blades.

**30. Elymus bakeri** (E.E. Nelson) Á. Löve BAKER'S WHEATGRASS [p. *381*, *516*]

*Elymus bakeri* grows in high, but not alpine, mountain meadows of Colorado and northern New Mexico. It differs from the awned phase of *Pseudoroegneria spicata* in having rather thicker culms and spikes, and stouter lemma awns.

**31. Elymus scribneri** (Vasey) M.E. Jones SCRIBNER'S WHEATGRASS [p. *381*, *516*]

*Elymus scribneri* grows in rocky areas in open subalpine and alpine regions, at 2500–3200 m, often in windswept locations, in southwestern Alberta and the western United States. It differs from *E. elymoides* in having only one spikelet per node, wider glumes, and more tardily disarticulating rachises, and from *E. sierrae* in its disarticulating rachises, denser spikes, and shorter anthers.

**32. Elymus sierrae** Gould SIERRA WHEATGRASS [p. *381*, *516*]

*Elymus sierrae* is best known from rocky slopes and ridgetops in the Sierra Nevada, at 2100–3400 m, and is also found in Washington and Oregon. It resembles *E. scribneri*, differing in its non-disarticulating rachises, longer rachis internodes, and longer anthers. Hybrids with *E. elymoides* have glumes with awns 15+ mm long, and some spikelets with narrower glume bases and shorter anthers. Specimens with wide-margined glumes suggest hybridization with *E. violaceus*.

**33. Elymus wawawaiensis** J.R. Carlson & Barkworth SNAKERIVER WHEATGRASS [p. *381*, *516*]

*Elymus wawawaiensis* grows primarily in shallow, rocky soils of slopes in coulees and reaches of the Salmon, Snake, and Yakima rivers of Washington, northern Oregon, and Idaho. There are also a few records from localities at some distance from the Snake River and its tributaries. These probably reflect deliberate introductions.

*Elymus wawawaiensis* resembles a vigorous version of *Pseudoroegneria spicata*, and was long confused with that species. It differs in its more imbricate spikelets and narrower, stiff glumes. In its primary range, *E. wawawaiensis* is often sympatric with *P. spicata*, but the two tend to grow in different habitats, *E. wawawaiensis* growing in shallow, rocky soils and *P. spicata* in medium- to fine-textured loess soil.

**34. Elymus albicans** (Scribn. & J.G. Sm.) Á. Löve MONTANA WHEATGRASS [p. *381*, *516*]

*Elymus albicans* grows primarily in the central Rocky Mountains and the western portion of the Great Plains. It tends to grow in shallow, rocky soils on wooded or sagebrush-covered slopes, rather than in deep loams. It is derived from hybrids between *Pseudoroegneria spicata* and *E. lanceolatus*.

**35. Elymus repens** (L.) Gould QUACKGRASS, COUCHGRASS, CHIENDENT, CHIENDENT RAMPANT [p. *381*, *516*]

*Elymus repens* is native to Eurasia; it is now established through most of the *Manual* region. It grows well in disturbed sites, spreading rapidly via its long rhizomes, as well as by seed. It is also drought tolerant. Although it is listed a noxious weed in several states, it provides good forage. It differs from *E. hoffmannii* in having widely spaced, unequally prominent leaf veins and, usually, shorter awns.

**36. Elymus hoffmannii** K.B. Jensen & Asay HOFFMANN'S WHEATGRASS [p. *381*]

*Elymus hoffmannii* was described from a breeding line of plants developed from seeds collected in Erzurum Province, Turkey. No information is available about its native distribution. It differs from *E. repens* primarily in its evenly prominent, closely spaced leaf veins and, usually, in having longer awns. Its distribution within the *Manual* region is not known. The cultivar 'NewHy' is often called *E. hoffmannii*.

**37. Elymus tsukushiensis** Honda [p. *381*, *516*]

*Elymus tsuskushiensis* is native to northeastern China, Japan, and Korea. It was collected from ballast dumps in Portland, Oregon, but is not established in the *Manual* region.

**38. Elymus ciliaris** (Trin.) Tzvelev [p. *382*, *516*]

*Elymus ciliaris* is native to northern China and Japan. It was collected from ballast dumps in Portland, Oregon, in 1899 and 1902; it is not established in the *Manual* region.

**39. Elymus semicostatus** (Nees *ex* Steud.) Melderis [p. *382*]

*Elymus semicostatus* is native to central Asia, from Afghanistan through Pakistan to northeastern India (Sikkim). Reports of its presence in the *Manual* region are based on misidentifications.

**Named hybrids**

*Elymus* is notorious for its ability to hybridize. Most of its interspecific hybrids are partially fertile, permitting introgression between the parents. The parentage of all hybrids is best determined in the field. Perennial hybrids, such as those in *Elymus*, can persist in an area after one or both parents have died out, but the simplest assumption is that both are present. Interspecific hybrids of *Elymus* that have disarticulating rachises presumably have *E. elymoides* or *E. multisetus* as one of their parents.

**40. Elymus ×cayouetteorum** (B. Boivin) Barkworth [p. *382*]

*Elymus ×cayouetteorum* consists of hybrids between *E. trachycaulus* and *E. canadensis*. The type specimen was collected on the Îlets Jérémie, Quebec. It is not known how widespread such hybrids are.

**41. Elymus ×palmerensis** (Lepage) Barkworth & D.R. Dewey [p. *382*]

*Elymus ×palmerensis* is the name for hybrids between *E. macrourus* and *E. sibiricus*. It is known from disturbed sites around Palmer, Alaska, and in south-central Alaska. It has also been reported from Fort Liard, in the MacKenzie District, Northwest Territories.

**42. Elymus ×pseudorepens** (Scribn. & J.G. Sm.) Barkworth & D.R. Dewey FALSE QUACKGRASS [p. *382*]

*Elymus ×pseudorepens* consists of hybrids between *E. lanceolatus* and *E. trachycaulus*. It appears to be fairly common, having been reported from Alberta to Michigan and south to Arizona, New Mexico, and Arkansas.

**43. Elymus ×yukonensis** (Scribn. & Merr.) Á. Löve [p. *382*]

The parents of *Elymus ×yukonensis* have not been identified. Morphological and geographic considerations suggest that they may be *E. lanceolatus* subsp. *psammophilus* and *E. alaskanus*.

### 44. Elymus ×saundersii Vasey [p. *382*]

*Elymus ×saundersii* comprises hybrids between *E. trachycaulus* and *E. elymoides*. Such hybrids are found throughout much of the western portion of the contiguous United States, mostly in disturbed areas. The hybrids are generally sterile and, as in all hybrids involving *E. elymoides* or *E. multisetus*, the rachises disarticulate at maturity.

### 45. Elymus ×hansenii Scribn. [p. *382*]

*Elymus ×hansenii* refers to hybrids between *E. glaucus* and either *E. elymoides* or *E. multisetus*. It is not clear which of the latter two species is involved. It is a fairly common hybrid in those parts of western North America where both parents grow. The glumes of the type specimen are as wide as those in *E. glaucus*, and some are divided longitudinally, as in *E. elymoides* and *E. multisetus*. As in other hybrids involving *E. elymoides* and *E. multisetus*, the rachis of *E. ×hansenii* disarticulates at maturity.

### 46. Elymus ×pinalenoensis (Pyrah) Barkworth & D.R. Dewey [p. *382*]

*Elymus ×pinalenoensis* consists of hybrids between *E. elymoides* subsp. *brevifolius* and *E. arizonicus*. It has been found in the Pinaleno and Santa Catalina mountains of Graham County, Arizona, in areas disturbed by logging, road building, summer home development, and recreation.

### 47. Elymus ×ebingeri G.C. Tucker [p. *382*]

*Elymus ×ebingeri* is the name for hybrids between *E. virginicus* and *E. hystrix*. It is frequently found where the two parental species grow together, often with later hybrid generations and introgressants to the two parents. It has been reported from southern Ontario, and from Wisconsin to New York and Illinois. Most published reports simply refer to the existence of these hybrids, the name itself not having been published until 1996.

# 13.14 ×ELYLEYMUS B.R. Baum[1]

**Pl** per; smt rhz. **Clm** 40–235 cm, erect. **Infl** usu spikes, smt spikelike rcm, 5–35 cm, erect, with 1–3 spklt per nd, ped, when present, to 3 mm. **Spklt** with 2–8 flt; **dis** usu above the glm and beneath the flt, smt below the glm, smt in the rchs, usu tardy. **Glm** linear to lanceolate, often awn-tipped; **lm** 6–25 mm, glab or hairy, usu awned, awns to 15 mm; **anth** 1.5–5 mm.

×*Elyleymus* consists of hybrids between *Elymus* and *Leymus*. So far as is known, they are completely sterile, having thin anthers (usually less than 0.5 mm thick) and failing to develop mature caryopses. Only the named hybrids are accounted for below. Each of the entities appears to be distinct, but identification of the parents is, in some instances, tentative. All the illustrations are based on type specimens.

The hybrids fall into two groups. Those with *Leymus mollis* as the *Leymus* parent (species 7–11) tend to have wider and flatter glumes than those with one of the inland species of *Leymus* as the *Leymus* parent.

### 1. ×Elyleymus turneri (Lepage) Barkworth & D.R. Dewey [p. *382*]

×*Elyleymus turneri* refers to hybrids between *Elymus lanceolatus* and *Leymus innovatus*. The type specimen is from the banks of the Saskatchewan River, 2 miles below Fort Saskatchewan, Alberta.

### 2. ×Elyleymus aristatus (Merr.) Barkworth & D.R. Dewey [p. *382*]

It has been argued that ×*Elyleymus aristatus* comprises hybrids between *Elymus elymoides* and *Leymus cinereus* or *L. triticoides*. It has been found at many locations where the parents are sympatric.

### 3. ×Elyleymus colvillensis (Lepage) Barkworth [p. *382*]

×*Elyleymus colvillensis* consists of hybrids between *Leymus innovatus* and, probably, *Elymus alaskanus*. The original collections were made on the banks of the Colville River at Umiat, Alaska. It is not known how widely it is distributed.

### 4. ×Elyleymus hirtiflorus (Hitchc.) Barkworth [p. *383*]

×*Elyleymus hirtiflorus* may consist of hybrids between *Elymus trachycaulus* and *Leymus innovatus*. The name, however, is based on collections from the banks of the Green River, Wyoming, where neither putative parent grows. The more likely parents are *E. lanceolatus* and *L. simplex*. Admittedly, the short anthers argue for *E. trachycaulus* rather than *E. lanceolatus* as the *Elymus* parent. Canadian specimens are here treated as belonging to ×*Elyleymus ontariensis*.

### 5. ×Elyleymus mossii (Lepage) Barkworth [p. *383*]

It has been suggested that *Elymus canadensis* is one parent of this hybrid, but *Elymus canadensis* is generally absent from the region around Lake Louise, Alberta, where the holotype was collected. The parents of ×*Elyleymus mossii* are probably *E. glaucus* and *Leymus innovatus*, both species that are common in the holotype area.

### 6. ×Elyleymus ontariensis (Lepage) Barkworth [p. *383*]

×*Elyleymus ontariensis* may comprise hybrids between *Elymus trachycaulus* and *Leymus innovatus*. It differs from ×*Elyleymus hirtiflorus* in having wider, more parallel-sided glumes and longer rachis internodes.

### 7. ×Elyleymus uclueletensis (Bowden) B.R. Baum [p. *383*]

×*Elyleymus uclueletensis* comprises hybrids between *Leymus mollis* and *Elymus glaucus*. It is known from two locations, near Ucluelet and along Gold River, both on the west coast of Vancouver Island, British Columbia.

### 8. ×Elyleymus aleuticus (Hultén) B.R. Baum [p. *383*]

×*Elyleums aleuticus* comprises hybrids between *Elymus hirsutus* and *Leymus mollis*. It is known only from the type locality, Atka, Alaska. It probably occurs at other locations where the two parents are sympatric.

### 9. ×Elyleymus hultenii (Melderis) Barkworth [p. *383*]

×*Elyleymus hultenii* consists of hybrids between *Elymus alaskanus* subsp. *alaskanus* and *Leymus mollis*. The original collection is from Deering, Alaska.

### 10. ×Elyleymus jamesensis (Lepage) Barkworth [p. *383*]

×*Elyleymus jamesensis* comprises hybrids between *Elymus trachycaulus* and *Leymus mollis*.

### 11. ×Elyleymus ungavensis (Louis-Marie) Barkworth [p. *383*]

×*Elyleymus ungavensis* is a northern hybrid, collected along the sandy banks of the Koksoak River, near Fort Chimo [= Kuujjuaq], at the southern end of Ungava Bay, Quebec. It consists of hybrids between *Elymus violaceus* and *Leymus mollis* subsp. *mollis*. The involvement of *E. violaceus* is suggested by the wide glume margins; that of *L. mollis* subsp. *mollis* by the relatively thick culms.

[1]Mary E. Barkworth

## 13.15 PASCOPYRUM Á. Löve[1]

**Pl** per; rhz. **Clm** 20–100 cm. **Lvs** bas concentrated; **shth** striate when dry, smooth, usu glab, rarely pilose; **aur** present; **lig** memb. **Infl** tml, distichous spikes, spklt usu 1 per nd, occ in pairs at the lo nd, spklt at the lo 4–6 nd often strl; **lowest intnd** to 26 mm, 2 times as long as the mid intnd. **Spklt** 12–26(30) mm, 1–3 times the length of the intnd, straight, usu ascending, not appressed, with 2–12 flt; **dis** above the glm, beneath the flt. **Glm** 5–15 mm, $^1/_2$–$^2/_3$ the length of the spklt, usu narrowly lanceolate, stiff, tapering from midlength or below, slightly curving to the side distally, not keeled, 3–5-veined bas, 1-veined distally, apc acuminate; **lm** lanceolate, rounded on the back, acute, mucronate to awned, awns to 5 mm, straight; **pal** slightly shorter than the lm; **anth** 3, 2.5–6 mm. **Car** 4–5 mm, falling with the lm and pal.

*Pascopyrum* is a North American allooctoploid genus with one species. It is the only species that combines the genomes of *Leymus* with those of *Elymus*.

### 1. Pascopyrum smithii (Rydb.) Barkworth & D.R. Dewey WESTERN WHEATGRASS [p. 383, 516]

*Pascopyrum smithii* is native to sagebrush deserts and mesic alkaline meadows, growing in both clay and sandy soils. *Pascopyrum smithii* is probably derived from a *Leymus triticoides* × *Elymus lanceolatus* cross; it is frequently confused with both. *Leymus triticoides* differs in usually having 2 spikelets per node and glumes that are narrower at the base. In *E. lanceolatus*, the leaves tend to be more evenly distributed and the glumes have straight midveins, become narrow beyond midlength, and tend to be wider at $^3/_4$ length. In addition, the first rachilla internodes of *E. lanceolatus* are often longer and narrower.

## 13.16 ×PASCOLEYMUS (B. Boivin) Barkworth[2]

**Pl** per; rhz. **Clm** to 120 cm, glab. **Lvs** evenly distributed; **shth** open; **aur** present; **lig** 0.5–1 mm, ciliate; **bld** 3–10 mm wide, slightly glaucous, long-tapering to the narrowly acute apc, abx surfaces scabridulous, glab, adx surfaces scabrous. **Infl** distichous spikes, with 2 spklt per nd. **Spklt** sessile, with 8–9 flt; **dis** above the glm, beneath the flt. **Glm** often unequal, linear-lanceolate to subulate; **lm** awned, awns shorter than the lm body; **pal** subequal to the lm, apc about 0.2 mm wide; **anth** about 3.5 mm, indehiscent.

*×Pascoleymus* consists of hybrids between *Pascopyrum* and *Leymus*. Only one such hybrid has been recognized.

### 1. ×Pascoleymus bowdenii (B. Boivin) Barkworth [p. 384]

The holotype of *×Pascoleymus bowdenii* was collected at Beaverlodge, Alberta. It has been considered to be a hybrid between *Agropyon smithii* [≡ *Pascopyrum smithii*] and *Elymus innovatus* [≡ *Leymus innovatus*]. The tapering leaf blades and hairy lemmas support the involvement of *L. innovatus*; support for the involvement of *P. smithii* is less evident. *×Pascoleymus bowdenii* differs it its wide blades from *×Elyleymus turneri*, another hybrid having *L. innovatus* as one of its parents, but these do not provide any particular support for identifying *P. smithii* as the other parent.

## 13.17 LEYMUS Hochst.[3]

**Pl** per; smt csp, often rhz. **Clm** 10–350 cm, erect, with exvag brchg. **Lvs** bas or evenly distributed; **shth** open; **aur** usu present; **lig** memb, truncate to rounded; **bld** often stiff, adx surfaces usu with subequal, closely spaced, prominently ribbed veins, smt with unequal, widely spaced, not prominently ribbed veins. **Infl** usu distichous spikes with 1–8 spklt per nd, smt pan with (2)3–35 spklt associated with each rchs nd; **rchs** with scabrous or ciliate edges; **intnd** 3.5–12(15) mm. **Spklt** $^1/_2$–3 $^3/_4$ times the length of the rchs intnd, usu sessile, smt pedlt, ped to 5 mm, appressed to ascending, with 2–12 flt, the tml flt usu rdcd; **dis** above the glm, beneath the flt. **Glm** usu 2, usu equal to subequal, the lo or both glm smt rdcd or absent, lanceolate and narrowing in the distal $^1/_4$, or lanceolate to subulate and tapering from below midlength, pilose or glab, smt scabrous, 0–3(7)-veined, veins evident at least at midlength, smt keeled, keels straight or almost so, apc acute, acuminate, or tapering to an awnlike tip, if distinctly awned, awns to 4 mm; **lm** glab or with hairs, smt scabrous distally, inconspicuously 5–7-veined, rounded over the back proximally, smt keeled distally, keels not conspicuously scabrous distally, apc acute, unawned or awned, awns usu to 7 mm, smt 16–33 mm, straight; **pal** slightly shorter than to slightly longer than the lm, keels usu scabrous or ciliate on the distal portion, smt throughout; **lod** 2, shortly hairy, lobed; **anth** 3, 2.5–10 mm. **Car** with hairy apc.

*Leymus* is a genus of approximately 50 species; all are native to temperate regions in the Northern Hemisphere. They are most abundant in eastern Asia, with North America being a secondary center. Of the 17 species treated, 11 are native to the *Manual* region, 4 are introduced, and 2 are naturally occurring hybrids.

Most species of *Leymus*, including most North American species, grow well in alkaline soils. They are used for soil stabilization and forage. All the species are self-incompatible, outcrossing polyploids. One of the haplomes present is the Ns genome; this genome is also found in *Psathyrostachys*, most species of which are diploids. There is disagreement concerning the second haplome. Morphologically,

[1,2,3]Mary E. Barkworth

*Psathyrostachys* and *Leymus* are very similar, the major differences being that *Psathyrostachys* is never rhizomatous, has disarticulating rachises, and, usually, distinctly awned lemmas.

*Leymus arenarius* and *L. mollis* are sometimes mistaken for *Ammophila*, which grows in the same habitats and has a similar habit. *Ammophila* differs from *Leymus*, however, in having only one floret per spikelet.

In most species of *Leymus*, at least some of the spikelets are on pedicels up to 2 mm long. Despite this, it is customary to identify the inflorescence of such species as a spike rather than a raceme, as is done in this treatment. Culm thicknesses are measured on the lower internodes.

1. Glumes absent or shorter than 1 mm; lemmas awned, awns 16–33 mm long . . . . . . . . . . . . . . . . . . . . . 17. *L. californicus*
1. Glumes developed, 3+ mm long, at least 1 on each spikelet; lemmas unawned or awned, awns to 7 mm long.
  2. Glumes flat or rounded on the back, tapering from midlength or above, flexible, the central portion scarcely thicker than the margins . . . . . . . . . . . . . . . . . . . . . . . . . . . . . . . . . . . . . . . . . . . . . 3. *L. mollis*
  2. Glumes keeled, at least distally, tapering from below midlength, stiff, the central portion thicker than the margins.
    3. Anthers usually indehiscent; plants rhizomatous, restricted to coastal regions from British Columbia to California.
      4. Glumes pubescent distally; lemmas awned, awns to 4 mm long; inflorescences spikes, not branched . . . . . . . . . . . . . . . . . . . . . . . . . . . . . . . . . . . . . . . . . . . . . . . . . . . . . . . 4. *L. ×vancouverensis*
      4. Glumes glabrous; lemmas acute to awned, awns to 1.8 mm long; inflorescences sometimes with strongly ascending branches . . . . . . . . . . . . . . . . . . . . . . . . . . . . . . . . . . . . . . . . . . . 11. *L. ×multiflorus*
    3. Anthers dehiscent; plants rhizomatous or cespitose, widespread, including coastal regions from British Columbia to California.
      5. Inflorescences with 2–4 branches to 6 cm long at the proximal nodes; culms 115–350 cm tall . . 10. *L. condensatus*
      5. Inflorescences without branches; culms 10–270 cm tall.
        6. Lemmas densely hairy, hairs 0.7–3 mm long, occasionally glabrate.
          7. Lemmas awned, awns 2–4 mm long; lemma hairs 0.7–2.5 mm long . . . . . . . . . . . . . . . . . 15. *L. innovatus*
          7. Lemmas unawned or the awns to 2 mm long; lemma hairs 2–3 mm long . . . . . . . . . . . . . . 16. *L. flavescens*
        6. Lemmas usually wholly or partly glabrous, or if hairy, the hairs shorter than 0.5(0.8) mm.
          8. Leaves equaling or exceeding the spikes; culms 10–30(60) cm tall; spikes 2–8 cm long, with 1–2 spikelets per node; plants of California coastal bluffs . . . . . . . . . . . . . . . . . . . . . . . . . 5. *L. pacificus*
          8. Leaves exceeded by the spikes; culms 35–270 cm tall; spikes 3–35 cm long, with 1–8 spikelets per node; plants widespread in the western part of the *Manual* region, including the coastal bluffs of California.
            9. Plants cespitose, not or weakly rhizomatous, culms several to many together.
              10. Spikes with 2–7 spikelets per node; blades 3–12 mm wide; culms (70)100–270 cm tall . . . . . . . . . . . . . . . . . . . . . . . . . . . . . . . . . . . . . . . . . . . . . . . . . . . . . . . . 12. *L. cinereus*
              10. Spikes with 1 spikelet at the distal nodes, often at all nodes, sometimes with 2(3) at the lower nodes; blades 1–6 mm wide; culms 35–140 cm tall.
                11. Blades with 5–9 adaxial veins; lemma awns to 2.5 mm long . . . . . . . . . . . . . . . . 13. *L. salina*
                11. Blades with (9)11–17 adaxial veins; lemma awns 1.3–7 mm long . . . . . . . . . . . 14. *L. ambiguus*
            9. Plants rhizomatous, culms solitary or few together.
              12. Culms 1–3 mm thick; glumes 4–16 mm long.
                13. Spikes with 1 spikelet at all or most nodes, sometimes with 2 at a few nodes; lemma awns 2.3–6.5 mm long; culms 35–55 cm tall . . . . . . . . . . . . . . . . . 6. *L. simplex*
                13. Spikes with 2+ spikelets at most nodes; lemma awns to 3 mm long; culms 45–125 cm tall.
                  14. Adaxial surfaces of the blades usually with closely spaced, prominently ribbed, subequal veins; calluses usually glabrous, occasionally with a few hairs about 0.1 mm long . . . . . . . . . . . . . . . . . . . 7. *L. triticoides*
                  14. Adaxial surfaces of the blades usually with widely spaced, not prominently ribbed veins, the primary veins evidently larger than the intervening secondary veins; calluses with hairs about 0.2 mm long . . . . . . 8. *L. multicaulis*
              12. Culms 2.5–12 mm thick; glumes 10–30 mm long.
                15. Spikelets 3–8 per node; lemmas hairy proximally, glabrous distally . . . . . . . . . . 1. *L. racemosus*
                15. Spikelets 2–3 per node; lemmas glabrous or hairy their whole length.
                  16. Anthers 6–9 mm long; blades 3–11 mm wide; glumes with hairs to 1.3 mm long; plants established around the Great Lakes and the coast of

Greenland, also found at a few other scattered locations, including western North America, sometimes cultivated . . . . . . . . . . . . . . . . . . . . . . 2. *L. arenarius*
16. Anthers 3–5 mm long; blades 5–7 mm wide; glumes glabrous, sometimes scabrous; plants cultivated . . . . . . . . . . . . . . . . . . . . . . . . . . 9. *L. angustus*

## 1. Leymus racemosus (Lam.) Tzvelev MAMMOTH WILDRYE [p. *384*, *516*]

*Leymus racemosus* is native to Europe and central Asia, where it grows on dry, sandy soils. It has been introduced into the *Manual* region, and collected at various locations, particularly in the northwestern contiguous United States; it is not clear how many of the populations represented by these specimens are still extant.

## 2. Leymus arenarius (L.) Hochst. EUROPEAN DUNEGRASS, LYMEGRASS, ÉLYME DES SABLES D'EUROPE [p. *384*, *516*]

*Leymus arenarius* is native to Europe. It has become established in sandy habitats around the Great Lakes and the coast of Greenland. It has also been found at a few other widely scattered locations. It is sometimes cultivated, forming large, attractive, blue-green clumps, but its tendency to spread may be undesirable.

## 3. Leymus mollis (Trin.) Pilg. AMERICAN DUNEGRASS, SEA LYMEGRASS, ÉLYME DES SABLES D'AMERIQUE, SEIGLE DE MER [p. *384*, *516*]

*Leymus mollis* is native to Asia and North America. It is treated here as having two very similar subspecies that have somewhat different ranges. The subspecies are sometimes treated as separate species, but they may be little more than environmentally induced variants. Both subspecies grow primarily on coastal beaches, close to the high tide line, and along some inland waterways, particularly in the arctic.

1. Spikes 12–34 cm long, with 12–33 nodes; basal blades 5–15 mm wide; culms 50–170 cm tall . . . . . subsp. *mollis*
1. Spikes 5–13(16) cm long, with 3–14 nodes; basal blades 3–8 mm wide; culms 12–70 cm tall . . subsp. *villosissimus*

## Leymus mollis (Trin.) Pilg. subsp. mollis [p. *384*]

In the *Manual* region, *Leymus mollis* subsp. *mollis* grows primarily on the west coast; on the east coast, it grows in New Brunswick and Nova Scotia, particularly along the St. Lawrence River, and on the coast of Greenland. Outside the *Manual* region, it is native in the coastal region of eastern Asia. *Leymus ×vancouverensis* is thought to be a hybrid between *L. mollis* subsp. *mollis* and *L. triticoides*, although its range extends beyond the current range of *L. triticoides*.

## Leymus mollis subsp. villosissimus (Scribn.) Á. Löve [p. *384*]

*Leymus mollis* subsp. *villosissimus* is an arctic taxon found primarily in eastern Siberia, Alaska, and northwestern Canada. It grows mostly on arctic coasts, but is also known from a few inland locations.

## 4. Leymus ×vancouverensis (Vasey) Pilg. VANCOUVER WILDRYE [p. *384*, *517*]

*Leymus ×vancouverensis* grows at scattered locations on beaches along the Pacific coast, from southern British Columbia to California. It is a sterile hybrid, probably between *L. mollis* and *L. triticoides*. The northern populations are outside the current range of *L. triticoides*.

## 5. Leymus pacificus (Gould) D.R. Dewey PACIFIC WILDRYE [p. *384*, *517*]

*Leymus pacificus* is found on coastal bluffs from Mendocino to Santa Barbara counties, California. It is poorly represented in herbaria. In some years it grows almost entirely vegetatively, often being represented by scattered innovations with somewhat curved leaves.

## 6. Leymus simplex (Scribn. & T.A. Williams) D.R. Dewey ALKALI WILDRYE [p. *384*, *517*]

*Leymus simplex* is found in meadows and drifting sand in southern Wyoming, and along the Green River in northeastern Utah.

1. Culms 55–75 cm tall; spikes 10–27 cm long; internodes 10–20 mm . . . . . . . . . . . . . . . . . . . . var. *luxurians*
1. Culms 35–55 cm tall; spikes 1.5–13 cm long; internodes 7–9 mm . . . . . . . . . . . . . . . . . . . . . var. *simplex*

## Leymus simplex var. luxurians (Scribn. & T.A. Williams) Beetle [p. *384*]

*Leymus simplex* var. *luxurians* grows at a few locations in Wyoming. It sometimes grows close to var. *simplex*. It may represent clones that have access to more water and/or more nutrients, but the absence of intermediate plants suggests a genetic distinction.

## Leymus simplex (Scribn. & T.A. Williams) D.R. Dewey var. simplex [p. *384*]

*Leymus simplex* var. *simplex* is found throughout the range of the species, sometimes in close proximity to var. *luxurians*. The two may be environmentally induced variants, but the lack of intermediates suggests a genetic distinction.

## 7. Leymus triticoides (Buckley) Pilg. BEARDLESS WILDRYE [p. *384*, *517*]

*Leymus triticoides* grows in dry to moist, often saline meadows. Its widely scattered populations extend from southern British Columbia to Montana, south to California, Arizona, and New Mexico. It is not known from Mexico. There is considerable variation within the species. It is very similar to *L. multicaulis*, strains of which were initially released as *L. triticoides* by the U.S. Department of Agriculture. The most consistent differences between them appear to be in the venation of the leaf blades and the vesture of the calluses. *Leymus triticoides* is also very similar to *L. simplex*, differing from it in the number of spikelets at the midspike nodes.

*Leymus triticoides* hybridizes with other species of *Leymus*; hybrids with *L. mollis* are called *L. ×vancouverensis*, those with *L. condensatus* are called *L. ×multiflorus*. Hybrids with *L. cinereus* are known, but have not been formally named. Plants identified as *Elymus arenicolus* Scribn. & J.G. Sm. are here included in *L. flavescens*, but may represent hybrids between *L. triticoides* and *L. flavescens*.

## 8. Leymus multicaulis (Kar. & Kir.) Tzvelev MANY-STEM WILDRYE [p. *384*]

*Leymus multicaulis* is native to Eurasia, extending from the Volga River delta in Russia to Xinjiang, China. In its native range, it grows in alkaline meadows and saline soils, and as a weed in fields, near roads, and around human habitations. It is very similar to *L. triticoides*, and hybrids with that species are highly fertile. A cultivar of *L. multicaulis*, 'Shoshone', that was originally thought to be a productive strain of *L. triticoides*, has been widely distributed for forage. *Leymus multicaulis* differs from *L. triticoides* primarily in having both primary and secondary veins in its blades, and small hairs on its calluses. Its distribution in North America is unknown.

## 9. Leymus angustus (Trin.) Pilg. ALTAI WILDRYE [p. *385*]

*Leymus angustus* is a Eurasian species that, in its native range, grows in alkaline meadows, and on sand and gravel in river and lake valleys. Several cultivars of *L. angustus* have been developed for use as forage, particularly in Canada. The distribution of *L. angustus* in the *Manual* region is not known.

## 10. Leymus condensatus (J. Presl) Á. Löve GIANT WILDRYE [p. *385*, *517*]

*Leymus condensatus* is found primarily on dry slopes and in open woodlands of the coastal mountains and offshore islands of California, at elevations of 0–1500 m. Both its large size and paniculate inflorescence tend to make it a distinctive species in the *Triticeae*. Hybrids between *L. condensatus* and *L. triticoides*, known as *Leymus ×multiflorus*, are relatively common where the parents are sympatric.

## 11. Leymus ×multiflorus (Gould) Barkworth & R.J. Atkins MANY-FLOWERED WILDRYE [p. *385*, *517*]

*Leymus ×multiflorus* is a sterile hybrid between *L. condensatus* and *L. triticoides* that grows near the coast of central and southern California.

## 12. Leymus cinereus (Scribn. & Merr.) Á. Löve GREAT BASIN WILDRYE [p. *385*, *517*]

*Leymus cinereus* grows along streams, gullies, and roadsides, and in gravelly to sandy areas in sagebrush and open woodlands. It is widespread and common in western North America. It differs from *Psathyrostachys juncea* in its non-disarticulating rachises, larger spikelets with more florets, and longer ligules. Spontaneous hybridization between *L. cinereus* and *L. triticoides* is known; the hybrids do not have a scientific name.

## 13. Leymus salina (M.E. Jones) Á. Löve [p. *385*, *517*]

The three subspecies of *Leymus salina* differ in their pubescence and geographic distribution, with subsp. *salina* being the most common of the three.

1. Basal sheaths and blades conspicuously hairy on the abaxial surfaces . . . . . . . . . . . . . . . . . . . . . . . . subsp. *salmonis*
1. Basal sheaths glabrous; blades usually glabrous on the abaxial surfaces.
    2. Blades strongly involute, usually densely hairy just above the ligules . . . . . . . . . . . . . . . . . . . subsp. *salina*
    2. Blades flat or almost flat, not densely hairy above the ligules . . . . . . . . . . . . . . . . . . subsp. *mojavensis*

### Leymus salina subsp. mojavensis Barkworth & R.J. Atkins MOJAVE WILDRYE [p. *385*]

*Leymus salina* subsp. *mojavensis* grows on steep, north-facing slopes of the New York, Providence, and Clark mountains in California, and the south rim of the Grand Canyon in Arizona.

### Leymus salina (M.E. Jones) Barkworth subsp. salina SALINA WILDRYE [p. *385*]

*Leymus salina* subsp. *salina* grows on rocky hillsides, primarily in eastern Utah and western Colorado, extending into southern Wyoming and northern Arizona and New Mexico.

### Leymus salina subsp. salmonis (C.L. Hitchc.) R.J. Atkins SALMON WILDRYE [p. *385*]

*Leymus salina* subsp. *salmonis* grows at scattered locations on rocky hillsides in the mountains of southern Idaho, Nevada, and western Utah.

## 14. Leymus ambiguus (Vasey & Scribn.) D.R. Dewey COLORADO WILDRYE [p. *385*, *517*]

*Leymus ambiguus* grows on steep, often boulder-strewn hillsides at scattered locations in Colorado and New Mexico.

## 15. Leymus innovatus (Beal) Pilg. DOWNY RYEGRASS, BOREAL WILDRYE [p. *385*, *517*]

*Leymus innovatus* is a North American species that grows in open woods and forests, riverbanks, open prairies, and rocky soils, and often in sandy, gravelly, or silty soils, primarily from northern Alaska to Hudson Bay, and south into the Black Hills region of Wyoming and South Dakota. Morphologically, the two subspecies show some overlap.

1. Spikes 8–16 cm long, 8–15 mm wide; lemma hairs 0.7–2.5 mm long . . . . . . . . . . . . . . . . . . . . . . subsp. *innovatus*
1. Spikes 3–8 cm long, 15–20 mm wide; lemma hairs 1.5–2.5 mm long . . . . . . . . . . . . . . . . . . . . . . subsp. *velutinus*

### Leymus innovatus (Beal) Pilg. subsp. innovatus [p. *385*]

*Leymus innovatus* subsp. *innovatus* is the more widespread of the two subspecies, extending across North America from the southern Yukon Territory to Ontario, south in the Rocky Mountains to northern Montana, and, as a disjunct, to the Black Hills region of Wyoming and South Dakota.

### Leymus innovatus subsp. velutinus (Bowden) Tzvelev [p. *385*]

*Leymus innovatus* subsp. *velutinus* is the more northern of the two subspecies, growing in Alaska, the Yukon Territory, and the western Northwest Territories.

## 16. Leymus flavescens (Scribn. & J.G. Sm.) Pilg. YELLOW WILDRYE [p. *385*, *517*]

*Leymus flavescens* grows on sand dunes and open sandy flats, and ditch- and roadbanks, of the Snake and Columbia river valleys. Plants identified as *Elymus arenicolus* Scribn. & J.G. Sm. are included here, but they may represent hybrids between *Leymus flavescens* and *L. triticoides*.

## 17. Leymus californicus (Bol. *ex* Thurb.) Barkworth CALIFORNIA BOTTLEBRUSH [p. *385*, *517*]

*Leymus californicus* is endemic to coniferous forests near the coast in western California, from Sonoma to Santa Cruz counties, at elevations from near sea level to 300 m. It used to be included in *Hystrix* Moench, a genus that was described as lacking glumes.

# 13.18 ×LEYDEUM Barkworth[1]

**Pl** per; rhz, smt shortly so. **Clm** to 140 cm tall, 1–3 mm thick. **Spikes** 10–15 cm long, 5–12 mm wide excluding the awns, erect, smt lax, nd with 2–3 spklt; **intnd** 3–5 mm; **dis** in the rchs, smt delayed. **Spklt** appressed, with 1–3 flt. **Glm** equal or unequal, 10–25 mm long, 0.2–1.5 mm wide, tapering from below midlength or subulate from the base; **lm** glab or hairy, awned, awns 1–10 mm; **anth** 1.8–3 mm long, 0.1–0.3 mm thick. **Car** not developed.

×*Leydeum* consists of hybrids between *Hordeum* and *Leymus*. Of the three species recognized, two involve the coastal species *L. mollis*, which has flat glumes; these hybrids have glumes intermediate between their parents in morphology, and rachises that tend to disarticulate. ×*Leydeum piperi*, the only hybrid involving one of the inland species of *Leymus*, differs from *Leymus* in its disarticulating rachises, and from *Hordeum* in having 2 spikelets per node and 2–3 florets in the larger spikelets.

[1]Mary E. Barkworth

1. ×**Leydeum piperi** (Bowden) Barkworth [p. *386*]

The parents of ×*Leydeum piperi* are probably *Hordeum jubatum* and *Leymus triticoides*. It is not known how common or widespread the hybrid is. It differs from ×*Elyhordeum macounii* in its subulate, rather than narrowly linear, glumes.

2. ×**Leydeum dutillyanum** (Lepage) Barkworth [p. *386*]

×*Leydeum dutillyanum* consists of hybrids between *Hordeum jubatum* and *Leymus mollis*. It has been reported only from Vieux-Comptoir, Quebec. It appears to disarticulate more readily than ×*L. littorale*.

3. ×**Leydeum littorale** (H.J. Hodgs. & W.W. Mitch.) Barkworth [p. *386*]

×*Leydeum littorale* consists of hybrids between *Hordeum brachyantherum* and *Leymus mollis*. It has been collected in the Matanuska Valley, Alaska, and on the coast of Vancouver Island, British Columbia; it may be more widespread.

## 13.19 PSATHYROSTACHYS Nevski[1]

Pl per; csp, forming dense to loose clumps, smt stln, smt rhz. Clm 15–120 cm, erect or decumbent. Shth of the bas lvs closed, becoming fibrillose, of the up cauline lvs open; aur smt present; lig 0.2–0.3 mm, memb; bld with prominently ribbed veins on the adx surfaces. Infl spikes, with 2–3 spklt per nd; dis in the rchs. Spklt appressed to ascending, with 1–2(3) flt, often with additional rdcd flt distally. Glm equal to unequal, (3.5)4.2–48.5(65) mm including the awns, subulate, stiff, scabrous to pubescent, obscurely 1-veined, not united at the base; lm 5.5–14.3 mm, narrowly elliptic, rounded, glab or pubescent, 5–7-veined, veins often prominent distally, apc sharply acute to awned, smt with a minute tooth on either side of the awn base, awns 0.8–34 mm, straight, ascending to slightly divergent, smt violet-tinged; pal equaling or slightly longer than the lm, memb, scabrous or pilose on and smt also between the keels, bifid; anth 3, 2.5–6.8(7) mm, yellow or violet; lod 2, acute, entire, ciliate. Car pubescent distally, tightly enclosed by the lm and pal at maturity.

*Psathyrostachys* has eight species, all of which are native to arid regions of central Asia, from eastern Turkey to eastern Siberia, Russia, and Xinjiang Province, China. One species, *P. juncea*, was introduced to North America as a potential forage species, and is now established in the *Manual* region.

*Psathyrostachys* is very similar to *Leymus*, particularly the cespitose species of *Leymus*. The major differences are that *Psathyrostachys* has disarticulating rachises and, usually, distinctly awned lemmas.

1. **Psathyrostachys juncea** (Fisch.) Nevski RUSSIAN WILDRYE [p. *386*, *517*]

*Psathyrostachys juncea* is native to central Asia, primarily to the Russian and Mongolian steppes. It has become established at various locations, from Alaska to Arizona and New Mexico. It is drought-resistant and tolerant of saline soils. In its native range, it grows on stony slopes and roadsides, at elevations to 5500 m. It differs from *Leymus cinereus* primarily in having shorter ligules and a rachis that breaks up at maturity. Immature plants can be identified by the more uniform appearance of the spikelets. *Psathyrostachys juncea* also tends to have smaller spikelets with fewer florets than *L. cinereus*.

## 13.20 THINOPYRUM Á. Löve[2]

Pl per; csp or not, smt rhz. Clm 10–250 cm, usu erect. Shth open, glab or ciliate; aur 0.2–1.8 mm or absent; lig memb; bld convolute or flat. Infl tml, distichous spikes, usu not disarticulating at maturity, with 1 spklt at all or most nd; intnd 5–30 mm. Spklt 1–3 times the length of the middle intnd, solitary, appressed to ascending, often diamond-shaped in outline and arching outwards at maturity; dis tardy, usu beneath the flt, smt in the rchs. Glm rectangular to lanceolate, narrowing beyond midlength, stiff, indurate to coriaceous, glab or with hairs, keeled or rounded at the base, usu more strongly keeled distally than proximally, 4–9-veined, midveins usu scabrous distally, mrg often hyaline, apc truncate to acute, smt mucronate, unawned, without lat teeth; lm 5-veined, coriaceous, glab or with hairs, truncate, obtuse, or acute, smt mucronate or awned, awns to 3 cm; anth 3, 2.5–12 mm.

*Thinopyrum* includes approximately ten species, most of which are alkaline tolerant. It is native from the Mediterranean region to western Asia. Four species are established in the *Manual* region; only *T. intermedium* and *T. ponticum* are common. The genus is sometimes included in *Elytrigia* Desv. or *Elymus*.

*Thinopyrum* differs from the other *Triticeae* in its thick, stiff glumes and lemmas.

In the key, measurements and comments about rachis internodes refer to the internodes at midspike. The lowest internodes of a spike are usually 2–4 times as long as those at midspike.

1. Plants not rhizomatous; glumes truncate, midveins about equal in length and prominence to the lateral veins . . . . . . . . . . . . . . . . . . . . . . . . . . . . . . . . . . . . . . . . . . . . . . . . . . . . . . . . . . . . . . . . . 4. *T. ponticum*
1. Plants rhizomatous; glumes obliquely truncate or obtuse to acute, midveins usually slightly longer and more prominent than the lateral veins.
   2. Glumes 9–18 mm long . . . . . . . . . . . . . . . . . . . . . . . . . . . . . . . . . . . . . . . . . . . . . . . . . . 3. *T. pycnanthum*
   2. Glumes 4.5–8.5 mm long.

---

[1]Claus Baden†  [2]Mary E. Barkworth

3. Lemmas 7.5–10 mm long, glabrous or hairy; rachis internodes 7–12 mm long; plants widespread in the *Manual* region, particularly in the western United States . . . . . . . . . . . . . . . . . . . . . . . . . . . . 1. *T. intermedium*
3. Lemmas 10–17 mm long, glabrous; rachis internodes 12–28 mm long; plants known only from a few coastal locations in the *Manual* region . . . . . . . . . . . . . . . . . . . . . . . . . . . . . . . . . . . . . . . . 2. *T. junceum*

### 1. Thinopyrum intermedium (Host) Barkworth & D.R. Dewey [p. 386, 517]

*Thinopyrum intermedium* is native to Europe and western Asia. It is widely established in western North America, having been introduced for erosion control, revegetation, forage, and hay. It also occurs in scattered locations further east. One of its advantages for erosion control and revegetation is that it establishes rapidly in many different habitats. In its native range, it grows in dry areas with sandy or stony soils.

Several subspecies have been recognized within *Thinopyrum intermedium*, usually based on differences in the vesture of the glumes and lemmas, the presence or absence of lemma awns, and the color of the plants. Because they have been found to be ecologically distinct, they are formally recognized here as subspecies. Plants with hairs only near the lemma margins are included under *T. intemedium* subsp. *intermedium*.

1. Lemmas and glumes glabrous . . . . . . . . . . . subsp. *intermedium*
1. Lemmas with hairs, sometimes only on the margins, hairs 1–1.5 mm long; glumes usually hairy throughout, sometimes glabrous but scabrous over the veins . . . . . . . . . . . . . . . . . . . . . . . . subsp. *barbulatum*

### Thinopyrum intermedium subsp. barbulatum (Schur) Barkworth & D.R. Dewey HAIRY WHEATGRASS [p. 386]

There is no known difference in geographic distribution between subsp. *barbulatum* and subsp. *intermedium* in the *Manual* region, but subsp. *barbulatum* is adapted to areas with 11–12 inches of rainfall per year.

### Thinopyrum intermedium (Host) Barkworth & D.R. Dewey subsp. intermedium INTERMEDIATE WHEATGRASS [p. 386]

There is no known difference in geographic distribution between subsp. *intermedium* and subsp. *barbulatum* in the *Manual* region, but subsp. *intermedium* is adapted to areas with 12–13 inches of rainfall per year.

### 2. Thinopyrum junceum (L.) Á. Löve RUSSIAN WHEATGRASS [p. 386, 517]

*Thinopyrum junceum* is native to the coast of Portugal, the Mediterranean, and the Black Sea. In the *Manual* region, it has been found on the coasts of southern California and Nova Scotia. In its native range, it grows on maritime rocky coasts, shifting beach sands, and, occasionally, by brackish water near river mouths.

### 3. Thinopyrum pycnanthum (Godr.) Barkworth TICK QUACKGRASS [p. 386, 517]

*Thinopyrum pycnanthum* is native to the coasts of western and southern Europe. It is reported from scattered locations in the western United States, and from Nova Scotia to Pennsylvania in eastern North America. In its native range, it grows in maritime sands and gravels, or river gravels.

### 4. Thinopyrum ponticum Barkworth & D.R. Dewey TALL WHEATGRASS, RUSH WHEATGRASS [p. 386, 517]

*Thinopyrum ponticum* is native to southern Europe and western Asia. In its native range, it grows in dry and/or saline soils. In the *Manual* region, *T. ponticum* is planted along roadsides for soil stabilization, and is spreading naturally in cooler areas because of its tolerance of the saline conditions caused by salting roads in winter.

## 14. POEAE R. Br.[1]

Pl ann or per; csp, rhz, or stln. Clm ann, not wd, not brchg above the base; intnd usu hollow. Shth usu open for most of their length, smt closed; col without tufts of hair on the sides; aur usu absent; lig memb to hyaline, smt ciliate, those of the up and lo cauline lvs usu similar; psdpet not developed; bld linear to narrowly lanceolate, venation parallel, cross venation not evident, without arm or fusoid cells, epdm without microhairs, not papillate, cross sections non-Kranz. Infl tml, usu pan, smt spikes, pan smt spikelike or rdcd to rcm in depauperate specimens; dis usu above the glm and beneath the flt, smt below the glm. Spklt 0.7–50 mm, lat compressed, smt weakly so, smt viviparous, usu with 2–22 flt, smt with 1, strl flt usu distal to the reproductively fnctl flt, smt with 1 or 2 stmt or strl flt below a bisx flt, strl flt often rdcd in size; rchl smt prolonged beyond the base of the distal flt. Glm (0, 1) 2, equal or unequal, shorter or longer than the adjacent flt, smt exceeding the distal flt; flt lat compressed; cal glab or hairy, not well developed; lm lanceolate to ovate, 1–7(9)-veined, unawned or awned, veins usu converging distally, smt parallel, awns from bas to tml on the lm, straight or bent; pal 2-keeled, from shorter than to longer than the lm, smt absent or minute; lod 2, memb, not or weakly veined; anth 3; ov usu glab, smt hairy distally; sty 2, bases free. Car longitudinally grooved or not, not beaked, prcp thin; hila punctate to linear; emb from $1/4$–$1/3$ as long as the car.

The *Poeae* constitute the largest tribe of grasses, encompassing around 115 genera and 2500 species. The species are primarily cool-temperate to arctic in their distribution. In the *Manual* region, there are 63 non-hybrid genera with 344 species, and 4 hybrid genera, each of which has one species. Many of the tribe's species are well known as lawn and pasture grasses, for example, *Poa pratensis* (Kentucky bluegrass), *Dactylis glomerata* (orchard grass), and *Phleum pratense* (timothy).

The tribe's circumscription and its infratribal taxonomy are unclear. It is interpreted here as including generic groups

[1]Mary E. Barkworth

that are, or have been, treated in other works as tribes (e.g., *Agrostideae* Dumort., *Aveneae* Dumort., *Hainardeae* Greut., and *Phalarideae* Dumort.).

The following key does not include these four hybrid genera: ×*Agropogon* (*Agrostis* × *Polypogon*, p. 156), ×*Arctodupontia* (*Arctophila* × *Dupontia*, p. 141), ×*Dupoa* (*Dupontia* × *Poa*, p. 141), and ×*Pucciphippsia* (*Puccinellia* × *Phippsia*, p. 110). They are described on the pages indicated. In the key that follows, branch measurements include spikelets, but not awns.

1. All or almost all spikelets viviparous, the spikelets producing plantlets [if sexual spikelets are common, take the alternate lead].
  2. Panicle branches smooth or slightly scabrous, the scabrules widely spaced; blades with a translucent line on either side of the midvein, apices usually prowlike . . . . . . . . . . . . . . . . . . . . . . . . . . . . . . . . . . . 14.13 *Poa* (in part)
  2. Panicle branches scabrous; blades without a translucent line on either side of the midvein, apices usually not prowlike.
    3. Sheaths closed for ¹/₂ or more of their length; ligules 0.1–0.6 mm long . . . . . . . . . . . . . . . . . 14.01 *Festuca* (in part)
    3. Sheaths open; ligules 1.5–13 mm long . . . . . . . . . . . . . . . . . . . . . . . . . . . . . . . . . . 14.26 *Deschampsia* (in part)
1. Some, usually all, spikelets sexually functional, with 1–25 bisexual or unisexual florets, sometimes with sterile and sexual spikelets mixed within an inflorescence.
  4. Inflorescences with 2 morphologically distinct forms of spikelets.
    5. Spikelets in pairs, the pedicels not fused at the base, smooth or slightly scabrous; disarticulation above the glumes and beneath the florets . . . . . . . . . . . . . . . . . . . . . . . . . . . . . . . . . . . . . 14.37 *Cynosurus*
    5. Spikelets in fascicles, the pedicels fused at the base, glabrous, hispid or strigose; disarticulation at the base of the fused pedicels.
      6. Secondary panicle branches sharply bent below the pedicels; glumes not winged . . . . . . . . . . . . . 14.11 *Lamarckia*
      6. Secondary panicle branches straight below the pedicels; glumes winged . . . . . . . . . . . . . . . 14.61 *Phalaris* (in part)
  4. Inflorescences with all spikelets morphologically alike.
    7. Glumes with pilose awns . . . . . . . . . . . . . . . . . . . . . . . . . . . . . . . . . . . . . . . . . . . . . . . . . . 14.30 *Lagurus*
    7. Glumes, if present, unawned or with glabrous awns.
      8. Inflorescences spikes with 1–2(4) spikelets per node, or spikelike racemes with 1 spikelet at all or most nodes.
        9. Spikelets with 1 functional floret, sometimes a reduced, sterile floret also present.
          10. Glumes membranous, flexible; all spikelets pedicellate, pedicels 0.5–1 mm long, 0.1–0.2 mm thick . . . . . . . . . . . . . . . . . . . . . . . . . . . . . . . . . . . . . . . . . . . . . . 14.59 *Mibora*
          10. Glumes coriaceous, stiff; lower spikelets sessile, upper spikelets sometimes pedicellate.
            11. Spikelets radial to the rachises, most spikelets with 1 glume, the terminal spikelets with 2 glumes . . . . . . . . . . . . . . . . . . . . . . . . . . . . . . . . . . . . . . . . . . . . 14.40 *Hainardia*
            11. Spikelets tangential to the rachises, all with 2 glumes.
              12. Lemmas unawned . . . . . . . . . . . . . . . . . . . . . . . . . . . . . . . . . . . . . . . . . 14.38 *Parapholis*
              12. Lemmas awned, awns 2–4 mm . . . . . . . . . . . . . . . . . . . . . . . . . . . . . . . . . 14.39 *Scribneria*
        9. Spikelets with 2–25 functional florets.
          13. Lemmas awned from about midlength, awns 8–26 mm long, twisted proximally.
            14. Adaxial surfaces of the leaves ribbed; rachillas pilose on all sides; ligules truncate to rounded, 0.5–1.5 mm long . . . . . . . . . . . . . . . . . . . . . . . . . . . . 14.47 *Helictotrichon* (in part)
            14. Adaxial surfaces of the leaves unribbed; rachillas glabrous on the side adjacent to the paleas, hairy elsewhere; ligules acute to truncate, 0.5–7 mm long . . . . . . . . 14.44 *Avenula* (in part)
          13. Lemmas unawned or apically awned, awns straight.
            15. Spikelets sessile; lemmas 2–12 mm long.
              16. Spikelets radial to the rachises, most spikelets with 1 glume, only the terminal spikelet with 2 glumes . . . . . . . . . . . . . . . . . . . . . . . . . . . . . . . . . . . . 14.05 *Lolium*
              16. Spikelets tangential to the rachises, all spikelets with 2 glumes . . . . . . . . . . . . . . 14.50 *Gaudinia*
            15. Spikelets subsessile to pedicellate, pedicels 0.5–3 mm long.
              17. Plants perennial . . . . . . . . . . . . . . . . . . . . . . . . . . . . . . . . . . . . . . . . . . 14.01 *Festuca* (in part)
              17. Plants annual.
                18. Inflorescences usually exceeded by the leaves; spikelets with (2)3–4(7) florets; lemmas (5)7–9-veined, apices round to emarginate, not bifid; culms usually prostrate or procumbent . . . . . . . . . . . . . . . . . . . . 14.09 *Sclerochloa* (in part)
                18. Inflorescences usually exceeding the leaves; spikelets with 4–25 florets; lemmas 5-veined, apices acute to obtuse, sometimes bifid; culms procumbent to erect . . . . . . . . . . . . . . . . . . . . . . . . . . . . . . . . . . . . 14.35 *Desmazeria* (in part)
      8. Inflorescences panicles or racemes, with more than 1 spikelet associated with each node.
        19. Inflorescences racemes or spikelike panicles, with all branches shorter than 1 cm [for opposite lead, see p. 84].

20. Leaves usually exceeding the inflorescences; culms usually prostrate to procumbent; lemmas indurate at maturity . . . . . . . . . . . . . . . . . . . . . . . . . . . . . . . . . . . . . . . . 14.09 *Sclerochloa* (in part)
20. Leaves usually exceeded by the inflorescences; culms usually erect or decumbent at the base; lemmas usually membranous or papery, sometimes coriaceous, not indurate.
  21. Spikelets disarticulating below the glumes or, if the spikelets are attached to stipes, at the base of the stipes; glume bases sometimes fused.
    22. Spikelets weakly laterally compressed, with stipes that fall with the spikelets; glume bases not fused; glumes usually awned . . . . . . . . . . . 14.28 *Polypogon* (in part)
    22. Spikelets strongly laterally compressed, without stipes; glume bases sometimes fused; glumes unawned or awned.
      23. Lemmas dorsally awned; spikelets oval in outline; glumes often connate at the base, often winged distally, keels sometimes ciliate, apices never abruptly truncate . . . . . . . . . . . . . . . . . . . . . . . 14.66 *Alopecurus* (in part)
      23. Lemmas usually unawned, occasionally subterminally awned; spikelets often U-shaped in outline, sometimes oval; glumes not connate at the base, not winged, often strongly ciliate on the keels and abruptly truncate to an awnlike apex . . . . . . . . . . . . . . . . . . . . . . 14.31 *Phleum* (in part)
  21. Spikelets disarticulating above the glumes; glume bases not fused.
    24. Spikelets with 2–25 bisexual florets, the sterile or staminate florets, if present, distal to the bisexual florets.
      25. Sheaths closed for at least ¹/₂ their length.
        26. Lemma midveins sometimes excurrent up to 2.2 mm, other veins not excurrent; plants native, arctic . . . . . . . . . . . . . . . . . . . 14.15 *Dupontia* (in part)
        26. Lemmas with 3–5 veins excurrent, forming awnlike teeth; plants cultivated . . . . . . . . . . . . . . . . . . . . . . . . . . . . . . . . . . . . . . . . . . . . . . 14.34 *Sesleria*
      25. Sheaths open for all or almost all of their length.
        27. Distal lemmas, sometimes all lemmas, awned from below midlength . . . . . . . . . . . . . . . . . . . . . . . . . . . . . . . . . . . . . . . . . . . . . 14.22 *Aira* (in part)
        27. All lemmas unawned or apically awned.
          28. Lemmas coriaceous at maturity, unawned, sometimes mucronate . . . . . . . . . . . . . . . . . . . . . . . . . . . . . . . . . . . 14.35 *Desmazeria* (in part)
          28. Lemmas membranous, apically awned, awns 0.3–22 mm long.
            29. Lemma margins involute, not scarious . . . . . . . . . . . . 14.04 *Vulpia* (in part)
            29. Lemma margins flat, scarious . . . . . . . . . . . . . . . . . 14.58 *Rostraria* (in part)
    24. Spikelets with 1 bisexual floret, sometimes with 1–2 sterile florets below the bisexual floret, the sterile florets sometimes reduced to lemmas, sometimes resembling tufts of callus hair.
      30. Spikelets with 1–2 sterile or staminate florets below the bisexual florets, these from larger than to much smaller than the bisexual florets, sometimes resembling tufts of hair; glumes sometimes winged distally.
        31. Fresh leaves not sweet-smelling when crushed; sterile lemmas unawned; bisexual lemmas usually hairy, sometimes sparsely so; glumes subequal, sometimes winged distally . . . . . . . . . . . . . . 14.61 *Phalaris* (in part)
        31. Fresh leaves sweet-smelling when crushed; sterile lemmas awned; bisexual lemmas glabrous; glumes unequal, not winged . . . . . . . . . . . . . . . . . . . . . . . . . . . . . . . . . . . . . . . . . . . . . 14.60 *Anthoxanthum* (in part)
      30. Spikelets without sterile or staminate florets below the bisexual floret; glumes not winged distally.
        32. Lemmas dorsally awned, awns geniculate; lateral lemma veins excurrent, forming 4 teeth, teeth sometimes awnlike . . . . . . 14.45 *Bromidium* (in part)
        32. Lemmas unawned or with only 1 awn, awns not strongly geniculate; lateral lemma veins not excurrent.
          33. Spikelets 8–15 mm long; lemmas more than ³/₄ as long as the glumes; plants strongly rhizomatous . . . . . . . . . . . . . . 14.64 *Ammophila* (in part)
          33. Spikelets 1.2–7 mm long; lemmas less than ³/₄ as long as the glumes; plants rhizomatous or not.
            34. Sheaths closed for at least ¹/₂ their length.

35. Calluses glabrous; exposed at maturity; lemmas 1–3-veined, unawned; [other genera may develop long caryopses when infected by nematodes or fungi; such caryopses are usually deformed and filled with eggs, larvae, or spores] . . . . . . . . . . . 14.08 *Phippsia* (in part)

35. Calluses with a ring of stiff hairs, hairs to about 1 mm long; lemmas 3–11-veined . . . . . . . . . . . . 14.15 *Dupontia* (in part)

34. Sheaths open for most of their length.

36. Spikelet bases usually U-shaped, sometimes cuneate; glumes equal, midveins usually strongly ciliate . . . . . . . . . . . . . . . . . . . . . . . . . . . . . 14.31 *Phleum* (in part)

36. Spikelet bases cuneate; glumes unequal, midveins not strongly ciliate.

37. Both glumes twice as long as the lemmas; lemmas pubescent . . . . . . . . . . . . . . . . . 14.32 *Gastridium* (in part)

37. Glumes from slightly shorter than to slightly longer than the lemmas; lemmas glabrous, sometimes scabridulous or scabrous.

38. Lemma awns 4–16 mm long; plants annual; paleas from $^3/_4$ as long as to slightly longer than the lemmas . . . . . . . . . . . . . . 14.67 *Apera*

38. Lemma awns to 10 mm, if longer than 4 mm, plants perennial and/or paleas less than $^1/_2$ as long as the lemmas . . . . . . . . 14.27 *Agrostis* (in part)

19. Inflorescences panicles, dense to open, sometimes compact, usually at least some branches longer than 1 cm [for opposite lead, see p. 82].

39. Caryopses usually as long as or longer than the lemmas, exposed at maturity; lemmas 1–3-veined, unawned; spikelets with 1 floret; sheaths of the flag leaves closed for at least $^1/_2$ their length; calluses glabrous [other genera may develop long caryopses when infected by nematodes or fungi; such caryopses are usually deformed or filled with eggs, larvae, or spores].

40. Lemmas 1-veined, narrowed to awnlike apices; sheaths strongly inflated; glumes absent; plants of temperate habitats . . . . . . . . . . . . . . . . . . . . . . . . . . 14.23 *Coleanthus*

40. Lemmas 1–3-veined, apices acute to rounded; sheaths not inflated; glumes developed, caducous or persistent; plants of arctic or alpine habitats . . . . . . . 14.08 *Phippsia* (in part)

39. Caryopses shorter than the lemmas, concealed at maturity; lemmas 3–11-veined; spikelets with 1 or more florets; leaf sheaths open or closed; calluses glabrous or with hairs.

41. Panicle branches secund, appearing 1-sided; spikelets strongly imbricate, subsessile.

42. Culms usually prostrate or procumbent; glumes obtuse to emarginate . . 14.09 *Sclerochloa* (in part)

42. Culms erect or ascending; glumes apiculate to awn-tipped.

43. Lemmas awned, awns of the lowest lemmas 0.3–22 mm long . . . . . . . 14.04 *Vulpia* (in part)

43. Lemmas unawned, sometimes awn-tipped.

44. Spikelets circular to ovate or obovate in outline, with 1–2 florets; glumes almost entirely concealing the sides of the florets; disarticulation below the glumes . . . . . . . . . . . . . . . . . . . . . . . . . 14.12 *Beckmannia*

44. Spikelets oval in outline, longer than wide, with 2–6 florets; glumes partially exposing the sides of the florets; disarticulation above the glumes . . . . . . . . . . . . . . . . . . . . . . . . . . . . . . . . . . . . . . 14.10 *Dactylis*

41. Panicle branches not secund; spikelets usually widely spaced to somewhat imbricate, usually clearly pedicellate, sometimes subsessile, sometimes on stipes.

45. All or most spikelets in an inflorescence with 1 bisexual floret, sometimes with 1–2 sterile or staminate florets below the bisexual floret, the sterile florets sometimes resembling tufts of hair . . . . . . . . . . . . . . . . . . . . . . . . . . . *Poeae* Subkey I

45. All or most spikelets in an inflorescence with 2–25 sexual florets, usually all florets bisexual or the distal florets sterile or unisexual, sometimes all florets unisexual, sometimes the plants unisexual . . . . . . . . . . . . . . . . . . . . . . *Poeae* Subkey II

## *POEAE* SUBKEY I

Synoecious or monoecious grasses with panicles having at least some branches longer than 1 cm and spikelets with 1 bisexual floret, sometimes with 1–2 sterile or staminate florets below the bisexual floret.

1. Spikelets with 1–2 staminate or sterile florets below the bisexual floret, sterile florets sometimes knoblike or resembling tufts of hair.
  2. Spikelets with 2 florets of similar size, the lower floret staminate; lower lemmas awned, the lemmas of the terminal floret unawned or awned  . . . . . . . . . . . . . . . . . . . . . . . . . . . . . . . . . . . . . . . . . . . .14.54 *Arrhenatherum*
  2. Spikelets with 2–3(4) florets, the lower 1–2 florets staminate or sterile, sometimes knoblike or resembling tufts of hair, sometimes larger than the bisexual floret; lemmas of the lower florets awned or unawned, the lemmas of the terminal floret unawned.
    3. Lower sterile florets 2, from shorter than to exceeding the bisexual floret; fresh leaves sweet-smelling when crushed  . . . . . . . . . . . . . . . . . . . . . . . . . . . . . . . . . . . . . . . . . . . . . . . .14.60 *Anthoxanthum* (in part)
    3. Lower sterile florets 1–2, varying from knoblike projections on the callus of the bisexual floret to linear or lanceolate lemmas up to ³/₄ as long as the bisexual floret; fresh leaves not sweet-smelling when crushed  . . . . . . . . . . . . . . . . . . . . . . . . . . . . . . . . . . . . . . . . . . . . . . . .14.61 *Phalaris* (in part)
1. Spikelets without staminate or sterile florets below the bisexual florets.
  4. Spikelets 15–50 mm long; lemmas usually dorsally awned, awns 20–90 mm long, sometimes unawned . . . . . . . . . . . . . . . . . . . . . . . . . . . . . . . . . . . . . . . . . . . . . . . . . . . . . . . . . .14.52 *Avena* (in part)
  4. Spikelets 1–15 mm long; lemmas unawned or awned, awns to 18 mm long, basal, dorsal, subterminal, or terminal.
    5. Glume bases gibbous and subcoriaceous; disarticulation above the glumes . . . . . . . . . . . . .14.32 *Gastridium* (in part)
    5. Glumes bases not gibbous, usually membranous; disarticulation above or below the glumes.
      6. Lemmas awned, awns longer than 2 mm.
        7. Glumes coriaceous, rigid, hispid or scabrous; lemmas awned, awns 5–14.5 mm long, subterminal . . . . . . . . . . . . . . . . . . . . . . . . . . . . . . . . . . . . . . . . . . . . . . . . . . . . . . . .14.63 *Limnodea*
        7. Glumes membranous, flexible, glabrous or with soft hairs, usually smooth; lemmas awned, awns 0.5–18 mm long, sometimes subterminal.
          8. Disarticulation below the glumes.
            9. Spikelets borne on stipes; disarticulation at the base of the stipes; lemmas 0.5–2 mm long; glumes usually awned, sometimes unawned  . . . . . . . . . . . . . .14.28 *Polypogon* (in part)
            9. Spikelets borne on pedicels; disarticulation immediately below the glumes; lemmas 1.5–7.5 mm long; glumes usually unawned.
              10. Paleas absent or greatly reduced; lemma awns attached at midlength or below; glume bases often fused; rachillas not prolonged beyond the floret base  . . 14.66 *Alopecurus* (in part)
              10. Paleas from ³/₄ to nearly as long as the lemmas; lemma awns subterminal; glume bases not fused; rachillas usually prolonged beyond the base of the distal floret as a minute stub or slender bristle  . . . . . . . . . . . . . . . . . . . . . . . . . . . . . . . . . . . . .14.62 *Cinna*
          8. Disarticulation above the glumes.
            11. Rachillas not prolonged beyond the base of the distal floret; paleas absent, minute, or subequal to the lemmas; lemmas 0.5–4 mm long.
              12. Lemmas usually glabrous, sometimes pubescent, unawned or awned, if awned, the awns usually shorter than 4.5 mm, sometimes to 10 mm long, basal, dorsal, subterminal, or terminal; veins usually not excurrent, if excurrent, not forming awnlike teeth; panicles often open, sometimes contracted and cylindrical . . . . . . . . . . . . . . . . . . . . . . . . . . . . . . . . .14.27 *Agrostis* (in part)
              12. Lemmas pilose and dorsally awned, awns 4.5–6 mm long; lateral lemma veins excurrent, forming 4 teeth, teeth sometimes awnlike; panicles dense . . . . 14.45 *Bromidium* (in part)
           11. Rachillas prolonged beyond the base of the distal floret; paleas at least ¹/₂ as long as the lemmas; lemmas 1–8 mm long.
              13. Plants annual; calluses glabrous or sparsely hairy; lemma apices entire; marginal veins not excurrent; awns subterminal  . . . . . . . . . . . . . . . . . . . . . . . .14.67 *Apera*
              13. Plants perennial; calluses usually abundantly, sometimes sparsely hairy, hairs 0.2–6.5 mm long; lemma apices denticulate or the marginal veins excurrent; awn attachment from nearly basal to subterminal.
                14. Lemma surfaces mostly glabrous; lemma apices denticulate; marginal lemma veins not excurrent . . . . . . . . . . . . . . . . . . . . . . . . . . . . . .14.49 *Calamagrostis* (in part)
                14. Lemma surfaces hairy; lemma apices erose or toothed; marginal lemma veins excurrent . . . . . . . . . . . . . . . . . . . . . . . . . . . . . . . . . . . . . . . . . . . . .14.43 *Lachnagrostis*

6. Lemmas unawned or, if awned, awns shorter than 2 mm.
 15. Disarticulation below the glumes.
  16. Glumes attached to stipes, disarticulation at the base of the stipes; glumes usually awned, awns flexuous ......................................... 14.28 *Polypogon* (in part)
  16. Glumes attached to pedicels, disarticulation immediately beneath the glumes; glumes unawned or with stiff awns.
   17. Lemma awns subterminal; glume bases not fused; paleas from ³/₄ to nearly as long as the lemmas; rachillas prolonged beyond the base of the distal floret for 0.1–1.3 mm ...................................................... 14.62 *Cinna*
   17. Lemma awns attached at midlength or below; glume bases often fused; paleas absent or greatly reduced; rachillas not prolonged beyond the base of the distal floret ........................................... 14.66 *Alopecurus* (in part)
 15. Disarticulation above the glumes.
  18. Glumes 8–15 mm long; plants strongly rhizomatous .................. 14.64 *Ammophila* (in part)
  18. Glumes 1–10 mm long; plants rhizomatous or not.
   19. Spikelets dorsally compressed; lemmas dark, coriaceous, lustrous, and glabrous ...... 14.65 *Milium*
   19. Spikelets laterally compressed, sometimes weakly so; lemmas not simultaneously dark, coriaceous, lustrous, and glabrous.
    20. Lower glumes exceeded by the florets, upper glumes exceeded by to exceeding the florets; sheaths usually closed for up to ¹/₅ their length ....... 14.33 *Arctagrostis*
    20. Both glumes subequal to or exceeding the florets; sheaths open to the base.
     21. Paleas absent or minute to subequal to the lemmas, not veined; rachillas not prolonged beyond the base of the distal florets; lemmas often unawned, sometimes awned, awn attachment basal to terminal ...................................................... 14.27 *Agrostis* (in part)
     21. Paleas more than ¹/₂ as long as the lemmas, 2-veined; rachillas prolonged beyond the base of the floret by at least 0.1 mm; lemmas often awned, awn attachment usually on the proximal ¹/₂ of the lemmas.
      22. Calluses hairy, hairs 0.5–4.5 mm long; lemmas usually awned, awns usually attached to the proximal ¹/₂, if the attachment higher, the callus hairs longer than 2 mm and/or the awns geniculate ................................... 14.49 *Calamagrostis* (in part)
      22. Calluses glabrous or with hairs to about 1 mm long; lemmas unawned or terminally awned, awns to 1(2.2) mm long.
       23. Glumes 4–9 mm long; sheaths closed for ¹/₂–²/₃ their length; plants of arctic and subarctic regions ............... 14.15 *Dupontia* (in part)
       23. Glumes 1.6–4.3 mm long; sheaths open; plants of western North America, from Alaska to California ................ 14.42 *Podagrostis*

## *POEAE* SUBKEY II

Synoecious, monoecious, or dioecious grasses with spikelets having 2–22 sexual florets, the lower florets sexual, the distal florets sometimes sterile.

1. One or both glumes exceeding the adjacent lemmas, sometimes exceeding the distal floret [for opposite lead, see p. 88].
 2. All lemmas within a spikelet unawned or with awns shorter than 2 mm.
  3. Spikelets usually with 2 florets, lemmas of the lower florets unawned, lemmas of the upper florets awned, the awns strongly curved or hooked ................................... 14.53 *Holcus* (in part)
  3. Spikelets with 2–22 florets, all lemmas unawned or if awned, the awns straight.
   4. Leaf sheaths closed for at least ¹/₂ their length; caryopses falling free of the lemma and palea; plants of arctic or subarctic regions.
    5. Lemma apices obtuse; paleas subequal to the lemmas ..................... 14.17 *Arctophila* (in part)
    5. Lemma apices acute to acuminate; paleas shorter than the lemmas ............... 14.15 *Dupontia* (in part)
   4. Leaf sheaths open for most of their length; caryopses usually falling with the lemma and palea attached; plants of temperate, arctic, or subarctic regions.
    6. Glumes 15–50 mm long; plants annual ................................. 14.52 *Avena* (in part)
    6. Glumes 0.4–9 mm long; plants annual or perennial.

7. Lemmas inflated, about as wide as long, with broadly rounded backs; calluses glabrous; spikelets pendulous . . . . . . . . . . . . . . . . . . . . . . . . . . . . . . . . . . . . . . . . . . . . . . . . . . . . . . . 14.21 *Briza* (in part)
7. Lemmas not inflated, longer than wide, keeled to rounded over the midvein; calluses usually with hairs, sometimes glabrous; spikelets not pendulous.
    8. Plants annual; spikelets with 2 florets; lemmas evenly hairy, 3-veined . . . . . . . . . . . 14.46 *Dissanthelium*
    8. Plants usually perennial, sometimes annual; spikelets with 2–10 florets; lemmas usually glabrous or with unevenly distributed hairs, never both annual and with evenly distributed hairs, 3–9-veined.
        9. Rachilla internodes hairy, hairs at least 1 mm long.
            10. Lemma apices truncate, erose to 2–4-toothed . . . . . . . . . . . . . . . . . . 14.26 *Deschampsia* (in part)
            10. Lemma apices acute, bifid . . . . . . . . . . . . . . . . . . . . . . . . . . . . . . . . . 14.56 *Trisetum* (in part)
        9. Rachilla internodes glabrous or with hairs shorter than 1 mm on the distal portion.
            11. Plants strongly rhizomatous; glumes 5–9 mm long . . . . . . . . . . . . . . . 14.51 *Scolochloa* (in part)
            11. Plants not or weakly rhizomatous; glumes 0.4–9 mm long.
                12. Panicle branches densely pubescent, hairs 0.1–0.2 mm long; lemma apices entire, sometimes mucronate; lemma veins converging distally . . 14.57 *Koeleria* (in part)
                12. Panicles branches glabrous, sometimes scabrous; lemma apices entire or serrate to erose, not mucronate; lemma veins more or less parallel distally . . . . . . . . . . . . . . . . . . . . . . . . . . . . . . . . . . . . . . . . . . . . . 14.06 *Puccinellia* (in part)
2. One or all lemmas within a spikelet awned, the awns at least 2 mm long.
   13. Lemmas 14–40 mm long; glumes 7–11-veined . . . . . . . . . . . . . . . . . . . . . . . . . . . . . . . . . . . . . . . 14.52 *Avena*
   13. Lemmas 1.3–16 mm long; glumes 1–9-veined.
      14. Lemmas 7–16 mm long.
         15. Adaxial surfaces of the leaves ribbed; rachillas pilose on all sides; ligules truncate to rounded, 0.5–1.5 mm long . . . . . . . . . . . . . . . . . . . . . . . . . . . . . . . . 14.47 *Helictotrichon* (in part)
         15. Adaxial surfaces of the leaves unribbed; rachillas glabrous on the side adjacent to the paleas, hairy elsewhere; ligules acute to truncate, 0.5–7 mm long . . . . . . . . . . . . . . 14.44 *Avenula* (in part)
      14. Lemmas 1.3–7 mm long.
         16. Lemmas 1-veined, awned, awns articulated near the middle, the proximal segment yellow-brown to dark brown, the distal segment pale green to whitish, the junction marked by a ring of minute, conical protuberances . . . . . . . . . . . . . . . . . . . . . . . . . . 14.55 *Corynephorus*
         16. Lemmas 3–7-veined, at least some lemmas awned, awns not articulated.
            17. Disarticulation below the glumes.
                18. Spikelets usually with 2 florets, the lower florets bisexual with unawned lemmas, the upper florets staminate or sterile with awned lemmas . . . . . . . 14.53 *Holcus* (in part)
                18. Spikelets with 2–5 florets, all florets bisexual or sometimes the distal florets sterile; all lemmas awned . . . . . . . . . . . . . . . . . . . . . . . . . . . . . . . . . . . . 14.56 *Trisetum* (in part)
            17. Disarticulation above the glumes.
                19. Lowest lemma within a spikelet unawned or with a straight awn up to 4 mm long, the distal lemmas within a spikelet always awned, awns 10–16 mm long, geniculate . . . . . . . . . . . . . . . . . . . . . . . . . . . . . . . . . . . . . . . . . . 14.36 *Ventenata* (in part)
                19. All lemmas within a spikelet similarly awned or the awns of the lower lemmas longer than those of the upper lemmas, or the upper lemmas with awns shorter than 10 mm.
                  20. Callus hairs about $^1\!/_2$ as long as the lemmas; rachillas not prolonged or prolonged about 0.5 mm or less beyond the base of the distal floret; plants loosely cespitose . . . . . . . . . . . . . . . . . . . . . . . . . . . . . . . . . . . . . . . . . 14.41 *Vahlodea*
                  20. Calluses usually glabrous or the hairs much shorter than $^1\!/_2$ the length of the lemmas, if about $^1\!/_2$ as long, the rachillas prolonged more than 0.5 mm beyond the base of the distal floret and the plants usually densely cespitose.
                      21. Plants annual; culms 1–60 cm tall; rachillas not prolonged beyond the base of the distal florets . . . . . . . . . . . . . . . . . . . . . . . . . . . . . . 14.22 *Aira* (in part)
                      21. Plants perennial or annual; culms 5–150 cm tall; rachillas prolonged beyond the base of the distal florets, the prolongations hairy.
                          22. Rachilla internodes glabrous or with hairs shorter than 1 mm on the distal portion; panicle branches densely pubescent, not scabrous . . . . . . . . . . . . . . . . . . . . . . . . . . . . . . . . . . . . . . 14.57 *Koeleria* (in part)

22. Rachilla internodes hairy, hairs at least 1 mm long; panicle branches usually glabrous, sometimes scabrous.
    23. Lemma apices truncate, erose or 2–4-toothed .... 14.26 *Deschampsia* (in part)
    23. Lemma apices acute, bifid ...................... 14.56 *Trisetum* (in part)
1. Both glumes shorter than or subequal to the adjacent lemmas [for opposite lead, see p. 86].
  24. Upper lemma(s) in a spikelet with hooked or geniculate awns, awns 2–16 mm long; lowest lemmas unawned or terminally awned, awns straight, to 4 mm long.
    25. Spikelets 9–15 mm long, with 2–20 florets; awns of the distal florets 10–16 mm long ... 14.36 *Ventenata* (in part)
    25. Spikelets 3–7 mm long, with 2 florets; awns of the distal floret 2–5 mm long ............ 14.53 *Holcus* (in part)
  24. Lemmas all similarly awned or unawned.
    26. Lower lemmas with awns longer than 2 mm.
      27. Calluses hairy; rachillas prolonged beyond the base of the distal florets.
        28. Glumes shorter than the adjacent lemmas; ligules 4.5–20 mm long .............. 14.48 *Amphibromus*
        28. Glumes subequal to the adjacent lemmas; ligules 0.5–6 mm long ............ 14.56 *Trisetum* (in part)
      27. Calluses glabrous or sparsely hairy; rachillas sometimes prolonged beyond the base of the distal florets.
        29. Panicles dense, spikelike; plants annual ............................. 14.59 *Rostraria* (in part)
        29. Panicles not both dense and spikelike; plants perennial or annual.
          30. Anthers 1; plants annual ............................. 14.04 *Vulpia* (in part)
          30. Anthers 3; plants perennial.
            31. Leaves without auricles; blades flat, conduplicate, involute, or convolute
                .................................................... 14.01 *Festuca* (in part)
            31. Lower leaves with auricles; blades flat ...................... 14.03 *Schedonorus* (in part)
    26. Lower lemmas unawned, mucronate, or with awns up to 2 mm long.
      32. Lemmas inflated, about as wide as long; spikelets pendulous ...................... 14.21 *Briza* (in part)
      32. Lemmas not inflated, longer than wide; spikelets appressed to divergent, not pendulous.
        33. Lemmas apices rounded, truncate, obtuse, or emarginate.
          34. Lemmas conspicuously 3-veined; lower glumes 0–3-veined.
            35. Lower glumes without veins; lemmas not keeled over the lateral veins ....... 14.19 *Catabrosa*
            35. Lower glumes 1–3-veined; lemmas keeled over each vein ............ 14.20 *Cutandia* (in part)
          34. Lemmas (3)5–9-veined, the veins often inconspicuous; lower glumes 1–5-veined.
            36. Inflorescences usually exceeded by the leaves; lemmas indurate at maturity; pedicels 0.5–0.8 mm thick; culms usually prostrate to procumbent, sometimes ascending; upper glumes 2.6–6.2 mm long ............ 14.09 *Sclerochloa* (in part)
            36. Inflorescences exceeding the leaves at maturity; lemmas usually membranous at maturity, sometimes coriaceous; pedicels less than 0.5 mm thick; culms usually erect; upper glumes 0.7–4.5(9) mm long.
              37. Lower glumes about as long as the upper glumes but no more than ¹/₂ as wide; disarticulation below the glumes ............... 14.25 *Sphenopholis* (in part)
              37. Lower glumes shorter than the upper glumes or subequal and more than ¹/₂ as wide; disarticulation above the glumes.
                38. Sheaths closed for more than ¹/₂ their length ............. 14.17 *Arctophila* (in part)
                38. Sheaths open their entire length.
                  39. Panicle branches stiff; lemmas coriaceous; plants annual; culms to 60 cm tall ......................... 14.35 *Desmazeria* (in part)
                  39. Panicle branches flexible; lemmas usually membranous, sometimes coriaceous; plants usually perennial, sometimes annual or biennial; culms 2–145 cm tall.
                    40. Lemma veins excurrent, lemma apices indistinctly 3-lobed or toothed; plants strongly rhizomatous, rhizomes succulent ................................. 14.51 *Scolochloa* (in part)
                    40. Lemma veins not excurrent, lemma apices entire, serrate, or erose; plants sometimes rhizomatous, rhizomes not succulent.
                      41. Lemma veins (5)7–9, prominent; plants of non-saline and non-alkaline habitats ........ 14.18 *Torreyochloa* (in part)
                      41. Lemma veins (3)5(7), inconspicuous or prominent; plants of saline and alkaline habitats ........ 14.06 *Puccinellia* (in part)
        33. Lemma apices acute to acuminate, sometimes mucronate or shortly awn-tipped.

42. Lemmas (3)5–9-veined, veins more or less parallel distally, conspicuous.
   43. Lemma veins (5)7–9; plants rhizomatous, growing in non-saline and non-alkaline habitats . . . . . . . . . . . . . . . . . . . . . . . . . . . . . . . . . . . . . 14.18 *Torreyochloa* (in part)
   43. Lemma veins (3)5(7); plants not truly rhizomatous, sometimes the culms rooting at buried lower nodes, growing in saline and alkaline habitats . . 14.06 *Puccinellia* (in part)
42. Lemmas 3–9-veined, veins converging distally, usually inconspicuous, sometimes conspicuous.
   44. Lemmas conspicuously 3-veined, keeled over each vein; panicle branches divaricate; plants annual . . . . . . . . . . . . . . . . . . . . . . . . . . . . . . . . . . 14.20 *Cutandia* (in part)
   44. Lemmas inconspicuously (3)5–9-veined, sometimes keeled over the midvein, not over the other veins; panicle branches divaricate or not; plants annual or perennial.
      45. Disarticulation below the glumes; lower glumes subequal to the upper glumes but no more than $^1/_2$ as wide . . . . . . . . . . . . . . . . . . . . 14.25 *Sphenopholis* (in part)
      45. Disarticulation above the glumes, sometimes above the basal floret; lower glumes shorter than the upper glumes or, if subequal, more than $^1/_2$ as wide.
         46. Panicle branches smooth, hairy, hairs soft . . . . . . . . . . . . . . . 14.57 *Koeleria* (in part)
         46. Panicle branches smooth or scabrous, glabrous or strigose, never covered with soft hairs.
            47. Rachillas pilose, hairs at least 2 mm long (see *Stipeae*, p. 110) . . . . . . . . . . . . . . . . . . . . . . . . . . . . . . . . . . . . . 10.01 *Ampelodesmos*
            47. Rachillas glabrous or with hairs shorter than 1 mm.
               48. Basal leaves with auricles . . . . . . . . . . . . . . . . . . 14.03 *Schedonorus* (in part)
               48. No leaves with auricles.
                  49. Lemma veins parallel distally; plants of saline and alkaline habitats . . . . . . . . . . . . . . . . . . . . . . . 14.06 *Puccinellia* (in part)
                  49. Lemma veins converging distally; plants of many habitats, including saline habitats.
                     50. Leaf blades with translucent lines on either side of the midvein, apices often prow-tipped; lemmas often with a tuft of hair at the base of the midvein; hila round to oval . . . . . . . . . . . . . 14.13 *Poa* (in part)
                     50. Leaf blades without translucent lines on either side of the midvein, apices not prow-tipped, often flat; lemmas without a tuft of hair below the midvein; hila usually linear, always linear in perennial species.
                        51. Plants perennial.
                           52. Plants and florets bisexual; glumes not translucent; caryopses obovoid-oblong . . . . . . . . . . . . . . . . . . . . . 14.01 *Festuca* (in part)
                           52. Plants unisexual; glumes translucent; caryopses fusiform . . . . . . . . . . . . . . . . . 14.02 *Leucopoa*
                        51. Plants annual.
                           53. Ligules up to 1 mm long; lemma apices mucronate or awned . . . . . . . . 14.04 *Vulpia* (in part)
                           53. Ligules 1–4 mm long; lemma apices never awned, sometimes mucronate.
                              54. Panicle branches up to 2 cm long, stiff, spikelet-bearing to the base; culms procumbent to erect . . . . . . . . . . . . . . . . . . . 14.35 *Desmazeria* (in part)
                              54. Panicles branches 2–10 cm long, flexible, spikelets confined to the distal portion; culms erect . . . . . . . . 14.24 *Eremopoa*

# 14.01 FESTUCA L.[1]

**Pl** per; bisx; usu densely to loosely csp, with or without rhz, occ stln. **Clm** 5–150(275) cm, usu glab and smooth throughout, smt scabrous or densely pubescent below the infl. **Shth** from open to the base to closed almost to the top, in some species shth of previous years persisting and the bld usu deciduous, in other species the senescent shth rapidly shredding into fibers and decaying between the veins and the bld not deciduous; **col** inconspicuous, usu glab; **aur** absent; **lig** 0.1–2(8) mm, memb, smt longest at the mrg, usu truncate, smt acute, usu ciliate, smt erose; **bld** flat, conduplicate, involute, or convolute, smt glaucous or pruinose, abx surfaces usu glab or scabrous, smt puberulent or pubescent, rarely pilose, adx surfaces usu scabrous, smt hirsute or puberulent, with or without ribs over the major veins; **abx sclerenchyma tissue** varying from longitudinal strands at the mrg and opposite the midvein to adjacent to some or all of the lat veins, longitudinal strands smt lat confluent with other strands into an interrupted or continuous band, smt reaching to the veins and forming *pillars*; **adx sclerenchyma tissue** smt present in strands opposite the veins at the epdm, the strands smt extending to the veins and, in combination with the abx sclerenchyma, forming *girders* of sclerenchyma tissue extending from one epdm to the other at some or all of the veins. **Infl** usu open or contracted pan, smt rdcd to rcm, usu with 1–2(3) br at the lo nd; **br** usu erect, spreading to widely spreading at anthesis, smt the lo br reflexed; **Spklt** with (1)2–10 mostly bisx flt, distal flt rdcd or abortive; **rchl** usu scabrous or pubescent, smt smooth and glab; **dis** above the glm, beneath the flt. **Glm** subequal or unequal, usu exceeded by the flt, ovate to lanceolate, acute to acuminate; **lo glm** from shorter than to about equal to the adjacent lm, 1(3)-veined; **up glm** 3(5)-veined; **cal** usu wider than long, usu glab and smooth, smt scabrous, occ pubescent; **lm** usu chartaceous, smt coriaceous, bases more or less rounded dorsally, slightly or distinctly keeled distally, veins 5(7), prominent or obscure, apc acute to attenuate, smt minutely bidentate, usu tml or subtml awned or mucronate; **pal** from shorter than to slightly longer than the lm, veins sparsely to densely scabrous-ciliate, intercostal region usu smooth and glab at the base, usu scabrous and/or puberulent distally, bidentate; **anth** 3; **ov** glab or with hispidulous apc, hairs persisting on the mature car. **Car** obovoid-oblong, adxly grooved, usu free of the lm and pal, smt adhering along the groove, smt adhering more broadly; **hila** linear, from ¹/₂ as long as to almost as long as the car.

*Festuca* is a widespread genus, probably having more than 500 species. The species grow in alpine, temperate, and polar regions of all continents except Antarctica. There are 37 species native to the *Manual* region, 2 introduced species that have become established, and 5 introduced species that are known only as ornamentals or waifs. One species, *F. rubra*, is represented by both native and introduced subspecies.

Many native species provide good forage in western North American grasslands and montane forests. Important cultivated species include *Festuca rubra*, grown for forage and as a turf grass, and *F. trachyphylla*, used as a turf grass and for erosion control. A number of species are cultivated as ornamentals; only *F. amethystina* and *F. glauca* are included in this account.

The distribution of some taxa that are grown for turf, revegetation, and, to a lesser extent, horticulture—such as *Festuca rubra* subsp. *rubra*, *F. trachyphylla*, *F. filiformis*, and *F. valesiaca*—is continually expanding because of their wide commercial availability. The occurrence of these in the *Manual* region is no doubt much more extensive than current herbarium collections indicate.

The taxonomy of the genus is problematic and contentious, and this treatment is far from definitive. Keying the species ultimately relies on characters that are sometimes difficult to detect on herbarium specimens, such as ovary pubescence and leaf blade sclerenchyma patterns. Because of the intraspecific variability in many characters, combinations of overlapping characters must be employed for identification.

The distribution of sclerenchyma tissue within the vegetative shoot leaves is often an important diagnostic character in *Festuca*. Taxa in a small region can often be identified reliably without resorting to consideration of these patterns but, for the *Manual* region as a whole, their use is essential. These patterns should be observed in cross sections made from mature, but not senescent, leaves of vegetative shoots, ¹/₄ to halfway up the blades; they can be made freehand, with a single-edged razor blade. Sections are best viewed at 40× or greater magnification, and with transmitted light (polarized if possible). Important features seen in the blade cross section are identified in Fig. 1, p. 91.

There are five main sclerenchyma distribution patterns in *Festuca* (Fig. 2). Almost all species have a strand of sclerenchyma tissue along the margins and opposite the midvein against the abaxial epidermis (Fig. 2C). Strands may be narrow (about as wide as the adjacent veins or narrower, Fig. 2D) to broad (wider than the adjacent veins, Fig. 2E). Additional strands are often present at the abaxial surface opposite the veins; these strands may be confluent (Figs. 1, 2E), sometimes combining to form a cylinder around the leaf and appearing as a continuous ring or band in the cross sections (Figs. 2F, I). Some species have additional strands on the adaxial surface opposite some or all of the veins (Figs. 2G, H). Another variant is for the abaxial sclerenchyma strands to extend inwards to some or all of the vascular bundles (veins), forming *pillars* in the cross sections (Figs. 2H, I). If both the abaxial and adaxial strands extend inward to the vascular bundles, they are said to form *girders* (Figs. 2A, B, I).

Some of the patterns described may co-occur within a leaf. For instance, some veins may be associated with pillars, others with girders; some sclerenchyma strands within a leaf may be confluent, whereas others are not. Although there may be considerable variation in the extent of sclerenchyma development, the general pattern within a species is usually

[1]Stephen J. Darbyshire and Leon E. Pavlick†

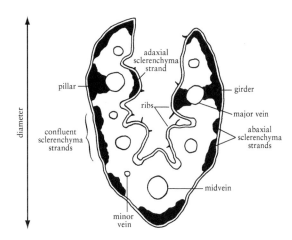

Figure 1. *Festuca* blade cross section.

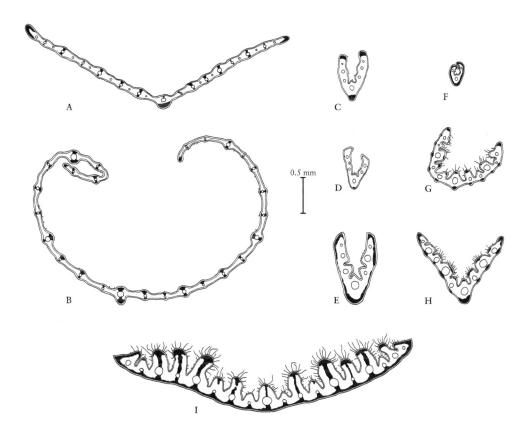

Figure 2. *Festuca* sclerenchyma distribution patterns.   A–Leaf blade flat, ribs indistinct, sclerenchyma girders at most veins (*F. subverticillata*);   B–Leaf blade loosely convolute, ribs indistinct, sclerenchyma girders at most veins (*F. subverticillata*);   C–Leaf blade conduplicate, ribs indistinct, sclerenchyma in broad strands at margins and midvein (*F. lenensis*);   D–Leaf blade conduplicate, ribs indistinct to distinct, sclerenchyma in narrow abaxial strands opposite veins (*F. hyperborea*);   E–Leaf blade conduplicate, ribs distinct, sclerenchyma in broad, mostly confluent strands (*F. trachyphylla*);   F–Leaf blade conduplicate, ribs indistinct, sclerenchyma in a continuous abaxial cylinder or band (*F. filiformis*);   G–Leaf blade loosely rolled, ribs distinct, sclerenchyma in narrow abaxial and adaxial strands (*F. dasyclada*);   H–Leaf blade loosely folded; ribs distinct, sclerenchyma in broad abaxial and adaxial strands forming pillars (*F. viridula*);   I–Leaf blade flat, ribs distinct, sclerenchyma in a continuous abaxial cylinder or band, forming girders at most veins (*F. californica*).

constant. These patterns have not, however, been examined for all species.

Descriptions of leaf blades are based on the leaves of the basal vegetative shoots, where present. For those without basal tufts of vegetative shoots, the cauline leaves are described. Width measurements are provided for leaves that are usually flat, or almost so, when encountered in the field or herbarium. "Diameter" is given for leaves that are usually folded or conduplicate when encountered; for leaves that are oval in cross section when folded, it is the largest diameter (or width).

Closure of the leaf sheaths should be checked on young leaves, because the sheaths often split with age, leading to underestimations of the extent of their closure. The fraction of the leaf sheath that is closed varies within and between species of *Festuca*, but the species can be divided into three categories in this regard: those such as *F. rubra*, in which the leaves are closed for at least $^3/_4$ their length; those such as *F.*

*saximontana*, in which they are closed from $^1/_3$ to slightly more than $^1/_2$ their length; and those such as *F. trachyphylla*, in which they are not closed or closed for less than $^1/_4$ their length. The descriptions indicate to which of these categories each species belongs. Lemma awns tend to be longer, and should be measured, on the distal florets within a spikelet.

Under adverse conditions, many species may *proliferate vegetatively*, where leafy bulbils or shoots form in place of some or all spikelets. Some populations of *Festuca* are largely (or completely) sterile, reproducing almost entirely through such bulbils, a process termed *pseudovivipary*. Pseudo-viviparous plants may be common or even abundant in certain areas and habitats. Since these stabilized forms are largely reproductively isolated, often of unusual ploidy, and largely morphologically distinct, they are treated as separate species. Although the lower bracts in pseudoviviparous spikelets are usually more or less normal in form, they are sometimes elongated or distorted, as are the upper bracts.

1. Blades usually flat, sometimes loosely conduplicate or convolute, 1.8–10 mm wide; sheaths closed to about $^1/_2$ their length, never about $^3/_4$ their length; sclerenchyma girders or pillars associated with at least some of the major veins; ovary apices usually pubescent, rarely glabrous; plants rarely pseudoviviparous [for opposite lead, see p. 93].
   2. Primary and secondary inflorescence branches stiffly and strongly divaricate at maturity, angles densely scabrous or ciliate; spikelets with 2(3) florets . . . . . . . . . . . . . . . . . . . . . . . . . . . . . 44. *F. dasyclada* (in part)
   2. Primary inflorescence branches lax or stiff, erect to ascending or spreading at maturity, sometimes lax, secondary branches not stiffly divaricate, angles smooth or scabrous; spikelets with 2–6(10) florets.
      3. Lemma awns (1.3)1.5–20 mm long, occasionally absent.
         4. Lemma awns usually less than $^1/_3$ the lemma length, rarely absent; anthers (3)4–8.5 mm long; plants densely cespitose . . . . . . . . . . . . . . . . . . . . . . . . . . . . . . . . . . . . . 13. *F. californica* (in part)
         4. Lemma awns usually more than $^1/_3$ the lemma length; anthers 1.5–5.7 mm long; plants usually loosely cespitose, rarely densely cespitose.
            5. Lemma calluses longer than wide, pubescent, at least basally; awns flexuous or kinked . . . . . . . 7. *F. subuliflora*
            5. Lemma calluses wider than long, glabrous, sometimes slightly scabrous; awns usually straight or slightly curved, bent or kinked.
               6. Anthers 1.5–3 mm long; lemmas entire, glabrous, sometimes sparsely scabrous; awns terminal, (2.5)5–15(20) mm long; leaf blades 3–10 mm wide . . . . . . . . . . . . . . . . . . . . . . . . . 4. *F. subulata*
               6. Anthers (3)3.4–5.7 mm long; lemmas bidentate or entire, puberulent or scabrous; awns subterminal or terminal, 1–5(8) mm long, occasionally absent; leaf blades usually less than 6 mm wide.
                  7. Upper glumes 3–4.6 mm long; lemma awns (1.5)2–5(8) mm long; blades 1.8–6 mm wide, blades of the vegetative shoots narrower than the cauline blades, usually flat or convolute or loosely conduplicate; plants of forests, usually below 500 m . . . . . . . . . . . . . . 6. *F. elmeri*
                  7. Upper glumes (4)5.5–7(8) mm long; lemma awns 1–3(3.5) mm long; blades 1.5–3 mm wide, the vegetative shoot and cauline blades similar in width, usually loosely conduplicate, or sometimes flat; plants of subalpine and low alpine habitats . . . . . . . . . . . . . . . . . . . . . . . . . . . . . . . . . . . . . . . . . . . 43. *F. washingtonica* (in part)
      3. Lemmas unawned, mucronate, or with awns shorter than 2 mm.
         8. Ligules 2–9 mm long; lemmas unawned, sometimes mucronate, mucros to 0.2 mm long.
            9. Rhizomes present; blades with (5)7–9 veins; lemmas 4–6.5 mm long, unawned; anthers 1.5–2.6 mm long . . . . . . . . . . . . . . . . . . . . . . . . . . . . . . . . . . . . . . . . . . . . . . 11. *F. ligulata* (in part)
            9. Rhizomes absent; blades with 9–15 veins; lemmas 6–10 mm long, unawned, sometimes with a mucro to 0.2 mm long; anthers 3–4.5 mm long . . . . . . . . . . . . . . . . . . . 12. *F. thurberi* (in part)
         8. Ligules 0.1–1.5(2) mm long; lemmas awned, mucronate, or unawned.
            10. Lemmas 3–5(5.2) mm long, unawned; anthers (0.7)1–2(2.5) mm long.
               11. Inflorescence branches usually reflexed at maturity; spikelets not or only slightly imbricate, elliptic to ovate; upper glumes 3–4(4.7) mm . . . . . . . . . . . . . . . . . . . . . 2. *F. subverticillata*
               11. Inflorescence branches ascending to spreading at maturity; spikelets closely imbricate, elliptic to obovate; upper glumes (3.5)4–5(5.5) mm . . . . . . . . . . . . . . . . . . . . . . 3. *F. paradoxa*

10. Lemmas (4.8)5–12 mm long, unawned or with awns to 2 mm long; anthers 1.6–6 mm long.

  12. Senescent sheaths not persistent, rapidly shredding into fibers; plants loosely cespitose; cauline nodes usually exposed.

    13. Lower glumes 4–7 mm long; lemmas smooth and glabrous, apices sometimes sparsely scabrous, unawned, sometimes mucronate . . . . . . . . . . . . . . . . . . . . . . . . . . 1. *F. versuta*

    13. Lower glumes 1.5–4.5 mm long; lemmas scabrous or puberulent, unawned or awned, awns to 2 mm long . . . . . . . . . . . . . . . . . . . . . . . . . . . . . . . . . . . . . . . . 5. *F. sororia*

  12. Senescent sheaths persistent or only slowly shredding into fibers; plants densely cespitose; cauline nodes usually not exposed.

    14. Panicle branches more or less erect, stiff; abaxial sclerenchyma forming continuous or interrupted bands; lower glumes from shorter than to about equal to the adjacent lemmas.

      15. Spikelets with 2–3(4) florets; glumes about equaling or slightly exceeding the upper florets; lemmas 5.5–8(9) mm long . . . . . . . . . . . . . . . . . . . . . . . 9. *F. hallii* (in part)

      15. Spikelets with (3)4–5(7) florets; glumes exceeded by the upper florets; lemmas (6.2)7–8.5(10) mm long . . . . . . . . . . . . . . . . . . . . . . . . . . . . 10. *F. campestris* (in part)

    14. Panicle branches lax, loosely erect, spreading, recurved, or reflexed; abaxial sclerenchyma in strands about the same width as the adjacent veins, not forming a continuous or interrupted band; lower glumes distinctly shorter than the adjacent lemmas.

      16. Senescent sheaths persistent, not shredding into fibers; blades deciduous; spikelets lustrous; ovary apices usually sparsely pubescent, rarely glabrous; plants densely cespitose . . . . . . . . . . . . . . . . . . . . . . . . . . . . . . . . . . . . . . 8. *F. altaica* (in part)

      16. Senescent sheaths persistent or slowly shredding into fibers; blades not deciduous; spikelets not lustrous; ovary apices usually densely pubescent, sometimes sparsely pubescent; plants loosely to densely cespitose.

        17. Blades of the lower cauline leaves much shorter and stiffer than those of the upper cauline leaves; lemmas smooth or slightly scabrous, unawned or the awns to 1.5(2) mm long . . . . . . . . . . . . . . . . . . . . . . . 42. *F. viridula* (in part)

        17. Blades of the lower cauline leaves similar in length and stiffness to those of the upper cauline leaves; lemmas scabrous or puberulent distally, the awns usually longer than 1 mm, rarely absent . . . 43. *F. washingtonica* (in part)

1. Blades usually conduplicate or folded and less than 2.5 mm in diameter, sometimes convolute or flat, sometimes the leaves of the vegetative shoots conduplicate and the cauline leaves more or less flat, up to 6(7) mm wide when flat; sheath closure varied, from completely open to closed for about ³⁄₄ their length, if the blades 2+ mm wide then the sheaths closed for about ³⁄₄ their length; sclerenchyma girders usually absent, present if the blades 3+ mm wide, pillars sometimes present; ovary apices pubescent or glabrous; plants sometimes pseudoviviparous [for opposite lead, see p. 92].

  18. Collars usually pubescent, at least at the margins, sometimes glabrous; lemmas (7)7.5–11 mm long, scabrous or pubescent; spikelets not pseudoviviparous; ovary apices densely pubescent . . 13. *F. californica* (in part)

  18. Collars glabrous; lemmas 2–10(11) mm long, smooth or scabrous, glabrous or with hairs; spikelets sometimes pseudoviviparous; ovary apices glabrous or pubescent or ovaries not developed.

    19. Ligules 2–9 mm long.

      20. Rhizomes present; blades with (5)7–9 veins; lemmas 4–6.5 mm long, unawned; anthers 1.5–2.6 mm long . . . . . . . . . . . . . . . . . . . . . . . . . . . . . . . . . . . . . 11. *F. ligulata* (in part)

      20. Rhizomes absent; blades with 9–15 veins; lemmas 6–10 mm long, unawned, sometimes with a mucro to 0.2 mm long; anthers 3–4.5 mm long . . . . . . . . . . . . . . . . . . . . . 12. *F. thurberi* (in part)

    19. Ligules to 1.5(2) mm long.

      21. Most or all spikelets pseudoviviparous; anthers and ovaries usually absent or abortive.

        22. Rhizomes present; sheaths closed for about ³⁄₄ their length or more, senescent sheaths rapidly shredding into fibers.

          23. Cauline blades 0.3–1 mm wide, conduplicate or folded; inflorescences sometimes racemose or subracemose, with 1–3 spikelets on the lower branches; plants of boreal and alpine eastern North America . . . . . . . . . . . . . . . . . . . . . . . . . . 15. *F. prolifera*

          23. Cauline blades 1.4–2.5 mm wide, flat; inflorescences paniculate, with 2–5 spikelets on the lower branches; plants known only from the Queen Charlotte Islands . . . . . . . . . . . . . . . . . . . . . . . . . . . . . . . . . . . . . . . . . . . . . . . . . . . . . . . . . 16. *F. pseudovivipara*

        22. Rhizomes absent; sheaths closed for less than ³⁄₄ their length, senescent sheaths sometimes persistent, sometimes shredding into fibers.

24. Abaxial sclerenchyma in broad, sometimes confluent strands that together cover ¹/₂ or more of the abaxial surface; glumes densely pubescent throughout; inflorescences 1.5–10 cm long . . . . . . . . . . . . . . . . . . . . . . . . . . . . . . . . . . . . 35. *F. frederikseniae*

24. Abaxial sclerenchyma in narrow strands that together cover less than ¹/₂ the abaxial surface; glumes glabrous or pubescent; inflorescences 1–4.8 cm long . . . . . 36. *F. viviparoidea*

21. No spikelets pseudovivaparous; anthers and ovaries well developed.

  25. Glumes about equaling or slightly exceeding the upper florets; lemma awns absent or 0.5–1.3 mm long; anthers 4–6 mm long . . . . . . . . . . . . . . . . . . . . . . . . . . . . 9. *F. hallii* (in part)

  25. Glumes distinctly exceeded by the upper florets; lemma awns various; anthers 0.3–6 mm long.

    26. Rhizomes usually present; sheaths of the vegetative shoots closed for about ³/₄ their length, glabrous or pubescent, hairs retrorse or antrorse, senescent sheaths rapidly shredding into fibers.

      27. Anthers 1.8–4.5 mm long; ovary apices glabrous . . . . . . . . . . . . . . . . . . . . . . 14. *F. rubra*

      27. Anthers 0.6–1.4 mm long; ovary apices densely pubescent . . . . . . . . . . . . . 17. *F. earlei* (in part)

    26. Rhizomes absent; sheaths of the vegetative shoots usually closed for less than ²/₃ their length, sometimes closed for about ³/₄ their length, usually glabrous and smooth, sometimes scabrous or puberulent, hairs rarely retrorse, senescent sheaths usually persistent for several years, sometimes slowly shredding into fibers.

      28. Primary and secondary inflorescence branches stiffly and strongly divaricate at maturity; spikelets with 2(3) florets . . . . . . . . . . . . . . . . . 44. *F. dasyclada* (in part)

      28. Primary inflorescence branches stiffly erect or laxly spreading at maturity, secondary branches not divaricate; spikelets with (1)2–10 florets.

        29. Culms densely scabrous or densely pubescent below the inflorescences; ligules 0.5–1.5(2) mm long; anthers (2)3–4(4.2) mm long . . . . . . . . . . . . . . 39. *F. arizonica*

        29. Culms usually smooth and glabrous below the inflorescences, sometimes scabrous or sparsely pubescent, if scabrous or pubescent then the ligules to 0.6 mm long; ligules 0.1–0.8(1) mm long; anthers 0.3–4.5(5) mm long.

          30. Lemmas unawned, mucronate, or with awns to 3.5 mm long, if awned then the leaf blades with adaxial sclerenchyma strands present.

            31. Blades 0.2–1.2(1.5) mm in diameter; lemmas usually unawned, sometimes mucronate, mucros to 0.4 mm long; ovary apices glabrous or sparsely pubescent; adaxial sclerenchyma absent, pillars and girders not formed; plants introduced, usually of disturbed habitats.

              32. Blades with (5)7–9 veins and 5–9 indistinct or distinct ribs; sclerenchyma in (5)7–9 sometimes partly confluent abaxial strands; inflorescences (3)8–18(25) cm long; spikelets (5)6–8.5(10) mm long; lemmas (3.5)4–5.6(6.6) mm long; anthers (2)3–4 mm long . . . . . . . . . . . . . . . . . . 20. *F. amethystina*

              32. Blades with 5(7) veins and 1 distinct rib; sclerenchyma in a continuous or almost continuous abaxial band; inflorescences 1–6(14) cm long; spikelets 3–6(6.5) mm long; lemmas 2.3–4(4.4) mm long; anthers (1)1.5–2.2 mm long . . . . . . . . . . . . . . . . . . . . . . . . . . . . . . . . . . . . . . . . . . . . 23. *F. filiformis*

            31. Blades 0.3–3 mm in diameter; lemmas unawned or with awns to 3.5 mm long; ovary apices pubescent, sometimes sparsely so; usually at least some veins associated with adaxial sclerenchyma and pillars or girders; plants native, of western alpine, subalpine, and montane habitats.

              33. Panicle branches erect to stiffly spreading; abaxial sclerenchyma forming continuous or interrupted bands; lower glumes from shorter than to about equal to the adjacent lemmas . . . . . . . . . . . . . . . . . . . . . . . . . . 10. *F. campestris* (in part)

33. Panicle branches lax, loosely erect, spreading, recurved, or reflexed; abaxial sclerenchyma in strands about the same width as the adjacent veins, not forming a continuous or interrupted band; lower glumes distinctly shorter than the adjacent lemmas.

    34. Senescent sheaths persistent, not shredding into fibers; blades deciduous; spikelets lustrous; ovary apices usually sparsely pubescent, rarely glabrous; plants densely cespitose; plants of alpine or arctic habitats from central British Columbia northward . . . . . . . . . . . . . . . . . . . . . . . . . . . . . . . . . . . . . . 8. *F. altaica* (in part)

    34. Senescent sheaths persistent or slowly shredding into fibers; blades not deciduous; spikelets not lustrous; ovary apices usually densely pubescent, sometimes sparsely pubescent; plants loosely to densely cespitose; plants of alpine and subalpine habitats from southern British Columbia southward.

        35. Blades of the lower cauline leaves shorter and stiffer than those of the upper cauline leaves; lemmas glabrous, smooth or slightly scabrous . . . . . . . . . . . . . . . . . . . . . . . . . . . . . . . 42. *F. viridula* (in part)

        35. Blades of the lower cauline leaves similar in length and stiffness to those of the upper cauline leaves; lemmas scabrous or puberulent distally . . . . . . . . . . . . . . . . . . . . . . 43. *F. washingtonica* (in part)

30. Lemmas usually awned, occasionally unawned, awns 0.3–12 mm long; leaf blades without adaxial sclerenchyma strands.

    36. Anthers (1.8)2–4.5 mm long; plants mostly not of arctic, subarctic, or alpine habitats (except *F. auriculata* and *F. lenensis*) . . . . . . . . . . . . . . . . . . . . . . . . . . . . . . . . . . . . . . . . . . . . . Subkey I

    36. Anthers 0.3–1.8(2) mm long; plants mostly of arctic, subarctic, or alpine habitats (except *F. occidentalis*, *F. ovina*, and *F. saximontana*) . . . . . . . . . . . . . . . . . . . . . . . . . . . . . . . . . . . . . . . Subkey II

## *Festuca* Subkey I

1. Ovary apices pubescent.

    2. Sheaths closed for about ³/₄ their length or more; vegetative shoot leaf blades narrow and conduplicate, the cauline blades broader and flat; anthers 2.5–4.5 mm long . . . . . . . . . . . . . . . . . . . . 18. *F. heterophylla*

    2. Sheaths closed for no more than ¹/₂ their length; vegetative shoot and cauline leaf blades similar, conduplicate; anthers 1–3.5 mm long.

        3. Lower inflorescence branches usually reflexed at maturity; lemma awns 3–12 mm long; ovary apices densely pubescent . . . . . . . . . . . . . . . . . . . . . . . . . . . . . . . . . . . . . . . . . . . . . . . . 37. *F. occidentalis*

        3. Lower inflorescence branches erect at maturity; lemma awns 1–2.5 mm long; ovary apices sparsely pubescent . . . . . . . . . . . . . . . . . . . . . . . . . . . . . . . . . . . . . . . 38. *F. calligera* (in part)

1. Ovary apices glabrous, or with up to 5 hairs in *F. calligera*.

    4. Abaxial sclerenchyma forming continuous or interrupted bands; lemmas (2.6)3–6(6.2) mm long; blades with 1–3(5) indistinct ribs; anthers 1.4–3 mm long; species used as ornamentals or for turf or soil stabilization, rarely spreading from cultivation.

        5. Plants usually not glaucous or pruinose; sheaths glabrous; plants used for turf . . . . . . . . . . . . . . 21. *F. ovina* (in part)

        5. Plants usually glaucous or pruinose; sheaths pubescent or glabrous; plants grown as ornamentals . . . . . . 22. *F. glauca*

    4. Abaxial sclerenchyma usually in 3–7 discrete or somewhat confluent strands, if in a continuous band, the lemmas 3.8–7(8.2) mm long; blades with 1–9 distinct ribs; anthers 0.4–4.5 mm long; most species native, a few introduced for turf.

        6. Abaxial sclerenchyma in 5–7 strands or in interrupted to continuous bands; blades with (1)3–9 well-defined ribs.

            7. Lemmas 3.8–5(6.5) mm long, usually scabrous or pubescent distally, especially on the margins, rarely entirely pubescent; lemma awns usually less than ¹/₂ the length of the lemma bodies . . . . . . . . . . . . . . . . . . . . . . . . . . . . . . . . . . . . . . . . . . . . . . . . . . 24. *F. trachyphylla*

7. Lemmas 5–10 mm long, scabrous distally; lemma awns usually more than ¹/₂ the length of the lemma bodies.

    8. Blades with (1)3–5 ribs; adaxial surfaces of the blades pubescent or scabrous; inflorescence branches usually somewhat spreading at maturity . . . . . . . . . . . . . . . . . . . . . . . . . . . . . . . . 40. *F. idahoensis*

    8. Blades with 5–9 ribs; adaxial surfaces of the blades glabrous or pubescent, sometimes scabrous; inflorescence branches erect to slightly spreading at maturity . . . . . . . . . . . . . . . . . . . . . 41. *F. roemeri*

6. Abaxial sclerenchyma strands usually restricted to the margins and midvein, rarely with additional strands in between; blades with 1–5 well-defined ribs.

    9. Sheaths closed distinctly less than ¹/₂ their length; culms 15–65 cm tall; inflorescences 5–15 cm long; plants native to the southwestern United States . . . . . . . . . . . . . . . . . . . . . . . . . . 38. *F. calligera* (in part)

    9. Sheaths closed for about ¹/₂ their length; culms 8–50(60) cm tall; inflorescences 1.5–10 cm long; plants introduced or native to the extreme northwest of the *Manual* region.

        10. Inflorescences (3)5–10 cm long, panicles; abaxial blade surfaces glabrous or pubescent, not pilose; lower glumes 2–3 mm long; anthers 2.2–2.6 mm long; plants introduced . . . . . . . . . 19. *F. valesiaca*

        10. Inflorescences 1.5–5(5.5) cm long, panicles or racemes; abaxial blade surfaces glabrous, pubescent, or pilose, varying within individual plants; lower glumes 2.5–3.4 mm long; anthers (2)2.4–3.5 mm long; plants native to the extreme northwest of the *Manual* region.

            11. Abaxial sclerenchyma strands distinctly narrower than the veins; spikelets 5–6.5(8) mm long . . . . . . . . . . . . . . . . . . . . . . . . . . . . . . . . . . . . . . . . . . . . . . . . . . . . . . . . . . . . 25. *F. auriculata*

            11. Abaxial sclerenchyma strands about the same width as or wider than the veins; spikelets (5)7–9(11) mm long . . . . . . . . . . . . . . . . . . . . . . . . . . . . . . . . . . . . . . 26. *F. lenensis* (in part)

## *Festuca* Subkey II

1. Ovary apices pubescent, sometimes with only a few hairs.

    2. Ovary apices densely and conspicuously pubescent.

        3. Lemma awns 0.3–1.5 mm long; lemmas 3–4.5 mm long . . . . . . . . . . . . . . . . . . . . . . . . . . . . . 17. *F. earlei* (in part)

        3. Lemma awns 3–12 mm long; lemmas (4)4.5–6.5(8) mm long . . . . . . . . . . . . . . . . . . . . . . . . . . . 37. *F. occidentalis*

    2. Ovary apices sparsely and inconspicuously pubescent.

        4. Culms densely pubescent or shortly pilose below the inflorescence; lemmas 3.5–6 mm long; anthers 0.3–0.7(1.1) mm long . . . . . . . . . . . . . . . . . . . . . . . . . . . . . . . . . . . . . . 32. *F. baffinensis* (in part)

        4. Culms glabrous below the inflorescence; lemmas (2)2.2–3.5(4) mm long; anthers (0.4)0.6–1.2 mm long . . . . . . . . . . . . . . . . . . . . . . . . . . . . . . . . . . . . . . . . . . . . . . . . . . . 34. *F. minutiflora* (in part)

1. Ovary apices glabrous.

    5. Abaxial sclerenchyma in a continuous band; anthers longer than 1.4 mm; plants persisting from historical use for turf and soil stabilization . . . . . . . . . . . . . . . . . . . . . . . . . . . . . . . . . . . . . . . . 21. *F. ovina* (in part)

    5. Abaxial sclerenchyma in 3+ discrete or confluent strands, rarely in a continuous band; anthers various; plants native in the *Manual* region.

        6. Abaxial sclerenchyma strands at least twice as wide as high, varying to more or less confluent, rarely forming a continuous band; plants of various habitats.

            7. Anthers (0.8)1.2–1.7(2) mm long; inflorescences 2–10(13) cm long; lemmas (3)3.4–4(5.6) mm long; spikelets (3)4.5–8.8(10) mm long; plants widespread in continental North America . . . . . . 29. *F. saximontana*

            7. Anthers 0.8–1.3 mm long; inflorescences 1.5–5 cm long; lemmas (2.5)3–3.5(4) mm long; spikelets 4–5.6(6) mm long; plants known only from Greenland . . . . . . . . . . . . . . . . . . . 33. *F. groenlandica*

        6. Abaxial sclerenchyma strands usually less than twice as wide as high, not confluent, never forming a continuous band; plants restricted to arctic or alpine habitats.

            8. Blades usually with 3 abaxial sclerenchyma strands, 2 in the margins and 1 opposite the midvein, occasionally with 2 additional abaxial strands opposite the lateral veins.

                9. Abaxial sclerenchyma strands 3(5), narrower than the veins; spikelets 5–6.5(8) mm long . . . . . . . . . . . . . . . . . . . . . . . . . . . . . . . . . . . . . . . . . . . . . . . . . . . . . . . . . . . . . 25. *F. auriculata* (in part)

                9. Abaxial sclerenchyma strands 3, about the same width as or wider than the veins; spikelets (5)7–9(11) mm long . . . . . . . . . . . . . . . . . . . . . . . . . . . . . . . . . . . . . . . . . . . 26. *F. lenensis* (in part)

            8. Blades with 3–7(9) abaxial sclerenchyma strands.

                10. Blades (0.2)0.3–0.4(0.6) mm in diameter; spikelets 2.5–5 mm long, with (1)2–3(5) florets; lemmas (2)2.2–3.5(4) mm long; lemma awns 0.5–1.5(1.7) mm long; plants of alpine habitats. . . . . . . . . . . . . . . . . . . . . . . . . . . . . . . . . . . . . . . . . . . . . . . . . . . . 34. *F. minutiflora* (in part)

                10. Blades (0.3)0.4–1.2 mm in diameter; spikelets (3)3.5–8.5 mm long, with 2–6 florets; lemmas 2.5–7 mm long; lemma awns (0.2)0.5–3.5 mm long; plants of arctic or alpine habitats.

11. Culms densely pubescent or pilose below the inflorescences; anthers 0.3–0.7(1.1)
    mm long . . . . . . . . . . . . . . . . . . . . . . . . . . . . . . . . . . . . . . . . . . . . . . . . . . 32. *F. baffinensis* (in part)
11. Culms usually glabrous and smooth below the inflorescences, occasionally slightly
    scabrous or sparsely puberulent; anthers (0.3)0.4–1.3 mm long.
    12. Inflorescences usually panicles; lower branches with 2+ spikelets; flag leaf
        sheaths not inflated; flag leaf blades (0.3)1–3 cm long . . . . . . . . . . . . . . . . . . . . . 28. *F. brachyphylla*
    12. Inflorescences often racemes; lower branches with 1–2(3+) spikelets; flag leaf
        sheaths usually slightly to distinctly inflated; flag leaf blades 0.2–5(8) cm long.
        13. Culms erect, more than twice as tall as the basal tuft of leaves; glumes
            ovate-lanceolate to lanceolate; lemmas (3)4–5.5(7) mm long; plants of
            Alaska to the western Northwest Territories . . . . . . . . . . . . . . . . . . . . . . . . . 27. *F. brevissima*
        13. Culms erect to prostrate, to twice as tall as the basal tuft of leaves; glumes
            ovate to ovate-lanceolate; lemmas 2.9–5.2 mm long; plants of Alaska to the
            eastern arctic.
            14. Culms usually erect, sometimes semi-prostrate; flag leaf blades
                0.5–5(8) mm long; blades often curved or somewhat falcate; spikelets
                (3)4–5.5(7) mm long; upper glumes 2.2–3.2 mm long; lemma apices
                usually minutely bidentate; awns usually slightly subterminal . . . . . . . . . 30. *F. hyperborea*
            14. Culms usually geniculate to prostrate, becoming erect at anthesis; flag
                leaf blades (0.3)0.5–2 cm long; most blades straight; spikelets 4.5–8.5
                mm long; upper glumes 2.9–4.3 mm long; lemma apices entire; awns
                usually terminal, sometimes slightly subterminal . . . . . . . . . . . . . . . . . . . .31. *F. edlundiae*

## Festuca subg. **Montanae** E.B. Alexeev

Three species have been placed in this section; one occurs in the *Manual* region, and the other two are Central American.

### 1. **Festuca versuta** Beal  TEXAS FESCUE [p. *387*, <u>517</u>]

*Festuca versuta* grows in moist, shaded sites on rocky slopes in open woods, from Oklahoma and Arkansas to Texas. It is an uncommon species.

## Festuca subg. **Obtusae** E.B. Alexeev

*Festuca* subg. *Obtusae* has been divided into two sections. The two species which occur in eastern North America belong to **Festuca** sect. **Obtusae** E.B. Alexeev. A third species, *F. japonica* Makino, belongs in **Festuca** sect. **Fauria** E.B. Alexeev.

### 2. **Festuca subverticillata** (Pers.) E.B. Alexeev
NODDING FESCUE [p. *387*, <u>517</u>]

*Festuca subverticillata* grows in moist to dry, deciduous or mixed forests with organic rocky soils, from Manitoba to Nova Scotia, south to eastern Texas, Florida, and northeastern Mexico. It resembles *F. paradoxa*, but its spikelets are less crowded on the branches. Plants that are sparsely pilose over the sheaths and blades have been named *F. subverticillata* f. *pilosifolia* (Dore) Darbysh. They frequently grow in mixed populations with *F. subverticillata* (Pers.) E.B. Alexeev f. *subverticillata*.

### 3. **Festuca paradoxa** Desv.  CLUSTER FESCUE [p. *387*, <u>517</u>]

*Festuca paradoxa* grows in prairies, open woods, thickets, and low open ground, from Wisconsin to Pennsylvania, south to northeastern Texas and northern Georgia. It resembles *F. subverticillata*, but its spikelets are more crowded on the branches.

## Festuca subg. **Subulatae** (Tzvelev) E.B. Alexeev

*Festuca* subg. *Subulatae* contains about 30–35 species. It is known from eastern Asia and western North America, as well as Central and South America. Three of the five sections that have been described occur in North America.

## Festuca sect. **Subulatae** Tzvelev

*Festuca* sect. *Subulatae* is the largest section in this subgenus and contains about 20–25 species. Its range includes eastern Asia, western North America, and Central and South America.

### 4. **Festuca subulata** Trin.  BEARDED FESCUE [p. *387*, <u>517</u>]

*Festuca subulata* grows on stream banks and in open woods, meadows, shady forests, and thickets, to about 2800 m. Its range extends from the southern Alaska panhandle eastward to southwestern Alberta and western South Dakota, and southward to central California and Colorado. It differs from *F. subuliflora* in having blunter, glabrous calluses and glabrous, often scabrous or puberulent leaf blades that are obscurely ribbed.

### 5. **Festuca sororia** Piper  RAVINE FESCUE [p. *387*, <u>517</u>]

*Festuca sororia* grows in open woods and on shaded slopes and stream banks, at 2000–3000 m. It is restricted to the United States, growing from central Utah and Colorado to Arizona and New Mexico.

## Festuca sect. **Elmera** E.B. Alexeev

One species has been placed in this section.

### 6. **Festuca elmeri** Scribn. & Merr.  COAST FESCUE, ELMER'S FESCUE [p. *387*, <u>517</u>]

*Festuca elmeri* grows on moist wooded slopes, usually below 300(500) m, from Oregon to south-central California. The more southerly populations, which have larger spikelets with 5–6, rather than 3–4, florets and a more compact inflorescence with more or less erect panicle branches, have been named *F. elmeri* subsp. *luxurians* Piper.

## Festuca sect. **Subuliflorae** (E.B. Alexeev) Darbysh.

Only one species has been placed in this section.

### 7. **Festuca subuliflora** Scribn.  CRINKLE-AWN FESCUE, COAST RANGE FESCUE [p. *387*, <u>517</u>]

*Festuca subuliflora* grows in shady sites in dry to moist forests, usually below 700 m. Its range extends from southwestern British Columbia to central California. It differs from *F. subulata* in having more elongated and distinctly hairy calluses, and often in having softly pubescent foliage and more strongly ribbed blades.

## Festuca L. subg. Festuca

*Festuca* subg. *Festuca* is most abundant in the Northern Hemisphere, but it is distributed on all continents except Antarctica. Estimating the number of species in this subgenus is difficult in the absence of adequate treatments for many parts of the world, but it probably exceeds 400.

## Festuca sect. Breviaristatae Krivot.

*Festuca* sect. *Breviaristatae* is distributed in Asia and North America. It contains about 15 species.

### 8. Festuca altaica Trin. NORTHERN ROUGH FESCUE, ALTAI FESCUE, FÉTUQUE D'ALTAI [p. 387, 517]

*Festuca altaica* is a plant of rocky alpine habitats, arctic tundra, and open boreal or subalpine forests. Its primary distribution extends from Alaska eastward to the western Northwest Territories, and south in the alpine regions of British Columbia and west-central Alberta. Disjunct populations occur in Quebec, western Labrador and Newfoundland, and in Michigan, where it may be introduced. From the Bering Sea it extends westward to the Altai Mountains of central Asia. The spikelets of *F. altaica* are lustrous and usually intensely purplish; plants with greenish spikelets have been named *F. altaica* f. *pallida* Jordal. A form producing pseudoviviparous spikelets, *F. altaica* f. *vivipara* Jordal, has been described from Alaska.

### 9. Festuca hallii (Vasey) Piper PLAINS ROUGH FESCUE [p. 387, 518]

*Festuca hallii* is a major component of grasslands in the northern Great Plains and the grassland-boreal forest transition zone, where it is an important source of forage. Its range extends from the Rocky Mountains of Canada east to western Ontario and south to Colorado. At the southern end of its range in Colorado, it grows in alpine meadows. It differs from *F. campestris* in usually having short rhizomes, stiffly erect panicles, and smaller spikelets. Where the two species are sympatric, as in the foothills of the Rocky Mountains, *F. hallii* is usually found at lower elevations.

### 10. Festuca campestris Rydb. MOUNTAIN ROUGH FESCUE [p. 387, 518]

*Festuca campestris* is a common species in prairies and montane and subalpine grasslands, at elevations to about 2000 m. Its range extends from southern British Columbia, Alberta, and southwestern Saskatchewan south through Washington, Oregon, Idaho, and Montana. It is highly palatable and provides nutritious forage. It differs from *F. hallii* in having larger spikelets, less stiffly erect panicles and, usually, in lacking rhizomes. Where the two are sympatric, *F. campestris* tends to grow at higher elevations.

### 11. Festuca ligulata Swallen GUADALUPE FESCUE [p. 387, 518]

*Festuca ligulata* grows on moist, shady slopes in the mountains of western Texas and north-central Mexico. It is listed as an endangered species under the Endangered Species Act of the United States.

### 12. Festuca thurberi Vasey THURBER'S FESCUE [p. 388, 518]

*Festuca thurberi* is a large bunchgrass of dry, rocky slopes and hills, open forests, and meadows in montane and subalpine regions, at (1000)2000–3500 m. Its range extends from southern Wyoming south through Utah and Colorado to New Mexico.

### 13. Festuca californica Vasey CALIFORNIA FESCUE [p. 388, 518]

*Festuca californica* grows on dry, open slopes and moist streambanks in thickets and open woods, from sea level to 2000 m. Its range extends from Clackamas County, Oregon, to the Sierra Nevada and southern California; it is not known to extend into Mexico. It is the largest species of *Festuca* in the *Manual* region.

1. Culms 30–80(100) cm tall, usually pubescent for more than 5 mm below the nodes; lower sheaths densely retrorsely pubescent; vegetative shoot blades with (3)5–9 ribs, the ribs to about ¹/₂ as deep as the blade thickness; abaxial sclerenchyma in small strands or forming continuous bands; adaxial sclerenchyma strands present or absent; sclerenchyma pillars rarely formed; girders not developed; spikelets with 3–4(5) florets . . . . . . . . subsp. *parishii*
1. Culms 60–150(200) cm tall, glabrous or pubescent for less than 5 mm below the nodes; lower sheaths glabrous or pubescent, if pubescent then usually not densely retrorsely hairy; vegetative shoot blades with 7–15(17) ribs, the ribs usually more than ¹/₂ as deep as the blade thickness; abaxial sclerenchyma forming a continuous band; adaxial sclerenchyma in strands; sclerenchyma pillars or girders usually associated with most of the veins; spikelets with (3)4–6(8) florets.
  2. Ligules 0.2–1(1.2) mm long, ciliate; spikelets (8)13–18(20) mm long . . . . . . . . . . . . . . . subsp. *californica*
  2. Ligules (1)1.5–5 mm long, ciliate or not; spikelets 8–12(17) mm long . . . . . . . . . subsp. *hitchcockiana*

#### Festuca californica Vasey subsp. californica [p. 388]

*Festuca californica* subsp. *californica* is the most widespread variety, growing from west-central Oregon to central California. The lower leaf sheaths are typically glabrous and scabrous, but sometimes have spreading hairs. This subspecies differs from subsp. *parishii* in having wider and longer leaf blades and more extensively developed sclerenchyma.

#### Festuca californica subsp hitchcockiana (E.B. Alexeev) Darbysh. [p. 388]

*Festuca californica* subsp. *hitchcockiana* is distinguished by its relatively long ligules. It is known only from Santa Clara and San Luis Obispo counties, California.

#### Festuca californica subsp. parishii (Piper) Darbysh. [p. 388]

*Festuca californica* subsp. *parishii* grows in southern California, in the San Bernardino, San Gabriel, and Palomar mountains. Its leaf blades tend to be narrower and shorter than in subsp. *californica* (10–30 cm long versus more than 30 cm long), and the sclerenchyma is less developed, with pillars only sometimes present and girders absent. The lower leaf sheaths are densely retrorsely pubescent.

## Festuca L. sect. Festuca

*Festuca* sect. *Festuca* is most abundant in the Northern Hemisphere. Its species are native to all continents except Antarctica. There are perhaps 400 or more species in this section, with new ones constantly described.

### 14. Festuca rubra L. RED FESCUE, FÉTUQUE ROUGE [p. 388, 518]

*Festuca rubra* is interpreted here as a morphologically diverse polyploid complex that is widely distributed in the arctic and temperate zones of Europe, Asia, and North America. Its treatment is complicated by the fact that Eurasian material has been introduced in other parts of the world. In addition, hundreds of forage and turf cultivars have been developed, many of which have also been widely distributed.

Within the complex, morphologically, ecologically, geographically, and/or cytogenetically distinct taxa have been described, named, and given various taxonomic ranks. In some cases these taxa represent extremes, and in other cases they are morphologically intermediate between other taxa. Moreover, hybridization and/or introgression between native taxa, and between native and non-native taxa, may be occurring.

Overlap in morphological characters between most taxa in the complex has led some taxonomists to ignore the variation within the complex, calling all its members *Festuca rubra* without qualification. This obscures what is known about the complex, and presents an extremely heterogeneous assemblage of plants as a single "species"— or a mega-species. The following account attempts to reflect the genetic diversity of the *F. rubra* complex in the *Manual* region. All the taxa are recognized as subspecies, but they are not necessarily equivalent in terms of their distinction and genetic isolation. Much more work on the taxonomy of the *F. rubra* complex is needed before the boundaries of individual taxa can be firmly established.

*Festuca earlei* is sometimes confused with *F. rubra*. It differs in having pubescent ovary apices.

1. Plants not rhizomatous, densely cespitose.
    2. Anthers 2.3–3.2 mm long; lemma awns 0.1–3 mm long; plants of natural habitats in coastal areas . . . . . . . . . . . . . . . . . subsp. *pruinosa*
    2. Anthers 1.8–2.2(3) mm long; lemma awns 1–3.3 mm long; plants of lawns, road verges, and other disturbed
       areas . . . . . . . . . . . . . . . . . . . . . . . . . . . . . . . . . . . . . . . . . . . . . . . . . . . . . . . . . . . . . . . . . . . . . . . . . . . . . subsp. *commutata*
1. Plants rhizomatous, usually loosely to densely cespitose, sometimes with solitary culms.
    3. Vegetative shoot blades usually flat or loosely conduplicate; plants strongly rhizomatous; adaxial sclerenchyma
       strands always present . . . . . . . . . . . . . . . . . . . . . . . . . . . . . . . . . . . . . . . . . . . . . . . . . . . . . . . . . . . . . . . . . . subsp. *fallax*
    3. Vegetative shoot blades usually conduplicate, sometimes flat; plants strongly or weakly rhizomatous; adaxial
       sclerenchyma strands sometimes present.
        4. Plants not or only loosely cespitose, the culms usually single and widely spaced; plants of moist meadows in
           montane and subalpine regions of the western cordillera, usually above 1000 m . . . . . . . . . . . . . . . . . . . . . . subsp. *vallicola*
        4. Plants loosely to densely cespitose, with several culms arising from the same tuft; plants of various habitats
           and elevations.
            5. Inflorescence branches scabrous or pubescent; lemmas usually moderately to densely pilose, sometimes
               only partially pilose, occasionally glabrous; lemma awns (0.2)0.5–1.6 (2.5) mm
               long; plants of subalpine, alpine, boreal, and arctic regions, both littoral and inland . . . . . . . . . . . . . . . . . . . . . subsp. *arctica*
            5. Inflorescence branches scabrous; lemmas usually glabrous, the lemmas of littoral plants sometimes hairy;
               lemma awns (0.1)0.4–5 mm long; plants of various habitats.
                6. Plants widely distributed, sometimes coastal.
                    7. Lower glumes 3–4.5 mm long; inflorescences 7–12 cm long, lanceolate; plants of disturbed
                       habitats throughout temperate and mesic regions . . . . . . . . . . . . . . . . . . . . . . . . . . . . . . . . . . subsp. *rubra*
                    7. Lower glumes 2.2–3.2(4.5) mm long; inflorescences 3–10 (20) cm long, linear to lanceolate; plants
                       of natural habitats in coastal areas . . . . . . . . . . . . . . . . . . . . . . . . . . . . . . . . . . . . . . . . . . . subsp. *pruinosa*
                6. Plants of the Pacific coast, often growing close to the littoral zone.
                    8. Cauline leaf sheaths tightly enclosing the culms; mature inflorescences usually completely exserted
                       from the sheaths.
                        9. Lemmas 4.5–6.5 mm long; sheaths glabrous or pubescent; plants of coastal rocks, cliffs, and
                           sands . . . . . . . . . . . . . . . . . . . . . . . . . . . . . . . . . . . . . . . . . . . . . . . . . . . . . . . . . . . subsp. *pruinosa*
                        9. Lemmas 6–9.5 mm long; sheaths pubescent; plants of maritime sands and gravels . . . . . . . . . . . . . . . subsp. *arenaria*
                    8. Cauline leaf sheaths loosely or tightly enclosing the culms; mature inflorescences usually partly
                       included in the uppermost sheaths.
                        10. Lemmas 4.5–6 mm long, acuminate in side view . . . . . . . . . . . . . . . . . . . . . . . . . . . . . . . . . . subsp. *mediana*
                        10. Lemmas 5.8–9 mm long, attenuate in side view.
                            11. Inflorescences 10–25 cm long; cauline leaf blades 2–4 mm wide, usually flat or loosely
                                conduplicate, not glaucous; lemmas 6–9 mm long, usually glabrous . . . . . . . . . . . . . . . . . . . . subsp. *aucta*
                            11. Inflorescences 7.5–12 cm long; cauline leaf blades to 2.5 mm wide when flat, usually
                                loosely to tightly conduplicate, sometimes glaucous; lemmas 5.8–6.6 mm long, glabrous
                                or hairy . . . . . . . . . . . . . . . . . . . . . . . . . . . . . . . . . . . . . . . . . . . . . . . . . . . . . . . . . . subsp. *secunda*

### Festuca rubra subsp. arctica (Hack.) Govor. ARCTIC
RED FESCUE, FÉTUQUE DE RICHARDSON [p. *388*]

*Festuca rubra* subsp. *arctica* grows in sands, gravels, silts, and stony soils of river banks, bars, and flats; in periglacial outwashes, beaches, sand dunes, muskegs, solifluction slopes, and scree slopes in tundra, subarctic forest, and barren regions; and subalpine areas in the mountains. It extends from Alaska, the southern part of the Canadian arctic archipelago, and Greenland to northwestern British Columbia, the coast of Hudson Bay and James Bay, and Quebec and Labrador, extending farthest south in the Rocky Mountains of Alberta. It also grows in arctic and subarctic Europe and Asia, and in the Ural Mountains.

### Festuca rubra subsp. arenaria (Osbeck) F. Aresch.
FÉTUQUE ROUGE DES SABLES [p. *388*]

*Festuca rubra* subsp. *arenaria* is a European taxon that grows in maritime sands and gravels. It is known in the *Manual* region only from one specimen collected on Vancouver Island; it is not known to have persisted. In the *Manual* region, the name has long been

misapplied to *F. richardsonii* Hook. [= *F. rubra* subsp. *arctica*], which also has hairy lemmas.

### Festuca rubra subsp. aucta (V.I. Krecz. & Bobrov)
Hultén ALEUT FESCUE [p. *388*]

*Festuca rubra* subsp. *aucta* is a coastal taxon, growing above the high tide line in the sand of stabilized sand dunes, beaches, etc., or in silt deposits. Its range extends along the Pacific coast from the Kamchatka Peninsula through the Aleutian Islands, Queen Charlotte Islands, and Vancouver Island and the adjacent continental coastline. *Festuca pseudovivipara* has been described as a form of *F. rubra* subsp. *aucta*, but differs from that taxon in having pseudoviviparous spikelets. It is also ecologically, altitudinally, and probably reproductively isolated from *F. rubra* subsp. *aucta*.

### Festuca rubra subsp. commutata Gaudin CHEWING'S
FESCUE [p. *388*]

*Festuca rubra* subsp. *commutata* is extensively used for lawns and road verges. It is native to Europe, growing from southern Sweden southward, but is widely introduced elsewhere in the world. In the

*Manual* region, it is common south of Alaska, Yukon Territory, and the Northwest Territories.

### Festuca rubra subsp. fallax (Thuill.) Nyman FLATLEAF

RED FESCUE, FÉTUQUE TROMPEUSE [p. *388*]

*Festuca rubra* subsp. *fallax* is a robust taxon that grows in damp, often disturbed places. It is native to northern and central Europe, but has been introduced widely in the *Manual* region, occurring from British Columbia to eastern Quebec and south to California.

### Festuca rubra subsp. mediana (Pavlick) Pavlick DUNE

RED FESCUE [p. *388*]

*Festuca rubra* subsp. *mediana* grows in sand beaches and dunes along exposed coasts, from Vancouver Island to Oregon.

### Festuca rubra subsp. pruinosa (Hack.) Piper ROCK

FESCUE, FÉTUQUE PRUINEUSE [p. *388*]

*Festuca rubra* subsp. *pruinosa* grows in the crevices of rocks, in pilings, and occasionally on pebble or sand beaches, extending upward from the upper littoral zone of the Pacific and Atlantic coasts of North America and Europe. Plants growing on coastal sands from California to Vancouver Island that are loosely cespitose and have abaxial sclerenchyma in large strands are sometimes distinguished as *F. rubra* subsp. *arenicola* E.B. Alexeev [= *F. ammobia* Pavlick]. The rhizomes are rarely present on herbarium specimens.

### Festuca rubra L. subsp. rubra RED FESCUE, FÉTUQUE

ROUGE TRAÇANTE [p. *388*]

*Festuca rubra* subsp. *rubra* grows in disturbed soil. It is often planted as a soil binder, or as turf or forage grass, in mesic temperate parts of the *Manual* region. Originally from Eurasia, it has been widely introduced elsewhere in the world, including most of the *Manual* region, from southern Alaska east to Newfoundland and Greenland and south to California and Georgia. It also grows in Mexico. Because *F. rubra* subsp. *rubra* has often been misunderstood, confounded, and lumped with other taxa of the *F. rubra* complex, statements about its distribution, including that given here, should be treated with caution. It is to be expected throughout the *Manual* region, in all but the coldest and driest habitats.

### Festuca rubra subsp. secunda (J. Presl) Pavlick SECUND

RED FESCUE [p. *388*]

*Festuca rubra* subsp. *secunda* grows on pebble beaches and in soil pockets on rocks, meadows, cliffs, banks, and stabilized sand dunes along seashores with high annual rainfall, on the Pacific coast of North America from Alaska south to Oregon.

### Festuca rubra subsp. vallicola (Rydb.) Pavlick

MOUNTAIN RED FESCUE [p. *388*]

*Festuca rubra* subsp. *vallicola* grows in moist meadows, lake margins, and disturbed soil, at 1000–2000 m, in montane and subalpine habitats from the Yukon Territory/British Columbia border area south to Wyoming.

### 15. Festuca prolifera (Piper) Fernald PROLIFEROUS

FESCUE [p. *388*, <u>*518*</u>]

*Festuca prolifera* is often abundant, and may be a dominant species in some habitats. The leafy bulbils or plantlets sometimes root when the top-heavy inflorescence is bent to the ground.

*Festuca prolifera* has two varieties: **Festuca prolifera** (Piper) Fernald var. **prolifera**, with glabrous lemmas; and **Festuca prolifera** var. **lasiolepis** Fernald, with pubescent lemmas. *Festuca prolifera* var. *prolifera* grows in arctic, alpine, or boreal rocky areas, in calcareous, basic or neutral soils, and is found in the James Bay area, Ungava Bay, western Newfoundland, Cape Breton, the Gaspé Peninsula, the White Mountains (New Hampshire), and Katahdin (Maine). *Festuca prolifera* var. *lasiolepis* is found in moist, sandy riverbanks, lake shores, rocky areas, and cliffs, often on limestone, from the

southeastern Northwest Territories to northern Quebec, Anticosti Island, and western Newfoundland.

### 16. Festuca pseudovivipara (Pavlick) Pavlick

PSEUDOVIVIPAROUS FESCUE [p. *388*, <u>*518*</u>]

*Festuca pseudovivipara* grows on coastal mountainsides, scree slopes, and other rocky areas, at 300–800 m. It is known only from the Queen Charlotte Islands, British Columbia. *Festuca pseudovivipara* has been described as a form of *F. rubra* subsp. *aucta*, but differs from that taxon in having pseudoviviparous spikelets. It is also ecologically, altitudinally, and probably reproductively isolated from *F. rubra* subsp. *aucta*.

### 17. Festuca earlei Rydb. EARLE'S FESCUE [p. *388*, <u>*518*</u>]

*Festuca earlei* grows in rich subalpine and alpine meadows, at 2800–3800 m, in Utah, Colorado, Arizona, and New Mexico. It often grows with the non-rhizomatous species *F. brachyphylla* subsp. *coloradensis* and *F. minutiflora*. It can be distinguished from the former by its pubescent ovary apices, and from the latter by its larger spikelets and lemmas. Because of its short rhizomes (which are often missing from herbarium specimens), *F. earlei* is sometimes confused with members of the *F. rubra* complex. It differs from them in having pubescent ovary apices and shorter anthers.

### 18. Festuca heterophylla Lam. VARIOUS-LEAVED FESCUE,

FÉTUQUE HÉTÉROPHYLLE [p. *389*]

*Festuca heterophylla* is native to open forests and forest edges in Europe and western Asia. In the *Manual* region, it used to be planted as a turf grass for shady areas, and sometimes persists in old lawns.

### 19. Festuca valesiaca Schleich. ex Gaudin VALAIS

FESCUE, FÉTUQUE DU VALAIS [p. *389*, <u>*518*</u>]

*Festuca valesiaca* is widely distributed through central Europe and northern Asia, where it grows in steppes, dry meadows, and open rocky or sandy areas. It is sold in the North American seed trade as *F. pseudovina* Hack. *ex* Wiesb., and has been collected at a few scattered localities in the *Manual* region, apparently having become established from deliberate seeding.

### 20. Festuca amethystina L. TUFTED FESCUE, FÉTUQUE À

COULEUR D'AMÉTHYSTE [p. *389*]

*Festuca amethystina* is sometimes cultivated as an ornamental species; it may occasionally escape.

### 21. Festuca ovina L. SHEEP FESCUE, FÉTUQUE DES OVINS

[p. *389*]

*Festuca ovina* was introduced from Europe as a turf grass. It is not presently used in the North American seed trade. The sporadic occurrences are mostly from old lawns and cemeteries, or sites seeded for soil stabilization. *Festuca ovina* used to be interpreted very broadly in North America, including almost any fine-leaved fescue that lacked rhizomes. Consequently, much of the information reported for *F. ovina*, and many of the specimens identified as such, belong to other species. The only confirmed recent reports are from Ontario; Piatt County, Illinois; and Okanogan County, Washington. Species in this treatment that have frequently been included in *F. ovina* are *F. arizonica*, *F. auriculata*, *F. baffinensis*, *F. brachyphylla*, *F. brevissima*, *F. calligera*, *F. edlundiae*, *F. frederikseniae*, *F. hyperborea*, *F. idahoensis*, *F. lenensis*, *F. minutiflora*, *F. saximontana*, *F. trachyphylla*, and *F. viviparoidea*.

### 22. Festuca glauca Vill. BLUE FESCUE, GRAY FESCUE,

FÉTUQUE GLAUQUE [p. *389*]

*Festuca glauca* is widely grown as an ornamental in the *Manual* region because of its attractive dense tufts of glaucous foliage. It is not known to have escaped cultivation. Several other Eurasian species of fescue with white or bluish foliage are also sold in the horticultural trade as "*Festuca glauca*". Providing a key to the species involved is beyond the scope of this treatment.

### 23. Festuca filiformis Pourr. HAIR FESCUE, FINE-LEAVED SHEEP FESCUE, FÉTUQUE CHEVELUE [p. 389, 518]

*Festuca filiformis* is a European species that has been introduced to the *Manual* region as a turf grass. It grows well on poor, dry soils and is becoming a ruderal weed in some areas. It is particularly common in the northeastern United States and southeastern Canada, but has been reported from scattered locations elsewhere.

### 24. Festuca trachyphylla (Hack.) Krajina HARD FESCUE, SHEEP FESCUE, FÉTUQUE DRESSÉE À FEUILLES SCABRES [p. 389, 518]

*Festuca trachyphylla* is native to open forests and forest edge habitats of Europe. It has been introduced and has become naturalized in many temperate regions. In the *Manual* region, *F. trachyphylla* is generally sold under the name 'Hard Fescue', and is popular as a durable turf grass and soil stabilizer. It is particularly common in the eastern United States and southeastern Canada, but is probably grown throughout the temperate parts of the region. Its naturalized distribution can be expected to expand.

### 25. Festuca auriculata Drobow LOBED FESCUE [p. 389, 518]

*Festuca auriculata* is an amphiberingian species that extends from the Ural Mountains of Russia through Alaska to the western continental Northwest Territories. It grows on dry, rocky cliffs and slopes, in low arctic and alpine regions. In the *Manual* region, this species seems to intergrade with, and is sometimes included in, *F. lenensis*. The two species tend to differ in their leaf surfaces as well as in the width of their sclerenchyma strands.

### 26. Festuca lenensis Drobow LENA FESCUE [p. 389, 518]

*Festuca lenensis* is an amphiberingean species of dry, eroding, rocky slopes in alpine and low arctic habitats. Its range extends from Siberia, Russia, and Mongolia to Alaska and the Yukon Territory. In North America, this species seems to intergrade with, and is sometimes treated as including, *F. auriculata*. The two species usually differ in their leaf surfaces as well as in the width of their sclerenchyma strands.

### 27. Festuca brevissima Jurtsev SHORT FESCUE [p. 389, 518]

*Festuca brevissima* is an amphiberingian diploid species that grows in rocky tundra habitats from the Russian Far East to Alaska and the western part of the Northwest Territories.

### 28. Festuca brachyphylla Schult. & Schult. f. ALPINE FESCUE, FÉTUQUE À FEUILLES COURTES [p. 389, 518]

*Festuca brachyphylla* is a variable, circumpolar, arctic, alpine, and boreal species of open, rocky places. It is palatable to livestock, and is important in some areas as forage for wildlife. The spikelets are usually tinged red to purple by anthocyanin pigments; plants which lack anthocyanins in the spikelets have been named *F. brachyphylla* f. *flavida* Polunin. *Festuca brachyphylla* may hybridize with *F. baffinensis* and/or other species to form *F. viviparoidea*. Three subspecies have been recognized in North America.

1. Culms usually more than twice as long as the vegetative shoot leaves; spikelets 4.4–7(8.5) mm long; lemmas (3)3.5–4.5(6) mm long; plants boreal, arctic, and alpine in the northern cordillera . . . . . . . subsp. *brachyphylla*
1. Culms up to twice as long as the vegetative shoot leaves; spikelets 3.5–5.5 mm long; lemmas (2.5)3.5–4 mm long; plants alpine in the southern cordillera.
    2. Culms usually twice as long as the vegetative shoot leaves; awns 2–3(3.2) mm long; spikelets 4.4–5.6(7) mm long; lemmas 3–4(4.5) mm long . . . . . . . . . . . . . . . . . . . . . . . subsp. *coloradensis*
    2. Culms usually less than twice as long as the vegetative shoot leaves; awns 1–2(2.2) mm long; spikelets 3.5–5(5.5) mm long; lemmas 2.5–4 mm long . . . . . . . . . . . . . . . . . . . . . . . . . . . subsp. *breviculmis*

### Festuca brachyphylla Schult. & Schult. f. subsp. brachyphylla [p. 389]

*Festuca brachyphylla* subsp. *brachyphylla* is circumpolar in its distribution. In the *Manual* region, it extends from Alaska to Newfoundland, south in the mountains to Washington in the west and in the high peaks of the Appalachian Mountains of eastern Quebec and New England in the east.

### Festuca brachyphylla subsp. breviculmis Fred. [p. 389]

*Festuca brachyphylla* subsp. *breviculmis* is endemic to California, where it grows in alpine habitats in the Sierra Nevada and White Mountains.

### Festuca brachyphylla subsp. coloradensis Fred. [p. 389]

*Festuca brachyphylla* subsp. *coloradensis* is a common species in alpine areas of Colorado, Utah, Wyoming, Arizona, and New Mexico. It often grows with *F. earlei*, from which it can be distinguished by its lack of rhizomes, and smaller spikelets and lemmas.

### 29. Festuca saximontana Rydb. ROCKY MOUNTAIN FESCUE, MOUNTAIN FESCUE, FÉTUQUE DES MONTAGNES ROCHEUSES, FÉTUQUE DES ROCHEUSES [p. 390, 518]

*Festuca saximontana* grows in grasslands, meadows, open forests, and sand dune complexes of the northern plains and boreal, montane, and subalpine regions in the *Manual* region, extending from Alaska to Greenland, south to southern California, northern Arizona, and New Mexico in the west and to the Great Lakes region in the east. It is also reported from the Russian Far East. *Festuca saximontana* provides good forage for livestock and wildlife. It is closely related to *F. brachyphylla*, and is sometimes included in that species as *F. brachyphylla* subsp. *saximontana* (Rydb.) Hultén.

Three weakly differentiated taxa have been recognized at the varietal level in North America.

1. Culms 25–50(60) cm tall, usually 3–5 times the height of the vegetative shoot leaves; abaxial surfaces of the blades usually scabrous; abaxial sclerenchyma in 3–5 strands, sometimes partly confluent or forming a continuous band; plants of lowland, montane, or boreal habitats . . . . . . . . var. *saximontana*
1. Culms (5)8–37 cm tall, usually 2–3 times the height of the vegetative shoot leaves; abaxial surfaces of the blades smooth or scabrous; abaxial sclerenchyma in 5–7 narrow strands; plants of subalpine or lower alpine habitats.
    2. Culms (5)8–20(25) cm tall, usually glabrous below the inflorescence; outer vegetative shoot sheaths mostly stramineous; blades with hairs shorter than 0.06 mm on the ribs; lemmas usually scabrous towards the apices and often along the margins . . . . . . . . . . . . . . . . . . . var. *purpusiana*
    2. Culms 16–37 cm tall, usually sparsely scabrous or pubescent below the inflorescence; outer vegetative shoot sheaths brownish on the lower 1/2; blades with hairs to 0.1 mm on the ribs; lemmas often scabrous on the distal 1/2 . . . . . var. *robertsiana*

### Festuca saximontana var. purpusiana (St.-Yves) Fred. & Pavlick [p. 390]

*Festuca saximontana* var. *purpusiana* grows in subalpine or lower alpine habitats. The distribution of this taxon is poorly known; it probably extends from Alaska south to northern California. It is also reported from the Chukchi Peninsula in eastern Russia.

### Festuca saximontana var. robertsiana Pavlick [p. 390]

*Festuca saximontana* var. *robertsiana* grows in subalpine or lower alpine habitats. It has only been reported from British Columbia.

**Festuca saximontana** Rydb. var. **saximontana** [p. *390*]

*Festuca saximontana* var. *saximontana* grows throughout the range of the species.

30. **Festuca hyperborea** Holmen *ex* Fred. Northern Fescue [p. *390*, 518]

*Festuca hyperborea* is a high arctic species that grows from Banks Island in the Canadian Arctic east to Greenland and south to Quebec. It differs from *F. brachyphylla* in its semi-prostrate habit, the loose sheaths and short blades of its flag leaves, the more pronounced ribs in its lower leaf blades, and its subterminal awn. It differs from *F. edlundiae* in having flag leaf blades shorter than 5 mm and smaller spikelets.

31. **Festuca edlundiae** S. Aiken, Consaul & Lefk. Edlund's Fescue [p. *390*, 518]

*Festuca edlundiae* is a high arctic species that is closely related to *F. brachyphylla*. It grows primarily on fine-grained and calcareous substrates in arctic regions of the Russian Far East, Alaska, the arctic islands of Canada, northern Greenland, and Svalbard. It differs from *F. hyperborea* in having flag leaf blades that are usually at least 5 mm long and larger spikelets.

32. **Festuca baffinensis** Polunin Baffin Island Fescue [p. *390*, 518]

*Festuca baffinensis* grows chiefly in damp, exposed, gravelly areas in calcareous and volcanic regions. It is circumpolar in distribution, growing in arctic and alpine habitats and extending southward in the Rocky Mountains to Colorado. It may hybridize with *F. brachyphylla* and/or other species to form *F. viviparoidea*.

33. **Festuca groenlandica** (Schol.) Fred. Greenland Fescue [p. *390*, 518]

*Festuca groenlandica* is endemic to Greenland. It differs from *F. brachyphylla* in having more extensive blade sclerenchyma, usually 7 broad abaxial strands rather than 5 narrow strands.

34. **Festuca minutiflora** Rydb. Little Fescue, Small-Flowered Fescue [p. *390*, 518]

*Festuca minutiflora* grows in alpine regions of the western mountains, from southeastern Alaska and the southwestern Yukon Territory to Arizona, New Mexico, and the Sierra Nevada of California. It has often been overlooked or included with *F. brachyphylla*, from which it differs in its laxer and narrower leaves, looser panicles, smaller spikelets, more pointed lemmas, shorter awns, and scattered hairs on the ovary. In the southern Rocky Mountains, it may grow with *F. earlei*, which has short rhizomes and larger spikelets and lemmas.

35. **Festuca frederikseniae** E.B. Alexeev Frederiksen's Fescue, Fétuque de Frederiksen [p. *390*, 518]

*Festuca frederikseniae* grows on cliffs, rocky or sandy barrens, and alpine regions in southern Quebec (Mingan and Anticosti islands), Newfoundland, southern Labrador, and southern Greenland. In Iceland and southern Greenland, putative hybrids between *Festuca frederikseniae* or *F. vivipara* and *F. rubra* have been reported, and named *F. villosa-vivipara* (Rosenv.) E.B. Alexeev. These plants are highly variable but, unlike *F. frederikseniae*, produce extravaginal shoots, have closed sheaths, and have blades about 1 mm wide, with 7-9 small strands of abaxial sclerenchyma. Such hybrids can be expected within the range of *F. frederikseniae* in North America.

36. **Festuca viviparoidea** Krajina *ex* Pavlick Viviparous Fescue [p. *390*, 518]

*Festuca viviparoidea* is circumboreal in distribution. It may consist of hybrids between *Festuca baffinensis* and *F. brachyphylla* and/or other species.

1. Plants loosely cespitose; culms usually glabrous and smooth throughout, rarely sparsely puberulent near the inflorescence; sheaths brownish, slowly shredding into fibers; abaxial sclerenchyma strands less than 2 times as wide as high; glumes and lemmas puberulent throughout or only near the apices .................................... subsp. *krajinae*
1. Plants densely cespitose; culms densely to sparsely puberulent below the inflorescence; sheaths stramineous, persistent; abaxial sclerenchyma strands 2-3 times wider than high; glumes and lemmas smooth or scabrous near the apices . . subsp. *viviparoidea*

**Festuca viviparoidea** subsp. **krajinae** Pavlick [p. *390*]

*Festuca viviparoidea* subsp. *krajinae* grows in alpine sites of the western cordillera, from southern Alaska and the Yukon Territory through British Columbia to southwestern Alberta.

**Festuca viviparoidea** Krajina *ex* Pavlick subsp. **viviparoidea** [p. *390*]

*Festuca viviparoidea* subsp. *viviparoidea* is circumpolar and found in the high arctic, including Alaska, Yukon Territory, Nunavut, Greenland, Svalbard, and Russia.

37. **Festuca occidentalis** Hook. Western Fescue [p. *390*, 518]

*Festuca occidentalis* grows in dry to moist, open woodlands, forest openings, and rocky slopes, up to 3100 m. It extends from southern Alaska and northern British Columbia to southwestern Alberta, south to southern California and eastward to Wyoming, and, as a disjunct, around the upper Great Lakes in Ontario, eastern Wisconsin, and Michigan. It is sometimes important as a forage grass, but is usually not sufficiently abundant.

38. **Festuca calligera** (Piper) Rydb. Callused Fescue [p. *390*, 519]

*Festuca calligera* is a poorly known, often overlooked species. It grows in grasslands and open montane forests, at 2500-3400 m, from southern Utah to south-central Wyoming and central Colorado, south to Arizona and New Mexico. It is often found with *F. arizonica*.

39. **Festuca arizonica** Vasey Arizona Fescue, Pinegrass [p. *390*, 519]

*Festuca arizonica* grows in dry meadows and openings of montane forests, in gravelly, rocky soil, at 2100-3400 m. Its range extends from southern Nevada and southern Utah east to Colorado and south to Arizona, western Texas, and northern Mexico. It is abundant and valuable forage in some parts of its range. It is often found with *F. calligera*. It differs from *F. idahoensis* in its prominently ribbed blades and pubescent ovary apices.

40. **Festuca idahoensis** Elmer Idaho Fescue, Blue Bunchgrass, Bluebunch Fescue [p. *391*, 519]

*Festuca idahoensis* grows in grasslands, open forests, and sagebrush meadow communities, mostly east of the Cascade Mountains, from southern British Columbia eastward to southwestern Saskatchewan and southward to central California and New Mexico. It extends up to 3000 m in the southern part of its range. It is often a dominant plant, and provides good forage. The young foliage is particularly palatable. *Festuca idahoensis* differs from *F. arizonica* in its less prominently ribbed blades and glabrous ovary apices.

41. **Festuca roemeri** (Pavlick) E.B. Alexeev Oregon Fescue, Roemer's Fescue [p. *391*, 519]

*Festuca roemeri* grows primarily west of the Cascade Mountains, from southeastern Vancouver Island southward to the coast of central California. Two varieties are now recognized (Wilson 2007)[1]. They tend to be more distinct north of Lake County, California.

1. Hairs on adaxial surfaces of leaves 8-14, 0.06-0.2 mm long ............................. var. *klamathensis*
1. Hairs on adaxial surfaces of leaves 2-9, 0.01-0.1 mm long .................................... var. *roemeri*

**Festuca roemeri var. klamathensis** B.L.Wilson [p. *391*]

*Festuca roemeri* var. *klamathensis* grows east of the coastal mountains from southern Douglas County, Oregon, to northwest California, in mesic to dry pine or oak savanna, grasslands, and edges of grassy balds, on a variety of substrates including serpentine.

**Festuca roemeri** (Pavlick) E.B. Alexeev var. **roemeri** [p. *391*]

*Festuca roemeri* var. *roemeri* grows throughout the range of the species in grasslands and open forests.

**42. Festuca viridula** Vasey  MOUNTAIN BUNCHGRASS, GREENLEAF FESCUE, GREEN FESCUE [p. *391*, *519*]

*Festuca viridula* grows in low alpine and subalpine meadows, forest openings, and open forests, at (900)1500–3000 m, from southern British Columbia east to Montana and south to central California and

Nevada. It is highly palatable to livestock, and is an important forage species in some areas.

**43. Festuca washingtonica** E.B. Alexeev  WASHINGTON FESCUE, HOWELL'S FESCUE [p. *391*, *519*]

*Festuca washingtonica* grows in subalpine to low alpine regions of British Columbia and Washington.

**44. Festuca dasyclada** Hack. *ex* Beal  OPEN FESCUE, INTERMOUNTAIN FESCUE [p. *391*, *519*]

*Festuca dasyclada* grows on rocky slopes in open forests and shrublands of western Colorado and central and southern Utah. For many years it was known only from the type collection. When the seeds are mature, the panicles break off the culms and are blown over the ground like a tumbleweed, shedding seeds as they travel.

## 14.02 LEUCOPOA Griseb.[1]

Pl per; unisx. Clm 30–120 cm. Shth closed only at the base; aur absent; lig memb; bld with sclerenchyma girders extending from the abx to adx surfaces. Infl open or contracted pan, usu erect to strongly ascending, not spikelike; br glab, smooth or somewhat scabrous, at least some br longer than 1 cm; ped smt longer than 3 mm, thinner than 1 mm. Spklt pedlt, somewhat dimorphic in unisx pl, lat compressed, with (2)3–5(6) flt; dis above the glm and beneath the flt. Glm subequal to unequal, shorter than the adjacent lm, more or less equally wide, glab, smt scabrous, mostly hyaline and thinner than the lm, memb adjacent to the midvein, unawned; lo glm 1-veined; up glm 1–3-veined; cal glab; lm memb to chartaceous, smooth or scabrous, smt

hirsute, 5-veined, veins converging distally, usu extending almost to the apc, apc entire, acute, usu unawned, smt awned, awns to 2 mm; pal about equaling the lm, scabrous on the veins, scarious or memb distally, veins terminating at the apex; lod 2, memb; anth 3; ov with glab or pubescent apc. Car shorter than the lm, concealed at maturity, fusiform, usu adhering at least to the pal; hila linear.

*Leucopoa* is a genus of about 10 species, most of which are Asian. One species is native to the *Manual* region. It is sometimes included in *Festuca*, but species of *Leucopoa* differ from those of *Festuca* in their dioecious habit, the hyaline glumes that are much thinner than the lemmas, and their differing ovary and caryopsis morphology.

**1. Leucopoa kingii** (S. Watson) W.A. Weber  SPIKE FESCUE, SPIKEGRASS, WATSON'S FESCUE-GRASS, KING'S FESCUE [p. *391*, *519*]

*Leucopoa kingii* grows from Oregon and Montana to Nebraska, south to southern California and northern New Mexico. It occurs in habitats from dry sagebrush plains to subalpine meadows, at

1700–3600 m. Although palatable to livestock in the early part of the season, *L. kingii* is only occasionally abundant enough to be an important forage species.

## 14.03 SCHEDONORUS P. Beauv.[2]

Pl per; csp, smt rhz. Clm to 2 m, slender to stout, erect to decumbent. Shth open, rounded, smooth or scabrous; aur present, usu falcate and clasping, smt an undulating flange; lig memb, glab; bld flat, linear. Infl tml pan, erect, not spikelike; br glab, smooth or scabrous, most br longer than 1 cm; ped smt longer than 3 mm, thinner than 1 mm. Spklt pedlt, lat compressed, with 2–22 flt; dis above the glm and between the flt. Glm 2, shorter than the adjacent lm, more or less equally wide, lanceolate to oblong, rounded on the back, memb, 3–9-veined, apc acute, unawned; cal glab or sparsely hairy; lm lanceolate, ovate or oblong, rounded on the back, memb, chartaceous, 3–7-veined, apc acute, smt hyaline,

unawned or awned, awns to 18 mm, tml or subtml, straight; pal narrower than the lm, memb, usu smooth, keels ciliolate, veins terminating at or beyond midlength; lod 2, lanceolate to ovate; anth 3; ov glab. Car shorter than the lm, concealed at maturity, dorsally compressed, oblong, broadly elliptic, or ovate, longitudinally sulcate, adherent to the pal; hila linear; emb $^1/_5$–$^1/_3$ as long as the car.

Three species of the Eurasian genus *Schedonorus* are established in North America, having been widely introduced as forage and ornamental grasses. *Schedonorus* has traditionally been included in *Festuca*, despite all the evidence pointing to its close relationship to *Lolium*.

1. Lemma awns 10–18 mm long, longer than the lemmas . . . . . . . . . . . . . . . . . . . . . . . . . . . . . . . . . . . . . . . . . . 2. *S. giganteus*
1. Lemmas unawned or the awns shorter than 4 mm, shorter than the lemmas.

[1,2]Stephen J. Darbyshire

2. Auricles glabrous; panicle branches at the lowest node 1 or 2, if paired the shorter with 1–2(3) spikelets, the longer with 2–6(9) spikelets; lemmas usually smooth, sometimes slightly scabrous distally, unawned or with a mucro to 0.2 mm long . . . . . . . . . . . . . . . . . . . . . . . . . . . . . . . . . . . . . . . . . . 1. *S. pratensis*
2. Auricles ciliate, having at least 1 or 2 hairs along the margins (check several leaves); panicle branches at the lowest node usually paired, the shorter with 1–13 spikelets, the longer with 3–19 spikelets; lemmas usually scabrous or hispidulous, at least distally, rarely smooth, unawned or with an awn up to 4 mm long . . . . . . . . . . . . . . . . . . . . . . . . . . . . . . . . . . . . . . . . . . . . . . . . . . 3. *S. arundinaceus*

### 1. **Schedonorus pratensis** (Huds.) P. Beauv. MEADOW FESCUE, FÉTUQUE DES PRÉS [p. *392*, <u>519</u>]

*Schedonorus pratensis* is a Eurasian species that is now widely established in the *Manual* region. It used to be a popular forage grass in the contiguous United States and southern Canada, but is now rarely planted.

### 2. **Schedonorus giganteus** (L.) Holub GIANT FESCUE [p. *392*, <u>519</u>]

*Schedonorus giganteus* is adventive from Europe. It is cultivated as an ornamental, and has escaped to woodland openings and edges and to shaded ravines, at isolated localities in Quebec, Ontario, Michigan, New York, and Connecticut.

### 3. **Schedonorus arundinaceus** (Schreb.) Dumort. TALL FESCUE, FÉTUQUE ÉLEVÉE [p. *392*, <u>519</u>]

*Schedonorus arundinaceus* is a Eurasian species that is grown in all but the coldest and most arid regions of the *Manual* region for forage, soil stabilization, and coarse turf, and often escapes. It is frequently infected with the endophytic fungi *Neotyphodium coenophialum*, which confers insect and drought resistance to the plant, as well as producing ergot alkaloids that are toxic to livestock.

## 14.04 VULPIA C.C. Gmel.[1]

Pl usu ann, rarely per. **Clm** 5–90 cm, erect or ascending from a decumbent base, usu glab. **Shth** open, usu glab; **aur** absent; **lig** usu shorter than 1 mm, memb, usu truncate, ciliate; **bld** flat or rolled, glab or pubescent. **Infl** pan or rcm, smt spikelike, usu with more than 1 spklt associated with each nd; **br** 1–3 per nd, appressed or spreading, usu glab, scabrous. **Spklt** pedlt, lat compressed, with 1–11(17) flt, distal flt rdcd; **dis** above the glm and beneath the flt, occ also at the base of the ped. **Glm** shorter than the adjacent lm, subulate to lanceolate, apc acute to acuminate, unawned or awn-tipped; **lo glm** much shorter than the up glm, 1-veined; **up glm** 3-veined; **rchl** terminating in a rdcd flt; **cal** blunt, glab; **lm** memb, lanceolate, 3–5-veined, veins converging distally, mrg involute over the edges of the car,

apc entire, acute to acuminate, mucronate or awned; **pal** usu slightly shorter than to equaling the lm, smt longer; **anth** usu 1, rarely 3 in chasmogamous specimens. **Car** shorter than the lm, concealed at maturity, elongate, dorsally compressed, curved in cross section, falling with the lm and pal.

*Vulpia*, a genus of 30 species, is most abundant in Europe and the Mediterranean region. The *Manual* region has three native and three introduced species. Most species, including ours, are weedy, cleistogamous annuals, usually having one anther per floret. *Festuca*, in which *Vulpia* is sometimes included, consists of chasmogamous species having three anthers per floret.

In the key, the spikelet and lemma measurements exclude the awns.

1. Lower glumes less than ¹/₂ the length of the upper glumes.
    2. Lemmas 5-veined, glabrous except the margins sometimes ciliate; rachilla internodes 0.75–1.9 mm long . . . . . . . . . . . . . . . . . . . . . . . . . . . . . . . . . . . . . . . . . . . . . . . . . . . . . . . . . . 1. *V. myuros*
    2. Lemmas 3(5)-veined, pubescent or glabrous, the margins ciliate; rachilla internodes 0.4–0.9 mm long . . . . . . . . . . . . . . . . . . . . . . . . . . . . . . . . . . . . . . . . . . . . . . . . . . . . . . . . . . 6. *V. ciliata*
1. Lower glumes ¹/₂ or more the length of the upper glumes.
    3. Lemmas 2.5–3.5 mm long, the apices more pubescent than the bases; caryopses 1.5–2.5 mm long . . . . . . . . . 2. *V. sciurea*
    3. Lemmas 2.7–9.5 mm long, if pubescent, the apices no more so than the bases but occasionally ciliate; caryopses 1.7–6.5 mm long.
        4. Panicle branches 1–2 per node; spikelets with 4–17 florets; rachilla internodes 0.5–0.7 mm long; awn of the lowermost lemma in each spikelet 0.3–9 mm long; caryopses 1.7–3.7 mm long . . . . . . . . . . 3. *V. octoflora*
        4. Panicle branches solitary; spikelets with 1–8 florets; rachilla internodes 0.6–1.2 mm long; awn of the lowermost lemma in each spikelet 2–20 mm long; caryopses 3.5–6.5 mm long.
            5. Panicle branches appressed to erect at maturity, without axillary pulvini; paleas equal to or shorter than the lemmas . . . . . . . . . . . . . . . . . . . . . . . . . . . . . . . . . . . . . . . . . . . . . 4. *V. bromoides*
            5. Panicle branches spreading to reflexed at maturity, with axillary pulvini; paleas usually slightly longer than the lemmas . . . . . . . . . . . . . . . . . . . . . . . . . . . . . . . . . . . . . . . . . . . 5. *V. microstachys*

### 1. **Vulpia myuros** (L.) C.C. Gmel. FOXTAIL FESCUE, RATTAIL FESCUE [p. *392*, <u>519</u>]

*Vulpia myuros* grows in well-drained, sandy soils and disturbed sites. It is native to Europe and North Africa. **Vulpia myuros** f. **megalura**

(Nutt.) Stace & R. Cotton differs from **Vulpia myuros** (L.) C.C. Gmel. f. **myuros** in having ciliate lemma margins. It was once thought to be native to North America, but it grows throughout the European and North African range of f. *myuros*, even in undisturbed areas.

[1]Robert I. Lonard

## 2. Vulpia sciurea (Nutt.) Henrard SQUIRRELTAIL FESCUE [p. 392, 519]

*Vulpia sciurea*, our most distinctive native species, is restricted to the *Manual* region. It can be recognized by its small spikelets and apically pubescent lemmas, and grows mostly in deep, sandy soils of open woodlands, old fields, roadside ditches, and sand hills in the southeastern *Manual* region. It is listed as endangered in New Jersey.

## 3. Vulpia octoflora (Walter) Rydb. SIXWEEKS FESCUE [p. 392, 519]

*Vulpia octoflora*, a widespread native species, tends to be displaced by the introduced *Bromus tectorum* in the Pacific Northwest. It grows in grasslands, sagebrush, and open woodlands, as well as in disturbed habitats and areas of secondary succession, such as old fields, roadsides, and ditches. Three varieties are recognized here, but their characterization is not completely satisfactory, e.g., plants of the southwestern United States with spikelets in the size range of var. *glauca* often have densely pubescent lemmas, the distinguishing characteristic of var. *hirtella*.

1. Spikelets usually 4–6.5 mm long; awn of the lowermost lemma in each spikelet 0.3–3 mm long . . . var. *glauca*
1. Spikelets usually 5.5–13 mm long; awn of the lowermost lemma in each spikelet 2.5–9 mm long.
    2. Lemmas scabrous to pubescent . . . . . . . . . . . . . var. *hirtella*
    2. Lemmas usually smooth, sometimes scabridulous distally and on the margins . . . . . . . . var. *octoflora*

## Vulpia octoflora var. glauca (Nutt.) Fernald [p. 392]

*Vulpia octoflora* var. *glauca* is most frequent in southern Canada and the northern half of the United States, and is the most common representative of *V. octoflora* from North Dakota to western Kansas, and east to Maine and Virginia.

## Vulpia octoflora var. hirtella (Piper) Henrard [p. 392]

*Vulpia octoflora* var. *hirtella* is most frequent from British Columbia south through the western United States and into Mexico. It is the most common variey of *V. octoflora* in the southwest.

## Vulpia octoflora (Walter) Rydb. var. octoflora [p. 392]

*Vulpia octoflora* var. *octoflora* is widespread throughout southern Canada, the United States, and Mexico, and has been introduced into temperate regions of South America, Europe, and Asia. It is most common from northern Oklahoma to Virginia, south to the Texas Gulf prairie and Florida.

## 4. Vulpia bromoides (L.) Gray BROME FESCUE [p. 392, 519]

*Vulpia bromoides* is a common European species that grows in wet to dry, open habitats. It is adventive and naturalized in North and South America. In North America, it is most common on the west coast, where it grows from British Columbia to northern Baja California; it occurs sparingly in other regions.

## 5. Vulpia microstachys (Nutt.) Munro SMALL FESCUE [p. 392, 519]

*Vulpia microstachys* is native to western North America, growing from British Columbia south through the western United States into Baja California. Four varieties are recognized here on the basis of spikelet indumentum, but they frequently occur together, and intergrading forms are known. No difference in their geographic or ecological distribution is known.

1. Glumes and lemmas smooth or scabrous . . . . . . . var. *pauciflora*
1. Glumes and/or lemmas pubescent.
    2. Glumes and lemmas pubescent . . . . . . . . . . . . . . var. *ciliata*
    2. Glumes or lemmas, but not both, pubescent.
        3. Glumes pubescent; lemmas glabrous . . . . . . . var. *confusa*
        3. Glumes glabrous; lemmas pubescent . . . var. *microstachys*

## Vulpia microstachys var. ciliata (A. Gray) Lonard & Gould EASTWOOD FESCUE [p. 392]

*Vulpia microstachys* var. *ciliata* grows in loose, sandy soils.

## Vulpia microstachys var. confusa (Piper) Lonard & Gould CONFUSING FESCUE [p. 392]

*Vulpia microstachys* var. *confusa* grows in sandy, open sites.

## Vulpia microstachys (Nutt.) Munro var. microstachys DESERT FESCUE [p. 392]

*Vulpia microstachys* var. *microstachys* grows most commonly in loose soil on open slopes and roadsides.

## Vulpia microstachys var. pauciflora (Scribn. *ex* Beal) Lonard & Gould PACIFIC FESCUE [p. 392]

*Vulpia microstachys* var. *pauciflora* grows in sandy, often disturbed sites, and is the most common and widespread variety of the complex. It is often intermingled with plants of the other varieties.

## 6. Vulpia ciliata Dumort. FRINGED FESCUE [p. 392, 519]

*Vulpia ciliata* is native to Europe, the Mediterranean area, and southwest and central Asia. It grows in open, dry habitats. It is easily distinguished from other members of the genus because its upper glumes have broadly membranous tips that break off, making the glumes appear truncate or blunt. In the *Manual* region, it was known until recently only from an old ballast dump record from Philadelphia. In May 2004, it was collected immediately north of the Odgen Bay Waterfowl Management Area, Weber County, Utah, in an upland area of the site.

# 14.05 LOLIUM L.[1]

Pl ann or per; csp, smt shortly rhz. Clm 10–150 cm, slender to stout, erect to decumbent, rarely prostrate. Shth open, rounded, glab, smt scabrous; lig to 4 mm, memb, glab; aur smt present; bld flat, linear. Infl distichous spikes, with solitary spklt oriented radial to the rchs, perpendicular to the rchs concavities. Spklt lat compressed, with 2–22 flt, distal flt rdcd; rchl glab; dis above the glm, beneath the flt. Glm usu 1, 2 in the tml spklt, lanceolate to oblong, rounded over the midvein, memb to indurate, 3–9-veined, unawned; lo glm absent from all but the tml spklt; up glm from shorter than to exceeding the distal flt; cal short, blunt, glab; lm lanceolate, ovate or oblong, rounded over the midvein, memb, chartaceous, 3–7-veined, apc smt hyaline, unawned or awned, awns subtml, more or less straight; pal memb, usu smooth, keels ciliolate; lod 2, free, lanceolate to ovate; anth 3; ov glab. Car dorsally compressed, oblong, broadly elliptic or ovate, longitudinally sulcate; hila linear, in the furrow; emb $1/5$–$1/3$ as long as the car.

[1]Edward E. Terrell

As interpreted here, *Lolium* comprises five species that are native to Europe, temperate Asia, and northern Africa. All have been introduced to the *Manual* region, often as forage grasses; most have become established. *Lolium* used to be included in the *Triticeae*, but evidence from genetics, morphology, and other studies shows its closest relationship to be to the species included here in *Schedonorus*.

1. Plants either long-lived perennials with 2–10 florets per spikelet, or annuals or short-lived perennials with 10–22 florets per spikelet.
    2. Plants long-lived perennials, with 2–10 florets per spikelet; lemmas unawned or awned, awns to about 8 mm long . . . . . . . . . . . . . . . . . . . . . . . . . . . . . . . . . . . . . . . . . . . . . . . . . . . . . . . . . . . . . . . 1. *L. perenne*
    2. Plants annuals or short-lived perennials, with 10–22 florets per spikelet; lemmas usually awned, awns to 15 mm long, rarely unawned . . . . . . . . . . . . . . . . . . . . . . . . . . . . . . . . . . . . . . . . . 2. *L. multiflorum*
1. Plants annuals, with 2–10(11) florets per spikelet.
    3. Spikelets somewhat sunken in the rachises and partly concealed by the glumes . . . . . . . . . . . . . . . . . . . . . . . 3. *L. rigidum*
    3. Spikelets not sunken in the rachises and not concealed by the glumes.
        4. Lemmas 3.5–8.5 mm long; paleas from 1.2 mm shorter than to 0.8 mm longer than the lemmas; mature florets and caryopses 2–3 times longer than wide . . . . . . . . . . . . . . . . . . . . . . . . . . . . . 4. *L. temulentum*
        4. Lemmas (5.2)7–12 mm long; paleas usually 0.5–1.8 mm longer than the lemmas; mature florets and caryopses 3.7–5 times longer than wide . . . . . . . . . . . . . . . . . . . . . . . . . . . . . . . . . . . . . . 5. *L. persicum*

### 1. Lolium perenne L. PERENNIAL RYEGRASS, ENGLISH RYEGRASS, IVRAIE VIVACE, RAY-GRASS ANGLAIS [p. *393*, 519]

*Lolium perenne*, a Eurasian species, is now established in disturbed areas throughout much of the *Manual* region. It is commercially important, being included in lawn seed mixtures as well as being used for forage and erosion prevention. It intergrades and is interfertile with *L. multiflorum*; it also intergrades with *L. rigidum*. Typical *L. perenne* differs from *L. multiflorum* in being a shorter, longer-lived perennial with narrower leaves that are folded, rather than rolled, in the bud. Hybrids between the two species are called **Lolium ×hybridum** Hausskn.

### 2. Lolium multiflorum Lam. ANNUAL RYEGRASS, ITALIAN RYEGRASS, IVRAIE MULTIFLORE, RAY-GRASS D'ITALIE [p. *393*, 519]

*Lolium multiflorum*, a European species, now grows in most of the *Manual* region. It is planted as a cover crop, as a temporary lawn grass, for roadside restoration, and for soil or forage enrichment; it often escapes from cultivation, becoming established in disturbed sites. *Lolium multiflorum* and *L. perenne* are interfertile and intergrade. *Lolium multiflorum* differs from *L. perenne* in being a taller, shorter-lived perennial or annual with wider leaves that are rolled, rather than folded, in the bud. Hybrids between the two species are called **Lolium ×hybridum** Hausskn. *Lolium multiflorum* also hybridizes with *L. rigidum*; those hybrids are called **Lolium ×hubbardii** Jansen & Wacht. *ex* B.K. Simon.

### 3. Lolium rigidum Gaudin STIFF RYEGRASS [p. *393*, 519]

*Lolium rigidum* is native to Europe, North Africa, and western Asia. It has been found as a weed of roadsides and waste places at scattered locations in the contiguous United States and Canada. It intergrades with *L. perenne*, *L. multiflorum*, and, occasionally, *L. temulentum*. Hybrids with *L. multiflorum* are called **Lolium ×hubbardii** Jansen & Wacht. *ex* B.K. Simon.

### 4. Lolium temulentum L. [p. *393*, 519]

*Lolium temulentum* is said to be the tares of the Bible. Its two subspecies differ mainly in quantitative characters.

1. Lemmas 3.5–5.5 mm long, 1.2–1.8 mm wide; glumes 5–16 mm long; caryopses 3.2–4.5 mm long, 1.2–1.8 mm wide; rachises slender . . . . . . . . . . subsp. *remotum*
1. Lemmas 4.5–8.5 mm long, 1.5–3 mm wide; glumes (5.5)7–28 mm long; caryopses (3.8)4–7 mm long, (1)1.5–3 mm wide; rachises rather stout . . . . subsp. *temulentum*

### Lolium temulentum subsp. remotum (Schrank) Á. Löve & D. Löve FLAX DARNEL, IVRAIE DU LIN [p. *393*]

*Lolium temulentum* subsp. *remotum* is native to Europe, Asia, and northern Africa. It originated as a weed in flax fields, through unintentional selection for seeds that could not be separated from flax seed using early harvesting techniques. It is a rare weed in the *Manual* region, being reported only from southern Ontario and California, where it grows in waste places and fields.

### Lolium temulentum L. subsp. temulentum L. DARNEL, IVRAIE ENIVRANTE [p. *393*]

*Lolium temulentum* subsp. *temulentum* is found occasionally in disturbed sites throughout much of the *Manual* region. It is native to the Eastern Hemisphere, where it is known only as a weed, especially of grain fields. Awn presence or absence and length vary, and have no taxonomic significance.

### 5. Lolium persicum Boiss. & Hohen. PERSIAN DARNEL [p. *393*, 519]

*Lolium persicum*, a native of southwest Asia, has been found as a weed in grain fields and waste places in southern Canada, Montana, North Dakota, and Wyoming and, as an adventive, in New York and Missouri. It is now one of the top ten weeds of western Canadian cereal crops.

## 14.06 PUCCINELLIA Parl.[1]

Pl ann, bien, or per; usu csp, smt weakly or strongly stln and mat-forming. **Clm** 2–100 cm, erect or decumbent, smt geniculate; **intnd** hollow. **Shth** open to the base or nearly so; **aur** absent; **lig** memb, acute to truncate, entire or erose; **bld** flat, folded, or involute. **Infl** tml pan, open to contracted; **br** smooth or scabrous, some br longer than 1 cm; **ped** usu longer than 3 mm, thinner than 0.5 mm. **Spklt** pedlt, subterete to weakly lat compressed, with 2–10 flt; **dis** above the glm, beneath the flt. **Glm** usu unequal, smt subequal to equal, usu distinctly shorter than the lowest lm in the spklt, smt only slightly shorter, rarely longer, memb, rounded or weakly keeled, veins obscure or prominent, apc unawned; **lo glm** 1(3)-veined; **up glm** (1)3(5)-veined; **cal** blunt, glab or pubescent; **lm** memb to slightly or distinctly coriaceous, glab or pubescent, pubescence smt restricted to the

[1]Jerrold I. Davis and Laurie L. Consaul

bases of the veins, rounded or weakly keeled, at least distally, (3)5(7)-veined, veins obscure to prominent, more or less parallel distally, usu not extending to the apc, lat veins smt rdcd, apical mrg with or without scabrules, apc usu acute to truncate, smt acuminate, entire or serrate to erose, unawned; **pal** subequal to the lm, scarious or memb distally, 2-veined, veins terminating at or beyond midlength; **lod** 2, free, glab; **anth** 3; **ov** glab. **Car** shorter than the lm, concealed at maturity, oblong, terete to dorsally flattened, falling free or with the pal or both the lm and pal attached; **hila** oblong, about ¹/₃ or less the length of the car.

*Puccinellia*, a genus of approximately 120 species, is most abundant in the middle and high latitudes of the Northern Hemisphere. There are 21 species in the *Manual* region, of which 3 are introduced. Ten are confined to the arctic, four are circumarctic and two are transberingian. Most species of *Puccinellia* are halophytes, either in coastal habitats or in saline or otherwise mineralized soils of interior habitats. Polyploidy, selfing, and hybridization are widespread in the genus, and many of the species boundaries are controversial.

The angle of the panicle branches (whether erect, ascending, etc.) refers to their position when the caryopses are mature. Lemma measurements should be made on the lowest lemma in the spikelets. Principal features of the lemmas are as follows. Scabrules (short, pointed hairs, similar in form to those that occur on the pedicels and inflorescence branches of many species of *Puccinellia*, and generally requiring magnification to observe) often occur along the distal margins of the lemmas. When present, they may be few and irregularly scattered, with gaps between them that are either wider than the individual scabrules (e.g., in some *P. pumila*), or arranged in a continuous palisade-like row that lacks gaps (e.g., in *P. distans*). Independent of the presence or absence of scabrules, the lemma margins may be entire (e.g., in *P. pumila* and *P. distans*) or serrate to erose (e.g., in *P. andersonii* and *P. vahliana*).

1. Plants stoloniferous perennials, forming low, often extensive mats; most plants lacking inflorescences, the spikelets, when present, usually not producing mature pollen or caryopses. . . . . . . . . . . . . . . . . . . . . . 1. *P. phryganodes*
1. Plants annual, biennial, or cespitose perennials, sometimes stoloniferous but not mat-forming; plants reproducing sexually, forming mature pollen and caryopses.
  2. Lemmas slightly to markedly coriaceous for most or all of their length; plants of temperate regions.
    3. Lemmas with hyaline apical margins; lemma midveins prominent . . . . . . . . . . . . . . . . . . . . . . . . . 2. *P. rupestris*
    3. Lemmas with coriaceous apical margins; lemma midveins obscure.
      4. Lemmas 1.8–3 mm long; lower branches of the panicles ascending to erect, spikelet-bearing nearly to the base; anthers 0.6–1 mm long . . . . . . . . . . . . . . . . . . . . . . . . . . . . . . . 3. *P. fasciculata*
      4. Lemmas 3–5 mm long; lower branches of the panicles erect to descending, spikelet-bearing from about midlength; anthers 1.5–2.6 mm long . . . . . . . . . . . . . . . . . . . . . . . . . . . . . . 4. *P. maritima*
  2. Lemmas mostly membranous or herbaceous, apical margins sometimes hyaline; plants of temperate and arctic regions.
    5. Plants annual, of temperate regions.
      6. Lemma apices acute; lemmas 2.5–4 mm long, veins glabrous or hairy, particularly on the basal ¹/₂, short (about 0.1 mm) hairs sparsely and evenly distributed between the veins . . . . . . . . . . . . . . . 5. *P. simplex*
      6. Lemma apices obtuse to truncate; lemmas 1.8–2.2 mm long, veins densely hairy on the basal ¹/₂–³/₄, glabrous between the veins . . . . . . . . . . . . . . . . . . . . . . . . . . . . . . . . . . . . . 6. *P. parishii*
    5. Plants perennial, of temperate and arctic regions.
      7. Palea veins with curly, intertwined hairs proximally, scabrous distally; plants of arctic regions.
        8. Pedicels smooth; apical margins of the lemmas smooth, veins obscure or distinct.
          9. Lower glumes ²/₃ to nearly as long as the adjacent lemmas; culms 5–15 cm; panicles 2–4 cm, anthers 0.8–1.5 mm . . . . . . . . . . . . . . . . . . . . . . . . . . . . . . . . . . . . . . . . 7. *P. vahliana*
          9. Lower glumes usually less than ²/₃ as long as the adjacent lemmas, culms 15–40 cm; panicles 5–8 cm, anthers 1.5–2.5 mm . . . . . . . . . . . . . . . . . . . . . . . . . . . . . . . . . . 8. *P. wrightii*
        8. Pedicels scabrous; apical margins of the lemmas scabrous, sometimes minutely so, veins obscure.
          10. Culms 50–65 cm tall; panicles 15–30 cm long . . . . . . . . . . . . . . . . . . . . . . . . . 9. *P. groenlandica*
          10. Culms 5–35 cm tall; panicles 1–13 cm long.
            11. Lemmas 3.5–5.2 mm long; panicles (4)5–13 cm long . . . . . . . . . . . . . . . . . . . . . 10. *P. angustata*
            11. Lemmas 2.8–3.8 mm long; panicles 1–4 cm long . . . . . . . . . . . . . . . . . . . . . . 11. *P. bruggemannii*
      7. Palea veins glabrous, shortly ciliate, or with fewer than 5 longer hairs proximally, never with curly intertwined hairs, scabrous or smooth distally; plants of temperate and arctic regions.
        12. Lemma margins smooth or with a few scabrules at and near the apices.
          13. Lemmas 2–2.5 mm long, usually purple with whitish margins, veins distinct, apices obtuse to truncate; lemmas and palea veins smooth and glabrous; pedicels smooth . . . . . . . . 12. *P. tenella*
          13. Lemmas 2.4–4.6 mm long, variously colored, margins not white, veins obscure to distinct, apices acute to truncate; lemmas and palea veins glabrous or hairy on the lower portion, often scabrous distally; pedicels smooth or scabrous.
            14. Lemmas glabrous or with a few hairs on the lower portion of the veins; lemma apices entire; plants of temperate regions or the low arctic, but not of the high arctic.

15. Pedicels scabrous; palea veins scabrous distally; anthers 1–2 mm long; plants not littoral . . . . . . . . . . . . . . . . . . . . . . . . . . . . . . . . . . . 13. *P. lemmonii* (in part)

15. Pedicels smooth or with a few scattered scabrules; palea veins smooth or with a few scabrules distally; anthers 0.5–1.2 mm long; plants littoral . . . . . . . . . . 14. *P. pumila*

14. Lemmas usually sparsely to moderately hairy, particularly on the vein bases, sometimes glabrous; lemma apices entire, irregularly serrate, or erose; plants of the low and high arctic.

16. Panicles with (2)3–5 branches at the lowest node; lemmas 2.5–3.7 mm long, veins obscure to distinct, apices entire or slightly erose; anthers 1.2–2.2 mm long . . . . . . . . . . . . . . . . . . . . . . . . . . . . . . . . . . . 15. *P. arctica* (in part)

16. Panicles usually with 2 branches at the lowest node; lemmas 3–4.5 mm long, veins obscure, apices irregularly serrate or erose; anthers 0.8–1.2 mm long . . . . . . . . . . . . . . . . . . . . . . . . . . . . . . . . . . . 16. *P. andersonii* (in part)

12. Lemma margins densely scabrous at and near the apices.

17. Lemmas 1.5–2.2 mm long, apices widely obtuse to truncate; anthers 0.4–0.8 mm long; lower panicle branches horizontal to descending . . . . . . . . . . . . . . . . . . . . . . . . . . 17. *P. distans*

17. Lemmas 2–5 mm long, apices usually acute to obtuse, occasionally acuminate or rounded; anthers 0.5–2.2 mm long; lower panicle branches erect to descending.

18. Lemma apices irregularly serrate or erose; lemmas 3–4.5 mm long; anthers 0.8–1.2 mm long . . . . . . . . . . . . . . . . . . . . . . . . . . . . . . . . . . . 16. *P. andersonii* (in part)

18. Lemma apices entire or slightly erose; lemmas 2–4.5(5) mm long; anthers 0.5–2.2 mm long.

19. Pedicels smooth or with a few scattered scabrules; lemmas glabrous or with a few hairs on the lower $^{1}/_{2}$, principally along the veins; anthers 1.5–2 mm long; plants restricted to mineralized springs in California . . . . . . . . . . . . . . . . . 18. *P. howellii*

19. Pedicels smooth to uniformly scabrous; lemmas glabrous or sparsely to moderately hairy on the lower $^{1}/_{2}$; anthers 0.5–2.2 mm long; plants of varied habitats, including hot springs.

20. Lemma midveins often extending to the apical margins; lemma apices acute; lemmas mostly smooth, midveins often slightly scabrous distally; leaf blades involute, 1.2–1.9 mm wide when flattened; leaves concentrated at the base of plant; plants of inland, temperate habitats . . . . . . . . . . . . . . . . . . . . . . . . . . . . . . . . . . . . . . . . . . . . . . . 13. *P. lemmonii* (in part)

20. Lemma midveins usually not extending to the margins; lemma apices usually acute to obtuse, occasionally acuminate; lemmas scabrous or smooth distally; leaf blades involute or flat, 0.5–6 mm wide when flat, leaves ranging from nearly all basal to evenly distributed along the culms; plants of coastal and inland habitats in temperate and arctic regions.

21. Culms 10–100 cm tall; lower glumes 0.5–1.6 mm long; plants usually growing south of 65° N latitude.

22. Pedicel epidermal cells not tumid, pedicels uniformly scabrous; lower branches of the panicles erect to descending; lemmas (2)2.2–3(3.5) mm long; plants usually of interior habitats, occasionally of coastal habitats . . . . . . . . . . . 19. *P. nuttalliana* (in part)

22. Pedicel epidermal cells often tumid, pedicels sparsely to densely scabrous; lower branches of the panicles usually erect to ascending, occasionally spreading to descending; lemmas (2.2)3–4.5(5) mm long; plants of coastal habitats . . . . . . . . . . . . 20. *P. nutkaensis*

21. Culms 6–30(40) cm tall; lower glumes 0.8–2.5 mm long; plants usually growing north of 65° N latitude.

23. Anthers 1.2–2.2 mm long; lateral margins of the lemmas often inrolled . . . . . . . . . . . . . . . . . . . . . . . . . . . . . . . . . . . 15. *P. arctica* (in part)

23. Anthers 0.6–1.2 mm long; lateral margins of the lemmas usually not inrolled.

24. Lemmas 2.8–4 mm long; panicles usually barely exserted from the sheaths . . . . . . . . . . . . . . . . . . . . . . . . . 21. *P. vaginata*

24. Lemmas 2–2.8 mm long; panicles usually distinctly exserted from the sheaths . . . . . . . . . . . . . . 19. *P. nuttalliana* (in part)

## 1. Puccinellia phryganodes (Trin.) Scribn. & Merr.
GOOSE GRASS, PUCCINELLIE RAMPANTE, PUCCINELLIE TROMPEUSE [p. *393*, 519]

*Puccinellia phryganodes* is a widespread and common circumpolar arctic species that grows on seashores at or near the high tide line, in wet saline meadows, and in saline or brackish marshes.

## 2. Puccinellia rupestris (With.) Fernald & Weath. STIFF SALTMARSH GRASS [p. *393*, 519]

*Puccinellia rupestris* grows in coastal and noncoastal habitats in Eurasia; North American collections were apparently introduced in ballast.

## 3. Puccinellia fasciculata (Torr.) E.P. Bicknell BORRER'S SALTMARSH GRASS [p. *393*, 519]

*Puccinellia fasciculata* is native to Europe. In the *Manual* region, it is found principally along the east coast, but it is also established at a few sites in Arizona and Utah, and has been reported from Nevada. All occurrences in the *Manual* region are probably the result of human introductions.

## 4. Puccinellia maritima (Huds.) Parl. COMMON SALTMARSH GRASS, PUCCINELLIE MARITIME [p. *393*, 520]

*Puccinellia maritima* grows in coastal environments in North America and Greenland. It is native to Europe; most or all occurrences in the *Manual* region are probably the result of human introduction.

## 5. Puccinellia simplex Scribn. WESTERN ALKALI GRASS [p. *393*, 520]

*Puccinellia simplex* is widespread in, and mostly confined to, saline soils of central California. The records from Utah probably reflect introductions.

## 6. Puccinellia parishii Hitchc. PARISH'S ALKALI GRASS [p. *393*, 520]

*Puccinellia parishii* grows in saline seepage areas in California, Arizona, and New Mexico.

## 7. Puccinellia vahliana (Liebm.) Scribn. & Merr.
VAHL'S ALKALI GRASS, PUCCINELLIE DE VAHL [p. *394*, 520]

*Puccinellia vahliana* is an arctic species that is circumpolar, except in the Beringian region. In the *Manual* region, it extends from Alaska through northern Canada to Greenland. It is generally non-halophytic, growing in calcareous gravel, sand, clay, or moss of imperfectly drained moist areas, and on seepage slopes from near sea level to 700 m, or, rarely, in seasonally dry, turfy sites. It is often a pioneering species in moist clay and silt by alpine brooks, ephemeral lakes, glacial runoff streams, and on snowbeds. The roots of this species and *P. wrightii* are characteristically thicker and more tightly curled than those of other *Puccinellia* species. It sometimes hybridizes with *Phippsia algida*.

## 8. Puccinellia wrightii (Scribn. & Merr.) Tzvelev
WRIGHT'S ALKALI GRASS [p. *394*, 520]

*Puccinellia wrightii* is an uncommon arctic species. Its range extends from the Chukotka Peninsula in the Russian Far East to western Alaska. Like *P. vahliana*, its roots are characteristically thicker and more tightly curled than those of other *Puccinellia* species.

## 9. Puccinellia groenlandica T.J. Sørensen GREENLAND ALKALI GRASS [p. *394*, 520]

*Puccinellia groenlandica* grows in littoral and nonlittoral environments. It is endemic to Greenland.

## 10. Puccinellia angustata (R. Br.) E.L. Rand & Redfield TALL ALKALI GRASS, PUCCINELLIE ÉTROITE [p. *394*, 520]

*Puccinellia angustata* is a common and widespread arctic species that grows in disturbed silty or sandy sediments. It is usually non-littoral, but when in coastal areas it grows above the influence of high tide.

## 11. Puccinellia bruggemannii T.J. Sørensen
BRUGGEMANN'S ALKALI GRASS [p. *394*, 520]

*Puccinellia bruggemannii* is restricted to arctic islands of Canada and northern Greenland. In Canada, it is a widespread yet local northern, western, and central arctic island species. It is found in calcareous, barren, gravelly, sandy, or silty sites, and is sometimes coastal. It sometimes superficially resembles *Poa abbreviata* or small *Poa glauca* in the field, because of its small, dense inflorescence.

## 12. Puccinellia tenella (Lange) Holmb. TUNDRA ALKALI GRASS [p. *394*, 520]

*Puccinellia tenella* is a halophytic, circumpolar subarctic and low arctic species. It is found above the high tide zone on sandy spits, in salt marshes, in silty soils, and among granitic rocks to 30 m above sea level.

## 13. Puccinellia lemmonii (Vasey) Scribn. LEMMON'S ALKALI GRASS [p. *394*, 520]

*Puccinellia lemmonii* grows in non-littoral saline environments in the western portion of the contiguous United States. Reports from Saskatchewan are probably based on depauperate specimens of *P. nuttalliana*.

## 14. Puccinellia pumila (Vasey) Hitchc. SMOOTH ALKALI GRASS, PUCCINELLIE NAINE [p. *394*, 520]

*Puccinellia pumila* is primarily North American, growing on the Pacific, Arctic, and Atlantic coasts. It also grows in Kamchatka, Russia. It generally grows in sand and among stones in protected intertidal environments.

*Puccinellia alaskana* Scribn. & Merr. is included here in *P. pumila*, but its status is currently under investigation. It differs morphologically from *P. pumila* mainly in its relatively distinct lemma veins, but also in having smaller lemmas (2.5–3 mm) and anthers (0.5–0.9 mm), and in being diploid. It represents the Aleutian Islands component of the geographic distribution given for *P. pumila*.

## 15. Puccinellia arctica (Hook.) Fernald & Weath.
ARCTIC ALKALI GRASS [p. *394*, 520]

*Puccinellia arctica* is restricted to the North American arctic, where it grows in silt, clay, and sandy substrates near the coast, and on alkaline, sparsely vegetated soils further inland. As treated here, it includes three entities that are sometimes treated as distinct species: *P. arctica sensu stricto*, *P. poacea* T.J. Sørensen, and *P. agrostidea* T.J. Sørensen. *Puccinellia arctica sensu stricto* is restricted to the southwestern arctic, *P. poacea* to the high arctic (Ellesmere and Axel Heiberg islands), and *P. agrostidea* to the southwestern arctic and possibly also Ellesmere Island. There are no morphological characters known for distinguishing these three entities.

## 16. Puccinellia andersonii Swallen ANDERSON'S ALKALI GRASS [p. *394*, 520]

*Puccinellia andersonii* is a widespread, coastal arctic species. It grows near the tideline and on otherwise barren, reworked marine sediments of eroded flood plains. Its decumbent growth form often gives it an unhealthy appearance. It is unique among *Puccinellia* species in the *Manual* region in having blunt, rather than pointed, scabrules in the apical region of its lemmas.

**17. Puccinellia distans** (Jacq.) Parl. EUROPEAN ALKALI GRASS, PUCCINELLIE À FLEURS DISTANTES [p. *394*, 520]

*Puccinellia distans* is a Eurasian native, reportedly introduced in North America, where it is widespread, particularly as a weed in nonlittoral environments, including the margins of salted roads. It is also found occasionally in coastal environments.

**18. Puccinellia howellii** J.I. Davis HOWELL'S ALKALI GRASS [p. *394*, 520]

*Puccinellia howellii* is known only from the type locality in Shasta County, California, where it is a dominant element of the vegetation associated with a group of three mineralized seeps. Isozyme profiles suggest that it is a polyploid.

**19. Puccinellia nuttalliana** (Schult.) Hitchc. NUTTALL'S ALKALI GRASS [p. *395*, 520]

*Puccinellia nuttalliana* is a widespread and variable species, restricted to the *Manual* region. It grows principally in the interior, but is also found in coastal settings, where it is difficult to distinguish from *P. nutkaensis*.

**20. Puccinellia nutkaensis** (J. Presl) Fernald & Weath. ALASKA ALKALI GRASS, PACIFIC ALKALI GRASS, PUCCINELLIE BRILLANTE [p. *395*, 520]

*Puccinellia nutkaensis* grows in coastal habitats of continental North America and Greenland, generally in sand and stones in protected intertidal environments. It is variable in form, ranging from diminutive plants that resemble *P. pumila* to tall, erect plants, often with dense or open inflorescences, resembling *P. nuttalliana*.

**21. Puccinellia vaginata** (Lange) Fernald & Weath. SHEATHED ALKALI GRASS, PUCCINELLIE ENGAINÉE [p. *395*, 520]

*Puccinellia vaginata* is a widespread North American arctic species. It is locally common in Greenland to about 78° N latitude; it is uncommon in Canada, where it is found primarily in the low arctic, rarely in the high arctic. One specimen is reported from eastern Siberia. It grows on coastal marine sediments and on eroding, raised marine sediments inland, and forms large tussocks near bird cliffs.

## 14.07 ×PUCCIPHIPPSIA Tzvelev[1]

Pl per; csp. Clm to 23 cm. Shth closed for $^1/_3$–$^2/_3$ their length, glab; aur absent; lig memb; bld flat or folded, glab. Infl compact pan. Spklt lat compressed, with (1)2–3(4) flt; rchl prolonged beyond the base of the distal flt; dis above the glm. Glm usu 2, occ 1, unequal, memb, hyaline, acute to obtuse; lm unawned; pal 2-veined, veins spiculose; anth indehiscent. Car absent.

×*Pucciphippsia* consists of hybrids between *Puccinellia* and *Phippsia*. Two different hybrids are known. Only one, ×*Pucciphippsia vacillans*, grows in the *Manual* region.

**1. ×Pucciphippsia vacillans** (Th. Fr.) Tzvelev [p. *395*, 520]

×*Pucciphippsia vacillans* is a sterile hybrid between *Puccinellia vahliana* and *Phippsia algida*. It grows in damp, open habitats such as wet meadows, wet tundra, pond margins, and imperfectly drained areas in silt, till, and moss. It is currently known in North America

×*Pucciphippsia czukczorum* Tzvelev is currently known only from the Chukotsk Peninsula in northeastern Russia, although its parents, *Puccinellia wrightii* and *Phippsia algida*, have both been found in Alaska on the Seward Peninsula, and in Greenland on the Hayes Peninsula and in the Ammassalik region. Both hybrids are completely sterile, as is indicated by their indehiscent anthers and lack of caryopses. They differ from *Puccinellia* in having small, compact panicles, and from *Phippsia* in usually having more than one floret per spikelet.

from Axel Hedberg, Bathurst, Cornwallis, Ellesmere, Devon, and Baffin islands in arctic Nunavut, and from Midgaardsormen, Greenland. Outside North America, it is found on Svalbard and Novaya Zemlya islands, where both parents are also present.

## 14.08 PHIPPSIA (Trin.) R. Br.[2]

Pl usu per, ann in some environments that have a sufficient growing season for sd production in a single year; csp or matlike. Clm 2–19 cm, erect or procumbent, not rooting at the lo nd, glab; prophylls 4–10 mm. Shth not inflated, those of the bas lvs usu fused only near the base, those of the flag lvs closed for at least $^1/_2$ their length; aur absent; lig memb, glab, acute; bld flat or conduplicate. Infl pan, dense or diffuse. Spklt pedlt, with 1 flt, lat compressed to nearly terete; rchl not prolonged beyond the flt; dis above the glm and beneath the flt. Glm to $^1/_3$ the length of the flt, ovate, without veins, caducous or persistent, unawned; lo glm highly rdcd; cal short, glab; lm 1–3-veined, not strongly keeled, apc acute to rounded, unawned; pal subequal to the lm; lod 2, free, glab; anth 1 or 2; ov glab. Car exceeding the lm and exposed at maturity.

*Phippsia* has two species, one of which is found in arctic Eurasia, Greenland, and the Canadian arctic islands; the other is circumpolar in the arctic, and is also known from the Rocky Mountains of North America and the high Andes of South America.

1. Panicles (0.5)1–2(3) cm long, 3–7 mm wide; spikelet length less than twice the width; glumes caducous, often colorless when present; lemmas often yellow-green, purple coloration, when present, not reaching the apices, glabrous or with a few soft hairs on the lower $^1/_3$; caryopses ellipsoid, widest at or just above the middle . . . . . . . . . . . . . . . . . . . . . . . . . . . . . . . . . . . . . . . . . . . . . . . . . . . . . . . . . . . . 1. *P. algida*
1. Panicles (1)3–9 cm long, (4)5–15 mm wide; spikelet length 2–3 times the width; glumes, particularly the upper glumes, not caducous, often still present on the previous season's growth, often with some

[1]Stephen J. Darbyshire  [2]Laurie L. Consaul and Susan G. Aiken

deep purple coloration; lemmas usually purplish-red with a strong coloration over the veins, the color over the midvein extending to the apex, veins with stiff hairs, soft or stiff hairs elsewhere over the lower $^1/_2-^2/_3$ of the surface; caryopses ovoid, widest below the middle . . . . . . . . . . . . . . . . . . . . . . . . . . . . . . . . . . . 2. *P. concinna*

**1. Phippsia algida** (Sol.) R. Br. ICEGRASS, PHIPPSIE FROIDE [p. *395*, 520]

*Phippsia algida* is a circumpolar species that also grows at high elevations in the Rocky Mountains and the Andes. It is one of the first grasses to flower in the high arctic, which may contribute to its success as an early colonizer of disturbed areas. Although highly nitrophilous, it can tolerate a wide range of soils, from highly alkaline to peat and imperfectly drained mud flats. It sometimes hybridizes with *Puccinellia vahliana*. Plants of *P. algida* with slightly hairy lemmas have been recognized as *P. algida* f. *vestita* Holmb.

**2. Phippsia concinna** (Th. Fr.) Lindeb. SNOWGRASS [p. *395*, 520]

The distribution of *Phippsia concinna* has previously been restricted to arctic Eurasia and Greenland, but is now thought to include the Canadian arctic archipelago. Specimens of *P. concinna* from latitudes near 60° N in Eurasia are relatively large plants, usually near 20 cm tall, with open inflorescences from spreading panicle branches. Specimens from northern Greenland latitudes near 80° N are usually smaller plants, with compact inflorescences and erect panicle branches. More thorough investigation is needed to determine the extent of *P. concinna* in North America.

## 14.09 SCLEROCHLOA P. Beauv.[1]

**Pl** ann. **Clm** 2–30 cm. **Shth** open to closed; **aur** absent; **lig** memb; **bld** flat or folded. **Infl** tml, usu rcm, smt rdcd pan, 1-sided, usu exceeded by the lvs. **Spklt** lat compressed, subsessile to pedlt; **ped** 0.5–1 mm long, 0.5–0.8 mm thick, stout, with 2–7 flt; **rchl** glab, lowest intnd thicker than those above; **dis** tardy, not strongly localized. **Glm** unequal, shorter than the lowest lm, glab, with wide hyaline mrg, apc obtuse to emgt, unawned; **lo glm** (1)3–5-veined; **up glm** (3)5–9-veined; **cal** blunt, glab; **lm** memb, with hyaline mrg, indurate at maturity, (5)7–9-veined, veins prominent, apc rounded to emgt, entire, unawned; **pal** shorter than to equaling the lm, dorsally compressed; **lod** 2, free, glab, entire to lacerate; **anth** 3. **Car** shorter than the lm, concealed at maturity, beaked from the persistent style base, falling free; **hila** round.

*Sclerochloa* is a genus of two species, both of which are native to southern Europe and western Asia. The species found in the *Manual* region, *S. dura*, is now a cosmopolitan weed.

**1. Sclerochloa dura** (L.) P. Beauv. HARDGRASS, FAIRGROUND GRASS [p. *395*, 520]

First collected in the United States in 1895, *Sclerochloa dura* is probably more widespread than indicated, because it is easily overlooked. It grows in lawns, campsites, roadsides, athletic fields, fairgrounds, and other disturbed sites. It is frequently found in severely compacted soils, because it can withstand heavy traffic by vehicles and pedestrians.

*Sclerochloa dura* and *Poa annua* are superficially similar, occupy similar habitats, and have a similar phenology, but *S. dura* has blunt, glabrous lemmas and racemose inflorescences, whereas *P. annua* has obtuse to acute lemmas that are smooth and usually sericeous or crisply puberulent over the veins, and paniculate inflorescences. Plants of *S. dura* become stramineous in age, making them easy to locate because areas dominated by this species change color.

## 14.10 DACTYLIS L.[2]

**Pl** per; csp, smt with short rhz. **Clm** to 2.1+ m, bases lat compressed; **intnd** hollow; **nd** glab. **Lvs** mostly bas, glab; **shth** closed for at least $^1/_2$ their length; **aur** absent; **lig** memb; **bld** flat to folded. **Infl** pan; **primary br** 1-sided, naked proximally, with dense clusters of subsessile spklt distally, at least some br longer than 1 cm. **Spklt** oval to elliptic in outline, lat compressed, with 2–6 flt; **rchl** glab, not prolonged beyond the distal flt; **dis** above the glm and beneath the flt. **Glm** shorter than the flt, lanceolate, 1–3-veined, ciliate-keeled, awn-tipped; **cal** short, blunt; **lm** 5-veined, scabrous to ciliate-keeled, tapering to a short awn; **pal** 2-keeled, tightly clasped by the lm, unawned, apc notched; **lod** 2, glab, toothed; **anth** 3; ov glab. **Car** shorter than the lm, concealed at maturity, oblong to ellipsoid, falling free or adhering to the lm and/or pal; **hila** round.

*Dactylis* is interpreted here as a variable unispecific genus, although five species are recognized by Russian taxonomists.

**1. Dactylis glomerata** L. ORCHARDGRASS, DACTYLE PELOTONNÉ [p. *396*, 520]

*Dactylis glomerata* grows in pastures, meadows, fence rows, roadsides, and similar habitats throughout North America. Native to Eurasia and Africa, it has been introduced throughout most of the cool-temperate regions of the world as a forage grass. It provides nutritious forage that is relished by all livestock, as well as by deer, geese, and rabbits. When abundant, the pollen can be a major contributor to hay fever.

## 14.11 LAMARCKIA Moench[3]

**Pl** ann; tufted. **Clm** 5–40 cm, erect or decumbent at the base, glab. **Shth** open for at least $^1/_3$ their length, glab, mrg memb and continuous with the lig; **aur** absent; **lig** memb, acute, glab, apc somewhat erose; **bld** flat, glab. **Infl** pan, dense, secund, golden-yellow to purplish; **primary br** appressed to the rchs; **secondary br** capillary,

[1]David M. Brandenburg  [2]Kelly W. Allred  [3]Lynn G. Clark

smooth, glab, flexuous, terminating in 1–4 fascicles of pedlt spklt, strongly bent below the junction with the fascicle base; **ped of spklt** in each fascicle fused at the base, strigose. **Spklt** dimorphic, the tml spklt of each fascicle ftl, the others strl; **dis** at the base of the fused ped. **Ftl spklt** terete to somewhat lat compressed, with 2 flt, 1 bisx, the other a rudiment, borne on long rchl intnd; **cal** short, blunt, glab; **glm** narrow, acuminate or short-awned, 1-veined; **bisx lm** scarcely veined, each with a delicate, straight, subapical awn; **pal** equal or subequal to the lm; **lod** 2, glab, toothed, **anth** 3; **ov** glab; **rdmt flt** highly rdcd lm, each with a delicate, straight awn. **Strl spklt** linear, lat compressed; **flt** 5–10; **glm** similar to those of the ftl spklt; **lm** imbricate, obtuse, unawned; **pal** absent. **Car** ovoid or ellipsoid, adhering to the lm and/or pal.

*Lamarckia* is a unispecific genus that has been introduced into North America from the Mediterranean and the Middle East.

### 1. Lamarckia aurea (L.) Moench GOLDENTOP [p. 396, 520]

*Lamarckia aurea* grows on open ground, rocky hillsides, and in sandy soil, at elevations from sea level to 700 m. Within the *Manual* region, it is known only from the southwest. It is an attractive, but rather weedy species.

## 14.12 BECKMANNIA Host[1]

**Pl** ann and tufted, or per and rhz. **Clm** 20–150 cm, smt tuberous at the base, erect. **Lvs** mostly cauline; **shth** open, glab, ribbed; **aur** absent; **lig** memb, acute; **bld** flat, glab. **Infl** dense, spikelike pan; **br** 1-sided, racemosely arranged, secondary br few, at least some br longer than 1 cm, with closely imbricate spklt; **dis** below the glm, the spklt falling entire. **Spklt** lat compressed, circular, ovate or obovate in side view, subsessile, with 1–2 flt; **rchl** not prolonged beyond the base of the distal flt. **Glm** subequal, slightly shorter than the lm, inflated, keeled, D-shaped in side view, unawned; **cal** blunt, glab; **lm** lanceolate, inconspicuously 5-veined, unawned; **pal** subequal to the lm; **lod** 2, free; **anth** 3; **ov** glab. **Car** shorter than the lm, concealed at maturity.

*Beckmannia* is a genus of two species: an annual species usually with one fertile floret per spikelet that is native to North America and Asia, and a perennial species with two fertile florets per spikelet that is restricted to Eurasia.

### 1. Beckmannia syzigachne (Steud.) Fernald AMERICAN SLOUGHGRASS, BECKMANNIE À ÉCAILLES UNIES [p. 396, 520]

*Beckmannia syzigachne* grows in damp habitats such as marshes, floodplains, the edges of ponds, lakes, streams, and ditches, and in standing water. It is a good forage grass, but frequently grows in easily damaged habitats.

## 14.13 POA L.[2]

**Pl** ann or per; usu synoecious, smt monoecious, gynodioecious, dioecious, and/or asexual; with or without rhz or stln, densely to loosely tufted or the clm solitary. **Bas brchg** invag, psinva, or exvag; **prophylls** of invag shoots 2-keeled and open, of psinva shoots not keeled and tubular, of exvag shoots scalelike. **Clm** 1–150 cm, hollow, usu unbrchd above the base. **Shth** from almost completely open to almost completely closed, terete or weakly to strongly compressed; **aur** absent; **lig** memb, truncate to acuminate; **bld** 0.4–12 mm wide, flat, folded, or involute, adx surfaces with a groove on each side of the midvein, other intercostal depressions shallow, indistinct, apc often prow-shaped. **Infl** usu tml pan, rarely rdcd and rcmlike. **Spklt** 2–12 mm, usu lat compressed, infrequently terete to subterete, usu lanceolate, smt ovate; **flt** (1)2–6(13), usu sx, smt bulb-forming; **rchl** usu terete, smt prolonged beyond the base of the distal flt; **dis** above the glm and beneath the flt. **Glm** usu shorter than the lowest lm in the spklt, usu keeled, 1–3(5)-veined, unawned; **cal** blunt, usu terete or slightly lat compressed, smt slightly dorsally compressed, glab or hairy, hairs often concentrated in 1(3) tufts or *webs*, smt distributed around the cal below the lm as a *crown of hairs*; **lm** usu keeled, infrequently weakly keeled or rounded, similar in texture to the glm, 5(7–11)-veined, lat veins smt faint, mrg scarious-hyaline distally, apc scarious-hyaline, truncate or obtuse to acuminate, unawned; **pal** from $^2/_3$ as long as to subequal to the lm, distinctly 2-keeled, mrg and intercostal regions milky white to slightly greenish; **lod** 2, broadly lanceolate, glab, lobed; **fnctl anth** (1–2)3, 0.1–5 mm; **ov** glab. **Car** 1–4 mm, ellipsoidal, often shallowly ventrally grooved, solid, with lipid; **hila** sub-bas, round or oval, to $^1/_6$ the length of the car.

*Poa* includes about 500 species. It grows throughout the world, principally in temperate and boreal regions. Sixty-one species and five hybrid species are native to the *Manual* region; nine species are introduced.

*Poa* is taxonomically difficult because most species are polyploid, many are apomictic, and hybridization is common. A variety of sexual reproductive systems are represented within the genus, but individual species are usually uniform in this regard.

[1]Stephan L. Hatch  [2]Robert J. Soreng

Vegetative characteristics that may be useful for distinguishing *Poa* from other morphologically similar genera are: the more or less straight, rather than curly, roots; two-grooved, prow-shaped blades; partially or wholly closed flag leaf sheaths; and isomorphic collar margins. Useful spikelet characteristics include: terete rachillas; multiple, relatively small, unawned florets; webbed calluses; well-developed palea keels; and the greenish or milky white intercostal regions of the paleas.

Callus hairs in *Poa* follow one of three patterns. In the most common pattern, there is an isolated dorsal tuft of crinkled or pleated hairs, the *web*, below the lemma keel. In a few species, additional webs may be present below the marginal veins. In the second pattern, crinkled hairs are distributed around the lemma base, but are somewhat concentrated and longer towards the back; this pattern is called a *diffuse web*. Webbed calluses are found only in *Poa*. In the third pattern, the hairs are straight to slightly sinuous, and more or less evenly distributed around the lemmas bases; calluses with such a pattern are described as having a *crown of hairs*.

Three named infrasectional hybrids are included in this treatment. One, *Poa arida*, is accounted for in the key. The other two are not. *Poa* ×*limosa* is too variable, and *P.* ×*gaspensis* is known from too few specimens to make their inclusion in the key helpful.

Unless stated otherwise, sheath closure is measured on the flag leaf, and ligule length on the upper 1–2 culm leaves; spikelet, floret, callus, lemma, and palea measurements are on non-bulb-forming florets; floret pubescence is evaluated on the lower florets within several spikelets; length of the callus hairs refers to their length when stretched out; anther measurements are based on functional anthers, i.e., those that produce pollen, as indicated by their being plump or, after the pollen is shed, by their open sacs. For hair lengths, puberulent is to about 0.15 mm long, short-villous to about 0.3 mm long, and long-villous from 0.3–0.4+ mm long, but these are only guidelines, not discrete categories; some species are only on one end of the range, and ranges have not been confirmed for every species. Many species key more than once, due in part to infraspecific variation.

1. Culms with bulbous bases; spikelets often bulbiferous (sect. *Arenariae*) . . . . . . . . . . . . . . . . . . . . . . . . . . . . .8. *P. bulbosa*
1. Culms with non-bulbous bases; spikelets sometimes bulbiferous.
    2. Some or all spikelets bulbiferous . . . . . . . . . . . . . . . . . . . . . . . . . . . . . . . . . . . . . . . . . . . . . . . . . . . . . . . . . . . . . Subkey I
    2. Spikelets not bulbiferous, florets developing normally.
        3. Anthers 0.1–1(1.2) mm long in all florets and well developed, or only the upper 1–2 florets with
            rudimentary anthers; plants annual or perennial . . . . . . . . . . . . . . . . . . . . . . . . . . . . . . . . . . . . . . . . . . Subkey II
        3. Some anthers (1.2)1.3–4 mm long, or the florets pistillate and all anthers vestigial and 0.1–0.2
            mm long, or longer and poorly developed; plants perennial.
            4. Plants rhizomatous or stoloniferous, rhizomes or stolons usually longer than 5 mm; basal
                leaves of the erect shoots with well-developed blades; plants densely to loosely cespitose or
                the culms solitary . . . . . . . . . . . . . . . . . . . . . . . . . . . . . . . . . . . . . . . . . . . . . . . . . . . . . . . . . . . . . . Subkey III
            4. Plants neither rhizomatous nor stoloniferous; basal leaves of the erect shoots sometimes
                without blades; plants densely cespitose . . . . . . . . . . . . . . . . . . . . . . . . . . . . . . . . . . . . . . . . . . . . Subkey IV

## *Poa* Subkey I

**Culms** without bulbous bases. **Leaf sheaths** not swollen at the base. **Spikelets** mostly bulbiferous, lower spikelets in each panicle frequently normal or subnormal, basal florets of the bulbiferous spikelets frequently normal or subnormal.

1. Sheaths closed for ¹/₁₀–¹/₅ their length; plants without rhizomes; calluses glabrous or with a crown of
   hairs, rarely webbed.
    2. Panicles 2–4 cm long, open; plants delicate, 15–20 cm tall; blades 0.7–1.5 mm wide; lemmas 2–3
       mm long, keels crisply puberulent; calluses glabrous; plants known only from the Brooks Range in
       Alaska . . . . . . . . . . . . . . . . . . . . . . . . . . . . . . . . . . . . . . . . . . . . . . . . . . . . .60. *P. pseudoabbreviata* (in part)
    2. Panicles 1–15 cm long, open or contracted; plants usually coarser, 10–60 cm tall; blades 1–7 mm
       wide; lemmas 3–8 mm long, keels villous; calluses often hairy; plants widespread.
        3. Basal branching all or nearly all extravaginal; calluses glabrous or webbed; plants strongly
            glaucous, rare, alpine and arctic . . . . . . . . . . . . . . . . . . . . . . . . . . . . . . . . . . . . . . . . . .57. *P. glauca* (in part)
        3. Basal branching both extravaginal and intravaginal; calluses glabrous or with a crown of hairs;
            plants glaucous or not, not rare, arctic to subarctic or coastal and boreal.
            4. Basal branching strictly intravaginal; calluses glabrous; plants densely tufted . . . . . . . . . . . . . .9. *P. alpina* (in part)
            4. Basal branching all or partly extravaginal; calluses glabrous or hairy; plants loosely to densely
               tufted.
               5. Panicle branches smooth or sparsely scabrous; blades folded and inrolled; lemmas loosely
                  villous between the veins; glumes somewhat shiny, not glaucous; plants of the high arctic
                . . . . . . . . . . . . . . . . . . . . . . . . . . . . . . . . . . . . . . . . . . . . . . . . . . . . . . . . . . . . . . .67. *P. hartzii* (in part)

  5. Panicle branches densely scabrous, at least distally; blades flat; lemmas glabrous between
   the veins; glumes not shiny, often glaucous; plants subarctic or boreal and coastal  . . . . 69. *P. stenantha* (in part)
1. Sheaths closed for ($^1$/$_5$)$^1$/$_4$–$^3$/$_4$ their length; plants with or without rhizomes; calluses usually webbed,
 sometimes glabrous.
  6. Basal branching intravaginal; plants densely tufted, neither rhizomatous nor stoloniferous; calluses
   glabrous . . . . . . . . . . . . . . . . . . . . . . . . . . . . . . . . . . . . . . . . . . . . . . . . . . . . . . . . . . . . . . . . . 9. *P. alpina* (in part)
  6. Basal branching completely or partly extravaginal; plants loosely or densely tufted, rhizomatous or
   stoloniferous; calluses usually webbed, sometimes glabrous.
    7. Panicles open, longest branches (5)8–15 cm long; sheaths closed for $^1$/$_2$–$^3$/$_4$ their length; plants
     green, not anthocyanic; plants of coastal regions of British Columbia . . . . . . . . . . . . . . . . . 24. *P. laxiflora* (in part)
    7. Panicles contracted to open, longest branches 1–8 cm long; sheaths closed for ($^1$/$_5$)$^1$/$_4$–$^1$/$_2$ their
     length; plants strongly anthocyanic; plants widespread.
     8. Paleas glabrous intercostally; spikelets 4–5.5 mm long, lemmas 2.5–3.5 mm long, intercostal
      regions glabrous; panicles contracted, sometimes opening eventually; plants slender, mostly
      high arctic . . . . . . . . . . . . . . . . . . . . . . . . . . . . . . . . . . . . . . . . . . . . . . 13. *P. pratensis* (in part)
     8. Paleas puberulent or hispidulous intercostally; spikelets 5.5–12 mm long; lemmas 4.5–8 mm
      long, intercostal regions hairy; panicles open or loosely contracted; plants slender to stout,
      subarctic and coastal to high arctic.
      9. Paleas, if recognizable, hispidulous intercostally; glumes distinctly keeled, keels scabrous;
       plants stout, of subarctic coastal regions . . . . . . . . . . . . . . . . . . . . . . . . . . . . 15. *P. macrocalyx* (in part)
      9. Paleas softly puberulent intercostally; glumes indistinctly keeled, keels smooth or nearly so;
       plants slender to stout, mainly in alpine and arctic habitats, rarely coastal . . . . . . . . . . . . 16. *P. arctica* (in part)

## *Poa* Subkey II

**Plants** annual or perennial. **Culms** not bulbous-based. **Basal leaf sheaths** not swollen at the base. **Spikelets**
not bulbiferous, florets developing normally. **Anthers** 0.1–1(1.2) mm long in all florets, anther sacs usually
well developed, the distal 1–2 fertile florets sometimes with rudimentary stamens.

1. Calluses glabrous; lemmas usually softly puberulent to long-villous on the keel and marginal veins,
 often also on the lateral veins, glabrous between the veins, non-alpine plants rarely glabrous
 throughout; palea keels smooth, usually short- to long-villous near the apices, rarely glabrous; panicle
 branches and glume keels smooth; plants annual, sometimes surviving for a second season, introduced,
 weedy species (sect. *Micrantherae*).
  2. Anthers 0.6–1(1.1) mm long, oblong prior to dehiscence; spikelets crowded or sparsely arranged on
   the branches; plants widespread . . . . . . . . . . . . . . . . . . . . . . . . . . . . . . . . . . . . . . . . . . . . . . . . 10. *P. annua*
  2. Anthers 0.1–0.5 mm long, round to elliptical prior to dehiscence; spikelets crowded on the branches;
   plants uncommon outside of California . . . . . . . . . . . . . . . . . . . . . . . . . . . . . . . . . . . . . . . . . . 11. *P. infirma*
1. Calluses webbed or glabrous, if glabrous, the lemma pubescence not as above or the palea keels at least
 slightly scabrous near the apices; panicle branches and glume keels smooth or scabrous; plants annual
 or perennial, native, sometimes growing in disturbed habitats.
  3. Calluses webbed; lemma keels glabrous throughout or, if hairy on the proximal $^1$/$_2$, the marginal
   veins glabrous.
    4. Culms 5–15(20) cm tall; plants alpine . . . . . . . . . . . . . . . . . . . . . . . . . . . . . . . . . . . . . 61. *P. abbreviata* (in part)
    4. Culms 20–126 cm tall; plants of shady forests and forest openings.
     5. Lemmas hairy only on the keels; branches in whorls of (2)3–5(7) . . . . . . . . . . . . . . . . . . . . . 2. *P. alsodes*
     5. Lemmas usually glabrous, marginal veins rarely sparsely hairy at the base, hairs to 0.15 mm
      long; branches 1–3 per node.
      6. Sheaths closed for at least $^9$/$_{10}$ their length . . . . . . . . . . . . . . . . . . . . . . . . . . . . . . . . 4. *P. marcida*
      6. Sheaths closed for $^1$/$_3$–$^3$/$_4$ their length.
       7. Plants perennial; panicles lax, less than $^1$/$_4$ the height of the plants; second rachilla
        internodes shorter than 1 mm; lemma apices obtuse to sharply acute or acuminate . . 1. *P. saltuensis* (in part)
       7. Plants usually annual, rarely longer-lived; panicles erect, $^1$/$_4$–$^1$/$_2$ the height of the plants;
        second rachilla internodes longer than (1)1.2 mm; lemma apices sharply acute . . . . . . . . . . 17. *P. bolanderi*
  3. Calluses webbed or glabrous, if webbed, the lemmas hairy on the keel and marginal veins.
    8. Plants annual, rarely persisting for a second season; calluses webbed.
     9. Anthers 1, 0.1–0.2(0.3) mm long; palea keels softly puberulent to long-villous at midlength;
      panicles eventually open; plants from east of the 100th meridian . . . . . . . . . . . . . . . . . . . . . 19. *P. chapmaniana*

9. Anthers (1–2)3 per floret, 0.2–1(1.2) mm long; palea keels scabrous or softly puberulent to short-villous at midlength; panicles contracted or eventually open; plants from west of the 100th meridian or Texas.
    10. Panicles eventually open; blade apices narrowly prow-shaped; lemmas with hairs of similar length over and between the veins; palea keels scabrous or softly puberulent at midlength . . . . . . . . . . . . . . . . . . . . . . . . . . . . . . . . . . . . . . . . . . . . . . . . . . . . . . 18. *P. howellii*
    10. Panicles contracted; blade apices broadly prow-shaped; lemmas long- to short-villous over the keels and marginal veins, glabrous or pilose between the veins; palea keels softly puberulent to short-villous at midlength . . . . . . . . . . . . . . . . . . . . . . . . . . . . . . . . . . . 20. *P. bigelovii*
8. Plants perennial; calluses glabrous or webbed.
  11. Panicles open, broadly rhomboidal to pyramidal, branches divaricately ascending to spreading, longest branches 1.5–5 cm long, pedicels often longer than the 3–5 mm long spikelets; calluses glabrous; lemmas crisply puberulent on the keel and marginal veins, glabrous elsewhere, rarely glabrous throughout . . . . . . . . . . . . . . . . . . . . . . 60. *P. pseudoabbreviata* (in part)
  11. Panicles contracted or open, if open and broadly rhomboidal to pyramidal, the branches not as above or, if approximately so, the calluses webbed or the lemma keels and marginal veins short- to long-villous.
    12. Sheaths closed from $^1/_{10}$–$^1/_4$($^1/_3$) their length; panicles 1–7(10) cm long, contracted, branches shorter than 1.5 cm; plants densely tufted, basal branching all or mainly intravaginal; lower glumes usually 3-veined, lanceolate to broadly lanceolate, upper glumes subequal to or longer than the lowest lemmas; culms 1–25(30) cm tall; plants of high alpine and tundra habitats (sect. *Abbreviatae*).
      13. Lemmas short- to long-villous on the marginal veins and distal $^3/_4$ of the keel, glabrous or softly puberulent between the veins; calluses webbed or glabrous . . 61. *P. abbreviata* (in part)
      13. Lemmas glabrous or the keel and marginal veins sparsely puberulent proximally, glabrous between the veins; calluses glabrous.
        14. Anthers 0.2–0.8 mm long; spikelets 3–4 mm long; lower glumes usually exceeding the lower lemmas; upper florets frequently exceeded by or only slightly exceeding the glumes; blades thin, flat, folded, or slightly inrolled . . . . . . . 59. *P. lettermanii*
        14. Anthers 0.8–1.2 mm long; spikelets 3.5–7 mm long; lower glumes shorter than to equaling the lower lemmas; upper florets exceeding both glumes; blades moderately thick, often folded and inrolled on the margins.
          15. Adaxial surfaces of the innovation blades smooth or sparsely scabrous, long cells papillate (at 100×); upper culm blades with 7–15 closely spaced ribs on the adaxial surface; plants of California . . . . . . . . . . . . . . . . . . 62. *P. keckii* (in part)
          15. Adaxial surfaces of the innovation blades densely and minutely hispidulous, puberulent, or scabrous, rarely smooth, long cells not papillate (at 100×); upper culm blades with 5–9 well-spaced ribs on the adaxial suface; plants of British Columbia, Washington, and Oregon . . . . . . . . . . . . . . . . . . . . . . . . . . . . . . . . . . . . . . . . . . 63. *P. suksdorfii* (in part)
    12. Sheaths closed for $^1/_{10}$–$^7/_8$ their length; panicles 1–40 cm long, loosely contracted to open, branches 0.5–20 cm long, or the panicles contracted to loosely contracted with the branches 0.5–2 cm long and the plants loosely tufted; basal branching mainly extravaginal or pseudointravaginal; lower glumes 1–3-veined, subulate to broadly lanceolate; upper glumes shorter than to subequal to the lowest lemmas; culms 5–150 cm tall; plants of various habitats.
      16. Sheaths closed for $^1/_{10}$–$^1/_5$ their length; basal branching mainly extravaginal; lower 1–3 leaves of the culms and innovations bladeless; anthers 0.8–1.2 mm long, sometimes poorly developed.
        17. Flag leaf nodes at or above midculm length . . . . . . . . . . . . . . . . . . . . . 55. *P. nemoralis* (in part)
        17. Flag leaf nodes usually in the basal $^1/_3$ of the culms.
          18. Anthers poorly developed, mature anther sacs about 0.1 mm wide and indehiscent; panicles dense to moderately dense, ovoid, 1.5–3.5 cm long; panicle branches not glaucous, angles smooth or sparsely scabrous; glumes broadly lanceolate, equal; upper glumes 3.7–4.7 mm long, the length 3.6–4.1 times the width; lemmas 3.7–4.5 mm long, glabrous between the veins, lateral veins usually glabrous, infrequently softly puberulent . . . . . . . . . . . . . . . . . . . . . . . . . . . . . . . . . . . . . . . . . . . 51. *P. laxa × glauca*
          18. Anthers well developed, mature anther sacs usually about 0.2 mm wide and dehiscent, rarely aborted; panicles dense to loose, ovoid to

lanceoloid, 1–10 cm long; panicle branches glaucous, the angles scabrous, at least below the spikelets; glumes narrowly to broadly lanceolate, subequal; upper glumes 2–3.8(5.2) mm long, the length usually more than 4.1 times the width; lemmas 2.5–4 mm long, glabrous or softly puberulent between the veins, lateral veins usually sparsely softly puberulent to short-villous . . . . . . . . . . . . . . . . . . . . . . . . . . . . . . . . 57. *P. glauca* (in part)

16. Sheaths closed for $^1/_5$–$^7/_8$ their length; basal branching extravaginal, mixed extra- and intravaginal, or pseudointravaginal; culms and innovations with or without bladeless leaves; anthers 0.2–1.2 mm long, well developed.

   19. Calluses usually glabrous, rarely sparsely and shortly webbed; palea keels softly puberulent to short-villous for much of their length; lemmas puberulent between the veins; panicles (5)8–20 cm long, broadly pyramidal; panicle branches moderately to densely scabrous on the angles, longest branches 5–12 cm long, with 3–8 spikelets . . . . . . . . . . . . . . . . . . . . . . . . . . . . . . . 6. *P. autumnalis* (in part)

   19. Calluses webbed or glabrous, if glabrous then the palea keels glabrous, lemmas glabrous between the veins, panicles 2–8 cm long, usually loosely contracted, infrequently contracted, panicle branches smooth or sparsely scabrous, longest branches 1–3(4) cm long, with 1–8 spikelets.

      20. Panicles 2–8 cm long, usually loosely contracted, infrequently contracted; panicle branches usually ascending or weakly divergent, infrequently erect, smooth or sparsely scabrous, sulcate, longest branches 1–3(4) cm long; calluses webbed or glabrous; anthers (0.6)0.8–1.1(1.3) mm long; sheaths closed for $^1/_5$–$^1/_3$ their length; plants alpine . . . . . . . . . . . . . . . . . . 50. *P. laxa* (in part)

      20. Panicles 2.5–40 cm long, open; panicle branches loosely ascending, spreading, or reflexed, smooth or scabrous, angled, sulcate, or terete, longest branches usually longer than 3 cm; calluses webbed; anthers 0.2–1.2 mm long; sheaths closed for $^1/_4$–$^3/_4$ their length; plants alpine or not.

         21. Panicle branches smooth or sparsely scabrous, usually terete or slightly sulcate; lower glumes subulate to broadly lanceolate; lemmas glabrous between the veins.

            22. Lower glumes subulate to narrowly lanceolate, keels usually scabrous; lemma keels short- to long-villous for $^1/_3$–$^2/_3$ their length; lateral veins glabrous; lemma apices narrowly acute; spikelets lanceolate to narrowly lanceolate, green to purple; panicle branches sparsely scabrous, longest branches with (3)4–15 spikelets; palea keels evenly pectinate-ciliate or scabrous at midlength . . . . . . . . . . . . . . . . . . . . . . . . . . . 53. *P. leptocoma* (in part)

            22. Lower glumes narrowly to broadly lanceolate, keels usually smooth; lemma keels long-villous for $^1/_2$–$^4/_5$ their length; lateral veins glabrous or hairy; lemma apices acute; spikelets lanceolate to ovate; panicle branches smooth or sparsely scabrous, if sparsely scabrous, with 1–3 ovate, dark purple spikelets on the lax to drooping capillary branches; longest branches with 1–18 spikelets; palea keels softly puberulent or scabrous at midlength.

               23. Longest panicle branches with (3)6–18 spikelets; palea keels scabrous or sparsely puberulent at midlength; lemmas usually sparsely puberulent on the lateral veins; lower branches usually reflexed . . . . . . . . . . . . . . . . . . . . . . . . . . . . . . 22. *P. reflexa*

               23. Longest panicle branches with 1–3(5) spikelets; palea keels sparsely scabrous at midlength; lemmas glabrous on the lateral veins; lower branches usually laxly ascending to spreading . . . . . . . . . . . . . . . . . . . . . . . . . . . . . . . . . . . . 23. *P. paucispicula*

         21. Panicle branches sparsely to densely scabrous, terete or angled; lower glumes subulate or broader; lemmas glabrous or puberulent between the veins.

            24. Panicles conical, the lower nodes with (2)3–10 branches; branches eventually reflexed; upper sheaths often ciliate on the overlapping margins near the point of fusion; intercostal regions

of the lemmas usually sparsely puberulent, lateral veins at least sparsely puberulent; palea keels puberulent . . . . . . . . . . . . . . . 3. *P. sylvestris* (in part)

24. Panicles not conical, the lower nodes with 1–7 branches; branches usually ascending to spreading, sometimes drooping or reflexed; upper sheaths not ciliate on the margins; lemmas glabrous or puberulent between the keel and marginal veins; palea keels puberulent or glabrous.

    25. Lemmas usually puberulent on the lateral veins and between the veins; lower cauline sheaths and ligules densely retrorsely scabrous; panicles (6)12–40 cm long, with 2–7 branches at the lower nodes; lower glumes 1-veined; plants densely tufted . . . . . . . . . . . . . . . . . . . . . . . . . . . . . . . . . . 21. *P. occidentalis*

    25. Lemmas glabrous on the keel, lateral veins, and between the veins, rarely puberulent on the lateral veins and between the veins; sheaths and ligules smooth or sparsely to moderately densely retrorsely scabrous; panicles 3–30 cm long, branches 1–3(5) per node; lower glumes 1–3-veined; plants densely to loosely tufted.

        26. Plants loosely tufted or with solitary culms, long-rhizomatous; lower glumes lanceolate; palea keels scabrous; panicles 14–30 cm long . . . . . . . . . . . . . . 24. *P. laxiflora* (in part)

        26. Plants densely to loosely tufted, sometimes shortly rhizomatous; lower glumes subulate to lanceolate; palea keels scabrous or puberulent; panicles 3–15(18) cm long.

            27. Ligules 1.5–4(6) mm long, obtuse to acute; lemmas often purple, keels pubescent for $^1/_3$–$^2/_3$ their length, apices usually bronze-colored, sharply acute to acuminate; palea keels evenly pectinate-ciliate to scabrous; lower glumes 1-veined . . . . . . . . . . . . . . . . . . . . . . . . . . . 53. *P. leptocoma* (in part)

            27. Ligules 0.3–2.1 mm long, truncate; lemmas green, keels pubescent for $^2/_3$ their length or more, apices white or faintly bronze, acute to obtuse; palea keels scabrous or puberulent; lower glumes 1–3-veined.

                28. Palea keels puberulent; anthers (0.5) 0.8–1.2 mm long; lemmas (2.5)3.2–4.7 mm long, lateral veins distinct . . . . . . . . . . . . . . . . . . . . . 7. *P. wolfii*

                28. Palea keels scabrous; anthers 0.2–0.8 mm long; lemmas 2.5–4 mm long, lateral veins faint . . . . . . . . . . . . . . . . . . . . . . . . . . . . . . . 52. *P. paludigena*

## *Poa* Subkey III

**Plants** rhizomatous or stoloniferous, densely to loosely tufted or the culms solitary. **Anthers** longer than 1.2 mm, or the florets pistillate and all anthers vestigial and 0.1–0.2 mm long, or longer and poorly developed.

1. Calluses usually with a crown of hairs, hairs 1–2 mm long, sinuous; lemmas 4.5–7 mm long, 5–7-veined, outer margins usually with hairs to 0.2 mm long, marginal veins usually glabrous, sometimes long-villous; bases of the basal sheaths densely retrorsely strigose, hairs 0.1–0.2 mm long, thick; plants of subsaline boreal to low arctic coastal beaches and meadows (*Poa* subg. *Arctopoa*) . . . . . . . . . . . . . . . . . . 72. *P. eminens*

1. Calluses usually glabrous or webbed, sometimes with a crown of hairs; lemmas 2–8 mm long, 5(7)-veined, outer margins glabrous, marginal veins glabrous or not; bases of the basal sheaths glabrous; plants of various habitats.

    2. Culms and nodes strongly compressed; culms usually geniculate; lower culm nodes usually exserted; panicle branches angled, scabrous on the angles; sheaths closed for $^1/_{10}$–$^1/_5$ their length . . . . . . . . . . . . . 58. *P. compressa*

    2. Culms terete to somewhat compressed, nodes not or only weakly compressed; culms geniculate or not; lower culm nodes exserted or not; panicle branches angled or terete, smooth or scabrous; sheath closure varied.

3. Panicles 3–12(18) cm long, narrowly cylindrical or lobed, congested, usually with over 100 spikelets; plants unisexual; spikelets sexually dimorphic; pistillate plants: calluses webbed dorsally and below the marginal veins, lemmas 4.2–6.4 mm long, keels and marginal veins densely long-villous, panicle branches usually moderately to densely coarsely scabrous; staminate plants: calluses glabrous or sparsely webbed dorsally, rarely also webbed below the marginal veins, lemmas 3.5–5 mm long, keels and marginal veins glabrous or moderately densely and shortly pubescent, panicle branches sparsely to moderately scabrous; all plants: blades flat or folded, adaxial surfaces glabrous; plants native to the southern Great Plains, infrequently introduced elsewhere . . . . . . . . . . . . . . . . . . . . . . . . . . . . . . . . . . . . . . . . . . . . . . . . . .48. *P. arachnifera*

3. Panicles 1–30 cm long, contracted to open, infrequently narrowly cylindrical or lobed and congested with over 100 spikelets; plants unisexual or bisexual; spikelets not sexually dimorphic; calluses glabrous, webbed, or with a crown of hairs, rarely with 3 webs; lemma keels and marginal veins glabrous or hairy; panicle branches smooth or scabrous; blades flat, folded, or involute, adaxial surfaces sometimes hairy in plants with contracted or loosely contracted panicles and unisexual spikelets; plants widespread.

    4. Basal branching extravaginal, branches initiated as pinkish to purplish, fleshy-scaled buds, the scales becoming brownish and flabelliform after shoot development; sheaths closed for at least $^9/_{10}$ their length; florets unisexual . . . . . . . . . . . . . . . . . . . . . . . . . . . . . . . . .34. *P. sierrae*

    4. Basal branching extra- or intravaginal or both, branches not initiated as persistent pinkish to purplish, fleshy-scaled buds; sheaths closed for at least $^1/_{15}$ their length; florets bisexual or unisexual.

        5. Lemmas totally glabrous, often scabrous; calluses webbed or diffusely webbed, hairs at least (1)2 mm long.

            6. Panicles 10–20 cm long, open, pyramidal, sparse; sheaths closed for $^1/_{15}$–$^1/_5$ their length; florets bisexual; plants of coastal redwood forests in northern California . . . . . . . . . . . . . . . . . .5. *P. kelloggii*

            6. Panicles 1–10.5 cm long, loosely contracted to open, lanceoloid to pyramidal, congested to sparse; sheaths closed for $^1/_3$–$^9/_{10}$ their length; florets unisexual or bisexual; plants of the Pacific coast states and provinces.

                7. Lemmas 4–7 mm long, smooth or sparsely scabrous between the veins.

                    8. Blades flat or folded, margins smooth, adaxial surfaces smooth or sparsely scabrous; blades of culm leaves gradually reduced in length upwards; collars smooth or sparsely scabrous . . . . . . . . . . . . . . . . . . . . . . . . . . . . . . . . . . .33. *P. chambersii* (in part)

                    8. Blades involute, margins scabrous, adaxial surfaces moderately to densely scabrous or pubescent, especially those of the innovations; blades of the culm leaves steeply reduced in length upwards, some collars usually sparsely hispidulous . . . . . . . . . . . . . . . . . . . . . . . . . . . . . . . . . . . . . . . . . . . . . . .39. *P. piperi*

                7. Lemmas 2.5–5 mm long, moderately to densely scabrous between the veins.

                    9. Lemmas 2.5–4(4.5) mm long; rachilla internodes 0.8–1.1 mm long; panicles 1–5(7) cm long, ovoid, loosely contracted, congested or moderately congested, branches erect to ascending, longest branches 0.5–3 cm; blades 0.5–1(1.5) mm wide . . . . . . . . . . . . . . . . . . . . . . . . . . . . . . . . . . . . . . . . . . . . . . . . . .37. *P. confinis* (in part)

                    9. Lemmas (3.2)4.2–5 mm long; rachilla internodes 1–1.3 mm long; panicles (4) 5.5–10.5(12.5) cm long, ovoid to broadly pyramidal, open, or eventually loosely contracted, sparse, branches laxly ascending, longest branches 2.1–4.5 (7) cm; blades 1.5–2.4 mm wide . . . . . . . . . . . . . . . . . . . . . . . . . . . . . . . . . . .38. *P. diaboli*

         5. Lemmas variously pubescent or glabrous; calluses glabrous or not, webbed or not, hairs long or short; florets never with both glabrous lemmas and long-webbed calluses.

            10. Plants 8–12(20) cm tall; panicles 2.5–5 cm long, erect, with 10–25(30) spikelets, branches smooth or sparsely scabrous; calluses glabrous; palea keels smooth, glabrous or softly puberulent to short-villous; glume keels smooth; leaf blades thin, flat, soft; plants stoloniferous . . . . . . . . . . . . . . . . . . . . . . . . . . . . . . . . . . . . . . . . . . . . . . .12. *P. supina*

            10. Plants (5)10–150 cm tall; panicles 1–30(41) cm long, erect or lax, with 10–100+ spikelets, branches smooth or scabrous; calluses glabrous or with hairs; palea keels sometimes partially scabrous; glume keels smooth or scabrous; leaf blades various; plants stoloniferous or not.

                11. Lemma keels softly puberulent for $^3/_5$ their length, hairs usually sparse, marginal veins glabrous or puberulent to $^1/_4$ their length, intercostal regions smooth and glabrous; lateral veins prominent; calluses webbed; palea keels smooth, muriculate, tuberculate, or scabridulous; lower glumes 1-veined, usually arched

to sickle-shaped; ligules 3–10 mm long, acute to acuminate; panicle branches angled, angles densely scabrous; plants usually weakly stoloniferous . . . . . . . . 49. *P. trivialis* (in part)

11. Lemmas glabrous or variously pubescent, if as above, the lateral veins faint or moderately prominent or the calluses glabrous or the palea keels distinctly scabrous or hairy or the lower glumes 3-veined; calluses glabrous or hairy; palea keels scabrous at least near the apices; lower glumes 1–3-veined, not arched, not sickle-shaped; ligules 0.5–18 mm long, truncate to acuminate; panicle branches terete or angled, smooth or scabrous; plants stoloniferous or not.

12. Culm leaf blades steeply reduced in length upward, flag leaf blades absent or to 1(3) cm long, less than $^1/_9(^1/_5)$ the length of the sheath; calluses glabrous; lemmas usually villous on the keel and marginal veins, glabrous elsewhere, sometimes glabrous throughout; sheaths closed for about $^1/_3$ their length; blades usually all involute and moderately firm, adaxial surfaces, at least those of the innovations, usually densely scabrous to puberulent; panicles contracted; spikelets laterally compressed; florets unisexual; plants of mountain slopes, never of low, wet ground . . . . . . . . . . . . . . . . . . . . . 41. *P. fendleriana* (in part)

12. Culm leaf blades gradually reduced in length upward or the midculm blades longer than those below; flag leaf blades usually over (0.5)1 cm long, usually more than $^1/_7$ the length of the sheath or, if as above, most or all blades flat and the panicles open or the sheaths closed for $^1/_{10}–^1/_5(^1/_4)$ their length and the blades folded and firm and the adaxial surfaces smooth or nearly so and the florets bisexual or the spikelet lengths 4–5 times the widths; calluses glabrous or with hairs; lemmas glabrous or variously pubescent; sheaths closed for $^1/_{10}–^9/_{10}$ their length; blades as above or not; panicles contracted or open; spikelets laterally compressed or subterete; florets bisexual or unisexual; plants of various habitats, sometimes of low, wet ground.

13. Calluses glabrous, diffusely webbed with hairs to $^1/_2$ the lemma length, or with a crown of hairs, or sparsely and dorsally webbed with hairs to $^1/_4$ the lemma length; lemmas glabrous or pubescent [for opposite lead, see p. 122].

14. Spikelets subterete to weakly laterally compressed, the lengths (3.8)4–5 times the widths; panicles usually contracted, sometimes open at anthesis; sheaths closed for $^1/_{10}–^1/_4$ their length; calluses glabrous or with a crown of hairs; adaxial surfaces of the innovation blades glabrous, smooth or scabrous, not densely scabrous between the veins, flat and soon withering or folded and somewhat firm; florets bisexual . . . . . . . . . . . . . . . . . . . . . . . 64. *P. secunda* (in part)

14. Spikelets laterally compressed, the lengths 2–3.5(3.8) times the widths; panicles contracted or open; sheaths closed for $^1/_{10}–^9/_{10}$ their length; calluses glabrous, diffusely webbed, with a crown of hairs, or dorsally webbed with hairs to $^1/_4$ the lemma length; adaxial surfaces of the innovation blades glabrous or with hairs, smooth or densely scabrous between the veins, flat and late withering, folded, soft and firm, or involute; florets bisexual or unisexual.

15. Sheaths closed for $^1/_{10}–^1/_5(^1/_4)$ their length, smooth or sparsely scabrous, glabrous; panicles usually contracted, sometimes loosely contracted or open; paleas usually glabrous between the keels, if hairy, the panicles contracted; lemma keels, marginal veins, and, often, lateral veins short- to long-villous, intercostal regions usually glabrous, if hairy, the panicles contracted; lemma apices often blunt; calluses usually glabrous, occasionally dorsally webbed, hairs to $^1/_4$ the lemma length; innovation blades usually folded and firm, infrequently flat and somewhat soft, adaxial surfaces glabrous, smooth or moderately scabrous, mainly over the veins; florets bisexual; plants usually of low, wet, somewhat alkaline or subsaline soils, from the valleys of the eastern foothills of the Rocky Mountains to the Great Plains, sometimes extending to timberline, rarely on slopes west of the continental divide . . . . . . . . . . 73. *P. arida*

15. Sheaths closed for $^1/_6$–$^9/_{10}$ their length, smooth or retrorsely scabrous, glabrous or with hairs; panicles contracted or open; paleas glabrous or hairy between the keels; lemmas glabrous or variously hairy, apices blunt or pointed; calluses glabrous, shortly webbed, or diffusely webbed; innovation blades flat, folded, or involute, soft or firm, adaxial surfaces glabrous or hairy, smooth or densely scabrous between the veins; florets bisexual or unisexual; plants widespread but not of subalkaline or subsaline soils from the eastern slope of the Rocky Mountains to the Great Plains.

  16. Sheaths closed for $^1/_3$–$^9/_{10}$ their length, sheaths of some leaves densely retrorsely scabrous or short- pubescent, at least on or near the collar margins; ligules of the lower culm leaves and innovations truncate, abaxial surfaces densely scabrous or softly puberulent; upper ligules 0.5–2 mm long; lemmas glabrous, or the keel and marginal veins softly puberulent to short-villous, intercostal region glabrous or hispidulous, infrequently softly puberulent; calluses usually glabrous, rarely shortly webbed.

    17. Sheaths hairy, hairs usually concentrated on and about the collars, collar margin hairs distinctly longer than those below the collar; sheaths closed for $^2/_3$–$^9/_{10}$ their length; blades flat or a few folded, adaxial surfaces smooth or sparsely scabrous, particularly over the veins; florets bisexual and unisexual; plants from west of the Cascade divide . . . . . . . . 30. *P. nervosa*

    17. Sheaths retrorsely scabrous or pubescent for $^1/_4$ or more of the length below the collars, collar and sheath vestiture not differing in length; sheaths closed for $^1/_3$–$^3/_4$ their length; blades of the innovations usually involute, adaxial surfaces usually densely scabrous to hispidulous on and between the veins; florets usually all pistillate, rarely bisexual or staminate; plants primarily from between the 100th meridian and the Cascade and Sierra Nevada mountains of western North America, rarely further west . . . . . . . . . . . . . . . . . . . . . . . . . . . . . . . . . . . . . . . 31. *P. wheeleri*

  16. Sheaths closed for ($^1/_6$)$^1/_5$–$^9/_{10}$ their length, glabrous, collars glabrous, smooth or infrequently moderately scabrous; ligules of the lower culm leaves and lateral shoots truncate to acuminate, smooth or scabrous abaxially, glabrous or softly puberulent; upper ligules 0.5–7 mm long; lemmas glabrous or variously pubescent; calluses glabrous, shortly webbed, or with a crown of hairs.

    18. Paleas pubescent between the keels; sheaths closed for ($^1/_6$)$^1/_5$–$^2/_5$ their length, smooth or slightly scabrous; lemma keels and marginal veins long-villous, intercostal regions usually short-villous, sometimes slightly softly puberulent on the lower back; panicles loosely contracted to open, branches smooth or sparsely scabrous; calluses glabrous or dorsally webbed; florets usually bisexual, anthers aborted late in development or 1.4–2.5 mm long; plants of subalpine to alpine and arctic habitats . . . . 16. *P. arctica* (in part)

    18. Paleas glabrous between the keels; sheaths closed for $^1/_3$–$^9/_{10}$ their length, smooth or scabrous; lemma keels and marginal veins glabrous or pubescent, intercostal regions usually glabrous, infrequently softly puberulent; panicles contracted to open, branches

smooth or sparsely to densely scabrous or hispidulous; calluses glabrous, diffusely webbed or with a crown of hairs; florets bisexual or unisexual; pistillate florets with anthers 0.1–0.2 mm long; plants coastal to subalpine.

19. Lemmas pubescent; calluses glabrous, diffusely webbed, or with a crown of hairs; blades involute, adaxial surfaces scabrous or pubescent, frequently densely so between the veins, or smooth and glabrous and the blades 0.5–1(1.5) mm wide; sheaths closed for $^1/_3$–$^2/_3$ their length; plants of sand dunes and sandy soils along the Pacific coast.

    20. Lemmas 2.5–4(4.5) mm long; panicles fairly tightly to loosely contracted; culms 0.4–0.9 mm thick; blades 0.5–1(1.5) mm wide, thin to moderately thick, soft, mostly filiform, adaxial surfaces sparsely scabrous; calluses diffusely webbed . . . . . . . . . . . . . 37. *P. confinis* (in part)

    20. Lemmas 5–11 mm long; panicles tightly contracted; culms 1–2 mm thick; blades 1–4 mm wide, thick, moderately firm to firm; adaxial surfaces densely scabrous or hispidulous; calluses glabrous, diffusely webbed, or with a crown of hairs.

        21. Panicle rachises and culms beneath the panicles densely hispidulous; lemmas 5–7.5 mm long . . . . . . . . . . . . . . . . . . . . . 35. *P. douglasii*

        21. Panicle rachises and culms beneath the panicles glabrous, smooth or sparsely to moderately scabrous; lemmas (6)7.5–11 mm long . . . . . . . . . . . . . . . . . 36. *P. macrantha*

19. Lemmas and calluses totally glabrous or, if the lemmas pubescent or the calluses dorsally webbed with hairs to $^1/_4$ the lemma length, then the blades flat or folded, 2–5 mm wide, smooth or sparsely scabrous adaxially; sheaths closed for $^1/_4$–$^9/_{10}$ their length; plants of inland regions, not growing in sand.

    22. Panicles 3–7 cm long, densely contracted, branches smooth or sparsely scabrous distally; spikelets 3.5–5.5 mm long, compact, rachilla internodes about 0.5 mm long; lemmas and calluses smooth, glabrous . . . . . . . . . . . . . . . . . . . . . . . . . . . . . . . .40. *P. atropurpurea*

    22. Panicles 2–22 cm long, densely contracted or open, if densely contracted, the branches sparsely to densely scabrous or spikelets 5.5–12 mm long; spikelets 3–12 mm long, looser, rachilla internodes 0.5–1.5 mm long; lemmas and calluses smooth or scabrous, glabrous or hairy.

        23. Basal branching nearly all intravaginal or mixed intra- and extravaginal; at least some innovation leaves with involute blades 0.5–2 mm wide and scabrous or pubescent on the adaxial surfaces; plants rarely rhizomatous, usually densely tufted; lemmas sparsely to densely scabrous, glabrous or

sparsely softly puberulent near the base of the keels and/or marginal veins . . . . . . . . . . . . . . . . . . . . . . . . . . . 42. *P. cusickii* (in part)

23. Basal branching all or mainly extravaginal; blades flat or folded, 2–5 mm wide, adaxial surfaces smooth or sparsely scabrous; plants shortly rhizomatous, loosely tufted or the culms solitary; lemmas smooth or sparsely scabrous, glabrous or the keel and marginal veins hairy, intercostal regions rarely pubescent.

24. Panicles (5)12–22 cm, open, branches spreading to eventually reflexed; calluses glabrous . . . . . . . . . . 28. *P. arnowiae*

24. Panicles 2–9 cm long, tightly to loosely contracted, branches erect to ascending or scarcely spreading; calluses of some lemmas usually shortly webbed . . . . . . . . . . . 33. *P. chambersii* (in part)

13. Calluses dorsally webbed, hairs over (¹/₃)¹/₂ the length of the lemmas, sometimes with additional webs below the marginal veins; lemma short- to long-villous on the keels and marginal veins [for opposite lead, see p. 119].

25. Sheaths closed for ¹/₁₀–¹/₅ their length; spikelets 3–5 mm long; lemmas 2–3 mm long, glabrous between the keels and marginal veins; panicle branches angled, angles densely scabrous; plants sometimes stoloniferous, sometimes branching above the culm bases; florets bisexual . . . . . . . . . . . . . . . . . . . . . . . . . . . 54. *P. palustris* (in part)

25. Sheaths closed for (¹/₆)¹/₅–⁹/₁₀ their length; spikelets 3.5–12 mm long; lemmas 2–8 mm long, glabrous or hairy between the keels and marginal veins; panicle branches terete or angled, smooth or scabrous; plants rarely stoloniferous, usually rhizomatous, never branching above the culm bases; florets bisexual or unisexual.

26. Sheaths closed for (²/₅)¹/₂–⁹/₁₀ their length, weakly to distinctly compressed, keels distinct, sometimes winged, wing to 0.5 mm wide, glabrous or the sides, collars, or throats pubescent; plants loosely tufted, shortly rhizomatous, never forming dense turf; culm blades flat or slightly folded, infrequently folded; innovations all or almost all extravaginal or a few intravaginal, with the intravaginal blades not involute and distinctly narrower than the culm blades; florets bisexual or unisexual, commonly some florets pistillate; anthers 0.1–0.2 mm or (1.3)2–4 mm long; plants mostly of forest openings and mountain thickets.

27. Blades steeply reduced in length up the culms, flag leaf blades 0.2–3(6) long; panicles broadly pyramidal; usually at least some upper lemmas within the spikelets pubescent between the veins . . . . . . . . . . . . . . . . . . . . . . . . . . . 29. *P. cuspidata*

27. Blades not steeply reduced in length up the culm, midculm blades sometimes longer than those below, flag leaf blades (1.4)3–20 cm long; panicles loosely contracted to narrowly pyramidal; lemmas glabrous or sparsely pubescent between the veins.

28. Panicles erect, usually narrowly pyramidal, (8)13–29 cm long, proximal internodes usually longer than 4 cm; usually some lemmas within the spikelets pubescent between the veins . . . . . . . . . . . . . . . . . . . . . . . . . 27. *P. tracyi*

28. Panicles nodding, ovoid, (2)4–10 cm long, proximal internodes 1.8–3 cm long; lemmas glabrous between the veins . . . . . . . . . . . . . . . . . . . . . . . . . . . . . . . . . . . 32. *P. rhizomata*

26. Sheaths closed for $(^1/_6)^1/_5$–$^1/_2(^3/_5)$ their length, terete to compressed, with or without distinct keels, usually glabrous, the sides infrequently retrorsely scabrous or pubescent; plants densely to loosely tufted or with solitary culms, sometimes forming dense turf; culm blades flat or folded; innovations all extravaginal or some intravaginal, blades of the intravaginal shoots sometimes involute and distinctly narrower than the culm blades; florets bisexual; anthers usually 1.2–2.5 mm long, sometimes some anthers aborting late in development and 1–1.5 mm long; plants widespread, sometimes of coastal belts and alpine and arctic habitats.

29. Glumes subequal in length and width, usually nearly equaling the adjacent lemmas, distinctly keeled, keels scabrous; lower glumes (4)4.5–7 mm long; lemmas (4)5–8 mm long, the intercostal regions usually moderately to densely scabrous or hispidulous, infrequently softly puberulent to short-villous near the base and moderately to densely scabrous to hispidulous in the middle $^1/_3$, rarely nearly smooth near the base and sparsely scabrous distally; intercostal regions of the paleas usually hispidulous, infrequently puberulent; blades (2)3–7 mm wide; culms usually stout, (20)30–120 cm tall; plants of coastal shores and low elevation wet meadows in Alaska and the low arctic . . . . . . . . . . . . . . . . . . . . . . . . . . 15. *P. macrocalyx* (in part)

29. Glumes unequal to subequal in length and width, distinctly shorter than to subequal to the adjacent lemmas, distinctly or weakly keeled, keels smooth or scabrous; lower glumes 1.5–5(6) mm long; lemmas 2–6(7) mm long, the intercostal regions smooth, glabrous or pilose to long-villous, and smooth or sparsely scabrous distally; intercostal regions of the paleas glabrous or pilose to short-villous; blades 0.4–6 mm wide; culms slender to stout, 10–70(100) cm tall; plants widespread.

30. Palea keels usually pubescent, rarely nearly glabrous, intercostal regions usually at least sparsely and softly puberulent near the base, sometimes glabrous; glumes weakly to distinctly keeled, the keels smooth or sparsely to moderately scabrous; upper glumes usually subequal to the lower lemmas or slightly shorter; lemma intercostal regions and lateral veins pubescent near the base; ligules smooth or sparsely scabrous, usually rounded or obtuse to acute, infrequently truncate, entire or lacerate, not ciliolate; panicle branches (1)2–5 per node, usually smooth or sparsely scabrous, infrequently moderately scabrous . . . . . . . . . . . . . . . . . . . . . . . . . . . . . . . . . . . . . . . . . . 16. *P. arctica* (in part)

30. Palea keels glabrous or pubescent, intercostal regions glabrous, rarely sparsely hispidulous; glumes distinctly keeled, the keels usually sparsely to densely scabrous distally, infrequently smooth; upper glumes usually distinctly shorter than the lower lemmas; lemma intercostal regions glabrous, lateral veins glabrous or pubescent; ligules smooth or scabrous, usually truncate or rounded, infrequently obtuse to acute, entire, glabrous, or ciliolate; panicle branches (1)2–7(9) per node, smooth or sparsely to fairly densely scabrous.

31. Intercostal surfaces of the lemmas visible, not or only partly concealed by hairs; lemma keels and marginal veins moderately to densely long-villous, more or less straight, lateral veins glabrous or softly puberulent, infrequently short-villous; panicle branches and ligules smooth or sparsely to fairly densely scabrous, longest branches 1–9 cm; plants widespread . . . . . . . . . . . . . . . . . . . . . . . . . . . . . . . . . . . . . . . . . . . . 13. *P. pratensis* (in part)
31. Intercostal surfaces of the lemmas concealed by the hairs over the keels and veins; lemma keels, marginal veins, and lateral veins copiously hairy, hairs of the keels and marginal veins cottony, those of the lateral veins somewhat shorter and sparser; panicle branches and ligules smooth or nearly so, longest branches 1–3 cm; plants of high arctic sands . . . . . . . . . . . . . . . . . . . . . . . . . . . 14. *P. sublanata*

## *Poa* Subkey IV

**Plants** perennial, not rhizomatous, not stoloniferous, loosely to densely tufted. **Culms** not bulbous-based. **Basal sheaths** not swollen. **Spikelets** not bulbiferous, florets developing normally. **Anthers** (1.2)1.3–4 mm long and dehiscent, or all rudimentary, having no or poorly formed pollen.

1. Calluses usually dorsally webbed, webs sometimes with 1 to few minute hairs, rarely the hairs somewhat diffuse [for opposite lead, see p. 125].
    2. Lemma lateral veins pronounced, keels pubescent, marginal veins glabrous or softly puberulent at the base, lemmas glabrous elsewhere; lower glumes 1-veined, subulate to narrowly lanceolate, usually arched to sickle-shaped; callus web well developed . . . . . . . . . . . . . . . . . . . . . . . . . . . 49. *P. trivialis* (in part)
    2. Lemma lateral veins obscure to pronounced, keels glabrous throughout or, if pubescent, the marginal veins distinctly pubescent for more than ¼ their length, lemma lateral veins and intercostal regions glabrous or pubescent, or, if pubescent as in *P. trivialis*, then the callus web short, scant, poorly developed and the lower glumes 3-veined and lanceolate or broader.
        3. Panicles open, conical, with whorls of (2)3–10, spreading to eventually reflexed, scabrous-angled branches at the lower nodes; lemmas hairy on the keel and veins, sometimes the intercostal regions also hairy; callus webs well developed . . . . . . . . . . . . . . . . . . . . . . . . . . . . . . . . . . 3. *P. sylvestris* (in part)
        3. Panicles contracted to open, if open then not conical and without whorls of (2)3–10, eventually reflexed, scabrous-angled branches at the lower nodes; branches smooth or scabrous-angled; lemmas glabrous or hairy; calluses glabrous, with diffuse hairs, or with a scanty or well-developed web.
            4. Sheaths closed for (⅕)⅓–¾ their length.
                5. Culms 8–35 cm tall, 0.5–0.8 mm thick; panicle branches smooth or sparsely scabrous; anthers to 1.3 mm long; plants alpine . . . . . . . . . . . . . . . . . . . . . . . . . . . . . . . . . . . . . . 50. *P. laxa* (in part)
                5. Culms 23–120 cm tall, 0.5–2 mm thick; panicle branches smooth or scabrous; anthers to 1.8 mm long; plants of many habitats, including alpine habitats.
                    6. Lemmas usually hairy on the keel and marginal veins, usually also on the intercostal regions; palea keels softly puberulent to short-villous at midlength; panicles open and erect, broadly pyramidal at maturity; callus webs sparse, poorly developed . . . . . . . 6. *P. autumnalis* (in part)
                    6. Lemmas glabrous or with a few hairs at the base of the keel or marginal veins; palea keels scabrous, glabrous; panicles contracted to loosely contracted or open and lax; callus webs scant and short or well developed.
                        7. Callus webs well developed; lemma keels glabrous; plants of eastern North America . . . . . . . . . . . . . . . . . . . . . . . . . . . . . . . . . . . . . . . . . . . . . . . . . . . 1. *P. saltuensis* (in part)
                        7. Callus webs minute, sometimes somewhat diffuse; lemma keels glabrous or sparsely softly puberulent near the base; plants of western North America . . . . . . . . . . . . . 42. *P. cusickii* (in part)
            4. Sheaths closed for 1/20–¼(⅓) their length.
                8. Basal branching all or mostly intravaginal; plants not stoloniferous.
                    9. Culms 30–90 cm tall; panicles open; plants of the mountains in and around the Chihuahuan Desert . . . . . . . . . . . . . . . . . . . . . . . . . . . . . . . . . 26 *P. strictiramea* (in part)

9. Culms 5–15(20) cm tall; panicles contracted; alpine plants of the Rocky Mountains
. . . . . . . . . . . . . . . . . . . . . . . . . . . . . . . . . . . . . . . . . . . . . . . . . . . . . . . . . . . . . . . . . .61. *P. abbreviata* (in part)

8. Basal branching all or mostly extravaginal, or extra- and intravaginal and the plants stoloniferous.

10. Flag leaf nodes usually in the lower $^1/_{10}$–$^1/_3$ of the culms; flag leaf blades usually distinctly shorter than their sheaths; lemmas sometimes softly puberulent between the veins, lateral veins usually with at least a few minute hairs; ligules 1–4(5) mm long . . 57. *P. glauca* (in part)

10. Flag leaf nodes usually in the upper $^2/_3$ of the culms; flag leaf blades shorter or longer than their sheaths; lemmas glabrous between the veins, lateral veins usually glabrous, rarely with 1 to several minute hairs; ligules 0.2–6 mm long.

11. Spikelets narrowly lanceolate to lanceolate; glumes subulate to narrowly lanceolate, gradually tapering to narrowly acuminate apices; lower glume lengths 6.4–11 times the widths; ligules 0.2–0.5(1) mm long, truncate; flag leaf nodes at or above the middle of the culms; flag leaf blades usually longer than their sheaths; rachillas usually hairy, hairs to 0.15 mm long; webs usually short, scanty
. . . . . . . . . . . . . . . . . . . . . . . . . . . . . . . . . . . . . . . . . . . . . . . . .55. *P. nemoralis* (in part)

11. Spikelets and glumes not as above or, if so, the ligules 1.5–6 mm long, truncate to acute, and the rachillas glabrous; flag leaf nodes at or above the lower $^1/_3$ of the culm; flag leaf blades longer or shorter than their sheaths; webs short or long, scanty or not.

12. Panicles (9)13–30(41) cm long, branches 4–15 cm long; culms closely spaced to isolated at the base; lower glumes tapering to the apices, lengths 6.4–10 times the widths; lemma keels abruptly inwardly arched beneath the scarious apices; lemma margins distinctly inrolled; rachillas usually muriculate, rarely sparsely hispidulous; web hairs usually longer than $^2/_3$ the length of the lemmas . . . . . . . . . . . . . . . . . . . . . . . . . . . . . . . . . . . . . . . .54. *P. palustris* (in part)

12. Panicles (1.5)3–15(17) cm long, branches 0.4–8(9) cm long; culms closely spaced at the base; lower glumes abruptly narrowing to the apices, lengths 4.5–6.3 times the widths; lemma keels not abruptly inwardly arched beneath the scarious apices; lemma margins not or slightly inrolled; rachillas usually muriculate or softly puberulent; web hairs shorter than $^1/_2($$^2/_3)$ the length of the lemmas . . . . . . . . . . . . . . . . . . . . . . . . . . . . . . . . . . . . . . . .56. *P. interior* (in part)

1. Callus glabrous or with a crown of hairs, hairs 0.1–2 mm long [for opposite lead, see p. 124].

13. Lemmas and calluses glabrous [for opposite lead, see p. 127].

14. Blades (4) 6–15 mm wide, flat or folded; sheaths closed for $^1/_2$–$^3/_4$ their length, strongly compressed, keeled, keels winged . . . . . . . . . . . . . . . . . . . . . . . . . . . . . . . . . . . . . . . . . . .25. *P. chaixii*

14. Blades 0.5–5 mm wide, flat, folded, or involute; sheaths closed for $^1/_{20}$–$^4/_5$ their length, not strongly compressed, if keeled, keels not winged.

15. Sheaths closed for $^2/_5$–$^4/_5$ their length; panicles 1–5(8) cm long, with (1)6–17(22) spikelets, nodes with 1–2 branches; branches appressed to spreading, smooth or sparsely scabrous; spikelets strongly compressed, lanceolate to broadly ovate; ligules hyaline, smooth, (1)2–4 mm long; blades 0.5–1 mm wide, thin, lax, filiform, soon withering; plants from the Columbia Plateau to southwestern Idaho and northwestern Nevada . . . . . . . . . 45. *P. leibergii*

15. Sheaths closed for $^1/_{20}$–$^3/_4$ their length, if for $^2/_5$–$^3/_4$, then the panicles longer than 8 cm or with more than 20 spikelets or the ligules of the innovations (and sometimes also the culms) 0.5–2.5 mm long and scabrous and often milky white; blades (0.5)1–5 mm wide, sometimes moderately thick and firm and holding their form; plants of many regions, including the range of *P. leibergii*.

16. Panicles (7)10–30 cm long, open, pyramidal, nodes with 2–5 moderately to densely scabrous branches; sheaths closed for $^1/_{20}$–$^1/_{10}$ their length; basal branching intravaginal; plants of the Chisos Mountains of Texas to Mexico . . . . . . . . . . 26. *P. strictiramea* (in part)

16. Panicles 1–25 cm long, contracted to loosely contracted or, if open, nodes with 1–3(5) smooth or scabrous branches; sheaths closed for $^1/_{20}$–$^3/_4$ their length; basal branching intravaginal, extravaginal, or both; plants of many regions, including the range of *P. strictiramea*.

17. Sheaths closed for $(^1/_4)$$^1/_3$–$^3/_4$ their length; florets often unisexual, anthers 2–3.5 mm long or nonfunctional and to 1.8 mm long; uppermost ligules of the innovation leaves 0.2–0.5(2.5) mm long, scabrous, usually truncate; innovation blades usually involute; panicles contracted, loosely contracted, or open; lower glumes distinctly shorter than the lowest lemmas.

18. Flag leaf blades usually absent or to 1 cm long; blades of the culm leaves sharply reduced in length upwards, similar in thickness and form to those of the innovations, moderately firm, usually involute; plants of southeastern Arizona and southwestern New Mexico . . . . . . . . . . . . . 41. *P. fendleriana* (in part)

18. Flag leaf blades usually present and 1+ cm long; blades of the culm leaves not sharply reduced in length upwards, sometimes differing in thickness or form from those of the innovations, soft, narrow and withering or broader and flat; plants from other parts of the *Manual* region.

   19. Panicles contracted or loosely contracted, branches smooth or sparsely to densely scabrous; innovation blades 0.5–2 mm wide, abaxial surfaces smooth or scabrous, adaxial surfaces usually densely scabrous or hispidulous; plants from southern Yukon Territory to California and Colorado . . . . . . . . . . . . . . . . . . . . . . . . . . . . . . . . . . 42. *P. cusickii* (in part)

   19. Panicles open or slightly contracted, branches smooth or sparsely scabrous; innovation blades 1–2 mm wide, abaxial surfaces smooth, adaxial surfaces usually smooth or sparsely scabrous; plants of Alaska, Yukon Territory, and Northwest Territories . . . . . . . . . . . . . . . . . . . 44. *P. porsildii*

17. Sheaths closed for $^1/_{20}$–$^2/_5$ their length; if sheaths closed for $^1/_4$–$^2/_5$ their length then all florets bisexual, or the functional anthers 1.2–1.8 mm long, or the ligules of the uppermost innovation leaves 2+ mm long and smooth or scabrous, or the lower glumes subequal to the lowest lemmas; blades of the innovation leaves involute or not; panicles contracted; lower glumes shorter than to equaling the lowest lemmas.

   20. Culms 5–40 cm tall; panicles 3–7 cm long, densely contracted, nearly cylindrical; culm blades to 5 mm wide, often a bit fleshy and broader than those of the innovations, those of the innovations usually thin and soon withering, infrequently all blades flat and a bit fleshy; florets bisexual; plants of the Pacific coast . . . . . . . . . . . . . . . . . . . . . . . . . . . 71. *P. unilateralis* (in part)

   20. Culms 2–120 cm tall; panicles 1–25 cm long, densely to loosely contracted, not cylindrical; culm blades 0.5–5 mm wide and soft, culm and innovation blades not much differentiated or, if differentiated, then the basal blades moderately firm and involute; florets unisexual or bisexual; plants of non-coastal regions.

      21. Culms 30–120 cm tall; panicles (4)5–25 cm long; spikelet lengths 3–5 times the widths; plants of saline or non-saline habitats, often below the subalpine zone, if of non-saline habitats, the spikelet lengths (3.8)4–5 times the widths and the panicles usually over 10 cm long.

         22. Spikelets (4)7–10 mm long, subterete, narrowly lanceolate, lengths usually (3.8)4–5 times the widths; plants of many habitats, widespread . . . . . . . . . . . . . . . . . . . . . . . . . . . . . . . 64. *P. secunda* (in part)

         22. Spikelets 4.5–7 mm long, compressed, lengths 3–3.5 times the widths; plants of mineralized soils around hot springs in Napa County, California . . . . . . . . . . . . . . . . . . . . . . . . . . . . . . . 70. *P. napensis* (in part)

      21. Culms 2–40 cm tall; panicles 1–8 cm long; spikelet lengths 2–4 times the widths; plants of non-saline, subalpine or alpine habitats.

         23. Ligules of the innovations 2.5–6 mm long, hyaline, smooth; panicles loosely contracted or contracted; basal branching extravaginal; lower glumes distinctly shorterthan the lowest lemmas; florets often unisexual . . . . . . . . . . . . . . . . . . . . . . . . . . . . . 46. *P. stebbinsii*

         23. Ligules of the innovations 0.5–2.5 mm long, usually milky, often scabrous; panicles contracted; some or all basal branching intravaginal; lower glumes distinctly shorter than or subequal to the lowest lemmas; florets bisexual or unisexual.

            24. Florets unisexual; anthers 2–4 mm long; blades involute . . . . . . . . . . 47. *P. pringlei*

            24. Florets bisexual; anthers 0.6–3.5 long; blades flat, folded, or involute.

               25. Anthers 2.2–3.5 mm long; culms 15–40 cm tall; longest culm blades 1–3 cm long and fairly firm, with thick

white margins and broadly prow-tipped apices, basal blades similar; plants of serpentine soils in Washington . . . . . . . . . . . . . . . . . . . . . . . . . . . . . . . . . . . . . . . 66. *P. curtifolia* (in part)

25. Anthers 0.6–1.2(2) mm long; culms 2–25 cm tall; basal and upper culm leaves not always similar, culm leaves without the above combination of characteristics; plants of non-serpentine soils from British Columbia to California.

26. Abaxial surfaces of the innovation blades smooth or sparsely scabrous, epidermes with papillae on the long cells (at 100×); abaxial surfaces of the flag leaf blades with 7–15 closely spaced ribs; culms 2–6(10) cm tall; plants of California . . . . . . . . . . . 62. *P. keckii* (in part)

26. Abaxial surfaces of the innovation blades densely hispidulous, scabrous, or softly puberulent, rarely smooth and glabrous, lacking papillae on the long cells (at 100×); abaxial surfaces of the flag leaf blades with 5–9 well-spaced ribs; culms 7–25 cm tall; plants of British Columbia, Washington, and Oregon . . . . . . . . . . . . . . . . . . . . . . . . . . . . 63. *P. suksdorfii* (in part)

13. Lemmas with hairs; calluses glabrous or hairy [for opposite lead, see p. 125].

27. Sheaths closed for ¹/₂–³/₄ their length; lemmas mostly glabrous, lower lemmas of some spikelets usually sparsely softly puberulent near the base of the keels and/or marginal veins, lemmas glabrous elsewhere; panicles 4–7 cm long, with 13–50 spikelets, branches smooth or sparsely scabrous; spikelets 7–10 mm long, strongly laterally compressed; florets pistillate; plants of subalpine to alpine habitats, from southern British Columbia to California . . . . . . 42. *P. cusickii* (in part)

27. Sheaths closed for ¹/₂₀–³/₄ their length, if closed for ¹/₂–³/₄ their length, the lemmas pilose between the veins or the panicle branches moderately to densely scabrous; lemmas variously pubescent, frequently pilose between the veins; panicles 1–30 cm long, with 9–100+ spikelets; branches smooth or sparsely to densely scabrous; spikelets 3–12 mm long, subterete to strongly laterally compressed; florets bisexual or unisexual; plants of various habitats, including subalpine to alpine habitats, widely distributed, including from British Columbia to California.

28. Sheaths closed for ¹/₃–¹/₂ their length; panicles (5)8–20 cm, erect or lax, broadly pyramidal at maturity, open, lower axils sometimes sparsely hairy; panicle branches spreading to reflexed, angled, longest branches 5–12 cm long, with 3–8 spikelets in the distal ¹/₃–¹/₄; paleas pilose; florets bisexual; plants of eastern North American woods . . . . . . . . . . . . . . . . . . . . . . . . . . . . . . . . . . . . . . . . . . . . . . . . . 6. *P. autumnalis* (in part)

28. Sheaths closed for ¹/₂₀–³/₄ their length, if closed for ¹/₃–³/₄ their length, then the panicles contracted or loosely contracted, or the branches smooth, or the longest branches shorter than 5 cm, or the paleas glabrous, or the spikelets unisexual; panicles 1–40 cm long, lower axils glabrous; panicle branches erect, ascending, or widely divergent, terete or angled; plants widely distributed, including eastern North American woods.

29. Basal branching mainly extravaginal, usually occurring late in the season; sheaths closed for ¹/₁₀–¹/₅ their length; blades usually flat, sometimes folded, thin, soft; panicle branches usually scabrous-angled; lemmas distinctly keeled; spikelets laterally compressed, lengths 2–3 times the widths.

30. Lemmas glabrous on the lateral veins and intercostal regions; culms usually with 1–2(3) nodes exserted, uppermost node usually at or above the lower ¹/₃ of the culm . . . . . . . . . . . . . . . . . . . . . . . . . . . . . . . . . . . . . . . . 56. *P. interior* (in part)

30. Lemmas usually at least sparsely softly puberulent on the lateral veins, intercostal regions with similar hairs or glabrous; culms with 0–1 nodes exserted, uppermost node usually in the lower ¹/₁₀–¹/₃ of the culms . . . . . . . . . 57. *P. glauca* (in part)

29. Basal branching intra- or extravaginal or mixed, if mostly extravaginal, branching often occurring early in the season; sheaths closed for ¹/₂₀–³/₄ their length; blades involute, flat, or folded, thin and soft to thick and firm; panicle branches terete or angled, smooth or scabrous; lemmas weakly to distinctly keeled; spikelets subterete or laterally compressed, lengths 1.5–5 times the widths.

31. Spikelets ovate, rachilla internodes 0.5–0.8 mm; panicles 2–6(8) cm long, open or loosely contracted; branches terete, usually smooth or sparsely scabrous,

rarely moderately densely scabrous, longest branches 1–3(4) cm; leaves mostly basal, blades 1–6(12) cm long, 2–4.5 mm wide, flat, soft; calluses glabrous; lemmas distinctly keeled, keels and marginal veins long- to short-villous, intercostal regions short-villous; paleas softly puberulent to short-villous at midlength; florets bisexual . . . . . . . . . . . . . . . . . . . . . . . . . . . . . . . . . . . . 9. *P. alpina* (in part)

31. Spikelets lanceolate to narrowly ovate, rachilla internodes 0.5–2 mm long; panicles 1–40 cm long, open or contracted; branches terete or angled, smooth or variously scabrous, longest branches 0.5–15 cm; leaves not as above; calluses glabrous or with a crown of hairs; lemmas weakly to distinctly keeled, variously hairy; paleas glabrous or softly puberulent; florets bisexual or unisexual.

32. Panicles contracted; florets usually unisexual, rarely bisexual, commonly pistillate; blades usually involute.

33. Sheaths closed for $^1/_7$–$^1/_3$ their length; lemmas weakly keeled; calluses usually with a crown of hairs around the base of the lemma; adaxial surfaces of the innovation blades smooth or somewhat scabrous; anthers late-aborted, 0.8–1.8 mm long; plants of the high arctic . . . . . 67. *P. hartzii* (in part)

33. Sheaths closed for $^1/_4$–$^3/_4$ their length; lemmas strongly keeled; calluses glabrous; adaxial surfaces of the innovation blades usually hispidulous to softly puberulent on and between the veins; anthers of pistillate plants rudimentary, 0.1–0.2 mm long; plants not arctic.

34. Sheaths closed for about $^1/_3$ their length; culm leaf blades sharply reduced in length upwards, the flag leaf blades absent or vestigial, commonly less than 1 cm long, always less than $^1/_5$ the sheath length, when present usually firm, not withering; innovation blades usually 1–3 mm wide; lemmas short- to long-villous on the keel and marginal veins . . . . . . . . . . . . . . . . . . . . . . . . . 41. *P. fendleriana* (in part)

34. Sheaths closed for $^1/_4$–$^3/_4$ their length; culm leaf blades gradually reduced in length upward along the culm or some midculm blades longer than the lower culm blades, culm blades narrow, thin and withering; innovation blades usually 0.5–1(2) mm wide; lemmas usually softly puberulent, sometimes short-villous on the keel and marginal veins . . . . . . . . . . . . . . . . . . . . . . . . . . . . . . . . . . . . 43. *P. ×nematophylla*

32. Panicles contracted or open; florets usually bisexual, if unisexual, the panicles open; blades flat, folded, or involute.

35. Panicles open or loosely contracted at maturity, 5–30 cm long, spikelets not crowded.

36. Spikelets laterally compressed, lengths usually 3–3.8 times the widths; lemmas distinctly keeled, intercostal regions glabrous or with hairs distinctly shorter than those over the keel and marginal veins.

37. Lemmas 2.5–3.5 mm long, usually glabrous throughout, infrequently with keels and marginal veins softly puberulent to short- or long-villous and/or intercostal regions sparsely softly puberulent; blades usually involute, rarely flat, scabrous; calluses usually glabrous, rarely sparsely and shortly webbed; panicles (7)10–30 cm long; plants of the Chisos Mountains in Texas and northern Mexico . . . . . 26. *P. strictiramea* (in part)

37. Lemmas 4–6 mm long, keels and marginal veins, sometimes also the lateral veins, short- to long-villous, intercostal regions glabrous or sparsely pilose or hispidulous near the bases; blades flat or folded, smooth or sparsely scabrous; calluses glabrous or with a crown of hairs, hairs 0.2–2 mm; panicles 5–18(25) cm long; plants of coastal Alaska, the Pacific Northwest, and Rocky Mountains . . . . . . . . . . . 69. *P. stenantha* (in part)

36. Spikelets subterete, lengths usually (3.8)4–5 times the widths; lemmas weakly keeled, usually at least sparsely softly puberulent, infrequently short-villous, between the veins, the hairs usually about the same length as those of the keel and marginal veins.

38. Ligules of the culm leaves usually 2–6 mm long, smooth or scabrous, truncate to acuminate; basal tuft of leaves narrow

or loosely clumped; basal leaves reaching 2–20+ cm, blades filiform or to 3 mm wide; panicle branches capillary or stouter, smooth or scabrous; plants widespread, sometimes on serpentine soils, often in wet habitats . . . . . . . . . . . . . . 64. *P. secunda* (in part)

38. Ligules of the culm leaves 0.5–1.5(2.5) mm long, scabrous, apices truncate to obtuse (acute); basal tuft of leaves narrow, tightly clumped; basal leaves reaching 2–8(13) cm, basal blades filiform; panicle branches capillary, distinctly scabrous; plants of thin, early drying serpentine soils in the Sierra Nevada foothills of California . . . . . . . . . . . . . . . . . . . . . 65. *P. tenerrima*

35. Panicles contracted at maturity, sometimes open during anthesis, 1–30 cm long, spikelets crowded or not.

39. Plants 2–6(10) cm tall; panicles 1–4(6) cm long; cauline blades soft, folded, 1–3.5(4.5) cm long, upper cauline blades 0.9–1.8 mm wide, abaxial surfaces with 7–15 ribs; spikelets 3.5–6 mm long; calluses glabrous; lemmas glabrous or the keel and marginal veins sparsely softly puberulent near the base; plants of high alpine habitats in the Sierra Nevada and adjacent ranges . . . . . . . 62. *P. keckii* (in part)

39. Plants 5–120 cm tall; panicles 2–30 cm long; leaves not as above in all respects; spikelets 3–10 mm long; lemmas nearly glabrous to copiously pubescent, if hairy, only so near the base on the keel and marginal veins, the hairs softly to crisply puberulent; calluses glabrous or with a crown of hairs; plants widespread.

40. Plants 5–40 cm tall; panicles 3–7 cm long, usually densely contracted, rarely loosely contracted, nearly cylindrical; culm blades 2–5 mm wide, flat or folded; innovation blades usually 1–1.5 mm wide, involute, infrequently similar to the culm blades; anthers fully developed; plants of the Pacific coast . . . . . . . . . . . . . . . . . . . . . . . . . . . . . . . . . . . . 71. *P. unilateralis* (in part)

40. Plants 10–120 cm tall; panicles 2–30 cm long, densely to loosely contracted, not nearly cylindrical; culm and innovation blades similar, 1–5 mm wide; anthers sometimes aborted late in development; plants of the high arctic or interior habitats of western North America.

41. Anthers usually sterile and to 1.5 mm long, infrequently well developed and 2–2.8 mm long; plants 10–33(45) cm tall; blades folded to involute, 1.5–3 mm wide, abaxial surfaces smooth or sparsely scabrous; spikelets lustrous; lemmas usually weakly keeled, more or less evenly and loosely short- to long-villous on the lower $^1/_3$–$^1/_2$, hairs mostly longer than 0.5 mm; calluses usually with a crown of hairs to 2 mm long; panicle branches smooth or sparsely to moderately scabrous; plants of the high arctic . . . . . . . . . . . . . . . . . . . . . . . . . . . . . . . . 67. *P. hartzii* (in part)

41. Anthers well developed, 1.2–3.5 mm long; plants 10–120 cm tall; blades flat, folded, or involute, 0.5–5 mm wide, abaxial surfaces smooth or scabrous; spikelets lustrous or not; lemmas weakly keeled or not, if the intercostal regions hairy, the hairs distinctly shorter than those on the keels or, if the lemmas more or less evenly hairy, then the hairs usually shorter than 0.5 mm; calluses usually glabrous, infrequently with a crown of hairs to 2 mm long; panicle branches smooth or sparsely to densely scabrous; plants of the high arctic or western North America, if of the high arctic, the lemma hairs shorter than 0.3 mm.

42. Lemmas usually evenly strigulose across the lower $^1/_3$–$^1/_2$, hairs 0.1–0.2 mm long, rarely to 0.3 mm on the keel and marginal veins; blades soft, involute; culms 10–30 cm tall; panicles 3–6 cm long, longest

branches 1–3(4) cm long, smooth or sparsely
scabrous ........................................ 68. *P. ammophila*

42. Lemmas variously hairy, if as above, the panicle
branches usually scabrous and/or the blades flat
and soon withering; culms (10)20–120 cm tall;
panicles 2–30 cm long; branches 1–15 cm long,
sparsely to densely scabrous.

43. Culm blades 1–3 cm long, flat, (1)1.5–3 mm
wide, with thick, white margins and broadly
prow-shaped apices; panicles 4–8 cm long,
narrowly lanceoloid; spikelets 7–9 mm long;
plants of serpentine slopes in the Wenatchee
Mountains of Washington ............. 66. *P. curtifolia* (in part)

43. Culm blades longer or narrower than above,
margins not thick and white, apices narrowly
prow-shaped; panicles 2–30 cm long, narrowly
lanceoloid to ovoid; spikelets (4)5–10 mm
long; plants widespread.

44. Spikelets subterete to weakly laterally
compressed, (4)5–10 mm long, lengths
3.5–5 times the widths; rachilla internodes
usually 1–2 mm long; lemmas usually
weakly keeled, 3.5–6 mm long, nearly
glabrous or hairy all over the basal $^2/_3$;
culms (10)15–120 cm tall; panicles
2–25(30) cm long ................. 64. *P. secunda* (in part)

44. Spikelets laterally compressed, (4)4.5–7
mm long, lengths 3–3.5 times the widths;
rachilla internodes usually shorter than 1
mm; lemmas distinctly keeled, 3–4 mm
long, usually glabrous, keels and marginal
veins rarely sparsely puberulent
proximally; culms 30–100 cm tall;
panicles 5–15 cm long ............. 70. *P. napensis* (in part)

## Poa L. subg. Poa

*Poa* subg. *Poa* is the largest subgenus of *Poa*. Its
distribution is essentially the same as that of the genus. It
includes all but one of the 70 species of *Poa* in the *Manual*
region; *P. eminens* is included in subg. *Arctopoa*.

## Poa sect. Sylvestres V.L. Marsh *ex* Soreng

*Poa* sect. *Sylvestres* includes seven species, all of which are
endemic to the *Manual* region.

### 1. Poa saltuensis Fernald & Wiegand OLDPASTURE BLUEGRASS [p. 396, 521]

*Poa saltuensis* grows in woodlands of the north-central and
northeastern United States and adjacent Canada, extending south to
Tennessee. The two subspecies are sometimes treated as species. The
variation between the two overlaps and is correlated to some extent
with ecology and geography. *Poa marcida*, a western species once
included in *P. saltuensis*, differs in having closed sheaths and attenuate
lemmas.

1. Anthers 0.4–1 mm long; lemma apices obtuse to
acute, firm or scarious for up to 0.25 mm ..... subsp. *languida*
1. Anthers 0.9–1.5 mm long; lemma apices acute to
acuminate, scarious for 0.25–0.5 mm ........ subsp. *saltuensis*

### Poa saltuensis subsp. languida (Hitchc.) A. Haines [p. 396]

*Poa saltuensis* subsp. *languida* grows in rich open woodlands and
thickets with dry to mesic soils of moderate pH, and, where soils are
thin, over limestone and marble substrates. It is most prevalent in the
southern portion of the species' range.

### Poa saltuensis Fernald & Wiegand subsp. saltuensis [p. 396]

*Poa saltuensis* subsp. *saltuensis* grows throughout the range of the
species, in open forests and woodlands with low to moderate pH soils.

### 2. Poa alsodes A. Gray GROVE BLUEGRASS [p. 396, 521]

*Poa alsodes* grows in mesic woodlands of eastern Canada and the
northeastern United States, extending south to Illinois, Tennessee, and
North Carolina, particularly in the Appalachian Mountains.

### 3. Poa sylvestris A. Gray WOODLAND BLUEGRASS [p. 396, 521]

*Poa sylvestris* grows in southeastern Canada and throughout much of
the eastern United States, mainly at low elevations in woodlands,
especially in riparian zones. It is distinguished from *P. wolfii* by its
smaller, more numerous spikelets and lemmas that are usually
sparsely hairy between the veins. Plants from the middle Appalachian
Mountains have been confused with *P. paludigena*; *P. sylvestris* is
usually larger, has more than 2 branches per panicle node, is
pubescent between the lemma veins and palea keels, and has larger
anthers.

## 4. Poa marcida Hitchc. WEEPING BLUEGRASS [p. 397, 521]

*Poa marcida* is an uncommon endemic of breaks in rich, mesic, generally old growth forests of the Pacific coast, from Vancouver Island through the western foothills of the northern Cascade Mountains to central Oregon. It differs from *P. saltuensis* in its closed sheaths and attenuate lemmas.

## 5. Poa kelloggii Vasey KELLOGG'S BLUEGRASS [p. 397, 521]

*Poa kelloggii* grows in rich coastal forests, especially redwood forests, in California. It is not common.

## 6. Poa autumnalis Muhl. *ex* Elliott AUTUMN BLUEGRASS [p. 397, 521]

*Poa autumnalis* grows primarily in the southeastern United States, being found in forests of the eastern and western Appalachian piedmont and coastal plain. It is readily distinguished from other perennial species of the eastern United States by its combination of glabrous calluses and pubescent palea keels.

## 7. Poa wolfii Scribn. WOLF'S BLUEGRASS [p. 397, 521]

*Poa wolfii* is an uncommon species that grows in boggy areas of eastern deciduous forests, primarily west of the Appalachian divide. It differs from *P. sylvestris* in having fewer branches, larger spikelets, and lemmas that are usually glabrous between the veins.

## Poa sect. Arenariae (Hegetschw.) Stapf

*Poa* sect. *Arenariae* is native to Eurasia and North Africa. It includes 14 species. These are easily recognized as members of the section by the bulbous bases of their new shoots. One species is established in the *Manual* region.

## 8. Poa bulbosa L. BULBOUS BLUEGRASS [p. 397, 521]

*Poa bulbosa* is a European species that is now established in the *Manual* region. In southern Europe and the Middle East, it is considered an important early spring forage.

1. Spikelets not bulbiferous . . . . . . . . . . . . . . . . . . subsp. *bulbosa*
1. All or some spikelets bulbiferous . . . . . . . . . . . . subsp. *vivipara*

## Poa bulbosa L. subsp. bulbosa [p. 397]

*Poa bulbosa* subsp. *bulbosa* is common in its native Europe. It is uncommon in the *Manual* region, with the only known collections being from Drake, Butler, and Preble counties, Ohio. Whether these collections represent independent introductions or reversion to reproduction by seed is not known.

## Poa bulbosa subsp. vivipara (Koel.) Arcang. [p. 397]

*Poa bulbosa* subsp. *vivipara* was introduced from Europe into the Pacific Northwest as a forage grass; it has since spread across temperate areas of the *Manual* region, particularly in the Pacific Northwest and northern Great Basin. It is highly tolerant of grazing and disturbance.

## Poa sect. Alpinae (Hegetschw. *ex* Nyman) Stapf

*Poa* sect. *Alpinae* includes seven species. They are all cespitose perennials with intravaginal branching and broad leaves. One species is circumboreal; the other six are native to Europe.

## 9. Poa alpina L. [p. 397, 521]

*Poa alpina* is a fairly common circumboreal forest species of subalpine to arctic habitats, extending south in the Rocky Mountains to Utah and Colorado in the west, and to the northern Great Lakes region in the east. It often grows in disturbed ground and is calciphilic. *Poa* ×*gaspensis* is a natural hybrid which seems to be between *P. alpina*

and *P. pratensis* subsp. *alpigena*; it differs from *P. alpina* in its extravaginal branching, rhizomatous habit, and webbed calluses.

1. Spikelets not bulbiferous . . . . . . . . . . . . . . . . . . . . subsp. *alpina*
1. Some or all spikelets bulbiferous . . . . . . . . . . . . subsp. *vivipara*

## Poa alpina L. subsp. alpina ALPINE BLUEGRASS [p. 397]

*Poa alpina* subsp. *alpina* is the more common of the two subspecies. In the *Manual* region, it grows throughout the range of the species.

## Poa alpina subsp. vivipara (L.) Arcang. [p. 397]

*Poa alpina* subsp. *vivipara* grows at scattered locations in Greenland, and has been reported for Alaska. It is common in alpine regions of northern and central Europe.

## Poa sect. Micrantherae Stapf

*Poa* sect. *Micrantherae* includes eight species, all of which are native to Eurasia and North Africa. They are gynomonoecious, with smooth or sparsely scabrous panicle branches. The calluses are glabrous in most species; the palea keels are usually hairy.

## 10. Poa annua L. ANNUAL BLUEGRASS [p. 397, 521]

*Poa annua* is one of the world's most widespread weeds. It thrives in anthropomorphic habitats outside of the arctic. A native of Eurasia, it is now well established throughout most of the *Manual* region. It differs from *P. infirma* in having larger anthers, and from *P. chapmaniana* in having glabrous calluses and three larger anthers, rather than one.

## 11. Poa infirma Kunth WEAK BLUEGRASS [p. 397, 521]

*Poa infirma* was introduced from Europe to the Americas, and was first described from Colombia. It is sporadically established along the Pacific coast and in the central valleys of California, and has been collected in Charleston, South Carolina. It is rare elsewhere in the *Manual* region. *Poa infirma* often resembles *P. infirma*, which is thought to be one of its parents, but *P. annua* is tetraploid and has anthers 0.6–1.1 mm long. Both species are gynomonoecious.

## 12. Poa supina Schrad. SUPINE BLUEGRASS [p. 397]

*Poa supina* is native to boreal and alpine regions of Eurasia. In the 1990s, the cultivar 'Supernova' was introduced for seeding in wet to moist, cool, shady areas subject to heavy traffic. It has been tested in both Canada and the United States, and is expected to gradually escape cultivation, probably becoming established throughout the cool-temperate portion of the *Manual* region. Its current distribution is not known. *Poa supina* differs from *P. annua*, of which it is thought to be one of the parents, in having longer anthers and a more stoloniferous habit, as well as in being diploid. It is gynomonoecious.

## Poa L. sect. Poa

*Poa* section *Poa* includes 32 species. All the species are synoecious perennials; most are strongly rhizomatous.

## 13. Poa pratensis L. KENTUCKY BLUEGRASS [p. 398, 521]

*Poa pratensis* is common, widespread, and well established in many natural and anthropogenic habitats of the *Manual* region. The only taxa that are clearly native to the region are the arctic and subarctic subspp. *alpigena* and *colpodea*. Outside the *Manual* region, *P. pratensis* is native in temperate and arctic Eurasia. It is now established in temperate regions around the world.

*Poa pratensis* is a highly polymorphic, facultatively apomictic species, having what is probably the most extensive series of polyploid chromosome numbers of any species in the world. Natural hybrids have been identified between *Poa pratensis* and *P. alpina*, *P. arctica*, *P. wheeleri*, and *P. secunda*.

*Poa rhizomata* resembles *P. pratensis*, but has acute ligules and sparse inflorescences, florets that are usually unisexual, and generally larger spikelets; *Poa macrocalyx* looks like a robust *P. pratensis* with large spikelets, its lemmas and paleas generally hispidulous between the veins and palea keels. *Poa confinis* also resembles *P. pratensis*, but differs in having glabrous or sparsely hairy lemmas, and diffusely webbed calluses.

1. At least some spikelets bulbiferous; plants of the high arctic tundra . . . . . . . . . . . . . . . . . . . . subsp. *colpodea*
1. Spikelets not bulbiferous; plants widely distributed.
  2. Panicle branches smooth or almost smooth.
    3. Basal branching primarily extravaginal; blades flat or folded, soft, adaxial surfaces usually glabrous, sometimes sparsely hairy; plants of alpine and tundra regions . . . . . subsp. *alpigena*
    3. Basal branching both intra- and extravaginal; blades folded or involute, somewhat firm, adaxial surfaces often sparsely hairy; plants widely distributed, but not in alpine or tundra regions . . . . . . . . . . . . . . . . . subsp. *agassizensis*
  2. Panicles branches more or less scabrous.
    4. Intravaginal innovation shoots present, intra- and extravaginal blades alike, 0.4–1 mm wide, folded to involute, somewhat firm, adaxial surfaces often sparsely and softly hairy; plants of dry meadows and forests . . . . . . . . . . . . . . . . . . . . . . . . . . . subsp. *angustifolia*
    4. Intravaginal innovation shoots present or absent, if present then differentiated or alike, at least some with blades 1.5–4.5 mm wide, flat or folded, adaxial surfaces rarely hairy; plants widespread, often of more mesic sites.
      5. Culms 8–30(50) cm tall, often somewhat glaucous, particularly the glumes; blades flat; intravaginal shoots absent or present and with blades similar to those of the extravaginal shoots; panicles with few spikelets per branch and 1–2(5) branches per node; plants of low, wet, often sandy ground . . . . . . . . . . . . . . . . . . . . . . . subsp. *irrigata*
      5. Culms to 100 cm tall, not glaucous; blades flat or folded; intravaginal shoots present, with blades similar to those of the extravaginal shoots or distinctly narrower; panicles with several to many spikelets per branch and 3–5(7) branches per node; plants of various habitats, including those of subsp. *irrigata* . . . . subsp. *pratensis*

## Cultivars of **Poa pratensis** L.

More than 60 cultivars of *Poa pratensis* have been released in the *Manual* region. Plants grown from commercially distributed seed have generally been placed in subsp. *pratensis* by North American authors, but they appear to include genetic contributions from at least three major subspecies, e.g., subspp. *angustifolia*, *pratensis*, and *irrigata*. These and intermediate forms, especially those favoring subspp. *irrigata* and *pratensis*, are best simply referred to as *Poa pratensis sensu lato* or labeled as cultivated material.

## **Poa pratensis** subsp. **agassizensis** (B. Boivin & D. Löve) Roy L. Taylor & MacBryde [p. 398]

*Poa pratensis* subsp. *agassizensis* grows throughout the drier, cool-temperate range of the species in North America. The least distinctive of the subspecies treated here, it closely approaches subsp. *angustifolia* in having involute leaves and small spikelets, but has shorter and broader leaves, and more condensed panicles.

## **Poa pratensis** subsp. **alpigena** (Lindm.) Hiitonen
ALPIGENE BLUEGRASS [p. 398]

*Poa pratensis* subsp. *alpigena* is a circumpolar, mesophytic to subhydrophytic, arctic and alpine subspecies that extends into boreal forests in northern parts of the *Manual* region. It is infrequent south of Canada. It also grows in southern Patagonia.

*Poa pratensis* subsp. *alpigena* differs from *P. arctica* in being glabrous between the lemma and palea veins, having somewhat more dense, later-opening panicles, and, usually, having smaller spikelets and more closely spaced palea keels. It differs from a likely hybrid with *P. alpina*, *P.* ×*gaspensis*, in its truncate to rounded ligules, lemmas that are glabrous between the veins, and the lack of a basal tuft of leaves. In this treatment, bulbiferous plants are placed in subsp. *colpodea*, a subspecies which is more common than subsp. *alpigena* in the high arctic.

## **Poa pratensis** subsp. **angustifolia** (L.) Lej. [p. 398]

*Poa pratensis* subsp. *angustifolia* is a western Eurasian subspecies that is known from scattered locations throughout the temperate North American distribution of the species. It is characterized by the predominance of fascicles of elongate, narrow, involute blades on the intravaginal vegetative shoots, and slender panicles with small spikelets.

## **Poa pratensis** subsp. **colpodea** (Th. Fr.) Tzvelev [p. 398]

*Poa pratensis* subsp. *colpodea* is circumpolar. In the *Manual* region, its range extends from Alaska and British Columbia to Greenland. It is more common than *P. pratensis* subsp. *alpigena* in the high arctic. The two sometimes grow together, and there is some evidence of a shift in dominance from year to year.

## **Poa pratensis** subsp. **irrigata** (Lindm.) H. Lindb. [p. 398]

*Poa pratensis* subsp. *irrigata* is poorly understood in the *Manual* region. As interpreted here, it includes subarctic to boreal, coastal and lowland plants that differ from those of subsp. *pratensis* in their primarily extravaginal branching, shorter stature, wide, flat blades, short, stout panicles with fewer branches, larger spikelets, and relatively longer glumes that are often pruinose.

## **Poa pratensis** L. subsp. **pratensis** [p. 398]

*Poa pratensis* subsp. *pratensis* grows throughout most of the range of the species, but is absent from the high arctic, and only sporadic in the low arctic. It usually has a few narrow, flat or involute, intravaginal shoot leaves, in addition to some broader, extravaginal shoot leaves, and is intermediate between subspp. *angustifolia* and *irrigata*.

## 14. **Poa sublanata** Reverd. COTTONBALL BLUEGRASS [p. 398, 521]

*Poa sublanata* grows in the high arctic of Alaska and Russia, usually on sandy ground. Bulbiferous plants are known from Russia; none have been found in the *Manual* region.

## 15. **Poa macrocalyx** Trautv. & C.A. Mey. LARGE-GLUME BLUEGRASS [p. 398, 521]

*Poa macrocalyx* grows mainly in coastal areas of boreal Alaska, and from the eastern coast of Russia to northern Japan. Bulbiferous plants are occasionally found. *Poa macrocalyx* resembles an exaggeratedly robust *P. pratensis*, with large spikelets and lemmas, proportionally longer glumes, and paleas that are generally hispidulous between the veins and palea keels.

## 16. **Poa arctica** R. Br. ARCTIC BLUEGRASS [p. 398, 521]

*Poa arctica* is a common circumboreal species of arctic and alpine regions, growing mainly in mesic to subhydric, acidic tundra and alpine meadows, and on rocky slopes. It extends south in the Rocky Mountains to New Mexico.

The most reliable way to distinguish *Poa arctica* from *P. pratensis*, particularly subsp. *alpigena*, is by the wider paleas and the presence of hairs between the palea keels. Bulbiferous forms of *P. arctica* differ from *P. stenantha* var. *vivipara* in not being glaucous, and in having rhizomes and terete, smooth panicle branches. *Poa* ×*gaspensis* also resembles *P. arctica*, but it has sharply keeled, more scabrous glumes and a spikelet shape that is intermediate between *P. pratensis* and *P. alpina*. *Poa arctica* forms natural hybrids with both *P. pratensis* and *P. secunda*.

1. Plants lacking well-developed rhizomes; anthers aborted late in development; plants of the high arctic . . . . . . . . . . . . . . . . . . . . . . . . . . . . . . . . . subsp. *caespitans*
1. Plants usually with well-developed rhizomes; anthers normal or plants not of the high arctic.
  2. Panicles erect, the branches relatively stout, fairly straight; longest branches of the lowest panicle nodes $^{1}/_{4}$–$^{1}/_{2}$ the length of the panicles; culms wiry, usually several together; calluses glabrous or shortly webbed; paleas sometimes glabrous; plants glaucous, growing in the southern Rocky Mountains and adjacent portions of the Intermountain region . . . . . . . . subsp. *aperta*
  2. Panicles lax to erect, the branches slender, flexuous to fairly stout and straight; longest branches of the lowest panicle nodes $^{2}/_{5}$–$^{3}/_{5}$ the length of the panicles; culms slender to stout, varying from solitary to several together; calluses glabrous or webbed, the hairs usually more than $^{1}/_{2}$ as long as the lemmas; paleas pubescent; plants sometimes glaucous, widespread in distribution.
    3. Calluses glabrous; spikelets not bulbiferous . . . . . . . . . . . . . . . . . . . . . subsp. *grayana*
    3. Calluses webbed, often copiously so, sometimes glabrous in bulbiferous spikelets; spikelets sometimes bulbiferous.
      4. Spikelets (5)6–8 mm long; lemmas 4–6 mm long; blades 2–6 mm wide; rachillas usually hairy; plants primarily of the western arctic, extending to northwestern British Columbia . . . . . . . . . . . . . . . . . subsp. *lanata*
      4. Spikelets (3.5)4–7 mm long; lemmas (2.7)3–4.5 mm long; blades 1.5–3 mm wide; rachillas commonly glabrous; plants widespread . . . . . . . . . . . . . . . . subsp. *arctica*

### Poa arctica subsp. aperta (Scribn. & Merr.) Soreng [p. 398]

*Poa arctica* subsp. *aperta* is restricted to subalpine and low alpine habitats on the Wasatch Escarpment and high mountains of the Colorado Plateau in southern Utah, and the Rocky Mountains of southern Colorado and northern New Mexico. It has softer leaves, and is more densely hairy between the lemma veins and the palea keels, than subsp. *arctica*. It can be distinguished from subsp. *grayana* by its more wiry culms, and less contracted panicles with straighter branches.

### Poa arctica R. Br. subsp. arctica [p. 398]

*Poa arctica* subsp. *arctica* is polymorphic and circumpolar. It grows in alpine and tundra habitats as far south as Wheeler Peak, New Mexico. Bulbiferous plants are known from alpine habitats in Alaska and British Columbia. It has tougher leaves, and is less densely hairy between the lemma veins and palea keels, than subsp. *aperta*. It often grows with subsp. *lanata*, but can be distinguished by its smaller and, usually, more numerous spikelets and narrower leaves.

### Poa arctica subsp. caespitans Simmons *ex* Nannf. [p. 398]

*Poa arctica* subsp. *caespitans* grows in moist tundra of the high arctic, or infrequently in the low arctic of northeastern Canada and Greenland. It also grows in Norway and in the Russian high arctic; it is rare in both regions. Some plants included here tend towards *Poa glauca*, while others are distinguished from other *P. arctica* subspecies only by their more tufted habit and sterile anthers. As interpreted here, subsp. *caespitans* includes *P. trichopoda* Lange [= *P. tolmatchewii* Roshev.].

### Poa arctica subsp. grayana (Vasey) Á. Löve, D. Löve & B.M. Kapoor [p. 398]

*Poa arctica* subsp. *grayana* grows only in the alpine regions of the middle and southern Rocky Mountains of Utah, Wyoming, Colorado, and New Mexico. It is characterized by its glabrous calluses, densely hairy lemmas, and paleas that are densely hairy between the keels. It has less wiry culms, and panicles with more flexuous branches, than subsp. *aperta* and, like that subspecies, can be difficult to distinguish from *P. arida*.

### Poa arctica subsp. lanata (Scribn. & Merr.) Soreng [p. 398]

*Poa arctica* subsp. *lanata* is amphiberingian in distribution, extending to northwestern British Columbia. It often grows with subsp. *arctica*, from which it differs in having larger and, usually, fewer spikelets and broader leaves. It intergrades with *P. macrocalyx* in one direction and *P. arctica* subsp. *arctica* in another. *Poa malacantha* Kom. is included here in *P. arctica* subsp. *lanata*.

## Poa sect. Homalopoa Dumort.

*Poa* sect. *Homalopoa* is the largest and most heterogeneous section of the genus, having at least 170 species, including many annuals and short-lived perennials. Most species are cespitose, have sheaths closed for $^{1}/_{4}$–$^{3}/_{4}$ their length and anthers up to 1 mm long. The section is widespread in its distribution, growing almost everywhere the genus is native.

### 17. Poa bolanderi Vasey BOLANDER'S BLUEGRASS [p. 399, 521]

*Poa bolanderi* grows mainly in pine to fir forest openings of mountain slopes in the western United States, from Washington to California and Utah. It differs from *P. howellii* in having smooth to scabrous, rather than puberulent, lemmas; it also grows at higher elevations, mostly at 1500–3000 m.

### 18. Poa howellii Vasey & Scribn. HOWELL'S BLUEGRASS [p. 399, 521]

*Poa howellii* grows primarily on rocky banks and wooded slopes, from the coastal ranges of southern British Columbia to southern California. It differs from *P. bolanderi* in having puberulent, rather than smooth or scabrous, lemmas, and in growing at lower elevations, mostly from near sea level to 1000 m.

### 19. Poa chapmaniana Scribn. CHAPMAN'S BLUEGRASS [p. 399, 521]

*Poa chapmaniana* is native from the central part of the Great Plains east and southward to the coast. It grows in dry to mesic forests, forest openings, and the margins of bottomlands, often in disturbed ground and on acidic substrates. Records from New York probably represent introductions. Its web and single short anther distinguish *P. chapmaniana* from *P. annua* and most plants of *P. bigelovii*. It also differs from *P. bigelovii*, probably its closest relative, in having narrower leaf blades, and panicle branches that are eventually spreading.

### 20. Poa bigelovii Vasey & Scribn. BIGELOW'S BLUEGRASS [p. 399, 521]

*Poa bigelovii* grows in arid upland regions, particularly on shady, rocky slopes of the southwestern United States and northern Mexico. Plants from southeastern Arizona eastwards are usually glabrous between the lemma veins, whereas more western plants are usually puberulent between the lemma veins. Plants with 1 or 2 small anthers are found in the eastern portion of the species' range; they differ from *P. chapmaniana* in their persistently contracted panicles and broader leaf blades.

### 21. **Poa occidentalis** Vasey  NEW MEXICAN BLUEGRASS [p. 399, <u>521</u>]

*Poa occidentalis* grows in natural openings and disturbed sites in mixed coniferous forests of the southwestern United States. It differs from *P. tracyi* in having shorter, well-developed anthers and in lacking rhizomes. It also usually has longer ligules relative to the blade width, and is shorter-lived. A few plants are intermediate in some characteristics. Small plants of *P. occidentalis* sometimes resemble *P. reflexa*.

### 22. **Poa reflexa** Vasey & Scribn.  NODDING BLUEGRASS [p. 399, <u>521</u>]

*Poa reflexa* grows in subalpine forests, meadows, and low alpine habitats, primarily in the central and southern Rocky Mountains. It usually grows on drier and more disturbed sites, and appears shorter-lived, than the frequently sympatric or parapatric *P. leptocoma*, from which it differs in usually having hairs on the palea keels and lateral veins of the lemmas, and smooth panicle branches. In addition, *P. reflexa* is tetraploid, whereas *P. leptocoma* is hexaploid. *Poa reflexa* may resemble small plants of *P. occidentalis* in habit.

### 23. **Poa paucispicula** Scribn. & Merr.  FEW-FLOWER BLUEGRASS [p. 399, <u>521</u>]

*Poa paucispicula* grows in arctic and alpine regions, from the north coast of Alaska and the western Northwest Territories south to Washington, Idaho, and Wyoming; it also grows in arctic far east Russia. It is a delicate species that prefers open, mesic, rocky slopes. It has sometimes been included in *P. leptocoma*, but it differs in having smoother branches, fewer spikelets, and broader glumes.

### 24. **Poa laxiflora** Buckley  LAX-FLOWER BLUEGRASS [p. 399, <u>521</u>]

*Poa laxiflora* is restricted to mesic, old growth, mixed conifer forests of the Pacific coast, from Alaska south through the western foothills of the northern Cascades to Oregon. It is not a common species.

### 25. **Poa chaixii** Vill.  CHAIX'S BLUEGRASS [p. 399, <u>521</u>]

*Poa chaixii* was introduced from Europe as an attractive ornamental, and has occasionally escaped. A population in southwestern Quebec has been extirpated.

### 26. **Poa strictiramea** Hitchc.  BIG BEND BLUEGRASS [p. 399, <u>521</u>]

*Poa strictiramea* grows on shady, upland mountain slopes, usually below north-facing cliffs, in and around the Chihuahuan Desert. In the United States, it is known only from the Chisos Mountains, Texas. Plants from the eastern part of its range, including the Chisos Mountains, commonly have short, truncate ligules, whereas westward in Mexico, plants with long, acute ligules are more common.

## Poa sect. **Madropoa** Soreng

*Poa* sect. *Madropoa* is confined to North America. Its 20 species exhibit breeding systems ranging from sequential gynomonoecy to gynodioecy and dioecy. The gyno-monoecious species usually grow in forests and have broad, flat leaves. The gynomonoecious and dioecious species grow mainly in more open habitats. They have normally developed anthers that are 1.3–4 mm long, and involute innovation blades that, in several species, are densely scabrous or hairy on the adaxial surfaces.

There are two subsections in the *Manual* region: subsects. *Madropoa* and *Epiles*.

## Poa nervosa Complex

The seven species of the *Poa nervosa* complex are typically forest species, with broad, flat leaf blades and short rhizomes. They exhibit breeding systems ranging from sequential gynomonoecy to dioecy. In populations of species with sequential gynomonoecy, plants with only bisexual florets exist in roughly the same number as plants that produce pistillate florets. In most of those producing pistillate florets, the number of pistillate florets increases as the growing season progresses. The pistillate florets are initially concentrated in the lower spikelets of the panicles and the upper florets within these spikelets; they later develop throughout the panicle.

### 27. **Poa tracyi** Vasey  TRACY'S BLUEGRASS [p. 400, <u>522</u>]

*Poa tracyi* grows primarily in coniferous forest openings, sometimes with gambel oak, and in subalpine mesic meadows. It is restricted to the front ranges of the southern Rocky Mountains; it is not common. It differs from *P. occidentalis* in having longer and/or rudimentary anthers, shorter ligules relative to the leaf blade width, and a loose, shortly rhizomatous habit. It is sequentially gynomonoecious.

### 28. **Poa arnowiae** Soreng  WASATCH BLUEGRASS [p. 400, <u>522</u>]

*Poa arnowiae* grows in openings within the coniferous forests of the mountain ranges in southeastern Idaho, northern Utah, and adjacent Wyoming. It is sequentially gynomonoecious.

### 29. **Poa cuspidata** Nutt.  EARLY BLUEGRASS [p. 400, <u>522</u>]

*Poa cuspidata* is a common species of forest openings in the Appalachian Mountains. It is an eastern counterpart of *P. arnowiae*, *P. tracyi*, and *P. nervosa*. Like those species, it is sequentially gynomonoecious.

### 30. **Poa nervosa** (Hook.) Vasey  VEINY BLUEGRASS [p. 400, <u>522</u>]

*Poa nervosa* occurs infrequently at low elevations in the western foothills of the northern Cascade Mountains and adjacent coast ranges, extending eastward up the Columbia Gorge as far as Multnomah Falls. It usually grows in wet habitats, such as mossy cliffs with seeps and around waterfalls, but it is also found in rich, old growth, mixed deciduous and conifer forests. It appears to be sexually reproducing and sequentially gynomonoecious. It differs from *P. wheeleri* in having densely pubescent leaf collar margins, and glabrous or more sparsely and shortly pubescent sheaths. It also differs in usually having well-developed anthers, and in being tetraploid. The two species are geographically isolated and ecologically distinct.

### 31. **Poa wheeleri** Vasey  WHEELER'S BLUEGRASS [p. 400, <u>522</u>]

*Poa wheeleri* is common at mid- to high elevations, generally on the east side of the coastal mountains from British Columbia to California, and from Manitoba to New Mexico. It generally grows in submesic coniferous forests to subalpine habitats. Most plants have densely retrorsely pubescent or scabrous sheaths, involute innovation blades that are pubescent adaxially, and pistillate florets. *Poa wheeleri* resembles *P. rhizomata* and *P. chambersii* more than *P. nervosa sensu stricto*. It differs from *P. chambersii* in having at least some proximal sheaths that are densely retrorsely scabrous or pubescent (sometimes obscurely so), and folded or involute innovation blades that are scabrous to hispidulous on the adaxial surfaces. Natural hybrids have been found between *P. wheeleri* and *P. pratensis*.

### 32. **Poa rhizomata** Hitchc.  RHIZOME BLUEGRASS [p. 400, <u>522</u>]

*Poa rhizomata* is a rare species that grows in upper elevation, mixed coniferous forests on ultramafic (gabro or peridotite) rocks of the

Klamath–Siskiyou region. It is subdioecious. It differs from *P. pratensis* in having acute ligules, sparse inflorescences, florets that are usually unisexual, and generally larger spikelets. It also resembles *P. chambersii*, but has more open sheaths, longer ligules, more pubescent lemmas, and a more well-developed web. It used to include *P. piperi*, which differs in having involute, adaxially hairy leaves and glabrous lemmas.

## 33. Poa chambersii Soreng CHAMBERS' BLUEGRASS [p. 400, 522]

*Poa chambersii* is known only from upland forest openings in the Cascades of western Oregon, where it is dioecious, and from high elevations on Steens Mountain in southeastern Oregon, where it is gynodioecious. It resembles *P. rhizomata*, but has more closed sheaths, shorter ligules, less pubescent or glabrous lemmas, and lacks a well-developed web. It approaches *P. cusickii* subsp. *purpurascens*, but is rhizomatous and sexually reproducing. It differs from *P. wheeleri* in having glabrous sheaths and flat or folded, glabrous innovation blades.

## 34. Poa sierrae J.T. Howell SIERRA BLUEGRASS [p. 400, 522]

*Poa sierrae*, a distinctive dioecious species, is a narrow endemic of mid-elevation canyon slopes on the west side of the Sierra Nevada, California. It can be distinguished from all other *Poa* species by the scaly, pink to purplish buds on the rhizomes, and by the entirely or almost entirely closed upper culm sheaths that are shorter than their blades.

## Poa subsect. Madropoa Soreng

The seven species of *Poa* subsect. *Madropoa* are strongly dioecious, usually rhizomatous, and usually have involute blades.

## 35. Poa douglasii Nees DOUGLAS' BLUEGRASS [p. 401, 522]

*Poa douglasii* is a dioecious endemic that grows on coastal sand dunes in California, a habitat that is being invaded by exotic species. It is rare north of Mendocino. Its hairy rachises distinguish *P. douglasii* from all other species of *Poa* in the *Manual* region. It differs from *P. macrantha*, which occupies similar habitats, in this and in its usually longer glumes and lemmas.

## 36. Poa macrantha Vasey DUNE BLUEGRASS [p. 401, 522]

*Poa macrantha* is a dioecious coastal sand dune species that grows from southern Alaska to northern California. It competes better than *P. douglasii* with the invasion of its habitat by *Ammophila* and other exotic species. It used to be treated as a subspecies of *P. douglasii*, but differs in its glabrous rachises and usually longer glumes and lemmas.

## 37. Poa confinis Vasey COASTAL BLUEGRASS [p. 401, 522]

*Poa confinis* grows on sandy beaches and forest margins of the west coast, a habitat that is being lost to invasion by exotic species and development. It is closely related to *P. diaboli*, from which it differs by a suite of characters, in addition to being ecologically and geographically distinct. *Poa confinis* differs from *P. pratensis* in having glabrous or sparsely hairy lemmas and diffusely webbed calluses. It is gynodioecious.

## 38. Poa diaboli Soreng & D.J. Keil DIABLO BLUEGRASS [p. 401, 522]

*Poa diaboli*, which is sequentially gynomonoecious, is endemic to upper shaly slopes, in soft coastal scrub and openings in Bishop Pine stands, in the coastal mountains of San Luis Obispo County, California. It differs from *P. confinis* by a suite of characters. The two species are also ecologically and geographically distinct.

## 39. Poa piperi Hitchc. PIPER'S BLUEGRASS [p. 401, 522]

*Poa piperi* grows in forest openings on serpentine rocks in the Coast Ranges of southwestern Oregon and northwestern California. It used to be included in *P. rhizomata*, from which it differs in its involute leaves and glabrous lemmas. It is dioecious.

## 40. Poa atropurpurea Scribn. SAN BERNARDINO BLUEGRASS [p. 401, 522]

*Poa atropurpurea* is a rare dioecious endemic of mesic upland meadows in southern California. It is federally listed as endangered.

## 41. Poa fendleriana (Steud.) Vasey VASEY'S MUTTONGRASS [p. 401, 522]

*Poa fendleriana* grows on rocky to rich slopes in sagebrush-scrub, interior chaparral, and southern (rarely northern) high plains grasslands to forests, and from desert hills to low alpine habitats. Its range extends from British Columbia to Manitoba and south to Mexico. It is one of the best spring fodder grasses in the eastern Great Basin, Colorado plateaus, and southern Rocky Mountains. It is dioecious. *Poa fendleriana* hybridizes with *Poa cusickii* subsp. *pallida*. The hybrids are called *P. ×nematophylla*.

1. Lemma keels and marginal veins glabrous or almost
   so . . . . . . . . . . . . . . . . . . . . . . . . . . . . . . . . . . . subsp. *albescens*
1. Lemma keels and marginal veins conspicuously hairy.
   2. Ligules of the middle cauline leaves 0.2–1.2 (1.5) mm long, not decurrent, usually scabrous, apices truncate to rounded, upper margins ciliolate or scabrous . . . . . . . . . . . . . . . . . . . . . . . . subsp. *fendleriana*
   2. Ligules of the middle cauline leaves (1.5)1.8–18 mm long, decurrent, usually smooth to sparsely scabrous, apices obtuse to acuminate, upper margins usually smooth, glabrous . . . . . . . . subsp. *longiligula*

## Poa fendleriana subsp. albescens (Hitchc.) Soreng [p. 401]

*Poa fendleriana* subsp. *albescens* is endemic to the northern Sierra Madre Occidental, extending from the southwestern United States to Chihuahua and Sonora, Mexico. It grows mainly in upland forest openings. It intergrades with subsp. *fendleriana* where sexual populations have come into contact.

## Poa fendleriana (Steud.) Vasey subsp. fendleriana [p. 401]

*Poa fendleriana* subsp. *fendleriana* grows chiefly in the southern and middle Rocky Mountains, and in the mountains surrounding the Colorado plateaus. It intergrades with subspp. *albescens* and *longiligula* where sexual or partially sexual populations have come into contact.

## Poa fendleriana subsp. longiligula (Scribn. & T. A. Williams) Soreng LONGTONGUE MUTTONGRASS [p. 401]

*Poa fendleriana* subsp. *longiligula* tends to grow to the west of the other two subspecies, in areas where winter precipitation is more consistent and summer precipitation less consistent. Apomixis is far more common and widespread than sexual reproduction in this subspecies.

## Poa subsect. Epiles Hitchc. ex Soreng

The five species of *Poa* subsect. *Epiles* are cespitose, gynodioecious or dioecious, and have involute or folded leaf blades.

## 42. Poa cusickii Vasey CUSICK'S BLUEGRASS [p. 402, 522]

*Poa cusickii* grows in rich meadows in sagebrush scrub to rocky alpine slopes, from the southwestern Yukon Territory to Manitoba and North Dakota, south to central California and eastern Colorado. It is gynodioecious or dioecious.

1. Panicle branches smooth or slightly scabrous, or the basal blades more than 1.5 mm wide and flat or folded; cauline blades more than 1.5 mm wide, often flat; some basal branching extravaginal; lemmas and calluses sometimes sparsely puberulent.
    2. Lemmas usually glabrous, rarely plants from the Rocky Mountains with puberulent keels and marginal veins; calluses glabrous; panicles erect, usually contracted; branches smooth to slightly scabrous . . . . . . . . . . . . . . . . . . . . . . . . . . . . . subsp. *epilis*
    2. Lemmas rarely completely glabrous, at least some florets with sparsely puberulent keels, the marginal veins glabrous or puberulent; calluses frequently with a sparse, short web; panicles somewhat lax and loosely contracted; branches smooth or sparsely to moderately scabrous . . . . . . . . . . . . . . . . . . . . . . . . . . . . . subsp. *purpurascens*
1. Panicle branches moderately to strongly scabrous; basal and cauline blades usually less than 1.5 mm wide, involute, rarely flat or folded; basal branching intravaginal; lemmas and calluses glabrous.
    3. Panicle branches longer than 1.7 cm in at least some panicles; panicles open or contracted . . . subsp. *cusickii*
    3. Panicle branches up to 1.7 cm long, stout; panicles contracted . . . . . . . . . . . . . . . . . . . . subsp. *pallida*

### Poa cusickii Vasey subsp. cusickii [p. 402]

*Poa cusickii* subsp. *cusickii* grows mainly in mesic desert upland and mountain meadows, on and around the Columbia plateaus of northern California, Oregon, southern Washington, and adjacent Idaho and Nevada. It is highly variable, with fairly open- to contracted-panicle populations, and from gynodioecious to dioecious populations. It appears to have hybridized with *P. pringlei* around Mount Shasta, California, and Mount Rose, Nevada. *Poa stebbinsii*, an endemic in the high Sierra Nevada, is easily distinguished from *P. cusickii* subsp. *cusickii* by its long hyaline ligules.

### Poa cusickii subsp. epilis (Scribn.) W.A. Weber SKYLINE BLUEGRASS [p. 402]

*Poa cusickii* subsp. *epilis* tends to grow around timberline. It is strictly pistillate. It is usually quite distinct from subspp. *cusickii* and *pallida*, and differs from subsp. *purpurascens* in having on average more and shorter spikelets, lemmas that are shorter and rarely pubescent, and both intra- and extravaginal branching. It occurs throughout most of the range of the species, but is absent from the Yukon Territory, and uncommon in the Cascade Mountains.

### Poa cusickii subsp. pallida Soreng [p. 402]

*Poa cusickii* subsp. *pallida* grows in forb-rich mountain grasslands to alpine habitats, from the southern Yukon Territory to California, across the Great Basin and through the Rocky Mountains to central Colorado. It is found mainly east and north of subsp. *cusickii*. The shorter branch length serves to distinguish it from the narrow-panicled subsp. *cusickii* forms in most cases. It hybridizes with *P. fendleriana*, forming *P. ×nematophylla*. The hybrids may have hairy lemmas or, less often, broader leaf blades and glabrous lemmas.

### Poa cusickii subsp. purpurascens (Vasey) Soreng [p. 402]

*Poa cusickii* subsp. *purpurascens* grows in subalpine habitats in the coastal mountains from southern British Columbia to southern Oregon, with sporadic occurrences eastward in British Columbia to the Rocky Mountains and south to the central Sierra Nevada. It tends to differ from subsp. *epilis* in having predominantly extravaginal branching, fewer and longer spikelets, and longer lemmas that are usually sparsely hairy on the keel and marginal veins. It differs from *P. chambersii* in lacking rhizomes and in being strictly pistillate; and from *P. porsildii* in its longer spikelets and in tending to have longer panicles with more spikelets.

### 43. Poa ×nematophylla Rydb. [p. 402, 522]

*Poa ×nematophylla* is believed to consist of hybrids between *P. cusickii* subsp. *pallida* and *P. fendleriana*. It is mostly pistillate and apomictic; few staminate plants have been found. It usually resembles *P. cusickii* most, but grades towards *P. fendleriana*. It tends to grow on drier slopes than either parent, mainly in and around sagebrush desert/forest interfaces.

### 44. Poa porsildii Gjaerev. PORSILD'S BLUEGRASS [p. 402, 522]

*Poa porsildii* is an alpine, calciphilic, mesophilic, dioecious species that grows from eastern Alaska to the western Northwest Territories. It differs from *P. cusickii* subsp. *purpurascens* in having panicles with laxer, smooth, and more slender branches, lemmas that are usually glabrous, and in having staminate plants.

### 45. Poa leibergii Scribn. LEIBERG'S BLUEGRASS [p. 402, 522]

*Poa leibergii* grows on mossy ledges and around vernal pools and the outer margins of *Camassia* swales, in sagebrush desert to low alpine habitats, especially where snow persists. It is found primarily on and around the basaltic Columbia plateaus, and is gynodioecious.

### 46. Poa stebbinsii Soreng STEBBINS' BLUEGRASS [p. 402, 522]

*Poa stebbinsii* is endemic to the high Sierra Nevada. It grows primarily in the outer margins of subalpine wet meadows, and is gynodioecious. It is easily recognized by its long hyaline ligules, thin glabrous lemmas, and the absence of intravaginal shoots.

### 47. Poa pringlei Scribn. PRINGLE'S BLUEGRASS [p. 402, 522]

*Poa pringlei* grows on rocky subalpine and alpine slopes in Oregon and California. Sexual populations, with approximately equal numbers of pistillate and staminate plants, are confined to the Klamath–Siskiyou region; Sierra Nevada populations are pistillate and apomictic. Hybrids of *Poa pringlei* with *P. cusickii* have been found on Mount Shasta, California, Mount Rose, Nevada, and near Crater Lake, Oregon. *Poa pringlei* differs from *P. curtifolia* in being dioecious and in having blades that are involute, soft to moderately firm, and abaxially pubescent.

### Poa sect. Dioicopoa E. Desv.

*Poa* sect. *Dioicopoa* includes 29 species; all except the North American *P. arachnifera* are native to South America. The North American species is strictly dioecious. All species appear to reproduce sexually; a few are also bulbiferous. Many, including *P. arachnifera*, are characterized by having 3 well-developed webs on the calluses of the pistillate florets.

### 48. Poa arachnifera Torr. TEXAS BLUEGRASS [p. 402, 522]

*Poa arachnifera* grows on moist, sandy to rich, black bottomlands of the southern Great Plains. At one time it was cultivated for winter pasture in the southeastern United States. It is strictly dioecious, with a 1:1 ratio of staminate to pistillate plants among herbarium samples.

### Poa sect. Pandemos Asch. & Graebn.

*Poa* sect. *Pandemos* includes two diploid species of European origin. One, *P. trivialis*, is now widespread around the world.

## 49. **Poa trivialis** L. ROUGH BLUEGRASS [p. *402*, <u>522</u>]

*Poa trivialis* is an introduced European species. Only **Poa trivialis** subsp. **trivialis** is present in the *Manual* region. Several cultivars have been planted for pastures and lawns, and have often escaped cultivation. *Poa trivialis* sometimes grows with *P. paludigena*, but has distinctly longer ligules and anthers. It is easily recognized by its flat blades, long ligules, sickle-shaped lower glumes, prominent callus webs, and lemmas with pubescent keels and pronounced lateral veins.

## Poa sect. **Oreinos** Asch. & Graebn.

*Poa* sect. *Oreinos* is circumboreal. It includes seven species: four strictly Eurasian, one amphiatlantic, one primarily western North American with isolated occurrences in the Russian Far East, and one restricted to North America. The species are boreal, alpine to low arctic, and grow in bogs and on alpine slopes. They are primarily slender perennials with extravaginal tillering.

## 50. **Poa laxa** Haenke LAX BLUEGRASS [p. *403*, <u>522</u>]

*Poa laxa* is a low arctic to high alpine amphiatlantic species. Its short anthers and smoother branches usually distinguish it from *P. glauca*, with which it can hybridize to form *P. laxa × glauca*.

*Poa laxa* has four subspecies, two of which are native to the *Manual* region; subsp. *laxa* grows in central Europe; and subsp. *flexuosa* (Sm.) Hyl. in northwestern Europe.

1. Innovations primarily extravaginal; panicle
   branches fairly straight; calluses glabrous . . . . . subsp. *banffiana*
1. Innovations primarily intravaginal; panicle branches
   flexuous, usually at least some florets having a
   webbed callus . . . . . . . . . . . . . . . . . . . . . . . . subsp. *fernaldiana*

## Poa laxa subsp. **banffiana** Soreng [p. *403*]

*Poa laxa* subsp. *banffiana* grows primarily in mesic alpine locations of the Rocky Mountains in Canada and the United States. It is sometimes difficult to distinguish from *P. glauca*.

## Poa laxa subsp. **fernaldiana** (Nannf.) Hyl. MOUNT WASHINGTON BLUEGRASS [p. *403*]

*Poa laxa* subsp. *fernaldiana* is native from Newfoundland south to New England.

## 51. **Poa laxa × glauca** [p. *403*]

*Poa laxa × glauca* is an eastern low arctic entity. It can be distinguished from *P. laxa* by its more open sheaths and poorly developed, indehiscent anthers. It differs from *P. glauca* in its broad, thin glumes and lemmas; compact panicles; smooth or nearly smooth, non-glaucous branches; and poorly developed, indehiscent anthers. It also grows in wetter habitats than *P. glauca*, often around seeps.

## 52. **Poa paludigena** Fernald & Wiegand EASTERN BOG BLUEGRASS [p. *403*, <u>522</u>]

*Poa paludigena* is an inconspicuous species restricted to the northeastern United States. It grows in shady bogs and fens, often underneath other plants. *Poa trivialis* sometimes grows with *P. paludigena*; the former has distinctly longer ligules and anthers. Plants from the middle Appalachian Mountains are sometimes confused with *P. sylvestris*. *Poa paludigena* is generally shorter and more slender, has shorter panicles with only 1–2 branches per node, is glabrous between the lemma veins and on the palea keels, has shorter anthers, and grows in colder habitats.

## 53. **Poa leptocoma** Trin. WESTERN BOG BLUEGRASS [p. *403*, <u>523</u>]

*Poa leptocoma* grows around lakes and ponds and along streams, in subalpine and alpine to low arctic habitats, in western North America

from Alaska to California and New Mexico, and on the Kamchatka Peninsula, Russia. It often grows with or near *P. reflexa*, from which it differs in its more scabrous panicle branches, shorter anthers, glabrous or pectinately ciliate palea keels, and preference for wet sites. The two also differ in their ploidy level, *P. leptocoma* being hexaploid, and *P. reflexa* tetraploid. It differs from *P. paucispicula* in its more scabrous panicle branches, narrower glumes and lemmas, and its more sparsely hairy calluses and lemmas.

## Poa sect. **Stenopoa** Dumort.

*Poa* sect. *Stenopoa* includes 30 species. Most are Eurasian; three are native in, and one is restricted to, the *Manual* region. The North American species are cespitose or weakly stoloniferous, and have sheaths open for much of their length, scabrous panicle branches, and faint lateral lemma veins. The new shoots for the following year are initiated late in the growing season, after flowering and fruiting; vegetative and flowering shoots are usually not present at the same time.

## 54. **Poa palustris** L. FOWL BLUEGRASS [p. *403*, <u>523</u>]

*Poa palustris* is native to boreal regions of northern Eurasia and North America, and is widespread in cool-temperate and boreal riparian and upland areas. European plants have also been introduced to other parts of North America. Plants in the Pacific Northwest and the southern United States are usually regarded as introduced, but some populations may be native. *Poa palustris* is used for soil stabilization and waterfowl feed. Plants from drier woods and meadows tends to resemble *P. interior*. The best features for recognizing *P. palustris* include its loose growth habit, more steeply ascending leaf blades, well-developed callus webs, narrowly hyaline lemma margins, and incurving lemma keels. It also has a tendency to branch at the nodes above the base.

## 55. **Poa nemoralis** L. WOODLAND BLUEGRASS [p. *403*, <u>523</u>]

Introduced from northern Eurasia, *Poa nemoralis* is established primarily at low elevations in deciduous and mixed conifer/deciduous forests. It is now common in southeastern Canada and the northeastern United States, and is spreading in the west. It can be distinguished from *P. glauca* and *P. interior* by its consistently short ligules, high top culm node, relatively long flag leaf blades, and narrow glumes and lemmas. It is usually hexaploid.

## 56. **Poa interior** Rydb. INTERIOR BLUEGRASS [p. *403*, <u>523</u>]

*Poa interior*, a native species, grows from Alaska to western Quebec and New York, south to Arizona and New Mexico. It is restricted to the *Manual* region. It is fairly common from boreal forests to low alpine habitats of the Rocky Mountains. It grows in subxeric to mesic habitats, such as mossy rocks and scree, usually in forests. It is usually tetraploid.

In alpine habitats, *Poa interior* is often quite short, and often sympatric with *P. glauca*, from which it differs in lemmas that are glabrous between the marginal veins and keels or, rarely, sparsely puberulent on the lateral veins. It usually also differs from *P. glauca* subsp. *rupicola* in having at least a few hairs on its calluses. It can be distinguished from *P. nemoralis* by its longer ligules, lower top culm node, and wider glumes and lemmas. It differs from *P. palustris* in having lemmas with wider hyaline margins and straight or gradually arched keels, a densely tufted habit, and scantly webbed calluses.

## 57. **Poa glauca** Vahl [p. *403*, <u>523</u>]

*Poa glauca* is a common, highly variable, circumboreal, boreal forest to alpine and high arctic species. It grows from Alaska to Greenland, south to California and New Mexico in the west, and through Canada and the northeastern United States in the east. It also grows at scattered locations in Patagonia. It generally favors dry habitats and tolerates disturbance well. It can be distinguished from *P. nemoralis* and *P. interior* by its longer ligules, lower top culm node, and wider glumes and lemmas. It can be difficult to distinguish from *P. laxa* subsp. *banffiana*. *Poa glauca* is often confused in herbaria with *P. abbreviata* subsp. *pattersonii*, from which it differs in having primarily extravaginal branching and, usually, longer anthers. It hybridizes with *P. laxa*, forming *P. laxa × glauca*. It is also known to hybridize with *P. hartzii*, and is suspected to hybridize with *P. arctica* and *P. secunda*. It is highly polyploid, and presumed to be highly apomictic.

1. All or some spikelets bulbiferous . . . . . . . . . . var. *pekulnejensis*
1. Spikelets not bulbiferous.
  2. Calluses usually webbed, sometimes glabrous; lemmas glabrous or hairy between the veins . . . subsp. *glauca*
  2. Calluses glabrous; lemmas hairy between the veins . . . . . . . . . . . . . . . . . . . . . . . . . subsp. *rupicola*

### Poa glauca Vahl subsp. glauca GLAUCOUS BLUEGRASS [p. 403]

*Poa glauca* subsp. *glauca* is the widespread and common subspecies in the Northern Hemisphere. It is also disjunct in South America. Plants with glabrous calluses are found only in the arctic. In the Rocky Mountains, *P. glauca* subsp. *glauca* often grows with subsp. *rupicola* and *P. interior*. It is often confused in herbaria with subsp. *rupicola*, but can sometimes be distinguished by its webbed calluses and lemmas that are glabrous between the veins.

### Poa glauca var. pekulnejensis (Jurtzev & Tzvelev) Prob. PEKULNEI BLUEGRASS [p. 403]

*Poa glauca* var. *pekulnejensis* is known only from sporadic records in Alaska and the Russian Far East. It can be difficult to distinguish from *P. hartzii* subsp. *vrangelica*, but differs in its tighter habit and thicker glumes.

### Poa glauca subsp. rupicola (Nash) W.A. Weber TIMBERLINE BLUEGRASS [p. 403]

*Poa glauca* subsp. *rupicola* is endemic to dry alpine areas of western North America. It is often confused in herbaria with subsp. *glauca* and *P. interior*, but its calluses lack even a vestige of a web, and its lemmas have at least a few hairs between the lemma veins. It is often sympatric with both taxa outside of California. It is not common in the northern Rocky Mountains.

### Poa sect. Tichopoa Asch. & Graebn.

*Poa* sect. *Tichopoa* has two species, both of which are native to Europe. They are similar to species of *Poa* sect. *Stenopoa*, differing in having strongly compressed culms and nodes, and in being rhizomatous.

### 58. Poa compressa L. CANADA BLUEGRASS [p. 403, 523]

*Poa compressa* is common in much of the *Manual* region. It is sometimes considered to be native, but this seems doubtful. It is rare and thought to be introduced in Siberia and only local in the Russian Far East, but is common in Europe. In the *Manual* region, it is often seeded for soil stabilization, and has frequently escaped. It grows mainly in riparian areas, wet meadows, and disturbed ground. Its distinctly compressed nodes and culms, exserted lower culm nodes, rhizomatous growth habit, and scabrous panicle branches make it easily identifiable.

### Poa sect. Abbreviatae Nannf. ex Tzvelev

*Poa* sect. *Abbreviatae* includes five North American species, two of which also grow in arctic regions of the Eastern Hemisphere. The species are principally high alpine to high arctic. Two of the species are known or reputed to be diploid.

### 59. Poa lettermanii Vasey LETTERMAN'S BLUEGRASS [p. 403, 523]

*Poa lettermanii* grows on rocky slopes of the highest peaks and ridges in the alpine zone, from northern British Columbia to western Alberta and south to California and Colorado, usually in the shelter of rocks or on mesic to wet, frost-scarred slopes. Its glabrous calluses and lemmas usually distinguish it from *P. abbreviata*; it also differs in having flat or folded leaf blades, and shorter spikelets with glumes that are longer than the adjacent florets.

### 60. Poa pseudoabbreviata Roshev. SHORT-FLOWERED BLUEGRASS [p. 403, 523]

*Poa pseudoabbreviata* is a low arctic to subarctic and alpine species of Alaska, northwestern Canada, Siberia, and the Russian Far East. It grows mainly on frost scars, rocky slopes, and ridges, often on open ground. It is easily distinguished from all other alpine and arctic species of *Poa* by its spreading, capillary branches, long pedicels, short stature, small spikelets, and glabrous calluses.

### 61. Poa abbreviata R. Br. [p. 404, 523]

*Poa abbreviata* is an alpine and circumarctic species which has two subspecies in the western cordilleras, and one in the high arctic. It grows mainly on frost scars and mesic rocky slopes, usually on open ground. In rare cases where the lemmas and calluses of *P. abbreviata* are glabrous, it can be confused with *P. lettermanii*, but that species has shorter spikelets and glumes that are longer than the adjacent florets.

1. Lemmas glabrous; calluses webbed . . . . . . . . . . . subsp. *marshii*
1. Lemmas usually with hairs over the veins; calluses glabrous or webbed, rarely both the lemmas and calluses glabrous.
  2. Anthers 0.2–0.8 mm long; lemma intercostal regions hairy . . . . . . . . . . . . . . . . . . . . . subsp. *abbreviata*
  2. Anthers 0.6–1.2(1.8) mm long; lemma intercostal regions glabrous or hairy . . . . . subsp. *pattersonii*

### Poa abbreviata R. Br. subsp. abbreviata DWARF BLUEGRASS [p. 404]

*Poa abbreviata* subsp. *abbreviata* is a common high arctic subspecies that also grows at scattered locations in the Brooks Range, Alaska, and in the Rocky Mountains of the Northwest Territories. It has a circumpolar distribution, but is not common in the Eurasian continental arctic.

### Poa abbreviata subsp. marshii Soreng MARSH'S BLUEGRASS [p. 404]

*Poa abbreviata* subsp. *marshii* is rather uncommon. It is known from scattered alpine peaks across the interior western United States: from the White Mountains of California, the Schell Creek Range of Nevada, the southern Rockies of Idaho, the Little Belt Mountains of Montana, and the Big Horn Mountains of Wyoming, mostly where the other subspecies are absent.

### Poa abbreviata subsp. pattersonii (Vasey) Á. Löve, D. Löve & B.M. Kapoor PATTERSON'S BLUEGRASS [p. 404]

*Poa abbreviata* subsp. *pattersonii* is an alpine taxon that extends from the Brooks Range, Alaska, to the Sierra Nevada, California, where it is rare, and through the Rocky Mountains to southern Colorado. It also grows in the Russian Far East. It is often confused in herbaria with *P. glauca*, but differs in having predominantly intravaginal branching, an abundance of vegetative shoots, and usually shorter anthers. Plants from northern British Columbia to Alaska and Russia have been called *P. abbreviata* subsp. *jordalii* (A.E. Porsild) Hultén. They have webbed calluses, very short (occasionally nonexistent) lemma hairs, panicles often exserted well above the basal tuft of leaves, and particularly slender culms.

**62. Poa keckii** Soreng KECK'S BLUEGRASS [p. *404*, <u>*523*</u>]

*Poa keckii* is endemic to high alpine frost scars and ledges, usually on open ground, in the Sierra Nevada and adjacent Sweetwater and White mountains of California. It is very similar to *Poa suksdorfii*, but is consistently distinct in its details.

**63. Poa suksdorfii** (Beal) Vasey *ex* Piper SUKSDORF'S BLUEGRASS [p. *404*, <u>*523*</u>]

*Poa suksdorfii* is a high alpine species of open rocky ground in the Pacific Northwest. It has narrow panicles like *P. pringlei* and *P. curtifolia*.

## Poa sect. Secundae V.L. Marsh *ex* Soreng

*Poa* sect. *Secundae* includes nine North American species. Two of the species also grow as disjuncts in South America. One species grows on high arctic islands in the Eastern Hemisphere. All the species tend to grow in arid areas, sometimes on wetlands within such areas. One species is confined to dry bluffs along the Pacific coast. All the species are primarily cespitose, but hybridization with members of *Poa* sect. *Poa* results in the formation of rhizomatous plants. Typically, members of sect. *Secundae* have sheaths that are closed for $^1/_{10}$–$^1/_4$ their length, contracted panicles, and anthers that are 1.2–3.5 mm long.

There are two subsections in the *Manual* region: subsects. *Secundae* and *Halophytae*.

## Poa subsect. Secundae Soreng

Species of *Poa* subsect. *Secundae* usually have elongated, weakly compressed spikelets.

**64. Poa secunda** J. Presl SECUND BLUEGRASS [p. *404*, <u>*523*</u>]

*Poa secunda* is one of the major spring forage species of temperate western North America. It is very common in high deserts, mountain grasslands, saline wetlands, meadows, dry forests, and on lower alpine slopes, primarily from the Yukon Territory east to Manitoba and south to Baja California, Mexico. Both subspecies are present, as disjuncts, in Patagonia. *Poa secunda* is known or suspected to hybridize with several other species, including *P. arctica*, *P. arida*, *P. glauca*, and *P. pratensis*. *Poa secunda* differs from *P. curtifolia* in having longer leaf blades that are sometimes folded or involute, and more spikelets per branch. Apomixis is common and facultative.

1. Lemmas usually glabrous, the keels and marginal veins infrequently sparsely puberulent at the base; basal branching mainly extravaginal; leaves slightly lax to firm, remaining intact through the growing season; ligules of the innovations to 2 mm long . . . . . . . . . . . . . . . . . . . . . . . . . . . . . . . . . . subsp. *juncifolia*
1. Lemmas sparsely to densely puberulent or short-villous on the basal $^2/_3$; basal branching mixed intra- or extravaginal or mainly intravaginal; leaves usually lax, withering with age; ligules of the innovations usually longer than 2 mm . . . . . . . . subsp. *secunda*

## Poa secunda subsect. juncifolia (Scribn.) Soreng ALKALI BLUEGRASS, BIG BLUEGRASS, NEVADA BLUEGRASS [p. *404*]

*Poa secunda* subsp. *juncifolia* is usually more robust than subsp. *secunda*, and generally inhabits moister and sometimes saline habitats. It comprises two fairly distinct variants: a robust upland variant that is frequently used for revegetation (*P. ampla* Merr., Big Bluegrass) that grows in deep, rich, montane soils; and a riparian and wet meadow variant (*P. juncifolia* Scribn., Alkali Bluegrass). Apart

from generally having glabrous lemmas, short ligules on the vegetative shoots, and leaf blades that hold their form better, *P. secunda* subsp. *juncifolia* differs anatomically in the predominance of sinuous-walled, rectangular long cells in the blade epidermis; smooth-walled, fusiform long cells are predominant in *P. secunda* subsp. *secunda*. Plants with glabrous lemmas and long ligules on the vegetative shoots have been called *P. nevadensis* Vasey *ex* Scribn.; they are intermediate between the subspecies.

## Poa secunda J. Presl subsp. secunda PACIFIC BLUEGRASS, PINE BLUEGRASS, SANDBERG BLUEGRASS, CANBY BLUEGRASS [p. *404*]

*Poa secunda* subsp. *secunda* comprises several forms or ecotypes which intergrade morphologically and overlap geographically. It generally grows in more xeric habitats than subsp. *juncifolia*; it is also common in alpine habitats. Some of the major variants, and the names that have been applied to them, are: scabrous plants, primarily from west of the Cascade/Sierra Nevada axis (*P. scabrella* (Thurb.) Benth. *ex* Vasey, Pine Bluegrass); smoother, large plants extending eastward (*P. canbyi* (Scribn.) Howell, Canby Bluegrass); tiny, early-spring-flowering plants of stony and mossy ground (*P. sandbergii* Vasey, Sandberg Bluegrass); and slender, sparse plants, generally of mesic shady habitats, with panicles that remain open (*P. gracillima* Vasey, Pacific Bluegrass). Alpine plants have been called *P. incurva* Scribn. & T.A. Williams.

*Poa secunda* subsp. *secunda* differs from *P. stenantha* var. *stenantha* in having more rounded lemma keels, hairs between the veins of the lemmas, and calluses that are glabrous or have hairs shorter than 0.2 mm. It resembles *P. tenerrima*, but lacks that species' combination of persistently wide, open panicles, very scabrous branches, short-truncate ligules, and very fine foliage.

**65. Poa tenerrima** Scribn. DELICATE BLUEGRASS [p. *404*, <u>*523*</u>]

*Poa tenerrima* is a rare species, endemic to serpentine barrens along the western base of the Sierra Nevada. It differs from *P. secunda* subsp. *secunda*) in combining consistently wide, open panicles, very scabrous branches, short-truncate ligules, and very fine foliage.

**66. Poa curtifolia** Scribn. WENATCHEE BLUEGRASS [p. *404*, <u>*523*</u>]

*Poa curtifolia* is endemic to upper serpentine slopes in the Wenatchee Mountains, Kittitas and Chelan counties, Washington. It has narrow panicles like *P. pringlei* and *P. suksdorfii*. It differs from *P. secunda* in having all blades short, flat, and firm, and few spikelets per branch.

**67. Poa hartzii** Gand. HARTZ'S BLUEGRASS [p. *404*, <u>*523*</u>]

*Poa hartzii* grows only in the high arctic. It generally grows on open ground, on sandy or clayey soils, or on slumping slopes of old marine terraces. Morphologically, it is closest to *P. secunda* and *P. ammophila*.

1. Spikelets bulbiferous . . . . . . . . . . . . . . . . . . . . subsp. *vrangelica*
1. Spikelets not bulbiferous.
   2. Lemmas 5.5–7 mm long; anthers well developed, 2–2.8 mm long . . . . . . . . . . . . . . . . . . . subsp. *alaskana*
   2. Lemmas 3.3–5.4 mm long; anthers usually aborted and shorter than 1.5 mm . . . . . . . . . . subsp. *hartzii*

## Poa hartzii subsp. alaskana Soreng [p. *404*]

*Poa hartzii* subsp. *alaskana* grows on the North Slope of Alaska, mainly in sandy places. It generally resembles robust plants of *P. hartzii* subsp. *hartzii*, but has a looser habit, longer lemmas, and well-developed anthers.

## Poa hartzii Gand. subsp. hartzii [p. *404*]

*Poa hartzii* subsp. *hartzii* is the common subspecies on high arctic islands and in Greenland. It grows at scattered locations on the continental margin, from the Mackenzie River Delta to the Ungava

Peninsula, Canada. Outside the *Manual* region, it is known from Wrangel Island in Russia and from Svalbard, Norway. It is apomictic, setting seed despite rarely forming anthers. Robust plants of subsp. *hartzii* resemble those of subsp. *alaskana*, but have a tighter habit and poorly developed anthers that are usually aborted.

## Poa hartzii subsp. vrangelica (Tzvelev) Soreng & L.J. Gillespie [p. 404]

*Poa hartzii* subsp. *vrangelica* can be difficult to distinguish from *P. glauca*, but it has a looser habit and thin glumes. It grows along the Sagavanirktok River, from the Franklin Hills to Prudhoe Bay, Alaska, as well as at scattered locations along the coast of the Beaufort Sea in Alaska, the Queen Elizabeth Islands in Nunavut, and Wrangell Island, Russia. It includes two varieties: **Poa hartzii** var. **vivipara** Polunin, which grows in the Canadian portion of its range; and **Poa hartzii** var. **vrangelica** (Tzvelev) Prob., which is more western and favors *P. glauca*.

## 68. Poa ammophila A.E. Porsild SAND BLUEGRASS [p. 404, 523]

*Poa ammophila* is endemic to the Mackenzie River Delta region, Northwest Territories. It grows primarily north of treeline and, as its name indicates, usually on sandy soils. Its close relative, *P. hartzii*, also reaches the continental coastline in this region.

## Poa subsect. Halophytae V.L. Marsh *ex* Soreng

Members of *Poa* subsect. *Halophytae* have spikelets that resemble those in other sections of *Poa* more closely than those of subsect. *Secundae*, being shorter and more compressed than those of that subsection.

## 69. Poa stenantha Trin. NARROW-FLOWER BLUEGRASS [p. 405, 523]

*Poa stenantha* grows in coastal meadows and on cliffs in subarctic and boreal forests; it is less common in moist, more southern subalpine and low alpine meadows and thickets. Its range extends from western Alaska to the northern Cascades and Rocky Mountains and, as a disjunct, to Patagonia.

1. Spikelets not bulbiferous . . . . . . . . . . . . . . . . . . . var. *stenantha*
1. Spikelets bulbiferous . . . . . . . . . . . . . . . . . . . . . var. *vivipara*

## Poa stenantha Trin. var. stenantha [p. 405]

*Poa stenantha* var. *stenantha* differs from *P. secunda* subsp. *secunda* in its strongly keeled lemmas with glabrous intercostal regions, and, when present, callus hairs longer than 0.2 mm. Plants with large panicles and glabrous calluses have been called *P. macroclada* Rydb. Such plants grow infrequently in the U.S. Rocky Mountain portion of the species' range. They intergrade with the more compact typical form.

## Poa stenantha var. vivipara Trin. [p. 405]

*Poa stenantha* var. *vivipara* is the common form of the species in the Aleutian Islands; it extends eastward to Sitka, Alaska. It differs from bulbiferous forms of *P. arctica* in its lack of rhizomes, more open sheaths, and usually glaucous and scabrous panicle branches.

## 70. Poa napensis Beetle NAPA BLUEGRASS [p. 405, 523]

*Poa napensis* is endemic to mineralized ground around hot springs in Napa County, California. It is listed as an endangered species by the United States Fish and Wildlife Service.

## 71. Poa unilateralis Scribn. SEA-BLUFF BLUEGRASS [p. 405, 523]

*Poa unilateralis* grows on grassy bluffs and cliffs near the Pacific coast of North America, from Washington to California.

1. Lemmas villous on the keels and marginal veins for more than $1/3$ the length of the lemmas; blades usually involute, the cauline and innovation blades similar . . . . . . . . . . . . . . . . . . . . . . . . . subsp. *pachypholis*
1. Lemmas glabrous or the keels and marginal veins villous or puberulent for less than $1/5$ the length of the lemmas; blades flat or folded, the cauline blades sometimes wider and thicker than the innovation blades . . . . . . . . . . . . . . . . . . . . . . . subsp. *unilateralis*

## Poa unilateralis subsp. pachypholis (Piper) D. D. Keck *ex* Soreng [p. 405]

*Poa unilateralis* subsp. *pachypholis* is known from populations in Lincoln County, Oregon, and Pacific County, Washington.

## Poa unilateralis Scribn. subsp. unilateralis [p. 405]

The range of *Poa unilateralis* subsp. *unilateralis* extends from northern Oregon to central California.

## Poa subg. Arctopoa (Griseb.) Prob.

*Poa* subg. *Arctopoa* includes five species in two sections; only one species grows in the *Manual* region. The species are generally robust, rhizomatous perennials of subsaline or subalkaline soils of wetlands. They have cilia along the lemma and sometimes the glume margins, and the bases of the basal sheaths have short, thick hairs. The subgenus is sometimes treated as a genus.

## Poa sect. Arctopoa (Griseb.) Tzvelev

*Poa* sect. *Arctopoa* has one species. It grows along boreal and low arctic coasts in North America, and from the Russian Far East to northern Japan. **Poa** sect. **Aphydris** (Griseb.) Tzvelev, the other section in the subgenus, is restricted to central and east Asia.

## 72. Poa eminens J. Presl EMINENT BLUEGRASS [p. 405, 523]

*Poa eminens* grows along low arctic and boreal coasts and estuaries, in subsaline meadows and beaches. It also grows along the Asian coast from Hokkaido Island, Japan, to the Chukchi Peninsula, Russia. It hybridizes with *Dupontia* (×*Dupoa*, p. 141).

## Named intersectional hybrids

## 73. Poa arida Vasey PLAINS BLUEGRASS [p. 405, 523]

*Poa arida* grows mainly on the eastern slope of the Rocky Mountains and in the northern Great Plains, primarily in riparian habitats of varying salinity or alkalinity. It is spreading eastward along heavily salted highway corridors. *Poa glaucifolia* Scribn. & T.A. Williams refers to specimens of the northern Great Plains that have a more lax growth form with broader leaves and occasionally somewhat open panicles, florets with a small web, and sometimes lacking hairs between the keel and marginal veins of the lemma.

## 74. Poa ×limosa Scribn. & T.A. Williams [p. 405, 523]

*Poa ×limosa* grows at scattered locations in western North America. It prefers wet to moist, often saline or alkaline meadows, primarily in the sagebrush zone. It is probably a hybrid between *P. pratensis* and *P. secunda* subsp. *juncifolia*.

## 75. Poa ×gaspensis Fernald [p. 405, 523]

*Poa ×gaspensis* is found in the coastal mountains of the Gaspé Peninsula. It seems to consist of hybrids between *P. pratensis* subsp. *alpigena* and *P. alpina*. *Poa ×gaspensis* differs from *P. alpina* in its extravaginal branching, rhizomatous habit, and webbed calluses; from *P. pratensis* in its acute ligules and more pubescent lemmas; and

from *P. arctica* in its sharply keeled, more scabrous glumes and its spikelet shape, which approaches those of *P. alpina* and *P. pratensis*.

## 14.14 ×DUPOA J. Cay. & Darbysh.[1]

Pl per; rhz. Clm 25–80 cm, glab, often glaucous, with 2–4 nd. Shth closed for $^1/_3$–$^2/_3$ their length, glab, not conspicuously glaucous; aur absent; lig memb, whitish- or yellowish-brown, acute to obtuse, lacerate to erose, ciliate; bld glab. Infl pan. Spklt with (1)2–3(4) flt. Glm usu not exceeding the flt, glab, often glaucous, acute to

acuminate; cal bearded, hairs often crinkled below the keel and/or absent from the sides; lm usu scabrous, smt smooth towards the base, acute; pal smooth or slightly scabrous; anth 3, indehiscent. Car absent.

×*Dupoa* is a sterile hybrid between *Dupontia* and *Poa*. Only one species is known.

### 1. ×Dupoa labradorica (Steudel) J. Cay. & Darbysh.
LABRADOR BLUEGRASS, PÂTURIN DU LABRADOR [p. 406, <u>523</u>]

×*Dupoa labradorica* is known from southeastern Hudson Bay, eastern James Bay, and northern Labrador. Its distribution is completely within the sympatric range of the parental species. Additional records are to be expected from regions where both parental species are found, such as Alaska and the Russian Far East. It grows on seashores, lagoon margins, and salt marshes, usually in relatively deep organic matter over clay layers sometimes mixed with sandy or gravelly deposits. It sometimes forms dense and almost pure stands.

×*Dupoa labradorica* is a sterile hybrid between *Dupontia fisheri* and *Poa eminens*. It is similar to *Poa eminens*, but has narrower leaves, stems, and rhizomes, and more open panicles. It differs from *Dupontia fisheri* in having glumes that usually do not exceed the florets, and some woolly or crinkly callus hairs. It differs from both parental species in having indehiscent anthers and no caryopses.

## 14.15 DUPONTIA R. Br.[2]

Pl per; rhz. Clm 5–80 cm. Shth closed for $^1/_2$–$^2/_3$ their length, glab; aur absent; lig memb, glab, truncate; bld flat or folded, usu glab, adx surfaces smt scabrous or shortly pubescent. Infl pan, diffuse to dense and spikelike, with few spklt; br 0.7–3.5(8.5) cm, smooth, stiff, erect to reflexed, sec br usu appressed. Spklt pedlt, slightly lat compressed, with 1–4(5) flt, distal flt strl; rchl prolonged beyond the upmost flt or terminating in a vestigial flt, glab; dis above the glm and beneath the ftl flt. Glm subequal, equaling or usu exceeding the distal flt, ovate and obtuse to lanceolate-attenuate, memb, glab, 1–3-veined, unawned; cal short, blunt, with a ring of stiff hairs, hairs to about 1 mm; lm ovate

to ovate-lanceolate, memb to coriaceous, glab or pubescent, with 3(5) fine veins, apc acute to acuminate, midveins smt excurrent as a mucro or awn to 1(2.2) mm; pal shorter than the lm, glab, smt scabrous on the veins; lod 2, memb, glab; anth 3; ov glab. Car shorter than the lm, concealed at maturity, falling free of the lm and pal; hila $^1/_6$–$^1/_3$ the length of the car, ovate.

*Dupontia* is a unispecific genus of arctic and subarctic wetlands, found throughout the holarctic region except in Scandinavia. Hybrids with *Arctophila* are referred to ×*Arctodupontia*; hybrids with *Poa eminens* are referred to ×*Dupoa*.

### 1. Dupontia fisheri R. Br. FISHER'S TUNDRAGRASS,
DUPONTIE DE FISHER [p. 406, <u>523</u>]

*Dupontia fisheri* grows in wet meadows, wet tundra, marshes, and along streams and the edges of lagoons, ponds, and lake shores, in sand, silt, clay, moss, and rarely in bogs. It hybridizes with *Poa eminens* to form ×*Dupoa labradorica*. The hybrids differ from *D.*

*fisheri* in having glumes that usually do not exceed the florets, and in having some woolly or crinkly callus hairs. The hybrid genus ×*Arctodupontia* is formed from *D. fisheri* and *Arctophila fulva*, and differs from *D. fisheri* in having lemmas with truncate, lacerate to dentate apices, rather than acute to acuminate apices.

## 14.16 ×ARCTODUPONTIA Tzvelev[3]

Pl per; rhz. Clm to 38 cm. Shth closed for $^1/_3$–$^3/_4$ their length; aur absent; lig memb, acute to truncate, lacerate; bld flat or folded, glab. Infl partly open pan; br stiff, glab, lo br smt reflexed. Spklt somewhat compressed, with 2–4 flt. Glm subequal, ovate, memb, acute; cal with scant, stiff hairs; lm ovate, memb to subcoriaceous,

1(3)-veined, apc truncate and lacerate-dentate; pal 2-keeled, glab; anth indehiscent; ov glab; lod 2, memb. Car apparently absent.

×*Arctodupontia* is a sterile hybrid between the two unispecific genera *Dupontia* and *Arctophila*. It is intermediate between the two parents.

### 1. ×Arctodupontia scleroclada (Rupr.) Tzvelev [p. 406, <u>524</u>]

×*Arctodupontia scleroclada* is known from Mansfield Island in northern Hudson Bay, and from Malaya Zemlya and the Chukchi Peninsula in Russia. Plants seemingly intermediate between *Dupontia fisheri* and *Arctophila fulva* are common, but caution must be

exercised in assigning them to ×*Arctodupontia*. The hybrids differ from their parents in being sterile and in having lemmas with truncate, lacerate to dentate apices. *Dupontia* has lemmas with acute to acuminate apices; *Arctophila* has lemmas with obtuse, entire apices.

[1,2]Jacques Cayouette and Stephen J. Darbyshire [3]Stephen J. Darbyshire and Jacques Cayouette

## 14.17 ARCTOPHILA (Rupr.) Andersson[1]

Pl per; rhz, smt producing aquatic lvs when submerged. Clm (5)10–100 cm, erect, glab, rooting at the lo nd; nd glab. Shth of aquatic lvs closed to near the apc, translucent, pale pinkish brown; col inconspicuous; lig acute; bld pinkish brown. Shth of aerial lvs usu closed for over ¹/₂ their length, smt open to the base, opaque, green to olive-green, glab; col conspicuous as a zone of contrasting color; aur absent; lig memb, truncate, lacerate; bld usu flat, glab, up bld conspicuously longer than the lo bld. Infl open pan; br stiff and ascending to pendulous, glab, some br longer than 1 cm. Spklt pedlt, somewhat lat compressed, with (1)2–7(9) flt; rchl prolonged beyond the base of the distal flt, glab; dis above the glm, beneath the flt. Glm subequal, broadly lanceolate to ovate, memb to subcoriaceous, glab, 1–3(5)-veined, acute to obtuse, unawned; cal short, blunt, shortly pubescent to almost glab; lm ovate, glab, memb to subcoriaceous, with 3(5) obscure veins, lat veins usu not reaching the lm apc, apc entire or somewhat erose, obtuse, unawned or rarely the cent vein extended as a short mucro to about 0.2 mm; pal subequal to the lm; lod 2, free, glab, toothed or entire; anth 3; ov glab. Car shorter than the lm, concealed at maturity, falling free; hila broadly ovate, ¹/₆–¹/₅ the length of the car.

*Arctophila* is a unispecific, but highly polymorphic, holarctic genus closely related to *Dupontia*.

### 1. Arctophila fulva (Trin.) Andersson PENDANT GRASS, ARCTOPHILE FAUVE [p. *406*, <u>524</u>]

*Arctophila fulva* grows as an emergent species in shallow, standing water, or along slow-moving streams, wet meadows, marshes, and saturated soils of low arctic and subarctic regions, where it often forms pure stands. It is one of the few grasses that develop aquatic leaves. In the *Manual* region, it grows from Alaska through the Yukon, Northwest Territories, Nunavut, Ontario, Quebec, and Labrador to Greenland. Its range extends across Eurasia to arctic Scandinavia. It forms a sterile hybrid, ×*Arctodupontia scleroclada*, with *Dupontia fisheri*. The hybrid differs from *Arctophila fulva* in having lemmas with truncate, lacerate to dentate apices, rather than obtuse, entire apices.

## 14.18 TORREYOCHLOA G.L. Church[2]

Pl per; rhz. Clm 18–145 cm, usu erect, smt decumbent and rooting at the lo nd; intnd hollow. Shth open to the base; aur absent; lig memb; bld flat. Infl tml pan; br scabrous, usu densely scabrid distally; ped less than 0.5 mm thick. Spklt pedlt, lat compressed to terete, with 2–8 flt; dis above the glm and beneath the flt. Glm unequal, shorter than the lowest lm, rounded to weakly keeled, memb, veins obscure to prominent, unawned; lo glm 1(3)-veined; up glm (1)3(5)-veined; cal blunt, glab; lm memb, rounded to weakly keeled, smt pubescent, particularly proximally, prominently (5)7–9-veined, veins more or less parallel, veins and intervens usu scabridulous, particularly distally, lat veins usu rdcd or absent, apc scabridulous and entire to serrate-erose, unawned; pal subequal to the lm, 2-veined; lod 2, free, glab, entire or toothed; anth usu 3; ov usu hairy, smt glab. Car shorter than the lm, concealed at maturity, oblong, flattened dorsally, falling free; hila oblong, about ¹/₃ the length of the car.

*Torreyochloa* grows in cold, wet, non-saline environments. It includes the two North American species treated below, plus two additional taxa in northeastern Asia. *Torreyochloa* is distinguished from *Glyceria* by its open leaf sheaths, and from *Puccinellia* by the 7–9 (occasionally 5) prominent, rather than faint, lemma veins.

1. Mature inflorescences linear to narrowly elliptic, 0.3–1 cm wide, 5.5–19 times as long as wide; widest cauline blades 3.4–7.2 mm wide . . . . . . . . . . . . . . . . . . . . . . . . . . . . . . . . . . . . . . . . . . . . . . . . . . . . . . . . . . . .1. *T. erecta*
1. Mature inflorescences conic, ovoid, or obovoid, 1–16 cm wide, 1–7.5 times as long as wide; widest cauline blades 1.5–18 mm wide . . . . . . . . . . . . . . . . . . . . . . . . . . . . . . . . . . . . . . . . . . . . . . . . . . . . .2. *T. pallida*

### 1. Torreyochloa erecta (Hitchc.) G.L. Church SPIKED FALSE MANNAGRASS [p. *407*, <u>524</u>]

*Torreyochloa erecta* grows at elevations above 2000 m, in the margins of subalpine and alpine lakes and streams of the Sierra Nevada and Cascade ranges.

### 2. Torreyochloa pallida (Torr.) G.L. Church [p. *407*, <u>524</u>]

All three varieties of *Torreyochloa pallida* grow in swamps, marshes, bogs, and the margins of lakes and streams. They are usually morphologically distinct, and tend to have different geographic ranges.

1. Widest cauline blades 1.5–3 mm wide; anthers of the lowest floret in each spikelet 0.3–0.6 mm long . . . . var. *fernaldii*
1. Widest cauline blades 2.8–18 mm wide; anthers of the lowest floret in each spikelet 0.5–1.5 mm long.
    2. Anthers of the lowest floret in each spikelet 0.7–1.5 mm long; basal diameter of the culms 1.2–3 mm; upper glumes 1.4–2.7 mm long; plants generally of eastern North America . . . . . . var. *pallida*
    2. Anthers of the lowest floret in each spikelet 0.5–0.7 mm long; basal diameter of the culms (1.3)1.8–4.8 mm; upper glumes 0.9–1.8 mm long; plants of western North America . . . . . . var. *pauciflora*

[1]Jacques Cayouette and Stephen J. Darbyshire  [2]Jerrold I. Davis

**Torreyochloa pallida var. fernaldii** (Hitchc.) Dore
FERNALD'S FALSE MANNAGRASS, GLYCÉRIE DE FERNALD [p. 407]

*Torreyochloa pallida* var. *fernaldii* grows from Saskatchewan through southern Ontario and the Great Lakes area to New Brunswick and the New England states. It is often difficult to distinguish from var. *pallida* where their ranges overlap.

**Torreyochloa pallida** (Torr.) G.L. Church var. **pallida**
PALE FALSE MANNAGRASS [p. 407]

The range of *Torreyochloa pallida* var. *pallida* extends from southeastern Manitoba to north of Lake Superior and Maine, and south to Missouri and Georgia. It can be difficult to distinguish from var. *fernaldii*.

**Torreyochloa pallida var. pauciflora** (J. Presl) J.I. Davis
WEAK MANNAGRASS [p. 407]

*Torreyochloa pallida* var. *pauciflora* grows in western North America, from sea level to 3500 m. Robust plants from Pacific lowland forests are often taller than 1 m; plants from farther east and higher elevations tend to be shorter.

## 14.19 CATABROSA P. Beauv.[1]

Pl per; not rhz, smt stln. Clm 5–70 cm, usu decumbent and rooting at the lo nd; nd glab. Shth closed; aur absent; lig memb; bld flat. Infl open pan, with at least some br longer than 1 cm. Spklt pedlt, lat compressed to terete, with (1)2(3) flt; rchl glab, prolonged beyond the base of the distal ftl flt, empty or with rdcd flt; dis above the glm and beneath the flt. Glm unequal, much shorter than the lm, scarious, veinless or the up glm with 1 vein at the base, apc rounded to truncate, unawned; cal short, blunt, glab; lm glab, conspicuously 3-veined, veins raised, rounded over the midvein, apc rounded to truncate, erose and scarious, unawned; pal subequal to the lm, 2-veined; lod 2, truncate, irregularly lobed; anth 3; ov glab. Car shorter than the lm, concealed at maturity, fusiform; hila ovoid.

*Catabrosa*, a genus of two species, grows in marshes and shallow waters of the Northern Hemisphere and South America. It resembles members of the *Meliceae* in its closed leaf sheaths, truncate, scarious lemma apices, and chromosome base number, but lacks the distinctive lodicule morphology of that tribe. One species is native to the *Manual* region.

1. **Catabrosa aquatica** (L.) P. Beauv. BROOKGRASS,
WATER WHORLGRASS, CATABROSA AQUATIQUE [p. 407, 524]

*Catabrosa aquatica* grows in wet meadows and the margins of streams, ponds, and lakes in the *Manual* region, Argentina, Chile, Europe, and Asia. It is listed as endangered in Wisconsin. Although palatable, it is never sufficiently abundant to be important as a forage species. The species is regarded here as being variable, but having no infraspecific taxa.

## 14.20 CUTANDIA Willk.[2]

Pl ann. Clm (4)10–42 cm. Shth open to the base; aur absent; lig memb; bld flat to convolute. Infl usu pan, rarely rcm; br divaricate, some br longer than 1 cm; ped of lat spklt 0.3–2 mm, ped of tml spklt on a br longer. Spklt lat compressed, with 2–9(12) flt, distal flt often rdcd; dis occurring variously at the base of the flt, the ped, or the br. Glm equal or unequal, usu shorter than the adjacent lm, memb, acuminate, rounded, or emgt, unawned or shortly awned, awns glab; lo glm 1–3-veined; up glm 1–5-veined; cal short, glab; lm memb, 3-veined, keeled on each vein, apc usu emgt or bifid, varying to rounded or acuminate, unawned, mucronate, or awned from the sinus or slightly below; pal subequal to the lm; lod 2, free, memb, ciliate; anth 3; ov glab. Car shorter than the lm, concealed at maturity, narrow, longitudinally grooved; hila oval.

*Cutandia* is a Mediterranean and western Asian genus of six species, one of which has been collected in California.

1. **Cutandia memphitica** (Spreng.) K. Richt. MEMPHIS
GRASS [p. 407, 524]

*Cutandia memphitica* is native to maritime sands, inland dunes, sandy gravel, and gypsum soils in the Mediterranean region and Middle East. In the *Manual* region, it was collected in sandy soil at a nursery at Devil's Canyon in the San Bernardino Mountains, California, in 1933. How it got there and whether it persists are not known.

## 14.21 BRIZA L.[3]

Pl ann or per; csp. Clm 5–100 cm, usu erect, unbrchd; intnd hollow; nd glab. Shth smt less than ¹/₂ as long as the intnd, open; aur absent; lig hyaline; bld flat, usu erect. Infl open pan; br sparsely strigose, capillary, spklt usu pendulous, some br longer than 1 cm. Spklt pedlt, pendulous, oval to triangular in side view, becoming light brown at maturity, lat compressed but the glm and lm with broadly rounded backs, glm and flt strongly divergent from the rchl, with 4–12(15) chartaceous flt, distal flt rdmt; rchl glab, not prolonged beyond base of the distal flt; dis above the glm and beneath the flt. Glm subequal, shorter than to longer than the adjacent lm, naviculate, faintly 3–7-veined, mrg more or less memb, apc obtuse, unawned; cal short, glab; lm inflated, about

[1,2]Mary E. Barkworth [3]Neil Snow

as wide as long, with broadly rounded backs, similar in shape to the glm but somewhat cordate, mrg becoming hyaline, frequently splitting perpendicular to the midveins, unawned; **pal** shorter than the lm, scarious or chartaceous; **lod** 2, joined or free, usu entire, smt toothed; **anth** 3; **ov** glab. **Car** shorter than the lm, concealed at maturity, usu falling with the lm and pal, ovoid to obovoid; **hila** round to elliptic.

*Briza*, a genus of about 20 species, is native to Eurasia and South America. Most species have little to no fodder value because of the scant foliage. The ornamental value of the genus is more significant; the species are often grown for use in dried floral arrangements. Three European species are now scattered in the more temperate parts of southern Canada and the United States, and will undoubtedly be collected in areas not indicated here. *Briza* species can become weedy where established.

1. Plants perennial; ligules about 0.5 mm long; sheaths open for about ¹/₂ their length . . . . . . . . . . . . . . . . . . . . 1. *B. media*
1. Plants annual; ligules 3–13 mm long; sheaths open to near the base.
   2. Spikelets 10–20 mm long . . . . . . . . . . . . . . . . . . . . . . . . . . . . . . . . . . . . . . . . . . . . . . . . . . . . . . . . . . . . . 2. *B. maxima*
   2. Spikelets 2–7 mm long . . . . . . . . . . . . . . . . . . . . . . . . . . . . . . . . . . . . . . . . . . . . . . . . . . . . . . . . . . . . . . 3. *B. minor*

### 1. **Briza media** L. PERENNIAL QUAKINGGRASS, AMOURETTE COMMUNE, AMOUR DU VENT [p. *407*, 524]

*Briza media* is native to chalk and clay grasslands of Europe. It grows in acid to calcareous soils in moist to somewhat dry, sunny conditions, in meadow floodplains, forest clearings, old meadows, and pastures. It is often grown as an ornamental, and can colonize artificial habitats such as roadsides, but does not appear to invade recently disturbed locations. In the *Manual* region, it is most abundant in eastern North America, and is found in a few widely scattered locations elsewhere.

### 2. **Briza maxima** L. BIG QUAKINGGRASS [p. *407*, 524]

*Briza maxima* is native to the Mediterranean region. Cultivated as an ornamental, it is possibly one of the earliest grasses grown for other than edible purposes. It occasionally becomes naturalized in dry to

somewhat moist but well-drained, fine or sandy soil on banks, rocky places, open woodlands, and cultivated areas such as roadsides and pastures. In the *Manual* region, it is known from scattered locations, mostly in Oregon and California, where it is an invader of coastal dune habitat.

### 3. **Briza minor** L. LITTLE QUAKINGGRASS [p. *407*, 524]

*Briza minor* is native to the Mediterranean region. It is the most widespread species of *Briza* in the *Manual* region, growing in many habitats: swamp margins, seasonal wetlands and around vernal pools, open woodlands, sandhills, roadsides, and pastures. It appears to be established from southern British Columbia south through western Oregon to California and Arizona, and in the east from the Atlantic states to the Gulf Coast states, inland to Oklahoma and Arkansas.

## 14.22 AIRA L.[1]

**Pl** ann; tufted. **Clm** 1–55 cm, erect to decumbent. **Lvs** cauline; **shth** open for most of their length, glab, usu scabridulous, occ smooth; **aur** absent; **lig** memb; **bld** of the upmost lvs greatly red. **Infl** open or contracted pan, smt spikelike, with more than 1 spklt associated with most nd; **br** longer than 5 mm, capillary, appressed to strongly divergent; **ped** capillary, appressed to divergent. **Spklt** 1.5–3.8 mm, lat compressed, with 2 bisx flt, both usu awned, the lo flt smt unawned, occ both unawned; **rchl** glab, lowest segments about 0.2 mm, flt appearing opposite, usu not or scarcely extended beyond the base of the distal flt; **dis** above the glm and beneath the flt. **Glm** equal to subequal, longer than the flt, memb, 1–3-veined, unawned; **cal** puberulent; **lm** subcoriaceous, glab, scabridulous, 5-veined, apc bifid, awned or

unawned, awns attached below midlength, usu geniculate, smt straight; **pal** memb, 2-veined; **lod** 2, free; **anth** 3; **ov** glab. **Car** shorter than the lm, concealed at maturity, adhering to the lm and/or pal, longitudinally grooved, dorsally compressed.

*Aira* has eight species; two have been introduced into the *Manual* region. All members of the genus are native to temperate Europe and the Mediterranean region, North Africa, and western Asia. Frequently adventive, they are now widespread outside of their native range as weeds, although they are not considered particularly troublesome. They have little forage value because most are delicate, with extremely small leaves. All the species grow in open, disturbed places on usually dry, occasionally mesic, sandy to rocky soils.

1. Panicle branches ascending to divergent; panicles 1.2–13.5 cm long, 1.5–10 cm wide . . . . . . . . . . . . . . . 1. *A. caryophyllea*
1. Panicle branches appressed to the rachises; panicles 0.5–4.1 cm long, 0.3–0.7 cm wide . . . . . . . . . . . . . . . . 2. *A. praecox*

### 1. **Aira caryophyllea** L. [p. *408*, 524]

*Aira caryophyllea* is native to Eurasia and Africa; it has become established in the *Manual* region, primarily on the Pacific, Gulf, and Atlantic coasts, and through much of the southeastern United States. It is usually found in disturbed areas, in vernally moist to dry, sandy to rocky, open sites, from sea level to subalpine elevations. It sometimes invades lawns or rock gardens.

1. Glumes subobtuse, usually denticulate, often mucronate; pedicels with abruptly thickened apices
   . . . . . . . . . . . . . . . . . . . . . . . . . . . . . . . var. *cupaniana*

1. Glumes acute; pedicels gradually thickening to the apices.
   2. Pedicels usually 2–8 times as long as the spikelets; spikelets 1.7–2.5 mm long . . . . . . . . var. *capillaris*
   2. Pedicels usually 1–2 times as long as the spikelets; spikelets 2–3.5 mm long . . . . . . . var. *caryophyllea*

### **Aira caryophyllea** var. **capillaris** (Mert. & W.D.J. Koch) Mutel DELICATE HAIRGRASS [p. *408*]

*Aira caryophyllea* var. *capillaris* is native to Europe, northern Africa, and western Asia. It usually grows in dry to somewhat moist, sandy loam soils of grassy banks, woodland openings, and disturbed sites such as pastures and roadsides.

**Aira caryophyllea** L. var. **caryophyllea** SILVER HAIRGRASS [p. 408]

*Aira caryophyllea* var. *caryophyllea* is native to the Mediterranean region. It usually grows in dry, sandy to rocky soil and on rock outcrops, in open and disturbed sites in woods, grassy flats, pastures, paths, and roadsides; it is occasionally found in damp ground at swamp or lagoon margins.

**Aira caryophyllea** var. **cupaniana** (Guss.) Fiori SILVERY HAIRGRASS [p. 408]

*Aira caryophyllea* var. *cupaniana* is native to southern Europe and northern Africa, growing in mesic, open habitats in disturbed areas or open woodland. It was discovered in a prescribed burn area of Mount Diablo State Park in Contra Costa County, California, in 1995, but was not relocated in 1999.

**2. Aira praecox** L. SPIKE HAIRGRASS [p. 408, 524]

*Aira praecox* is native to Europe. In the *Manual* region, it grows mainly along or near the Pacific and Atlantic coasts, in dry to vernally moist sand dunes or in sandy to rocky soils, on rock faces and ledges, and in disturbed areas such as the edges of roads, railways, and airports. It is usually found in lowland areas, though it occasionally grows at montane to subalpine elevations.

## 14.23 COLEANTHUS Seidl[1]

**Pl** ann, tufted. **Clm** 2–7(10) cm, ascending or decumbent. **Shth** closed, inflated; **aur** absent; **lig** memb, ovate; **bld** flat. **Infl** pan; **br** distant, verticillate. **Spklt** slightly lat compressed, pedlt, ped hairy, with 1 flt; **rchl** not prolonged beyond the base of the flt; **dis** below the flt. **Glm** absent; **cal** short, glab; **lm** 1-veined, narrowed to awnlike apc; **pal** 2–4-toothed, 2-keeled; **lod** absent; **anth** 2; **ov** glab. **Car** elliptic, exceeding the lm and pal, falling free; **hila** oval.

*Coleanthus* is a unispecific genus that is native to Eurasia and North America.

**1. Coleanthus subtilis** (Tratt.) Seidl MOSS GRASS, MUD GRASS [p. 408, 524]

*Coleanthus subtilis* is an ephemeral pioneer species of wet, open habitats. It grows on muddy to sandy calcium-deficient soils on the shores of lakes, sandbars, and islands. In the *Manual* region, it is known from the Columbia River, and around Hatzic, Arrow and Shuswap lakes in British Columbia. It also grows in Europe, Russia, and China. It is easily overlooked because of its diminutive size, and because it flowers in early spring or late fall. It is not clear whether it is native or introduced in the *Manual* region.

## 14.24 EREMOPOA Roshev.[2]

**Pl** ann. **Clm** 5–60 cm. **Shth** open for most of their length; **aur** absent; **lig** glab; **bld** flat or convolute. **Infl** pan; **br** glab, scabrous, some br longer than 1 cm, some ped longer than 3 mm, usu less than 1 mm thick, flexible. **Spklt** pedlt, terete to slightly lat compressed, with (1)2 to many flt, ftl flt (1)2–6, distal flt vestigial; **rchl** straight or slightly bowed, scabrous or puberulent; **dis** above the glm and beneath the flt. **Glm** unequal in length, subequal in width, usu shorter than the adjacent lm, unawned; **lo glm** 1-veined; **up glm** 3-veined; **cal** short, blunt, glab; **lm** lanceolate, glab or the bases slightly pilose, 5-veined, veins converging distally, apc not mucronate, unawned; **pal** shorter than the lm; **lod** 2, free, glab; **anth** 3; **ov** glab. **Car** shorter than the lm, concealed at maturity, linear, adhering to the lm and/or pal.

*Eremopoa* is a genus of two to six species that are native from the eastern Mediterranean to western China.

**1. Eremopoa altaica** (Trin.) Roshev. ALTAI GRASS [p. 408, 524]

*Eremopoa altaica* is native to the high deserts and mountain steppes of Asia, from Turkey through Afghanistan and Pakistan to the Altai region, the Himalayas, and western China. It was once found in railway yards and roadsides at Brandon, Manitoba, but misidentified as *Eremopoa persica* (Trin.) Roshev. Although the population persisted for several years, recent efforts to relocate it have failed.

## 14.25 SPHENOPHOLIS Scribn.[3]

**Pl** usu per, rarely winter ann; usu csp, smt the clm solitary. **Clm** (5)20–130 cm, lvs evenly distributed. **Shth** open; **aur** absent; **lig** memb, erose; **bld** flat or involute, glab or pubescent. **Infl** pan, open or contracted, nodding to erect; **dis** below the glm, the distal flt smt disarticulating first. **Spklt** pedlt, 2.1–9.5 mm, lat compressed, with 2–3 flt; **rchl** glab or pubescent, prolonged beyond the base of the distal flt as a slender bristle. **Glm** almost equaling the lowest flt, dissimilar in width, memb to subcoriaceous, mrg scarious, apc unawned; **lo glm** narrower than the up glm, 1(3)-veined, strongly keeled, apc acute; **up glm** elliptical to oblanceolate, obovate, or subcucullate, 3(5)-veined, strongly to slightly keeled, apc acuminate, acute, rounded, or truncate; **cal** glab; **lm** hrb, not indurate, rounded on the lo back, smooth or partly or wholly scabrous, usu keeled near the apc, 3(5)-veined, veins usu not visible, unawned or awned from just below the apc,

[1]Sandy Long  [2]Stephen J. Darbyshire  [3]Thomas F. Daniel

awns straight or geniculate; **pal** hyaline, shorter than the lm; **lod** 2, free, memb, glab, toothed or entire; **anth** 3; ov glab. **Car** shorter than the lm, concealed at maturity, linear-ellipsoid, glab; **endosperm** liquid.

*Sphenopholis* includes six species, all of which are native to the *Manual* region. Its greatest diversity is in the southeastern United States. One species, *Sphenopholis obtusata*, extends outside the region to southern Mexico and the Caribbean.

Glume widths are measured in side view, from the lateral margin to the midvein.

1. Distal lemmas awned, awns 3–9 mm long; spikelets 4.5–9.5 mm long . . . . . . . . . . . . . . . . . . . . . . . . . . . 1. *S. pensylvanica*
1. Distal lemmas unawned or awned, awns 0.1–3 mm long; spikelets 2.1–5.2 mm long.
  2. Distal lemmas scabrous on the sides, unawned or awned; anthers (0.5)1–2 mm long.
    3. Lower glumes at least ⅓ as wide as the upper glumes; blades (1)2–5(7) mm wide, flat to slightly involute; awns absent or shorter than 1 mm . . . . . . . . . . . . . . . . . . . . . . . . . . . . . . . . 2. *S. nitida*
    3. Lower glumes less than ⅓ as wide as the upper glumes, rarely slightly wider; blades 0.3–1.5(2) mm wide, involute to filiform; awns absent or (0.1)1–3 mm long . . . . . . . . . . . . . . . . . . . . . 3. *S. filiformis*
  2. Distal lemmas usually smooth on the sides, sometimes scabrous or scabridulous distally, unawned; anthers 0.2–1 mm long.
    4. Upper glumes subcucullate, the width/length ratio 0.3–0.5; panicles usually erect, often spikelike, the spikelets usually densely arranged . . . . . . . . . . . . . . . . . . . . . . . . . . . . . . . . . . . 4. *S. obtusata*
    4. Upper glumes not subcucullate, the width/length ratio 0.17–0.35; panicles usually nodding, not spikelike, the spikelets usually loosely arranged.
      5. Spikelets 2.1–4 mm long; lowest lemmas 2.1–3 mm long; upper glumes 1.9–2.9 mm long . . . . . . . . 5. *S. intermedia*
      5. Spikelets 4–5.2 mm long; lowest lemmas 3.1–4.2 mm long; upper glumes 3–3.9 mm long . . . . . . . . 6. *S. longiflora*

### 1. Sphenopholis pensylvanica (L.) Hitchc. SWAMP OATS [p. 408, 524]

*Sphenopholis pensylvanica* grows in springheads, seepage areas, swamps, marshes, and other moist to wet places, at 0–1100 m, in the eastern and southeastern United States. It hybridizes with *S. obtusata*. The hybrids, which are called **Sphenopholis ×pallens** (Biehler) Scribn., differ from *S. pensylvanica* in having generally shorter (3–5 mm) spikelets and shorter (0.1–4 mm), straight or bent awns on the distal lemmas. They differ from *S. obtusata* in having awns on the distal lemmas.

### 2. Sphenopholis nitida (Biehler) Scribn. SHINY WEDGEGRASS [p. 408, 524]

*Sphenopholis nitida* grows in moist to dry, deciduous and coniferous forests, on clay or silt banks and slopes, at 0–1200 m, in southern Ontario and the eastern United States. It can be confused with occasional forms of *S. obtusata* that have somewhat scabrous distal lemmas, but *S. nitida* has broader lower glumes.

### 3. Sphenopholis filiformis (Chapm.) Scribn. SOUTHERN WEDGEGRASS [p. 408, 524]

*Sphenopholis filiformis* grows in sandy soils of pine and mixed pine forests, at 0–500 m, in the southeastern United States. It is found primarily in the coastal plain, but extends to the piedmont. *Sphenopholis filiformis* differs from occasional forms of *S. obtusata* with somewhat scabrous distal lemmas in having narrower leaves.

### 4. Sphenopholis obtusata (Michx.) Scribn. PRAIRIE WEDGEGRASS, SPHENOPHOLIS OBTUS [p. 409, 524]

*Sphenopholis obtusata* grows in prairies, marshes, dunes, forests, and waste places, at 0–2500 m. Its range extends from British Columbia to New Brunswick, through most of the United States, to southern Mexico and the Caribbean. The distal lemmas of *S. obtusata* are occasionally somewhat scabrous. Such plants can be distinguished from *S. nitida* by their narrower lower glumes, from *S. filiformis* by their wider leaves, and from *S. pensylvanica* by their shorter, unawned spikelets. Hybrids with *S. pensylvanica*, called **Sphenopholis ×pallens**, have short (0.1–4 mm) awns on the distal lemmas.

### 5. Sphenopholis intermedia (Rydb.) Rydb. SLENDER WEDGEGRASS, SPHENOPHOLIS INTERMÉDIAIRE [p. 409, 524]

*Sphenopholis intermedia* grows at 0–2500 m in wet to damp sites, sites that dry out after the growing season, and sites with clay soils that retain moisture. Restricted to the *Manual* region, it is found in forests, meadows, and waste places throughout most of the region other than the high arctic. It differs from *Koeleria macrantha* in its more open panicles and in having spikelets that disarticulate below the glumes.

### 6. Sphenopholis longiflora (Vasey *ex* L.H. Dewey) Hitchc. BAYOU WEDGEGRASS [p. 409, 524]

*Sphenopholis longiflora* grows in forest bottoms along bayous and streams, from 0–50 m, in Texas, Arkansas, and Louisiana.

## 14.26 DESCHAMPSIA P. Beauv.[1]

**Pl** usu per, smt ann; csp or tufted. **Clm** 5–140 cm, hollow, erect. **Lvs** usu mainly bas, often forming a dense tuft; **shth** open; **aur** absent; **lig** memb, decurrent, rounded to acuminate; **bld** often all or almost all tightly rolled or folded and some flat, smt most flat, others rolled or folded. **Infl** tml pan, open or contracted; **dis** above the glm, beneath the flt. **Spklt** 3–9 mm, with 2(3) flt in all or almost all spklt, flt usu bisx, smt viviparous; **rchl** hairy, usu prolonged more than 0.5 mm beyond the base of the distal flt, smt terminating in a highly rdcd flt. **Glm** subequal to unequal, usu exceeding the adjacent flt, often exceeding all flt, 1- or 3-veined, acute to acuminate; **cal** antrorsely strigose; **lm** obscurely (3)5–7-veined, rounded over the back, apc truncate-erose to 2–4-toothed, awned, awns usu attached on the lo ½ of the lm, occ subapical, straight to strongly geniculate, slightly to strongly twisted proximally, straight distally; **pal** shorter than the lm, 2-keeled, keels often scabrous;

[1]Mary E. Barkworth

lod 2, lanceolate to ovate-lanceolate, usu entire; **anth 3;**
**ov** glab; **sty** 2. **Car** oblong; **emb** about ¹/₄ the length of
the car.

*Deschampsia* includes 20–40 species. It is best represented
in the Americas and Eurasia, but it grows in cool, damp
habitats throughout the world. Seven species are native to the
*Manual* region; none of the remaining species have been
introduced.

*Deschampsia* differs from *Vahlodea*, which it used to
include, in having primarily basal, rather than primarily
cauline, leaves, and hairy rachillas that extend more than 0.5
mm beyond the base of the distal floret in a spikelet. *Trisetum*

differs from *Deschampsia* primarily in its more acute, bifid
lemmas, and in having awns that are inserted at or above the
midpoint of the lemmas. In *Deschampsia*, the awns are
usually inserted near the base.

Because the treatments of *Deschampsia brevifolia* and *D.*
*sukatschewii* were revised shortly before going to press, the
maps are preliminary, particularly with respect to the southern
portion of the distribution of these two species.

Lemma length, awn attachment, and awn length should be
examined on the lower florets within the spikelets. The upper
florets often have shorter lemmas, and shorter awns that are
attached higher on the back than those of the lower florets.

1. All or most spikelets viviparous; panicle branches smooth . . . . . . . . . . . . . . . . . . . . . . . . . . . . . . . . . . . 5. *D. alpina*
1. All or most spikelets bisexual or, if viviparous, the panicle branches scabrous.
  2. Plants annual; awns strongly geniculate . . . . . . . . . . . . . . . . . . . . . . . . . . . . . . . . . . . . . . . 7. *D. danthonioides*
  2. Plants perennial; awns straight to strongly geniculate.
    3. Lemmas scabridulous or puberulent, dull . . . . . . . . . . . . . . . . . . . . . . . . . . . . . . . . . . . . 8. *D. flexuosa*
    3. Lemmas glabrous, shiny.
      4. Glumes mostly green, apices purple; spikelets strongly appressed, overlapping each other by
        less than ¹/₂ their length . . . . . . . . . . . . . . . . . . . . . . . . . . . . . . . . . . . . . . . . . . . . 6. *D. elongata*
      4. Glumes usually purplish proximally, sometimes purplish over more than ¹/₂ the surface,
        whitish to golden distally; spikelets usually divergent, and/or overlapping each other by more
        than ¹/₂ their length. (*D. cespitosa* complex).
        5. Spikelets 6–7.5 mm long; culms sometimes decumbent and rooting at the lower nodes;
          plants of sandy areas around lakes in the Northwest Territories and northern
          Saskatchewan . . . . . . . . . . . . . . . . . . . . . . . . . . . . . . . . . . . . . . . . . . . . . . 2. *D. mackenzieana*
        5. Spikelets 2–7.6 mm; culms erect, not rooting at the lower nodes; plants of gravels, wet
          meadows, and bogs, widely distributed in cooler regions of North America.
          6. Spikelets strongly imbricate, often rather densely clustered on the ends of the branches,
            sometimes evenly distributed on the branches; glumes and lemmas dark purple
            proximally for over more than ¹/₂ their surface; lemmas 2.2–4 mm long . . . . . . . . . . . . . . . 4. *D. brevifolia*
          6. Spikelets usually not or only moderately imbricate, not in dense clusters at the ends of
            the branches; glumes usually purple over less than ¹/₂ their surface, often with a green
            base, a distal purple band, and pale apices; lemmas 2–5(7)mm long.
            7. Basal blades with 5–11 ribs, usually most or all ribs scabridulous or scabrous, outer
              ribs often more strongly so, sometimes the ribs only papillose or puberulent, usually
              at least some blades flat and 1–4 mm wide, the majority folded or rolled and 0.5–1
              mm in diameter; lower glumes often scabridulous distally over the midvein; lower
              panicle branches often scabridulous or scabrous, sometimes smooth . . . . . . . . . . . . . . . 1. *D. cespitosa*
            7. Basal blades with 3–5 ribs, ribs usually smooth or papillose, sometimes puberulent
              or the outer ribs scabridulous, all blades of the current year usually strongly involute
              and hairlike, 0.3–0.5(0.8) in diameter; lower glumes smooth over the midvein; lower
              panicle branches usually smooth, sometimes sparsely scabridulous; . . . . . . . . . . . . . . 3. *D. sukatschewii*

### 1. Deschampsia cespitosa (L.) P. Beauv. [p. 409, 524]

*Deschampsia cespitosa* is circumboreal in the Northern Hemisphere,
and also grows in New Zealand and Australia. It is an attractive taxon
that grows in wet meadows and bogs, and along streams and lakes,
from sea level to over 3000 m in cool-temperate, but not arctic,
habitats.

There are widely varying opinions concerning the taxonomic
treatment of *Deschampsia cespitosa*. The circumscription adopted
here is narrower than has been customary in North America. Some of
the distribution records shown, particularly those from the northern
part of the region, may reflect the broad interpretation of the species.
In western North America, *Deschampsia cespitosa* exhibits both
ecotypic differentiation and a high degree of plasticity. The following
three subspecies intergrade.

1. Panicles contracted at anthesis, the branches
  appressed to ascending; glumes 4.5–5.8 mm long,

  midvein of the lower glumes scabrous distally
  . . . . . . . . . . . . . . . . . . . . . . . . . . . . . . . . . . . subsp. *holciformis*
1. Panicles open at anthesis, the branches strongly
  divergent to drooping; glumes 2–7.5 mm long;
  midvein of the lower glumes smooth or scabridulous
  distally.
  2. Plants often glaucous; glumes 4.4–7.5 mm long;
    awns usually exceeding the lemmas; plants of the
    northwest coast of North America . . . . . . . subsp. *beringensis*
  2. Plants not glaucous; glumes 2–6 mm long; awns
    exceeded by or exceeding the lemmas; plants
    widespread in North America . . . . . . . . . . . subsp. *cespitosa*

### Deschampsia cespitosa subsp. beringensis (Hultén) W.E. Lawr. Beringian Hairgrass [p. 409]

*Deschampsia cespitosa* subsp. *beringensis* is primarily a coastal
species, growing up to 800 m along the Aleutian chain and the
southern coast of Alaska south to Sonoma County, California, and

west to the Kamchatka Peninsula, Russia. Typical plants are tall, glaucous, have long ligules and spikelets, and long, narrow glumes, but in the Pribiloff Islands and at scattered locations elsewhere, they intergrade with plants that are only 15–25 cm tall and also have smaller spikelet parts. *Deschampsia cespitosa* subsp. *beringensis* differs from *D. mackenzieana* primarily in its coastal distribution and lower chromosome number. It supposedly differs from *D. cespitosa* subsp. *cespitosa* in having long glumes but there is considerable overlap in this and other characters. The morphological, geographic, and ecological boundaries between the two subspecies need further study.

### Deschampsia cespitosa (L.) P. Beauv. subsp. cespitosa

TUFTED HAIRGRASS, DESCHAMPSIE CESPITEUSE [p. *409*]

*Deschampsia cespitosa* subsp. *cespitosa* is treated here as a circumboreal taxon that is most prevalent in boreal and temperate North America, growing at 0–3000 m; many reports from arctic and alpine North America refer to what are treated here as *D. sukatschewii* or *D. brevifolia*. Even with this narrower interpretation, *D. cespitosa* is highly polymorphic. Larger plants are difficult to distinguish from *D. cespitosa* subsp. *beringensis*. Many cultivars of *Deschampsia cespitosa* subsp. *cespitosa* have been developed. They are treated here as one part of the spectrum of variation in subsp. *cespitosa*.

### Deschampsia cespitosa subsp. holciformis (J. Presl) W.E. Lawr. [p. *409*]

*Deschampsia cespitosa* subsp. *holciformis* grows in coastal marshes and sandy soils, from the Queen Charlotte Islands, British Columbia, to central California. It intergrades and is interfertile with *D. cespitosa* subsp. *beringensis*, differing in its closed panicles, scabrous veins on the lower glumes, and more strongly imbricate spikelets. There are relatively few collections; it is not clear whether this reflects lack of collecting or rarity.

### 2. Deschampsia mackenzieana Raup MACKENZIE HAIRGRASS [p. *409*, *524*]

*Deschampsia mackenzieana* grows on the sandy shores and dunes around Great Slave Lake, Northwest Territories, and Lake Athabasca, Saskatchewan. The decumbent culms of some plants may be a response to shifting substrate.

### 3. Deschampsia sukatschewii (Popl.) Roshev.

DESCHAMPSIE NAINE [p. *409*, *524*]

*Deschampsia sukatschewii* is a circumboreal species that extends from northern Russia through Alaska, northern Canada, and Greenland to Svalbard, and southward in the Rocky Mountains to Nevada and Utah. It ranges from short plants that form dense, mossy tufts on the Arctic coast to larger plants in subalpine and alpine habitats of the Rocky Mountains that have frequently been included in *D. cespitosa*.

Arctic taxonomists recognize two subspecies of *Deschampsia sukatschewii* in arctic North America: the amphiberingian subsp. *orientalis* (Hultén) Tzvelev that extends to the northern coast of Alaska and the western Northwest Territories; and subsp. *borealis*

(Trautv.) Tzvelev, which is circumpolar in the arctic. Efforts to circumscribe infraspecific taxa of *D. sukatschewii* for this treatment failed.

### 4. Deschampsia brevifolia R. Br. [p. *409*, *524*]

*Deschampsia brevifolia* is a circumboreal taxon that grows in wet places in the tundra, often in disturbed soils associated with riverbanks, frost-heaving, etc. It is interpreted here as extending southward through the Rocky Mountains to Colorado, where it grows at elevations up to 4300 m. It is to be expected from high elevations in British Columbia and Alberta. In its typical appearance, *Deschampsia brevifolia* is quite distinctive because of its dark, narrow panicles. Culm height can vary substantially from year to year, probably in response to the environment.

### 5. Deschampsia alpina (L.) Roem. & Schult. ALPINE HAIRGRASS, DESCHAMPSIE ALPINE [p. *409*, *524*]

*Deschampsia alpina* grows in damp, rocky places, on calcareous substrates with low organic content, in Greenland and northeastern Canada and, outside the *Manual* region, in the mountains of Scandinavia and Russia in the Kola Peninsula and Novaya Zemlya. Plants of *D. alpina* differ from viviparous plants of *D. cespitosa* in having smooth, rather than scabrous, panicle branches.

### 6. Deschampsia elongata (Hook.) Munro SLENDER HAIRGRASS [p. *409*, *524*]

*Deschampsia elongata* grows in moist to wet habitats, from near sea level to alpine elevations, from Alaska and the Yukon south to northern Mexico and east to Montana, Wyoming, and Arizona. It also grows, as a disjunct, in Chile. The records from Maine and Colorado probably represent introductions.

### 7. Deschampsia danthonioides (Trin.) Munro ANNUAL HAIRGRASS [p. *409*, *525*]

*Deschampsia danthonioides* grows in temperate and cool-temperate regions, usually in open, wet to dry habitats and often in disturbed ground. Its primary range extends from southern British Columbia, through Washington and Idaho, to Baja California, Mexico. It also grows, as a disjunct, in Chile and Argentina. Records from the Yukon Territory date from the late 1800s and early 1900s; it has not been seen since in the region. Records from east of the primary region are also probably introductions; it is not known whether the species has persisted at these locations.

### 8. Deschampsia flexuosa (L.) Trin. CRINKLED HAIR-GRASS, WAVY HAIRGRASS, DESCHAMPSIE FLEXUEUSE [p. *409*, *525*]

*Deschampsia flexuosa* grows on dry, often rocky slopes, and in woods and thickets, often in disturbed sites. In the *Manual* region, it is primarily eastern in distribution, with records from west of the Great Lakes and Appalachians probably being introductions. It is also known from Mexico, Central America, South America, Borneo, the Philippines, and New Zealand.

---

# 14.27 AGROSTIS L.[1]

Pl usu per; usu csp, smt rhz or stln. **Clm** (3)5–120 cm, usu erect. **Shth** open, usu smooth and glab, smt scabrous to scabridulous, rarely hairy; **col** not strongly developed; **aur** absent; **lig** memb, smooth or scabridulous dorsally, apc truncate, obtuse, rounded, or acute, usu erose to lacerate, the lacerations smt obscuring the shape, or entire; **bld** flat, folded, or involute, usu smooth and glab, smt scabridulous, adx surfaces somewhat ridged. **Infl** tml pan, narrowly cylindrical and dense to open and diffuse; **br** usu in

whorls, usu more or less scabrous, rarely smooth, some br longer than 1 cm; **sec pan** smt present in the leaf axils. **Spklt** 1.2–7 mm, pedlt, lat compressed, lanceolate to narrowly oblong or ovate, with 1(2) flt; **rchl** not prolonged beyond the base of the flt(s); **dis** above the glm, beneath the flt, smt initially at the panicle base. **Glm** (1)1.3–2(4) times longer than the lm, 1(3)-veined, glab, usu mostly smooth, vein(s) often scabrous to scabridulous, backs keeled or rounded, apc acute to acuminate or awn-tipped; **lo glm** usu 0.1–0.3 mm

[1]M.J. Harvey

longer than the up glm, rarely equal; **cal** poorly developed, blunt, glab or hairy, hairs to about $^1/_2$ as long as the lm; **lm** thinly memb to hyaline, usu smooth and glab, smt scabridulous, occ pubescent, rarely warty-tuberculate, 3–5-veined, veins not convergent, smt excurrent as 2–5 teeth, apc acute to obtuse or truncate, smt erose, unawned or awned, smt varying within an inflorescence, awns arising from near the lm bases to near the apc, usu geniculate, smt straight; **pal** absent, or minute to subequal to the lm, usu thin, veins not or only weakly developed; **lod** 2, free; **anth** (1)3, 0.1–2 mm, not penicillate; **sty** 2, free to the base, white; **ov** glab. **Car** with a hard, soft, or liquid endosperm, the latter resulting from the substitution of lipids for starch.

*Agrostis* in the older, broad sense is a genus comprised of species with the spikelets reduced to single florets. As such, it is found in all inhabited continents, is presumably of ancient origins, and many of the 150–200 species may be only distantly related.

*Agrostis* usually differs from both *Podagrostis* and *Lachnagrostis* in having no, or very reduced, paleas, and rachillas that are not prolonged beyond the base of the floret. *Agrostis* also differs from *Lachnagrostis* in certain features of the lemma epidermes. *Agrostis* can be distinguished from *Apera* by its lemmas that are less firm than the glumes, paleas that are often absent or minute, and its lack of a rachilla prolongation. There is no single character that distinguishes all species of *Agrostis* from those of *Calamagrostis*. In general, *Agrostis* has smaller plants with smaller, less substantial lemmas and paleas than *Calamagrostis*, and tends to occupy drier habitats. It differs from *Polypogon* in having spikelets that disarticulate above the glumes.

Species of *Agrostis* growing in the temperate regions of the Northern Hemisphere and on tropical mountains are mostly perennials, with the annual species predominantly in warmer climates, such as the Mediterranean and the Southern Hemisphere. Of the 26 species known from the *Manual* region, 21 are native and 5 are introductions. Two additional species, *A. tolucensis* and *A. anadyrensis*, have been reported; the reports are dubious.

Species with awns on the lemmas frequently exhibit a developmental gradient within the inflorescence. Upper florets may possess a well-developed geniculate awn inserted at the base or on the lower half of the lemma; mid-inflorescence spikelets may have a shorter, possibly non-geniculate awn inserted high on the lemma, while basal spikelets may possess only a terminal bristle on the lemma. This phenomenon is particularly sharply shown in *Agrostis castellana*, where a single side branch of only a dozen or so spikelets can show the whole sequence. When using the key, it is advised to examine spikelets from the upper parts of an inflorescence. Many species key more than once, due to the potential for awns to be either present or absent.

1. Paleas at least $^2/_5$ as long as the lemmas.
  2. Lemmas 0.5–0.8 mm long, transparent; paleas similar to the lemmas and almost as long; panicles usually over $^1/_2$ the length of the culms, extremely diffuse . . . . . . . . . . . . . . . . . . . . . . . . . . . . . . . . . . 28. *A. nebulosa*
  2. Lemmas 1.2–2.5 mm long, opaque to translucent; paleas shorter than the lemmas; panicles less than $^1/_2$ the length of the culms, diffuse or not.
    3. Panicles narrowly contracted, sometimes open at anthesis, 0.5–4(6) cm wide; branches ascending to appressed.
      4. Stolons present, plants mat-forming, rhizomes absent; paleas 0.7–1.4 mm long; anthers 0.9–1.4 mm long . . . . . . . . . . . . . . . . . . . . . . . . . . . . . . . . . . . . . . . . . 4. *A. stolonifera* (in part)
      4. Stolons absent, plants usually cespitose, rhizomes sometimes present; paleas to 0.5 mm long; anthers 0.3–0.6 mm long . . . . . . . . . . . . . . . . . . . . . . . . . . . . . . . . . . . . . . . . 15. *A. exarata* (in part)
    3. Panicles open at maturity, sometimes somewhat contracted after anthesis, (1)2–15 cm wide; branches spreading to ascending.
      5. Ligules of the upper leaves longer than wide, 2–7.5 mm long; usually at least some lower panicle branches with spikelets to the base.
        6. Stolons absent, rhizomes present; culms 20–120 cm tall; panicles 8–30 cm long; longest lower panicle branches 4–9 cm long . . . . . . . . . . . . . . . . . . . . . . . . . . . . . . . . . . . . . . . . . 3. *A. gigantea*
        6. Stolons present, rhizomes absent; culms 8–60 cm tall; panicles 3–20 cm long; longest lower panicle branches 2–6 cm long . . . . . . . . . . . . . . . . . . . . . . . . . . . . . . . . . . . . . . 4. *A. stolonifera* (in part)
      5. Ligules of the upper leaves usually shorter than wide, 0.3–3 mm long; lower panicle branches with spikelets confined to the distal $^1/_3$–$^1/_2$.
        7. Calluses glabrous or with a few hairs to 0.1 mm long; adjacent pedicels divergent, giving well-separated spikelets; panicles stiffly erect, 3–20 cm long; awns rarely present, to 2 mm; lemmas glabrous . . . . . . . . . . . . . . . . . . . . . . . . . . . . . . . . . . . . . . . . . . . . . . 1. *A. capillaris*
        7. Calluses abundantly hairy, hairs to 0.6 mm long; adjacent pedicels not divergent, spikelets appearing clustered; panicles somewhat lax, 10–30 cm long; awns, if present, to 5 mm long on the terminal spikelet of a cluster; lemmas occasionally with hairs on the lower $^1/_2$ . . . . . . . . 2. *A. castellana*
1. Paleas absent or less than $^2/_5$ as long as the lemmas.
  8. Lemmas awned [for opposite lead, see p. 151].
    9. Panicles dense, often spikelike, 0.2–4 cm wide; lower branches usually shorter than 2 cm, appressed to ascending, usually hidden by the spikelets.

10. Lemma awns 3.5–10 mm long; calluses with hairs to 1 mm long; plants annual.
  11. Lemmas 2.5–4 mm long, teeth to 1.5 mm long; awns (5)8–10 mm long; blades 1–4.5 cm long ...................................................... 25. *A. hendersonii*
  11. Lemmas 1.5–2.3 mm long, teeth to 0.5 mm long; awns 3.5–8 mm long; blades 3–15 cm long ...................................................... 26. *A. microphylla*
10. Lemma awns to 3.5 mm long; calluses glabrous or with hairs to 0.3 mm long; plants perennial.
  12. Blades less than 2 mm wide, usually involute or folded.
    13. Calluses glabrous; panicles often partly enclosed by the upper sheaths at maturity; lemma awns to 0.7 mm long ............................ 22. *A. blasdalei* (in part)
    13. Calluses hairy; panicles exserted from the upper sheaths at maturity; lemma awns to 3.5 mm long.
      14. Lemma apices acute, entire; lemma awns to 2.8 mm long, usually not exserted from the spikelets .......................................... 21. *A. variabilis* (in part)
      14. Lemma apices truncate, denticulate; lemma awns 2–3.5 mm long, exserted from the spikelets ........................................... 23. *A. tolucensis* (in part)
  12. Blades 2–10 mm wide, usually flat, sometimes folded.
    15. Lemma apices truncate to acute; blades to 4 mm wide, flat or involute; ligules 2–6.2 mm long; panicles 0.5–1.5 cm wide ................. 23. *A. tolucensis* (in part)
    15. Lemma apices acute to obtuse; blades to 10 mm wide, flat; ligules 1–11.2 mm long; panicles 0.5–4 cm wide.
      16. Anthers 0.3–0.6 mm long; paleas usually absent, rarely to 0.5 mm long and about $^1/_5$ the length of the lemmas; lemmas entire or with teeth to 0.12 mm long ...................................................... 15. *A. exarata* (in part)
      16. Anthers 0.5–2 mm long; paleas 0.3–0.7 mm long, to about $^1/_3$ the length of the lemmas; lemmas usually with teeth to 0.3 mm long ................. 16. *A. densiflora* (in part)
9. Panicles open or diffuse, or somewhat contracted but not spikelike, 0.4–20 cm wide; lower branches 1.5–12 cm long, erect to spreading, readily visible.
  17. Leaves usually involute or becoming so, sometimes only the basal leaves involute, less than 1 mm in diameter when involute, 0.5–2 mm wide when flat; plants without rhizomes or stolons.
    18. Anthers 1, 0.1–0.2 mm long, usually persistent at the apices of the caryopses; awns attached just below the apices of the lemmas, flexuous but not geniculate, deciduous ...................................................... 27. *A. elliottiana* (in part)
    18. Anthers 3, 0.4–1.5 mm long, usually shed at anthesis; awns attached below midlength on the lemmas, usually geniculate, persistent.
      19. Basal leaves usually withered at anthesis; lower sheaths finely tomentose; callus hairs abundant; plants endemic to coastal California ......................... 19. *A. hooveri*
      19. Basal leaves persistent; lower sheaths smooth or scabrous; callus hairs sparse; plants widespread, especially in northern and montane parts of the *Manual* region, including California.
        20. Panicles (2)3–10 cm long; branches not capillary, fairly stiff, smooth or sparsely scabridulous; callus hairs to 0.4 mm long; caryopses 1.4–2 mm long, endosperm solid ........................................... 7. *A. mertensii* (in part)
        20. Panicles (4)8–25(50) cm long; branches capillary, flexible, scabrous; callus hairs to 0.2 mm long; caryopses 0.9–1.4 mm long; endosperm liquid ....... 10. *A. scabra* (in part)
  17. Leaves usually remaining flat, 0.5–6 mm wide; plants with or without rhizomes or stolons.
    21. Lemmas with 4 teeth up to 0.5 mm long, lemmas 2.5–3 mm long; awns 4–6 mm long ...... 20. *A. howellii*
    21. Lemmas usually entire, sometimes minutely toothed or erose, teeth to 0.4 mm long, lemmas 1–3 mm long; awns to 5 mm long.
      22. Rhizomes and stolons absent; blades to 30 cm long; anthers 0.4–1.2 mm long.
        23. Panicle branches widely divergent, the whole panicle often detaching at the base at maturity, forming a tumbleweed; cauline nodes usually 1–3; blades 1–2 mm wide; glume apices acuminate ............................. 10. *A. scabra* (in part)
        23. Panicle branches usually erect to ascending, if widely divergent then the panicle not forming a tumbleweed; cauline nodes 2–10; blades 0.5–5 mm wide; glume apices acute to acuminate.
          24. Lemma awns 1–4.4 mm long, geniculate, exserted; blades to 13 cm long.

25. Leaf blades 0.5–3 mm wide, flat to involute; panicles 2–10 cm long, usually open; awns 2–4.4 mm long, inserted just below midlength on the lemmas . . . . . . . . . . . . . . . . . . . . . . . . . . . . . . . . . . . . . . . . . . . . 7. *A. mertensii* (in part)

25. Leaf blades 3–4 mm wide, flat; panicles 6–20 cm long, somewhat contracted; awns 1–1.5 mm long, inserted just above midlength on the lemmas . . . . . . . . . . . . . . . . . . . . . . . . . . . . . . . . . . . . . . . . . . . . . . . 8. *A. anadyrensis*

24. Lemma awns minute or to 2 mm long, straight, usually not exserted; blades 6–30 cm long.

26. Basal leaves usually withered by anthesis; culm leaves 3–10, as broad and substantial as the lower leaves; callus hairs dense; plants primarily from east of the 100th Meridian . . . . . . . . . . . . . . . . 12. *A. perennans* (in part)

26. Basal leaves persisting; culm leaves 5 or fewer, usually less substantial than the lower leaves; callus hairs sparse; plants primarily western . . . . . . . . . . . . . . . . . . . . . . . . . . . . . . . . . . . . . . 14. *A. oregonensis* (in part)

22. Rhizomes or stolons present; blades 1–10 cm long; anthers 0.7–1.8 mm long.

27. Rhizomes absent; stolons present, to about 25 cm long, producing tufts of shoots at the nodes; glumes 1.7–3 mm long; panicles open, branches erect to spreading . . . . . . . . . . . . . . . . . . . . . . . . . . . . . . . . . . . . . . . . . . . . . 5. *A. canina* (in part)

27. Rhizomes present, to about 10 cm long; stolons absent; glumes 2–4 mm long; panicles open to constricted, branches more or less erect to ascending.

28. Lemma apices blunt, entire; lemmas usually awned from near the base, rarely unawned, awns 2–4.5 mm long, geniculate . . . . . . . . . . . . . . . . . 6. *A. vinealis* (in part)

28. Lemma apices acute, entire or toothed; lemmas usually unawned, rarely awned from below the apices, awns to 0.5(2.7) mm long, straight . . . . . 17. *A. pallens* (in part)

8. Lemmas unawned [for opposite lead, see p. 149].

29. Mature panicles dense; lower panicle branches to 3(4) cm long, often hidden by the spikelets; spikelets crowded.

30. Blades 0.5–2 mm wide, in dense basal tufts; panicles 0.2–2 cm wide; culms 5–30 cm tall.

31. Lemma veins not excurrent; anthers 0.4–1 mm long; plants of western alpine and subalpine zones . . . . . . . . . . . . . . . . . . . . . . . . . . . . . . . . . . . . 21. *A. variabilis* (in part)

31. Lemma veins excurrent to 0.2 mm; anthers 0.7–2 mm long; plants of western coastal cliffs, dunes, and shrublands . . . . . . . . . . . . . . . . . . . . . . . . . . . . . . 22. *A. blasdalei* (in part)

30. Blades 2–10 mm wide, not basally concentrated; panicles 0.5–4 cm wide; culms 8–100 cm tall.

32. Anthers 0.3–0.6 mm long; paleas usually absent, rarely to 0.5 mm long and about ¹/₅ the length of the lemmas; lemmas entire or with teeth to 0.12 mm long . . . . . . . . . . 15. *A. exarata* (in part)

32. Anthers 0.5–2 mm long; paleas 0.3–0.7 mm long, to about ¹/₃ the length of the lemmas; lemmas usually toothed, teeth to 0.3 mm long . . . . . . . . . . . . . . . . . . . 16. *A. densiflora* (in part)

29. Mature panicles open to diffuse; lower panicle branches often longer than 3 cm, usually not hidden by the spikelets; spikelets crowded or not.

33. Blades 0.5–14 cm long, 0.5–2 mm wide, usually involute or becoming so; anthers 0.1–0.9 mm long.

34. Anthers 1, 0.1–0.2 mm long; callus hairs dense, to 0.6 mm long; plants annual . . 27. *A. elliottiana* (in part)

34. Anthers 3, 0.2–0.9 mm long; callus hairs sparse, to 0.3 mm long; plants perennial or annual.

35. Panicles (4)8–50 cm long; lower panicle branches 4–15 cm long.

36. Lemmas 1.4–2 mm long, exceeding the ripe caryopses by 0.3+ mm; anthers 0.4–0.8 mm long; pedicels to 9.6 mm long, spikelets not appearing clustered . . . . . . . . . . . . . . . . . . . . . . . . . . . . . . . . . . . . . . . . . . . . . . . . . . 10. *A. scabra* (in part)

36. Lemmas 0.8–1.2 mm long, exceeding the ripe caryopses by no more than 0.2 mm; anthers 0.2–0.5 mm long; pedicels to 3.5 mm long, spikelets appearing clustered . . . . . . . . . . . . . . . . . . . . . . . . . . . . . . . . . . . . . . . . . . . . . . . 11. *A. hyemalis*

35. Panicles 1.5–13 cm long; lower panicle branches 1–4 cm long.

37. Anthers 0.3–0.6 mm long; upper culm sheaths not inflated; plants perennial, of western seepage areas and bogs . . . . . . . . . . . . . . . . . . . . . . . . . . . 13. *A. idahoensis* (in part)

37. Anthers 0.5–0.9 mm long; upper culm sheaths inflated; plants annual, near hot springs . . . . . . . . . . . . . . . . . . . . . . . . . . . . . . . . . . . . . . . . . . . . . . . . 24. *A. rossiae*

33. Blades 1–30 cm long, 1–7.5 mm wide, usually flat; anthers 0.3–2.3 mm long.

38. Rhizomes present, to 50 cm long, stolons absent; panicle branches branching from midlength or to near the base; lower panicle branches 1–5 cm long; anthers 0.7–2.3 mm long.
    39. Lemma apices blunt, entire; callus hairs sparse, to 0.1 mm long; panicles 2–15 cm long; pedicels 0.5–2 mm long; blades 1–3 mm wide . . . . . . . . . . . . . . . . . . . . . . 6. *A. vinealis* (in part)
    39. Lemma apices usually acute, entire or toothed, teeth to about 0.2 mm long; callus hairs sparse or abundant, to 2 mm long; panicles 5–22 cm long; pedicels 0.5–7 mm long; blades 1–6 mm wide.
        40. Anthers 0.7–1.8 mm long; callus hairs to 0.3(1) mm long, sparse; leaf blades 1.5–11.5 cm long; caryopses 1–1.5 mm long . . . . . . . . . . . . . . . . . . . . . 17. *A. pallens* (in part)
        40. Anthers 1.5–2.3 mm long; callus hairs 0.8–2 mm long, abundant; leaf blades 6–20 cm long; caryopses 1.5–2 mm long . . . . . . . . . . . . . . . . . . . . . . . . . . . . . . 18. *A. hallii*
38. Rhizomes absent, stolons sometimes present, to 25 cm long; panicle branches mostly branching at or beyond midlength; lower panicle branches 1–12 cm long; anthers 0.3–1.5 mm long.
    41. Stolons present, to about 25 cm long, producing tufts of shoots at the nodes; anthers 1–1.5 mm long; blades 1–3 mm wide; glume apices acute . . . . . . . . . . . . 5. *A. canina* (in part)
    41. Stolons absent; anthers 0.3–1.2 mm long; blades 0.5–7 mm wide; glume apices acute to acuminate.
        42. Blades to 2 mm wide, 1–14 cm long, flat or involute; leaves mostly basal.
            43. Lower panicle branches 4–12 cm long; whole panicle often detaching at the base at maturity, forming a tumbleweed; blades 4–14 cm long . . . . 10. *A. scabra* (in part)
            43. Lower panicle branches 1–4 cm long; panicle not detaching at maturity and forming a tumbleweed; blades 1–7 cm long . . . . . . . . . . . . . . 13. *A. idahoensis* (in part)
        42. Blades to 7 mm wide, usually at least some wider than 2 mm, 5–30 cm long, flat; leaves mostly cauline to mostly basal.
            44. Plants annual or short-lived perennials; glumes 1.5–2.8 mm long, subequal; lemmas smooth and glabrous; anthers 0.3–0.6 mm long; caryopses 0.9–1.3 long . . . . . . . . . . . . . . . . . . . . . . . . . . . . . . . . . . . . . . . . . . . . . 9. *A. clavata*
            44. Plants perennial; glumes 1.8–3.6 mm long, unequal; lemmas smooth or scabridulous, sometimes pubescent; anthers 0.4–1.2 mm long; caryopses 1–1.9 mm long.
                45. Basal leaves usually withered by anthesis; cauline nodes 3–10; blades of the upper leaves as broad and substantial as those of the lower leaves; callus hairs abundant; plants primarily from cast of the 100th Meridian . . . . . . . . . . . . . . . . . . . . . . . . . . . . . . 12. *A. perennans* (in part)
                45. Basal leaves persisting to anthesis; cauline nodes 5 or fewer; blades of the upper leaves usually less substantial than those of the lower leaves; callus hairs sparse; plants primarily western . . . . . . . . 14. *A. oregonensis* (in part)

**1. Agrostis capillaris** L. BROWNTOP, RHODE ISLAND BENT, COLONIAL BENT, AGROSTIDE FINE [p. *410*, <u>525</u>]

*Agrostis capillaris* grows along roadsides and in disturbed areas. It was introduced from Europe, and is now well established in western and eastern North America. It is often used for fine-leaved lawns; commercial seed sold as *Agrostis tenuis* 'Highland' usually contains *A. capillaris*.

*Agrostis capillaris* differs from *A. gigantea* in its short ligules, especially on the vegetative shoots, and the open panicles that lack spikelets near the base of the branches. It differs from *A. castellana* in having diffuse rather than clustered spikelets, fewer rhizomes, divaricate panicle branches after anthesis, calluses that are glabrous or with hairs up to 0.1 mm long, and glabrous lemmas. It also tends to flower somewhat earlier than *A. castellana*. *Agrostis capillaris* readily hybridizes with *A. vinealis*, the hybrids being somewhat intermediate between the two parents.

**2. Agrostis castellana** Boiss. & Reut. HIGHLAND BENT, DRYLAND BROWNTOP [p. *410*, <u>525</u>]

*Agrostis castellana* is native to southern Europe. It was introduced to North America in the 1930s for use in lawns and golf greens, under the name *Agrostis tenuis* 'Highland'; commercial samples of 'Highland' often contain *A. capillaris*. Recorded habitats have ranged from sunny gravel roadsides to moist ground alongside cranberry bogs, at elevations from near sea level to over 600 m. In view of its extensive commercial use for over 70 years and its drought tolerance, it is likely that it is more widespread than shown.

*Agrostis castellana* belongs to a Eurasian group that includes *A. gigantea*, *A. stolonifera*, and *A. capillaris*. It differs from *A. gigantea* and *A. stolonifera* in having shorter, truncate ligules about as short as wide, and in not possessing extensive rhizomes and stolons. It differs from *A. capillaris* in having clustered rather than diffuse spikelets, more abundant rhizomes, somewhat constricted panicle branches after anthesis, abundantly hairy calluses with hairs up to 0.3(0.6) mm long, and lemmas that are sometimes dorsally pubescent. It also tends to flower somewhat later than *A. capillaris*.

**3. Agrostis gigantea** Roth REDTOP, AGROSTIDE BLANCHE [p. *410*, <u>525</u>]

*Agrostis gigantea* grows in fields, roadsides, ditches, and other disturbed habitats, mostly at lower elevations. It is a serious agricultural weed, as well as a valuable soil stabilizer. In the *Manual* region, its range extends from the subarctic to Mexico; it is considered to be native to Eurasia. It is more heat tolerant than most species of *Agrostis*.

*Agrostis gigantea* differs from *A. stolonifera* in having rhizomes and a more open panicle. *Agrostis stolonifera* has elongated leafy stolons, mainly all above the surface, that root at the nodes, and the panicles are condensed and often less strongly pigmented than in *A. gigantea*. Its distribution tends to be more northern and coastal where ditches and pond margins are common habitats, and its stolons enable it to form loose mats. *Agrostis gigantea* is ecologically adapted to a more extreme climate—hot summers/cold winters and drought—than *A. stolonifera*. It is also similar to *A. capillaris* and *A. castellana*; it differs from both in its longer ligules, from *A. capillaris* in its less open panicles with spikelets near the base of the branches, and from *A. castellana* in being more extensively rhizomatous.

When *Agrostis gigantea* grows in damp hollows under trees it becomes more like *A. stolonifera*, particularly when the inflorescence is young, not expanded, and pale. If the rootstock is not collected, identification is a major problem.

### 4. Agrostis stolonifera L. CREEPING BENT, AGROSTIDE STOLONIFÈRE [p. *410*, <u>525</u>]

*Agrostis stolonifera* grows in areas that are often temporarily flooded, such as lakesides, marshes, salt marshes, lawns, and damp fields, as well as moist meadows, forest openings, and along streams. It will also colonize disturbed sites such as ditches, clearcuts, and overgrazed pastures. Its North American range extends from the subarctic into Mexico, mostly at low to middle elevations.

*Agrostis stolonifera* has been confused with *A. gigantea*. It is considered to be Eurasian, but some northern salt marsh and lakeside populations may be native. *Agrostis stolonifera* differs from *A. castellana* in having longer, acute to truncate ligules that are longer than wide, and in possessing extensive stolons. A hybrid between *A. stolonifera* and *Polypogon monspeliensis*, ×*Agropogon lutosus*, has been found in the *Manual* region. It differs from *A. stolonifera* in having awned glumes and lemmas. *Agrostis stolonifera* readily hybridizes with *A. vinealis*, the hybrids being somewhat intermediate between the two parents.

### 5. Agrostis canina L. VELVET BENT, AGROSTIDE DES CHIENS [p. *410*, <u>525</u>]

*Agrostis canina* is a Eurasian species that is now established in both western and eastern North America, where it grows on roadsides and open ground in summer-cool climates. It is used for fine-textured lawns and golf greens. Although similar to *A. vinealis*, it differs from that species in having creeping, leafy stolons that form a dense carpet, and leaves with a finer, softer texture.

### 6. Agrostis vinealis Schreb. BROWN BENT [p. *410*, <u>525</u>]

*Agrostis vinealis* is native to Eurasia; it is not clear if populations in Greenland and Alaska represent a circumboreal distribution, or are introductions. It forms a fine, compact turf. It is similar to *A. canina* in its habitat, except that it appears to be more heat tolerant and drought resistant. It used to be included in *A. canina*, but differs from that species in its subterranean rhizomes and lack of leafy stolons. *Agrostis vinealis* readily hybridizes with *A. capillaris* and *A. stolonifera*, the hybrids being somewhat intermediate between the two parents.

### 7. Agrostis mertensii Trin. NORTHERN BENT, AGROSTIDE DE MERTENS [p. *410*, <u>525</u>]

*Agrostis mertensii* grows on banks and gravel bars in river and lake valleys, and on open grasslands and rocky slopes of mountains and cliffs. It has a circumboreal distribution. In the *Manual* region, it extends from Alaska across Canada to Newfoundland and Greenland, south in the mountains to Wyoming and Colorado in the west, and West Virginia, Tennessee, and North Carolina in the east. It also grows in arctic Europe, Scandinavia, the mountainous regions of Mexico, and northwestern South America.

*Agrostis mertensii* differs from dwarf, awned forms of *A. scabra* in its larger spikelets, more culm nodes, larger anthers, slightly wider, flatter leaves, and panicles that are less expanded and less than $^1/_3$ the culm length. It tends to grow in better-drained habitats than *A. idahoensis*.

It differs from *A. anadyrensis* in being less robust, having narrower, less abundant basal leaves, smaller panicles, and minor differences in the insertion of the awns on the lemmas. In addition, the panicle branches are smooth to weakly scabrous, contrasting with the branches of *A. anadyrensis*, which are strongly scabrous, with long acicules throughout their length.

### 8. Agrostis anadyrensis Soczava ANADYR BENT [p. *410*]

*Agrostis anadyrensis* grows in sand and gravel shores of rivers and lakes, and in meadows and shrubby valleys in eastern Siberia. It has been reported from arctic Russia and southern Alaska. Specimens from Alaska purporting to be *A. anadyrensis* so far have proven to be *A. mertensii*. *Agrostis anadyrensis* differs from *A. mertensii* in being more robust, with wider, more abundant basal leaves, larger panicles, and minor differences in the insertion of the awns on the lemmas. In addition, the panicle branches are strongly scabrous, with long acicules throughout their length. *A. mertensii* has smooth to weakly scabrous branches.

### 9. Agrostis clavata Trin. CLAVATE BENT [p. *410*, <u>525</u>]

*Agrostis clavata* grows in disturbed ground on sandbars and gravelbars, and in wet meadows and coniferous forests, from Sweden across northern Asia to Kamchatka. It was recently found in Alaska, the Yukon, and the Northwest Territories, and appears to be native there. It differs from the similarly large-panicled *A. scabra* in its much broader, flat leaves.

### 10. Agrostis scabra Willd. TICKLEGRASS, ROUGH BENT, FOIN FOU, AGROSTIDE SCABRE [p. *411*, <u>525</u>]

*Agrostis scabra* grows in a wide variety of habitats, including grasslands, meadows, shrublands, woodlands, marshes, and stream and lake margins, as well as disturbed sites such as roadsides, ditches, and abandoned pastures. It occurs throughout much of the *Manual* region, but is not common in the Canadian high arctic or the southeastern United States. It extends south into Mexico; it is also native to the Pacific coast from Kamchatka to Japan and Korea, and has been introduced elsewhere.

Plants in the *Agrostis scabra* aggregate are variable. Awned and unawned plants often occur together, the difference presumably being caused by a single gene. At least three groups may be distinguished within the species as treated here: widespread, lowland, rather weedy plants capable of producing very large panicles that have been introduced into the southern United States; smaller, short-leaved, slow-growing plants of rocks and screes, which are widespread in the Rockies, the Appalachians, and much of Alaska, Canada, and Greenland; and luxuriant, broad-leaved plants that are characteristically found in sheltered, frost-free canyons of the southwestern United States. The second group has sometimes been called *A. scabra* var. *geminata* (Trin.) Swallen or *A. geminata* Trin.

*Agrostis scabra* is often confused with a number of other species: *A. mertensii*, *A. clavata*, *A. hyemalis*, *A. perennans*, and *A. idahoensis*.

### 11. Agrostis hyemalis (Walter) Britton, Sterns & Poggenb. WINTER BENT, AGROSTIDE D'HIVER [p. *411*, <u>525</u>]

*Agrostis hyemalis* is most abundant along roadsides and in open pastures, scrub, and rocky areas. It is centered in the southeastern United States; historically it extended north to coastal Maine, where it may be extinct, west to Wisconsin and Texas, and south into the Caribbean, Mexico, and Ecuador. Records from further north and west in North America are confused; many reflect the former inclusion of the generally more northern *A. scabra* in *A. hyemalis*. *Agrostis hyemalis* differs from *A. scabra* in its smaller spikelets and anthers, more conspicuous culm leaves, and more clustered spikelets.

### 12. Agrostis perennans (Walter) Tuck. AUTUMN BENT, UPLAND BENT, AGROSTIDE PERENNANT [p. *411*, <u>525</u>]

*Agrostis perennans* grows along roadsides and in fields, fens, woodlands, and periodically inundated stream banks. It is widespread and common in eastern North America; it also grows from central Mexico to central South America. There are old records from Oregon and Washington, but *A. perennans* does not appear to be established in western North America. It is more tolerant of shade and moisture than *Agrostis scabra*, from which it differs in its later flowering, leafier culms, and its basal leaves that usually wither by anthesis.

### 13. Agrostis idahoensis Nash IDAHO REDTOP [p. *411*, 525]

*Agrostis idahoensis* grows in western North America, from British Columbia to California and New Mexico, in alpine and subalpine meadows along wet seepage areas and bogs, and in wet openings with *Sphagnum* in coniferous forests. It was recently discovered in Chile and Argentina; it is not known whether it is native or introduced there. *Agrostis idahoensis* is often confused with *A. mertensii* and dwarf forms of *A. scabra*, both of which tend to grow in better-drained habitats.

### 14. Agrostis oregonensis Vasey OREGON REDTOP [p. *411*, 525]

*Agrostis oregonensis* grows in wet habitats, such as stream and lake margins, damp woods, and meadows, in western North America, primarily in the Pacific Northwest from British Columbia to California and Wyoming. It has not been found in Mexico.

### 15. Agrostis exarata Trin. SPIKE BENT [p. *411*, 525]

*Agrostis exarata* is common and widely distributed in western North America, usually growing in moist ground in open woodlands, river valleys, tidal marshes, and swamp and lake margins; it also grows in dry habitats such as grasslands and shrublands. It extends from Alaska into Mexico, and is also found in Kamchatka and the Kuril Islands. Eastern North American records probably reflect introductions. It readily colonizes roadsides and bare soil, and exhibits ecological and developmental flexibility. *Agrostis exarata* is recognized here as a single, variable species that includes what others have treated as distinct species or varieties.

### 16. Agrostis densiflora Vasey CALIFORNIA BENT [p. *411*, 525]

*Agrostis densiflora* is endemic to coastal Oregon and California. It grows in sandy soils, on cliffs, and in scrublands. It appears to be related to *A. exarata*, and hybridizes with *A. blasdalei*.

### 17. Agrostis pallens Trin. DUNE BENT [p. *411*, 525]

*Agrostis pallens* grows on coastal sands and cliffs, in meadows, and in open, xeric woodlands to subalpine woodlands at 3500 m. It extends from British Columbia south into Baja California, Mexico, and east to western Montana and Utah.

### 18. Agrostis hallii Vasey HALL'S BENT [p. *411*, 525]

*Agrostis hallii* is primarily coastal, growing in open areas of oak and coniferous forests in Oregon and California.

### 19. Agrostis hooveri Swallen HOOVER'S BENT [p. *412*, 525]

*Agrostis hooveri* is an uncommon species, endemic to dry, sandy soils, open chaparral, and oak woodlands of San Luis Obispo and Santa Barbara counties, California.

### 20. Agrostis howellii Scribn. HOWELL'S BENT [p. *412*, 525]

*Agrostis howellii* is a rare Washington and Oregon endemic, growing in shady woodlands and at the base of cliffs.

### 21. Agrostis variabilis Rydb. MOUNTAIN BENT [p. *412*, 525]

*Agrostis variabilis* grows in alpine and subalpine meadows and forests and on talus slopes, at elevations up to 4000 m, from British Columbia and Alberta south to California and New Mexico. It differs from dwarf forms of *Podagrostis humilis* in not having paleas.

### 22. Agrostis blasdalei Hitchc. BLASDALE'S BENT [p. *412*, 525]

*Agrostis blasdalei* is a xerophytic species that is known only from Mendocino to Santa Cruz counties, California, where it grows on coastal cliffs and dunes and in shrublands. It hybridizes with *A. densiflora*.

### 23. Agrostis tolucensis Kunth [p. *412*]

*Agrostis tolucensis* grows in alpine meadows, usually in damp areas by lakes or streams. It is native from Mexico to Chile, Bolivia, and Argentina, growing in the Andes at 1800–4900 m. Its presence in the *Manual* region is dubious; two specimens in the S.M. Tracy herbarium, from Brewster and Brown counties in Texas, are listed in the Flora of Texas online database. Attempts to locate the specimens in 2005 were unsuccessful, suggesting the records may have been based upon misidentifications which have since been rectified.

### 24. Agrostis rossiae Vasey ROSS' BENT [p. *412*, 525]

*Agrostis rossiae* is a rare species, known only from alkaline soils near hot springs in Yellowstone National Park, Wyoming.

### 25. Agrostis hendersonii Hitchc. HENDERSON'S BENT [p. *412*, 525]

*Agrostis hendersonii* is a rare species that grows below 600 m in clay or adobe, sometimes rocky, soils around the edges of vernal pools in Oregon and California.

### 26. Agrostis microphylla Steud. SMALL-LEAF BENT [p. *412*, 526]

*Agrostis microphylla* grows in thin, rocky soils, sandy areas, cliffs, vernal pools, and serpentine areas, mostly along the Pacific coast from British Columbia to northern Baja California, Mexico. It is a winter annual, flowering in late winter to spring, adapted to low-competition habitats with summer drought. It may be related to, or conspecific with, *Agrostis hendersonii*.

### 27. Agrostis elliottiana Schult. ELLIOTT'S BENT [p. *413*, 526]

*Agrostis elliottiana* grows in fields and scrublands and along roadsides. It has a disjunct distribution, occurring in western North America in northern California and southern Arizona and New Mexico; in eastern North America from Kansas and Texas east to Pennsylvania and northern Florida; and in Yucatan, Mexico. It resembles *A. scabra* and *A. hyemalis* in its diffuse panicle, but differs in its flexible awn and single anther.

### 28. Agrostis nebulosa Boiss. & Reut. CLOUD GRASS [p. *413*, 526]

*Agrostis nebulosa* is native to Spain and Portugal. It is cultivated as an ornamental and for dried flower arrangements, but occasional escapes have been found on roadsides, ditches, and in fields in widely scattered locations in the *Manual* region.

---

## 14.28 POLYPOGON Desf.[1]

**Pl** ann or per; not rhz. **Clm** 4–120 cm, erect to decumbent, rooting at the lo nd, sparingly brchd near the base. **Lvs** usu no more than 5 per clm, bas and cauline; **shth** open, smooth or scabridulous; **aur** absent; **lig** memb or hyaline, acute to broadly rounded, erose, ciliate; **bld** flat to convolute. **Infl** tml pan, dense, continuous or interrupted below; **br** flexible, usu some longer than 1 cm; **ped** absent and the spklt borne on a

---

[1]Mary E. Barkworth

stipe, or present and terminating in a stipe; **stipes** scabrous, flaring distally; **dis** at the base of the stipes. **Spklt** 1–5 mm, weakly lat compressed, with 1 bisx flt; **rchl** not prolonged beyond the base of the flt. **Glm** exceeding the flt, lanceolate, bases not fused, apc entire to emgt or bilobed, usu awned from the sinuses or apc, awns flexuous, glab, smt unawned; **lm** 1–3(5)-veined, often awned, awns usu tml or subtml, smt arising from just above midlength; **pal** from $^1/_3$ as long as to equaling the lm; **lod** 2, oblong-lanceolate to lanceolate; **anth** 3;

ov glab; **sty** separate. **Car** slightly flattened, broadly ellipsoid to oblong-ellipsoid; **hila** $^1/_6$–$^1/_4$ as long as the car, ovate.

*Polypogon* is a pantropical and warm-temperate genus of about 18 species. There are eight species in the *Manual* region; one species, *P. interruptus*, is native.

*Polypogon* is similar to *Agrostis*, and occasionally hybridizes with it. It differs from *Agrostis* in having spikelets that disarticulate below the glumes, often at the base of a stipe.

1. Glumes with awns 3–12 mm long.
    2. Glumes deeply lobed, the lobes more than $^1/_6$ the length of the glume body . . . . . . . . . . . . . . . . . . . . . . 7. *P. maritimus*
    2. Glumes not lobed or the lobes $^1/_{10}$ or less the length of the glume body.
        3. Plants annual; glume apices rounded, lobed, the lobes 0.1–0.2 mm long; ligules 2.5–16 mm long . . 6. *P. monspeliensis*
        3. Plants perennial; glume apices acute to truncate, unlobed or the lobes shorter than 0.1 mm; ligules 1–6 mm long.
            4. Glume awns (3)4–6 mm long; longest blades 13–17 cm long . . . . . . . . . . . . . . . . . . . . . . . . . . . . . 5. *P. australis*
            4. Glume awns 1.5–3.2 mm long; longest blades 5–9 cm long . . . . . . . . . . . . . . . . . . . 2. *P. interruptus* (in part)
1. Glumes unawned or with awns to 3.2 mm long.
    5. Glumes unawned . . . . . . . . . . . . . . . . . . . . . . . . . . . . . . . . . . . . . . . . . . . . . . . . . . . . . . . . . . . . . . . . . . . 1. *P. viridis*
    5. Glumes awned, the awns 0.2–3.2 mm long.
        6. Stipes 1.5–2.5 mm long; glumes tapering from about midlength to the acute, unlobed apices . . . . . . . . 3. *P. elongatus*
        6. Stipes 0.2–1.5 mm long; glumes not tapering to the apices, the apices usually rounded to truncate, sometimes acute, often lobed.
            7. Lemmas 1–2 mm long; paleas about $^1/_2$ as long as the lemmas; the lower glumes longer than the upper glumes . . . . . . . . . . . . . . . . . . . . . . . . . . . . . . . . . . . . . . . . . . . . . . . . . . . . . . . . . 8. *P. imberbis*
            7. Lemmas 0.7–1.5 mm long; paleas from $^3/_4$ as long as to equaling the lemmas; glumes of each spikelet subequal to equal.
                8. Plants annual; glumes acute to rounded, lobed, the lobes 0.1–0.2 mm long . . . . . . . . . . . . . . . . . . . 4. *P. fugax*
                8. Plants perennial, often flowering the first year; glumes acute to truncate, if lobed, the lobes to 0.1 mm long . . . . . . . . . . . . . . . . . . . . . . . . . . . . . . . . . . . . . . . . . . . . . . . . . 2. *P. interruptus* (in part)

### 1. Polypogon viridis (Gouan) Breistr. WATER BEARDGRASS [p. *413*, *526*]

*Polypogon viridis* grows in mesic habitats associated with rivers, streams, and irrigation ditches. It is native from southern Europe to Pakistan, but is now established in the *Manual* region, particularly the southwestern United States. Records from the Atlantic coast are based on plants found on ballast dumps; there have been no recent collections from these locations.

### 2. Polypogon interruptus Kunth DITCH BEARDGRASS [p. *413*, *526*]

*Polypogon interruptus* grows in moist soil at lower elevations. It is native to the Western Hemisphere, extending south from the western United States into northern Mexico, and through the American tropics to Argentina and Bolivia. The more eastern records may indicate introductions; it is not known whether or not the species persists at these locations.

### 3. Polypogon elongatus Kunth SOUTHERN BEARDGRASS [p. *413*, *526*]

*Polypogon elongatus* is native from Mexico to Argentina. It now grows at scattered locations in the *Manual* region, primarily in California.

### 4. Polypogon fugax Nees *ex* Steud. ASIAN BEARDGRASS [p. *413*, *526*]

*Polypogon fugax* is native from Iraq to Myanmar [Burma]. It was collected in Santa Barbara, California, and from salt marshes around Oakland, California, in the nineteenth century, and from Portland,

Oregon, in the early twentieth century. There are no recent collections from the *Manual* region.

### 5. Polypogon australis Brongn. CHILEAN BEARDGRASS [p. *413*, *526*]

*Polypogon australis* is native to South America. It has become established in western North America, where it grows along ditches and streams. The records from Washington and Oregon are from ballast dumps; it is not known from recent collections in those states.

### 6. Polypogon monspeliensis (L.) Desf. RABBITSFOOT GRASS [p. *413*, *526*]

*Polypogon monspeliensis* is native to southern Europe and Turkey. It is now a common weed throughout the world, including much of the *Manual* region. It grows in damp to wet, often alkaline soils, particularly in disturbed areas.

In Europe, *Polypogon monspeliensis* hybridizes with *Agrostis stolonifera*, producing the sterile ×*Agropogon lutosus*. The hybrid differs from *P. monspeliensis* in having more persistent spikelets, less blunt short-awned glumes, and lemmas with subterminal rather than terminal awns.

### 7. Polypogon maritimus Willd. MEDITERRANEAN BEARDGRASS [p. *413*, *526*]

*Polypogon maritimus* grows in disturbed, moist places, from sea level to 700 m. It is a Mediterranean species that now occurs at scattered locations in North America, being particularly common in, or possibly just well-reported from, California. There are two varieties.

Plants from the *Manual* region belong to **P. maritimus** Willd. var. **maritimus**, having stipes about as long as they are wide, glumes that never become strongly indurate at the base, and uninflated, acute hairs on the glume bases. Plants of **P. maritimus** var. **subspathaceus** (Req.) Bonnier & Layens have stipes that are 3–4 times as long as wide, glumes that become strongly indurate at maturity, and hairs on the glume bases that are strongly inflated and subobtuse.

## 14.29 ×AGROPOGON P. Fourn.[1]

**Pl** per; loosely csp to spreading, rhz. **Clm** 8–60 cm, usu brchd below, ascending from a decumbent base or the lo nd geniculate. **Shth** smooth, open; **aur** absent; **lig** memb, puberulent, acute to obtuse, often bifid or denticulate with age; **bld** usu glab, smt scabrous. **Infl** pan, moderately dense to very dense, often interrupted; **dis** tardy, below the glm. **Spklt** lat compressed, with 1 flt.

### 1. ×Agropogon lutosus (Poir.) P. Fourn. PERENNIAL BEARDGRASS [p. *414*]

×*Agropogon lutosus* is a sterile hybrid between *Agrostis stolonifera* and *Polypogon monspeliensis* that sometimes grows in locations where both parents occur, such as damp to wet, often alkaline soils on

## 14.30 LAGURUS L.[2]

**Pl** ann; tufted. **Clm** 6–80 cm, erect or ascending, pilose or villous. **Lvs** pilose or villous; **shth** open nearly to the base; **aur** absent; **lig** truncate, erose, ciliate, abx surfaces densely pubescent; **bld** flat at maturity, convolute in the bud. **Infl** tml pan, dense, ovoid. **Spklt** lat compressed, with 1 flt; **rchl** prolonged beyond the base of the flt, pubescent; **dis** above the glm, beneath the flt. **Glm** 2, equal, exceeding the flt, lanceolate, memb, pilose, hairs 2–3 mm, 1-veined, veins acuminate and awned, awns pilose; **cal** blunt, pubescent; **lm** lanceolate, pilose or scabridulous, 3-veined, 3-awned, apc bifid, lat veins extending as slender awns, not twisted below, cent awn

### 1. Lagurus ovatus L. HARETAIL GRASS [p. *414*, 526]

*Lagurus ovatus* grows in disturbed sites in full sun. It is a waif, or sparingly naturalized, in North Carolina and western Florida; it is

## 14.31 PHLEUM L.[3]

**Pl** ann or per; csp, smt rhz, occ stln. **Clm** 2–150 cm, erect or decumbent; **nd** glab. **Shth** open; **aur** absent or inconspicuous; **lig** memb, not ciliate; **bld** usu flat. **Infl** dense, spikelike pan, more than 1 spklt associated with each nd; **br** often shorter than 2 mm, always shorter than 7 mm, stiff; **ped** shorter than 1 mm, smt fused; **dis** above the glm or, late in the season, beneath the glm. **Spklt** strongly lat compressed, bases usu U-shaped, smt cuneate, with 1 flt; **rchl** glab, smt prolonged beyond the base of the flt. **Glm** equal, longer and firmer than the flt, stiff, bases not connate, strongly keeled, keels usu strongly ciliate, smt glab, smt scabrous, 3-veined, apc truncate to tapered, midveins often extending into

### 8. Polypogon imberbis (Phil.) Johow SHORTHAIRED BEARDGRASS [p. *413*, 526]

*Polypogon imberbis* is a South American species that has been collected at two locations in California, one from Oceano Beach, San Luis Obispo County, and the other near Martines, Contra Costa County. It does not appear to be established there, the last collections having been made before 1950. In South America, it grows in moist, sandy soils near streams, lagoons, and the coast.

**Glm** similar, narrowly oblong to elliptic, 1-veined, apc notched, midveins extending into a short awn; **cal** glab; **lm** shorter than the glm, inconspicuously 5-veined, awned from just below the minutely toothed apc; **pal** about ³⁄₄ as long as the lm; **anth** usu indehiscent.

×*Agropogon* comprises hybrids between *Agrostis* and *Polypogon*; one hybrid grows in the *Manual* region.

lakesides, marshes, ditches, and intermittently flooded fields. It differs from *Polypogon* in having more persistent spikelets, less blunt short-awned glumes, and lemmas with subterminal rather than terminal awns; and from *Agrostis* in having awned glumes and awned lemmas.

arising from the distal ¹⁄₃ of the lm, twisted below; **pal** nearly as long as the lm, veins scabridulous, extending as awnlike points; **lod** 2, narrowly elliptic, glab, apc minutely bilobed; **anth** 3; **ov** glab; **sty** fused or separate. **Car** stipitate, subterete, ellipsoid; **hila** ¹⁄₅–¹⁄₄ as long as the car, ovate.

*Lagurus* is a unispecific genus endemic to the Mediterranean region. *Lagurus ovatus* has been introduced in North America, as well as South America, southern Africa, and Australia. It is sometimes cultivated as an ornamental, and the dried flowering culms are used in floral arrangements.

established and plentiful in central California. It may not have persisted at the other sites shown.

short, stiff, awnlike apc; **cal** blunt, glab; **lm** white, often translucent, not keeled, 5–7-veined, unawned, bases not connate, apc acute, entire, smt with a weak, subapical awn; **pal** subequal to the lm, 2-veined; **lod** 2, free, glab, toothed; **anth** 3; **ov** glab. **Car** elongate-ovoid; **emb** ¹⁄₆–¹⁄₄ the length of the car.

*Phleum* is a genus of approximately 15 species, most of which are native to Eurasia. One species, *P. alpinum*, is native to the *Manual* region and six are introduced. One of the introduced species, *P. pratense*, is widely cultivated as a fodder grass, both in the *Manual* region and in other parts of the world. *Phleum exaratum* has been reported from the United States. No specimens supporting the report have been seen. It

[1]Mary E. Barkworth  [2]Gordon C. Tucker  [3]Mary E. Barkworth

resembles *P. arenarium*, but has anthers 1.5–2 mm long rather than 0.3–1.2 mm, and an inflorescence that is rounded at the base.

Species of *Phleum* are sometimes mistaken for *Alopecurus*, but *Alopecurus* has obtuse to acute glumes that are unawned or taper into an awn, lemmas that are both awned and keeled, and paleas that are absent or greatly reduced. The species of *Phleum* that are most abundant in the *Manual* region are easily recognized by their strongly ciliate, abruptly truncate, awned glumes and adnate panicle branches. In addition, in *Phleum* the lemmas are not keeled, and the paleas are always subequal to the lemmas.

1. Plants perennial.
    2. Panicles tapering distally; panicle branches free from the rachis; glumes scabrous to shortly ciliate
       on the keels; plants known, in the *Manual* region, only from British Columbia . . . . . . . . . . . . . . . . . . . . . 3. *P. phleoides*
    2. Panicles not tapering distally; panicle branches adnate to the rachises; glumes conspicuously ciliate
       on the keels; plants widespread in the *Manual* region.
        3. Sheaths of the flag leaves not inflated; panicles 2–14(17) cm long, 5–20 times as long as wide;
           lower internodes of the culms frequently enlarged or bulbous; widespread in the *Manual* region . . . . . . . 1. *P. pratense*
        3. Sheaths of the flag leaves inflated; panicles 1–6 cm long, usually 1.5–3 times as long as wide;
           lower internodes of the culms not enlarged or bulbous . . . . . . . . . . . . . . . . . . . . . . . . . . . . . . . . . 2. *P. alpinum*
1. Plants annual.
    4. Glume apices abruptly truncate . . . . . . . . . . . . . . . . . . . . . . . . . . . . . . . . . . . . . . . . . . . . . . 4. *P. paniculatum*
    4. Glume apices gradually narrowed to tapered.
        5. Glumes semi-elliptical in outline, apices pointing towards each other; keels usually glabrous,
           sometimes scabrous . . . . . . . . . . . . . . . . . . . . . . . . . . . . . . . . . . . . . . . . . . . . . . . . . . . . 5. *P. subulatum*
        5. Glumes oblong-lanceolate, apices parallel or divergent; keels ciliate . . . . . . . . . . . . . . . . . . . . 6. *P. arenarium*

### 1. Phleum pratense L. TIMOTHY, FLÉOLE DES PRÉS, PHLÉOLE DES PRÉS, MIL [p. *414*, <u>526</u>]

*Phleum pratense* grows in pastures, rangelands, and disturbed sites throughout most of the mesic, cooler regions of North America. Originally introduced from Eurasia as a pasture grass, it is now well established in many parts of the world, including the *Manual* region. North American plants belong to the polyploid **Phleum pratense** L. subsp. **pratense**, which differs from the diploid **P. pratense** subsp. **bertolonii** (DC.) Bornm. in having obtuse ligules. Depauperate specimens of *P. pratense* are hard to distinguish from *P. alpinum*.

### 2. Phleum alpinum L. ALPINE TIMOTHY, FLÉOLE ALPINE, PHLÉOLE ALPINE [p. *414*, <u>526</u>]

*Phleum alpinum* grows along stream banks, on moist prairie hillsides, and in wet mountain meadows. It is a circumboreal species extending, in the *Manual* region, from northern North America southward through the mountains to Mexico and South America. It is also widespread in northern Eurasia. Isolated, depauperate plants of *P. pratense* may be difficult to distinguish from *P. alpinum*; there is never any difficulty in the field. North American plants belong to **P. alpinum** L. subsp. **alpinum** and are tetraploid.

### 3. Phleum phleoides (L.) H. Karsten PURPLE-STEM CAT'S TAIL [p. *414*, <u>526</u>]

*Phleum phleoides* is native to dry grasslands from Europe through central Asia. It was collected, in 1990, beside railroad tracks in Coquitlam, British Columbia.

### 4. Phleum paniculatum Huds. BRITISH TIMOTHY [p. *414*, <u>526</u>]

*Phleum paniculatum* is native to dry habitats in southern and south-central Europe. In the *Manual* region, it is known only from old ballast dump records.

### 5. Phleum subulatum (Savi) Asch. & Graebn. ITALIAN TIMOTHY [p. *414*, <u>526</u>]

*Phleum subulatum* is native to the grasslands of southern Europe. In the *Manual* region, it is known only from old ballast dump records.

### 6. Phleum arenarium L. SAND TIMOTHY [p. *414*, <u>526</u>]

*Phleum arenarium* is native to maritime sands and shingles of southern and western Europe. It is known only from old ballast dump records in the *Manual* region.

## 14.32 GASTRIDIUM P. Beauv.[1]

**Pl** ann; tufted or the clm solitary. **Clm** 7–70 cm, erect to decumbent. **Lvs** cauline; **shth** open; **aur** vestigial or absent; **lig** memb; **bld** flat. **Infl** pan, contracted or spikelike, more than 1 spklt associated with each nd; **br** appressed to ascending. **Spklt** pedlt, lat compressed, not U-shaped at the base, solitary, with 1 bisx flt; **rchl** prolonged as bristles beyond the base of the flt or absent; **dis** above the glm, beneath the flt. **Glm** unequal, exceeding the flt, 1-veined, subcoriaceous and gibbous proximally, memb distally, gradually narrowing to the apc, apc acuminate, unawned; **cal** short, blunt, glab; **lm** memb, pubescent or glab, 5-veined, lat veins usu faint, not forming awns, apc more or less truncate and denticulate, unawned or awned, unawned and awned lm present in the same panicle, awns arising in the up $^1/_3$, smt subtml, shorter or longer than the lm, geniculate; **pal** subequal to the lm, hyaline, 2-veined, bifid; **lod** 2, memb, glab, entire, unlobed; **anth** 3; **ov** glab. **Car** shorter than the lm, concealed at maturity, slightly adhering to the lm and/or pal.

*Gastridium* is a genus of two species that are native from Europe and North Africa east to Iran. They grow in grassy or disturbed sites. One species is established in the *Manual* region.

[1]J.K. Wipff

## 1. Gastridium phleoides (Nees & Meyen) C.E. Hubb.
NIT GRASS [p. *415*, *526*]

Native to southwest Asia and northeast Africa, *Gastridium phleoides* now grows in Australia, South Africa, North America, and South America, in dry, often disturbed areas. In the *Manual* region, it is established in Oregon and California; it has also been collected, but may not be established, in southwestern British Columbia, Arizona,

western Texas, Massachusetts, and South Carolina. In North America, *Gastridium phleoides* has been mistakenly placed in *G. ventricosum* (Gouan) Schinz & Thell. It differs from that species in having densely pubescent lemmas and well-developed, densely pubescent rachilla prolongations.

## 14.33 ARCTAGROSTIS Griseb.[1]

Pl per; rhz. Clm 10–150 cm, erect, glab. Shth usu closed for up to ¹/₅ their length; aur absent; lig memb, obtuse, erose or lacerate; bld usu flat, glab, scabrous. Infl tml pan, loose and open to narrow and contracted; br 0.5–27 cm, scabridulous. Spklt 3–6.5 mm, pedlt, lat compressed, usu with 1 flt, rarely with 2; rchl hairy or scabrous, smt prolonged beyond the base of the single flt or terminating in a well-formed or vestigial flt; dis above the glm, beneath the flt(s). Glm usu shorter than the lm, up glm smt equaling the lm(s), ovate to lanceolate, glab, smt scabridulous, 1–3-veined, acute to

obtuse, unawned; cal weakly developed, glab; lm scabrous or puberulent, 3–5-veined, lat veins smt obscure, not reaching the apc, apc obtuse to acute, smt mucronate, unawned; pal about as long as the lm, folded along 1 prominent vein, the second vein faint; lod 2, glab; anth 3; ov glab. Car 1.7–3 mm, shorter than the lm, concealed at maturity; hila ¹/₄–¹/₃ the length of the car.

*Arctagrostis* is a circumpolar genus with one species and two subspecies.

### 1. Arctagrostis latifolia (R. Br.) Griseb. POLARGRASS [p. *415*, *526*]

*Arctagrostis latifolia* grows in a wide range of habitats, from sea level to 2000 m, and exhibits considerable phenotypic plasticity.

1. Panicles (4)10–35(44) cm long; longest branches (1.4)3–27 cm long, with (5)50–140 spikelets per branch; secondary branches usually spreading; uppermost sheaths shorter than the blades; spikelets 3–5.2 mm long; glumes unequal, the upper glumes varying from shorter than to longer than the lemmas . . . . . . . . . . . . . . . . . . . . . . . . . . . . subsp. *arundinacea*
1. Panicles 2.5–10(17) cm long; longest branches 0.5–4(6.5) cm long, with 3–15(40) spikelets per branch; secondary branches usually appressed, spreading in warmer habitats; uppermost sheaths

usually longer than the blades; spikelets (3)4–6.5 mm long; glumes subequal, about 1 mm shorter than the lemmas . . . . . . . . . . . . . . . . . . . . . . subsp. *latifolia*

**Arctagrostis latifolia subsp. arundinacea** (Trin.) Tzvelev REED POLARGRASS [p. *415*]

*Arctagrostis latifolia* subsp. *arundinacea* is the predominant subspecies from Alaska east to the western Northwest Territories, northern British Columbia, and Alberta. Cultivars for hay and revegetation mixes have been developed.

**Arctagrostis latifolia subsp. latifolia** POLARGRASS, ARCTAGROSTIS À LARGES FEUILLES [p. *415*]

*Arctagrostis latifolia* subsp. *latifolia* is the common subspecies from the arctic coast of Alaska to Nunavut, northern Quebec, and northern Greenland.

## 14.34 SESLERIA Scop.[2]

Pl per; more or less csp, smt rhz, rarely stln. Clm 3–80(100) cm. Lvs mostly bas; shth closed for most of their length, often shredding when mature; aur absent; lig 0.1–1 mm, memb, truncate to obtuse, usu ciliolate; bld flat, plicate, or involute. Infl single tml pan, cylindrical to globose, usu dense, smt subtended by ovate to round, scarious or hyaline, erose bracts, smt by 1–2 pubescent scales, smt without subtending scales or bracts, more than 1 spklt associated with each nd; br shorter than 10 mm. Spklt 3–9 mm, lat compressed, with 2–5 flt, upmost flt rdcd; rchl usu glab, rarely sparsely pilose, smt prolonged beyond the base of the distal flt, smt terminating in a rdcd flt; dis above the

glm, beneath the flt. Glm unequal, usu shorter than the lowest lm, scarious to memb, 1–3-veined, apc awned, awns glab; cal very short, broadly rounded, glab or with scattered hairs; lm memb, 5–7-veined, 3–5 veins usu extending into awnlike teeth, cent teeth longer than the lat teeth; pal equaling or exceeding the lm; lod 2, free, glab, toothed; anth 3; ov pubescent. Car 1.5–3 mm; emb ¹/₄–¹/₃ the length of the car.

*Sesleria* has approximately 30 species. Most abundant in the Balkans, it extends from Iceland, Great Britain, and southern Sweden through central and southern Europe into northwest Asia. Four species are cultivated in North America.

1. Panicles cylindrical, 4.5–10 cm long, 4–7 mm wide . . . . . . . . . . . . . . . . . . . . . . . . . . . . . . . . . . . 1. *S. autumnalis*
1. Panicles ovoid, spherical, or cylindrical, 0.9–4(8) cm long, 5–15 mm wide.
  2. Glumes 5–6.5 mm long, lanceolate; anthers about 2.2 mm long; lemmas glabrous . . . . . . . . . . . . . . . . . . . . . 4. *S. nitida*
  2. Glumes 3–5 mm long, ovate to ovate-lanceolate; anthers 2.3–4 mm long; lemmas pubescent.
    3. Glumes 3–4 mm long, ovate-lanceolate, coriaceous; anthers about 4 mm long . . . . . . . . . . . . . . . 2. *S. heufleriana*
    3. Glumes 4–5 mm long, ovate, hyaline; anthers 2.3–3.2 mm long . . . . . . . . . . . . . . . . . . . . . . . . . . . 3. *S. caerulea*

[1]Susan G. Aiken  [2]Mary E. Barkworth

**1. Sesleria autumnalis** (Scop.) F.W. Schultz AUTUMN MOORGRASS [p. *415*]

*Sesleria autumnalis* is native from northern and eastern Italy to central Albania, where it grows primarily on limestone-derived soils in deciduous, often littoral forests. It sometimes forms intermediates with *S. nitida*. It is now popular in the *Manual* region as an ornamental, being an attractive, low-maintenance plant that provides excellent ground cover.

**2. Sesleria heufleriana** Schur BLUE-GREEN MOORGRASS [p. *415*]

*Sesleria heufleriana* is native to east-central Europe, where it grows in woods and on rocks, usually over calcareous substrates. It is grown as an ornamental in the *Manual* region.

**3. Sesleria caerulea** (L.) Ard. BLUE MOORGRASS [p. *415*]

*Sesleria caerulea* is native to Europe, ranging from central Sweden to northwestern Russia and central Bulgaria. It usually grows in moist to wet, calcareous pastures and bogs. It is grown as an ornamental in the *Manual* region.

**4. Sesleria nitida** Ten. GRAY MOORGRASS [p. *415*]

*Sesleria nitida* is native to the mountains of central Italy and Sicily, where it grows in broken, rocky, calcareous habitats, sometimes forming intermediates with *S. autumnalis*. It is grown as an ornamental in the *Manual* region.

## 14.35 DESMAZERIA Dumort.[1]

Pl ann. **Clm** to 60 cm, procumbent to erect, sparingly brchd at the base. **Lvs** bas and cauline; **shth** open, glab; **aur** absent; **lig** longer than wide, acute; **bld** linear, usu flat, smt convolute when dry, glab. **Infl** tml, rcm or pan, usu with 1 br per nd; **br** stiff, not secund, ped 0.5–3 mm. **Spklt** subsessile, tangential to the rchs, lanceolate to ovate, lat compressed, with 4–25 flt, distal flt rdcd; **dis** above the glm, beneath the flt; **rchl** not prolonged beyond the base of the distal flt. **Glm** unequal to subequal, shorter than or subequal to the adjacent lm, 1–5-veined, unawned; **cal** blunt, rounded, glab; **lm** narrowly elliptic, coriaceous at maturity, incon-spicuously 5-veined, glab, smt scabridulous towards the apc, apc acute to obtuse, smt bifid, often mucronate, unawned; **pal** about as long as the lm, 2-veined; **lod** 2, free, lanceolate; **anth** 3, only slightly exserted at anthesis; **ov** glab. **Car** shorter than the lm, concealed at maturity, ellipsoid-oblong, dorsally flattened, falling with the pal; **hila** about ¹⁄₁₀ as long as the car, ovate.

*Desmazeria* has six or seven species, all of which are native around the Mediterranean. There are two species in the *Manual* region, one established as a weed and one introduced as an ornamental.

1. Lemmas 2–3 mm long, rounded on the back or weakly keeled distally; anthers 0.4–0.6 mm long; inflorescences usually panicles, sometimes racemes . . . . . . . . . . . . . . . . . . . . . . . . . . . . . . . . . . . . . . . . . . . . 1. *D. rigida*
1. Lemmas 3.5–4.5 mm long, strongly keeled; anthers 0.8–1.4 mm long; inflorescences racemes . . . . . . . . . . . . . . 2. *D. sicula*

**1. Desmazeria rigida** (L.) Tutin FERN GRASS [p. *415*, 526]

*Desmazeria rigida* is native to Europe, and appears to have no distinctive habitat preferences. In the *Manual* region, it is now established as a weed in disturbed sites such as roadsides, ditches, and the edges of fields. It is probably more widespread than indicated on the map, because herbarium records of weed distributions are often poor.

**2. Desmazeria sicula** (Jacq.) Dumort. SPIKE GRASS [p. *415*]

*Desmazeria sicula* used to be cultivated as an ornamental in the *Manual* region, but it is not included in contemporary treatments of ornamental grasses.

## 14.36 VENTENATA Koeler[2]

Pl ann; tufted. **Clm** 10–75 cm, erect, puberulent below the nd; **nd** glab. **Lvs** mostly on the lo ¹⁄₂ of the clm; **shth** open, glab or sparsely pubescent; **aur** absent; **lig** hyaline, acute or obtuse, usu lacerate; **bld** flat initially, involute with age. **Infl** open or contracted pan, with spklt borne near the ends of the br on clavate ped. **Spklt** lat compressed, with 2–10 flt; **dis** above the first flt and between the distal flt; **rchl** smt prolonged beyond the base of the distal flt, smt terminating in a rdcd flt. **Glm** unequal, lanceolate, hispidulous, similar in texture to the lm, mrg scarious, apc acuminate; **lo glm** 3–7-veined; **up glm** 3–9-veined; **cal** of the lo flt shorter than those of the up flt, sparsely hairy, cal of the distal flt with a dense tuft of white hairs; **lm** lanceolate, chartaceous, 5-veined, mrg scarious, apc entire or bifid, awned or unawned; **lowest lm** within a spklt awned or unawned, awns straight, tml; **distal lm** within a spklt awned, awns dorsal, geniculate; **pal** shorter than the lm, memb, keels ciliate distally; **lod** 2, memb, glab, toothed or not toothed; **anth** 3; **ov** glab. **Car** shorter than the lm, concealed at maturity, glab.

*Ventenata* is native from central and southern Europe and north Africa to Iran. It has five species, all of which grow in dry, open habitats. Only one species is established in the *Manual* region.

[1]Gordon C. Tucker  [2]William J. Crins

## 1. Ventenata dubia (Leers) Coss. VENTENATA, NORTH-AFRICA GRASS [p. *416*, <u>526</u>]

*Ventenata dubia* is established in crop and pasture lands of eastern Washington and western Idaho, and has been found, but has not necessarily become established, at scattered locations elsewhere. Mature specimens can be confusing because the first, straight-awned floret remains after the distal, bisexual florets have disarticulated.

# 14.37 CYNOSURUS L.[1]

Pl ann or per; smt rhz. **Clm** 1.5–90 cm, erect. **Cauline lvs** 1–3; **shth** open to the base; **aur** absent; **lig** truncate, entire, erose, or ciliolate; **bld** flat. **Infl** tml pan, condensed, often spikelike, linear to almost globose, more or less unilat; **br** and **ped** stiff, straight, smooth and glab or almost so, smt slightly scabridulous, scabrules/hairs to 0.1 mm. **Spklt** dimorphic, usu paired, subsessile to shortly pedlt, lat compressed, proximal spklt of each pair strl, almost completely concealing the ftl spklt, distal spklt on each br smt solitary; **dis** above the glm, beneath the flt. **Strl spklt** persistent, with 6–18 flt, flt rdcd to narrow, linear-lanceolate lm, smt awned, awns tml; **glm** narrow, linear. **Ftl spklt** adx to the strl spklt, with 1–5 flt; **rchl** glab, prolonged beyond the base of the distal flt; **glm** 2, subequal and shorter than the spklt, thin, lanceolate, 1-veined, acute, smt awned; **cal** short, blunt, glab; **lm** glab or pubescent, 5-veined, acute or bidentate, unawned to conspicuously awned, awns tml; **pal** about as long as the lm, bifid; **lod** 2, free, glab, ovate, bilobed; **anth** 3; **ov** broadly ellipsoid, glab; **sty** separate. **Car** oblong-ellipsoid, subterete, slightly dorsally compressed, smt adherent to the pal; **hila** $^1/_5$–$^1/_2$ as long as the car, oblong to linear.

*Cynosurus* is a genus of eight species that grow in open, grassy, often weedy habitats. It is native around the Mediterranean and in western Asia. The affinities of the genus are obscure. Two species are established in the *Manual* region.

1. Plants perennial; panicles linear; fertile lemmas unawned or with awns shorter than 3 mm . . . . . . . . . . . . . . 1. *C. cristatus*
1. Plants annual; panicles ovoid to almost globose; fertile lemmas with awns 5–25 mm long . . . . . . . . . . . . . . 2. *C. echinatus*

## 1. Cynosurus cristatus L. CRESTED DOGTAIL, CYNOSURE ACCRÊTÉ, CRÉTELLE DES PRÉS [p. *416*, <u>526</u>]

*Cynosurus cristatus* is a European native that is now established in North America. It grows in a wide range of soils in dry or damp habitats. In Europe it is used for fodder and pasture, especially for sheep, but in North America it is regarded as a weedy species. It is self-incompatible.

## 2. Cynosurus echinatus L. BRISTLY DOGTAIL [p. *416*, <u>526</u>]

*Cynosurus echinatus* is native to southern Europe. It is now established in dry, open habitats in North America, South America, and Australia.

# 14.38 PARAPHOLIS C.E. Hubb.[2]

Pl ann; tufted. **Clm** 2–45 cm, erect to prostrate, smt brchd above the base; **nd** glab, usu purple; **intnd** hollow. **Lvs** not bas concentrated; **shth** not keeled; **aur** smt present; **lig** 0.3–2.3 mm, memb, truncate; **bld** 0.3–3.5 mm wide, linear, flat to convolute. **Infl** tml and ax spikes, often curved, with solitary spklt sunk into the rchs; **dis** in the rchs below each spklt. **Spklt** sessile, cylindrical, straight to strongly curved, tangential to the rchs, with 1 flt; **rchl** smt prolonged beyond the base of the distal flt. **Glm** 2, subequal, lying side by side, usu exceeding or smt slightly shorter than the flt and covering the rchs cavities, smt asymmetric and winged, coriaceous, stiff, veins conspicuous, mrg translucent, apc unawned; **lm** 3.5–5.5 mm, memb, translucent, glab, rounded on the back, 1–3-veined, unawned; **pal** more or less equal to the lm, translucent; **lod** ovate-acute; **anth** 3; **sty** 2, free to the base, white. **Car** 3–3.6 mm, narrowly ovoid to ellipsoid.

*Parapholis* is a genus of six species that grow in coastal habitats and salt marshes. Its native range extends from western Europe to India. Two species are established in the *Manual* region.

1. Spikes curved and twisted; upper sheath margins expanded, enclosing the lowest spikelets; anthers 0.5–1.3 mm long . . . . . . . . . . . . . . . . . . . . . . . . . . . . . . . . . . . . . . . . . . . . . . . . . . . . . . . . . . . . . . . . . . . . . . . . 1. *P. incurva*
1. Spikes straight, not twisted; sheath margins usually all alike, the lowest spikelets generally exserted beyond the sheaths; anthers 1.5–4 mm long . . . . . . . . . . . . . . . . . . . . . . . . . . . . . . . . . . . . . . . . . . . . . . . . . . . . 2. *P. strigosa*

## 1. Parapholis incurva (L.) C.E. Hubb. CURVED SICKLEGRASS, SICKLEGRASS [p. *416*, <u>526</u>]

*Parapholis incurva* is established at various locations on the coasts of the contiguous United States. It grows in both poorly drained and well-drained disturbed soils, at and above the high tide mark. It tends to grow in more saline soils, and at lower elevations with respect to the tide, than *P. strigosa*.

## 2. Parapholis strigosa (Dumort.) C.E. Hubb. HAIRY SICKLEGRASS, STRIGOSE SICKLEGRASS [p. *416*, <u>527</u>]

*Parapholis strigosa* has been found in disturbed areas of Humboldt Bay, California, and has also been reported from Del Norte, Mendicino, and Sonoma counties. It grows on moist soils above normal high tides, usually in well-compacted sandy loams. In general, it is found in less saline soils and at higher elevations than *P. incurva*.

[1]Sandy Long  [2]Thomas Worley

# 14.39 SCRIBNERIA Hack.[1]

**Pl** ann; tufted. **Clm** 3–35 cm, ascending to erect, often brchd at the lowest nd, glab; **nd** purple. **Shth** open; **aur** absent; **lig** memb; **bld** involute, nearly filiform. **Infl** tml distichous spikes, with 1 spklt at all or most nd, occ 2 at some nd, very rarely 3 or 4 spklt per nd, lo spklt sessile, up spklt pedlt; **ped** shorter than 3 mm. **Spklt** tangential to and partially embedded in the rchs, lat compressed, with 1 flt; **rchl** prolonged beyond the base of the flt; **dis** above the glm, beneath the flt. **Glm** 2, exceeding the flt, glab, coriaceous, stiff, reddish- or purplish-tinged, 2-keeled, unawned; **lo glm** longer and narrower than the up glm, 2–3-veined; **up glm** 3–4-veined; **cal** pubescent; **lm** memb, inconspicuously 5-veined, shortly bifid, awned from the sinus, awns 2–4 mm; **pal** tightly clasped by the lm; **anth** 1; **ov** glab. **Car** about 2.5 mm, fusiform; **hila** punctiform; **emb** about ¼ the length of the car.

*Scribneria* is a unispecific genus native to North America.

### 1. Scribneria bolanderi (Thurb.) Hack. SCRIBNER GRASS [p. *416*, *527*]

*Scribneria bolanderi* grows between 500–3000 m, from Washington to Baja California, Mexico. It grows in diverse habitats, ranging from dry, sandy or rocky soils to seepages and vernal pools. It is often overlooked because it is relatively inconspicuous.

# 14.40 HAINARDIA Greuter[2]

**Pl** ann. **Clm** 5–45 cm, brchd above the base; **intnd** solid. **Upmost shth** open, often partially enclosing the infl; **aur** absent; **lig** memb, truncate; **bld** flat or convolute. **Infl** single, tml spikes, cylindrical, with solitary spklt embedded in and radial to the rchs, the abx surface of the up glm exposed; **dis** at the rchs nd. **Spklt** dorsiventrally compressed, with 1–2 flt, second flt rdcd and strl; **rchl** smt prolonged beyond the base of the distal flt. **Lo glm** absent from all but the tml spklt; **up glm** coriaceous, stiff, longer and firmer than the lm, rigid, 3–7(9)-veined, acute, unawned, smt mucronate; **lo lm** memb, lanceolate, 3(5)-veined, unawned; **pal** hyaline; **anth** 1–3; **lod** 2, oblique, glab, fleshy bas. **Car** shorter than the lm, oblong, somewhat dorsally compressed, with an apical appendage, concealed at maturity; **emb** about ⅕ the length of the car; **hila** short, linear.

*Hainardia* is a unispecific European genus that grows in saline and alkaline soils. It resembles *Parapholis*, which also occupies coastal salt marshes, but *Parapholis* differs in having spikelets with 2 glumes and culms with hollow internodes.

### 1. Hainardia cylindrica (Willd.) Greuter HARDGRASS, THINTAIL [p. *417*, *527*]

*Hainardia cylindrica* is now established in California, along the Gulf coasts of Texas and Louisiana, in northern Baja California, Mexico, and in South Carolina. It grows in coastal salt marshes and alkaline soils below 300 m.

# 14.41 VAHLODEA Fr.[3]

**Pl** per; loosely csp. **Clm** 15–80 cm. **Shth** open nearly to the base; **aur** absent; **lig** memb; **bld** rolled in the bud, flat. **Infl** open or closed pan; **br** often flexuous, capillary, spklt distal, some br longer than 1 cm. **Spklt** pedlt, usu with 2 flt, smt with additional distal rdmt flt; **rchl** prolonged beyond the base of the distal flt about 0.5 mm or less, usu glab, smt with a few hairs; **first intnd** about 0.5 mm; **dis** above the glm, beneath the flt. **Glm** subequal to equal, equaling or exceeding the flt, memb, acute to acuminate, unawned; **lo glm** 1(3)-veined; **up glm** 3-veined; **cal** obtuse, pilose, hairs about ½ as long as the lm; **lm** ovate, with 5(7) obscure veins, awned, awns attached near the middle of the lm, twisted, geniculate, visible between the glm; **pal** subequal to the lm; **lod** 2, memb, toothed or not toothed; **anth** 3; **ov** glab. **Car** shorter than the lm, concealed at maturity, ellipsoid and irregularly triangular or ovate in cross section, deeply grooved, smt adhering to the lm and pal; **hila** linear to oblong, ¼–½ the length of the car.

*Vahlodea* is a unispecific genus with a discontinuous circumboreal distribution. It also grows in southern South America. The genus is sometimes included in *Deschampsia*, from which it differs in having loosely cespitose shoots and leaves that are usually mostly cauline, rather than mostly basal as in the perennial species of *Deschampsia*. The rachilla is prolonged beyond the upper floret for less than 0.5 mm in *Vahlodea* and is usually glabrous or has only a few hairs. In *Deschampsia*, the rachilla is usually prolonged more than 0.5 mm and is usually densely pubescent with long hairs. The caryopses of *Vahlodea* are ellipsoid, irregularly triangular or ovate in cross section, deeply grooved, and with a hilum ¼–½ the length of the caryopsis. In *Deschampsia*, the caryopses are narrowly ellipsoid to fusiform, elliptic or ovate in cross section, grooved or not, and with a hilum ¹⁄₁₀–⅓ the length of the caryopsis.

[1,2]James P. Smith, Jr. [3]Jacques Cayouette and Stephen J. Darbyshire

## 1. Vahlodea atropurpurea (Wahlenb.) Fr. *ex* Hartm.

MOUNTAIN HAIRGRASS, DESCHAMPSIE POURPRE [p. *417*, 527]

*Vahlodea atropurpurea* grows in moist to wet, open woods, forest edges, streamsides, snowbeds, and meadows, in montane to alpine and subarctic habitats.

## 14.42 PODAGROSTIS (Griseb.) Scribn. & Merr.[1]

Pl per; csp, smt rhz. Clm 5–90 cm, erect or decumbent at the base. Lvs bas concentrated; **shth** open to the base, smooth, glab; **aur** absent; **lig** memb, scabridulous dorsally, truncate to subacute, entire to lacerate; **bld** flat or involute. Infl pan, exserted at maturity, not disarticulating; **br** ascending to erect. Spklt pedlt, weakly lat compressed, with 1 flt; **rchl** usu prolonged 0.1–1.9 mm beyond the base of the flt, smt absent, especially from the lo spklt within a panicle, apc glab or with hairs, hairs to 0.3 mm; **dis** above the glm, beneath the flt. Glm equal or the lo glm longer than the up glm, flexible, acute to acuminate, smt apiculate, unawned; **cal** glab or hairy, hairs to 0.5 mm; **lm** memb, (3)5-veined, veins mostly obscure, smt prominent distally, apc truncate to rounded or acute, unawned or awned, awns to about 1.3 mm, usu subapical, occ attached near midlength; **pal** more than $^1/_2$ as long as the lm, 2-veined, thinner than the lm; **anth** 3. Car shorter than the lm, concealed at maturity.

*Podagrostis* is a genus of six or more species that grow in cool, wet areas. Four or more species are native to Central and South America, and two to the *Manual* region. *Podagrostis* differs from *Agrostis*, in which it has formerly been included, in its combination of a relatively long palea and, usually, the prolongation of the rachilla beyond the base of the floret. It differs from *Calamagrostis* in the poorly developed callus hairs and awns.

1. Glumes 2.3–4.3 mm long, generally equal; rachilla prolongations 0.5–1.9 mm long . . . . . . . . . . . . . . . . . 1. *P. aequivalis*
1. Glumes 1.6–2.3 mm long, lower glumes equal to or longer than the upper glumes; rachilla prolongations 0.1–0.6 mm long . . . . . . . . . . . . . . . . . . . . . . . . . . . . . . . . . . . . . . . . . . . . . 2. *P. humilis*

### 1. Podagrostis aequivalis (Trin.) Scribn. & Merr.

ARCTIC BENT [p. *417*, 527]

*Podagrostis aequivalis* grows along lake, bog, and stream margins, and in forest fens. It is common in the coastal regions of Alaska and British Columbia, and occurs less frequently inland, as well as to about 1500 m in the Cascade Mountains south to Oregon.

### 2. Podagrostis humilis (Vasey) Björkman   ALPINE BENT

[p. *417*, 527]

*Podagrostis humilis* is a western North American species that grows in undisturbed alpine and subalpine meadows and screes at over 3500 m, down to meadows, fens, and open woodlands at less than 200 m. In the field, dwarf forms of *P. humilis* mimic *Agrostis variabilis*; they differ from that species in having paleas.

## 14.43 LACHNAGROSTIS Trin.[2]

Pl ann, or short-lived per; csp, smt rhz. Clm 10–80 cm, erect or geniculately ascending. Shth open, rounded over the midvein; **aur** absent; **lig** memb; **bld** flat or folded, mrg smt involute. Infl pan, lax. Spklt pedlt, with 1(2) flt(s), lat compressed; **rchl** prolonged beyond the base of the flt, smt equaling the pal and the apc hairy, or minute and glab; **dis** above the glm, beneath the flt, in per species the pan detaching with a portion of the upmost cauline nd at maturity, in ann species the pan persistent. Glm equal to subequal, exceeding the flt, ovate-elliptic to lanceolate, memb, 1(3)-veined, lat veins much shorter than the midveins, keels scabridulous to scabrous, apc unawned; **cal** minute, blunt, usu hairy, smt glab, hairs $^1/_5$–$^2/_3$ the length of the lm; **lm** usu shorter and more flexible than the glm, rarely as long as and firmer than the glm, usu hairy, smt glab, rarely scabrous, (3)5-veined, rounded over the midveins, apc often denticulate, mrg veins slightly excurrent, apc erose or toothed, usu awned, smt unawned, awns not or only slightly exceeding the glm, attached near midlength or subapically, dorsal awns straight or geniculate, subapical awns straight; **pal** from $^1/_2$ as long as to equaling the lm, hyaline, weakly 2-keeled; **lod** 2, linear to lanceolate, glab; **anth** 3, not penicillate; **ov** glab; **sty** 2. Car shorter than the lm, concealed at maturity, fusiform, endosperm doughy or dry.

*Lachnagrostis* includes about 20 species. It is native to the Southern Hemisphere, having its greatest concentration in Australasia. One species is established in the *Manual* region.

*Lachnagrostis* has usually been included in *Agrostis*, but differs from that genus in its combination of sometimes disarticulating panicles, paleas at least half as long as the lemmas, well-developed, sometimes hairy rachilla prolongations, and smooth lemma epidermes in which the walls of the long cells are wavy and more or less flush with the surface rather than raised. Both genera have individual species that resemble the other species in one of these features, but there is usually no difficulty in placing them in the correct genus.

[1,2]M.J. Harvey

**1. Lachnagrostis filiformis** (G. Forst.) Trin. PACIFIC
BENT [p. *417*, <u>527</u>]

*Lachnagrostis filiformis* is native to New Guinea, Australia, New Zealand, and Easter Island. In North America, it grows in open, disturbed sites such as roadsides and burned areas, and has been spreading into vernal pools around San Diego, California. It was introduced to North America in the late nineteenth century, but is only known to be established in California.

## 14.44 AVENULA (Dumort.) Dumort.[1]

**Pl** per; **csp**, smt **stln**. **Clm** 10–110 cm. **Shth** usu open, smt closed for most of their length; **aur** absent; **lig** memb, acute to truncate; **bld** flat or folded, adx surfaces unribbed, with a furrow on either side of the midveins, mrg sclerenchymatous. **Infl** rdcd pan, many br (all br in depauperate specimens) with a single spklt. **Spklt** with 2–7 flt; **rchl** glab on the side adjacent to the pal, hairy elsewhere; **dis** above the glm, beneath the flt. **Glm** as long as or longer than the adjacent lm, 1–3-veined, unawned; **cal** acute; **lm** 5–7-veined, obtuse, bifid, awned from about midlength, awns geniculate, flattened or terete and twisted below the bend; **pal** with lat wings less than ½ as wide as the intercostal region, apc shallowly bifid; **lod** 2, entire, unlobed; **anth** 3. **Car** more than twice as long as the hila, shorter than the lm, concealed at maturity; **endosperm** liquid or semi-liquid.

*Avenula* is a genus of approximately 30 species, most of which are European. One species is native to the *Manual* region, and one has been introduced. The genus is frequently included in *Helictotrichon*, from which it differs in having acute cauline ligules, unribbed leaves, rachillas glabrous on one side, unlobed lodicules, short hila, liquid to semi-liquid endosperm, and no sclerenchyma ring in its roots.

1. Sheaths closed for less than ⅓ their length, sheaths and blades smooth to scabridulous; panicles 4–13 cm long, usually 0.8–2.5 cm wide; awns 10–17 mm long, flattened below the bend . . . . . . . . . . . . . . . . . . . . 1. *A. hookeri*
1. Sheaths closed to near the top, sheaths and blades usually pubescent; panicles 6–20 cm long, 2–6 cm wide; awns 12–26 mm long, terete below the bend . . . . . . . . . . . . . . . . . . . . . . . . . . . . . . . . . . . . 2. *A. pubescens*

**1. Avenula hookeri** (Scribn.) Holub SPIKE OATGRASS [p. *418*, <u>527</u>]

*Avenula hookeri* grows on mesic to dry, open prairie slopes, hillsides, forest openings, and meadows, in montane to subalpine zones, from the Yukon and Northwest Territories to northern New Mexico.

**2. Avenula pubescens** (Huds.) Dumort. DOWNY ALPINE OATGRASS [p. *418*, <u>527</u>]

*Avenula pubescens* is native to Eurasia, where it grows in meadows, pastures, and woodland clearings. The most widespread taxon is **Avenula pubescens** (Huds.) Dumort. subsp. **pubescens**, which differs from **Avenula pubescens** subsp. **laevigata** (Schur) Holub in having smaller spikelets. *Avenula pubescens* subsp. *pubescens* has been collected in southern Ontario, Anticosti Island in Quebec, and in New England, but is not known to be established in Canada.

## 14.45 BROMIDIUM Nees & Meyen[2]

**Pl** ann or per; **csp**, smt **rhz**. **Clm** 5–50 cm, erect. **Shth** open almost to the base; **aur** absent; **lig** memb; **bld** flat or subconvolute, smooth or scabrous. **Infl** pan, spikelike or subspikelike, dense, usu interrupted towards the base. **Spklt** with 1 flt; **rchl** not prolonged beyond the base of the flt; **dis** above the glm, beneath the flt. **Glm** 2, exceeding the flt, equal or unequal, memb, 1-veined, keeled, keels scabrous or scabrous-ciliate, apc unawned; **cal** short, pilose on all sides or with 2 tufts of lat hairs, hairs not reaching past the lo ⅓ of the lm; **lm** memb, glab or pilose, mrg incurved, 5-veined, veins scab-ridulous on the distal ⅓, lat and mrg veins extending beyond the lm mrg as awns or 0.1–2 mm awnlike teeth, midvein forming an awn from the lo or up ⅓ of the lm, awns exceeding the glm, stout, geniculate, bases twisted, hygroscopic; **pal** absent or much rdcd; **lod** 2, memb, acute; **anth** 3, anth sacs separated in the distal ⅓ after dehiscence; **ov** oblong; **sty** 2; **stigmas** plumose. **Car** shorter than the lm, concealed at maturity, fusiform, grooved, usu falling free of the lm and pal; **hila** punctiform or oval; **emb** small; **endosperm** lipid, liquid, doughy, or starchy.

*Bromidium* includes five species, all of which are native to South America. One species, *B. tandilense*, is now established in California. None of the species is important for forage. Four of the five species are annual. *Bromidium* is sometimes included in *Agrostis*, but differs in its combination of dense, contracted, spikelike panicles with a relatively large number of spikelets, unawned glumes, and 4-awned or -toothed lemmas.

**1. Bromidium tandilense** (Kuntze) Rúgolo [p. *418*, <u>527</u>]

*Bromidium tandilense* is native to Argentina, Brazil, Uruguay, and Paraguay. It now grows around vernal pools in Solano, Monterey, and San Diego counties, California

[1]Gordon C. Tucker  [2]Zulma E. Rúgolo de Agrasar

## 14.46 DISSANTHELIUM Trin.[1]

**Pl** ann or per; csp, smt rhz. **Clm** to 10(25) cm. **Shth** open, usu glab, lo shth shorter than the up shth; **aur** absent; **lig** 2–6 mm, memb. **Infl** contracted pan, some br longer than 1 cm. **Spklt** pedlt, 2.5–5 mm, lat compressed, with 2(3) flt, all flt bisx or some flt bisx and others pist; **dis** above the glm, beneath the flt. **Glm** equal or subequal, usu exceeding all the flt, smt subequal to them, ovate or acuminate, mrg scarious, apc unawned; **lo glm** 1-veined; **up glm** 3-veined; **cal** poorly developed, glab; **lm** oval or elliptic, 3(5)-veined, lat veins near the mrg, glab, scabrous, or pilose, apc

acute or obtuse, smt denticulate, unawned; **pal** slightly shorter than the lm; **lod** 2; **anth** 3. **Car** shorter than the lm, concealed at maturity.

*Dissanthelium* contains about 20 species, and has an amphi-neotropical distribution. Most species grow in South America; two grow in North America. One of the two North American species, *D. mathewsii* (Ball) R.C. Foster & L.B. Sm., has a disjunct distribution, growing in both central Mexico and South America. The other, *D. californicum*, is discussed below.

### 1. Dissanthelium californicum (Nutt.) Benth.
CALIFORNIA DISSANTHELIUM [p. *418*, <u>527</u>]

*Dissanthelium californicum* is known only from Santa Catalina and San Clemente islands, California, and Guadalupe Island, Baja California, Mexico. It grows in coastal sage-scrub. Until its

rediscovery on Santa Catalina Island in 2005, it was thought to be extinct in California, not having been reported from the state since 1903.

## 14.47 HELICTOTRICHON Besser *ex* Schult. & Schult. f.[2]

**Pl** per; csp. **Clm** 5–150 cm, erect. **Shth** open nearly to the base; **aur** absent; **lig** about as long as wide, memb, truncate to rounded, ciliate-erose; **bld** convolute or involute, adx surfaces ribbed over the veins. **Infl** narrow pan or rcm, some br longer than 1 cm. **Spklt** lat compressed, with (1)2–8 flt; **rchl** pilose on all sides, terminating in rdcd flt; **dis** above the glm, beneath the flt. **Glm** equaling or exceeding the adjacent lm, exceeded by the distal flt, 1–3(5)-veined; **cal** acute, strigose; **lm** pilose or glab, 3–5-veined, apc acute, toothed, usu awned from about midlength, awns geniculate, twisted and terete below the bend, distal lm smt unawned; **pal** shorter than

the lm, wings more than $^{1}/_{2}$ as wide as the intercostal region; **lod** 2, lobed; **anth** 3; **ov** pubescent distally. **Car** shorter than the lm, concealed at maturity, with a solid endosperm, longitudinally grooved, with a tml tuft of hairs; **hila** more than $^{1}/_{2}$ as long as the car, linear.

*Helictotrichon* has about 15 species. Most are native to Europe; one is endemic to the *Manual* region, and one has been introduced as an ornamental. The genus is sometimes interpreted as including *Avenula*, from which it differs in having truncate to rounded ligules, ribbed leaves, rachillas that are pilose on all sides, lobed lodicules, long hila, solid endosperm, and sclerenchyma rings in its roots.

1. Culms 5–20 cm tall; panicles 2–8 cm long, most branches with 1 spikelet; plants native . . . . . . . . . . . . 1. *H. mortonianum*
1. Culms 30–150 cm tall; panicles 8–20 cm long, most branches with 3–10 spikelets; plants cultivated as ornamentals . . . . . . . . . . . . . . . . . . . . . . . . . . . . . . . . . . . . . . . . . . . . . . . . . . . . . . . . . . . . . 2. *H. sempervirens*

### 1. Helictotrichon mortonianum (Scribn.) Henrard
ALPINE OATGRASS [p. *418*, <u>527</u>]

*Helictotrichon mortonianum* grows in alpine and subalpine meadows and summits, at 3000–4200 m. It is restricted to the central and southern Rocky Mountains of the contiguous United States.

### 2. Helictotrichon sempervirens (Vill.) Pilg. BLUE
OATGRASS [p. *418*]

*Helictotrichon sempervirens* is a native of the southwestern Alps in Europe, where it grows on rocky soils and in stony pastures. In the *Manual* region, it is frequently grown as an ornamental species; it is not established in the region.

## 14.48 AMPHIBROMUS Nees[3]

**Pl** ann or per; csp, smt rhz or stln. **Clm** to 180 cm, erect or geniculate. **Shth** open, lo shth often enclosing cleistogamous pan with unawned spklt having fewer flt and smaller anth than the aerial spklt; **aur** absent; **lig** elongate, memb, becoming lacerate; **bld** flat or inrolled. **Tml infl** pan, open to spikelike. **Spklt** lat compressed, with 2–10 flt, cleistogamous flt occ intermixed with the chasmogamous flt, distal flt often rdcd, stmt; **rchl** pubescent, prolonged beyond the base of the most distal pist flt, empty or terminating in a rdcd flt; **dis** above the

glm, beneath the flt. **Glm** subequal to unequal, shorter than the adjacent lm, ovate to lanceolate, scarious, acute to obtuse, often erose, unawned; **lo glm** shorter and narrower than the up glm, 1–5-veined; **up glm** 3–7-veined; **cal** blunt, pubescent; **lm** chartaceous, smooth or scabrous, with 5–9 prominent veins, awned, apc with 2–4 teeth or lobes, outer lobes often smaller than the inner lobes, all lobes aristate to obtuse, awns arising from below midlength to near the apc, smt straight when young, geniculate at maturity, spreading or

[1]Nancy F. Refulio  [2]Gordon C. Tucker  [3]Surrey W.L. Jacobs

recurved; **pal** subequal to or much shorter than the lm, bilobed; **lod** 2, free, glab, not lobed; **anth** 3; **ov** glab. **Car** shorter than the lm, concealed at maturity, terete, apc often with a few hairs; **hila** to ¹/₂ the length of the car.

*Amphibromus* is a genus of 12 species, two native to South America, one to both New Zealand and Australia, and the remainder endemic to Australia; two have been introduced into the *Manual* region. Most species grow in open, damp habitats such as floodplains and other areas that are periodically flooded, and on the banks of, and sometimes in, inland and coastal rivers, marshes, lagoons, waterholes, and swamps.

1. Pedicels absent or to 10 mm long; lowest internodes usually swollen; awns 8–17 mm long . . . . . . . . . . . . 1. *A. scabrivalvis*
1. Pedicels usually longer than 10 mm; lowest internodes not swollen; awns 12–26 mm long.
   2. Awns arising from the lower ²/₅–³/₅ of the lemmas; lemma apices not appearing constricted . . . . . . . . . . . 2. *A. nervosus*
   2. Awns arising from the upper ²/₃–³/₄ of the lemmas; lemma apices appearing constricted . . . . . . . . . . . . . . . . 3. *A. neesii*

### 1. Amphibromus scabrivalvis (Trin.) Swallen ROUGH AMPHIBROMUS [p. *419*, 527]

*Amphibromus scabrivalvis* is native to open grasslands of South America. It was discovered growing in strawberry patches in Tangipahoa Parish, Louisiana, in the late 1950s. Despite efforts to eradicate it, the species persists there; it is not known to have spread elsewhere. North American plants belong to **Amphibromus scabrivalvis** (Trin.) Swallen var. **scabrivalvis**.

### 2. Amphibromus nervosus (Hook. f.) Baill. COMMON SWAMP WALLABYGRASS [p. *419*, 527]

*Amphibromus nervosus* is the most common species in the genus. It has frequently been misidentified as *A. neesii*, but has a lower awn insertion. Such misidentification is the basis of the 1990 report of *A.*

*neesii* from a vernal pool in Sacramento County, California; examination of the voucher specimens showed them to be *A. nervosus*. The discovery of living plants is of particular concern, because of their ability to invade and survive in vernal pools.

### 3. Amphibromus neesii Steud. SOUTHERN SWAMP WALLABYGRASS [p. *419*]

*Amphibromus neesii* is an Australian species that grows on floodplains and river banks, and in marshes and lagoons. It was first reported as growing in North America in 1990; examination of the voucher specimens showed them to be *A. nervosus*, which differs from *A. neesii* in having a lower lemma awn insertion. Both species are included in this treatment to help prevent future misidentification of the two species in North America.

## 14.49 CALAMAGROSTIS Adans.[1]

**Pl** per; often csp, usu rhz. **Clm** 10–210 cm, unbrchd or brchd, more or less smooth, **nd** 1–8. **Shth** open, smooth or scabrous; **aur** absent; **lig** memb, usu truncate to obtuse, smt acute, entire or lacerate, lacerations often obscuring the shapes; **bld** flat to involute, smooth or scabrous, rarely with hairs. **Infl** pan, open or contracted, smt spikelike; **br** appressed to more or less drooping, some br longer than 1 cm. **Spklt** pedlt, weakly lat compressed, with 1(2) flt; **rchl** prolonged beyond the base of the distal flt(s), usu hairy; **dis** above the glm. **Glm** memb, subequal, equal to, or longer than the lm, rounded or keeled, backs smooth or scabrous, rarely long-scabrous with bent projections, veins obscure to prominent, apc acute to acuminate, rarely awn-tipped or attenuate; **lo glm** 1(3)-veined; **up glm** 3-veined; **cal** hairy, hairs 0.2–6.5 mm, sparse to abundant; **lm** 3(5)-veined, smooth or scabrous, apc usu tapering into 4 teeth, awned; **awns** arising from near the base to near the apc, straight or bent, smt delicate and indistinct from the callus hairs, smt exserted beyond the lm mrg; **pal** well developed, almost as long as to slightly longer than the lm, thin, 2-veined; **anth** 3, smt strl. **Car** shorter than the lm, concealed at maturity, oblong, usu glab.

*Calamagrostis* grows in cool-temperate regions and is especially diverse in mountainous regions. Its species grow in both moist and xeric habitats. There are about 100 species of *Calamagrostis*. Twenty-five species of *Calamagrostis* grow in the *Manual* region; one, *C. epigejos*, is introduced. Some species of *Calamagrostis* are rangeland forage grasses, but most occur too sparsely to be important for livestock.

This treatment includes one cultivar, *Calamagrostis* ×*acutiflora* 'Karl Foerster', that is becoming increasingly popular in horticulture. A cultivar of *C. canadensis* has been registered for use in revegetation in arctic Alaska.

Vivipary and agamospermy occur in some species. Interspecific hybridization, polyploidy, and apomixis contribute to the taxonomic difficulty of the genus. There is a high degree of misidentification of taxa within this genus, and species distributions should be taken as a guide only.

*Calamagrostis* is sometimes confused with *Agrostis*; there is no single character that distinguishes all species of *Calamagrostis* from those of *Agrostis*. In general, *Calamagrostis* has larger plants with larger, more substantial lemmas and paleas than *Agrostis*, and tends to occupy wetter habitats.

Measurements of the rachilla and callus hairs reflect the longest hairs present. Panicle widths refer to pressed specimens. The following key will enable typical specimens to be identified readily, but atypical specimens are common. For this reason, most leads require observation of a combination of characters, notably awn length, length of callus hairs relative to the lemma, glume length and scabrosity, panicle size, and leaf width.

1. Callus hairs more than 1.3 times as long as the lemmas; lemmas at least 2 mm shorter than the glumes, long-acuminate . . . . . . . . . . . . . . . . . . . . . . . . . . . . . . . . . . . . . . . . . . . . . . . . . . . . . . . . . . . . . . . . . . . . . . 1. *C. epigejos*
1. Callus hairs usually less than 1.2 times as long as the lemmas; if the callus hairs longer than the lemmas, then the lemmas less than 2 mm shorter than the glumes and not long-acuminate.

[1]Kendrick L. Marr, Richard J. Hebda, and Craig W. Greene†

2. Blades usually densely hairy on the adaxial surfaces; glumes keeled, scabrous; awns 4.5–9 mm long .. 2. *C. purpurascens*
2. Blades glabrous or sparsely hairy on the adaxial surfaces; glumes keeled or rounded, scabrous or smooth; awns 0.5–17 mm long.
  3. Leaves with abundant white glands between the veins, visible at about 10×; awns 5–8 mm long; plants of California . . . . . . . . . . . . . . . . . . . . . . . . . . . . . . . . . . . . . . . . . . . . . . . . . . . . . . . . . . 4. *C. ophitidis*
  3. Leaves without abundant white glands between the veins; awns 0.5–17 mm long; plants from throughout the *Manual* region, including California.
    4. Awns 5–17 mm long, always exserted and bent; if the awns 5–6 mm long, either some blades wider than 2 mm or the abaxial blade surfaces scabrous.
      5. Panicles open, (2)3.5–6.5(8) cm wide when pressed, branches spikelet-bearing only beyond midlength; awns 10–16 mm long . . . . . . . . . . . . . . . . . . . . . . . . . . . . . . . . . . . . . . . 3. *C. howellii*
      5. Panicles usually contracted, (0.5)0.8–3 cm wide if open, the branches spikelet-bearing to below midlength, usually to the base; awns 5–17 mm long.
        6. Some leaf blades 6–13 mm wide; culms (47)60–120(150) cm tall . . . . . . . . . . . . . . . . . . . . . . . 5. *C. tweedyi*
        6. All leaf blades (1.5)2–7 mm wide; culms (15)30–60(95) cm tall, if the culms taller than 60 cm, then the blades less than 4 mm wide.
          7. Awns 12–14(17) mm long; plants of California . . . . . . . . . . . . . . . . . . . . . . . . . . . . . . . . 6. *C. foliosa*
          7. Awns (5.4)7–11(13) mm long; plants of Alaska, British Columbia, Washington, and Oregon.
            8. Glume apices long-acuminate, usually twisted distally; glume keels usually scabrous for their whole length . . . . . . . . . . . . . . . . . . . . . . . . . . . . . . . . . . . . 7. *C. sesquiflora*
            8. Glume apices usually acute, if acuminate, not twisted distally; glume keels smooth or sparsely scabrous on the distal ¹/₂ . . . . . . . . . . . . . . . . . . . . . . . . . 8. *C. tacomensis*
    4. Awns 0.5–6 mm long, exserted or not, bent or straight; if the awns 5–6 mm long, then either all blades less than 2 mm wide or the abaxial blade surfaces smooth or nearly so.
      9. Awns attached on the distal ²/₅ of the lemmas, 0.5–2 mm long, straight; blades flat; panicles contracted, 0.7–2.5(3) cm wide.
        10. Lateral veins of the glumes prominent; rachillas hairy only distally . . . . . . . . . . . . . . . . . . . 9. *C. cinnoides*
        10. Lateral veins of the glumes obscure; rachillas hairy throughout their length . . . . . . . . . . 10. *C. scopulorum*
      9. Awns attached on the lower ¹/₂(⁷/₁₀) of the lemmas, 0.9–6 mm long, straight or bent; blades flat or involute; panicles open or contracted, 0.4–5.5(9) cm wide.
        11. Blades 0.2–1.7 mm wide; panicles (1.5)1.9–8.5 cm long; callus hairs sparse.
          12. Blades involute, 0.2–0.4 mm in diameter; ligules 1–2.5 mm long . . . . . . . . . . . . . . . 11. *C. muiriana*
          12. Blades flat, 0.9–1.7 mm wide, sometimes involute and 0.4–0.6 mm in diameter when dry; ligules 1.7–6 mm long . . . . . . . . . . . . . . . . . . . . . . . . . . . . . . . . . . . 12. *C. breweri*
        11. Blades (1)1.5–20 mm wide, most wider than 2 mm; panicles (2)3–30(40) cm long; callus hairs sparse to abundant.
          13. Awns usually exserted, (2.8)3–6 mm long; callus hairs 0.1–0.7 times as long as the lemmas; leaf collars hairy or glabrous.
            14. Culms 10–55(60) cm tall; panicles open; blades (1)1.5–3(4) mm wide.
              15. Blades 2–8(15) cm long; panicles erect; plants of brackish arctic and subarctic coastal habitats . . . . . . . . . . . . . . . . . . . . . . . . . . . . . . . . . 13. *C. deschampsioides*
              15. Blades (5)15–39 cm long; panicles often drooping; plants of rocky soils and disturbed sites in the mountains of Tennessee and North Carolina . . . . . . . . 14. *C. cainii*
            14. Culms (26)50–210 cm tall; panicles open or contracted; blades (1)1.5–8(12) mm wide, most blades wider than 3 mm.
              16. Panicles open, 2.5–6 cm wide; panicle branches with the spikelets confined to the distal ¹/₄–¹/₂; leaf collars glabrous . . . . . . . . . . . . . . . . . . . . . 15. *C. bolanderi*
              16. Panicles usually contracted, 0.7–3(7) cm wide; panicle branches usually with the spikelets confined to the distal ¹/₂–²/₃, sometimes spikelet-bearing to the base; leaf collars hairy or glabrous.
                17. Culms 135–210 cm tall; plants cultivated ornamentals . . . . . . . . . . . 16. *C. ×acutiflora*
                17. Culms 26–120 cm tall; plants native.
                  18. Awns 4–5.5 mm long; leaf collars glabrous; plants often densely cespitose; rhizomes usually 2–4 mm thick . . . . . . . . . . . 17. *C. koelerioides*
                  18. Awns 2–4.5 mm long; leaf collars sometimes hairy; plants loosely cespitose; rhizomes 0.5–2 mm thick.

19. Blades (2)3–8(12) mm wide; panicle branches usually
with the spikelets restricted to the distal $^1/_2$–$^2/_3$, sometimes
spikelet-bearing to the base; plants from east of the 100th
meridian . . . . . . . . . . . . . . . . . . . . . . . . . . . . . . . . . . . . . . 18. *C. porteri*

19. Blades (1)2–5(8) mm wide; panicle branches spikelet-
bearing to the base; plants from west of the 100th
meridian . . . . . . . . . . . . . . . . . . . . . . . . . . . . . . . . . . . . 19. *C. rubescens*

13. Awns usually not exserted, or if exserted, then barely so, 0.9–3.1(4) mm long;
callus hairs (0.1)0.2–1.2(1.5) times as long as the lemmas; leaf collars glabrous
or hairy, if hairy, then the callus hairs more than 0.7 times as long as the lemmas.

20. Callus hairs shorter than 1 mm, 0.2–0.3 times as long as the lemmas; awns
bent . . . . . . . . . . . . . . . . . . . . . . . . . . . . . . . . . . . . . . . . . 21. *C. pickeringii*

20. Callus hairs longer than 1 mm, (0.2)0.3–1.2(1.5) times as long as the
lemmas; awns straight or bent.

21. Culms usually scabrous, rarely smooth; awns slightly bent; callus hairs
0.4–0.8 times as long as the lemmas; blades usually involute, 1–4 mm
wide, the abaxial surfaces scabrous; nodes 1–2 . . . . . . . . . . . . . . . . . . . 22. *C. montanensis*

21. Culms smooth to slightly scabrous; awns usually straight, rarely bent;
callus hairs (0.2)0.5–1.2(1.5) times as long as the lemmas; blades flat or
involute, 1–20 mm wide, the abaxial surfaces smooth or scabrous;
nodes 1–8.

22. Lemmas (3)4–5 mm long; glumes keeled; blades flat . . . . . . . . . . . . . 20. *C. nutkaensis*

22. Lemmas 2–4(5) mm long, if the lemmas longer than 4 mm, then the
glumes rounded and the abaxial blade surfaces smooth; glumes
keeled or rounded; blades flat or involute.

23. Panicle branches 2.7–6.5(12) cm long; ligules lacerate; glumes
scabrous on the keels, often throughout; blades flat, the
abaxial surfaces scabridulous or scabrous; nodes (2)3–7(8);
panicles open.

24. Awns bent, stout, readily distinguished from the callus
hairs; collars densely hairy; plants of New York State . . . . . . . . 23. *C. perplexa*

24. Awns usually straight, delicate, often difficult to
distinguish from the callus hairs; collars rarely hairy;
plants of northern and western North America . . . . . . . . . . . 24. *C. canadensis*

23. Panicle branches (1)1.4–5(9.5) cm long; if the panicle branches
longer than 3.7 cm, then the ligules usually entire; glumes
smooth or scabrous only on the keels; blades flat or involute,
the abaxial surfaces smooth or scabrous; nodes 1–3(4);
panicles loosely contracted.

25. Glume lengths usually more than 3 times the widths,
smooth, the keels rarely slightly scabrous, the lateral veins
obscure; spikelets 3.5–5.5 mm long; awns usually slender
and similar to the callus hairs . . . . . . . . . . . . . . . . . . . . . . . 25. *C. lapponica*

25. Glume lengths usually less than 3 times the widths,
usually smooth, rarely scabrous, the keels smooth or
scabrous, the lateral veins prominent or obscure; spikelets
2–5 mm long; awns stout, usually readily distinguished
from the callus hairs . . . . . . . . . . . . . . . . . . . . . . . . . . . . . 26. *C. stricta*

**1. Calamagrostis epigejos** (L.) Roth Bushgrass, Chee
Reedgrass, Feathertop, Calamagrostide Épigéios,
Calamagrostide Commune [p. *419*, <u>527</u>]

*Calamagrostis epigejos* is an introduced Eurasian species. It grows in
waste places, along roadsides, in juniper swamps, sandy woods, and
thickets, and on rehabilitated tailings and cinders of railway beds. It
is known from scattered locations in southern Canada and the
contiguous United States. It is probably more widespread than shown.
Hybrids of *C. epigejos* with *C. arundinacea* (L.) Roth are called *C.
×acutiflora*.

**2. Calamagrostis purpurascens** R. Br. Purple
Reedgrass, Calamagrostide Pourpre [p. *419*, <u>527</u>]

*Calamagrostis purpurascens* grows in alpine tundra, on subalpine
slopes, in grasslands, sand dunes, meadows, coniferous and deciduous
forests, and disturbed soils, usually on rocky ridgetops and slopes
and, infrequently, on valley floors. It prefers well- to moderately-
drained, medium- to coarse-textured substrates, including scree and
talus, that are often calcareous, at elevations from 15–4000 m. Its
range extends from Alaska through Canada to Greenland and
Newfoundland, including the islands of the Canadian arctic, and
south in the western mountains to California and northern New
Mexico. It does not occur near the open coast except in the Aleutian

Islands, the Arctic, and the Olympic Peninsula in Washington. In Asia, it ranges from eastern and central arctic Siberia south to the Kamchatka Peninsula and Sakhalin Island.

The hairy adaxial leaf surfaces are a reliable diagnostic characteristic for *Calamagrostis purpurascens*. It differs from *C. tacomensis* in its leaf vestiture, shorter awns and panicle branches, and more scabrous glumes. Plants of *C. purpurascens* that have short awns barely projecting beyond the lemma margins have been mistaken for *C. montanensis*, but that species does not have hairy adaxial leaf surfaces.

### 3. Calamagrostis howellii Vasey HOWELL'S REEDGRASS [p. *419*, 527]

*Calamagrostis howellii* grows on dry rocky slopes, banks, ledges, and in cliff crevices, sometimes on basalt, from 100–500 m. It grows only in the Columbia River Gorge of Washington and Oregon.

### 4. Calamagrostis ophitidis (J.T. Howell) Nygren SERPENTINE REEDGRASS [p. *419*, 527]

*Calamagrostis ophitidis* grows in meadows, seeps, grasslands, and chaparral, as well as in coniferous forests, on serpentine outcrops and soils, at 50–1100 m. It is known only from Sonoma, Marin, Mendocino, Lake, and Napa counties in California.

### 5. Calamagrostis tweedyi (Scribn.) Scribn. CASCADE REEDGRASS, TWEEDY'S REEDGRASS [p. *419*, 527]

*Calamagrostis tweedyi* grows in montane to subalpine moist meadows and coniferous forests, often in association with *Carex geyeri*, at 900–2000 m. Its range extends from Washington and Oregon to western Montana.

### 6. Calamagrostis foliosa Kearney LEAFY REEDGRASS [p. *419*, 527]

*Calamagrostis foliosa* grows in coastal scrub and forest, and on rocks and crevices of bluffs and cliffs, from sea level to 1200 m. It is known only from Del Norte, Humboldt, Mendocino, and Sonoma counties in California.

### 7. Calamagrostis sesquiflora (Trin.) Tzvelev ONE-AND-A-HALF-FLOWERED REEDGRASS [p. *419*, 527]

*Calamagrostis sesquiflora* grows at 0–1000 m in open heath, meadows, and forest openings, on or at the base of open rocky cliffs and knolls, as well as in moist talus. It grows in strictly maritime habitats along the west coast of North America, from the Aleutian Islands in Alaska to the coast of British Columbia. In northeast Asia, it ranges into the Kamchatka Peninsula and Kuril Archipelago. It has sometimes included *C. tacomensis*, but differs in preferring moister habitats, having wider leaves, callus hairs that are shorter relative to the lemmas, shorter panicle branches, and glumes that are often twisted at the apices.

### 8. Calamagrostis tacomensis K.L. Marr & Hebda RAINIER REEDGRASS [p. *419*, 527]

*Calamagrostis tacomensis* grows on montane to alpine slopes in dry or wet meadows, seeps, rocky talus slopes, and cliff crevices, at 400–2200 m. It grows only in the mountains of western Washington and in the Steens Mountains of southeastern Oregon. It differs from *C. purpurascens* in having glabrous leaves, generally longer awns and inflorescence branches, and smoother glumes. It differs from *C. sesquiflora* in having narrower leaves, callus hairs that are longer relative to the lemmas, longer inflorescence branches, and glume apices that are not twisted, as well as in often preferring drier habitats.

### 9. Calamagrostis cinnoides (Muhl.) W.P.C. Barton SMALL REEDGRASS [p. *419*, 527]

*Calamagrostis cinnoides* is found on roadsides, in ditches, pond edges, and boggy streamhead seepages, and along streams in oak or oak-pine woods on sandy to peaty soils, at 5–1100 m. Its range extends throughout eastern North America, from Nova Scotia and Maine to Georgia and Louisiana. It is adventive in Ohio.

### 10. Calamagrostis scopulorum M.E. Jones JONES' REEDGRASS, DITCH REEDGRASS [p. *420*, 527]

*Calamagrostis scopulorum* grows on canyon slopes and wash bottoms, and in dry to moist montane to alpine habitats, often on rocky, sandy to silty soil, at 1000–3550 m. It grows from western Montana and Wyoming south to Arizona and New Mexico.

### 11. Calamagrostis muiriana B.L. Wilson & Sami Gray MUIR'S REEDGRASS [p. *420*, 527]

*Calamagrostis muiriana* grows in moist to dry, subalpine and alpine floodplain meadows, lake margins, and stream banks, at 2400–3900 m, in the Sierra Nevada south of Sonora Pass in central California. It differs from *C. bolanderi* in having basally concentrated leaves.

### 12. Calamagrostis breweri Thurb. SHORTHAIR REEDGRASS [p. *420*, 528]

*Calamagrostis breweri* grows in moist subalpine and alpine meadows, lake margins, and stream banks, at 1700–2600 m, from Mount Hood in Oregon south to north of the Carson Pass area in Alpine and Amador counties, California. It differs from *C. bolanderi* in having basally concentrated leaves.

### 13. Calamagrostis deschampsioides Trin. CIRCUMPOLAR REEDGRASS, CALAMAGROSTIDE FAUSSE-DESCHAMPSIE [p. *420*, 528]

*Calamagrostis deschampsioides* is a halophyte that grows, in the *Manual* region, on coastal dunes and beach ridges, gravel beaches, and in brackish coastal marshes, sometimes with *Carex lyngbyei*, at or near sea level. Its distribution is circumboreal, extending in North America from the islands of the Bering Sea and coastal Alaska, across the arctic coast to Hudson Bay and northern Labrador. It also extends from the arctic coast of Europe to Siberia and Japan.

### 14. Calamagrostis cainii Hitchc. CAIN'S REEDGRASS [p. *420*, 528]

*Calamagrostis cainii* grows on bouldery substrates and in soil pockets, landslides, and disturbed sites, at 1200–2100 m. It has been found in only three locations: the slopes of Mount LeConte, Sevier County, Tennessee; and in North Carolina on Craggy Pinnacle, Buncombe County, and the summit of Mount Craig, Yancey County. The species is of conservation concern because of its limited distribution.

### 15. Calamagrostis bolanderi Thurb. BOLANDER'S REEDGRASS [p. *420*, 528]

*Calamagrostis bolanderi* grows in marshes, swamps, bogs, fens, seeps, moist meadows, open and closed coniferous and broadleaf forests, prairies, and coastal scrub, from sea level to 500 m. It is known only from sites near the coast in Humboldt, Mendocino, and Sonoma counties, California. It differs from *C. breweri* and *C. muiriana* in having leaves evenly distributed along the culms.

### 16. Calamagrostis ×acutiflora (Schrad.) D.C. 'Karl Foerster' FOERSTER'S REEDGRASS, FEATHER REEDGRASS [p. *420*]

*Calamagrostis ×acutiflora* is a hybrid of European origin that is now widely planted as an ornamental, especially in dry sites and gardens throughout northern North America. The parents are *C. arundinacea* (L.) Roth and *C. epigejos*; the hybrids are seed-sterile.

### 17. Calamagrostis koelerioides Vasey DENSE-PINE REEDGRASS [p. *420*, 528]

*Calamagrostis koelerioides* grows in mountain meadows, chaparral, and Jeffrey pine and blue spruce forests, and on talus slopes, dry hills, and ridges, occasionally on serpentine soils, at 50–2100 m. It extends from Washington south to southern California and east to Montana and western Wyoming. It differs from *C. rubescens* in its longer lemmas, glumes, and awns.

## 18. Calamagrostis porteri A. Gray PORTER'S REEDGRASS [p. 420, 528]

*Calamagrostis porteri* grows in dry chestnut/oak forests, often on rocky ridgetops, piedmont bluffs, and slopes, at (100)600–1300 m. It is now restricted to the northeastern and central United States. Historically, its range extended from Missouri and Arkansas east to New York and Alabama. Flowering appears to be a response to disturbance; plants in undisturbed habitats remain vegetative and may go unnoticed. Thus the species may be more widespread and abundant than reported.

1. Leaf blades glaucous on both surfaces; leaf collars glabrous . . . . . . . . . . . . . . . . . . . . . . . . . . . . . subsp. *insperata*
1. Leaf blades light green and glaucous on the adaxial surfaces, darker green on the abaxial surfaces; leaf collars usually with prominent tufts of hair, rarely glabrous . . . . . . . . . . . . . . . . . . . . . . . . . . . . subsp. *porteri*

## Calamagrostis porteri subsp. insperata (Swallen) C.W. Greene BARTLEY'S REEDGRASS [p. 420]

Isolated populations of *Calamagrostis porteri* subsp. *insperata* grow from southern Missouri and Arkansas to Ohio and Kentucky. It is listed as possibly extirpated or of conservation concern in several states.

## Calamagrostis porteri A. Gray subsp. porteri [p. 420]

*Calamagrostis porteri* subsp. *porteri* grows from New York to Virginia, Tennessee, and Kentucky. It appears to form hybrids with the nearly sympatric *C. canadensis* in rocky wooded sites in central Virginia. These putative hybrids have hairy collars, relatively long callus hairs, and short awns.

## 19. Calamagrostis rubescens Buckley PINEGRASS, PINE REEDGRASS [p. 420, 528]

*Calamagrostis rubescens* grows at 50–2800 m, usually in open montane pine or aspen forests and parklands, infrequently in sagebrush steppes, chaparral, and meadows. It is primarily a species of interior western North America, although it reaches the Pacific coast in southern California. It is considered threatened in Saskatchewan. It differs from *C. koelerioides* in its shorter lemmas, glumes, and awns.

## 20. Calamagrostis nutkaensis (J. Presl) Steud. PACIFIC REEDGRASS [p. 420, 528]

*Calamagrostis nutkaensis* grows in wetlands and openings in coniferous forests, on marine and freshwater beaches and dunes, and, sometimes, on cliffs. It is usually found within a few kilometers of the marine shoreline at or near sea level, but it sometimes occurs as high as 1100 m. It grows along the Pacific coast of North America from the Aleutian Islands in Alaska to San Luis Obispo County, California, and also in the Kamchatka Peninsula of Russia.

## 21. Calamagrostis pickeringii A. Gray PICKERING'S REED BENTGRASS, CALAMAGROSTIDE DE PICKERING [p. 420, 528]

*Calamagrostis pickeringii* grows in bogs, open white spruce scrub, wet meadows, coastal peatlands and lake shores, heaths, frost pockets (hollows), pitch pine barrens, and on sandy beaches, at 0–1600 m. It is found from Newfoundland and Nova Scotia south to the mountains of New Hampshire, New York, and New Jersey.

## 22. Calamagrostis montanensis (Scribn.) Vasey PLAINS REEDGRASS [p. 421, 528]

*Calamagrostis montanensis* inhabits prairie grasslands and sagebrush flats, benchlands, valley bottoms, and occasionally woodlands, at 200–2600 m. It grows in the continental interior from eastern British Columbia and adjacent Alberta, south to southern Wyoming and east to Manitoba and western Minnesota. It may be mistaken for *C. purpurascens*, but the latter species has hairy adaxial leaf surfaces and longer awns.

## 23. Calamagrostis perplexa Scribn. WOOD REEDGRASS [p. 421, 528]

*Calamagrostis perplexa* grows on wet rocks and in dry woods at "Thatcher's Pinnacle" [Pinnacle Rock], Tompkins County, New York. There is also an unverified report of this species in Columbia County, New York. This apparently sterile species is intermediate between *C. porteri* and *C. canadensis*. It is of conservation concern because of its limited distribution.

## 24. Calamagrostis canadensis (Michx.) P. Beauv. BLUEJOINT, CALAMAGROSTIDE DU CANADA [p. 421, 528]

*Calamagrostis canadensis* is a species of moist meadows, thickets, bog edges, and forest openings. It grows from sea level to 3400 m. It occurs widely throughout the *Manual* region. Its range also extends from northern Asia to northeastern China and Japan, with additional scattered populations elsewhere in Asia. It hybridizes with *Ammophila breviligulata* in Grand Island, Michigan and on the adjacent mainland, forming ×**Calammophila don-hensonii** Reznicek & Judz. *Calamagrostis canadensis* also appears to form hybrids with the nearly sympatric *C. porteri* in rocky wooded sites in central Virginia. These putative hybrids have hairy collars, relatively long callus hairs, and short awns. The apparently sterile *C. perplexa* is intermediate between *C. canadensis* and *C. porteri*.

1. Spikelets (3.5)4–4.5(5.2) mm long; glumes usually scabrous over the entire surface, the prickles on the keels hairlike, often bent; glume apices distinctly acuminate . . . . . . . . . . . . . . . . . . . . . . . . var. *langsdorffii*
1. Spikelets 2–4 mm long; glumes smooth or scabrous, often scabrous only on the keels, prickles straight; glume apices acute, rarely acuminate.
    2. Spikelets 2.5–4 mm long, lemmas usually shorter than the glumes; glumes rounded to broadly keeled, with raised midveins; glume apices usually acute, rarely acuminate . . . . . . . . . . . var. *canadensis*
    2. Spikelets 2–3 mm long; lemmas usually about as long as the glumes; glumes rounded, midveins not raised; glume apices acute . . . . . . . . . . var. *macouniana*

## Calamagrostis canadensis (Michx.) P. Beauv. var. canadensis [p. 421]

*Calamagrostis canadensis* var. *canadensis* is widespread throughout the range of the species, except on the arctic islands, Greenland, and the adjacent mainland.

## Calamagrostis canadensis var. langsdorffii (Link) Inman [p. 421]

A circumboreal taxon, *Calamagrostis canadensis* var. *langsdorffii* grows from Alaska to Greenland and south to northern California, Colorado, Minnesota, New Hampshire, and New York.

## Calamagrostis canadensis var. macouniana (Vasey) Stebbins MACOUN'S REEDGRASS [p. 421]

*Calamagrostis canadensis* var. *macouniana* is common in marshes and mud flats. It ranges eastward from Saskatchewan to Prince Edward Island and Nova Scotia, and from North Dakota and Missouri to New Jersey and Virginia.

## 25. Calamagrostis lapponica (Wahlenb.) Hartm. LAPLAND REEDGRASS, CALAMAGROSTIDE DE LAPPONIE [p. 421, 528]

*Calamagrostis lapponica* grows in northern and alpine tundra, particularly on ridgecrests and upper slopes, often with low shrubs including heathers, dwarf willows, and dwarf birch, usually on well-drained and coarse-textured soils, infrequently in meadows beside streams and lakeshores, very rarely in standing water, at 30–2300 m. It is circumboreal and circumpolar, ranging from Alaska to western Greenland and Labrador, including the islands of the high arctic, south into the mountains of northern British Columbia and the west-central Rocky Mountains of Alberta. In Europe it extends south to about 60° N latitude, and in Asia south to North Korea. It can be confused with *C. stricta*, but the two grow in different habitats. In addition, the glumes of *C. lapponica* have a smoother, more glossy appearance than those of *C. stricta* and are typically purple for most of their length, including the apices; the glumes of *C. stricta* are generally brown at the apices.

### 26. Calamagrostis stricta (Timm) Koeler SLIMSTEM REEDGRASS [p. *421*, <u>528</u>]

*Calamagrostis stricta* grows throughout northern North America; it also is found in Europe and northeastern Asia. It grows in habitats ranging from meadows and grassland to wetlands, sandy shorelines, and sand dunes, from sea level to 3400 m. Primarily a species of open settings, it is frequently found in association with shrubs. Both subspecies have a notable but not exclusive association with alkaline to saline substrates. *Calamagrostis stricta* differs from *C. lapponica* in having glumes that are not as smooth and glossy, and are generally brown rather than purple at the tips.

1. Spikelets 3–4(5) mm long; callus hairs 2–4.5 mm long; rachilla prolongations 1–1.5 mm long; panicle branches 1.5–9.5 cm long; culms usually scabrous, sometimes smooth . . . . . . . . . . . . . . . . . . . . subsp. *inexpansa*

1. Spikelets 2–2.5(3) mm long; callus hairs 1–3 mm long; rachilla prolongations 0.5–1 mm long; panicle branches 1.4–4 cm long; culms usually smooth, sometimes slightly scabrous . . . . . . . . . . . . . . . . . subsp. *stricta*

### Calamagrostis stricta subsp. inexpansa (A. Gray) C.W. Greene NORTHERN REEDGRASS, CALAMAGROSTIDE CONTRACTÉE [p. *421*]

*Calamagrostis stricta* subsp. *inexpansa* differs from subsp. *stricta* in its more robust growth and coarser habit. In North America, it extends from Alaska to Labrador and Newfoundland and south to California, Arizona, Minnesota, Iowa, Ohio, and New York. It also grows in northeastern Asia. This subspecies usually grows in moist meadows, sphagnum bogs, and grasslands associated with rivers and streams, and less frequently on grassy slopes, in open woods, and beside sand dunes. It is noted to grow at the edge of, rather than in, wetlands.

### Calamagrostis stricta (Timm) Koeler subsp. stricta CALAMAGROSTIDE DRESSÉE [p. *421*]

A circumboreal taxon, *Calamagrostis stricta* subsp. *stricta* favors moist meadows and fens, occurring less frequently in marshes and bogs, and sometimes grows near sand dunes. It is usually associated with fine-textured substrates. In the *Manual* region, its range extends from Alaska to Newfoundland and Labrador and Greenland, and south to California, Utah, North Dakota, Minnesota, New Hampshire, and New York. In Europe, it extends from 50° N latitude to Switzerland and the southern Caucasus.

## 14.50 GAUDINIA P. Beauv.[1]

Pl ann or per. Clm 15–120 cm. Shth open; aur absent; lig memb; bld flat. Infl solitary, distichous spikes; dis in the rchs, immediately above the spklt. Spklt lat compressed, sessile, tangential, more or less appressed to the rchs, with 3–11 flt; rchl glab, prolonged beyond the base of the distal ftl flt, terminating in a rdcd flt. Glm 2, unequal, from shorter than to about as long as the spklt, unawned; cal blunt, glab; lo glm 3(5)-veined; up glm 5–7(11)-veined; lm coriaceous, obscurely 7–9-veined,

unawned or awned near the apc; pal shorter than the lm; lod 2, free, memb, glab, toothed; anth 3; ov pubescent. Car with a tml tuft of hairs; hila round.

*Gaudinia* is a weedy genus of four species that are native to the Mediterranean, the Azores, and the Canary Islands. Its inflorescence is reminiscent of some *Triticeae*; it differs from members of that tribe in its manner of disarticulation and in having compound starch grains in its endosperm. One species has become established in North America.

### 1. Gaudinia fragilis (L.) P. Beauv. FRAGILE OAT [p. *421*, <u>528</u>]

*Gaudinia fragilis* is the most widespread species of the genus in the Mediterranean region. The first record of its presence in the United States dates from a collection made on ballast ground in Alabama in

1885. In 1991, it was discovered in Sonoma County, California, on an open, grassy hilltop in the thin, rocky soil of open oak woodlands, in a region that has long been used for agriculture.

## 14.51 SCOLOCHLOA Link[2]

Pl per; strongly rhz, rhz succulent. Clm 70–200 cm, smt rooting at the lo nd. Shth open; aur absent; lig memb, truncate to rounded; bld 4–12 mm wide, flat. Infl tml, open pan, some br longer than 1 cm. Spklt pedlt, lat compressed, with 3–4 flt; rchl prolonged beyond the base of the distal flt, intnd glab or with a few hairs to 0.2 mm long; dis above the glm, between the flt. Glm unequal, acute to acuminate, unawned; lo glm shorter than the adjacent lm, 1–5-veined; up glm usu exceeding, smt equaling, the distal flt, 3–7-veined; cal short, blunt, hairy; lm glab, rounded, 3–9-veined, veins excurrent, apc indistinctly 3-lobed or toothed, unawned; pal about

as long as the lm; lod free, glab; anth 3; ov with pubescent apc. Car shorter than the lm, concealed at maturity, about 2 mm, dorsiventrally compressed.

*Scolochloa* is a unispecific genus that grows in shallow water and marshes of the temperate regions of North America and Eurasia. In the *Manual* region, it grows primarily in the northern Great Plains and prairie pothole region, from British Columbia to Manitoba and south through Nebraska and Iowa, with disjunct populations in Alaska, eastern Oregon, Montana, Idaho, and Wyoming. Other outlying populations probably reflect introductions.

[1]Thomas F. Daniel   [2]Mary E. Barkworth

**1. Scolochloa festucacea (Willd.) Link** COMMON
RIVERGRASS, WHITETOP [p. *421*, <u>528</u>]

*Scolochloa festucacea* grows in ponds, marshes, seasonally flooded basins, and the shallow margins of freshwater to moderately saline lakes and streams. It provides good nesting cover for some waterfowl and shorebirds, and can provide valuable forage for livestock and wildlife.

# 14.52 AVENA L.[1]

**Pl** ann or per. **Clm** 8–200 cm, erect or decumbent. **Shth** open; **aur** absent; **lig** memb; **bld** usu flat, smt involute, lax. **Infl** pan, diffuse, smt 1-sided, some br longer than 1 cm. **Spklt** 15–50 mm, pedlt, lat compressed, with 1–6(8) flt; **rchl** not prolonged beyond the base of the distal flt; **dis** above the glm, usu also beneath the flt, cultivated forms not disarticulating. **Glm** usu exceeding the flt, memb, glab, 3–11-veined, acute, unawned; **cal** rounded to pointed, with or without hairs; **lm** usu indurate and enclosing the car at maturity, 5–9-veined, often with twisted, strigose hairs below midlength, apc dentate to bifid or biaristate, usu awned, smt unawned, awns dorsal, usu once-geniculate and strongly twisted in the bas portion; **pal** bifid or entire, keels ciliate; **lod** 2, free, glab, toothed or not toothed; **anth** 3; **ov** hairy. **Car** shorter than the lm, concealed at maturity, terete, ventrally grooved, pubescent; **hila** linear.

*Avena*, a genus of 29 species, is native to temperate and cold regions of Europe, North Africa, and central Asia; it has become nearly cosmopolitan through the cultivation of cereal oats, and the inadvertent introduction of the weedy species. Six species have been introduced into the *Manual* region.

Reports of *Avena strigosa* Schreb. from California are based on misidentifications. The specimens involved belong to *A. barbata*.

1. Florets not disarticulating from the glumes, remaining attached to the plant even at maturity; calluses glabrous . . . . . . . . . . . . . . . . . . . . . . . . . . . . . . . . . . . . . . . . . . . . . . . . . . . . . . . . . . . . . . . . 5. *A. sativa*
1. Florets disarticulating at maturity, only the glumes remaining attached; calluses bearded.
  2. Florets falling from the glumes as a unit . . . . . . . . . . . . . . . . . . . . . . . . . . . . . . . . . . . . . . . . . 6. *A. sterilis*
  2. Florets falling separately.
    3. Lemma apices biaristate, 2 veins extending 2–4 mm beyond the apices . . . . . . . . . . . . . . . . . . . . . . 1. *A. barbata*
    3. Lemma apices erose to bifid, the veins not extending beyond the apices.
      4. Spikelets with 2(3) florets; disarticulation scar of the lower florets in a spikelet round to oval or triangular, that of the third floret, if present, similar . . . . . . . . . . . . . . . . . . . . . . 2. *A. fatua*
      4. Spikelets with 2–4(5) florets; disarticulation scar of the lower florets in a spikelet round to elliptic, those of the third and fourth florets (and sometimes the second) heart-shaped.
        5. Glumes 15–23 mm long . . . . . . . . . . . . . . . . . . . . . . . . . . . . . . . . . . . . . . . . . . . . . 3. *A. hybrida*
        5. Glumes 28–40 mm long . . . . . . . . . . . . . . . . . . . . . . . . . . . . . . . . . . . . . . . . 4. *A. occidentalis*

**1. Avena barbata** Pott *ex* Link SLENDER OATS, SLENDER WILD OATS [p. *422*, <u>528</u>]

*Avena barbata* is native to the Mediterranean region and central Asia. It has become naturalized in western North America, particularly California, displacing native grasses. It was collected once in Vancouver, British Columbia.

**2. Avena fatua** L. WILD OATS, FOLLE AVOINE [p. *422*, <u>528</u>]

*Avena fatua* is native to Europe and central Asia. It is known as a weed in most temperate regions of the world; it is considered a noxious weed in some parts of Canada and the United States. It differs from *A. occidentalis* in having shorter, wider spikelets, fewer florets, and a distal floret which does not have a heart-shaped disarticulation scar. Hybrids between *A. fatua* and *A. sativa* are common in plantings of cultivated oats. The hybrids resemble *A. sativa*, but differ in having the *fatua*-type lodicule; some also have a weak awn on the first lemma. They are easily confused with fatuoid forms of *A. sativa*.

**3. Avena hybrida** Peterm. [p. *422*, <u>528</u>]

*Avena hybrida* is native to western and central Asia; it grows as a weed in Europe. It has been reported from Essex County, Massachusetts, and Prince Edward Island.

**4. Avena occidentalis** Durieu WESTERN OATS [p. *422*, <u>528</u>]

*Avena occidentalis* is native to the Canary Islands, coastal North Africa, and Saudi Arabia; it is now established in western North America, from California to northern Mexico. It differs from *A. fatua* in its longer, narrower spikelets, greater number of florets, and the heart-shaped disarticulation scars of the distal florets.

**5. Avena sativa** L. OATS, CULTIVATED OATS, NAKED OATS, AVOINE, AVOINE CULTIVÉE [p. *422*, <u>528</u>]

*Avena sativa*, a native of Eurasia, is widely cultivated in cool, temperate regions of the world, including North America. Fall-sown oats are planted in the Pacific and southern states in the United States; spring-sown oats are more important elsewhere in North America. It is sometimes planted as a fast-growing soil stabilizer along roadsides. It hybridizes readily with *A. fatua*, forming hybrids with the *fatua*-type lodicule. The hybrids are easily confused with fatuoid forms of *A. sativa*, which differ in having the *sativa*-type lodicule.

**6. Avena sterilis** L. ANIMATED OATS, AVOINE STÉRILE [p. *422*, <u>528</u>]

*Avena sterilis* is native from the Mediterranean region to Afghanistan; it now grows on all continents. It has become naturalized in California, where it can be found in fields, vineyards, orchards, and on hillsides. It has been reported from Ontario. It is listed as a noxious weed by the U. S. Department of Agriculture.

[1]Bernard R. Baum

# 14.53 HOLCUS L.[1]

Pl usu per, rarely ann; csp or rhz, rarely both csp and rhz. Clm (8)20–200 cm, glab or pubescent; nd glab or retrorsely pubescent. Shth open; aur absent; lig 1–5 mm, memb, entire or erose-ciliate, glab or puberulent; bld flat, pubescent. Infl tml pan, contracted to open. Spklt lat compressed, with 2(3) flt, lo flt bisx, up flt(s) stmt or strl; rchl curved below the lowest flt, smt prolonged beyond the base of the distal flt; dis below the glm. Glm equaling to exceeding the flt, strongly keeled, unawned; lo glm 1-veined; up glm 3-veined; cal glab or pubescent; lm firm, shiny, glab or pubescent, obscurely 3–5-veined,

often bidentate; lo lm unawned; up lm awned from below the apc, awns hooked or geniculate; pal thin, subequal to the lm; lod 2, glab, toothed or not; anth 3; ov glab. Car shorter than the lm, concealed at maturity, glab.

*Holcus*, a genus of eight species, is native to Europe, North Africa, and the Middle East. One species, *H. lanatus*, has become widely naturalized in the Americas, Japan, and Hawaii; a second, *H. mollis*, has become a troublesome weed in some areas of the *Manual* region.

1. Awns 1–2 mm long, forming a curved hook at maturity; culms densely pilose adjacent to the lower
   nodes; plants cespitose, not rhizomatous . . . . . . . . . . . . . . . . . . . . . . . . . . . . . . . . . . . . . . . . . . . . . . . . . 1. *H. lanatus*
1. Awns 3–5 mm long, straight or geniculate at maturity; culms glabrous or sparsely pubescent adjacent
   to the lower nodes; plants not cespitose, rhizomatous . . . . . . . . . . . . . . . . . . . . . . . . . . . . . . . . . . . . . . 2. *H. mollis*

### 1. Holcus lanatus L. Velvetgrass, Yorkshire Fog, Houlque Laineuse [p. 422, 528]

*Holcus lanatus*, a native of Europe, grows in disturbed sites, moist waste places, lawns, and pastures, in a wide range of edaphic conditions and at elevations from 0–2300 m.

### 2. Holcus mollis L. Creeping Velvetgrass, Houlque Molle [p. 422, 528]

*Holcus mollis* grows in moist soil and disturbed sites, including lawns and damp pastures. It is a European introduction that has persisted in

the *Manual* region, becoming a problematic weed in ungrazed pastures, prairie remnants, and oak savannahs in portions of the Pacific Northwest. It is also sold as an ornamental. There are two subspecies: Holcus mollis L. subsp. *mollis* (stems not thickened and tuberous at the base; panicles lax, brownish or purplish) and H. mollis subsp. *reuteri* (Boiss.) Malag. (stems thickened and tuberous at the base; panicles narrow, whitish). North American introductions belong to subsp. *mollis*.

# 14.54 ARRHENATHERUM P. Beauv.[2]

Pl per; csp, smt rhz. Clm 30–200 cm, bas intnd occ globose. Shth open, not overlapping; aur absent; lig memb, smt ciliate; bld flat or convolute. Infl tml, narrow pan; br spreading until after anthesis, then becoming loosely appressed to the rchs. Spklt pedlt, lat compressed, with 2 flt, lo flt stmt, up flt pist or bisx, a rdmt flt occ present distally; rchl pubescent; dis above the glm, the flt usu falling together, rarely falling separately. Glm unequal, hyaline, unawned; lo glm less than 3/4 the length of the up glm, 1- or 3-veined; up glm 3-veined; cal short, blunt, pubescent; lo lm memb, 3–7-veined, acute, awned below midlength, awns twisted

and geniculate; up lm memb to subcoriaceous, glab or hairy, 7-veined, acute, usu unawned, smt awned from near the apc, awns short, straight, rarely awned similarly to the lo lm; pal subequal to the lm, 2-veined, 2-keeled, keels scabrous or hairy, apc notched; lod 2, free, linear, memb, glab, entire; anth 3, 3.4–6.5 mm; ov pubescent. Car shorter than the lm, concealed at maturity, not grooved, dorsally compressed to terete, hairy; hila long-linear.

*Arrhenatherum* is a Mediterranean and eastern Asian genus of six species; one has become established in North America.

### 1. Arrhenatherum elatius (L.) P. Beauv. *ex* J. Presl & C. Presl Tall Oatgrass, Fenasse, Fromental [p. 422, 528]

*Arrhenatherum elatius* is grown as a forage grass, and yields palatable hay. It readily escapes from cultivation, and can be found in mesic to dry meadows, the edges of woods, streamsides, rock outcrops, and disturbed areas such as fields, pastures, fence rows, and roadsides. Variegated forms are cultivated as ornamentals. There are two subspecies, both of which have been found in the *Manual* region. Arrhenatherum elatius (L.) P. Beauv. *ex* J. Presl & C. Presl subsp.

elatius has glabrous nodes and basal internodes 2–4 mm thick. It is more common than Arrhenatherum elatius subsp. bulbosum (Willd.) Schübl. & G. Martens, which has densely hairy nodes, and swollen basal internodes 5–10 mm thick. While both can be weedy, the latter subspecies is especially difficult to control in cultivated fields, as tilling the soil spreads the swollen internodes, which then propagate vegetatively.

# 14.55 CORYNEPHORUS P. Beauv.[3]

Pl ann or per. Clm to 60 cm, erect. Shth open; aur absent; lig memb; bld flat or involute. Infl tml pan. Spklt pedlt, lat compressed, with 2 bisx flt; rchl prolonged beyond the base of the distal flt, pilose; dis

above the glm, beneath the flt. Glm subequal, exceeding the flt, lanceolate, memb, acute, unawned; lo glm 1-veined; up glm 3-veined at the base; cal pilose; lm about 1/2 as long as the glm, ovate-lanceolate, memb, 1-veined,

[1]Lisa A. Standley  [2]Stephan L. Hatch  [3]John W. Thieret†

acute, awned from just above the base, awns geniculate, articulated near the middle, with a ring of conical protuberances near the joint, proximal segment yellow-brown to dark brown, smooth, thicker than distal segment, distal segment pale green to whitish, clavate; **pal** shorter than the lm, memb; **lod** 2, free, glab, toothed; **anth** 3; **ov** glab. **Car** shorter than lm, concealed

### 1. Corynephorus canescens (L.) P. Beauv. GRAY HAIRGRASS [p. *423*, *529*]

*Corynephorus canescens* is native to Europe. It grows on coastal sand dunes and on sandy soils inland, and in disturbed areas. It has been

at maturity, usu adhering to the lm and/or pal, longitudinally grooved.

*Corynephorus* is a Eurasian and North African genus of five species. One has been introduced into the *Manual* region. Flowering specimens are easily recognized, but sterile plants, with their involute, glaucous leaves, resemble involute-leaved species of *Festuca*, such as *F. trachyphylla*.

recorded from scattered locations in North America, but its current status in these locations is not known.

### 14.56 TRISETUM Pers.[1]

**Pl** ann or per; smt rhz, smt csp. **Clm** 5–150 cm, glab or pubescent, bas brchg exvag. **Shth** open the entire length or fused at the base; **aur** absent; **lig** memb, often erose to lacerate, smt ciliolate; **bld** rolled in the bud. **Infl** tml pan, open and diffuse to dense and spikelike; **br** antrorsely scabrous. **Spklt** 2.5–12 mm, usu subsessile to pedlt, rarely sessile, lat compressed, with 2–5 flt, rdcd flt smt present distally; **rchl** hairy, intnd evident, prolonged beyond the base of the distal bisx flt; **dis** usu initially above the glm and beneath the flt, subsequently below the glm, in some species initially below the glm. **Glm** subequal or unequal, keels scabrous, apc usu acute, unawned, often apiculate; **lo glm** 1(3)-veined; **up glm** 3(5)-veined, lat veins less than ¹/₂ the glm length; **cal** hairy; **lm** 3–7-veined, mrg hyaline, unawned or awned from above the middle with a single awn, apc usu bifid,

smt entire; **pal** from subequal to longer than the lm, memb, 2-veined, veins usu extending as bristlelike tips; **lod** 2, shallowly and usu slenderly lobed to fimbriate; **anth** 3; **ov** glab or pubescent; **sty** 2. **Car** shorter than the lm, concealed at maturity, elongate-fusiform, compressed, brown; **emb** elliptic, to ¹/₃ the length of the car; **endosperm** milky.

*Trisetum* has about 75 species and grows in temperate, subarctic, and alpine regions. ten are native to the *Manual* region; two have been introduced. Barkworth has modified this treatment to reflect the treatment of *Trisetum* in Finot et al. (2005). *Trisetum* differs from *Sphenopholis* in its combination of longer awns and spikelets that disarticulate above the glumes, and from *Deschampsia* in its more acute, distally awned, bifid lemmas. In *Deschampsia*, the awns are usually inserted at or below midlength, often near the base.

1. Lemmas entire or slightly bilobed, unawned or with straight awns up to 2 mm long that scarcely exceed the lemma apices [*Graphephorum* Finot].
  2. Panicles usually 2–4 cm wide, open, nodding; callus hairs 1.3–2.5 mm long . . . . . . . . . . . .1. *T. melicoides*
  2. Panicles usually 1–1.5 cm wide, contracted, erect; callus hairs up to 1 mm long . . . . . . . . . . . . . .2. *T.wolfii*
1 Lemma apices bidentate, lateral veins extending into 2(4) teeth, at least some lemmas with a dorsal awn 1.3–16 mm long [*Trisetum*].
  3. Plants annual, without sterile shoots.
    4. Disarticulation below the glumes; lower glumes 3-veined; spikelets 3–6 mm; plants native . . . . . . . . . . . . . . . . . . . . . . . . . . . . . . . . . . . . . . . . . . . . .10. *T. interruptum*
    4. Disarticulation above the glumes; lower glumes 1-veined; spikelets 2.5–3.5 mm; plants introduced, not established . . . . . . . . . . . . . . . . . . . . . . . . . . . . . . . . . . . . . . .12. *T. aureum*
    5. Plants rhizomatous, culms usually solitary.
      6. Lemma teeth 3–6 mm; ligules 0.5–1(1.2) mm . . . . . . . . . . . . . . . . . . . . . .11. *T. flavescens*
      6. Lemma teeth usually shorter than 1 mm; ligules 1–5 mm.
        7. Awns 4–6 mm, straight to curved; culms 80–110 cm . . . . . . . . . . . . . . .3. *T. orthochaetum*
        7. Awns 6–8 mm, usually geniculate and twisted; culms 15–65 cm . . . . . . . . . . . . . .6. *T. sibiricum*
    5. Plants not rhizomatous; culms clumped.
      8. Culms densely hairy below the panicles, hairs (0.2)0.5–1 mm; anthers 0.5–1 mm . . . .7. *T. spicatum*
      8. Culms glabrous below the panicles or, if hairy, hairs about 0.1 mm; anthers 0.5–3 mm.
        9. Awns 3.5–4.5 mm; anthers 0.8–1.2 mm; ovaries glabrous . . . . . . . . . . . . . . . . . .9. *T. montanum*
        9. Awns 5–16 mm; anthers 0.5–3 mm; ovaries glabrous or hairy near the apices.
          10. Leaf blades densely pilose on both surfaces; upper glumes equaling or exceeding the florets; panicles dense, individual branches not apparent; ligules densely pilose on dorsal surface . . . . . . . . . . . . . . . . . . . . . . . .8. *T. projectum*

[1]Mary E. Barkworth and John H. Rumely

10. Leaf blades glabrous, canescent, or sparsely pilose on the adaxial surface,
glabrous or canescent on the abaxial surface; ligules glabrous or pilose, but
not densely pilose, on the dorsal surface.
  11. Anthers 2–3 mm; ovaries glabrous . . . . . . . . . . . . . . . . . . . . . 11. *T. flavescens* (in part)
  11. Anthers 0.5–1.8 mm; ovaries with hairs near the apices.
    12. Most panicle branches spikelet-bearing their full length, ascending
    to somewhat divergent; lower glumes 3–5 mm . . . . . . . . . . . . . . . . . 5. *T. canescens*
    12. Most panicle branches spikelet-bearing only distally, branches at
    lowest 1–3 whorls strongly divergent to reflexed . . . . . . . . . . . . . . . 5. *T. cernuum*

### 1. Trisetum melicoides (Michx.) Scribn. FALSE MELIC, TRISÈTE FAUSSE-MÉLIQUE [p. 423, 529]

*Trisetum melicoides* is a native species that grows on moist, cool stream banks, gravelly shores, shaded rock ledges (especially calcareous ones), and in damp woods. It grows only in southeastern Canada and the northeastern United States. It is listed as endangered in Wisconsin, New York, and Maine. Finot et al. (2005) treat this species as *Graphephorum melicoides* (Michx.) Desv.

### 2. Trisetum wolfii Vasey WOLF'S TRISETUM [p. 423, 529]

*Trisetum wolfii* grows in moist meadows and marshes, and on stream banks in aspen groves and parks in the spruce-fir forest zone, at medium to high, but usually not alpine, elevations. It is restricted to southwestern Canada and the western United States. Finot et al. (2005) treat this species as *Graphephorum wolfii* (Vasey) Vasey *ex* Coult.

### 3. Trisetum orthochaetum Hitchc. BITTERROOT TRISETUM [p. 423, 529]

*Trisetum orthochaetum* is known only from Montana, in or near the edges of marshes, seeps, and creeksides, at about 1465 m.

### 4. Trisetum canescens Buckley TALL TRISETUM [p. 423, 529]

*Trisetum canescens* grows on or near stream banks and in forest margins or interiors, in moist to dry areas in the western *Manual* region. It is especially abundant in ponderosa pine stands and spruce-fir forests. The vestiture of different parts varies throughout the species range.

### 5. Trisetum cernuum Trin. NODDING TRISETUM [p. 423, 529]

*Trisetum cernuum* grows in moist woods, on stream banks, lake and pond shores, and floodplains of the western *Manual* region. The hairiness of the leaf sheaths varies, often within a plant.

### 6. Trisetum sibiricum Rupr. SIBERIAN TRISETUM [p. 423, 529]

*Trisetum sibiricum* grows on coastal beaches and creek banks, and in moist meadows and open forests, from sea level to 300+ m. It is often abundant and has significant value as a pasture plant. Circumpolar in distribution, in the *Manual* region it grows in Alaska and the Yukon Territory. Most North American plants belong to **Trisetum sibiricum** subsp. **litorale** Rupr. *ex* Roshev., having culms 15–30 cm tall, leaf blades 2.5–4 mm wide, panicles 3–5 cm long and 2–3 cm wide, branches to 2 cm long, and lemma awns 5–8 mm long. **Trisetum sibiricum** Rupr. subsp. **sibiricum** grows in the Yukon Territory and Eurasia. It differs from *T. spicatum* in its smooth culms and leaves, and its broad, less dense panicles.

### 7. Trisetum spicatum (L.) K. Richt. SPIKE TRISETUM, TRISÈTE À ÉPI [p. 423, 529]

*Trisetum spicatum* grows from Alaska to Greenland and south to California, New Mexico, and the northern United States at 150–3750 m. Plants with glabrous glumes belong to **T. spicatum** (L.) K. Richt. var. **spicatum**. They grow in western North America. Plants with hairy glumes belong to **T. spicatum** var. **pilosiglume** Fernald. They grow in eastern Canada, Greenland, and the northeastern United States.

### 8. Trisetum projectum Louis Marie SIERRAN TRISETUM [p. 423, 529]

*Trisetum projectum* is restricted to the western United States, usually in dry woods, at 1200–2900 m. Its pilose leaf sheaths, glabrous culms, and yellowish panicles distinguish it from *T. spicatum*. Unlike *T. canescens* and *T. cernuum*, it has glabrous ovaries.

### 9. Trisetum montanum Vasey MOUNTAIN TRISETUM [p. 424, 529]

*Trisetum montanum* grows in moist to wet, loam to rocky soils, in valleys and on mountain slopes at 1900–4500 m from Yukon Territory to Arizona and New Mexico. Its culms are usually glabrous, but occasionally canescent, beneath the panicle and it usually has a less dense panicle than *T. spicatum*.

### 10. Trisetum interruptum Buckley PRAIRIE TRISETUM [p. 424, 529]

*Trisetum interruptum* grows in open, dry or moist soil in deserts, plains, arid shrublands, and riparian woodlands, from the southern United States into Mexico. It is often weedy. Finot et al. (2004) treat this species as *Sphenopholis interrupta* (Buckley) Scribn. subsp. *interrupta*.

### 11. Trisetum flavescens (L.) P. Beauv. YELLOW OATGRASS, AVOINE JAUNÂTRE [p. 424, 529]

*Trisetum flavescens* grows in seeded pastures, roadsides, and as a weed in croplands. Native to Europe, west Asia, and north Africa, it was introduced into the *Manual* region because of its drought resistance, wide soil tolerance, and high palatability to domestic livestock. It contains calcinogenic glycosides, which can lead to vitamin D toxicity in grazing animals.

### 12. Trisetum aureum (Ten.) Ten. GOLDEN OATGRASS [p. 424, 529]

*Trisetum aureum* is native to the Mediterranean region. It was collected from a ballast dump in Camden, New Jersey, in 1896. It has not been reported since from the *Manual* region.

## 14.57 KOELERIA Pers.[1]

Pl per; usu csp, smt weakly rhz. Clm 5–130 cm, erect. Shth open; aur absent; lig memb; bld flat to involute, pubescent or glab. Infl pan, erect, usu dense and spikelike, smt lax, stiffly and narrowly pyramidal at anthesis; main rchs and br smooth, softly hairy. Spklt lat compressed, with 2–4 flt; rchl to 1 mm, glab or pubescent, usu prolonged beyond the base of the distal flt, or with a vestigial flt; dis above the glm, beneath the flt. Glm subequal to or slightly exceeding the lm, memb, scabrid to tomentose, keels smt ciliate, unawned; lo glm 1-veined, somewhat narrower and shorter than the up glm; up glm obscurely 3(5)-veined; cal glab or hairy; lm

[1]Lisa A. Standley

thin, memb, 5-veined, mrg shiny, scarious, apc acute, smt mucronate or awned; **pal** equaling or subequal to the lm, hyaline; **lod** 2, glab, toothed; **anth** 3; **ov** glab. Car glab.

*Koeleria* is a cosmopolitan genus of about 35 species that grow in dry grasslands and rocky soils; two are native to the

*Manual* region. *Koeleria pyramidata* is sometimes said to be in the *Manual* region; such reports seem to reflect a different interpretation of the species. *Koeleria* sometimes includes the genus *Rostraria*, which differs in its annual growth habit, and in having awned lemmas and paleas.

1. Lemmas pubescent to densely tomentose, purple to almost black; panicles 1–5 cm long; culms 5–35 cm tall, finely pubescent throughout . . . . . . . . . . . . . . . . . . . . . . . . . . . . . . . . . . . . . . . . . . . . . . . . 1. *K. asiatica*
1. Lemmas usually glabrous, usually green when young, sometimes purple-tinged, stramineous at maturity; panicles 4–27 cm long; culms 20–130 cm tall, mostly glabrous, pubescent near the nodes and sometimes below the panicle.
   2. Spikelets 2.5–6.5 mm long; old sheaths usually breaking off with age or, if disintegrating, the fibers straight or nearly so; margins of the basal leaf blades glabrous or with hairs usually shorter than 1 mm near the base . . . . . . . . . . . . . . . . . . . . . . . . . . . . . . . . . . . . . . . . . . . . . . . . 2. *K. macrantha*
   2. Spikelets 6–10 mm long; old sheaths weathering to wavy, curled, or arched fibers; margins of the basal leaf blades frequently with hairs longer than 2 mm near the base . . . . . . . . . . . . . . . . . . . 3. *K. pyramidata*

### 1. Koeleria asiatica Domin EURASIAN JUNEGRASS [p. 424, 529]

*Koeleria asiatica* grows on the gravel bars of creeks, dry tundra, and scree slopes. Its range extends from the Ural Mountains through the Kamchatka Peninsula to northern Alaska and northwestern Canada.

### 2. Koeleria macrantha (Ledeb.) Schult. JUNEGRASS, KOELERIE ACCRÊTÉ, KOELERIE À CRÊTES [p. 424, 529]

*Koeleria macrantha* is widely distributed in temperate regions of North America and Eurasia. In North America, it grows in semi-arid to mesic conditions, on dry prairies or in grassy woods, generally in

sandy soil, from sea level to 3900 m. It is treated here as a polymorphic, polyploid complex. It differs from *Sphenopholis intermedia* in its less open panicles, and in having spikelets that disarticulate above the glumes.

### 3. Koeleria pyramidata (Lam.) P. Beauv. CRESTED HAIRGRASS [p. 424]

*Koeleria pyramidata*, as interpreted here, is confined to Europe. Some North American records for *K. pyramidata* are based on robust specimens of *K. macrantha*; others reflect an interpretation of *K. pyramidata* that includes *K. macrantha*.

## 14.58 ROSTRARIA Trin.[1]

**Pl** ann; tufted or with solitary clm. **Clm** 3–60 cm, erect. **Lvs** mostly cauline; **shth** open, glab, pubescent, or pilose; **aur** absent; **lig** memb; **bld** flat or involute, stiff. **Infl** spikelike pan, dense; **br** scabrous to pubescent. **Spklt** lat compressed, with 2–7 flt; **rchl** sparsely to moderately pubescent, prolonged or not beyond the base of the distal flt; **dis** above the glm, beneath the flt. **Glm** unequal, memb, glab or hirsute, keels ciliate, apc unawned; **lo glm** shorter and narrower than the up glm, 1-veined; **up glm** subequal to the lowest lm, 3-veined;

cal usu glab, occ sparsely hairy; **lm** thin, memb, 5-veined, glab or hirsute, mrg shiny, scarious, apc acute, bifid, subtml awned or unawned; **pal** subequal to the lm, hyaline, veins smt extended into awnlike apc; **anth** 3; **ov** glab. Car glab.

*Rostraria* is a genus of approximately 10 species, all of which are native to the Mediterranean, southeastern Europe, and western Asia, where they grow in dry, disturbed sites. The genus is sometimes included in *Koeleria*; it differs in its annual growth habit and awned lemmas and paleas.

### 1. Rostraria cristata (L.) Trin. MEDITERRANEAN HAIRGRASS [p. 424, 529]

*Rostraria cristata* is native to Europe. In the *Manual* region, it is found in California, around the Gulf of Mexico, and in New York State. Plants with hirsute glumes and lower lemmas belong to **Rostraria**

cristata (L.) Trin. var. **cristata**; those with glumes and lemmas that are almost glabrous apart from their ciliate keels belong to **R. cristata** var. **glabriflora** (Trautv.) Doğan.

## 14.59 MIBORA Adans.[2]

**Pl** ann; tufted. **Clm** 2–15 cm tall, to 0.3 mm wide, erect. **Lvs** primarily bas; **shth** closed almost to the top; **aur** absent; **lig** about 1 mm, hyaline, truncate; **bld** 0.3–1 mm wide, flat or involute. **Infl** single, tml, spikelike rcm; **rchs** smooth, glab, with 1 spklt per nd; **ped** 0.2–1 mm long, 0.1–0.2 mm thick. **Spklt** imbricate, in 2 rows on 1 side of the rchs, slightly lat compressed, with 1 flt; **rchl** not prolonged beyond the base of the flt; **dis** above the glm, beneath the flt. **Glm** subequal, exceeding the flt, memb, glab, flexible, smooth, 1-veined, rounded on the back,

unawned; **cal** glab; **lm** about $2/3$ the length of the spklt, elliptical in side view, thinner than the glm, 5-veined, shortly and densely pubescent throughout, unawned, apc truncate, often denticulate; **pal** about as long as the lm, 2-veined, shortly and densely pubescent between the inconspicuous keels; **lod** 2; **anth** 3; **ov** glab; **sty** fused at the base, dividing into 2 feathery stigmas. **Car** elliptical, terete, $2/3$ the length of the lm, enclosed by, but not fused to, the lm and pal at maturity; **emb** about $1/5$ the length of the car, elliptic; **hila** punctate, bas.

[1]Lisa A. Standley  [2]Hans J. Conert

*Mibora* is a genus of two species, both of which grow in damp, sandy soils. *Mibora minima* has been introduced to the *Manual* region, and also grows throughout much of western Europe. The second species, *M. maroccana* (Maire) Maire, is restricted to northwest Africa.

### 1. Mibora minima (L.) Desv. EARLY SANDGRASS [p. 425, 529]

*Mibora minima* is native to western Europe, where it grows on sandy and other light, damp soils in places with mild winters. It was collected from ballast dumps in Plymouth, Massachusetts, in the first part of the twentieth century, and from an experimental farm in Sydney, British Columbia, in 1914. It has also been collected in Monroe County, New York. None of these populations appear to have led to establishment of the species in North America.

## 14.60 ANTHOXANTHUM L.[1]

Pl per or ann; densely to loosely csp, smt rhz; fragrant. Clm 4–100 cm, erect or geniculate, smt brchd; intnd hollow. Lvs cauline or bas concentrated, glab or softly hairy; **shth** open; **aur** absent or present; **lig** memb, smt shortly ciliate or somewhat erose; **bld** flat or rolled, glab or sparsely pilose. Infl open or contracted pan, smt spikelike. Spklt pedlt or sessile, 2.5–10 mm, lat compressed, stramineous to brown at maturity, with 3 flt, lowest 2 flt stmt or rdcd to dorsally awned lm subequal to or exceeding distal flt, distal flt bisx, unawned; **rchl** not prolonged beyond the base of the distal flt; **dis** above the glm, the flt falling together. Glm unequal or subequal, equaling or exceeding the flt, lanceolate to ovate, glab or pilose, keeled; **cal** blunt, glab or hairy. Lowest 2 flt: lm strongly compressed, 3-veined, strigose, hairs brown, apc bilobed, unawned or dorsally awned. Distal flt: lm somewhat indurate, glab or pubescent, shiny, inconspicuously 3–7-veined, unawned; pal 1-veined, enclosed by the lm; **lod** 2 or absent; **anth** 2 or 3. Car shorter than the lm, concealed at maturity, tightly enclosed in the flt; **hila** less than 1/3 the length of the car, oval.

*Anthoxanthum* is a cool-season genus of about 50 species that grow in temperate and arctic regions throughout the world. There are seven species in the *Manual* region, five of which are native.

This treatment merges what have traditionally be treated as two genera, *Anthoxanthum* and *Hierochloë*. *Phalaris* resembles *Anthoxanthum sensu lato* in its spikelet structure, differing only in the greater reduction of the lower florets.

1. Glumes unequal, the lower glumes shorter than the upper glumes; lowest 2 florets sterile.
   2. Plants annual; ligules 1–3 mm long; blades 1–5 mm wide; panicles 1–4 cm long . . . . . . . . . . . . . . . . . . 1. *A. aristatum*
   2. Plants perennial; ligules 2–7 mm long; blades 3–10 mm wide; panicles 3–14 cm long . . . . . . . . . . . . . . . 2. *A. odoratum*
1. Glumes subequal; lowest 2 florets staminate.
   3. Staminate lemmas awned, the awns of the upper staminate florets 4.5–10.5 mm long; plants densely to loosely tufted, with rhizomes rarely more than 2 cm long . . . . . . . . . . . . . . . . . . . . . . . . . 3. *A. monticola*
   3. Staminate lemmas unawned or with an awn no more than 1 mm long; plants long-rhizomatous.
      4. Panicles spikelike, 0.3–0.5 cm wide, with 1–2 spikelets per branch; rhizomes 0.3–1 mm thick; plants of the high arctic . . . . . . . . . . . . . . . . . . . . . . . . . . . . . . . . . . . . . . . . . . 4. *A. arcticum*
      4. Panicles not spikelike, 1–10 cm wide, the longer branches usually with 3+ spikelets; rhizomes 0.7–3 mm thick; plants non-arctic or arctic.
         5. Lower staminate lemmas in each spikelet narrowly elliptic, lengths more than 5 times widths; glumes equaling or slightly exceeded by the apices of the bisexual florets; blades 3–15 mm wide . . . . . . . . . . . . . . . . . . . . . . . . . . . . . . . . . . . . . . . . . . . . . . . . . . . . 5. *A. occidentale*
         5. Lower staminate lemmas in each spikelet elliptic, lengths usually no more than 4 times widths; glumes exceeding the bisexual florets; blades 2–8 mm wide.
            6. Hairs on the distal portion of the bisexual florets mostly shorter than 0.5 mm, longer hairs, if present, concentrated near the midvein . . . . . . . . . . . . . . . . . . . . . . . . . . . . . . 6. *A. nitens*
            6. Hairs on the distal portion of the bisexual florets 0.5–1 mm long, evenly distributed around the apices . . . . . . . . . . . . . . . . . . . . . . . . . . . . . . . . . . . . . . . . . . . 7. *A. hirtum*

### 1. Anthoxanthum aristatum Boiss. VERNALGRASS [p. 425, 529]

*Anthoxanthum aristatum* is native to Europe. It is now established but not common in the *Manual* region, being found in mesic to dry, open, disturbed habitats of western and eastern North America. North American plants belong to **Anthoxanthum aristatum** Boiss. subsp. **aristatum**, which differs from **Anthoxanthum aristatum** subsp. **macranthum** Valdés in having well-exserted awns and deeply bifid, sterile lemmas. Another annual species of *Anthoxanthum*, *A. gracile* Biv., is occasionally cultivated for dry bouquets, but it does not appear to be widely available at present. It differs from *A. aristatum* in having longer (10–12 mm) spikelets and simple or sparingly branched culms.

### 2. Anthoxanthum odoratum L. SWEET VERNALGRASS, FLOUVE ODORANTE, FOIN D'ODEUR [p. 425, 529]

*Anthoxanthum odoratum* is native to southern Europe. In the *Manual* region, it grows in meadows, pastures, grassy beaches, old hay fields, waste places, and openings in coniferous forests, occasionally in dense shade or as a weed in lawns. It is most abundant on the western and eastern sides of the continent, and is almost absent from the central region.

[1]Kelly W. Allred and Mary E. Barkworth

### 3. Anthoxanthum monticola (Bigelow) Veldkamp

ALPINE SWEETGRASS, HIÉROCHLOÉ ALPINE [p. *425*, *529*]

*Anthoxanthum monticola* is circumpolar, usually growing above or north of the tree line, occasionally in open forests. It occurs sporadically on well-drained, weakly acidic to neutral sand, gravel, and rocky barrens in most of arctic North America; it is not common to the south, even at high elevations. It is listed as threatened or endangered in several parts of its range. There are two subspecies in the *Manual* region.

1. Awns of the upper staminate florets 5–10.5 mm long, attached from near the base to about midlength; awn usually strongly geniculate, the lower portion usually twisted, with 2–4 gyres . . . subsp. *alpinum*
1. Awns of the upper staminate florets 4.5–7 mm long, attached at or above midlength, not or only weakly geniculate, the lower portion not twisted or twisted with 1–2 gyres . . . . . . . . . . . . . . . . . . . . . . . . subsp. *monticola*

### Anthoxanthum monticola subsp. alpinum (Sw. *ex* Willd.) Soreng [p. *425*]

*Anthoxanthum monticola* subsp. *alpinum* is the common subspecies in the *Manual* region, extending from western Alaska to eastern Greenland. It extends south to the Canadian border in the Rocky Mountains. It usually grows above or north of the tree line, in places that are strongly exposed to the wind and have little snow cover during the winter.

### Anthoxanthum monticola (Bigelow) Veldkamp subsp. monticola [p. *425*]

*Anthoxanthum monticola* subsp. *monticola* grows from Greenland to the eastern side of Hudson Bay, through Labrador, and south to northern New England. It usually grows in similar, but wetter and more exposed, habitats than those occupied by subsp. *alpinum*.

### 4. Anthoxanthum arcticum Veldkamp ARCTIC

SWEETGRASS, HIÉROCHLOÉ PAUCIFLORE [p. *425*, *529*]

*Anthoxanthum arcticum* is a coastal and lowland circumpolar species of the Alaskan, Canadian, and Russian arctic; it is absent from Greenland. It generally grows in wet tundra on acidic, peaty soils. In the warmest sectors of the western high arctic, it is rooted in mats of moss that are growing over carbonate substrates.

### 5. Anthoxanthum occidentale (Buckley) Veldkamp

CALIFORNIA SWEETGRASS [p. *425*, *529*]

*Anthoxanthum occidentale* grows in moist to fairly dry forested areas, from Klickitat County, Washington, south to the coastal mountains of San Luis Obispo County, California. Its long flag leaf blades and more elongate spikelet parts make it easier to distinguish from *A. hirtum* than the key suggests.

### 6. Anthoxanthum nitens (Weber) Y. Schouten & Veldkamp VANILLA SWEETGRASS, HIÉROCHLOÉ ODORANTE, FOIN D'ODEUR, HERBE SAINTE [p. *425*, *529*]

*Anthoxanthum nitens* is primarily a European species. In the *Manual* region, it grows along the coast from northern Labrador to New England. It is not known from Greenland, although it grows in Iceland and northwestern Europe. It grows in wet meadows and at the edges of sloughs, marshes, roadsides, and fields. Only **A. nitens** (Weber) Y. Schouten & Veldkamp subsp. **nitens** is present in the region; it is also present in Europe. It differs from **A. nitens** subsp. **balticum** (G. Weim.) G.C. Tucker in being almost always awned.

### 7. Anthoxanthum hirtum (Schrank) Y. Schouten & Veldkamp HAIRY SWEETGRASS [p. *425*, *529*]

*Anthoxanthum hirtum* is the most widely distributed species of *Anthoxanthum* in the *Manual* region, extending from Alaska to northeastern Quebec and south to Washington and Colorado, South Dakota, Illinois, Ohio, and New York. Outside the *Manual* region, it extends from Scandinavia south to Germany and east to Asiatic Russia. It grows in wet meadows and marshes with good water, not in salt- or brackish water. Because much of its native habitat has been drained, it is becoming less common. Its short flag leaf blades and more circular spikelets distinguish it from *A. occidentale*, and the relative abundance and even distribution of hairs longer than 0.5 mm distinguish it from *A. nitens*.

# 14.61 PHALARIS L.[1]

Pl ann or per; smt csp, smt rhz. **Clm** 4–230 cm tall, erect or decumbent, smt swollen at the base, not brchg above the base. **Lvs** more or less evenly distributed, glab; **shth** open for most of their length, upmost shth often somewhat inflated; **aur** absent; **lig** hyaline, glab, truncate to acuminate, entire or lacerate; **bld** usu flat, smt revolute. **Infl** tml pan, smt spikelike, ovoid to cylindrical, dense, smt interrupted, with 10–200 spklt borne singly or in clusters, spklt homogamous in species with single spklt, heterogamous in species with spklt in clusters, lo spklt in the clusters usu stmt, rarely strl, tml spklt bisx or pist. **Spklt** pedlt, lat compressed, with 1–3(4) flt, the tml or only flt usu sexual, lo flt(s), if present, strl; **dis** above the glm, beneath the strl flt in species with solitary spklt, in species with clustered spklt usu at the base of the spklt clusters, smt beneath the bisx or pist spklt. **Glm** subequal, exceeding the flt, 1–5-veined, keeled, keels often conspicuously winged; **lo (strl) flt** rdcd, varying from knoblike projections on the cal of the bisx flt to linear or lanceolate lm less than $^3/_4$ as long as the bisx flt; **tml flt** usu bisx, in the lo spklt of a spklt cluster the tml flt pist or stmt, rarely strl; **lm of tml flt** coriaceous to indurate, shiny, glab or hairy, inconspicuously 5-veined, acute to acuminate or beaked, unawned; **pal** similar to the lm in length and texture, enclosed by the lm at maturity, 1-veined, mostly glab, veins shortly hairy; **lod** absent or 2 and rdcd; **anth** 3; **ov** glab; **sty** 2, plumose. **Car** shorter than the lm, concealed at maturity, with a reticulate prcp, falling free of the lm and pal; **hila** long-linear.

*Phalaris* has 22 species, most of which grow primarily in temperate regions. It is found in a wide range of habitats, although most species prefer somewhat mesic, disturbed areas. There are 11 species in the *Manual* region, 5 native and 6 introduced.

The sterile florets of *Phalaris* are frequently mistaken for tufts of hair at the base of a solitary functional floret. Close examination will reveal that the hairs are actually growing from linear to narrowly lanceolate pieces of tissue. Developmental studies have shown that these structures are reduced lemmas.

[1]Mary E. Barkworth

1. Spikelets in clusters, heterogamous, the lower 4–7 spikelets in each cluster with a staminate (rarely sterile) terminal floret, only the terminal spikelet in the clusters with a pistillate or bisexual terminal floret; disarticulation usually at the base of the spikelet clusters, sometimes beneath the bisexual or pistillate spikelets.

    2. Glumes of the bisexual or pistillate spikelets winged, the wings with 1 prominent tooth; plants annual; culms not swollen at the base . . . . . . . . . . . . . . . . . . . . . . . . . . . . . . . . . . . . . . . . . . . . 1. *P. paradoxa*

    2. Glumes of the bisexual or pistillate spikelets winged, the wings entire or irregularly dentate to crenate distally; plants perennial; culms with swollen bases . . . . . . . . . . . . . . . . . . . . . . . . 2. *P. coerulescens*

1. Spikelets borne singly, homogamous, all spikelets with a bisexual terminal floret; disarticulation above the glumes, beneath the sterile florets.

    3. Glume keels not winged or with wings no more than 0.2 mm wide.

        4. Plants perennial; bisexual florets with acute to somewhat acuminate apices.

            5. Panicles ovoid to cylindrical, 1.5–6 cm long, branches not evident; sterile florets usually more than $^1/_2$ as long as the bisexual florets . . . . . . . . . . . . . . . . . . . . . . . . . . . . . . 8. *P. californica*

            5. Panicles elongate, 5–40 cm long, evidently branched towards the base; sterile florets less than $^1/_2$ as long as the bisexual florets . . . . . . . . . . . . . . . . . . . . . . . . . . . . . . . . . . 9. *P. arundinacea*

        4. Plants annual; bisexual florets with beaked or strongly acuminate apices.

            6. Apices of the bisexual florets glabrous; glumes scabrous over the lateral veins and keels, and adjacent to the keels . . . . . . . . . . . . . . . . . . . . . . . . . . . . . . . . . . . . . . . . . . . . . 7. *P. lemmonii*

            6. Apices of the bisexual florets hairy; glumes smooth or scabridulous over the lateral veins and keels, the wing surface smooth . . . . . . . . . . . . . . . . . . . . . . . . . . . . . 10. *P. caroliniana* (in part)

    3. Glume keels broadly winged, the wings 0.2–1 mm wide.

        7. Sterile florets usually 1, if 2, the lower floret up to 0.7 mm long and the upper floret 1–3 mm long.

            8. Plants annual; sterile florets 1, glabrous or almost so; wings of the glume keels irregularly dentate to crenate, varying within a panicle . . . . . . . . . . . . . . . . . . . . . . . . . . . . . 3. *P. minor*

            8. Plants perennial; sterile florets usually 1, sometimes 2, hairy; wings of the glume keels usually entire . . . . . . . . . . . . . . . . . . . . . . . . . . . . . . . . . . . . . . . . . . . . . . . . . 4. *P. aquatica*

        7. Sterile florets 2, equal to subequal, 0.5–4.5 mm long.

            9. Panicles cylindrical, sometimes lobed; anthers 0.5–1.3 mm long . . . . . . . . . . . . . . . . . . . . . . . . . . . . 11. *P. angusta*

            9. Panicles usually ovoid to ellipsoid or oblong-ovoid, occasionally cylindrical, not lobed; anthers 1.5–4 mm long.

                10. Sterile florets 0.6–1.2 mm long, about $^1/_5$ the length of the bisexual florets . . . . . . . . . . . . 5. *P. brachystachys*

                10. Sterile florets 1.5–4.5 mm long, $^1/_3$ or more the length of the bisexual florets.

                    11. Glumes 7–10 mm long, 2–2.5 mm wide; bisexual florets 4.5–6.8 mm long; anthers 2–4 mm long . . . . . . . . . . . . . . . . . . . . . . . . . . . . . . . . . . . . . . . . . . . 6. *P. canariensis*

                    11. Glumes 3.8–6(8) mm long, 0.8–1.5 mm wide; bisexual florets 2.9–4.7 mm long; anthers 1.5–2 mm long . . . . . . . . . . . . . . . . . . . . . . . . . . . . . . . . . . . . 10. *P. caroliniana* (in part)

### 1. Phalaris paradoxa L. HOODED CANARYGRASS [p. *425*, *529*]

*Phalaris paradoxa* is native to the Mediterranean region; it is now found throughout the world, primarily in harbor areas and near old ballast dumps. It is an established weed in parts of Arizona and California. Within an inflorescence, the most reduced sterile spikelets are located near the base, and the most nearly normal spikelets are near the top.

### 2. Phalaris coerulescens Desf. SUNOLGRASS [p. *425*, *530*]

*Phalaris coerulescens* is native around the Mediterranean; it is now established in northern Europe and South America. It was found in Contra Costa County, California, in 2000.

### 3. Phalaris minor Retz. LESSER CANARYGRASS [p. *426*, *530*]

*Phalaris minor* is native around the Mediterranean and in northwestern Asia, but is now found throughout the world. Even where it is native, it usually grows in disturbed ground, often around harbors and near refuse dumps. Although it has been found at numerous locations in the *Manual* region, it is only established in the southern portion of the region.

The compact panicle with its truncate to rounded base, and the rather variable edges of the glume wings, usually distinguish *Phalaris minor* from other species in the genus.

### 4. Phalaris aquatica L. BULBOUS CANARYGRASS [p. *426*, *530*]

A native of the Mediterranean region, *Phalaris aquatica* now grows in many parts of the world, frequently having been introduced because of its forage value. Even where it is native, it usually grows in disturbed areas, often those subject to seasonal flooding. It is now established in western North America, being most common along the coast, and as an invasive in disturbed wet prairies with clay soils.

### 5. Phalaris brachystachys Link SHORTSPIKE CANARYGRASS [p. *426*, *530*]

*Phalaris brachystachys* is native to the Mediterranean region and the Canary Islands, where it grows on waste ground, at the edges of cultivated fields, and on roadsides. It is adventive in northern Europe, Australia, and North America. It is known from a few locations in the *Manual* region, most of them in California.

**6. Phalaris canariensis** L. ANNUAL CANARYGRASS, PHALARIS DES CANARIES, ALPISTE DES CANARIES [p. 426, 530]

*Phalaris canariensis* is native to southern Europe and the Canary Islands, but is now widespread in the rest of the world, frequently being grown for birdseed. The exposed ends of the glumes are almost semicircular in outline, making this one of our easier species of *Phalaris* to identify.

**7. Phalaris lemmonii** Vasey LEMMON'S CANARYGRASS [p. 426, 530]

*Phalaris lemmonii* is native to California, but it has also been found in Victoria, Australia. It grows in moist areas, and appears to hybridize with both *P. caroliniana* and *P. angusta*. The strongly beaked tips of the bisexual florets are a useful distinguishing feature.

**8. Phalaris californica** Hook. & Arn. CALIFORNIA CANARYGRASS [p. 426, 530]

*Phalaris californica* is native to California and southwestern Oregon. It grows in ravines and on open, moist ground. Records from further north probably represent introductions. The relatively long, sterile florets of *P. californica* distinguish it from other species of *Phalaris* in the *Manual* region.

**9. Phalaris arundinacea** L. REED CANARYGRASS, ALPISTE ROSEAU, PHALARIS ROSEAU, ROSEAU [p. 426, 530]

*Phalaris arundinacea* is a circumboreal species, native to north-temperate regions; it occurs, as an introduction, in the Southern Hemisphere. It grows in wet areas such as the edges of lakes, ponds, ditches, and creeks, often forming dense stands; in some areas it is a problematic weed. *Phalaris arundinacea* hybridizes with other species of *Phalaris*. One hybrid, **P.** ×monspeliensis Daveau [= *P. arundinacea* × *P. aquatica*] is grown for forage.

**10. Phalaris caroliniana** Walter CAROLINA CANARYGRASS [p. 426, 530]

*Phalaris caroliniana* grows in wet, marshy, and swampy ground. It is a common species in suitable habitats through much of the southern portion of the *Manual* region and in northern Mexico. It has also been found in Puerto Rico, where it may be an introduction, and in Europe and Australia, where it is undoubtedly an introduction. It appears to hybridize with *P. lemmonii* and *P. caroliniana*.

**11. Phalaris angusta** Nees *ex* Trin. NARROW CANARYGRASS [p. 426, 530]

*Phalaris angusta* grows in the contiguous United States, primarily in the south. In South America, it is most abundant in a band from Chile to Argentina; it also grows in Ecuador, Peru, and Bolivia. Throughout its distribution, it tends to grow in open grasslands and prairies.

## 14.62 CINNA L.[1]

**Pl** per; csp, smt rhz. **Clm** 20–203 cm, solitary or clustered, often rooting at the lo nd, usu glab. **Shth** open, glab; **aur** absent; **lig** scarious; **bld** flat, mrg scabrous, surfaces scabrous or smooth. **Infl** pan; **br** spreading to ascending, some br longer than 1 cm; **ped** slightly flared, scabrous or smooth; **dis** below the glm. **Spklt** lat compressed, with 1 flt, rarely with a second rdmt or ftl flt; **rchl** usu prolonged beyond the base of the flt as a minute stub or bristle, smooth or scabridulous, smt not prolonged. **Glm** from slightly shorter than to slightly longer than the flt, 1- or 3-veined, mrg hyaline, keeled, keels scabrous, apc acute, smt minutely awn-tipped; **lo glm** from somewhat shorter than to equaling the up glm, flt sessile or stipitate; **cal** short, glab; **lm** 3- or 5-veined, smt obscurely so, apc acute, minutely bifid, usu awned, awns subtml, smt unawned; **pal** 3/4 to nearly as long as the lm, 1-veined or with 2 closely spaced veins; **anth** 1 or 2. **Car** shorter than the lm, concealed at maturity, often beaked.

*Cinna* is a genus of four species, all of which generally grow in damp woods, along streams, or in wet meadows. One species, *C. latifolia*, is northern temperate and circumboreal. The other three species are restricted to the Western Hemisphere. *Cinna poaeformis* (Kunth) Scribn. & Merr. extends from Mexico to Venezuela and Bolivia.

1. Anthers 2; lemmas 5-veined; florets more or less sessile . . . . . . . . . . . . . . . . . . . . . . . . . . . . . . . . . . . 3. *C. bolanderi*
1. Anthers 1; lemmas 3(5)-veined; florets on a 0.1–0.65 mm stipe.
    2. Upper glumes prominently 3-veined; spikelets (3.5)4–6(7.5) mm long . . . . . . . . . . . . . . . . . . . . . . . 1. *C. arundinacea*
    2. Upper glumes usually 1-veined, rarely 3-veined; spikelets (2)2.5–4(5) mm long . . . . . . . . . . . . . . . . . . . . . . 2. *C. latifolia*

**1. Cinna arundinacea** L. STOUT WOODREED, SWEET WOODREED, CINNA ROSEAU [p. 426, 530]

*Cinna arundinacea* grows in southeastern Canada and throughout most of the eastern United States, at 0–900 m. It is most common in moist woodlands and swamps, depressions, along streams, and in floodplains and upland woods. It is less frequent in wet meadows, marshes, and disturbed sites. It flowers in late summer to fall. *Cinna arundinacea* is most easily distinguished from *C. latifolia* by its 3-veined upper glumes and larger spikelets.

**2. Cinna latifolia** (Trevir. *ex* Göpp.) Griseb. DROOPING WOODREED, SLENDER WOODREED, CINNA À LARGES FEUILLES [p. 426, 530]

*Cinna latifolia* is a circumboreal species, extending from Alaska to Newfoundland in North America, and across Eurasia from Norway to the Kamchatka Peninsula, Russia. It grows in moist to wet soil in open coniferous or mixed forests, swamps, thickets, bogs, and streamsides, at 0–2600 m. It flowers in late summer and fall. *Cinna latifolia* differs from *C. arundinacea* in its 1(3)-veined upper glumes and its smaller spikelets; and from *C. bolanderi* in having 1 anther, shorter anthers and spikelets, and stipitate florets.

**3. Cinna bolanderi** Scribn. SIERRAN WOODREED, BOLANDER'S WOODREED [p. 426, 530]

*Cinna bolanderi* is endemic to meadows and streamsides at 1900–2400 m in Sequoia, Kings Canyon, and Yosemite national parks. It flowers from late summer to fall. It used to be included in *C. latifolia*, but it differs from that species in having 2 anthers, longer anthers and spikelets, and sessile florets. The two species do not overlap in distribution.

[1]David M. Brandenburg

## 14.63 LIMNODEA L.H. Dewey[1]

Pl ann; tufted. Clm to 60 cm, often prostrate and brchg at the base, glab; intnd solid. Shth open or closed, rounded on the back, frequently hispid; aur absent; lig memb, usu lacerate and minutely ciliate; bld flat, mostly ascending, abx and/or adx surfaces glab or hispid. Infl pan, loosely contracted; br spklt-bearing to the base or nearly so, some br longer than 1 cm; dis below the glm. Spklt pedlt, subterete, with 1 flt; rchl prolonged beyond the base of the flt as a slender bristle, glab. Glm equal, rigid, coriaceous, hispid or scabrous throughout, rounded over the midvein, veinless or obscurely 3–5-veined, acute, unawned; cal blunt, glab; lm equaling the glm, chartaceous, smooth, veinless or inconspicuously 3-veined, minutely bifid or acute, awned, awns subtml, exceeding the flt, geniculate near midlength, bas segment twisted; pal shorter than the lm, veinless or the base 2-veined, hyaline; lod 2, glab, toothed or not; anth 3; ov glab. Car about 2.5 mm, shorter than the lm, concealed at maturity, linear.

*Limnodea* is a unispecific genus of the southern United States and adjacent Mexico.

### 1. Limnodea arkansana (Nutt.) L.H. Dewey
OZARKGRASS [p. 427, 530]

*Limnodea arkansana* grows in dry, usually sandy soils of prairies, open woodlands, disturbed areas, and riverbanks. Along the Gulf coast, it grows on maritime shell mounds and middens, and upper beaches where shells accumulate. It has also been found, as an introduction, around a wool-combing mill in Jamestown, South Carolina.

## 14.64 AMMOPHILA Host[2]

Pl per; strongly rhz. Clm 20–130 cm, erect, glab. Lvs mostly bas; shth open; aur absent; lig memb, smt ciliolate; bld 0.5–8 mm wide, involute or convolute. Infl tml pan, dense, cylindrical; br strongly ascending and overlapping. Spklt pedlt, lat compressed, with 1 flt; rchl prolonged beyond the flt, glab or hairy; dis above the glm, beneath the flt. Glm equaling or exceeding the flt, subequal, linear-lanceolate, papery, keeled, acute to acuminate; lo glm 1-veined; up glm 3-veined; cal short, pilose; lm chartaceous, linear-lanceolate, obscurely 3–5-veined, keeled, smt slightly rounded at the base, apc entire or minutely bifid, unawned or awned, awns 0.2–0.5 mm, subtml; pal equaling the lm, often appearing 1-keeled, 2- or 4-veined, cent veins close together; lod 2, free, memb, ciliate or glab, not toothed; anth 3, 3–7 mm; ov glab. Car enclosed by the hardened lm and pal, ellipsoid, longitudinally grooved; hila about 2/3 as long as the car.

*Ammophila* has two species, one native to the coast of Europe and northern Africa, and one to eastern North America. Both species are effective sand binders and dune stabilizers. They are sometimes mistaken for *Leymus arenarius* and *L. mollis*, which grow in the same habitats and have a similar habit, but species of *Leymus* have more than 1 floret per spikelet.

1. Ligules 1–4.6 mm long, truncate to obtuse . . . . . . . . . . . . . . . . . . . . . . . . . . . . . . . . . . . . . . . . . . . . . . . . . . . . . . . . 1. *A. breviligulata*
1. Ligules 10–35 mm long, acute and bifid or lacerate . . . . . . . . . . . . . . . . . . . . . . . . . . . . . . . . . . . . . . . . . . . . . . . . . . 2. *A. arenaria*

### 1. Ammophila breviligulata Fernald [p. 427, 530]

The two subspecies overlap in their morphological characteristics and occupy similar habitats, the primary difference between them being their flowering time.

1. Flowering late July to September; panicles (9.5) 13–40 cm long; anthers 3.5–7 mm long . . . . subsp. *breviligulata*
1. Flowering late June to early July; panicles 9–17 cm long; anthers 3–4.5 mm long . . . . . . . . . subsp. *champlainensis*

### Ammophila breviligulata Fernald subsp. breviligulata
AMERICAN BEACHGRASS, AMMOPHILE À LIGULE COURTE [p. 427]

*Ammophila breviligulata* subsp. *breviligulata* grows on sand dunes and dry, sandy shores from around the Great Lakes to the Atlantic coast from Newfoundland to South Carolina and, as an introduction, on the west coast. Anthesis is from late July to September. It hybridizes with *Calamagrostis canadensis* on Grand Island, Michigan, and the adjacent mainland, to form ×**Calammophila don-hensonii** Reznicek & Judz. The hybrid is morphologically intermediate between its parents. Most florets appear to be sterile; some appear normal.

### Ammophila breviligulata subsp. champlainensis (F. Seym.) P.J. Walker, C.A. Paris & Barrington *ex* Barkworth CHAMPLAIN BEACHGRASS [p. 427]

*Ammophila breviligulata* subsp. *champlainensis* differs from subsp. *breviligulata* primarily in its earlier anthesis, from June to early July rather than late July to September. It also tends to be smaller, but the two subspecies overlap in all their morphological characteristics.

### 2. Ammophila arenaria (L.) Link EUROPEAN BEACHGRASS, MARRAMGRASS [p. 427, 530]

*Ammophila arenaria* is a European species that has become naturalized in most temperate countries. It was introduced along the Pacific coast and in the interior of western North America as a sand binder. It is known from a single 1941 collection on a sand dune in Erie County, Pennsylvania. *Ammophila arenaria* flowers from the end of June to August.

North American plants belong to **Ammophila arenaria** (L.) Link subsp. **arenaria**, in which the glumes exceed the lemma and the callus hairs are about 2–3 mm long. It is native from northern and western Europe to northwestern Spain. **Ammophila arenaria** subsp. **arundinacea** H. Lindb. has glumes that equal the lemma, and callus hairs 4–6 mm long. It is native around the Mediterranean.

[1]Neil Snow  [2]Mary E. Barkworth

# 14.65 MILIUM L.[1]

Pl ann or per; csp, smt rhz. **Clm** 10–180 cm, glab or hispidulous; **nd** 2–5, glab. **Shth** open, smooth or scabrous; **aur** absent; **lig** hyaline, glab, obtuse to acute; **bld** flat, smooth or scabrous over the veins. **Infl** open pan; **br** drooping to ascending, smooth or scabrous, some br longer than 1 cm. **Spklt** pedlt, dorsally compressed, with 1 flt; **rchl** not prolonged beyond the base of the flt; **dis** above the glm, beneath the flt. **Glm** equal, equaling or exceeding the lm, memb, smooth or scabrous, unawned; **cal** blunt, glab; **lm** dark, coriaceous, glab, lustrous, obscurely 5-veined, mrg involute, apc unawned; **pal** similar to the lm and partly enfolded by them; **lod** 2, free, glab, toothed or not; **anth** 3; **ov** glab. **Car** shorter than the lm, concealed at maturity, glab; hila $^1/_5$ to nearly $^1/_2$ the length of the car.

*Milium* is a circumtemperate genus of four species. All the species grow in mesic to dry mixed woods and dry open habitats. *Milium effusum* is native to the *Manual* region; *M. vernale* has become established.

1. Plants perennial; blades 8–17 mm wide; panicles 10–27 cm long; lemmas 2.3–3 mm long . . . . . . . . . . . . . . . 1. *M. effusum*
1. Plants annual; blades 1.9–5 mm wide; panicles 4–11.5 cm long . . . . . . . . . . . . . . . . 2. *M. vernale*

### 1. Milium effusum L. WOOD MILLET, MILLET DIFFUS [p. 427, 530]

*Milium effusum* is widespread in temperate to subarctic regions in the Northern Hemisphere. North American plants belong to **M. effusum** var. **cisatlanticum** Fernald, an elegant native grass that grows in woodlands in eastern North America. It differs from **M. effusum** L. var. **effusum**, which grows from Europe to Asia and Japan, in having 2–3 panicle branches at most nodes and spikelets 2.5–5 mm long, rather than 4–5 panicle branches at most nodes and spikelets about 3 mm long.

### 2. Milium vernale M.-Bieb. EARLY MILLET, SPRING MILLETGRASS [p. 427, 530]

Native to Eurasia, *Milium vernale* was first detected in North America in 1987, when it was found infesting winter wheat and other crops in north-central Idaho. The infested area has since increased.

# 14.66 ALOPECURUS L.[2]

Pl ann or per; smt csp, smt shortly rhz. **Clm** 5–110 cm, clumped or solitary, erect or decumbent, occ cormlike at the base; **nd** glab. **Lvs** inserted mostly on the lo $^1/_2$ of the clm; **shth** open, up shth smt inflated; **aur** absent; **lig** 0.6–6.5 mm, truncate to acute, memb, puberulent or glab, entire to lacerate; **bld** 0.7–12 mm wide, flat or involute, glab or scabrous, bld of upmost lvs smt short or absent. **Infl** tml pan, spikelike, capitate to cylindrical; **br** usu shorter than 5 mm, lo br smt to 2 cm; **dis** below the glm. **Spklt** 1.8–7 mm, pedlt, strongly lat compressed, oval in outline, with 1 flt; **rchl** not prolonged beyond the base of the flt. **Glm** equaling or exceeding the flt, memb or coriaceous, free or connate in at least the lo $^1/_2$, narrowing from above midlength, 3-veined, keeled, keels ciliate, at least bas, apc obtuse to acute or shortly awned; **cal** blunt, glab; **lm** memb, mrg often connate in the lo $^1/_2$, keeled, indistinctly 3–5-veined, apc truncate to acute, awned dorsally from just above the base to about midlength, geniculate or straight; **pal** absent or greatly rdcd; **lod** absent; **anth** 3, 0.3–4.1 mm; **ov** glab; **sty** fused, with 2 br. **Car** shorter than the lm, concealed at maturity, glab; hila short.

*Alopecurus* has 36 species. They grow primarily in open, mesic habitats, and are native to the northern temperate zone and South America. Four species are native to the *Manual* region, four were introduced and have become established, and two were introduced and are not known to persist.

Some species of *Alopecurus* appear similar to *Phleum*, which has truncate glumes that are abruptly awned or mucronate, lemmas without awns or keels, and well-developed paleas; *Alopecurus* has glumes that are obtuse to acute and gradually awned or unawned, lemmas with both awns and keels, and paleas that are absent or greatly reduced.

1. Glume keels winged; glumes glabrous, pubescent over the veins.
   2. Glumes 4.5–7.5 mm long, connate in the lower $^1/_2$, the apices acute, convergent to parallel; lemma apices acute . . . . . . . . . . . . . . . . . . . . . . . . . . . . . . . . . . . . . . . . . . . . . . 9. *A. myosuroides*
   2. Glumes 3–4.5 mm long, connate in the lower $^1/_2$–$^4/_5$, the apices obtuse, mucronate, divergent; lemma apices truncate . . . . . . . . . . . . . . . . . . . . . . . . . . . . . . . . 10. *A. creticus*
1. Glume keels not winged; glumes usually sparsely to densely pubescent, sometimes glabrous.
   3. Plants annual, without rhizomes, not rooting at the lower nodes; blades 1–16 cm long, 0.9–4 mm wide; culms 5–50 cm tall.
      4. Glumes 5–6.4 mm long, coriaceous and dilated in the lower $^1/_2$; glume and lemma apices acute to acuminate; anthers about 3 mm long . . . . . . . . . . . . . . . . . . . . . . . . . . . . . . 6. *A. rendlei*
      4. Glumes 2.1–5 mm long, membranous, not dilated below; glume and lemma apices obtuse; anthers 0.3–1.8 mm long.

[1,2]William J. Crins

5. Upper sheaths conspicuously inflated; glumes 3–5 mm long; lemmas 3–5 mm long, awns
exceeding the lemmas by 3–6 mm; panicles 5.5–13 mm wide, excluding the awns . . . . . . . . . . . . . 7. *A. saccatus*
5. Upper sheaths not or only slightly inflated; glumes 2.1–3.1 mm long; lemmas 1.9–2.7 mm
long, awns exceeding the lemmas by 1.6–4 mm; panicles 3–6 mm wide, excluding the awns . . . . 8. *A. carolinianus*
3. Plants perennial, often rhizomatous, sometimes rooting at the lower nodes; blades 2–40 cm long,
1–12 mm wide; culms 5–110 cm tall.
6. Glumes 1.8–3.7 mm long, the apices obtuse; anthers 0.5–2.2 mm long.
7. Awns geniculate, exceeding the lemmas by 1.2–4 mm; anthers (0.9)1.4–2.2 mm long . . . . . . . . . . . 4. *A. geniculatus*
7. Awns straight, not exceeding the lemmas or exceeding them by less than 2.5 mm; anthers
0.5–1.2 mm long . . . . . . . . . . . . . . . . . . . . . . . . . . . . . . . . . . . . . . . . . . . . . . . . . . . . . . . . . . . . . . 5. *A. aequalis*
6. Glumes 3–6 mm long, the apices acute; anthers 2–4 mm long.
8. Glume margins connate in the lower ¹/₈; glumes densely pilose throughout . . . . . . . . . . . . . . . . 3. *A. magellanicus*
8. Glume margins connate in the lower ¹/₅–¹/₃; glumes with long hairs mainly restricted to the
veins.
9. Lemma apices acute; glume apices parallel or convergent . . . . . . . . . . . . . . . . . . . . . . . . . . 1. *A. pratensis*
9. Lemma apices obtuse to truncate; glume apices divergent . . . . . . . . . . . . . . . . . . . . . . . . 2. *A. arundinaceus*

## 1. Alopecurus pratensis L. MEADOW FOXTAIL, VULPIN DES PRÉS [p. 427, 530]

*Alopecurus pratensis* is native from temperate northern Eurasia south to North Africa. It is now widely naturalized in temperate regions throughout the world. It grows in poorly to somewhat drained soils in meadows, riverbanks, lakesides, ditches, roadsides, and fence rows. It has been widely introduced as a pasture grass and is now established throughout much of the *Manual* region.

## 2. Alopecurus arundinaceus Poir. CREEPING MEADOW FOXTAIL, VULPIN ROSEAU [p. 427, 530]

*Alopecurus arundinaceus* is native to Eurasia, extending north of the Arctic Circle and south to the Mediterranean. It grows in wet, moderately acid to moderately alkaline soils, on flood plains, near vernal ponds, and along rivers, streams, bogs, potholes, and sloughs. It was introduced for pasture in North Dakota and now occurs more widely, having been promoted as a forage species. It is sometimes used in seed mixtures for revegetation projects.

## 3. Alopecurus magellanicus Lam. ALPINE FOXTAIL, BOREAL FOXTAIL, VULPIN ALPIN, VULPIN BORÉALE [p. 427, 530]

*Alopecurus magellanicus* has an arctic-alpine to subalpine circumpolar distribution, but it has not been found in Scandinavia or Iceland. It grows primarily in wet soils in tundra, meadows, along streams, shorelines, gravelbars, and floodplains, and occasionally in somewhat drier forest openings, in fine or silty to stony soils or moss. It is sometimes co-dominant with *Dupontia fisheri* in the arctic and subarctic portion of its range. The anthocyanic tint of the plant as a whole greatly increases to the north.

## 4. Alopecurus geniculatus L. WATER FOXTAIL, VULPIN GÉNICULÉ [p. 428, 530]

*Alopecurus geniculatus* is native to Eurasia and parts of North America, growing in shallow water, ditches, open wet meadows, shores, and streambanks, from lowland to montane zones. It has been naturalized in eastern North America. The status of populations in the west is less certain. Many occur in moist sites within native rangeland, but these areas have also been affected by European settlement.

*Alopecurus* ×*haussknechtianus* Asch. & Graebn. is a hybrid between *A. geniculatus* and *A. aequalis*, which occurs fairly frequently in areas of sympatry, particularly in drier midcontinental areas from Alberta to Saskatchewan, south to Arizona and New Mexico.

## 5. Alopecurus aequalis Sobol. [p. 428, 530]

*Alopecurus aequalis* is native to temperate zones of the Northern Hemisphere. It generally grows in wet meadows, forest openings, shores, springs, and along streams, as well as in ditches, along roadsides, and in other disturbed sites, from sea level to subalpine

elevations. It is the most widespread and variable species of *Alopecurus* in the *Manual* region.

*Alopecurus* ×*haussknechtianus* Asch. & Graebn. is a hybrid between *A. aequalis* and *A. geniculatus*, which occurs fairly frequently in areas of sympatry, particularly in drier midcontinental areas from Alberta to Saskatchewan, south to Arizona and New Mexico.

1   Panicles 3–6 mm wide; glumes 1.8–3 mm long; awns
    not exceeding the lemmas or exceeding them by less
    than 1 mm; anthers 0.5–0.9 mm long . . . . . . . . . . var. *aequalis*
1   Panicles 4–9 mm wide; glumes to 3.7 mm long;
    awns exceeding the lemmas by 1–2.5 mm; anthers
    1–1.2 mm long . . . . . . . . . . . . . . . . . . . . . . . var. *sonomensis*

## Alopecurus aequalis Sobol. var. aequalis SHORTAWN FOXTAIL, VULPIN À COURTES ARÊTES [p. 428]

*Alopecurus aequalis* var. *aequalis* is the widespread variety in the *Manual* region. It is listed as threatened in the state of Connecticut.

## Alopecurus aequalis var. sonomensis P. Rubtzov SONOMA SHORTAWN FOXTAIL [p. 428]

*Alopecurus aequalis* var. *sonomensis* is endemic to Marin and Sonoma counties, California, where it grows in shallow water and marshy or moist ground, usually in the open. It is listed as endangered by the U.S. Fish and Wildlife Service.

## 6. Alopecurus rendlei Eig RENDLE'S MEADOW FOXTAIL [p. 428, 530]

*Alopecurus rendlei* is native to wet meadows, and adventive in roadsides and waste places, in southern and western Europe. It was found growing on ballast in Philadelphia, Pennsylvania in 1880; it has not been collected in North America since then.

## 7. Alopecurus saccatus Vasey PACIFIC MEADOW FOXTAIL [p. 428, 530]

*Alopecurus saccatus* is a native annual that inhabits moist, open meadows, valley plains, and vernal pools, at elevations below 700 m, from Washington to California.

## 8. Alopecurus carolinianus Walter TUFTED FOXTAIL [p. 428, 531]

*Alopecurus carolinianus* is native to the central plains, Mississippi valley, and southeastern United States, where it is common in wet meadows, ditches, wetland edges, and other moist, open habitats; it is occasionally a weed of rice fields. At the northern limit of its range it is clearly adventive, growing in gardens and nurseries. It also occurs in arid areas of the prairies and southwest, growing sporadically along sloughs and in ditches and vernal pools. Whether such populations are native or naturalized is not clear.

**9. Alopecurus myosuroides** Huds. BLACKGRASS, SLENDER MEADOW FOXTAIL [p. *428*, 531]

*Alopecurus myosuroides* is native to Eurasia, and grows in moist meadows, deciduous forests, and cultivated or disturbed ground. A significant weed species in temperate cereal crops, it is one of the most damaging weeds of winter cereals in England. It has been introduced repeatedly as a weed of cultivation into many parts of the *Manual* region, but apparently has not spread to a large degree outside of cultivation. *Alopecurus myosuroides* has been listed as a noxious weed in the state of Washington, one of the states where winter wheat is a major crop.

**10. Alopecurus creticus** Trin. CRETAN MEADOW FOXTAIL [p. *428*, 531]

*Alopecurus creticus* is native to marshes and wet places in the southern part of the Balkan Peninsula. It was discovered on ballast dumps in Philadelphia, Pennsylvania in the nineteenth century; it has not persisted in the *Manual* region.

## 14.67 APERA Adans.[1]

Pl ann; tufted or the clm solitary. **Clm** 5–120 cm, erect or geniculate, glab. **Lvs** mostly cauline; **shth** open, rounded to slightly keeled; **col** glab, midveins continuous; **aur** absent; **lig** memb, often lacerate to erose; **bld** flat or weakly involute, glab. **Infl** tml pan; **br** strongly ascending to divergent. **Spklt** pedlt, slightly lat compressed, with 1 flt, rarely more, distal flt, if present, vestigial; **rchl** prolonged beyond the base of the flt as a bristle, rarely terminating in a vestigial flt; **dis** above the glm, beneath the flt. **Glm** unequal, lanceolate, scabrous on the distal ¹/₂, unawned; **lo glm** 1-veined; **up glm** slightly shorter than to slightly longer than the flt, 3-veined; **cal** blunt, glab or sparsely hairy; **lm** firmer than the glm, folded to nearly terete, obscurely 5-veined, mrg veins not excurrent, apc entire, awned, awns subtml; **pal** about ³/₄ as long as to equaling the lm, hyaline, 2-veined; **lod** 2, free, glab, usu toothed; **anth** 3; **ov** glab. **Car** shorter than the lm, concealed at maturity, 1.2–2 mm, ellipsoidal, slightly sulcate; **hila** broadly ovate, ¹/₅ the length of the car.

*Apera* is genus of three species, native to Europe and western Asia. It is similar to *Agrostis*. It differs in its firm lemmas; paleas that are always present and equal to the lemma, or nearly so; and prolonged rachillas. In North America, two species have been introduced, growing as weeds in lawns and disturbed ground, and in grain fields.

1. Anthers 0.3–0.5 mm long; panicles contracted, 0.4–3 cm wide; most branches spikelet-bearing to
within 2 mm of the base . . . . . . . . . . . . . . . . . . . . . . . . . . . . . . . . . . . . . . . . . . . . . . . . . . . . . 1. *A. interrupta*
1. Anthers 1–2 mm long; panicles pyramidal, 2–15 cm wide; branches naked at the base for 5 mm or
more . . . . . . . . . . . . . . . . . . . . . . . . . . . . . . . . . . . . . . . . . . . . . . . . . . . . . . . . . . . . . . . 2. *A. spica-venti*

**1. Apera interrupta** (L.) P. Beauv. INTERRUPTED WINDGRASS [p. *428*, 531]

*Apera interrupta* grows as a weed in lawns, grain fields (especially winter wheat), sandy open ground, and roadsides. Introduced from Europe, it now grows from British Columbia south to Arizona and New Mexico, as well as in Ontario and at a few scattered locations in the eastern part of the *Manual* region.

**2. Apera spica-venti** (L.) P. Beauv. COMMON WINDGRASS, LOOSE SILKYBENT [p. *428*, 531]

*Apera spica-venti* grows as a weed in lawns, waste places, grain fields, sandy ground, and roadsides. Introduced from Europe, it is found in scattered locations in the *Manual* region.

# 5. ARUNDINOIDEAE Burmeist.

The *Arundinoideae* are interpreted here as including one tribe, the *Arundineae*.

## 15. ARUNDINEAE Dumort.[2]

Pl usu per; csp or not, smt rhz, smt stln. **Clm** 15–1000 cm, ann, hrb to somewhat wd, intnd usu hollow. **Lvs** usu mostly cauline, often conspicuously distichous; **shth** usu open; **aur** usu absent; **abx lig** usu absent (of hairs in *Hakonechloa*); **adx lig** memb or of hairs, if memb, often ciliate; **bld** without psdpet, smt deciduous at maturity; **mesophyll** usu non-radiate (radiate in *Arundo*); **adx palisade layer** absent; **fusoid cells** absent; **arm cells** usu absent (present in *Phragmites*); **Kranz anatomy** absent; **midribs** simple; **adx bulliform cells** present; **stomatal subsidiary cells** low dome-shaped or triangular; **bicellular microhairs** usu present, usu with long, narrow tml cells; **papillae** usu absent. **Infl** usu tml, ebracteate, usu pan, occ spicate or rcm. **Spklt** lat compressed, with 1–several bisx flt or all flt unisx and the species dioecious; **flt** 1–several, terete or lat compressed, distal flt often rdcd; **dis** above the glm. **Glm** 2, from shorter than the adjacent lm to exceeding the distal flt; **lm** (3)5–7-veined, lanceolate to elliptic, acute to acuminate, smt awned; **awns** 1 or 3, if 3 not fused into a single bas column; **pal** subequal to the lm; **lod** 2, usu free, occ joined at the base, fleshy, usu glab, not, scarcely, or heavily vascularized; **anth** (1)2–3; **ov** glab; **sty** 2, usu free, bases close together. **Car** usu punctate (long-linear in *Molinia*); **endosperm** hard, without lipid; **starch grains** compound; **haustorial synergids** absent; **emb** usu large compared to the car,

[1]Kelly W. Allred    [2]Grass Phylogeny Working Group and Kelly W. Allred

waisted or not; **epiblasts** absent; **scutellar cleft** present; **mesocotyl intnd** elongate; **emb lf mrg** usu meeting (overlapping in *Hakonechloa*).

The morphological circumscription of the *Arundineae* is difficult. The most abundant genera in North America, *Phragmites* and *Arundo*, have tall culms bearing numerous, conspicuously distichous, broad leaves and large, plumose panicles, a habit frequently described as "reedlike", but not all members of the tribe have this habit. *Arundo*, *Phragmites*, and *Molinia* have hollow culm internodes, punctate hila, and convex sides to the adaxial ribs in the leaf blades, but these characters have not been examined in all genera of the tribe.

Members of the *Arundineae* are found in tropical and temperate areas around the world. The reedlike species are found in marshy to damp soils, but some of the other species grow in xeric habitats.

1. Plants cespitose, not rhizomatous; rachillas and lemmas glabrous . . . . . . . . . . . . . . . . . . . . . . . . . . . . . . 15.01 *Molinia*
1. Plants rhizomatous or stoloniferous, sometimes also loosely cespitose; rachillas or lemmas hairy.
    2. Lemmas glabrous . . . . . . . . . . . . . . . . . . . . . . . . . . . . . . . . . . . . . . . . . . . . . . . . . . . . . . . . 15.03 *Phragmites*
    2. Lemmas hairy.
        3. Rachilla internodes hairy; lemmas with papillose-based hairs on the margins . . . . . . . . . . . . . . . . . 15.02 *Hakonechloa*
        3. Rachilla internodes glabrous; lemmas pilose, the hairs not papillose-based . . . . . . . . . . . . . . . . . . . 15.04 *Arundo*

## 15.01 MOLINIA Schrank[1]

**Pl** per; csp, not rhz. **Clm** 15–250 cm, often disarticulating at the first nd, bas intnd persistent, often swollen and clavate. **Lvs** mostly bas; **lig** of hairs; **bld** flat or convolute, eventually disarticulating from the shth. **Infl** tml, pan, not plumose. **Spklt** lat compressed, with (1)2–5 flt; **rchl** prolonged beyond the distal flt, terminating in a rdmt flt, intnd ⅓–½ as long as the flt, glab; **dis** beneath the flt. **Glm** exceeded by the flt, 1- or 3-veined; **cal** 0.1–0.3 mm, blunt, glab or sparsely strigose, hairs to 0.5 mm; **lm** glab, inconspicuously 3(5)-veined, rounded over the back, acute to obtuse, unawned; **pal** subequal to the lm; **anth** 3. **Car** falling free from the lm and pal; **prcp** loosely adherent.

*Molinia* is a genus of two to five species, all of which are native to temperate Eurasia. One species is established in the *Manual* region.

**1. Molinia caerulea** (L.) Moench Purple Moorgrass, Moline Bleue [p. *428*, <u>531</u>]

*Molinia caerulea* is established at scattered locations in the *Manual* region, but not at all the locations where it has been found. Most records are from southeastern Canada and the northeastern United States, but it has also been reported as being established in western Oregon.

## 15.02 HAKONECHLOA Makino *ex* Honda[2]

**Pl** per; loosely csp, rhz and stln. **Clm** 30–90 cm, erect or geniculate at the base. **Shth** open; **aur** absent; **abx lig** present, composed of a line of hairs across the collar; **adx lig** memb and sparsely ciliate, smt lacerate, cilia subequal to the base; **bld** flat, linear-lanceolate, resupinate, in living pl the glaucous-green adx surface facing downwards and the bright green abx surface facing upwards. **Pan** not plumose. **Spklt** pedlt, somewhat lat compressed, with 5–10 flt; **rchl intnd** conspicuously pilose; **dis** at the base of the rchl segment and below each spklt. **Glm** unequal, lanceolate, unawned; **cal** 1.5–2 mm, strigose, hairs 1–1.5 mm; **lm** chartaceous, 3-veined, mrg with papillose-based hairs near the base, apc inconspicuously bidentate, awned from between the teeth; **awns** 3–5 mm, straight; **pal** 2-keeled. **Car** glab.

*Hakonechloa* is a unispecific genus, endemic to Japan, but grown as an ornamental in the *Manual* region. The resupination of the blades is not evident on herbarium specimens.

**1. Hakonechloa macra** (Munro) Makino Japanese Forest Grass, Hakone Grass [p. *429*]

In Japan, *Hakonechloa macra* grows on rocks along rivers. Although rhizomatous, it is not an invasive species and is recommended for mass planting. The form most commonly cultivated in the *Manual* region is forma *aureola* Makino *ex* Ohwi, with yellow leaves having narrow green stripes.

## 15.03 PHRAGMITES Adans.[3]

**Pl** per; rhz or stln, often forming dense stands. **Clm** 1–4 m tall, 0.5–1.5 cm thick, lfy; **intnd** hollow. **Lvs** cauline, mostly glab; **shth** open; **lig** memb, ciliate; **bld** flat or folded. **Infl** tml, plumose pan. **Spklt** with 2–8 flt, weakly lat compressed, lo 1–2 flt stmt, distal 1–2 flt rdmt, remaining flt bisx; **rchl intnd** sericeous; **dis** above the glm and below the flt. **Glm** unequal, shorter than the flt, 1–3-veined, glab; **lo glm** much shorter than the up glm; **cal** pilose, hairs 6–12 mm; **lm** 3-veined, glab, unawned; **anth** 1–3. **Car** rarely maturing.

*Phragmites australis* grows throughout the world and has been put to many uses. In the Americas, native peoples have

[1,2]Mary E. Barkworth  [3]Kelly W. Allred and Kristin Saltonstall

used it for thatch, arrow shafts, pipe stems, clothing, reed boats, and medicine. Nowadays it is used extensively in bio-remediation facilities. Saltonstall (2002) demonstrated that there are three strains present in North America. Some strains approved for use in bioremediation are imported from Europe. It is not known whether they belong to the invasive strain identified here.

1. Ligules 1–1.7 mm long; lower glumes 3–6.5 mm long; upper glumes 5.5–11 mm long; leaf sheaths caducous with age; culms exposed in the winter, smooth and shiny; rarely forming a monoculture . . . . . . . . . . . . . . . . . . . . . . . . . . . . . . . . . . . . . . . . . . . . . . . . . . . . . . . . . . . . . . . . *P. australis* subsp. *americanus*
1. Ligules 0.4–0.9 mm long; lower glumes 2.5–5 mm long; upper glumes 4.5–7.5 mm long; leaf sheaths not caducous with age; culms not exposed in the winter, smooth and shiny or ridged and not shiny; usually forming a monoculture.
    2. Culms smooth and shiny . . . . . . . . . . . . . . . . . . . . . . . . . . . . . . . . . . . . . . . . . . . . . . . *P. australis* subsp. *berlandieri*
    2. Culms ridged and not shiny . . . . . . . . . . . . . . . . . . . . . . . . . . . . . . . . . . . . . . . . . . . . . . . . . . . Invasive plants

### Phragmites australis subsp. americanus Saltonstall, P.M. Peterson & Soreng COMMON REED, PHRAGMITE COMMUN, ROSEAU COMMUN [p. *429*, <u>531</u>]

*Phragmites australis* subsp. *americanus* is the native North American strain. It grows across much of Canada and in the United States, from New England and the Mid-Atlantic states across to the Pacific coast and into the southwest. Regional structuring can be found within this lineage, with east coast, midwestern, and western populations showing different genetic profiles (Saltonstall 2003a,b).

### Phragmites australis subsp. berlandieri (E. Fourn.) Saltonstall & Hauber [p. *429*]

*Phragmites australis* subsp. *berlandieri* is the strain that grows from the Atlantic coast of Florida, around the Gulf of Mexico, southwestern Arizona, northern Mexico, and extends into Central America and South America. It is not yet clear whether it is native or introduced. Some taxonomists treat it as *P. karka* (Retz.) Trin. *ex* Steud., a species that was described from India.

### Phragmites australis (Cav.) Trin. ex Steud. [p. *429*] (Invasive)

The appropriate name for the introduced plants is not clear although they probably originate in Europe. The name *Phragmites australis*, and hence the name *P. australis* subsp. *australis*, is based on plants collected in Australia. A number of subspecific names are available for European plants but, until their taxonomy is better understood, it is impossible to determine whether the name for the invasive strain that has been introduced to North America should be subsp. *australis* or one of the subspecies names applied to European plants. Nevertheless, because of the management issues involved, it is important to provide guidance on distinguishing the introduced plants from the two native strains.

## 15.04 ARUNDO L.[1]

**Pl** per; **rhz**, rhz short, usu more than 1 cm thick. **Clm** 2–10 m tall, 1–3.5 cm thick, usu erect, occ pendant from cliffs; **nd** glab; **intnd** hollow. **Lvs** cauline, conspicuously distichous, glab; **shth** open, longer than the intnd; **lig** memb, shortly ciliate; **bld** flat or folded, mrg scabrous. **Pan** tml, plumose, silvery to purplish. **Spklt** lat compressed, with 1–several flt; **rchl intnd** glab; **dis** above the glm and between the flt. **Glm** longer than the flt, 3–5-veined; **lm** pilose, hairs not papillose-based, 3–7-veined, apc entire or minutely awned; **pal** shorter than the lm, 2-veined; **anth** 3.

*Arundo*, a genus of three species, grows throughout the tropical and warm-temperate regions of the world. Only one species has been introduced to the Western Hemisphere. *Arundo* is similar to, but usually larger than, *Phragmites*, a much more common genus in North America. In addition, *Arundo*, but not *Phragmites*, has a wedge-shaped, light to dark brown area at the base of its blades.

### 1. Arundo donax L. GIANT REED [p. *429*, <u>531</u>]

Within the *Manual* region, *Arundo donax* grows in the southern half of the contiguous United States, being found along ditches, culverts, and roadsides where water accumulates. It has been used extensively as a windbreak, and planted for erosion control on wet dunes. It is also grown for the ornamental value of its tall, leafy culms and large panicles, but its tendency to spread is sometimes a disadvantage.

## 6. CHLORIDOIDEAE Kunth ex Beilschm.[2]

The subfamily *Chloridoideae* is most abundant in dry, tropical and subtropical regions. In the *Manual* region, it reaches its greatest diversity in the southwestern United States. Almost all its members, and all those in the *Manual* region, have $C_4$ photosynthesis.

There is considerable disagreement concerning the tribal treatment within the *Chloridoideae*, the number of tribes recognized varying from two to eight. The treatment presented here is conservative in recognizing the *Orcuttieae* and *Pappophoreae* as distinct tribes. It departs from most other treatments in merging all other North American taxa into a single tribe, the *Cynodonteae*.

1. Leaves with little or no distinction between the sheath and blade; ligules not present; plants annual, viscid . . . . . . . . . . . . . . . . . . . . . . . . . . . . . . . . . . . . . . . . . . . . . . . . . . . . . . . . . . . . . . . . . . . . 18. *Orcuttieae*
1. Leaves clearly differentiated into sheath and blade; ligules present; plants annual or perennial, not viscid.

[1]Kelly W. Allred [2]Grass Phylogeny Working Group

2. Lemmas 5–13-veined, all the veins extending into awns, often alternating with hyaline lobes or teeth . . . . . . . . . . . . . . . . . . . . . . . . . . . . . . . . . . . . . . . . . . . . . . . . . . . . . . . . . . . . . . . . 17. *Pappophoreae*

2. Lemmas 1–11-veined, unawned or with 1 or 3 awns, sometimes with hyaline lobes on either side of the central awns . . . . . . . . . . . . . . . . . . . . . . . . . . . . . . . . . . . . . . . . . . . . . . . . . . . . . . . . . . 16. *Cynodonteae*

## 16. CYNODONTEAE Dumort.[1]

Pl ann or per. **Clm** 1–500 cm, not wd, usu not brchd above the base. **Shth** usu open, often with coarse hairs at the top; **aur** rarely present; **lig** of hairs or memb, if memb, often ciliate, cilia smt longer than the memb base; **bld** often with stiff, coarse mrgl hairs adjacent to the lig, glab or variously pubescent elsewhere. **Infl** tml, smt also ax, simple pan, pan of 1–many spikelike br, spikelike rcm, spikes, or, in 1 genus, a solitary spklt, in dioecious taxa the stmt and pist infl smt morphologically distinct; **dis** usu beneath the ftl flt or the glm but, particularly if the pan br are short, smt at the base of the br. **Spklt** usu lat compressed, with 1–60 flt, strl or rdcd flt, if present, usu distal to the bisx flt. **Glm** from shorter than the adjacent flt to exceeding the distal flt; **lm** 1–3-veined or 7–13-veined, rarely 5-veined, if with 7–13 veins, the veins often in 3 groups; **lod** 2, or absent.

Most members of the *Cynodonteae* in the *Manual* region can be recognized by their possession of two or more of the following characteristics: 1–3- or 7–13-veined lemmas, laterally compressed spikelets, spikelike inflorescence branches, and the presence of coarse hairs near the junction of the sheath and blade. All tend to grow in hot, dry areas. Having said this, it must be acknowledged that each of these characteristics can be found in other tribes and, within the *Cynodonteae*, there are genera that lack one or more of them.

1. Inflorescences clearly exceeded by the upper leaves, often completely or almost completely enclosed in the upper leaf sheaths; culms 1–30(75) cm tall.
   2. Lemmas 3-lobed, the lobes ciliate; spikelets with 4 florets . . . . . . . . . . . . . . . . . . . . . . . . . . . 16.17 *Blepharidachne*
   2. Lemmas not 3-lobed or the lobes not ciliate; spikelets with 1–60 florets.
      3. Spikelets (and often the plants) unisexual.
         4. Leaves strongly distichous; lemmas 9–11-veined; plants unisexual, growing in saline and alkaline soils.
            5. Pistillate and staminate inflorescences consisting of a single spikelet . . . . . . . . . . . . . . . 16.07 *Monanthochloë*
            5. Pistillate and staminate inflorescences consisting of more than 1 spikelet . . . . . . . . . . . 16.04 *Distichlis* (in part)
         4. Leaves not strongly distichous; lemmas 1–5-veined; plants unisexual or, if bisexual, with separate pistillate and staminate inflorescences, growing in a variety of soils.
            6. Spikelets 5–26 mm long; pistillate and staminate inflorescences similar, simple panicles; glumes and lemmas unawned, mucronate, or 1-awned . . . . . . . . . . . . . . . . . . . . . . . 16.23 *Eragrostis* (in part)
            6. Spikelets 2.5–7 mm long; pistillate and staminate inflorescences strongly dimorphic; staminate inflorescences with pectinate, spikelike branches; pistillate spikelets with conspicuously 3-awned glumes or distal florets.
               7. Upper glumes of the pistillate spikelets white, rigid, globose structures terminating in 3 awnlike teeth; pistillate spikelets with 1 floret, without rudimentary florets, the floret unawned or shortly 3-awned; staminate spikelets 4–6 mm long; anthers 2.5–3 mm long, brownish, red, or orange; widespread species of the central plains . . . . . . . . . . . . . . 16.48 *Buchloë* (in part)
               7. Upper glumes of the pistillate spikelets membranous, unawned; pistillate spikelets with one 3-awned floret and a distal 3-awned rudiment; staminate spikelets 3–4 mm long; anthers 2–2.5 mm long, pale; known, within the *Manual* region, only from Florida . . . 16.47 *Opizia* (in part)
      3. Spikelets bisexual, usually at least the lowest floret in each spikelet bisexual, in *Dasyochloa* the third floret in each spikelet bisexual or pistillate, if pistillate, the lowest 2 florets staminate.
         8. Lemma margins with a tuft of hairs at midlength, glabrous elsewhere; blades with white, thickened margins and sharply pointed . . . . . . . . . . . . . . . . . . . . . . . . . . . . . . . . . . . 16.18 *Munroa*
         8. Lemma margins glabrous, or with hairs but the hairs not forming a tuft at midlength; blades without white cartilaginous margins, not sharply pointed.
            9. Lemmas awned, the awns 1–11 mm long.
               10. Plants stoloniferous; inflorescences 1–2.5 cm long, dense panicles; lemmas bilobed, the lobes about ½ as long as the lemmas; ligules of hairs . . . . . . . . . . . . . . . . . . . . . 16.15 *Dasyochloa*
               10. Plants not stoloniferous; inflorescences 1.5–76 cm long, not dense; lemmas entire or minutely bilobed; ligules membranous, sometimes ciliate . . . . . . . . . . . . . . . . . 16.19 *Leptochloa* (in part)
            9. Lemmas unawned, sometimes mucronate, with mucros less than 1 mm long.
               11. Spikelets with 2–20 florets; inflorescences panicles of 2–120 spikelike branches . . . 16.19 *Leptochloa* (in part)
               11. Spikelets with 1(3) florets; inflorescences simple panicles, often highly contracted, without spikelike branches.

[1]Mary E. Barkworth

12. Inflorescences 0.3–7.5 cm long, dense, spikelike or capitate panicles 1–8 times longer than wide; glumes strongly keeled; plants annual . . . . . . . . . . . . . . . . . 16.31 *Crypsis* (in part)

12. Inflorescences 1–60 cm long, sometimes dense and spikelike but, if less than 8 cm long, more than 8 times longer than wide; glumes rounded or weakly keeled; plants annual or perennial . . . . . . . . . . . . . . . . . . . . . . . . . . . . . . . . . . 16.30 *Sporobolus* (in part)

1. Inflorescences usually equaling or exceeding the upper leaves; culms 1–500 cm tall.

13. Inflorescences with disarticulating branches, disarticulation at the base of the branches or (in *Lycurus*) the fused pedicels; branches 0.04–7 cm long, often globose or spikelike (fused to the rachis and not evident in *Lycurus*), usually with fewer than 15 spikelets per branch.

14. Upper glumes with straight or uncinate spinelike projections; spikelets crowded, the branches condensed into burs . . . . . . . . . . . . . . . . . . . . . . . . . . . . . . . . . . . . . . . . . . . . 16.52 *Tragus*

14. Upper glumes without spinelike projections; spikelets sometimes crowded, but not forming burlike clusters.

15. Branches not fused to the rachises; spikelets more than 3 per branch, usually all alike, sometimes the proximal spikelet sterile or replaced with short secondary branches.

16. Both glumes much longer than the florets, usually exceeding the distal florets; plants annual . . . . . . . . . . . . . . . . . . . . . . . . . . . . . . . . . . . . . . . . . . . 16.22 *Dinebra*

16. Lower glumes shorter than or subequal to the lower florets, 1 or both glumes usually exceeded by the distal florets; plants usually perennial.

17. Spikelets with 1–2(3) florets; branches disarticulating promptly, before the spikelets disarticulate . . . . . . . . . . . . . . . . . . . . . . . . . . . . 16.46 *Bouteloua* (in part)

17. Spikelets with 2–8 florets; branches disarticulating tardily, initially the spikelets disarticulating above the glumes . . . . . . . . . . . . . . . . . . . . 16.25 *Pogonarthria* (in part)

15. Branches sometimes fused to the rachises; spikelets 1–3 per branch, usually some spikelets on each branch sterile or staminate and 1 pistillate or bisexual.

18. Axes of the branches extending beyond the base of the distal florets . . . . . . . . . 16.46 *Bouteloua* (in part)

18. Axes of the branches terminating at the base of the distal spikelet.

19. Spikelets in pairs; glumes awned, the lower glumes (1)2(3)-awned, the upper glumes 1-awned; panicle branches often fused to the rachises . . . . . . . . . . . . 16.34 *Lycurus* (in part)

19. Spikelets in triplets; glumes unawned, 1-awned, or 3-awned; panicle branches sometimes appressed, but not fused, to the rachises.

20. Branches straight at the base; central spikelets sessile . . . . . . . . . . . . . . . . . . . . . . 16.51 *Hilaria*

20. Branches sharply curved at the base; central spikelets sessile or pedicellate.

21. Central spikelets with 1 bisexual floret; lateral spikelets with 1 floret, varying to rudimentary . . . . . . . . . . . . . . . . . . . . . . . . . . . . . . . . . 16.50 *Aegopogon*

21. Central spikelets with 3–4 florets, the lowest floret pistillate, bisexual, or staminate, the distal florets staminate or sterile; lateral spikelets usually with 2 florets . . . . . . . . . . . . . . . . . . . . . . . . . . . . . 16.49 *Cathestecum*

13. Inflorescences without disarticulating branches; branches, if present, often more than 4.5 cm long, variously shaped, including spikelike but not globose, often with more than 16 spikelets per branch.

22. Inflorescences spikes or racemes.

23. Spikelets with 1 bisexual or staminate floret and no additional florets.

24. Rachises falcate or curved; both glumes exceeding the florets . . . . . . . . . . . . . 16.43 *Microchloa* (in part)

24. Rachises straight; lower glumes exceeded by the florets, sometimes absent.

25. Spikelets solitary at each node; disarticulation below the glumes or the spikelets not disarticulating . . . . . . . . . . . . . . . . . . . . . . . . . . . . . . . . . . . . . . . . 16.53 *Zoysia*

25. Spikelets paired, terminal on branches that are fused to the rachises; disarticulation at the base of the fused pedicel pairs . . . . . . . . . . . . . . . . . 16.34 *Lycurus* (in part)

23. Spikelets with more than 1 floret but sometimes only 1 floret bisexual, the additional florets sterile or staminate.

26. All spikelets unisexual, the functional florets either staminate or pistillate; plants either unisexual or with both pistillate and staminate spikelets.

27. Lemmas 9–11-veined; plants of saline habitats . . . . . . . . . . . . . . . . . . 16.04 *Distichlis* (in part)

27. Lemmas 3-veined; plants of various habitats.

28. Lemmas of the pistillate florets with awns 3.4–6.8 mm long; branches of staminate inflorescences pectinate; staminate spikelets with 1 floret . . . . . . 16.47 *Opizia* (in part)

28. Lemmas of the pistillate florets with awns 30–150 mm long; branches of staminate inflorescences not pectinate; staminate spikelets with 5–20 florets . . . . . . . . . . . . . . . . . . . . . . . . . . . . . . . . . . . . . . . . . . . 16.13 *Scleropogon* (in part)

26. Some or all spikelets bisexual, the florets bisexual or unisexual, but both staminate and pistillate florets present within an individual spikelet.
    29. Lemmas of the pistillate or bisexual florets with awns 30–150 mm long; bisexual florets rarely found . . . . . . . . . . . . . . . . . . . . . . . . . . . . . . . . . 16.13 *Scleropogon* (in part)
    29. Lemmas of the bisexual florets unawned or with awns less than 10 mm long; pistillate florets not present.
        30. Inflorescences 5–15 cm long, apparently a pectinate spike, actually a solitary, pectinate, spikelike branch . . . . . . . . . . . . . . . . . . . . . . . . . . 16.42 *Ctenium* (in part)
        30. Inflorescences 1.5–10 cm long, spikes, spikelike racemes, or panicles, linear or densely cylindrical to ovoid.
            31. Inflorescences linear, (1.5)4–10 cm long; rachises not concealed by the spikelets . . . . . . . . . . . . . . . . . . . . . . . . . . . . . . . . . . . . . . . . . . . . . . . . . . . . 16.20 *Tripogon*
            31. Inflorescences cylindrical to ovoid, 1.5–5 cm long; rachises concealed by the spikelets . . . . . . . . . . . . . . . . . . . . . . . . . . . . 16.01 *Fingerhuthia* (in part)
22. Inflorescences simple panicles (sometimes highly condensed) or panicles of 1–120 spikelike branches.
    32. Inflorescences simple panicles, sometimes highly contracted, even spikelike in appearance; spikelike branches not evident [for opposite lead, see p. 190].
        33. Spikelets usually with only 1 floret, occasionally with 2–3 florets.
            34. Ligules membranous, hyaline, or coriaceous, sometimes ciliate; lemmas 3-veined (occasionally appearing 5-veined), usually awned, sometimes unawned or mucronate.
                35. Lemmas and paleas densely sericeous over the veins and margins, glabrous between the veins . . . . . . . . . . . . . . . . . . . . . . . . . . . . . . . . . . . . 16.16 *Blepharoneuron*
                35. Lemmas and paleas glabrous to variously hairy but not densely sericeous over the veins and margins.
                    36. Lemmas usually awned or mucronate; spikelets usually with 1 floret . . . . . . . . . . . . . . . . . . . . . . . . . . . . . . . . . . . . . . . . . 16.33 *Muhlenbergia* (in part)
                    36. Lemmas unawned or mucronate; spikelets frequently with 2–3 florets . . . . . . . . . . . . . . . . . . . . . . . . . . . . . . . . . . . . . . . . . . . . . 16.23 *Eragrostis* (in part)
             34. Ligules of hairs; lemmas 1(3)-veined, unawned, sometimes mucronate.
                37. Panicles 0.3–4(7.5) cm long, 3–15 mm wide, spikelike or capitate, 1–8 times longer than wide; plants annual . . . . . . . . . . . . . . . . . . . . . . . . . . . . 16.31 *Crypsis* (in part)
                37. Panicles 1–80 cm long, 2–600 mm wide, dense to open, if less than 8 cm long, often 10 or more times longer than wide; plants annual or perennial.
                    38. Calluses usually glabrous or almost so; paleas glabrous; fruits falling free of the lemma and palea . . . . . . . . . . . . . . . . . . . . . . . . . . . . . 16.30 *Sporobolus* (in part)
                    38. Calluses evidently hairy, the hairs ¼–⅞ as long as the lemmas; paleas hairy; fruits falling with the lemma and palea . . . . . . . . . . . . . . . . . . . . . 16.32 *Calamovilfa*
        33. Spikelets with more than 1 floret.
            39. Lemmas with (5)9–11 veins (the lateral veins obscure in *Allolepis*).
                40. Spikelets unisexual; plants almost always unisexual, occasionally bisexual.
                    41. Lemmas 9–11-veined; glumes 2–7 veined; plants rhizomatous and/or stoloniferous, found in saline or alkaline soils . . . . . . . . . . . . . . . 16.04 *Distichlis* (in part)
                    41. Lemmas 1–6-veined; lower glumes of the staminate spikelets 1-veined, those of the pistillate spikelets 1–5-veined; plants stoloniferous or rooting at the lower nodes, not rhizomatous, not found in saline or alkaline soils.
                        42. Plants perennial, stoloniferous; paleas of the pistillate florets completely surrounding the ovaries, the intercostal region coriaceous . . . . . . . . . . . . . . . . . . . . . . . . . . . . . . . . . . . . . . . 16.06 *Allolepis* (in part)
                        42. Plants annual, rooting at the lower nodes; paleas of the pistillate florets not completely surrounding the ovaries, the intercostal region membranous or hyaline . . . . . . . . . . . . . . . . . . . . . 16.23 *Eragrostis* (in part)
                40. All spikelets with at least 1 bisexual floret.
                43. Glumes longer than the adjacent lemmas.
                    44. Glumes exceeded by the distal florets; spikelets with 3–7 bisexual florets plus reduced distal florets; lemmas of the bisexual florets 5–7-veined throughout . . . . . . . . . . . . . . . . . . . . . . . . . . . . . . . . . . . . 16.03 *Swallenia*

44. Glumes exceeding the distal florets; spikelets with 2–4 florets, only the lowest floret bisexual; lemmas of the bisexual florets 3-veined basally, 5–7-veined distally . . . . . . . . . . . . . . . . . . . . . . 16.01 *Fingerhuthia* (in part)
43. Glumes shorter than the adjacent lemmas.
45. Spikelets ovate-elliptical to ovate-triangular, 15–50 mm long, 6–16 mm wide; lower florets sterile, without paleas . . . . . . . . . . . . . . . 16.02 *Uniola*
45. Spikelets usually elliptical to lanceolate, 1–26 mm long, 0.6–9 mm wide; lower florets in each spikelet bisexual, with paleas.
46. Calluses glabrous or sparsely pubescent; lemmas (1)3(5)-veined; spikelets 1–26 mm long, 0.6–9 mm wide . . . . . . 16.23 *Eragrostis* (in part)
46. Calluses densely pubescent; lemmas 5-, 7-, or 9-veined; spikelets 10–16 mm long, 2.5–5 mm wide . . . . . . . . . . . . . . . 16.26 *Vaseyochloa*
39. Lemmas with 1–3 veins (occasionally with scabrous lines that may be mistaken for additional veins).
47. Florets unisexual.
48. Staminate and pistillate florets strongly dimorphic; plants unisexual or bisexual, bisexual plants with unisexual or bisexual spikelets; pistillate spikelets with 3–5 functional florets and lemma awns (30)50–150 mm long; staminate spikelets with 5–10(20) florets and unawned or shortly awned (to 3 mm) lemmas . . . . . . . . . . . . . . . . . . . . . . . . . . . . . 16.13 *Scleropogon* (in part)
48. Staminate and pistillate florets similar; plants unisexual; spikelets with 4–60 florets, all or almost all functional; lemmas 1.5–10.5 mm, unawned, sometimes mucronate.
49. Plants perennial, stoloniferous; paleas of the pistillate florets completely surrounding the ovaries, the intercostal region coriaceous . . . . . . . . . . . . . . . . . . . . . . . . . . . . . . . . . . . 16.06 *Allolepis* (in part)
49. Plants annual, rooting at the lower nodes; paleas of the pistillate florets not completely surrounding the ovaries, the intercostal region membranous or hyaline . . . . . . . . . . . . . . . . . . . . . 16.23 *Eragrostis* (in part)
47. At least 1 floret in each spikelet bisexual.
50. Lemmas, including the calluses, glabrous or inconspicuously hairy; lemma apices usually entire, sometimes minutely toothed.
51. Spikelets with (1)2–60 florets; lemmas unawned, sometimes mucronate; ligules usually membranous and ciliate or ciliolate, sometimes of hairs . . . . . . . . . . . . . . . . . . . . . . . . . . . . . . 16.23 *Eragrostis* (in part)
51. Spikelets with 1(2–3) florets; lemmas often awned, sometimes unawned or mucronate; ligules membranous, sometimes ciliolate, not ciliate . . . . . . . . . . . . . . . . . . . . . . . . . . . . . . . . 16.33 *Muhlenbergia* (in part)
50. Lemma bodies conspicuously hairy over the veins and/or calluses conspicuously hairy; lemma apices usually with emarginate, bilobed, or trilobed apices, sometimes entire.
52. Leaf margins evidently cartilaginous . . . . . . . . . . . . . . . . . . . . . . 16.14 *Erioneuron*
52. Leaf margins not cartilaginous.
53. Palea keels long-hairy distally, the distal hairs 0.5–2 mm long . . . . 16.12 *Triplasis*
53. Palea keels glabrous or with hairs less than 0.5 mm long.
54. Lemmas unawned, the midveins sometimes excurrent up to 0.5 mm.
55. Lemmas rounded to truncate, emarginate to bilobed; all 3 lemma veins often pilose basally . . . . . . . . . . 16.10 *Tridens*
55. Lemmas acute, entire or with 3 minute teeth, glabrous or shortly pubescent on the distal ⅔, the pubescence not confined to the veins . . . . . . . . . . . . . . 16.11 *Redfieldia*
54. Lemmas awned, the awns 1–7 mm long.
56. Plants 80–500 cm tall; panicles 35–73 cm long, plumose; lemma margins pilose; lemma apices bifid, awned from between the teeth; awns about 3 mm long . . . . . . . . . . . . . . . . . . . . . . . . . . . . . . . . . . . . 16.08 *Neyraudia*
56. Plants 2–90 cm tall; panicles 6–30 cm long, not plumose; lemma margins sparsely pilose; lemma

apices 3–4-lobed or -toothed and 3-awned; central
awns 5–7 mm long, lateral awns 6–7 mm long . . . . . . . . . . 16.09 *Triraphis*

32. Inflorescences panicles of spikelike branches, the branches digitately or racemosely
arranged on the rachises [for opposite lead, see p. 188].

  57. Inflorescence branches 1 or more, if more than 1, arranged in terminal, digitate
clusters, sometimes with additional branches or whorls below the terminal cluster.

    58. Inflorescences with 1(2) falcate branches.

      59. Spikelets with 2 well-developed sterile or staminate florets below the
bisexual florets; additional sterile or staminate florets present distal to the
bisexual floret . . . . . . . . . . . . . . . . . . . . . . . . . . . . . . . . . . . . . 16.42 *Ctenium* (in part)

      59. Spikelets usually with 1, rarely 2, florets, the lowest or only floret bisexual
. . . . . . . . . . . . . . . . . . . . . . . . . . . . . . . . . . . . . . . . . . . . 16.43 *Microchloa* (in part)

    58. Inflorescences with more than 1 branch or, if only 1, the branch not strongly
falcate.

      60. Plants unisexual.

        61. Staminate spikelets 4–6 mm long, with 2 florets; upper glumes of the
pistillate spikelets indurate, white . . . . . . . . . . . . . . . . . . . . . . . 16.48 *Buchloë* (in part)

        61. Staminate spikelets 3–4 mm long, with 2 florets; upper glumes of the
pistillate spikelets membranous . . . . . . . . . . . . . . . . . . . . . . . . . 16.47 *Opizia* (in part)

      60. Plants bisexual, all spikelets with at least 1 bisexual floret.

        62. Spikelets with more than 1 bisexual floret.

          63. Panicle branches 0.4–7 cm long, terminating in a point . . . . . . . . . 16.29 *Dactyloctenium*

          63. Panicle branches 1–22 cm long, terminating in a functional or
rudimentary spikelet.

            64. Disarticulation eventually below the glumes, initially below
the lemmas and caryopses, the paleas persistent; panicle
branches terminating in a rudimentary spikelet . . . . . . . . . . . . . . 16.28 *Acrachne*

            64. Disarticulation above the glumes, usually also below the
florets; panicle branches terminating in a functional spikelet.

              65. Lemmas 3-awned, the central awns 8–12 mm long . . 16.38 *Trichloris* (in part)

              65. Lemmas not 3-awned.

                66. Lemmas usually with hairs over the veins, at least
basally, the apices often toothed, sometimes
mucronate or awned . . . . . . . . . . . . . . . . . . 16.19 *Leptochloa* (in part)

                66. Lemmas glabrous, the apices entire, neither
mucronate nor awned . . . . . . . . . . . . . . . . . . . . . . . . . 16.27 *Eleusine*

        62. Spikelets usually with only 1 bisexual floret (occasionally 2 in some
genera), often with additional staminate, sterile, or modified florets.

          67. Spikelets usually without sterile or modified florets; lemmas
unawned . . . . . . . . . . . . . . . . . . . . . . . . . . . . . . . . . . . . . . . . . . . . 16.44 *Cynodon*

          67. Spikelets with 1 or more sterile florets distal to the bisexual floret;
lemmas of the bisexual florets often awned.

            68. Lowest lemmas in the spikelets 3-awned, the central awns
8–12 mm long, the lateral awns 0.5–12 mm long . . . . . . . 16.38 *Trichloris* (in part)

            68. Lowest lemmas in the spikelets usually unawned or with a
single awn, if 3-awned, the lateral awns less than 0.5 mm
long.

              69. Spikelets dorsally compressed . . . . . . . . . . . . . . . 16.37 *Enteropogon* (in part)

              69. Spikelets laterally compressed or terete.

                70. Upper glumes truncate or bilobed; lowest lemmas
unawned or with an awn to 1.2 mm long . . . . . . . . . . . . . 16.36 *Eustachys*

                70. Upper glumes acute to acuminate; lowest lemmas
usually awned, the awns to 37 mm long . . . . . . . . . . . . . . 16.35 *Chloris*

  57. Inflorescence branches more than 1, racemosely arranged on the rachises.

    71. All spikelets unisexual.

      72. Staminate spikelets 4–6 mm long, the anthers 2.5–3 mm long; pistillate
spikelets with 1 unawned or shortly awned floret; widespread species of the
central plains . . . . . . . . . . . . . . . . . . . . . . . . . . . . . . . . . . . . . . . . . . 16.48 *Buchloë* (in part)

72. Staminate spikelets 3–4 mm long, the anthers 2–2.5 mm long; pistillate lemmas with awns 3.4–6.8 mm long; in the *Manual* region, known only as an occasional escape from lawns and experimental plots in Florida . . . . . . 16.47 *Opizia* (in part)
71. All spikelets with at least 1 bisexual floret.
    73. Inflorescence branches woody, terminating in hard, sharp points . . . . . . . . . . . 16.24 *Cladoraphis*
    73. Inflorescence branches not particularly stiff or rigid, terminating in spikelets or points.
        74. Spikelets with more than 1 bisexual floret, sometimes also with reduced florets.
            75. Lemmas 7–11-veined, mucronate, the mucros 0.1–0.3 mm long; not established in the *Manual* region . . . . . . . . . . . . . . . . . . . . . . . 16.05 *Aeluropus*
            75. Lemmas 3-veined, unawned, mucronate, or awned, the awns often much more than 1 mm long; established in the *Manual* region.
                76. Lower glumes exceeding the lowest lemmas, sometimes exceeding the distal lemmas . . . . . . . . . . . . . . . . . . . . . . . . . 16.21 *Trichoneura*
                76. Lower glumes not or only slightly exceeding the lowest lemmas.
                    77. Inflorescences with 50 or more closely spaced, arcuate, tardily deciduous branches . . . . . . . . . . . . . . . . 16.25 *Pogonarthria* (in part)
                    77. Inflorescences with 2–120 straight, non-disarticulating branches.
                        78. Spikelets with (2)3–12(20) bisexual florets . . . 16.19 *Leptochloa* (in part)
                        78. Spikelets with 2–4 florets, but only the lowest 1(2) florets bisexual . . . . . . . . . . . . . . . . . . . . . 16.41 *Gymnopogon* (in part)
        74. Spikelets with 1 bisexual floret, sometimes with sterile, rudimentary, or modified florets distal to the bisexual floret.
            79. Functional spikelets with sterile, rudimentary, or modified florets distal to the bisexual floret.
                80. Spikelets widely spaced to slightly imbricate, appressed to the branch axes . . . . . . . . . . . . . . . . . . . . . . . . . . . . 16.41 *Gymnopogon* (in part)
                80. Spikelets densely imbricate, varying from appressed to strongly divergent.
                    81. Inflorescence branches usually solitary at each node (sometimes only 1 per panicle); spikelets laterally compressed or terete . . . . . . . . . . . . . . . . . . . . . . . 16.46 *Bouteloua* (in part)
                    81. Inflorescence branches usually more than 1 at the lower nodes; spikelets dorsally compressed . . . . . . . . . . 16.37 *Enteropogon* (in part)
            79. Functional spikelets with only 1 floret, lacking sterile, rudimentary, or modified florets.
                82. Spikelets distant to slightly imbricate, appressed to the branches; branches strongly divergent.
                    83. Blades with thick, white margins and a well-developed midrib . . . . . . . . . . . . . . . . . . . . . . . . . . . . . . . . . . . . 16.40 *Schedonnardus*
                    83. Blades lacking both thick, white margins and well-developed midribs . . . . . . . . . . . . . . . . . . . . . . 16.41 *Gymnopogon* (in part)
                82. Spikelets clearly imbricate, appressed to strongly divergent; branches appressed to strongly divergent.
                  84. Spikelets laterally compressed, appressed to divergent . . . . . . . . . 16.45 *Spartina*
                  84. Spikelets dorsally compressed, appressed . . . . . . . . . . . . . 16.39 *Willkommia*

## 16.01 FINGERHUTHIA Nees[1]

Pl usu per, occ ann in desert areas; csp and shortly rhz. Clm 5–120 cm, unbrchd. Lvs mostly bas; **lig** of hairs; **bld** 2–5 mm wide. Infl tml, exceeding the up lvs, dense, cylindrical to ovoid pan, occ rdcd to rcm; **br** short, non-disarticulating; **rchs** concealed by the spklt; **dis** beneath the glm. Spklt lat compressed, with 2–4 flt, only the bas flt bisx, the next 2 flt usu stmt, the fourth flt, if present, strl. **Glm** subequal, clearly exceeding the flt, awned or unawned; **lm** firmly memb, 3-veined bas, 5–7-veined distally, mucronate to shortly awned, awns shorter than 10 mm; **anth** 3; **ov** glab.

*Fingerhuthia* has two species, one native to southern Africa and western Asia, the other endemic to southern Africa. One species has been grown in the *Manual* region.

[1]Mary E. Barkworth

**1. Fingerhuthia africana** Lehm. THIMBLEGRASS, ZULU FESCUE [p. *429*]

*Fingerhuthia africana* is native to southern Africa and western Asia. It has been grown at the Santa Rita Experimental Range in Pima County, Arizona, but is not established in the *Manual* region.

## 16.02 UNIOLA L.[1]

Pl per; rhz or stln. **Clm** to 2.5 m, erect, glab, unbrchd. **Lig** of hairs; **bld** flat, becoming involute when dry, mrg scabrous, tapering to an attenuate apex. **Infl** tml, simple pan, exceeding the lvs; **dis** below the glm. **Spklt** 8–50 mm long, 6–16 mm wide, ovate-elliptical to ovate-triangular, strongly lat compressed, with 3–34 flt, lowest 2–8 flt strl, remaining flt(s) bisx. **Glm** subequal, shorter than the adjacent lm, midveins keeled, serrate to serrulate, apc unawned; **lm** 3–9-veined, midveins keeled, serrate to serrulate, apc somewhat blunt to acute or mucronate, unawned; **pal**, if present, from slightly shorter than to exceeding the lm, 2-keeled, keels winged, serrulate or ciliate; **anth** 3; **ovary** glab; **sty** 1, with 2 style br. **Car** linear; **emb** less than ½ as long as the car.

*Uniola* has two species, both of which grow on coastal sand dunes. There is one species native to the *Manual* region; the second, *U. pittieri* Hack., extends from northern Mexico to Ecuador, primarily along the Pacific coast.

**1. Uniola paniculata** L. SEA OATS [p. *430*, 531]

*Uniola paniculata* grows on the beaches and sand dunes of the Atlantic and Gulf coastal plains from Maryland to Veracruz, Mexico, and on the Florida Keys, the Bahama Islands, and Cuba. Seed production is generally poor; the reason is not known.

## 16.03 SWALLENIA Soderstr. & H.F. Decker[2]

Pl per; clumped, rhz wd. **Clm** 10–60 cm, brchd above the base. **Lvs** mostly bas; **aur** absent; **lig** of hairs; **bld** flat, strongly veined, sharply pointed. **Infl** tml, usu exceeding the up lvs, contracted pan; **br** ascending to erect. **Spklt** lat compressed, unawned, with 3–7 bisx flt, distal flt rdcd; **dis** beneath the car. **Glm** subequal, longer than the adjacent lm but exceeded by the distal flt, acuminate; **lo** glm 5–7-veined; **up glm** 7–11-veined; **cal** hairy; **lm** memb to papery, 5–7-veined, densely villous on the mrg, smt also between the veins, unawned to mucronate; **pal** equaling or exceeding the lm; **anth** 3. **Car** falling free from the lm and pal.

*Swallenia* is a unispecific genus, endemic to California. It is unusual in that only its caryopses break off the plant.

**1. Swallenia alexandrac** (Swallen) Soderstr. & H.F. Decker EUREKA VALLEY DUNEGRASS [p. *430*, 531]

*Swallenia alexandrae* grows on sand dunes in Inyo County, California. It is only known from four sites, all between 900–1200 m, in the Eureka Valley of northern Inyo County. At these sites, it forms dense colonies 1–2 m across. It is state-listed as rare and federally listed as endangered because of off-road vehicle activity.

## 16.04 DISTICHLIS Raf.[3]

Pl per; usu unisx, occ bisx; strongly rhz and/or stln. **Clm** to 60 cm, usu erect, glab. **Lvs** conspicuously distichous; lo lvs rdcd to scalelike shth; up lf shth strongly overlapping; **lig** shorter than 1 mm, memb, serrate; up bld stiff, glab, ascending to spreading, usu equaling or exceeding the pist pan. **Infl** tml, contracted pan or rcm, smt exceeding the up lvs. **Spklt** lat compressed, with 2–20 flt; **dis** of the pist spklt above the glm and below the flt, stmt spklt not disarticulating. **Glm** 3–7-veined; lm coriaceous, stmt lm thinner than the pist lm, 9–11-veined, unawned; **pal** 2-keeled, keels narrowly to broadly winged, serrate to toothed, smt with excurrent veins; **anth** 3. **Car** glab, free from the pal at maturity, brown.

*Distichlis* has about five species that grow in saline soils of the coasts and interior deserts of the Western Hemisphere and Australia. All the species grow in South America, but only one, *D. spicata*, is found in North America.

**1. Distichlis spicata** (L.) Greene SALTGRASS [p. *430*, 531]

*Distichlis spicata* grows in saline soils of the Western Hemisphere and Australia.

## 16.05 AELUROPUS Trin.[4]

Pl per; usu strongly rhz or stln, rhz and stln with persistent shth. **Clm** 5–40 cm, prostrate to erect, solitary or not; intnd numerous, short. **Shth** overlapping; **lig** of hairs; **bld** usu stiffly spreading, flat or convolute below, folded distally, apc often cartilaginous and sharp. **Infl** tml, dense pan of non-disarticulating spikelike br racemosely

[1]H. Oliver Yates  [2]James P. Smith, Jr.  [3,4]Mary E. Barkworth

arranged on elongate rchs, exceeding the up lvs; **br** 2.5–5 mm wide, axes triquetrous, spklt closely packed, subsessile, in 2 rows. **Spklt** bisx, dorsally compressed, with 2–14 flt. **Glm** unequal, exceeded by the flt, keeled; **lo glm** 1–5(7)-veined; **up glm** 3–9-veined; **lm** 7–11-veined, mucronate, mucros 0.1–0.3 mm; **lod** 2; **anth** 3, 0.8–1.6

mm; **sty br** 2. **Car** ovoid-ellipsoid; **hila** punctate, bas; **emb** about ½ as long as the car.

*Aeluropus* is a genus of five species that extends from Portugal to China and northern India. One species has been cultivated in the *Manual* region.

### 1. Aeluropus littoralis (Gouan) Parl. [p. *430*]

*Aeluropus littoralis* is native to Eurasia, where it grows in sandy, often saline habitats. It has been cultivated experimentally in Pima County, Arizona; it is not established in the *Manual* region.

## 16.06 ALLOLEPIS Soderstr. & H.F. Decker[1]

**Pl** per; dioecious, stmt and pist pl similar; stln. **Clm** 10–70 cm. **Lvs** mostly cauline; **lig** memb, ciliate. **Infl** tml pan, exceeding the up lvs; **br** appressed or ascending, usu spklt-bearing to the base. **Spklt** not disarticulating. **Pist spklt** with 5–10 flt; **lo glm** 1–5-veined, only the midveins conspicuous; **up glm** with 3 conspicuous veins, smt with 2–4 inconspicuous lat veins; **lm** 3–5(6)-veined, unawned; **pal** shorter than the lm, completely surrounding the ov, saccate bas, narrowing distally and surrounding the sty,

intercostal region coriaceous, veins keeled, scabrous and puberulent, mrg scarious. **Stmt spklt** with 4–21 flt; **lo glm** 1-veined; **up glm** 1–3-veined; **lm** conspicuously 3-veined, unawned; **pal** equaling or slightly longer than the lm, linear, 2-veined, veins ciliolate; **lod** 2, cuneate.

*Allolepis* is a unispecific North American genus. It used to be included in *Distichlis*, but it differs in its paleal morphology, in never producing rhizomes, and in its soil preferences.

### 1. Allolepis texana (Vasey) Soderstr. & H.F. Decker
FALSE SALTGRASS [p. *431*, <u>531</u>]

*Allolepis texana* grows in the Big Bend region of Texas and northern Mexico. It is not common, but it may be locally abundant, growing on sandy or silty, but not alkaline, soils.

## 16.07 MONANTHOCHLOË Engelm.[2]

**Pl** per; dioecious; mat-forming, with long, wiry stln. **Clm** 5–20 cm, erect, with numerous short, lfy, lat br. **Lvs** distinctly distichous, clustered on distant to closely-spaced, short, lat shoots; **lig** thickly memb ciliate rims; **bld** stiff, subulate. **Infl** tml, composed of a single glab spklt, this enclosed, and almost concealed, by the upmost lf shth. **Pist spklt** subterete, with 3–5 flt, distal flt rdmt; **dis** tardy, below the lowest flt; **glm** absent; **lm** coriaceous, glab, 9-veined, acute; **pal** coriaceous, keels prominently

winged, wings overlapping and enclosing the car. **Stmt spklt** similar to the pist spklt, but smaller and the lm and pal thinner.

*Monanthochloë* is a genus of two species, one growing along the southern coastlines of North America, the other on inland salt pans in Argentina. It is probably related to *Distichlis*, but differs from that genus, and all others in the *Manual* region, in its highly reduced inflorescence.

### 1. Monanthochloë littoralis Engelm. SHOREGRASS [p. *431*, <u>531</u>]

*Monanthochloë littoralis* grows in moist, sandy, saline soils along the coast of southern California and the southeastern United States, northeastern Mexico, and the Caribbean islands.

## 16.08 NEYRAUDIA Hook. f.[3]

**Pl** per; csp, with short, thick, scaly rhz. **Clm** 80–500 cm tall, to 1.5 cm thick, reedlike, almost wd; **intnd** solid. **Lvs** cauline, evenly distributed, distichous; **shth** with tightly overlapping mrg; **lig** a cartilaginous ridge subtending a line of hairs; **bld** broad, flat, deciduous at maturity, mrg not cartilaginous. **Infl** tml, plumose pan, 30–80 cm, exceeding the up lvs; **dis** above the glm and between the flt. **Spklt** lat compressed; **flt** (2)4–10, bisx, lowest flt smt

only an empty lm; **rchl intnd** glab. **Glm** 2, shorter than the lm, subequal, narrowly ovate, glab, memb, 1-veined, acute; **cal** obtuse, pilose; **lm** narrowly ovate, thinly memb to hyaline, 3-veined, midveins glab, mrg and lat veins long-pilose on the distal ⅔, apc bifid, shortly awned from between the teeth; **pal** shorter than the lm, hyaline, 2-keeled.

[1]J.K. Wipff [2]John W. Thieret† [3]Gerald F. Guala

*Neyraudia* is a genus of 2–4 species from the tropics of the Eastern Hemisphere. One Asian species is established in Florida. It differs from the other species in the genus in having a sterile lemma at the base of the spikelets. *Neyraudia* is often mistaken for *Phragmites*, but that genus has villous rachilla internodes and glabrous lemmas.

### 1. **Neyraudia reynaudiana** (Kunth) Keng *ex* Hitchc.
BURMA REED, SILK REED [p. *431*, <u>531</u>]

*Neyraudia reynaudia* is an Asian species that was introduced at Chapman Field, U.S. Department of Agriculture, Coral Gables, Florida, in 1915. It is now a troublesome weed in that state, growing in a variety of habitats from marshy areas to dry pinelands.

## 16.09 TRIRAPHIS R. Br.[1]

**Pl** ann or per; csp. **Clm** (1)4–140 cm, erect or geniculate at the lo nd. **Lvs** cauline; **aur** absent; **lig** of hairs or memb and long-ciliate; **bld** narrow, often involute. **Infl** tml, open or contracted (occ spikelike) simple pan, exceeding the lvs. **Spklt** lat compressed, with 3–9 bisx flt, rdcd flt (if present) distal; **dis** above the glm and beneath the flt. **Glm** subequal, exceeded by the flt, 1-veined; **cal** short, bearded; **lm** 3-veined, 3–4-lobed or toothed, lat veins pilose or ciliate, midveins glab or sparsely pubescent, all 3 veins extending into awns; **pal** shorter than the lm, 2-veined; **lod** 2; **anth** 3. **Car** trigonous, falling free of the lm and pal; **emb** large relative to the car.

*Triraphis* is a genus of seven species that are often found in dry, open habitats in sandy or rocky soil. Most of its species are native to Africa and Arabia, but one species native to Australia is established in the *Manual* region.

### 1. **Triraphis mollis** R. Br. PURPLE NEEDLEGRASS [p. *431*, <u>531</u>]

*Triraphis mollis* usually flowers in response to rain. It is common on sandy soils in New South Wales, Australia. In the *Manual* region it is currently known only from Dimmit County, Texas; it will probably spread.

## 16.10 TRIDENS Roem. & Schult.[2]

**Pl** per; usu csp, often with short, knotty rhz, occ with elongate rhz, never stln. **Clm** 5–180 cm, erect, mostly glab, lo nd smt with hairs. **Shth** shorter than the intnd, open; **lig** memb and ciliate or of hairs; **bld** 6–25 cm long, 1–8 mm wide, flat or involute, mrg not thick and cartilaginous. **Infl** tml, usu pan (smt rdcd to rcm), 5–40 cm, exceeding the up lvs, exserted. **Spklt** 4–10(13) mm, lat compressed, with 4–11(16) flt, more than 1 flt bisx; **strl flt** distal to the ftl spklt; **dis** above the glm. **Glm** from shorter than to equaling the distal flt; **lo glm** 1(3)-veined; **up glm** shorter than or about equal to the lo glm, 1–3(9)-veined, unawned; **cal** usu glab, smt pilose; **lm** hyaline or memb, 3-veined, veins usu shortly hairy below, apc rounded to truncate, emgt to bilobed, midvein often excurrent to 0.5 mm, lat veins not or more shortly excurrent; **pal** glab or shortly pubescent on the lo back and mrg, veins glab or ciliolate; **lod** 2, free or adnate to the pal; **anth** 3, reddish-purple. **Car** dorsiventrally compressed and reniform in cross section, dark brown; **emb** about 2/5 as long as the car.

*Tridens*, a genus of 14 species, is native to the Americas; all ten species described here are native to the the *Manual* region.

1. Primary panicle branches appressed to strongly ascending; panicles 0.3–4 cm wide, dense and spikelike.
    2. Lateral veins of the lemmas glabrous or pubescent only at the base . . . . . . . . . . . . . . . . . . . . . . . . . . . . .1. *T. albescens*
    2. Lateral veins of the lemmas pilose to well above the base.
        3. Glumes evidently longer than the adjacent lemmas, often twice as long, usually equaling or exceeding the distal florets . . . . . . . . . . . . . . . . . . . . . . . . . . . . . . . . . . . . . . . . .2. *T. strictus*
        3. Glumes from shorter than to equaling the adjacent lemmas, often exceeded by the distal florets.
            4. All 3 lemma veins shortly excurrent; calluses pilose . . . . . . . . . . . . . . . . . . . . . . . . . . . .3. *T. carolinianus*
            4. Lateral lemma veins not excurrent, often terminating before the distal margin, the midvein sometimes excurrent; calluses glabrous or shortly pilose.
                5. Panicles 7–25 cm long, 0.3–0.8 cm wide; lemma midveins rarely excurrent . . . . . . . . . . . . . . . . . .4. *T. muticus*
                5. Panicles 5–8(10) cm long, 1.2–2.5 cm wide; lemma midveins always shortly excurrent . . . . . . . .5. *T. congestus*
1. Primary panicle branches ascending to reflexed or drooping; panicles 1–20 cm wide, open, not spikelike.
    6. All pedicels shorter than 1 mm . . . . . . . . . . . . . . . . . . . . . . . . . . . . . . . . . . . . . . . . . . . . . . .6. *T. ambiguus*
    6. Some pedicels longer than 1 mm.
        7. Lateral veins of the lemmas rarely excurrent.
            8. Lemmas 4–6 mm; ligules 0.4–1 mm . . . . . . . . . . . . . . . . . . . . . . . . . . . . . . .7. *T. buckleyanus*
            8. Lemmas 2–3.2 mm; ligules 1.2–3 mm . . . . . . . . . . . . . . . . . . . . . . . . . . . . . .8. *T. eragrostoides*
        7. Lateral veins of the lemmas commonly excurrent as short points.
            9. Blades 1–5 mm wide; panicles 5–16 cm long . . . . . . . . . . . . . . . . . . . . . . . . . . . . .9. *T. texanus*
            9. Blades mostly 3–10 mm wide; panicles 15–40 cm long . . . . . . . . . . . . . . . . . . . . . . .10. *T. flavus*

[1]J.K. Wipff  [2]Jesús Valdés-Reyna

### 1. Tridens albescens (Vasey) Wooton & Standl. WHITE TRIDENS [p. *432*, 531]

*Tridens albescens* grows in plains and open woods, often in clay soils that periodically receive an abundance of water. Its range extends into northern Mexico.

### 2. Tridens strictus (Nutt.) Nash LONGSPIKE TRIDENS [p. *432*, 531]

*Tridens strictus* grows in open woods, old fields, right of ways, and coastal grasslands. It is endemic to the United States.

### 3. Tridens carolinianus (Steud.) Henrard CREEPING TRIDENS [p. *432*, 531]

*Tridens carolinianus* grows in pinelands and open sandy woods along the coastal plain from North Carolina to Louisiana.

### 4. Tridens muticus (Torr.) Nash SLIM TRIDENS [p. *432*, 531]

1. Upper glumes 4–5(6) mm long, 1-veined . . . . . . . . . var. *muticus*
1. Upper glumes usually 5.5–10 mm long, 3–7-veined . . . . . . . . . . . . . . . . . . . . . . . . . . . var. *elongatus*

### Tridens muticus var. elongatus (Buckley) Shinners [p. *432*]

*Tridens muticus* var. *elongatus* grows on well-drained, clayey and sandy soils from Colorado to Missouri and from Arizona to Louisiana. It is not known from Mexico.

### Tridens muticus (Torr.) Nash var. muticus [p. *432*]

*Tridens muticus* var. *muticus* is a common species on dry, sandy or clay soils in the arid southwestern United States and adjacent Mexico.

### 5. Tridens congestus (L.H. Dewey) Nash PINK TRIDENS [p. *432*, 531]

*Tridens congestus* grows in moist depressions, ditches, and low flats of otherwise dry hills in Texas. It resembles *T. albescens*, but usually has shorter panicles, spikelets that are more or less evenly pink rather than purple-tipped, and more deeply cleft lemma apices.

### 6. Tridens ambiguus (Elliott) Schult. PINE-BARREN TRIDENS [p. *432*, 531]

*Tridens ambiguus* grows on the southeastern coastal plain, from North Carolina to Texas. It is usually found in mesic to perennially moist soils of pine flatwoods and pine-oak savannahs, in seasonally inundated depressions, and at the margins of pitcher plant bogs, often in disturbed sites.

### 7. Tridens buckleyanus (L.H. Dewey) Nash BUCKLEY'S TRIDENS [p. *432*, 531]

*Tridens buckleyanus* is endemic to the southeastern portion of the Edwards Plateau, Texas. It grows on rocky slopes along shaded stream banks and the borders of woodlands.

### 8. Tridens eragrostoides (Vasey & Scribn.) Nash LOVEGRASS TRIDENS [p. *432*, 531]

*Tridens eragrostoides* grows in brush grasslands, generally in partial shade. Its range extends from the southern United States into Mexico and Cuba.

### 9. Tridens texanus (S. Watson) Nash TEXAS TRIDENS [p. *432*, 531]

*Tridens texanus* grows in clayey and sandy loam soils, often in the protection of shrubs and along fenced road right of ways. Its range extends from southern Texas into northern Mexico.

### 10. Tridens flavus (L.) Hitchc. PURPLETOP TRIDENS [p. *432*, 531]

1. Panicles nodding; pulvini inconspicuously or conspicuously hairy, the hairs confined to the adaxial side of the branches . . . . . . . . . . . . . . . . . . . . . . . . var. *flavus*
1. Panicles erect throughout; pulvini always conspicuously hairy, the hairs extending around the base of the branches . . . . . . . . . . . . . . . . . . . . . . . . var. *chapmanii*

### Tridens flavus var. chapmanii (Small) Shinners [p. *432*]

*Tridens flavus* var. *chapmanii* grows in pine and oak woods of the southeastern United States from Missouri to Virginia and south from eastern Texas to Florida.

### Tridens flavus (L.) Hitchc. var. flavus [p. *432*]

*Tridens flavus* var. *flavus* grows in old fields and open woods. Its range extends to Nuevo Léon, Mexico. It was discovered for the first time in Canada in 1976, growing along a railway track in southern Ontario.

## 16.11 REDFIELDIA Vasey[1]

Pl per; with extensive, often deep, horizontal or vertical rhz. Clm 50–130 cm, erect, bases usu buried and rooting at the nd. Lvs cauline; shth shorter than the intnd, open, ribbed; lig memb, ciliate; aur absent; bld loosely involute, smt scabridulous, apc attenuate. Infl tml, conical to oblong pan, open to diffuse, exceeding the up lvs; br slender, widely spreading. Ped longer than the spklt, capillary, flexible. Spklt ovate to obovate, olive-green to brownish, with (1)2–6 flt; strl flt distal to the bisx flt; dis above the glm and below the flt. Glm unequal, usu exceeded by the flt, glab, acute; lo glm 1-veined; up glm 1- or 3-veined; cal with a tuft of soft hairs; lm lanceolate to falcate, glab or shortly pubescent, at least on the distal ⅔, 3-veined, lat veins converging distally, apc acute to awn-tipped, entire or with 3 minute teeth; anth 3.

*Redfieldia* is a unispecific genus that is endemic to the *Manual* region.

### 1. Redfieldia flexuosa (Thurb. *ex* A. Gray) Vasey BLOWOUT-GRASS [p. *433*, 532]

*Redfieldia flexuosa* grows on sandhills and dunes. It is a common and important soil binder in blowout areas. The only Arizona collection was made in 1896; the Washington population was introduced for erosion control.

[1]Stephan L. Hatch

## 16.12 TRIPLASIS P. Beauv.[1]

**Pl** ann or per; csp, occ rhz. **Clm** 14–100 cm, ascending to erect; **nd** pubescent to hirsute. **Shth** open; **aur** absent; **lig** of hairs or memb and ciliate; **bld** 1–5 mm wide, flat or involute. **Pri infl** tml, open pan, with few spklt, exserted or partially included in the up shth, apc exceeding the up lf bld, ax pan smt also present; **cleistogamous infl** also present in the up shth. **Spklt** lat compressed, purplish, with 2–5 flt; **strl flt** above the ftl flt; **rchl** prolonged; **dis** above the glm and beneath the flt and, subsequently, at the cauline nd. **Glm** equal or unequal, shorter than the first lm, 1-veined, keeled; **cal** hairy; **lm** 3-veined, veins villous, apc bilobed to incised, midveins smt extending into an awn, awns to 11 mm; **pal** bowed-out, keels hairy, distal hairs 0.5–2 mm, longer than those below; **lod** 2, truncate; **anth** 3, yellow or reddish-purple; **stigmas** pink to purple. **Car** dorsiventrally compressed.

*Triplasis* is an American genus of two species that is probably related to *Tridens*. The disarticulating culm, which helps disperse the cleistogenes, aids in distinguishing *Triplasis* from most other genera.

1. Lemmas with lobes 4.5–8 mm long, tapering to acute tips; lemma awns 5–11 mm long; culm internodes puberulent to pilose . . . . . . . . . . . . . . . . . . . . . . . . . . . . . . . . . . . . . . . . . . . . . . . . . . . . . . . . . . . . . . . 1. *T. americana*
1. Lemmas with lobes about 1 mm long, rounded; lemma awns less than 2 mm long; culm internodes glabrous . . . . . . . . . . . . . . . . . . . . . . . . . . . . . . . . . . . . . . . . . . . . . . . . . . . . . . . . . . . . . . . . . . . . . . . . . . . . 2. *T. purpurea*

### 1. Triplasis americana P. Beauv. PERENNIAL SANDGRASS [p. 433, 532]

*Triplasis americana* is endemic to the southeastern United States. It grows on sandy soils in prairies and woods, being less common in maritime dunes than *Triplasis purpurea*.

### 2. Triplasis purpurea (Walter) Chapm. PURPLE SANDGRASS [p. 433, 532]

*Triplasis purpurea* grows in sandy soils throughout the eastern and central portion of the *Manual* region, extending southward through Mexico to Costa Rica. It has been found in Washington and Oregon, but may not persist there. It is far more common in maritime dunes than *T. americana*. Plants in the *Manual* region belong to **Triplasis purpurea (Walter) Chapm. var. purpurea**.

## 16.13 SCLEROPOGON Phil.[2]

**Pl** per; usu monoecious, less frequently dioecious, occ synoecious; bearing wiry, often arching, stln with 5–15 cm intnd, smt also weakly rhz. **Lvs** mostly bas; **shth** short, strongly veined, bas lvs commonly hispid or villous; **lig** of hairs; **bld** firm, flat or folded. **Infl** tml, usu exceeding the up lvs, spikelike rcm or contracted pan with few spklt, in bisx pl stmt and pist flt in the same spklt with the stmt flt below the pist flt or in separate spklt, bisx flt occ produced; **br** not pectinate; **dis** above the glm and below the lowest pist flt in a spklt, flt falling together, lowest flt with a bearded, sharp-pointed callus.

**Stmt spklt** with 5–10(20) flt; **glm** memb, pale, 1–3-veined, acuminate; **lm** 3-veined, similar to the glm, unawned or awned, awns to 3 mm; **pal** shorter than the lm, often conspicuously so. **Pist spklt** appressed to the br axes, usu the 3–5 lo flt fnctl, up flt rdcd to awns; **glm** acuminate, strongly 3-veined, occ with a few fine accessory veins; **lm** narrow, 3-veined, veins extending into awns, awns (30)50–100(150) mm, spreading or reflexed at maturity.

*Scleropogon* is a unispecific American genus with a disjunct distribution.

### 1. Scleropogon brevifolius Phil. BURROGRASS [p. 433, 532]

*Scleropogon brevifolius* grows on grassy plains and flats, generally being most abundant on disturbed or overgrazed land. Its North American range extends from the southwestern United States to central Mexico; its South American range is from Chile to northwestern Argentina.

## 16.14 ERIONEURON Nash[3]

**Pl** per; usu csp, occ stln. **Clm** 6–65 cm, erect. **Lvs** mostly bas; **shth** smooth, glab, striate, mrg hyaline, col with tufts of 1–3 mm hairs; **bld** usu folded, pilose bas, mrg white, cartilaginous, apc acute but not sharp. **Infl** tml, simple pan (rcm in depauperate specimens), exserted well above the lvs. **Spklt** lat compressed, with 3–20 flt, distal flt stmt or strl; **dis** above the glm and between the flt. **Glm** thin, memb, 1-veined, acute to acuminate; **cal** with hairs; **lm** rounded on the back, 3-veined, veins conspicuously pilose, at least bas, apc toothed or obtusely 2-lobed, midveins often extended into awns, awns to 4 mm, lat veins smt extended as small mucros; **pal** shorter than the lm, keels ciliate, intercostal regions pilose bas; **lod** 2, adnate to the bases of the pal; **anth** 1 or 3. **Car** glossy, translucent; **emb** more than ½ as long as the car.

*Erioneuron* is an American genus of three species. Its seedlings appear to have a shaggy, white-villous indumentum, but this is composed of a myriad of small, water-soluble crystals.

Stoloniferous plants are unusual in the region covered by the *Manual*, but they are quite common in populations of *Erioneuron nealley* and *E. avenaceum* from central Mexico.

[1]Stephan L. Hatch  [2]John R. Reeder  [3]Jesús Valdés-Reyna

1. Lemmas entire or with teeth to 0.5 mm long, the awns 0.5–2.5 mm long; both glumes shorter than the lowest floret . . . . . . . . . . . . . . . . . . . . . . . . . . . . . . . . . . . . . . . . . . . . . . . . . . . . . . . . . . . . . . . . . . . . 1. *E. pilosum*
1. Lemmas 2-lobed, the lobes 1–2.5 mm long, the awns 1–4 mm long; upper glumes equaling or exceeding the lowest floret.
   2. Lemma lobes obtuse to broadly acute, 1–2 mm long; lateral veins not forming mucros; plants 7–40 cm tall . . . . . . . . . . . . . . . . . . . . . . . . . . . . . . . . . . . . . . . . . . . . . . . . . . . . . . . . . . . . . . . 2. *E. avenaceum*
   2. Lemma lobes rounded to truncate, 1.5–2.5 mm long; lateral veins forming mucros to 1 mm long; plants 15–65 cm tall . . . . . . . . . . . . . . . . . . . . . . . . . . . . . . . . . . . . . . . . . . . . . . . . . . . . . . . . . . . . 3. *E. nealleyi*

### 1. Erioneuron pilosum (Buckley) Nash HAIRY TRIDENS, HAIRY WOOLYGRASS [p. *433*, <u>532</u>]

*Erioneuron pilosum* grows on dry, rocky hills and mesas, often in oak and pinyon-juniper woodlands. In North America, it is represented by *E. pilosum* var. **pilosum**. This variety differs from the other two varieties, both of which are restricted to Argentina, in its longer, less equal glumes and shorter awns.

### 2. Erioneuron avenaceum (Kunth) Tateoka LARGE-FLOWERED TRIDENS, SHORTLEAF WOOLYGRASS [p. *433*, <u>532</u>]

*Erioneuron avenaceum* is common in rocky areas from the southwestern United States to central Mexico; it also grows in Bolivia and Argentina. North American plants belong to **E. avenaceum** (Kunth) Tateoka var. **avenaceum**. Stoloniferous plants occur in the *Manual* region, but they are most common in central Mexico.

### 3. Erioneuron nealleyi (Vasey) Tateoka NEALLEY'S ERIONEURON, NEALLEY'S WOOLYGRASS [p. *433*, <u>532</u>]

*Erioneuron nealleyi* is found on rocky slopes in the southwestern United States and central Mexico. Stoloniferous plants are known only from central Mexico.

## 16.15 DASYOCHLOA Willd. *ex* Rydb.[1]

**Pl** per; stln, smt mat-forming. **Clm** (1)4–15 cm, initially erect, eventually bending and rooting at the base of the infl. **Lvs** not bas aggregated on the pri clm; **shth** with a tuft of hairs to 2 mm at the throat; **lig** of hairs; **bld** involute. **Infl** tml, short, dense pan of spikelike br, each subtended by lfy bracts and exceeded by the up lvs; **br** with 2–4 subsessile to shortly pedlt spklt. **Spklt** lat compressed, with 4–10 flt; **dis** above the glm. **Glm** subequal to the adjacent lm, glab, 1-veined, rounded or weakly keeled, shortly awned to mucronate; **flt** bisx; **lm** rounded or weakly keeled, densely pilose on the lo ½ and on the mrg, thinly memb, 3-veined, 2-lobed, lobes about ½ as long as the lm and obtuse, midveins extending into awns as long as or longer than the lobes, lat veins not excurrent; **pal** about as long as the lm; **anth** 3. **Car** oval in cross section, translucent; **emb** more than ½ as long as the car.

*Dasyochloa* is a unispecific genus that is restricted to the United States and Mexico. *Dasyochloa* differs from *Triodia*, *Tridens*, and *Erioneuron*, but resembles *Munroa*, in its leafy-bracteate inflorescence. Seedlings of *Dasyochloa*, like those of *Erioneuron*, are shaggy-white-villous. This indumentum is composed of myriads of hairlike, water soluble crystals that wash off in water. They are the product of transpiration and evaporation.

### 1. Dasyochloa pulchella (Kunth) Willd. *ex* Rydb. FLUFFGRASS [p. *434*, <u>532</u>]

*Dasyochloa pulchella* grows in rocky soils of arid regions. It extends from the United States to central Mexico and is the most common grass in the *Larrea-Flourensia* scrub of the southwestern United States and adjacent Mexico. It was also found on Maryland ore piles in 1953.

## 16.16 BLEPHARONEURON Nash[2]

**Pl** ann or per. **Clm** 10–70 cm. **Shth** open, glab, usu longer than the intnd; **lig** memb or hyaline, truncate to obtuse, often decurrent; **bld** flat to involute, abx surfaces glab, smt scabrous, adx surfaces shortly pubescent. **Infl** tml, pan, exceeding the lvs; **br** spreading to ascending; **ped** capillary, lax, minutely glandular just below the spklt. **Spklt** with 1 flt, slightly lat compressed, grayish-green; **dis** above the glm. **Glm** subequal, ovate to obtuse, faintly 1-veined, glab; lm slightly longer and firmer than the glm, 3-veined, veins and mrg densely sericeous, hairs 0.1–0.7(1) mm, apc acute to obtuse, occ mucronate; **pal** 2-veined, densely villous between the veins; **anth** 3, purplish.

*Blepharoneuron* is a genus of two species: a slender annual, *B. shepherdii* (Vasey) P.M. Peterson & Annable, which is known only from northern Mexico, and *B. tricholepis*, which is native to the *Manual* region.

### 1. Blepharoneuron tricholepis (Torr.) Nash HAIRY DROPSEED [p. *434*, <u>532</u>]

*Blepharoneuron tricholepis* grows in dry, rocky to sandy slopes, dry meadows, and open woods in pine-oak-madrone forests from Utah and Colorado to the state of Puebla, Mexico, at elevations of 700–3660 m. It flowers from mid-June through November.

[1]Jesús Valdés-Reyna  [2]Paul M. Peterson and Carol R. Annable

## 16.17 BLEPHARIDACHNE Hack.[1]

Pl per (rarely ann); csp, from a knotty base, often mat-forming. Clm 3–8(20) cm, often decumbent and rooting at the lo nd, frequently brchd above the bases, forming short spur shoots at the ends of long intnd; intnd minutely pubescent. Lvs clustered at the bases of the pri and spur shoots; bas shth shorter than the intnd; lig of hairs or absent; bld linear to triangular, convolute to conduplicate, or flat to plicate, sharp, those of the up lvs usu exceeding the infl. Infl tml, compact pan, exserted or partially included in the up shth(s). Spklt lat compressed, subsessile or pedlt, with 4 flt per spklt; dis above the glm but not between the flt. Glm subequal to each other and the lowest lm, rounded or weakly keeled, 1-veined, awn-tipped or unawned; lowest 2 flt in each spklt stmt or strl; third flt pist or bisx; lm rounded on the back, 3-veined, mostly glab but pilose across the bases and on the mrg, strongly 3-lobed, lat lobes wider than the cent lobes, all

lobes ciliate on 1 or both mrg, lo lm with the lat lobes rounded or mucronate to awned, cent lobes awned; third lm 3-lobed, lobes awned; pal from slightly shorter to slightly longer than the lm; lod absent; anth 2 or 3 (rarely 1); sty br 2. Distal flt rdmt, 3-awned, plumose, or hairy. Car lat compressed.

The four species comprising *Blepharidachne* are restricted to the Americas, growing in arid and semi-arid regions of the United States, Mexico, and Argentina. *Blepharidachne bigelovii* and *B. kingii* are endemic to North America, whereas *B. benthamiana* (Hack.) Hitchc. and *B. hitchcockii* Lahitte are native to Argentina. *Blepharidachne* differs from all other genera in the tribe in having four florets per spikelet, with the first two florets being sterile or staminate, the third bisexual or pistillate, and the fourth a rudimentary 3-awned structure.

1. Glumes subacute, exceeded by the distal florets . . . . . . . . . . . . . . . . . . . . . . . . . . . . . . . . . . . . . . . . . . . . . . 1. *B. bigelovii*
1. Glumes acuminate or awn-tipped, exceeding the florets . . . . . . . . . . . . . . . . . . . . . . . . . . . . . . . . . . . . . . . . . 2. *B. kingii*

### 1. Blepharidachne bigelovii (S. Watson) Hack.
BIGELOW'S DESERTGRASS [p. *434*, 532]

*Blepharidachne bieglovii* grows on rocky slopes in western Texas and adjacent areas of New Mexico, and Coahuila and Zacatecas, Mexico.

### 2. Blepharidachne kingii (S. Watson) Hack. KING'S
EYELASH GRASS [p. *434*, 532]

*Blepharidachne kingii* grows at scattered locations in arid regions of the Great Basin, sometimes being locally abundant.

## 16.18 MUNROA Torr.[2]

Pl ann; stln, mat-forming; stln 2–8 cm, terminating in fascicles of lvs from which new clm arise. Clm 3–15(30) cm. Lvs mostly bas, smt with a purple tint; shth with a tuft of hairs at the throat; aur absent; lig of hairs; bld linear, usu involute, smt flat or folded, with white, thickened mrg, apc sharply pointed. Infl tml, capitate pan of spikelike br; br almost completely hidden in a subtending lfy bract, bearing 2–4 subsessile or pedlt spklt. Spklt lat compressed, with 2–10 flt; lo flt bisx or pist; tml flt strl; dis above the glm or beneath the lvs subtending the br. Glm shorter than the spklt, keeled, 1-veined, unawned; lo glm usu present on all spklt (absent from all spklt in *M. mendocina*); up glm absent or rdcd

on the tml spklt; lm with a pilose tuft of hairs along the mrg at midlength, memb or coriaceous, 3-veined, lat veins occ shortly excurrent, apc emgt or 2-lobed; pal glab, smooth; lod present or absent, truncate; anth 2 or 3, yellow; sty br elongate, 2(3), barbellate. Car dorsally compressed.

*Munroa*, a genus of five species, is endemic to the Western Hemisphere. One species occurs in the *Manual* region, the remainder being confined to South America. Its closest relatives are thought to be *Blepharidachne* and *Dasyochloa*, both of which are stoloniferous, mat-forming species with leafy-bracteate panicles. *Munroa* differs from both in its annual habit.

### 1. Munroa squarrosa (Nutt.) Torr. FALSE BUFFALOGRASS
[p. *434*, 532]

*Munroa squarrosa* grows in dry, open areas, usually in sandy soil or disturbed sites, from Saskatchewan and Alberta south to Chihuahua, Mexico.

## 16.19 LEPTOCHLOA P. Beauv.[3]

Pl ann or per; csp. Clm (3)10–250(300) cm, usu ascending to erect, often geniculate at the lo nd, occ prostrate and rooting at the lo nd, often brchg at the aerial nd; nd usu glab; intnd usu hollow. Lvs usu primarily cauline, occ in bas rosettes; shth open; lig 0.2–10(15) mm, obtuse to attenuate, usu memb, smt

ciliate; bld flat, involute when dry, usu ascending to erect, apc attenuate. Pri infl tml, pan of 2–150 non-disarticulating, spikelike br, usu exceeding the lvs; br 1–22 cm, digitate, subdigitate, or rcm on the rchs, 1-sided, usu spklt-bearing throughout their length, spklt in 2 rows, axes terminating in a fnctl spklt, lo br occ

[1,2]Jesús Valdés-Reyna
[3]Neil Snow

with sec brchg; **sec pan** smt present, ax to and concealed by the lo shth, their flt not disarticulating; **dis** in the pri pan beneath the flt. **Spklt** rounded to slightly keeled, distant to tightly imbricate, not conspicuously pubescent, with (2)3–12(20) bisx flt; **rchl** rarely prolonged. **Glm** usu unequal, smt subequal, exceeded by the flt, memb, rounded to weakly keeled, 1-veined, veins scabrous, apc unawned (rarely mucronate); **lo glm** 0.5–4.9 mm; **up glm** 0.9–6 mm; **flt** usu bisx; **cal** distinct or poorly developed, glab or pubescent; **lm** memb, usu pubescent at least over the lo portion of the veins, 3(5)-veined, apc entire or minutely bilobed, unawned, mucronate, or awned; **pal** usu subequal to the lm, memb or hyaline; **anth** 1–3, 0.1–2.7 mm. **Car** obovate to elliptic, falling free of the lm and pal.

*Leptochloa* is a pantropical, warm-temperate genus of 32 species. Eight of the ten species in this treatment are native to the *Manual* region. Of the other two, *Leptochloa chloridiformis* was introduced over 60 years ago but has not become established. *Leptochloa chinensis* is not yet known from the region; it is included here because of its potential threat as an invasive weed.

*Leptochloa* tends to grow in somewhat basic soils. Many of the species, particularly the annual species, are poor ecological competitors and grow in relatively open, seasonally inundated soils, such as are found along rivers. In disturbed areas, they are associated with roadside ditches, the margins of reservoirs, and mesic agricultural lands. A few species, primarily perennial, grow on well-drained soils. The vegetative vigor of all species is greatly influenced by soil moisture availability.

1. Panicle branches digitate or subdigitate; plants perennial.
    2. Lemma apices obtuse to truncate and often emarginate; lemmas membranous; plants often with secondary panicles concealed in the lower leaf sheaths . . . . . . . . . . . . . . . . . . . . . . . . . . . . . . . 1. *L. dubia* (in part)
    2. Lemma apices usually acute; lemmas chartaceous; plants without secondary panicles concealed in the lower leaf sheaths.
        3. Panicle branches always digitate, 7–17 cm long; lemmas mucronate but not awned; in the *Manual* region, known only from a few old collections in Cameron County, Texas . . . . . . . . . . . . 2. *L. chloridiformis*
        3. Panicle branches subdigitate, 1.5–18 cm long; lemmas unawned, mucronate, or awned; native in much of the southeastern United States, including parts of Texas . . . . . . . . . . . . . . . . . . . . . . 3. *L. virgata* (in part)
1. Panicle branches racemose; plants annual or perennial.
    4. Ligules 2–8 mm long, attenuate, becoming lacerate at maturity . . . . . . . . . . . . . . . . . . . . . . . . . . . . . . 4. *L. fusca*
    4. Ligules 0.3–5.4 mm long, truncate to obtuse, erose, ciliate, or lacerate.
        5. Sheaths sparsely to densely hairy, the hairs papillose-based . . . . . . . . . . . . . . . . . . . . . . . . 5. *L. panicea*
        5. Sheaths lacking hairs or with hairs that are not papillose-based.
            6. Panicles with 25–150 branches.
                7. Lemmas 2.4–3 mm long; anthers 0.6–0.8 mm long; spikelets 4–5 mm long . . . . . . . . 10. *L. panicoides* (in part)
                7. Lemmas 1.2–2.4 mm long; anthers 0.2–0.6 mm long; spikelets 2.5–4.5 mm long.
                    8. Leaf blades 8–16 mm wide; lemmas 2.1–2.4 mm long . . . . . . . . . . . . . . . . . . . . . . . . 7. *L. scabra*
                    8. Leaf blades 4–8 mm wide; lemmas 1–2 mm long.
                        9. Lower glumes 0.7–0.8 mm long; anthers 0.2–0.4 mm long; plants native to the southern United States . . . . . . . . . . . . . . . . . . . . . . . . . . . . . . . . . . . . . . . 6. *L. nealleyi*
                        9. Lower glumes 1.1–1.7 mm long; anthers 0.4–0.6 mm long; plants aggressive weeds, currently not known from the *Manual* region . . . . . . . . . . . . . . . . . . . . . . . 8. *L. chinensis*
            6. Panicles with 2–25 branches.
                10. Plants perennial.
                    11. Lemmas 4–5 mm long, membranous, their apices broadly acute, obtuse, or truncate, unawned; panicles with 2–15 branches; caryopses 1.9–2.3 mm long, strongly dorsally compressed; secondary panicles often present in, and concealed by, the lower leaf sheaths . . . . . . . . . . . . . . . . . . . . . . . . . . . . . . . . 1. *L. dubia* (in part)
                    11. Lemmas 1.5–3.6 mm long, chartaceous, their apices usually acute, rarely obtuse, unawned, mucronate, or awned, the awns to 11 mm long; panicles with 9–25 branches; caryopses 1.3–1.8 mm long, somewhat laterally compressed; secondary panicles not present in the lower leaf sheaths . . . . . . . . . . . . . . . . . . . 3. *L. virgata* (in part)
                10. Plants annual.
                    12. Panicles 2–17 cm long, with 5–23 branches; anthers 0.4–0.5 mm long; caryopses 0.4–0.5 mm wide . . . . . . . . . . . . . . . . . . . . . . . . . . . . . . . . . . . . . . . . . . . . . . . . 9. *L. viscida*
                    12. Panicles 20–35 cm long, with 20–90 branches; anthers 0.6–0.8 mm long; caryopses about 0.7 mm wide . . . . . . . . . . . . . . . . . . . . . . . . . . . . . . . . . . . . 10. *L. panicoides* (in part)

## 1. Leptochloa dubia (Kunth) Nees GREEN SPRANGLETOP [p. *435*, 532]

*Leptochloa dubia* grows from the southwestern United States and Florida through Mexico to Argentina, often in well-drained, sandy or rocky soils. It provides fair to good forage, but is seldom abundant.

## 2. Leptochloa chloridiformis (Hack.) Parodi ARGENTINE SPRANGLETOP [p. *435*, 532]

*Leptochloa chloridiformis* is native to Uruguay, southern Paraguay, and northern Argentina. It was introduced in the early part of the twentieth century but has not become established in the *Manual* region. The only known collections are from Cameron County, Texas, the most recent having been made in the 1940s.

## 3. Leptochloa virgata (L.) P. Beauv. TROPICAL SPRANGLETOP [p. *435*, 532]

*Leptochloa virgata* is a common neotropical species that extends from the southeastern United States through the West Indies to Argentina.

## 4. Leptochloa fusca (L.) Kunth [p. *435*, 532]

*Leptochloa fusca* grows in warm areas throughout the world. The two American subspecies, subsp. *uninervia* and subsp. *fascicularis*, are usually distinct, but they intergrade repeatedly with *L. fusca* subsp. *fusca*.

1. Uppermost leaf blades exceeding the panicles; panicles usually partially enclosed in the uppermost leaf sheaths; mature lemmas often smoky white with a dark spot in the basal ½ . . . . . . . . . . . . . . . . . . . . . . . . . . subsp. *fascicularis*
1. Uppermost leaf blades exceeded by the panicles; panicles usually completely exserted; mature lemmas usually lacking a dark spot.
    2. Anthers 0.5–2.7 mm long; spikelets 6–14 mm long; lemmas obtuse, acute, or acuminate, sometimes bifid, light brown to dark green . . . . . . . . . . . . subsp. *fusca*
    2. Anthers 0.2–0.6(1) mm long; spikelets 5–10 mm long; lemmas obtuse to truncate, usually notched and mucronate, often dark green or lead-colored . . . . . . . . . . . . . . . . . . . . . . . . . . . . . subsp. *uninervia*

### Leptochloa fusca subsp. fascicularis (Lam.) N. Snow BEARDED SPRANGLETOP [p. *435*]

*Leptochloa fusca* subsp. *fascicularis* extends from southern British Columbia and Ontario to Argentina. It differs from *L. viscida*, which grows in the same region, in its longer panicles, frequently unawned or mucronate lemmas, and whitish florets.

### Leptochloa fusca (L.) Kunth subsp. fusca BEETLEGRASS SPRANGLETOP [p. *435*]

*Leptochloa fusca* subsp. *fusca* is the most variable of the subspecies. In North America, it is known only from a few specimens collected at scattered locations in California; it may no longer be in the *Manual* region.

### Leptochloa fusca subsp. uninervia (J. Presl) N. Snow MEXICAN SPRANGLETOP [p. *435*]

*Leptochloa fusca* subsp. *uninervia* is native from the southern United States to Argentina. It differs from *L. scabra* in its truncate or obtuse lemmas.

## 5. Leptochloa panicea (Retz.) Ohwi [p. *435*, 532]

*Leptochloa panicea* is a cosmopolitan species that somewhat resembles *L. chinensis*, an aggressive weed that has not yet been found in the *Manual* region. It differs in its sparsely to densely hairy, rather than glabrous or almost glabrous, sheaths and blades. Two of its three subspecies grow in the *Manual* region.

1. Glumes linear to narrowly lanceolate, exceeding the florets; lemmas 0.9–1.2 mm long; caryopses without a ventral groove, often somewhat coarsely rugose, the apices broadly obtuse . . . . . . . . . . . . . . . . . . . . . . . . . . . . . subsp. *mucronata*
1. Glumes lanceolate to narrowly elliptic, not or only slightly exceeding the florets; lemmas 1.3–1.7 mm long; caryopses usually with a narrow, shallow ventral groove, smooth, the apices broadly obtuse to acute . . . . . . . . . . . . . subsp. *brachiata*

### Leptochloa panicea subsp. brachiata (Steud.) N. Snow RED SPRANGLETOP [p. *435*]

*Leptochloa panicea* subsp. *brachiata* extends from the southern half of the United States to Argentina. It is common in disturbed and mesic agricultural sites, and is considered a noxious weed by the U.S. Department of Agriculture.

### Leptochloa panicea subsp. mucronata (Michx.) Nowack MISSISSIPPI SPRANGLETOP [p. *435*]

*Leptochloa panicea* subsp. *mucronata* grows in the southern portion of the United States, primarily from Kansas and Missouri through Texas and Louisiana.

## 6. Leptochloa nealleyi Vasey NEALLEY'S SPRANGLETOP [p. *435*, 532]

*Leptochloa nealleyi* is native to coastal Texas, Louisiana, Florida, and Mexico; it also grows, but rarely, in Cuba. The numerous, short, stiffly ascending or erect panicle branches make *Leptochloa nealleyi* easy to identify.

## 7. Leptochloa scabra Nees ROUGH SPRANGLETOP [p. *435*, 532]

*Leptochloa scabra* is a neotropical species that extends into Louisiana and southwestern Alabama. It differs from *L. panicoides* in having more, flexuous to arcuate panicle branches, shorter spikelets, and less prominent lemma veins; from *L. fusca* subsp. *uninervia* in its acute lemmas; and from *L. virgata* in its hollow, flattened culms and the complete lack of lemma awns.

## 8. Leptochloa chinensis (L.) Nees ASIAN SPRANGLETOP [p. *435*]

*Leptochloa chinensis* is not yet known from the *Manual* region but, if introduced, it could become an aggressive weed because it competes well in undisturbed mesic sites. Although it resembles *L. panicea*, *L. chinensis* differs in its glabrous, or nearly glabrous, sheaths and blades.

## 9. Leptochloa viscida (Scribn.) Beal SONORAN SPRANGLETOP [p. *435*, 532]

*Leptochloa viscida* is a Sonoran Desert species that occurs from southern California to southwestern New Mexico and south into adjacent Mexico. It differs from *L. fusca* subsp. *fascicularis*, which grows in the same region, in its consistently short-awned lemmas, smaller panicles, often prostrate and much-branched growth habit, and often reddish florets.

## 10. Leptochloa panicoides (J. Presl) Hitchc. AMAZON SPRANGLETOP [p. *435*, 532]

*Leptochloa panicoides* is native from the central Mississippi and Ohio river drainages south through Mesoamerica to Brazil. It usually grows in somewhat mesic habitats. It differs consistently from *L. scabra* in the number of panicle branches, spikelet length, and prominence of the lemma veins.

## 16.20 TRIPOGON Roem. & Schult.[1]

Pl per or ann; csp or tufted. Clm 4–65 cm, erect, slender. Lvs linear, flat, usu becoming folded and filiform; lig memb, ciliate. Infl tml, unilat linear spikes or spikelike rcm, with 1 spklt per nd, exceeding the lvs; rchs visible, not concealed by the spklt. Spklt appressed, in 2 rows along 1 side of the rchs, with 3–20 bisx flt, distal flt strl or stmt; dis above the glm and between the flt. Glm unequal, 1(3)-veined; lm 1–3-veined, backs slightly keeled or rounded, apc lobed or bifid, mucronate or awned from between the lobes, lat veins smt also excurrent, awns usu straight; anth 1–3.

*Tripogon* is a genus of approximately 30 species, most of which are native to the tropics of the Eastern Hemisphere, especially Africa and India. One species, *Tripogon spicatus*, is native to the Western Hemisphere.

### 1. Tripogon spicatus (Nees) Ekman AMERICAN TRIPOGON [p. *435*, <u>532</u>]

*Tripogon spicatus* grows in shallow rocky soils, usually on granite outcroppings, occasionally on limestone. The flowering period, April–July(October, November), apparently depends on rainfall. Its range includes the West Indies, Mexico, and South America, in addition to central Texas.

## 16.21 TRICHONEURA Andersson[2]

Pl ann or per. Clm 12–155 cm, nd glab, intnd solid. Lig memb; bld linear, narrow, usu flat. Infl tml, pan of 5–40 racemosely arranged, spikelike br, exceeding the lvs; br spreading to appressed, persistent, unilat, with 1 spklt per nd. Spklt 5.3–14 mm long, with 2 or more flt, typically with 2–8 bisx flt, strl or stmt flt smt present distal to the bisx flt; rchl intnd pilose bas, apc oblique; dis above the glm and below the flt. Glm from shorter than to greatly exceeding the flt, equal or subequal to each other, narrow, apc acuminate and mucronate, awnlike, or awned; cal well-developed, strigose; lm 3-veined, conspicuously hairy adjacent to and on the lat veins, apc cleft, midveins excurrent from the sinuses, smt forming awns.

*Trichoneura* is a genus of seven species that grow in dry, sandy, or stony soils. Five species are native to the Eastern Hemisphere and two to the Western Hemisphere, one of which is native to the *Manual* region.

### 1. Trichoneura elegans Swallen SILVEUS GRASS [p. *436*, <u>532</u>]

*Trichoneura elegans* usually grows in dry, deep, sandy soil. Its range extends from south central Texas to northern Tamaulipas, Mexico.

## 16.22 DINEBRA Jacq.[3]

Pl ann. Clm 13–120 cm, not wd. Lig memb, truncate, lacerate, smt ciliate; bld linear, flat. Infl tml, pan of 1–70, 1-sided, spikelike br, irregularly disposed on elongate rchs, clearly exceeding the up lvs; br with 2 rows of 1 or more closely imbricate, sessile spklt, proximal spklt smt replaced by short, tardily deciduous, sec br; dis at the base of the br or at the base of the sec br and (eventually) beneath the flt. Spklt lat compressed, cuneate, with 1–3 flt. Glm subequal, much longer than the flt, usu exceeding the distal flt, coriaceous or memb, strongly keeled, acuminate-aristate; lm thinly memb, weakly keeled, 3-veined, pilose over the veins, apc acute to 2-lobed, cent veins excurrent, forming mucros. Car elliptic-oblong, trigonous.

*Dinebra*, a genus of three species, is native from Africa to Madagascar and India. One species has been reported from the *Manual* region.

### 1. Dinebra retroflexa (Vahl) Panz. VIPER GRASS [p. *436*, <u>532</u>]

*Dinebra retroflexa* is native from southern Africa through tropical Africa to Egypt, Iraq, Pakistan, and India. It has reportedly been found on chrome ore piles in Canton, Maryland, as well as in Mecklenberg County, North Carolina, and Riverside County, California. It is a common weed of rich soils in moist, tropical regions.

## 16.23 ERAGROSTIS Wolf[4]

Pl ann or per; usu synoecious, smt dioecious; csp, stln, or rhz. Clm 2–160 cm, not wd, erect, decumbent, or geniculate, smt rooting at the lo nd, simple or brchd; intnd solid or hollow. Lvs not strongly distichous; shth open, often with tufts of hairs at the apc, hairs 0.3–8 mm; lig usu memb and ciliolate or ciliate, cilia smt longer than the memb base, occ of hairs or memb and non-ciliate; bld flat, folded, or involute. Infl tml, smt also ax, simple pan, open to contracted or spikelike, tml pan usu exceeding the up lvs; pulvini in the axils of the

[1,2]J.K. Wipff [3]Mary E. Barkworth [4]Paul M. Peterson

pri br glab or not; **br** not spikelike, not disarticulating. Spklt 1–27 mm long, 0.5–9 mm wide, lat compressed, with (1)2–60 flt; **dis** below the ftl flt, smt also below the glm, acropetal with deciduous glm and lm but persistent pal, or basipetal with the glm often persistent and the flt usu falling intact. **Glm** usu shorter than the adjacent lm, 1(3)-veined, not lobed, apc obtuse to acute, unawned; **cal** glab or sparsely pubescent; **lm** usu glab, obtuse to acute, (1)3(5)-veined, usu keeled, unawned or mucronate; **pal** shorter than the lm, longitudinally bowed-out by the car, 2-keeled, keels usu ciliate, intercostal region memb or hyaline; **anth** 2–3; **ov** glab; **sty** free to the bases. **Cleistogamous spklt** occ present, smt on the ax pan, smt on the tml pan. **Car** variously shaped.

*Eragrostis*, a genus of approximately 350 species, grows in tropical and subtropical regions throughout the world. About 110 species are native or adventive in the Western Hemisphere; 25 species are native in the *Manual* region, 24 are introduced. In most taxa native to the Western Hemisphere, disarticulation is acropetal and the lemmas fall with the caryopses, leaving the paleas attached to the rachilla.

1. Plants annual, tufted or mat-forming, without innovations [for opposite lead, see p. 203].
    2. Palea keels prominently ciliate, the cilia 0.2–0.8 mm long.
        3. Spikelets 1–3.6 mm long, 0.9–2 mm wide, with 4–12 florets; lemmas 0.7–1.3 mm long.
            4. Anthers 2; pedicels 0.1–1 mm long, mostly shorter than the spikelets, straight . . . . . . . . . . . . . . . . 1. *E. ciliaris*
            4. Anthers 3; pedicels 1–4(7) mm long, as long as or longer than the spikelets, mostly curved . . . . . . . . 3. *E. amabilis*
        3. Spikelets 5–20 mm long, 1.4–4 mm wide, with 10–42 florets; lemmas 1.3–2.8 mm long.
            5. Lemmas and culms without glands; anthers 0.1–0.2 mm long, purplish . . . . . . . . . . . . . . . 2. *E. cumingii* (in part)
            5. Lemmas with 1–3 crateriform glands on the keels, similar glands also often present below the cauline nodes; anthers 0.2–0.5 mm long, yellow . . . . . . . . . . . . . . . . . . . . . . . . . . 18. *E. cilianensis* (in part)
    2. Palea keels smooth or scabrous, the scabridities less than 0.2 mm long.
        6. Plants mat-forming; panicles 1–3.5 cm long; erect portion of culms (2)5–20 cm, the basal portion prostrate and rooting at the nodes.
            7. Spikelets bisexual; anthers 2, 0.2–0.3 mm long . . . . . . . . . . . . . . . . . . . . . . . . . . . . . . . . 4. *E. hypnoides*
            7. Spikelets and plants unisexual; anthers 3, 1.4–2.2 mm long . . . . . . . . . . . . . . . . . . . . . . . . . 5. *E. reptans*
        6. Plants usually not forming mats; panicles 3–55 cm long; culms (2)6–130 cm tall, not prostrate or rooting at the lower nodes.
            8. Ligules membranous, neither ciliolate nor ciliate . . . . . . . . . . . . . . . . . . . . . . . . . . . . . . . . . 6. *E. japonica*
            8. Ligules membranous and ciliolate to ciliate, the cilia often longer than the basal membrane.
                9. Caryopses with a shallow or deep ventral groove, ovoid to rectangular-prismatic or dorsally compressed, if dorsally compressed, the surface striate or smooth.
                    10. Bases of the caryopses greenish; caryopses dorsally compressed, the distal ⅔ translucent, the surface smooth; leaf sheaths with oblong glands; in the *Manual* region, known from a single collection at Canton, Maryland . . . . . . . . . . . . . . . . . . . . . 7. *E. cylindriflora*
                    10. Bases of the caryopses reddish-brown or brownish; caryopses laterally compressed or rectangular-prismatic to ovoid, the distal ⅔ opaque, the surface striate; sheaths without oblong glands, sometimes with glandular pits; plants found at many locations in the *Manual* region.
                        11. Spikelets 4–11 mm long, with 5–15 florets; pedicels somewhat divergent to almost appressed . . . . . . . . . . . . . . . . . . . . . . . . . . . . . . . . . . . . . . . . . . . . . . . . . . 12. *E. mexicana*
                        11. Spikelets 1.4–5 mm long, with 2–7 florets; pedicels divergent.
                            12. Panicles 4–20 cm long, less than ½ the height of the plant; pedicels 1.5–5 mm long; glandular pits often present below the cauline nodes, on the rachises, and on the panicle branches . . . . . . . . . . . . . . . . . . . . . . . . . . . . . . . 13. *E. frankii* (in part)
                            12. Panicles 10–45(55) cm long, ⅔ or more the height of the plant; pedicels 4–25 mm long; plants without glandular pits . . . . . . . . . . . . . . . . . . . . . . . . . . . . . 14. *E. capillaris*
                9. Caryopses without a ventral groove, usually globose, rarely flattened, pyriform, obovoid, ellipsoid, or rectangular-prismatic, the surface smooth to faintly striate.
                    13. Plants with glandular pits or bands somewhere, the location(s) various, including any or all of the following: below the cauline nodes, on the sheaths, blades, rachises, panicle branches, or pedicels, or on the keels of the lemmas and paleas.
                        14. Panicles 0.5–2 cm wide, contracted; primary panicle branches usually appressed, occasionally diverging up to 30° from the rachises; spikelets light yellowish, occasionally with reddish-purple markings . . . . . . . . . . . . . . . . . . . . . . . . . . 15. *E. lutescens*
                        14. Panicles 2–18 cm wide, open to somewhat contracted; primary panicle branches diverging 20–110° from the rachises; spikelets plumbeous, greenish, or reddish-purple.

15. Spikelets 1.7–5.6 mm long, with 3–6 florets . . . . . . . . . . . . . . . . . . . . . . . 13. *E. frankii* (in part)
15. Spikelets (2)3.5–20 mm long, with (3)5–40 florets.
    16. Spikelets 0.6–1.4 mm wide; pedicels 1–10 mm long, lax, appressed or
        divergent . . . . . . . . . . . . . . . . . . . . . . . . . . . . . . . . . . . . . . . . . . . . . . . . . 16. *E. pilosa* (in part)
    16. Spikelets 1.1–4 mm wide; pedicels 0.2–4 mm long, stiff, straight, usually
        divergent.
        17. Lemmas 2–2.8 mm long, with 1–3 crateriform glands along the
            keels; spikelets 6–20 mm long, 2–4 mm wide, with 10–40 florets;
            disarticulation below the florets, the rachillas persistent; anthers
            yellow . . . . . . . . . . . . . . . . . . . . . . . . . . . . . . . . . . . . . . 18. *E. cilianensis* (in part)
        17. Lemmas 1.4–1.8 mm long, rarely with 1 or 2 crateriform glands
            along the keels; spikelets 4–7(11) mm long, 1.1–2.2 mm wide, with
            7–12(20) florets; disarticulation below the lemmas, both the paleas
            and rachillas usually persistent; anthers reddish-brown.
            18. Panicles with glandular regions below the nodes, the glandular
                tissue forming a ring or band, often shiny or yellowish; anthers
                3; blade margins without crateriform glands; pedicels without
                glandular bands . . . . . . . . . . . . . . . . . . . . . . . . . . . . . . . . . . . . 19. *E. barrelieri*
            18. Panicles sometimes with areas, but rarely rings, of glandular
                spots or crateriform pits below the nodes, the glands usually
                dull greenish-gray to stramineous; anthers 2; blade margins
                sometimes with crateriform glands; pedicels usually with
                glandular bands . . . . . . . . . . . . . . . . . . . . . . . . . . . . . . . . . . . . . . 20. *E. minor*
13. Plants without glandular pits or bands.
    19. Spikelets (1.6)2–4 mm wide; florets disarticulating intact from the persistent
        rachillas . . . . . . . . . . . . . . . . . . . . . . . . . . . . . . . . . . . . . . . . . . . . . . . . . . . . . 21. *E. unioloides*
    19. Spikelets 0.6–2.5 mm wide; lemmas disarticulating separately from the paleas,
        sometimes both the paleas and the rachillas persistent.
        20. Spikelets with 3–6 florets; plants of the central and northeastern United
            States and southern Ontario, Canada . . . . . . . . . . . . . . . . . . . . . . . . 13. *E. frankii* (in part)
        20. Spikelets with (3)5–42 florets; plants from throughout the contiguous United
            States and southern Ontario, Canada.
            21. Lemmas 1.6–3 mm long; caryopses 0.7–1.3 mm long, obovoid, smooth,
                light brown to white; plants cultivated, occasionally escaping . . . . . . . . . . . . . . . . 22. *E. tef*
            21. Lemmas 1–2.2 mm long; caryopses 0.3–1.1 mm long, subglobose,
                pyriform, or obovoid to prism-shaped, smooth or faintly striate,
                brownish; plants native species or established introductions, variously
                distributed.
                22. Lemmas with conspicuous, often greenish lateral veins; caryopses
                    0.3–0.6 mm long, ovoid, subglobose to obovoid.
                  23. Spikelets 5–12(18) mm long, with 12–42 florets; primary
                    branches 6–10 per culm; lemmas 1.3–2 mm long; anthers 3 . . 2. *E. cumingii* (in part)
                23. Spikelets 2–4.6 mm long, with 5–15 florets; primary branches
                    (12)15–20 per culm; lemmas 1–1.3 mm long; anthers 2 . . . . . . . . . 23. *E. gangetica*
                22. Lemmas with inconspicuous to moderately conspicuous lateral veins,
                  the veins usually not greenish; caryopses 0.5–1.1 mm long, pyriform
                  or obovoid to prism-shaped.
                  24. Lower glumes 0.5–1.5 mm long, at least ½ as long as the lowest
                    lemmas; spikelets 1.2–2.5 mm wide; panicle branches solitary
                    or paired at the lowest 2 nodes; lemmas with moderately
                    conspicuous lateral veins . . . . . . . . . . . . . . . . . . . . . . . . . . . . . . 17. *E. pectinacea*
                24. Lower glumes 0.3–0.6(0.8) mm long, usually less than ½ as
                    long as the lowest lemmas; spikelets 0.6–1.4 mm wide; panicle
                    branches usually whorled at the lowest 2 nodes; lemmas with
                    inconspicuous lateral veins . . . . . . . . . . . . . . . . . . . . . . . . . 16. *E. pilosa* (in part)
1. Plants perennial, sometimes rhizomatous, forming innovations at the basal nodes [for opposite lead,
   see p. 202].
    25. Paleas with a broad lower portion forming a wing or tooth on each side, these often projecting
       beyond the lemmas.

26. Spikelets 5.5–16 mm long, 2.7–9 mm wide; lemmas 3–5 mm long, the keels without crateriform glands; pedicels with a narrow band or abscission line just below the apices; anthers 1.4–2.8 mm long . . . . . . . . . . . . . . . . . . . . . . . . . . . . . . . . . . . . . . . . . . . . . . . 24. *E. superba*
26. Spikelets 2–5 mm long, 2–3.5 mm wide; lemmas 1.8–2.3 mm long, the keels with a few crateriform glands; pedicels without a narrow band or abscission line just below the apices; anthers 0.5–0.9 mm long . . . . . . . . . . . . . . . . . . . . . . . . . . . . . . . . . . . . . . . . . . . 25. *E. echinochloidea*
25. Paleas without a broad lower portion forming a wing or tooth, the bases never projecting beyond the lemmas.
  27. Plants rhizomatous; disarticulation always below the florets, the paleas falling with the lemmas and caryopses.
    28. Plants with long, scaly rhizomes, 4–8 mm thick; spikelets 8–14 mm long; lemmas 3.8–4.5 mm long, 3–5-veined, the apices acute to obtuse, usually erose; caryopses 1.6–2 mm long . . . . . . . . . . . . . . . . . . . . . . . . . . . . . . . . . . . . . . . . . . . . . . . . . . . . . . . . . . . . . . . . .26. *E. obtusiflora*
    28. Plants with short, knotty rhizomes less than 4 mm thick, often stout but never elongated; spikelets 2.5–7.6 mm long; lemmas 1–2.5 mm long, 3-veined, the apices acute, usually entire; caryopses 0.5–0.8 mm long.
      29. Sheaths, blades, and culms not viscid or glandular; caryopses strongly flattened, the ventral surface with 2 prominent ridges separated by a groove; anthers 0.3–0.5 mm long; lemmas leathery . . . . . . . . . . . . . . . . . . . . . . . . . . . . . . . . . . . . . . . . 27. *E. spectabilis*
      29. Sheaths, blades, and/or culms often viscid, sometimes glandular; caryopses terete, the ventral surfaces without 2 ridges separated by a groove; anthers 0.2–0.4 mm long; lemmas membranous.
        30. Pedicels 0.2–1.2 mm long, appressed; lemmas 1.5–2.2 mm long; caryopses 0.6–0.8 mm long . . . . . . . . . . . . . . . . . . . . . . . . . . . . . . . . . . . . . 28. *E. curtipedicellata*
        30. Pedicels (1)1.5–12 mm long, divergent or appressed; lemmas 1.1–1.4 mm long; caryopses 0.5–0.6 mm long . . . . . . . . . . . . . . . . . . . . . . . . . . . . . . . . . . 29. *E. silveana*
  27. Plants not rhizomatous; disarticulation often below the lemmas, the paleas persistent, sometimes below the florets and the paleas falling with the lemmas and caryopses.
    31. Panicles 0.3–0.6 cm wide, spicate, dense; spikelets with 2–3 florets . . . . . . . . . . . . . . . . . . . . . . 30. *E. spicata*
    31. Panicles 1–45 cm wide, ovate to obovate or elliptic, open to somewhat condensed and glomerate; spikelets with 1–45 florets.
      32. Caryopses with shallowly to deeply grooved adaxial surfaces, rectangular-prismatic to ellipsoid, ovoid, or obovoid in overall shape [for opposite lead, see p. 205].
        33. Caryopses strongly dorsally compressed, translucent, mostly light brown, bases sometimes greenish.
          34. Lemmas 1.8–3 mm long; panicles 16–35(40) cm long, (4)8–24 cm wide; blades 12–50(65) cm long; caryopses 1–1.7 mm long; ligules 0.6–1.3 mm long . . . . . . . . . . . . . . . . . . . . . . . . . . . . . . . . . . . . . . . . . . . . . 9. *E. curvula* (in part)
          34. Lemmas 1.4–1.7 long; panicles 6–18 cm long, 2–8 cm wide; blades 2–12 cm long; caryopses 0.4–0.8 mm long; ligules 0.3–0.5 mm long.
            35. Plants without woolly hairs at the base; glumes unequal; lateral lemma veins not green, inconspicuous throughout; spikelets 0.8–1.2 mm wide; naturalized in the southwestern United States . . . . . . . . . . . . . . . 10. *E. lehmanniana* (in part)
            35. Plants with conspicuous, woolly hairs at the base; glumes subequal; lateral lemma veins green, conspicuous basally, obscure near the lemma apices; spikelets 1.3–2 mm wide; in the *Manual* region, known only from waste areas near a woolen mill in South Carolina . . . . . . . . . . . . . . . . . . . . 11. *E. setifolia* (in part)
        33. Caryopses laterally compressed, terete, or slightly dorsally compressed, usually opaque, usually reddish-brown.
          36. Lateral veins of the lemmas conspicuous, often greenish, the lemmas strongly keeled.
            37. Panicles 2–8 cm wide, contracted to somewhat open, narrowly oblong to narrowly lanceolate; primary branches appressed or diverging up to 30° from the rachises; lemmas with punctate glands along the keels; pedicels 1–7 mm long, appressed; plants native to Africa, in the *Manual* region, known only from waste areas near sheep and cattle lots in South Carolina and Alabama . . . . . . . . . . . . . . . . . 31. *E. plana*
            37. Panicles 4–30 cm wide, open, ovate to oblong; primary branches diverging 10–90° from the rachises; lemmas without punctate glands

on the keels; pedicels 0.4–22 mm long, usually diverging, occasionally
appressed; plants native to the southern United States.

38. Pedicels with a glandular band; culms with a glandular band
below the nodes; anthers 0.3–0.5 mm long; restricted to southern
Texas . . . . . . . . . . . . . . . . . . . . . . . . . . . . . . . . . . . . . . . . . . . . . . . . . . . . 32. *E. swallenii*

38. Pedicels and culms without glandular bands; anthers 0.6–1.6 mm
long; often found outside southern Texas.

39. Glumes 1.8–4 mm long, the upper glumes generally equaling
or exceeding the lower lemmas; spikelets 1.5–3.6 mm wide,
greenish-yellow with a reddish-purple tinge; lemmas 2.2–3.5
mm long; caryopses 0.8–1.3 mm long . . . . . . . . . . . . . . . . . . . . . . . 33. *E. trichodes*

39. Glumes 1.1–2.2 mm long, the upper glumes exceeded by the
lower lemmas; spikelets 1–2 mm wide, plumbeous; lemmas
2–2.6 mm long; caryopses 0.6–0.8 mm long . . . . . . . . . . . . 34. *E. palmeri* (in part)

36. Lateral veins of the lemmas inconspicuous, the lemmas sometimes only
weakly keeled.

40. Lemmas 1.2–1.8 mm long; culms 30–70 cm tall.

41. Culms with a glandular ring below the nodes; bases of primary
panicle branches with a glandular band; panicles 2–7 cm wide;
pedicels glandular; known, in the *Manual* region, from only a few
collections at Canton, Maryland . . . . . . . . . . . . . . . . . . . . . . . . . . . . 8. *E. trichophora*

41. Culms without a glandular ring below the nodes; bases of primary
panicle branches without a glandular band; pedicels not glandular
at the base; panicles 5–27 cm wide; plants known from many parts
of the southern United States.

42. Spikelets 1.1–1.6 mm wide, uniformly plumbeous; sheaths
sometimes densely pilose dorsally and on the collars; distal
margins of the lemmas not hyaline . . . . . . . . . . . . . . . . . . . . . . 35. *E. polytricha*

42. Spikelets 0.5–1(1.3) mm wide, plumbeous to reddish-purple;
sheaths usually glabrous dorsally and on the collars; distal
margins of the lemmas hyaline . . . . . . . . . . . . . . . . . . . . . . . 36. *E. lugens* (in part)

40. Lemmas 1.6–3 mm long; culms (30)40–110(120) cm tall.

43. Spikelets greenish with a purplish tinge, with 2–6 florets; blades
25–60 cm long, 3–11 mm wide, flat to loosely involute; sheaths
densely hirsute with papillose-based hairs on the collar, back,
and base . . . . . . . . . . . . . . . . . . . . . . . . . . . . . . . . . . . . . . . . . . . . . 37. *E. hirsuta* (in part)

43. Spikelets olivaceous to plumbeous, with (3)5–12 florets; blades
(4)10–35 cm long, 1–3.8 mm wide, involute or flat; sheaths never
with papillose-based hairs, sometimes villous over the back.

44. Lemmas 1.6–2.2 mm long; anthers 0.5–0.8 mm long,
purplish . . . . . . . . . . . . . . . . . . . . . . . . . . . . . . . . . . . . . . . . . . . . . . 38. *E. intermedia*

44. Lemmas 2–3 mm long; anthers 0.6–1.7 mm long, purplish
to yellowish.

45. Caryopses 0.8–1.6 mm long; lemmas 2.4–3 mm long . . . . . . . . . 39. *E. erosa*

45. Caryopses 0.6–0.8 mm long; lemmas 2–2.6 mm long . . 34. *E. palmeri* (in part)

32. Caryopses not grooved on the adaxial surfaces, ellipsoid, subellipsoid, ovoid,
obovoid, globose, to pyriform in overall shape [for opposite lead, see p. 204].

46. Anthers 2.

47. Panicles 15–45 cm wide, open, diffuse, broadly ovate to obovate; primary
branches lax; pedicels 0.5–35(50) mm long, the lower pedicels longer or
shorter than the spikelets.

48. Spikelets with appressed pedicels; only the terminal pedicels of each
branch longer than the spikelets; disarticulation usually in the rachilla
beneath the florets . . . . . . . . . . . . . . . . . . . . . . . . . . . . . . . . . . . . . . . . . . 40. *E. refracta*

48. Spikelets with divergent pedicels; all pedicels usually longer than the
spikelets; disarticulation below the lemmas, the paleas persistent . . . . . . . . . . 41. *E. elliottii*

47. Panicles (1)2–17 cm wide, contracted to open, narrowly ovate to oblong;
primary branches stiff; pedicels absent or 0.3–6 mm long, always shorter
than the spikelets.

49. Spikelets 2.4–5 mm wide; glumes 1.7–4 mm long; lemmas 2–6 mm long, the apices usually acuminate or attenuate . . . . . . . . . . . . . . . . . . . 42. *E. secundiflora*
49. Spikelets 1–2.4 mm wide; glumes 1–2.2 mm long; lemmas 1.1–2.5 mm long, the apices usually acute, occasionally acuminate.
  50. Spikelets 0.7–1.4 mm wide; anthers 0.2–0.3 mm long; caryopses flattened ventrally . . . . . . . . . . . . . . . . . . . . . . . . . . . . . . . . . . . . . . . . . . . 43. *E. prolifera*
  50. Spikelets 1.3–2.4 mm wide; anthers (0.2)0.3–0.7 mm long; caryopses rounded, not flattened ventrally.
    51. Terminal panicles 1–3.5 cm wide, contracted, condensed into glomerate lobes; primary branches 0.8–3 cm long . . . . . . . . . . . . . 44. *E. elongata*
    51. Terminal panicles (1)2–17 cm wide, open to contracted; primary branches 1–15 cm long.
      52. Plants without axillary panicles; terminal panicles 15–45 cm long; blades (8)12–40 cm long, 2–5 mm wide, flat to involute; caryopses 0.6–0.8 mm long, striate, obovoid to ellipsoid . . . . . . . . . . . . . . . . . . . . . . . . . . . . . . 45. *E. bahiensis*
      52. Plants usually with axillary panicles, these contracted and partially to completely enclosed by the subtending sheaths; terminal panicles 5–15 cm long; blades 4–8(18) cm long, 1–2 mm wide, usually involute; caryopses 0.5–0.6 mm long, smooth, globose . . . . . . . . . . . . . . . . . . . . . 46. *E. scaligera*
46. Anthers 3.
  53. Primary panicle branches not rebranched; proximal spikelets on each branch sessile or subsessile, the pedicels shorter than 0.4 mm . . . . . . . . . . . . . . . . . . . 47. *E. sessilispica*
  53. Primary panicle branches usually with secondary branches; proximal spikelets on each branch usually pedicellate, the pedicels longer than 0.4 mm.
    54. Spikelets 1.3–2 mm long, with 1–3 florets; lemmas 0.8–1.2 mm long . . . . . . 48. *E. airoides*
    54. Spikelets 2–19 mm long, with 2–22 florets; lemmas 1.2–2.4 mm long.
      55. Spikelets 2–4.5(5) mm long.
        56. Blades 25–60 cm long, 3–11 mm wide; lemmas 1.6–2.4 mm long; spikelets 1–1.7 mm wide; sheaths densely hirsute, with papillose-based hairs on the base, back, and collar . . . . . . . . 37. *E. hirsuta* (in part)
        56. Blades 4–22 cm long, 1–3.5 mm wide; lemmas 1.2–1.8 mm long; spikelets 0.5–1.3 mm wide; sheaths sometimes hirsute, at least partially, but the hairs never papillose-based . . . . . . . . . . 36. *E. lugens* (in part)
      55. Spikelets 4–19 mm long.
        57. Spikelets with 10–22 florets; caryopses terete to laterally compressed, opaque, uniformly reddish brown . . . . . . . . . . . . . . . 49. *E. atrovirens*
        57. Spikelets with 3–12(14) florets; caryopses dorsally compressed, translucent, greenish over the embryo.
          58. Lemmas 1.8–3 mm long; panicles 16–35(40) cm long, (4)8–24 cm wide; blades 12–50(65) cm long; caryopses 1–1.7 mm long; ligules 0.6–1.3 mm long . . . . . . . . . . . 9. *E. curvula* (in part)
          58. Lemmas 1.4–1.7 long; panicles 6–18 cm long, 2–8 cm wide; blades 2–12 cm long; caryopses 0.4–0.8 mm long; ligules 0.3–0.5 mm long.
            59. Plants without woolly hairs on the base; glumes unequal; lateral lemma veins not green, inconspicuous throughout; spikelets 0.8–1.2 mm; naturalized in the southwestern United States . . . . . . . . . . . . . . . . . . . . . . . . . . . . 10. *E. lehmanniana* (in part)
            59. Plants with conspicuous, woolly hairs on the base; glumes subequal; lateral lemma veins green, conspicuous basally, obscure near the lemma apices; spikelets 1.3–2 mm wide; in the *Manual* region, known only from waste areas near a woolen mill in South Carolina . . . . . . . . . . . . . . . . . . 11. *E. setifolia* (in part)

## 1. Eragrostis ciliaris (L.) R. Br. GOPHERTAIL LOVEGRASS [p. *436*, <u>532</u>]

*Eragrostis ciliaris* is native to the paleotropics. It is naturalized in parts of the United States, growing along roadsides, on waste sites, in xerothermic vegetation, and sometimes in saline habitats, at 0–200 m. It may be more widespread than indicated.

1. Panicles 0.2–1.5 cm wide, contracted, the branches mostly appressed to the rachises, congested, forming glomerate lobes; spikelets densely packed . . . . . . . . . var. *ciliaris*
1. Panicles 1.5–5 cm wide, open, the branches spreading 20–50° from the rachises; spikelets widely separated from each other . . . . . . . . . . . . . . . . . . . . . . var. *laxa*

## Eragrostis ciliaris (L.) R. Br. var. **ciliaris** [p. *436*]

*Eragrostis ciliaris* var. *ciliaris* is more common than *E. ciliaris* var. *laxa* in the *Manual* region.

## Eragrostis ciliaris var. laxa Kuntze [p. *436*]

*Eragrostis ciliaris* var. *laxa* grows in five counties of Florida, the Caribbean Islands, and the Yucatan Peninsula, Mexico.

## 2. Eragrostis cumingii Steud. CUMING'S LOVEGRASS [p. *436*, <u>533</u>]

*Eragrostis cumingii* is native to southeast Asia and Australia. Within the *Manual* region, it has become established in Florida, growing in waste places and along roadsides in sandy or gravelly soils, at 0–150 m.

## 3. Eragrostis amabilis (L.) Wight & Arn. *ex* Nees JAPANESE LOVEGRASS [p. *436*, <u>533</u>]

*Eragrostis amabilis* is native to the Eastern Hemisphere. It is now naturalized in the southeastern United States, growing in open areas such as cultivated fields, forest margins, and roadsides at 0–200 m.

## 4. Eragrostis hypnoides (Lam.) Britton, Sterns & Poggenb. TEEL LOVEGRASS, ÉRAGROSTIDE HYPNOÏDE [p. *436*, <u>533</u>]

*Eragrostis hypnoides* grows along muddy or sandy shores of lakes and rivers and in moist, disturbed sites, at 10–1600 m. It is native to the Americas, extending from southern Canada to Argentina.

## 5. Eragrostis reptans (Michx.) Nees CREEPING LOVEGRASS [p. *436*, <u>533</u>]

*Eragrostis reptans* grows in wet sand, gravel, and clay soils along rivers and lake margins from the United States to northern Mexico, at 0–400 m, frequently with *Cynodon dactylon* and *Heliotropium*. It flowers from April through November.

## 6. Eragrostis japonica (Thunb.) Trin. POND LOVEGRASS [p. *436*, <u>533</u>]

*Eragrostis japonica* is native to the tropics of the Eastern Hemisphere; it is now established in moist areas along rivers and streams in the southern portion of the contiguous United States, usually in sandy soils, at 0–200 m.

## 7. Eragrostis cylindriflora Hochst. [p. *437*, <u>533</u>]

*Eragrostis cylindriflora* is native to Africa. It is not established in the *Manual* region, but has been collected from a disturbed site in Canton, Maryland.

## 8. Eragrostis trichophora Coss. & Durieu [p. *437*]

*Eragrostis trichophora* is native to Africa, where it often grows in moist, disturbed or overgrazed sites. It has been collected from disturbed sites at Canton, Maryland.

## 9. Eragrostis curvula (Schrad.) Nees WEEPING LOVEGRASS [p. *437*, <u>533</u>]

*Eragrostis curvula* is native to southern Africa. It is often used for reclamation because it provides good ground cover but, once introduced, it easily escapes. In the *Manual* region, it grows on rocky slopes, at the margins of woods, along roadsides, and in waste ground, at 20–2400 m, usually in pine-oak woodlands, and yellow pine and mixed hardwood forests.

## 10. Eragrostis lehmanniana Nees LEHMANN'S LOVEGRASS [p. *437*, <u>533</u>]

*Eragrostis lehmanniana* is native to southern Africa, where it grows in sandy, savannah habitats. It was introduced for erosion control in the southern United States, where it often displaces native species. In the *Manual* region, it grows in sandy flats, along roadsides, on calcareous slopes, and in disturbed areas, at 200–1830 m. It is commonly found in association with *Larrea tridentata*, *Opuntia*, *Quercus*, *Juniperus*, and *Bouteloua gracilis*.

## 11. Eragrostis setifolia Nees NEVERFAIL LOVEGRASS [p. *437*]

*Eragrostis setifolia* is an Australian species that was collected around the Santee Wool Combing Mill, Jamestown, Berkeley County, South Carolina, in 1958. It is not known to have spread from that location.

## 12. Eragrostis mexicana (Hornem.) Link MEXICAN LOVEGRASS [p. *437*, <u>533</u>]

*Eragrostis mexicana* grows along roadsides, near cultivated fields, and in disturbed open areas, at 100–3000 m. It is native to the Americas, its native range extending from the southwestern United States through Mexico, Central and northern South America, to Argentina. Within the *Manual* region, it has been introduced beyond its native range, often becoming an established part of the flora.

1. Spikelets ovate to oblong in outline, 1.5–2.4 mm wide; lower glumes 1.2–2.3 mm long; sum of the spikelet width and lower glume length 2.7–4.7 mm; culms and sheaths sometimes with glandular depressions . . . . . . . . . . . . . . . . . . . . . . . . . subsp. *mexicana*
1. Spikelets linear to linear-lanceolate, 0.7–1.4 wide; lower glumes 0.7–1.7 mm long; sum of the spikelet width and lower glume length 1.5–3.1 mm; culms and sheaths without glandular depressions . . . . . . . . . . . . . . . . . . . . . . . . . subsp. *virescens*

## Eragrostis mexicana (Hornem.) Link subsp. **mexicana** [p. *437*]

*Eragrostis mexicana* subsp. *mexicana* grows from Ontario through the midwestern United States to California, South Carolina, and Texas and southwards to Mexico.

## Eragrostis mexicana subsp. virescens (J. Presl) S.D. Koch & Sánchez Vega [p. *437*]

*Eragrostis mexicana* subsp. *virescens* has a disjunct distribution, growing in California and western Nevada and, in South America, from Ecuador to Chile, southern Brazil, and northern Argentina. It has also been found, as an introduction, at various other locations in North America, including eastern North America.

## 13. Eragrostis frankii C.A. Mey. *ex* Steud. SANDBAR LOVEGRASS, ÉRAGROSTIDE DE FRANK [p. *437*, <u>533</u>]

*Eragrostis frankii* is native in the central and eastern United States, but it has been found, as an introduction, in southern Ontario, and appears to be increasingly common in the northeastern United States. It grows in moist meadows, along streams and sand bars, in forest openings, and along roadsides, at 5–1500 m, usually in association with *Pinus*, *Quercus*, *Acer*, and *Fagus grandiflora*. *Eragrostis frankii* differs from *E. capillaris* in its frequent possession of glandular pits, its flat or more shallowly grooved caryopses, shorter pedicels, and glabrous sheath margins, and in having panicles that are usually less than half as long as the culms.

## 14. **Eragrostis capillaris** (L.) Nees Lacegrass [p. *437*, 533]

*Eragrostis capillaris* is native to the eastern portion of the *Manual* region. It grows in open, dry, sandy riverbanks, floodplains, rocky roadsides, and gravel pits, at 150–1500 m, usually in association with *Pinus, Quercus, Carrya,* and *Liquidambar styraciflua.* Its range extends into northeastern Mexico. It differs from *E. frankii* in its lack of glandular pits, deeply grooved caryopses, longer pedicels, pilose sheath margins, and larger panicles. The two species are sympatric over much of the eastern United States.

## 15. **Eragrostis lutescens** Scribn. Sixweeks Lovegrass [p. *437*, 533]

*Eragrostis lutescens* grows on the sandy banks of streams and lakes and in moist alkaline flats of the western United States at 300–2000 m. It has not been reported from Mexico.

## 16. **Eragrostis pilosa** (L.) P. Beauv. India Lovegrass, Éragrostide Poilue [p. *437*, 533]

*Eragrostis pilosa* is native to Eurasia but has become naturalized in many parts of the world. In the *Manual* region, it grows in forest margins and disturbed sites such as roadsides, railroad embankments, gardens, and cultivated fields, at 0–2500 m.

1. Plants with numerous glandular pits scattered over the whole plant, especially on the midribs of the sheaths and blades; lemmas 1.8–2 mm long . . . . . . var. *perplexa*
1. Plants with a few glandular pits scattered on the culms or without any glandular pits; lemmas 1.2–1.8 mm long . . . . . . . . . . . . . . . . . . . . . . . . . . . . . var. *pilosa*

## **Eragrostis pilosa** var. **perplexa** (L.H. Harv.) S.D. Koch [p. *437*]

*Eragrostis pilosa* var. *perplexa* is known from widely scattered locations in Wyoming, North Dakota, Nebraska, Colorado, and northwestern Texas.

## **Eragrostis pilosa** (L.) P. Beauv. var. **pilosa** [p. *437*]

*Eragrostis pilosa* var. *pilosa* is more common than *E. pilosa* var. *perplexa* in the *Manual* region. Most of the records shown on the map are for this variety.

## 17. **Eragrostis pectinacea** (Michx.) Nees Tufted Lovegrass, Éragrostide Pectinée [p. *438*, 533]

*Eragrostis pectinacea* is native from southern Canada to Argentina. In the *Manual* region, it grows in disturbed sites such as roadsides, railroad embankments, gardens, and cultivated fields, at 0–1200 m.

1. Anthers 0.5–0.7 mm long . . . . . . . . . . . . . . . . . . . . . . var. *tracyi*
1. Anthers 0.2–0.4 mm long.
    2. Pedicels appressed, rarely diverging to 20° from the branches . . . . . . . . . . . . . . . . . . . . var. *pectinacea*
    2. Pedicels widely divergent, usually diverging 20–60° from the branches . . . . . . . . . . . . . . . var. *miserrima*

## **Eragrostis pectinacea** var. **miserrima** (E. Fourn.) Reeder [p. *438*]

*Eragrostis pectinacea* var. *miserrima* grows in the southern United States, from Texas to Florida, and south through the lowland tropics to Brazil. It usually flowers from July–November in the *Manual* region.

## **Eragrostis pectinacea** (Michx.) Nees var. **pectinacea** [p. *438*]

*Eragrostis pectinacea* var. *pectinacea* grows throughout the range of the species, including most of the contiguous United States. Within the *Manual* region, it is most common in the eastern states and usually flowers from July–November.

## **Eragrostis pectinacea** var. **tracyi** (Hitchc.) P.M. Peterson [p. *438*]

*Eragrostis pectinacea* var. *tracyi* is known only from Lee, Manatee, and Sarasota counties, Florida. It flowers from March–May and August–December in the *Manual* region.

## 18. **Eragrostis cilianensis** (All.) Vignolo *ex* Janch. Stinkgrass, Éragrostide Fétide [p. *438*, 533]

*Eragrostis cilianensis* is an introduced European species that now grows in disturbed sites such as pastures and roadsides, at 0–2300 m, through most of the contiguous United States and southern Canada.

## 19. **Eragrostis barrelieri** Daveau Mediterranean Lovegrass [p. *438*, 533]

*Eragrostis barrelieri* is a European species that is now naturalized in the *Manual* region, primarily in the southwestern United States. It grows on gravelly roadsides, in gardens, and other disturbed, sandy sites, especially near railroad yards, at 10–2000 m. The ring of glandular tissue is most conspicuous below the upper cauline nodes.

## 20. **Eragrostis minor** Host Little Lovegrass, Éragrostide Faux-Pâturin [p. *438*, 533]

*Eragrostis minor* is a European species that now grows in gravelly roadsides and disturbed sites, especially near railroad yards, at 20–1600 m in southern Canada and the contiguous United States.

## 21. **Eragrostis unioloides** (Retz.) Nees *ex* Steud. Chinese Lovegrass [p. *438*, 533]

*Eragrostis unioloides* is an Asian species that is now established in the southeastern United States, growing along roadsides and in disturbed ground, at 20–150 m.

## 22. **Eragrostis tef** (Zucc.) Trotter Teff [p. *438*]

*Eragrostis tef* is native to northern Africa. In Ethiopia, it is used both as a grain and as fodder for cattle. It is also grown, but not commonly, for these purposes in the *Manual* region and is occasionally found as an escape from cultivation.

## 23. **Eragrostis gangetica** (Roxb.) Steud. Slimflower Lovegrass [p. *438*, 533]

*Eragrostis gangetica* is an Asian species that now grows in the southeastern United States. It can be found in the sandy margins of ponds, roadsides, and ditches, at 0–100 m, usually in association with *Pinus, Taxodium distichum, Rynchospora,* and *Steinchisma hians.* It differs from *E. bahiensis* in its annual habit and shorter spikelets, lemmas, anthers, and caryopses.

## 24. **Eragrostis superba** Peyr. Sawtooth Lovegrass [p. *438*, 533]

*Eragrostis superba* is native to Africa, where it is grown for hay, being fairly palatable and drought resistant. It is also used for erosion control and revegetation. In the *Manual* region, it grows on rocky slopes, in sandy flats, and along roadsides, at 480–1650 m, often with *Acacia, Prosopis, Fouquieria splendens, Juniperus,* and *Quercus.*

## 25. **Eragrostis echinochloidea** Stapf Tickgrass [p. *438*, 533]

*Eragrostis echinochloidea* is native to southern Africa. It is now established in Arizona, growing in gravel soils, often along roadsides and in sidewalks, from 700–1000 m. It has also been found in Prince George's County, Maryland.

## 26. Eragrostis obtusiflora (E. Fourn.) Scribn. ALKALI LOVEGRASS [p. *438*, 533]

*Eragrostis obtusiflora* is native to the southwestern United States and Mexico. It grows in dry or wet alkali flats, often in association with *Distichlis* and *Sarcobatus*, at 900–1400 m.

## 27. Eragrostis spectabilis (Pursh) Steud. PURPLE LOVEGRASS, ÉRAGROSTIDE BRILLANTE [p. *438*, 533]

*Eragrostis spectabilis* is native in the eastern portion of the *Manual* region, extending from southern Canada through the United States, Mexico, and Central America to Belize. It grows in fields and on the margins of woods, along roadsides, and in other disturbed sites, usually in sandy to clay loam soils, at 0–1830 m, and is associated with hardwood forests, *Prosopis-Acacia* grasslands, and shortgrass prairies. A showy species, *E. spectabilis* is available commercially for planting as an ornamental.

## 28. Eragrostis curtipedicellata Buckley GUMMY LOVEGRASS [p. *438*, 533]

The range of *Eragrostis curtipedicellata* extends from southern Colorado, Kansas, and Missouri to northeastern Mexico. It grows near fields, along roadsides, and in the margins of woods, at 10–1525 m.

## 29. Eragrostis silveana Swallen SILVEUS' LOVEGRASS [p. *439*, 533]

*Eragrostis silveana* grows in various open habitats, from sandy prairies to clay loam flats, near roadsides, railroads, and fields at 0–100 m. Its range is limited to the coastal plain of Texas and northern Mexico. Morphologically, *E. silveana* is somewhat intermediate between *E. spectabilis* and *E. curtipedicellata*, and grows where the distribution of these two species overlaps.

## 30. Eragrostis spicata Vasey SPIKE LOVEGRASS [p. *439*, 534]

*Eragrostis spicata* grows in moist areas in prairies, usually in deep, sandy, clay loam soils, at 0–70 m. It is native from southern Texas to Mexico and in Paraguay and Argentina. In North America, it grows with *Andropogon*, *Quercus stellata*, *Prosopis glandulosa*, and *Acacia*.

## 31. Eragrostis plana Nees [p. *439*]

*Eragrostis plana* is native to southern Africa. It is known from two locations in the *Manual* region, both waste areas near sheep and cattle lots in Florence County, South Carolina.

## 32. Eragrostis swallenii Hitchc. SWALLEN'S LOVEGRASS [p. *439*, 534]

*Eragrostis swallenii* grows in sandy sites along coastal grasslands and roadsides, often with *Andropogon* and *Spartina*, at 30–150 m. Its range extends around the Gulf Coast from Texas to Mexico.

## 33. Eragrostis trichodes (Nutt.) Alph. Wood SAND LOVEGRASS [p. *439*, 534]

*Eragrostis trichodes* grows in sandy to gravelly prairies, open sandy woods, rocky slopes, and roadsides, at 100–2150 m, often in associations with *Quercus marilandica*, *Q. stellata*, *Juniperus*, and *Redfieldia flexuosa*. It is endemic to the contiguous United States, and is available commercially as an ornamental.

## 34. Eragrostis palmeri S. Watson RIO GRANDE LOVEGRASS [p. *439*, 534]

*Eragrostis palmeri* grows on rocky slopes and hills between 300–2150 m, generally in association with *Pinus edulis*, *Juniperus monosperma*,

*Bouteloua gracilis*, and *Prosopis*. Its range extends from the southwestern United States into Mexico. It resembles *E. erosa*, but differs in its shorter lemmas and caryopses.

## 35. Eragrostis polytricha Nees HAIRYSHEATH LOVEGRASS [p. *439*, 534]

*Eragrostis polytricha* grows in sandy and rocky areas, at 0–30 m, usually in open pinelands. It is native to Florida but its primary range lies to the south of the *Manual region*, from southern Mexico through Central America to Venezuela, Chile, and Argentina.

## 36. Eragrostis lugens Nees MOURNING LOVEGRASS [p. *439*, 534]

*Eragrostis lugens* grows on sandy dunes and along river banks, at 1–300 m. Its range extends from the southern United States to Peru and Argentina.

## 37. Eragrostis hirsuta (Michx.) Nees BIGTOP LOVEGRASS [p. *439*, 534]

*Eragrostis hirsuta* grows in sandy clay loams on the coastal plain and along roadsides, at 0–150 m, usually in association with *Pinus palustris* and *Quercus*. Its range extends from the southeastern United States through eastern Mexico to Guatemala and Belize.

## 38. Eragrostis intermedia Hitchc. PLAINS LOVEGRASS [p. *439*, 534]

*Eragrostis intermedia* grows in clay, sandy, and rocky soils, often in disturbed sites, at 0–1850 m. Its range extends from the United States through Mexico and Central America to South America. *Eragrostis intermedia* differs from the more widespread *E. lugens* in having wider spikelets, longer lemmas, and caryopses with a prominent adaxial groove.

## 39. Eragrostis erosa Scribn. ex Beal CHIHUAHUA LOVEGRASS [p. *439*, 534]

*Eragrostis erosa* grows on rocky slopes and hills, at 1200–2300 m, often in association with *Pinus edulis*, *Juniperus monosperma*, and *Bouteloua gracilis*. Its range extends from New Mexico and western Texas to northern Mexico.

## 40. Eragrostis refracta (Muhl.) Scribn. COASTAL LOVEGRASS [p. *439*, 534]

*Eragrostis refracta* grows in sandy pinelands, savannahs, marshes, and woodlands on the coastal plain of the southeastern United States, at 0–150 m. It is not known from Mexico.

## 41. Eragrostis elliottii S. Watson ELLIOTT'S LOVEGRASS [p. *440*, 534]

*Eragrostis elliottii* grows in sandy pinelands and live-oak woodlands on the coastal plain, at 0–150 m. Its range extends from the southeastern United States through the West Indies and Gulf coast of Mexico to Central and South America.

## 42. Eragrostis secundiflora J. Presl RED LOVEGRASS [p. *440*, 534]

There are two subspecies of *E. secundiflora*; plants from the *Manual* region belong to **E. secundiflora** subsp. **oxylepis** (Torrey) S.D. Koch. They grow in sandy soils, dunes, grasslands, beaches, and roadsides of the southern United States and northern Mexico, at 0–1000 m. **Eragrostis secundiflora** J. Presl subsp. **secundiflora** grows in Mexico and Central and South America.

## 43. Eragrostis prolifera (Sw.) Steud. DOMINICAN LOVEGRASS [p. *440*, 534]

*Eragrostis prolifera* grows on beaches, in brackish water, and along roadsides, at elevations below 5 m in Florida. Its range extends southward from Florida through Mexico and Central America to Colombia.

### 44. Eragrostis elongata (Willd.) Jacq. LONG LOVEGRASS [p. 440]

*Eragrostis elongata* is native to southeastern Asia and Australia, where it grows in disturbed, sandy soils at 0–50 m. It was collected once near Washington, D.C., probably as an escape from the U.S. Department of Agriculture's experimental grass garden; it has not become established in the *Manual* region.

### 45. Eragrostis bahiensis (Schrad.) Schult. BAHIA LOVEGRASS [p. 440, 534]

*Eragrostis bahiensis* grows in sandy soils near river banks, lake shores, and roadsides, at 0–200 m. Its range extends south from the Gulf Coast of the United States through Mexico to Peru, Bolivia, Paraguay, and Argentina.

### 46. Eragrostis scaligera Salzm. *ex* Steud. [p. 440, 534]

*Eragrostis scaligera* is known from Lee and Collier counties, Florida, where it grows in sandy areas in the coastal scrub zone and along adjacent roadsides, at 0–10 m. It is native to French Guiana and Brazil.

### 47. Eragrostis sessilispica Buckley TUMBLE LOVEGRASS [p. 440, 534]

*Eragrostis sessilispica* grows in prairies, limestone mesas, partial forest openings, and grasslands, generally in sandy soils, at 0–1220 m, often in association with *Prosopis* and *Quercus*. Its range extends into northern Mexico.

### 48. Eragrostis airoides Nees [p. 440, 534]

*Eragrostis airoides* is a South American species that, in the *Manual* region, is known from roadsides and disturbed sites in Brazos County, Texas, and Lamar County, Alabama. It is often treated as *Sporobolus brasiliensis* (Raddi) Hack., but its frequent possession of spikelets with more than 1 floret and its mode of spikelet disarticulation argue for its retention in *Eragrostis*.

### 49. Eragrostis atrovirens (Desf.) Trin. *ex* Steud. THALIA LOVEGRASS [p. 440, 534]

*Eragrostis atrovirens* is native to northern Africa, but it is now established in southeastern United States, where it grows along railways and roads, on beaches and in ditches, often in wet sandy soils and in association with *Pinus*, *Taxodium*, and *Sabal*.

## EXCLUDED SPECIES

The following species have been reported from the *Manual* region, but no specimens supporting their presence, other than in experimental plots, have been found: *Eragrostis acutiflora* (Kunth) Nees, *Eragrostis leptostachya* (R. Br.) Steud., and *Eragrostis suaveolens* Becker *ex* Claus.

## 16.24 CLADORAPHIS Franch.[1]

**Pl** per; synoecious; **rhz**, occ also **stln. Clm** 2–80 cm, hard, persistent, brchd above the base. **Lig** memb, ciliate, cilia as long as or longer than the bas membrane; **bld** linear-lanceolate, becoming rolled, hard, and sharp-pointed. **Infl** tml, exceeding the up lvs, pan of racemosely arranged spikelike pri br; **pri br** wd, not disarticulating, apc hard, sharp; **sec br** shorter than 1 cm, otherwise similar to the pri br, smt clustered. **Spklt** 7–16 mm, lat compressed, with 3–16(20) flt; **flt** bisx; **dis** above the glm and beneath the flt. **Glm** more or less equal, markedly exceeded by the flt; **lm** 3-veined, unawned; **lod** 2; **anth** 3. **Car** glab.

*Cladoraphis* is a southern African genus of two species, both of which grow in open, xeric, sandy habitats.

### 1. Cladoraphis cyperoides (Thunb.) S.M. Phillips
BRISTLY LOVEGRASS [p. 440]

*Cladoraphis cyperoides* was once collected on a ballast dump at Linnton (near Portland, Oregon). It is not known to have persisted in North America.

## 16.25 POGONARTHRIA Stapf[2]

**Pl** ann or per; csp. **Clm** 13–100(250) cm, not wd. **Shth** open; **lig** of hairs or memb and ciliate; **bld** flat or loosely involute. **Infl** tml, pan of numerous spikelike br on elongate rchs. **Spklt** in 2 rows on 1 side of the flat or trigonous br axes, with 2–8 flt, additional rdcd flt smt present distal to the fnctl flt; **rchl intnd** tipped with a few short hairs; **dis** initially above the glm and between the flt or the lm falling and the pal persistent, subsequently at the bases of the pan br. **Glm** unequal, shorter than the spklt, keeled, acute to acuminate, unawned; **lm** 3-veined, keeled, memb, acute, acuminate, or shortly awned; **pal** shorter than the lm. **Car** ellipsoid to fusiform.

*Pogonarthria* includes four species, all of which are native to tropical and southern Africa. One species has become established in Arizona.

### 1. Pogonarthria squarrosa (Licht.) Pilg. HERRINGBONE GRASS, SEKELGRAS [p. 441, 534]

*Pogonarthria squarrosa* is native to eastern and southern Africa. In the *Manual* region, *P. squarrosa* grows spontaneously only in a small area in the foothills of the Huachuca Mountains, Cochise County, Arizona, at an elevation of about 1450 m, where it seems to be competing well with native grasses and *Eragrostis lehmanniana*, another African introduction. The plants tend to grow in rather dense colonies of a few square meters, scattered through the area. It turns reddish-brown as it matures, causing it to stand out among its associates.

[1]Mary E. Barkworth   [2]John R. Reeder

## 16.26 VASEYOCHLOA Hitchc.[1]

Pl per; csp, occ rhz. **Clm** 60–110 cm, erect, glab, unbrchd. **Shth** glab; **lig** memb, ciliate, cilia subequal to or longer than the memb base. **Infl** tml, lanceolate or lance-ovate pan, exceeding the lvs; **br** 1 per nd. **Spklt** bisx, lanceolate, with 5–10 flt; **dis** above the glm and between the flt. **Glm** shorter than the adjacent lm, about ¼ as long as the spklt, lanceolate, glab, unawned; **cal** densely pubescent; **lm** 5-, 7-, or 9-veined, lanceolate, densely pubescent below and glab above, apc truncate to rounded or obtuse, smt retuse, unawned; **pal** shorter than the lm, glab, splitting down the midline as the car matures. **Car** to 3 mm, suborbicular, concave, glab, amber, with 2 persistent hornlike style br.

*Vaseyochloa* is a unispecific genus endemic to the coastal zone of southern Texas.

### 1. Vaseyochloa multinervosa (Vasey) Hitchc.
TEXASGRASS [p. *441*, <u>*534*</u>]

*Vaseyochloa multinervosa* grows in islands of live oaks within rolling sand dunes on the Texas mainland, North Padre Island, and on naturally occurring islands in the Laguna Madre of Texas.

## 16.27 ELEUSINE Gaertn.[2]

Pl ann or per; csp. **Clm** 10–150 cm, hrb, glab, brchg both at and above the base. **Shth** open; **lig** memb, ciliate. **Infl** tml, pan of (1)2–20 non-disarticulating, spikelike br, exceeding the up lvs; **br** 1–17 cm, all or most in a digitate cluster, smt 1(2) br attached immediately below the tml whorl, axes flattened, terminating in a fnctl spklt. **Spklt** 3.5–11 mm, lat compressed, with 2–15 bisx flt; **dis** above the glm and between the flt (*E. coracana* not disarticulating). **Glm** unequal, shorter than the lo lm; **lo glm** 1–3-veined; **up glm** 3–5(7)-veined; **lm** 3-veined, glab, keeled, apc entire, neither mucronate nor awned; **pal** smt with winged keels; **anth** 3, 0.5–1 mm; **ov** glab. **Car** modified, prcp thin, separating from the sd at an early stage in its development; **sd** usu obtusely trigonous, the surfaces ornamented.

Eight of the nine species of *Eleusine* are native to Africa, where they grow in mesic to xeric habitats; the exception, *E. tristachya*, is native to South America. Three species have become established in the *Manual* region. When moistened, the seeds of all species are easily freed from the thin pericarp.

1. Panicles with 1–3 oblong branches 1–6(8) cm long, attached in a single digitate cluster . . . . . . . . . . . . . . . . 3. *E. tristachya*
1. Panicles with 4–20 linear branches 3.5–17 cm long, 1(2) of the branches attached below the terminal, digitate cluster.
   2. Lower glumes 1-veined; panicle branches 3–5.5 mm wide; surface of the seeds striate . . . . . . . . . . . . . . . . . . 1. *E. indica*
   2. Lower glumes 2- or 3-veined; panicle branches 5–15 mm wide; surface of the seeds granular . . . . . . . . . . 2. *E. coracana*

### 1. Eleusine indica (L.) Gaertn. GOOSEGRASS, ELEUSINE DES INDES, ELEUSINE D'INDE [p. *441*, <u>*534*</u>]

*Eleusine indica* is a common weed in the warmer regions of the world. In the *Manual* region, it usually grows in disturbed areas and lawns, and has been found in most states of the contiguous United States.

### 2. Eleusine coracana (L.) Gaertn. [p. *441*]

*Eleusine coracana* is an allotetraploid, one of its genomes being derived from *E. indica*.

Two subspecies are recognized; only subsp. *coracana* is known from North America.

1. Seeds almost globose, the surface granular to smooth; florets not disarticulating . . . . . . . . . . . subsp. *coracana*
1. Seeds oblong, the surface shallowly ridged and uniformly granular; florets disarticulating at maturity . . . . . . . . . . . . . . . . . . . . . . . . subsp. *africana*

### Eleusine coracana subsp. africana (Kenn.-O'Byrne) Hilu & de Wet AFRICAN FINGER MILLET [not illustrated]

This weedy subspecies hybridizes freely with the cultivated subsp. *coracana*. It tends to have more slender branches than subsp. *coracana* (5–7 mm wide rather than 7–15 mm), which led to its previous inclusion in *E. indica*.

### Eleusine coracana (L.) Gaertn. subsp. coracana FINGER MILLET, RAGI [p. *441*]

*Eleusine coracana* subsp. *coracana* is the domesticated variant of *E. coracana*. It is cultivated at various agricultural experiment stations and occasionally escapes.

### 3. Eleusine tristachya (Lam.) Lam. THREESPIKE GOOSEGRASS [p. *441*, <u>*534*</u>]

*Eleusine tristachya* is native to tropical America. In the 1800s and early 1900s, it was found on ballast dumps at various ports and transportation centers in the United States. More recently, it has been found as a weed in the Imperial Valley of California. All records of collections outside of California appear to be historical, with no populations persisting.

[1]Robert I. Lonard  [2]Khidir W. Hilu

## 16.28 ACRACHNE Wight & Arn. *ex* Chiov.[1]

Pl ann; tufted. Clm to approximately 50 cm, erect or geniculate, not wd. Shth open; lig memb, ciliate; bld broadly linear. Infl tml, pan of spikelike br, exceeding the up lvs; br 1.5–10 cm, subdigitate or in whorls along elongate rchs, axes flattened, with imbricate, subsessile spklt, terminating in a rdmt spklt. Spklt lat compressed, with 3–25 flt; dis of the spklt below the glm, of the lm within the spklt acropetal, spklt falling wholly or in part after only a few lm have fallen, pal persistent. Glm 1-veined, keeled, exceeded by the flt; lm 3-veined, strongly keeled, firmly memb to cartilaginous, glab, cuspidate or awn-tipped. Car modified, prcp hyaline, rupturing at maturity; sd deeply sulcate, ornamented.

*Acrachne* has four species, all of which are native to the Eastern Hemisphere. One species, *Acrachne racemosa*, which is widely distributed in the tropics, was recently found in southern California. The genus resembles *Eleusine* and *Dactyloctenium* in its fruits and ornamented seeds, but differs from both in its mode of disarticulation.

### 1. Acrachne racemosa (B. Heyne *ex* Roem. & Schult.) Ohwi  [p. *441*, *534*]

*Acrachne racemosa* grows in areas of seasonal rainfall in tropical regions of Africa, Asia, and Australia. It has been found in Riverside County, California and may become established there.

## 16.29 DACTYLOCTENIUM Willd.[2]

Pl ann or per; tufted, stln, or rhz. Clm 5–115(160) cm, erect or decumbent, often rooting at the lo nd, not brchg above the base. Shth not overlapping, open, keeled; aur absent; lig memb, memb and ciliate, or of hairs; bld flat or involute. Infl tml, pan of 2–11, digitately arranged spicate br; br with axes 0.8–11 cm long, extending beyond the spklt, terminating in a point, the spklt imbricate in 2 rows on the lo sides. Spklt with 3–7 bisx flt, additional strl flt distally; dis usu above the glm, the flt falling as a unit. Glm unequal, shorter than the adjacent lm, 1-veined, keeled; lo glm acute, mucronate; up glm subapically awned, awns curved; cal glab; lm memb, glab, 3-veined (lat veins smt indistinct), strongly keeled, apc entire, mucronate, or awned; pal glab; anth 3, yellow; ov glab; sty fused. Sd falling free of the hyaline prcp, transversely rugose or granular.

*Dactyloctenium* is primarily an African and Australian genus of 10–13 species. Three species have been introduced in the *Manual* region, two of which have become established. *Dactyloctenium aegyptium* is widespread throughout the warmer areas of the world.

1. Panicle branches 0.4–1.5 cm long; most spikelets touching those of an adjacent branch . . . . . . . . . . . . . . . . . 2. *D. radulans*
1. Panicle branches 1.5–7 cm long; only the first few proximal spikelets on each branch in contact with
   those on an adjacent branch.
   2. Anthers 0.5–0.9 mm long; upper glume awns 1–2.5 mm long . . . . . . . . . . . . . . . . . . . . . . . . . 1. *D. aegyptium*
   2. Anthers 1.1–1.7 mm long; upper glume awns 4.5–10 mm long . . . . . . . . . . . . . . . . . 3. *D. geminatum*

### 1. Dactyloctenium aegyptium (L.) Willd. DURBAN CROWFOOT [p. *442*, *534*]

*Dactyloctenium aegyptium* is a widely distributed weed of disturbed sites in the *Manual* region.

### 2. Dactyloctenium radulans (R. Br.) P. Beauv. BUTTONGRASS [p. *442*, *534*]

*Dactyloctenium radulans* has been found at few locations in the *Manual* region, most of which were associated with wool waste. It is native to Australia, where it is regarded as a valuable ephemeral pasture grass in the drier inland areas but also as a garden weed.

### 3. Dactyloctenium geminatum Hack. DOUBLE COMBGRASS [p. *442*]

*Dactyloctenium geminatum* is native to tropical eastern Africa. It was found at one time on ballast dumps in Maryland, but has not survived in North America.

## 16.30 SPOROBOLUS R. Br.[3]

Pl ann or per; usu csp, smt rhz, rarely stln. Clm 10–250 cm, usu erect, rarely prostrate, glab. Shth open, usu glab, often ciliate at the apc; lig of hairs; bld flat, folded, involute, smt terete. Infl tml, open or contracted pan, smt partially included in the upmost shth. Spklt rounded to lat compressed, with 1(–3) flt(s) per spklt; dis above the glm. Glm 0–1-veined; cal poorly developed, usu glab; lm memb or chartaceous, 1(3)-veined, unawned; pal glab, 2-veined, often splitting between the veins at maturity; anth (2)3. Car ellipsoid, obovoid, fusiform, or quadrangular, prcp free from the sd, becoming mucilaginous when moist in most species, remaining dry and partially adherent to the sd in *S. heterolepis* and *S. clandestinus*. Cleistogamous spklt occ present in the lo lf shth.

[1]Sylvia M. Phillips  [2]Stephan L. Hatch  [3]Paul M. Peterson, Stephan L. Hatch, and Alan S. Weakley

*Sporobolus* is a cosmopolitan genus of more than 160 species that grow in tropical, subtropical, and warm-temperate regions throughout the world. Seventy-four species are native to the Western Hemisphere; 27 are native to the *Manual* region, three are established introductions, one was introduced but has not persisted, and the status of one is uncertain. Two genera of the Western Hemisphere, *Calamovilfa* and *Crypsis*, resemble *Sporobolus* in having hairy ligules, spikelets with 1 floret, 1-veined lemmas, and fruits with a free pericarp.

1. Plants annuals or short-lived perennials flowering in the first year.
  2. Lower panicle nodes with 7–20 branches.
    3. Pedicels 0.1–0.5(1) mm long, appressed . . . . . . . . . . . . . . . . . . . . . . . . . . . . . . . . . . . . . . 2. *S. pyramidatus* (in part)
    3. Pedicels (2)3–6(8) mm long, widely spreading . . . . . . . . . . . . . . . . . . . . . . . . . . . . . . . . . . . 3. *S. coahuilensis*
  2. Lower panicle nodes with 1–3 branches.
    4. Spikelets 0.7–1.1 mm long; anthers 0.2–0.3 mm long . . . . . . . . . . . . . . . . . . . . . . . . . . . . . . 1. *S. tenuissimus*
    4. Spikelets 1.6–6 mm long; anthers 0.3–3.2 mm long.
      5. Mature panicles 10–35 cm long, 4.5–30 cm wide, open; secondary branches spreading; pedicels usually 6–25 mm long, spreading . . . . . . . . . . . . . . . . . . . . . . . . . . . . . . . . . . . . . . . . . 20. *S. texanus* (in part)
      5. Mature panicles 1–5 cm long, 0.2–0.5 cm wide, contracted; secondary branches appressed; pedicels usually 0.1–4 mm long, appressed.
        6. Lemmas strigose; spikelets 2.3–6 mm long; mature fruits (1.1)1.8–2.7 mm long . . . . . . . . . . . 4. *S. vaginiflorus*
        6. Lemmas glabrous; spikelets 1.6–3 mm long; mature fruits 1.2–1.8 mm long . . . . . . . . . . . . . . . 5. *S. neglectus*
1. Plants perennial.
  7. Plants with rhizomes.
    8. Spikelets 1.4–3.2 mm long.
      9. Panicles 0.4–1.6 cm wide, spikelike, blades usually conspicuously distichous . . . . . . . . . . . . . . . . . 6. *S. virginicus*
      9. Panicles 2.4–8 cm wide, somewhat contracted to lax and open, blades not obviously distichous . . . . . . . . . . . . . . . . . . . . . . . . . . . . . . . . . . . . . . . . . . . . . . . . . . . . . . . . . . . . . . . 13. *S. fimbriatus*
    8. Spikelets 4–10 mm long.
      10. Panicles (0.6)1–8 cm wide, open to somewhat contracted, narrowly pyramidal, well-exserted from the uppermost sheath; branches without spikelets on the lower ⅓ . . . . . 26. *S. interruptus* (in part)
      10. Panicles 0.04–1.6 cm wide, narrow or spikelike, partially to wholly included in the uppermost sheath; branches spikelet-bearing to the base.
        11. Fruits 1–2 mm long; pericarp gelatinous, slipping from the seed when wet; panicles 5–30 cm long, 0.4–1.6 cm wide; lemmas glabrous, smooth . . . . . . . . . . . . . . . . . . . . . . . . 7. *S. compositus* (in part)
        11. Fruits (1.5)2.4–3.5 mm long; pericarp loose but neither gelatinous nor slipping from the seed when wet; panicles 5–11 cm long, 0.04–0.3 cm wide; lemmas minutely pubescent or scabridulous . . . . . . . . . . . . . . . . . . . . . . . . . . . . . . . . . . . . . . . . . . . . . . 8. *S. clandestinus* (in part)
  7. Plants without rhizomes.
    12. Upper glumes usually less than ⅔ as long as the florets.
      13. Lower panicle branches much shorter than the adjacent internodes, appressed to strongly ascending . . . . . . . . . . . . . . . . . . . . . . . . . . . . . . . . . . . . . . . . . . . . . . . . . . . . . . . . . . . . 11. *S. creber*
      13. Lower panicle branches usually as long as or longer than the adjacent internodes, appressed or ascending.
        14. Spikelets 2–2.7 mm long; upper glumes usually ½–⅔ as long as the florets, acute to obtuse, entire . . . . . . . . . . . . . . . . . . . . . . . . . . . . . . . . . . . . . . . . . . . . . . . . . . . . . . . . 9. *S. indicus*
        14. Spikelets 1.3–1.8(2) mm long; upper glumes usually less than ½ as long as the florets, rarely longer; truncate, erose to denticulate.
          15. Anthers 0.9–1.1 mm long, usually 3, rarely 2; branches spikelet-bearing to the base . . . . . . . . . . . . . . . . . . . . . . . . . . . . . . . . . . . . . . . . . . . . . . . . . . . . . . . . 10. *S. jacquemontii*
          15. Anthers 0.5–0.8 mm long, usually 2, rarely 3; branches without spikelets on the lower ¼ . . . . . . . . . . . . . . . . . . . . . . . . . . . . . . . . . . . . . . . . . . . . . . . . . . . . . . . . . 12. *S. diandrus*
    12. Upper glumes at least ⅔ as long as the florets, often longer.
      16. Spikelets 1–2.5(2.8) mm long [for opposite lead, see p. 214].
        17. Lower sheaths keeled and flattened below . . . . . . . . . . . . . . . . . . . . . . . . . . . . . . . . 14. *S. buckleyi*
        17. Lower sheaths rounded below.
          18. Panicles 12–35 cm wide, open.
            19. Sheath apices with a conspicuous tuft of white hairs; flag blades nearly perpendicular to the culms . . . . . . . . . . . . . . . . . . . . . . . . . . . . . . . . . 18. *S. cryptandrus* (in part)
            19. Sheath apices glabrous or with a few scattered hairs; flag blades ascending.
              20. Secondary panicle branches spikelet-bearing to the base; pedicels mostly appressed, mostly 0.2–0.5 mm long; panicles 20–60 cm long . . . . . . . . . . . . . 16. *S. wrightii*

20. Secondary panicle branches without spikelets on the lower ¼–½; pedicels mostly spreading, mostly 0.5–25 mm long; panicles 10–45 cm long.
  21. Pedicels 0.5–2 mm long; anthers 1.1–1.8 mm long . . . . . . . . . . . . . . . . . 17. *S. airoides*
  21. Pedicels 6–25 mm long; anthers 0.3–1 mm long . . . . . . . . . . . . . 20. *S. texanus* (in part)
18. Panicles 0.2–12(14) cm wide, contracted to open.
  22. Mature panicles 0.2–5 cm wide, contracted, often spikelike, the panicle branches appressed or diverging no more than 30° from the rachises.
    23. Primary panicle branches without spikelets on the lower ⅛–½ of their length.
      24. Leaf blades 1–1.5 mm wide; ligules 0.2–0.4 mm long . . . . . . . . . 21. *S. nealleyi* (in part)
      24. Leaf blades 2–6 mm wide; ligules 0.3–1 mm long.
        25. Lower panicle nodes with 7–12(15) branches; anthers 0.2–0.4 mm long . . . . . . . . . . . . . . . . . . . . . . . . . . . . . . . . . . . . . . 2. *S. pyramidatus* (in part)
        25. Lower panicle nodes with 1–3 branches; anthers 0.5–1 mm long . . . . . . . . . . . . . . . . . . . . . . . . . . . . . . . . . . . . . . . 18. *S. cryptandrus* (in part)
    23. Primary panicle branches spikelet-bearing to the base.
      26. Lower glumes usually 1-veined; mature panicles 0.2–0.8(1) cm wide; lemmas 2–3.2 mm long, linear-lanceolate; upper glumes 2–3.2 mm long; anthers 3, 0.3–0.5 mm long; plants primarily from west of the Mississippi River . . . . . . . . . . . . . . . . . . . . . . . 19. *S. contractus* (in part)
      26. Lower glumes usually without veins; mature panicles 1–5 cm wide; lemmas 1.1–2 mm long, ovate; upper glumes 1.1–2 mm long; anthers 2 or 3, 0.5–1 mm long; plants primarily from east of the Mississippi River . . . . . . . . . . . . . . . . . . . . . . . . . . . . . . . . . 15. *S. domingensis*
  22. Mature panicles 4.5–30 cm wide, open, pyramidal to subovate or oblong, the panicle branches diverging more than 10° from the rachises, sometimes reflexed.
    27. Lower panicle nodes with 7–12(15) branches; anthers 0.2–0.4 mm long . . . . . . . . . . . . . . . . . . . . . . . . . . . . . . . . . . . . . . . . . . 2. *S. pyramidatus* (in part)
    27. Lower panicle nodes with 1–2(3) branches; anthers 0.4–1 mm long.
      28. Pedicels 6–25 mm long, spreading; panicles 4.5–30 cm wide, about as long as wide, diffuse . . . . . . . . . . . . . . . . . . . . . . . . . . . . 20. *S. texanus* (in part)
      28. Pedicels 0.1–3 mm long, appressed or spreading; panicles 0.3–14 cm wide, longer than wide, open and/or drooping.
        29. Culms 10–50(60) cm tall, 0.7–1.2 mm thick near the base; plants with hard, knotty bases; blades (0.6)1.5–6(7) cm long, 1–1.5 mm wide, involute, spreading at right angles to the culms . . . . . . . . . . . . . . . . . . . . . . . . . . . . . . . . . . . . . . . . . . 21. *S. nealleyi* (in part)
        29. Culms 30–120 cm tall, 1–3.5 mm thick near the base; plant bases not hard and knotty; blades (2)5–26 cm long, 2–6 mm wide, flat to involute, ascending or at right angles to the culms.
          30. Pedicels appressed to the secondary branches; primary branches appressed, spreading, or reflexed; pulvini glabrous; rachises straight, erect; mature panicles narrowly pyramidal, lower branches longer than the middle branches . . . . . . . . . . . . . . . . . . . . . . . . 18. *S. cryptandrus* (in part)
          30. Pedicels spreading from the secondary branches; primary branches reflexed; pulvini pubescent; rachises drooping or nodding; mature panicles subovate to oblong, lower branches no longer than those in the middle . . . . . . . . . . . . . . 22. *S. flexuosus*
16. Spikelets 2.5–10 mm long [for opposite lead, see p. 213].
  31. Lower panicle nodes with 3 or more branches.
    32. Mature panicles 2–6 cm wide, pyramidal; panicle branches diverging 20–100° from the rachises; blades 0.8–2 mm wide; fruits 1.4–1.8 mm long . . . . . . . . . . . . . . . . 24. *S. junceus*
    32. Mature panicles 0.4–1.6 cm wide, narrow, contracted; panicle branches appressed or diverging to 20° from the rachises; blades 2–5 mm wide; fruits 1.8–2.3 mm long . . . . . . . . . . . . . . . . . . . . . . . . . . . . . . . . . . . 25. *S. purpurascens*
  31. Lower panicle nodes with 1–2(3) branches.
    33. Mature panicles 0.04–4 cm wide, spikelike; panicle branches appressed.

34. Spikelets 4–6(10) mm long, stramineous to purplish-tinged; panicles terminal and axillary; sheaths without a conspicuous apical tuft of hairs.
  35. Lemmas minutely pubescent or scabridulous, chartaceous and opaque; pericarps loose but neither gelatinous nor slipping off the seeds when wet; fruits (1.5)2.4–3.5 mm long . . . . . . . . . . . . . . . . . . . . . . . . . 8. *S. clandestinus* (in part)
  35. Lemmas usually glabrous and smooth, membranous to chartaceous and hyaline; pericarps gelatinous, slipping off the seeds when wet; fruits 1–2 mm long . . . . . . . . . . . . . . . . . . . . . . . . . . . . . . . . . . . . . . . . . . . . . 7. *S. compositus* (in part)
34. Spikelets 1.7–3.5(4) mm long, whitish to plumbeous; panicles all terminal; sheaths with a conspicuous apical tuft of hairs.
  36. Culms 40–100(120) cm tall, 2–4(5) mm thick near the base; mature panicles 0.2–0.8(1) cm wide; anthers 0.3–0.5 mm long . . . . . . . . . . 19. *S. contractus* (in part)
  36. Culms 100–200 cm tall, (3)4–10 mm thick near the base; mature panicles 1–4 cm wide; anthers 0.6–1 mm long . . . . . . . . . . . . . . . . . . . . . . . 23. *S. giganteus*
33. Mature panicles (0.6)1–30 cm wide, usually open, narrowly pyramidal to pyramidal or ovate; panicle branches appressed or spreading.
  37. Spikelets 2.3–3 mm long; panicles 4.5–30 cm wide, diffuse, about as long as wide; branches capillary; anthers 0.3–1 mm long . . . . . . . . . . . . . . . . . . . . 20. *S. texanus* (in part)
  37. Spikelets 3–7.2 mm long; panicles 0.6–15 cm wide, longer than wide, not diffuse; branches not capillary; anthers 1.5–5 mm long.
    38. Mature spikelets plumbeous; sheath bases dull, fibrous.
      39. Anthers 3–4.2 mm long; ligules 0.2–0.7 mm long; plants from Arizona . . . . . . . . . . . . . . . . . . . . . . . . . . . . . . . . . . . . . . . 26. *S. interruptus* (in part)
      39. Anthers 1.7–3 mm long; ligules 0.1–0.3 mm long; plants not known from Arizona . . . . . . . . . . . . . . . . . . . . . . . . . . . . . . . . . . . 27. *S. heterolepis*
    38. Mature spikelets purplish-brown to purplish; sheath bases shiny, indurate.
      40. Blades 0.5–1.2 mm wide, subterete to terete in cross section, at least at the base, sometimes channeled for portions of their length, sometimes becoming tightly involute distally, senescing or turning tan in late fall, the margins smooth; pedicels with scattered ascending hairs . . . . . . . . . . . . . . . . . . . . . . . . . . . . . . . . . . . . . . . . 28. *S. teretifolius*
      40. Blades 0.8–10 mm wide, flat or V-shaped in cross section, flat, folded, or involute when dry, remaining green well into winter or yellowing at maturity, the margins usually scabridulous, occasionally smooth; pedicels glabrous, sometimes scabridulous or scabrous.
        41. Lower glumes from 0.9 times as long as to longer than the upper glumes; culms 30–80(90) cm tall; panicles 10–25 cm long; pedicels 0.5–4(8) mm long, usually shorter than the spikelets, appressed . . . . . . . . . . . . . . . . . . . . . . . . . . . . . . . . . . . . . . 29. *S. curtissii*
        41. Lower glumes from 0.6–0.9 (0.94) times as long as the upper glumes; culms (30)45–250 cm tall; panicles 15–50 cm long; pedicels 2–22 mm long, spreading or appressed.
          42. Pedicels appressed; lemmas 4.4–6.5 mm long; anthers 3.5–5 mm long; spikelets purplish . . . . . . . . . . . . . . . . . . . . . . . 30. *S. silveanus*
          42. Pedicels spreading; lemmas 3–4.3 mm long; anthers 2–3.4 mm long; spikelets purplish-brown.
            43. Blades (2)3–10 mm wide, pale bluish-green, yellowing at maturity; panicles (18)30–50 cm long, 4–15 cm wide; lower glumes (0.6)0.75–0.94 times as long as the upper glumes . . . . . . . . . . . . . . . . . . . . . . . . . . . . . . . . 31. *S. floridanus*
            43. Blades 1.2–2(3) mm wide, dark green, remaining green well into winter; panicles 15–30 cm long, 2–6 cm wide; lower glumes 0.6–0.83 times as long as the upper glumes . . . . . . . . . . . . . . . . . . . . . . . . . . . . . . . . 32. *S. pinetorum*

## 1. Sporobolus tenuissimus (Mart. *ex* Schrank) Kuntze TROPICAL DROPSEED [p. *442*, <u>*535*</u>]

*Sporobolus tenuissimus* is native to the Western Hemisphere, and introduced to Africa and Asia. Its native distribution in the Americas is tropical, extending from southern Mexico to Brazil and Paraguay. It has been found at a few locations in the southeastern United States, at 0–100 m. It grows in disturbed areas, often occurring as a weed in gardens and cultivated fields.

## 2. Sporobolus pyramidatus (Lam.) Hitchc. WHORLED DROPSEED [p. *442*, <u>*535*</u>]

*Sporobolus pyramidatus* is native to the Americas, extending from the southern United States to Argentina. It grows in disturbed soils, roadsides, railways, coastal sands, and alluvial slopes in many plant communities, at elevations from 0–1500 m.

## 3. Sporobolus coahuilensis Valdés-Reyna [p. *442*, <u>*535*</u>]

*Sporobolus coahuilensis* is primarily known from central Coahuila in Mexico. It was first found in Brewster County, Texas, in 1966, and it has been collected there as recently as 2003. It was also found in Hudspeth County, Texas, in 1980. It appears to be closely related to the widespread species *S. pyramidatus*, from which it differs in its long capillary pedicels and usually wider panicles.

## 4. Sporobolus vaginiflorus (Torr. *ex* A. Gray) Alph. Wood POVERTY GRASS, SPOROBOLE ENGAINÉ [p. *442*, <u>*535*</u>]

*Sporobolus vaginiflorus* is a North American species, native to the eastern portion of the *Manual* region and probably introduced in the west. It grows in disturbed sites within many plant communities, commonly in sandy to sandy-clay soils, these often derived from calcareous parent materials. Its elevational range is 1–1250 m.

1. Sheath bases sparsely hairy; glumes usually longer
   than the florets; lemmas always faintly 3-veined . . var. *ozarkanus*
1. Sheath bases usually glabrous; glumes usually shorter
   than the florets; lemmas usually 1-veined  . . . . . . var. *vaginiflorus*

### Sporobolus vaginiflorus var. ozarkanus (Fernald) Shinners OZARK DROPSEED [p. *442*]

*Sporobolus vaginiflorus* var. *ozarkanus* grows primarily in the central and southeastern United States.

### Sporobolus vaginiflorus (Torr. *ex* A. Gray) Alph. Wood var. vaginiflorus [p. *442*]

*Sporobolus vaginiflorus* var. *vaginiflorus* is the most wide-ranging of the two varieties, extending north into Canada.

## 5. Sporobolus neglectus Nash PUFFSHEATH DROPSEED, SPOROBOLE NÉGLIGÉ [p. *442*, <u>*535*</u>]

*Sporobolus neglectus* is native to the *Manual* region, and grows at 0–1300 m in sandy soils, on river shores, and in dry, open areas within many plant communities, often in disturbed sites. It appears to have been extirpated from Maine and Maryland and is considered endangered or of special concern in Connecticut, Massachusetts, New Hampshire, and New Jersey. It differs from *S. neglectus* in having strigose lemmas, sheaths that are sparsely hairy towards the base and, usually, longer spikelets.

## 6. Sporobolus virginicus (L.) Kunth SEASHORE DROPSEED [p. *442*, <u>*535*</u>]

*Sporobolus virginicus* grows on sandy beaches, sand dunes, and in saline habitats, primarily along the southeastern coast, occasionally inland. Its range extends through Mexico and Central America to Peru, Chile, and Brazil. No fruits of this species have been found despite examination of several natural populations and over 200 herbarium specimens.

## 7. Sporobolus compositus (Poir.) Merr. ROUGH DROPSEED, SPOROBOLE RUDE [p. *442*, <u>*535*</u>]

*Sporobolus compositus* grows along roadsides and railroad right of ways, on beaches, and in cedar glades, pine woods, live oak-pine forests, prairies, and other partially disturbed, semi-open sites at 0–1600 m. Its range lies entirely within the *Manual* region.

The *Sporobolus compositus* complex is a difficult assemblage of forms, perhaps affected by their primarily autogamous breeding system. Asexual proliferation via rhizomes adds to the species' ability to maintain local population structure and to perpetuate unique character combinations.

1. Rhizomes present . . . . . . . . . . . . . . . . . . . . . . . . . . . . var. *macer*
1. Rhizomes absent.
   2. Culms slender, 1–2(2.5) mm thick; upper sheaths
      usually less than 2.5 mm wide; panicles with
      16–36 spikelets per cm² when pressed . . . . . var. *drummondii*
   2. Culms stout, 2–5 mm thick; upper sheaths usually
      2.6–6 mm wide; panicles with 30–90 spikelets per
      cm² when pressed . . . . . . . . . . . . . . . . . . . . . var. *compositus*

### Sporobolus compositus (Poir.) Merr. var. compositus [p. *442*]

*Sporobolus compositus* var. *compositus* is the most widespread of the three varieties, being found throughout most of the range shown for the species, but not in South Carolina or Florida.

### Sporobolus compositus var. drummondii (Trin.) Kartesz & Gandhi [p. *442*]

*Sporobolus compositus* var. *drummondii* is most abundant in Kansas, Oklahoma, and Texas.

### Sporobolus compositus var. macer (Trin.) Kartesz & Gandhi [p. *442*]

*Sporobolus compositus* var. *macer* is known only from the south-central United States.

## 8. Sporobolus clandestinus (Biehler) Hitchc. HIDDEN DROPSEED [p. *442*, <u>*535*</u>]

*Sporobolus clandestinus* grows primarily in sandy soils along the coast and, inland, along roadsides. In the southeastern United States, it is found in dry to mesic longleaf pine-oak-grass communities and cedar glades. Its range lies entirely within the *Manual* region.

## 9. Sporobolus indicus (L.) R. Br. SMUTGRASS [p. *443*, <u>*535*</u>]

*Sporobolus indicus* is a pantropical species. It commonly grows in disturbed places and open areas such as roadsides, pastures, and lake shores. In the *Manual* region, it is found on sandy or clay soils and is associated with many plant communities.

## 10. Sporobolus jacquemontii Kunth RATSTAIL [p. *443*, <u>*535*</u>]

*Sporobolus jacquemontii*, like *S. indicus*, is native to North America. It is not a common species in the *Manual* region, being known only from coastal and low elevation sites in Florida. It is sometimes included in *S. indicus* or *S. pyramidalis* P. Beauv.

## 11. Sporobolus creber De Nardi [p. *443*, <u>*535*</u>]

*Sporobolus creber* is an Australian species that was found in 1995 growing spontaneously on a ranch in Glenn County, California. It differs from *S. indicus* in its widely spaced, closely appressed, and densely spikeleted branches.

## 12. Sporobolus diandrus (Retz.) P. Beauv. [p. *443*, <u>*535*</u>]

*Sporobolus diandrus* is native from India to southeast Asia and Australia. It is not common in North America, being known only from a few counties in Florida, Mississippi, and Texas.

### 13. Sporobolus fimbriatus (Trin.) Nees [p. *443*]

*Sporobolus fimbriatus* is an African species that has only been found in waste areas near the sites of old wool mills in Berkeley and Florence counties, South Carolina.

### 14. Sporobolus buckleyi Vasey BUCKLEY'S DROPSEED [p. *443*, *535*]

*Sporobolus buckleyi* grows between 0–150 m, in loamy soils near the margins of woods or thorn scrub, sometimes in partial sunlight. Its range extends from southeastern Texas to Belize.

### 15. Sporobolus domingensis (Trin.) Kunth CORAL DROPSEED [p. *443*, *535*]

*Sporobolus domingensis* grows in sandy, rocky, or alkaline soils, often in disturbed sites adjacent to the coast and below 20 m. Its range extends to the Antilles and the Yucatan Peninsula, Mexico.

### 16. Sporobolus wrightii Munro *ex* Scribn. BIG ALKALI SACATON [p. *443*, *535*]

*Sporobolus wrightii* grows in moist clay flats and on rocky slopes near saline habitats, from 5–1800 m. Its range extends to central Mexico.

### 17. Sporobolus airoides (Torr.) Torr. ALKALI SACATON [p. *443*, *535*]

*Sporobolus airoides* grows on dry, sandy to gravelly flats or slopes, at elevations from 50–2350 m. It is usually associated with alkaline soils. Its range extends into northern Mexico.

### 18. Sporobolus cryptandrus (Torr.) A. Gray SAND DROPSEED, SPOROBOLE À FLEURS CACHÉES [p. *443*, *535*]

*Sporobolus cryptandrus* is a widespread North American species, extending from Canada into Mexico. It grows in sandy soils and washes, on rocky slopes and calcareous ridges, and along roadsides in salt-desert scrub, pinyon-juniper woodlands, yellow pine forests, and desert grasslands. Its elevational range is 0–2900 m.

### 19. Sporobolus contractus Hitchc. SPIKE DROPSEED [p. *443*, *535*]

*Sporobolus contractus* grows in dry to moist, sandy soils, at elevations from 300–2300 m. It is found occasionally in salt-desert scrub, desert grasslands, and pinyon-juniper woodlands. Its range extends to the states of Baja California and Sonora in Mexico.

### 20. Sporobolus texanus Vasey TEXAS DROPSEED [p. *443*, *535*]

*Sporobolus texanus* grows along rivers, ponds, and in wet alkaline habitats, at 100–3300 m. It is known only from the United States.

### 21. Sporobolus nealleyi Vasey GYPGRASS [p. *444*, *535*]

*Sporobolus nealleyi* grows in sandy and gravelly soils, usually in those derived from gypsum, or near alkaline habitats associated with desert grasslands. It is known only from the southwestern United States, where it grows at 700–3000 m.

### 22. Sporobolus flexuosus (Thurb. *ex* Vasey) Rydb. MESA DROPSEED [p. *444*, *535*]

*Sporobolus flexuosus* grows on sandy to gravelly slopes, flats, and roadsides in the southwestern United States and northern Mexico. It is associated with desert scrub, pinyon-juniper woodlands, and yellow pine forests. Its elevational range is 800–2100 m.

### 23. Sporobolus giganteus Nash GIANT DROPSEED [p. *444*, *535*]

*Sporobolus giganteus* grows in sand dunes and sandy areas along rivers and roadsides, from 100–1830 m. Its range includes the southwestern United States and northern Mexico.

### 24. Sporobolus junceus (P. Beauv.) Kunth PINEY WOODS DROPSEED [p. *444*, *535*]

*Sporobolus junceus* grows in openings in pine and hardwood forests, coastal prairies, and pine barrens, usually in sandy to loamy soils, at 2–400 m. Its range lies entirely within the southern United States.

### 25. Sporobolus purpurascens (Sw.) Ham. PURPLE DROPSEED [p. *444*, *535*]

*Sporobolus purpurascens* grows in oak scrub, prairie grasslands, and sandy sites near railroad crossings and roadsides, at elevations from 2–300 m. It extends from southern Texas through eastern Mexico, the West Indies, and Central America to Brazil.

### 26. Sporobolus interruptus Vasey BLACK DROPSEED [p. *444*, *535*]

*Sporobolus interruptus* grows on rocky slopes and in dry meadows of open yellow pine and oak-pine forests and pinyon-juniper woodlands, at elevations from 1500–2300 m. It is an Arizona endemic that is morphologically similar to *S. heterolepis*, but the range of the latter lies to the north and east of Arizona. The only reliable morphological difference between them is anther length (3–4.2 mm long in *S. interruptus*, 1.7–3 mm long in *S. heterolepis*).

### 27. Sporobolus heterolepis (A. Gray) A. Gray PRAIRIE DROPSEED, SPOROBOLE À GLUMES INÉGALES [p. *444*, *536*]

*Sporobolus heterolepis* grows at elevations of 40–2250 m, in lowland and upland prairies, along the borders of woods, roadsides, and swamps, and in north-facing swales. It is associated with many plant communities, and is also available commercially as an ornamental. It is restricted to the *Manual* region.

### 28. Sporobolus teretifolius R.M. Harper WIRELEAF DROPSEED [p. *444*, *536*]

*Sporobolus teretifolius* is restricted to the southeastern United States, where it grows in wet to moist flatwoods and savannahs, at elevations of 10–150 m.

### 29. Sporobolus curtissii Small *ex* Kearney CURTISS' DROPSEED [p. *444*, *536*]

*Sporobolus curtissii* is restricted to the southeastern United States, where it grows in dry-mesic to moist flatwoods, in soils seasonally saturated at the surface or rather well-drained throughout the year. Its elevational range is 0–100 m.

### 30. Sporobolus silveanus Swallen SILVEUS' DROPSEED [p. *444*, *536*]

*Sporobolus silveanus* is restricted to the southeastern United States. It grows in wet to mesic pine woodlands and adjoining glades and barren openings, and in blackland prairies, at elevations of 5–200 m.

### 31. Sporobolus floridanus Chapm. FLORIDA DROPSEED [p. *444*, *536*]

*Sporobolus floridanus* grows in wet to mesic pine woodlands, seepage bogs, and treeless swales, in soils semi-permanently to seasonally saturated at the surface, and in places where water may pond for weeks, at elevations of 0–100 m. It is endemic to the southeastern United States.

### 32. Sporobolus pinetorum Weakley & P.M. Peterson CAROLINA DROPSEED [p. *444*, *536*]

*Sporobolus pinetorum* grows in wet to moist pine woodlands, in soils seasonally to semi-permanently saturated, at elevations of 0–160 m. It is endemic to the southeastern United States.

## 16.31 CRYPSIS Aiton[1]

Pl ann; synoecious. **Clm** 1–75 cm, erect to geniculately ascending, smt brchg above the base; **nd** usu exposed. **Shth** open, often becoming inflated, jnct with the bld evident; **lig** of hairs; **aur** absent; **bld** often disarticulating. **Infl** tml or tml and ax, spikelike or capitate pan subtended by, and often partially enclosed in, 1 or more of the upmost lf shth, additional pan often present in the axils of the lvs below. **Spklt** 2–6 mm, strongly lat compressed, with 1 flt; **flt** bisx; **dis** above or below the glm. **Glm** 1-veined, strongly keeled; **lm** memb, glab, 1-veined, strongly keeled, not lobed, unawned, smt mucronate; **pal** hyaline, 1–2-veined; **lod** absent; **anth** 2 or 3; **ov** glab. **Car** oblong, prcp loosely enclosing the sd and easily removed when wet; **hila** punctate.

*Crypsis*, a genus of eight species, is native from the Mediterranean region to northern China. Its species tend to occur in moist soils, often in areas subject to winter flooding. The three species found in the *Manual* region are very plastic in the lengths of their culms and leaves.

1. Spikelets 1.5–2.8 mm long; panicles 7–8 times longer than wide, usually completely exserted from the uppermost sheath at maturity . . . . . . . . . . . . . . . . . . . . . . . . . . . . . . . . . . . . . . . . . . . .1. *C. alopecuroides*
1. Spikelets 2.5–4 mm long; panicles 1–5 times longer than wide, the bases usually enclosed in the uppermost sheath at maturity.
    2. Collars glabrous; glumes unequal, the margins glabrous; anthers 0.7–1.1 mm long . . . . . . . . . . . . . . . .2. *C. schoenoides*
    2. Collars pilose; glumes subequal, at least the lower glumes pilose on the margin; anthers 0.5–0.9 mm long . . . . . . . . . . . . . . . . . . . . . . . . . . . . . . . . . . . . . . . . . . . . . . . . . . 3. *C. vaginiflora*

### 1. Crypsis alopecuroides (Piller & Mitterp.) Schrad. FOXTAIL PRICKLEGRASS [p. 445, 536]

*Crypsis alopecuroides* is common to abundant in sandy soils around drying lake margins in Oregon and southern Washington, and within the last forty years has become widespread in northern California; it is also known from several other western states. It was collected in the late 1800s from shipyard areas in and around Philadelphia, but has not been collected since in the eastern United States. In the Eastern Hemisphere, it extends from France and northern Africa to the Urals and Iraq.

### 2. Crypsis schoenoides (L.) Lam. SWAMP PRICKLEGRASS [p. 445, 536]

*Crypsis schoenoides* is common to abundant in clay or sandy clay soils around drying lake margins and vernal pools. In the *Manual*

region, it is most abundant in California, but also appears to be established in a few other western states. It is known from a few collections in several eastern states, though apparently none more recent than 1955. Its native range extends from southern Europe and northern Africa through western Asia to India.

### 3. Crypsis vaginiflora (Forssk.) Opiz MODEST PRICKLEGRASS [p. 445, 536]

*Crypsis vaginiflora* is common to abundant in clay or sandy clay soil in California, where it was first introduced in the late 1800s. It has since been found at a few locations in Washington, Idaho, and Nevada, and will probably spread to additional sites with suitable habitat. It is native to Egypt and southwestern Asia.

## 16.32 CALAMOVILFA Hack.[2]

Pl per; synoecious; rhz, rhz short or elongate. **Clm** 50–250 cm, solitary or few. **Lvs** cauline; **shth** open; **lig** of hairs, dense, short; **bld** elongate, long-tapering. **Infl** tml, simple pan, usu exserted and exceeding the up lvs, open to contracted; **pan** 8–80 cm long, to 60 cm wide, simple, flexible, br not spikelike, not disarticulating. **Spklt** with 1 flt, lat compressed, unawned; **rchl** not prolonged; **dis** above the glm, achenes falling with the lm and pal. **Glm** subequal to unequal, 1-veined, acute; **cal** evidently hairy, hairs ¼–⅞ as long as the lm; **lm** similar to the glm, from shorter than to longer than the up glm, 1-veined, acute, unawned; **pal** longitudinally grooved; **anth** 3, 2.4–5.5 mm; **ov** glab. **Prcp** free from the sd.

*Calamovilfa* is a genus of five species, all of which are endemic to temperate portions of the *Manual* region.

1. Rhizomes elongate; ligules 0.7–2.5 mm (sect. *Interior*).
    2. Lemmas or paleas (or both) pubescent, although sometimes sparsely so; spikelets 7–10.8 mm long . . . . . . .1. *C. gigantea*
    2. Lemmas and paleas glabrous; spikelets 5–8.5 mm long . . . . . . . . . . . . . . . . . . . . . . . . . . . . . . . . . . . . . .2. *C. longifolia*
1. Rhizomes short; ligules to 0.7 mm (sect. *Calamovilfa*).
    3. Panicles contracted, to 3.5 cm wide . . . . . . . . . . . . . . . . . . . . . . . . . . . . . . . . . . . . . . . . . . . . .3. *C. curtissii*
    3. Panicles open, 4–40 cm wide.
        4. Spikelets 6–7.4 mm; glumes acute to acuminate, usually arcuate; lemmas 5.5–7 mm, arcuate, attenuate . . . . . . . . . . . . . . . . . . . . . . . . . . . . . . . . . . . . . . . . . . . . . . . . . . . . . . . . . . . . .4. *C. arcuata*
        4. Spikelets 4–5.8 mm; glumes acute, straight; lemmas 4–5.4 mm, straight, acuminate . . . . . . . . . . . . . . .5. *C. brevipilis*

[1]Barry E. Hammel and John R. Reeder  [2]John W. Thieret†

**1. Calamovilfa gigantea** (Nutt.) Scribn. & Merr. GIANT
SANDREED [p. *445*, <u>*536*</u>]

*Calamovilfa gigantea* grows on sand dunes, prairies, river banks, and flood plains in the Rocky Mountains and central plains from Utah and Nebraska to Arizona and Texas.

**2. Calamovilfa longifolia** (Hook.) Scribn. PRAIRIE
SANDREED [p. *445*, <u>*536*</u>]

*Calamovilfa longifolia* usually grows in sand or sandy soils, but is occasionally found in clay soils or loess. Two geographically contiguous varieties exist. They differ as shown in the following key; the differences between the two are more striking in the field.

1. Most spikelets overlapping no more than 1 other
   spikelet, usually with a brownish cast . . . . . . . . . . . . . . var. *magna*
1. Most spikelets overlapping 2–3 other spikelets,
   usually without a brownish cast . . . . . . . . . . . . . . var. *longifolia*

**Calamovilfa longifolia** (Hook.) Scribn. var. **longifolia**
[p. *445*]

*Calamovilfa longifolia* var. *longifolia* is a characteristic grass on the drier prairies of the interior plains, from southern Canada to northern New Mexico, with reports from southern Arizona. It also grows, as an adventive, in Washington, Wisconsin, Michigan, and Missouri.

**Calamovilfa longifolia** var. **magna** Scribn. & Merr. [p. *445*]

*Calamovilfa longifolia* var. *magna* grows on dunes and sandy shores around lakes Superior, Michigan and Huron, with outlying stations in sand or sandy soils.

**3. Calamovilfa curtissii** (Vasey) Scribn. FLORIDA
SANDREED [p. *445*, <u>*536*</u>]

*Calamovilfa curtissii* is a rare species, although sometimes locally common. It is restricted to two disjunct regions in Florida. Most Gulf Coast populations grow in moist flatwoods or adjacent to wet cypress depressions; Atlantic coast populations occur in interdune swales.

**4. Calamovilfa arcuata** K.E. Rogers CUMBERLAND
SANDREED [p. *445*, <u>*536*</u>]

*Calamovilfa arcuata* is known only from a few scattered locations in the south-central United States. It grows along streams and rivers.

**5. Calamovilfa brevipilis** (Torr.) Scribn. PINE-BARREN
SANDREED [p. *445*, <u>*536*</u>]

*Calamovilfa brevipilis* grows in moist to dry pine barrens, savannahs, bogs, swamp edges, and pocosins. It is a common grass on the New Jersey pine barrens and locally common across the coastal plain of North Carolina, but rare at present in Virginia and South Carolina.

# 16.33 **MUHLENBERGIA** Schreb.[1]

**Pl** ann or per; usu rhz, often csp, smt mat-forming, rarely stln. **Clm** 2–300 cm, erect, geniculate, or decumbent, usu hrb, smt becoming wd. **Shth** open; **lig** memb or hyaline (rarely firm or coriaceous), acuminate to truncate, smt minutely ciliolate, smt with lat lobes longer than the cent portion; **bld** narrow, flat, folded, or involute, smt arcuate. **Infl** tml, smt also ax, open to contracted or spikelike pan; **dis** usu above the glm, occ below the ped. **Spklt** with 1(2–3) flt. **Glm** usu (0)1(2–3)-veined, apc entire, erose, or toothed, truncate to acuminate, smt mucronate or awned; **lo glm** smt rdmt or absent, occ bifid; **up glm** shorter than to longer than the flt; **cal** poorly developed, glab or with hairs; **lm** glab, scabrous, or with short hairs, 3-veined (occ appearing 5-veined), apc awned, mucronate, or unawned; **awns**, if present, straight, flexuous, sinuous, or curled, smt borne between 2 minute teeth; **pal** shorter than or equal to the lm, 2-veined; **anth** (1–2)3, purple, orange, yellow, or olivaceous. **Car** elongate, fusiform or elliptic, slightly dorsally compressed. **Cleistogamous pan** smt present in the axils of the lo cauline lvs, enclosed by a tightly rolled, somewhat indurate shth.

*Muhlenbergia* is primarily a genus of the Western Hemisphere. It has approximately 155 species. Sixty-nine of the 70 species treated here are native to the *Manual* region. The other species, *M. diversiglumis*, is included because it was recently reported from Texas, but no specimens supporting the report have been found. It may be based on a misidentification. Within the *Manual* region, *Muhlenbergia* is represented best in the southwestern United States.

In the key and descriptions, "puberulent" refers to having hairs so small that they can only be seen with a 10× hand lens.

1. Plants annual [for opposite lead, see p. 221].
   2. Lemmas unawned or mucronate, the mucros to 1 mm long.
      3. Glumes strigulose, at least on the margins or towards the apices, the hairs 0.1–0.3 mm long.
         4. Pedicels of most spikelets strongly curved below the spikelets, often through 90° or more; anthers olivaceous, 0.6–1.2 mm long . . . . . . . . . . . . . . . . . . . . . . . . . . . . . . . . . . . . . . . . . . . . . 65. *M. sinuosa*
         4. Pedicels of most spikelets straight to somewhat curved below the spikelets, rarely curved through 90°; anthers purplish, 0.2–0.7 mm long.
            5. Sheaths and culm internodes strigulose; lemmas 1.3–2 mm long, the apices acute to acuminate, mucronate or shortly awned . . . . . . . . . . . . . . . . . . . . . . . . . . . . . . . 67. *M. texana* (in part)
            5. Sheaths and culm internodes glabrous, sometimes scabridulous; lemmas 0.8–1.5 mm long, the apices obtuse to subacute, unawned . . . . . . . . . . . . . . . . . . . . . . . . . . . . . . . . . . . 66. *M. minutissima*
      3. Glumes glabrous.
         6. Panicles contracted, less than 0.5 cm wide; branches closely appressed at maturity; culms often rooting at the lower nodes . . . . . . . . . . . . . . . . . . . . . . . . . . . . . . . . . . . . . 38. *M. filiformis* (in part)
         6. Panicles open or diffuse, 0.6–11 cm wide; branches spreading at maturity; culms not rooting at the lower nodes.

[1]Paul M. Peterson

7. Primary panicle branches 0.5–3.2 cm long; pedicels stout, 1–3 mm long, 0.5–1.5 mm thick; lemmas mottled, with greenish-black and greenish-white areas . . . . . . . . . . . . . . . . . . . . . . . . 70. *M. ramulosa*
7. Primary panicle branches 0.4–6.2 cm long; pedicels delicate, 0.2–10 mm long, about 0.02 mm thick; lemmas not mottled, purplish, plumbeous, or brownish.
  8. Primary panicle branches diverging 80–100° from the rachises; branches not developing below the lower leaf nodes; ligules with lateral lobes (vertical extensions of the sheath margins); plants truly annual . . . . . . . . . . . . . . . . . . . . . . . . . . . . . . . . . . . . . . . . 69. *M. fragilis*
  8. Primary panicle branches diverging less than 80° from the rachises; branches developing below the lower leaf nodes; ligules without lateral lobes; plants perennial but often appearing annual . . . . . . . . . . . . . . . . . . . . . . . . . . . . . . . . . . . . 42. *M. uniflora* (in part)
2. Lemmas awned, the awns 1–32 mm long.
  9. Upper glumes 2- or 3-veined, apices 2- or 3-toothed, not awned.
    10. Lemma awns olive-green, sinuous to curled; lemmas widest near the middle, 1.7–2.2 mm long  . . 47. *M. crispiseta*
    10. Lemma awns purplish, flexuous; lemmas widest near the base, 1.4–4.2 mm long . . . . . . . . . . . 48. *M. peruviana*
  9. Upper glumes 1-veined or veinless, entire, erose, or awned.
    11. Lower glumes 2-veined, minutely to deeply bifid, the teeth aristate or with awns to 1.8 mm long; spikelets often in sessile-pedicellate pairs; disarticulation at the base of the pedicels.
      12. Glumes subequal to the lemmas; lemmas 2.5–4.5 mm long, those of the upper spikelets with awns 6–15 mm long . . . . . . . . . . . . . . . . . . . . . . . . . . . . . . . . . . . . 63. *M. depauperata*
      12. Glumes up to ⅔ as long as the lemmas; lemmas 3.5–6 mm long, those of the upper spikelets with awns 10–20 mm long . . . . . . . . . . . . . . . . . . . . . . . . . . . . . . . . . 64. *M. brevis*
    11. Lower glumes, if present, veinless or 1-veined, not bifid, unawned or with a single awn; spikelets borne singly; disarticulation above the glumes.
      13. Lemma awns 1–5 mm long.
        14. Glumes 0.1–0.3 mm long, the lower glumes often almost absent; ligules 0.2–0.5 mm long . . . . . . . . . . . . . . . . . . . . . . . . . . . . . . . . . . . . 13. *M. schreberi* (in part)
        14. Glumes 0.8–1.8 mm long; lower glumes 0.8–1.6 mm long; ligules 0.9–2.5 mm long.
          15. Pedicels 2–7 mm long, usually longer than the florets, usually divergent; lemmas 1.3–2 mm long; lemma awns 0.1–1(2) mm long; caryopses 0.8–1 mm long; paleas minutely appressed-pubescent on the lower ½, 1.3–2 mm long . . . . . . . . 67. *M. texana* (in part)
          15. Pedicels 1–2(3) mm long, usually shorter than the florets and appressed to the branches; lemmas (1.7)1.9–2.5 mm long; lemma awns 1.2–3.5 mm long; caryopses 1.3–2.3 mm long; paleas glabrous, 1.8–2.4 mm long . . . . . . . . . . . . . . . . . . 68. *M. eludens*
      13. Lemma awns 10–32 mm long.
        16. Panicles secund; primary branches with 2–5 spikelets; secondary branches not developed; spikelets dimorphic with respect to the glumes, the glumes of the proximal spikelet on each branch subequal, to 0.7 mm long, orbicular and unawned, those of the distal spikelets evidently unequal, the lower glumes to 8 mm long and usually awned, the upper glumes orbicular, sometimes awned . . . . . . . . . . . . . . . . . . 18. *M. diversiglumis*
        16. Panicles not secund; primary branches always with more than 2 spikelets, usually with more than 5; secondary branches well-developed; spikelets monomorphic with respect to the glumes.
          17. Ligules 0.3–0.9 mm long, membranous and ciliate; distal portion of the sheath margins with hairs to 1 mm long; lemmas subulate to lanceolate, with a scabrous line between the midvein and lateral veins, giving the lemmas a 5-veined appearance . . . . . . . . . . . . . . . . . . . . . . . . . . . . . . . . . . . . . . . . . 17. *M. pectinata*
          17. Ligules 1–3(5) mm long, hyaline to membranous, often lacerate; sheath margins glabrous, even distally; lemmas lanceolate, smooth over most of their length, scabridulous to scabrous distally, not appearing 5-veined.
            18. Cleistogamous panicles not present in the axils of the lower cauline leaves; upper glumes 1.5–2.8 mm long, acute . . . . . . . . . . . . . . . . . . . . . . 14. *M. tenuifolia* (in part)
            18. Cleistogamous panicles of 1–3 florets present in the axils of the lower cauline leaves; upper glumes 0.6–2 mm long, obtuse to subacute.
              19. Lemmas 2.5–3.8(5.3) mm long; glumes 0.4–1.3 mm long; ligules 1–2 mm long . . . . . . . . . . . . . . . . . . . . . . . . . . . . . . . . . . . . . . . . . 15. *M. microsperma*
              19. Lemmas 4–6.2(7.5) mm long; glumes 1–2 mm long; ligules 1.5–3 mm long . . . . . . . . . . . . . . . . . . . . . . . . . . . . . . . . . . . . . . . . . . . 16. *M. appressa*

1. Plants perennial [for opposite lead, see p. 219].
  20. Plants rhizomatous, usually not cespitose; rhizomes scaly and creeping [for opposite lead, see p. 223].
    21. Panicles open, (2)4–20 cm wide; panicle branches capillary (0.05–0.1 mm thick), diverging 30–100° from the rachises at maturity.
      22. Lemmas awned, awns 1–12(20) mm long.
        23. Blades stiff and pungent; lemma awns 1–1.5(2) mm long, straight; primary branches of the panicles appearing fascicled in immature plants . . . . . . . . . . . . . . . . . . . . . . . . . . 32. *M. pungens*
        23. Blades not stiff and pungent; lemma awns 4–12(20) mm long, flexuous; primary branches of the panicles not appearing fascicled . . . . . . . . . . . . . . . . . . . . . . 24. *M. arsenei* (in part)
      22. Lemmas unawned, sometimes mucronate with mucros to 0.3 mm long.
        24. Culms compressed-keeled; blades conduplicate; panicles cylindrical, 4–8 cm wide . . . . . 39. *M. torreyana*
        24. Culms terete to somewhat compressed-keeled near the base; blades usually flat, occasionally conduplicate; panicles ovoid, 4–16 cm wide.
          25. Ligules 0.5–2 mm long, hyaline, with well-developed lateral lobes; blade margins and midveins prominent, white, thick . . . . . . . . . . . . . . . . . . . . . . . . . . . . . . . . . . . . 41. *M. arenacea*
          25. Ligules 0.2–1 mm long, ciliate, without lateral lobes; blade margins and midveins not prominent, greenish, not particularly thick . . . . . . . . . . . . . . . . . . . 40. *M. asperifolia*
    21. Panicles contracted, 0.1–2(3) cm wide; panicle branches more than 0.1 mm thick, appressed or diverging up to 30(40)° from the rachises at maturity.
      26. Culms 100–300 cm tall, 3–6 mm thick, woody and bamboolike . . . . . . . . . . . . . . . . . . . . . . 33. *M. dumosa*
      26. Culms 4–100(140) cm tall, 0.5–2(3) mm thick, herbaceous, not bamboolike.
        27. Blades 0.2–2(2.6) mm wide, flat, involute, or folded at maturity.
          28. Lemmas awned, the awns 1–25 mm long.
            29. Lemmas pubescent for ¾ their length, the hairs about 1.5 mm long . . . . . . . . . 21. *M. curtifolia*
            29. Lemmas scabridulous or pubescent for no more than ½ their length, the hairs often less than 1.5 mm long.
              30. Lemma awns generally less than 4(6) mm long . . . . . . . . . . . . . . . . . 19. *M. glauca* (in part)
              30. Lemma awns 4–25 mm long.
                31. Lemmas and paleas mostly glabrous, the calluses with a few short hairs; ligules with lateral lobes, the lobes 1.5–3 mm longer than the central portion; culms erect; plants tightly cespitose at the base; sheaths and blades commonly with dark brown necrotic spots . . . . . . . . . . . . . . . . . . . . . . . . . . . . . . . . . . . . 22. *M. pauciflora* (in part)
                31. Lemma midveins and margins and paleas appressed-pubescent on the lower ⅓–⅔; ligules without lateral lobes or with lobes less than 1.5 mm longer than the central portion; culms decumbent; plants loosely cespitose at the base; sheaths and blades without necrotic spots.
                  32. Anthers orange, 1.5–2 mm long; lemmas elliptic, 2–3.5 mm long, the awns 10–20(25) mm long; panicles 0.3–1.8 cm wide . . . . . . . . . . . . . . . . . . . . . . . . . . . . . . . . . . . . . . 23. *M. polycaulis* (in part)
                  32. Anthers purple, 1.3–3 mm long; lemmas lanceolate, 3.5–5 mm long, the awns 4–12(20) mm long; panicles 1–3(5) cm wide . . . . . . . . . . . . . . . . . . . . . . . . . . . . . . . . . . . . . . 24. *M. arsenei* (in part)
        28. Lemmas unawned, mucronate, or shortly awned, the awns to 1 mm long.
          33. Lemmas and paleas pubescent, the hairs 0.4–1.2 mm long.
            34. Glumes acuminate, usually awned, the awns to 1.5 mm long; anthers orange; blades flat to involute distally, never arcuate . . . . . . . . . . . . . . 19. *M. glauca* (in part)
            34. Glumes acute, neither mucronate nor awned; anthers yellow, dark green, or purple; blades tightly involute, often arcuate.
              35. Lemmas 2.6–4 mm long; glumes nearly as long as the lemmas; anthers 2.1–2.3 mm long . . . . . . . . . . . . . . . . . . . . . . . . . . . . . . 20. *M. thurberi*
              35. Lemmas 1.4–2.5 mm long; glumes ½–⅔ as long as the lemmas; anthers 0.9–1.4 mm long . . . . . . . . . . . . . . . . . . . . . . . . . . . . . . . 34. *M. villiflora*
          33. Lemmas and paleas scabrous, glabrous, or with hairs less than 0.3 mm long.
            36. Glumes more than ½ as long as the lemmas; lemmas 2.6–4.2 mm long, attenuate . . . . . . . . . . . . . . . . . . . . . . . . . . . . . . . . . . . . . . . . . . . . . . . 35. *M. repens*

36. Glumes ½ as long as the lemmas or less; lemmas 1.3–3.1 mm long; not
    attenuate.
    37. Ligules 0.2–0.8 mm long; panicles usually partially included, the
        rachises usually visible between the branches . . . . . . . . . . . . . . . . . . . . . . . 36. *M. utilis*
    37. Ligules 0.8–3 mm long; panicles exserted, the rachises usually
        hidden by the branches . . . . . . . . . . . . . . . . . . . . . . . . . . . . . 37. *M. richardsonis* (in part)
27. Blades (1.5)2–15 mm wide, flat at maturity.
    38. Glumes awn-tipped, 3–8 mm long (including the awns), about 1.3–2 times
        longer than the lemmas.
        39. Internodes dull, puberulent, usually terete, rarely keeled; culms seldom
            branched above the base; ligules 0.2–0.6 mm long; anthers 0.8–1.5 mm long . . . . . . 2. *M. glomerata*
        39. Internodes smooth and polished for most of their length, elliptic in cross
            section and strongly keeled; culms much branched above the base; ligules
            0.6–1.7 mm long; anthers 0.4–0.8 mm long . . . . . . . . . . . . . . . . . . . . . . . . . . . 1. *M. racemosa*
    38. Glumes unawned or awn-tipped, 0.4–4 mm long (including the awns), from
        shorter than to about 1.2 times longer than the lemmas.
        40. Lemmas usually completely glabrous, rarely with a few appressed hairs.
            41. Culms 30–100 cm tall, bushy and much branched above; axillary
                panicles common, partly included or exserted from the sheaths; lemmas
                shiny, stramineous or purplish, not mottled; anthers 0.3–0.5 mm long . . . . . . 5. *M. glabrifloris*
            41. Culms 5–30 cm tall, often mat-forming; axillary panicles not present;
                lemmas dark greenish or plumbeous, sometimes mottled; anthers 0.9–1.6
                mm long . . . . . . . . . . . . . . . . . . . . . . . . . . . . . . . . . . . . 37. *M. richardsonis* (in part)
        40. Lemmas with hairs, these sometimes restricted to the callus.
            42. Lemma bases with hairs about as long as the florets, usually 2–3.5 mm
                long . . . . . . . . . . . . . . . . . . . . . . . . . . . . . . . . . . . . . . . . . . . . . . . . . . . . . . . 6. *M. andina*
            42. Lemma bases glabrous or with hairs shorter than the florets, usually
                shorter than 1.5 mm.
                43. Glumes unequal in length, the lower glumes 0.4–1.5 mm long, the
                    upper glumes 0.8–1.9 mm long . . . . . . . . . . . . . . . . . . . . . . . 7. *M. ×curtisetosa* (in part)
                43. Glumes subequal, 1–4 mm long.
                    44. Axillary panicles often present, always partly included in the
                        sheaths; internodes smooth, shiny.
                        45. Ligules 0.2–0.6 mm long; leaves of the lateral branches
                            often shorter and narrower than those of the main
                            branches; glumes 1.4–2 mm long, ⅓–⅔ as long as the
                            lemmas . . . . . . . . . . . . . . . . . . . . . . . . . . . . . . . . . . . . . . . . . . . . . 8. *M. bushii*
                        45. Ligules 0.7–1.7 mm long; leaves of the lateral branches
                            similar in length and width to those of the main
                            branches; glumes 2–4 mm long, about as long as the
                            lemmas . . . . . . . . . . . . . . . . . . . . . . . . . . . . . . . . . . . . . . . . . . 9. *M. frondosa*
                    44. Axillary panicles, if present, exserted on elongated
                        peduncles; internodes smooth, scabrous, or pubescent,
                        sometimes smooth and shiny.
                        46. Glumes much shorter than the lemmas, acute.
                            47. Anthers 0.4–1 mm long; lemmas unawned, or
                                with an awn less than 1 mm long; internodes
                                and sheaths glabrous . . . . . . . . . . . . . . . . . . . . . . . . 10. *M. sobolifera*
                            47. Anthers 1.1–2.2 mm long; lemmas usually awned,
                                occasionally unawned, the awns to 12 mm long;
                                internodes, and usually the base of the sheaths,
                                pubescent . . . . . . . . . . . . . . . . . . . . . . . . . . . . . . . . . . 11. *M. tenuiflora*
                        46. Glumes nearly as long as or longer than the lemmas,
                            acuminate.
                            48. Anthers 1–1.7 mm long; sheaths scabrous; lemma
                                awns 0.2–2.2 mm long; restricted to the Transverse
                                Ranges of southern California . . . . . . . . . . . . . . . . . . . 4. *M. californica*
                            48. Anthers 0.3–0.8 mm long, sheaths smooth for most
                                of their length; lemma awns to 18 mm long;

widespread species, but not known from southern California.

    49. Ligules 0.4–1 mm long; panicles dense; pedicels up to 2 mm long; anthers 0.3–0.5 mm long . . . . . . . . . . . . . . . . . . . . . . . . . . . . . . . . . . . . . . 3. *M. mexicana*

    49. Ligules 1–2.5 mm long; panicles not dense, pedicels 0.8–3.5 mm long; anthers 0.4–0.8 mm long . . . . . . . . . . . . . . . . . . . . . . . . . . . . . . . . . . . . . . 12. *M. sylvatica*

20. Rhizomes absent; plants cespitose or bushy [for opposite lead, see p. 221].

  50. Upper glumes usually 3-veined and 3-toothed; old sheaths flattened, ribbonlike or papery, sometimes spirally coiled.

    51. Lemmas unawned, mucronate, or with awns to 5 mm long.

      52. Lower glumes awned, the awns to 1.6 mm long; blades 2–6 cm long, filiform, tightly involute, sharp-tipped . . . . . . . . . . . . . . . . . . . . . . . . . . . . . . . . . 43. *M. filiculmis*

      52. Lower glumes unawned; blades (5)6–12 cm long, flat or loosely involute to subfiliform, not sharp-tipped . . . . . . . . . . . . . . . . . . . . . . . . . . . . . . 44. *M. jonesii*

    51. Lemmas awned, the awns (2)6–27 mm long.

      53. Upper glumes as long as or longer than the lemmas, the apices acuminate to acute, occasionally minutely 3-toothed; old sheaths conspicuously spirally coiled . . . . . . . . . . 45. *M. straminea*

      53. Upper glumes ⅓–⅔ as long as the lemmas, the apices truncate to acute, 3-toothed; old sheaths occasionally spirally coiled . . . . . . . . . . . . . . . . . . . . . . . . . . . . . . 46. *M. montana*

  50. Upper glumes usually 1-veined, (rarely 2- or 3-veined), rounded, obtuse, acute, or acuminate, entire or erose; old sheaths not flattened or papery, never spirally coiled.

    54. Panicles loosely contracted, open, or diffuse, (1)2–40 cm wide; panicle branches usually not appressed at maturity, often naked basally [for opposite lead, see p. 225].

      55. Culms arising from the bases of old depressed culms; plants loosely matted, delicate; lemmas 1.2–2 mm long, unawned . . . . . . . . . . . . . . . . . . . . . . . . . . . . . 42. *M. uniflora* (in part)

      55. Culms not arising from the bases of old depressed culms; plants cespitose, not delicate; lemmas 2–5 mm long, awned or unawned.

        56. Lemmas unawned or with awns to 4(6) mm long.

          57. Basal sheaths laterally compressed, usually keeled.

            58. Glumes longer than the lemmas; ligules 10–25 mm long, membranous throughout; lemmas 2–3 mm long, usually awned, the awns to 15 mm long, occasionally unawned . . . . . . . . . . . . . . . 49. *M. emersleyi* (in part)

            58. Glumes shorter than the lemmas; ligules 3–12 mm long, firm and brown near the base, membranous distally; lemmas 3–4.2 mm long, awned, the awns 0.5–4 mm long . . . . . . . . . . . . . . . . . . . . . . . 50. *M. ×involuta*

          57. Basal sheaths rounded to somewhat flattened, not keeled.

            59. Culms 10–60(70) cm tall, somewhat decumbent; blades 0.5–10(16) cm long.

              60. Blade margins and midveins conspicuous, thick, white, and cartilaginous; ligules 1–2 mm long . . . . . . . . . . . . . . . . . . . . . . . . . 29. *M. arizonica*

              60. Blade margins not conspicuously thickened, greenish; ligules 2–9 mm long.

                61. Blades arcuate, 0.3–0.9 mm wide, 1–3(5) cm long; usually no culm nodes exposed; most leaf blades reaching no more than ⅛ of the plant height . . . . . . . . . . . . . . . . . . . . . . . . . . . . . 30. *M. torreyi*

                61. Blades not arcuate, 1–2.2 mm wide, 4–10(16) cm long; 1 or more culm nodes exposed; leaf blades reaching ¼–½ of the plant height . . . . . . . . . . . . . . . . . . . . . . . . . . . . . 31. *M. arenicola*

            59. Culms 40–160 cm tall, erect from the base; blades (8)10–100 cm long.

              62. Pedicels 0.1–2.5 mm long, usually shorter than the florets; glumes usually longer than the florets; ligules 10–30 mm long; panicle branches not capillary, appressed or spreading up to 60° from the rachises, the lower branches with 30–60 spikelets . . . . . . . . 51. *M. longiligula* (in part)

              62. Pedicels 3–50 mm long, longer than the florets; glumes usually shorter than the florets; ligules 1.8–10 mm long; panicle branches capillary, spreading 30–100° from the rachises, the lower branches with 5–20 spikelets.

63. Panicles about as long as wide, not diffuse, 10–20(30) cm long; branches and pedicels stiff or flexible . . . . . . . . . . . . . . . . 55. *M. reverchonii* (in part)
63. Panicles longer than wide, diffuse, 15–60 cm long; branches and pedicels flexible.
     64. Glumes more than ½ as long as the lemmas; lemmas usually unawned, if awned, the awns no more than 3 mm long . . . . . . . . . . . . . . . . . . . . . . . . . . . . . . . . . . . . . 52. *M. expansa*
     64. Glumes less than ½ as long as the lemmas; lemmas usually awned, the awns to 18 mm long . . . . . . . . . . . . . . . 53. *M. capillaris* (in part)
56. Lemma awns 6–35 mm long.
     65. Plants conspicuously branched, bushy in appearance; culms wiry, with geniculate, stiff, widely divergent branches . . . . . . . . . . . . . . . . . . . . . . . . . . . . . . . 26. *M. porteri*
     65. Plants not conspicuously branched, not bushy in appearance, usually typical bunchgrasses; culms not wiry, when branched, the branches not both geniculate and widely divergent.
         66. Basal sheaths laterally compressed, commonly keeled; glumes (excluding any awns) exceeding the florets; culms (50)80–150 cm tall . . . . . . . . . . . . . . . . . . . . . . . . . . . . . . . . . . . . . . . . . . . . . . . . . . . . . .49. *M. emersleyi* (in part)
         66. Basal sheaths rounded; glumes (excluding any awns) exceeded by the florets; culms 30–160 cm tall.
             67. Glume apices puberulent or scabridulous, acuminate to acute; lemmas long-acuminate, the demarcation of the lemma body and awn not evident . . . . . . . . . . . . . . . . . . . . . . . . . . . . . . . . . . . . . . . . . . 57. *M. elongata*
             67. Glume apices glabrous or sparsely hirtellous, obtuse to acute, sometimes awned; lemmas acute to acuminate, the demarcation of the lemma body and awn evident.
                 68. Glumes neither awned nor mucronate; spikelets dark purple; lemmas scabrous distally.
                     69. Panicles 8–41 cm wide, open, diffuse, the branches strongly divergent; pedicels 10–40(50) mm long; in the *Manual* region, restricted to the eastern United States, growing from Connecticut south to Kansas, Oklahoma, eastern Texas, and Florida . . . . . . . . . . . . . . . . . . . 53. *M. capillaris* (in part)
                     69. Panicles 2–5 (12) cm wide, loosely contracted to open, most branches appressed to ascending, occasionally a few diverging up to 80°; pedicels 1–10 mm long; in the *Manual* region, restricted to the southwestern United States, growing from Arizona to western Texas . . . . . . . . . . . . . . . 56. *M. rigida*
                 68. Glumes usually awn-tipped or mucronate; spikelets stramineous, brown, or purplish; lemmas smooth or scabrous distally.
                     70. Lemmas shiny, smooth; blades tightly involute, 0.2–1.2 mm wide; panicles 2–5 cm wide, loosely contracted; branches diverging up to 70° from the rachises . . . . . . . . . . . . 58. *M. setifolia*
                     70. Lemmas not shiny, usually scabrous, at least distally; blades flat or involute, 1–4 mm wide; panicles 5–40 cm wide, loosely contracted to open; branches diverging 30°–100° from the rachises.
                         71. Panicles 10–20(30) cm long, about equally wide, not diffuse; branches spreading up to 80° from the rachises at maturity; spikelets stramineous, brownish or purple-tinged; lemmas awned, the awns 0.5–6 mm long . . . . . . . . . . . . . . . . . . . 55. *M. reverchonii* (in part)
                         71. Panicles 15–50(60) cm long, narrower than long, diffuse; branches spreading 30–100° from the rachises at maturity; spikelets usually purplish; lemmas unawned or, if awned, the awns to 35 mm long.
                           72. Upper glumes unawned or with awns to 5 mm long; lemmas without setaceous teeth or the

teeth no more than 1 mm long; lemma awns
2–13(18) mm long . . . . . . . . . . . . . . . . . . 53. *M. capillaris* (in part)

72. Upper glumes awned, the awns 2–25 mm long;
lemmas with setaceous teeth 1–5 mm long;
lemma awns 8–35 mm long . . . . . . . . . . . . . . . . . . 54. *M. sericea*

54. Panicles narrow, 0.2–3(5) cm wide; branches appressed to ascending at maturity, usually
spikelet-bearing their whole length [for opposite lead, see p. 223].

73. Panicles spikelike, 0.1–1.2 cm wide, sometimes interrupted near the base; branches
appressed, 0.3–1.2(4) cm long.

74. Culms 3–20(35) cm tall, often decumbent and rooting at the lower nodes; blades
1–4(6) cm long . . . . . . . . . . . . . . . . . . . . . . . . . . . . . . . . . . . . . . . . . . . . 38. *M. filiformis* (in part)

74. Culms (15)20–150 cm tall, stiffly erect, not rooting at the lower nodes; blades
1.4–50 cm long, at least some more than 6 cm long.

75. Lemmas awned, awns 3–10 mm long; lemmas scabridulous, the veins and
margins glabrous . . . . . . . . . . . . . . . . . . . . . . . . . . . . . . . . . . . . 59. *M. palmeri* (in part)

75. Lemmas unawned, sometimes mucronate, the mucros up to 1 mm long;
lemmas with hairs on the lower ⅙, the hairs sometimes extending to ¾ of
the lemma length over the midvein and margins.

76. Basal leaf sheaths rounded on the back; panicles 15-60 cm long . . . . . . . . . . . 62. *M. rigens*

76. Basal leaf sheaths compressed, the backs keeled; panicles 4-16 cm
long.

77. Ligules 0.2–0.8 mm long; paleas glabrous; glume apices gradually
acute to acuminate, mucronate, the mucros to 0.3 mm long . . . . . . . . 27. *M. cuspidata*

77. Ligules 1–3(5) mm long; paleas pubescent between the veins; glumes
abruptly narrowed, acute or obtuse, awned, the awns 0.5–1 mm
long . . . . . . . . . . . . . . . . . . . . . . . . . . . . . . . . . . . . . . . . . . . . 28. *M. wrightii*

73. Panicles not spikelike, 0.6–5 cm wide; branches appressed, ascending, or diverging
up to 70°, 0.2–13 cm long.

78. Ligules 10–35 mm long, firm and brown basally, membranous distally; glumes
as long as or longer than the florets.

79. Basal sheaths rounded; lemmas unawned or with awns to 2 mm long; plants
of Arizona and New Mexico . . . . . . . . . . . . . . . . . . . . . . . . . . . 51. *M. longiligula* (in part)

79. Basal sheaths compressed-keeled, at least basally; lemmas unawned or with
awns to 4 mm long; endemic to south central Texas, sometimes grown as an
ornamental . . . . . . . . . . . . . . . . . . . . . . . . . . . . . . . . . . . . . . . . . . . . 60. *M. lindheimeri*

78. Ligules 0.2–10 mm long, usually membranous throughout, sometimes firmer
basally, never brownish; glumes shorter than the florets.

80. Lemma awns 0.5–6 mm long.

81. Blades 10–60 cm long.

82. Glumes acute, unawned; ligules 4–10 mm long, acute, lacerate;
spikelets grayish-green . . . . . . . . . . . . . . . . . . . . . . . . . . . . . . . 61. *M. dubia*

82. Glumes acute to acuminate, awned, the awns to 1.5 mm long;
ligules 1–3 mm long, truncate, ciliolate; spikelets yellowish-
brown to purplish . . . . . . . . . . . . . . . . . . . . . . . . . . . . . . 59. *M. palmeri* (in part)

81. Blades to 10 cm long.

83. Glumes 2–4 mm long; lemmas 3.5–5 mm long, the awns flexuous;
ligules 1–2 mm long, with lateral lobes less than 1.5 mm longer
than the central portion; anthers 1.3–3 mm long, purple . . . . . . . . . . . . 24. *M. arsenei*

83. Glumes less than 2 mm long; lemmas 1.8–3(3.4) mm long, the
awns straight; ligules 0.2–1.1 mm long, without lateral lobes;
anthers 0.2–0.9 mm long, yellow.

84. Upper glumes veinless, 0.1–0.3 mm long; lower glumes
rudimentary or lacking, veinless . . . . . . . . . . . . . . . . . . . 13. *M. schreberi* (in part)

84. Upper glumes 1(2)-veined, 0.8–1.9 mm long; lower glumes
0.4–1.5 mm long, veinless or 1-veined . . . . . . . . . . . . . 7. *M. ×curtisetosa* (in part)

80. Lemma awns 6–40 mm long.

85. Glumes 0.3–1 mm long, obtuse to acute, sometimes erose, not awned;
anthers 0.9–1.6 mm long . . . . . . . . . . . . . . . . . . . . . . . . . . . . . . . . . . . . 25. *M. spiciformis*

85. Glumes (1)1.2–3.5 mm long, acute to acuminate, usually awn-
tipped; anthers 0.9–3 mm long.

86. Blades 15–50 cm long; lemma awns 3–10 mm long; plants of
southern Arizona and Chihuahua, Mexico . . . . . . . . . . . . . . . 59. *M. palmeri* (in part)
86. Blades 1–15 cm long; lemma awns 4–30 mm long; plants of the
southwestern United States, including southern Arizona.
   87. Lemmas and paleas almost glabrous, with only a few short
      hairs on the calluses; ligules with lateral lobes, the lobes 1.5–3
      mm longer than the central portion; culms erect and plants
      tightly cespitose at the base; sheaths and blades usually with
      dark brown necrotic spots . . . . . . . . . . . . . . . . . . . . . . 22. *M. pauciflora* (in part)
   87. Lemmas and paleas pubescent on the lower ⅓–⅔ of the
      midveins and margins; ligules without lateral lobes or with
      lobes less than 1.5 mm longer than the central portion; culms
      decumbent and plants loosely cespitose; sheaths and blades
      without necrotic spots.
      88. Anthers 0.9–1.5 mm long, yellowish; panicles usually
         7–20 cm long . . . . . . . . . . . . . . . . . . . . . . . . . . . . 14. *M. tenuifolia* (in part)
      88. Anthers 1.3–2 mm long, purplish or orange; panicles 2–13
         cm long.
         89. Anthers orange, 1.5–2 mm long; lemmas elliptic,
            2–3.5 mm long, with awns 10–20(25) mm long;
            panicles 0.3–1.8 cm wide . . . . . . . . . . . . . . . . 23. *M. polycaulis* (in part)
         89. Anthers purple, 1.3–3 mm long; lemmas lanceolate,
            3.5–5 mm long, with awns 4–12(20) mm long;
            panicles 1–5 cm wide . . . . . . . . . . . . . . . . . . . . . 24. *M. arsenei* (in part)

## 1. Muhlenbergia racemosa (Michx.) Britton, Sterns & Poggenb. MARSH MUHLY [p. *445*, 536]

*Muhlenbergia racemosa* grows on rocky slopes, beside irrigation ditches, in seasonally wet meadows, on the margins of cultivated fields, railways and roadsides, in prairies, on sandstone outcrops, on stream banks, and in forest ecotones at elevations of 30–3400 m. It is most common in the north-central United States, but can be found at scattered locations throughout the western United States, and extends into northern Mexico.

## 2. Muhlenbergia glomerata (Willd.) Trin. SPIKE MUHLY, MUHLENBERGIE AGGLOMÉRÉE [p. *445*, 536]

*Muhlenbergia glomerata* grows in meadows, marshes, bogs, alkaline fens, lake margins, stream banks, beside irrigation ditches and hot springs, and on gravelly slopes, in many different plant communities, at elevations of 30–2300 m. It is most common in southern Canada and the northeastern United States, but grows sporadically throughout the western United States. It is not known from Mexico.

## 3. Muhlenbergia mexicana (L.) Trin. WIRESTEM MUHLY, MUHLENBERGIE DU MEXIQUE, MUHLENBERGIE MEXICAINE [p. *445*, 536]

*Muhlenbergia mexicana* usually grows in mesic to wet areas such as moist prairies and woodlands, stream banks, roadsides, ditch banks, lake margins, swamps, bogs, and hot springs, at elevations 50–3300 m, and is found in many different plant communities. Despite its name, *M. mexicana* grows only in Canada and the United States.

Plants with awns 3–10 mm long belong to **Muhlenbergia mexicana** var. **filiformis** (Torr.) Scribn., and those without an awn or with awns less than 3 mm long to **Muhlenbergia mexicana** (L.) Trin. var. **mexicana**. Early in the flowering season, *M. mexicana* may be confused with plants of *M. bushii* in which the axillary panicles are poorly developed, but they differ in their dull internodes and the fact that the blades on the secondary branches are usually similar in length and width to those of the main branches.

## 4. Muhlenbergia californica Vasey CALIFORNIA MUHLY [p. *446*, 536]

*Muhlenbergia californica* grows in canyons, along moist ditches, and on sandy slopes, at elevations of 100–2150 m. It is endemic to the Transverse Ranges of southern California.

## 5. Muhlenbergia glabrifloris Scribn. INLAND MUHLY [p. *446*, 536]

*Muhlenbergia glabrifloris* grows at the edge of dry forests, in prairies, thickets, and along roadsides in pine and oak associations, at elevations of 20–400 m. It is restricted to the southern portion of the central contiguous United States. It differs from *M. frondosa* in its glabrous lemmas and shorter caryopses (1.2–1.4 mm rather than 1.6–1.9 mm).

## 6. Muhlenbergia andina (Nutt.) Hitchc. FOXTAIL MUHLY [p. *446*, 536]

*Muhlenbergia andina* grows in damp places such as stream banks, gravel bars, marshes, lake margins, damp meadows, around springs, and in canyons, at elevations of 700–3000 m. It grows only in the western part of southern Canada and the contiguous United States.

## 7. Muhlenbergia ×curtisetosa (Scribn.) Bush [p. *446*, 536]

*Muhlenbergia* ×*curtisetosa* grows in abandoned fields and forest openings, often near bogs, at elevations of 20–300 m. It may be a hybrid between *M. schreberi* (which contributes the short glumes) and either of two rhizomatous species, *M. frondosa* and *M. tenuiflora*.

## 8. Muhlenbergia bushii R.W. Pohl NODDING MUHLY [p. *446*, 536]

*Muhlenbergia bushii* grows in sandy alluvium, open thickets, dry woodlands, and flood plains, at elevations of 10–250 m in the central portion of the contiguous United States. Early season plants, in which the axillary panicles are poorly developed, can be distinguished from those of *M. mexicana* by their shiny internodes and the tendency of the blades on the secondary branches to be shorter and narrower than those on the main branches.

### 9. Muhlenbergia frondosa (Poir.) Fernald WIRESTEM MUHLY, MUHLENBERGIE FEUILLÉE [p. 446, 536]

*Muhlenbergia frondosa* grows in forest borders, thickets, clearings, alluvial plains, and disturbed sites within deciduous forests, at elevations of 20–1000 m. It grows only in southern Canada and the contiguous United States. Plants with unawned lemmas or with awns shorter than 4 mm can be called *M. frondosa* forma *frondosa*, and those with lemma awns 4–13 mm long, *M. frondosa* forma *commutata* (Scribn.) Fernald.

### 10. Muhlenbergia sobolifera (Muhl. *ex* Willd.) Trin. ROCK MUHLY [p. 446, 536]

*Muhlenbergia sobolifera* grows in dry upland forests, oak woodlands, and on rock outcrops of sandstone, chert, or limestone formations, at elevations of 0–1200 m. It is restricted to the *Manual* region.

### 11. Muhlenbergia tenuiflora (Willd.) Britton, Sterns & Poggenb. SLIMFLOWERED MUHLY, MUHLENBERGIE TÉNUE [p. 446, 536]

*Muhlenbergia tenuiflora* grows only in the *Manual* region, usually being found on sandy or rocky slopes derived from sandstone, chert, or limestone formations, in mixed hardwood and oak-hickory forests, at elevations of 40–1500 m.

### 12. Muhlenbergia sylvatica (Torr.) Torr. *ex* A. Gray WOODLAND MUHLY, MUHLENBERGIE DES BOIS [p. 446, 537]

*Muhlenbergia sylvatica* grows in upland forests, along creeks and hollows, on rocky ledges derived from sandstone, shale, or calcareous parent materials, moist prairies, and swamps, at elevations from 30–1500 m. It is restricted to the *Manual* region, its primary range being southeastern Canada and the midwestern and eastern United States.

### 13. Muhlenbergia schreberi J.F. Gmel. NIMBLEWILL [p. 446, 537]

*Muhlenbergia schreberi* grows in moist to dry woods and prairies on rocky slopes, in ravines, and along sandy riverbanks, at elevations of 60–1600 m. It is also common in disturbed sites near cultivated fields, pastures, and roads at these elevations. Its geographic range includes central, but not northern, Mexico. Records from the western United States probably reflect recent introductions. The species is considered a noxious, invasive weed in California.

### 14. Muhlenbergia tenuifolia (Kunth) Trin. SLENDER MUHLY [p. 446, 537]

*Muhlenbergia tenuifolia* grows in grama grasslands and pine-oak woodlands on rocky slopes, limestone rock outcrops, gravelly roadsides, and in sandy drainages, at elevations of 1200–2200 m. Its range extends through Mexico to northern South America.

### 15. Muhlenbergia microsperma (DC.) Trin. LITTLESEED MUHLY [p. 446, 537]

*Muhlenbergia microsperma* grows on sandy slopes, drainages, cliffs, rock outcrops, and disturbed roadsides, at elevations of 0–2400 m. It is usually found in creosote scrub, thorn-scrub forest, sarcocaulescent desert, and oak-pinyon woodland associations. Its range extends from the southwestern United States through Central America to Peru and Venezuela. Morphological variation among and within its populations is marked.

### 16. Muhlenbergia appressa C.O. Goodd. DEVIL'S-CANYON MUHLY [p. 447, 537]

*Muhlenbergia appressa* grows in sandy drainages, canyon bottoms, rocky road cuts, and sandy slopes, at elevations of 20–1750 m. Its range extends from Arizona to Baja California, Mexico. It grows in grama grasslands, oak-juniper woodlands, and chaparral associations.

### 17. Muhlenbergia pectinata C.O. Goodd. COMBTOP MUHLY [p. 447, 537]

*Muhlenbergia pectinata* grows on rock outcrops, rocky cliffs, canyon walls, steep slopes, and road cuts, at elevations of 45–2400 m in thorn-scrub forests, grama grasslands, and pine-oak woodlands. It is almost entirely restricted to vertical surfaces that are seasonally wet. Its range extends from southeastern Arizona to Oaxaca, Mexico.

### 18. Muhlenbergia diversiglumis Trin. MIXEDGLUME MUHLY [p. 447, 537]

*Muhlenbergia diversiglumis* has been collected from Galveston County, Texas. The species is native from Mexico to Peru and Venezuela, where it grows on moist cliffs, along water courses, sandy slopes, and road cuts, primarily in moist shaded environments of broadleaf evergreen forests and pine-oak forests, at elevations of 600–2500 m.

### 19. Muhlenbergia glauca (Nees) B.D. Jacks. DESERT MUHLY [p. 447, 537]

*Muhlenbergia glauca* grows on calcareous rocky slopes, cliffs, canyon walls, table rocks, and volcanic rock outcrops, at elevations of 1200–2780 m. Its range extends from the southwestern United States to central Mexico. *M. glauca* resembles *M. polycaulis*, but differs in its shorter lemma awns and strongly rhizomatous habit.

### 20. Muhlenbergia thurberi (Scribn.) Rydb. THURBER'S MUHLY [p. 447, 537]

*Muhlenbergia thurberi* usually grows in moist soil in seeps near canyon cliffs, sandstone slopes, and rocky ledges, at elevations of 1350–2300 m. It appears to be restricted to the southwestern United States. It flowers from July to September. It resembles *M. curtifolia*, but differs in its tightly involute blades, and longer anthers and ligules. The two species have been found growing within 50 m of each other in Apache County, Arizona, but in different habitats, *M. curtifolia* growing in a damp drainage areas whereas *M. thurberi* grew near a moist but dryer canyon cliff.

### 21. Muhlenbergia curtifolia Scribn. UTAH MUHLY [p. 447, 537]

*Muhlenbergia curtifolia* grows on damp ledges and in rock crevices of vertical cliffs, and beneath large calcareous boulders above the canyon floor, at elevations of 1600–2750 m, in the southwestern United States. It resembles *M. thurberi*, differing in its flatter leaf blades and shorter ligules and anthers. It also tends to grow in more mesic habitats than *M. thurberi*.

### 22. Muhlenbergia pauciflora Buckley NEW MEXICAN MUHLY [p. 447, 537]

*Muhlenbergia pauciflora* grows in open or closed forests on rocky slopes, cliffs, canyons, and rock outcrops of granitic or calcareous origin, at elevations of 1200–2500 m. Its range extends from the southwestern United States to central Mexico.

### 23. Muhlenbergia polycaulis Scribn. CLIFF MUHLY [p. 447, 537]

*Muhlenbergia polycaulis* grows in open vegetation on steep rocky slopes, canyon walls, cliffs, table rocks, and volcanic rock outcrops, at elevations of 1200–2400 m. Its range extends from the southwestern United States to central Mexico. It differs from *M. glauca* in its longer lemma awns, shorter rhizomes, and loosely tufted habit.

### 24. Muhlenbergia arsenei Hitchc. NAVAJO MUHLY [p. 447, 537]

*Muhlenbergia arsenei* grows among granitic boulders, on rocky slopes, limestone rock outcrops, and in arroyos, at elevations of 1400–2850 m.

Its range extends from the southwestern United States into Baja California, Mexico. It flowers from August to September.

### 25. Muhlenbergia spiciformis Trin. LONGAWN MUHLY [p. *447*, <u>537</u>]

*Muhlenbergia spiciformis* grows on rocky slopes, cliffs, and calcareous rock outcrops, often in thorn-scrub and open woodland communities. Its elevational range is 450–2800 m; its geographic range extends from the southwestern United States to northern Mexico.

### 26. Muhlenbergia porteri Scribn. *ex* Beal BUSH MUHLY [p. *447*, <u>537</u>]

*Muhlenbergia porteri* grows among boulders on rocky slopes and on cliffs, and in dry arroyos, desert flats, and grasslands, frequently in the protection of shrubs, at elevations of 600–1700 m. Its geographic range extends from the southwestern United States to northern Mexico.

### 27. Muhlenbergia cuspidata (Torr. *ex* Hook.) Rydb. PLAINS MUHLY [p. *447*, <u>537</u>]

*Muhlenbergia cuspidata* grows in dry, gravelly prairies, on gentle rocky slopes, rocky limestone outcrops, and in sandy drainages, at elevations of 300–1400 m, primarily in the central portion of the *Manual* region. It flowers from June to October. It is often confused with *M. richardsonis*, but that species has rhizomes and longer ligules.

### 28. Muhlenbergia wrightii Vasey *ex* J.M. Coult. SPIKE MUHLY [p. *447*, <u>537</u>]

*Muhlenbergia wrightii* grows in gravelly prairies, on rocky slopes, and in meadows on granitic, sandstone, or limestone-derived soils, at elevations of 1100–3000 m. Its range extends from the southwestern United States to northern Mexico.

### 29. Muhlenbergia arizonica Scribn. ARIZONA MUHLY [p. *448*, <u>537</u>]

*Muhlenbergia arizonica* grows in sandy drainages and gravelly canyons, and on plateaus and rocky slopes in open desert grasslands, at elevations of 1220–2230 m. Its range extends from the southwestern United States into northwestern Mexico. Flowering is from August to October.

### 30. Muhlenbergia torreyi (Kunth) Hitchc. *ex* Bush RING MUHLY [p. *448*, <u>537</u>]

*Muhlenbergia torreyi* grows in desert grasslands and open woodlands on sandy mesas, calcareous rock outcrops, and rocky slopes, at elevations of 1000–2450 m. Its range extends from the southwestern United States to northern Mexico. It also grows, as a disjunct, in northwestern Argentina.

### 31. Muhlenbergia arenicola Buckley SAND MUHLY [p. *448*, <u>537</u>]

*Muhlenbergia arenicola* grows on sandy mesas, limestone benches, and in valleys and open desert grasslands, at elevations of 600–2135 m. Its range extends from the southwestern United States to central Mexico. It also grows, as a disjunct, in northwestern Argentina.

### 32. Muhlenbergia pungens Thurb. *ex* A. Gray SANDHILL MUHLY [p. *448*, <u>537</u>]

*Muhlenbergia pungens* grows in loose sandy soils near sand dunes to sandy clay loam slopes and flats in desert shrub and open woodlands, at elevations of 600–2500 m. It is known only from the western and central contiguous United States.

### 33. Muhlenbergia dumosa Scribn. *ex* Vasey BAMBOO MUHLY [p. *448*, <u>537</u>]

*Muhlenbergia dumosa* grows on rocky slopes, canyon ledges, and cliffs, in areas protected from grazing animals in oak-pine and thorn-scrub forests and oak-grama savannahs, at elevations of 600–1800 m, from Arizona to southern Mexico. It is also grown as an ornamental in the southwestern United States.

### 34. Muhlenbergia villiflora Hitchc. HAIRY MUHLY [p. *448*, <u>537</u>]

In the United States, *Muhlenbergia villiflora* grows in open ground with alkaline to calcareous soils and on gypsum rock flats, at elevations of 600–1200 m. It usually forms small, isolated populations. Plants in the United States belong to **Muhlenbergia villiflora** var. **villosa** (Swallen) Morden. This variety differs from **M. villiflora** Hitchc. var. **villiflora**, which grows in Mexico, in its longer spikelets (1.8–2.5 mm versus 1.4–2.3 mm) and preference for calcareous, rather than gypsiferous, soils.

### 35. Muhlenbergia repens (J. Presl) Hitchc. CREEPING MUHLY [p. *448*, <u>537</u>]

*Muhlenbergia repens* grows in open, sandy meadows, canyon bottoms, calcareous rocky flats, gypsum flats, and on rolling slopes and roadsides, at elevations of 100–3120 m. Its range extends from the southwestern United States to southern Mexico.

### 36. Muhlenbergia utilis (Torr.) Hitchc. APAREJOGRASS [p. *448*, <u>537</u>]

*Muhlenbergia utilis* grows in wet soils along streams, ponds, and depressions in grasslands and alkaline or gypsiferous plains, at elevations of 200–1800 m. Its range extends from the southern United States through Mexico to Costa Rica.

### 37. Muhlenbergia richardsonis (Trin.) Rydb. MAT MUHLY, MUHLENBERGIE DE RICHARDSON [p. *448*, <u>538</u>]

*Muhlenbergia richardsonis* grows in open sites in alkaline meadows, prairies, sandy arroyo bottoms, talus slopes, rocky flats and the shores of rivers, at elevations of 60–3300 m. It is the most widespread species of *Muhlenbergia* in the *Manual* region, extending from the Yukon Territory to Quebec in the north and to northern Baja California, Mexico, in the south. It is often confused with *M. cuspidata*, which differs in lacking rhizomes and having shorter ligules, and sometimes with *M. filiformis*, which differs in being a weak annual with glabrous internodes and obtuse, erose glumes.

### 38. Muhlenbergia filiformis (Thurb. *ex* S. Watson) Rydb. PULL-UP MUHLY [p. *448*, <u>538</u>]

*Muhlenbergia filiformis* grows in open, moist meadows, on gravelly lake shores, along stream banks, and in moist humus near thermal springs, at elevations of 1060–3050 m. It is usually associated with yellow pine forests, but also grows in many other plant communities. Its range extends into northern Mexico. It differs from *M. richardsonis* in having glabrous internodes and subacute apices.

### 39. Muhlenbergia torreyana (Schult.) Hitchc. NEW JERSEY MUHLY [p. *448*, <u>538</u>]

*Muhlenbergia torreyana* grows in perennially wet or moist, usually seasonally inundated habitats such as the sphagnous margins of shallow ponds and seasonally wet depressions, often within pine-oak or oak barrens and at elevations of 0–150 m. It is rare even in those states where it is still growing. It resembles the western *M. asperifolia* but differs in its strigose, strongly compressed, keeled culms and less strongly divergent panicle branches.

## 40. **Muhlenbergia asperifolia** (Nees & Meyen *ex* Trin.) Parodi SCRATCHGRASS [p. 448, <u>538</u>]

*Muhlenbergia asperifolia* grows in moist, often alkaline meadows, playa margins, and sandy washes, on grassy slopes, and around seeps and hot springs, at elevations of 55–3000 m. Its geographic range includes northern Mexico. It differs from the southeastern *M. torreyana* in having glabrous, weakly compressed culms and more widely divergent panicle branches.

## 41. **Muhlenbergia arenacea** (Buckley) Hitchc. EAR MUHLY [p. 449, <u>538</u>]

*Muhlenbergia arenacea* grows in sandy flats, plains, alluvial fans, washes, depressions, and alkaline mesas in open grasslands, at elevations of 1000–2200 m. Its range extends from the southwestern United States into northern Mexico.

## 42. **Muhlenbergia uniflora** (Muhl.) Fernald BOG MUHLY, MUHLENBERGIE UNIFLORE [p. 449, <u>538</u>]

*Muhlenbergia uniflora* grows in bogs, wet meadows, and lake shores in sandy or peaty, often acidic, soils, at elevations of 0–650 m. It is native to eastern North America, but was collected once in British Columbia, probably having been introduced from ship ballast, and was recently collected from a commercial cranberry bog in Oregon. The collection from Texas may also be an introduction.

## 43. **Muhlenbergia filiculmis** Vasey SLIMSTEM MUHLY [p. 449, <u>538</u>]

*Muhlenbergia filiculmis* grows on rocky slopes, dry meadows, and dry gravelly flats in forest openings and grasslands, at elevations of 2500–3500 m in the southern Rocky Mountains and northern Arizona. It differs from *M. montana* in its shorter spikelets and lemma awns and tightly involute or filiform, sharp blades.

## 44. **Muhlenbergia jonesii** (Vasey) Hitchc. MODOC MUHLY [p. 449, <u>538</u>]

*Muhlenbergia jonesii* is endemic to northern California. It grows on open slopes, pumice flats, and in openings in pine forests, at elevations of 1130–2130 m.

## 45. **Muhlenbergia straminea** Hitchc. SCREWLEAF MUHLY [p. 449, <u>538</u>]

*Muhlenbergia straminea* grows on rolling, rocky slopes, volcanic tuffs, canyon bottoms, and ridges, usually in open pine forests, at elevations of 1800–2600 m. It is known only from the southwestern United States.

## 46. **Muhlenbergia montana** (Nutt.) Hitchc. MOUNTAIN MUHLY [p. 449, <u>538</u>]

*Muhlenbergia montana* grows on rocky slopes and ridge tops and in dry meadows and open grasslands, at elevations of 1400–3500 m. Its range extends from the western United States to Guatemala. It differs from *M. filiculmis* in its longer spikelets and lemma awns, and flatter leaf blades that are not sharply tipped.

## 47. **Muhlenbergia crispiseta** Hitchc. MEXICALI MUHLY [p. 449, <u>538</u>]

*Muhlenbergia crispiseta* grows on rock outcrops, in rocky drainages, and on white tablelands, on soils derived from calcareous parent materials in pine-oak and pinyon-juniper woodlands, at elevations of 1900–2600 m. It is basically a Mexican species, with a disjunct population in Brewster County, Texas.

## 48. **Muhlenbergia peruviana** (P. Beauv.) Steud. PERUVIAN MUHLY [p. 449, <u>538</u>]

*Muhlenbergia peruviana* grows in open gravelly flats, meadows, rock outcrops, sandy washes, gravelly drainages, rocky slopes, disturbed road cuts, and volcanic flats, in yellow pine forest associations, at elevations of 2000–4600 m. Its primary distribution is to the south of the *Manual* region, extending from the southwestern United States through Mexico to Ecuador, Peru, Bolivia, and Argentina. As treated here, *Muhlenbergia peruviana* includes some names identified as *M. pulcherrima* Scribn. *ex* Beal, *M. pusilla* Steud., and *M. peruviana s. s.* There are, however, numerous intermediates among the three extremes represented by these names.

## 49. **Muhlenbergia emersleyi** Vasey BULLGRASS [p. 449, <u>538</u>]

*Muhlenbergia emersleyi* grows on rocky slopes, gravelly washes, canyons, cliffs, and arroyos, often in soils derived from limestone, at elevations of 1200–2500 m, and is also grown as an ornamental. Its range extends from the southwestern United States through Mexico to Panama. It differs from the closely related *M. longiligula* in its compressed-keeled sheaths, pubescent florets, and membranous ligules.

## 50. **Muhlenbergia ×involuta** Swallen CANYON MUHLY [p. 449, <u>538</u>]

*Muhlenbergia ×involuta* grows on rocky, calcareous slopes in openings and along canyons, at elevations of 150–500 m. It has only been found growing naturally in Texas, but it is also available commercially as an ornamental. *Muhlenbergia reverchonii* and *M. lindheimeri* may be its parents, but *M. rigida* is another plausible possibility.

## 51. **Muhlenbergia longiligula** Hitchc. LONGTONGUE MUHLY [p. 449, <u>538</u>]

*Muhlenbergia longiligula* grows on rocky slopes, canyons, and rock outcrops derived from volcanic or calcareous parent materials, at elevations of 1220–2500 m. It is a common species in Arizona and southwestern New Mexico, and extends into northwestern Mexico. It differs from *M. emersleyi* in its rounded basal leaf sheaths, glabrous lemmas, and panicle branches that are spikelet-bearing to the base. It differs from *M. lindheimeri* in its rounded basal sheaths.

## 52. **Muhlenbergia expansa** (Poir.) Trin. SAVANNAH HAIRGRASS [p. 449, <u>538</u>]

*Muhlenbergia expansa* grows in perennially moist to wet soils in pitcher plant bogs, pine savannahs, and flatwoods, usually in sandy soils and at elevations of 0–300 m. Its primary range is the coastal plain of the southeastern United States.

## 53. **Muhlenbergia capillaris** (Lam.) Trin. HAIRY-AWN MUHLY [p. 450, <u>538</u>]

In the southeastern United States, *Muhlenbergia capillaris* usually grows in rocky or clay soils in open woodlands and savannahs and on calcareous outcrops, at elevations of 0–500 m. In the northeastern states, it is also found on diabase and sandstone outcrops and ridges. Its native range includes the southeastern United States, Bahamas, and possibly various Caribbean islands. It is also grown as an ornamental. It differs from *M. reverchonii* in its dull, apically scabrous lemmas.

## 54. **Muhlenbergia sericea** (Michx.) P.M. Peterson DUNE HAIRGRASS, PURPLE MUHLY [p. 450, <u>538</u>]

*Muhlenbergia sericea* grows in sandy maritime habitats on the barrier islands and in coastal woodlands of the southeastern United States, at elevations of 0–50 m. It is available as an ornamental, sometimes under the name 'Purple Muhly'.

## 55. **Muhlenbergia reverchonii** Vasey & Scribn. SEEP MUHLY [p. 450, <u>538</u>]

*Muhlenbergia reverchonii* grows on calcareous rocky slopes, flats, and limestone rock outcrops, at elevations of 150–650 m. It is restricted to Oklahoma and Texas. It differs from *M. capillaris* in its smooth and shiny lemmas, and from *M. setifolia* in its wider panicles, spreading panicle branches, acute glumes, and more shortly awned lemmas.

### 56. Muhlenbergia rigida (Kunth) Trin. PURPLE MUHLY [p. *450*, 538]

*Muhlenbergia rigida* grows on rocky slopes, ravines, and sandy, gravelly slopes derived from granitic and calcareous substrates, at elevations of 1200–2200 m, in two disjunct areas: the southwestern United States south to Chiapas, Mexico, and in Ecuador, Peru, Bolivia, and Argentina. It is often a common upland bunchgrass, and is also grown as an ornamental. It differs from *M. setifolia* and *M. reverchonii* in its purplish, scabridulous to scabrous lemmas.

### 57. Muhlenbergia elongata Scribn. *ex* Beal SYCAMORE MUHLY [p. *450*, 538]

*Muhlenbergia elongata* grows on rock outcrops, cliffs, canyon walls, and moist rock walls, on rhyolitic and volcanic conglomerates, at elevations of 850–2100 m. It extends south from Arizona into northern Mexico.

### 58. Muhlenbergia setifolia Vasey CURLYLEAF MUHLY [p. *450*, 538]

*Muhlenbergia setifolia* grows on calcareous rocky slopes, rock outcrops, and in desert grasslands, at elevations of 1000–2250 m. Its range extends from the southwestern United States into northern Mexico. It differs from *M. reverchonii* in its narrower panicles, less widespread panicle branches, truncate to obtuse glumes, and longer lemma awns.

### 59. Muhlenbergia palmeri Vasey PALMER'S MUHLY [p. *450*, 538]

*Muhlenbergia palmeri* grows in rocky drainages and in sandy soil along creeks, at elevations of 1000–2100 m. Its range extends from southern Arizona into northern Mexico.

### 60. Muhlenbergia lindheimeri Hitchc. LINDHEIMER'S MUHLY [p. *450*, 538]

*Muhlenbergia lindheimeri* grows in sandy draws to rocky, calcareous soils, generally in open areas, at elevations of 150–500 m. It is uncommon throughout its range, which includes northern Mexico in addition to southern Texas, but it is also grown as an ornamental. It differs from the closely related *M. longiligula* in its compressed-keeled basal sheaths, grayish spikelets, and, when present, bifid glume apices.

### 61. Muhlenbergia dubia E. Fourn. PINE MUHLY [p. *450*, 538]

*Muhlenbergia dubia* grows on steep slopes, ridge tops, limestone rock outcrops, and along draws, at elevations of 1500–2300 m. Its range extends into northern Mexico. It resembles *M. rigens*, but differs in having looser, contracted (but not spikelike) panicles, longer ligules, olivaceous anthers, and generally longer lemmas.

### 62. Muhlenbergia rigens (Benth.) Hitchc. DEERGRASS [p. *450*, 539]

*Muhlenbergia rigens* grows in sandy washes, gravelly canyon bottoms, rocky drainages, and moist, sandy slopes, often along small streams, at elevations of 90–2500 m. Its geographic range extends to central Mexico. It is available commercially as an ornamental. It differs from *M. dubia* in having tighter, spikelike panicles, shorter ligules, yellow or purplish anthers, and shorter lemmas.

### 63. Muhlenbergia depauperata Scribn. SIXWEEKS MUHLY [p. *450*, 539]

*Muhlenbergia depauperata* grows in gravelly flats, rock outcrops, exposed bedrock, and sandy banks, in grama grassland associations, usually on soils derived from calcareous parent materials, at elevations of 1530–2400 m. Its range extends from the southwestern United States to southern Mexico. *Muhlenbergia depauperata* and *M. brevis* share several features with *Lycurus*: spikelets borne in pairs, 2-veined and 2-awned lower glumes, 1-veined and awned upper glumes, acuminate, awned lemmas with short pubescence along the margins, and pubescent paleas.

### 64. Muhlenbergia brevis C.O. Goodd. SHORT MUHLY [p. *450*, 539]

*Muhlenbergia brevis* grows on rocky slopes, gravelly flats, and rock outcrops, particularly those derived from calcareous parent materials, at elevations of 1700–2500 m, in grama grasslands, pinyon-juniper woodlands, and pine-oak woodlands. Its range extends from the southwestern United States to central Mexico. Like *Muhlenbergia depauperata*, *M. brevis* shares several features with *Lycurus*, notably the paired spikelets with 2-veined and 2-awned lower glumes, 1-veined and awned upper glumes, acuminate, awned lemmas with shortly pubescent margins, and pubescent paleas.

### 65. Muhlenbergia sinuosa Swallen MARSHLAND MUHLY [p. *451*, 539]

*Muhlenbergia sinuosa* grows in sandy soil along washes, on open slopes and rocky ledges, and in roadside ditches, at elevations of 1650–2300 m. It is usually found in oak-pine forests, pinyon-juniper woodlands, oak-grama savannahs, and riverine woodlands. Its range extends from the southwestern United States into northern Mexico.

### 66. Muhlenbergia minutissima (Steud.) Swallen ANNUAL MUHLY [p. *451*, 539]

*Muhlenbergia minutissima* grows in sandy and gravelly drainages, rocky slopes, flats, road cuts, and open sites. It is usually found in yellow pine and oak-pine forests, pinyon-juniper woodlands, thorn-scrub forests, and oak-grama savannahs, at elevations of 1200–3000 m. Its range extends from the western United States to southern Mexico.

### 67. Muhlenbergia texana Buckley TEXAS MUHLY [p. *451*, 539]

*Muhlenbergia texana* grows on open slopes, in sandy, gravelly drainages, and on rock outcrops, at elevations of 1200–2750 m. Its range extends from the southwestern United States into northwestern Mexico.

### 68. Muhlenbergia eludens C. Reeder GRAVELBAR MUHLY [p. *451*, 539]

*Muhlenbergia eludens* grows in open sandy gullies, washes, rocky slopes, and roadsides. It is found at elevations of 1700–2450 m in the southwestern United States and northwestern Mexico.

### 69. Muhlenbergia fragilis Swallen DELICATE MUHLY [p. *451*, 539]

*Muhlenbergia fragilis* grows on rocky talus slopes, cliffs, canyon walls, road cuts, and sandy slopes, often over calcareous parent materials, at elevations of 480–2200 m. It is usually found in oak-grama savannahs, thorn scrub forests, oak-yellow pine forests, and pinyon-juniper woodlands. Its range extends from the southwestern United States to southern Mexico.

### 70. Muhlenbergia ramulosa (Kunth) Swallen GREEN MUHLY [p. *451*, 539]

*Muhlenbergia ramulosa* grows in open, well-drained areas including slopes, sandy meadows, washes, gravelly road cuts, and rock outcrops in yellow pine-oak forests and in open meadows of pine-fir forests, at elevations of 2100–3500 m. Its range includes the southwestern United States, Mexico, Guatemala, Costa Rica, and Argentina.

## 16.34 LYCURUS Kunth[1]

Pl per; csp. **Clm** 10–60 cm, erect to somewhat decumbent, usu brchd. **Shth** open, compressed-keeled, glab, smooth or scabridulous, mostly shorter than the intnd, a 2-veined prophyllum often present; **lig** hyaline, strongly decurrent, truncate or rounded to elongate and acuminate, smt with narrow triangular lobes extending from the edges of the shth on either side; **bld** folded or flat, rather stiff, with prominent, firm, scabrous mrg, midveins smt extending as short mucros or fragile, scabrous, awnlike apc. **Infl** tml and ax, dense, bristly, spikelike pan; **br** short, fused to the rachis, terminating in a pair of unequally pedlt spklt or a pedlt spklt and a short sec br bearing two spklt, occ in a solitary spklt, usu the lo spklt in a pair stmt or strl and the up spklt bisx, smt vice versa, or both spklt bisx; **dis** at the fused base of the peds or ped and br, paired spklt falling as a unit, leaving a cuplike tip. **Spklt** with 1 flt. **Glm** subequal, awned; **lo glm** with (1)2(3) awns, usu unequal, awns commonly longer than the body; **up glm** 1-veined, with a single flexuous awn that is usu longer than the glm body, rarely a finer second awn present; **lm** lanceolate, 3-veined, pubescent on the mrg, mostly glab over the back, tapering to a scabrous awn that is usu shorter than the lm body; **pal** about equal to the lm, acute or occ the 2 veins extending as very short mucros, pubescent between the veins and on the sides, except for the narrow, glab, hyaline mrg; **anth** 3. **Car** fusiform, brownish.

*Lycurus* is a genus of three species of open rocky slopes and mesas. It is native to two disjunct regions, one extending from Colorado and southern Utah to southern Mexico and Guatemala, the other from Colombia through western South America to west-central Argentina. Two species are native to the *Manual* region. They can be reliably distinguished only by their vegetative characters.

1. Upper leaves terminating in a fragile, awnlike tip (3)4–7(12) mm long; ligules (2)3–10(12) mm long, elongate, acute or acuminate, sometimes with a small cleft on either side; culms erect . . . . . . . . . . . . . . . . . . . . 1. *L. setosus*
1. Upper leaves acute or with a mucro or bristle 1–3 mm long; ligules 1.5–3 mm long, with evident narrow triangular lobes 1.5–3(4)mm long on the sides; culms erect to ascending, often geniculate . . . . . . . . 2. *L. phleoides*

### 1. Lycurus setosus (Nutt.) C. Reeder BRISTLY WOLFSTAIL [p. *451*, 539]

*Lycurus setosus* grows on rocky slopes and open mesas, at elevations of 570–3400 m. Its range extends from the southwestern United States to northern Mexico, and, as a disjunct, in Argentina and Bolivia. It was found as an adventive (associated with wool waste) in North Berwick, York County, Maine, in 1902, but it has not been reported from there since. Its flowering time is July–October. It is sometimes confused with *Muhlenbergia wrightii*, which has a somewhat similar aspect but is normally found in moist habitats.

Also, in *M. wrightii* the first glume is 1-veined with a very short awn, and the lemma is acuminate and unawned or with an awn no more than 1 mm long.

### 2. Lycurus phleoides Kunth COMMON WOLFSTAIL [p. *451*, 539]

*Lycurus phleoides* grows on rocky hills and open slopes, at elevations of 670–2600 m. It grows from the southwestern United States to southern Mexico, and in northern South America. It flowers from July–October.

## 16.35 CHLORIS Sw.[2]

Pl ann or per; habit various, rhz, stln, or csp. **Clm** 10–300 cm; **intnd** pith-filled. **Shth** strongly keeled, glab, scabrous, or pubescent; **lig** memb, erose to lacerate or ciliate, occ absent; **bld**, particularly those of the bas lvs, often with long, coarse hairs near the base of the adx surface and mrg. **Infl** tml, pan with (1)5–30 spikelike br, these usu borne digitately, occ in 2–several whorls, smt with a few isolated br below the pri whorl(s), all br usu exceeding the up lvs; **br** with spklt in 2 rows on 1 side of the br axes. **Spklt** solitary, sessile to pedlt, lat compressed, with 2–3(5) flt, usu only the lowest flt bisx, rarely the lo 2 flt bisx, remaining flt(s) strl or stmt; **flt** lat compressed or terete, cylindrical to obovoid, awned or unawned, strl and stmt flt progressively rdcd distally if more than 1 present; **dis** usu beneath the lowest flt in the spklt, all flt falling as a unit, smt at the uppermost cauline nd, pan falling intact. **Glm** unequal, exceeded by the flt, lanceolate, acute to acuminate, usu unawned, occ awned, awns to 0.3 mm; **cal** bearded; **lm of bisx flt** 3-veined, mrgl veins pubescent, midveins usu glab, smt scabrous, usu extending into an awn, smt merely mucronate, awns to 37 mm, lm apc truncate or obtuse, entire or bilobed, lobes, when present, smt awn-tipped, awns to 0.6 mm; **pal** shorter than the lm, 2-veined, veins scabrous; **anth** 3; **lod** 2. **Car** ovoid, elliptic, or obovoid.

As interpreted here, *Chloris* is a tropical to subtropical genus of 55–60 species. It is most abundant in the Southern Hemisphere. Of the 18 species treated here, 11 are native, five have been collected in the *Manual* region but seem not to have persisted, one is an introduced species that has become established, and one is known only from cultivation.

1 Panicle branches tightly entangled for most of their length, separable only with difficulty and forming a cylindrical, spikelike inflorescence, the individual branches visibly distinct only at the tip . . . . . . . . . . . . . . . . . 1. *C. berroi*
1. Panicle branches sometimes appressed to each other but easily separable, inflorescences evidently composed of multiple spikelike branches borne in 1–several whorls.

[1]Charlotte G. Reeder  [2]Mary E. Barkworth

2. Panicles with 1–3 branches.
  3. Lowest (bisexual) lemmas usually glabrous or scabrous, occasionally with a few scattered hairs
     on the margins.
    4. Lemma of the first sterile floret in each spikelet entire; panicle branches 5–11 cm long, with
       about 10 spikelets per cm . . . . . . . . . . . . . . . . . . . . . . . . . . . . . . . . . . . .9. *C. ventricosa* (in part)
    4. Lemma of the first sterile floret in each spikelet bilobed for ⅓–½ of its length; panicle branches
       4–17 cm long, with 3–7 spikelets per cm on the distal portion . . . . . . . . . . . . . . . . . . . .11. *C. divaricata* (in part)
  3. Lowest (bisexual) lemmas conspicuously hairy on the margins and keels.
    5. Lemma of the lowest (bisexual) floret in each spikelet 1.8–2.8 mm long . . . . . . . . . . . . . . . . . .2. *C. ciliata* (in part)
    5. Lemma of the lowest (bisexual) floret in each spikelet 2.7–4.2 mm long . . . . . . . . . . . . . .3. *C. canterae* (in part)
2. Panicles with 4 or more branches.
  6. Spikelets with 3 or more florets (the third often concealed), the lowest (first) floret bisexual, the
     remainder sterile or staminate, the terminal floret sometimes represented only by a clavate
     rachilla segment.
    7. Second sterile florets ovoid to subspherical, strongly inflated; first sterile florets 0.9–1.3 mm
      long . . . . . . . . . . . . . . . . . . . . . . . . . . . . . . . . . . . . . . . . . . . . . . . . . . . . . . . . . . . . . . . . . . . . .5. *C. barbata*
    7. Second sterile florets widened or not distally, not inflated or inflated only near the apices; first
      sterile florets 1–3.5 mm long.
      8. All sterile or staminate florets awned.
        9. Culms to 300 cm tall; panicle branches averaging 10 spikelets per cm; spikelets tawny
          . . . . . . . . . . . . . . . . . . . . . . . . . . . . . . . . . . . . . . . . . . . . . . . . . . . . . . . .8. *C. gayana* (in part)
        9. Culms 30–50 cm tall; panicle branches averaging 6 spikelets per cm; spikelets dark
          brown to black . . . . . . . . . . . . . . . . . . . . . . . . . . . . . . . . . . . .10. *C. truncata* (in part)
      8. At least 1 of the sterile or staminate florets in each spikelet unawned.
        10. Spikelets barely imbricate, averaging 5–7 per cm; plants annual or short-lived
          perennials . . . . . . . . . . . . . . . . . . . . . . . . . . . . . . . . . . . . . . . . . . . . . . . . . . . .6. *C. pilosa* (in part)
        10. Spikelets strongly imbricate, averaging 10–14 per cm; plants perennial or annual.
          11. Margins of the lowest lemma in each spikelet scabrous, glabrous, or appressed
            pubescent . . . . . . . . . . . . . . . . . . . . . . . . . . . . . . . . . . . . . . . . . . . . . .8. *C. gayana* (in part)
          11. Margins of the lowest lemma in each spikelet conspicuously hairy, at least
            distally, the hairs 0.5–3 mm long.
            12. Plants annual; third florets, if present, shorter than the subtending rachilla
              segment . . . . . . . . . . . . . . . . . . . . . . . . . . . . . . . . . . . . . . . . . . . . . . . . . . .7. *C. virgata* (in part)
            12. Plants perennial; third florets as long as or longer than the subtending rachilla
              segment.
              13. Lowest lemmas 2.7–4.2 mm long . . . . . . . . . . . . . . . . . . . . . . . . . . . .3. *C. canterae* (in part)
              13. Lowest lemmas 1.5–2.8 mm long.
                14. Panicles with 2–7 branches; branches 3.5–8 cm long; awns of the
                  first sterile florets 0.9–1.4 mm long . . . . . . . . . . . . . . . . . . . . . . . .2. *C. ciliata* (in part)
                14. Panicles with (4)8–30 branches; branches (5)8–20 cm long; awns of
                  the first sterile florets 0.8–4 mm long.
                  15. First sterile florets 1–1.6 mm long . . . . . . . . . . . . . . . . . . . . . . .4. *C. elata* (in part)
                  15. First sterile florets 2.2–3.2 mm long . . . . . . . . . . . . . . . . . . . . .8. *C. gayana* (in part)
  6. Spikelets with 2 florets, 1 bisexual and 1 sterile or staminate.
    16. Lemmas of the sterile or staminate florets bilobed for at least ⅓ of their length; lateral lobes
      sometimes awned.
      17. Spikelets pectinate, diverging from the branch axes, crowded, averaging 10–14 per cm;
        plants annual . . . . . . . . . . . . . . . . . . . . . . . . . . . . . . . . . . . . . . . . . . . . . . . . . . . . .12. *C. pectinata*
      17. Spikelets appressed to the branch axes, not crowded, averaging 3–7 per cm; plants
        perennial . . . . . . . . . . . . . . . . . . . . . . . . . . . . . . . . . . . . . . . . . . . . . . . . . . . .11. *C. divaricata* (in part)
    16. Lemmas of the sterile or staminate florets not bilobed or lobed less than ¼ of their length;
      lateral lobes, if present, not awned.
      18. Lowest (bisexual) lemmas unawned, mucronate, or shortly awned, the awns less than
        1.5 mm long.
        19. Plants annuals or short-lived perennials, sometimes shortly stoloniferous . . . . . . . . .6. *C. pilosa* (in part)
        19. Plants perennial, sometimes stoloniferous.
          20. Lowest (bisexual) lemmas unawned, sometimes mucronate, the mucros to 1
          mm long.
            21. Sterile or staminate florets inflated, 1–1.5 mm wide and about equally
              long . . . . . . . . . . . . . . . . . . . . . . . . . . . . . . . . . . . . . . . . . . . . . . . . . .13. *C. cucullata* (in part)

21. Sterile or staminate florets 0.3–0.9 mm wide, usually at least twice as long　..17. *C. submutica*
20. Lowest (bisexual) lemmas always awned, the awns 1–11 mm long.
　22. Panicles with 2–9 branches; branches 5–11 cm long . . . . . . . . . . . . . .9. *C. ventricosa* (in part)
　22. Panicles with 4–28 branches (usually more than 8); branches 2–20 cm
　　long.
　　23. Panicle branches 10–20, 2–5 cm long; spikelets with only 1 sterile
　　　floret . . . . . . . . . . . . . . . . . . . . . . . . . . . . . . . . . . . . . . . . . . . .13. *C. cucullata* (in part)
　　23. Panicle branches 4–28 (usually more than 8), 5–20 cm long, spikelets
　　　usually with 2 sterile florets, the lowest often enclosing and concealing
　　　the second . . . . . . . . . . . . . . . . . . . . . . . . . . . . . . . . . . . . . . . . . .4. *C. elata* (in part)
18. Lowest (bisexual) lemmas always awned, the awns 1.5–16 mm long.
　24. Lowest lemmas with a conspicuous glabrous or pubescent groove on each side;
　　plants annual or short-lived perennials . . . . . . . . . . . . . . . . . . . . . . . . . . . . .6. *C. pilosa* (in part)
　24. Lowest lemmas not conspicuously grooved on the sides; plants perennial or
　　annual.
　　25. Margins of the lowest (bisexual) lemmas conspicuously hairy along most of
　　　their length, the hairs strongly divergent, at least some longer than 1 mm.
　　　26. Plants rarely stoloniferous; second (first sterile) florets 1–1.6 mm long　. . . .4. *C. elata* (in part)
　　　26. Plants usually stoloniferous; second (first sterile) florets 2.2–3.2 mm long
　　　　. . . . . . . . . . . . . . . . . . . . . . . . . . . . . . . . . . . . . . . . . . . . . . . . .8. *C. gayana* (in part)
　　25. Margins of the lowest (bisexual) lemmas glabrous, scabrous, or appressed
　　　pubescent, sometimes with a few scattered longer hairs, or with strongly
　　　divergent hairs distally but not basally.
　　　27. Panicle branches borne in 2 or more, clearly distinct whorls.
　　　　28. Second (first sterile) florets 0.4–0.7 mm long . . . . . . . . . . . . . . . .18. *C. radiata* (in part)
　　　　28. Second (first sterile) florets 0.9–2.5 mm long.
　　　　　29. Panicle branches spikelet-bearing to the base; panicles with 10–16
　　　　　　branches, these usually in several, well separated whorls below a
　　　　　　solitary, vertical branch . . . . . . . . . . . . . . . . . . . . . . . . . . . . . .14. *C. verticillata*
　　　　　29. Panicle branches naked on the basal 2–5 cm; panicles with 8–10
　　　　　　branches, these usually digitate, sometimes a second whorl
　　　　　　present below the terminal whorl . . . . . . . . . . . . . . . . . . . . . .16. *C. texensis* (in part)
　　　27. Panicle branches usually digitate, in a single terminal cluster, sometimes
　　　　with a poorly-developed second whorl just below the terminal cluster.
　　　　30. Margins of the lowest (bisexual) lemmas with conspicuously longer
　　　　　hairs distally than basally, the distal hairs usually more than 1.5 mm
　　　　　long.
　　　　　31. Plants annual; third floret, if present, shorter than its
　　　　　　subtending rachilla segment . . . . . . . . . . . . . . . . . . . . . . . . .7. *C. virgata* (in part)
　　　　　31. Plants perennial; third floret, if present, longer than its
　　　　　　subtending rachilla segment . . . . . . . . . . . . . . . . . . . . . . . . .8. *C. gayana* (in part)
　　　　30. Margins of the lowest (bisexual) lemmas glabrous or with appressed
　　　　　hairs less than 1 mm long, occasionally with a few scattered longer
　　　　　hairs.
　　　　　32. Panicle branches without spikelets on the basal 2–5 cm . . . . . .16. *C. texensis* (in part)
　　　　　32. Panicle branches spikelet-bearing to the base or naked for less
　　　　　　than 2 cm.
　　　　　　33. Second florets (first sterile florets) 0.1–0.5 mm wide.
　　　　　　　34. Plants annual, sometimes rooting at the lower nodes,
　　　　　　　　not stoloniferous; second florets 0.4–0.7　mm long . . .18 *C. radiata* (in part)
　　　　　　　34. Plants perennial, not rooting at the lower nodes but
　　　　　　　　sometimes stoloniferous; second florets 0.7–2.6 mm
　　　　　　　　long.
　　　　　　　　35. Culms 10–40 cm; lowest lemmas 1.9–2.7 mm long
　　　　　　　　　with awns 1.9–5.2 mm long; second florets 0.9–1.7
　　　　　　　　　mm long . . . . . . . . . . . . . . . . . . . . . . . . . . . . .15. *C. andropogonoides*
　　　　　　　　35. Culms to 100 cm; lowest lemmas 2–5.4 mm long,
　　　　　　　　　with awns 1–11 mm long; second florets 1–2.6 mm
　　　　　　　　　long . . . . . . . . . . . . . . . . . . . . . . . . . . . . . . . .9. *C. ventricosa* (in part)

33. Second florets (first sterile florets) 0.5–1 mm wide.
    36. Lemmas of the second florets inconspicuously bilobed; spikelets with (1)2–4 staminate or sterile florets . . . . . . 8. *C. gayana* (in part)
    36. Lemmas of the second florets usually not bilobed; spikelets with 1(2) staminate or sterile florets.
        37. Panicle branches 5–23 cm long; margins of the lowest lemmas appressed pubescent, occasionally sparsely so; second florets awned, the awns 3.1–12.5 mm long; blades without basal hairs . . . . . . . . . . . . . . . . . . . . . . . . . . . . . . . . . 10. *C. truncata* (in part)
        37. Panicle branches 5–11 cm long; margins of the lowest lemmas usually glabrous or scabrous, occasionally sparsely pubescent; second florets awned, the awns 0.5–7.5 mm long; blades with basal hairs up to 3 mm long . . . . . . . . . . . . . . . 9. *C. ventricosa* (in part)

## 1. Chloris berroi Arechav. [p. *451*]

*Chloris berroi* is native to the Rio de la Plata region of Argentina and Uruguay. It has been cultivated at scattered locations in the United States, but is not known to be established in the *Manual* region.

## 2. Chloris ciliata Sw. FRINGED WINDMILL-GRASS [p. *451*, *539*]

*Chloris ciliata* is a native species of grasslands from the Gulf Coast of Texas, through the Caribbean islands and Mexico to Central America, then, as a disjunct, in Argentina. It has been found, as an introduction, in New York.

## 3. Chloris canterae Arechav. PARAGUAYAN WINDMILL-GRASS [p. *451*, *539*]

*Chloris canterae* is native to South America. Both of its varieties are found in the coastal plain of Texas and Louisiana. In South America, they are essentially sympatric, but occupy different habitats.

1. Leaves primarily cauline, 2.5–6 mm wide, flat; panicle branches 4–14 cm long . . . . . . . . . . . . . . . . . var. *canterae*
1. Leaves primarily basal, 1–1.5 mm wide, involute; panicle branches 3–6 cm long . . . . . . . . . . . . . . var. *grandiflora*

### Chloris canterae Arechav. var. canterae [p. *451*]

*Chloris canterae* var. *canterae* has been collected in Texas and Louisiana. In South America, it grows on moist soils of the *campo* in Paraguay, southern Brazil, and northeastern Argentina.

### Chloris canterae var. grandiflora (Roseng. & Izag.) D.E. Anderson [p. *451*]

*Chloris canterae* var. *grandiflora* was collected around woolen mills in southeastern North America in the nineteenth and early twentieth centuries. It has also been found in Texas, but less frequently than var. *canterae*. In South America, it grows in drier, more rocky areas than var. *canterae*.

## 4. Chloris elata Desv. TALL WINDMILL-GRASS [p. *452*, *539*]

The range of *Chloris elata* lies primarily to the south of the *Manual* region, extending from southern Florida and the Caribbean islands to Peru and Argentina.

## 5. Chloris barbata Sw. SWOLLEN WINDMILL-GRASS [p. *452*, *539*]

*Chloris barbata* grows in subtropical and tropical coastal regions on loams, limestone-derived soils, and along beaches. The main portion of its range lies to the south of the *Manual* region, through the Caribbean and the east coast of Mexico, Central America, and South America. It is a weedy species, often growing in waste areas, but also in cultivated fields.

## 6. Chloris pilosa Schumach. [p. *452*]

*Chloris pilosa* is native to equatorial Africa, but it is sometimes planted for forage. It has been collected in Kleberg County, Texas, possibly from an experimental forage planting; it is not known to be established in the *Manual* region.

## 7. Chloris virgata Sw. FEATHER WINDMILL-GRASS, FEATHER FINGERGRASS [p. *452*, *539*]

*Chloris virgata* is a widespread species that grows in many habitats, from tropical to temperate areas with hot summers, including much of the United States. It is a common weed in alfalfa fields of the southwestern United States.

## 8. Chloris gayana Kunth RHODESGRASS [p. *452*, *539*]

*Chloris gayana* grows in warm-temperate to tropical regions throughout the world, including the southern United States. It is cultivated as a meadow grass in irrigated regions of the southwest.

## 9. Chloris ventricosa R. Br. PLUMP WINDMILL-GRASS [p. *452*, *539*]

*Chloris ventricosa*, an Australian species, has been found near old woolen mills in South Carolina and has been cultivated. It usually differs from *C. truncata* in its shorter panicle branches and tawny bisexual lemmas with glabrous margins.

## 10. Chloris truncata R. Br. BLACK WINDMILL-GRASS [p. *452*, *539*]

*Chloris truncata*, like the rather similar *C. ventricosa*, is an Australian native that has been found near woolen mills in South Carolina and beside a road near Lake Skinner in Riverside County, California. It usually differs from *C. ventricosa* in having longer panicle branches, and very dark, almost black, bisexual lemmas with appressed pubescent margins.

## 11. Chloris divaricata R. Br. SPREADING WINDMILL-GRASS [p. *452*, *539*]

*Chloris divaricata* is an Australian species that was collected around woolen mills of South Carolina in the first half of the twentieth century. It has since become established in Texas and New Mexico.

## 12. Chloris pectinata Benth. COMB WINDMILL-GRASS [p. *452*, *539*]

*Chloris pectinata* is an Australian species that was collected around woolen mills in South Carolina in the first half of the twentieth century.

## 13. Chloris cucullata Bisch. HOODED WINDMILL-GRASS [p. 452, 539]

*Chloris cucullata* is common along roadsides and in waste areas throughout much of Texas and adjacent portions of New Mexico and Mexico. Records from outside this area probably represent introductions. *Chloris cucullata* hybridizes with both *C. andropogonoides* and *C. verticillata*.

## 14. Chloris verticillata Nutt. TUMBLE WINDMILL-GRASS [p. 452, 539]

*Chloris verticillata* is a common weed of roadsides, lawns, and waste areas in the central United States. Prior to disruption of the native vegetation, it grew in low areas of the central prairies. It also grows in northern Mexico. *Chloris verticillata* hybridizes with both *C. cucullata* and *C. andropogonoides*.

## 15. Chloris andropogonoides E. Fourn. SLIMSPIKE WINDMILL-GRASS [p. 452, 539]

*Chloris andropogonoides* grows along grassy roadsides and prairie relicts of the coastal plain of southern Texas and northeastern Mexico.

## Hybridization and introgression between *Chloris cucullata*, *Chloris verticillata*, and *Chloris andropogonoides*.

*Chloris cucullata*, *C. verticillata*, and *C. andropogonoides* are sympatric in southern and central Texas, and often form mixed populations that include many apparent hybrids and introgressants. These plants combine the morphological features of their parents. In some populations, no 'pure' parental plants are found, having been eliminated either through competition or hybridization.

## 16. Chloris texensis Nash TEXAS WINDMILL-GRASS [p. 453, 539]

*Chloris texensis* appears to be rare. It is endemic to Texas.

## 17. Chloris submutica Kunth MEXICAN WINDMILL-GRASS [p. 453, 540]

*Chloris submutica* grows from the southwestern United States through Mexico, Guatemala, and Colombia to Venezuela. In Mexico, it is generally found between 1000–2100 m.

## 18. Chloris radiata (L.) Sw. RADIATE WINDMILL-GRASS [p. 453, 540]

*Chloris radiata* is a weedy species of the eastern Caribbean, Central America, and northern South America. It may be native to Florida, but the record from Linton, Oregon, was from a ballast dump. The species is no longer found in Oregon.

## 16.36 EUSTACHYS Desv.[1]

Pl per; csp, shortly rhz and often stln. Clm 20–150 cm, erect or decumbent, flattened, glab; **intnd** hollow. **Lvs** bas and cauline; **shth** open, keeled, strongly compressed, distinctly distichous, equitant; **lig** to 0.5 mm, scarious, densely short ciliate; **bld** flat or folded, erect to spreading, both surfaces usu glab, smt scabrous or scabridulous. **Infl** tml, exceeding the up lvs, pan of 1–36 non-disarticulating spikelike br; **br** digitately arranged, axes triquetrous, with spklt in 2 rows on the abx sides of the br. **Spklt** solitary, diverging strongly from the br axes, lat compressed, sessile or subsessile, with 2–3 flt; **lowest flt** bisx; **second flt** usu rdcd to a stipitate empty lm, occ with a pal, stmt if a third flt is present; **third flt**, if present, strl and usu rdmt, stipitate; **dis** above the glm. **Glm** unequal, 1-veined; **lo glm** somewhat smaller than the up glm, narrow, acuminate; **up glm** almost as long as the spklt, flattened, scarious, glab, green, pale, or purplish, veins antrorsely scabrous, apc truncate, bilobed, or bifid, often mucronate; **lowest lm** cartilaginous, light to dark brown, scabridulous distally, 3-veined, unawned, mucronate, or with a single awn, awns to 1.2 mm; **pal** equaling or slightly shorter than the lm, glab, 2-veined, veins keeled, shortly ciliate or scabridulous; **anth** 3, deep purple to purple-red. **Car** 1–1.7 mm, trigonous-ellipsoid, glab, translucent when fresh, pale to slightly reddish-purple-tinged.

*Eustachys*, as treated here, is a genus of approximately 12 species, most of which are native to the Western Hemisphere; four are native to the *Manual* region and three have been introduced. It is, in many ways, morphologically similar to *Chloris*, with the placement of a few species being problematic.

1. Lateral veins of the lowest lemma in each spikelet usually glabrous, occasionally with a few short, stiff hairs . . . . . . . . . . . . . . . . . . . . . . . . . . . . . . . . . . . . . . . . . . . . . . . . . . . . . . . . . 1. *E. glauca*
1. Lateral veins of the lowest lemma in each spikelet pubescent.
    2. Keels of the lowest lemma in each spikelet glabrous.
        3. Spikelets 1.5–2.1 mm long; panicles with 6–15 branches; branches 4–10 cm long, straight, somewhat stiff; leaf blades 5–10 mm wide . . . . . . . . . . . . . . . . . . . . . . . . . . . . . . . . . . . 6. *E. retusa*
        3. Spikelets 2.4–3 mm long; panicles with 10–36 branches; branches 6–15 cm long, flexible; leaf blades 10–15 mm wide . . . . . . . . . . . . . . . . . . . . . . . . . . . . . . . . . . . . . . 7. *E. distichophylla*
    2. Keels of the lowest lemma in each spikelet pubescent.
        4. Spikelets 1.5–2.5 mm long; lowest lemma in each spikelet mucronate.
            5. Lowest lemma in each spikelet dark brown, the lateral veins with appressed hairs shorter than 0.5 mm . . . . . . . . . . . . . . . . . . . . . . . . . . . . . . . . . . . . . . . . . . . . . . . . . . . 2. *E. petraea*

[1]Cynthia Aulbach

5. Lowest lemma in each spikelet tawny to reddish-brown, the lateral veins with spreading hairs longer than 0.5 mm . . . . . . . . . . . . . . . . . . . . . . . . . . . . . . . . . . . . . . . . . . . . . . 5. *E. caribaea*
4. Spikelets 2.6–3.7 mm long; lowest lemma in each spikelet awned, awns 0.4–1.2 mm.
6. Panicle branches 1–3; awns of lowest lemma in each spikelet 0.4–0.6 mm long; spikelets 3–3.7 mm long . . . . . . . . . . . . . . . . . . . . . . . . . . . . . . . . . . . . . . . . . . . . . 3. *E. floridana*
6. Panicle branches (3)4–9; awns of lowest lemma in each spikelet 0.7–1.2 mm long; spikelets 2.6–3 mm long . . . . . . . . . . . . . . . . . . . . . . . . . . . . . . . . . . . . . . . . . 4. *E. neglecta*

### 1. Eustachys glauca Chapm. Saltmarsh Fingergrass [p. 453, <u>540</u>]

*Eustachys glauca* grows in the margins of low woods and in ditches and brackish marshes. It is endemic to the southeastern United States.

### 2. Eustachys petraea (Sw.) Desv. Pinewoods Fingergrass [p. 453, <u>540</u>]

*Eustachys petraea* grows on dunes and open sandy areas and along roadsides and salt and brackish marshes. Its range extends south from the United States through Mexico to Panama.

### 3. Eustachys floridana Chapm. Florida Fingergrass [p. 453, <u>540</u>]

*Eustachys floridana* grows in dry, sandy woods and old fields. It is endemic to the southeastern United States.

### 4. Eustachys neglecta (Nash) Nash Four-spike Fingergrass [p. 453, <u>540</u>]

*Eustachys neglecta* grows in sandy fields, roadsides, and pinelands. It is endemic to the southeastern United States.

### 5. Eustachys caribaea (Spreng.) Herter Chickenfoot Grass [p. 453, <u>540</u>]

*Eustachys caribaea* has been introduced to the United States from South America. It is now established in North America, growing along roadsides at a few locations within the *Manual* region.

### 6. Eustachys retusa (Lag.) Kunth [p. 453, <u>540</u>]

*Eustachys retusa* is native to South America, but it is now established along roadsides, sandy fields, and waste areas in the southeastern United States.

### 7. Eustachys distichophylla (Lag.) Nees [p. 453, <u>540</u>]

*Eustachys distichophylla* is native to South America, but is now established along sandy roadsides, fields, and waste areas in the southern United States.

## 16.37 ENTEROPOGON Nees[1]

Pl ann or per; csp, smt stln or rhz. Clm to 120 cm, not wd. Shth open; lig memb, ciliate; bld flat. Infl tml, pan of 1–20 non-disarticulating, spikelike br, exceeding the up lvs; br digitately or racemosely arranged, if rcm, the lo nd usu with more than 1 br. Spklt solitary, strongly imbricate, appressed to somewhat divergent, dorsally compressed, with 2–6 flt, lowest flt bisx, elongate, remaining flt progressively rdcd, usu the distal flt rdmt and strl, occ stmt; dis beneath the glm. Glm unequal, subulate to lanceolate, memb; up glm much shorter than the lowest flt, acute or shortly awned; cal strigose; lowest lm stiff, 3-veined, ridged over the midveins, apc acute or bidentate, usu awned from between the teeth, without lat awns; pal almost as long as the lm, 2-keeled and bidentate; distal lm awned; lod 2; anth 1–3. Car sulcate; emb ¼–⅓(½) as long as the car.

*Enteropogon* is a tropical genus of 17 species. It differs from *Chloris*, in which it has sometimes been included, in its dorsally compressed, indurate lemmas that are conspicuously ridged over the midvein. The caryopses also differ, being dorsally flattened with a shallow ventral groove in *Enteropogon*; in *Chloris*, the caryopses are basically triangular in cross section although there may be a shallow ventral groove. The embryos also tend to be shorter relative to the caryopses in *Enteropogon* than in *Chloris*, but there is some overlap.

There is one species native to the *Manual* region; two others have been found at various locations but have not persisted.

1. Panicles with 3–15 branches racemosely arranged; plants rhizomatous . . . . . . . . . . . . . . . . . . . . . . . . . . . 1. *E. chlorideus*
1. Panicles with 1–10 branches in a single, digitate cluster; plants not rhizomatous, sometimes stoloniferous.
2. Panicle branches 6–11 cm long, erect to slightly diverging; spikelets with 5–6 florets, the distal 4–5 sterile; plants stoloniferous . . . . . . . . . . . . . . . . . . . . . . . . . . . . . . . . . . . . . . 2. *E. prieurii*
2. Panicle branches 7–25 cm long, divergent to drooping; spikelets with 2 florets, the distal floret sterile; plants not stoloniferous . . . . . . . . . . . . . . . . . . . . . . . . . . . . . . . . . 3. *E. dolichostachyus*

### 1. Enteropogon chlorideus (J. Presl) Clayton Buryseed Umbrellagrass [p. 453, <u>540</u>]

*Enteropogon chlorideus* is native from the southwestern United States through Mexico to Honduras. The spikelet-bearing rhizomes distinguish *Enteropogon* from most other grasses, but they are often missing from herbarium specimens.

### 2. Enteropogon prieurii (Kunth) Clayton Prieur's Umbrellagrass [p. 453]

*Enteropogon prieurii* is native to the tropics of the Eastern Hemisphere. It was found near wharves in Alabama and North Carolina at the beginning of the twentieth century, but it is not known to be established in the *Manual* region.

[1]Mary E. Barkworth

### 3. Enteropogon dolichostachyus (Lag.) Keng *ex* Lazarides [p. 453]

*Enteropogon dolichostachyus* grows from Afghanistan through southeast Asia to Australia. It has been collected from near a woolen mill in South Carolina, but is not known to be established in the *Manual* region.

## 16.38 TRICHLORIS E. Fourn. *ex* Benth.[1]

Pl per; csp, smt stln. Clm to 150 cm, hrb, solid, glab. Shth open, rounded; lig memb, ciliate, cilia longer than the memb base, conspicuous tufts of stiff hairs present on either side of the lig; bld linear, flat or folded. Infl tml, pan of non-disarticulating spikelike br, exceeding the lvs; br in 1 or more whorl(s), spklt in 2 rows on the abx side of the br, axes terminating in a fnctl spklt. Spklt lat compressed, with 2–5 flt, lowest 1–2 flt bisx, distal 1–3 flt progressively rdcd and strl; dis above the glm, all the flt falling as a unit. Glm much shorter than the spklt, memb; lo glm linear, acuminate; up glm lanceolate-ovate, awned; cal bearded; lowest lm linear-lanceolate, 3-veined, veins prolonged into 3 awns; cent awns 8–12 mm; lat awns 0.5–12 mm; pal 2-keeled, acute; distal flt(s) 1–3-awned; lod 2; anth 2 or 3. Car sulcate; emb ½ as long as the car; hila punctate.

*Trichloris* has two species, both of which are native to the *Manual* region. It differs from *Chloris* in its 3-awned lemmas. Both species of *Trichloris* have a disjunct distribution, with populations in both North and South America.

1. Lowest lemma awns subequal, the central awns 8–12 mm long, equaling or slightly longer than the lateral awns . . . . . . . . . . . . . . . . . . . . . . . . . . . . . . . . . . . . . . . . . . . . . . . . . . . . . . . . . . . . . . . . . 1. *T. crinita*
1. Lowest lemma awns unequal, the central awns 8–12 mm long, the lateral awns 0.5–1.5 mm long . . . . . . . . 2. *T. pluriflora*

### 1. Trichloris crinita (Lag.) Parodi FALSE RHODESGRASS [p. 454, 540]

*Trichloris crinita* is a native species that grows in the southwestern United States and northern Mexico, and, as a disjunct, in northern Argentina.

### 2. Trichloris pluriflora (E. Fourn.) Clayton MULTIFLOWER FALSE RHODESGRASS [p. 454, 540]

*Trichloris pluriflora* is native from southern Texas to Guatemala and, as a disjunct, from Ecuador to Argentina.

## 16.39 WILLKOMMIA Hack.[2]

Pl ann or per; csp or tufted, often stln. Clm 12–80 cm, hrb, unbrchd distally. Lvs involute to flat; lig memb, ciliate. Infl tml, pan of several, racemosely arranged, spikelike br, mostly exceeding the lvs, lo br smt partially included in the up lf shth at maturity; br not wd, unilateral, with 2 rows of appressed, imbricate, solitary spklt, terminating in a rdmt, strl spklt, strl spklt smt consisting of 1 or 2 small scales. Spklt dorsally compressed, with 1 flt; dis below the glm. Lo glm to ⅔ as long as the spklt, veinless; up glm equaling the flt, 1-veined; cal acute to pointed; lm thinly memb, 3-veined, apc rounded to acute, mucronate, shortly awned, or unawned. Car ellipsoid.

*Willkommia* is a genus of four species, three native to southern tropical Africa and one to the Americas, including the *Manual* region.

### 1. Willkommia texana Hitchc. WILLKOMMIA [p. 454, 540]

*Willkommia texana* grows on clay pans, alkaline flats, and sandy soils, in open or bare areas of Texas and (rarely) Oklahoma and, as a disjunct, in Argentina. The Oklahoma record may represent a recent introduction. North American plants belong to **Willkommia texana** Hitchc. var. **texana**. They differ from the Argentinean variety, **W. texana** var. **stolonifera** Parodi, in being cespitose rather than stoloniferous.

## 16.40 SCHEDONNARDUS Steud.[3]

Pl per; csp. Clm 8–55 cm, smt geniculate and brchd bas, usu curving distally; intnd minutely retrorsely pubescent, mostly solid. Lvs mostly bas; shth compressed-keeled, closed, glab, mrg scarious; lig 1–3.5 mm, memb, lanceolate; bld 1–12 cm long, 0.7–2 mm wide, stiff, usu folded, often spirally twisted, midrib well-developed, mrg thick and whitish. Infl tml, pan of widely spaced, racemosely arranged, spikelike br, exceeding the up lvs; br strongly divergent, with distant to slightly imbricate, closely appressed spklt. Spklt 3–5.5 mm, mostly sessile, lat compressed, with 1 flt; flt bisx; dis at the base of the pan and above the glm. Glm lanceolate, unequal, 1-veined; lm usu exceeding the glm, 3-veined, unawned or shortly awned; pal subequal to the lm; anth 3; sty 2. Car fusiform.

*Schedonnardus* is a unispecific North American genus that grows in the prairies and central plains of Canada, the United States, and northwestern Mexico. It has also been found, as a recent introduction, in California and Argentina. It is not known if it is established in California.

[1]Mary E. Barkworth  [2]J.K. Wipff  [3]Neil Snow

### 1. Schedonnardus paniculatus (Nutt.) Trel.
TUMBLEGRASS [p. *454*, <u>540</u>]

*Schedonnardus paniculatus* is frequently found in disturbed areas. At maturity, the panicle breaks at the base and functions as a tumbleweed for seed dispersal.

## 16.41 GYMNOPOGON P. Beauv.[1]

Pl usu per; often csp in appearance, rhz. **Clm** 10–100 cm, erect to decumbent, simple or sparingly brchd. **Lvs** cauline, evidently distichous; **shth** often strongly overlapping; **aur** absent; **lig** 0.1–0.5 mm, memb, ciliate; **bld** linear to ovate-lanceolate, lacking midribs. **Infl** tml, pan of spikelike br, these subdigitately or racemosely arranged, usu strongly divergent to reflexed, smt naked bas, spklt borne singly. **Spklt** widely spaced to slightly imbricate, appressed to the br, shortly pedlt, lat compressed, with 1–2(4) flt, only the lowest 1(2) flt(s) bisx; **rchl extensions** present, usu with a highly rdcd, strl flt(s); **dis** above the glm, flt falling together. **Glm** subequal, usu exceeding the bisx flt, narrow, acuminate, 1-veined; **lm of bisx flt** 3-veined, midveins prominent, apc minutely bidentate, usu awned from between the teeth, rarely unawned; **anth** (2)3.

*Gymnopogon*, a genus of around 15 species, extends from the United States to South America, with one additional species ranging from India to Thailand. Three species are native to the *Manual* region. *Gymnopogon* is most likely to be confused with *Chloris*, but differs in having a more highly reduced, sterile floret at the end of the rachilla extension, and distichous leaves.

1. Plants with elongate rhizomes; panicle branches naked for at least ⅓ of their length . . . . . . . . . . . . . . . . . . 1. *G. brevifolius*
1. Plants with short, knotty rhizomes or cespitose with a knotty base; panicle branches naked for less than
⅓ of their length, often spikelet-bearing to the base.
  2. Lemma awns 4–12.2 mm long . . . . . . . . . . . . . . . . . . . . . . . . . . . . . . . . . . . . . . . . 2. *G. ambiguus*
  2. Lemma awns 0–2.2 mm long . . . . . . . . . . . . . . . . . . . . . . . . . . . . . . . . . . . . . . . 3. *G. chapmanianus*

### 1. Gymnopogon brevifolius Trin. SHORTLEAF
SKELETONGRASS [p. *454*, <u>540</u>]

*Gymnopogon brevifolius* grows in dry to somewhat moist sandy pine woodlands of the southeastern United States, usually in loamy soils. It generally has rather weak, decumbent culms that tend to be obscured by the surrounding vegetation. Plants with stiffer culms tend to be confused with *Gymnopogon ambiguus*.

### 2. Gymnopogon ambiguus (Michx.) Britton, Sterns & Poggenb. BEARDED SKELETONGRASS [p. *454*, <u>540</u>]

*Gymnopogon ambiguus* grows in sandy pine woodlands of the southeastern United States, Haiti, and the Dominican Republic. It often grows with *G. brevifolius*, from which it differs in being more robust, having long, wider leaves, longer lemma awns, and, usually, having panicle branches that are spikelet-bearing to the base.

### 3. Gymnopogon chapmanianus Hitchc. CHAPMAN'S
SKELETONGRASS [p. *454*, <u>540</u>]

*Gymnopogon chapmanianus* grows in sandy pine barrens and sites inhabited by dwarf palmetto, *Serenoa repens*. As interpreted here, *G. chapmanianus* includes *G. floridanus* Swallen.

## 16.42 CTENIUM Panz.[2]

Pl usu per, smt ann; csp or rhz. **Clm** 10–150 cm, simple. **Lvs** often aromatic; **lig** memb, smt ciliate; **bld** flat or convolute, up bld rdcd. **Infl** tml, pan of 1–3 strongly pectinate br, usu exceeding the up lvs; **br** usu falcate, if more than 1, digitately arranged, axes crescentic in cross section, with 2 rows of solitary, subsessile spklt. **Spklt** strongly divergent, lat compressed, with 2 well-developed strl or stmt flt below the single bisx flt, rdcd strl or stmt flt also present beyond the bisx flt; **dis** above the glm. **Glm** unequal; **lo glm** shorter than the up glm, 1-veined, keeled; **up glm** 2-3-veined, awned dorsally; **lm** thin, 3-veined, entire or bidentate, awned, awns dorsal, attached just below the lm apc, or tml; **lod** 2, glab; **anth** 3 in bisx flt, 2 in stmt flt; **sty** 2. **Car** ellipsoid.

*Ctenium* has 17–20 species that are native to tropical areas of Africa and the Americas, generally being found in savannah associations. The awned upper glumes and the presence of sterile or staminate florets both below and above the fertile floret set it apart from other genera. The two native North American species are highly fire-adapted, flourishing in communities that regularly burn on a 1–5 year basis.

1. Plants without rhizomes, forming dense tufts; upper glumes with a row of prominent glands on each
side of the midvein; awns of the upper glumes strongly diverging at maturity . . . . . . . . . . . . . . . . . . . . 1. *C. aromaticum*
1. Plants with slender, scaly rhizomes; upper glumes without glands, or the glands inconspicuous; awns of
the upper glumes straight to ascending . . . . . . . . . . . . . . . . . . . . . . . . . . . . . . . . . . . . . . 2. *C. floridanum*

[1]James P. Smith, Jr.  [2]Mary E. Barkworth

## 1. Ctenium aromaticum (Walter) Alph. Wood
TOOTHACHE GRASS [p. 455, 540]

*Ctenium aromaticum* is a common species that grows in wet to moist pine flatwoods, savannahs, prairies, pitcher plant bogs, and ecotones between pine uplands and wet streamheads of the southeastern coastal plain.

## 2. Ctenium floridanum (Hitchc.) Hitchc. FLORIDA ORANGEGRASS [p. 455, 540]

*Ctenium floridanum* is an uncommon endemic of Georgia and Florida, where it grows in dry to mesic pine-oak uplands and pine flatwoods. It is also cultivated, the graceful curve of its spikes making it an attractive addition to gardens.

## 16.43 MICROCHLOA R. Br.[1]

**Pl** per (rarely ann); erect and csp, or decumbent and mat-forming. **Clm** 5–60 cm; **nd** glab. **Lvs** mostly bas; **lig** memb and ciliate or of hairs; **bld** often stiff, convolute. **Infl** tml, pan with a solitary (rarely 2), slender, spikelike **br**; **rchs** curved or falcate, semi-circular or crescentic in cross section. **Spklt** solitary, with 1(2) flt, terete to dorsally compressed; **flt** bisx; **dis** above the glm. **Glm** subequal, exceeding the flt, 1-veined; **lo glm**

asymmetric, keeled, keels somewhat twisted; **up glm** symmetric, flat, midveins straight; **lm** memb or hyaline, 3-veined, veins hairy; **pal** with 2 pubescent keels. **Car** 0.9–1.5 mm, glab.

*Microchloa* includes three African and one pan-tropical species. The species usually grow in open mesic to xeric habitats in tropical regions, often in shallow, hard soils. One species has become established in the *Manual* region.

### 1. Microchloa kunthii Desv. KUNTH'S SMALLGRASS [p. 455, 540]

*Microchloa kunthii* grows on granitic outcrops on rocky slopes. Its range extends southward from Carr Canyon, Huachuca Mountains,

Arizona and the Big Bend region of Texas through Mexico to Guatemala.

## 16.44 CYNODON Rich.[2]

**Pl** per; smt stln, smt also rhz, often forming dense turf. **Clm** 4–100 cm. **Shth** open; **aur** absent; **lig** of hairs or memb; **bld** flat, conduplicate, convolute, or involute, smt disarticulating. **Infl** tml, digitate or subdigitate pan of spikelike **br**; **br** (1)2–20, 1-sided, with 2 rows of solitary, subsessile, appressed, imbricate spklt. **Spklt** lat compressed, with 1(–3) flt, only the lowest flt fnctl; **rchl extension** usu present, smt terminating in a rdcd flt; **dis** above the glm. **Glm** usu shorter than the lm, memb, keeled, usu muticous; **lo glm** 1-veined; **up glm** 1–3-veined, occ shortly awned; **lm** memb to cartilaginous, 3-veined, keeled, keels with hairs, occ winged, apc

mucronate or muticous; **pal** about as long as the lm, 2-keeled; **anth** 3; **style br** 2, plumose; **lod** 2.

*Cynodon* has nine species, all of which are native to tropical regions of the Eastern Hemisphere. Several species are used as lawn and forage grasses in tropical and warm-temperate regions. The status of several species in the *Manual* region is unclear. Species other than *C. dactylon* usually grow only under cultivation, but there are scattered records of populations of other species from the southern United States that appear to have become established. Many cultivars of *Cynodon* have been developed that may exhibit combinations of features not found in the wild species, making it difficult to accommodate them in a key.

1. Lemma keels winged; panicle branches with flattened axes (subg. *Pterolemma*) . . . . . . . . . . . . . . . . . . . . . 7. *C. incompletus*
1. Lemma keels not winged; panicle branches with triquetrous axes (subg. *Cynodon*).
   2. Glumes 0.1–0.6 mm long . . . . . . . . . . . . . . . . . . . . . . . . . . . . . . . . . . . . . . . . . . . . . . . . . . . 1. *C. plectostachyus*
   2. Glumes 1.1–2.6 mm long.
      3. Panicles with 1–3(4) branches; culms 5–30 cm tall; blades 1–1.5 mm wide . . . . . . . . . . . . . . . . 2. *C. transvaalensis*
      3. Panicles with (2)4–20 branches; culms 5–100 cm tall; blades (1)2–7 mm wide.
         4. Panicles with 2–6(9) branches in a single whorl; culms 5–40(50) cm tall.
            5. Panicles with (2)4-6(9) branches; anthers dehiscent at maturity . . . . . . . . . . . . . . . . . . . . . . . 3. *C. dactylon*
            5. Panicles with 2-4 branches; anthers indehiscent at maturity . . . . . . . . . . . . . . . . . . . . . . . 4. *C. ×magennisii*
         4. Panicles with 4–20 branches in 1–5 whorls; culms 20–100 cm tall.
            6. Lemma keels glabrous or with a few scattered hairs; panicle branches usually in 2–5 whorls, stiff, frequently red or purple; culms 25–100 cm tall, woody . . . . . . . . . . . . . . . . . . . . 5. *C. aethiopicus*
            6. Lemma keels shortly pubescent; panicle branches usually in 1 whorl, lax, usually green; culms 20–60 cm tall, not woody . . . . . . . . . . . . . . . . . . . . . . . . . . . . . . . . . . . . . . 6. *C. nlemfuënsis*

### 1. Cynodon plectostachyus (K. Schum.) Pilg. STARGRASS [p. 455, 540]

*Cynodon plectostachyus* is native to tropical Africa. Its status in the *Manual* region is unclear. The records shown are from non-cultivated plants, but it is not known whether they represent established populations.

### 2. Cynodon transvaalensis Burtt Davy AFRICAN DOGSTOOTH GRASS, FLORIDAGRASS [p. 455]

*Cynodon transvaalensis* is native to southern Africa. Strains of the species have been crossed with strains of *C. dactylon*, and cultivars developed from these crosses are sometimes used as turf grasses in the southern United States and in similar climates throughout the world.

[1]Evangelina Sánchez [2]Mary E. Barkworth

### 3. Cynodon dactylon (L.) Pers. BERMUDAGRASS [p. 455, 540]

*Cynodon dactylon* is a variable species. Several varieties have been described, in addition to which numerous cultivars have been developed, some as turf grasses for lawns or putting greens, others as pasture or forage grasses. The most commonly encountered variety, both in the *Manual* region and in other parts of the world, is *C. dactylon* var. *dactylon*, largely because it thrives in severely disturbed, exposed sites; it does not invade natural grasslands or forests. Determining how many other varieties are established in the *Manual* region is almost impossible, because there has been no global study of variation in the species. For most purposes, it is probably neither necessary nor feasible to identify the variety of *C. dactylon* encountered.

1. Rhizomes near the surface (sometimes surfacing for a short distance before submerging again), the tips eventually surfacing and, like the lateral buds, producing tillers . . . . . . . . . . . . . . . . . . . . . . . . var. *dactylon*
1. Rhizomes growing up to 1 m deep, the tips remaining below ground, only the lateral buds producing tillers . . . . . . . . . . . . . . . . . . . . . . . . var. *aridus*

### Cynodon dactylon var. aridus J.R. Harlan & de Wet [not illustrated]

A cultivar of this variety has been introduced to the Yuma region of Arizona.

### Cynodon dactylon (L.) Pers. var. dactylon [p. 455]

As noted above, this is by far the most common variety of *Cynodon dactylon*.

### 4. Cynodon ×magennisii Hurcombe MAGENNIS' DOGSTOOTH GRASS [not illustrated]

*Cynodon ×magennisii* is a natural triploid hybrid between *C. dactylon* and *C. transvaalensis*. Several cultivars have been developed for lawns

and golf courses in the southern United States. These exhibit differing mixes of the characteristics of the two parent species.

### 5. Cynodon aethiopicus Clayton & J.R. Harlan ETHIOPIAN DOGSTOOTH GRASS [p. 455, 540]

*Cynodon aethiopicus* is native to the East African rift. It is now established along the canal bank in the Santa Ana National Wildlife Refuge in Texas, and is expected to spread.

### 6. Cynodon nlemfuënsis Vanderyst AFRICAN BERMUDAGRASS [p. 455, 540]

*Cynodon nlemfuënsis* is native to east and central Africa, but it is now established in southern Texas, and may be present in other parts of the southern United States. It differs from *C. dactylon* in being larger and lacking rhizomes. It is also less hardy, not becoming established where temperatures fall below -4°C. Plants in the *Manual* region belong to **Cynodon nlemfuënsis var. nlemfuënsis**, which differs from *C. nlemfuënsis* var. **robustus** Clayton & J.R Harlan in having shorter inflorescence branches and thinner culms.

### 7. Cynodon incompletus Nees [p. 455]

*Cynodon incompletus* is native to southern Africa. A hybrid between the two varieties identified below, *Cynodon ×bradleyi* Stent, is used as a lawn grass in North America.

1. Blades glabrous or sparsely hirsute; spikelets 2.5–3 mm long, narrowly ovate . . . . . . . . . . . . . . . . . var. *incompletus*
1. Blades densely hirsute; spikelets 2–2.5 mm long, broadly ovate . . . . . . . . . . . . . . . . . . . . . . . . . . var. *hirsutus*

---

## 16.45 SPARTINA Schreb.[1]

Pl per; csp from knotty bases or rhz. **Clm** 10–350 cm, erect, terete, solitary or in small to large clumps. **Lvs** mostly cauline; **shth** open, smooth, smt striate; **lig** memb, ciliate, cilia longer than the memb bases; **bld** flat or involute. **Infl** tml, usu exceeding the up lvs, 3–70 cm, pan of 1–75 spikelike br attached to an elongate rachis; **br** racemosely arranged, alternate, opposite, or whorled, appressed to strongly divergent, axes 3-sided, spklt usu sessile on the 2 lo sides, usu divergent to strongly divergent; **dis** beneath the glm. **Spklt** lat compressed, with 1 flt. **Glm** unequal, strongly keeled; **lo glm** shorter than the flt, 1-veined; **up glm** usu longer than the flt, 1–6-veined; **lm** shorter than the pal, 1–3-

veined, midveins keeled, lat veins usu obscure; **pal** thin, papery, 2-veined, obscurely keeled; **anth** 3; **lod** smt present, truncate, vascularized; **sty** 2, plumose. **Car** rarely produced.

*Spartina* has 15–17 species, most of which grow in moist to wet, saline habitats, both coastal and interior. Reproduction of all the species is almost entirely vegetative. There are nine native and two introduced species in the *Manual* region, plus three hybrids, one of which is native, the other two being deliberate introductions. One of the introduced species, *S. maritima*, grows in both Europe and at a few locations in Africa; the African populations may also represent introductions.

1. Leaf blades with smooth or slightly scabrous margins.
    2. Panicle branches 2–8 cm long, usually closely appressed and often twisted, the lower branches evidently less closely imbricate than the upper branches; glumes usually curved; plants of California and Baja California, Mexico . . . . . . . . . . . . . . . . . . . . . . . . . . . . . . . . . . . . . . . . . . . . . . . . . . . . . . . . . . . . 3. *S. foliosa*
    2. Panicle branches 2–24 cm long, usually loosely appressed or divergent, usually not twisted, lower and upper branches more or less equally imbricate; glumes straight; plants of varied distribution, including California and Baja California, Mexico.
        3. Glumes usually mostly glabrous on the sides, sometimes with appressed hairs; panicles with 3–25 branches . . . . . . . . . . . . . . . . . . . . . . . . . . . . . . . . . . . . . . . . . . . . . . . . . . . . . . . . . . . . . . . . . . . 2. *S. alterniflora*
        3. Glumes usually with appressed hairs on the sides, the margins sometimes glabrous; panicles with 1–12 branches.
            4. Ligules 2–3 mm long; anthers 5–13 mm long, well-filled, dehiscent at maturity . . . . . . . . . . . . . . . 6. *S. anglica*

[1]Mary E. Barkworth

4. Ligules 0.2–1.8 mm long; anthers 3–10 mm long, sometimes poorly filled and indehiscent at maturity.
   5. Ligules 0.2–0.6 mm long; leaf blades 6–12 cm long; anthers 3–6.5 mm long, well-filled, dehiscent at maturity . . . . . . . . . . . . . . . . . . . . . . . . . . . . . . . . . . . . . . . . . . . . . 4. *S. maritima*
   5. Ligules 1–1.8 mm long; leaf blades 6–30 cm long; anthers 5–10 mm long, poorly filled and indehiscent at maturity . . . . . . . . . . . . . . . . . . . . . . . . . . . . . . . . . . . . . . . . . . . . 5. *S.* ×*townsendii*
1. Leaf blades with strongly scabrous margins.
   6. Panicles smooth in outline, with (6)15–75 tightly appressed panicle branches; branches 0.5–4(7) cm long; plants without rhizomes . . . . . . . . . . . . . . . . . . . . . . . . . . . . . . . . . . . . . . . . . . . . . . . . . . . . 1. *S. spartinae*
   6. Panicles not smooth in outline, with 2–67 tightly appressed to strongly divergent branches; branches 1–15 cm; plants with more than 15 panicle branches always strongly rhizomatous, those with less than 16 branches with or without rhizomes.
      7. Plants without rhizomes or the rhizomes short; culms usually clumped; panicle branches 2–16.
         8. Upper glumes 1-veined . . . . . . . . . . . . . . . . . . . . . . . . . . . . . . . . . . . . . . . 9. *S. densiflora*
         8. Upper glumes 3–4-veined.
            9. Spikelets 6–9 mm long; culms to 200 cm tall; plants of the southeastern United States . . . . . . . . . . 7. *S. bakeri*
            9. Spikelets 10–17 mm long; culms to 120 cm tall; plants of the northeastern United States . . . . . . . . . . . . . . . . . . . . . . . . . . . . . . . . . . . . . . . . . . . 12. *S.* ×*caespitosa* (in part)
      7. Plants with well-developed rhizomes; culms usually solitary, sometimes a few together; panicle branches 3–67.
         10. Rhizomes whitish; upper glumes 1-veined or with all lateral veins on the same side of the keels; panicles with 2–15 branches, the branches 1–9 cm long.
            11. Spikelets 6–11 mm long, ovate to lanceolate; inland plants of western North America, rarely found east of Lake Winnipeg and the Mississippi Valley . . . . . . . . . . . . . . . . . . . . . . . 10. *S. gracilis*
            11. Spikelets 7–17 mm long, linear-lanceolate to ovate-lanceolate; usually coastal, also known from a few inland sites in northeastern North America.
               12. Spikelets 7–12 mm long; blade of the second leaf below the panicles 0.5–4(7) mm wide; plants of disturbed and undisturbed coastal habitats from the Gulf of St. Lawrence to the Gulf of Mexico and, as an introduction, on the west coast of North America . . . . . . . . . . . . . . . . . . . . . . . . . . . . . . . . . . . . . . . . . . . 11. *S. patens*
               12. Spikelets 10–17 mm long; blade of the second leaf below the panicles 2–7 mm wide; plants of disturbed habitats and artificial wetlands from Maine to Maryland . . . . . . . . . . . . . . . . . . . . . . . . . . . . . . . . . . . . 12. *S.* ×*caespitosa* (in part)
         10. Rhizomes light brown to brownish-purple; upper glumes 1-veined or with lateral veins on either side of the keels; panicles with 3–67 branches, the branches 1.5–15 cm long.
            13. Blade of the second leaf below the panicles 2–5(7) mm wide, usually involute even when fresh; panicles with 3–9 branches, the branches 3–9 cm long . . . . . . . . . . . . . . . . 12. *S.* ×*caespitosa* (in part)
            13. Blade of the second leaf below the panicles 5–14 mm wide, flat when fresh; panicles with 5–67 branches, the branches 1.5–15 cm long.
               14. Lower glumes ¾ as long as to equaling the adjacent lemmas; upper glumes awned, the awns 3–8 mm, with glabrous, rarely hispid, lateral veins . . . . . . . . . . . . . . . . . . . . 13. *S. pectinata*
               14. Lower glumes less than ½ as long to ⅔ as long as the adjacent lemmas; upper glumes unawned or with awns up to 2 mm long, usually with hispid lateral veins . . . . . . . . . 8. *S. cynosuroides*

### 1. Spartina spartinae (Trin.) Merr. *ex* Hitch. GULF CORDGRASS [p. 456, 540]

*Spartina spartinae* grows from the Gulf Coast through Mexico to Costa Rica in North America and, in South America, in Paraguay and northern Argentina. In the United States, it grows in sandy beaches, roadsides, ditches, wet meadows, and arid pastures near the coast, the most inland collection being 60 miles from the coast. In other parts of its range it sometimes grows well inland in saline soils where *Pinus palustris* (longleaf pine) is dominant or co-dominant.

### 2. Spartina alterniflora Loisel. SMOOTH CORDGRASS, SPARTINE ALTERNIFLORE [p. 456, 541]

*Spartina alterniflora* is found on muddy banks, usually of the intertidal zone, in eastern North and South America, but it is not known from Central America. In addition, it has become established on the west coast of North America, and in England and southeastern France. It hybridizes with *S. pectinata* in Massachusetts, and with *S. foliosa* in California. *Spartina alterniflora* is considered a serious threat to coastal ecosystems in Washington and California. It out-

competes many of the native species in these habitats and frequently invades mud flats and channels, converting them to marshlands.

### 3. Spartina foliosa Trin. CALIFORNIA CORDGRASS [p. 456, 541]

*Spartina foliosa* grows in the intertidal zone from northern California to Baja California, Mexico. Populations in San Francisco Bay are threatened by various introduced species of *Spartina*. Of particular concern is *S. alterniflora*, which forms hybrids with *S. foliosa* that have a broader ecological amplitude than either parent. *Spartina foliosa* differs from *S. densiflora*, which is also established in some regions, in being rhizomatous and having softer culms and wider leaf blades.

### 4. Spartina maritima (Curtis) Fernald SMALL CORDGRASS [p. 456]

*Spartina maritima* is a European species that has been reported as growing in Mississippi; the record has not been verified for this treatment. It also grows in Africa, possibly as an introduction.

### 5. Spartina ×townsendii H. Groves & J. Groves

TOWNSEND'S CORDGRASS [p. 456]

*Spartina ×townsendii* is a sterile hybrid between the European *S. maritima* and the American *S. alterniflora*. It seems to have formed spontaneously at several locations in Europe, often taking over the areas formerly occupied by its progenitors. At some locations it has given rise to the fertile amphidiploid *S. anglica*, from which it differs morphologically in its narrower, less divergent upper blades, shorter ligules, shorter, less hairy spikelets, and poorly filled, indehiscent anthers. *Spartina ×townsendii* has been used throughout the world for tideland reclamation because it is easy to establish, but it displaces native species.

### 6. Spartina anglica C.E. Hubb. ENGLISH CORDGRASS [p. 456, 541]

*Spartina anglica* is a naturally formed amphidiploid, derived from *Spartina ×townsendii*. Like *S. ×townsendii*, it has been introduced for reclamation of tidal mudflats, but differs from that species in its wider and more widely divergent upper blades, longer ligules, longer, more hairy spikelets, and longer, well-filled anthers.

### 7. Spartina bakeri Merr. SAND CORDGRASS [p. 456, 541]

*Spartina bakeri* grows on sandy maritime beaches and other salt water sites in the southeastern coastal states and on the shores of inland, freshwater lakes in Florida. Its inflorescence is similar to that of *S. patens*, but the branches of *S. patens* usually diverge from the rachises at maturity, whereas those of *S. bakeri* remain appressed. *Spartina bakeri* is distinct from most other species of *Spartina* in North America in forming dense clumps and in being able to grow in freshwater habitats.

### 8. Spartina cynosuroides (L.) Roth BIG CORDGRASS [p. 456, 541]

*Spartina cynosuroides* grows in brackish estuaries, tidal lagoons and bays, and in maritime habitats bordering the strand and intertidal zones. It grows primarily on the eastern and Gulf coasts of the United States, but has also been found in Michigan, possibly introduced by shipping.

### 9. Spartina densiflora Brongn. DENSELY-FLOWERED CORDGRASS [p. 456, 541]

*Spartina densiflora* is native to South America, where it grows in coastal marshes and at inland sites. It was introduced to Humboldt Bay, Humboldt County, California, possibly during the nineteenth century. It is now established there and in several locations around San Francisco Bay and in Washington, Oregon, and Texas, as well as the Mediterranean coast of Europe. It differs from *S. foliosa* in its indurate culms, narrow, inrolled leaves, and cespitose growth habit and tendency to grow among *Salicornia* in the upper intertidal zone or in open mud.

### 10. Spartina gracilis Trin. ALKALI CORDGRASS [p. 456, 541]

*Spartina gracilis* is found on the margins of alkaline lakes and along stream margins and river bottoms. Its range extends from the southern portion of the Northwest Territories, Canada, to central Mexico.

### 11. Spartina patens (Aiton) Muhl. SALTMEADOW CORDGRASS, SPARTINE ÉTALÉE [p. 456, 541]

*Spartina patens* grows in coastal salt and brackish waters. It is native to the east coast of North and Central America, extending through the Caribbean Islands to the north coast of South America, but is now established at scattered locations on the west coast of Canada and the United States. On the east coast, it is usually one of the dominant components of coastal salt marshes, frequently extending from the dry, sandy beach above the intertidal zone well up into the drier portions of the marshes. The older inland collections are from areas associated with brine deposits or saline soils, but there is some indication that the species' range is increasing inland because of the use of salt to de-ice roads in winter.

The inflorescence of *Spartina patens* is similar to that of *S. bakeri* when young, but its inflorescence branches usually diverge at maturity, whereas those of *S. bakeri* remain appressed. *Spartina patens* is probably one of the parents of *S. ×caespitosa*, *S. pectinata* being the other. Unlike *S. ×caespitosa*, *S. patens* grows in both disturbed and undisturbed habitats.

### 12. Spartina ×caespitosa A. A. Eaton MIXED CORDGRASS [p. 456, 541]

*Spartina ×caespitosa* is found in disturbed areas of the drier portions of salt and brackish marshes, at some distance above the intertidal zone. It occurs sporadically along the coast from Maine to Maryland, a region where its putative parents, *S. pectinata* and *S. patens*, are sympatric.

### 13. Spartina pectinata Link PRAIRIE CORDGRASS, SPARTINE PECTINÉE [p. 456, 541]

*Spartina pectinata* is native to Canada and the United States, but it has been introduced at scattered locations on other continents. On the Atlantic coast, it grows in marshes, sloughs, and flood plains, being a common constituent of ice-scoured zones of the northeast and growing equally well in salt and fresh water habitats. In western North America, it grows in both wet and dry soils, including dry prairie habitats and along roads and railroads. It is thought to be one of the parents of *S. ×caespitosa*, the other parent being *S. patens*.

---

## 16.46 BOUTELOUA Lag.[1]

**Pl** ann or per; synoecious; habit various, csp, stln, or rhz. **Clm** 1–80 cm. **Lvs** usu mostly bas; **shth** open; **lig** of hairs, memb, or memb and ciliate. **Infl** tml, pan of 1–80 solitary, spikelike br, exceeding the up lvs; **br** 4–50(75) mm, not wd, 1-sided, usu rcm on elongate rchs, smt digitate or subdigitate, with 1–130+ sessile to subsessile spklt in 2 rows, axes terminating in a spklt or extending beyond the base of the distal spklt. **Spklt** closely imbricate, appressed to pectinate, lat compressed or terete, with 1–2(3) flt, lowest flt in each spklt bisx, distal flt stmt or strl; **dis** at the base of the br or above the glm. **Glm** unequal or subequal, 1 or both glm equaled or exceeded by the distal flt, 1-veined, acute or acuminate, smt shortly awned; **lo glm** usu shorter than the lowest flt; **lm of lowest flt** entire, bilobed, trilobed, or 4-lobed, 3-veined, veins usu extended into 3 short awns; **pal of lowest flt** 2-veined, veins smt excurrent; **distal flt(s)** stmt or strl, varying from similar to the lowest flt in shape, size, and venation to strl and rdcd to an awn column with well-developed awns or to a flabellate scale.

*Bouteloua*, a genus of the Western Hemisphere with its center of diversity in Mexico, has about 40 species; all 19 species treated here are native to the *Manual* region. Several of its taxa are important forage grasses, and some are important constituents of the native North American grasslands.

---

[1]J.K. Wipff

1. Panicle branches deciduous, disarticulation occurring at their bases; spikelets usually 1–15 per branch, usually appressed rather than pectinate (subg. *Bouteloua*).
    2. All or most panicle branches with 1 spikelet . . . . . . . . . . . . . . . . . . . . . . . . . . . . . . . . . . . . . . . . . 3. *B. uniflora*
    2. All or most panicle branches with 2–15 spikelets.
        3. First (proximal) spikelet on each branch with 1 floret, the remaining spikelets with 2 florets; plants annual; panicles with 1–15 branches . . . . . . . . . . . . . . . . . . . . . . . . . . . . . . . . . . . 4. *B. aristidoides*
        3. Spikelets all alike or with 2 or more florets; plants annual or perennial; panicles with 1–80 branches.
            4. Central awns of lemmas flanked by 2 membranous lobes at maturity, the lobes 0.5–1.5 mm.
                5. Upper glumes bilobed, awned, the awns arising from between the teeth; inflorescence branch axes with deeply bi- or trifurcate apices; second florets sterile, rudimentary . . . . . . . . . . . . . . . . . . 7. *B. rigidiseta*
                5. Upper glumes acute, unawned or awn-tipped; inflorescence branch axes with apices entire; second florets usually staminate.
                    6. Base of plants dense, hard and knotty; culms straight, unbranched; panicle branches (15)20–30 mm long; plants rhizomatous . . . . . . . . . . . . . . . . . . . . . . . . . . . . . . . 9. *B. radicosa* (in part)
                    6. Base of plants usually not dense, hard, or knotty; culms straight or geniculate, branching; panicle branches 10–20 mm long; plants not rhizomatous . . . . . . . . . . . . . . . 8. *B. repens* (in part)
            4. Central awns of lemmas, if present, not flanked by membranous lobes or the lobes less than 0.3 mm long.
                7. Upper glumes with hairs, at least over the midveins.
                    8. Upper glumes with hairs only over the veins . . . . . . . . . . . . . . . . . . . . . . . . . . . . . . 8. *B. repens* (in part)
                    8. Upper glumes with hairs over the veins and elsewhere.
                        9. Panicles 6–10 cm long; branches with 2–6 spikelets . . . . . . . . . . . . . . . . . . . . . . . . . 5. *B. eludens*
                        9. Panicles 2.5–6 cm long; branches with 8–12 spikelets . . . . . . . . . . . . . . . . . . . . . . 6. *B. chondrosoides*
                7. Upper glumes glabrous, sometimes scabrous.
                    10. Second florets sterile, usually rudimentary, usually without paleas; central awns rarely to 7 mm long; panicles with 9–80 branches.
                        11. At least some leaf blades more than 2.5 mm wide, flat or folded when dry; ligules 0.3–0.5 mm long; anthers yellow, orange, red, or purple . . . . . . . . . . . . . . . . . . . . . 1. *B. curtipendula*
                        11. Leaf blades 1–1.5(2.5) mm wide, involute when dry; ligules 1–1.5 mm long; anthers dark purple . . . . . . . . . . . . . . . . . . . . . . . . . . . . . . . . . . . . . . . . . . . . . . . . . . 2. *B. warnockii*
                    10. Second florets bisexual, pistillate or staminate, with well-developed paleas; central awns 4–10 mm long; panicles with 2–17 branches.
                        12. Base of plants dense, hard and knotty; culms straight, unbranched; panicle branches (15)20–30 mm long; plants rhizomatous . . . . . . . . . . . . . . . . . . . . . . . 9. *B. radicosa* (in part)
                        12. Base of plants usually not dense, hard, or knotty; culms straight or geniculate, branching; panicle branches 10–20 mm long; plants not rhizomatous . . . . . . . . . . 8. *B. repens* (in part)
1. Panicle branches persistent; disarticulation above the glumes; spikelets 6–130 or more per branch, pectinate (subg. *Chondrosum*).
    13. Upper glumes of at least some spikelets with papillose-based hairs.
        14. Panicle branches extending beyond the base of the terminal spikelets . . . . . . . . . . . . . . . . . . . . . . . 11. *B. hirsuta*
        14. Panicle branches terminating in a spikelet.
            15. Plants tufted annuals or short-lived stoloniferous perennials; panicle branches 4–8, the axes with papillose-based hairs; lowest lemmas 3–4 mm long . . . . . . . . . . . . . . . . . . . . . . . 17. *B. parryi*
            15. Plants perennial, often shortly rhizomatous; panicle branches 1–3(6), the axes scabrous, never with papillose-based hairs; lowest lemmas 3.5–6 mm long . . . . . . . . . . . . . . . . 10. *B. gracilis* (in part)
    13. Upper glumes glabrous, scabrous, or hairy, but the hairs not papillose-based.
        16. Lower cauline internodes woolly-pubescent . . . . . . . . . . . . . . . . . . . . . . . . . . . . . . . . . . . . . . 12. *B. eriopoda*
        16. Lower cauline internodes glabrous or mostly so, sometimes pubescent immediately below the nodes.
            17. Central awns of lemmas not flanked by membranous lobes . . . . . . . . . . . . . . . . . . . . . . . . . 13. *B. trifida*
            17. Central awns of lemmas flanked by 2 membranous lobes.
                18. Lowest paleas in the spikelets awned, awns 1–2 mm long; panicles with 2–20 branches.
                    19. Lowest lemmas glabrous, with awns 3–4 mm long; panicle branches with 6–20 spikelets; plants perennial . . . . . . . . . . . . . . . . . . . . . . . . . . . . . . . . . . . . . . . . . 14. *B. kayi*
                    19. Lowest lemmas densely pilose, with awns 0.5–3 mm long; panicle branches with 20–50 spikelets; plants annual or short-lived perennials . . . . . . . . . . . . . . . . . . . . . . 15. *B. barbata*
                18. Lowest paleas in the spikelets unawned, but the veins sometimes excurrent for less than 1 mm; panicles with 1–6 branches.

20. Plants annual . . . . . . . . . . . . . . . . . . . . . . . . . . . . . . . . . . . . . . . . . . . . . . . . . . . . 16. *B. simplex*
20. Plants perennial.
    21. Culms usually with 2–3 nodes, not woody at the base; caryopses 2.5–3 mm
        long; lower paleas shallowly bilobed, the veins sometimes excurrent . . . . . 10. *B. gracilis* (in part)
    21. Culms usually with 4–5 nodes, somewhat woody at the base; caryopses
        1–1.2 mm long; lower paleas acute to acuminate, the veins not excurrent.
        22. Lower culm internodes with a thick, white, chalky bloom distally; panicle
            branches stramineous, mostly appressed, usually straight to slightly
            arcuate; plants rhizomatous, growing on gypsum soils . . . . . . . . . . . . . . . . . . . 18. *B. breviseta*
        22. Lower culm internodes without a conspicuous bloom; panicle branches
            dark, mostly ascending to widely divergent, usually becoming arcuate;
            plants not rhizomatous, growing on limestone soils . . . . . . . . . . . . . . . . . . . . . . 19. *B. ramosa*

## Bouteloua Lag. subg. Bouteloua

### 1. Bouteloua curtipendula (Michx.) Torr. SIDEOATS GRAMA [p. 456, <u>541</u>]

*Bouteloua curtipendula* is a common, often dominant or co-dominant species in open grasslands and wetlands of the drier portions of the central grasslands of North America. It is highly regarded as a forage species and is also an attractive ornamental. Its range extends from the *Manual* region through Mexico and Central America to western South America. There are three varieties. Two of the three grow in the *Manual* region; the third, *B. curtipendula* var. *tenuis* Gould & Kapadia, is endemic to Mexico.

1. Plants long-rhizomatous; culms solitary or in small
   clumps . . . . . . . . . . . . . . . . . . . . . . . . . . . . . . var. *curtipendula*
1. Plants not long-rhizomatous, bases sometimes
   knotty with short rhizomes; culms in large or small
   clumps . . . . . . . . . . . . . . . . . . . . . . . . . . . . . . . var. *caespitosa*

### Bouteloua curtipendula var. caespitosa Gould & Kapadia [p. 456]

*Bouteloua curtipendula* var. *caespitosa* grows on loose, sandy or rocky, well drained limestone soils at 200–2500 m in the southwestern United States, Mexico, and South America. It frequently grows, and may hybridize, with *B. warnockii*.

### Bouteloua curtipendula (Michx.) Torr. var. curtipendula [p. 456]

*Bouteloua curtipendula* var. *curtipendula* is the common variety of *B. curtipendula* in most of the *Manual* region. It grows on rich, loamy, well-drained prairie soils. Its elevational range extends from below 100 m to 2500 m.

### 2. Bouteloua warnockii Gould & Kapadia WARNOCK'S GRAMA [p. 456, <u>541</u>]

*Bouteloua warnockii* grows on limestone ledges and dry slopes below limestone outcrops. Its range extends from the southwestern United States to the state of Coahuila in northern Mexico. It frequently grows, and may hybridize, with *B. curtipendula* var. *caespitosa*.

### 3. Bouteloua uniflora Vasey NEALLY'S GRAMA [p. 456, <u>541</u>]

*Bouteloua uniflora* grows primarily in fertile, rocky, limestone soils of Texas and adjacent Coahuila, Mexico at 300–1000 m. A disjunct collection has been reported from Zion National Park, Utah. Plants in the *Manual* region belong to **Bouteloua uniflora** Vasey var. **uniflora**, which differs from **B. uniflora** var. **coahuilensis** Gould & Kapadia in having taller, leafy (rather than scapose) culms, longer leaf blades, and more panicle branches.

### 4. Bouteloua aristidoides (Kunth) Griseb. [p. 457, <u>541</u>]

There are two varieties, both of which grow in the *Manual* region.

1. Panicle branches with 2–5 spikelets, usually 5–16
   mm to the base of the terminal spikelets, axes usually
   extending an additional 6–10 mm . . . . . . . . . . . var. *aristidoides*
1. Panicle branches with 6–10 spikelets, usually 15–35
   mm to the base of the terminal spikelets, axes
   extending an additional 2–5(7) mm . . . . . . . . . . . var. *arizonica*

### Bouteloua aristidoides (Kunth) Griseb. var. aristidoides NEEDLE GRAMA [p. 457]

*Bouteloua aristidoides* var. *aristidoides* grows in dry mesas, plains, and washes from near sea level to about 2000 m. It matures rapidly following summer rains, and can be abundant over large areas within its range, which extends from California to western Texas and Mexico.

### Bouteloua aristidoides var. arizonica M.E. Jones ARIZONA NEEDLE GRAMA [p. 457]

*Bouteloua aristidoides* var. *arizonica* grows in the same kind of habitats as var. *aristidoides*, but only from 500–800 m. It has a more restricted range than *B. aristidoides* var. *aristidoides* (which extends into northern Mexico), being known only from New Mexico, Arizona, and Chihuahua, Mexico. In its extreme form, var. *arizonica* is very different from var. *aristidoides*, but the two varieties do intergrade.

### 5. Bouteloua eludens Griffiths ELUSIVE GRAMA [p. 457, <u>541</u>]

*Bouteloua eludens* grows on dry, rocky slopes and rolling desert flats at 1200–1800 m. It is only known from Cochise, Santa Cruz, and eastern Pima counties in Arizona, adjacent portions of New Mexico and Sonora, Mexico. Although its range is small, *B. eludens* is not rare. It resembles *B. chrondrosoides* in having pubescent panicle branches, but *B. eludens* usually has 12–16 branches 5–11 mm long with 2–6 spikelets, whereas *B. chrondrosoides* usually has 3–8 branches 10–15 mm long with 8–12 spikelets per branch. *Bouteloua rigidiseta* differs from *B. eludens* in its glume pubescence and geographic distribution, being found only in Oklahoma, Texas, and northeastern Mexico.

### 6. Bouteloua chondrosioides (Kunth) Benth. *ex* S. Watson SPRUCETOP GRAMA [p. 457, <u>541</u>]

*Bouteloua chondrosioides* grows on dry, rocky slopes and grassy plateaus at 200–2500 m. Its range extends from southern Arizona and western Texas to Costa Rica. It resembles *B. eludens* in having pubescent panicle branches, but *B. eludens* usually has 12–16 branches 5–11 mm long with 2–6 spikelets whereas *B. chrondrosioides* usually has 3–8 branches 10–15 mm long with 8–12 spikelets per branch.

### 7. Bouteloua rigidiseta (Steud.) Hitchc. TEXAS GRAMA [p. 457, <u>541</u>]

*Bouteloua rigidiseta* grows in grassy pastures and openings in woods, usually in clay or sandy clay soils, from near sea level to approximately 700 m. It is both widespread and abundant within its range, which extends from the southern United States to northern Mexico, but has little value as a forage grass. It is one of the earliest flowering warm season grasses. Although similar to *B. eludens*, *B.*

*rigidiseta* differs in its geographic distribution and glume pubescence, so the two taxa are unlikely to be confused in the field.

## 8. Bouteloua repens (Kunth) Scribn. & Merr. SLENDER GRAMA [p. 457, <u>541</u>]

*Bouteloua repens* grows in open, usually hilly terrain on many soil types, from sandy ocean shores to montane slopes, reaching elevations of 2500 m. Its native range extends from the southwestern United States through the Caribbean islands, Mexico, and Central America to Colombia and Venezuela. It has been found in Maine, where it is not established, and in Massachusetts, where it may not persist.

## 9. Bouteloua radicosa (E. Fourn.) Griffiths PURPLE GRAMA [p. 457, <u>541</u>]

*Bouteloua radicosa* grows on dry, rocky slopes at 1000–3000 m, from Arizona and southern New Mexico to southern Mexico. It has also become established in Maine, growing in disturbed habitats, but is not common there. It frequently grows with *B. repens* at lower elevations, but extends higher than that species. Like *B. repens*, *B. radicosa* exhibits great variation in spikelet and inflorescence characters.

# Bouteloua subg. Chondrosum (Desv.) A. Gray

## 10. Bouteloua gracilis (Kunth) Lag. ex Griffiths BLUE GRAMA, EYELASH GRASS [p. 457, <u>541</u>]

*Bouteloua gracilis* grows in pure stands in mixed prairie associations and disturbed habitats, usually on rocky or clay soils and mainly at elevations of 300–3000 m. Its native range extends from Canada to central Mexico; records from the eastern portion of the *Manual* region represent introductions. It is an important native forage species and also an attractive ornamental.

## 11. Bouteloua hirsuta Lag. [p. 457, <u>541</u>]

*Bouteloua hirsuta* is a widespread species, with two subspecies that frequently hybridize in areas of sympatry.

1. Rachilla internodes subtending second florets with a distal tuft of hairs; culms erect from the base, usually unbranched . . . . . . . . . . . . . . . . . . . . . . subsp. *pectinata*
1. Rachilla internodes subtending second florets without a distal tuft of hairs; culms usually decumbent and branched basally . . . . . . . . . . . . . subsp. *hirsuta*

### Bouteloua hirsuta Lag. subsp. hirsuta HAIRY GRAMA [p. 457]

*Bouteloua hirsuta* subsp. *hirsuta* grows from the open plains to slightly shaded openings in woods and brush on well-drained, often rocky soils at 50–300 m. It is morphologically, ecologically, and cytologically more variable than subsp. *pectinata*. Its range extends from North Dakota and Minnesota to central Mexico. In the northern portion of its range, it is not densely tufted and the culms are decumbent and branched; in the southwestern United States and northern Mexico, it grows in isolated, dense clumps, with erect, stout, unbranched culms and mostly basal leaves.

### Bouteloua hirsuta Lag. subsp. pectinata (Feath.) Wipff & S.D. Jones TALL GRAMA [p. 457]

*Bouteloua hirsuta* subsp. *pectinata* grows in well-drained, relatively undisturbed, calcareous soils, usually on thin-soiled limestone outcrops, at 60–500 m. Its range extends from southern Oklahoma to central Texas. At sites where subsp. *pectinata* is sympatric with subsp. *hirsuta*, swarms of morphologically intermediate plants are found.

## 12. Bouteloua eriopoda (Torr.) Torr. BLACK GRAMA [p. 457, <u>541</u>]

*Bouteloua eriopoda* grows on dry plains, foothills, and open forested slopes, often in shrubby habitats, and also in waste ground. It is usually found between 1000–1800 m, but extends to 2500 m. Once dominant in much of its range, under heavy grazing *B. eriopoda* persists only where protected by shrubs or cacti because it is highly palatable. Its range extends from the southwestern United States to northern Mexico.

## 13. Bouteloua trifida Thurb. ex S. Watson RED GRAMA [p. 457, <u>541</u>]

*Bouteloua trifida* grows on dry open plains, shrubby hills, and rocky slopes, at 2200–2500 m. Its range extends from the southwestern United States to central Mexico. It is a drought-resistant species that is sometimes mistaken for *Aristida* because of its delicate, cespitose growth habit and purplish, 3-awned spikelets. Juvenile plants may also be confused with *B. barbata* but that species is annual, with the central awn flanked by two membranous lobes and the lowest paleas 4-lobed and 2-awned.

1. Lower lemmas densely appressed pubescent; awns 2.2–4.5 mm long; anthers 0.2–0.3 mm long . . . . . . . var. *burkii*
1. Lower lemmas glabrous or sparsely appressed pubescent along both sides of the veins; awns (3.2)4–6.6 mm long; anthers 0.3–0.4 mm long . . . . . . var. *trifida*

### Bouteloua trifida var. burkii (Scribn. ex S. Watson) Vasey ex L.H. Dewey [p. 457]

*Bouteloua trifida* var. *burkii* grows in southern New Mexico, southern Texas, and adjacent Mexico.

### Bouteloua trifida Thurb. ex S. Watson var. trifida [p. 457]

*Bouteloua trifida* var. *trifida* grows in dry plains and rocky slopes, mostly at 300–1500 m, from southern California, Nevada, and Utah to Texas and Mexico.

## 14. Bouteloua kayi Warnock KAY'S GRAMA [p. 457, <u>541</u>]

*Bouteloua kayi* is known only from the mountainous limestone terrain along the Rio Grande River in southwestern Brewster County, Texas, at 2200–2500 m. Superficially, it resembles *B. trifida*.

## 15. Bouteloua barbata Lag. [p. 457, <u>541</u>]

The range of *Bouteloua barbata* extends from the southwestern United States to southern Mexico. It is often confused with juvenile plants of the perennial *B. trifida*, but in *B. barbata* the central awn is flanked by two membranous lobes and the lowest paleas are 4-lobed and 2-awned. Three varieties are recognized. The two that grow in the *Manual* region are often sympatric, but are usually easily distinguished in the field by their growth habit. The third variety, **B. barbata** var. **sonorae** (Griffiths) Gould, is usually stoloniferous; it is known only from the states of Sonora and Sinaloa, Mexico.

1. Plants annual; culms usually decumbent and geniculate, occasionally rooting at the lower nodes . . var. *barbata*
1. Plants short-lived perennials; culms erect from the base . . . . . . . . . . . . . . . . . . . . . . . . . . . . . . . . var. *rothrockii*

### Bouteloua barbata Lag. var. barbata SIXWEEKS GRAMA [p. 457]

*Bouteloua barbata* var. *barbata* grows in loose sands, rocky slopes, and washes, often on disturbed soils, usually at elevations below 2000 m. Its range extends from the southwestern United States to northwestern Mexico.

### Bouteloua barbata var. rothrockii (Vasey) Gould ROTHROCK'S GRAMA [p. 457]

*Bouteloua barbata* var. *rothrockii* grows on dry slopes and sandy flats, mostly at 750–1700 m. It grows throughout the southwestern United States and Mexico, sometimes covering large areas. It used to be the most important forage grass in southern Arizona and neighboring regions. It differs from *B. parryi* var. *parryi* in its lack of papillose-based hairs on the keels of the upper glumes.

### 16. **Bouteloua simplex** Lag. MAT GRAMA [p. *458*, <u>542</u>]

*Bouteloua simplex* grows on rocky, open slopes in grassy and open shrub vegetation at 1200–2500 m. Its native range extends from the southwestern United States through Mexico and Central America to western South America. It is adventive in Maine, where it grows in disturbed places, but it is not common there.

### 17. **Bouteloua parryi** (E. Fourn.) Griffiths PARRY'S GRAMA [p. *458*, <u>542</u>]

*Bouteloua parryi* grows on sandy slopes and flats at elevations from near sea level to 2000 m. Its range extends from the southwestern United States to central Mexico. Plants in the *Manual* region belong to **B. parryi** (E. Fourn.) Griffiths var. **parryi**, which differs from **B. parryi** var. **gentryi** (Gould) Gould in comprising tufted annuals rather than stoloniferous perennials. *Bouteloua parryi* var. *parryi* resembles *B. barbata* var. *rothrockii*, but differs in the papillose-based hairs on the keels of its upper glumes.

### 18. **Bouteloua breviseta** Vasey GYPSUM GRAMA [p. *458*, <u>542</u>]

*Bouteloua breviseta* is locally abundant on gypsum soils in southeastern New Mexico and the northern portion of the Trans Pecos region in Texas. It also grows in the state of Chihuahua, Mexico.

### 19. **Bouteloua ramosa** Scribn. *ex* Vasey CHINO GRAMA [p. *458*, <u>542</u>]

*Bouteloua ramosa* is locally common on rocky limestone slopes and flats among shrubs and *Agave lecheguilla*. Its range extends from the Trans Pecos region of western Texas to adjacent northern Mexico, particularly the state of Coahuila.

## 16.47 OPIZIA J. Presl[1]

Pl ann, per, or of indefinite duration; monoecious or dioecious, infl unisx and dimorphic; stln and mat-forming. Clm 1–15(30) cm, not wd. Lvs not clustered, not strongly distichous; shth open, keeled; lig memb, not ciliate; bld flat. Infl tml, with spikes or spikelike br on elongate rchs. Stmt infl pan of 1–6 spikelike pectinate br on elongate rchs, exserted well above the uppermost lvs; br 0.5–2 cm, terminating in a point; stmt spklt glab, with 1 flt; glm unequal, much shorter than the flt; lm 3-veined, unawned. Pist infl 1-sided spikes, with 6–12 spklt, often partially enclosed by the subtending shth, smt with br to 6 mm long at the lo nd; dis below the glm, the spklt falling intact; pist spklt lat compressed, with 1 bisx flt and a conspicuously 3-awned rdmt; lo glm rdmt or absent; up glm subequal to the lm, memb, flat, not enclosing the flt, unawned; cal blunt, with hairs; lm coriaceous, keeled, 3-veined, 3-awned, awns emanating from between 4 short, hyaline teeth; pal keels adnate to the rchl bas, widely winged distally.

*Opizia* is a North American genus of two species. Both species are probably native to Mexico, but one of them now grows, possibly from introductions, in Florida and the West Indies.

### 1. **Opizia stolonifera** J. Presl ACAPULCO GRASS [p. *458*, <u>542</u>]

*Opizia stolonifera* grows along dry roadsides in Florida. No pistillate plants have been found in the *Manual* region.

## 16.48 BUCHLOË Engelm.[2]

Pl per; usu dioecious; strongly stln, smt mat-forming. Clm 1–30 cm, erect, solid, mostly unbrchd, those of the pist infl much shorter than those of the stmt infl; nd mostly glab. Lvs bas tufted, not clustered or strongly distichous; shth open, rounded, often sparsely pilose near the collar; lig memb or of hairs; bld usu flat bas, curling when dry, glab or sparsely pilose, apc involute. Stmt infl tml, usu exceeding the up lvs, pan of 1–3(4) racemosely arranged, unilat, pectinate br; br not enclosed at maturity, spklt densely crowded in 2 rows. Stmt spklt with 2 flt; glm unequal, glab, 1- or 2-veined; lm 3-veined, glab, unawned; anth brownish to red or orange. Pist infl tml, pan, partially hidden within bracteate lf shth; br 2–3(4), 2.5–4.5 mm, burlike, with 3–5(7) spklt; dis at the base of the pan br. Pist spklt with 1 flt, almost completely enclosed by the up glm; lo glm irregular and rdcd; br axes and lo portion of up glm globose, white, indurate, terminating in 3 awnlike teeth; lm firmly memb, glab, 3-veined, unawned or shortly 3-awned.

*Buchloë* is a unispecific genus of the central plains of North America. It is usually dioecious, infrequently monoecious, or rarely synoecious.

### 1. **Buchloë dactyloides** (Nutt.) Engelm. BUFFALOGRASS [p. *458*, <u>542</u>]

*Buchloë dactyloides* is a frequent dominant on upland portions of the semi-arid, shortgrass component of the Great Plains, ranging from the southern prairie provinces of Canada through the desert southwest of the United States to much of northern Mexico. Collections from east of the Mississippi River and south of the Ohio River probably represent recent introductions. *Buchloë dactyoides* may be confused in the southern portion of its range with *Hilaria belangeri*, which consistently has pilose nodes, or in the Big Bend region of Texas with *Cathestecum erectum*, which has three spikelets per node and distinctly awned lemmas.

[1]Mary E. Barkworth  [2]Neil Snow

## 16.49 CATHESTECUM J. Presl[1]

Pl ann or per; monoecious or dioecious; often stln. **Clm** 5–40 cm, smt decumbent. **Lvs** mostly bas; **shth** open; **lig** of hairs; **bld** 0.5–2 mm wide. **Infl** tml, racemelike pan of short br, usu exceeding the up lvs, smt appearing 1-sided; **br** about 0.5 cm, not appressed to the rchs, with 3 spklt, bases strongly curved, shortly strigose, axes not extending beyond the distal spklt; **dis** at the base of the br. **Spklt** pedlt or sessile, lat spklt with shorter ped than the cent spklt or sessile; **lat spklt** with 2 flt, lo flt usu stmt or strl, rarely pist, up flt much rdcd, usu strl; **cent spklt** pedlt, with 3 flt, lowest flt bisx or pist (stmt in stmt pl), distal flt stmt or strl. **Glm** 2, very unequal, shorter than

the spklt, unawned; **lo glm** shorter than the up glm, those of the cent spklt flabellate, glab, veinless; **up glm** approximately equal to the adjacent lm, 1-veined, acuminate; **lm** thinner than the glm, pilose, 3-veined, 4-lobed, all 3 veins excurrent, forming awns that equal or exceed the lobes, lobes of the strl flt usu deeper than those of the other flt; **pal** 2-veined, veins often extending as awns.

*Cathestecum* is a genus of six species that extends from the southern United States to Guatemala; two are native to the *Manual* region.

1. Stolon internodes to 12 cm long, straight or almost so; culms 5–15 cm tall; spikelets frequently reddish or purplish; lateral spikelets of branches with bisexual florets poorly-developed, sterile (rarely staminate) florets; central spikelet on staminate branches with sparsely pilose lemmas . . . . . . . . . . . . . . . . 1. *C. brevifolium*
1. Stolon internodes usually 15 cm or longer, strongly arching; culms 10–30 cm tall; spikelets frequently pale green; lateral spikelets of branches with bisexual florets usually well developed, usually staminate or sterile; central spikelet on staminate branches with glabrous lemmas . . . . . . . . . . . . . . . . . . . . . . . . . . . 2. *C. erectum*

### 1. Cathestecum brevifolium Swallen [p. 458, 542]

The range of *Cathestecum brevifolium* extends from Ragged Top Mountain, Arizona, to El Salvador and Honduras.

### 2. Cathestecum erectum Vasey & Hack. FALSE GRAMA [p. 458, 542]

*Cathestecum erectum* grows on dry hills in the Great Bend region of western Texas and in northern Mexico.

## 16.50 AEGOPOGON Humb. & Bonpl. *ex* Willd.[2]

Pl ann or per; synoecious; tufted, csp, or sprawling. **Clm** 2–30 cm. **Shth** open; **lig** memb; **bld** flat. **Infl** tml, racemelike, 1-sided pan, usu exceeding the up lvs; **br** 0.4–0.8 mm, not appressed to the rachis, with 3 spklt, bases sharply curved, strigose, axes not extending beyond the distal spklt; **dis** at the base of the br. **Spklt** with 1 flt; **lat spklt** pedlt, stmt or strl, varying from rdmt to as large as the cent spklt; **cent spklt** sessile or pedlt, lat compressed, bisx. **Glm** exceeded by the flt, cuneate, truncate to bilobed, 1-veined, awned from the midveins,

smt also from the lat lobes; **lm** 3-veined, cent veins and smt the lat veins extended into awns, cent awns always the longest; **pal** almost as long as the lm, 2-keeled, 2-awned; **lod** 2; **anth** 3; **sty** 2. **Car** about twice as large as the emb; **hila** punctate.

*Aegopogon* is an American genus of four species that extends from the southwestern United States to Peru, Bolivia, and northern Argentina. One species is native in the *Manual* region.

### 1. Aegopogon tenellus (DC.) Trin. FRAGILE GRASS [p. 459, 542]

*Aegopogon tenellus* usually grows between 1550–2150 m in shady habitats of moist canyons, but it is sometimes found along roadsides

and in other open areas. Its range extends from southern Arizona into northern South America.

## 16.51 HILARIA Kunth[3]

Pl per or ann; tufted or csp, smt stln, per species smt rhz. **Clm** 5–250 cm, erect or decumbent; **nd** usu villous or pilose, particularly the up nd. **Shth** open, glab or pilose, lo shth often glab bas and pilose distally, mrg smt villous or pilose, up shth often glab even if the lo shth are pilose; **lig** 0.5–5 mm, memb, lacerate or ciliate. **Infl** tml, spikelike pan of red, disarticulating br, exceeding the up lvs; **br** with 3 spklt, appressed to the rchs, bases straight, seated in a ciliate, cuplike structure, smt with a 0.5–2 mm cal, cal pilose, axes not extending past the distal flt; **dis** at the base of the br, leaving the zig-zag

rchs. **Lat spklt** of each br shortly pedlt, with 1–4(5) strl or stmt flt; **glm** almost as long as the flt, deeply cleft into 2 or more lobes, with 1 or more dorsal awns; **lm** memb, hyaline. **Cent spklt** sessile, with 1 pist or bisx flt; **glm** shorter than the flt, rigid, indurate and fused bas, apc with 2 or more lobes; **lm** memb, awned or unawned.

*Hilaria* is a genus of 10 species that ranges from the southwestern United States to northern Guatemala, growing primarily in dry grasslands and desert areas. Most of the species are important forage species. *Hilaria* is interpreted here as having two groups, the *Hilaria* group and the

[1,2,3]Mary E. Barkworth

*Pleuraphis* group [= *Pleuraphis* Torr.]. These are sometimes treated as separate genera but, although they differ consistently in some morphological characters, their overall similarity is striking.

In the key and descriptions below, the term "fascicle" refers to a branch and its spikelet. Actual branch lengths are much shorter and harder to measure.

1. Glumes thickened, indurate, and conspicuously fused at the base; central spikelets with 1 pistillate floret (*Hilaria* group).
    2. Glumes pale to purplish, those of the lateral spikelets with dark glands confined to the base or lacking, awned from below midlength . . . . . . . . . . . . . . . . . . . . . . . . . . . . . . . . . . . . . . . . . . . . . . 4. *H. belangeri*
    2. Glumes gray to dark brown, those of the lateral spikelets evenly covered with dark glands, awned from above midlength . . . . . . . . . . . . . . . . . . . . . . . . . . . . . . . . . . . . . . . . . . . . . . . . . . . 5. *H. swallenii*
1. Glumes papery or membranous throughout, not conspicuously fused at the base; central spikelets with 1 bisexual floret (*Pleuraphis* group).
    3. Glumes of the lateral spikelets flabellate, the awns not exceeding the apical lobes; cauline nodes usually only shortly pubescent, sometimes glabrous . . . . . . . . . . . . . . . . . . . . . . . . . . . . . . 1. *H. mutica*
    3. Glumes of the lateral spikelets lanceolate or parallel-sided, the awns exceeding the apical lobes; cauline nodes pilose, villous, or glabrous.
        4. Lower cauline internodes tomentose . . . . . . . . . . . . . . . . . . . . . . . . . . . . . . . . . . . 2. *H. rigida*
        4. Lower cauline internodes glabrous . . . . . . . . . . . . . . . . . . . . . . . . . . . . . . . . . . . . . 3. *H. jamesii*

### 1. Hilaria mutica (Buckley) Benth. TOBOSAGRASS [p. 459, 542]

*Hilaria mutica* grows in level upland areas and desert valleys subject to occasional flooding but lacking permanent streams. Its range extends into northern Mexico.

### 2. Hilaria rigida (Thurb.) Benth. *ex* Scribn. BIG GALLETA [p. 459, 542]

*Hilaria rigida* grows in deserts and open juniper stands, at low elevations, from the southwestern United States to central Mexico.

### 3. Hilaria jamesii (Torr.) Benth. GALLETA [p. 459, 542]

*Hilaria jamesii* is endemic to the southwestern United States, and grows in deserts, canyons, and dry plains. It is usually less pubescent than *H. rigida*, the difference being most marked on the lower cauline nodes.

### 4. Hilaria belangeri (Steud.) Nash CURLY MESQUITE [p. 459, 542]

Both varieties of *Hilaria belangeri* are found on mesas and plains within the regions indicated.

1. Plants stoloniferous; blades 3–10 cm long, 1–2 mm wide; ligules about 1–1.5 mm long . . . . . . . . var. *belangeri*
1. Plants not stoloniferous; blades 3–15 cm long, to 3.5 mm wide; ligules 2.5–3 mm long . . . . . . . . . . . . . var. *longifolia*

### Hilaria belangeri (Steud.) Nash var. belangeri [p. 459]

*Hilaria belangeri* var. *belangeri* grows from Arizona to Texas, and south through Mexico. It was the dominant grass on Texas' shortgrass prairies.

### Hilaria belangeri var. longifolia (Vasey) Hitchc. [p. 459]

*Hilaria belangeri* var. *longifolia* is more restricted than var. *belangeri* in its distribution, growing from Arizona to Texas, and south to northwestern Mexico.

### 5. Hilaria swallenii Cory SWALLEN'S CURLY MESQUITE [p. 459, 542]

*Hilaria swallenii* grows on dry plains and rocky mesas in New Mexico, Texas, and northern Mexico.

## 16.52 TRAGUS Haller[1]

Pl ann or per; csp. Clm (2)5–65 cm, hrb, usu rooting at the lo nd; nd and intnd glab. Lvs cauline; shth open, usu shorter than the intnd, mostly glab but long-ciliate at the edges of the collar; lig memb, truncate, ciliate; bld usu flat, mrg ciliate. Infl tml, exceeding the up lvs, narrow, cylindrical pan; br 0.5–5 mm, resembling burs, with 2–5 spklt; dis at the base of the br. Spklt crowded, attached individually to the br, with 1 flt; proximal spklt(s) bisx, larger than the distal spklt(s); tml spklt often strl. Glm unequal; lo glm absent or minute, veinless, memb; up glm usu exceeding the flt, 5–7-veined, with 5–7 longitudinal rows of straight or uncinate spinelike projections; lm 3-veined; pal 2-veined, hyaline, memb.

*Tragus* has seven species, all of which are native to the tropics and subtropics of the Eastern Hemisphere; four have been introduced into the *Manual* region. The genus is easily recognized by the spinelike projections on the upper glumes. The number of glume veins should be determined by examining the adaxial surface, where they appear as green lines.

1. Upper glumes with 5 longitudinal rows of spinelike projections, 5-veined.
    2. Proximal internodes of the primary branches not longer than the second internode . . . . . . . . . . . . . . 1. *T. berteronianus*
    2. Proximal internodes of the primary branches 2–3 (or more) times longer than the second internode . . . . 2. *T. australianus*
1. Upper glumes with (5)6–7 longitudinal rows of spinelike projections, 7-veined.
    3. Panicle branches with 2 (rarely 3) spikelets; proximal spikelets on the branches 3–3.5 mm long . . . . . . 3. *T. heptaneuron*
    3. Panicle branches with 3–5 (rarely 2) spikelets; proximal spikelets on the branches 3.8–6.6 mm long . . . . . 4. *T. racemosus*

[1]J.K. Wipff

**1. Tragus berteronianus** Schult. SPIKE BURGRASS [p. *459*, <u>542</u>]

*Tragus berteronianus* is native to Africa and Asia, and is now established in Arizona, New Mexico, and Texas. It was collected in Maine, Massachusetts, New York, and Virginia in the nineteenth century, and Virginia in 1959.

**2. Tragus australianus** S.T. Blake AUSTRALIAN BURGRASS [p. *459*, <u>542</u>]

*Tragus australianus* is native to Australia, where it becomes established rapidly on disturbed or bare soil after summer rains. In the Western Hemisphere, it is known from Berkeley and Florence counties, South Carolina, and Argentina.

**3. Tragus heptaneuron** Clayton SEVEN-VEINED BURGRASS [p. *459*, <u>542</u>]

*Tragus heptaneuron* is native to tropical Africa, but it has been collected in Florence County, South Carolina.

**4. Tragus racemosus** (L.) All. STALKED BURGRASS [p. *459*, <u>542</u>]

*Tragus racemosus* is native from the Mediterranean region to southwest Asia, but now grows in the United States, primarily in Cochise County, Arizona. It was collected in several eastern states in the late nineteenth century, but does not appear to be established in these locations.

## 16.53 ZOYSIA Willd.[1]

**Pl** per, rhz, mat-forming. **Clm** 5–40 cm tall. **Lig** to 0.3 mm, of hairs, often with longer hairs at the base of each bld immediately behind the lig; **bld** usu glab abxly, smt with a ciliate cal at the base of each collar, adx surfaces glab, scabrous, or sparsely pilose, apc often sharply pointed. **Infl** tml, exceeding the lvs, solitary spikelike rcm (a single spklt in *Z. minima*), spklt solitary, shortly pedlt, lat appressed to the rchs; **dis** beneath the glm or not occurring. **Spklt** lat compressed, with 1 flt; **flt** bisx. **Lo glm** usu absent; **up glm** enclosing the flt, chartaceous to coriaceous, awned, awns to 2.5 mm; **lm** thin, lanceolate or linear, acute to emgt, 1-veined; **pal** thin, rarely present.

*Zoysia* has 11 species. They are native to coastal sands between 42°N and 42°S and from Mauritius to Polynesia. Some species also grow in disturbed inland areas. Three species are used as lawn grasses in mesic tropical and subtropical areas, including parts of the *Manual* region. Because lawn grasses are rarely collected, the distribution maps for *Zoysia* show the states in which each species can be successfully grown rather than individual county records.

As is common with commercially important species, several cultivars of *Zoysia* have been developed, sometimes from hybrids. Whether or not they are hybrids, cultivars often exceed the normal range of variation for a species in one or more respects. This makes it difficult to account for them in a key. The following key is written for specimens that fall within the normal range of the species concerned. In employing it, caution must be used when interpreting blade widths because blades of *Zoysia* become involute or convolute under drought or salinity stress. In addition, hybridization has resulted in cultivars with vegetative characteristics more like those of one species and reproductive characteristics more like those of another species.

1. Blades to 0.5 mm in diameter; racemes with 3–12 spikelets; peduncles included or extending to 1 cm beyond the sheaths of the flag leaves . . . . . . . . . . . . . . . . . . . . . . . . . . . . . . . . . . . . . . . . . . . . . . . 1. *Z. pacifica*
1. Blades 0.5–5 mm wide; racemes with 10–50 spikelets; peduncles extending (0.3)1–6.5 cm beyond the sheaths of the flag leaves.
  2. Pedicels 1.6–3.5 mm long; spikelets ovate, 1–1.4 mm wide; culm internodes 2–10 mm long; blades ascending . . . . . . . . . . . . . . . . . . . . . . . . . . . . . . . . . . . . . . . . . . . . . . . . . . . . . . . . . . . . 2. *Z. japonica*
  2. Pedicels 0.6–1.6 mm long; spikelets lanceolate, 0.6–1 mm wide; culm internodes 5–40 mm long, all plants with at least some internodes more than 14 mm long; blades patent . . . . . . . . . . . . . . . . . . . . . . . 3. *Z. matrella*

**1. Zoysia pacifica** (Goudsw.) M. Hotta & Kuroki MASCARENEGRASS, KOREAN VELVETGRASS [p. *460*, <u>542</u>]

*Zoysia pacifica* is less cold-tolerant than either *Z. matrella* or *Z. japonica*. It is not a common lawn grass in the *Manual* region. The cultivar 'Cashmere' has many of the characteristics of *Z. pacifica*; it is probably derived from a hybrid between *Z. matrella* and *Z. pacifica*.

**2. Zoysia japonica** Steud. JAPANESE LAWNGRASS, KOREAN LAWNGRASS [p. *460*, <u>542</u>]

*Zoysia japonica* was the first species of *Zoysia* introduced into cultivation in the United States. It is the most cold-tolerant and coarsely textured of the three species that have been introduced to the *Manual* region, and is the only species that is currently available as seed in the United States.

**3. Zoysia matrella** (L.) Merr. MANILAGRASS [p. *460*, <u>542</u>]

Many of the *Zoysia* lawn grasses grown in the southern and eastern United States are derived from hybrids between *Z. matrella* and *Z. pacifica* or *Z. japonica*, and have retained many of the characteristics of *Z. matrella*. They are used as lawn grasses from Connecticut southwards, and have occasionally been found as escapes in that region.

## 17. PAPPOPHOREAE Kunth[2]

**Pl** per; rarely ann. **Clm** hrb. **Lig** of hairs; **bld** epdm with bicellular microhairs. **Infl** tml, dense, narrow to somewhat open pan; **dis** above the glm but not between the flt (except in *Cottea*). **Spklt** with 3–10 flt, smt only the lowest flt bisx. **Lm** 5–13-veined, veins extending into awns, often with intermixed hyaline lobes. **Car** with punctate hila; **emb** ½ or more as long as the car.

[1]Sharon J. Anderson [2]John R. Reeder

The tribe *Pappophoreae* includes five genera and approximately 40 species. It is represented in tropical and warm regions around the world.

1. Spikelets disarticulating above the glumes and between the florets; florets 6–10, the lemmas with both awns and awned teeth, these not forming a pappuslike crown . . . . . . . . . . . . . . . . . . . . . . . . . . . . . . . . . . . . .17.03 *Cottea*
1. Spikelets disarticulating above the glumes but not between the florets, these falling as a unit; florets 3–6, the lemmas awned but without awned teeth, the awns forming a pappuslike crown.
   2. Lower glumes 1-veined; lemma awns scabridulous, not plumose . . . . . . . . . . . . . . . . . . . . . . . . . .17.01 *Pappophorum*
   2. Lower glumes 5–7-veined; lemma awns plumose, not scabridulous . . . . . . . . . . . . . . . . . . . . . . . . .17.02 *Enneapogon*

## 17.01 PAPPOPHORUM Schreb.[1]

Pl per; csp, essentially glab throughout. Clm 30–130 cm. Shth open; lig of hairs; microhairs of bld with an inflated tml cell similar in length to the bas cell. Infl tml, narrow, spikelike or somewhat open pan; dis above the glm but not between the flt, these falling together. Spklt with 3–5 flt, only the lo 1–3 bisx. Glm subequal, thin, memb, 1-veined, acute; lm rounded on the back, hairy at least bas, obscurely (5)7(9)-veined, veins extending into scabridulous awns of unequal length, several additional narrow awnlike lobes usu also present, awns and lobes together forming a pappuslike crown; pal textured like the lm, 2-veined, 2-keeled, keels scabrous or hairy; anth 3; sty 2. Car elliptical, plump, slightly dorsally flattened or nearly terete; emb about ½ as long as the car.

*Pappophorum* is an American genus with about eight species. It grows in warm regions of North and South America. Two species are native to the *Manual* region.

1. Panicles purple-tinged, narrow, but usually with some slightly spreading branches; lemma bodies 3–4 mm, the awns mostly not more than 1.5 times as long, these rarely spreading at right angles . . . . . . . . . . . . . . .1. *P. bicolor*
1. Panicles white or tawny, rarely slightly purplish, tightly contracted; lemma bodies 3–3.2 mm, the awns about twice as long, commonly spreading at right angles when mature . . . . . . . . . . . . . . . . . . . . . . . . . . .2. *P. vaginatum*

### 1. Pappophorum bicolor E. Fourn. PINK PAPPUSGRASS [p. 460, 542]

*Pappophorum bicolor* grows in open valleys, road right of ways, and grassy plains in Texas and northern Mexico.

### 2. Pappophorum vaginatum Buckley WHIPLASH PAPPUSGRASS [p. 460, 542]

*Pappophorum vaginatum* grows in similar habitats to *P. bicolor*, the two species sometimes growing together. Its range extends from southern Arizona to Texas and northern Mexico and from Uruguay to Argentina.

## 17.02 ENNEAPOGON Desv. *ex* P. Beauv.[2]

Pl per or ann; csp, more or less hairy throughout. Clm 3–100 cm; nd hairy; intnd hollow. Shth open; lig of hairs; microhairs of bld each with an elongated bas cell and an inflated tml cell. Infl tml, spikelike to somewhat open pan, bases often included within the upmost lf shth; dis above the glm but not between the flt, flt falling as a unit. Spklt with 3–6 flt, frequently only the lowest flt bisx, distal flt progressively rdcd. Glm subequal, as long as or slightly shorter than the flt (including the awns), more or less pubescent; lo glm 5–7-veined; lm firm, rounded on the backs, villous below the mid, strongly 9-veined, veins extending into equal, plumose awns 3–5 times as long as the lm bodies and forming a pappuslike crown; pal longer than the lm, entire, thinly memb, 2-veined, 2-keeled, keels hairy; anth 3, 0.2–1.5 mm; sty 2, free to the base, white.

*Enneapogon* includes about 28 species. It is found in tropical and warm regions of the world, especially in Africa and Australia. Two species are found in the *Manual* region: one is native, and one is an introduction that has become established in the region.

1. Culms 50–100 cm tall, about 2 mm thick; panicles loosely contracted to somewhat open, up to 3 cm wide at maturity; plants annual . . . . . . . . . . . . . . . . . . . . . . . . . . . . . . . . . . . . . . . . . . . . . . . . . . . . .1. *E. cenchroides*
1. Culms 20–45 cm tall, up to 1 mm thick; panicles usually tightly cylindrical, rarely more than 1 cm wide; plants perennial . . . . . . . . . . . . . . . . . . . . . . . . . . . . . . . . . . . . . . . . . . . . . . . . . . . . . . . . . .2. *E. desvauxii*

### 1. Enneapogon cenchroides (Licht.) C.E. Hubb. SOFTFEATHER PAPPUSGRASS [p. 460, 542]

*Enneapogon cenchroides* has been introduced and is persisting in the Ajo, Santa Catalina, Tucson, and Galiuro mountains of southern Arizona. Outside the Americas, its range includes Ascension Island, and extends from Sudan southward to the Cape Provinces of South Africa, through Arabia to India.

### 2. Enneapogon desvauxii P. Beauv. NINEAWN PAPPUSGRASS [p. 460, 542]

*Enneapogon desvauxii* grows in open areas of the southwestern United States and in much of Mexico. It also grows in Peru, Bolivia, Argentina, and most of Africa, from which it extends eastward through Arabia and India to China.

[1,2]John R. Reeder

## 17.03 COTTEA Kunth[1]

**Pl** per; csp, softly pilose throughout. **Clm** 25–70 cm, ascending or erect from hard knotty bases. **Shth** pilose, rounded on the backs; **bld** flat, linear; **microhairs of bld** each with an elongated bas cell and an inflated tml cell. **Infl** tml, open, rather narrow pan; **dis** above the glm and between the flt. **Spklt** with 6–10 flt, distal flt rdcd. **Glm** subequal, about as long as the lowest lm, pilose, 7–13-veined, midveins smt prolonged as short awns, apc acuminate or 3-toothed; **lm** rounded on the backs, with 9–13 prominent veins, some of these extending into antrorsely barbed awns of various lengths, others into awned teeth, awns and teeth not forming a pappuslike crown; **pal** slightly longer than the lm, 2-veined, 2-keeled, keels hairy; **anth** 3. Cleistogamous spklt, usu consisting of a single flt, produced in the lo shth.

*Cottea* is a unispecific American genus.

### 1. Cottea pappophoroides Kunth COTTA GRASS [p. 460, 543]

*Cottea pappophoroides* grows on open hillsides from Arizona and Texas south to central Mexico, and from Ecuador to Argentina.

## 18. ORCUTTIEAE Reeder[2]

**Pl** ann; viscid, aromatic. **Clm** with solid, pithy interiors, often with 5 or more nd. **Lvs** with little or no distinction between shth and bld; **lig** absent; **bld** of up cauline lvs similar in length to those below; **siliceous cells of the lf epdm** absent or irregular to dumbbell-shaped; **microhairs of the bld** bicellular, small, sunken, "mushroom-button" shaped. **Infl** tml, narrow, cylindrical to clavate or capitate, usu dense pan or spikes, spklt sessile or shortly pedlt; **rchl** tardily disarticulating between the flt. **Spklt** with 4–25(40) flt. **Glm** entire, denticulate, toothed, or absent; **lm** with 5–17 conspicuous veins; **pal** subequal to or slightly shorter than the lm, well-developed, keels glab; **lod** absent or 2, obscure, rounded, truncate to slightly emgt; **anth** 3. **Car** lat compressed; **hila** large, punctate, bas; **emb** ¾ or more as long as the car.

The tribe *Orcuttieae* includes three genera and nine species, all of which are restricted to vernal pools and similar habitats in California and Baja California, Mexico.

1. Lemmas deeply cleft into 5 mucronate or awn-tipped teeth, the teeth ⅓ as long as to equaling the lemma bodies; spikelets distichously arranged . . . . . . . . . . . . . . . . . . . . . . . . . . . . . . . . . . . . . . . . . 18.01 *Orcuttia*
1. Lemmas entire or denticulate, often with a central mucro; spikelets spirally arranged.
    2. Inflorescences clavate, often partially enclosed at maturity; spikelets laterally compressed, with glumes; lemmas rectangular, not translucent between the veins, the apices mucronate, otherwise entire or denticulate; caryopses without a viscid exudate, the embryo visible through the pericarp . . . . . . . 18.02 *Tuctoria*
    2. Inflorescences cylindrical, usually completely exposed at maturity; spikelets dorsally compressed, without glumes; lemmas flabellate, translucent between the veins, the apices ciliolate; caryopses covered with a viscid exudate, the embryos obscured by the pericarp . . . . . . . . . . . . . . . . . . . . . . . . . . 18.03 *Neostapfia*

## 18.01 ORCUTTIA Vasey[3]

**Pl** ann; viscid-aromatic, pilose, smt sparsely so, producing long, juvenile, floating bas lvs. **Clm** 3–35 cm, erect, ascending, or decumbent, smt becoming prostrate, not breaking apart at the nd, usu brchg only at the lo nd. **Lvs** without lig, smt with a "collar" line visible at the junction of the shth and bld, especially when dry; **bld** flat or becoming involute in drying. **Infl** tml, clavate to capitate spikes, exserted at maturity, spklt distichously arranged; **dis** tardy, above the glm and between the flt. **Spklt** lat compressed, with 4–40 flt. **Glm** irregularly 2–5-toothed; **lm** deeply cleft and strongly 5-veined, veins terminating in prominent mucronate or awn-tipped teeth ⅓–½ or more as long as the lm bodies, each tooth with an additional weaker vein on either side of a strong cent vein, these extending about halfway to the base of the lm; **pal** well-developed, 2-veined; **lod** absent; **anth** 3, white or pinkish, exserted on long, slender, ribbonlike filaments at anthesis; **sty** 2, apical, elongate, filiform, stigmatic for ⅓–½ of their length; **stigmatic hairs** short, often sparse. **Car** slightly compressed lat, oblong to elliptic; **emb** ¾ as long as to equaling the car; **epiblast** absent.

*Orcuttia* is a genus of five species, all of which are restricted to vernal pools and similar habitats in California and northern Baja California, Mexico.

1. Lemma teeth unequal, the central tooth the longest.
    2. Lemmas 6–7 mm long, the teeth terminating in awns at least 1 mm long; caryopses 2.3–2.5 mm long . . . . . 1. *O. viscida*
    2. Lemmas 4–5 mm long, the teeth sharp-pointed or with awns to 0.5 mm long; caryopses 1.3–1.8 mm long.

[1,2,3]John R. Reeder

3. Plants sparingly hairy; culms usually prostrate; spikes clavate . . . . . . . . . . . . . . . . . . . . . . . . 2. *O. californica* (in part)
    3. Plants conspicuously hairy, grayish; culms erect or decumbent; spikes somewhat capitate . . . . . . . 3. *O. inaequalis*
1. Lemma teeth essentially equal in length.
    4. Culms usually prostrate; caryopses 1.5–1.8 mm long . . . . . . . . . . . . . . . . . . . . . . . . . . 2. *O. californica* (in part)
    4. Culms erect, ascending, or decumbent; caryopses 2–3 mm long.
        5. Culms 1–2 mm thick, branching only at the lower nodes; spikes congested, crowded towards the
            top; leaf blades 3–5 mm wide . . . . . . . . . . . . . . . . . . . . . . . . . . . . . . . . . . . . . . 4. *O. pilosa*
        5. Culms 0.5–1 mm thick, often branching from the upper nodes; spikes not congested, even
            towards the top; leaf blades 1.5–2 mm wide . . . . . . . . . . . . . . . . . . . . . . . . . . . . . . . 5. *O. tenuis*

### 1. Orcuttia viscida (Hoover) Reeder SACRAMENTO ORCUTTGRASS [p. *461*, <u>543</u>]

*Orcuttia viscida* grows at elevations below 120 m in Sacramento County, California. Its awn-tipped lemma teeth curve outward at maturity, giving the spikes a distinctive, bristly appearance. It is listed as an endangered species by the U.S. Fish and Wildlife Service.

### 2. Orcuttia californica Vasey CALIFORNIA ORCUTTGRASS [p. *461*, <u>543</u>]

*Orcuttia californica* grows at elevations below 625 m in Los Angeles, Riverside, and San Diego counties, California and northern Baja California, Mexico. It is listed as an endangered species by the U.S. Fish and Wildlife Service.

### 3. Orcuttia inaequalis Hoover SAN JOAQUIN ORCUTTGRASS [p. *461*, <u>543</u>]

*Orcuttia inaequalis* grows at elevations below 575 m in Fresno, Madera, Merced, Stanislaus, and Tulare counties, California. It is listed as a threatened species by the U.S. Fish and Wildlife Service.

### 4. Orcuttia pilosa Hoover HAIRY ORCUTTGRASS [p. *461*, <u>543</u>]

*Orcuttia pilosa* grows at elevations below 150 m in Madera, Merced, Stanislaus, and Tehama counties, California. It is listed as an endangered species by the U.S. Fish and Wildlife Service.

### 5. Orcuttia tenuis Hitchc. SLENDER ORCUTTGRASS [p. *461*, <u>543</u>]

*Orcuttia tenuis* grows at 25–100 m in Shasta and Tehama counties of the Central Valley of California, with outlying populations in Sacramento County and the lower montane regions of Lake, Shasta, and Siskiyou counties, California. It is listed as a threatened species by the U.S. Fish and Wildlife Service.

## 18.02 TUCTORIA Reeder[1]

**Pl** ann; viscid-aromatic, more or less hairy throughout, not producing juvenile floating lvs. **Clm** 5–15(30) cm, simple or brchg at the up nd, erect or ascending, often rather fragile, readily breaking apart at the nd. **Lvs** without lig, with little or no distinction between shth and bld; **bld** flat, becoming involute when dry. **Infl** tml, clavate spikes, partially included or exserted at maturity, spklt spirally arranged; **dis** tardy, above the glm and between the flt. **Spklt** lat compressed, with 5–40 flt. **Glm** irregularly short-toothed or entire; **lm** (3)4–7 mm, 11–17-veined, not translucent between the veins, entire or denticulate, usu with a cent mucro; **pal** subequal to or slightly shorter than the lm; **lod** 2, 0.1–0.5 mm, smt fused to the pal; **anth** 3, exserted on long, slender, ribbonlike filaments at anthesis; **sty** 2, apical, long, filiform, stigmatic for $\frac{1}{3}$–$\frac{1}{2}$ of their length; **stigmatic hairs** short, often sparse. **Car** lat compressed, pyriform to oblong, prcp not viscid; **emb** visible through the prcp, brown, from $\frac{3}{4}$ as long as to nearly equaling the car; **epiblasts** present.

*Tuctoria* has three species, all of which grow in vernal pools or similar habitats, two in the Central Valley of California and one, *T. fragilis* (Swallen) Reeder, in Baja California Sur, Mexico. Both species found in the *Manual* region are endangered by loss of habitat to urbanization and agriculture.

1. Spikes exserted from the upper leaf sheaths at maturity; lemmas more or less truncate; caryopses about 2
    mm long, minutely rugose . . . . . . . . . . . . . . . . . . . . . . . . . . . . . . . . . . . . . . . . . . . . . 1. *T. greenei*
1. Spikes partially included in the upper leaf sheaths at maturity; lemmas tapering gradually to a
    mucronate apex; caryopses about 3 mm long, smooth . . . . . . . . . . . . . . . . . . . . . . . . . . . . 2. *T. mucronata*

### 1. Tuctoria greenei (Vasey) Reeder AWNLESS SPIRALGRASS [p. *461*, <u>543</u>]

*Tuctoria greenei* grows at elevations below 150 m in the Central Valley of California.

### 2. Tuctoria mucronata (Crampton) Reeder PRICKLY SPIRALGRASS [p. *461*, <u>543</u>]

*Tuctoria mucronata* is known from only two locations in Solano County, California; both locations are at elevations below 10 m.

## 18.03 NEOSTAPFIA Burtt Davy[2]

**Pl** ann; glab or sparsely pilose, not producing juvenile floating lvs. **Clm** 10–30 cm, simple, ascending or decumbent, often geniculate, not breaking apart at the nd. **Lvs** without lig, with little or no distinction between shth and bld. **Infl** tml, dense cylindrical spikes, usu completely exposed at maturity, spklt spirally arranged,

[1,2]John R. Reeder

rchs often extending beyond the spklt as a naked or scale-covered axis; **dis** below the spklt. **Spklt** dorsally compressed, usu with 5 flt. **Glm** absent; **lm** about 5 mm long, flabellate, to 3 mm wide distally, 7–11-veined, translucent between the veins, apc entire, ciliolate; **pal** slightly shorter and much narrower than the lm, hyaline, 2-veined; **lod** 2, about 0.2 mm, truncate or slightly emgt; **anth** 3, exserted at anthesis, **sty** 2. **Car** lat compressed, obovoid, prcp thick and covered with a viscid exudate, obscuring the emb; **epiblast** present.

*Neostapfia* is a unispecific genus endemic to the Central Valley of California.

## 1. Neostapfia colusana (Burtt Davy) Burtt Davy
COLUSAGRASS [p. *461*, <u>*543*</u>]

*Neostapfia colusana* grows in vernal pools of Colusa, Merced, Solano, and Stanislaus counties, California, at elevations below 125 m. It is listed by the U.S. Fish and Wildlife Service as a threatened species because of its restricted habitat, much of which has been destroyed. The stout, cylindrical spikes emerging from the sheathing leaves resemble miniature ears of maize. This and the abundant viscid secretion make *N. colusana* a particularly distinctive species.

# 7. DANTHONIOIDEAE N.P. Barker & H.P. Linder[1]

The *Danthonioideae* include one tribe, the *Danthonieae*, which used to be included in the *Arundinoideae*. The combination of haustorial synergids, ciliate ligules, elongated embryo mesocotyls, and $C_3$ photosynthesis distinguishes the *Danthonioideae* from other subfamilies of the *Poaceae*.

# 19. DANTHONIEAE Zotov[2]

**Pl** usu per, smt ann; when per, csp, rhz, or stln. **Clm** usu solid, rarely hollow. **Lvs** distichous; **shth** usu open; **abx lig** usu absent; **aur** usu absent; **adx lig** of hairs or memb and ciliate; **bld** not psdpet; **mesophyll** non-radiate; **adx palisade layer** absent; **fusoid cells** absent; **arm cells** absent; **Kranz anatomy** absent; **midrib** simple, usu with 1 vascular bundle (an arc of bundles in *Cortaderia*); **adx bulliform** cells present or not; **stomata** usu with dome-shaped or parallel-sided subsidiary cells (rarely slightly triangular or high dome-shaped); **bicellular microhairs** usu present, distal cell long, narrow; **papillae** usu absent. **Infl** ebracteate (subtending lf somewhat spatheate in *Urochlaena* Nees), usu pan, smt rcm or spicate, occ a single spklt; **dis** usu above the glm and between the flt, smt below the glm or in the clm. **Spklt** bisx (smt with unisx flt) or unisx, with 1–7(20) bisx or pist flt, distal flt in the bisx spklt often strl or stmt; **rchl extension** present. **Glm** 2, usu equal, (1)3–7-veined, usu exceeding the distal flt; **flt** lat compressed; **lm** firmly memb to coriaceous, 3–9-veined, rounded across the back, glab or with non-uncinate hairs, these smt in tufts or fringes, lm apc shortly to deeply bilobed, lobes often setaceous, midveins often extended as awns, awns usu geniculate, bas segment often flat and twisted; **pal** well-developed, smt short relative to the lm; **lod** 2, usu free, usu fleshy, rarely with a memb apical flap, glab or ciliate, often with microhairs, smt heavily vascularized; **anth** 3; **ov** usu glab, rarely with apical hairs; **haustorial synergids** present, smt weakly developed; **sty** 2, bases usu widely separated. **Car** separate from the lm and pal; **hila** punctate or long-linear; **emb** large or small relative to the car; **endosperm** hard; **stch g** usu compound; **epiblasts** absent; **scutellar cleft** present; **mesocotyl intnd** elongated; **embryonic lf mrg** usu meeting, smt overlapping.

The *Danthonieae*, the only tribe in the *Danthonioideae*, include approximately 13 genera and 290 species, most of which grow in mesic to xeric, open habitats such as grasslands, heaths, and open woods. It is most abundant in the Southern Hemisphere, with only *Danthonia* being native in the Northern Hemisphere.

Two of the genera recognized here, *Karroochloa* and *Austrodanthonia*, are frequently included in *Danthonia*, from which they can be distinguished by the tufts of hairs on their lemmas. It is much more difficult to identify a character, or combination of characters, that will consistently distinguish them from each other. Glume length works in this *Manual* because of the species involved, but it is not generally reliable.

1. Culms 200–700 cm tall; inflorescences plumose, 30–130 cm long . . . . . . . . . . . . . . . . . . . . . . . . . . . . . . . . . . . . . . . . . . 19.01 *Cortaderia*
1. Culms 2–100 cm tall; inflorescences not plumose, 0.5–12 cm long.
  2. Lemmas with hairs in 1 or more transverse row(s) above the callus and/or at midlength.
    3. Panicles subcapitate; glumes 3.5–7 mm long . . . . . . . . . . . . . . . . . . . . . . . . . . . . . . . . . . . . . . . . 19.04 *Karroochloa*
    3. Panicles narrow; glumes 8–24 mm long . . . . . . . . . . . . . . . . . . . . . . . . . . . . . . . . . . . . . . . 19.05 *Austrodanthonia*
  2. Lemmas glabrous or, if with hairs, the hairs not in transverse rows.
    4. Plants annual . . . . . . . . . . . . . . . . . . . . . . . . . . . . . . . . . . . . . . . . . . . . . . . . . . . . . . . . . . . . . . . 19.03 *Schismus*
    4. Plants perennial.
      5. Lemma apices entire, acute to acuminate . . . . . . . . . . . . . . . . . . . . . . . . . . . . . . . . . . . . . . . 19.06 *Tribolium*
      5. Lemma apices bifid, obtuse, acute, or acuminate . . . . . . . . . . . . . . . . . . . . . . . . . . . . . . . . . 19.02 *Danthonia*

[1]Grass Phylogeny Working Group  [2]Grass Phylogeny Working Group and Mary E. Barkworth

## 19.01 CORTADERIA Stapf[1]

**Pl** per; often dioecious or monoecious; csp. **Clm** 2–7 m, erect, densely clumped. **Lvs** primarily bas; **shth** open, often overlapping, glab or hairy; **aur** absent; **lig** of hairs; **bld** to 2 m, flat to folded, arching, edges usu sharply serrate. **Infl** tml, plumose pan, 30–130 cm, subtended by a long, ciliate bract; **br** stiff to flexible. **Spklt** somewhat lat compressed, usu unisx, smt bisx, with 2–9 unisx flt; **dis** above the glm and below the flt. **Glm** unequal, nearly as long as the spklt, hyaline, 1-veined; **cal** pilose; **lm** 3–5(7)-veined, long-acuminate, bifid and awned or entire and mucronate; **lm of pist** and **bisx flt** usu long-sericeous; **lm of stmt flt** less hairy or glab; **lod** 2, cuneate and irregularly lobed, ciliate; **pal** about ½ as long as the lm, 2-veined; **anth** of bisx flt 3, 1.5–6 mm, those of the pist flt smaller or absent. **Car** 1.5–3 mm; **hila** linear, about ½ as long as the car; **emb** usu shorter than 1 mm.

*Cortaderia*, a genus of about 25 species, is native to South America and New Zealand, with the majority of species being South American. Both of the species that are found in North America were originally introduced as ornamentals; both are now considered aggressive weeds in parts of the *Manual* region.

1. Sheaths hairy; panicles elevated well above the foliage; culms 4–5 times as long as the panicles . . . . . . . . . . . . . . . 1. *C. jubata*
1. Sheaths glabrous or sparsely hairy; panicles elevated only slightly, if at all, above the foliage; culms 2–4 times as long as the panicles . . . . . . . . . . . . . . . . . . . . . . . . . . . . . . . . . . . . . . . . . . . . . . . . . . . . . . . . . . . . . . . . . . . . . . . . 2. *C. selloana*

### 1. Cortaderia jubata (Lemoine *ex* Carrière) Stapf
PURPLE PAMPAS GRASS [p. *461*, <u>543</u>]

*Cortaderia jubata* is found on the west coast of the conterminous United States, growing in disturbed, open ground such as brushy slopes, eroded banks and cliffs, road cuts, cut-over timber areas, and sand dunes. It is native to mountainous areas of Ecuador, Peru, and Bolivia. It was grown in the past as an ornamental because of its attractive panicles, but is now a serious weed in California, reproducing apomictically and invading many open habitats.

### 2. Cortaderia selloana (Schult. & Schult. f.) Asch. & Graebn. PAMPAS GRASS [p. *461*, <u>543</u>]

*Cortaderia selloana* is native to central South America. It is cultivated as an ornamental in the warmer parts of North America. It was thought that it would not become a weed problem because most plants sold as ornamentals are unisexual, but it is now considered an aggressive weed in California and Bendigo, Australia.

## 19.02 DANTHONIA DC.[2]

**Pl** per; csp, smt shortly rhz. **Clm** 7–130 cm, erect. **Shth** open to the base, with tufts of hairs at the aur position, smt with a line of hairs around the col; **aur** absent; **lig** of hairs; **bld** rolled in the bud, flat or involute when dry. **Infl** tml; pan, rcm, or a solitary spklt, to 12 cm; **rchs, br,** and **ped** scabrous or hirsute. **Spklt** terete or lat compressed, with 3–12 flt, tml flt rdcd; **dis** beneath the flt, also at the cauline nd in some species. **Glm** subequal or the lo glm a little longer than the up glm, usu exceeding the flt (excluding the awns and lm teeth), lanceolate, chartaceous, 1–7-veined, keels glab or sparsely scabrous; **rchl** glab; **cal** densely strigose on the sides; **lm bodies** obscurely (5)7–11-veined, backs glab or pilose, mrg usu densely pilose proximally, apc with 2 acute to aristate lobes, mucronate or awned between the lobes; **awns,** when present, geniculate and twisted below the geniculation; **pal** about as long as the lm bodies, 2-veined, veins scabrous, apc obtuse, smt bifid; **lod** 2, glab or with a few hairs; **anth** 3, their size depending on whether the flt are chasmogamous or cleistogamous; **ov** glab. **Car** 1.5–5.5 mm, ovate to obovate, dorsally flattened, brown; **hila** linear, ⅓–¾ as long as the car. **Clstgn** usu present in the lo shth, with 1(–10) flt, not disarticulating; **rchl intnd** about as long as or longer than the adjacent flt; **lm** coriaceous, glab or scabrous near the apex, entire, unawned; **pal** smt slightly longer than the lm; **anth** 3, minute; **ov** glab; **car** more linear than in the aerial flt.

*Danthonia* is interpreted here as a genus of about 20 species that are native in Europe, North Africa, and the Americas. Of the eight species found in the *Manual* region, seven are native and one is an introduction that is now established.

NOTE: In the key and descriptions, lemma lengths do not include the apical teeth. Callus characteristics are best seen on the middle to upper florets in the spikelet.

1. Lemmas mucronate, not awned . . . . . . . . . . . . . . . . . . . . . . . . . . . . . . . . . . . . . . . . . . . . . . . . . . . 1. *D. decumbens*
1. Lemmas not mucronate, with a twisted, geniculate awn.
  2. Calluses of the middle florets from shorter to slightly longer than wide, convex abaxially; lemma bodies 2.5–6 mm long, the backs usually pilose, occasionally glabrous or sparsely pilose.
    3. Awns 10–17 mm; hairs of the lemma margins evidently increasing in length distally, longest hairs 2.5–4 mm long . . . . . . . . . . . . . . . . . . . . . . . . . . . . . . . . . . . . . . . . . . . . . . . . . . . . . . . . . . . . . . . . . . . . . . . 2. *D. sericea*
    3. Awns 5–10 mm; hairs of the lemma margins not evidently increasing in length distally, longest hairs 0.5–2 mm long.
      4. Lemma lobes 2–4 mm long, usually ⅔ or more as long as the lemma bodies, aristate; lower inflorescence branches usually flexible, divergent after anthesis; pedicels on the lowest inflorescence branch as long as or longer than the spikelets; leaves not curling at maturity . . . . . . . 3. *D. compressa*

[1]Kelly W. Allred  [2]Stephen J. Darbyshire

4. Lemma lobes 0.5–2 mm long, less than ⅔ as long as the lemma bodies, acute to aristate; inflorescence branches stiff, appressed to strongly ascending after anthesis; pedicels on the lowest inflorescence branch from shorter than to equaling the spikelets; blades usually becoming curled at maturity . . . . . . . . . . . . . . . . . . . . . . . . . . . . . . . . . . . . . . . . . . . 4. *D. spicata*
2. Calluses of the middle florets longer than wide, concave abaxially; lemma bodies 3–11 mm long, the backs usually glabrous or sparsely pilose (pilose in *D. parryi*).
  5. Lower inflorescence branches (pedicels if the inflorescence racemose) stiff, erect; pedicels from shorter than to as long as the spikelets.
    6. Spikelets 1(–3), if 2–3, the inflorescence a raceme; lemma bodies 5.5–11 mm; mature culms disarticulating at the nodes . . . . . . . . . . . . . . . . . . . . . . . . . . . . . . . . . . . . . . 8. *D. unispicata*
    6. Spikelets (4)5–10; lower inflorescence branches usually with 2–3 spikelets; lemma bodies 3–6 mm; mature culms not disarticulating at the nodes . . . . . . . . . . . . . . . . . . . . 5. *D. intermedia*
  5. Lower inflorescence branches (pedicels if the inflorescence racemose) flexible, slightly to strongly divergent; pedicels usually as long as or longer than the spikelets (sometimes shorter in *D. parryi*).
    7. Uppermost cauline blades usually strongly divergent or reflexed; inflorescences usually racemose; pedicels usually much longer than the spikelets and usually strongly divergent; lemmas glabrous or sparsely hairy over the back; mature culms disarticulating at the nodes . . . . . . 7. *D. californica*
    7. Uppermost cauline blades usually erect to ascending; inflorescences usually paniculate; pedicels shorter than to as long as the spikelets; lemmas pilose over the back, at least basally; mature culms not disarticulating at the nodes . . . . . . . . . . . . . . . . . . . . . . . . . . . . . . . . 6. *D. parryi*

## 1. Danthonia decumbens (L.) DC. MOUNTAIN HEATH-GRASS, DANTHONIE DÉCOMBANTE [p. 462, 543]

*Danthonia decumbens* grows throughout most of Europe, the Caucasus, and northern Turkey, and is now established on the west and east coasts of North America. It grows in heathlands, sandy or rocky meadows, clearings, and sometimes along roadsides.

## 2. Danthonia sericea Nutt. DOWNY OATGRASS [p. 462, 543]

*Danthonia sericea* is restricted to the eastern United States. It grows mostly on sand barrens and in open woods on dry soils. A less common form, with glabrous foliage and lemma backs, is found in bogs, seepage areas, and low moist areas adjacent to lakes and rivers and has been called *D. sericea* var. *epilis* (Scribn.) Gleason or *D. epilis* Scribn.

## 3. Danthonia compressa Austin FLATTENED OATGRASS, DANTHONIE COMPRIMÉE [p. 462, 543]

*Danthonia compressa* grows in open and semi-shaded areas, including meadows, open woods, and woodland openings. Although not a true pioneer species, it may sometimes occur as a weed in perennial crops. It is restricted to eastern North America.

## 4. Danthonia spicata (L.) P. Beauv. *ex* Roem. & Schult. POVERTY OATGRASS, DANTHONIE À ÉPI [p. 462, 543]

*Danthonia spicata* grows in dry rocky, sandy, or mineral soils, generally in open sunny places. Its range includes most of boreal and temperate North America and extends south into northeastern Mexico.

Phenotypically, *Danthonia spicata* is quite variable, expressing different growth forms under different conditions. Plants of shady or moist habitats often lack the distinctive curled or twisted blades usually found on plants growing in open habits. Such plants tend to have smaller spikelets and pilose foliage. The terminal inflorescence is usually primarily cleistogamous, but plants with chasmogamous inflorescences are found throughout the range of the species. Chasmogamous plants differ in having divergent inflorescence branches at anthesis, larger anthers, and well-developed lodicules.

## 5. Danthonia intermedia Vasey TIMBER OATGRASS, DANTHONIE INTERMÉDIAIRE [p. 462, 543]

*Danthonia intermedia* grows in boreal and alpine meadows, open woods, and on rocky slopes and northern plains. Its range extends from Kamchatka, Russia, to North America, south along the cordillera, and east, through boreal and alpine regions, to Quebec and Newfoundland and Labrador.

## 6. Danthonia parryi Scribn. PARRY'S OATGRASS [p. 462, 543]

*Danthonia parryi* is endemic to western North America and is often a major component of grasslands on the eastern foothills of the Rocky Mountains. It grows in open grassland, open woods, and rocky slopes, at elevations up to 4000 m. It rarely produces caryopses in the terminal inflorescences.

## 7. Danthonia californica Bol. CALIFORNIA OATGRASS [p. 462, 543]

*Danthonia californica* grows in prairies, meadows, and open woods. It has a disjunct distribution, one portion of its range being located in western North America, the other in Chile. An introduced population has been found at Mansfield, Massachusetts.

## 8. Danthonia unispicata (Thurb.) Munro *ex* Vasey ONE-SPIKE OATGRASS [p. 462, 543]

*Danthonia unispicata* is restricted to western North America, where it grows in prairies and meadows, on rocky slopes, and in dry openings up to timberline in the mountains. It differs from *D. californica* in its shorter stature, usually densely pilose foliage, short, erect pedicels, and the usually erect cauline leaf blades. It is closely related to *D. californica*.

## 19.03 SCHISMUS P. Beauv.[1]

Pl ann or short-lived per; tufted. **Clm** 2–30 cm, smt decumbent, glab. **Shth** open, usu shorter than the intnd, with tufts of 1.5–4 mm hairs on the mrg of the col; **aur** absent; **lig** memb, ciliate; **bld** flat or folded, becoming involute on drying. **Infl** tml, dense pan, 1–7 cm long, 0.5–2(3) cm wide, br 1–2 per nd; **dis** initially above the

[1]Elizabeth A. Kellogg

glm, glm and ped smt falling together later. **Spklt** with (4)5–7(10) flt. **Glm** subequal, exceeding or exceeded by the distal flt, 3–7-veined, mrg hyaline; **lm** 7–9-veined, mrg and intercostal regions usu pubescent, varying to glab, mrg hyaline, apc bifid or merely notched, sinuses smt mucronate, mucros to 1.5 mm; **pal** spatulate, memb, 2-veined, 2-keeled; **anth** 3, 0.2–0.5 mm. **Car** ovoid.

*Schismus* is a genus of five species that is native to Africa and Asia. Two species are established in the *Manual* region. In using the key and descriptions, the lowest floret in a spikelet should be examined. Succeeding florets tend to have shorter, less acute or acuminate lobes, and a shallower sinus.

1. Lower glumes equaling or exceeding the distal florets; lemma lobes longer than wide, acute to acuminate; paleas always shorter than the lemmas . . . . . . . . . . . . . . . . . . . . . . . . . . . . . . . . . . . . . . . . . 1. *S. arabicus*
1. Lower glumes exceeded by the distal florets; lemma lobes as wide as or wider than long, acute to obtuse; paleas of the lower florets in the spikelets as long as or longer than the lemmas . . . . . . . . . . . . . . . . . . . . . . 2. *S. barbatus*

**1. Schismus arabicus** Nees ARABIAN SCHISMUS [p. 462, 543]

*Schismus arabicus* is native to southwestern Asia, but it is now established in the southwestern United States, growing in open and disturbed sites.

**2. Schismus barbatus** (Loefl. *ex* L.) Thell. COMMON MEDITERRANEAN GRASS [p. 462, 543]

*Schismus barbatus* is native to Eurasia, but it is now established in the southwestern United States. It grows in sandy, disturbed sites along roadsides and fields and in dry riverbeds.

## 19.04 KARROOCHLOA Conert & Türpe[1]

**Pl** ann or per; csp, stln or rhz. **Clm** 4–40 cm. **Shth** open; **lig** of hairs; **bld** to 2 mm wide, flat, rolled, or involute. **Infl** tml, contracted pan, 0.5–6 cm. **Spklt** 4–8 mm, lat compressed, with 3–7 flt; **dis** above the glm. **Glm** 3.5–7 mm, subequal, equaling or exceeding the flt, 3–5(7)-veined; **cal** with lat tufts of hairs; **lm** pubescent and/or with fringes, rows, or tufts of hair, 9-veined, awned from between the 2 apical lobes or teeth, awns twisted and geniculate; **pal** 2-veined; **lod** 2, pubescent; **anth** 3; **ov** glab. **Car** about 1 mm; **hila** short.

*Karroochloa* is a southern African genus of four species.

**1. Karroochloa purpurea** (L. f.) Conert & Türpe [p. 462]

*Karroochloa purpurea* was grown in the grass garden of the University of California, Berkeley. There is no evidence that it has become established in the *Manual* region.

## 19.05 RYTIDOSPERMA[2]

**Pl** per; csp to somewhat spreading, smt shortly rhz. **Clm** (1.5)30–90(140) cm. **Shth** open, glab or hairy, apc with tufts of hair, these smt extending across the col; **lig** of hairs; **bld** persistent or disarticulating. **Infl** tml, rcm or pan. **Spklt** with 3–10 flt; **flt** bisx, tml flt rdcd; **dis** above the glm and between the flt. **Glm** (2)8–24 mm, subequal or equal, usu exceeding the flt, stiffly memb; **cal** with lat tufts of stiff hairs; **lm** ovate to lanceolate, with 2 complete or incomplete transverse rows of tufts of hairs, smt rdcd to mrgl tufts, 5–9-veined, apc bilobed, lobes usu at least as long as the body, acute, acuminate, or aristate, awned from between the lobes, awns longer than the lobes, twisted, usu geniculate; **lod** 2, fleshy, with hairs or glab. **Car** 1.2–3 mm, obovate to elliptic. **Clstgn** absent.

*Rytidosperma* is a genus of about 45 species that are native to Australia and Malesia. Several species have been cultivated in research plots or forage trials in North America. The species treated here have escaped cultivation and persisted in California and Oregon. Other species that have been grown experimentally in both the United States and Canada include *R. setaceum* (R. Br.) Connor & Edgar, and *R. tenuius* (Steud.) A. Hansen & P. Sunding. They are not known to have escaped or persisted in North America.

1. Lemmas with a transverse row of hairs 4–5 mm long below the sinus and additional hairs about 2.5 mm long evenly distributed below . . . . . . . . . . . . . . . . . . . . . . . . . . . . . . . . . . . . . . . . . . . . . . . 5. *R. richardsonii*
1. Lemmas with two transverse rows of hairs in addition to the callus hairs or with one transverse row and two marginal tufts below the sinus.
    2. Distal hairs on the lemmas forming a complete row of tufts of hairs.
        3. Lemma bodies 1.8–2.2(2.8) mm long . . . . . . . . . . . . . . . . . . . . . . . . . . . . . . . . . . . . . . . . . . 2. *R. biannulare*
        3. Lemma bodies 4–6 mm long . . . . . . . . . . . . . . . . . . . . . . . . . . . . . . . . . . . . . . . . . . . . . . . . 4. *R. caespitosum*
    2. Distal hairs on the lemmas restricted to marginal tufts or marginal tufts plus two dorsal tufts.

[1]Stephen J. Darbyshire  [2]Darbyshire and Henry E. Connor

4. Lower row of lemma hairs reaching upper row; calluses (0.6)09.–1.5 mm long; callus hairs
usually not overlapping lower lemma hairs . . . . . . . . . . . . . . . . . . . . . . . . . . . . . . . . . . . . . . . . . . . 3. *R. racemosum*
4. Lower row of lemma hairs not reaching upper row; calluses 0.–0.8(1) mm long; callus hairs
usually overlapping lower lemma hairs . . . . . . . . . . . . . . . . . . . . . . . . . . . . . . . . . . . . . . . . . . . 1. *R. penicillatum*

## 1. Rytidosperma penicillatum (Labill.) Connor & Edgar HAIRY DANTHONIA, HAIRY OATGRASS, POVERTY GRASS [p. 463, 543]

*Rytidosperma penicillatum* is endemic to Australia and has been introduced to New Zealand as well as North America. Although considered a poor quality forage, it was introduced and grown experimentally in several states under the name *Danthonia pilosa* R. Br. [= *R. pilosa* (R. Br.) Connor & Edgar]. It has become well-established in northern California and southwestern Oregon, mainly in coastal areas. Since it does well on dry, nutrient depleted soils and competes well with more desirable species, it is considered a troublesome pest.

## 2. Rytidosperma biannulare (Zotov) Connor & Edgar [p. 463, 543]

*Rytidosperma biannulare* is endemic to New Zealand. As early as 1905, it was grown experimentally under the name *Danthonia semiannularis* (Labill.) R. Br. in several states. Only a few specimens document its persistence in Oregon and California.

## 3. Rytidosperma racemosum (R. Br.) Connor & Edgar [p. 463, 543]

*Rytidosperma racemosum* is endemic to Australia and has been introduced to New Zealand. Grown experimentally in several places in North America, including Berkeley, California, it seems to have become established in only a few places in central California.

## 4. Rytidosperma caespitosum (Gaudich.) Connor & Edgar [p. 463, 543]

*Rytidosperma caespitosum* is native to Australia and naturalized in New Zealand. It has been found as an escape in Alameda and San Mateo counties, California.

## 5. Rytidosperma richardsonii (Cashmore) Connor & Edgar [p. 463, 544]

*Rytidosperma richardsonii* is native to Australia. It has been found as an escape in Alameda and San Mateo counties, California.

## 19.06 TRIBOLIUM Desv.[1]

Pl ann or per; csp, smt stln, occ with rhz. Clm 2–60 cm. Shth open, throats glab or long-ciliate; lig of hairs, or memb and ciliate, or ciliolate; bld flat or rolled, glab or villous. Infl tml, smt also ax, pan, spikes, or rcm. Spklt with 2–5(10) flt, distal flt rdcd; dis above the glm and between the flt. Glm exceeding or exceeded by the flt, hrb, 3–5-veined, glab or scabridulous, smt with stiff, papillose-based hairs, glm apc acute or acuminate; cal glab; lm hrb, with acute or clavate hairs in mrgl rows or variously scattered, lm apc unlobed, acute to long-acuminate; pal smt with tufts of hairs on the mrg; lod 2, obtriangular, glab or with bristles, 1–3-veined, lod apc at least as thick as the base; anth 3, 0.3–3.1 mm. Car 0.7–2.2 mm long, 0.4–1.1 mm wide; prcp poorly separable, dull, smooth or rugulose, glab; emb ⅓–½ as long as the car; hila punctiform.

*Tribolium* is a southern African genus of 10 species. It is unusual in the tribe in having unlobed lemmas, but has the haustorial synergids, bilobed or bi-awned prophylls, and stalked ovaries characteristic of the tribe.

## 1. Tribolium obliterum (Hemsley) Renvoize [p. 463, 544]

*Tribolium obliterum* is native to Cape Province, South Africa, where it usually grows in gravelly, well-drained soils at elevations below 600 m. It has been introduced into Australia and St. Helena and has been discovered in a roadside ditch near Fort Ord, Monterey County, California. It appears to be naturalized there.

# 8. ARISTIDOIDEAE Caro

The subfamily *Aristidoideae* includes one tribe, the *Aristideae*.

# 20. ARISTIDEAE C.E. Hubb.[2]

Pl ann or per; usu csp. Clm ann, erect, solid or hollow, usu unbrchd. Lvs distichous; shth usu open; aur absent; abx lig absent or of hairs; adx lig memb and ciliate or of hairs; bld without psdpet; mesophyll cells radiate or non-radiate; adx palisade layer absent; fusoid cells absent; arm cells absent; Kranz anatomy absent or present, when present, with 1 or 2 parenchyma shth; midribs simple; adx bulliform cells present; stomatal subsidiary cells dome-shaped or triangular; bicellular microhairs present, with long, slender, thin-walled tml cells. Infl tml, not lfy, usu pan, smt spikes or rcm. Spklt bisx, with 1 flt; rchl extension absent; dis above the glm. Glm 2, usu longer than the flt, usu acute or acuminate; flt terete or lat compressed, with well-developed cal; lm 1- or 3-veined, more or less coriaceous, with a germination flap, lm mrg overlapping at maturity and concealing the pal, apc evidently 3-awned; awn bases often forming a column, lat awns occ rdcd or absent; pal less than ½ as long as the lm; lod usu present, 2, free, memb, glab, heavily vascularized; anth 1–3; ov

[1]Gerrit Davidse  [2]Grass Phylogeny Working Group and Kelly W. Allred

glab; **haustorial synergids** absent; **sty** 2, free to the base but close. **Car** usu fusiform, falling with the lm and pal attached; **hila** short or long, linear; **endosperm** hard, without lipid; **stch g** compound; **emb** small or large relative to the car; **epiblasts** absent; **scutellar cleft** present or absent; **mesocotyl intnd** elongated; **embryonic lf mrg** meeting.

The tribe *Aristideae* has three genera and 300–350 species, and is primarily pan-tropical in its distribution. Its members are usually readily recognized by their terete, 3-awned lemmas with overlapping margins. *Aristida*, which has many more species than the other two genera combined, is the only genus found in the Americas.

## 20.01 ARISTIDA L.[1]

**Pl** usu per; hrb, usu csp, occ rhz. **Clm** 10–150 cm, not wd, smt brchd above the base; **intnd** usu pith-filled, smt hollow. **Lvs** smt predominantly bas, smt predominantly cauline; **shth** open; **aur** lacking; **lig** of hairs or very shortly memb and long-ciliate, the 2 types generally indistinguishable. **Infl** tml, usu pan, smt rcm, occ spikes; **pri br** without ax pulvini and usu appressed to ascending, or with ax pulvini and ascending to strongly divergent or divaricate. **Spklt** with 1 flt; **rchl** not prolonged beyond the flt; **dis** above the glm. **Glm** often longer than the flt, thin, usu 1–3-veined, acute to acuminate; **flt** terete or weakly lat compressed; **cal** well-developed, hirsute; **lm** fusiform, 3-veined, convolute, usu glab or scabridulous, usu enclosing the pal at maturity, usu with 3 tml awns, lat awns rdcd or obsolete in some species, lm apc smt narrowed to a straight or twisted beak below the awns; **awns** ascending to spreading, usu straight, bases smt twisted together into a column or the bases of the individual awns coiled, twisted, or otherwise contorted, occ disarticulating at maturity; **pal** shorter than the lm, 2-veined, occ absent; **anth** 1 or 3. **Car** fusiform; **hila** linear.

*Aristida* is a tropical to warm-temperate genus of 250–300 species. It grows throughout the world in dry grasslands and savannahs, sandy woodlands, arid deserts, and open, weedy habitats and on rocky slopes and mesas. All 29 species in this treatment are native to the *Manual* region.

NOTE: Lemma lengths are measured from the base of the callus to the divergence of the awns.

1. Lower glumes 3–7-veined.
    2. Awns nearly equal, the lateral awns 8–66 mm long and at least ¾ as long as the central awns . . . . . . . . . 13. *A. oligantha*
    2. Awns markedly unequal, the lateral awns 1–4 mm long, no more than ½ as long as the central awns, sometimes absent.
        3. Plants annual; inflorescences 5–12 cm long, 2–4 cm wide . . . . . . . . . . . . . . . . . . . . . . . 12. *A. ramosissima* (in part)
        3. Plants perennial; inflorescences 10–30 cm long, 4–26 cm wide . . . . . . . . . . . . . . . . . . . . 7. *A. schiedeana* (in part)
1. Lower glumes 1–2(3)-veined.
    4. Central awns spirally coiled at the base.
        5. Lateral awns 1–4 mm long, erect . . . . . . . . . . . . . . . . . . . . . . . . . . . . . . . . . . . . . . . 15. *A. dichotoma*
        5. Lateral awns 5–13 mm long, spreading . . . . . . . . . . . . . . . . . . . . . . . . . . . . . . . . . . . . 14. *A. basiramea*
    4. Central awns straight to curved, sometimes loosely contorted but not spirally coiled, at the base.
        6. Lateral awns markedly reduced, usually ⅓ or less as long as the central awns.
            7. Panicles 1–6 cm wide, the branches erect-appressed to strongly ascending, without axillary pulvini or the pulvini only weakly developed.
                8. Plants annual; culms often highly branched above the base.
                    9. Awns flattened at the base . . . . . . . . . . . . . . . . . . . . . . . . . . . . . . 17. *A. adscensionis* (in part)
                    9. Awns terete at the base.
                        10. Lemmas 2.5–10 mm long; central awns curving up to 100° at the base . . . . . . 16. *A. longespica* (in part)
                        10. Lemmas 8–22 mm long; central awns with a semicircular bend at the base . . . . 12. *A. ramosissima* (in part)
                8. Plants perennial; culms rarely branched above the base.
                    11. Collars densely pilose, the hairs 1–3 mm, often densely tangled and deflexed; blades usually tightly involute, about 0.5 mm in diameter . . . . . . . . . . . . . . . . . . . . . 11. *A. gypsophila* (in part)
                    11. Collars mostly glabrous or with straight hairs, often with long hairs at the sides; blades usually flat to loosely involute, sometimes tightly involute.
                        12. Lateral awns absent; panicle branches spikelet-bearing to the base; plants of the Florida keys . . . . . . . . . . . . . . . . . . . . . . . . . . . . . . . . . . . . . . . 5. *A. floridana* (in part)
                        12. Lateral awns usually present, varying from much shorter than to equaling the central awns; panicle branches sometimes naked near the base; plants rarely found east of the Mississippi and not known at all from Florida.
                            13. Primary panicle branches 3–6 cm long; lateral awns (1)8–140 mm long . . . 19. *A. purpurea* (in part)
                            13. Primary panicle branches 6–16 cm long; lateral awns absent or to 1(3) mm long . . . . . . . . . . . . . . . . . . . . . . . . . . . . . . . . . . . . . . . 7. *A. schiedeana* (in part)

[1]Kelly W. Allred

7. Panicles 6–45 cm wide, at least the lower branches spreading and having well-developed axillary pulvini.
  14. Lateral awns absent or no more than 3 mm long.
    15. Central awns often deflexed at a sharp angle when mature; lemma apices often twisted at maturity.
      16. Blades usually flat, sometimes folded, 1–2 mm wide; plants of juniper, oak, or pine woodlands . . . . . . . . . . . . . . . . . . . . . . . . . . . . . . . . . . . . . . . . . . . 7. *A. schiedeana* (in part)
      16. Blades usually tightly involute, about 0.5 mm in diameter; plants of thorn-scrub deserts . . . . . . . . . . . . . . . . . . . . . . . . . . . . . . . . . . . . . . . . . . . . . . . . 11. *A. gypsophila* (in part)
    15. Central awns usually straight or arcuate; lemma apices not twisted.
      17. Panicle branches spikelet-bearing from the base; lower glumes longer than the upper glumes . . . . . . . . . . . . . . . . . . . . . . . . . . . . . . . . . . . . . . . . . . . . . . 5. *A. floridana* (in part)
      17. Panicle branches usually naked at the base; lower glumes about equal to the upper glumes . . . . . . . . . . . . . . . . . . . . . . . . . . . . . . . . . . . . . . . . . . . . . . . 6. *A. ternipes* (in part)
  14. Lateral awns 3–23 mm long.
    18. Anthers 0.8–1 mm long.
      19. Spikelets usually divergent and the pedicels with axillary pulvini; secondary branches usually absent; primary branches 2–6 cm long; lemma apices with 0–2 twists when mature . . . . . . . . . . . . . . . . . . . . . . . . . . . . . . . . . . . . 9. *A. havardii* (in part)
      19. Spikelets usually appressed and the pedicels without axillary pulvini; secondary branches usually well-developed; primary branches 5–13 cm long; lemma apices with 4 or more twists when mature . . . . . . . . . . . . . . . . . . . . . . . . 8. *A. divaricata* (in part)
    18. Anthers 1.2–3 mm long.
      20. Collars glabrous or strigillose; blades with scattered hairs 1.5–3 mm long above the ligule on the adaxial surface; lower glumes about equal to or slightly shorter than the upper glumes . . . . . . . . . . . . . . . . . . . . . . . . . . . . . . . . . 6. *A. ternipes* (in part)
      20. Collars pubescent, with hairs 0.2–0.8 mm long; blades glabrous, sometimes scabridulous, above the ligule on the adaxial surface; lower glumes slightly longer than the upper glumes . . . . . . . . . . . . . . . . . . . . . . . . . . . . . . . . 4. *A. patula* (in part)
6. Lateral awns well-developed, usually at least ½ as long as the central awns.
  21. Blades tightly involute, the adaxial surfaces densely scabrous or densely short-pubescent . . . . . . . . 21. *A. stricta*
  21. Blades flat or folded and lax, or, if involute, the adaxial surfaces neither densely scabrous nor densely short pubescent.
    22. Rachis nodes and leaf sheaths usually lanose or floccose, sheaths occasionally glabrous . . . . . . 25. *A. lanosa*
    22. Rachis nodes glabrous, scabrous, or with straight hairs; leaf sheaths glabrous, pilose, or floccose.
      23. Junction of the lemma and awns evident; awns disarticulating at maturity.
        24. Plants perennial.
          25. Culms 45–100 cm tall; culms unbranched or sparingly branched; blades 12–28 cm long . . . . . . . . . . . . . . . . . . . . . . . . . . . . . . . . . . . . . 18. *A. spiciformis* (in part)
          25. Culms 10–40 cm tall; culms much branched; blades usually less than 6 cm long . . . . . . . . . . . . . . . . . . . . . . . . . . . . . . . . . . . . . . . . . . 3. *A. californica* (in part)
        24. Plants annual.
          26. Awns divergent but not arcuate or entwined above the column; cauline internodes pubescent or glabrous . . . . . . . . . . . . . . . . . . . . . . . . . . . . . . 3. *A. californica*
          26. Awns strongly arcuate, often entwined above the column or no column present; cauline internodes glabrous.
            27. Glumes 10–17 mm long; lemmas beaked, the beak 2–7 mm long; awns not forming a column; calluses 1–2.5 mm long . . . . . . . . . . . . . . . . . . . . . 1. *A. desmantha*
            27. Glumes 20–30 mm long; lemmas not beaked; awns forming a column 8–15 mm long; calluses 3–4 mm long . . . . . . . . . . . . . . . . . . . 2. *A. tuberculosa*
      23. Junction of the lemma and awns not evident; awns not disarticulating at maturity.
        28. Lemmas terminating in a beak 7–30 mm long; upper glumes awned, the awns 10–12 mm long . . . . . . . . . . . . . . . . . . . . . . . . . . . . . . . . . . . . . 18. *A. spiciformis* (in part)
        28. Lemmas not beaked or with a beak less than 7 mm long; upper glumes unawned or with an awn to 6 mm long.
          29. At least the lower primary panicle branches divergent and with axillary pulvini.
            30. Lateral awns about ½ as thick as the central awns . . . . . . . . . . . . . . 4. *A. patula* (in part)
            30. Lateral awns nearly as thick as the central awns.

31. Panicles narrow and contracted above, usually only the lower
1–2 branches spreading and with a pulvinus; lemma apices
0.2–0.3 mm wide ................................. 19. *A. purpurea* (in part)
31. Almost all panicle branches spreading and with axillary pulvini;
lemma apices 0.1–0.2 mm wide.
  32. Anthers 0.8–1 mm long.
    33. Spikelets usually divergent and the pedicels with
axillary pulvini; secondary branches absent or nearly
so; primary branches 2–6 cm long; lemma apices
straight or with 1 or 2 twists ................. 9. *A. havardii* (in part)
    33. Spikelets usually appressed and the pedicels without
axillary pulvini; secondary branches usually well-
developed; primary branches 5–13 cm long; lemma
apices with 4 or more twists at maturity .......... 8. *A. divaricata* (in part)
  32. Anthers 1–3 mm long.
    34. Base of the blades with scattered hairs 1.5–3 mm long
on the adaxial surfaces ..................... 6. *A. ternipes* (in part)
    34. Base of the blades glabrous or puberulent on the
adaxial surface, the hairs, if present, less than 0.5 mm
long.
      35. Glumes reddish, the lower glumes often shorter
than the upper glumes; awns ascending to
divaricate, (8)13–140 mm long; terminal spikelets
usually appressed and without axillary pulvini .. 19. *A. purpurea* (in part)
      35. Glumes brownish, equal or unequal; awns
spreading to horizontal, 5–15 mm long;
terminal spikelets often spreading from axillary
pulvini ..................................... 10. *A. pansa*
29. Lower primary panicle branches (pedicels in racemose species) appressed,
without axillary pulvini.
  36. Plants with well-developed rhizomes; basal sheaths shredding into
threadlike segments at maturity .......................... 22. *A. rhizomophora*
  36. Plants tufted, without rhizomes; basal sheaths not fibrous, not
shredding into threadlike segments even when old.
    37. Lower inflorescence nodes with only 1 spikelet; inflorescences
spicate or racemose ........................................ 23. *A. mohrii*
    37. Lower inflorescence nodes with 2 or more spikelets;
inflorescences racemose or paniculate.
      38. Lower glumes usually ⅓–¾ as long as the upper glumes.
        39. Plants annual ......................... 17. *A. adscensionis* (in part)
        39. Plants perennial.
          40. Lemma awns 8–15 mm long; lemmas 5–7 mm long
........................................ 29. *A. gyrans* (in part)
          40. Lemma awns (8)15–140 mm long; lemmas 6–16
mm long ........................... 19. *A. purpurea* (in part)
      38. Lower glumes usually more than ¾ as long as the upper
glumes.
        41. Plants annual.
          42. Awns flat at the base ................. 17. *A. adscensionis* (in part)
          42. Awns terete at the base ................. 16. *A. longespica* (in part)
        41. Plants perennial.
          43. Lemma apices prominently twisted for 3–6 mm;
blades usually curled at maturity; leaves forming a
basal tuft ................................. 20. *A. arizonica*
          43. Lemma apices straight or only slightly twisted;
blades usually not curled at maturity; leaves
variously distributed.
            44. Lower glumes prominently 2-keeled,
(7.5)9–13 mm long; central awns 15–40 mm
long ..................................... 26. *A. palustris*

44. Lower glumes usually 1-keeled, if 2-keeled, 5–9
   mm long; central awns 8–25 mm long.
   [⇐ revert to left, Ed.]
45. Central awns about twice as thick as the lateral awns, divergent to arcuate-reflexed.
   46. All 3 awns divergent to reflexed and contorted at the base; lower rachis nodes usually associated
     with 2 spikelets (occasionally 1 or 3), 1 pedicellate and 1 sessile . . . . . . . . . . . . . . . . . . . . . . . . 24. *A. simpliciflora*
   46. Lateral awns usually erect to ascending and not contorted at the base; lower rachis nodes
     usually associated with more than 2 spikelets, pedicellate to subsessile . . . . . . . . . . . 27. *A. purpurascens* (in part)
45. Central awns about the same thickness as the lateral awns, erect to spreading.
   47. Lower glumes 1–4 mm longer than the upper glumes . . . . . . . . . . . . . . . . . . . . . . . . 27. *A. purpurascens* (in part)
   47. Lower glumes from shorter than to 1 mm longer than the upper glumes.
      48. Culms usually 3–6 mm thick at the base; primary panicle branches 4–20 cm long; lower
         glumes 1-veined . . . . . . . . . . . . . . . . . . . . . . . . . . . . . . . . . . . . . . . . . . . . . . . . . 28. *A. condensata*
      48. Culms usually 1–4 mm thick at the base; primary panicle branches 1–5 cm long; lower
         glumes 1–2-veined.
         49. Calluses 0.4–0.8 mm long . . . . . . . . . . . . . . . . . . . . . . . . . . . . . . . 27. *A. purpurascens* (in part)
         49. Calluses 1–2 mm long . . . . . . . . . . . . . . . . . . . . . . . . . . . . . . . . . . . . . 29. *A. gyrans* (in part)

## 1. Aristida desmantha Trin. & Rupr. CURLY THREEAWN [p. 463, <u>544</u>]

*Aristida desmantha* grows in sandy fields, dry pine woods, and waste places in the United States. It has shorter glumes, calluses, and awns than *A. tuberculosa*.

## 2. Aristida tuberculosa Nutt. SEASIDE THREEAWN [p. 463, <u>544</u>]

*Aristida tuberculosa* grows in sandy fields, hills, pinelands, and disturbed areas. Along the Atlantic coastal fringe, it grows on maritime dunes; inland it is associated with xeric pine-oak sandhills. Like *A. desmantha*, *A. tuberculosa* is restricted to the United States, but has longer glumes, calluses, and awns.

## 3. Aristida californica Thurb. MOJAVE THREEAWN [p. 463, <u>544</u>]

The range of both varieties of *Aristida californica* extends from the southwestern United States into northwestern Mexico.

1. Cauline internodes puberulent to nearly lanose . . var. *californica*
1. Cauline internodes glabrous . . . . . . . . . . . . . . . . . var. *glabrata*

### Aristida californica Thurb. var. californica [p. 463]

*Aristida californica* var. *californica* grows in dry, sandy plains, dunes, and flats of the Sonoran and Mojave deserts at elevations of 0–700 m.

### Aristida californica var. glabrata Vasey [p. 463]

*Aristida californica* var. *glabrata* grows in sandy to rocky soils of desert grassland and desert thorn-scrub communities in the Sonoran and Mojave deserts at elevations of 500–1400 m, generally to the east of var. *californica*.

## 4. Aristida patula Chapm. ex Nash TALL THREEAWN [p. 463, <u>544</u>]

*Aristida patula* grows in sandy fields, low pinelands, and roadsides. It is endemic to Florida.

## 5. Aristida floridana (Chapm.) Vasey FLORIDA THREEAWN [p. 463, <u>544</u>]

*Aristida floridana* grows in waste places, along roadsides, and on railroad embankments. It is rare in the United States, being known only from Key West and Ramrod Key, Florida. It is more common in the Yucatan Peninsula, Mexico.

## 6. Aristida ternipes Cav. [p. 464, <u>544</u>]

1. Lateral awns 2–23 mm long . . . . . . . . . . . . . . . . . . var. *gentilis*
1. Lateral awns 0–2 mm long . . . . . . . . . . . . . . . . . . var. *ternipes*

### Aristida ternipes var. gentilis (Henrard) Allred HOOK THREEAWN [p. 464]

*Aristida ternipes* var. *gentilis* grows on dry slopes and plains and along roadsides from Californica to Texas and south through Mexico to Guatemala.

### Aristida ternipes Cav. var. ternipes SPIDERGRASS [p. 464]

*Aristida ternipes* var. *ternipes*, like var. *gentilis*, grows on dry slopes and plains and along roadsides, but its range extends from Arizona to Texas south through Mexico and Central America to South America.

## 7. Aristida schiedeana Trin. & Rupr. SINGLE THREEAWN [p. 464, <u>544</u>]

*Aristida schiedeana* grows on rocky slopes and plains, generally in pinyon-juniper, oak, or ponderosa pine communities. Plants from the southwestern United States and northern Mexico belong to **A. schiedeana** var. **orcuttiana** (Vasey) Allred & Valdés-Reyna, in which the lower glumes are usually glabrous and longer than the upper glumes, and the collar and throat are usually glabrous. **Aristida schiedeana** var. **schiedeana** grows in Mexico, Guatemala, Nicaragua, and Honduras, and has puberulent, equal glumes and pilose collars and throats.

## 8. Aristida divaricata Humb. & Bonpl. ex Willd. POVERTY GRASS [p. 464, <u>544</u>]

*Aristida divaricata* grows on dry hills and plains, especially in pinyon-juniper-grassland zones, from the southwestern United States through Mexico to Guatemala. It occasionally intergrades with *A. havardii*, but that species has lemma beaks that are straight or have only 1–2 twists, shorter primary branches, usually no secondary branches, and pedicels that more frequently have axillary pulvini so the spikelets are more frequently divergent than in *A. divaricata*.

## 9. Aristida havardii Vasey HAVARD'S THREEAWN [p. 464, <u>544</u>]

*Aristida havardii* grows on dry hills and plains in desert grassland to pinyon-juniper zones, and in sandy to rocky ground from the southwestern United States to northern Mexico. It occasionally intergrades with *A. divaricata*, but that species differs in having more twisted lemma beaks, longer primary branches, well-developed secondary branches, and, usually, appressed spikelets.

## 10. Aristida pansa Wooton & Standl. WOOTON'S THREEAWN [p. 464, <u>544</u>]

*Aristida pansa* grows in desert scrub, commonly in the Chihuahuan Desert of the southwestern United States and Mexico, but its ecological range extends into the lower juniper zone and its geographic range to southern Mexico. It prefers cobbly to sandy, often gypsiferous soil. It is very similar to the single-awned *A. gypsophila*, but it has also been confused with *A. purpurea* var. *perplexa*, which differs in having reddish glumes of unequal length and longer ascending awns.

## 11. Aristida gypsophila Beetle GYPSUM THREEAWN [p. 464, 544]

*Aristida gypsophila* grows on rocky limestone or gypsum hills in thorn-scrub communities of the Chihuahuan Desert, almost always in the protection of shrubs. It is similar to *A. pansa*, which differs in having three well-developed awns and being, usually, shorter in stature. Both species have involute blades with a characteristic tuft of cobwebby hairs at the collar. Plants from the United States have spreading primary branches with axillary pulvini and appressed spikelets. Mexican plants sometimes have primary branches with no axillary pulvini.

## 12. Aristida ramosissima Engelm. *ex* A. Gray S-CURVE THREEAWN [p. 464, 544]

*Aristida ramosissima* grows in open, dry, sterile ground, fallow fields, and roadsides. It is restricted to the United States.

## 13. Aristida oligantha Michx. OLDFIELD THREEAWN [p. 464, 544]

*Aristida oligantha* grows in waste places, dry fields, roadsides, along railroads, and in burned areas, usually in sandy soil. It has been reported from Coahuila, Mexico, but is otherwise unknown outside southern Canada and the United States.

## 14. Aristida basiramea Engelm. *ex* Vasey FORKTIP THREEAWN, ARISTIDE À RAMEAUX BASILAIRES [p. 464, 544]

*Aristida basiramea* grows in open, sandy, often barren ground in southern Ontario and in the United States. It differs from *A. dichotoma* in its longer lateral awns.

## 15. Aristida dichotoma Michx. CHURCHMOUSE THREEAWN [p. 464, 544]

*Aristida dichotoma* grows in sandy fields and clearings, disturbed sites and sterile ground, pine woods, and on granitic outcrops of the United States and southern Ontario. It differs from *A. basiramea* in its shorter lateral awns.

1. Glumes unequal; lemmas smooth or scabridulous, 6–11 mm long . . . . . . . . . . . . . . . . . . . . . . . . . var. *curtissii*
1. Glumes equal or subequal; lemmas sparsely appressed-pubescent, 3–8 mm long . . . . . . . . . . . var. *dichotoma*

### Aristida dichotoma var. curtissii A. Gray [p. 464]

*A. dichotoma* var. *curtissii* has similar ecological preferences and extensive overlap with var. *dichotoma*, but is somewhat more western in its distribution.

### Aristida dichotoma Michx. var. dichotoma [p. 464]

*A. dichotoma* var. *dichotoma* has similar ecological preferences and extensive overlap with var. *curtissii*, but var. *dichotoma* is somewhat more eastern in its distribution.

## 16. Aristida longespica Poir. [p. 464, 544]

*Aristida longespica* grows along roadsides and in waste places, sandy fields, and clearings in pine and oak woods of southern Ontario and the eastern and central United States.

1. Central awns 8–27 mm long, lateral awns usually 6–18 mm long . . . . . . . . . . . . . . . . . . . . . . . . var. *geniculata*
1. Central awns 1–14 mm long and/or lateral awns usually 0–5 mm long . . . . . . . . . . . . . . . . . . . . . .var. *longespica*

### Aristida longespica var. geniculata (Raf.) Fernald KEARNEY'S THREEAWN [p. 464]

*Aristida longespica* var. *geniculata* has a similar geographic range to var. *longespica*. The two varieties are often found growing together.

### Aristida longespica Poir. var. longespica SLIMSPIKE THREEAWN [p. 464]

*Aristida longespica* var. *longespica* has a similar geographic range to var. *geniculata*. The two varieties often grow together.

## 17. Aristida adscensionis L. SIXWEEKS THREEAWN [p. 464, 544]

*Aristida adscensionis* grows in waste ground, along roadsides, and on degraded rangelands and dry hillsides, often in sandy soils. It is associated with woodland, prairie, and desert shrub communities. Its range extends from the United States south through Mexico and Central America to South America. It is highly variable in height, panicle size, and awn development, but most of the variation appears to be environmentally induced.

## 18. Aristida spiciformis Elliott BOTTLEBRUSH THREEAWN [p. 464, 544]

*Aristida spiciformis* grows in pine savannahs, pine flatwoods, pine-oak sandhills, and oak woods, frequently being associated with *Pinus palustris*. It is a primary fire carrier in these habitats. Its range includes Cuba and Puerto Rico as well as the southeastern United States.

## 19. Aristida purpurea Nutt. [p. 465, 544]

*Aristida purpurea* is composed of several intergrading varieties.

1. Lower or all primary panicle branches stiff, divergent to divaricate from the base, with axillary pulvini; awns 13–30 mm.
    2. Lower glumes ½–⅔ as long as the upper glumes . . . . . . . . . . . . . . . . . . . . . . . . . . . . . . . . . . . var. *perplexa*
    2. Lower glumes from ¾ as long as to equaling the upper glumes . . . . . . . . . . . . . . . . . . . . . . . var. *parishii*
1. Primary panicle branches appressed or ascending at the base, sometimes drooping distally, without axillary pulvini; awns 8–140 mm.
    3. Awns 35–140 mm long.
        4. Lemmas apices 0.1–0.3 mm wide; awns 0.1–0.2(0.3) mm wide at the base, 35–60 mm long; upper glumes usually shorter than 16 mm . . . . . . . . . . . . . . . . . . . . . . . . . . .var. *purpurea*
        4. Lemma apices 0.3–0.8 mm wide; awns 0.2–0.5 mm wide at the base, 40–140 cm long; upper glumes 14–25 mm long . . . . . . . .var. *longiseta*
    3. Awns 8–35 mm long.
        5. Lemma apices 0.1–0.3 mm wide distally; awns 0.1–0.3 mm wide at the base.
            6. All or most of the panicle branches straight (lower branches sometimes lax); pedicels straight, appressed to ascending . . var. *nealleyi*
            6. All or most of the panicle branches and pedicels drooping to sinuous distally . . . var. *purpurea*
        5. Lemma apices 0.2–0.3 mm wide; awns stout, 0.2–0.3 mm wide at the base.
            7. Mature panicle branches and pedicels flexible, lax or drooping distally . . . . . . . . var. *purpurea*
            7. Mature panicle branches and pedicels usually stiff, straight.
                8. Panicles usually 3–15 cm long; blades 4–10 cm long . . . . . . . . . . . . . . . . var. *fendleriana*
                8. Panicles usually 15–30 cm long; blades 10–25 cm long.

9. Glumes and lemmas reddish or dark-colored at anthesis or earlier (fading to stramineous), usually in marked contrast with the current foliage; panicles dense, the lower nodes with 8–18 spikelets; flowering March to May, after winter rains . . . . . . . . . . . . . . . . . var. *parishii*

9. Glumes and lemmas tan to brown (also fading to stramineous), giving the panicle a brownish appearance; old growth gray-green, not in marked contrast with the current foliage; panicles less dense, the lower nodes with 2–10 spikelets . . . . var. *wrightii*

## Aristida purpurea var. fendleriana (Steud.) Vasey

FENDLER'S THREEAWN [p. *465*]

*Aristida purpurea* var. *fendleriana* grows on open slopes, hills, and sandy flats, at low to medium elevations, from the western United States into northern Mexico. It differs from var. *longiseta* in having narrower lemma apices and thinner, shorter awns.

## Aristida purpurea var. longiseta (Steud.) Vasey RED

THREEAWN [p. *465*]

*Aristida purpurea* var. *longiseta* grows on sandy or rocky slopes and plains, and in barren soils of disturbed ground from western Canada to northern Mexico. It is the most variable variety of *A. purpurea*, ranging from short plants with basal leaves and short panicles suggestive of var. *fendleriana*, to tall plants with long cauline leaves and long, drooping panicles resembling var. *purpurea*. The length of its glumes, width of its lemma apices, and the length and thickness of its awns distinguish it from all the other varieties.

## Aristida purpurea var. nealleyi (Vasey) Allred

NEALLEY'S THREEAWN [p. *465*]

*Aristida purpurea* var. *nealleyi* grows on dry slopes and plains at lower elevations than the other varieties, frequently in desert grassland vegetation. Its range extends from the southwestern United States into Mexico. Although var. *nealleyi* is more distinct than the other varieties, having tight tufts of foliage exceeded by narrow, straw-colored panicles, it grades into var. *purpurea*, and the panicles resemble those of var. *wrightii*. It differs from *A. arizonica* in having involute blades that are usually straight, and shorter awns.

## Aristida purpurea var. parishii (Hitchc.) Allred PARISH'S

THREEAWN [p. *465*]

*Aristida purpurea* var. *parishii* grows on sandy plains and hills of the southwestern United States and Baja California, Mexico. In many respects it is intermediate between *A. purpurea* and other species of *Aristida* with spreading panicle branches, especially *A. ternipes* var. *gentilis*. Its spikelets are indistinguishable from those of var. *wrightii*, but var. *parishii* frequently has axillary pulvini associated with the lower branches. The two also differ in their phenology: var. *parishii* flowers from March through May in response to winter rains, whereas var. *wrightii* flowers from May through October in response to summer rains.

## Aristida purpurea var. perplexa Allred & Valdés-Reyna

JORNADA THREEAWN [p. *465*]

*Aristida purpurea* var. *perplexa* grows in sandy to rocky plains and on mesas in desert grassland and scrub communities, often in calcareous soils, in both the *Manual* region and Mexico. It is sometimes confused with *A. pansa*, which differs in having cobwebby hairs at the collar, equal glumes, and shorter awns.

## Aristida purpurea Nutt. var. purpurea PURPLE

THREEAWN [p. *465*]

*Aristida purpurea* var. *purpurea* grows in sandy to clay soils, along right of ways, or on dry slopes and mesas. Its range extends from the *Manual* region to Mexico and Cuba. As treated here var. *purpurea* is a broadly defined taxon, incorporating slender plants with small spikelets that used to be referred to as *A. roemeriana* Scheele, but also occasional plants with somewhat flexible branches that are intermediate to var. *wrightii* and var. *nealleyi*.

## Aristida purpurea var. wrightii (Nash) Allred WRIGHT'S

THREEAWN [p. *465*]

*Aristida purpurea* var. *wrightii* grows on sandy to gravelly hills and flats from the southwestern United States to southern Mexico. It is the most robust variety of *A. purpurea*, and has dark, stout awns and long panicles. It may be confused with var. *nealleyi*, which has narrower lemmas and awns and a light-colored panicle, but it also intergrades with var. *purpurea* and var. *parishii*. *Aristida purpurea* forma **brownii** (Warnock) Allred & Valdés-Reyna refers to plants with short central awns and lateral awns that are only 1–3 mm long.

## 20. Aristida arizonica Vasey ARIZONA THREEAWN [p. *465*, <u>*544*</u>]

*Aristida arizonica* grows in pine, pine-oak, and pinyon-juniper woodlands from the southwestern United States to southern Mexico. It differs from *A. purpurea* var. *nealleyi* in having flat, curly leaf blades and longer awns.

## 21. Aristida stricta Michx. WIREGRASS, PINELAND THREEAWN [p. *465*, <u>*544*</u>]

*Aristida stricta* grows in pine barrens and sandy fields of the coastal plain from Louisiana to North Carolina.

## 22. Aristida rhizomophora Swallen [p. *465*, <u>*544*</u>]

*Aristida rhizomophora* is not well-collected. It is endemic to Florida, where it grows in moist to wet pine flatwoods, and on the borders of ponds and bald-cypress depressions.

## 23. Aristida mohrii Nash MOHR'S THREEAWN [p. *465*, <u>*544*</u>]

*Aristida mohrii* is endemic to the southeastern United States, growing on dry, sandy pinelands and oak barrens, and occasionally in waste places. It is sometimes confused with *A. simpliciflora* because both have reduced, spikelike inflorescences, but *A. simpliciflora* has lateral awns that are only about half as thick as the central awn, and its spikelets are borne in pairs.

## 24. Aristida simpliciflora Chapm. SOUTHERN THREEAWN [p. *465*, <u>*545*</u>]

*Aristida simpliciflora* grows in wet savannahs, the upper portion of seepage bogs, and the moister portion of ecotones between such bogs and the surrounding dry uplands. It is restricted to the southeastern United States. It is sometimes confused with *A. mohrii* because both have reduced, spikelike inflorescences, but *A. mohrii* has lateral awns that are about as thick as the central awn, and its spikelets are solitary.

## 25. Aristida lanosa Muhl. *ex* Elliott WOOLY THREEAWN [p. *465*, <u>*545*</u>]

*Aristida lanosa* is restricted to the eastern United States, where it grows in dry fields, pine-oak woods, and uplands, chiefly in sandy soil. It differs from *A. palustris* in several reproductive, vegetative, and habitat characteristics.

**26. Aristida palustris** (Chapm.) Vasey LONGLEAF THREEAWN [p. 465, <u>545</u>]

*Aristida palustris* is endemic to the southeastern United States, where it grows in seepage bogs, pitcher plant savannahs, wet pine flatwoods, bald-cypress depressions, and wet prairies. It is a distinctive species of the southeastern coastal plain region that differs from *A. lanosa* in several reproductive, vegetative, and habitat characteristics.

**27. Aristida purpurascens** Poir. [p. 465, <u>545</u>]

*Aristida purpurascens* is composed of three intergrading varieties.

1. Central awns divaricate to reflexed, about twice as thick at the base as the lateral awns . . . . . . . . . . . . . . .var. *virgata*
1. Central and lateral awns divergent, all about the same thickness at the base.
  2. Lower glumes usually longer than the upper glumes; awns straight or only slightly contorted at the base; blades 1–3 mm wide, often curling. . . . . . . . . . . . . . . . . . . . . . . . . . . . . . . . . . var. *purpurascens*
  2. Lower glumes shorter than or equal to the upper glumes; awns spirally contorted at the base; blades mostly about 1 mm wide, usually not curling. . . . . . . . . . . . . . . . . . . . . . . . . . . . . . var. *tenuispica*

**Aristida purpurascens** Poir. var. **purpurascens**
ARROWFEATHER THREEAWN [p. 465]

*Aristida purpurascens* var. *purpurascens* grows in waste places, glades, fields, and pine savannahs in sandy or clay soils. Its range extends into northern Mexico.

**Aristida purpurascens** var. **tenuispica** (Hitchc.) Allred [p. 465]

*Aristida purpurascens* var. *tenuispica* grows in pine and oak woods, prairies, and along roadsides, at low elevations and usually in sandy soils. Within the *Manual* region, it grows on the coastal plain from North Carolina to Mississippi. It also grows in Mexico and Honduras.

**Aristida purpurascens** var. **virgata** (Trin.) Allred [p. 465]

*Aristida purpurascens* var. *virgata* grows in wet or moist areas such as seepage bogs, sandy pinelands, and wet prairies of the southeastern United States.

**28. Aristida condensata** Chapm. BIG THREEAWN [p. 465, <u>545</u>]

*Aristida condensata* grows on sandy hills, and in pine and oak barrens in the southeastern United States.

**29. Aristida gyrans** Chapm. CORKSCREW THREEAWN [p. 465, <u>545</u>]

*Aristida gyrans* is endemic to the southeastern United States, growing in sandy pine woods and oak scrub. It differs from other species in the genus by its combination of narrow blades, unequal glumes, long calluses, and contorted awns.

# 9. CENTOTHECOIDEAE Soderstr.[1]

The subfamily *Centothecoideae* is one of the subfamilies that cannot be characterized by a suite of morphological characteristics, but anatomical, micromorphological, and nucleic acid data all support its recognition. It is most abundant in warm-temperate woodlands and tropical forests.

1. Spikelets 4–50 mm long, with 1–15 florets, the lowest florets sometimes sterile, the upper florets bisexual; disarticulation at the base of the florets or the base of the spikelets; leaves not pseudopetiolate; culms 35–150 cm tall; plants not reedlike . . . . . . . . . . . . . . . . . . . . . . . . . . . . . . . . . . . . . . . . . 21. *Centotheceae*
1. Spikelets 1.2–1.8 mm long, with 2(3–4) florets, the lower florets sterile, the upper florets bisexual; disarticulation at the pedicel bases, subsequently below the spikelets; leaves pseudopetiolate; culms 150–400 cm tall; plants reedlike . . . . . . . . . . . . . . . . . . . . . . . . . . . . . . . . . . . . . . . . . 22. *Thysanolaeneae*

# 21. CENTOTHECEAE Ridl.[2]

**Pl** ann or per; csp, hrb, delicate to smt reedlike. **Clm** to 4 m tall, not wd, glab; **intnd** solid or hollow. **Shth** open; **aur** smt present; **lig** scarious or memb, truncate, smt ciliolate; **psdpet** present in some genera; **bld** flat, relatively wide, usu with evident cross venation (not in *Chasmanthium*), not disarticulating from the shth. **Infl** tml, pan, smt with spicate br. **Spklt** solitary, pedlt or subsessile, lat compressed, with 1–many flt, bisx, pist or stmt, lowest and distal flt often strl; **rchl** terminating in a rdmt flt; **dis** variable, at the base of the ped, below the glm, above the glm, beneath the flt, between the flt, or some combination of these. **Glm** smt lacking, if present, memb, shorter than the flt, 2–9-veined, acute to obtuse; **lowest flt** smt strl, with or without pal; **up lm** memb, 3–15-veined, unawned or awned; **up pal** nearly as long as the lm, apc entire or notched; **lod** 2, free (rarely fused in some genera), cuneate, truncate or somewhat lobed; **anth** 1, 2, or 3. **Car** ellipsoid to circular, or trigonous; **emb** about $\frac{1}{3}$ or less as long as the car; **hila** subbas to bas, punctate.

The tribe *Centotheceae* has approximately 10 genera and 30 species, most of which grow in tropical forests. *Chasmanthium*, the only member of the tribe represented in the *Manual* region, is also the only genus to extend into temperate regions. Most members of the tribe are easily recognized by the evident cross venation of their wide blades, but *Chasmanthium* lacks this trait.

[1]Grass Phylogeny Working Group   [2]J. Gabriel Sánchez-Ken

## 21.01 CHASMANTHIUM Link[1]

Pl per; csp or loosely colonial, rhz. **Clm** 35–150 cm, simple or brchd. **Lvs** cauline; **lig** memb, ciliate; **bld** not psdpet, flat. **Pan** open or contracted, smt becoming rcm distally; **dis** above the glm and between the flt. **Spklt** 4–50 mm, lat compressed, with 2–many flt, lo 1–4 flt strl. **Glm** 2, subequal, shorter than the spklt, glab, (2)3–9-veined, acute to acuminate; **lm** glab, 3–15-veined, compressed-keeled, keels serrate or ciliate, apc acuminate to acute, entire (rarely bifid); **pal** glab, gibbous bas, 2-keeled, keels winged, wings glab, scabrous, or pilose; **lod** 2, fleshy, cuneate, 2–4-veined, lobed-truncate; **anth** 1; **ov** glab; **sty** 2; **sty br** 2, plumose, reddish-purple at anthesis. **Car** 1.9–5 mm, lat compressed, brown to reddish-black or black.

*Chasmanthium*, a genus of five species endemic to North America, grows primarily in the southeastern and south-central parts of the United States.

1. Panicle branches nodding or drooping; pedicels 10–30 mm long; calluses pilose; lower glumes 4.2–9.1 mm long; keels of fertile lemmas winged, the wings scabrous to pilose their full length; caryopses 2.9–5 mm long . . . . . . . . . . . . . . . . . . . . . . . . . . . . . . . . . . . . . . . . . . . . . . . . . . . . . . . . 1. *C. latifolium*
1. Panicle branches erect or ascending; pedicels 0.5–2.5(5) mm long; calluses glabrous; lower glumes 1.2–5 mm long; keels of fertile lemmas not winged, scabridulous toward or at the apices; caryopses 1.9–3 mm long.
  2. Spikelets 9.5–24 mm long; fertile lemmas 9–13-veined; caryopses enclosed at maturity; blades 7–16(33) cm long, lanceolate-fusiform; culms leafy for 80% of their height.
    3. Axils of the panicle branches scabrous; fertile florets diverging to 45° from the rachilla; sterile florets (0)1(2); lower glumes 3.1–5 mm long, 7–9-veined; ligules entire . . . . . . . . . . . . . . . . . . . . 2. *C. nitidum*
    3. Axils of the panicle branches pilose; fertile florets diverging to 85° from the rachilla; sterile florets 2–4; lower glumes 2.5–2.9 mm long, 2–3-veined; ligules irregularly laciniate . . . . . . . . . . . . . 3. *C. ornithorhynchum*
  2. Spikelets 4–10 mm long; fertile lemmas 3–9-veined; caryopses exposed at maturity; blades (8)20–50 cm long, linear-lanceolate; culms leafy for 40–50% of their height.
    4. Collars and sheaths pilose; culms (1)2–3.5 mm thick at the nodes; fertile lemmas 7–9-veined, usually curved or irregularly contorted . . . . . . . . . . . . . . . . . . . . . . . . . . . . . . . . . . . . . . 4. *C. sessiliflorum*
    4. Collars and sheaths glabrous; culms to 1 mm thick at the nodes; fertile lemmas 3–7-veined, straight . . . . . . . . . . . . . . . . . . . . . . . . . . . . . . . . . . . . . . . . . . . . . . . . . . . . . . . . . . . . . . 5. *C. laxum*

### 1. Chasmanthium latifolium (Michx.) H.O. Yates
BROADLEAF CHASMANTHIUM [p. 466, <u>545</u>]

*Chasmanthium latifolium* grows along stream and river banks and in rich deciduous woods. It is the most widespread species of the genus, extending further west and east than any of the other four species. The map shows its verifiable range. Flowering in *C. latifolium* is sometimes cleistogamous.

### 2. Chasmanthium nitidum (Baldwin) H. O. Yates
SHINY CHASMANTHIUM [p. 466, <u>545</u>]

*Chasmanthium nitidum* grows along stream and river banks, roadside ditches, and the margins of low, moist woods in the southeastern United States.

### 3. Chasmanthium ornithorhynchum Nees BIRDBILL
CHASMANTHIUM [p. 466, <u>545</u>]

*Chasmanthium ornithorhynchum* grows along stream and river banks in low woods, and on hummocks in swamps. It is most common along the coastal plain from eastern Louisiana to western Florida, but is also found at a few other locations in the southeastern United States.

### 4. Chasmanthium sessiliflorum (Poir.) H.O. Yates
LONGLEAF CHASMANTHIUM [p. 466, <u>545</u>]

*Chasmanthium sessiliflorum* grows in rich woods, meadows, and swamps, especially on the coastal plain. It grows throughout most of the southeastern United States.

### 5. Chasmanthium laxum (L.) H.O. Yates SLENDER
CHASMANTHIUM [p. 466, <u>545</u>]

*Chasmanthium laxum* is almost completely sympatric with *C. sessiliflorum* in the southeastern United States, growing in similar habitats but extending farther into sphagnous stream heads, pine flatwoods, and pine savannahs. Putative, naturally occurring hybrids between *Chasmanthium ornithorhynchum* and *C. laxum* have been reported along streams of the outer coastal plain of Mississippi and Louisiana. In general appearance, the hybrids resemble *C. laxum*, their most striking difference being the enlarged, sterile spikelets.

## 22. THYSANOLAENEAE C.E. Hubb.[2]

Pl per; csp, shortly rhz, reedlike. **Clm** 1.5–4 m, wd, persistent, glab, usu not brchd above the base; **intnd** solid. **Shth** open; **aur** smt present; **lig** memb to somewhat leathery, entire, minutely erose, or ciliolate; **psdpet** poorly developed; **bld** flat, disarticulating from the shth when old, cross venation not evident. **Infl** tml, open diffuse pan; **dis** at the ped bases, subsequently below the spklt. **Spklt** solitary, pedlt, terete, with 2(3–4) flt, lowest flt strl; **rchl** prolonged, often terminating in a rdmt flt. **Glm** memb, shorter than the flt, 0–1-veined, obtuse; **lowest flt** equaling the up flt, strl; **lowest lm** memb, 1–3-veined; **lowest pal** not present; **up flt** bisx; **up lm** memb, 3-veined, unawned, mrgl veins with papillose-based hairs, hairs 0.8–1.5 mm, strongly diverging at maturity; **up pal** about ½ as long as the lm, apc notched; **lod** 2, free, broadly cuneate, truncate to irregularly lobed; **anth**

[1]J. Gabriel Sánchez-Ken and Lynn G. Clark   [2]J. Gabriel Sánchez-Ken

2(3). **Car** nearly spherical to broadly ovoid; **hila** sub-bas, punctate; **emb** large, about ¾ as long as the car.

The tribe *Thysanolaeneae* is native to tropical Asia and includes only the unispecific genus *Thysanolaena*.

## 22.01 THYSANOLAENA Nees[1]

**Pl** per; csp, shortly rhz, reedlike. **Clm** 1.5–4 m, wd, persistent, glab, usu not brchd above the base; **intnd** solid. **Lvs** cauline, distichous; **shth** open; **aur** smt present; **lig** memb to somewhat leathery, entire, minutely erose, or ciliolate; **psdpet** poorly developed; **bld** flat, disarticulating from the shth when old, cross venation not evident. **Infl** tml, open diffuse pan. **Spklt** solitary, pedlt, terete, with 2(3–4) flt, lowest flt in each spklt strl; **rchl extension** prolonged, often terminating in a rdmt flt; **dis** at the ped bases, subsequently below the spklt. **Glm** memb, shorter than the flt, 0–1-veined, obtuse; **lowest flt** equaling the up flt, strl; **lowest lm** memb, 1–3-veined; **lowest pal** not present; **up flt** bisx; **up lm** memb, 3-veined, unawned, mrgl veins with papillose-based hairs, hairs 0.8–1.5 mm, strongly diverging at maturity; **up pal** about ½ as long as the lm, apc notched; **lod** 2, free, broadly cuneate, truncate to irregularly lobed; **anth** 2(3). **Car** nearly spherical to broadly ovoid; **emb** about ¾ as long as the car; **hila** sub-bas, punctate.

*Thysanolaena* is a unispecific genus of tropical Asia, where it grows in open habitats, generally in mountainous areas. Its only species is grown as an ornamental in the *Manual* region.

### 1. Thysanolaena latifolia (Roxb. *ex* Hornem.) Honda
ASIAN BROOMGRASS [p. 466]

*Thysanolaena latifolia* is an important pasture species in Asia. It is grown in the United States as an ornamental, recommended for frost-free areas with full sunlight or light shade. It is not known to be established in the *Manual* region.

## 10. PANICOIDEAE Link[2]

The subfamily *Panicoideae* is most abundant in tropical and subtropical regions, particularly mesic portions of such regions, but several species grow in temperate regions of the world. Within the *Manual* region, the *Panicoideae* are represented by 63 genera and 363 species. They are most abundant in the eastern United States. Photosynthesis may be either $C_3$ or $C_4$.

Spikelets with two florets are found in many other subfamilies, but rarely do they follow the pattern of the lower floret being sterile or staminate and the upper floret bisexual.

## 23. GYNERIEAE Sánchez-Ken & L.G. Clark[3]

**Pl** per; dioecious. **Clm** 2–15 m; **intnd** solid. **Lvs** cauline; **shth** longer than the intnd, persistent, distal shth strongly overlapping; **bld** leathery, not psdpet, articulated with the shth, midveins conspicuous, those below midclm length disarticulating at maturity. **Infl** tml, pan. **Pist spklt** lat compressed, with 2 flt; **dis** above the glm and between the flt; **glm** unequal, exceeding the flt, lo glm shorter and less firm than the up glm; **cal** linear, glab; **lm** long silky-pilose distally, apc elongate, narrow, unawned; **sty br** 2, free; **rchl extension** absent. **Stmt spklt** with 2–4 flt; **dis** below the tml flt, glm and lo flt(s) remaining attached; **glm** subequal, memb; **lm** memb, glab or sparsely short pilose, (0)1(3)-veined; **anth** 2. **Car** oblong; **hila** punctate.

The tribe *Gynerieae* includes only one genus, the neotropical genus *Gynerium*.

## 23.01 GYNERIUM Willd. *ex* P. Beauv.[4]

**Pl** per; dioecious, stmt and pist pl similar in gross morphology; rhz. **Shth** overlapping, mostly glab or the young and bldless lvs sparsely pilose, mrg usu pilose distally; **col** pilose in young lvs with rdmt bld; **lig** pilose; **bld** to 2 m, leathery, glab, lo bld disarticulating at maturity, midribs 0.5–1.5 cm wide, smt with a line of

[1]J. Gabriel Sánchez-Ken   [2]Grass Phylogeny Working Group   [3,4]J. Gabriel Sánchez-Ken

scattered hairs on either side. **Pan** large, the pist pan plumose. **Pist spklt** with 2 flt; **dis** above the glm and between the flt; **glm** 1–3-veined, tardily disarticulating; **up glm** longer and firmer than the lo glm, exceeding the flt; **lm** shortly pilose below, long-pilose above, apc elongated and narrowed; **lod** 2, memb, faintly 2–3-veined, smt with long hairs, truncate; **stmd** 2; **sty** 2. **Stmt spklt** with 2–4 flt; **dis** below the tml flt; **glm** 1-veined; **lm** hyaline, glab, occ puberulent below; **lod** 2, free, faintly veined, truncate; **anth** 2; **ov** abortive.

*Gynerium* is a unispecific genus that ranges from Mexico and the West Indies to northeastern Argentina.

### 1. Gynerium sagittatum (Aublet) P. Beauv. WILDCANE [p. 466]

*Gynerium sagittatum* is grown as an ornamental in subtropical portions of the *Manual* region. Even when vegetative, it can be identified by its height, the absence of blades on the lower leaves, the strongly distichous, fan-shaped arrangement of the distal leaf blades, and the wide midveins of the blades. It does not flower when grown outdoors in the *Manual* region.

## 24. PANICEAE R. Br.[1]

**Pl** ann or per; habit various. **Clm** 3–800 cm, ann, usu not wd. **Lvs** bas and/or cauline; **shth** usu open; **lig** of hairs or memb, memb lig often ciliate, cilia smt longer than the memb base; **bld** occ psdpet, seldom disarticulating at maturity. **Infl** tml, smt also ax, occ subterranean pan; **br** smt spikelike and secund, smt less than 1 cm; **dis** usu below the glm, smt at the base of the pan br, occ below the flt. **Spklt** usu dorsally compressed, varying to terete or lat compressed, with 2(3) flt, lo flt stmt, strl, or red, up flt usu bisx; **cal** not developed. **Glm** usu memb; **lo glm** usu less than ½ as long as the spklt, smt absent; **up glm** usu subequal to the up flt, occ absent; **lo lm** similar to the up glm in length and texture; **up lm** indurate, coriaceous, or cartilaginous, with a germination flap at the base, mrg usu widely separated and involute at maturity, smt flat and hyaline; **up pal** similar to the up lm in length and texture; **lod** short; **anth** usu 3; **stigmas** usu red. **Car** usu dorsally compressed or terete; **emb** ½ or more the length of the car.

The tribe *Paniceae*, which includes about 100 genera and 2000 species, is primarily tropical in distribution. Within the *Manual* region, it is represented by 31 genera and 261 species, with its greatest representation being in the eastern portion of the contiguous United States.

The *germination flap* is a small area of soft tissue at the base of the upper lemma through which the primary root of the seedling grows.

1. Plants developing both subterranean and aerial inflorescences, only the subterranean spikelets setting seed . . . . . . . . . . . . . . . . . . . . . . . . . . . . . . . . . . . . . . . . . . . . . . . . . . . . . 24.04 *Amphicarpum* (in part)
1. Plants developing only aboveground inflorescences, the spikelets setting seed [*Amphicarpum* is also keyed out here to accommodate situations in which looking for subterranean inflorescences is not permitted or specimens have no underground parts].
    2. Inflorescences spikelike panicles, with the branches partially embedded in the flattened rachises; plants perennial, stoloniferous . . . . . . . . . . . . . . . . . . . . . . . . . . . . . . . . . . . . . . . . . . 24.24 *Stenotaphrum*
    2. Inflorescences panicles, sometimes spikelike, but the branches not embedded in the rachises or the rachises not flattened; plants annual or perennial, sometimes stoloniferous.
        3. Most spikelets or groups of 2–11 spikelets subtended by 1–many, distinct to more or less connate, stiff bristles or bracts.
            4. Spikelets in groups of 2–11, subtended by 4 flat, narrowly elliptic, coriaceous bracts; terete bristles not present . . . . . . . . . . . . . . . . . . . . . . . . . . . . . . . . . . . . . . . . . . . . . 24.18 *Anthephora*
            4. Spikelets solitary or in groups, subtended by 1–many stiff, terete bristles, sometimes appearing as an extension of the branch; flat, connate bristles sometimes present distal to the terete bristles.
                5. Bristles falling with the spikelets at maturity; disarticulation at the base of the reduced panicle branches (*fascicles*).
                    6. Bristles plumose or antrorsely scabrous, free or fused no more than ½ their length . . . . . . . 24.16 *Pennisetum*
                    6. Bristles glabrous, smooth, retrorsely scabrous, or strigose, usually at least some bristles fused for more than ½ their length . . . . . . . . . . . . . . . . . . . . . . . . . . . . . . . . 24.17 *Cenchrus*
                5. Bristles persistent; disarticulation below the spikelets.
                    7. Upper glumes indurate at maturity; lower lemmas somewhat indurate at the base; pedicels subtended by a single bristle . . . . . . . . . . . . . . . . . . . . . . . . . . . . . . . . . . 24.20 *Setariopsis*
                    7. Upper glumes membranous to herbaceous at maturity; lower lemmas neither constricted nor indurate at the base; pedicels subtended by 1–many bristles.
                        8. Spikelets subtended by 1–many bristles; paleas of the lower florets usually hyaline to membranous at maturity, rarely absent or reduced; paleal veins not keeled . . . . . . . 24.21 *Setaria* (in part)
                        8. Spikelets subtended by 1 bristle; paleas of the lower florets coriaceous to indurate at maturity, the keels thickened . . . . . . . . . . . . . . . . . . . . . . . . . . . . . . . . . . 24.19 *Ixophorus*

[1]Mary E. Barkworth

3. All or most spikelets not subtended by stiff bristles, sometimes the terminal spikelet on each branch subtended by a single bristle, and occasionally other spikelets with a single subtending bristle.

9. Terminal spikelet on each branch subtended by a single bristle; other spikelets occasionally with a single stiff subtending bristle . . . . . . . . . . . . . . . . . . . . . . . . . . . . . . . . . . . . . . 24.21 *Setaria* (in part)

9. None of the spikelets subtended by a stiff bristle.

10. Inflorescences of spikelike branches 1–3.7 cm long, the branch axes extending as a 2.5–4 mm bristle beyond the base of the distal spikelets . . . . . . . . . . . . . . . . . . . . . . . . . 24.22 *Paspalidium*

10. Inflorescences various but, if of spikelike branches, these terminating in a well-developed or rudimentary spikelet.

11. Lower glumes or lower lemmas awned, sometimes shortly so (the awn reduced to a point in *Echinochloa colona*).

12. Upper florets laterally compressed; spikelets also laterally compressed . . . . . . . 24.12 *Melinis* (in part)

12. Upper florets dorsally compressed; spikelets usually dorsally compressed or terete, sometimes laterally compressed.

13. Blades linear to linear-lanceolate, usually more than 10 times longer than wide, with prominent midribs; at least the upper leaves, often all leaves, without ligules; ligules usually absent, particularly from the upper leaves, of hairs when present . . . . . . . . . . . . . . . . . . . . . . . . . . . . . . . 24.07 *Echinochloa* (in part)

13. Blades triangular to lanceolate, less than 10 times longer than wide, the midribs not particularly prominent, at least distally; ligules present, of hairs or membranous.

14. Lower glumes awned, the awns exceeding the florets; upper glumes not ciliate-margined; culms trailing on the ground, frequently rooting and branching at the nodes . . . . . . . . . . . . . . . . . . . . . . . . 24.06 *Oplismenus*

14. Lower glumes unawned or shortly awned, the awns exceeded by the florets; upper glumes ciliate-margined; culms erect or decumbent below, sometimes rooting and branching at the lower nodes . . . . . . . . . 24.03 *Alloteropsis*

11. Lower glumes and lower lemmas unawned.

15. Upper florets laterally compressed . . . . . . . . . . . . . . . . . . . . . . . . . . . . . . . . 24.12 *Melinis* (in part)

15. Upper florets dorsally compressed or terete.

16. Upper lemmas and paleas cartilaginous and flexible at maturity; lemma margins flat, hyaline; lower glumes absent or to ¼ the length of the spikelets.

17. Aerial inflorescences with elongate rachises and glabrous spikelets; spikelets of the aerial panicles rarely setting seed; subterranean spikelets developed, seed-forming . . . . . . . . . . . . . . . . . . . . 24.04 *Amphicarpum* (in part)

17. Aerial inflorescences of digitate or subdigitate clusters of spikelike branches with glabrous or pubescent spikelets or with elongate rachises and conspicuously pubescent spikelets; aerial spikelets seed-forming; subterranean spikelets not developed.

18. Spikelets ellipsoid to obovoid; inflorescences simple panicles with erect to ascending branches on elongate rachises; branches ascending, not conspicuously spikelike . . . . . . . . . . . . . . . . . . . . . . . 25.02 *Anthenantia*

18. Spikelets lanceoloid to ellipsoid; inflorescences usually panicles with digitate or subdigitate clusters of spikelike branches, sometimes simple panicles with strongly divergent branches . . . . . . . . . . . 24.01 *Digitaria*

16. Upper lemmas and paleas chartaceous to indurate and rigid at maturity; lemma margins not hyaline, frequently involute; lower glumes varying from absent to subequal to the spikelets or extending beyond the distal floret.

19. Spikelets subtended by a cuplike callus . . . . . . . . . . . . . . . . . . . . . . . . . . . . 24.15 *Eriochloa*

19. Spikelets not subtended by a cuplike callus.

20. At least the upper leaves, often all leaves, without ligules; ligules, when present, of hairs . . . . . . . . . . . . . . . . . . . . . . . . . 24.07 *Echinochloa* (in part)

20. All leaves with ligules, ligules membranous or of hairs.

21. Paleas of the lower florets inflated and indurate at maturity; lower and upper florets standing apart from each other when mature . . . . . . . . . . . . . . . . . . . . . . . . . . . . . . . . . . . . . . . . . . . . 24.26 *Steinchisma*

21. Paleas of the lower florets neither inflated nor indurate at
    maturity; lower and upper florets closely appressed to each
    other when mature.
    [⇐ revert to left, Ed.]

22. Panicle branches 1-sided; secondary branches usually absent or few and clearly shorter than the
    primary branches, sometimes many and almost as long as the primary branches; panicles never
    spikelike.
    23. Lower glumes, sometimes both glumes, absent from all or almost all spikelets.
        24. Both glumes absent from all or most spikelets, upper glume sometimes present on the
            terminal spikelet on a branch; spikelets 3.8–5.2 mm . . . . . . . . . . . . . . . . . . . . . . . . . . . 24.31 *Reimarochloa*
        24. Upper glumes usually present on all spikelets, if absent the spikelets 1.8–2 mm.
            25. Axes of panicle branches flattened, usually winged; spikelets with upper glumes and
                lemmas appressed to the branches . . . . . . . . . . . . . . . . . . . . . . . . . . . . . . . . 24.28 *Paspalum* (in part)
            25. Axes of panicles branches triquetrous; spikelets with lower lemmas appressed to the
                branch axes . . . . . . . . . . . . . . . . . . . . . . . . . . . . . . . . . . . . . . . . . . . . . . . . . . . . 24.27 *Axonopus*
    23. Lower glumes present on all or most spikelets.
        26. Spikelets 5.5–7 mm long, lanceoloid; lower glumes nearly as long as the lower lemmas,
            exceeding the upper florets . . . . . . . . . . . . . . . . . . . . . . . . . . . . . . . . . . . . . . . . . . . . . . 24.30 *Phanopyrum*
        26. Spikelets 1–6 mm long, ovoid to obovoid; lower glumes shorter than the upper glumes.
            27. Lower glumes ¾ as long as to equaling the spikelets; upper glumes 7–11-veined; upper
                florets mostly glabrous but hairy at the apices . . . . . . . . . . . . . . . . . . . . . . . . . . . . . . . 24.23 *Hopia*
            27. Lower glumes usually less than ¾ the length of the spikelets; upper florets usually
                glabrous, sometimes hairy all over, never just at the apices; upper glumes 0–13-veined.
                28. Upper lemmas rugose at maturity; upper glumes 5–13-veined   . . . . . . . . . . . . . . . . . 24.14 *Urochloa*
                28. Upper lemmas smooth at maturity; upper glumes usually 0–5-veined, sometimes 7-
                    veined.
                    29. Spikelet lengths at least twice as long as their widths; lower glumes and upper
                        lemmas glabrous . . . . . . . . . . . . . . . . . . . . . . . . . . . . . . . . . . . . . . . . 24.10 *Panicum* (in part)
                    29. Spikelet lengths less than twice as long as their widths and/or lower glumes
                        and upper lemmas hairy.
                        30. Lower lemmas and glumes appressed to the panicle branches, villous . . . . . . 24.11 *Moorochloa*
                        30. Upper lemmas and glumes appressed to the panicle branches, usually
                            glabrous, sometimes puberulent or sparsely pubescent . . . . . . . . . . . . 24.28 *Paspalum* (in part)
22. Panicle branches not 1-sided, usually with well-developed secondary branches; inflorescences
    sometimes spikelike.
    31. Inflorescences dense, the spikelets concealing at least the distal ½ of the rachises.
        32. Upper glumes slightly to strongly saccate, 5–13-veined; panicle branches often fused to the
            rachises; blades 1.5–22 mm wide; culm internodes hollow . . . . . . . . . . . . . . . . . . . . . . . 24.08 *Sacciolepis*
        32. Upper glumes not saccate, 3–7-veined; panicle branches not fused to the rachises; blades
            12–28 mm wide; culm internodes filled with aerenchyma . . . . . . . . . . . . . . . . . . . . . . . . . 24.25 *Hymenachne*
    31. Inflorescences more or less open panicles, the spikelets not concealing the distal portion of the
        rachises.
        33. Lower glumes saccate at the base; glumes and lemmas with woolly pubescent apices; culms
            weakly lignified, rooting at the nodes . . . . . . . . . . . . . . . . . . . . . . . . . . . . . . . . . . . . . . . . 24.05 *Lasiacis*
        33. Lower glumes not saccate at the base; glumes and lemmas glabrous or with short, straight
            hairs, apices sometimes with a tuft of hairs but never woolly pubescent; culms usually not
            lignified, if lignified, not rooting at the nodes.
            34. Lemmas of upper florets transversely rugose.
                35. Panicles branches verticillate at lowest nodes . . . . . . . . . . . . . . . . . . . . . . . . . . 24.13 *Megathyrsus*
                35. Panicle branches solitary at the lowest nodes . . . . . . . . . . . . . . . . . . . . . . . . . . . . 24.29 *Zuloagaea*
            34. Lemmas of the upper florets usually smooth, sometimes verrucose, not rugose.
                36. Aerial spikelets lanceoloid, often without lower glumes; upper lemmas with flat
                    margins; plants forming subterranean spikelets . . . . . . . . . . . . . . . . . . . . . 24.04 *Amphicarpum* (in part)
                36. Spikelets ovoid to ellipsoid or lanceoloid; lower glumes present; upper lemmas
                    with involute margins; plants not forming subterranean spikelets.
                    37. Blades of the basal leaves clearly distinct from the cauline leaves; basal leaves
                        ovate to lanceolate, cauline leaves with longer and narrower blades; basal
                        leaves forming a distinct winter rosette . . . . . . . . . . . . . . . . . . . . . . 24.09 *Dichanthelium* (in part)
                    37. Blades of the basal and cauline leaves similar, usually linear to lanceolate,
                        varying from filiform to ovate; basal leaves not a distinct winter rosette.

## 24.01 DIGITARIA Haller[1]

Pl ann, per, or of indefinite duration. Clm 5–250 cm, erect or decumbent, brchg bas or at aerial clm nd, when ann or of indefinite duration usu decumbent and rooting at the lo nd. Shth open; lig memb, smt ciliate; bld usu flat. Infl tml, smt also ax, usu pan of 1-sided spikelike br (smt only 1 br) attached digitately or racemosely to a rachis, smt simple pan of solitary, pedlt spklt; spikelike br, if present, smt with sec br, pri br axes triquetrous, bearing spklt abxly, in 2 rows, usu in unequally pedlt groups of 2–5, occ borne singly. Spklt 1.2–8.2 mm, lanceoloid to ellipsoid, dorsally compressed, apc obtuse to acuminate, unawned, with 2 flt; dis beneath the glm. Lo glm absent or to ¼ as long as the spklt; up glm usu from ⅙ as long as to equaling the spklt, occ absent, 0–5-veined, usu pubescent; lo flt strl; lo lm memb, usu as long as the up lm, usu pubescent, (3)5–7(13)-veined; lo pal absent or rdcd; up lm mostly stiffly chartaceous to cartilaginous, obscurely veined, with 0.5–1 mm hyaline mrg that embrace the up pal; up pal similar to the up lm in texture and size; lod 3, cuneate; anth 3. Car plano-convex; emb ⅕–½ as long as the car; hila punctiform to ellipsoid.

*Digitaria* has approximately 200 species and grows primarily in tropical and warm-temperate regions, often in disturbed, open sites. There are 29 species known to occur in the *Manual* region; 18 are native to the region. Most annual species of *Digitaria* will survive several years in regions without a pronounced cold season; such species are described as being of indefinite duration.

NOTE: The pubescence of the lower lemma may be mistaken for two white lines between the veins because the individual hairs are not visible, being both tightly packed and closely appressed.

1. Inflorescences simple open panicles, with well-developed primary and secondary branches; branches and pedicels divergent; spikelets solitary.
  2. Upper glumes absent or to 0.6 mm long, veinless . . . . . . . . . . . . . . . . . . . . . . . . . . . . . . . . . . . . 4. *D. tomentosa*
  2. Upper glumes 1.8–3.8 mm long, 3–7-veined.
    3. Spikelets 3.5–4.6 mm long; upper glumes 5–7-veined . . . . . . . . . . . . . . . . . . . . . . . . . . . . 1. *D. arenicola*
    3. Spikelets 2.2–3.3 mm long; upper glumes 3(5)-veined.
      4. Lower lemmas 7-veined, veins not equally spaced . . . . . . . . . . . . . . . . . . . . . . . . . . . . . . . . . .2. *D. cognata*
      4. Lower lemmas 5-veined, veins equidistant . . . . . . . . . . . . . . . . . . . . . . . . . . . . . . . . . . . . . . .3. *D. pubiflora*
1. Inflorescences panicles of spikelike branches; secondary branches rarely present; spikelets appressed to the branches, in groups of 2–5 on the middle portion of the primary branches.
  5. Spikelets in groups of 3–5 on the middle portions of the primary branches, the longer pedicels in each group often adnate to the branch axes for part of their length.
    6. Upper lemmas pale yellow or gray when immature, light brown to brown when mature, sometimes purple-tinged.
      7. Upper glumes ⅙–⅓ long as the spikelets; sheaths and blades pubescent . . . . . . . . . . . . . . . . . . . . . . 13. *D. serotina*
      7. Upper glumes equaling or almost equaling the spikelets; sheaths and blades usually glabrous.
        8. Upper glumes 5-veined; spikelets elliptic to obovate; plants stoloniferous . . . . . . . . . . . . . . . . 14. *D. longiflora*
        8. Upper glumes 3-veined; spikelets lanceolate; plants not stoloniferous . . . . . . . . . . . . . 15. *D. floridana* (in part)
    6. Upper lemmas brown when immature, becoming dark brown when mature.
      9. Primary panicle branches wing-margined, the wings at least ½ as wide as the midribs.
        10. Plants always with axillary panicles in the lower leaf sheaths, these panicles sometimes completely concealed by the sheaths; spikelets 1.7–2.3 mm long . . . . . . . . . . . . . . . . . . . . . 16. *D. ischaemum*
        10. Plants without axillary panicles; spikelets 1.2–1.7 mm long.
          11. Primary panicle branches, if more than 2, racemose, the terminal branches erect, the other branches usually divergent; upper lemmas light brown to brown at maturity; upper glumes almost as long as the upper lemmas . . . . . . . . . . . . . . . . . . . . . . 15. *D. floridana* (in part)
          11. Primary panicle branches usually all digitate, sometimes 1 below the others, all the branches erect to ascending; upper lemmas dark brown at maturity; upper glumes ½ as long as to almost equaling the upper lemmas . . . . . . . . . . . . . . . . . . . . . . . . . . . . . . . 17. *D. violascens*
      9. Primary panicle branches not wing-margined or the wings not as wide as the midribs.
        12. Lower lemmas pubescent.

[1]J.K. Wipff

13. Plants annual, or short-lived perennials, branching at the lower nodes; cauline nodes 3–6 . . . . . . . . . . . . . . . . . . . . . . . . . . . . . . . . . . . . . . . . . . . . 7. *D. filiformis* (in part)
13. Plants perennial, not branching at the lower nodes; cauline nodes 1–2.
    14. Upper glumes and lower lemmas with long, glandular-tipped hairs along their margins and intercostal regions . . . . . . . . . . . . . . . . . . . . . . . . . . . . . 8. *D. leucocoma*
    14. Upper glumes and lower lemmas glabrous over most of their length, sparsely pubescent near the apices, the hairs short, not glandular-tipped . . . . . . . . . . . . . 6. *D. bakeri* (in part)
  12. Lower lemmas glabrous.
    15. Upper glumes less than ½ as long as the spikelets . . . . . . . . . . . . . . . . . . . . . . . . . . . 5. *D. gracillima*
    15. Upper glumes more than ½ as long as the spikelets.
      16. Plants annual, branching at the lower nodes; culm nodes 3–6; upper glumes obtuse . . . . . . . . . . . . . . . . . . . . . . . . . . . . . . . . . . . . . . . . . . . . . . 7. *D. filiformis* (in part)
      16. Plants perennial, not branching at the lower nodes; culm nodes 1–2; upper glumes acute . . . . . . . . . . . . . . . . . . . . . . . . . . . . . . . . . . . . . . . . . . . . . . 6. *D. bakeri* (in part)
5. Spikelets paired on the middle portions of the primary branches; pedicels not adnate to the branch axes.
  17. Upper lemmas brown when immature, almost always dark brown when mature; primary branches not wing-margined or with wings less than ½ as wide as the midribs.
    18. Spikelets (including pubescence) 1.3–3.1 mm long; lower lemmas (including pubescence) shorter than or no more than 0.5 mm longer than the upper florets; ligules 0.1–1.5 mm long, ciliate; lower lemmas sparsely to densely pubescent, the hairs less than 1 mm long, appressed, not spreading at maturity.
      19. Lower glumes 0.3–1 mm long; plants perennial, with hard, knotty, shortly rhizomatous bases; culms erect, not rooting at the lower nodes; lower lemmas 5-veined, the veins equally spaced . . . . . . . . . . . . . . . . . . . . . . . . . . . . . . . . . . . . . . . 9. *D. hitchcockii*
      19. Lower glumes absent or to 0.1 mm long; plants annual, or short-lived perennials; culms erect or decumbent, sometimes rooting at the lower nodes; lower lemmas 5–7-veined, the veins unequally spaced, the outer veins closely spaced.
        20. Upper lemmas dark brown at maturity; lower primary panicle branches without secondary branches; upper glumes with clavate to capitate hairs . . . . . . . . . . . . . 7. *D. filiformis* (in part)
        20. Upper lemmas usually gray, sometimes brown, at maturity; lower primary panicle branches with strongly divergent secondary branches; upper glumes with tapering or parallel-sided hairs . . . . . . . . . . . . . . . . . . . . . . . . . . . . . . . . . 25. *D. velutina* (in part)
    18. Spikelets (including pubescence) 3.7–7.5 mm long; lower lemmas (including pubescence) exceeding the upper florets by 0.8 mm or more; ligules 1–6 mm long, not ciliate; lower lemmas densely pubescent, the hairs 1–6 mm long, usually spreading at maturity.
      21. Terminal pedicels of primary branches 7.4–20 mm long; primary branches usually divergent, sometimes ascending, at maturity, the middle internodes (4.5)6–15 mm long . . . . . . . . . . . . . . . . . . . . . . . . . . . . . . . . . . . . . . . . . . . . . . . . . . . . 10. *D. patens*
      21. Terminal pedicels of primary branches 1.7–6(7) mm long; primary branches appressed to ascending at maturity, the middle internodes 2–6 mm long.
        22. Lower lemmas pubescent between most, sometimes all, of the veins and on the margins . . . . . . . . . . . . . . . . . . . . . . . . . . . . . . . . . . . . . . . . . . . . 12. *D. insularis*
        22. Lower lemmas glabrous between the veins, pubescent on the margins, sometimes also on the lateral veins . . . . . . . . . . . . . . . . . . . . . . . . . . . . . . . . . 11. *D. californica*
  17. Upper lemmas pale yellow, tan, or gray, sometimes purple-tinged, when immature; gray, yellow, tan, light brown, or purple at maturity; primary branches sometimes wing-margined, the margin widths various.
    23. Primary panicle branches not or only narrowly wing-margined, the wings no more than ½ as wide as the midribs.
      24. Spikelets 1.5–2.5 mm long.
        25. Upper glumes glabrous; plants rhizomatous; culms decumbent but usually not rooting at the lower nodes . . . . . . . . . . . . . . . . . . . . . . . . . . . . . . . 18. *D. abyssinica*
        25. Upper glumes shortly villous on the margins, sometimes also elsewhere; plants not rhizomatous; culms erect or decumbent and rooting at the lower nodes . . . . 21. *D. texana* (in part)
      24 Spikelets 2.5–3.6 mm long.
        26. Culms usually branching at the aerial nodes, not rooting at the lower nodes; leaf blades 2–2.2 mm wide, flat or folded; upper glumes glabrous . . . . . . . . . . . . . . . . . . . . . 19. *D. pauciflora*
        26. Culms not branching at the aerial nodes, often rooting at the lower nodes; leaf blades 2–7 mm wide, flat; upper glumes villous or glabrous.

27. Upper glumes 7–9-veined, glabrous or obscurely pubescent; plants of indefinite duration . . . . . . . . . . . . . . . . . . . . . . . . . . . . . . . . . . . . . . . . . . . . . . . 20. *D. simpsonii*
27. Upper glumes (3)5-veined, shortly villous on the margins and sometimes between the margins; plants perennial . . . . . . . . . . . . . . . . . . . . . . . . . . . . . . 21. *D. texana* (in part)
23. Primary branches winged, the wings at least ½ wide as the midribs.
   28. Plants perennial, usually stoloniferous, sometimes also rhizomatous.
      29. Leaf blades 1–3 mm wide; panicles with 2–4 primary branches, the branches 2–7 cm long . . . . . . . . . . . . . . . . . . . . . . . . . . . . . . . . . . . . . . . . . . . . . . . . . . . . . 22. *D. didactyla*
      29. Leaf blades 3–13 mm wide; panicles with 2–18 primary branches, the branches 5–25 cm long.
         30. Midveins of the lower lemmas scabrous, at least on the distal ½ . . . . . . . . . . . . 24. *D. milanjiana*
         30. Midveins of the lower lemmas smooth throughout . . . . . . . . . . . . . . . . . . . . . . . 23. *D. eriantha*
   28. Plants annual or of indefinite duration, usually neither rhizomatous nor stoloniferous.
      31. Lateral veins of the lower lemmas scabrous for the distal ⅔ of their length, sometimes scabrous throughout (use 20× magnification); leaf blades usually with papillose-based hairs on both surfaces . . . . . . . . . . . . . . . . . . . . . . . . . . . . . . . . 28. *D. sanguinalis*
      31. Lateral veins of the lower lemmas smooth throughout or scabrous only on the distal ⅓; leaf blades with or without papillose-based hairs.
         32. Lower lemmas of the lower spikelets of each pair 7-veined, the 2 lateral veins on each side crowded together near the margins, the 3 central veins equally spaced.
            33. Lower primary panicle branches with strongly divergent secondary branches . . . . . . . . . . . . . . . . . . . . . . . . . . . . . . . . . . . . . . . . . . . . . . . . . . 25. *D. velutina* (in part)
            33. Lower primary panicle branches without secondary branches.
               34. Spikelets 2.6–3.7 mm long; spikelets dimorphic with respect to their pubescence; lower lemmas of the upper spikelets of each spikelet pair with marginal hairs that become widely divergent at maturity; lower lemmas of the lower spikelets in each pair glabrous or with hairs that remain appressed at maturity; lowest panicle nodes glabrous or with hairs less than 0.4 mm long . . . . . . . . . . . . . . . . . . . . . . . . . . . . . . . . 29. *D. bicornis*
               34. Spikelets 1.7–2.8 mm long; spikelets homomorphic with respect to their pubescence; lowest panicle nodes with hairs more than 0.4 mm long.
                  35. Adaxial surfaces of the blades evenly, sometimes densely, hairy; leaf sheaths usually with scattered hairs; upper glumes ⅓–½ as long as the spikelets . . . . . . . . . . . . . . . . . . . . . . 26. *D. horizontalis* (in part)
                  35. Adaxial surfaces of the blades glabrous or with a few long hairs near the base; leaf sheaths glabrous or with a few long hairs near the base; upper glumes ⅖–⅘ as long as the spikelets . . . . . . . . . 27. *D. nuda* (in part)
         32. Lower lemmas of the lower spikelets of each pair 5- or 7-veined, the 2 or 3 lateral veins on each side crowded together near the margins, well separated from the midvein.
            36. Lower glumes absent or no more than 0.1 mm long.
               37. Anthers 0.6–1.3 mm long; upper glumes 0.2–1.3 mm long, less than 0.4 times as long as the spikelets . . . . . . . . . . . . . . . . . . . . . . . . . . . . . 31. *D. setigera*
               37. Anthers 0.3–0.6 mm; upper glumes 1–2.2 mm long, 0.4–0.8 times as long as the spikelets . . . . . . . . . . . . . . . . . . . . . . . . . . . . . . . . . 27. *D. nuda* (in part)
            36. Lower glumes 0.1–0.8 mm long.
               38. Lower glumes 0.2–0.8 mm long; primary branches glabrous or with hairs shorter than 1 mm; spikelets 2.7–4.1 mm long . . . . . . . . . . . . . . . . . . . . 30. *D. ciliaris*
               38. Lower glumes 0.1–0.2 mm long; primary branches often with scattered hairs 1–4 mm long; spikelets 1.7–2.8 mm long.
                  39. Adaxial surfaces of the blades evenly hairy, sometimes densely so; leaf sheaths usually with scattered hairs; upper glumes 0.3–0.5 times as long as the spikelets . . . . . . . . . . . . . . . . 26. *D. horizontalis* (in part)
                  39. Adaxial surfaces of the blades glabrous or with a few long hairs near the base; leaf sheaths glabrous or with a few long hairs near the base; upper glumes 0.4–0.8 as long as the spikelets . . . . . . . . 27. *D. nuda* (in part)

### 1. Digitaria arenicola (Swallen) Beetle SAND WITCHGRASS [p. 466, <u>545</u>]

*Digitaria arenicola* is endemic to deep sands along the coast of Texas, a very restricted habitat and one that is being lost to the development of coastal parks and housing.

### 2. Digitaria cognata (Schult.) Pilg. FALL WITCHGRASS [p. 466, <u>545</u>]

*Digitaria cognata* grows in dry, sandy soils in the eastern portion of the *Manual* region from southern Ontario and Vermont through the United States to southern Mexico.

### 3. Digitaria pubiflora (Vasey) Wipff WESTERN WITCHGRASS [p. 466, <u>545</u>]

*Digitaria pubiflora* grows in dry, sandy or rocky soils from Arizona to central Texas and south to central Mexico.

### 4. Digitaria tomentosa (J. König *ex* Rottler) Henrard [p. 467]

A native of southern India and Ceylon, *Digitaria tomentosa* is a noxious weed that is not known to occur in the *Manual* region. It is included here to help ensure that any introduction is correctly identified.

### 5. Digitaria gracillima (Scribn.) Fernald [p. 467, <u>545</u>]

*Digitaria gracillima* is a rare species, endemic to scrub and dry pinelands of peninsular Florida. It was formerly interpreted as including *D. bakeri*, but differs from that species both morphologically and ecologically.

### 6. Digitaria bakeri (Nash) Fernald [p. 467, <u>545</u>]

*Digitaria bakeri* grows in pastures, particularly horse pastures, from Florida through Mexico to Panama. It is probably more widespread in Florida than the map suggests but, because of its inclusion in *D. gracillima*, little information is available at present.

### 7. Digitaria filiformis (L.) Koeler SLENDER CRABGRASS [p. 467, <u>545</u>]

*Digitaria filiformis* grows throughout the warmer parts of the eastern United States, var. *filiformis* the most widespread of its varieties, extending into Mexico.

1. Lower lemmas glabrous . . . . . . . . . . . . . . . . . . var. *laeviglumis*
1. Lower lemmas pubescent.
   2. Basal leaf sheaths glabrous; cauline blades about 1 mm wide, folded or involute . . . . . . . . . . . . . . . . . . . . . . . . . . . . . . . . . . var. *dolichophylla*
   2. Basal leaf sheaths with papillose-based hairs; cauline blades 1–6 mm wide, flat.
      3. Spikelets 1.3–1.9 mm long; panicle branches 3–13 cm long; culms 10–80 cm tall . . . . . . . var. *filiformis*
      3. Spikelets 2–2.8 mm long; panicle branches 10–25 cm long; plants 75–150 cm tall . . . . . . . . var. *villosa*

### Digitaria filiformis var. dolichophylla (Henrard) Wipff [p. 467]

*Digitaria filiformis* var. *dolichophylla* is an uncommon species of moist pine barrens and open ground in southern Florida.

### Digitaria filiformis (L.) Koeler var. filiformis [p. 467]

*Digitaria filiformis* var. *filiformis* is a weed of sandy fields and open, disturbed ground in the southeastern United States and Mexico.

### Digitaria filiformis var. laeviglumis (Fernald) Wipff [p. 467]

*Digitaria filiformis* var. *laeviglumis* is endemic to sandy soils in New England.

### Digitaria filiformis var. villosa (Walter) Fernald [p. 467]

*Digitaria filiformis* var. *villosa* has essentially the same geographic range as var. *filiformis* and grows in similar habitats.

### 8. Digitaria leucocoma (Nash) Urb. [p. 467, <u>545</u>]

*Digitaria leucocoma* is known only from high pinelands near Lake Ella, Lake County, Florida. It has been treated in the past as a synonym of *D. filiformis* var. *villosa*.

### 9. Digitaria hitchcockii (Chase) Stuck. [p. 467, <u>545</u>]

*Digitaria hitchcockii* is an uncommon species of open, dry, gravelly slopes in southwestern Texas and northern Mexico.

### 10. Digitaria patens (Swallen) Henrard [p. 467, <u>545</u>]

*Digitaria patens* is endemic to southwestern and southern Texas and adjacent Mexico. It grows in well-drained, usually sandy, soils, often in disturbed habitats.

### 11. Digitaria californica (Benth.) Henrard [p. 467, <u>545</u>]

*Digitaria californica* grows on plains and open ground from California, Arizona, southern Colorado, and Oklahoma through Mexico and Central America to South America. Plants in the *Manual* region belong to **D. californica** (Benth.) Henrard var. **californica**. They differ from those of **D. californica** var. **villosissima** Henrard in having densely villous, rather than densely tomentose, leaves.

### 12. Digitaria insularis (L.) Mez *ex* Ekman [p. 467, <u>545</u>]

*Digitaria insularis* grows in low, open ground of the southern United States, and extends to the West Indies, Mexico, and through Central America to Argentina.

### 13. Digitaria serotina (Walter) Michx. [p. 467, <u>545</u>]

*Digitaria serotina* is native to the coastal plain of the southeastern United States. It has also been found in Cuba, possibly as an introduction, and on a ballast dump in Philadelphia, Pennsylvania. Its densely hairy sheaths and short, densely hairy blades make this one of the more distinctive species of *Digitaria* in the *Manual* region.

### 14. Digitaria longiflora (Retz.) Pers. [p. 468, <u>545</u>]

*Digitaria longiflora* is native to Africa and Asia. It is now established in disturbed areas of Florida, growing on railroad grades and in pastures and lawns.

### 15. Digitaria floridana Hitchc. [p. 468, <u>545</u>]

*Digitaria floridana* is a rare species that is known only from sandy pine woods in Hernando County, Florida.

### 16. Digitaria ischaemum (Schreb.) Muhl. SMOOTH CRABGRASS, DIGITAIRE ASTRINGENTE [p. 468, <u>546</u>]

*Digitaria ischaemum* is a Eurasian weed that is now common in lawns, gardens, fields, and waste ground in warm-temperate regions throughout the world, including much of the *Manual* region.

### 17. Digitaria violascens Link [p. 468, <u>546</u>]

*Digitaria violascens* is a weedy species that is native to tropical regions of the Eastern Hemisphere. It is now established in the *Manual* region, primarily in the southeastern United States, and in Mexico and Central America. It grows in disturbed sites.

### 18. Digitaria abyssinica (Hochst. *ex* A. Rich.) Stapf [p. 468]

Introduced from Africa, *Digitaria abyssinica* is not known to be established in the *Manual* region, although it has occasionally been cultivated in the southern United States. It is considered a potentially serious weed threat by the U.S. Department of Agriculture.

### 19. Digitaria pauciflora Hitchc. [p. *468*, 546]

*Digitaria pauciflora* is known only from pinelands in Dade, Collier, and Monroe counties in Florida.

### 20. Digitaria simpsonii (Vasey) Fernald [p. *468*, 546]

*Digitaria simpsonii* is a rare species, known only from sandy fields in Florida.

### 21. Digitaria texana Hitchc. [p. *468*, 546]

*Digitaria texana* grows in sandy oak woods and prairies of southern Texas and Florida.

### 22. Digitaria didactyla Willd. BLUE COUCH [p. *468*]

A native of Africa, *Digitaria didactyla* is often cultivated as a lawn grass in tropical and subtropical regions. It has been grown experimentally in Florida, but is not otherwise known from the *Manual* region.

### 23. Digitaria eriantha Steud. [p. *468*, 546]

*Digitaria eriantha* is an African species that is widely cultivated in warm climates as a pasture grass. The appearance of the spikelets varies considerably with the length of the hairs, those of subsp. *eriantha* usually being longer than those of subsp. *pentzii*. Several cultivars have been released for forage and hay use. One cultivar, 'Survenola', that has been released for use in the tropics and on well-fertilized upland soils in Florida, has much wider leaf blades than any other cultivars that have been released so far (usually 10–13 mm wide, rather than usually less than 8 mm) and glabrous leaf sheaths.

1. Plants cespitose . . . . . . . . . . . . . . . . . . . . . . . . subsp. *eriantha*
1. Plants stoloniferous . . . . . . . . . . . . . . . . . . . . . subsp. *pentzii*

### Digitaria eriantha Steud. subsp. eriantha [not illustrated]

*Digitaria eriantha* subsp. *eriantha* is seed-producing, but it is not clear whether it is being grown in the *Manual* region.

### Digitaria eriantha subsp. pentzii (Stent) Kok [p. *468*]

In the Western Hemisphere, *D. eriantha* subsp. *pentzii* has traditionally been referred to as *D. decumbens* Stent. It is widely cultivated as a forage grass throughout the tropics at low and intermediate altitudes (up to 2000 m). It does not set seed; propagation is by sprigging the stolons.

### 24. Digitaria milanjiana (Rendle) Stapf [p. *469*, 546]

*Digitaria milanjiana* is native to tropical and subtropical Africa. It has been found as an escape from experimental plantings in Florida.

### 25. Digitaria velutina (Forssk.) P. Beauv. [p. *469*]

*Digitaria velutina* is an African species, appearing on the noxious weed list of the U.S. Department of Agriculture. It has been erroneously reported as occurring in Texas.

### 26. Digitaria horizontalis Willd. [p. *469*, 546]

*Digitaria horizontalis* is native to tropical regions of the Americas. It has been found in hammocks and disturbed areas in central and southern Florida and at a few other locations in the southeastern United States, including ballast dumps in Mobile, Alabama.

### 27. Digitaria nuda Schumach. [p. *469*, 546]

*Digitaria nuda* is an African species that is now established in tropical regions throughout the world, including the Americas. So far as is known, it has only been collected once in the *Manual* region, in Columbia County, Florida.

### 28. Digitaria sanguinalis (L.) Scop. HAIRY CRABGRASS, DIGITAIRE SANGUINE [p. *469*, 546]

*Digitaria sanguinalis* is a weedy Eurasian species that is now found in waste ground of fields, gardens, and lawns throughout much of the world, including the *Manual* region.

### 29. Digitaria bicornis (Lam.) Roem. & Schult. [p. *469*, 546]

*Digitaria bicornis* is a common species on the sandy coastal plain of the southeastern United States. Its range extends through Mexico to Costa Rica and northern South America, as well as to the West Indies. The Californian record reflects a 1926 collection; the species is not known to be established in the state.

### 30. Digitaria ciliaris (Retz.) Koeler SOUTHERN CRABGRASS [p. *469*, 546]

*Digitaria ciliaris* is a weedy species, found in open, disturbed areas in most warm-temperate to tropical regions, primarily in the eastern United States. It is particularly abundant in the southeast. So far as is known, the two varieties distinguished in the following key do not differ in any other characters.

1. Lower lemmas without glassy yellow hairs . . . . . . . . var. *ciliaris*
1. Lower lemmas with glassy yellow hairs . . . . . . . var. *chrysoblephara*

*Digitaria ciliaris* (Retz.) Koeler var. *ciliaris* is the more common of the two varieties in the *Manual* region. **Digitaria ciliaris** var. **chrysoblephara** (Fig. & De Not.) R.R. Stewart is more common in southeast Asia, but it has been found in the northeastern United States.

### 31. Digitaria setigera Roth [p. *469*, 546]

*Digitaria setigera* is native to southeastern Asia. It is now established in tropical America, growing in disturbed habitats in Florida and Central America, and probably in tropical South America. It has often been confused with *D. sanguinalis*. Plants in the *Manual* region belong to **Digitaria setigera** Roth var. **setigera**. Unlike plants of **D. setigera** var. **calliblepharata** (Henrard) Veldkamp, they do not have large, glassy hairs on their lower lemmas.

---

## 24.02 ANTHENANTIA P. Beauv.[1]

**Pl** per; rhz. **Clm** 60–120 cm, stiffly erect, clustered. **Shth** open; **lig** 0.1–0.3 mm, memb, ciliate; **bld** flat, stiff, up bld much rdcd. **Infl** tml, simple pan with elongate rchs; **br** ascending to erect, not spikelike, lo br usu with 8 or more spklt; **dis** beneath the glm. **Spklt** 3–4 mm, globose, neither subtended by bristles nor sunken into the rchs, elliptic, obovoid, hirsute, unawned, with 2 flt. **Lo glm** absent; **up glm** as long as the spklt, obovate, densely hirsute, 5-veined; **lo flt** strl or stmt; **lo lm** densely hirsute, similar to the up glm; **lo pal** present or absent; **up lm** and **up pal** cartilaginous, glab, dark brown, separating slightly at maturity, exposing the car, lm 3-veined, mrg flat, hyaline, 0.5–1 mm wide, pal 2-veined.

*Anthenantia* has two species, both of which are endemic to the southeastern United States. It is very similar to *Leptocoryphium* Nees, a unispecific genus that extends from Mexico to Argentina, but differs in having laterally, rather than dorsally, compressed spikelets.

[1]J.K. Wipff

1. Junction of the sheath and blade not evident abaxially, the sheath and blades in line with each other; blades 2–5 mm wide, the margins scabrous; leaves 30–60 cm long . . . . . . . . . . . . . . . . . . . . . . . . . . . . . . . . . . . . . . 1. *A. rufa*
1. Junction of the sheath and blade evident, the blades not in line with the sheaths; blades 5–10 mm wide, the margins papillose-hispid; leaves mostly less than 30 mm long . . . . . . . . . . . . . . . . . . . . . . . . . . . . . . . . 2. *A. villosa*

### 1. Anthenantia rufa (Elliott) Schult. PURPLE SILKYSCALE [p. 469, 546]

*Anthenantia rufa* grows in wet pine flatwoods and savannahs, sphagnous streamhead ecotones, and pitcher plant bogs on the southeastern coastal plain from eastern Texas to North Carolina.

### 2. Anthenantia villosa (Michx.) P. Beauv. GREEN SILKYSCALE [p. 469, 546]

*Anthenantia villosa* grows in dry, usually sandy soil on the southeastern coastal plain from eastern Texas to North Carolina. It usually grows in wetter habitats in southeastern Texas than in other portions of its range.

## 24.03 ALLOTEROPSIS J. Presl[1]

**Pl** ann or per; csp. **Clm** 15–150 cm, erect or decumbent and rooting at the lo nd, erect distally; **nd** glab or pubescent. **Lvs** mostly bas; **shth** stiffly pubescent; **lig** of hairs or memb and ciliate; **bld** linear to lanceolate. **Infl** tml, pan of 4–11 digitate or subdigitate, spikelike br; **br** naked at the base, spklt in pairs or triplets, lo lm and lo glm closest to the br axes; **dis** below the glm. **Spklt** 2.5–7 mm, dorsally compressed, with 2 flt. **Glm** acute to shortly awned; **lo glm** about ½ as long as the spklt, 1–2-veined; **up glm** equaling the spklt, 3–5-veined, mrg ciliate; **lo flt** stmt; **lo lm** papery; **lo pal** rdcd, deeply bifid, 1-veined; **up flt** bisx; **up lm** stiffly memb, mrg involute, apc attenuate into a mucro or curved awn; **anth** 3. **Car** elliptic.

*Alloteropsis* has five to eight species that are native to tropical Asia and Australia. One species is now established in the *Manual* region.

### 1. Alloteropsis cimicina (L.) Stapf BUGSEED GRASS [p. 470, 546]

*Alloteropsis cimicina* is native to southeast Asia but has been collected in Alachua and Columbia counties, Florida, and Baltimore, Maryland. Being a weedy species, it should be sought at other disturbed locations along the east coast of the United States.

## 24.04 AMPHICARPUM Kunth[2]

**Pl** ann or per; rhz, rhz slender, terminating in a rdcd pan of cleistogamous spklt. **Clm** 30–100 cm, erect or decumbent. **Shth** open; **aur** absent; **lig** of hairs; **bld** flat. **Infl** subterranean and aerial, only the subterranean infl forming mature car; **subterranean pan** with 1–5 spklt; **aerial pan** tml, simple, with elongate rchs bearing erect to ascending br, usu with 15 or more spklt. **Spklt** glab, unawned, with 2 flt. **Subterranean spklt** setting sd, with 1 glm; **lo glm** absent; **up glm** and **lo lm** similar in size and texture, exceeded by the up flt; **up flt** turgid, ellipsoidal; **up lm** mostly indurate, mrg thin, flat, apc acuminate; **up pal** similar in texture to the lm; **anth** 3; **car** well-developed. **Aerial spklt** not setting sd, smt forming immature car, lanceoloid, dorsally compressed to terete; **glm** unequal or the lo glm absent; **up glm** and **lo lm** similar in size and texture; **up lm** mostly indurate, mrg thin, flat, apc acute; **lo flt** stmt or strl; **up flt** with pistils but fruit not developed.

*Amphicarpum* has two species, both endemic to the southeastern United States. It differs from all other North American grass genera in its production of subterranean, cleistogamous spikelets.

1. Leaf blades conspicuously hirsute; plants annual; culms erect . . . . . . . . . . . . . . . . . . . . . . . . . . . . . . . . . . . 1. *A. amphicarpon*
1. Leaf blades glabrous or almost glabrous; plants perennial; culms usually decumbent . . . . . . . . . . . . 2. *A. mühlenbergianum*

### 1. Amphicarpum amphicarpon (Pursh) Nash PURSH'S BLUE MAIDENCANE, HAIRY MAIDENCANE [p. 470, 546]

*Amphicarpum amphicarpon* grows in sandy pinelands of the eastern United States.

### 2. Amphicarpum mühlenbergianum (Schult.) Hitchc. BLUE MAIDENCANE [p. 470, 546]

*Amphicarpum mühlenbergianum* grows in damp areas, such as dried pond bottoms, ditches, flatwoods, and swampy pinewoods of the southeastern United States.

## 24.05 LASIACIS (Griseb.) Hitchc.[3]

**Pl** per (rarely ann); csp. **Clm** 0.5–8 m, weakly lignified, erect, arching, climbing, or decumbent, rooting at the nd. **Shth** open; **lig** memb, smt ciliate; **psdpet** smt present; **bld** linear to ovate, bases slightly to strongly asymmetric. **Infl** open or contracted pan, rchs usu visible, even distally, spklt attached obliquely to the ped; **dis** below the glm. **Spklt** subglobose to globose, with 2 flt. **Glm** memb, apc lanate pubescent, abruptly apiculate; **lo glm** ⅓–⅔ as long as the spklt, 5–13-veined, bases saccate, mrg overlapping; **up glm** about as long as the up flt, not

[1]David W. Hall  [2]J.K. Wipff  [3]Gerrit Davidse

saccate, 7–15-veined; **lo flt** strl or stmt; **lo lm** memb, apc lanate pubescent, abruptly apiculate; **lo pal** present, smt rdcd; **up flt** stipitate, bisx, appearing to be mucronate or acuminate; **up lm** indurate, usu broadly elliptic to obovate, mrg enclosing the edges of the pal, apc obtuse, somewhat woolly pubescent, usu dark brown at maturity; **up pal** similar to the lm, but saccate below and gibbous above. **Car** plano-convex, ovoid, or nearly orbicular; **emb** about ½ as long as the car; **hila** oblong to nearly round.

*Lasiacis* is a neotropical genus of 16 species that extends from southern Florida to Peru and Argentina. Two species are native to the *Manual* region. The shiny black color of its mature florets and the oil-filled cells of the inner epidermes of the glumes and sterile lemmas distinguish *Lasiacis* from all other grasses. The upper florets of *Lasiacis* species appear to be mucronate or acuminate. The mucro or acuminate apex is formed by the tuft of hairs at the apex of the upper floret.

1. Leaf blades linear-lanceolate to narrowly lanceolate, 3–16 cm long, 3–30 mm wide . . . . . . . . . . . . . . . . . . . . 1. *L. divaricata*
1. Leaf blades ovate to broadly lanceolate, 2–16 cm long, 8–56 mm wide . . . . . . . . . . . . . . . . . . . . . . . . . . . 2. *L. ruscifolia*

### 1. Lasiacis divaricata (L.) Hitchc. [p. 470, 546]

*Lasiacis divaricata* is a Caribbean species. Its range extends from Florida through the West Indies to Mexico, Panama, and northern Venezuela. In Florida, it usually grows in hammocks, but occasionally in pinelands. The whitish to brown upper florets are unusual in the genus. Plants in the *Manual* region belong to **Lasiacis divaricata** var. **divaricata**, which differs from the other two varieties in having panicles with fewer spikelets and panicle branches that are usually reflexed.

### 2. Lasiacis ruscifolia (Kunth) Hitchc. [p. 470, 546]

The range of *Lasiacis ruscifolia* extends from southern Florida to Peru. Plants in the *Manual* region belong to **L. ruscifolia** (Kunth) Hitchc. var. **ruscifolia**, which differs from **L. ruscifolia** var. **velutina** (Swallen) Davidse in being scabrous or puberulent, rather than velutinous, on the panicle branches.

## 24.06 OPLISMENUS P. Beauv.[1]

**Pl** ann or per. **Clm** 10–100 cm, weak, trailing on the ground, brchg. **Lvs** cauline; **lig** memb and ciliate, or of hairs; **bld** lanceolate. **Infl** tml, pan of unilat br, spklt paired (but the first spklt smt rdcd), rchs and br terminating in a spklt; **br** 0.1–7 cm, persistent; **dis** below the glm. **Spklt** dorsally compressed, not sunken into the rchs, lacking subtending bristles, with 2 flt. **Lo glm** awned; **up glm** not ciliate on the mrg, unawned or with awns shorter than those of the lo glm, awns of

both glm often becoming viscid; **lo flt** strl or stmt; **lo lm** acute to shortly awned; **lo pal** present or absent; **up flt** bisx; **up lm** papery to leathery, glab, smooth, unawned, white or yellow at maturity; **up pal** similar to the up lm.

*Oplismenus* is a genus of five closely related species that grow in shady, mesic forests of tropical and subtropical regions. One species is native to the *Manual* region. The awns of most species become viscid at maturity, aiding in fruit dispersal.

### 1. Oplismenus hirtellus (L.) P. Beauv. BRISTLE BASKETGRASS [p. 470, 546]

*Oplismenus hirtellus* grows at scattered locations in the southeastern United States, extending south in subtropical and tropical habitats to Argentina. The key below includes the three subspecies attributed to the *Manual* region.

*Oplismenus hirtellus* subsp. *fasciculatus* U. Scholz is restricted to southern Louisiana and eastern Mexico; *O. hirtellus* subsp. *setarius* (Lam.) Mez ex Ekman is more widely distributed, growing both in the southeastern United States and Mexico. *Oplismenus hirtellus* subsp. *undulatifolius* (Ard.) U. Scholz was found recently in Maryland; it is native to the Eastern Hemisphere.

1. Sheaths and culms noticeably pilose, the hairs 1–3
   mm long; lemmas 7-veined . . . . . . . . . . . . subsp. *undulatifolius*
1. Sheaths and culms glabrous or with a few scattered
   hairs less than 1 mm long; lemmas (7)9–11-veined.
   2. Spikelets 2.2–2.7(3.3) mm long; lowest panicle
      branches 0.1–0.5 cm long . . . . . . . . . . . . . . subsp. *setarius*
   2. Spikelets 3–3.4(4.5) mm long; lowest panicle
      branches 0.5–0.7 cm long . . . . . . . . . . . . subsp. *fasciculatus*

## 24.07 ECHINOCHLOA P. Beauv.[2]

**Pl** ann or per; with or without rhz. **Clm** 10–460 cm, prostrate, decumbent or erect, distal portions smt floating, smt rooting at the lo nd; **nd** usu glab; **intnd** hollow or solid. **Shth** open, compressed; **aur** absent; **lig** usu absent but, if present, of hairs; **bld** linear to linear-lanceolate, usu more than 10 times longer than wide, flat, with a prominent midrib. **Infl** tml, pan of simple or compound spikelike br attached to elongate rchs, axes

not terminating in a bristle, spklt subsessile, densely packed on the angular br; **dis** below the glm (cultivated taxa not or tardily disarticulating). **Spklt** plano-convex, with 2(3) flt; **lo flt** strl or stmt; **up flt** bisx, dorsally compressed. **Glm** memb; **lo glm** usu ¼–⅖ as long as the spklt (varying to more than ½ as long), unawned to minutely awn-tipped; **up glm** unawned or shortly awned; **lo lm** similar to the up glm in length and texture,

[1]J.K. Wipff  [2]P.W. Michael

unawned or awned, awns to 60 mm; **lo pal** vestigial to well-developed; **up lm** coriaceous, dorsally rounded, mostly smooth, apc short or elongate, firm or memb, unawned; **up pal** free from the lm at the apc; **lod** absent or minute; **anth** 3. **Car** ellipsoid, broadly ovoid or spheroid; **emb** usu 0.7–0.9 times as long as the car.

*Echinochloa* is a tropical to warm-temperate genus of 40–50 species that are usually associated with wet or damp places. Many of the species are difficult to distinguish because they tend to intergrade. Some of the characters traditionally used for distinguishing taxa, e.g., awn length, are affected by the amount of moisture available; others reflect selection by cultivation, e.g., non-disarticulation in grain taxa, mimicry of rice as weeds of rice fields. There are 13 species in the *Manual* region: five native and one possibly native, four established, two grown as commercial crops, and one in research.

NOTE: In this treatment, measurements of the spikelet do not include the awns. The color of the caryopses is based on fully ripe caryopses.

1. Ligules of stiff hairs present on the lower leaves; lower florets staminate; plants perennial.
  2. Plants without scaly rhizomes, sometimes rooting at the lower nodes; lower lemmas usually awned, sometimes merely apiculate; known outside of experimental plantings . . . . . . . . . . . . . . . . . . . . . . . . 1. *E. polystachya*
  2. Plants with short, scaly rhizomes; lower lemmas unawned, sometimes long-cuspidate; in the *Manual* region, known only from experimental plantings . . . . . . . . . . . . . . . . . . . . . . . . . . . . . . . . . . . . . . . . 2. *E. pyramidalis*
1. Ligules almost always absent from all leaves, the ligule region sometimes pubescent; lower florets sterile or staminate; plants usually annual, sometimes short-lived perennials.
  3. Lower lemmas usually unawned; spikelets, particularly those near the base of the panicles, not disarticulating at maturity; upper lemmas wider and longer than the upper glumes at maturity and, hence, exposed at maturity.
    4. Spikelets always green and pale at maturity, their apices usually obtuse, varying to acute; rachis nodes not or only sparsely hispid with papillose-based hairs; caryopses whitish . . . . . . . . . . . . . . . 9. *E. frumentacea*
    4. Spikelets purplish to blackish-brown at maturity, their apices obtuse to shortly acute; rachis nodes densely hispid with papillose-based hairs; caryopses brownish . . . . . . . . . . . . . . . . . . . . . . . . 11. *E. esculenta*
  3. Lower lemmas often awned; spikelets disarticulating at maturity; upper lemmas not or scarcely exceeding the upper glumes in length and width at maturity.
    5. Plants essentially obligate weeds of rice, growing in the fields; culms erect, densely tufted; spikelets 3.7–7 mm long; plants resembling rice in their vegetative growth.
      6. Panicles horizontal or drooping at maturity; spikelets broadly ovate to ovate; lower lemmas usually awned; caryopses 1.9–3 mm long, the embryos 70–85% as long as the caryopses . . . . . . . . 12. *E. oryzoides*
      6. Panicles erect to slightly drooping; spikelets ovate-elliptical; lower lemmas awned or not; caryopses 1.7–2.6 mm long, the embryos 89–98% as long as the caryopses . . . . . . . . . . . . . . . . 13. *E. oryzicola*
    5. Plants not obligate weeds of rice, found in summer crops and wet places, and often in rice fields; culms sprawling, decumbent or erect; spikelets 2–5 mm long; plants occasionally resembling rice vegetatively but, if so, the spikelets less than 3 mm long.
      7. Lower florets staminate; anthers of the upper florets 1.2–1.7 mm long . . . . . . . . . . . . . . . . . . . . . 3. *E. paludigena*
      7. Lower florets sterile; anthers of the upper florets 0.5–1.2 mm long.
        8. Panicle branches 0.7–2(4) cm long, without secondary branches; spikelets 2–3 mm long, unawned . . . . . . . . . . . . . . . . . . . . . . . . . . . . . . . . . . . . . . . . . . . . . . . . . . . . . . . . . . . . . . . 8. *E. colona*
        8. Panicle branches 1–14 cm long, usually rebranched, the secondary branches often short and inconspicuous; spikelets 2.5–5 mm long, awned or unawned.
          9. Upper lemmas broadly ovate to elliptical, if elliptical, each with a line of minute (need 25× magnification) hairs across the base of the early-withering tips.
            10. Upper lemmas with rounded or broadly acute coriaceous apices that pass abruptly into a membranous tip, a line of minute hairs present at the base of the tip . . . . . . . . . . 10. *E. crus-galli*
            10. Upper lemmas with acute or acuminate coriaceous apices that extend into the membranous tip, without hairs at the base of the tip . . . . . . . . . . . . . . . . . . . . . . . . . 4. *E. muricata*
          9. Upper lemmas narrowly ovate to elliptical, never with a line of minute hairs across the base of the early-withering, membranous tips.
            11. Spikelets 2.5–3.4 mm long; lower lemmas unawned or with awns 3–10(15) mm long, curved . . . . . . . . . . . . . . . . . . . . . . . . . . . . . . . . . . . . . . . . . . . . . . . . . . . . . . . . . 6. *E. crus-pavonis*
            11. Spikelets 3–5 mm long; lower lemmas usually with awns 8–25 mm long, typically straight.
              12. Blades 10–35(60) mm wide; sheaths usually hispid and the margins ciliate with prominent papillose-based hairs, sometimes the sheaths only papillose; lower lemmas awned, the awns 8–25(60) mm long; common in the eastern portion of the *Manual* region . . . . . . . . . . . . . . . . . . . . . . . . . . . . . . . . . . . . . . . . . . . . . . . 5. *E. walteri*

12. Blades 5–10 mm wide; sheaths glabrous or with papillose-based hairs; lower
    lemmas unawned or awned, the awns 8–16(50) mm long; in the *Manual*
    region, known only from southern Arizona . . . . . . . . . . . . . . . . . . . . . . . . . . . 7. *E. oplismenoides*

## 1. Echinochloa polystachya (Kunth) Hitchc. Creeping River Grass [p. *471*, 546]

*Echinochloa polystachya* grows in coastal marshes, often in standing water, from Texas and Louisiana south through Mexico and the Caribbean islands to Argentina. Two varieties exist. **Echinochloa polystachya** var. **polystachya** has glabrous culms and leaf sheaths; **Echinochloa polystachya** var. **spectabilis** (Nees *ex* Trin.) Mart. Crov. has swollen, pubescent cauline nodes and pubescent leaf sheaths.

## 2. Echinochloa pyramidalis (Lam.) Hitchc. & Chase Antelope Grass [p. *471*]

*Echinochloa pyramidalis* is native to Africa, where it is used both as a cereal and a pasture grass. It has been grown experimentally in Gainesville, Florida, but it is not established in North America.

## 3. Echinochloa paludigena Wiegand Florida Barnyard Grass [p. *471*, 546]

*Echinochloa paludigena* is native to swamps, riverbanks, and other wet habitats. Reports from Texas and Louisiana appear to be based on misidentifications; *E. paludigena* may be a Florida endemic.

## 4. Echinochloa muricata (P. Beauv.) Fernald American Barnyard Grass [p. *471*, 546]

*Echinochloa muricata* is native to North America, growing from southern Canada to northern Mexico in moist, often disturbed sites (but not rice fields). It resembles *E. crus-galli* in gross morphology and ecology, but differs consistently by the characters used in the key. The two varieties tend to be distinct, but there is some overlap in both morphology and geography.

1. Spikelets 2.5–3.8 mm long; lower lemmas unawned
   or awned, the awns to 10 mm long . . . . . . . . var. *microstachya*
1. Spikelets 3.5–5 mm long; lower lemmas usually
   awned, the awns 6–16 mm long . . . . . . . . . . . . . . var. *muricata*

## Echinochloa muricata var. microstachya Wiegand Échinochloa Piquant [p. *471*]

*Echinochloa muricata* var. *microstachya* is the common variety in the western part of North America, extending east to the Missouri River and the Texas panhandle.

## Echinochloa muricata (P. Beauv.) Fernald var. muricata Échinochloa de l'Ouest [p. *471*]

*Echinochloa muricata* var. *muricata* is the common variety in eastern North America.

## 5. Echinochloa walteri (Pursh) A. Heller Coast Barnyard Grass, Échinochloa de Walter [p. *471*, 546]

*Echinochloa walteri* grows in wet places, often in shallow water and brackish marshes. It is a native species that extends through Mexico to Guatemala. It is found in both disturbed and undisturbed sites although not in rice fields. Occasional specimens of *E. walteri* with glabrous lower sheaths and short awns differ from *E. crus-pavonis* in their less dense panicles.

## 6. Echinochloa crus-pavonis (Kunth) Schult. Gulf Barnyard Grass [p. *471*, 547]

*Echinochloa crus-pavonis* is a native species found in scattered locations from British Columbia to Arizona, east to Florida, and south into South America. It favors marshes and wet places at lower elevations, often being found in the water.

1. Lower paleas more than ½ as long as the lemmas;
   panicles usually drooping . . . . . . . . . . . . . . . . var. *crus-pavonis*
1. Lower paleas absent or much less than ½ as long as
   the lemmas; panicles usually stiffly erect . . . . . . . . . . var. *macra*

## Echinochloa crus-pavonis (Kunth) Schult. var. crus-pavonis [p. *471*]

This is generally the more southern of the two varieties, extending through Mexico and the Caribbean to Bolivia and Argentina. It appears, presumably as an adventive species, as far north as Humboldt County, California.

## Echinochloa crus-pavonis var. macera (Wiegand) Gould [p. *471*]

This variety extends south only as far as northern Mexico. On further review, it has been decided that the correct spelling of the epithet is 'macera'. See FNA 24:792 for further discussion.

## 7. Echinochloa oplismenoides (E. Fourn.) Hitchc. Chihuahuan Barnyard Grass [p. *471*, 547]

*Echinochloa oplismenoides* was first found in the United States, in southern Arizona, in 1993. It was previously known only from Mexico, with a range that extends from northwestern Mexico to Guatemala. The southern Arizonan plants were found near a cattle tank in wet grasslands.

## 8. Echinochloa colona (L.) Link Awnless Barnyard Grass [p. *471*, 547]

*Echinochloa colona* is widespread in tropical and subtropical regions. It is adventive and weedy in North America, growing in low-lying, damp to wet, disturbed areas, including rice fields. The unbranched, rather widely spaced panicle branches make this one of the easier species of *Echinochloa* to recognize.

## 9. Echinochloa frumentacea Link Siberian Millet, White Panic [p. *471*]

*Echinochloa frumentacea* originated in India, and possibly also in Africa. It is found occasionally in the contiguous United States and southern Canada, the primary source being birdseed mixes. It differs from *E. esculenta* in its whitish caryopses and proportionately smaller embryos.

## 10. Echinochloa crus-galli (L.) P. Beauv. Barnyard Grass, Échinochloa Pied-de-Coq [p. *472*, 547]

*Echinochloa crus-galli* is a Eurasian species that is now widely established in the *Manual* region, where it grows in moist, disturbed sites, including rice fields. Some North American taxonomists have interpreted *E. crus-galli* much more widely; others treat it as here, but recognize several infraspecific taxa based on such characters as trichome length and abundance, and awn length. There are several ecological and physiological ecotypes within the species, but the correlation between these and the species' morphological variation has not been established, so no infraspecific taxa are recognized here.

## 11. Echinochloa esculenta (A. Braun) H. Scholz Japanese Millet [p. *472*]

*Echinochloa esculenta* was derived from *E. crus-galli* in Japan, Korea, and China. It is cultivated for fodder, grain, or birdseed. It has sometimes been included in *E. frumentacea*, from which it differs in its brownish caryopses and longer pedicels.

## 12. Echinochloa oryzoides (Ard.) Fritsch EARLY BARNYARD GRASS [p. 472, 547]

*Echinochloa oryzoides* is a common weed of rice fields throughout the world, growing in the flooded portions of the fields. It differs from *E. oryzicola* in its shorter embryo, lax, strongly drooping panicle, and earlier (June–July) flowering period. This flowering period is also earlier than that of *Oryza*. In addition, *E. oryzoides* is usually conspicuously awned, having longer awns than even the awned variants of *E. oryzicola*, and it is rarely obviously pubescent on the cauline nodes, leaf sheaths, and collars..

## 13. Echinochloa oryzicola (Vasinger) Vasinger LATE BARNYARD GRASS [p. 472, 547]

Like *Echinochloa oryzoides*, *E. oryzicola* is an introduced weed of rice fields, where it grows in the flooded portion, with the rice. The two are quite distinct, with *E. oryzicola* flowering after *Oryza* and having a longer embryo and an erect panicle. It is also more likely to have evidently pubescent cauline nodes, leaf sheaths, and collars than *E. oryzoides* and is never conspicuously awned.

## 24.08 SACCIOLEPIS Nash[1]

Pl ann or per; rhz, stln, or csp. Clm 5–150 cm, not wd, brchd above the base; intnd hollow. Lvs cauline; aur smt present; lig memb, smt ciliate; bld flat or rolled, with or without cross venation. Infl tml, usu contracted, dense pan, distal ½ of the rchs concealed by the spklt; br fused to the rchs or free and appressed to ascending; ped with discoid apc; dis below the glm and below the up flt. Spklt bisx, with 2 flt, rounded to acute; rchl intnd not swollen. Glm unequal, prominently veined, unawned; lo glm 3–7-veined; up glm as long as or exceeding the up flt, distinctly saccate or gibbous, 5–13-veined; lo flt 0.8–1.9 mm, strl or stmt, less than ½ as long as the spklt; lo lm resembling the up glm but not saccate, smt with a transverse row of hairs, 5–9-veined, unawned; lo pal present or absent, 0–2-veined; up lm subcoriaceous to subindurate, dorsally compressed, glab, smooth, mrg inrolled or flat, never hyaline, faintly 3–5-veined; up pal similar to the lm, 2-veined; lod 2, fleshy, glab.

*Sacciolepis* is a genus of 30 species. It is represented throughout the tropics and subtropics, primarily in Africa. Two species grow in the *Manual* region. One is native; the other is an introduction that has become established. Most species grow along and in ponds, lakes, streams, ditches, and other moist areas. The prominently multi-veined, saccate upper glumes and contracted panicles distinguish *Sacciolepis* from all other grasses in the *Manual* region.

1. Primary branches fused to the rachises for at least ¾ of their length; lower branches 0.1–0.5 cm long; upper glumes 9-veined; paleas of the lower florets 0.5–1 mm long, to ½ as long as the lower lemmas . . . . . . . . . . 1. *S. indica*
1. Primary branches ascending, free from the rachises; lower branches 0.4–11.5 cm long; upper glumes 11(12)-veined; paleas of the lower florets 2–4 mm long, ¾ to almost as long as the lower lemmas . . . . . . . . . . 2. *S. striata*

## 1. Sacciolepis indica (L.) Chase CHASE'S GLENWOODGRASS [p. 472, 547]

*Sacciolepis indica* is native to the Eastern Hemisphere tropics. It is now established in the coastal states of the southeastern United States, where it grows in and along streams, ponds, lakes, ditches, and other moist places. It flowers from late summer to fall.

## 2. Sacciolepis striata (L.) Nash AMERICAN CUPSCALE [p. 472, 547]

*Sacciolepis striata* is native to the southeastern United States and the West Indies, and from the Guianas to Venezuela and Amapá, Brazil. It grows along and in ponds, lakes, streams, and ditches, and flowers in late summer to fall.

## 24.09 DICHANTHELIUM (Hitchc. & Chase) Gould[2]

Pl per; csp, smt rhz, smt with hard, cormlike bases, often with bas winter rosettes of lvs having shortly ovate to lanceolate bld, these often sharply distinct from the bld of the cauline lvs. Clm 5–150 cm, hrb, hollow, usu erect or ascending, rarely sprawling, in the spring often spreading, smt decumbent in the fall, usu brchg from the mid- or lo clm nd in summer and fall; br rebrchg 1–4 times, terminating in small sec pan that are usu partly included in the shth. Cauline lvs 3–14, usu distinctly longer and narrower than the rosette bld; lig of hairs, memb, or memb and ciliate, smt absent; psdlig of 1–5 mm hairs often present at the bases of the bld immediately behind the true lig; bld usu distinctly longer and narrower than those of the bas rosette, cross sections with non-Kranz anatomy; photosynthesis C$_3$. Infl pan, tml on the clm and br; strl br and bristles absent; dis below the glm. Pri pan terminating the clm, developing April–June(July), smt also in late fall, usu at least partially chasmogamous, often with a lo sd set than the sec pan; sec pan terminating the br, produced from (May)June to fall, usu partially or totally cleistogamous. Spklt 0.8–5.2 mm, not subtended by bristles, dorsally compressed, surfaces unequally convex, apc unawned. Glm apc not or only slightly gaping at maturity; lo glm ⅕–¾ as long as the spklt, 1–5-veined, truncate, acute, or acuminate; up glm slightly shorter than the spklt or exceeding the up flt by up to 1 mm, 5–11-veined, not saccate, apc rounded to attenuate. Lo flt strl or stmt; lo lm similar to the up glm; lo pal smt present, thin, shorter than the lo lm; up flt bisx, sessile, plump, usu apiculate to mucronate, smt minutely so, or subacute to (rarely) acute; up lm striate, chartaceous-indurate, shiny, usu glab, mrg involute; up pal striate; lod 2; anth 3. Car smooth; prcp thin; endosperm hard; hila round or oval.

[1]J.K. Wipff   [2]Robert W. Freckmann and Michel G. Lelong

*Dichanthelium* is a genus of approximately 72 species, 34 of which are native to the *Manual* region. It is often included in *Panicum*, the two taxa being similar in gross morphology.

When the branches of *Dichanthelium* develop, in late summer or fall, the culms acquire a very different aspect; comments about the 'fall phase' refer to the appearance of the plant or its culms following this branching. Unless stated otherwise, measurements refer to structures of the culms and primary panicles, not those of the branches and secondary panicles. Ligule measurements usually include the hairs of the pseudoligule, if present, because the two are often difficult to distinguish with less than 30× magnification.

1. Basal leaf blades similar in shape to those of the lower cauline leaves, usually erect to ascending, clustered at the base, sometimes small or vestigial; culms branching from near the base in the fall, with 2–4 leaves, only the upper 2–4 internodes elongated.
    2. Blades soft, 3–12 mm wide, usually ciliate; upper blades less than 20 times as long as wide; fall phase with short panicle-bearing branches, without sterile shoots (sect. *Strigosa*).
        3. Leaf sheaths with retrorse or spreading hairs; upper blades 4–17 cm long, at least ¾ as long as the basal blades; blade margins usually finely short ciliate, the cilia not papillose-based; spikelets with papillose-based hairs . . . . . . . . . . . . . . . . . . . . . . . . . . . . . . . . . . . . . . . . . . . . . . . . 29. *D. laxiflorum*
        3. Leaf sheaths glabrous or with ascending hairs; upper blades 1.5–6 cm long, less than ¾ as long as the basal blades; blade margins with papillose-based cilia; spikelets glabrous or pubescent, hairs not papillose-based . . . . . . . . . . . . . . . . . . . . . . . . . . . . . . . . . . . . . . . . . 30. *D. strigosum*
    2. Blades stiff, 1–5 mm wide, not ciliate; most upper blades at least 20 times as long as wide; fall phase with basal panicles and sterile shoots (sect. *Linearifolia*).
        4. Upper glumes and lower lemmas forming a beak extending 0.2–1 mm beyond the upper florets; spikelets 3.2–4.3 mm long; primary panicles with 7–25 spikelets . . . . . . . . . . . . . . . . . . . . . . 34. *D. depauperatum*
        4. Upper glumes and lower lemmas equaling or exceeding the upper florets by no more than 0.3 mm, not forming a beak; spikelets 2–3.4 mm long; primary panicles with 12–70 spikelets.
            5. Cauline blades 4–8 cm long, all alike; basal blades ascending to spreading . . . . . . . . . . . . . . . 31 *D. wilcoxianum*
            5. Uppermost cauline blades 10–20 cm long, distinctly longer than the lower blades; basal blades erect to ascending.
                6. Panicles 1–3 cm wide, with ascending branches and appressed pedicels; spikelets turgid, 2.6–3.4 mm long, 1–1.7 mm wide, upper florets obovoid . . . . . . . . . . . . . . . . . . . . . . . . . 32. *D. perlongum*
                6. Panicles 2–6 cm wide, with spreading branches and pedicels; spikelets not turgid, 2–3.2 mm long, 0 8–1.4 mm wide, upper florets ellipsoid . . . . . . . . . . . . . . . . . . . . . . . . 33. *D. linearifolium*
1. Basal leaf blades usually well differentiated from the cauline blades, ovate to lanceolate, spreading, forming a rosette, or basal blades absent; culms usually branching from the midculm nodes in the fall, with 3–14 leaves, usually all internodes elongated.
    7. Bases of the culms hard, cormlike; basal rosettes absent; spikelets with papillose-based hairs and attenuate basally (sect. *Pedicellata*).
        8. Culms erect in the spring; cauline leaves 4–7, with thin, glabrous or sparsely hirsute blades that widen distal to the rounded to subcordate bases; lower glumes not encircling the pedicels, subadjacent to the upper glumes . . . . . . . . . . . . . . . . . . . . . . . . . . . . . . . . . . . . . . . . . . 1. *D. pedicellatum*
        8. Culms decumbent to ascending in the spring; cauline leaves 8–14, with thick, firm, puberulent blades that are parallel-sided distal to the rounded to truncate bases; lower glumes almost to completely encircling the pedicels, attached about 0.2 mm below the upper glumes . . . . . . . . . . . . . . 2. *D. nodatum*
    7. Bases of the culms not cormlike; basal rosettes usually present; spikelets not both with papillose-based hairs and attenuate basally.
        9. Blades cordate, thick, with white, cartilaginous margins; spikelets usually spherical to broadly obovoid or broadly ellipsoid, 1–1.8 mm long (sect. *Sphaerocarpa*).
            10. Spikelets 1–1.4 mm long; lower glumes 0.2–0.4 mm long; cauline blades 5–10 mm wide . . . . 23. *D. erectifolium*
            10. Spikelets 1.3–1.8 mm long; lower glumes 0.4–0.8 mm long; cauline blades 5–25 mm wide.
                11. Cauline blades 4–7, 10–25 cm long, 14–25 mm wide, with evident veins; culms nearly erect; panicles less than ½ as wide as long . . . . . . . . . . . . . . . . . . . . . . . . . . . 24. *D. polyanthes*
                11. Cauline blades 3–4(6), 1.5–10 cm long, 5–14 mm wide, with obscure veins; culms decumbent or ascending; panicles more than ½ as wide as long . . . . . . . . . . . . . . . . . . 25. *D. sphaerocarpon*
        9. Blades not cordate or the spikelets not both spherical and less than 1.9 mm long; blade margins usually not white and cartilaginous.
            12. Lower glumes thinner and more weakly veined than the upper glumes, attached about 0.2 mm below the upper glumes, the bases clasping the pedicels; spikelets attenuate basally.
                13. Blades 2–7 cm long, about 10 times as long as wide, not or slightly involute, spreading, without raised veins, not longitudinally wrinkled; spikelets obovoid-obpyriform, planoconvex in side view (sect. *Lancearia*) . . . . . . . . . . . . . . . . . . . . . . . . . . . 26. *D. portoricense*

13. Blades 4–16 cm long, more than 14 times as long as wide, or involute, stiffly erect or ascending, with prominently raised veins, the lower blades usually longitudinally wrinkled; spikelets ellipsoid to obovoid, biconvex in side view (sect. *Angustifolia*).

    14. Culms densely villous; nodes densely bearded; spikelets densely pubescent . . . 28. *D. consanguineum*

    14. Culms glabrous, puberulent, or pilose with papillose-based hairs; nodes glabrous, puberulent to lightly bearded; spikelets glabrous or pubescent . . . . . . . . . . . . . . . . . . . 27. *D. aciculare*

12. Lower glumes similar in texture and vein prominence to the upper glumes, attached immediately below the upper glumes, the bases not clasping the pedicels; spikelets usually not attenuate basally.

    15. Culms arising from rhizomes 3–5 mm thick, with (5)7–14 cauline blades; sheaths strongly hispid or viscid, mottled with pale spots, constricted at the top (sect. *Clandestina*).

        16. Nodes densely bearded above a viscid glabrous ring, often swollen; blades densely soft pubescent . . . . . . . . . . . . . . . . . . . . . . . . . . . . . . . . . . . . . . . . . . . . . 10. *D. scoparium*

        16. Nodes glabrous or sparsely pubescent, not swollen; blades glabrous or sparsely pubescent.

            17. Cauline blades 7–15 mm wide, apices involute, long-tapering; spikelets glabrous or sparsely puberulent . . . . . . . . . . . . . . . . . . . . . . . . . . . . . . . . . 8. *D. scabriusculum*

            17. Cauline blades 15–30 mm wide, apices flat, acuminate; spikelets sparsely pubescent . . . . . . . . . . . . . . . . . . . . . . . . . . . . . . . . . . . . . . . . . . . . . . . . . . . . . . . . . 9. *D. clandestinum*

    15. Culms arising from caudices or from rhizomes to 2 mm thick, with 3–7(9) cauline blades; sheaths not viscid, rarely hispid, not mottled with pale spots or constricted at the top.

        18. Ligules with a membranous base, ciliate distally; culms usually arising from slender rhizomes; lower florets often staminate; cauline blades 5–40 mm wide, often with a cordate base (sect. *Macrocarpa*).

            19. Spikelets ellipsoid, not turgid, with pointed apices; cauline blades 4–6, cordate at the base; sheaths without papillose-based hairs.

                20. Spikelets 2.2–3.2 mm long; ligules about 0.3 mm long; blades 5–25 mm wide; lower floret sterile . . . . . . . . . . . . . . . . . . . . . . . . . . . . . . . . . . . 5. *D. commutatum*

                20. Spikelets 2.9–5.2 mm long; ligules 0.4–0.9 mm long; blades 15–40 mm wide; at least some lower florets staminate.

                    21. Nodes glabrous or slightly bearded; spikelets 2.9–3.9 mm long . . . . . . . . . . 3. *D. latifolium*

                    21. Nodes densely retrorsely bearded; spikelets 3.8–5.2 mm long . . . . . . . . . . . . . . 4. *D. boscii*

            19. Spikelets obovoid, turgid, with rounded apices; cauline blades 3–4, tapered, rounded or truncate to cordate at the base; sheaths with papillose-based hairs.

                22. Blades and spikelets with papillose-based hairs; panicles usually slightly longer than wide, with spreading to ascending branches . . . . . . . . . . . . . . . . . . . 6. *D. leibergii*

                22. Blades glabrous; spikelets puberulent to almost glabrous; panicles usually more than twice as long as wide, with nearly erect branches . . . . . . . . . . 7. *D. xanthophysum*

        18. Ligules of hairs (except for *D. nudicaule*); culms arising from caudices; lower florets sterile; cauline blades 1–18 mm wide, bases usually tapered, rounded, or truncate at the base, sometimes cordate.

            23. Lower internodes short, upper nodes elongated; flag leaves distant and much reduced; culms rarely branching in the fall; branches, if present, few, developing from basal and sub-basal nodes, erect (sect. *Nudicaulia*) . . . . . . . . 19. *D. nudicaule*

            23. Lower internodes about as long as the upper internodes; flag leaves usually not much reduced; culms branching in the fall; branches often many, developing from mid- or upper culm nodes, often spreading.

                24. Spikelets 2.5–4.3 mm long, usually obovoid, turgid; upper glumes usually with an orange or purple spot at the base, the veins prominent (sect. *Oligosantha*).

                    25. Nodes glabrous or sparsely pubescent; abaxial surfaces of the blades glabrous or pubescent, but not velvety pubescent . . . . . . . . . . . . . . . . . 11. *D. oligosanthes*

                    25. Nodes bearded with spreading to retrorse hairs; abaxial surfaces of the blades softly velvety pubescent.

                        26. Spikelets 3.7–4.3 mm long; culms 2–3 mm thick, stiffly erect; ligules 2–5 mm long, without pseudoligules; blades glabrous or sparsely pilose on the adaxial surfaces . . . . . . . . . . . . . . . . . . . . . . . 12. *D. ravenelii*

26. Spikelets 2.5–3.2 mm long; culms usually 1–2 mm thick, erect; ligules 0.5–1 mm long, with the adjacent pseudoligules 1–3 mm long; blades densely velvety pubescent on both surfaces . . . . . 13. *D. malacophyllum*
24. Spikelets 0.8–3 mm long, ellipsoid or obovoid, not turgid; upper glumes lacking an orange or purple spot at the base and the veins not prominent.
    27. Ligules and adjacent pseudoligules 1–5 mm long, or the culms and sheaths with long hairs and also puberulent; spikelets variously pubescent to subglabrous (sect. *Lanuginosa*).
        28. Spikelets 0.8–1.1 mm long, puberulent to subglabrous; culms delicate, 0.3–0.8 mm thick . . . . . . . . . . . . . . . . . . . . . . . . . . . 16. *D. wrightianum*
        28. Spikelets 1.1–3 mm long, variously pubescent; culms not delicate, usually more than 1 mm thick.
            29. Spikelets 1.1–2.1 mm long; sheaths glabrous or pubescent with hairs no more than 3 mm long . . . . . . . . . . . . . . . . . . 14. *D. acuminatum*
            29. Spikelets 1.8–3 mm long; sheaths with hairs to 4 mm long . . . . . . . 15. *D. ovale*
    27. Ligules absent or to 1.8 mm long, without adjacent pseudoligules; culms and at least the upper sheaths glabrous or sparsely pubescent with hairs of 1 length only; spikelets glabrous or pubescent.
        30. Culms (18)40–100 cm tall, rarely delicate, usually more than 1 mm thick; spikelets 1.5–2.7 mm long; blades 3.5–14 cm long, 5–14 mm wide (sect. *Dichanthelium*).
            31. Spikelets glabrous or, if pubescent, either the nodes bearded or the culms weak and prostrate; blade of the flag leaf usually spreading . . . . . . . . . . . . . . . . . . . . . . . . . . . . . . . . . . 17. *D. dichotomum*
            31. Spikelets pubescent; nodes glabrous; culms erect or ascending; blade of the flag leaf erect or ascending . . . . . . . . . . . . . 18. *D. boreale*
        30. Culms 5–40(55) cm tall, delicate, usually less than 1 mm thick; spikelets 1.1–1.7 mm long; longest blades 1.5–6 cm long, 1.5–6 mm wide (sect. *Ensifolia*).
            32. Culms reclining or weakly erect; cauline blades 4–9, usually without prominent white, cartilaginous margins; ligules often more than 1 mm long . . . . . . . . . . . . . . . . . . . . . . . . . 20. *D. ensifolium*
            32. Culms erect, sometimes geniculate basally; cauline blades 3–5, with prominent white, cartilaginous margins; ligules 0.2–0.7 mm long.
                33. Culms few per clump; the fall phase branching sparingly; cauline blades flat, the bases rounded; blades of the flag leaves much shorter than those of the lower leaves . . . . . . . . . . . . . . . . . . . . . . . . . . . . . . . . . . . . . 21. *D. tenue*
                33. Culms many per clump; the fall phase branching extensively; cauline blades often involute, the bases subcordate; blades of the flag leaves only slightly shorter than those of the lower leaves . . . . . . . . . . . . . . . 22. *D. chamaelonche*

## Dichanthelium sect. **Pedicellata** (Hitchc. & Chase) Freckmann & Lelong

### 1. Dichanthelium pedicellatum (Vasey) Gould CORM-BASED PANICGRASS [p. 472, <u>547</u>]

*Dichanthelium pedicellatum* grows on limestone outcroppings and in dry, open oak woodlands. Its range extends from Texas into Mexico and Guatemala. Primary panicles develop from late March into June (and sometimes from late August to November) and are open-pollinated; secondary panicles develop from May into fall and are at least partly cleistogamous.

### 2. Dichanthelium nodatum (Hitchc. & Chase) Gould SARITA PANICGRASS [p. 472, <u>547</u>]

*Dichanthelium nodatum* grows in oak savannahs near the Gulf Coast from Texas to northeastern Mexico. The primary panicles are produced from April into June (sometimes late August to November)

and are at least partly open-pollinated; the secondary panicles are produced from May into fall and are at least partly cleistogamous.

## Dichanthelium sect. **Macrocarpa** Freckmann & Lelong

### 3. Dichanthelium latifolium (L.) Harvill BROADLEAVED PANICGRASS, PANIC À LARGES FEUILLES [p. 472, <u>547</u>]

*Dichanthelium latifolium* grows in rich deciduous woods, often in slightly open areas within eastern North America. The primary panicles are open-pollinated and develop in May and June (and sometimes in September and October), the secondary panicles, which are produced from July through September, are rarely open-pollinated.

## 4. Dichanthelium boscii (Poir.) Gould & C.A. Clark
BOSC'S PANICGRASS, PANIC DE BOSC [p. 472, 547]

*Dichanthelium boscii* usually grows in semi-open areas in dry oak-hickory woods of the eastern United States. The primary panicles are open-pollinated and are produced from late April through June (and sometimes again in the fall); the secondary panicles are partly open-pollinated, and are produced from July through September.

## 5. Dichanthelium commutatum (Schult.) Gould
VARIABLE PANICGRASS [p. 473, 547]

*Dichanthelium commutatum* is fairly common in dry to wet, semi-open woodlands. Its range extends from the eastern United States to South America. The primary panicles are open-pollinated and are produced from April through June; the secondary panicles are primarily cleistogamous and are produced from June through fall. The four subspecies are fairly distinct in some parts of their ranges, but subsp. *commutatum* intergrades with the other three where they occur together.

1. Culms densely crisp-puberulent; spikelets 2.2–2.7 mm long; cauline blades usually 5–8 cm long, 5–10 mm wide, thick, the bases symmetrical; rosette blades usually less than 3 cm long and to 6 mm wide . . . . . . . . . . . . . . . . . . . . . . . . . . . . . . subsp. *ashei*
1. Culms usually glabrous or sparsely pubescent; spikelets 2.6–3.2 mm long; cauline blades usually more than 8 cm long and 10 mm wide, thin, bases sometimes asymmetrical; rosette blades large, some more than 4 cm long and 10 mm wide.
    2. Cauline blades nearly linear, 5–14 mm wide, about 10 times as long as wide; spikelets 3–3.2 mm long; lower glumes about ½ as long as the spikelets . . . . . . . . . . . . . . . . . . . . . . . subsp. *equilaterale*
    2. Cauline blades ovate-lanceolate, 6–25 mm wide, about 4–8 times as long as wide; spikelets 2.6–3.2 mm long; lower glumes about ¼ as long as the spikelets.
        3. Culms decumbent or sprawling, with loose caudices or rhizomes; blades strongly asymmetric-falcate, often; spikelets 2.9–3.2 mm long; lower lemmas pointed . . . . . . . . . . . . . subsp. *joorii*
        3. Culms more or less erect, with caudices; blades almost symmetrical, green, sometimes glaucous; spikelets 2.6–2.9 mm long; lower lemmas rounded . . . . . . . . . . . . . . . . subsp. *commutatum*

## Dichanthelium commutatum subsp. ashei (T.G. Pearson *ex* Ashe) Freckmann & Lelong [p. 473]

*Dichanthelium commutatum* subsp. *ashei* grows in open woodlands. It sometimes resembles, and may hybridize with, *D. dichotomum*.

## Dichanthelium commutatum (Schult.) Gould subsp. commutatum [p. 473]

*Dichanthelium commutatum* subsp. *commutatum* grows in wet to dry woodlands. Its range extends to South America.

## Dichanthelium commutatum subsp. equilaterale (Scribn.) Freckmann & Lelong [p. 473]

*Dichanthelium commutatum* subsp. *equilaterale* grows in sandy pine and oak woodlands. Its range extends to southeastern Mexico and Nicaragua.

## Dichanthelium commutatum subsp. joorii (Vasey) Freckmann & Lelong [p. 473]

*Dichanthelium commutatum* subsp. *joorii* grows in wet woodlands and swamps. Its range extends into Mexico.

## 6. Dichanthelium leibergii (Vasey) Freckmann
LEIBERG'S PANICGRASS [p. 473, 547]

*Dichanthelium leibergii* grows primarily on prairie relics, but is occasionally found in sandy woodlands. It is restricted to the *Manual* region. The primary panicles are produced from mid-May through July, the secondary panicles from late June to September. Sterile putative hybrids with *D. acuminatum* and *D. xanthophysum* are occasionally found.

## 7. Dichanthelium xanthophysum (A. Gray) Freckmann
PALE PANICGRASS, PANIC JAUNÂTRE [p. 473, 547]

*Dichanthelium xanthophysum* usually grows on sandy or rocky soils in semi-open pine, oak, or aspen woodlands. It extends from eastern Saskatchewan and northeastern Montana to Quebec, New England, and West Virginia. Plants from Minnesota and western Quebec approach *D. leibergii* in having cauline blades narrower than 10 mm, and papillose-based hairs. Sterile putative hybrids with *D. leibergii* and *D. boreale* are rare.

## Dichanthelium sect. Clandestina Freckmann & Lelong

## 8. Dichanthelium scabriusculum (Elliott) Gould & C.A. Clark TALL-SWAMP PANICGRASS [p. 473, 547]

*Dichanthelium scabriusculum* usually grows in wet, sandy, open sites, including shores, stream banks, swamps, and bogs. It is restricted to the eastern United States. The primary panicles develop from May to July, the secondary panicles, which are usually concealed within the sheaths, from July through November.

## 9. Dichanthelium clandestinum (L.) Gould DEER-TONGUE GRASS, PANIC CLANDESTIN [p. 473, 547]

*Dichanthelium clandestinum* usually grows in semi-open areas in damp or sandy woodlands, thickets, or on banks. It is restricted to the eastern part of the *Manual* region. The primary panicles are open-pollinated for a brief period, and produced from late May to early July; the secondary panicles, which are cleistogamous and usually concealed within the sheaths, are produced from July through September.

## 10. Dichanthelium scoparium (Lam.) Gould VELVETY PANICGRASS [p. 473, 547]

*Dichanthelium scoparium* grows in moist, sandy, open, often disturbed areas of the southeastern United States. It is also present in the West Indies. The primary panicles are open-pollinated, produced from May to early August; the secondary panicles are cleistogamous and are produced from July through October.

## Dichanthelium sect. Oligosantha (Hitchc.) Freckmann & Lelong

## 11. Dichanthelium oligosanthes (Schult.) Gould FEW-FLOWERED PANICGRASS [p. 473, 547]

*Dichanthelium oligosanthes* grows throughout the southern portion of the *Manual* region and extends into northern Mexico. The primary panicles are briefly open-pollinated, then cleistogamous, from late May to early June; the secondary panicles, which are produced from June to November, are cleistogamous. Specimens of *D. oligosanthes* that have few elongated internodes, but those elongated more than usual, are often mistaken for *D. wilcoxianum*. Unlike that species, however, they have turgid spikelets with an orange spot at the base of the lemma. Such specimens seem to be most common among collections made in the southern and southwestern states during November, February, or March. Apparent hybrids with *D. malacophyllum*, *D. ovale*, and *D. acuminatum* subsp. *columbianum* are occasionally found.

The subspecies intergrade in areas of overlapping range, but they are usually distinct elsewhere.

1. Spikelets ellipsoid to oblong-obovoid, usually 3.4–4.2 mm long, 1.7–2 mm wide, usually sparsely pubescent; blades usually 4–9 mm wide, more than 10 times longer than wide, often partly involute; ligules 2–3 mm . . . . . . . . . . . . . . . . . subsp. *oligosanthes*
1. Spikelets broadly obovoid-ellipsoid, 2.7–3.5 mm long, 2–2.4 mm wide, usually glabrous; blades usually 6–15 mm wide, less than 10 times longer than wide, flat; ligules 1–1.5 mm long . . . . subsp. *scribnerianum*

### Dichanthelium oligosanthes (Schult.) Gould subsp. oligosanthes [p. 473]

*Dichanthelium oligosanthes* subsp. *oligosanthes* grows in dry, open, sandy, oak or pine woodlands. Its range extends from southern Ontario and New Hampshire to the Texas Gulf Coast. It has not yet been reported from Mexico.

### Dichanthelium oligosanthes subsp. scribnerianum (Nash) Freckmann & Lelong SCRIBNER'S PANICGRASS [p. 473]

*Dichanthelium oligosanthes* subsp. *scribnerianum* grows in sandy or clayey banks and prairies. Its range extends from southern British Columbia to the east coast of the United States, and south into northern Mexico. It is the most widespread of the two varieties.

### 12. Dichanthelium ravenelii (Scribn. & Merr.) Gould RAVENEL'S PANICGRASS [p. 473, 547]

*Dichanthelium ravenelii* grows in dry, sandy woodlands of the southeastern United States. The primary panicles develop from early May through June, and are at least partly open-pollinated. The secondary panicles, which are produced from July through September, are cleistogamous.

### 13. Dichanthelium malacophyllum (Nash) Gould SOFT-LEAVED PANICGRASS [p. 473, 547]

*Dichanthelium malacophyllum* usually grows in cedar glades, on dry limestone soils. It is restricted to the United States. The primary panicles are briefly open-pollinated from late May to early June; the secondary panicles, which are produced from June to November, are cleistogamous. The species occasionally intergrades, and perhaps hybridizes, with *D. oligosanthes* and *D. acuminatum*.

### Dichanthelium sect. Lanuginosa (Hitchc.) Freckmann & Lelong

Hybridization, often followed by segregation in autogamous lines, produces a reticulate pattern of intergradation between members of this section. In the descriptions, no distinction is made between the ligules and pseudoligules because of the difficulty of distinguishing the two at less than 30×.

### 14. Dichanthelium acuminatum (Sw.) Gould & C.A. Clark HAIRY PANICGRASS, PANIC LAINEUX [p. 474, 547]

*Dichanthelium acuminatum* is common and ubiquitous in dry to wet, open, sandy or clayey woods, clearings, bogs, and swamps, or in saline soil near hot springs, growing in much of the *Manual* region and extending into northern South America. It is probably the most polymorphic and troublesome species in the genus. The treatment presented here attempts to delimit the major variants present, but does not fully reflect the intricate reticulate pattern of morphological variation that exists. There is considerable overlap among the nine subspecies recognized and, in addition, there appears to be widespread introgression from other *Dichanthelium* species, such as *D.*

*dichotomum*, *D. sphaerocarpon*, *D. ovale*, and *D. aciculare* into the *D. acuminatum* complex, contributing to the taxonomic difficulties.

1. Lower portion of the culms and lower sheaths usually glabrous or sparsely pubescent.
   2. Primary panicles congested, more than twice as long as wide; spikelets ascending to appressed . . . . . . . . . . . . . . . . . . . . . . . . . . . . . . . . . subsp. *spretum*
   2. Primary panicles open, less than twice as long as wide; spikelets diverging to ascending.
      3. Blades green or purplish, the margins not conspicuously ciliate at the base; spikelets 1.1–1.5 mm long, usually ellipsoid . . . . . . . . . . . . . . . . . . . . . . . subsp. *longiligulatum*
      3. Blades often yellowish-green, the margins usually with long, papillose-based cilia at the base; spikelets 1.3–1.6 mm long, usually obovoid . . . . . . . . . . . . . . . . . . . . . . subsp. *lindheimeri*
1. Lower portion of the culms and lower sheaths densely and variously pubescent or puberulent.
   4. Culms 15–30 cm tall; midculm sheaths nearly as long as the internodes; blades usually 2–6.5 cm long, less than 8 times longer than wide . . . . . . . . . . . . . . . . . . . . . . . . . . . . . . . . . subsp. *sericeum*
   4. Culms usually 30–100 cm tall; midculm sheaths about ½ as long as the internodes; blades usually 6–12 cm long, more than 8 times longer than wide.
      5. Culms and lower sheaths densely covered with spreading, villous hairs or soft, thin, papillose-based hairs, often with shorter hairs underneath; blades softly pubescent to velvety on the abaxial surfaces.
         6. Primary panicles usually poorly exserted, on peduncles less than 6 cm long; blades suberect, the margins lacking cilia on the distal ½ . . . . . . . . . . . . . . subsp. *thermale*
         6. Primary panicles usually well-exserted, on peduncles more than 8 cm long; blades ascending to spreading, the margins ciliate along most of their length . . . . . . . . . . . . . . . . . subsp. *acuminatum*
      5. Culms and sheaths pilose with papillose-based hairs to hispid with mostly ascending hairs, or densely puberulent with a few longer, ascending hairs also present; blades appressed-pubescent or puberulent abaxially, not velvety to the touch.
         7. Sheaths and culms densely puberulent, scattered long hairs often present also . . . . . . . . subsp. *columbianum*
         7. Sheaths and culms pilose with papillose-based hairs, the hairs mostly ascending, occasionally with inconspicuous, shorter hairs underneath.
            8. Blades usually 6–12 mm wide, spreading to ascending, the adaxial surfaces nearly glabrous or with hairs shorter than 3 mm long; spikelets 1.5–2 mm . . . . . . . . . . . . . . . subsp. *fasciculatum*
            8. Blades usually 2–6 mm wide, erect to ascending, spreading or reflexed, the adaxial surfaces glabrous or with hairs 3–6 mm long; spikelets 1.1–1.6 mm long.

9. Blades erect to ascending, the adaxial surfaces long-pilose; spikelets 1.3–1.6 mm long, usually broadly obovoid .............. subsp. *implicatum*
9. Blades ascending, spreading, or reflexed, the adaxial surfaces glabrous or sparsely pubescent; spikelets 1.1–1.5 mm long, usually ellipsoid .......... subsp. *leucothrix*

## Dichanthelium acuminatum (Sw.) Gould & C.A. Clark subsp. acuminatum [p. 474]

*Dichanthelium acuminatum* subsp. *acuminatum* grows primarily in moist, open, sandy areas on the Atlantic and Gulf coastal plains. Its range extends through Mexico, the West Indies, and Central America to northern South America.

## Dichanthelium acuminatum subsp. columbianum (Scribn.) Freckmann & Lelong PANIC DU DISTRICT DE COLUMBIA [p. 474]

*Dichanthelium acuminatum* subsp. *columbianum* grows in sandy woods or clearings in the northeastern portion of the species range. It is much less common than the other eastern subspecies of *D. acuminatum*. Occasionally, it resembles the more widespread subsp. *fasciculatum*, subsp. *implicatum*, and subsp. *lindheimeri*. The culms and sheaths of *D. acuminatum* subsp. *columbianum* are always puberulent with very short hairs. This puberulence should not be confused with the slightly longer hairs that develop on the secondary branches of other taxa.

## Dichanthelium acuminatum subsp. fasciculatum (Torr.) Freckmann & Lelong PANIC LAINEUX [p. 474]

*Dichanthelium acuminatum* subsp. *fasciculatum* grows primarily in disturbed areas, open or cut-over woods, thickets, and grasslands, in dry to moist soils, including river banks, lake margins, and marshy areas. It is widespread in temperate North America, growing from Canada to Mexico, but it is somewhat less common in the western part of its range, where it often occurs on moister areas. It includes probably the most widespread, ubiquitous, and variable assemblages of forms in the species. It is not always clearly separable from the other subspecies of *D. acuminatum*, especially subsp. *acuminatum*, subsp. *implicatum*, and subsp. *lindheimeri*.

## Dichanthelium acuminatum subsp. implicatum (Scribn.) Freckmann & Lelong [p. 474]

*Dichanthelium acuminatum* subsp. *implicatum* usually grows in low, moist areas, including open woodlands, meadows, bogs, and cedar and hemlock swamps, and also in drier, sandy areas. Its range extends from south-central Canada to the midwestern and northeastern United States. It intergrades occasionally with the more widespread subsp. *fasciculatum*.

## Dichanthelium acuminatum subsp. leucothrix (Nash) Freckmann & Lelong [p. 474]

*Dichanthelium acuminatum* subsp. *leucothrix* grows in low, sandy or peaty pine savannahs of the coastal plain. Its range extends through Mexico, the West Indies, and Central America to northern South America. It is closely related to, and often sympatric with, the more common, glabrous subsp. *longiligulata*.

## Dichanthelium acuminatum subsp. lindheimeri (Nash) Freckmann & Lelong [p. 474]

*Dichanthelium acuminatum* subsp. *lindheimeri* grows in dry or moist, sandy or clayey, open, often disturbed areas, open woodlands,

limestone glades, and roadsides, primarily in the eastern portion of the species range. It intergrades occasionally with the pubescent subsp. *fasciculatum* and subsp. *implicatum*.

## Dichanthelium acuminatum subsp. longiligulatum (Nash) Freckmann & Lelong [p. 474]

*Dichanthelium acuminatum* subsp. *longiligulatum* is common, especially in moist pine savannahs and bogs of the coastal plain; it also grows inland to Tennessee, and in Mexico, the West Indies, Central America, and South America. It is similar to subsp. *leucothrix*, which grows in the same habitat, often at the same sites.

## Dichanthelium acuminatum subsp. sericeum (Schmoll) Freckmann & Lelong [p. 474]

*Dichanthelium acuminatum* subsp. *sericeum* grows in warm or hot ground around geysers and hot springs in the Rocky Mountains from Banff, Alberta south to Yellowstone National Park and east to Bighorn County, Wyoming.

## Dichanthelium acuminatum subsp. spretum (Schult.) Freckmann & Lelong [p. 474]

*Dichanthelium acuminatum* subsp. *spretum* grows in wet to moist, sandy or peaty soil, pine savannahs, and bogs. It is not a common taxon, but is most frequent on the coastal plain and around the Great Lakes. It is very similar to the more common, southern subsp. *longiligulatum*. It also resembles *D. dichotomum* in size and overall habit.

## Dichanthelium acuminatum subsp. thermale (Bol.) Freckmann & Lelong GEYSER PANICGRASS [p. 474]

*Dichanthelium acuminatum* subsp. *thermale* grows on the mineralized crust of warm, moist soil at the Geysers, Sonoma County, California; it is listed as endangered in that state.

## 15. Dichanthelium ovale (Elliott) Gould & C.A. Clark STIFF-LEAVED PANICGRASS [p. 474, <u>547</u>]

*Dichanthelium ovale* grows in dry, open, sandy or rocky woodland borders, sand barrens, dunes, and dry prairies in southeastern Canada, the eastern United States, the West Indies, Mexico, and Central America. The growth form and certain morphological features of *D. ovale* resemble those of the widespread *D. laxiflorum*, which usually grows in more mesic habitats. Occasional specimens exhibit traits of *D. acuminatum*, *D. oligosanthes*, and *D. commutatum*.

The four subspecies often intergrade, especially subsp. *villosissimum* and subsp. *pseudopubescens* in the southeastern United States, and subsp. *villosissimum* and subsp. *praecocius* in the western part of their range.

1. Lower sheaths and lower culm internodes with soft, spreading or retrorse, papillose-based hairs, the longer hairs often longer than 4 mm; spikelets 1.8–2.5 mm long.
   2. Spikelets 2.1–2.5 mm long; culms usually more than 1 mm thick, stiff; largest blades usually 6–10 mm wide ................... subsp. *villosissimum*
   2. Spikelets 1.8–2.1 mm long; culms usually less than 1 mm thick, wiry; largest blades usually 2–6 mm wide ........................ subsp. *praecocius*
1. Lower sheaths and lower culm internodes with ascending or appressed, non-papillose-based hairs shorter than 4 mm or nearly glabrous; spikelets 2.1–3 mm long.
   3. Spikelets 2.5–3 mm long; basal blades with long hairs on or near the margins and bases ....... subsp. *ovale*
   3. Spikelets 2.1–2.6 mm long; basal blades usually without long hairs on or near the margins and bases ..................... subsp. *pseudopubescens*

**Dichanthelium ovale** (Elliott) Gould & C.A. Clark subsp. **ovale** [p. *474*]

*Dichanthelium ovale* subsp. *ovale* grows in dry, open, sandy woods, pinelands, and sandhills along the east coast of the United States from New Jersey southwards, extending into the coastal plain from eastern Texas to South Carolina, and in Mexico, Honduras, Guatemala, and Nicaragua. It intergrades somewhat with subsp. *pseudopubescens*. Occasional long-spikelet specimens exhibit morphological characteristics of *D. oligosanthes* and *D. commutatum*.

**Dichanthelium ovale** subsp. **praecocius** (Hitchc. & Chase) Freckmann & Lelong [p. *474*]

*Dichanthelium ovale* subsp. *praecocius* is most common in the midwest and in the tallgrass prairie states. It intergrades with subsp. *villosissimum*, especially in the western parts of the latter's range, and to a lesser extent, with *D. acuminatum* subsp. *fasciculatum* in the northern part of its range.

**Dichanthelium ovale** subsp. **pseudopubescens** (Nash) Freckmann & Lelong [p. *474*]

*Dichanthelium ovale* subsp. *pseudopubescens* grows in dry, sandy, open woods, sandhills, and sand dunes, over the same geographic range and in the same habitats as subsp. *villosissimum*, and often intergrades morphologically with that subspecies.

**Dichanthelium ovale** subsp. **villosissimum** (Nash) Freckmann & Lelong [p. *474*]

*Dichanthelium ovale* subsp. *villosissimum* grows in dry, sandy, open pine and oak woodlands. It and subsp. *pseudopubescens* are the most common and widespread subspecies throughout the eastern United States. The range of subsp. *villosissimum* extends to Mexico, Honduras, Guatemala, and Nicaragua. It grades into the less pubescent subsp. *pseudopubescens*, and occasional specimens with smaller spikelets approach *D. acuminatum* subsp. *acuminatum*, which is usually densely grayish, velvety-pubescent.

### 16. **Dichanthelium wrightianum** (Scribn.) Freckmann WRIGHT'S PANICGRASS [p. *474*, <u>547</u>]

*Dichanthelium wrightianum* grows in moist, sandy or peaty areas, low pine savannahs, bogs, the margins of ponds, and cypress swamps, in the coastal plain from Massachusetts to Texas and Florida, extending to Cuba, Mexico, Central America, and northern South America. Occasional specimens of *D. wrightianum*, particularly those with subglabrous spikelets, closely resemble *D. chamaelonche*. Others suggest *D. ensifolium*, and a few unusually robust specimens closely approach *D. acuminatum* subsp. *longiligulatum*. All of these taxa often grow together in the same habitats.

## **Dichanthelium** (Hitchc. & Chase) Gould sect. **Dichanthelium**

Gene exchange between the subspecies of *Dichanthelium dichotomum*, and between *D. dichotomum* and other species in the genus, appears to be rather common.

### 17. **Dichanthelium dichotomum** (L.) Gould FORKED PANICGRASS [p. *474*, <u>547</u>]

*Dichanthelium dichotomum* grows in dry, sandy, clayey, or rocky ground, often in woods, or (more commonly) in moist or wet places, including marshes, bogs, low woods, swamps, and the moist borders of lakes and ponds. Its range extends south from the *Manual* region into the Caribbean. It is a polymorphic and ubiquitous species, with many of its intergrading subspecies exhibiting traits of other widespread and variable species such as *D. commutatum*, *D. laxiflorum*, and *D. sphaerocarpon*, which often grow at the same sites.

1. Lower nodes hairy.
  2. Spikelets 1.5–1.8 mm long, upper floret 0.6–0.8 mm wide . . . . . . . . . . . . . . . . . . . . . . subsp. *microcarpon*
  2. Spikelets 1.8–2.5 mm long; upper floret 0.7–1.0 mm wide.
    3. Spikelets usually glabrous; midculm blades usually 5–7 mm wide . . . . . . . . . . . . subsp. *dichotomum*
    3. Spikelets pubescent; midculm blades usually 7–14 mm wide.
      4. Lower sheaths and blades glabrous or sparsely pubescent . . . . . . . . . . . . . . . subsp. *nitidum*
      4. Lower sheaths and blades more or less densely velvety pubescent . . . subsp. *mattamuskeetense*
1. Lower nodes glabrous.
  5. Larger blades more than 1 cm wide; sheaths often with pale glandular spots between the prominent veins; spikelets 1.9–2.6 mm long, acute to beaked . . . . . . . . . . . . . . . . . . . subsp. *yadkinense*
  5. Larger blades less than 1 cm wide; sheaths without glandular spots; spikelets 1.5–2.3 mm long, obtuse to subacute.
    6. Culms weak, ultimately reclining or sprawling, often flattened . . . . . . . . . . . . . . . subsp. *lucidum*
    6. Culms erect, terete.
      7. Blades usually spreading; spikelets ellipsoid, 1.8–2.3 mm long, rarely purplish at the base . . . . . . . . . . . . . . . . . . . . . subsp. *dichotomum*
      7. Blades usually ascending or erect; spikelets broadly ellipsoid or obovoid, 1.5–1.8 mm long, often purplish at the base . . . . . . . . . . . . . . . . . . . . . . subsp. *roanokense*

**Dichanthelium dichotomum** (L.) Gould subsp. **dichotomum** [p. *474*]

*Dichanthelium dichotomum* subsp. *dichotomum* usually grows in dry to mesic woods. Its range extends from southern Ontario to Maine and south through Illinois and Missouri to eastern Texas and to the east coast of the United States and central Florida.

**Dichanthelium dichotomum** subsp. **lucidum** (Ashe) Freckmann & Lelong [p. *474*]

*Dichanthelium dichotomum* subsp. *lucidum* grows in wet woods, the margins of cypress swamps, sphagnum bogs, and other similar, wet habitats. It is primarily a species of the coastal plain, ranging from New Jersey to Florida, southeastern Texas, and up the Mississippi embayment to western Tennessee and, as a disjunct, on the Indiana Dunes of Lake Michigan.

**Dichanthelium dichotomum** subsp. **mattamuskeetense** (Ashe) Freckmann & Lelong [p. *474*]

*Dichanthelium dichotomum* subsp. *mattamuskeetense* grows in low, moist, often sandy or peaty, ground and bogs. A relatively uncommon subspecies, it grows on the Atlantic coastal plain from Massachusetts to Florida.

**Dichanthelium dichotomum** subsp. **microcarpon** (Muhl. *ex* Elliott) Freckmann & Lelong [p. *474*]

*Dichanthelium dichotomum* subsp. *microcarpon* grows in wet woods, swamps, and wetland borders. It is a widespread subspecies, extending from southern Michigan to Massachusetts and south to eastern Oklahoma and Texas and throughout the southeast to central Florida.

**Dichanthelium dichotomum** subsp. **nitidum** (Lam.) Freckmann & Lelong [p. *474*]

*Dichanthelium dichotomum* subsp. *nitidum* grows in moist to wet areas, and the borders of swamps. It is primarily a coastal plain taxon, ranging from Virginia to southeastern Texas and Florida. It is very similar to both subsp. *microcarpon* and subsp. *mattamuskeetense*, and intergrades with each occasionally.

## Dichanthelium dichotomum subsp. roanokense (Ashe) Freckmann & Lelong [p. 474]

*Dichanthelium dichotomum* subsp. *roanokense* grows in marshes, wet pinelands, wet woods, and the borders of swamps. A relatively uncommon subspecies, it grows on the coastal plain from Delaware to southeastern Texas and in the West Indies. It is very similar to subsp. *dichotomum* and also exhibits traits of *D. sphaerocarpon* and *D. erectifolium*.

## Dichanthelium dichotomum subsp. yadkinense (Ashe) Freckmann & Lelong [p. 474]

*Dichanthelium dichotomum* subsp. *yadkinense* grows in rich, moist or wet woods. A relatively uncommon subspecies, its range extends from Pennsylvania to Maryland and south through southern Illinois and southeastern Missouri to Georgia and Louisiana. It exhibits traits of *D. laxiflorum* and *D. commutatum*.

## 18. Dichanthelium boreale (Nash) Freckmann
NORTHERN PANICGRASS, PANIC BOREAL [p. 475, 548]

*Dichanthelium boreale* grows in open woodlands and thickets, wet meadows, and fields. It is restricted to the *Manual* region. The primary panicles are mostly open-pollinated and are produced in May and June; the secondary panicles are predominantly cleistogamous and are produced from mid-June into October. It occasionally hybridizes with *D. acuminatum* and *D. xanthophysum*.

## Dichanthelium sect. Nudicaulia Freckmann & Lelong

*Dichanthelium nudicaule* is the only member of sect. *Nudicaulia* present in the *Manual* region.

## 19. Dichanthelium nudicaule (Vasey) B.F. Hansen & Wunderlin NAKED-STEMMED PANICGRASS [p. 475, 548]

*Dichanthelium nudicaule* is a rare species that grows in wet pine savannas, bogs (including *Sphagnum* mats), and the margins of cypress swamps in eastern Louisiana, southern Mississippi and Alabama, and western Florida. Vegetatively, it exhibits traits of *D. laxiflorum*, but its spikelets resemble those of small plants of *D. scabriusculum*, which are fairly widespread in similar habitats of the Gulf coastal plain. *Dichanthelium nudicaule* is protected by U.S. federal law.

## Dichanthelium sect. Ensifolia (Hitchc.) Freckmann & Lelong

## 20. Dichanthelium ensifolium (Baldwin *ex* Elliott) Gould SWORD-LEAF PANICGRASS [p. 475, 548]

*Dichanthelium ensifolium* grows in wet to moist, sandy pinelands, savannahs, and bogs, often on *Sphagnum* mats, primarily on the coastal plain. It extends south into Mesoamerica, and has been reported from Venezuela. Occasional specimens grade towards the larger *D. tenue*, and are usually found on somewhat drier sites. It also resembles *D. chamaelonche*, but that species is usually more densely cespitose, has slightly smaller, glabrous spikelets, and generally occupies drier, disturbed sites.

1. Sheaths sparsely spreading-pilose; ligules usually 1–1.8 mm long; blades sparsely pilose or glabrous on both surfaces .................................... subsp. *curtifolium*
1. Sheaths glabrous; ligules 0.2–1 mm long; blades usually puberulent abaxially, usually glabrous, occasionally pubescent adaxially .......... subsp. *ensifolium*

## Dichanthelium ensifolium subsp. curtifolium (Nash) Freckmann & Lelong [p. 475]

The two subspecies are sympatric, often growing together at the same sites.

## Dichanthelium ensifolium (Baldwin *ex* Elliott) Gould subsp. ensifolium [p. 475]

The two subspecies are sympatric, often growing together at the same sites.

## 21. Dichanthelium tenue (Muhl.) Freckmann & Lelong SLENDER PANICGRASS [p. 475, 548]

*Dichanthelium tenue* grows in moist to dry, sandy woods, savannahs, and disturbed sites. It also grows in Chiapas, Mexico. It exhibits features of *D. sphaerocarpon* and *D. dichotomum*. It is also closely related to *D. ensifolium*, and occasional specimens are intermediate between them.

## 22. Dichanthelium chamaelonche (Trin.) Freckmann & Lelong SMALL-SEEDED PANICGRASS [p. 475, 548]

*Dichanthelium chamaelonche* grows in low, open, sandy, coastal pine woods, savannahs, and moist depressions in sand dunes. It is restricted to the southeastern United States.

1. Culms 5–20 cm tall, glabrous or puberulent; spikelets 1.3–1.5 mm long, puberulent .......... subsp. *breve*
1. Culms 10–45 cm tall, glabrous; spikelets 1.1–1.4 mm long, glabrous .................. subsp. *chamaelonche*

## Dichanthelium chamaelonche subsp. breve (Hitchc. & Chase) Freckmann & Lelong [p. 475]

*Dichanthelium chamaelonche* subsp. *breve* grows only in peninsular Florida.

## Dichanthelium chamaelonche (Trin.) Freckmann & Lelong subsp. chamaelonche [p. 475]

*Dichanthelium chamaelonche* subsp. *chamaelonche* grows from North Carolina to Florida and Louisiana.

## Dichanthelium sect. Sphaerocarpa (Hitchc. & Chase) Freckmann & Lelong

*Dichanthelium* sect. *Sphaerocarpa* extends from the southeastern United States through Central America and Cuba to northern South America. Pairs of species often grow together, with infrequent apparent hybridization.

## 23. Dichanthelium erectifolium (Nash) Gould & C.A. Clark FLORIDA PANICGRASS, ERECT-LEAF PANICGRASS [p. 475, 548]

*Dichanthelium erectifolium* grows in sand and peat in wet pinelands, bogs, and the shores of ponds. Its range extends from the southeastern *Manual* region into the Caribbean.

## 24. Dichanthelium polyanthes (Schult.) Mohlenbr.
MANY-FLOWERED PANICGRASS [p. 475, 548]

*Dichanthelium polyanthes* grows in woods, stream banks, and ditches, and is restricted to the eastern United States. It occasionally hybridizes with *D. sphaerocarpon*.

## 25. Dichanthelium sphaerocarpon (Elliott) Gould
ROUND-FRUITED PANICGRASS [p. 475, 548]

*Dichanthelium sphaerocarpon* grows in dry, open woods and roadsides. Its range extends from eastern North America to Ecuador and Venezuela. It occasionally hybridizes with other species, including *D. polyanthes*, *D. acuminatum*, and *D. laxiflorum*.

## Dichanthelium sect. **Lancearia** (Hitchc.) Freckmann & Lelong

Only one species of sect. *Lancearia*, *Dichanthelium portoricense*, grows in the *Manual* region.

### 26. **Dichanthelium portoricense** (Desv. *ex* Ham.) B.F. Hansen & Wunderlin BLUNT-GLUMED PANICGRASS [p. 475, 548]

*Dichanthelium portoricense* grows in sandy woods, low pinelands, savannahs, and coastal sand dunes, usually in moist places. Its range extends south from the *Manual* region into Mexico, the Caribbean, and Mesoamerica. It is a highly variable species with numerous intergrading forms, some possibly resulting from hybridization with other widespread species in the same region, such as *D. sphaerocarpon* and *D. commutatum*.

1. Spikelets 1.8–2.6 mm long, usually densely pubescent or puberulent (rarely glabrous); cauline blades 4–7 cm long, 3.5–8 mm wide . . . . . . . . . . . . . . . . . . . . . . . subsp. *patulum*
1. Spikelets 1.5–2.0 mm long, puberulent to nearly glabrous; cauline blades 2–5 cm long, 2.5–4.5 mm wide . . . . . . . . . . . . . . . . . . . . . . . . . . . . . subsp. *portoricense*

### Dichanthelium portoricense subsp. **patulum** (Scribn. & Merr.) Freckmann & Lelong [p. 475]

*Dichanthelium portoricense* subsp. *patulum* is more common in moist, sandy pinelands and savannahs than subsp. *portoricense*. It also grows in coastal sand dunes, but is less abundant there than subsp. *portoricense*. It is the more variable of the two subspecies, grading into subsp. *portoricense* as well as *D. commutatum*. Occasional specimens resemble the widespread *D. sphaerocarpon*.

### Dichanthelium portoricense (Desv. *ex* Ham.) B.F. Hansen & Wunderlin subsp. **portoricense** [p. 475]

*Dichanthelium portoricense* subsp. *portoricense* is more common than subsp. *patulum* in coastal sand dunes. It also grows in sandy pinelands and savannahs. It resembles *D. aciculare* somewhat, but that species usually has ascending-pilose culms, strongly involute or acicular blades, and longer spikelets.

## Dichanthelium sect. **Angustifolia** (Hitchc.) Freckmann & Lelong

*Dichanthelium* sect. *Angustifolia* grows from the southeastern United States through Central America and the West Indies to northern South America.

### 27. **Dichanthelium aciculare** (Desv. *ex* Poir.) Gould & C.A. Clark NARROW-LEAVED PANICGRASS [p. 475, 548]

*Dichanthelium aciculare* grows in sandy, open areas in the southeastern United States, the West Indies and the Caribbean, southern Mexico, Central America, and northern South America. It has not been reported from northern Mexico. The primary panicles are open-pollinated (sometimes briefly) and develop from April to June; the secondary panicles are cleistogamous and develop from May into late fall.

The subspecies are often distinct when growing together, perhaps maintained by the predominant autogamy, but they are more difficult to separate over wider geographic areas. Rare, partly fertile putative hybrids with *Dichanthelium consanguineum*, *D. acuminatum*, *D. ovale*, *D. portoricense*, and (possibly) *D. dichotomum* apparently lead to some intergradation with these species.

1. Primary panicles usually contracted; branches appearing 1-sided; culms sparsely pubescent to almost glabrous . . . . . . . . . . . . . . . . . . . . subsp. *neuranthum*

1. Primary panicles not contracted; branches not appearing 1-sided; culms usually pubescent, at least on the lower internodes.
   2. Spikelets 1.7–2.3 mm long, with blunt apices . . . . . . . . . . . . . . . . . . . . . . . . . . . . . . . . . . subsp. *aciculare*
   2. Spikelets 2.4–3.6 mm long, with pointed or beaked apices.
      3. Spikelets 2.4–3 mm long, not strongly attenuate at the base; lower glumes attached less than 0.2 mm below the upper glumes . . . . . . . . . . . . . . . . . . . . . . . . . . . subsp. *angustifolium*
      3. Spikelets 2.9–3.6 mm long, strongly attenuate at the base; lower glumes attached 0.3–0.5 mm below the upper glumes . . . . subsp. *fusiforme*

### Dichanthelium aciculare (Desv. *ex* Poir.) Gould & C.A. Clark subsp. **aciculare** [p. 475]

*Dichanthelium aciculare* subsp. *aciculare* is common in sterile, open sands on the coastal plain. It is restricted to the eastern United States and Mexico.

### Dichanthelium aciculare subsp. **angustifolium** (Elliott) Freckmann & Lelong [p. 475]

*Dichanthelium aciculare* subsp. *angustifolium* grows in open pine woodlands, often in sandy soil with needle duff. It is restricted to the southeastern United States.

### Dichanthelium aciculare subsp. **fusiforme** (Hitchc.) Freckmann & Lelong [p. 475]

*Dichanthelium aciculare* subsp. *fusiforme* grows in sandy pine or oak savannahs. It tends to replace *D. aciculare* subsp. *angustifolium* from southern Florida through Central America and the Antilles.

### Dichanthelium aciculare subsp. **neuranthum** (Griseb.) Freckmann & Lelong [p. 475]

*Dichanthelium aciculare* subsp. *neuranthum* grows in moist, sandy, open ground and savannahs, primarily on the outer coastal plain and in Cuba and various other Caribbean islands.

### 28. **Dichanthelium consanguineum** (Kunth) Gould & C.A. Clark KUNTH'S PANICGRASS [p. 475, 548]

*Dichanthelium consanguineum* grows in sandy woodlands and low, boggy pinelands. It is restricted to the southeastern United States. The primary panicles are open-pollinated and produced from April to June; the secondary panicles are cleistogamous and produced from June into fall. Some specimens of *D. consanguineum* suggest that hybridization occasionally occurs with *D. aciculare* or *D. ovale*.

## Dichanthelium sect. **Strigosa** Freckmann & Lelong

### 29. **Dichanthelium laxiflorum** (Lam.) Gould SOFT-TUFTED PANICGRASS [p. 476, 548]

*Dichanthelium laxiflorum* is a widespread, common species that grows in mesic deciduous woods, and occasionally in drier, more open woodlands. Its range extends south from the *Manual* region into Mexico. The density of the pubescence on the blade surfaces varies greatly. The primary (spring) panicles are apparently chasmogamous; the secondary panicles are largely cleistogamous and are produced from late spring to winter.

### 30. **Dichanthelium strigosum** (Muhl. *ex* Elliott) Freckmann CUSHION-TUFTED PANICGRASS [p. 476, 548]

*Dichanthelium strigosum* extends from the southeastern *Manual* region south into Mexico, the Caribbean, and into northern South America. The primary panicles are briefly open-pollinated in April or

May; the secondary panicles, which are produced from May through November, are cleistogamous. The three subspecies are mostly sympatric and sometimes grow together, with occasional intergradation.

1. Spikelets pubescent, broadly ellipsoid, 1.6–2.1 mm long; lower glumes about ½ as long as the spikelets; blades glabrous . . . . . . . . . . . . . . . . . . . subsp. *leucoblepharis*
1. Spikelets glabrous, obovoid, 1.1–1.8 mm long; lower glumes about ⅓ as long as the spikelets; blades pilose or glabrous.
   2. Blades pilose; spikelets 1.1–1.6 mm long . . . . . . . . . . . . . . . . . . . . . . . . . . . . . . . subsp. *strigosum*
   2. Blades glabrous or sparsely pilose near the base; spikelets 1.4–1.8 mm long . . . . . . . . . . . . subsp. *glabrescens*

## Dichanthelium strigosum subsp. glabrescens (Griseb.) Freckmann & Lelong [p. 476]

*Dichanthelium strigosum* subsp. *glabrescens* grows in sandy, open pine woods and bogs. Its range extends from Mississippi along the coast to Florida and south through the West Indies.

## Dichanthelium strigosum subsp. leucoblepharis (Trin.) Freckmann & Lelong [p. 476]

*Dichanthelium strigosum* subsp. *leucoblepharis* grows in low, moist, sandy pinelands and bogs. Its range extends from North Carolina along the coastal plain to Florida and eastern Texas and into Mexico.

## Dichanthelium strigosum (Muhl. *ex* Elliott) Freckmann subsp. strigosum [p. 476]

*Dichanthelium strigosum* subsp. *strigosum* grows in sandy, low, open pine woods and bogs. It is the most widespread of the three subspecies, extending from southeastern Virginia through the coastal plain to eastern Texas, Florida, Cuba, and the West Indies to Colombia.

## Dichanthelium sect. Linearifolia Freckmann & Lelong

### 31. Dichanthelium wilcoxianum (Vasey) Freckmann
WILCOX'S PANICGRASS [p. 476, 548]

*Dichanthelium wilcoxianum* grows in dry prairies, especially in sandy or gravelly openings. It is restricted to the *Manual* region. The primary panicles, which are produced from mid-May to early June, are partially open-pollinated; the secondary panicles, which are produced in June, and occasionally also in September, are cleistogamous. Some specimens of *D. oligosanthes* subsp. *scribnerianum* from the southern Great Plains that have prematurely

elongating upper internodes resemble *D. wilcoxianum*, but they have greenish spikelets that are 1.7–2.4 mm wide, an orange spot at the base of the glumes, and larger basal rosettes.

### 32. Dichanthelium perlongum (Nash) Freckmann
LONG-STALKED PANICGRASS [p. 476, 548]

*Dichanthelium perlongum* grows in dry to mesic prairies, and is restricted to the *Manual* region. It appears to hybridize occasionally with *D. depauperatum* and *D. linearifolium*. The primary panicles are briefly open-pollinated and develop from May to early June; the secondary panicles are cleistogamous and are produced from mid-June through mid-July. It differs from *D. wilcoxianum* in having only the upper 1 or 2 blades greatly elongated (usually more than 20 times longer than wide), narrow, erect basal blades, and a contracted panicle with ascending branches. *Dichanthelium acuminatum* may also be confused with *D. perlongum* only if its upper internodes elongate, as tends to be the case after a spring fire, but *D. acuminatum* has less turgid spikelets and hairs in the ligule area that are 3–5 mm long.

### 33. Dichanthelium linearifolium (Scribn.) Gould
LINEAR-LEAVED PANICGRASS, PANIC À FEUILLES LINÉAIRES [p. 476, 548]

*Dichanthelium linearifolium* grows in dry, open woodlands, rock outcroppings, and sandy areas. It is restricted to the *Manual* region. The primary panicles are briefly open-pollinated, produced from May to early June; the secondary panicles are cleistogamous, produced from late June through July (rarely in fall). Plants in the northern United States and Canada tend to be shorter and more spreading, subglabrous, and to have spikelets 2–2.6 mm long. In the southwestern part of its range, especially in the Ozarks, most plants of *D. linearifolium* are tall, erect, densely pilose, with very elongated blades and spikelets often 2.6–3 mm long; they may hybridize with *D. perlongum*.

### 34. Dichanthelium depauperatum (Muhl.) Gould
STARVED PANICGRASS, PANIC APPAUVRI [p. 476, 548]

*Dichanthelium depauperatum* grows in dry, open woodlands and open, disturbed areas, especially on sand. It is restricted to the *Manual* region. The primary panicles, which are rarely open-pollinated, are produced from May to early June; the secondary, cleistogamous panicles are produced from late June through July (rarely in fall). The species is linked with *D. perlongum* and *D. linearifolium* by occasional hybrids and hybrid derivatives. In the northern United States and Canada, 80–90% of the plants are glabrous. The frequency of pilose plants increases southward, where some populations are entirely pilose.

# 24.10 PANICUM L.[1]

Pl ann or per; their habit variable. **Clm** 2–300 cm, hrb, smt hard and almost wd, or wd, simple or brchd; **intnd** solid, spongy, or hollow. **Lvs** cauline, bas, or both, bas lvs not forming a winter rosette; **lig** memb, usu ciliate; **bld** filiform to ovate, flat to involute, glab or pubescent, cross sections with Kranz anatomy and 1 or 2 bundle shth or with non-Kranz anatomy; **photosynthesis** $C_4$ with NAD-me or NADP-me pathways, or, in pl with non-Krantz anatomy, $C_3$. **Infl** tml on the clm and br, often also ax, tml pan typically appearing after midsummer; **strl br** and **bristles** absent; **dis** usu below the glm, smt at the base of the up flt, if at the base of the up flt, then the flt not very plump at maturity. **Spklt** 1–8 mm, usu dorsally compressed, smt subterete or lat compressed, unawned. **Glm** usu unequal, hrb, glab or pubescent, rarely tuberculate or glandular, apc not or only slightly gaping at maturity; **lo glm** minute to ⅔ as long as the spklt, 1–9-veined, truncate, acute, or acuminate; **up glm** slightly shorter to much longer than the spklt, 3–13(15)-veined, bases rarely slightly sulcate, apc rounded to attenuate; **lo flt** strl or stmt; **lo lm** similar to the up glm; **lo pal** absent, or shorter than the lo lm and hyaline; **up flt** bisx, sessile or stipitate, ⅖ as long as to almost equaling the spklt, apc acute, puberulent, or with a tuft of hairs; **up lm** usu more or less rigid and chartaceous-indurate, usu shiny, glab or (rarely) pubescent, usu smooth or striate, smt verrucose or inconspicuously rugose, mrg involute, usu clasping the pal, rarely with bas wings or lunate scars, apc obtuse, acute, apiculate, or with small green crests; **up pal** striate; **lod** 2; **anth** usu 3. **Car** smooth; **prcp** thin; **endosperm**

[1]Robert W. Freckmann and Michel G. Lelong, amended by Mary E. Barkworth to reflect research completed since publication of FNA 25

hard, without lipid, starch grains simple or compound, or both; **hila** round or oval.

*Panicum* is a large genus, but just how large is difficult to estimate because its limits are not yet clear. This treatment excludes *P. plenum* and *P. obtusum* (now placed in *Hopia*), *P. bulbosum* (now in *Zuloagaea*), and *P. gymnocarpon* (now in *Phanopyrum*).

Most species of *Panicum* are tropical, but many grow in warm, temperate regions. Of the 30 species occurring in the *Manual* region, 21 are native, seven are established introductions, and two are not established within the region. Within the *Manual* region, *Panicum* is most abundant in the southeastern United States.

1. Panicle branches 1-sided; spikelets usually subsessile, the longest pedicels usually less than 2 mm long, rarely 3 mm long.
    2. Lower florets staminate; lower paleas subequal to the lower lemmas; upper lemmas thin, flexible, clasping the paleas only at the base (sect. *Hemitoma*) . . . . . . . . . . . . . . . . . . . . . . . . . . . . . . . . . . . 26. *P. hemitomon*
    2. Lower florets sterile; lower paleas no more than ⅔ as long as the lower lemmas; upper lemmas thick, stiff, clasping the paleas throughout their length.
        3. Glumes and lower lemmas without keeled midveins; upper florets with glabrous apices; plants tufted, from knotty rhizomes; panicles with a few spikelets; pedicels with slender hairs near the apices (sect. *Tenera*) . . . . . . . . . . . . . . . . . . . . . . . . . . . . . . . . . . . . . . . . . . . 24. *P. tenerum*
        3. Glumes and lower lemmas with keeled midveins; upper florets with a tuft of small hairs at the apices; plants often with scaly rhizomes; panicles with many spikelets; pedicels glabrous (sect. *Agrostoidea*).
            4. Plants without conspicuous rhizomes, cespitose; culms and sheaths strongly compressed; spikelets usually 1.6–3.8 mm long, lanceolate, not falcate . . . . . . . . . . . . . . . . . . . . . . . . 22. *P. rigidulum*
            4. Plants with conspicuous, stout, short or elongate, scaly rhizomes; culms and sheaths slightly compressed; spikelets 2.3–3.9 mm long, rarely lanceolate, often falcate . . . . . . . . . . . . . . . . . . . . 23. *P. anceps*
1. Panicle branches usually not 1-sided; spikelets not subsessile, the longest pedicels 2–20 mm long.
    5. Upper glumes and lower lemmas warty-tuberculate (sect. *Verrucosa*).
        6. Lower lemmas verrucose with hemispheric warts; spikelets 1.7–2.2 mm long, about 1 mm wide, subacute or obtuse, glabrous; plants of wetlands . . . . . . . . . . . . . . . . . . . . . . . . . . . . . . . . 29. *P. verrucosum*
        6. Lower lemmas tuberculate-hispid; spikelets 3.2–4 mm long, about 1.5 mm wide, acute or acuminate; plants of dry, sandy or clayey areas . . . . . . . . . . . . . . . . . . . . . . . . . . . . . . . . 30. *P. brachyanthum*
    5. Upper glumes and lower lemmas glabrous, villous, or scabridulous, but not warty-tuberculate.
        7. Plants with rhizomes about 1 cm thick and with large, pubescent, scalelike leaves; culms hard, almost woody (sect. *Antidotalia*) . . . . . . . . . . . . . . . . . . . . . . . . . . . . . . . . . . . . . . 25. *P. antidotale*
        7. Plants without rhizomes or with rhizomes less than 0.5 cm thick and with small, glabrous, scalelike leaves; culms clearly not woody, except at the base of *P. hirsutum*.
            8. Glumes, lower lemmas, and upper lemma margins villous, with whitish hairs (sect. *Urvilleana*) . . 21. *P. urvilleanum*
            8. Glumes and lemmas usually glabrous, sometimes the lower lemmas sparsely pilose on the margins and near the apices.
                9. Plants perennial, usually with vigorous scaly rhizomes; lower florets staminate (sect. *Repentia*).
                    10. Lower glumes 0.5–1.5 mm long, less than ½ as long as the spikelet, 1–5-veined; upper glumes and lower lemmas extending 0.1–0.5 mm beyond the upper florets and scarcely separated (gaping); lower paleas oblong, not hastate-lobed.
                        11. Lower glumes subtruncate to broadly acute, faintly veined; upper florets widest at or above the middle, with rounded apices; plants not cespitose, with long, scaly rhizomes . . . . . . . . . . . . . . . . . . . . . . . . . . . . . . . . . . . . . . . . . . . . . . . 17. *P. repens*
                        11. Lower glumes acute, with evident veins; upper florets widest below the middle, with lightly beaked apices; plants cespitose, with short knotty rhizomes . . . . . . . . . . 18. *P. coloratum*
                    10. Lower glumes 1.8–4 mm long, more than ½ as long as the spikelets, with at least 5 veins; upper glumes and lower lemmas extending 0.4–3 mm beyond the upper florets, stiffly separated (gaping); lower paleas hastate-lobed.
                        12. Panicles contracted; branches appressed to strongly ascending; plants glabrous throughout . . . . . . . . . . . . . . . . . . . . . . . . . . . . . . . . . . . . . . . . . . . . . . . . 19. *P. amarum*
                        12. Panicles open; branches ascending to spreading; plants often pilose, at least at the base of the leaf blades . . . . . . . . . . . . . . . . . . . . . . . . . . . . . . . . . . . . . . . . 20. *P. virgatum*
                9. Plants annual, or perennials usually without rhizomes, sometimes rooting at the lower nodes; lower florets sterile.
                    13. Lower glumes truncate to subacute, ⅕–⅓ as long as the spikelets; sheaths more or less compressed, glabrous or sparsely pubescent; plants slightly succulent or spongy (sect. *Dichotomiflora*).

14. Plants usually annual, usually terrestrial, rooting at the lower nodes if in water, but not floating; blades 3–25 mm wide ................................... 15. *P. dichotomiflorum*
14. Plants perennial or of indefinite duration, usually aquatic, sometimes floating, rooting at the lower nodes; blades 2–15 mm wide.
    15. Spikelets 2–2.2 mm long; blades 2–4 mm wide; lower paleas absent; culms succulent ................................................... 14. *P. lacustre*
    15. Spikelets 3–4 mm long; blades 5–15 mm wide; lower paleas present; culms spongy .................................................. 16. *P. paludosum*
13. Lower glumes acute to attenuate, usually ⅓–¾ as long as the spikelets; sheaths rounded, usually hirsute or hispid; plants not succulent (sect. *Panicum*, in part).
    16. Spikelets 4–6.5 mm long.
        17. Upper glumes and lower lemmas only slightly exceeding the upper florets; upper florets 2–2.5 mm wide; plants annual; lower paleas truncate to bilobed ....... 1. *P. miliaceum*
        17. Upper glumes and lower lemmas exceeding the upper florets by 3–4 mm; upper florets 1–1.1 mm wide; plants perennial; lower paleas acute .......... 8. *P. capillarioides*
    16. Spikelets 1–4.2 mm long.
        18. Plants annual; panicle branches usually with secondary branches and pedicels attached to the distal ⅔.
            19. Blades 2–7 cm long, 5–20 mm wide, lanceolate, 4–6 times longer than wide (sect. *Monticola*, in part) .................................... 27. *P. trichoides*
            19. Blades 5–40 cm long, 1–18 mm wide, linear, more than 10 times longer than wide (sect. *Panicum*, in part).
                20. Panicles more than 2 times longer than wide at maturity; branches ascending to somewhat divergent; spikelets narrowly ovoid, usually about 3 times longer than wide ................. 4. *P. flexile*
                20. Panicles less than 1.5 times longer than wide at maturity; branches diverging; spikelets variously shaped, less than 3 times longer than wide.
        18. Plants perennial; panicle branches usually with all or most secondary branches confined to the distal ⅓.
            21. Lower panicle branches whorled; culms 2–10 mm thick, 50–300 cm tall.
                22. Sheaths with fragile, prickly hairs causing skin irritation; panicles not breaking at the base and becoming tumbleweeds; lower paleas 1.3–1.7 mm long ..................................................... 9. *P. hirsutum*
                22. Sheaths glabrous or sparsely to densely pubescent but without fragile, prickly hairs; panicles breaking at the base and becoming tumbleweeds; lower paleas 1.4–2.2 mm long ........................ 10. *P. bergii*
             21. Lower panicle branches solitary; culms 0.5–10 mm thick, 15–100 cm tall.
                23. Blades glabrous and glaucous on the adaxial surface; nodes sericeous or pilose, sometimes almost glabrous ................................ 13. *P. hallii*
                23. Blades sparsely to densely hirsute and not glaucous on the adaxial surface; nodes sericeous.
                    24. Spikelets 2.1–2.9 mm long; culms spreading to weakly ascending; blades spreading, 1–5 mm wide, without a prominent white midrib ................................ 11. *P. diffusum*
                    24. Spikelets 2.6–3.4 mm long; culms erect to decumbent; blades ascending to erect, 0.5–14 mm wide, with a prominent white midrib .................................................. 12. *P. ghiesbreghtii*
                [⇐ revert to left, Ed.]
25. Spikelets 2.1–4 mm long, upper glumes and lower lemmas with prominent veins; lower glumes ⅖–¾ as long as the spikelets; lower paleas 0.4–2 mm long, from ⅓ as long as the lower lemmas to equaling them; ligules 0.2–0.4 mm or 1–3.5 mm long.
    26. Lower glumes 0.7–1.1 mm long, about ⅖ as long as the spikelets; lower paleas 1–2 mm long; leaf blades 2–8 mm wide, usually completely glabrous, sometimes with a few marginal cilia near the base .................................................................. 7. *P. psilopodium*
    26. Lower glumes 1.2–2.4 mm long, ½–¾ as long as the spikelets; lower paleas 0.2–0.9 mm long; leaf blades 1–30 mm wide, hairs papillose-based.

27. Primary panicle branches appressed to the main axis; culms 2–8 cm long; spikelets 2–2.2 mm long . . . . . . . . . . . . . . . . . . . . . . . . . . . . . . . . . . . . . . . . . . . . . . . . . . . . . . . 6. *P. mohavense*
27. Primary panicle branches divergent; culms 11–110 cm long; spikelets 1.9–4 mm long . . . . . . . . . . 5. *P. hirticaule*
25. Spikelets 1.4–4 mm long, upper glumes and lower lemmas without prominent veins; lower glumes usually less than ½ as long as the spikelets; lower paleas usually small or absent; ligules 0.5–1.5 mm long.
28. Plants mostly glabrous, but the sheaths ciliate on the margins and the blades sometimes sparingly pilose adaxially (sect. *Monticola*, in part) . . . . . . . . . . . . . . . . . . . . . . . . . . . . . . . . . . . . 28. *P. bisulcatum*
28. Plants mostly hairy, even the sheaths hairy throughout.
29. Panicles usually more than ½ the total height of the plant, breaking at the base of the peduncle at maturity and becoming a tumbleweed; spikelets 1.9–4 mm long; mature upper florets stramineous or nigrescent (sect. *Panicum*, in part) . . . . . . . . . . . . . . . . . . . . . . . . . . . . . . . 2. *P. capillare*
29. Panicles usually less than ½ the total height of the plant, the base of the peduncle usually not breaking at maturity; spikelets 1.4–2.4 mm long; mature upper florets often dark brown . . . . 3. *P. philadelphicum*

## Panicum L. sect. Panicum

*Panicum* sect. *Panicum* includes approximately 22 species and extends from the southern United States to Argentina. Most species grow in dry, open places, but a few grow in moist sites such as river banks.

### 1. Panicum miliaceum L. BROOMCORN, PROSO MILLET, HOG MILLET, PANIC MILLET, MILLET COMMUN [p. 476, 548]

*Panicum miliaceum* is native to Asia, where it has been cultivated for thousands of years. In the *Manual* region, it is grown for bird seed and is occasionally planted for game birds. It is also found in corn fields and along roadsides.

1. Mature upper florets blackish, disarticulating at maturity; culms 70–210 cm tall; panicles erect, exserted at maturity, about twice as long as wide; panicle branches ascending to spreading; pulvini well-developed . . . . . . . . . . . . . . . . . . . . . . . . . . . subsp. *ruderale*
1. Mature upper florets stramineous to orange, not disarticulating; culms 20–120 cm tall; panicles usually nodding, not fully exserted, more than twice as long as wide; panicle branches ascending to appressed; pulvini almost absent . . . . . . . . . . . . . . . . . . . . . . . . . . . subsp. *miliaceum*

### Panicum miliaceum L. subsp. miliaceum [p. 476]

*Panicum miliaceum* subsp. *miliaceum* is the subspecies used in bird seed. It probably rarely persists because of the retention of the upper florets on the plant and, in northern states, poor seed survival over winter.

### Panicum miliaceum subsp. ruderale (Kitag.) Tzvelev [p. 476]

*Panicum miliaceum* subsp. *ruderale* is now naturalized over much of the *Manual* region. It may become a major weed, especially in corn fields.

### 2. Panicum capillare L. WITCHGRASS, PANIC CAPILLAIRE [p. 476, 548]

*Panicum capillare* grows in open areas, particularly in disturbed sites such as fields, pastures, roadsides, waste places, ditches, sand, and rock crevices, etc. It grows throughout temperate North America, including northern Mexico. It also grows in Bermuda, the Virgin Islands, and sporadically in South America, and has become naturalized in much of Europe and Asia. It appears to hybridize with *P. philadelphicum*.

1. Upper florets without a lunate scar, usually stramineous; lower paleas absent; pedicels and secondary branches strongly divergent . . . . . . . . . subsp. *capillare*
1. Upper florets with a lunate scar at the base, usually nigrescent; lower paleas present; pedicels and secondary branches often appressed, varying to narrowly divergent . . . . . . . . . . . . . . . . . . . . . . subsp. *hillmanii*

### Panicum capillare L. subsp. capillare [p. 476]

*Panicum capillare* subsp. *capillare* is the common subspecies, growing in weedy and dry habitats throughout the range of the species. Plants in the western United States and Canada have spikelets over 2.6 mm long more often than those in the east. Robust plants germinating early in the season and growing on better soils tend to spread more, and have wider, shorter blades and more exserted panicles than plants in the eastern United States and Canada growing under comparable conditions.

### Panicum capillare subsp. hillmanii (Chase) Freckmann & Lelong [p. 476]

*Panicum capillare* subsp. *hillmanii* grows in weedy habitats in California, New Mexico, Iowa, Kansas, Oklahoma, and Texas. It may be a southern Great Plains extension of the western plants of subsp. *capillare*, but it differs from subsp. *capillare* in more characters than such plants.

### 3. Panicum philadelphicum Bernh. ex Trin. PHILADELPHIA WITCHGRASS [p. 476, 548]

*Panicum philadelphicum* grows in open areas such as fallow fields, roadside ditches, receding shores, and rock crevices. It is restricted to the eastern part of the *Manual* region. It intergrades with *P. capillare*, possibly as a result of hybridization, especially in the southeastern United States.

1. Spikelets less than ½ as wide as long; plants purplish . . . . . . . . . . . . . . . . . . . . . . . . . . . . . . . . . . . . . . . . subsp. *lithophilum*
1. Spikelets usually more than ½ as wide as long; plants green or yellow-green.
2. Spikelets 1.9–2.4 mm long; apices of the upper glumes and lower lemmas straight; secondary branches and pedicels divergent; blades often 6–12 mm wide, those of the flag leaves usually more than ½ as long as the panicles . . . . . . . subsp. *gattingeri*
2. Spikelets 1.4–2.1 mm long; apices of the upper glumes and lower lemmas curving over the upper florets at maturity; secondary panicle branches and pedicels appressed; blades usually 2–6 mm wide, those of the flag leaves usually less than ½ as long as the panicles . . . . . . . . . . . . . subsp. *philadelphicum*

### Panicum philadelphicum subsp. gattingeri (Nash) Freckmann & Lelong PANIC DE GATTINGER [p. 476]

*Panicum philadelphicum* subsp. *gattingeri* is commonly found in fields, roadsides, and wet clay on receding shores. This subspecies seems to be more common in the warmer parts of the northeastern United States.

**Panicum philadelphicum** subsp. **lithophilum** (Swallen) Freckmann & Lelong [p. 476]

*Panicum philadelphicum* subsp. *lithophilum* is endemic to wet depressions in granitic outcroppings of Georgia and North and South Carolina.

**Panicum philadelphicum** Bernh. *ex* Trin. subsp. **philadelphicum** PANIC DE PHILADELPHIE [p. 476]

*Panicum philadelphicum* subsp. *philadelphicum* grows in meadows, open woods, sand, and on receding shores. Plants with decumbent culms, glabrous pulvini, flexuous pedicels without hairs over 0.2 mm long, spikelets 1.4–1.7 mm long, and the mature floret not disarticulating have been called *Panicum tuckermanii* Fernald. They are often fairly distinct on receding lake shores in New England and the Great Lakes area, but intergrade with subsp. *philadelphicum* elsewhere.

**4. Panicum flexile** (Gatt.) Scribn. WIRY WITCHGRASS, PANIC FLEXIBLE [p. 476, 548]

*Panicum flexile* grows in fens and other calcareous wetlands, in dry, calcareous or mafic rock barrens, and in open woodlands, especially on limestone derived soils. It is restricted to the *Manual* region.

**5. Panicum hirticaule** J. Presl ROUGHSTALKED WITCHGRASS [p. 477, 548]

*Panicum hirticaule* grows in rocky or sandy soils in waste places, roadsides, ravines, and wet meadows along streams. Its range extends from southeastern California and southwestern Texas southward through Mexico, Central America, Cuba, and Hispaniola to western South America and Argentina.

1. Blades rounded at the base, 3–16 mm wide; lower paleas less than ½ as long as the upper florets; panicles erect . . . . . . . . . . . . . . . . . . . . . . . subsp. *hirticaule*
1. Blades cordate, clasping at the base, 4–30 mm wide; lower paleas more than ½ as long as the upper florets; panicles often nodding.
    2. Nodes, sheaths, and blades glabrous or sparsely pilose, hairs papillose-based; culms usually less than 70 cm tall; spikelets 3.2–4 mm long . . . . . . . . . . . . . . . . . . . . . . . . . . . . . subsp. *stramineum*
    2. Nodes, sheaths, and blades hirsute, hairs papillose-based; culms robust, usually more than 70 cm tall; spikelets 3–3.3 mm long . . . subsp. *sonorum*

**Panicum hirticaule** J. Presl subsp. **hirticaule** [p. 477]

*Panicum hirticaule* subsp. *hirticaule* is the most common of the subspecies, growing throughout the range of the species but occurring more often in arid habitats.

**Panicum hirticaule** subsp. **sonorum** (Beal) Freckmann & Lelong [p. 477]

*Panicum hirticaule* subsp. *sonorum* has been collected only a few times. Its range extends from southern Arizona to Chiapas, Mexico. It may have originated through selection and cultivation.

**Panicum hirticaule** subsp. **stramineum** (Hitchc. & Chase) Freckmann & Lelong [p. 477]

*Panicum hirticaule* subsp. *stramineum* grows in rich bottomlands in southern Arizona, New Mexico, and western Mexico.

**6. Panicum mohavense** Reeder MOHAVE WITCHGRASS [p. 477, 548]

*Panicum mohavense* is known only from arid limestone terraces in Arizona and New Mexico.

**7. Panicum psilopodium** Trin. [p. 477, 548]

*Panicum psilopodium* is native to eastern Asia. It has been reported from chrome ore piles in Canton, Maryland, but no voucher specimens have been seen. In its native range it grows in open habitats, such as roadsides and waste places.

**8. Panicum capillarioides** Vasey LONG-BEAKED WITCHGRASS [p. 477, 548]

*Panicum capillarioides* grows in sandy grasslands, oak savannahs, and rangelands from southern Texas to northern Mexico.

**9. Panicum hirsutum** Sw. GIANT WITCHGRASS [p. 477, 549]

*Panicum hirsutum* grows along river banks or in ditches, often among shrubs in partial shade. Its range extends from southern Texas through eastern Mexico, Central America, Cuba, and the West Indies to Ecuador, Brazil, and Argentina.

**10. Panicum bergii** Arechav. BERG'S WITCHGRASS [p. 477, 549]

*Panicum bergii* is an eastern South American species that now grows in southeastern Texas. It occurs in ditches and shallow, and sporadically flooded depressions in grasslands.

**11. Panicum diffusum** Sw. SPREADING WITCHGRASS [p. 477, 549]

*Panicum diffusum* grows along river banks, ditches, and disturbed areas in wet, loamy or clayey soils. Its range extends from Texas to the Caribbean and northern South America.

**12. Panicum ghiesbreghtii** E. Fourn. GHIESBREGHT'S WITCHGRASS [p. 477, 549]

*Panicum ghiesbreghtii* grows in low, moist ground, wet thickets, and savannahs, from southern Texas through Mexico, Central America, Cuba, and the West Indies to northern South America.

**13. Panicum hallii** Vasey HALL'S WITCHGRASS [p. 477, 549]

*Panicum hallii* grows on sandy, gravelly, or rocky land, including roadsides, pastures, rangeland, oak and pine savannahs, chaparral, and moist areas in deserts and on mesas. Its range extends from the southwestern United States to southern Mexico.

1. Spikelets 3–4.2 mm long; panicles usually greatly exceeding the blades, with a few spikelets; blades clustered near the base of the plants, ascending, often curling at maturity . . . . . . . . . . . . . . . . . . . . . . . . . . . subsp. *hallii*
1. Spikelets 2.1–3 mm long; panicles scarcely exceeding the blades, with relatively crowded spikelets; blades not clustered near the base of the plants, lax, spreading, not curled . . . . . . . . . . . . . . subsp. *filipes*

**Panicum hallii** subsp. **filipes** (Scribn.) Freckmann & Lelong [p. 477]

*Panicum hallii* subsp. *filipes* often grows in moist soil. Its range extends from Arizona, Texas, and Louisiana to southern Mexico.

**Panicum hallii** Vasey subsp. **hallii** [p. 477]

*Panicum hallii* subsp. *hallii* usually grows on drier sites than subsp. *filipes*. Its range extends from southern Colorado and Kansas to north-central Mexico.

## Panicum sect. Dichotomiflora (Hitchc.) Honda

Members of sect. *Dichotomiflora* usually grow in wet, open areas, some as emergents in shallow water. They are often found on disturbed ground. There are about seven species in the Western Hemisphere, but only three grow in the *Manual* region.

### 14. Panicum lacustre Hitchc. & Ekman CYPRESS-SWAMP PANICUM [p. 477, <u>549</u>]

*Panicum lacustre* grows in shallow water or wet soil at the edge of cypress ponds in the Everglades of southern Florida. It also grows in Cuba.

### 15. Panicum dichotomiflorum Michx. FALL PANICUM, PANIC D'AUTOMNE [p. 477, <u>549</u>]

*Panicum dichotomiflorum* grows in open, often wet, disturbed areas such as cultivated and fallow fields, roadsides, ditches, open stream banks, receding shores, clearings in floodplain woods, and sometimes in shallow water. It is probably native throughout the eastern United States and adjacent Canada, but introduced elsewhere, including in the western United States. Its size and habit may be partly under genetic control, but these features also seem to be strongly affected by moisture levels, soil richness, competition, and the time of germination.

1. Spikelets 1.8–2.2 mm long, widest at the middle, acute; upper glumes and lower lemmas sub-membranaceous; pedicels often over 3 mm long . . . . . . . . . . . . . . . . . . . . . . . . . . . . . . subsp. *puritanorum*
1. Spikelets 2.2–3.8 mm long, widest below the middle, acuminate; upper glumes and lower lemmas subcoriaceous; most pedicels less than 3 mm long.
    2. Sheaths glabrous or sparsely pilose, hairs not papillose-based . . . . . . . . . . . . . . . . subsp. *dichotomiflorum*
    2. Sheaths hispid, hairs papillose-based . . . . . subsp. *bartowense*

### Panicum dichotomiflorum subsp. bartowense (Scribn. & Merr.) Freckmann & Lelong [p. 477]

*Panicum dichotomiflorum* subsp. *bartowense* grows in Florida, Cuba, and the Bahamas. Reports from more northerly areas may represent introductions or misidentifications.

### Panicum dichotomiflorum Michx. subsp. dichotomiflorum PANIC D'AUTOMNE DRESSÉ, PANIC À FLEURES DICHOTOM [p. 477]

*Panicum dichotomiflorum* subsp. *dichotomiflorum* is the most common of the three subspecies and is found throughout the range of the species.

### Panicum dichotomiflorum subsp. puritanorum (Svenson) Freckmann & Lelong [p. 477]

*Panicum dichotomiflorum* subsp. *puritanorum* has a sporadic distribution on receding shores along the Atlantic coast from Nova Scotia to Virginia, and around southern Lake Michigan.

### 16. Panicum paludosum Roxb. AQUATIC PANICUM [p. 477, <u>549</u>]

*Panicum paludosum* is an Asian species that grows in shallow water. It has been found in Baltimore, Maryland, but may not be established there.

### Panicum sect. Repentia Stapf

There are approximately 12 species of *Panicum* sect. *Repentia* in the Western Hemisphere, four of which grow in the *Manual* region. The species generally inhabit wet sites, such as coastal dunes, sea beaches, or the margins of rivers.

### 17. Panicum repens L. TORPEDO GRASS [p. 478, <u>549</u>]

*Panicum repens* grows on open, moist, sandy beaches and the shores of lakes and ponds, occasionally extending out into or onto the water. It is mostly, but not exclusively, coastal. It grows on tropical and subtropical coasts throughout the world and may have been introduced to the Americas from elsewhere.

### 18. Panicum coloratum L. KLEINGRASS [p. 478, <u>549</u>]

*Panicum coloratum* is an African species that has been widely introduced into tropical and subtropical regions around the world. It is now established in the *Manual* region, growing in open, usually wet ground; it is also occasionally cultivated as a forage grass.

### 19. Panicum amarum Elliott BITTER BEACHGRASS [p. 478, <u>549</u>]

*Panicum amarum* grows in the coastal dunes, wet sandy soils, and the margins of swamps, along the Atlantic Ocean and the Gulf of Mexico from Connecticut to northeastern Mexico. It is also known, as an introduction, from a few inland locations in New Mexico, North Carolina, and West Virginia, as well as in the Bahamas and Cuba.

1. Rhizomes short or ascending; culms often bunched and decumbent, usually more than 120 cm tall; lower glumes with 3–5 less evident veins, the midvein smooth distally; spikelet density high; panicles with 2 or more main branches per node; spikelets 4–5.9 mm long . . . . . . . . . . . . . . . . . . . . . . . subsp. *amarulum*
1. Rhizomes horizontally elongate; culms mostly solitary, less than 150 cm tall; lower glumes with 7–9 prominent veins, the midvein scabridulous distally; spikelet density moderate; panicles with 1 or 2 main branches per node; spikelets 4.7–7.7 mm long . . . . subsp. *amarum*

### Panicum amarum subsp. amarulum (Hitchc. & Chase) Freckmann & Lelong [p. 478]

*Panicum amarum* subsp. *amarulum* grows in swales behind the first dune and on sandy borders of wet areas. It extends as far north as northern New Jersey and extends southward into Mexico. It has been introduced to Massachusetts, West Virginia, Cuba, and the Bahamas. It is a fertile tetraploid, and possibly a progenitor of subsp. *amarum*, with which it intergrades in the Gulf region. Plants that intergrade with *P. virgatum* are evident in some coastal areas; they may represent hybrids.

### Panicum amarum Elliott subsp. amarum [p. 478]

*Panicum amarum* subsp. *amarum* grows on foredunes, where its longer rhizomes probably permit it to respond quickly to being buried under shifting sand, and occasionally in the swales with subsp. *amarulum*. It ranges farther north (into Connecticut) than subsp. *amarulum*, but apparently not into Mexico, Cuba, or the Bahamas.

### 20. Panicum virgatum L. SWITCHGRASS, PANIC RAIDE [p. 478, <u>549</u>]

*Panicum virgatum* grows in tallgrass prairies, especially mesic to wet types where it is a major component of the vegetation, and on dry slopes, sand, open oak or pine woodlands, shores, river banks, and brackish marshes. Its range extends, primarily on the eastern side of the Rocky Mountains, from southern Canada through the United States to Mexico, Cuba, Bermuda, and Costa Rica, and, possibly as an introduction, in Argentina. It has also been introduced as a forage grass to other parts of the world. It is not always readily separable from *P. amarum*, particularly *P. amarum* subsp. *amarulum*.

### Panicum sect. Urvilleana (Hitchc. & Chase) Pilg.

*Panicum* sect. *Urvilleana* consists of three species that grow on coastal dunes and sand in South America. One species also grows in the *Manual* region.

### 21. Panicum urvilleanum Kunth SILKY PANICGRASS [p. 478, <u>549</u>]

*Panicum urvilleanum* grows on desert sand dunes and in creosote bush scrubland in the Mojave and Colorado desert regions of southern California, southern Nevada, and western Arizona. It also grows in Peru, Chile, and Argentina.

## Panicum sect. Agrostoidea (Nash) C.C. Hsu

*Panicum* sect. *Agrostoidea* includes about four species, one of which is found in the *Manual* region.

### 22. Panicum rigidulum Bosc *ex* Nees REDTOP PANICUM [p. 478, <u>549</u>]

*Panicum rigidulum* grows in swamps, wet woodlands, floodplain forests, wet pine savannahs, marshy shores of rivers, ponds, and lakes, drainage ditches, and other similar wet to moist places; it is rarely found in dry sites. Its range extends from southern Canada to Mexico, Guatemala, and the Antilles.

1. Sheaths truncate or broadly auriculate; blade bases much narrower than the subtending sheaths . . . subsp. *abscissum*
1. Sheaths not truncate or broadly auriculate; blade bases about as wide as the subtending sheaths.
    2. Blades usually 5–12 mm wide, flat, mostly glabrous or scabridulous; ligules membranous, 0.3–1 mm long.
        3. Spikelets 1.6–2.5 mm long, usually over 0.6 mm wide, green or purplish-tinged . . . . . subsp. *rigidulum*
        3. Spikelets 2.4–3 mm long, usually less than 0.6 mm wide, conspicuously stipitate, usually purple . . . . . . . . . . . . . . . . . . . . . . . . subsp. *elongatum*
    2. Blades usually 2–7 mm wide, often folded or involute, usually pilose adaxially, at least near the base; ligules membranous, the cilia usually fimbriate, 0.5–3 mm long.
        4. Spikelets 2–2.7 mm long, green or purplish-stained, often obliquely set on the pedicels . . . . . . . . . . . . . . . . . . . . . . . . . . . . . subsp. *pubescens*
        4. Spikelets 2.6–3.8 mm long, usually purple, slender, erect on the pedicels . . . . . . . . . . subsp. *combsii*

### Panicum rigidulum subsp. abscissum (Swallen) Freckmann & Lelong [p. 478]

*Panicum rigidulum* subsp. *abscissum* is endemic to central Florida. It usually grows in marshy, sandy ground, but is occasionally found in dry, sandy sites. It is very similar vegetatively and reproductively to subsp. *pubescens* and subsp. *combsii*. In addition, its spikelets suggest those of *P. virgatum*.

### Panicum rigidulum subsp. combsii (Scribn. & C.R. Ball) Freckmann & Lelong [p. 478]

*Panicum rigidulum* subsp. *combsii* is restricted to the United States, where it grows in the same moist habitats as subsp. *pubescens*, but is much less common. Its long, narrow, purple, often gaping spikelets somewhat resemble those of *P. virgatum*, which often grows in the same habitats.

### Panicum rigidulum subsp. elongatum (Scribn.) Freckmann & Lelong [p. 478]

*Panicum rigidulum* subsp. *elongatum* is most common in the piedmont and mountain regions of the eastern United States.

### Panicum rigidulum subsp. pubescens (Vasey) Freckmann & Lelong [p. 478]

*Panicum rigidulum* subsp. *pubescens* grows primarily in moist pine savannahs, bogs, and other similar open, sandy habitats on the Atlantic and Gulf coastal plains of the United States.

### Panicum rigidulum Bosc *ex* Nees subsp. rigidulum [p. 478]

*Panicum rigidulum* subsp. *rigidulum* is the most common, most variable, and widest ranging of the five subspecies.

### 23. Panicum anceps Michx. BEAKED PANICGRASS [p. 478, <u>549</u>]

*Panicum anceps* grows in low, moist, primarily sandy areas, pine savannahs, the borders of floodplain swamps, mesic woodlands,

roadsides, and upland pine-hardwood forests. It is restricted to the United States.

1. Spikelets 2.7–3.9 mm long, often clearly falcate; rhizomes relatively short and stout . . . . . . . . . . . . . subsp. *anceps*
1. Spikelets 2.3–2.8 mm long, not clearly falcate; rhizomes relatively long and slender . . . . . . . . subsp. *rhizomatum*

### Panicum anceps Michx. subsp. anceps [p. 478]

*Panicum anceps* subsp. *anceps* is widespread in all physiographic provinces within its range.

### Panicum anceps subsp. rhizomatum (Hitchc. & Chase) Freckmann & Lelong [p. 478]

*Panicum anceps* subsp. *rhizomatum* grows in the Atlantic and Gulf coastal plains. Its small, crowded, often purplish spikelets often closely resemble those of *Panicum rigidulum*.

## Panicum sect. Tenera (Hitchc. & Chase) Pilg.

### 24. Panicum tenerum Beyr. *ex* Trin. BLUE-JOINT PANICGRASS [p. 478, <u>549</u>]

*Panicum tenerum* grows in wet or moist, sandy (often peaty) soil, depressions in pine savannahs, bogs, marshes, pond margins, and interdunal swales. Its range includes the Atlantic and Gulf coastal plains of the United States, the Antilles, Bahamas, and Central America. It exhibits numerous features of the widespread and polymorphic *P. rigidulum*, particularly *P. rigidulum* subsp. *pubescens*.

## Panicum sect. Antidotalia Freckmann & Lelong

In *Panicum* sect. *Antidotalia*, the culms become almost woody at maturity; even a hammer blow fails to flatten them.

### 25. Panicum antidotale Retz. BLUE PANICGRASS [p. 478, <u>549</u>]

*Panicum antidotale* is native to India. It is grown in the *Manual* region as a forage grass, primarily in the southwestern United States. It is now established in the region, being found in open, disturbed areas and fields.

## Panicum sect. Hemitoma (Hitchc.) Freckmann & Lelong

*Panicum* sect. *Hemitoma* includes only *P. hemitomon*.

### 26. Panicum hemitomon Schult. MAIDENCANE [p. 478, <u>549</u>]

*Panicum hemitomon* forms extensive, nearly pure stands in water or wet soils such as marshes, swamps, and along the shores of streams, canals, ditches, lakes, and ponds. It is restricted to the United States.

## Panicum sect. Monticola Stapf

Three species of *Panicum* sect. *Monticola* grow in the Western Hemisphere. Two species grow in the *Manual* region, one of which is native to Asia.

### 27. Panicum trichoides Sw. SMALL-FLOWERED PANICGRASS [p. 479, <u>549</u>]

*Panicum trichoides* grows in moist, often weedy fields, woodlands, and savannahs of Mexico, Central and tropical America, and the Caribbean. It has been found as a weed in Brownsville and Austin, Texas, and is probably introduced to the *Manual* region. It has also been introduced into Africa, tropical Asia, and the Pacific islands. In the *Manual* region, it flowers from August through October.

### 28. Panicum bisulcatum Thunb. [p. 479, <u>549</u>]

*Panicum bisulcatum* has been found at scattered locations on the coastal plain of Georgia and South Carolina, but has rarely become established.

### Panicum sect. Verrucosa (Nash) C.C. Hsu

### 29. Panicum verrucosum Muhl. WARTY PANICGRASS [p. 479, <u>549</u>]

*Panicum verrucosum* grows primarily in open, moist or wet sandy areas bordering swamps, marshes, or lakes or on roadside ditches; it also grows occasionally in open, drier woodlands. It is restricted to the eastern United States and is mostly, but not exclusively, coastal.

### 30. Panicum brachyanthum Steud. PRAIRIE PANICGRASS [p. 479, <u>549</u>]

*Panicum brachyanthum* grows in dry, sandy or clayey soils of open areas, remnant prairies, woodland borders, and roadsides and, less commonly, along the margins of bogs and on grassy shores in the western portion of the Gulf Coast plain. It is restricted to the southern United States. It resembles *P. verrucosum* in its growth habit, but is more restricted in its distribution.

## 24.11 MOOROCHLOA Veldkamp[1]

Pl ann. Clm 10–60 cm, hrb, not wd, often creeping. Lvs cauline; shth open, glab or pubescent; lig memb, with a ciliate fringe, fringe longer than the memb base. Infl tml, secund pan of 1-sided br; br erect to ascending, axes triquetrous, terminating in a well-developed spklt; sec br, when present, shorter than the pri br; dis below the glm and beneath the up flt. Spklt solitary, subsessile, dorsally compressed, unequally convex, in 2 rows, the lo glm and lm appressed or adjacent to the br axes, with 2 flt; lo flt strl or stmt; up flt stipitate, bisx, usu glab, readily disarticulating, acuminate. Lo glm to 0.5 mm, less than ½ as long as the spklt, glab, adjacent to the br axes, 0–1-veined; up glm and lo lm subequal, villous, 3–5-veined; up glm subequal to or slightly exceeding the up flt, not saccate; lo pal present; anth (if present) 3; up lm equaling the second glm, glab, indurate, smooth, shiny to lustrous, 5- or 7-veined, mrg involute, apc round to muticous; up pal similar to the up lm; anth 3. Car ovoid, dorsally compressed.

*Moorochloa*, as now interpreted, includes three species, all native to the Eastern Hemisphere. One species is established in the *Manual* region. *Moorochloa* differs from *Urochloa* in its smooth, rounded, distal floret and from *Panicum* in its secund panicle and stipitate, shiny to lustrous, disarticulating distal floret. It used to be known as *Brachiaria* (Trin.) Griseb.

### 1. Moorochloa eruciformis (Sm.) Veldkamp. SWEET SIGNALGRASS [p. 479]

*Moorochloa eruciformis* is native from the Mediterranean to tropical Africa and India. It tends to be weedy in many parts of the world, growing in moist, disturbed sites. It has been grown for evaluation as a forage crop at various experimental stations in the United States. A few of these plantings have resulted in escapes that have persisted for a short time, but the species has not become established in the *Manual* region.

## 24.12 MELINIS P. Beauv.[2]

Pl ann or per; habit various. Clm 20–150 cm, erect, decumbent, or prostrate. Shth open; lig of hairs or memb and ciliate. Infl tml, simple pan or pan of spikelike pri br, usu with capillary sec br and ped; dis below the glm, smt also below the up flt, the up flt then falling first. Spklt with 2 flt. Lo glm present or absent, 0–1-veined, unawned; up glm equaling or exceeding the flt, smt gibbous bas, 5–7-veined, emgt to bilobed, awned or unawned; lo flt stmt or strl; lo lm similar to the up glm, but not gibbous; up flt bisx, lat compressed; up lm subcoriaceous, glab, smooth, unawned; up pal resembling the up lm; lod 2, fleshy or memb.

*Melinis* is an African and western Asian genus of 22 species that grow in savannahs, open grasslands, and disturbed places. Two species have become established in the *Manual* region. A third species, *M. nerviglumis*, is sold as an ornamental.

1. Glumes and pedicels glabrous, scabridulous; spikelets 1–2 mm long . . . . . . . . . . . . . . . . . . . . . . . . . . . . . . . . 1. *M. minutiflora*
1. Glumes and pedicels conspicuously hairy; spikelets 2.2–5.7 mm long.
    2. Leaves not basally concentrated, blades flat, 2–13 mm wide; plants usually annual . . . . . . . . . . . . . . . . . . . . 2. *M. repens*
    2. Leaves strongly basally concentrated, blades rolled, 0.5–1.5 mm in diameter, plants usually perennial . . 3. *M. nerviglumis*

### 1. Melinis minutiflora P. Beauv. MOLASSES GRASS [p. 479, <u>549</u>]

*Melinis minutiflora* is native to Africa, but has been introduced throughout the tropics as a forage crop. It is now regarded as a serious weed in many places. In the *Manual* region, it is only known to be established in southern Florida.

### 2. Melinis repens (Willd.) Zizka NATAL GRASS [p. 479, <u>549</u>]

*Melinis repens* is probably native to Africa and western Asia. It is now established throughout the subtropics, including the southern portion of the *Manual* region. It has been grown as an ornamental, but it is now established and often weedy in warmer portions of the region. Plants in

[1]J.K. Wipff and Rahmona A. Thompson  [2]J.K. Wipff

the *Manual* region belong to **Melinis repens** (Willd.) Zizka subsp. **repens**.

### 3. Melinis nerviglumis (Franch.) Zizka [p. 479]

In its native southern Africa, *Melinis nerviglumis* flowers from November to September [sic]. It differs from *M. repens* subsp. *repens*

primarily in its strongly overlapping leaf sheaths and rolled leaf blades. The cultivar being marketed is 'Pink Crystal'.

## 24.13 MEGATHYRSUS (Pilg.) B.K. Simon & S.W.L. Jacobs[1]

**Pl** per; ces, with short, thick rhz. **Clm** usu erect, smt geniculate and rooting at the lower nd; **nd** pubescent or glab. **Shth** glab or pubescent, smt with papillose-based hairs, mrg smt ciliate; **lig** of hairs; **bld** flat, erect or ascending, glab or with hairs, smt with appressed papillose-based hairs, mrg smt ciliate bas. **Pan** 10–65 cm, rchs smooth or scabrous, br single or whorled at the lowest nd, lo axils with a tuft of hairs, base, up axils glab; **br** usu spikelike, sec and tertiary br usu appressed when present; ped unequal, straight or curved, glab or with a single setaceous hair near the apex. **Spklt** 2.7–4.3 mm long, usu glab, rarely densely covered with papillose-based hairs, solitary, paired or in triplets, usu appressed to the br axes. **Glm** appressed to the flt, rchl

between the glm not pronounced; **lo glm** to ⅓ the length of the spklt, 1–3-veined, obtuse or truncate, glab; **up glm** subequal to the up flt, 5-veined, glab; **lo lm** subequal to the up flt, glab, 5-veined, without cross venation, acute, muticous or mucronate; lo flt stmt; **up lm** ellipsoid, pale, glab, transversely rugose, apices acute or mucronulate.

Simon and Jacobs (2003) transferred *Urochloa maxima* to a new genus, *Megathyrsus*, citing its morphological distinction from the taxa indicated as its close relatives in molecular-trees. The genus includes two species, *M. maximus* and *M. infestus* (Peters) B.K. Simon & S.W.L. Jacobs. Only *M. maximus* is present in the *Manual* region. North America.

### 1. Megathyrsus maximus (Jacq.) B.K. Simon & S.W.L. Jacobs GUINEA GRASS [p. 479, 549]

*Megathyrsus maximus* is an important forage grass that is native to Africa. In the *Manual* region, it grows in fields, waste places, stream banks, and hammocks. It is cultivated widely as a forage grass at low elevations, especially near the coast, and often escapes.

There are usually two varieties recognized. Only **Megathyrsus maximus** (Jacq.) B.K. Simon & S.W.L. Jacobs var. **maximus**, which has glabrous spikelets, is known from the *Manual* area. Plants with densely pubescent spikelets belong to **M. maximus** var. **pubiglumis** (K. Schum.) B.K. Simon & S.W.L. Jacobs.

## 24.14 UROCHLOA P. Beauv.[2]

**Pl** ann or per; usu csp, smt mat-forming, smt stln. **Clm** 5–500 cm, hrb, erect, geniculate, or decumbent and rooting at the lo nd. **Shth** open; **aur** rarely present; **lig** apparently of hairs, the bas memb portion inconspicuous; **bld** ovate-lanceolate to lanceolate, flat. **Infl** tml or tml and ax, usu pan of spikelike pri br in 2 or more ranks, rchs not concealed by the spklt; **pri br** usu alternate or subopposite, spikelike, and 1-sided, axes flat or triquetrous, usu terminating in a rdmt spklt; **sec br** present or absent, axes flat or triquetrous; **dis** beneath the spklt. **Spklt** solitary, paired, or in triplets, subsessile or pedlt, divergent or appressed, ovoid to ellipsoid, dorsally compressed, in 1–2(4) rows, with 2 flt, lo or up glm adjacent to the br axes. **Glm** not saccate bas; **lo glm** usu ⅕–⅔ as long as the spklt, occ equaling the up flt, (0)1–11-veined; **up glm** 5–13-veined; **lo flt** strl or stmt; **lo lm** similar to the up glm, 5–9-veined; **lo pal** if present, usu hyaline, 2-veined; **up flt** bisx, sessile, ovoid to

ellipsoid, usu plano-convex, usu glab, not dis-articulating, mucronate or acuminate; **up lm** indurate, transversely rugose and verrucose, 5-veined, mrg involute, apc round to mucronate, or aristate; **up pal** rugose, shiny or lustrous; **lod** 2, cuneate, truncate; **anth** 3. **Car** ovoid to elliptic, dorsally compressed; **emb** ½–¾ as long as the car; **hila** punctate to linear.

*Urochloa* is a genus of approximately 100 tropical and subtropical species. Nineteen species are found in the *Manual* region. Eight species are established introductions, six are native, three are cultivated as grain or forage crops, and two have been found in the region but are not known to be established. *Urochloa* differs from *Moorochloa* in its two or more ranks of panicles and rugose, non-disarticulating, distal florets. The rugose, often mucronate or aristate, distal florets also distinguish it from most species of *Panicum*.

1. Spikelets paired at mid-branch length, sometimes solitary distally.
    2. Axes of the primary panicle branches flat; lower glumes (0)1–3-veined.
       3. Plants annual; culms 10–35 cm tall; spikelets 1.8–2.2 mm long . . . . . . . . . . . . . . . . . . . . . . . . . . . . . 2. *U. reptans*
       3. Plants perennial; culms 20–500 cm tall; spikelets 2.5–5 mm long.
          4. Upper lemmas awned, the awns 0.5–1.2 mm; lower glumes 3-veined, with 1–3 conspicuous,
            rigid hairs . . . . . . . . . . . . . . . . . . . . . . . . . . . . . . . . . . . . . . . . . . . . . . . . . . . . . . . . . . . 8. *U. mosambicensis*
          4. Upper lemmas unawned; lower glumes 1–3-veined, glabrous . . . . . . . . . . . . . . . . . . . . . . . . . . . . 1. *U. mutica*

[1]Mary E. Barkworth, modified from treatment of *Urochloa* by J.K. Wipff and R. Thompson.  [2]J.K. Wipff and Rahmona A. Thompson

2. Axes of the primary panicle branches triquetrous; lower glumes 3–7-veined.
  5. Spikelets 4.8–6 mm long; hila linear, about ½ as long as the caryopses . . . . . . . . . . . . . . . . . . . . . . . . . . . 3. *U. texana*
  5. Spikelets 2–4.2 mm long; hila punctate.
    6. Branch axes densely hairy, the hairs papillose-based . . . . . . . . . . . . . . . . . . . . . . . . . . . . . . . . . . . 4. *U. arizonica*
    6. Branch axes sometimes densely hairy, with few or no papillose-based hairs.
      7. Upper glumes with evident cross venation extending from near the bases to the apices;
        spikelets obovoid; upper glumes and lower lemmas usually glabrous; lower lemmas 7-
        veined . . . . . . . . . . . . . . . . . . . . . . . . . . . . . . . . . . . . . . . . . . . . . . . . . . . . . . . . . . . . . . 5. *U. fusca*
      7. Upper glumes without evident cross venation, or the cross venation confined to the distal
        ½; spikelets ellipsoid; upper glumes and lower lemmas glabrous or pubescent; lower
        lemmas 5-veined.
        8. Cauline nodes glabrous; plants 20–120 cm tall; spikelet apices abruptly acuminate . . . . . . . . . 6. *U. adspersa*
        8. Cauline nodes pubescent; plants 10–70 cm tall, spikelet apices broadly acute to acute.
          9. Upper glumes 5-veined; upper lemmas 1.8–2.1 mm long; glumes separated by an
            internode about 0.3 mm long . . . . . . . . . . . . . . . . . . . . . . . . . . . . . . . . . . . 11. *U. villosa* (in part)
          9. Upper glumes 7–9-veined; upper lemmas 2.3–3.3 mm long; glumes not separated by
            a conspicuous internode . . . . . . . . . . . . . . . . . . . . . . . . . . . . . . . . . . . . . . . . . . 7. *U. ramosa*
1. Spikelets solitary at mid-branch length.
  10. Panicle branches triquetrous, 0.2–0.4 mm wide.
    11. Plants perennial; upper glumes (9)11–13-veined; upper lemmas 2.4–2.8 mm long; glumes not
      separated by a conspicuous internode; lower florets staminate . . . . . . . . . . . . . . . . . . . . . . . 17. *U. ciliatissima*
    11. Plants annual; upper glumes 5-veined; upper lemmas 1.8–2.1 mm long; glumes separated by
      an internode about 0.3 mm long; lower florets sterile . . . . . . . . . . . . . . . . . . . . . . . . . . . 11. *U. villosa* (in part)
  10. Panicle branches flat or crescent-shaped in cross section, 0.5–2.5 mm wide.
    12. Upper lemmas awned, the awns 0.3–1.2 mm long, apices rounded.
      13. Plants perennial; lower glumes with 1–3 conspicuous, rigid hairs, the lower glumes ½–¾
        as long as the spikelet; lower florets staminate . . . . . . . . . . . . . . . . . . . . . . . . . . . 8. *U. mosambicensis*
      13. Plants annual; lower glumes without conspicuous, rigid hairs; the lower glumes ¼–⅓(½)
        as long as the spikelet; lower florets sterile . . . . . . . . . . . . . . . . . . . . . . . . . . . . . . 14. *U. panicoides*
    12. Upper lemmas unawned; apices variable, with or without a mucronate tip.
      14. Spikelets in a single row along the branches; spikelets 4–6 mm long; lower florets
        staminate; panicle branch axes crescent-shaped in cross section . . . . . . . . . . . . . . . . . . . . . 10. *U. brizantha*
      14. Spikelets in 2 rows along the branches; spikelets 2.5–6 mm long; lower florets sterile or
        staminate; panicle branches flat.
        15. Lower glumes 5–7-veined; glumes scarcely separated, the internode between them
          shorter than 0.3 mm.
          16. Upper glumes and lower lemmas pubescent, the hairs often long in the distal ⅓
            . . . . . . . . . . . . . . . . . . . . . . . . . . . . . . . . . . . . . . . . . . . . . . . . . . . . . 9. *U. piligera* (in part)
          16. Upper glumes and lower lemmas glabrous.
            17. Plants perennial, stoloniferous; lower florets staminate . . . . . . . . . . . . . . . . 16. *U. arrecta*
            17. Plants annual; lower florets sterile . . . . . . . . . . . . . . . . . . . . . . . . . . . . 15. *U. platyphylla*
        15. Lower glumes (7)9–11-veined; glumes separated by a conspicuous, 0.3–0.5 mm
          internode.
          18. Lower paleas absent; spikelets usually pubescent, sometimes glabrous . . . . . . . . . 9. *U. piligera* (in part)
          18. Lower paleas present; spikelets glabrous.
            19. Spikelets 3.3–3.7 mm long; base of blades rounded to subcordate, not
              clasping the stem . . . . . . . . . . . . . . . . . . . . . . . . . . . . . . . . . . . . . . . 12. *U. subquadripara*
            19. Spikelets 4–6 mm long; base of blades subcordate to cordate, clasping the
              stem . . . . . . . . . . . . . . . . . . . . . . . . . . . . . . . . . . . . . . . . . . . . . . . 13. *U. plantaginea*

## 1. Urochloa mutica (Forssk.) T.Q. Nguyen PARAGRASS [p. 480, 550]

An African species, *Urochloa mutica* is grown as a forage crop throughout the tropics, but it tends to become weedy. It grows on moist, disturbed soils and is established in the southeastern United States.

## 2. Urochloa reptans (L.) Stapf SPRAWLING SIGNALGRASS [p. 480, 550]

*Urochloa reptans* is widely distributed in tropical and subtropical regions of the world, growing in disturbed habitats. In the *Manual* region, it is found primarily in Texas and Louisiana.

### 3. Urochloa texana (Buckley) R.D. Webster TEXAS SIGNALGRASS, TEXAS MILLET [p. *480*, 550]

*Urochloa texana* grows in sandy, moist soils from the southern United States to northern Mexico. Populations in the United States outside of Texas may represent relatively recent introductions.

### 4. Urochloa arizonica (Scribn. & Merr.) Morrone & Zuloaga ARIZONA SIGNALGRASS [p. *480*, 550]

*Urochloa arizonica* is native to the southwestern United States and northern Mexico, but has been introduced to, and appears to be established in, the southeastern United States. It grows in open, dry areas with rocky or sandy soils.

### 5. Urochloa fusca (Sw.) B.F. Hansen & Wunderlin BROWNTOP SIGNALGRASS [p. *480*, 550]

*Urochloa fusca* grows from the southern United States to Peru, Paraguay, and Argentina, usually in moist, often disturbed areas at low elevations. It frequently occurs as a weed, but is occasionally grown for forage and grain. Plants having smaller, more compact panicles and larger (2.4–3.4 mm), mostly yellowish spikelets have been referred to as *Urochloa fusca* var. *reticulata* (Torr.) B.F. Hansen & Wunderlin. This variety is mainly found in the southwestern United States, but has been introduced into other areas, including Australia. *Urochloa fusca* (Sw.) B.F. Hansen & Wunderlin var. *fusca* has generally larger, more open panicles and smaller (2–2.5 mm), reddish-brown or bronze-colored spikelets. Much intergradation is reported between the two varieties.

### 6. Urochloa adspersa (Trin.) R.D. Webster DOMINICAN SIGNALGRASS [p. *480*, 550]

*Urochloa adspersa* grows in southern Florida, the West Indies, and Argentina. It prefers moist, open areas, often on coral limestone. It has also been found on ballast dumps in Mobile, Alabama; Philadelphia, Pennsylvania; and Camden, New Jersey; it has not persisted at these locations.

### 7. Urochloa ramosa (L.) T.Q. Nguyen BROWNTOP MILLET [p. *480*, 550]

A weedy species of tropical Africa and Asia, *Urochloa ramosa* has spread throughout the tropics and subtropics, including the southeastern United States. It is considered a weed in the *Manual* area, but it is cultivated in India as a grain and forage crop. The grain is sometimes used for birdseed.

### 8. Urochloa mosambicensis (Hack.) Dandy SABI GRASS [p. *480*, 550]

*Urochloa mosambicensis*, native to Africa, has been found in southern Texas and is expected to spread. It is grown for forage and hay in Africa.

### 9. Urochloa piligera (F. Muell. *ex* Benth.) R.D. Webster WATTLE SIGNALGRASS [p. *480*, 550]

*Urochloa piligera* is an Australian species that has been found in Florida. Glabrous and pubescent forms of *U. piligera* are identical except in the presence or absence of spikelet vestiture. Currently, only the pubescent form has been found in the *Manual* area. *Urochloa piligera* differs from *U. subquadripara* in lacking a lower palea, and having larger, usually closely overlapping spikelets.

### 10. Urochloa brizantha (Hochst. *ex* A. Rich.) R.D. Webster PALISADE SIGNALGRASS [p. *480*, 550]

*Urochloa brizantha*, a native of tropical Africa, was first reported from the *Manual* region in 1993, and is considered a sporadic introduction in the *Manual* area. It is reported to intergrade with U. **decumbens** (Stapf) R.D. Webster, and intermediates are often difficult to separate, although they do not seem to be very common in the wild.

Although *U. decumbens* has not been reported in the *Manual* area, it is expected in Florida, because it is widely used as forage in the tropics. The two taxa can be distinguished by their panicle branches: *U. decumbens* has flat, ribbonlike panicle branches 1–1.8 mm wide, whereas *U. brizantha* has crescentric panicle branches 0.5–1.2 mm wide.

### 11. Urochloa villosa (Lam.) T.Q. Nguyen HAIRY SIGNALGRASS [p. *480*, 550]

*Urochloa villosa* is a tropical African and Asian species that was reported from chrome and iron ore piles in Newport News, Virginia in 1959. No additional collections have been reported in the *Manual* area.

### 12. Urochloa subquadripara (Trin.) R.D. Webster ARMGRASS MILLET [p. *480*, 550]

*Urochloa subquadripara*, native to tropical Asia and Australasia, is established in Florida and, reportedly, Georgia, although no specimens documenting its presence in Georgia have been located. Its weediness and drought tolerance suggest that it might become a troublesome weed in some parts of the *Manual* region. It is similar to U. **distachya** (L.) T.Q. Nguyen, but that species supposedly has shorter (2.4–3 mm) spikelets and shorter (1.9–2.3 mm) upper florets. *Urochloa subquadripara* differs from the glabrous form of *U. piligera* in having a well-developed palea and smaller (3.3–3.8 mm), well-separated spikelets.

### 13. Urochloa plantaginea (Link) R.D. Webster PLANTAIN SIGNALGRASS [p. *481*, 550]

*Urochloa plantaginea*, native to western and central Africa, is found from the southeastern United States to Argentina. It is now established in the southeastern United States, growing in loose sand and loam soils. It is considered a weed in the *Manual* area, though it may provide good forage. A similar species, **Urochloa oligobrachiata** (Pilg.) Kartesz, has been reported from Florida, but it differs from *U. plantaginea* in having acute to acuminate lower glumes and shortly awned upper lemmas. It is native to western Africa.

### 14. Urochloa panicoides P. Beauv. LIVERSEED GRASS [p. *481*, 550]

*Urochloa panicoides* is native to Africa, and is considered a noxious weed by the U.S. Department of Agriculture. In the Western Hemisphere, it has been introduced into the southern United States, Mexico, and Argentina. Within the *Manual* region, it has been reported from disturbed sites in the southern and Gulf Coast areas of Texas, but it is expected to spread. Populations from New Mexico have been destroyed. *Urochloa panicoides* has morphological forms with glabrous, pubescent, or setosely-fringed spikelets. Only the glabrous form is known to occur in the *Manual* area.

### 15. Urochloa platyphylla (Munro *ex* C. Wright) R.D. Webster BROADLEAF SIGNALGRASS [p. *481*, 550]

*Urochloa platyphylla* is a weedy species found in open, sandy soil in the southeastern United States, West Indies, and South America.

### 16. Urochloa arrecta (Hack. *ex* T. Durand & Schinz) Morrone & Zuloaga AFRICAN SIGNALGRASS [p. *481*, 550]

*Urochloa arrecta* is native to Africa, but it has been introduced into Florida and Brazil as a forage grass. It is reported to be established in Collier County, Florida.

### 17. Urochloa ciliatissima (Buckley) R.D. Webster FRINGED SIGNALGRASS, SANDHILL GRASS [p. *481*, 550]

*Urochloa ciliatissima* is endemic to Texas, Oklahoma, and Arkansas, and grows on sandy soils.

## 24.15 ERIOCHLOA Kunth[1]

**Pl** ann or per; **csp**, smt with short rhz or stln, not producing subterranean spklt. **Clm** 20–250 cm, erect or decumbent, usu with 2–5 nd. **Shth** open; **aur** absent; **lig** memb, ciliate. **Infl** tml, pan of spikelike br on elongate rchs; **br** with many pedlt, loosely appressed spklt, terminating in a spklt, without stiff bristles or flat bracts, spklt in pairs, triplets, or solitary, often solitary distally when in pairs or triplets at the mid of the br; **ped** terminating in a well-developed disk; **dis** below the glm(s). **Spklt** with 2 flt, lo flt usu strl, up flt bisx. **Lo glm** typically rdcd (smt absent) and fused with the glab cal to form a cuplike structure; **up glm** lanceolate to ovate, glab or variously pubescent, 3–9-veined, unawned or awned; **lo lm** similar to the up glm in length, shape, venation, and pubescence, unawned; **lo pal** absent to fully developed; **up lm** lanceolate to ovate, indurate, rugose, dull, glab, rounded on the back, veins not pronounced, mrg involute; **anth** 3; **lod** 2, papery; **sty** with 2 br, purple, plumose. **Car** not longitudinally grooved; **endosperm** solid.

*Eriochloa* has 20–30 species that grow in tropical, subtropical, and warm-temperate areas of the world. Eight species of *Eriochloa* are native to the *Manual* region and three are introduced. Of the three introduced species only two have become naturalized.

1. Spikelets solitary at the middle of the branches, sometimes in unequally pedicellate pairs near the base.
    2. Pedicels with more than 12 long (1.5–3 mm) hairs near the apices, densely hirsute or villous below, the hairs mostly about 0.1 mm long, but some longer hairs interspersed among the short hairs.
        3. Blades 0.5–4 mm wide; spikelets 1.4–1.9 mm wide; plants perennial . . . . . . . . . . . . . . . . . . . . . . . . . .1. *E. sericea*
        3. Blades 5–12 mm wide; spikelets 2–2.5 mm wide; plants annual . . . . . . . . . . . . . . . . . . . . . . . . . . . . .2. *E. villosa*
    2. Pedicels with fewer than 10 long (1.5–3 mm) hairs near the apices; variously hirsute below.
        4. Lower floret of each spikelet with a palea . . . . . . . . . . . . . . . . . . . . . . . . . . . . . . . . . . . .3. *E. michauxii* (in part)
        4. Lower floret of each spikelet without a palea.
            5. Rachises hairy, the longer hairs 0.1–0.8 mm long; spikelets 3.1–5 mm long, 1.2–1.7 mm wide . . . . .4. *E. contracta*
            5. Rachises glabrous or scabrous, not hairy; spikelets 2.7–3.6 mm long, 0.8–1.5 mm wide . . .5. *E. fatmensis* (in part)
1. Spikelets in unequally pedicellate pairs or triplets at the middle of the branches, sometimes solitary distally.
    6. Adaxial surfaces of the blades velvety to the touch; cauline internodes pubescent to pilose . . . . . . . . . . . .6. *E. lemmonii*
    6. Adaxial surfaces of the blades glabrous or hairy, but not velvety to the touch; cauline internodes glabrous or pubescent.
        7. Upper lemmas unawned or the awns shorter than 0.2(0.3) mm.
            8. Plants annual; upper glumes acute to acuminate, often terminating in awnlike apices up to 1.5 mm long; lower florets without paleas.
                9. Longer pedicels of the spikelet pairs or triplets to 1 mm long; upper glumes acute to acuminate, unawned or awned, the awns up to 1.2 mm long . . . . . . . . . . . . . . . . . .10. *E. acuminata* (in part)
                9. Longer pedicels of the spikelet pairs to 3 mm long; upper glumes always acuminate, awned, the awns 0.5–3.5 mm long . . . . . . . . . . . . . . . . . . . . . . . . . . . . . . . .8. *E. aristata* (in part)
            8. Plants perennial; upper glumes acute, unawned; lower floret of each spikelet usually with a palea.
                10. Culms erect, not rooting at the lower nodes; spikelets 3.7–5.7 mm long, 1.3–1.8 mm wide; lower floret of each spikelet always with a palea . . . . . . . . . . . . . . . . . . . . . . .3. *E. michauxii* (in part)
                10. Culms decumbent, rooting at the lower nodes; spikelets 3.2–3.9 mm long, 1.1–1.3 mm wide; lower floret of each spikelet with or without a palea . . . . . . . . . . . . . . . . . . . . . . . .11. *E. polystachya*
        7. Upper lemmas awned, the awns 0.2–1.5 mm long.
            11. Spikelets 2.7–3.6 mm long, 0.8–1.5 mm wide . . . . . . . . . . . . . . . . . . . . . . . . . . . . . . . . . .5. *E. fatmensis* (in part)
            11. Spikelets 3.6–8.8 mm long, 0.9–1.6 mm wide.
                12. Pedicels uniformly hirsute, the hairs about 0.1 mm long; plants rhizomatous . . . . . . . . . . . . . .7. *E. punctata*
                12. Pedicels with some hairs 0.5–2.5 mm long, at least distally; plants not rhizomatous.
                    13. Plants perennial . . . . . . . . . . . . . . . . . . . . . . . . . . . . . . . . . . . . . . . . . . . . .9. *E. pseudoacrotricha*
                    13. Plants annual.
                        14. Longer pedicels of the spikelet pairs or triplets to 1 mm long; upper glumes acute to acuminate, unawned or awned, the awns up to 1.2 mm long . . .10. *E. acuminata* (in part)
                        14. Longer pedicels of the spikelet pairs to 3 mm long; upper glumes always acuminate, awned, the awns 0.5–3.5 mm long . . . . . . . . . . . . . . . . . . . . . . . . . . .8. *E. aristata* (in part)

[1]Robert B. Shaw, Robert D. Webster, and Christine M. Bern

**1. Eriochloa sericea** (Scheele) Munro ex Vasey TEXAS CUPGRASS [p. *481*, 550]

*Eriochloa sericea* usually grows on clay or clay-loam soils in prairies, roadsides, or protected areas. It is widespread in the blackland prairie and Edwards Plateau of Texas, but extends into Kansas, Nebraska, and Oklahoma, and onto the coastal prairie and rolling plains of Texas and northern Mexico.

**2. Eriochloa villosa** (Thunb.) Kunth ÉRIOCHLOÉ VELUE [p. *481*, 550]

*Eriochloa villosa* is a weedy species of eastern Asia that has been found at scattered locations in the *Manual* region.

**3. Eriochloa michauxii** (Poir.) Hitchc. LONGLEAF CUPGRASS [p. *481*, 550]

*Eriochloa michauxii* is endemic to the southeastern United States. Plants intermediate between the two varieties have been collected in Lee and Monroe counties, Florida.

1. Lower florets staminate; blades generally flat, usually 8–15 mm wide . . . . . . . . . . . . . . . . . . . var. *michauxii*
1. Lower floret sterile; blades involute to conduplicate, 5–8 mm wide . . . . . . . . . . . . . . var. *simpsonii*

*Eriochloa michauxii* (Poir.) Hitchc. var. **michauxii** grows in brackish or fresh water marshes, hammocks, and prairies of the southeastern United States, including the whole of Florida.

*Eriochloa michauxii* var. **simpsonii** (Hitchc.) Hitchc. is a rare variety that grows in low wet areas, roadsides, or on washed sand and shell beaches of southwestern Florida.

**4. Eriochloa contracta** Hitchc. PRAIRIE CUPGRASS [p. *481*, 550]

*Eriochloa contracta* grows in fields, ditches, and other disturbed areas. It is known only from the United States, being native and common in the central United States, and adventive to the east and southwest. It differs from *E. acuminata* in its tightly contracted, almost cylindrical panicles and longer lemma awns, but intermediate forms can be found. It can also be confused with first-year plants of the perennial *E. punctata*, which have glabrous leaves, narrower and more tapering spikelets, and longer lemma awns.

**5. Eriochloa fatmensis** (Hochst. & Steud.) Clayton [p. *481*, 550]

*Eriochloa fatmensis* is native to tropical Africa, Arabia, and India, where it usually grows in wet areas or grasslands. It has been found in Tucson, Arizona, and Biloxi, Mississippi, but is probably not established in the *Manual* region.

**6. Eriochloa lemmonii** Vasey & Scribn. CANYON CUPGRASS [p. *482*, 550]

*Eriochloa lemmonii*, a rare species, grows in canyons and on rocky slopes in Pima County, Arizona, Hidalgo County, New Mexico, and adjacent Mexico. The record from Tennessee reflects an introduction. It is not known if the species has persisted there. *Eriochloa lemmonii* may hybridize with *E. acuminata*, from which it differs in the frequent presence of lower paleas, raised veins of the upper glumes and lower lemmas, broad, velvety pubescent leaf blades, and blunt spikelets. Reports of *E. lemmonii* from Texas may be based on hybrids between the two species.

**7. Eriochloa punctata** (L.) Desv. *ex* Ham. LOUISIANA CUPGRASS [p. *482*, 550]

*Eriochloa punctata* grows in coastal marshes, along water courses, and in moist swales and ditches of the coastal plain from Texas and Louisiana south through Mexico to Central and South America. It has not been possible to verify the identification of the specimen from Georgia for this treatment. If correct, it suggests that the species may be more widespread than generally thought.

**8. Eriochloa aristata** Vasey BEARDED CUPGRASS [p. *482*, 550]

*Eriochloa aristata* is a weed of moist swales, roadsides, and irrigated fields of the southwestern United States. Its range extends through Mexico and Central America to Colombia. There are three specimens from Oktibbeha County, Mississippi, and the species may be established there.

**9. Eriochloa pseudoacrotricha** (Stapf *ex* Thell.) J.M. Black VERNAL CUPGRASS [p. *482*, 551]

*Eriochloa pseudoacrotricha* is an Australian species that has been introduced into south Texas and Mississippi. It grows in waste areas.

**10. Eriochloa acuminata** (J. Presl) Kunth SOUTHWESTERN CUPGRASS [p. *482*, 551]

*Eriochloa acuminata* is native to the southern United States and northern Mexico, but has become established outside this region. It may hybridize with *E. lemmonii*, from which it differs in its lack of lower paleas, upper glumes and lower lemmas with level veins, and narrower, glabrous or sparsely pubescent leaf blades.

Both varieties grow in Mexico as well as the United States.

1. Spikelets 4–6 mm long, long-acuminate or taper-ing to a short awn . . . . . . . . . . . . . . . . . . . . . . var. *acuminata*
1. Spikelets 3.8–4 mm long, acute . . . . . . . . . . . . . . . . var. *minor*

*Eriochloa acuminata* (J. Presl) Kunth var. **acuminata** generally grows in ditches, fields, right of ways, and other disturbed areas of the southern United States.

*Eriochloa acuminata* var. **minor** (Vasey) R.B. Shaw is common in irrigated fields, orchards and disturbed areas of the southwestern United States. It is adventive in Maryland.

**11. Eriochloa polystachya** Kunth CARIBBEAN CUPGRASS [p. *482*, 551]

*Eriochloa polystachya* is native to the West Indies, Costa Rica, Honduras, and South America. It was introduced into the United States as a forage crop and is now established at some locations in Florida and Texas.

## 24.16 PENNISETUM Rich.[1]

Pl ann or per; habit various. **Clm** 3–800 cm, not wd, smt brchg above the base; **intnd** solid or hollow. **Lig** memb and ciliate, or of hairs, rarely completely memb; **bld** smt psdpet. **Infl** spicate pan with highly rdcd br termed *fascicles*; pan 1–many per plant, tml on the clm or on both the clm and the sec br, or tml and ax, or only ax, usu completely exposed at maturity; **rchs** usu terete, with (1)5–many fascicles; **fascicle axes** 0.2–7.5(28) mm, with (1)3–130+ bristles and 1–12 spklt. **Bristles** free or fused at the base, disarticulating with the spklt at maturity; of 3 kinds, *outer*, *inner*, and *primary*, in some species with all 3 kinds present below each spklt, in others 1 or more kinds missing from some or all of the spklt; **outer** (lo) **bristles** antrorsely scabrous, terete; **inner** (up) **bristles** antrorsely scabrous or long-ciliate, usu flatter and wider than the outer bristles; **pri** (tml)

[1]J.K. Wipff

**bristles** located immediately below the spklt, solitary, antrorsely scabrous or long-ciliate, often longer than the other bristles associated with the spklt; **dis** usu at the base of the fascicles, smt also beneath the up flt. **Spklt** with 2 flt; **lo glm** absent or present, 0–5-veined; **up glm** longer, 0–11-veined; **lo flt** strl or stmt; **lo lm** usu as long as the spklt, memb, 3–15-veined, mrg usu glab; **lo pal** present or absent; **up lm** memb to coriaceous, 5–12-veined; **up pal** shorter than the lm but similar in texture; **lod** 0 or 2, glab; **anth** 3, if present.

*Pennisetum* has 80–130 species, most of which grow in the tropics and subtropics, and occupy a wide range of habitats. Twenty-five species are native to the Western Hemisphere, but none to the *Manual* region. Most of the species treated here are cultivated for food, forage, or as ornamentals. Four are classified as noxious weeds by the U.S. Department of Agriculture. Records known to be based on cultivated plants are not included in the distribution maps but, in many cases, it is not possible to determine whether a record is based on a cultivated plant or an escape.

The placement of the boundary between *Pennisetum* and *Cenchrus* is contentious. As treated here, *Pennisetum* has antrorsely scabrous bristles that are not spiny, and fascicle axes that terminate in a bristle. *Cenchrus* has retrorsely (rarely antrorsely) scabrous, spiny bristles, and fascicle axes that are terminated by a spikelet.

NOTE: Pedicel length is the distance from the base of the primary bristles to the base of the terminal spikelet. Fascicle axis lengths and fascicle densities are measured in the middle of the panicle; spikelet measurements refer to the largest spikelet in the fascicles.

1. Plants stoloniferous; panicles axillary, partially or wholly hidden in the leaf sheaths at maturity, the rachises flattened in cross section, with 1–6 fascicles; spikelets 10–22 mm long, bristles mostly shorter than the spikelet . . . . . . . . . . . . . . . . . . . . . . . . . . . . . . . . . . . . . . . . . . . . . . . . . . . . . . . . 1. *P. clandestinum*
1. Plants not stoloniferous; panicles terminal or terminal and axillary, fully exserted at maturity, the rachises terete, with 10–many fascicles; spikelets 2.5–12 mm long, the majority of the bristles as long as or longer than the spikelets.
   2. Fascicles with only 1 bristle and 1 spikelet . . . . . . . . . . . . . . . . . . . . . . . . . . . . . . . . . . . . . . . . 18. *P. petiolare*
   2. Fascicles with 6 or more bristles and 1–12 spikelets.
      3. Most or all bristles scabrous, the primary bristles sometimes sparsely and inconspicuously long-ciliate.
         4. Primary bristles not noticeably longer than all the other bristles in the fascicles.
            5. Terminal panicle erect; fascicles with a stipelike base 1.5–5.6 mm long . . . . . . . . . . 7. *P. alopecuroides* (in part)
            5. Terminal panicle drooping; fascicles subsessile, the bases 0.4–0.7 mm long.
               6. Plants green; most of the bristles only slightly longer than the spikelets; upper glumes (7)9-veined, about as long as the spikelets . . . . . . . . . . . . . . . . . . . . . . . . . . . . . . . 4. *P. nervosum*
               6. Plants purplish; bristles at least twice as long as the spikelets; upper glumes 1–3-veined, usually about ½ as long as the spikelets . . . . . . . . . . . . . . . . . . . . . . . . . . 5. *P. macrostachys*
         4. Primary bristles noticeably longer than all the other bristles in the fascicles.
            7. Panicles dense; rachises with 21–40 fascicles per cm.
               8. Rachises pubescent; bristles yellow or purple; leaf blades (4)12–40 mm wide; paleas of lower florets present . . . . . . . . . . . . . . . . . . . . . . . . . . . . . . . . . . . . . . . . 2. *P. purpureum* (in part)
               8. Rachises scabrous; bristles white to stramineous; leaf blades 4–12 mm wide; paleas of lower florets absent . . . . . . . . . . . . . . . . . . . . . . . . . . . . . . . . . . . . . . . . . . . . 6. *P. macrourum*
            7. Panicles less dense; rachises with 5–16 fascicles per cm.
               9. Panicles drooping, terminal and axillary; leaf blades 19–45 mm wide . . . . . . . . . . . . . . . . . 8. *P. latifolium*
               9. Panicles erect, all terminal; leaf blades 2–12 mm wide.
                  10. Plants not rhizomatous; lower part of rachises pubescent . . . . . . . . . . . . . 7. *P. alopecuroides* (in part)
                  10. Plants rhizomatous; lower part of rachises scabrous . . . . . . . . . . . . . . . . . . . . 14. *P. flaccidum* (in part)
      3. Bristles, at least the primary bristles, conspicuously long-ciliate.
         11. Spikelets 9–12 mm long . . . . . . . . . . . . . . . . . . . . . . . . . . . . . . . . . . . . . . . . . . . . . . . . . . 11. *P. villosum*
         11. Spikelets 2.5–7 mm long.
            12. Fascicles not disarticulating from the rachises; panicles 4–200 cm long; upper lemmas with pubescent margins; caryopses protruding from the florets at maturity . . . . . . . . . . . . . . . 3. *P. glaucum*
            12. Fascicles disarticulating from the rachises at maturity; panicles 2–37.5 cm long; upper lemmas with glabrous margins; caryopses concealed by the lemmas and paleas at maturity.
               13. Upper florets readily disarticulating at maturity; upper lemmas smooth and shiny, conspicuously different in texture from the lower lemmas.
                  14. Fascicles with 6–14 long-ciliate inner bristles and 13–30 scabrous outer bristles; fascicle axes 0.2–0.5 mm long; spikelets sessile . . . . . . . . . . . . . . . . . . . 9. *P. polystachion*
                  14. Fascicles with 40–90 long-ciliate inner bristles and 10–20 scabrous outer bristles; fascicle axes 1.5–2.5 mm long; spikelets pedicellate, the pedicels 1–3.5 mm long . . . . . . . . . . . . . . . . . . . . . . . . . . . . . . . . . . . . . . . . . . . . . . . 10. *P. pedicellatum*

13. Upper florets not disarticulating at maturity; lower and upper lemmas similar in texture.
  15. Lower portion of the rachises glabrous, sometimes scabrous.
    16. Inner bristles neither grooved nor fused, even at the base; spikelets 5.2–6.7 mm long, pedicellate, the pedicels 0.1–0.5 mm long . . . . . . . . 14. *P. flaccidum* (in part)
    16. Inner bristles grooved and fused, at least at the base; spikelets 2.5–5.6 mm long, sessile.
      17. Inner bristles fused for up to ¼ their length; many outer bristles exceeding the spikelets; terminal bristles 10.5–23 mm, noticeably longer than the other bristles in the fascicles . . . . . . . . . . . . . . . . . . . . . . . . . . 12. *P. ciliare*
      17. Inner bristles fused for ⅓–½ their length; outer bristles not exceeding the spikelets; terminal bristles 2.9–6.5 mm, usually not noticeably longer than other bristles in the fascicles . . . . . . . . . . . . . . . . . . . . . . . 13. *P. setigerum*
  15. Lower portion of the rachises pubescent.
    18. Plants 200–800 cm tall; midculm leaves (4)12–40 mm wide; panicles golden-yellow or dark purple; rachises with 30–40 fascicles per cm . . . . . . . . . . 2. *P. purpureum* (in part)
    18. Plants 50–200 cm tall; midculm leaves 2–11 mm wide; panicles white, burgundy, light purple, or pink; rachises with 5–17 fascicles per cm.
      19. Midculm leaves 2–3.5 mm wide, convolute or folded, green, the midvein noticeably thickened; lower florets of the spikelets usually sterile, sometimes staminate . . . . . . . . . . . . . . . . . . . . . . . . . . . . . . . . 15. *P. setaceum*
      19. Midculm leaves 3–11 mm wide, flat, green or burgundy, the midvein not noticeably thickened; lower florets of the spikelets staminate.
        20. Plants shortly rhizomatous; nodes pubescent; panicles erect to slightly arching, white or purple-tinged; leaves green; ligules 1–1.7 mm long; fascicles with 0–24 terete, scabrous outer bristles . . . . . . . . . . . . . . . . . . . . . . . . . . . . . . . . . . . . . . . . 16. *P. orientale*
        20. Plants not rhizomatous; nodes glabrous; panicles conspicuously drooping, burgundy (rarely whitish-green); leaves burgundy (rarely green); ligules 0.5–0.8 mm long; fascicles with 43–68 terete, scabrous outer bristles . . . . . . . . . . . . . . . . . . . . . . . . . . . . . 17. *P. advena*

## 1. Pennisetum clandestinum Hochst. *ex* Chiov. KIKUYU GRASS [p. 482, 551]

*Pennisetum clandestinum* is native to Africa. It now grows in many parts of the world, often as a forage or lawn grass. The U.S. Department of Agriculture considers it a noxious weed. In parts of the *Manual* region, it is well-established in lawns.

## 2. Pennisetum purpureum Schumach. ELEPHANT GRASS [p. 482, 551]

*Pennisetum purpureum* is native to Africa but now grows in tropical areas throughout the world, frequently becoming naturalized. It is grown as an ornamental in the *Manual* region, and, less commonly, for forage.

## 3. Pennisetum glaucum (L.) R. Br. PEARL MILLET [p. 482, 551]

*Pennisetum glaucum*, a native of Asia, is cultivated in the United States for grain, forage, and birdseed.

## 4. Pennisetum nervosum (Nees) Trin. BENTSPIKE FOUNTAINGRASS [p. 482, 551]

*Pennisetum nervosum* is native to South America. It has been introduced into the *Manual* region, being known from populations adjacent to the Rio Grande River in Cameron and Hidalgo counties, Texas, and San Diego County, California.

## 5. Pennisetum macrostachys (Brongn.) Trin. PACIFIC FOUNTAINGRASS [p. 482]

*Pennisetum macrostachys* is native to the South Pacific. It is grown in the *Manual* region as an ornamental, being sold as 'Burgundy Giant'.

## 6. Pennisetum macrourum Trin. WATERSIDE REED [p. 483, 551]

*Pennisetum macrourum* is native to Africa, where it grows along rivers and lake margins. In the *Manual* region, it is known only from one location in Monterey County, California. Although sometimes recommended as an ornamental, the U.S. Department of Agriculture considers it a noxious weed.

## 7. Pennisetum alopecuroides (L.) Spreng. FOXTAIL FOUNTAINGRASS [p. 483]

*Pennisetum alopecuroides* is native to southeast Asia. It is frequently grown as an ornamental in the *Manual* region.

## 8. Pennisetum latifolium Spreng. URUGUAY FOUNTAINGRASS [p. 483]

*Pennisetum latifolium* is native to South America. It is occasionally grown as an ornamental in the *Manual* region.

## 9. Pennisetum polystachion (L.) Schult. MISSION GRASS [p. 483, 551]

*Pennisetum polystachion* is a polymorphic, weedy African species that has become established in the tropics and subtropics, including Florida. The U.S. Department of Agriculture considers it a noxious weed. Only **Pennisetum polystachion** subsp. **setosum** (Sw.) Brunken has been found in the *Manual* region. It differs from **P. polystachion** (L.) Schutt. subsp. **polystachion** as indicated:

1. Plants annual, usually profusely branching; fascicles white, pink, red, or deep purple . . . . . . . . . subsp. *polystachion*
1. Plants perennial, usually sparingly branched; fascicles yellow, light brown, or purplish . . . . . . . subsp. *setosum*

**10. Pennisetum pedicellatum** Trin. HAIRY FOUNTAINGRASS [p. 483, 551]

*Pennisetum pedicellatum* is native to Africa. It now grows in many other areas, including Florida. The U.S. Department of Agriculture considers it a noxious weed.

**11. Pennisetum villosum** R. Br. *ex* Fresen. FEATHERTOP [p. 483]

*Pennisetum villosum* is native to Ethiopia, northern Somalia, and the Arabian Peninsula. It is grown as an ornamental in the *Manual* region.

**12. Pennisetum ciliare** (L.) Link BUFFEL GRASS [p. 483, 551]

*Pennisetum ciliare* is native to Africa, western Asia, and India. It now grows throughout the warmer, drier regions of the world, often as a forage crop, and is established in much of the southeastern United States. It is sometimes included in *Cenchrus*, based solely on the fusion of its bristles.

**13. Pennisetum setigerum** (Vahl) Wipff BIRDWOOD GRASS [p. 483, 551]

*Pennisetum setigerum* is grown as a forage grass in the southern United States, but is not known to be established in the *Manual* region. It is native to Africa, Arabia, and India. It is sometimes included in *Cenchrus*, based solely on the fusion of its bristles.

**14. Pennisetum flaccidum** Griseb. HIMALAYAN FOUNTAINGRASS [p. 483]

*Pennisetum flaccidum* is native to central Asia. Although grown primarily as an ornamental, it is reportedly used for forage in the

*Manual* region, but only one record, from Brazos County, Texas, is known.

**15. Pennisetum setaceum** (Forssk.) Chiov. TENDER FOUNTAINGRASS [p. 483, 551]

*Pennisetum setaceum* is a desert grass native to the eastern Mediterranean region. It is a popular ornamental throughout the southern United States, but it is also an invasive weed.

**16. Pennisetum orientale** Willd. *ex* Rich. WHITE FOUNTAINGRASS [p. 483, 551]

*Pennisetum orientale* is native from North Africa to India. It is grown as an ornamental in the *Manual* region, but has potential as a forage species.

**17. Pennisetum advena** Wipff & Veldkamp PURPLE FOUNTAINGRASS [p. 483]

The origin of *Pennisetum advena* is uncertain. It is frequently cultivated as an ornamental, usually being sold as *P. setaceum* 'Rubrum'.

**18. Pennisetum petiolare** (Hochst.) Chiov. [p. 483, 551]

*Pennisetum petiolare* is native to northern Africa, where it grows in disturbed habitats. The only collection in the *Manual* region is from Ames, Iowa, where it grew from fallen bird seed. It is not known to be established anywhere in the region.

## 24.17 CENCHRUS L.[1]

Pl ann or per. Clm 5–200 cm, erect or decumbent, usu geniculate; nd and intnd usu glab. Shth open, usu glab; lig memb, ciliate, cilia as long as or longer than the bas membrane; bld flat or folded, mrg cartilaginous, scabridulous. Infl tml, spikelike pan of highly rdcd br termed *fascicles* ("burs"); fascicles consisting of 1–2 series of many, stiff, partially fused, usu retrorsely scabridulous to strigose, sharp bristles surrounding, smt almost concealing, 1–4 spklt; outer (lo) bristles, if present, in 1 or more whorls, terete or flattened; inner (up) bristles usu strongly flattened, fused at least at the base and forming a disk, frequently to more than ½ their length and forming a cupule; dis at the base of the fascicles. Spklt sessile, with 2 flt; lo flt usu strl; up flt bisx. Lo glm ovate, scarious, glab, 1-veined, acute to acuminate; up glm and lo lm ovate, 3–9-veined; lo pal equaling the lm, tawny or purplish; up lm and pal subequal, indurate, ovate, obscurely veined, acuminate. Car obtrulloid.

*Cenchrus* has about 16 species. They are primarily tropical and most easily (and painfully) recognized by their spiny fascicles. Seven species are native to the *Manual* region; the eighth species, *C. biflorus*, is an introduction that does not appear to have become established. Species of *Cenchrus* are generally considered to be undesirable weeds. Most species of *Cenchrus* differ from those of *Pennisetum* in having retrorsely scabrous or strigose inner bristles that are fused to well above their bases.

1. All bristles terete, fused only at the base; fascicles not burlike . . . . . . . . . . . . . . . . . . . . . . . . . . . . . . . . . . . 7. *C. myosuroides*
1. Inner bristles flattened, variously fused, forming a shallow disk or distinct cupule; fascicles burlike.
    2. Fascicles with 1 whorl of fused, flattened inner bristles subtended by 5–25 free, terete, outer bristles.
        3. Inner bristles fused only at the base and forming a shallow disk, their abaxial surfaces with 1–3 grooves . . . . . . . . . . . . . . . . . . . . . . . . . . . . . . . . . . . . . . . . . . . . . . . . . . . . . . . . . . . . . . . . . . . . 8. *C. biflorus*
        3. Inner bristles fused for ⅓–½ their length or more, forming a globose cupule, their abaxial surfaces not grooved.
            4. Rachis internodes 0.8–1.7 mm long; the majority of the outer bristles equaling or slightly exceeding the inner, flattened bristles . . . . . . . . . . . . . . . . . . . . . . . . . . . . . . . . . . . . . . . . . . . 1. *C. brownii*
            4. Rachis internodes 2–4 mm long; the majority of the outer bristles about ½ as long as the inner, flattened bristles . . . . . . . . . . . . . . . . . . . . . . . . . . . . . . . . . . . . . . . . . . . . . . . . . . . . . . . . . . 2. *C. echinatus*
    2. Fascicles with more than 1 whorl of flattened inner bristles, these originating at irregular intervals throughout the body of the cupule, sometimes subtended by terete outer bristles.

[1]Michael T. Stieber and J.K. Wipff

5. Plants perennial, long-lived; fascicles not imbricate, usually glabrous; leaf blades 1–3.5 mm wide . . . . 3. *C. gracillimus*
5. Plants annual or perennial but short-lived; fascicles imbricate, usually pubescent; leaf blades (1)3–14.2 mm wide.
    6. Inner bristles 0.5–0.9(1.4) mm wide at the base; fascicles with 45–75 bristles . . . . . . . . . . . . . .5. *C. longispinus*
    6. Inner bristles 1–3 mm wide at the base; fascicles with 8–43 bristles.
        7. Fascicles densely pubescent, 9–16 mm long, with 1(2) spikelets; inner bristles 4–8 mm long . . . . 6. *C. tribuloides*
        7. Fascicles glabrous or sparsely to moderately pubescent, 5.5–10.2 mm long, with 2–4 spikelets; inner bristles 2–5.8 mm long . . . . . . . . . . . . . . . . . . . . . . . . . . . . . . . . . . . . .4. *C. spinifex*

## 1. Cenchrus brownii Roem. & Schult. SLIMBRISTLE SANDBUR, GREEN SANDBUR [p. 484, <u>551</u>]

*Cenchrus brownii* is native to sandy waste places and forest borders. It occurs infrequently on the coastal plain of the southeastern United States, but is common through the Caribbean, Central America, and the northern coast of South America. It has also been introduced to other parts of the world. The record from Texas may represent an introduction; only one specimen is known from the state.

## 2. Cenchrus echinatus L. SOUTHERN SANDBUR [p. 484, <u>551</u>]

*Cenchrus echinatus* grows in disturbed areas throughout the coastal plain and piedmont of the southern United States, Mexico, Central and South America, and, as an unwelcome introduction, elsewhere.

## 3. Cenchrus gracillimus Nash SLENDER SANDBUR [p. 484, <u>551</u>]

*Cenchrus gracillimus* grows in sandy soils of open pinelands, wet prairies, and river flats of the southeastern United States and the West Indies.

## 4. Cenchrus spinifex Cav. COASTAL SANDBUR, COMMON SANDBUR [p. 484, <u>551</u>]

*Cenchrus spinifex* is common in sandy woods, fields, and waste places throughout the southern United States and southwards into South America. It may be more widespread than shown in the northern portion of the contiguous United States because it has often been confused with *C. tribuloides*. It differs from *C. tribuloides* in its glabrous or less densely pubescent fascicles, narrower inner bristles, and larger number of bristles. It differs from *C. longispinus* in having shorter spikelets, fewer bristles overall, wider inner bristles, and outer bristles that are usually flattened rather than usually terete.

## 5. Cenchrus longispinus (Hack.) Fernald MAT SANDBUR, LONGSPINE SANDBUR, CENCHRUS À ÉPINES LONGUES [p. 484, <u>551</u>]

*Cenchrus longispinus* grows in sandy woods, fields, and waste ground in southern Canada and the contiguous United States. Its range extends south to Venezuela. It is often confused with *C. spinifex* and *C. tribuloides*; see discussion under those species.

## 6. Cenchrus tribuloides L. SANDDUNE SANDBUR, DUNE SANDBUR [p. 484, <u>551</u>]

*Cenchrus tribuloides* grows in moist, sandy dunes and is restricted to the eastern United States. It differs from *C. spinifex* in its larger spikelets and smaller number of spikelets per fascicle, and from *C. longispinus* in its densely pubescent fascicles, fewer bristles, and wider inner bristles.

## 7. Cenchrus myosuroides Kunth BIG SANDBUR [p. 484, <u>551</u>]

*Cenchrus myosuroides* is a native species that grows mostly along roadsides and in other waste places. Its native range extends through the Caribbean and Central America to northern South America.

## 8. Cenchrus biflorus Roxb. INDIAN SANDBUR [p. 484, <u>551</u>]

*Cenchrus biflorus* is widely distributed from Africa to India. It was collected once in Westchester County, New York, but has not become established in the *Manual* region.

# 24.18 ANTHEPHORA Schreb.[1]

Pl ann or per; tufted or csp. Clm 15–50 cm, not wd. Shth open; lig memb. Infl tml, spikelike pan, each nd supporting a highly rdcd br or *fascicle* of spklt; fascicles imbricate, with a short, thick, bas stipe subtending 4 thick, rigid, coriaceous, many-veined, flat, narrowly elliptic to ovate bracts; bracts fused at the base, enclosing 2–11 spklt, 1–2 of the spklt strl or rdcd; dis beneath the fascicles. Spklt with 2 flt. Lo glm absent; up glm acicular, 1-veined, awned; lo flt strl, smt rdcd; lo lm 7-veined; lo pal subequal to the lo lm; up flt bisx, strl, or rdcd; up lm faintly 3-veined. Car ellipsoidal.

*Anthephora* has 12 species, most of which are native to Africa and Arabia. One species is native to tropical America and has become established in the *Manual* region. The fascicle bracts are often interpreted as the lower glumes of the spikelets, but the developmental studies needed to evaluate this interpretation have not been conducted.

## 1. Anthephora hermaphrodita (L.) Kuntze OLDFIELD GRASS [p. 484, <u>551</u>]

*Anthephora hermaphrodita* is a weedy species, native to maritime beaches, lowland pastures, and disturbed areas from Mexico and the Caribbean Islands to Peru and Brazil. It is now established in Alachua County, Florida, having escaped from plantings at the Experiment Station of the University of Florida, Gainesville.

[1]Mary E. Barkworth

# 24.19 **IXOPHORUS** Schltdl.[1]

**Pl** ann or short-lived per; tufted. **Clm** 15–150 cm tall, 1–10 mm thick, dry to somewhat succulent, longitudinally grooved, glab. **Lvs** linear, vernation conduplicate; **shth** open, glab, compressed lat, often purple-streaked at the base; **lig** memb, long-ciliate; **bld** flat, midvein often white, adx surfaces puberulent immediately distal to the ligule, glab elsewhere, mrg strigillose. **Infl** tml, open, pyramidal pan; **rchs** scabridulous, bearing 4–50 alternate, spikelike pri br; **pri br** to 7 cm, flexuous, axes scabridulous, bearing spklt in 2 abx rows; **ped** shorter than 1 mm, cuplike, each with a single, smooth (occ scabridulous), terete bristle; **bristles** 4–12 mm, pale brown to black; **dis** below the glm. **Spklt** dorsally compressed, with 2 flt; **lo flt** stmt; **up flt** pist. **Lo glm** ¼–⅓ as long as the up glm, orbicular to triangular, 3-veined; **up glm** slightly shorter than the lo lm, often purple or green, 11-veined, acute; **lo lm** 5-veined, acute; **lo pal** hyaline, about as long as the up glm, accrescent, thinly memb at anthesis, becoming thicker and stiffer and about 3 times as wide as the lm in fruit, keels clasping the up flt at maturity; **anth** 3, about 2 mm, orange; **up lm** dorsally compressed, indurate, rugose, papillate, bases with a prominent germination flap, mrg enclosing the edges of the up pal; **up pal** flat, indurate, papillate; **stigmas** bright red, plumose. **Car** oblong obtuse, dorsally compressed; **emb** about ⅓ as long as car.

*Ixophorus* is native from central Mexico through Central America to northern South America. It is treated here as consisting of a single species, but it has been treated in the past as having as many as three species.

1. **Ixophorus unisetus** (J. Presl) Schltdl. Turkey Grass, Crane Grass [p. 485]

*Ixophorus unisetus* has been collected in Kleberg County, Texas, where it was being evaluated for its forage potential. It is not known to be established in the *Manual* region.

# 24.20 **SETARIOPSIS** Scribn.[2]

**Pl** ann. **Clm** 20–80 cm, to about 1 mm thick, solid, brchg above the base. **Shth** open; **lig** of hairs; **bld** flat. **Infl** tml pan, 8–23 cm long, 0.8–2 cm wide, with pilose rchs; **br** 0.5–1.5 cm, spklt congested, shortly pedlt, the ped subtended by a 3–10 mm, terete bristle; **dis** below the glm. **Spklt** dorsally compressed, with 2 flt, lo flt usu strl, up flt bisx. **Lo glm** about ¼ as long as the spklt, 5–7-veined, subclasping; **up glm** slightly shorter than the spklt, 11–19-veined, indurate at maturity, constricted at the base, auriculate above the point of constriction; **lo lm** longer than the glm, memb but somewhat indurate at the base; **lo pal** usu present, short; **up lm** indurate, finely transversely rugose, apiculate, mrg clasping the pal; **up pal** similar to the lm in length and texture; **lod** 2; **anth** 3, purple; **ov** glab; **sty br** 2, free to the base. **Car** ovate, plano-convex; **emb** about ½ as long as the car.

*Setariopsis* includes two species, both of which grow primarily south of the *Manual* region southward through Central America.

1. **Setariopsis auriculata** (E. Fourn.) Scribn. [p. 485, 551]

*Setariopsis auriculata* was found to be established in Arizona in 2001, growing in moist, shady habitats. Its range extends to Colombia and Venezuela.

# 24.21 **SETARIA** P. Beauv.[3]

**Pl** ann or per; csp, rarely rhz. **Clm** 10–600 cm, erect or decumbent. **Lig** memb and ciliate or of hairs; **bld** flat, folded, or involute, or plicate and petiolate (subg. *Ptychophyllum*). **Infl** tml, pan, usu dense and spikelike, occ loose and open; **dis** usu below the glm, spklt falling intact, bristles persistent. **Spklt** 1–5 mm, usu lanceoloid-ellipsoid, rarely globose, turgid, subsessile to short pedlt, in fascicles on short br or single on a short br, some or all subtended by 1–several, terete bristles (strl brlets). **Lo glm** memb, not saccate, less than ½ as long as the spklt, 1–7-veined; **up glm** memb to hrb at maturity, ½ as long as to nearly equaling the up lm, 3–9-veined; **lo flt** stmt or strl; **lo lm** memb, equaling or rarely exceeding the up lm, rarely absent, not constricted or indurate bas, 5–7-veined; **lo pal** usu hyaline to memb at maturity, rarely absent or rdcd, veins not keeled; **up flt** bisx; **up lm** and **pal** indurate, transversely rugose, rarely smooth; **anth** 3, not penicillate; **sty** 2, free or fused bas, white or red. **Car** small, ellipsoid to subglobose, compressed dorsiventrally.

*Setaria* has about 140 species that grow predominantly in tropical and warm-temperate regions. It is particularly well-represented in Africa, Asia, and South America. There are 27 species in the *Manual* region; fifteen are native, nine are established introductions, one is cultivated, and two are not established or have been collected only at scattered locations.

[1]Ken M. Hiser  [2]John R. Reeder  [3]James M. Rominger

1. Terminal spikelet of each panicle branch subtended by a single bristle, single bristles also present occasionally below the other spikelets.
    2. Blades not plicate, less than 10 mm wide; bristles present only below the terminal spikelets.
        3. Panicles nodding; spikelets 2-ranked on the branch axes (subg. *Paurochaetium*) .................. 4. *S. chapmanii*
        3. Panicles erect; spikelets randomly distributed on the branch axes (subg. *Reverchoniae*) ............ 5. *S. reverchonii*
    2. Blades plicate, more than 10 mm wide; a single bristle sometimes present below the non-terminal spikelets (subg. *Ptychophyllum*).
        4. Plants annual; blades 10–25 mm wide; rachises villous ................................... 1. *S. barbata*
        4. Plants perennial; blades 20–80 mm wide; rachises scabrous or puberulent.
            5. Panicles loosely open, branches lax, 6–10 cm long ...................................... 2. *S. palmifolia*
            5. Panicles lanceoloid, branches stiff, 2–5 cm long ...................................... 3. *S. megaphylla*
1. All spikelets subtended by 1–several bristles (subg. *Setaria*).
    6. Bristles 4–12 below each spikelet.
        7. Plants annual.
            8. Panicles erect; bristles 3–8 mm long; spikelets 2–3.4 mm long; blades 4–10 mm wide ............. 27. *S. pumila*
            8. Panicles arching and drooping from near the base; bristles about 10 mm long; spikelets 2.5–3 mm long; blades 10–20 mm wide ..................................... 24. *S. faberi* (in part)
        7. Plants perennial.
            9. Panicles 3–8(10) cm long, yellow to purple; knotty rhizomes present; native ................. 25. *S. parviflora*
            9. Panicles 5–25 cm long, usually orange to purple; stout rhizomes present; introduced .......... 26. *S. sphacelata*
    6. Bristles 1–3 (rarely 6) below each spikelet.
        10. Bristles retrorsely scabrous.
            11. Margins of sheaths glabrous; blades strigose on the abaxial surfaces; subtropical ............. 19. *S. adhaerans*
            11. Margins of sheaths ciliate distally; blades scabrous on the abaxial surfaces; temperate ....... 20. *S. verticillata*
        10. Bristles antrorsely scabrous.
            12. Plants perennial.
                13. Spikelets 2.8–3.2 mm long.
                    14. Blades scabrous; plants of Florida and Georgia ......................... 11. *S. macrosperma*
                    14. Blades pubescent; plants of Texas and possibly Arizona ................. 10. *S. villosissima*
                13. Spikelets 1.9–2.8(3) mm long.
                    15. Panicles 2–6 cm long; spikelets 1.9–2.1 mm long; culms branching at the upper nodes ......................................................... 6. *S. texana*
                    15. Panicles 5–30 cm long; spikelets 2–2.8(3) mm long; culms seldom branching at the upper nodes.
                        16. Lower paleas narrow, ½–¾ as long as the lemmas; spikelets elliptical.
                            17. Blades usually less than 5 mm wide; panicles 6–15 cm, columnar; bristles ascending .................................... 8. *S. leucopila*
                            17. Blades usually more than 5 mm wide; panicles 15–25 cm, tapering to the apex; bristles diverging ................................ 9. *S. scheelei*
                        16. Lower paleas broad, subequal to the lemmas in length; spikelets subspherical to ovate-lanceolate.
                            18. Panicles dense, cylindrical; spikelets subspherical ................... 7. *S. macrostachya*
                            18. Panicles interrupted, attenuate; spikelets ovate-lanceolate.
                                19. Blades 6–12 mm wide; lower glumes ½ as long as the spikelets ......... 12. *S. setosa*
                                19. Blades mostly less than 5 mm wide; lower glumes about ⅓ as long as the spikelets .......................................... 13. *S. rariflora*
            12. Plants annual.
                20. Upper glumes and lower lemmas with 7 veins, the outer pair of veins not coalescing with the inner 5; lower paleas absent ..................................... 15. *S. liebmannii*
                20. Upper glumes and lower lemmas with 5–7 veins, all of which coalesce near the apices; lower paleas present, sometimes reduced or absent.
                    21. Upper lemmas smooth and shiny, occasionally obscurely transversely rugose.
                        22. Spikelets about 2 mm long, lower paleas equal to the lower lemmas .............. 18. *S. magna*
                        22. Spikelets about 3 mm long, lower paleas absent or up to ½ as long as the lower lemmas .................................................. 23. *S. italica*
                    21. Upper lemmas distinctly transversely rugose, dull.
                        23. Upper lemmas coarsely rugose.
                            24. Panicles densely spicate; rachises sparsely villous; plants of the southeastern United States ................................ 17. *S. corrugata*
                            24. Panicles loosely spicate; rachises scabrous; plants of southern Arizona ...... 16. *S. arizonica*

23. Upper lemmas finely rugose.
    25. Panicles verticillate or loosely spicate; rachises visible, scabrous or hispid.
        26. Panicles verticillate; rachises scabrous; cauline nodes glabrous . . . . 21. *S. verticilliformis*
        26. Panicles loosely spicate, interrupted; rachises hispid; cauline nodes pubescent . . . . . . . . . . . . . . . . . . . . . . . . . . . . . . . . . . . . . . . . . . 14. *S. grisebachii*
    25. Panicles densely spicate; rachises not visible, villous.
        27. Blades softly pilose on the upper surface; spikelets 2.5–3 mm long; panicles nodding from the base . . . . . . . . . . . . . . . . . . . . . . . . . 24. *S. faberi* (in part)
        27. Blades scabrous; spikelets 1.8–2.2 mm long; panicles nodding only from near the apex . . . . . . . . . . . . . . . . . . . . . . . . . . . . . . . . . . . . . . 22. *S. viridis*

## Setaria subg. **Ptychophyllum** (A. Braun) Hitchc.

*Setaria* subg. *Ptychophyllum* is primarily an American taxon, but is also found in the Eastern Hemisphere. Three species, all introduced, have been found in the *Manual* region.

### 1. Setaria barbata (Lam.) Kunth MARY GRASS, CORN GRASS [p. 485, 552]

*Setaria barbata* is an African species that was apparently introduced to the Western Hemisphere from Asia. It is now common throughout the West Indies, but rare in the *Manual* region.

### 2. Setaria palmifolia (J. König) Stapf PALMGRASS [p. 485, 552]

*Setaria palmifolia* is primarily an Asiatic species. It is common in Jamaica, and has been reported from scattered locations around the southern coast of the United States. In the *Manual* region it is occasionally cultivated as an ornamental for its conspicuous, plicate leaves and large panicles.

### 3. Setaria megaphylla (Steud.) T. Durand & Schinz BIGLEAF BRISTLEGRASS [p. 485, 552]

*Setaria megaphylla* is a species of tropical Africa and tropical America that has become established in Florida.

## Setaria subg. **Paurochaetium** (Hitchc. & Chase) Rominger

*Setaria* subg. *Paurochaetium* includes seven taxa and extends from southern Florida through the West Indies into the Yucatan region of Mexico and Belize. One species, *S. chapmanii*, grows in the *Manual* region. *Setaria* subg. *Paurochaetium* usually differs from subg. *Reverchoniae* in its 2-ranked and smaller spikelets and in the presence of a palea in the lower floret. It is exceptional within subgenus *Paurochaetium* in lacking a lower palea.

### 4. Setaria chapmanii (Vasey) Pilg. CHAPMAN'S BRISTLEGRASS [p. 485, 552]

*Setaria chapmanii* is native to soils of coral or shell origin in the Florida Keys, the Bahamas, Cuba, and the Yucatan Peninsula, Mexico.

## Setaria subg. **Reverchoniae** W.E. Fox

*Setaria* subg. *Reverchoniae* is a unispecific subgenus. It usually differs from subg. *Paurochaetium* in the random disposition and larger size of the spikelets and, usually, in the absence of the lower palea.

### 5. Setaria reverchonii (Vasey) Pilg. [p. 485, 552]

*Setaria reverchonii* grows in sandy prairies and limestone hills from eastern New Mexico, southwestern Oklahoma, and Texas to northern Mexico.

1. Blades usually more than 15 cm long; spikelets 3.5–4.5 mm long . . . . . . . . . . . . . . . . . . . . . . subsp. *reverchonii*
1. Blades usually less than 15 cm long; spikelets 2.1–3.2 mm long.
    2. Blades 2–4 mm wide; spikelets about 2.5 mm long . . . . . . . . . . . . . . . . . . . . . . . . . . . subsp. *ramiseta*
    2. Blades 4–7 mm wide; spikelets about 3–3.2 mm long . . . . . . . . . . . . . . . . . . . . . . . . . . . subsp. *firmula*

### Setaria reverchonii subsp. firmula (Hitchc. & Chase) W.E. Fox KNOTGRASS [p. 485]

*Setaria reverchonii* subsp. *firmula* is endemic to the sandy prairies of southeastern Texas.

### Setaria reverchonii subsp. ramiseta (Scribn.) W.E. Fox RIO GRANDE BRISTLEGRASS [p. 485]

*Setaria reverchonii* subsp. *ramiseta* grows in the sandy plains and prairies of southeastern New Mexico, southern Texas, and northern Mexico.

### Setaria reverchonii (Vasey) Pilg. subsp. reverchonii REVERCHON'S BRISTLEGRASS [p. 485]

*Setaria reverchonii* subsp. *reverchonii* grows in sandy prairies and limestone hills from southwestern Texas to northern Mexico.

## Setaria P. Beauv. subg. **Setaria**

*Setaria* subg. *Setaria* is represented in subtropical and temperate regions throughout the world. It is the best represented subgenus in the *Manual* region.

### 6. Setaria texana Emery TEXAS BRISTLEGRASS [p. 486, 552]

*Setaria texana* grows in shaded habitats on sandy loam soils of the Rio Grande plain of south Texas and northeastern Mexico.

### 7. Setaria macrostachya Kunth PLAINS BRISTLEGRASS [p. 486, 552]

*Setaria macrostachya* is abundant in the desert grasslands of the southwestern United States, particularly in southern Arizona and Texas. It extends south through the highlands of central Mexico. It also grows in the West Indies, but is not common there. It is a valuable forage grass in the *Manual* region.

**8. Setaria leucopila** (Scribn. & Merr.) K. Schum. STREAMBED BRISTLEGRASS [p. 486, <u>552</u>]

*Setaria leucopila* grows in the southwestern United States and northern Mexico. It is the most common of the perennial "Plains bristlegrasses."

**9. Setaria scheelei** (Steud.) Hitchc. SOUTHWESTERN BRISTLEGRASS [p. 486, <u>552</u>]

*Setaria scheelei* grows in alluvial soils of canyons and river bottoms of New Mexico and Texas. Within the *Manual* region, it is particularly abundant in the limestone canyons of the Edwards Plateau of central Texas. Its range extends into central Mexico.

**10. Setaria villosissima** (Scribn. & Merr.) K. Schum. HAIRYLEAF BRISTLEGRASS [p. 486, <u>552</u>]

*Setaria villosissima* is a rare species that grows on granitic soils in southwestern Texas and northern Mexico. The villous sheaths and blades and large spikelets of *S. villosissima* aid in its identification.

**11. Setaria macrosperma** (Scribn. & Merr.) K. Schum. CORAL BRISTLEGRASS [p. 486, <u>552</u>]

*Setaria macrosperma* grows on shell or coral islands, and occasionally in old fields or hammocks. It is most frequent in Florida, but has been collected in both South Carolina and Georgia. It also grows in the Bahamas and Mexico.

**12. Setaria setosa** (Sw.) P. Beauv. WEST INDIES BRISTLEGRASS [p. 486, <u>552</u>]

*Setaria setosa* is native to the West Indies and Mexico. It is probably a recent introduction to Florida, but appears to be established there. The specimen from New Jersey was from a ballast dump; the species is not established in that state.

**13. Setaria rariflora** J.C. Mikan *ex* Trin. BRAZILIAN BRISTLEGRASS [p. 486, <u>552</u>]

*Setaria rariflora* has its center of distribution in South America. It is probably only recently adventive in North America, where it is known from Florida and the West Indies.

**14. Setaria grisebachii** E. Fourn. GRISEBACH'S BRISTLEGRASS [p. 486, <u>552</u>]

*Setaria grisebachii* is the most widespread and abundant native annual species of *Setaria* in the southwestern United States. It grows in open ground and extends along the central highlands of Mexico to Guatemala, usually at elevations of 750–2500 m. The specimens from Maryland were collected on chrome ore piles; the species is not established in the state.

**15. Setaria liebmannii** E. Fourn. LIEBMANN'S BRISTLEGRASS [p. 486, <u>552</u>]

Within the *Manual* region, *Setaria liebmannii* is known only from southern Arizona, but it is common along the Pacific slope from northern Mexico to Nicaragua, usually growing at elevations below 750 m. The five apically coalescing veins and the additional free pair at the periphery are unique among the *Setaria* species in the *Manual* region.

**16. Setaria arizonica** Rominger ARIZONA BRISTLEGRASS [p. 486, <u>552</u>]

*Setaria arizonica* is locally abundant in sandy washes on both sides of the Arizona-Sonora border, southwest of Tucson.

**17. Setaria corrugata** (Elliott) Schult. COASTAL BRISTLEGRASS [p. 486, <u>552</u>]

*Setaria corrugata* grows in pinelands and cultivated fields along the southeastern coast of the United States. It is also found in Cuba and the Dominican Republic. It differs from *S. viridis* in its coarsely rugose ("corrugated") lower lemmas.

**18. Setaria magna** Griseb. GIANT BRISTLEGRASS [p. 486, <u>552</u>]

*Setaria magna* grows in saline marshes along the eastern coast of the United States. There are also disjunct populations in brackish swamps in Arkansas, and in Texas and southeastern New Mexico as well as in Jamaica, Puerto Rico, Bermuda, Mexico, and Costa Rica. It may have been recently introduced to some of these regions, including inland areas of the *Manual* region.

**19. Setaria adhaerans** (Forssk.) Chiov. BUR BRISTLEGRASS, TROPICAL BARBED BRISTLEGRASS [p. 486, <u>552</u>]

*Setaria adhaerans* grows in subtropical regions throughout the world. In North America, it is known from the southern United States, northeastern Mexico, Guatemala, Cuba, and the Bahamas. The California record may represent a recent introduction. *Setaria adhaerans* differs from the temperate *S. verticillata* in having shorter panicles, shorter spikelets, glabrous sheath margins, and papillose-based strigose hairs on the blades.

**20. Setaria verticillata** (L.) P. Beauv. HOOKED BRISTLEGRASS, SÉTAIRE VERTICILLÉE [p. 486, <u>552</u>]

*Setaria verticillata* is a European adventive that is now common throughout the cooler regions of the contiguous United States and in southern Canada. It is an aggressive weed in the vineyards of central California. *Setaria verticillata* differs from *S. adhaerans* in having longer panicles and spikelets, sheath margins that are ciliate distally, and blades that are scabrous, not hairy. *Setaria verticillata* is a more northern species than *S. adhaerans*, but their ranges overlap in the *Manual* region.

**21. Setaria verticilliformis** Dumort. BARBED BRISTLEGRASS [p. 486, <u>552</u>]

*Setaria verticilliformis* is a European adventive that has been found at scattered, mostly urban, locations in the United States.

**22. Setaria viridis** (L.) P. Beauv. GREEN BRISTLEGRASS, SÉTAIRE VERTE [p. 487, <u>552</u>]

*Setaria viridis* resembles *S. italica* but differs in its shorter spikelets and rugose upper florets, and mode of disarticulation. It is also a more aggressive weed. It is native to Eurasia but is now widespread in warm temperate regions of the world.

1. Culms 100–250 cm tall; blades 10–25 mm wide; panicles 10–20 cm long . . . . . . . . . . . . . . . . . . . . . . . . var. *major*
1. Culms 20–100 cm tall; blades 4–12 mm wide; panicles 3–8 cm long . . . . . . . . . . . . . . . . . . . . . . . . . var. *viridis*

**Setaria viridis** var. **major** (Gaudin) Peterm. GIANT GREEN FOXTAIL [p. 487]

*Setaria viridis* var. *major* is a major adventive weed in corn and bean fields of the midwestern United States, dwarfing var. *viridis* in stature.

**Setaria viridis** (L.) P. Beauv. var. **viridis** GREEN FOXTAIL [p. 487]

*Setaria viridis* var. *viridis* is an aggressive adventive weed throughout temperate North America. It is the most common annual representative of *Setaria* in the *Manual* region.

**23. Setaria italica** (L.) P. Beauv. FOXTAIL MILLET,
SÉTAIRE ITALIENNE, SÉTAIRE D'ITALIE, MILLET DES OISEAUX
[p. 487, <u>552</u>]

*Setaria italica* is grown in China mostly for hay or as a pasture grass.
It is sometimes cultivated in North America, but it is better known as
a weed in moist ditches, mostly in the northeastern United States. It
differs from *S. viridis* in having longer (3 mm) spikelets and smooth,
shiny upper florets which readily disarticulate above the lower florets.
It exhibits considerable variation in seed and bristle color, bristle
length, and panicle shape.

**24. Setaria faberi** R.A.W. Herrm. CHINESE FOXTAIL,
SÉTAIRE GÉANTE [p. 487, <u>552</u>]

*Setaria faberi* spread rapidly throughout the North American corn
belt after being accidentally introduced from China in the 1920s. It
has become a major nuisance in corn and bean fields in the
midwestern United States.

**25. Setaria parviflora** (Poir.) Kerguélen KNOTROOT
BRISTLEGRASS [p. 487, <u>552</u>]

*Setaria parviflora* is a common, native species of moist ground. It is most
frequent along the Atlantic and Gulf coasts, but it also grows from the
Central Valley of California east through the central United States and
southward through Mexico to Central America, as well as in the West
Indies. The plant from Oregon was found on a ballast dump; the species
is not established in that state. *Setaria parviflora* is the most
morphologically diverse and widely distributed of the indigenous
perennial species of *Setaria*.

**26. Setaria sphacelata** (Schumach.) Stapf & C.E.
Hubb. AFRICAN BRISTLEGRASS [p. 487, <u>553</u>]

*Setaria sphacelata* is native to tropical Africa, but it has been found at
a few scattered locations in the *Manual* region, often near a port. Two
of the five varieties most likely to be introduced into the United States
are **Setaria sphacelata** (Schumach.) Stapf & C.E. Hubb. var.
**sphacelata** and *S. sphacelata* var. **aurea** (Hochst. *ex* A. Braun)
Clayton, with var. *aurea* differing from var. *sphacelata* in having
fibrous basal leaf sheaths and upper glumes that are often 3-veined.

**27. Setaria pumila** (Poir.) Roem. & Schult. YELLOW
FOXTAIL, PIGEON GRASS [p. 487, <u>553</u>]

1. Spikelets 3–3.4 mm long; bristles yellow . . . . . . . . . subsp. *pumila*
1. Spikelets 2–2.5 mm long; bristles reddish . . . . . subsp. *pallidefusca*

**Setaria pumila** subsp. **pallidefusca** (Schumach.) B.K.
Simon [p. 487]

*Setaria pumila* subsp. *pallidefusca* is native to tropical Africa. It is
now established as a weed in southeastern Louisiana, but it has also
been collected in the past on ballast dumps in Portland, Oregon.

**Setaria pumila** (Poir.) Roem. & Schult. subsp.
**pumila** SÉTAIRE GLAUQUE [p. 487]

*Setaria pumila* subsp. *pumila* is a European adventive that has become
a common weed in lawns and cultivated fields throughout temperate
North America.

## 24.22 PASPALIDIUM Stapf[1]

**Pl** ann or per. **Clm** to 100 cm, not brchg above the base.
**Aur** absent; **lig** memb and ciliate or of hairs. **Infl** pan of
racemosely arranged, spikelike br, br smt highly rdcd,
each pan appearing spikelike; **br** 1-sided, terminating in
a more or less inconspicuous bristle, bristles 2.5–4 mm;
**dis** beneath the spklt. **Spklt** subsessile, in 2 rows on 1
side of the br, lacking subtending bristles, dorsally compressed, with 2 flt, up glm and up flt appressed to the br
axes. **Glm** memb; **lo glm** much shorter than the spklt;
**up glm** subequal to the up flt; **lo flt** strl or stmt; **lo lm**
similar to the up glm in size and texture; **up flt** bisx; **up**
lm indurate, rugose, unawned, yellow or brown; **up pal**
similar to their lm; **anth** 3.

*Paspalidium* is a genus of approximately 40 species, one of
which is native to the *Manual* region. The genus grows in
tropical regions throughout the world. Most of its species
have an inflorescence of well-spaced, unilateral, spicate
branches and the resemblance to *Paspalum* is evident, but
species with closely crowded, highly reduced branches are
easily mistaken for species of *Setaria*, the terminal bristle
resembling a single, subtending bristle.

**1. Paspalidium geminatum** (Forssk.) Stapf EGYPTIAN
PASPALIDIUM, WATER PASPALIDIUM [p. 487, <u>553</u>]

*Paspalidium geminatum* grows in moist to wet, fresh to brackish
areas. It is native to the southeastern United States, the West Indies,
and tropical regions of the Americas.

## 24.23 HOPIA Zuloaga & Morrone[2]

**Pl** per; shortly rhz and long stln. **Clm** not cormous at the
base, not brchd above the base; **nd** swollen, villous,
particularly on the stln. **Lig** membranous, papery; **bld**
linear-lanceolate, cross-sections with Kranz anatomy
with a single mestome shth surrounding the vascular
bundles and in contact with the metaxylem vessels. **Pan**
contracted; **br** spikelike, appressed, 1-sided, spklt borne
in pairs. **Spklt** ellipsoid to obovoid. **Lo glm** ¾–⅘ as long
as the spklt, 5–7-veined; **up glm** subequal to the lo lm,
7–11-veined, blunt; **lo flt** stmt; **lo pal** well developed; **up**
flt obovoid, indurate, smooth, with microhairs and
simple papillae near the base and tip.

*Hopia* is a unispecific genus that extends from the
southwestern United States into northern Mexico. Its only
species used to be included in *Panicum*. Zuloaga et al. (2007)
recognized it as a distinct genus based on its unique
combination of characteristics: microhairs and papillae on the
upper floret, possession of the XYMS- subtype of C4
photosynthesis, chromosome base number of 10, and
molecular sequence data.

[1]Charles M. Allen  [2]Mary E. Barkworth

## 1. Hopia obtusa (Kunth) Zuloaga & Morrone VINE MESQUITE [p. 487, 553]

*Hopia obtusa* grows in seasonally wet sand or gravel, especially on stream banks, ditches, roadsides, wet pastures, and rangeland. Its range extends from the southwestern United States to central Mexico. Flowering is from May through October.

## 24.24 STENOTAPHRUM Trin.[1]

Pl ann or per; smt rhz or stln. Clm 10–60 cm, usu compressed; intnd solid. Lvs cauline; shth shorter than the intnd, compressed; lig memb and ciliate or of hairs; bld flat or folded. Infl spikelike pan; br very short, with fewer than 10 spklt, appressed to and partially embedded in the flattened, corky rchs; dis below the glm, often with a segment of the br. Spklt lanceolate to ovate, unawned, lo glm oriented away from the br axes. Glm memb; lo glm scalelike, usu without veins; up glm 5–7-veined; lo flt stmt or strl, lm 3–9-veined; up flt bisx; up lm longer than the glm, papery to subcoriaceous, 3–5-veined; up pal generally indurate, 2-veined; anth 3. Car lanceolate to ovate, often failing to develop.

*Stenotaphrum* has seven species that usually grow on the seashore or near the coast, primarily along the Indian Ocean rim. Three species are endemic to Madagascar, and one species is thought to be native to the *Manual* region.

## 1. Stenotaphrum secundatum (Walter) Kuntze ST. AUGUSTINE GRASS [p. 488, 553]

*Stenotaphrum secundatum* grows on sandy beaches, at the edges of swamps and lagoons, and along inland streams and lakes. It may be native to the southeastern United States, being known from the Carolinas prior to 1800, but it has become naturalized in most tropical and subtropical regions of the world. It is planted for turf in the southern United States and is now established from California to North Carolina and Florida. Numerous cultivars have been developed.

## 24.25 HYMENACHNE P. Beauv.[2]

Pl per. Clm 50–350 cm, decumbent, often rooting from the lo nd; nd glab; intnd glab, filled with spongy aerenchyma. Lvs evenly distributed; lig memb. Infl narrow, condensed, cylindrical or spikelike pan; br appressed but not fused to the rchs, densely and evenly short ciliate on the ridges. Spklt dorsally compressed, narrowly lanceolate, with 2 flt. Glm unequal, memb, with a definite rchl intnd between the 2 glm; lo glm no more than ½ as long as the spklt, 1–3-veined; up glm subequal to or longer than the lo lm, memb, not saccate, 3–7-veined; lo flt strl; lo lm similar to the up glm; up flt bisx, much shorter than the up glm and lo lm; up lm memb to leathery, white at maturity, mrg clasping the edges of the pal bas but not distally, apc acute; pal similar in texture to the lm; lod 2; anth 3; sty 2, fused. Car falling free from the lm and pal.

*Hymenachne* is a pantropical genus of approximately five species. It is unusual in the *Paniceae* in that all its members grow in aquatic or swampy habitats. It differs from *Sacciolepis* in its aerenchyma-filled internodes, a character that is probably an adaptation to its aquatic habitat. One species is native to the *Manual* region.

## 1. Hymenachne amplexicaulis (Rudge) Nees WEST INDIAN MARSH GRASS [p. 488, 553]

In the *Manual* region, *Hymenachne amplexicaulis* is known only from low, wet pastures in southern Florida and it is rare even in that state. It is more abundant in the remainder of its range which extends through Mexico to Argentina.

## 24.26 STEINCHISMA Raf.[3]

Pl per; csp, rhz, rhz short, slender. Clm slender, often compressed. Shth usu keeled; lig minute, memb, often erose or ciliate; bld exhibiting Kranz anatomy, with few organelles in the external shth and 5–7 isodiametric mesophyll cells between the vascular bundles. Infl tml, open to contracted pan; pri br few, slender; ped short, to 1 mm. Spklt ellipsoid or lanceolate, initially somewhat compressed, ultimately expanding greatly. Glm glab; lo glm ⅓–½ as long as the spklt, usu 3(5)-veined, acute; up glm and lo lm subequal, 3–5(7)-veined; lo flt strl or stmt, often standing apart from the up flt at maturity; lo pal longer than the lo lm, greatly inflated at maturity, indurate; up flt ovoid or ellipsoid; up lm usu dull-colored, minutely papillose, papillae in longitudinal rows, apc acute.

*Steinchisma* is a genus of 5–6 species that grow in moist or wet, usually open, sandy areas in warm-temperate and tropical regions of the Western Hemisphere. A single species is native the *Manual* region. It is sometimes included in *Panicum*.

[1]Kelly W. Allred   [2]Mary E. Barkworth   [3]Robert W. Freckman and Michel G. Lelong

### 1. Steinchisma hians (Elliott) Nash GAPING PANICGRASS [p. 488, <u>553</u>]

*Steinchisma hians* grows in moist or wet, usually open areas, and in moist pinelands, low woods, and ditches. Its range extends from the southeastern United States through Mexico and Central America to Colombia, Brazil, and Argentina.

## 24.27 AXONOPUS P. Beauv.[1]

**Pl** per, rarely ann; **csp**, loosely tufted, or mat-forming, smt rhz or stln. **Clm** 7–300 cm, not wd, often decumbent at the base, erect to ascending. **Shth** open; **lig** memb, truncate, ciliate; **bld** flat or convolute, usu obtuse. **Infl** tml, smt also ax, pan of 2–many, digitately, subdigitately, or racemosely arranged spikelike br; **br** triquetrous, spklt subsessile or sessile, solitary, in 2 rows, lo lm appressed to the br axes; **dis** below the glm. **Spklt** dorsally compressed, with 2 flt; **lo flt** strl or stmt; **up flt** sessile, bisx. **Lo glm** absent; **up glm** and **lo lm** equal, memb; **lo pal** absent; **up lm** indurate, usu glab, smt with an apical tuft of hairs, mrg slightly involute, clasping the pal, apc acute to obtuse; **up pal** similar to the up lm in texture. **Car** ellipsoid.

*Axonopus* is a genus of approximately 100 tropical and subtropical species, most of which are native to the Western Hemisphere. Three species are native to the *Manual* region; one additional species has been grown experimentally in Florida.

1. Panicles with 30–100+ branches; lower branches 10–24 cm long; culms (50)100–300 cm tall . . . . . . . . . . . . . . 4. *A. scoparius*
1. Panicles with 2–7 branches; lower branches 1–15 cm long; culms 7–100 cm tall.
   2. Spikelets 3.5–5.5 mm long; upper glumes glabrous; lower lemmas glabrous or sparsely pilose over the veins . . . . . . . . . . . . . . . . . . . . . . . . . . . . . . . . . . . . . . . . . . . . . . . . . . . . . . . . . . . 3. *A. furcatus*
   2. Spikelets 1.6–3.5 mm long; upper glumes and lower lemmas sparsely pilose on the margins or marginal veins.
      3. Upper glumes and lower lemmas extending beyond the upper florets, forming acute to acuminate apices; blades 3–20 mm wide . . . . . . . . . . . . . . . . . . . . . . . . . . . . . . . . . . . . . . . . . . 2. *A. compressus*
      3. Upper glumes and lower lemmas not or scarcely extending beyond the upper florets, forming obtuse to subacute apices; blades 1.5–6 mm wide . . . . . . . . . . . . . . . . . . . . . . . . . . . . . 1. *A. fissifolius*

### 1. Axonopus fissifolius (Raddi) Kuhlm. COMMON CARPETGRASS [p. 488, <u>553</u>]

*Axonopus fissifolius* is sometimes used as a lawn or pasture grass, but it is also an invasive weedy species, often growing in moist, disturbed sites. It is native in the southeastern United States and from central Mexico south to Bolivia and Argentina. It has also been introduced into tropical and subtropical regions of the Eastern Hemisphere.

### 2. Axonopus compressus (Sw.) P. Beauv. BROADLEAF CARPETGRASS [p. 488, <u>553</u>]

*Axonopus compressus* is native from the southeastern United States to Bolivia, Brazil, and Uruguay, and has become established in the Eastern Hemisphere. It is used as a lawn and forage grass but is also weedy, readily growing in moist, disturbed habitats.

### 3. Axonopus furcatus (Flüggé) Hitchc. BIG CARPETGRASS [p. 488, <u>553</u>]

*Axonopus furcatus* is endemic to the southeastern United States. It grows in moist pine barrens, marshes, river banks, wet ditches, pond margins, and other such damp areas.

### 4. Axonopus scoparius (Humb. & Bonpl. *ex* Flüggé) Kuhlm. CARPETGRASS [p. 488]

*Axonopus scoparius* is native from southern Mexico to Peru, Bolivia, and Brazil. It has been grown experimentally in Florida, but it is not winter hardy even there. Not surprisingly, *A. scoparius* is not established in the *Manual* region.

## 24.28 PASPALUM L.[2]

**Pl** ann or per; csp, rhz, or stln. **Clm** 3–400 cm, erect, spreading or prostrate, smt trailing for 200+ cm. **Shth** open; **aur** smt present; **lig** memb. **Infl** tml, smt also ax, pan of 1–many spikelike br, these digitate or rcm on the rchs, spreading to erect, 1 or more br completely or partially hidden in the shth in some species; **br axes** flattened, usu narrowly to broadly winged, usu terminating in a spklt, smt extending beyond the distal spklt but never forming a distinct bristle; **dis** below the glm. **Spklt** subsessile to shortly pedlt, plano-convex, rounded to acuminate, dorsally compressed, not subtended by bristles or a ringlike cal, solitary or paired (1 spklt of the pair rdcd in some species), in 2 rows along 1 side of the br, with 2 flt, first rchl segment not swollen, up glm and up lm adjacent to the br axes; **lo flt** strl; **up flt** sessile or stipitate, bisx, acute or rounded. **Lo glm** absent or present only on some spklt of each br, without veins or 1-veined, unawned; **up glm** and **lo lm** subequal, memb, apc rounded, unawned; **lo pal** absent or rdmt; **up lm** convex, indurate, smooth to slightly rugose, stramineous to dark brown, mrg scarious, involute, clasping the pal; **up pal** indurate, smooth to slightly rugose, stramineous to dark brown. **Car** orbicular to elliptical, plano-convex or flattened, white, yellow, or brown.

*Paspalum* includes 300–400 species, most of which are native to the Western Hemisphere. Forty-three species are found in the *Manual* region; twenty-four are native. Nineteen of the

[1]Mary E. Barkworth  [2]Charles M. Allen and David W. Hall

species growing in the *Manual* region are introduced, and some of them are weedy. Because weeds are under-represented in most herbaria, the distribution maps of such species probably understate their prevalence.

1. Spikelets solitary, not associated with a naked pedicel or rudimentary spikelets.
    2. Panicles with 1–70 branches, if more than 1, the branches racemosely arranged.
        3. Branches 7–70, disarticulating at maturity, the axes extending beyond the distal spikelets . . . . . . . . . . . . . . 1. *P. repens*
        3. Branches 1–6, persistent, terminating in a spikelet.
            4. Upper florets olive to dark brown . . . . . . . . . . . . . . . . . . . . . . . . . . . . . . . . . . . . . . 2. *P. scrobiculatum*
            4. Upper florets pale to stramineous.
                5. Axes of panicle branches 0.6–1.3 mm wide . . . . . . . . . . . . . . . . . . . . . . . . . . . . . . . . 3. *P. laeve*
                5. Axes of panicle branches 1.8–3.3 mm wide.
                    6. Spikelets 1.7–2.1 mm long; upper lemmas glabrous throughout . . . . . . . . . . . . . . . . . . . . 4. *P. dissectum*
                    6. Spikelets 3.2–4 mm long; upper lemmas with a few short hairs at the apices . . . . . . . . . . . 5. *P. acuminatum*
    2. Panicles usually composed of a terminal pair of branches, sometimes with 1(–5) additional branches below the terminal pair.
        7. Upper glumes pilose on the margins or shortly pubescent on the back.
            8. Spikelets 1.3–1.9 mm long; upper glumes pilose along the margins . . . . . . . . . . . . . . . . . . . . . 6. *P. conjugatum*
            8. Spikelets 2.4–3.2 mm long; upper glumes sparsely short-pubescent on the back . . . . . . . . . . . . . . 7. *P. distichum*
        7. Upper glumes glabrous.
            9. Spikelets elliptic, their apices acute to acuminate.
                10. Plants rhizomatous, not appearing cespitose; usually in brackish to salt marsh habitats . . . . . . 8. *P. vaginatum*
                10. Plants shortly rhizomatous but appearing cespitose; usually in disturbed inland habitats . . . . . . . . 9. *P. almum*
            9. Spikelets ovate to broadly elliptic, their apices obtuse to broadly acute.
                11. Spikelets 2.5–4 mm long; leaf blades flat or conduplicate . . . . . . . . . . . . . . . . . . . . . . . . . . 10. *P. notatum*
                11. Spikelets 1.9–2.3 mm long; leaf blades flat . . . . . . . . . . . . . . . . . . . . . . . . . . . . . . . . . . 11. *P. minus*
1. Spikelets paired, if only 1 spikelet functional, a naked pedicel or rudimentary, non-functional spikelet present.
    12. Spikelets 1–1.3 mm long.
        13. Panicle branches 2–6; spikelets elliptic to elliptic-obovate, appressed to the branch axes . . . . . . . . 12. *P. blodgettii*
        13. Panicle branches 18–50; spikelets ovate, diverging from the branch axes . . . . . . . . . . . . . . . . . . 13. *P. paniculatum*
    12. Spikelets 1.3–4.1 mm long.
        14. Margins of upper glumes and lower lemmas ciliate-lacerate and winged or pilose.
            15. Upper glumes and lower lemmas ciliate-lacerate, winged . . . . . . . . . . . . . . . . . . . . . . . . . . 14. *P. fimbriatum*
            15. Upper glumes and lower lemmas pilose.
                16. Panicle branches 2–7; spikelets 2.3–4 mm long . . . . . . . . . . . . . . . . . . . . . . . . . . . . . 15. *P. dilatatum*
                16. Panicle branches (4)10–30; spikelets 1.8–2.8 mm long . . . . . . . . . . . . . . . . . . . . . . . . . 16. *P. urvillei*
        14. Margins of upper glumes and lower lemmas neither ciliate-lacerate nor winged, glabrous or pubescent, if pubescent then the hairs not pilose, often glandular, papillose-based, or wrinkled.
            17. Upper florets olive to dark brown.
                18. Plants aquatic, the culms decumbent, rooting at the nodes; lower glumes often present . . . . . . . . . . . . . . . . . . . . . . . . . . . . . . . . . . . . . . . . . . . . . . . . . 17. *P. modestum* (in part)
                18. Plants not aquatic or, if aquatic, the culms erect; lower glumes absent.
                    19. Panicle branches 10–28 or more.
                        20. Plants annual; axes of panicle branches broadly winged, wings about as wide as the central portion . . . . . . . . . . . . . . . . . . . . . . . . . . . . . . . . . 18. *P. boscianum* (in part)
                        20. Plants perennial; axes of panicle branches narrowly winged, wings narrower than the central portion.
                            21. Axes of panicle branches 1–1.7 mm wide; spikelets 1.8–2.4 mm wide . . . . . . . 19. *P. virgatum*
                            21. Axes of panicle branches 0.5–1.2 mm wide; spikelets 1.1–1.8 mm wide . . . . . . . . . . . . . . . . . . . . . . . . . . . . . . . . . . . . . . . . . . . 20. *P. conspersum* (in part)
                    19. Panicle branches 1–10(28).
                        22. Plants annual.
                            23. Spikelets 1.3–1.8 mm wide, broadly elliptical to orbicular, glabrous; panicles with 1–10(28) branches, the axes 0.7–2.3 mm wide . . . . . 18. *P. boscianum* (in part)
                            23. Spikelets 1.7–2.4 mm wide, broadly obovate, shortly pubescent; panicles with 1–5 branches, the axes 0.8–1.3 mm wide . . . . . . . . . . . . . . . . 21. *P. convexum*
                        22. Plants perennial.
                            24. Plants cespitose, rhizomes sometimes present but not well-developed; culms 100–200 cm tall, stout; panicle branches ascending, divaricate, or reflexed.

25. Leaf blades 7–18 mm wide . . . . . . . . . . . . . . . . . . . . . . . . . 20. *P. conspersum* (in part)
25. Leaf blades 2.5–4 mm wide . . . . . . . . . . . . . . . . . . . . . . 22. *P. plicatulum* (in part)
24. Plants not cespitose, rhizomatous; culms 10–150 cm tall, varying in thickness; panicle branches ascending.
   26. Rhizomes short, indistinct . . . . . . . . . . . . . . . . . . . . . . . . . 22. *P. plicatulum* (in part)
   26. Rhizomes long, evident.
      27. Plants aquatic; upper florets chestnut brown . . . . . . . . . . . . . . . . . 23. *P. wrightii*
      27. Plants not aquatic; upper florets dark brown . . . . . . . . . . . . . . . . . 24. *P. nicorae*
17. Upper florets white, stramineous, or golden brown.
  28. Lower lemmas with well-developed ribs over the veins; upper glumes absent . . . . . . 25. *P. malacophyllum*
  28. Lower lemmas not ribbed over the veins; upper glumes present.
    29. Panicles with 15–100 branches.
      30. Plants annual; upper glumes and lower lemmas rugose . . . . . . . . . . . . . . . . . 26. *P. racemosum*
      30. Plants perennial; upper glumes and lower lemmas smooth.
        31. Plants rhizomatous, not cespitose; branch axes 0.9–1.2 mm wide; panicle branches often arcuate . . . . . . . . . . . . . . . . . . . . . . . . . . . . . . . . 27. *P. intermedium*
        31. Plants cespitose, not rhizomatous; branch axes 0.3–0.6 mm wide; panicle branches straight.
          32. Panicle branches spreading to reflexed (rarely ascending); leaf blades 10–23 mm wide; axes of panicle branches 0.3–0.4 mm wide . . . . . . . . . . . . . . . . . . . . . . . . . . . . . . . . . . . . . . 28. *P. coryphaeum*
          32. Panicle branches erect to ascending; leaf blades 4.9–6.1 mm wide; axes of panicle branches 0.5–0.6 mm wide . . . . . . . . . . . . . . . . . 29. *P. quadrifarium*
    29. Panicles with 1–15 branches.
      33. Spikelet pairs not imbricate; lower glumes usually present . . . . . . . . . . . . . . . . . 30. *P. bifidum*
      33. Spikelet pairs imbricate; lower glumes absent or present.
        34. Spikelets 1.3–2.5 mm long.
          35. Upper glumes, usually also the lower lemmas, shortly pubescent.
            36. Lower glumes present . . . . . . . . . . . . . . . . . . . . . . . . . . . . . 31. *P. langei* (in part)
            36. Lower glumes absent.
              37. Panicles both terminal and axillary, the axillary panicles partially or completely enclosed by the subtending leaf sheath . . . . . . . . . . . . . . . . . . . . . . . . . . . . . . . 32. *P. setaceum* (in part)
              37. Panicles all terminal.
                38. Leaf blades involute; culms 80–110 cm tall . . . . . . . . . . . . . . 33. *P. laxum*
                38. Leaf blades flat; culms 20–75 cm tall.
                  39. Spikelets 1.3–2 mm long, 0.7–1 mm wide, elliptic; upper glumes and lower lemmas 5-veined; culm bases swollen . . . . . . . . . . 34. *P. caespitosum* (in part)
                  39. Spikelets 2–2.5 mm long, 1.4–1.6 mm wide, ovate; upper glumes and lower lemmas 3-veined; culm bases not swollen . . . . . . . . . . . . . . . . . . . 35. *P. virletii* (in part)
          35. Upper glumes and lower lemmas glabrous.
            40. Panicles both terminal and axillary, the axillary panicles partially or completely enclosed by the subtending leaf sheath
. . . . . . . . . . . . . . . . . . . . . . . . . . . . . . . . . . . . . . . . . . . . . . . . . . . . 32. *P. setaceum* (in part)
            40. Panicles all terminal.
              41. Upper panicle branches erect . . . . . . . . . . . . . . 36. *P. monostachyum* (in part)
              41. Upper panicle branches spreading to ascending.
                42. Leaf blades mostly involute; plants of sandy or rocky areas, usually on the coast . . . . . . . . . . . . . . . 37. *P. pleostachyum*
                42. Leaf blades mostly flat; plants of inland areas or, if coastal, then in marshy areas.
                  43. Upper glumes and lower lemmas 3-veined.
                    44. Leaf blades usually conduplicate, 2.2–8.3 mm wide . . . . . . . . . . . . . . . . . . . . . . . 39. *P. praecox* (in part)
                    44. Leaf blades usually flat, 5–10 mm wide . . . 35. *P. virletii* (in part)
                  43. Upper glumes and lower lemmas 5-veined.
                    45. Axes of panicle branches 0.2–0.5 mm wide; ligules 0.2–0.4 mm long . . . . . 34. *P. caespitosum* (in part)

45. Axes of panicle branches 1.5–2 mm wide;
     ligules 2.2–4.7 mm long . . . . . . . . . . . . . 38. *P. lividum* (in part)
34. Spikelets 2.5–4.1 mm long.
  46. Upper glumes, and usually lower lemmas, pubescent.
    47. Lower glumes present . . . . . . . . . . . . . . . . . . . . . . . . . . . . 31. *P. langei* (in part)
    47. Lower glumes absent.
      48. Leaf blades 2–5 mm wide; upper glumes and lower
         lemmas abundantly pubescent, most hairs longer than
         0.1 mm; spikelets elliptic . . . . . . . . . . . . . . . . . . . . . . . 40. *P. hartwegianum*
      48. Leaf blades 4–18 mm wide; upper glumes and lower
         lemmas glabrous or sparsely pubescent, the hairs shorter
         than 0.1 mm; spikelets obovate to elliptic . . . . . . . . 41. *P. pubiflorum* (in part)
  46. Upper glumes, and usually lower lemmas, glabrous.
    49. Upper florets golden brown.
      50. Plants not rhizomatous; culms decumbent and rooting
         at the lower nodes; spikelets 1.3–1.6 mm wide; lower
         lemmas 5–7-veined; lower glumes often present . . . . 17. *P. modestum* (in part)
      50. Plants rhizomatous; culms erect, not rooting at the
         lower nodes; spikelets 1.9–3.1 mm wide; lower lemmas
         3-veined.
        51. Panicle branches 1–6; upper glumes 5-veined; leaf
             blades 3–18 mm wide . . . . . . . . . . . . . . . . . . . 42. *P. floridanum* (in part)
        51. Panicle branches 1–3; upper glumes 3-veined; leaf
             blades 3–4 mm wide . . . . . . . . . . . . . . . . . . . 43. *P. unispicatum* (in part)
    49. Upper florets stramineous to pale, but not golden brown.
      52. Terminal panicle branches erect.
        53. Blades involute; upper glumes 1-veined . . . 36. *P. monostachyum* (in part)
        53. Blades flat; upper glumes 3-veined . . . . . . . . 43. *P. unispicatum* (in part)
      52. Terminal panicle branches spreading to ascending.
        54. Spikelets 2.2–2.6 mm long.
          55. Spikelets 1.2–1.5 mm wide, elliptic to obovate
                 . . . . . . . . . . . . . . . . . . . . . . . . . . . . . . . . . 38. *P. lividum* (in part)
          55. Spikelets 2–2.8 mm wide, orbicular to
                 suborbicular . . . . . . . . . . . . . . . . . . . . . . . 39. *P. praecox* (in part)
        54. Spikelets 2.6–4.1 mm long.
          56. Plants decumbent, rooting at the lower nodes,
                 not rhizomatous; spikelets obovate to elliptic
                 . . . . . . . . . . . . . . . . . . . . . . . . . . . . . . . . . 41. *P. pubiflorum* (in part)
          56. Plants rhizomatous, neither decumbent nor
                 rooting at the lower nodes; spikelets orbicular
                 to elliptic.
            57. Spikelets 2.1–3.1 mm long, 2–2.8 mm
                     wide, orbicular to suborbicular; upper
                     glumes 3-veined; leaf blades conduplicate
                     . . . . . . . . . . . . . . . . . . . . . . . . . . . . . 39. *P. praecox* (in part)
            57. Spikelets 2.9–4.1 mm long, 1.9–3.1 mm
                     wide, suborbicular to elliptic; upper glumes
                     5-veined; leaf blades flat . . . . . . . . . . . . . 42. *P. floridanum* (in part)

## 1. Paspalum repens P.J. Bergius WATER PASPALUM [p. 489, 553]

*Paspalum repens* is a native species that grows along the edges of lakes, streams, and roadside ditches in the southeastern United States. Its range extends through tropical America to Peru, Bolivia, and Argentina.

## 2. Paspalum scrobiculatum L. INDIAN PASPALUM [p. 489, 553]

*Paspalum scrobiculatum* is native to India. It has been found growing in widely scattered disturbed areas of the southeastern United States, possibly as an escape from cultivation.

## 3. Paspalum laeve Michx. FIELD PASPALUM [p. 489, 553]

*Paspalum laeve* is restricted to the eastern United States. It grows at the edges of forests and in disturbed areas.

## 4. Paspalum dissectum (L.) L. MUDBANK PASPALUM [p. 489, 553]

*Paspalum dissectum* grows at the edges of lakes, ponds, rice fields, and wet roadside ditches. It is native to the eastern portion of the contiguous United States and to Cuba.

**5. Paspalum acuminatum** Raddi Brook Paspalum, Canoegrass [p. *489*, 553]

*Paspalum acuminatum* grows at the edges of lakes, ponds, rice fields, and wet roadside ditches. It is native to the Americas, with a range that extends from the southern United States to Argentina.

**6. Paspalum conjugatum** P.J. Bergius Sour Paspalum [p. *489*, 553]

*Paspalum conjugatum* is native to tropical and subtropical regions of both the Western and Eastern hemispheres, including the *Manual* region. It grows in disturbed areas and at the edges of forests, and is sometimes used as a lawn grass.

**7. Paspalum distichum** L. Knotgrass, Thompsongrass [p. *489*, 553]

*Paspalum distichum* grows on the edges of lakes, ponds, rice fields, and wet roadside ditches. It is native in warm regions throughout the world, being most abundant in humid areas. In the Western Hemisphere, it grows from the United States to Argentina and Chile.

**8. Paspalum vaginatum** Sw. Seashore Paspalum [p. *489*, 553]

*Paspalum vaginatum* grows in brackish and salt marshes. It is native to warm, coastal regions around the world, including the Americas. It has been grown for turf and in lawn trials, but is not yet widely used for these purposes.

**9. Paspalum almum** Chase Comb's Paspalum [p. *489*, 553]

*Paspalum almum* was probably introduced to North America as a forage species. Its native range is Brazil, Paraguay, Uruguay, and eastern Argentina. In the *Manual* region, it is found along roadsides and in pastures of southeastern Texas and southern Louisiana.

**10. Paspalum notatum** Flügge Bahiagrass [p. *489*, 553]

*Paspalum notatum* is native from Mexico through the Caribbean and Central America to Brazil and northern Argentina. It was introduced to the United States for forage, turf, and erosion control. It is now established, generally being found in disturbed areas and at the edges of forests in the southeastern United States. A number of cultivars have been developed for use as turf grasses.

**11. Paspalum minus** E. Fourn. Matted Paspalum [p. *489*, 553]

*Paspalum minus* grows in disturbed areas and on the edges of forests. It grows from southern Texas to Florida in the *Manual* region; outside the region, it extends through Mexico and the West Indies to Peru, Bolivia, Brazil, and Paraguay.

**12. Paspalum blodgettii** Chapm. Coral Paspalum [p. *489*, 553]

*Paspalum blodgettii* grows in hammocks, low pinelands, and along roadsides in southern peninsular Florida, the Bahamas, the Greater Antilles, southeastern Mexico, and Belize.

**13. Paspalum paniculatum** L. Arrocillo [p. *489*, 553]

*Paspalum paniculatum* is native from Mexico and the West Indies to Argentina. It is now established in Mississippi and southern Florida, growing in disturbed areas.

**14. Paspalum fimbriatum** Kunth Winged Paspalum, Panama Crowngrass [p. *490*, 553]

*Paspalum fimbriatum* has probably been introduced into the United States. Its primary range extends from southern Mexico to Colombia, Venezuela, and French Guiana. In the *Manual* region, it grows in disturbed areas of Florida.

**15. Paspalum dilatatum** Poir. Dallisgrass [p. *490*, 553]

*Paspalum dilatatum* is native to Brazil and Argentina. It is now well established in the *Manual* region, generally as a weed in waste places. It is also used as a turf grass.

**16. Paspalum urvillei** Steud. Vaseygrass [p. *490*, 554]

*Paspalum urvillei* has been introduced to the United States from South America. In the *Manual* region it grows in disturbed, moist to wet areas, primarily in the southeastern United States.

**17. Paspalum modestum** Mez Water Paspalum [p. *490*, 554]

*Paspalum modestum* grows in wet roadside ditches and rice fields of Texas and southern Louisiana. It was introduced to the United States from South America. Plants with pale florets may key to *P. lividum*, which differs from *P. modestum* in having shorter ligules.

**18. Paspalum boscianum** Flügge Bull Paspalum [p. *490*, 554]

*Paspalum boscianum* grows in moist to dry, disturbed areas, and at the edges of forests. It is native from the southeastern United States through the West Indies and Mexico to Brazil. The California record came from a weed in a rice field.

**19. Paspalum virgatum** L. Talquezal [p. *490*, 554]

*Paspalum virgatum* is native from Mexico to South America. It has been introduced to the southeastern United States, where it grows primarily in disturbed areas and cultivated fields.

**20. Paspalum conspersum** Schrad. Scattered Paspalum [p. *490*, 554]

*Paspalum conspersum* is native from Mexico to Argentina, but it has been introduced to the southern United States. It is grown for its forage value, and has become established at scattered locations from Texas to Florida, growing along roadsides and in other disturbed areas.

**21. Paspalum convexum** Humb. & Bonpl. *ex* Flügge Mexican Paspalum [p. *490*, 554]

*Paspalum convexum* grows in disturbed areas in the southern United States. It is native from Mexico and the Caribbean Islands to Brazil.

**22. Paspalum plicatulum** Michx. Brownseed Paspalum [p. *490*, 554]

*Paspalum plicatulum* grows in prairies, along forest margins, and in disturbed areas. Its range extends from the southeastern United States through the Caribbean and Mexico to Bolivia, Paraguay, and Argentina.

**23. Paspalum wrightii** Hitchc. & Chase Wright's Paspalum [p. *490*, 554]

The range of *Paspalum wrightii* extends from Cuba and Campeche, Mexico, to Bolivia, Paraguay, and Argentina. It is now established in the *Manual* region, growing along wet roadside ditches, primarily on the Gulf Coast of Texas.

**24. Paspalum nicorae** Parodi Brunswickgrass [p. *490*, 554]

*Paspalum nicorae* is native to Brazil, Uruguay, and Argentina. It was introduced to the United States for use in pastures and as a cover crop in waterways. It is now established in the southeastern United States, growing as a weed in pastures, turf, and other disturbed areas.

## 25. Paspalum malacophyllum Trin. RIBBED PASPALUM [p. 490, <u>554</u>]

*Paspalum malacophyllum* is native from Mexico to Bolivia and Argentina. It was introduced to the southern United States for forage and soil conservation, and is now established in the southeastern United States, growing in disturbed sites at scattered locations.

## 26. Paspalum racemosum Lam. PERUVIAN PASPALUM [p. 491, <u>554</u>]

*Paspalum racemosum* is native to Colombia, Ecuador, and Peru. Within the *Manual* region, it is known from disturbed sites at a few widely scattered locations.

## 27. Paspalum intermedium Munro *ex* Morong & Britton INTERMEDIATE PASPALUM [p. 491, <u>554</u>]

*Paspalum intermedium* is an introduced roadside weed in the *Manual* region. It is found in Mexico and South America, but not in Central America.

## 28. Paspalum coryphaeum Trin. EMPEROR PASPALUM [p. 491, <u>554</u>]

*Paspalum coryphaeum* is native from Costa Rica and the Caribbean south to northern South America. In the *Manual* region, it grows in disturbed habitats at scattered southeastern locations.

## 29. Paspalum quadrifarium Lam. PAJA MANSE, PAJA COLORADA, TUSSOCK PASPALUM [p. 491, <u>554</u>]

*Paspalum quadrifarium* is native to Uruguay, Paraguay, Brazil, and Argentina. It is grown as an ornamental in Florida, but has also become established in disturbed habitats of the southeastern United States. It is considered a noxious weed in New South Wales, Australia.

## 30. Paspalum bifidum (Bertol.) Nash PITCHFORK PASPALUM [p. 491, <u>554</u>]

*Paspalum bifidum* is restricted to the southeastern United States. It grows at the edges of forests in longleaf pine-oak-grass ecosystems, usually in dry to mesic loamy sandy soils. It grows vigorously following fire.

## 31. Paspalum langei (E. Fourn.) Nash RUSTYSEED PASPALUM [p. 491, <u>554</u>]

*Paspalum langei* is native from Texas to Florida, and extends through Mexico to Venezuela and the Antilles. It grows at the edges of moist woods and in disturbed areas.

## 32. Paspalum setaceum Michx. [p. 491, <u>554</u>]

*Paspalum setaceum* is a variable species that grows east of the Rocky Mountains in the contiguous United States and Mexico. The following treatment summarizes the major patterns of variation within the species. Some specimens will be hard to place, particularly old herbarium specimens that have lost their color. Nine varieties grow in the *Manual* region.

1. Leaf blades conspicuously basal, recurved, 3–10 mm wide; lower lemmas without evident midveins.
　2. Leaf blades yellowish-green, usually glabrous; spikelets usually glabrous, not spotted . . . . . . . . . . . . . . . . . . . . . . . . . . . . . . . . . . . var. *longepedunculatum*
　2. Leaf blades grayish-green, hirsute; spikelets short pubescent, often spotted . . . . . . . . . . . . . var. *villosissimum*
1. Leaf blades more evenly distributed, lax to straight, 1.5–20 mm wide; lower lemmas with or without evident midveins.
　3. Leaf blades glabrous (or almost so) on the surfaces, sometimes ciliate on the margins.
　　4. Leaf blades 2.4–6.1 mm wide, stiff; spikelets 2–2.6 mm long . . . . . . . . . . . . . . . . . . var. *rigidifolium*

　　4. Leaf blades 3–18 mm wide, lax to somewhat stiff but, if somewhat stiff, more than 6 mm wide; spikelets 1.7–2.4 mm long.
　　　5. Lower lemmas without evident midveins; blades yellowish-green to dark green . . . . . . . . . . . . . . . . . . . . . . var. *stramineum* (in part)
　　　5. Lower lemmas with evident midveins; blades dark green to purplish . . . . . . . . var. *ciliatifolium*
　3. Leaf blades evidently hirsute on the surfaces as well as on the margins.
　　6. Plants widely spreading to prostrate.
　　　7. Leaf blades grayish-green; lower lemmas without evident midveins . . . . . . . var. *psammophilum*
　　　7. Leaf blades yellowish-green; lower lemmas with or without evident midveins . . . . . . . . . . . . . . . . . . . . . . . . . . . . . var. *supinum*
　　6. Plants erect to spreading.
　　　8. Lower lemmas usually with evident midveins; spikelets 1.8–2.5 mm long, usually glabrous, light green to green . . . . . . . var. *muhlenbergii*
　　　8. Lower lemmas usually without evident midveins; spikelets 1.4–2.4 mm long, usually pubescent, pale yellow to light green.
　　　　9. Leaves grayish-green; blades 1.5–7 mm wide, always conspicuously hirsute . . . . . . . . . . . . . . . . . . . . . var. *setaceum*
　　　　9. Leaves yellowish-green to dark green; blades 3.3–13.5 mm wide, almost glabrous or conspicuously hirsute . . . . . . . . . . . . . . . . . . var. *stramineum* (in part)

## Paspalum setaceum var. ciliatifolium (Michx.) Vasey FRINGELEAF PASPALUM [p. 491]

*Paspalum setaceum* var. *ciliatifolium* is the most variable and widespread of the nine varieties of *P. setaceum*. It usually grows in sandy soil in open areas, including disturbed areas, of prairies and forest margins. Its range extends from Louisiana and the eastern United States to Panama, the West Indies, and Bermuda.

## Paspalum setaceum var. longepedunculatum (Leconte) Alph. Wood BARESTEM PASPALUM [p. 491]

*Paspalum setaceum* var. *longepedunculatum* grows on open ground, usually in moist areas such as along ditches and roadsides, as well as in flatwoods. It is found primarily in the coastal plain of the southeastern United States, but has also been found in Ohio, Kentucky, and Tennessee. It is similar to var. *villosissimum*, differing in its glabrous leaves, more delicate habit, and more poorly developed rhizomes. Both varieties grow in peninsular Florida, but var. *longepedunculatum* also grows along the coast as far west as the Mississippi delta and as far north as southern North Carolina.

## Paspalum setaceum var. muhlenbergii (Nash) D.J. Banks HURRAHGRASS [p. 491]

*Paspalum setaceum* var. *muhlenbergii* is endemic to the *Manual* region, extending from southern Ontario to the Gulf Coast of Texas and northern Florida. It grows in disturbed areas and on the margins of forests. It resembles var. *supinum*, differing in its erect habit and, usually, in its spikelet shape and presence of a midvein on the lower lemma.

## Paspalum setaceum var. psammophilum (Nash) D.J. Banks SAND PASPALUM [p. 491]

*Paspalum setaceum* var. *psammophilum* grows in sandy, maritime habitats and, inland, along sandy roadsides and in dry fields, from Massachusetts to the District of Columbia. The combination of its spreading to prostrate habit and densely puberulent foliage distinguishes it from other varieties of *P. setaceum*.

**Paspalum setaceum var. rigidifolium** (Nash) D.J. Banks
STIFF PASPALUM [p. *491*]

*Paspalum setaceum* var. *rigidifolium* grows on hammocks, sand barrens, high pinelands, and flatwoods of Georgia, Florida, and Cuba.

**Paspalum setaceum** Michx. var. **setaceum** THIN
PASPALUM [p. *491*]

*Paspalum setaceum* var. *setaceum* grows in open areas and sandy soils, often at the edges of forests, primarily on the southeastern coastal plain of the United States, from southern New England to eastern Mexico, but extending inland to western Virginia, Missouri, and Arkansas. It also grows in Cuba.

**Paspalum setaceum var. stramineum** (Nash) D.J. Banks
YELLOW SAND PASPALUM [p. *491*]

*Paspalum setaceum* var. *stramineum* grows at the edges of forests and in disturbed areas with sandy soil. Its range extends from the central plains and eastern United States to Mexico, Bermuda, and the West Indies.

**Paspalum setaceum var. supinum** (Bosc *ex* Poir.) Trin.
SUPINE THIN PASPALUM [p. *491*]

*Paspalum setaceum* var. *supinum* grows at the edges of forests and in disturbed areas. Within the *Manual* region, its range extends from Texas, Arkansas, and Louisiana to South Carolina, Georgia, and Florida. It resembles var. *muhlenbergii*, differing in its spreading habit and, usually, in its spikelet shape and lack of a midvein on the lower lemma.

**Paspalum setaceum var. villosissimum** (Nash) D.J.
Banks HAIRY PASPALUM [p. *492*]

*Paspalum setaceum* var. *villosissimum* grows in sandy fields and flatwoods of Florida and Cuba. It resembles var. *longepedunculatum*, differing in its pubescent leaves, more robust habit, and more developed rhizomes.

**33. Paspalum laxum** Lam. COCONUT PASPALUM [p. *492*, *554*]

*Paspalum laxum* grows in hammocks and along roads, often in sandy or limestone soils. It grows in southern Florida, the Antilles, and Belize.

**34. Paspalum caespitosum** Flüggé BLUE PASPALUM [p. *492*, *554*]

*Paspalum caespitosum* grows in hammocks and sandy pinelands. It is native in southern Alabama, Florida, the West Indies, Mexico, and Central America.

**35. Paspalum virletii** E. Fourn. VIRLET'S PASPALUM [p. *492*, *554*]

*Paspalum virletii* grows in dry, sandy soils in disturbed habits. It is known only from Arizona, where it is considered a rare species, and from Mexico, where it also appears to be either rare or poorly collected.

**36. Paspalum monostachyum** Vasey GULFDUNE
PASPALUM [p. *492*, *554*]

*Paspalum monostachyum* grows in sand and muck soils on coastal sand dunes, wet prairie, marshes, and disturbed habitats of the southern coastal plain from Florida to eastern Mexico.

**37. Paspalum pleostachyum** Döll TROPICAL PASPALUM [p. *492*, *554*]

*Paspalum pleostachyum* grows in sandy soil or rocky areas in Florida, the West Indies, and from northern South America to Brazil. It is usually found along the coast.

**38. Paspalum lividum** Trin. *ex* Schltdl. LONGTOM [p. *492*, *554*]

*Paspalum lividum* grows in fresh and brackish marshes and ditches. It is native from the Gulf Coast of the United States southward through Mexico and Central America to Cuba and Argentina. Plants of *P. modestum* with pale upper florets may be mistaken for *P. lividum*, but will have ligules that are only 1–2.3 mm long.

**39. Paspalum praecox** Walter EARLY PASPALUM [p. *492*, *554*]

*Paspalum praecox* grows in pitcher plant bogs, wet pine flatwoods, wet savannahs, prairies, and wet streamhead ecotones. It is restricted to the United States, growing predominantly on the southeastern coastal plain.

**40. Paspalum hartwegianum** E. Fourn. HARTWEG'S
PASPALUM [p. *492*, *554*]

*Paspalum hartwegianum* grows in wet prairies, ditches, and swales from southern Texas through Mexico and Central America to Paraguay and Argentina.

**41. Paspalum pubiflorum** Rupr. *ex* E. Fourn.
HAIRYSEED PASPALUM [p. *492*, *555*]

*Paspalum pubiflorum* grows on the edges of forests and in disturbed areas. It is native to the southeastern United States, Mexico, and Cuba.

**42. Paspalum floridanum** Michx. FLORIDA PASPALUM [p. *492*, *555*]

*Paspalum floridanum* grows along the edges of forests, flatwoods, and pinewoods and in open areas. It is a frequent component of dry-mesic soils in longleaf pine-oak-grass ecosystems, and is restricted to the eastern United States.

**43. Paspalum unispicatum** (Scribn. & Merr.) Nash
ONE-SPIKE PASPALUM [p. *492*, *555*]

*Paspalum unispicatum* grows in sandy soil in the coastal plain of Texas and extends southward through Mexico and Central America to Cuba and Paraguay, Uruguay, and Argentina. It has not been reported from Brazil.

---

# 24.29 ZULOAGAEA Bess[1]

**Pl** per; ces, rhz, rhz short, thin. **Clm** clumped or solitary, often with hard, cormlike bases, slightly compressed, erect or geniculate at the lo nd. **Shth** shorter than the intnd, keeled, often pilose, hairs papillose-based near the throat; **lig** membranous-based, ciliate; **bld** (6)20–75 cm long, 1.5–15 mm wide, flat, adaxial surfaces scabrous. **Pan** pyramidal, open, bearing 4–40 br, bas nd with 1 br; pri br to 30 cm, with 3–6 orders of brchg, straight or flexible, ascending to reflexed; **ped** scabridulous, divergent. **Spklt** 2.5–4.2(5.5) mm long, ellipsoid or lanceoloid, purplish or greenish, glab, acute or obtuse. **Lo glm** 1–3.7 mm, about ⅔ as long as the up

[1]Mary E. Barkworth

glm, 3–5-veined; **up glm** slightly shorter to subequal to the lo lm, glab, 5–7-veined; lo flt strl or stmt; **lo lm** 2.9–3.3 mm, glab, 5-veined, acute; **lo pal** 3–4 mm, hyaline; **up flt** 2.1–5 mm long, exceeding the up glm, mrg embracing the the lo lm, dull, pale, finely transversely rugose, apc acute, puberulent; **anth** about 2 mm, yellow-brown; **stigmas** pale purple, plumose. **Car** oblong, compressed; **emb** about ⅓ the length of the car.

*Zuloagaea* is a unispecific genus that is native to western North America. It used be included in *Panicum*, but molecular

phylogenetic studies consistently place it in the bristle clade of the *Paniceae*, although it never develops bristles. Bess et al. (2005) concluded that it is best treated as a unispecific genus. They noted that it is recognizable by its "open, loosely flowered pyramidal panicle with small spikelets that are purple in color if the plant is growing in the sun, or green in color if it is growing in the shade." Its distinctive vegetative characters include the thickened, often cormous, culm bases, elongate blades, and rather short, often pilose sheaths.

### 1. Zuloagaea bulbosa ( Kunth) Bess BULBOUS PANICGRASS [p. 493, 555]

*Zuloagaea bulbosa* grows in roadside ditches, on gravelly river banks and moist mountain slopes, often in ponderosa pine and oak woodlands, from southern Nevada and Arizona to western Texas and central Mexico. It is considered an important forage grass and is sometimes cut for hay but is not known to be cultivated. Flowering is from July to mid-October. Plants growing in sunlight tend to have

purple spikelets, those growing in the shade tend to have green spikelets.

In the past, three species have been recognized within *Panicum* sect. *Obtusum* but the variation among them is continuous and highly influenced by environment (Bess et al. 2005).

## 24.30 PHANOPYRUM Raf.[1]

**Pl** per; **stln. Clm** thick, decumbent, succulent, rooting profusely at the lo nd, without cormous bases. **Lig** memb; bld with non-Kranz anatomy. **Pan** open; **br** spikelike, stiffly ascending, with secund clusters of shortly pedlt to sessile spklt. **Spklt** 6–7 mm, lanceoloid, lat compressed, glab. **Glm** exceeding the up flt, apc separated at maurity; **lo glm** ⅘ as long as the spklt, nearly as long as the lo lm, 3-veined; **up glm** 3-veined; **lo flt** strl; **lo lm** 3-veined; **lo pal** small, lanceolate; **up flt** about ⅓ as long as the spklt, obovoid, stipitate; **up lm** indurate, smooth.

*Phanopyrum* differs from other members of the *Paniceae* in its one-sided panicle branches, large spikelets with spreading glumes, its long lower glumes, and the short, stipitate upper floret. It includes only one species and is restricted to the southeastern United States.

*Phanopyrum* used to be treated as *Panicum* subg. *gymnocarpon*. Aliscioni et al. (2003) showed that it is only distantly related to *Panicum sensu stricto* and not closely related to other members of the subgenus, such as *P. hemitomon* and *P. trichoides*. Because the appropriate treatment of other members of the subgenus is not yet evident, they have been left in *Panicum* for now.

### 1. Phanopyrum gymnocarpon Elliott SAVANNAH PANICGRASS [p. 493, 555]

*Phanopyrum gymnocarpon* grows in swamps, wet woodlands, and the marshy shores of lakes and streams. It is also found occasionally

in shallow water, often in the shade. It is restricted to the United States.

## 24.31 REIMAROCHLOA Hitchc.[2]

**Pl** ann; smt stln. **Clm** 10–100 cm, erect to ascending, brchg above the base. **Lvs** mostly cauline; **lig** of hairs; bld linear. **Infl** numerous, tml and ax, subdigitate or rcm pan of spikelike br, spklt borne singly in 2 rows on the abx sides of the br; **dis** at the base of the spklt. **Spklt** dorsally compressed, with 2 flt, up flt appressed to the br axes. **Glm** usu absent, up glm smt present on the tml spklt of a br; **lo flt** strl; **up flt** bisx; **lo lm** subequal to the

up lm; **up lm** memb to coriaceous, mrg narrow; **up pal** similar in texture to the lm, their bases enclosed by the lm; **anth** 1 or 2.

*Reimarochloa* is a genus of three species, all of which grow in damp habitats. The range of the genus extends from the southern United States to Argentina. One species is native to the *Manual* region.

### 1. Reimarochloa oligostachya (Munro *ex* Benth.) Hitchc. FLORIDA REIMARGRASS [p. 493, 555]

*Reimarochloa oligostachya* grows in water or wet soil of hammocks, riverbanks, ditches, and disturbed areas. Although not common in the *Manual* region, it grows in peninsular Florida, Mobile County,

Alabama, and Cuba. The Alabama record is probably an introduction.

[1,2]Mary E. Barkworth

# 25. ANDROPOGONEAE Dumort.[1]

Pl usu per. Clm 7–600 cm, ann, not wd, often reddish or purple, particularly at the nd, often brchd above the base. Shth open; lig usu scarious to memb, ciliate or not; bld mostly well-developed, lvs subtending an infl or an infl unit often with rdcd bld. Photosynthetic pathway NADP-ME; bundle shth single. Infl tml, frequently on both the clm and their br, smt also ax, usu of 1–many spikelike br, these in digitate clusters of 1–13+ on a peduncle or attached, directly or indirectly, to elongate rchs, often partially to almost completely enclosed by the subtending lf shth at maturity, in some taxa ax infl composed of multiple-stalked pedunculate clusters of infl br subtended by a modified lf; dis usu in the br axes beneath the sessile flt, the dispersal unit being a sessile flt, the intnd to the next sessile flt, the ped, and the pedlt spklt (branches with disarticulating axes are termed *rames* in the following accounts), smt beneath the glm, the br axes remaining intact. Spklt in unequally pedlt pairs, sessile-pedlt pairs, or triplets, or apparently solitary and sessile, pedlt spklt and smt the ped rdcd or absent, triplets usu with 1 sessile and 2 pedlt spklt, tml spklt units on the br often with 2 pedlt spklt even if the others have only 1 (all spikelet units with 2 sessile and 1 pedicellate spikelet in *Polytrias*). Spklt pairs or triplets *homogamous* (spikelets in the unit sexually alike) or *heterogamous* (spikelets in the unit sexually dissimilar); spklt of unequally pedlt pairs usu homogamous and homomorphic; spklt in sessile-pedlt pairs or triplets usu heterogamous and heteromorphic; sessile spklt usu bisx; pedlt spklt usu smaller than the sessile spklt, often stmt or strl, smt absent. Spklt usu with 2 flt (1 in *Polytrias*). Glm exceeding and usu concealing the flt (excluding the awns), rounded or dorsally compressed, usu tougher than the lm; lo flt in bisx or pist spklt strl or stmt, often rdcd to a hyaline scale; up flt bisx or pist, lm often hyaline, smt with an awn that exceeds the glm; lod cuneate; anth usu 3. Ped free or fused to the rchs intnd. Pedlt spklt variable, smt similar to the sessile spklt, smt differing in sexuality and shape, smt missing.

The tribe *Andropogoneae* includes about 87 genera and 1060 species, of which 31 genera and 102 species have been found in the *Manual* region; some of these have not become established. The tribe is common in tropical and subtropical regions, particularly in areas with significant summer rains, such as the central plains of North America.

Members of the *Andropogoneae* differ from those of *Paniceae* in the reduced lemmas and paleas of their florets and, usually, in their paired, unequally pedicellate spikelets, disarticulating inflorescence branches (*rames*), and the manner in which these branches are aggregated into inflorescences. Unequally pedicellate spikelet pairs are found in many other tribes, but they are more common, and the pedicels more strikingly unequal in length, in the *Andropogoneae*.

## Inflorescence Structures

Describing inflorescence structures in the *Andropogoneae* is not simple. There is a basic pattern, but its many modifications have resulted in great structural diversity. The following paragraphs provide an overview of this diversity and explain the words and phrases used in describing it. Diagrammatic representations of many of the structures mentioned are presented on page 321.

### Spikelets

Members of the *Andropogoneae*, like those of the *Paniceae*, generally have two florets per spikelet, the lower floret usually being reduced in size and sterile or staminate, and the upper floret bisexual (p. 321). Despite this similarity, spikelets of the two tribes are easy to distinguish. In the *Paniceae*, the lowest glume is usually much shorter than the floret, and the upper florets usually have lemmas that are thicker and tougher than the glume and lower lemma. In the *Andropogoneae*, the glumes usually exceed and enclose both florets, and are thicker and tougher than the lemmas. The florets of the *Andropogoneae* contrast strongly with the glumes, having hyaline or thinly membranous lemma bodies and hyaline paleas, or, in many cases, no paleas. They are almost always completely concealed by the glumes, except that the upper floret often has an awn that projects beyond the glume.

In some *Andropogoneae*, the glumes are merely thickly membranous, but most genera have coriaceous or indurate glumes. The lower glumes are sometimes tougher and larger than the upper glumes, and may even conceal the upper glumes as, for example, in *Heteropogon* (p. 502). In such genera, the lower glumes may be mistaken for lemmas. In dioecious species, or monoecious species with strongly differentiated staminate and pistillate spikelets, the staminate spikelets usually have softer glumes than the pistillate spikelets.

### Spikelet Units

The basic element of the inflorescence structure in the *Andropogoneae* is the *spikelet unit*. These units usually consist of pairs of spikelets, one sessile and one pedicellate (e.g., *Saccharum bengalense*, p. 494), but they may consist of a pair of unequally pedicellate spikelets (e.g., *Miscanthus sacchariflorus*, p. 494) or of three spikelets (e.g., *Chysopogon fulvus*, p. 496). If there are three spikelets in the unit, one is usually sessile and the other two pedicellate, but a few genera, such as *Polytrias*, have two sessile spikelets and one pedicellate spikelet.

Unequally pedicellate spikelet pairs or triplets are found in other tribes, but in the *Andropogoneae* they usually differ in size, shape, and sexuality. Spikelet units with spikelets that differ in their sexuality are described as *heterogamous*; those with sexually similar spikelets are said to be *homogamous*. Spikelet units with morphologically dissimilar spikelets are *heteromorphic* (e.g., *Andropogon longiberbis*, p. 500); those with morphologically similar spikelets are *homomorphic* (e.g., *Chrysopogon zizanioides*, p. 496). In most *Andropogoneae*, the spikelet units are heterogamous and heteromorphic. The

---

[1]Mary E. Barkworth

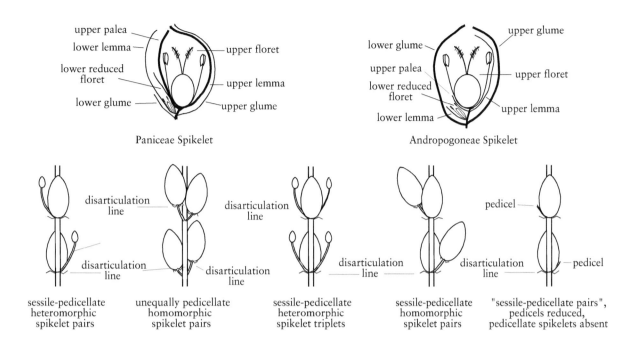

SPIKELETS and SPIKELET UNITS

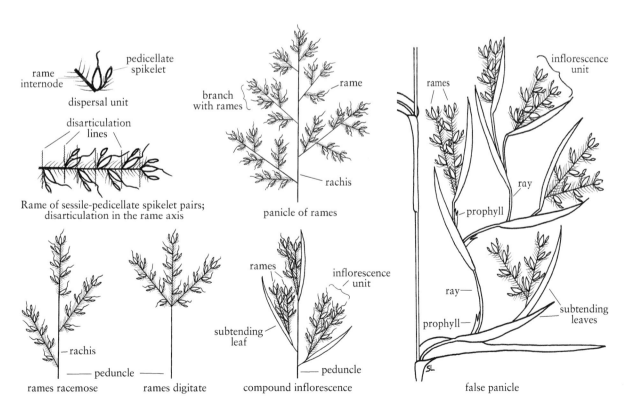

INFLORESCENCE STRUCTURES

sessile spikelets usually contain a bisexual or pistillate floret, and often exhibit features such as awns and calluses that are related to seed dispersal and establishment; the pedicellate spikelets are usually staminate, sterile, vestigial, or even absent. In some genera the situation is reversed, the pedicellate spikelets being bisexual or pistillate, and the sessile spikelets staminate or sterile. Sterile and staminate spikelets are sometimes morphologically similar to the pistillate or bisexual spikelets, but usually lack the features associated with seed dispersal and establishment.

In a few genera, such as *Sorghastrum* (p. 496), there is only an empty pedicel associated with each bisexual sessile spikelet, or even, as in *Arthraxon* (p. 501), only a stump where the pedicel and its spikelet would be.

## Inflorescence Structure

Further complexity is introduced to the *Andropogoneae* inflorescence structure by the manner in which the spikelet units are aggregated and the mode of disarticulation. Three patterns can be identified. The simplest pattern consists of inflorescences similar to those common in other tribes, in which neither the rachis nor the inflorescence branches break up at maturity. Genera with such inflorescences [e.g., *Miscanthus* (p. 494) and *Imperata* (p. 495)] have unequally pedicellate spikelets, and disarticulation is below the glumes. Such inflorescences are, however, in the minority within the *Andropogoneae*.

A more common situation is for the spikelets to be in sessile-pedicellate pairs and disarticulation to be in the branch axes, immediately below the attachment of the sessile spikelets. The resulting dispersal unit consists of the spikelet pair plus the internode that extends from the sessile spikelet to the next most distal sessile spikelet. These disarticulating inflorescence branches, termed *rames* in this *Manual*, form the basic unit of the typical *Andropogoneae* inflorescence. In other publications, the rames are often called *racemes*, a word that is restricted in this *Manual* to an entire inflorescence, not just an inflorescence branch.

Rames are usually composed of several spikelet units, but sometimes of only one. The spikelets may be evenly distributed, or the base of the rame axis may be naked. Individual plants may bear few to many rames, and the rames themselves may be aggregated in a wide array of primary and secondary arrangements; they may also be branched.

One or more rames may be borne on a single stalk. If this stalk is attached to a rachis, the unit formed by the stalk and its rame(s) constitutes an inflorescence branch. Such a pattern is seen, for example, in *Sorghum halepense* (p. 496) and *Bothriochloa bladhii* (p. 498). A more common situation is for one or more rames to be attached digitately to a common stalk, the peduncle. This peduncle may terminate a culm (as in *Dichanthium annulatum* [p. 497] or *Elionurus* [p. 502]) or be axillary to a subtending leaf (as in *Andropogon hallii* [p. 498] and *Hemarthria altissima* [p. 502]). Each peduncle and its associated rame(s) constitutes an *inflorescence unit*.

*False panicles* represent a further level of complexity. In these, the inflorescence units terminate *rays*, each of which has a *prophyll*, a 2-veined structure, in its axil. Several rays may develop within the axil of a single leaf sheath, and rays may themselves give rise to subtending leaves with multiple rays in their axils. The result is a complex, tiered inflorescence in which only the ultimate units are easily described. Such inflorescences are found, for example, in *Andropogon glomeratus* and *Cymbopogon citratus* (p. 500). Fortunately, identification of the *Andropogoneae* does not require analyzing false panicles, merely their ultimate inflorescence units.

In another inflorescence pattern, the rame axes are thick and the pedicels are either closely appressed or even fused to the rame axes. In these genera, the pedicellate spikelets are often highly reduced or absent. Pistillate rames of *Tripsacum* and wild taxa of *Zea* (p. 504) represent an extreme example of this pattern. In these genera, the sessile spikelets are completely embedded in the rame axes, the lower glumes being indurate and completely concealing the florets. Less extreme examples are seen in *Coelorachis* and *Hackelochloa* (p. 503).

1. Leaves smelling of lemon oil or citronella, the sheaths without glandular depressions on the keel; plants perennial, not reaching reproductive maturity in the *Manual* region when grown outdoors . . . . . . . . . . . 25.16 *Cymbopogon*
1. Leaves usually not aromatic or, if aromatic and smelling of citronella, the sheaths with glandular depressions along the keel and plants annual; plants reaching reproductive maturity in the *Manual* region.
   2. All spikelets unisexual, the pistillate and staminate spikelets in separate inflorescences or the pistillate spikelets below the staminate spikelets in the same inflorescence.
      3. Pistillate spikelets completely concealed within a hard, globose, beadlike structure (a modified leaf sheath) from which the staminate rames protrude . . . . . . . . . . . . . . . . . . . . . . . . . . . . . . . 25.30 *Coix*
      3. Pistillate spikelets exposed or enclosed by 1 or more subtending leaf sheaths and a hyaline prophyll; staminate spikelets either distal on the same branch or in a separate inflorescence on the same plant.
         4. Staminate and pistillate spikelets in the same inflorescence and on the same branch, the staminate spikelets distal to the pistillate spikelets . . . . . . . . . . . . . . . . . . . . . . . . . . . . . . . 25.28 *Tripsacum*
         4. Staminate and pistillate inflorescences usually separate; staminate inflorescences terminal on the culms and branches; pistillate inflorescences terminal on axillary peduncles, sometimes aggregated in false panicles . . . . . . . . . . . . . . . . . . . . . . . . . . . . . . . . . . . . . . . . . . . . . . 25.29 *Zea*
   2. Some spikelets bisexual (usually the sessile or more shortly pedicellate spikelet of each spikelet pair or triplet).
      5. Spikelets apparently solitary and sessile, the pedicellate spikelets absent; pedicels absent or present.

6. Culms decumbent, scrambling; leaf blades ovate to ovate-lanceolate; pedicels absent or shorter than 3 mm . . . . . . . . . . . . . . . . . . . . . . . . . . . . . . . . . . . . . . . . . . . . . . . . . . . . 25.18 *Arthraxon*
6. Culms erect; leaf blades lanceolate to linear-lanceolate; pedicels always present, usually longer than 3 mm.
   7. Inflorescences terminal and axillary, composed of digitate clusters of 1–13 rames on a common peduncle; peduncles subtended by, and often partially included in, a modified leaf
   . . . . . . . . . . . . . . . . . . . . . . . . . . . . . . . . . . . . . . . . . . . . . . . . . . . . . . 25.15 *Andropogon*
   7. Inflorescences terminal, with elongate rachises and branches with several to many rames; peduncles and branches not subtended by a modified leaf . . . . . . . . . . . . . . . . . . . . . . . . 25.09 *Sorghastrum*
5. Spikelets in sessile-pedicellate or unequally pedicellate pairs or triplets, the pedicellate spikelets often smaller than the sessile spikelets, sometimes rudimentary.
  8. Pedicels strongly appressed or fused to the thick rame axes, or the rames with only 1 spikelet unit, this a triplet with 2 unequally pedicellate spikelets; bisexual spikelets usually unawned; inflorescences of rames.
    9. Lower glumes of the sessile spikelets rugose, pitted, tuberculate, or alveolate or the keels winged or with spinelike projections at the base.
      10. Keels of the lower glumes with spinelike projections on the base, sometimes winged distally, the surface between the keels smooth; spikelets unawned . . . . . . . . . . . . . . . . . . 25.25 *Eremochloa*
      10. Keels of the lower glumes winged throughout or not winged, the surface between the keels rough, rugose, pitted, tuberculate, or alveolate; spikelets unawned or awned.
        11. Sessile spikelets awned . . . . . . . . . . . . . . . . . . . . . . . . . . . . . . . . . . . . . . 25.13 *Ischaemum*
        11. Sessile spikelets unawned.
          12. Plants perennial; sessile spikelets ovate, the lower glumes smooth, rugose, or pitted . . . . . . . . . . . . . . . . . . . . . . . . . . . . . . . . . . . 25.24 *Coelorachis* (in part)
          12. Plants annual; sessile spikelets hemispherical, the lower glumes alveolate . . . 25.27 *Hackelochloa*
    9. Lower glumes of the sessile spikelets smooth or scabrous, not sculptured, the keels without spinelike projections.
      13. Inflorescences false panicles; individual rames to 1 cm long, with 1 spikelet unit; spikelet units composed of 1 sessile and 2 unequally pedicellate and dissimilar spikelets . . . . . . . . . . . . 25.14 *Apluda*
      13. Inflorescences usually solitary rames, sometimes with 2 rames in a digitate cluster; individual rames 2–15 cm long, with more than 1 spikelet unit; spikelet units composed of sessile pedicellate pairs, the pedicellate spikelets often rudimentary or absent.
        14. Pedicels appressed, but not fused, to the rame axes.
          15. Pedicellate spikelets 1–3 mm long . . . . . . . . . . . . . . . . . . . . . . . . 25.24 *Coelorachis* (in part)
          15. Pedicellate spikelets 4–8 mm long . . . . . . . . . . . . . . . . . . . . . . . . . . . . 25.22 *Elionurus*
        14. Pedicels at least partially fused to the rame axes.
          16. Plants perennial; sheaths mostly glabrous, sparsely ciliate basally . . . . . . . . . . . 25.23 *Hemarthria*
          16. Plants annual; sheaths with stiff, papillose-based hairs 1–3 mm long . . . . . . . . . . . 25.26 *Rottboellia*
  8. Pedicels free; rame or branch internodes slender, sometimes thickened distally; bisexual spikelets usually awned; inflorescences of rames with the spikelets in sessile-pedicellate pairs or of non-disarticulating branches with the spikelets in unequally pedicellate pairs.
   17. All spikelet units homogamous, frequently also homomorphic.
    18. Terminal inflorescences a single rame or a digitate or subdigitate cluster of rames.
      19. Terminal inflorescences a digitate or subdigitate cluster of (1)2–6 rames; rames 3–7 cm long . . . . . . . . . . . . . . . . . . . . . . . . . . . . . . . . . . . . 25.06 *Microstegium* (in part)
      19. Terminal inflorescences solitary rames; rames 2–3 cm long . . . . . . . . . . . . . . . . . . . 25.05 *Polytrias*
    18. Terminal inflorescences with elongated rachises.
      20. Spikelets in unequally pedicellate pairs; disarticulation below the glumes, the branches remaining intact at maturity.
        21. Spikelets usually awned; inflorescence branches usually 7–35 cm long . . . . . . . 25.03 *Miscanthus*
        21. Spikelets unawned; inflorescence branches 1–7 cm long . . . . . . . . . . . . . . . . . . 25.04 *Imperata*
      20. Spikelets in sessile-pedicellate pairs or triplets; disarticulation in the rames, below the sessile spikelets.
        22. Culms to 100 cm tall, often decumbent and straggling; terminal inflorescences with 2–6 subdigitately to racemosely arranged rames . . . . . . . . . . . . . . . . . . . . . . . . . . . . . . . . . . . . . . . . . . 25.06 *Microstegium* (in part)
        22. Culms 40–600 cm tall, erect; terminal inflorescences panicles, with more than 6 primary branches; branches usually with 2 or more rames.

## 25.01 SPODIOPOGON Trin.[1]

Pl usu per; smt rhz. **Clm** 40–150 cm, erect, simple or brchg. **Lvs** not aromatic; **lig** memb; **bld** lanceolate to broadly linear, smt psdpet. **Infl** tml, open or contracted pan, with evident rchs with numerous subverticellate br that terminate in 1–3 short rames; **rames** with slender intnd and 2–5 sessile-pedlt homogamous spklt pairs; **dis** in the rames, below the sessile spklt. **Spklt** usu lanceolate. **Glm** equal, chartaceous, often pilose, scarcely keeled, with several raised veins, acute; **cal** glab or densely hairy; **lo flt** usu stmt, unawned; **up flt** bisx; **up lm** bilobed, with a geniculate awn; **anth** 3.

*Spodiopogon* has 10–15 species, most of which grow in subtropical regions of the Eastern Hemisphere, although *S. sibiricus* extends north to Irkutsk, Russia. One species is cultivated in the *Manual* region.

[1]Mary E. Barkworth

**1. Spodiopogon sibiricus** Trin. SILVER SPIKE [p. *493*]

*Spodiopogon sibiricus* is native to the grasslands of the montane regions that extend from central China to northeastern Siberia. It is grown as an ornamental in Canada and the contiguous United States.

# 25.02 SACCHARUM L.[1]

Pl per; csp, often with a knotty crown, smt rhz, rhz usu short but elongate in some species, rarely stln. Clm 0.8–6 m, erect. Lvs cauline, not aromatic; **shth** usu glab, smt ciliate at the throats; **lig** memb, ciliate; **bld** flat, lax, smooth, usu glab. Infl tml, large, often plumose, fully exserted pan with evident rchs and numerous, ascending to appressed br terminating in multiple rames, br alternate, smt naked below; **rames** with numerous sessile-pedlt spklt pairs and a tml triad of 1 sessile and 2 pedlt spklt, intnd slender, without a translucent median groove; **dis** beneath the pedlt spklt and in the rames beneath the sessile spklt, sessile spklt falling with the adjacent intnd and ped. Spklt pairs homogamous and homomorphic, or almost so, not embedded in the rame axes, dorsally compressed. Sessile spklt: cal truncate, usu with silky hairs; **glm** subequal, chartaceous to coriaceous, glab or villous, 2-keeled, veins not raised; **lo flt** strl; **lo lm** hyaline or memb; **lo pal** absent or vestigial, entire; **up flt** bisx; **up lm** entire or bidentate, muticous or awned; **lod** 2, truncate; **anth** 2 or 3. Ped neither appressed nor fused to the rame axes. Pedlt spklt well developed, from slightly shorter than to equaling the sessile spklt.

*Saccharum* has 35–40 species that grow throughout the tropics and subtropics. Nine species can be found in the *Manual* region; five are native, two are grown as ornamentals, one is grown for agriculture, and one for research. Some species of *Saccharum* hybridize naturally with other, presumably closely related, genera such as *Miscanthus*, *Imperata*, and *Sorghum*.

1. Spikelets unawned, or with awns less than 5 mm long; anthers 3.
    2. Spikelets with visible awns, the awns 2–5 mm long . . . . . . . . . . . . . . . . . . . . . . . . . . . . . . . 6. *S. ravennae*
    2. Spikelets unawned, or the awns concealed by the glumes.
        3. Lower glumes of sessile spikelets pubescent . . . . . . . . . . . . . . . . . . . . . . . . . . . . . 9. *S. bengalense*
        3. Lower glumes of sessile spikelets mostly glabrous, sometimes ciliate distally.
            4. Culms clumped, 2–5 cm thick; rhizomes short; blades 20–60 mm wide . . . . . . . . . . . . . . . . . . . . 8. *S. officinarum*
            4. Culms solitary or few together, 0.6–2 cm thick; rhizomes elongate; blades 10–25 mm wide . . . . . 7. *S. spontaneum*
1. Spikelets awned, the awns 10–26 mm long; anthers 2.
    5. Awns spirally coiled at the base.
        6. Callus hairs 3–7 mm long, equal to or shorter than the spikelets, white to brown; rachises glabrous or sparsely pilose . . . . . . . . . . . . . . . . . . . . . . . . . . . . . . . . . . . . . . . . . . . 3. *S. brevibarbe* (in part)
        6. Callus hairs 9–14 mm long, exceeding the spikelets, silvery or tinged with purple; rachises densely pubescent . . . . . . . . . . . . . . . . . . . . . . . . . . . . . . . . . . . . . . . . . . . 2. *S. alopecuroides*
    5. Awns straight to curved at the base.
        7. Callus hairs longer than the spikelets; lowest panicle nodes densely pilose . . . . . . . . . . . . . . . . . . . . 1. *S. giganteum*
        7. Callus hairs absent or no more than equaling the spikelets; lowest panicle nodes glabrous or sparsely pilose.
            8. Calluses glabrous or with hairs to 2 mm long and exceeded by the spikelets; panicles 1–2.5 cm wide . . . . . . . . . . . . . . . . . . . . . . . . . . . . . . . . . . . . . . . . . . . . . . . . . . . . . . 5. *S. baldwinii*
            8. Callus hairs 3–7 mm long, often equaling the spikelets; panicles 3–10 cm wide.
                9. Awns flat basally; lower lemmas of sessile spikelets not or indistinctly veined; upper lemmas 0.9–1 times as long as the lower lemmas . . . . . . . . . . . . . . . . . . . . . . . 3. *S. brevibarbe* (in part)
                9. Awns terete basally; lower lemmas of the sessile spikelets typically 3-veined; upper lemmas 0.7–0.8 times as long as the lower lemmas . . . . . . . . . . . . . . . . . . . . . . . . . . 4. *S. coarctatum*

**1. Saccharum giganteum** (Walter) Pers. SUGARCANE PLUMEGRASS [p. *494*, *555*]

*Saccharum giganteum* grows in wet soils of bogs, swales, and swamps. Its range extends from the eastern and southeastern United States to Central America. The combination of long callus hairs and straight awns distinguishes it from all other species of *Saccharum* in the *Manual* region.

**2. Saccharum alopecuroides** (L.) Nutt. SILVER PLUMEGRASS [p. *494*, *555*]

*Saccharum alopecuroides* grows in damp woods, open areas, and field margins. It is restricted to the southeastern United States. It is rare or non-existent on the sandy coastal plain, and there are few specimens from southern Florida and the higher elevations of the Appalachian Mountains. The combination of long rhizomes and long silvery callus hairs distinguishes *S. alopecuroides* from all other species in the region.

**3. Saccharum brevibarbe** (Michx.) Pers. SHORTBEARD PLUMEGRASS [p. *494*, *555*]

*Saccharum brevibarbe* grows only in the southeastern United States.

1. Awns 15–22 mm long, straight or sinuous at the base; upper lemmas of the sessile spikelets entire at maturity . . . . . . . . . . . . . . . . . . . . . . . . . . . . . . var. *brevibarbe*

[1]Robert D. Webster

1. Awns 10–18 mm long, spirally coiled at the base, usually with 2–4 coils; upper lemmas of the sessile spikelets bifid at maturity, teeth about 2–2.5 mm long . . . . . . . . . . . . . . . . . . . . . . . . . . . . . . . . . var. *contortum*

### Saccharum brevibarbe (Michx.) Pers. var. **brevibarbe** [p. 494]

*Saccharum brevibarbe* var. *brevibarbe* grows in the southeastern coastal states and is common in central and southern Arkansas, eastern Oklahoma, the piney woods region of eastern Texas, and northern Louisiana.

### Saccharum brevibarbe var. **contortum** (Baldwin) R.D. Webster [p. 494]

*Saccharum brevibarbe* var. *contortum* grows in moist, sandy pinelands and open ground of the coastal plain, from Maryland to Florida and inland to Tennessee and Oklahoma. Initially, the awns in var. *contortum* are not coiled and the lemmas are entire but, as the spirals develop, they tear the lemmas, creating the bifid apices.

### 4. Saccharum coarctatum (Fernald) R.D. Webster COMPRESSED PLUMEGRASS [p. 494, 555]

*Saccharum coarctatum* is common in wet, peaty or sandy soils of swales, pond margins, and meadows of the coastal plain of the southeastern United States. It is unusual in having lodicule veins that extend into hairlike projections up to 0.6 mm long.

### 5. Saccharum baldwinii Spreng. NARROW PLUMEGRASS [p. 494, 555]

*Saccharum baldwinii* commonly grows in sandy, shaded river and stream bottoms. It occurs throughout the southeastern United States,

but it is not as common as other members of the genus, and is rare or completely absent from higher elevations of the Appalachian Mountains.

### 6. Saccharum ravennae (L.) L. RAVENNAGRASS [p. 494]

*Saccharum ravennae* is native to southern Europe and western Asia. It is grown as an ornamental in the *Manual* region, occasionally escaping and persisting.

### 7. Saccharum spontaneum L. WILD SUGARCANE [p. 494]

*Saccharum spontaneum* is a weedy species, native to tropical Africa and Asia, that is now established in Mesoamerica but not, so far as is known, in the *Manual* region. It is listed as a noxious weed by the U.S. Department of Agriculture, but it is grown in breeding programs as a source of potentially useful genes for *S. officinarum* (sugar cane), with which it readily hybridizes. Because of the potential economic damage of uncontrolled hybridization between *S. spontaneum* and *S. officinarum*, the U.S. Department of Agriculture should be notified of plants found growing outside a controlled planting.

### 8. Saccharum officinarum L. SUGARCANE [p. 494, 555]

*Saccharum officinarum* is native to tropical Asia and the Pacific islands. It is cultivated for sugar production in various parts of the world, including Texas, Louisiana, and Florida. It is also becoming popular as an ornamental plant for gardens in warmer parts of the contiguous United States, and appears to be established in some parts of the southeastern United States. It hybridizes with *S. spontaneum* (see discussion above).

### 9. Saccharum bengalense Retz. TALL CANE [p. 494]

*Saccharum bengalense* is native from Iran to northern India. It is sometimes cultivated as an ornamental in the *Manual* region.

## 25.03 MISCANTHUS Andersson[1]

**Pl** per; csp, smt rhz. **Clm** 40–400 cm, erect. **Lvs** not aromatic; **shth** open; **lig** memb, truncate, ciliate; **bld** flat. **Infl** tml, ovoid or corymbose pan, with elongate rchs and numerous ascending, spikelike br; **br** usu more than 10 cm long, with unequally pedlt spklt pairs, spklt homogamous and homomorphic; **dis** below the glm. **Cal** short, blunt, pilose, with fine hairs, hairs often exceeding the spklt. **Glm** memb to coriaceous; **lo glm** broadly convex to weakly 2-keeled, without raised veins; **lo flt** strl; **up flt** bisx; **up lm** entire and unawned

or bidentate and awned from the sinuses; **anth** 2 or 3. **Ped** free.

   *Miscanthus* is a genus of approximately 25 species. Most of the species are native to southeast Asia; a few extend into Africa. Some species hybridize with *Saccharum*, from which *Miscanthus* differs in its non-disarticulating branches and unequally pedicellate, rather than sessile-pedicellate, spikelets. The five species found in the *Manual* region are all grown as ornamentals because of their large, plumose panicles and striking growth habit.

1. Callus hairs 2–4 times as long as the spikelets.
   2. Spikelets 4–6 mm long; upper lemmas unawned or the awns not exceeding the glumes . . . . . . . . . . 4. *M. sacchariflorus*
   2. Spikelets 2–2.8 mm long; upper lemmas awned, the awns 9–13 mm long, exceeding the glumes . . . . . . . 3. *M. nepalensis*
1. Callus hairs from shorter than to twice as long as the spikelets.
   3. Culms few together or solitary; basal leaves with reduced blades, only the cauline leaves with long blades; panicles loose, with 2–5 branches . . . . . . . . . . . . . . . . . . . . . . . . . . . . . . . . . . . . . . 5. *M. oligostachyus*
   3. Culms densely tufted, forming large clumps; many basal leaves with long blades; panicles usually with more than 15 branches.
      4. Spikelets 3.5–7 mm long; blades 6–20 mm wide; rachises ⅓–⅔ as long as the panicles . . . . . . . . . . . . . . 2. *M. sinensis*
      4. Spikelets 3–3.5 mm long; blades 15–40 mm wide; rachises ¾–⅘ as long as the panicles . . . . . . . . . . 1. *M. floridulus*

### 1. Miscanthus floridulus (Labill.) Warb. *ex* K. Schum. & Lauterb. GIANT CHINESE SILVERGRASS [p. 494]

*Miscanthus floridulus* is the most widespread species of *Miscanthus* in southeast Asia. In North America it is grown as an ornamental.

### 2. Miscanthus sinensis Andersson EULALIA [p. 494, 555]

*Miscanthus sinensis* is native to southeastern Asia. It is frequently cultivated in the United States and southern Canada, and is now established in some parts of the United States.

[1]Mary E. Barkworth

**3. Miscanthus nepalensis** (Trin.) Hack. HIMALAYA FAIRYGRASS [p. *494*]

*Miscanthus nepalensis* is native from Pakistan through the Himalayas to Myanmar. It is cultivated occasionally in the *Manual* region.

**4. Miscanthus sacchariflorus** (Maxim.) Benth. AMUR SILVERGRASS, MISCANTHUS [p. *494, 555*]

*Miscanthus sacchariflorus* is native to the margins of rivers or marshes in temperate to north-temperate regions of eastern Asia. It has escaped from cultivation in various parts of the *Manual* region.

**5. Miscanthus oligostachyus** Stapf SMALL JAPANESE SILVERGRASS [p. *494*]

*Miscanthus oligostachyus* is a native of Japanese and Korean forests that is sold as an ornamental species in the United States.

## 25.04 IMPERATA Cirillo[1]

Pl per; strongly rhz. Clm 10–150(217) cm, mostly erect and unbrchd, usu with 3–4 nd. Lvs not aromatic; shth open, ciliate at the mrg of the col; lig memb; bld of the bas lvs linear to lanceolate, smt ciliate bas, those of the cauline lvs rdcd. Infl tml, cylindrical to conical pan with an evident rchs; rchs often with numerous long hairs; infl br 1–7 cm, usu shorter than the rchs, with spklt in unequally pedlt pairs; dis below the glm. Spklt homogamous and homomorphic, unawned; cal very short, hairy, hairs 7–16 mm. Glm equal to subequal, memb, 3–9-veined, with hairs longer than the flt over at least the lo ½; lo flt rdcd to hyaline or memb lm; up flt bisx, lm, if present, hyaline, unawned; anth 1–2, yellow to brown; stigmas elongate, purple to brown; sty connate or free. Ped not fused to the br axes, terminating in cuplike tips. Car ovate to obovate, light to dark brown.

*Imperata* has nine species and is widely distributed in warm regions of both hemispheres. One species is native to the *Manual* region; two have been introduced.

1. Stamens 2, filaments not dilated at the base . . . . . . . . . . . . . . . . . . . . . . . . . . . . . . . . . . . . . . . . . . . . . 3. *I. cylindrica*
1. Stamens 1, filaments dilated at the base.
  2. Panicles 7.5–14(17) cm long; lower branches 1–3.5 cm long, appressed; upper florets usually without lemmas; southeastern United States . . . . . . . . . . . . . . . . . . . . . . . . . . . . . . . . . . . . . . . . . . 1. *I. brasiliensis*
  2. Panicles 16–34 cm long; lower branches 2–5 cm long, divergent; both florets with lemmas; southwestern United States . . . . . . . . . . . . . . . . . . . . . . . . . . . . . . . . . . . . . . . . . . . . . . . . . . 2. *I. brevifolia*

**1. Imperata brasiliensis** Trin. BRAZILIAN BLADYGRASS, BRAZILIAN SATINTAIL [p. *495, 555*]

The range of *Imperata brasiliensis* includes South America and Central America, Mexico, and Cuba. It is now thought to be established in the southeastern United States. *Imperata brasiliensis* and *I. cylindrica* differ in the number of their stamens and the frequent absence of the lower lemma in *I. brasiliensis*. *Imperata brasiliensis* is listed as a noxious weed by the U.S. Department of Agriculture.

**2. Imperata brevifolia** Vasey SATINTAIL [p. *495, 555*]

Once known from wet or moist sites in the southwestern deserts from southern California, Nevada, and Utah to western Texas, *Imperata*

*brevifolia* is currently known only from populations in Grand Canyon National Park.

**3. Imperata cylindrica** (L.) Raeusch. COGONGRASS, BLADYGRASS [p. *495, 555*]

*Imperata cylindrica* is the most variable species in the genus, and is one of the world's 10 worst weeds. It is listed as a noxious weed by the U.S. Department of Agriculture. It was introduced to Alabama by 1912, and has spread considerably through the southeastern United States since then.

## 25.05 POLYTRIAS Hack.[2]

Pl per; stln. Clm 10–40 cm, often decumbent and rooting at the lo nd. Lvs not aromatic; lig memb, ciliate or fimbriate. Infl tml, solitary rames, spklt in homomorphic sessile-pedlt triplets of 2 sessile spklt and 1 pedlt spklt; intnd without a median translucent line; dis in the rames below the sessile spklt, smt also beneath the pedlt spklt. Spklt dorsally compressed, with 1 flt; sessile spklt bisx; pedlt spklt bisx, unisx, or strl. Glm equal, oblong, truncate, memb; lo glm with the mrg incurved over the up glm; up glm keeled; flt bisx; lm hyaline, bifid almost to the base, awned from the cleft; awns twisted, geniculate; anth 3. Ped not fused to the rame axes.

*Polytrias* is a unispecific genus of the Asian tropics that has become naturalized in Africa and the Western Hemisphere. It is unusual within the *Andropogoneae* in having only one floret, rather than two, in its spikelets.

**1. Polytrias amaura** (Büse) Kuntze JAVA GRASS [p. *495, 555*]

*Polytrias amaura* is native to southeastern Asia. It is used as a lawn grass in tropical and subtropical regions, including Florida. It gives a purplish cast to a lawn.

[1]Mark L. Gabel  [2]Mary E. Barkworth

## 25.06 MICROSTEGIUM Nees[1]

Pl ann or per; straggling. **Clm** to 100 cm, often decumbent. **Lvs** not aromatic; **lig** memb; **bld** narrowly-elliptic to lanceolate, often psdpet. **Infl** tml, subdigitate to rcm clusters of 1–few rames; **rame intnd** slender, without a translucent longitudinal groove; **dis** in the rames beneath the sessile spklt, and below the pedlt spklt. **Spklt** in homogamous, homomorphic, sessile-pedlt pairs, with 1 or 2 flt. **Lo glm** hrb to cartilaginous,

longitudinally grooved, mrg inflexed, 4–6-veined, usu keeled; **up glm** 3-veined, mucronate or shortly awned; **lo flt** absent, or rdcd and strl; **up flt** bisx; **up lm** usu awned; **anth** (2)3.

*Microstegium* is a genus of approximately 15 species, most of which are native to southeastern Asia; one is established in the *Manual* region.

### 1. Microstegium vimineum (Trin.) A. Camus NEPALESE BROWNTOP [p. *495*, *555*]

*Microstegium vimineum* was introduced to Tennessee from Asia around 1919 and is now established in much of the eastern United States. Although often associated with forested and wetland areas, it also does well in many disturbed areas. In suitable habitats it quickly spreads by rooting from its prostrate culms, forming dense, unispecific stands. It differs from *Leersia virginica* in its glabrous cauline nodes and the presence of hairs at the summit of the leaf sheaths. In addition, *M. vimineum* flowers in late September and October and is clearly a member of the *Andropogoneae*, whereas *L. viriginica* flowers in June through July and is a member of the *Oryzeae*.

## 25.07 TRACHYPOGON Nees[2]

Pl ann or per; csp or shortly rhz. **Clm** 30–200 cm, unbrchd; **intnd** semi-solid. **Lvs** cauline, not aromatic; **shth** shorter than the intnd, rounded; **lig** memb; **bld** flat to involute. **Infl** tml, solitary rcm of heterogamous subsessile-pedlt spklt pairs (rarely of 2 digitate spikelike br), axes slender, without a translucent median groove; **dis** beneath the pedlt spklt. **Subsessile spklt** stmt or strl, without a cal and unawned, otherwise similar to the pedlt spklt. **Ped** slender, not fused to the rame axes. **Pedlt spklt** bisx; **cal** sharp, strigose; **glm** firm, enclosing

the flt; **lo glm** several-veined, encircling the up glm; **up glm** 3-veined; **lo flt** strl; **up flt** bisx, lm firm but hyaline at the base, tapering to an awn; **awns** (4)6–15 cm, twisted, pubescent to plumose; **pal** absent; **anth** 3.

*Trachypogon* is a tropical or warm-temperate genus that is native to Africa and tropical to subtropical America. Estimates of the number of species included range from one to ten. One species, *T. secundus*, is native to the *Manual* region.

### 1. Trachypogon secundus (J. Presl) Scribn. CRINKLE-AWN [p. *495*, *555*]

*Trachypogon secundus* is found in sandy prairies, woodlands, rocky hills, and canyons, in well-drained soils at 500–2000 m. Statements about its range are difficult to make because of disagreement as to whether northern plants, such as those found in the *Manual* region, belong to the same species as those found elsewhere. *Trachypogon secundus* resembles *Heteropogon*, but differs in the longer, non-disarticulating inflorescence and shorter, pale awns.

## 25.08 SORGHUM Moench[3]

Pl ann or per. **Clm** 50–500+ cm; **intnd** solid. **Lvs** not aromatic, bas and cauline; **aur** absent; **lig** memb and ciliate or of hairs; **bld** usu flat. **Infl** tml, pan with evident rchs; **pri br** whorled, compound, the ult units rames; **rames** with most spklt in heterogamous sessile-pedlt spklt pairs, tml spklt unit on each rame usu a triplet of 1 sessile and 2 pedlt spklt, rame axes without a translucent median line; **dis** in the rames below the sessile spklt, smt also below the pedlt spklt (cultivated taxa not or only tardily disarticulating). **Sessile spklt** dorsally compressed, **cal** blunt or pointed; **lo glm** dorsally compressed and

rounded bas, 2-keeled or winged distally, 5–15-veined, usu unawned; **up glm** 2-keeled, smt awned; **lo flt** rdcd to hyaline lm; **up flt** pist or bisx, lm hyaline, smt awned. **Ped** slender, neither appressed nor fused to the rame axes. **Pedlt spklt** stmt or strl, well-developed, often subequal to the sessile spklt in size.

Most of the approximately 25 species of *Sorghum* are native to tropical and subtropical regions of the Eastern Hemisphere, but one is native to Mexico. Two have been introduced into the *Manual* region.

1. Plants perennial, rhizomatous; spikelets disarticulating at maturity; caryopses not exposed at maturity . . . . . 1. *S. halepense*
1. Plants usually annual, sometimes short-lived perennials; spikelets either not disarticulating or doing so tardily; caryopses often exposed at maturity . . . . . . . . . . . . . . . . . . . . . . . . . . . . . . . . . . . . . . . . . . . . . . . . . . . . 2. *S. bicolor*

[1]John W. Thieret †   [2]Kelly W. Allred   [3]Mary E. Barkworth

## 1. Sorghum halepense (L.) Pers. JOHNSON GRASS [p. 496, <u>555</u>]

*Sorghum halepense* is native to the Mediterranean region. It is sometimes grown for forage in North America, but it is considered a serious weed in warmer parts of the United States. It hybridizes readily with *S. bicolor*, and derivatives of such hybrids are widespread. The annual **Sorghum ×almum** Parodi, which has wider (2–2.8 mm) sessile spikelets with more veins in the lower glumes (13–15 versus 10–13) than *S. halepense*, is one such derivative.

## 2. Sorghum bicolor (L.) Moench SORGHUM [p. 496, <u>555</u>]

*Sorghum bicolor* was domesticated in Africa 3000 years ago, and is now an important crop in the United States and Mexico. Numerous cultivated strains exist, some of which have been formally named. They are all interfertile with each other and with other wild species of *Sorghum*. *Sorghum bicolor* subsp. *arundinaceum* is the wild progenitor of the cultivated strains, all of which are treated as *S. bicolor* subsp. *bicolor*. These strains tend to lose their distinguishing characteristics if left to themselves. They will also hybridize with subsp. *arundinaceum*, and these hybrids can backcross to either parent, resulting in plants that may strongly resemble one parent while having some characteristics of the other. All such hybrids and backcrosses are treated here as *S. bicolor* subsp. ×*drummondii*.

1. Inflorescence branches remaining intact at maturity; caryopses exposed at maturity; sessile spikelets 3–9 mm long, elliptic to oblong . . . . . . . . . . . . . . . . . . subsp. *bicolor*

1. Inflorescence branches rames, disarticulating at maturity, sometimes tardily; caryopses not exposed at maturity; sessile spikelets 5–8 mm long, lanceolate to elliptic.
    2. Rames readily disarticulating . . . . . . . . subsp. *arundinaceum*
    2. Rames disarticulating tardily . . . . . . . . subsp. ×*drummondii*

### Sorghum bicolor subsp. arundinaceum (Desv.) de Wet & J.R. Harlan *ex* Davidse [p. 496]

*Sorghum bicolor* subsp. *arundinaceum* is native to, and most common, in Africa, but some strains have been introduced into the Western Hemisphere.

### Sorghum bicolor (L.) Moench subsp. bicolor SORGHUM, BROOMCORN, SORGO [p. 496]

All the cultivated sorghums are placed in *Sorghum bicolor* subsp. *bicolor*. 'Grain sorghums' have short panicles and panicle branches, 'broomcorns' have elongate panicles and panicle branches, and 'sweet sorghums' or 'sorgo' produce an abundance of sweet juice in their stems.

### Sorghum bicolor subsp. ×drummondii (Steud.) de Wet *ex* Davidse CHICKEN CORN, SUDANGRASS [p. 496]

The hybrids treated here as *Sorgum bicolor* subsp. ×*drummondii* are most common in the Eastern Hemisphere, but a few are cultivated in the United States. Among these are the plants known as 'chicken corn' and 'Sudangrass' [= *S. sudanense* (Piper) Stapf].

## 25.09 SORGHASTRUM Nash[1]

**Pl** ann or per; csp, smt rhz. **Clm** 50–300+ cm, erect, nodding or clambering, unbrchd; **nd** densely pubescent, particularly in young pl. **Lvs** not aromatic; **lig** memb, glab or pubescent; **bld** flat, involute, or folded. **Infl** tml, secund or equilat pan with evident rchs and numerous br, not subtended by modified lvs; **br** capillary, rebrchg, with many rames, not subtended by modified lvs; **dis** in the rames, beneath the sessile spklt. **Spklt** sessile, subtending a hairy ped (2 ped in the tml spklt units), dorsally compressed. **Cal** blunt or sharp; **glm** coriaceous; **lo glm** pubescent, 5–9-veined, acute; **up glm** slightly longer, usu glab, 5-veined, truncate; **lo flt** rdcd to hyaline lm; **up flt** bisx, lm hyaline, bifid, awned from the sinuses; **awns** usu once- or twice-geniculate, often spirally twisted, shortly strigose, brownish; **anth** 3; **ov** glab. **Car** flattened. **Ped** 3–6.5 mm, slender, not fused to the rame axes; **pedlt spklt** absent.

*Sorghastrum* includes about 18 species. Most are native to tropical or subtropical America, two are African, and four are native to the *Manual* region. Absence of the pedicellate spikelet, while confusing at first, makes *Sorghastrum* a readily recognizable genus.

1. Awns 10–22(30) mm long, once-geniculate; plants rhizomatous . . . . . . . . . . . . . . . . . . . . . . . . . . . . . . . . . . 3. *S. nutans*
1. Awns 21–40 mm long, twice-geniculate; plants not rhizomatous.
    2. Pedicels sharply curved to recurved; panicles secund; sessile spikelets 0.8–1.2 mm wide . . . . . . . . . . . . . . 2. *S. secundum*
    2. Pedicels flexuous; panicles not secund; sessile spikelets 1.1–1.8 mm wide . . . . . . . . . . . . . . . . . . . . . . . . . 1. *S. elliottii*

### 1. Sorghastrum elliottii (C. Mohr) Nash SLENDER INDIANGRASS [p. 496, <u>555</u>]

*Sorghastrum elliottii* usually grows in dry, open woods on sandy terraces of the lowlands in the southeastern United States, often over a clay subsoil.

### 2. Sorghastrum secundum (Elliott) Nash LOPSIDED INDIANGRASS [p. 496, <u>555</u>]

*Sorghastrum secundum* grows in woodlands, sandy soils, and occasionally at the edges of marshes, at elevations below 1000 m. Its native range extends north and west from Florida to the Appalachian Mountains; other records probably reflect introductions. The

mountains may have effectively prevented its further spread to the northwest. *Sorghastrum secundum* is easily confused with plants of *S. elliottii* that are not at anthesis, because both species have straight to slightly arching panicles with ascending branches, but the rachis nodes of *S. secundum* are glabrous or almost glabrous.

### 3. Sorghastrum nutans (L.) Nash INDIANGRASS, FAUX-SORGHO PENCHÉ [p. 496, <u>555</u>]

*Sorghastrum nutans* grows in a wide range of habitats, from prairies to woodlands, savannahs, and scrubland vegetation. It is native from Canada to Mexico. It is frequently used for forage, for erosion control on slopes and along highways, and in restoration work. There are several cultivars on the market.

## 25.10 CHRYSOPOGON Trin.[2]

**Pl** ann or per; if per, smt csp, smt rhz or stln. **Clm** 15–300 cm, erect, smt decumbent. **Lvs** not aromatic; mostly bas; **aur** absent; **lig** shortly memb and ciliolate to ciliate or of hairs; **bld** often rough and glaucous. **Infl** tml

[1]Patricia D. Dávila Aranda and Stephan L. Hatch  [2]David W. Hall and John W. Thieret†

pan with elongate rchs and numerous br, br often naked for a considerable distance before terminating in a rame; **rames** often with only a single heterogamous triplet of 1 sessile and 2 pedlt spklt, smt with 1(–3) heterogamous sessile-pedlt spklt pairs below the tml triplet, intnd without a translucent median groove; **dis** oblique, below the sessile spklt. **Sessile spklt** terete or lat compressed; **cal** usu sharp, setose, hairs white or yellow to brown; **glm** leathery to stiff, involute or folded and keeled above; **lo glm** rounded or lat compressed; **lo flt**

strl; **up flt** bisx, unawned or awned. **Ped** slender, not fused to the rame axes, without a translucent groove. **Pedlt spklt** dorsally compressed or absent, if present, lo flt strl and unawned, up flt strl or stmt, awned or unawned.

*Chrysopogon* is a tropical and subtropical genus of 26 species. All but one species are native to the Eastern Hemisphere tropics, the majority to India. *Chrysopogon pauciflorus* is native to Florida and Cuba; four species have been introduced in the *Manual* region.

1. Upper lemmas of sessile spikelets awned, the awns 2–16 cm long.
    2. Plants annual; pedicellate spikelets 7.2–15 mm long . . . . . . . . . . . . . . . . . . . . . . . . . . . . . . . . . . . . . . . 1. *C. pauciflorus*
    2. Plants perennial; pedicellate spikelets 2.5–8 mm long . . . . . . . . . . . . . . . . . . . . . . . . . . . . . . . . . . . . . 2. *C. fulvus*
1. Upper lemmas of sessile spikelets unawned or the awns no more than 8 mm long.
    3. Calluses of the sessile spikelets 0.6–0.8 mm long, rounded, with white hairs; plants not stoloniferous . . . . 4. *C. zizanioides*
    3. Calluses of the sessile spikelets 3–6.4 mm long, sharp, with golden-yellow hairs; plants extensively
       stoloniferous . . . . . . . . . . . . . . . . . . . . . . . . . . . . . . . . . . . . . . . . . . . . . . . . . . . . . . . . . . 3. *C. aciculatus*

### 1. Chrysopogon pauciflorus (Chapm.) Benth. *ex* Vasey FLORIDA RHAPHIS [p. 496, 556]

*Chrysopogon pauciflorus* is native, but infrequently encountered, in the southeastern United States, primarily in Florida; it also occurs in Cuba. It grows in flatwoods, abandoned fields, pinelands, marsh edges, and various disturbed sites.

### 2. Chrysopogon fulvus (Spreng.) Chiov. GOLDEN BEARDGRASS [p. 496, 556]

*Chrysopogon fulvus* is native from southern India to Thailand, where it is considered a good forage grass. It was grown at the experiment station in Gainesville, Florida, and subsequently found in adjacent flatwoods as an escape.

### 3. Chrysopogon aciculatus (Retz.) Trin. MACKIE'S PEST, LOVEGRASS [p. 496]

*Chrysopogon aciculatus* is native to tropical Asia, Australia, and Polynesia. In the contiguous United States, it is known only from

controlled plantings at the experiment station in Gainesville, Florida. It is a vigorous colonizer of bare ground that can withstand heavy grazing and trampling, and is difficult to eradicate once established. The sharp calluses are injurious to grazing animals. The U.S. Department of Agriculture considers *C. aciculatus* a noxious weed, and should be informed if the species is found growing in other than a controlled planting.

### 4. Chrysopogon zizanioides (L.) Roberty VETIVER, KHUS-KHUS, KHAS-KHAS [p. 496, 556]

*Chrysopogon zizanioides*, which used to be included in *Vetiveria*, is native to river banks and flood plains in the south Asian tropics and subtropics, but it has been deliberately established in the warmer areas of the United States. It grows in a variety of soils, from heavy clays to dune sand, and will tolerate windy coastal conditions. Hedges of *C. zizanioides* can control soil erosion or restore eroded land.

## 25.11 DICHANTHIUM Willemet[1]

**Pl** ann or per; csp, smt with extensive creeping stln. **Clm** 15–200 cm. **Lvs** usu not aromatic; **lig** memb, smt ciliate; **bld** 2-4 mm wide. **Infl** tml, smt also ax but the ax infl not numerous; **peduncles** with 1–many rames in digitate or subdigitate clusters; **rames** smt naked bas, axes terete to slightly flattened, without a translucent, longitudinal groove, bearing 1–many sessile-pedlt spklt pairs and a tml triplet of 1 sessile and 2 pedlt spklt, bas pair(s) homomorphic and homogamous, stmt or strl, unawned, persistent, distal spklt pairs homomorphic but heterogamous, sessile spklt bisx and awned, pedlt spklt stmt or strl and unawned; **dis** in the rames, beneath the

bisx sessile spklt. **Sessile spklt** often imbricate, dorsally compressed, with blunt cal; **lo glm** chartaceous to cartilaginous, broadly convex to slightly concave, smt pitted; **lo flt** rdcd, strl; **up flt** strl or stmt and unawned in the homogamous pairs, bisx and awned in the heterogamous pairs; **awns** 1–3.5 cm, usu glab; **anth** (2)3. **Ped** free of the rame axes, terete to somewhat flattened, slender, not grooved. **Pedlt spklt** strl or stmt.

*Dichanthium*, a genus of 20 species, grows in habitats ranging from subdeserts to marshlands in tropical Asia and Australia. It is frequently found in disturbed areas. Three species have been introduced to the *Manual* region.

1. Lower glume of the sessile spikelets with a subapical arch of hairs; pedicellate spikelets usually sterile . . . . . . . . . 1. *D. sericeum*
1. Lower glume of the sessile spikelets without a subapical arch of hairs; pedicellate spikelets usually
   staminate.
    2. Rame bases pilose; lower glume of the sessile spikelets more or less obovate . . . . . . . . . . . . . . . . . . . . . . 2. *D. aristatum*
    2. Rame bases glabrous; lower glume of the sessile spikelets elliptic to oblong . . . . . . . . . . . . . . . . . . . . . 3. *D. annulatum*

[1]Mary E. Barkworth

## 1. Dichanthium sericeum (R. Br.) A. Camus
QUEENSLAND BLUEGRASS [p. 497, *556*]

*Dichanthium sericeum* is an Australian species. There are two subspecies: D. sericeum (R. Br.) A. Camus subsp. sericeum is a perennial with sessile spikelets 4–4.5 mm long and to 1–1.4 mm wide, 9–10-veined lower glumes, and rames more than 4 cm long; D. sericeum subsp. humilius (J.M. Black) B.K. Simon is an annual, with sessile spikelets up to 4 mm long and about 1 mm wide, 5–7-veined lower glumes, and rames less than 4 cm long. *Dichanthium sericeum* subsp. *sericeum* is established in Texas and Florida.

## 2. Dichanthium aristatum (Poir.) C.E. Hubb. AWNED
DICHANTHIUM [p. 497, *556*]

*Dichanthium aristatum* was introduced to the Americas from southern Asia. It is sometimes used as a lawn grass in Texas, Louisiana, and Florida.

## 3. Dichanthium annulatum (Forssk.) Stapf RINGED
DICHANTHIUM [p. 497, *556*]

*Dichanthium annulatum* is native to southeastern Asia. It is now established at scattered locations in Arizona, Texas, Louisiana, and Florida.

## 25.12 BOTHRIOCHLOA Kuntze[1]

Pl per; csp or stln. Clm 30–250 cm, with pithy intnd. Lvs bas or cauline, not aromatic; shth open; aur absent; lig memb, smt also ciliate; bld usu flat, convolute in the bud. Infl tml, pan of subdigitate to racemosely arranged br, each br with (1)2–many rames, br not subtended by modified lvs; rames with spklt in heterogamous sessile-pedlt pairs, intnd with a translucent, longitudinal groove, often villous on the mrg; dis in the rames, beneath the sessile spklt. Spklt dorsally compressed; sessile spklt with 2 flt; lo glm rounded, several-veined, smt with a dorsal pit, mrg clasping the up glm; up glm somewhat keeled, 3-veined; lo flt hyaline scales, unawned; up flt bisx; up lm with a midvein that usu extends into a twisted, geniculate awn, occ unawned; anth 3. Ped similar to the intnd. Pedlt spklt rdcd or well-developed, strl or stmt, unawned. Car lanceolate to oblong, somewhat flattened; hila punctate, bas; emb about ½ as long as the car.

*Bothriochloa* is a genus of about 35 species that grow in tropical to warm-temperate regions. Nine are native to the *Manual* region; three Eastern Hemisphere species have been introduced into the southern United States for forage and range rehabilitation.

1. Pedicellate spikelets about as long as the sessile spikelets.
2. Sessile spikelets 5.5–7 mm long . . . . . . . . . . . . . . . . . . . . . . . . . . . . . . . . . . . . . . . . . . . .1. *B. wrightii*
2. Sessile spikelets 3–4.5 mm long.
3. Rachises longer than the branches . . . . . . . . . . . . . . . . . . . . . . . . . . . . . . . .10. *B. bladhii* (in part)
3. Rachises shorter than the branches.
4. Lower glumes of the sessile spikelets with a dorsal pit . . . . . . . . . . . . . . . . . . . . . . . . .12. *B. pertusa*
4. Lower glumes of the sessile spikelets without a dorsal pit . . . . . . . . . . . . . . . . . . . .11. *B. ischaemum*
1. Pedicellate spikelets much shorter than the sessile spikelets.
5. Sessile spikelets 2.5–4.5 mm long; awns absent or less than 17 mm long.
6. Sessile spikelets unawned or with awns less than 6 mm long . . . . . . . . . . . . . . . . . . . .4. *B. exaristata*
6. Sessile spikelets with awns 8–17 mm long.
7. Panicles reddish when mature; hairs below the sessile spikelets about ¼ as long as the spikelets, sparse, not obscuring the spikelets . . . . . . . . . . . . . . . . . . . . . . . . .10. *B. bladhii* (in part)
7. Panicles silvery-white or light tan; hairs below the sessile spikelets at least ½ as long as the spikelets, copious, at least somewhat obscuring the spikelets.
8. Panicles 9–20 cm long; sessile spikelets narrowly ovate to lanceolate; glumes acute; leaves evenly distributed on the culms; culms 2–4 mm thick . . . . . . . . . . . . . . . . . . . .3. *B. longipaniculata*
8. Panicles 4–12(14) cm long; sessile spikelets ovate; glumes blunt; leaves often clustered at the base of the culms; culms usually less than 2 mm thick . . . . . . . . . . . . . . . . . .2. *B. laguroides*
5. Sessile spikelets 4.5–8.5 mm long; awns 18–35 mm long.
9. Rachises 5–20 cm long, with numerous branches.
10. Panicles of the larger shoots 14–25 cm long; culms 130–250 cm tall, 2–4 mm thick, stiffly erect, little-branched distally, glaucous below the nodes; nodes with spreading hairs, the hairs 2–6 mm long . . . . . . . . . . . . . . . . . . . . . . . . . . . . . . . . . . . . . . . . . . . . . . . . . . . .5. *B. alta*
10. Panicles of the larger shoots 5–14(20) cm long; culms usually 60–120 cm tall, usually less than 2 mm thick, tending to be bent at the base and often branched at maturity, not glaucous below the nodes; nodes with ascending hairs less than 3 mm long . . . . . . . . . . . . . . . . . . . .6. *B. barbinodis*
9. Rachises usually less than 5 cm long, with 2–9 branches.
11. Cauline nodes densely pubescent, the hairs 3–7 mm long, white, spreading . . . . . . . . . . . . . .7. *B. springfieldii*
11. Cauline nodes glabrous or puberulent, the hairs always less than 3 mm long, usually off-white and ascending.

[1]Kelly W. Allred

12. Lower branches of the inflorescences rebranched; sessile spikelets 4.5–6.5 mm long; lower glumes sparsely hairy near the base; leaves primarily cauline, the blades 2–5 mm wide . . . . . . . . . . . . . . . . . . . . . . . . . . . . . . . . . . . . . . . . . . . . . . . . . . . . . . . . . . . . . . . 8. *B. hybrida*
12. Lower branches of the inflorescences simple, not rebranched; sessile spikelets 5–8 mm long; lower glumes glabrous; leaves primarily basal, the blades usually less than 2 mm wide . . . . . . . . . . . . . . . . . . . . . . . . . . . . . . . . . . . . . . . . . . . . . . . . . . . . . . . 9. *B. edwardsiana*

## 1. Bothriochloa wrightii (Hack.) Henrard WRIGHT'S BLUESTEM [p. 497, 556]

*Bothriochloa wrightii* grows in rocky grasslands and shrubby slopes of the pine-oak woodlands of southern Arizona, New Mexico, Texas, and northern Mexico, at 1200–1800 m. It was last collected in the United States in 1930. It differs from *B. barbinodis* in its glaucous foliage, short, fan-shaped panicles, and large, pedicellate spikelets.

## 2. Bothriochloa laguroides (DC.) Herter SILVER BLUESTEM [p. 497, 556]

*Bothriochloa laguroides* grows in well-drained soils of grasslands, prairies, roadsides, river bottoms, and woodlands, often on limestone, usually at 20–2100 m. Plants from the United States and northern Mexico belong to **B. laguroides** subsp. **torreyana** (Steud.) Allred & Gould, which differs from **B. laguroides** (DC.) Herter subsp. **laguroides** in its glabrous, or almost glabrous, nodes, long internode hairs, and pilose throat region. *Bothriochloa laguroides* differs from the more southern *B. saccharoides* (Sw.) Rydb. in having pilose leaves, a narrow central groove in the internodes and pedicels, and panicle branches with axillary pulvini.

## 3. Bothriochloa longipaniculata (Gould) Allred & Gould LONGSPIKE SILVER BLUESTEM [p. 497, 556]

*Bothriochloa longipaniculata* grows at 2–200 m, along roadsides and in fields, open woodlands, disturbed ground, and swales of the Gulf coastal prairie, often in heavy clay soil. Its range extends from southern Texas and Louisiana to northeastern Mexico and possibly Panama.

## 4. Bothriochloa exaristata (Nash) Henrard AWNLESS BLUESTEM [p. 497, 556]

*Bothriochloa exaristata* grows in heavy soils of fields and roadsides of the Gulf coastal prairie, at 2–150 m, as well as in coastal areas of southern Brazil and adjacent Argentina, and inland along the Rio Pilcomayo to Paraguay. It has been reported from Los Angeles County, California. When growing in dense grassland thickets, *B. exaristata* has rather spindly basal growth, but branches abundantly from the middle and upper nodes.

## 5. Bothriochloa alta (Hitchc.) Henrard TALL BLUESTEM [p. 497, 556]

*Bothriochloa alta* grows along roads, drainage ways, and gravelly slopes in the desert grasslands of the southwestern United States, at 600–1200 m, and extends south to Bolivia and Argentina. It is not a common species in the *Manual* region. It often grows with and is mistaken for *B. barbinodis*, but differs from that species in having longer culms, panicles, and nodal hairs.

## 6. Bothriochloa barbinodis (Lag.) Herter CANE BLUESTEM [p. 497, 556]

*Bothriochloa barbinodis* is a common species, growing at 500–1200 m along roadsides, drainage ways, and on gravelly slopes in desert grasslands, from the southwestern United States through Mexico and Central America to Bolivia and Argentina. It has also been found in the Hawaiian Islands. Plants with a pit on the back of their lower glumes occur sporadically; they do not differ in any other respect from those without pits. The species is sometimes used as an ornamental. It differs from *B. wrightii* in not being glaucous and in having oblong to merely somewhat fan-shaped panicles with pedicellate spikelets that are definitely shorter than the sessile spikelets; from *B. alta* in having

shorter culms, panicles, and nodal hairs; and from *B. springfieldii* in having taller culms, wider leaves, shorter nodal hairs, and more, less hairy panicle branches.

## 7. Bothriochloa springfieldii (Gould) Parodi SPRINGFIELD BLUESTEM [p. 497, 556]

*Bothriochloa springfieldii* grows in rocky uplands, ravines, plains, sandy areas, and roadsides, from southern Utah to western Texas and Mexico at 900–2500 m, and as a disjunct in northwest Louisiana. It differs from *B. barbinodis* in its less robust habit, narrower blades, longer nodal hairs, and fewer, more hairy panicle branches, and from *B. edwardsiana* in its pubescent nodes and wider, non-ciliate leaf blades.

## 8. Bothriochloa hybrida (Gould) Gould HYBRID BLUESTEM [p. 497, 556]

*Bothriochloa hybrida* grows in open grasslands, rangeland pastures, disturbed ground, and roadsides, often on calcareous soil, usually at 50–500 m. Its range extends from southern Texas and Louisiana to central Mexico. It resembles *B. edwardsiana* in some respects, but the latter species has a less robust habit, more predominantly basal foliage, and narrower leaf blades.

## 9. Bothriochloa edwardsiana (Gould) Parodi MERRILL'S BLUESTEM [p. 497, 556]

*Bothriochloa edwardsiana* grows in the rocky plains and prairies of the Edwards Plateau of Texas, on calcareous soil, at 300–600 m. It also grows in northern Mexico and Uruguay. It resembles *B. hybrida* in some respects, but that species has a more robust habit, predominantly cauline foliage, and wider leaf blades.

## 10. Bothriochloa bladhii (Retz.) S.T. Blake AUSTRALIAN BLUESTEM [p. 498, 556]

*Bothriochloa bladhii* grows along roadsides and in rangeland pastures, waste ground, and open disturbed areas, at 150–1800 m. It is native to subtropical Asia and Africa and was introduced to the *Manual* region as a forage grass. It is now established in the southern and central United States. A similar species, *B. decipiens* (Hack.) C.E. Hubb., has been grown at some experiment stations in the United States. It is not known to be established in North America. It differs from *B. bladhii* in having longer (4.7–5.3 mm) sessile spikelets and a single anther.

## 11. Bothriochloa ischaemum (L.) Keng [p. 498, 556]

*Bothriochloa ischaemum* grows along roadsides and in waste ground and rangeland pastures, at 50–1200 m. It is native to southern Europe and Asia. It was introduced to the United States for erosion control along right of ways and for livestock forage in the southwest. It is now established in the region and has spread along roadsides into other central and southern states.

## 12. Bothriochloa pertusa (L.) A. Camus PITTED BLUESTEM [p. 498, 556]

*Bothriochloa pertusa* is native to the Eastern Hemisphere, and was introduced to the southern United States as a warm-season pasture grass. It now grows in disturbed, moist, grassy places and pastures in the region, at elevations of 2–200 m. It has not persisted at all locations shown on the map.

## 25.13 ISCHAEMUM L.¹

Pl ann or per. Clm 10–350 cm, often decumbent, smt brchd above the base. **Lvs** not aromatic; **shth** open; **lig** memb, glab or ciliate, sides often higher than the mid. **Infl** tml, smt also ax; **infl units** with (1)2–many rames on a common peduncle; **rames** secund, ascending, members of a cluster smt so closely appressed as to appear as one; **intnd** stoutly linear to clavate. **Spklt** in homogamous or heterogamous sessile-pedlt or unequally pedlt pairs; **dis** in the rames, below both the sessile and pedlt spklt. **Sessile spklt** dorsally compressed; **glm** subequal; **lo glm** 2-keeled, keels smt winged; **up glm** keeled, smt awned; **lo flt** stmt; **up flt** bisx, lm usu bifid and awned from the sinus. **Ped** fused to the rame axes, clavate or inflated, smt as wide as the spklt. **Pedlt spklt** morphologically and sexually similar to the sessile spklt or stmt and rdcd.

*Ischaemum* includes approximately 65 species, all of which are native to tropical regions of the Eastern Hemisphere. None of the species is known to be established in North America. The genus is included in this treatment because two of its species are considered serious weed threats by the U.S. Department of Agriculture.

1. Lower glumes of the sessile spikelets not winged, rugose, with 4–5 ridges; plants annuals or short-lived perennials . . . . . . . . . . . . . . . . . . . . . . . . . . . . . . . . . . . . . . . . . . . . . . . . . . . . . . . . . . . . . . . 1. *I. rugosum*
1. Lower glumes of the sessile spikelets winged on the keels, not rugose; plants perennial . . . . . . . . . . . . . . . . . 2. *I. indicum*

### 1. Ischaemum rugosum Salisb. RIBBED MURAINAGRASS [p. 498]

*Ischaemum rugosum* is native to southern Asia, and is now established in moist, tropical habitats around the world, including Mexico. It has been found in southern Texas and on chrome ore piles in Canton, Maryland, but is thought to have been eliminated from both areas. The U.S. Department of Agriculture considers it a noxious weed; plants found growing within the continental United States should be promptly reported to that agency.

### 2. Ischaemum indicum (Houtt.) Merr. INDIAN MURAINAGRASS [p. 498, 556]

*Ischaemum indicum* was reported on chrome ore piles in Canton, Maryland. This reported introduction has not been verified.

## 25.14 APLUDA L.²

Pl per; often scrambling. **Clm** to 3 m, decumbent. **Lvs** not aromatic; **shth** open; **lig** memb; **bld** linear, often psdpet. **Infl** false pan, individual infl units with solitary rames; **rames** to 1 cm, often enclosed by the subtending lf shth, with 1 sessile and 2 unequally pedlt spklt; **dis** at the base of the sessile spklt, smt also at the base of the pedlt spklt. **Sessile spklt** lat compressed, with a large, bulbous cal; **lo glm** coriaceous, without keels or wings, smooth, bidentate; **up glm** unawned; **up lm** awned or unawned. **Ped** flat, wide, adjacent to each other, appressed but not fused to the rame axes. **Pedlt spklt** usu unequal, unawned, 1 stmt or bisx and as large as the sessile spklt, the other strl and usu smaller.

*Apluda* is treated here as consisting of a single weedy species that is native to tropical Asia and Australia, where it grows primarily in thickets and forest margins. It is not known to be established in the *Manual* region.

### 1. Apluda mutica L. MAURITIAN GRASS [p. 498, 556]

*Apluda mutica* was reported on chrome ore piles in Canton, Maryland. The report has not been verified.

## 25.15 ANDROPOGON L.³

Pl per; usu csp, smt rhz. **Clm** 20–310 cm, erect, much-brchd distally. **Lvs** not aromatic; **lig** memb, smt ciliate; **bld** linear, flat, folded, or convolute. **Infl** tml and ax or a false pan; **infl units** 1–600+ per clm; **peduncles** initially concealed by the subtending lf shth, smt exserted beyond the shth at maturity, with (1)2–5(13) rames; **rames** not reflexed at maturity, axes slender, terete to flattened, not longitudinally grooved, usu conspicuously pubescent, with spklt in heterogamous sessile-pedlt pairs (the tml spklt smt in triplets of 1 sessile and 2 pedlt spklt), apc of the intnd neither cupulate nor fimbriate; **dis** in the rames, below the sessile spklt. **Sessile spklt** bisx, awned, with short, blunt cal; **lo glm** 2-keeled, flat or concave, usu not veined between the keels, smt 2–9-veined; **anth** 1, 3(2). **Ped** usu longer than 3 mm, similar to the rame intnd in shape, length, and pubescence color, not fused to the rame axes. **Pedlt spklt** usu vestigial or absent, smt well-developed and stmt.

*Andropogon* is a cosmopolitan genus of tropical and temperate zones, comprising approximately 120 species. Thirteen species are native to the *Manual* region. All but *A. hallii* grow in the southeastern United States.

Species of *Andropogon* with solitary rames are easily confused with *Schizachyrium* but, in *Andropogon*, the lower glumes of the sessile spikelets are flat or concave and the rame internodes are not cupulate, whereas *Schizachyrium* has convex glumes and rame internodes with strongly cupulate apices. Successful identification of species in *Andropogon* sect. *Leptopogon* (numbers 3–14) requires mature, complete specimens and careful field study.

¹,²Mary E. Barkworth ³Christopher S. Campbell

1. Pedicellate spikelets usually well-developed, (3.5)6–12 mm long, usually staminate; sessile spikelets 5–12 mm long (sect. *Andropogon*).
  2. Sessile spikelets with awns 8–25 mm long; ligules 0.4–2.5 mm long; hairs of the rame internodes 2.2–4.2 mm long, sparse to dense; rhizomes sometimes present, the internodes usually less than 2 cm . . . . . . . 1. *A. gerardii*
  2. Sessile spikelets unawned or with awns less than 11 mm long; ligules (0.9)2.5–4.5 mm long; hairs of the rame internodes 3.7–6.6 mm long, usually dense; rhizomes always present, the internodes often more than 2 cm long . . . . . . . . . . . . . . . . . . . . . . . . . . . . . . . . . . . . . . . . . . . . . . . . . . . . 2. *A. hallii*
1. Pedicellate spikelets usually vestigial or absent, those of the terminal spikelet units occasionally well-developed and staminate; sessile spikelets 2.6–8.4 mm long (sect. *Leptopogon*).
  3. Peduncles with solitary rames; plants of southern Florida . . . . . . . . . . . . . . . . . . . . . . . . . . . 3. *A. gracilis*
  3. Peduncles with (1)2–13 rames; plants of varied distribution, including southern Florida.
    4. Rames not or scarcely exserted at maturity; peduncles mostly less than 15 mm long at maturity.
      5. Culms 30–140 (usually about 80) cm tall; blades 0.8–5 (usually about 2.5) mm wide; inflorescence units 2–31 per culm . . . . . . . . . . . . . . . . . . . . . . . . . . . . . . . . . . . . . . . . 9. *A. gyrans*
      5. Culms 20–250 (usually more than 90) cm tall; blades 1.7–9.5 (usually more than 3) mm wide; inflorescence units 3–600 per culm.
        6. Blades pubescent, most hairs appressed; callus hairs 1.5–5 mm long . . . . . . . . . . 13. *A. longiberbis*
        6. Blades glabrous or with spreading (rarely appressed) hairs; callus hairs 1–3 mm long.
          7. Blades 11–52 cm long; sheaths smooth, rarely somewhat scabrous; ligules 0.2–1 mm long; keels of the lower glumes usually smooth below midlength, scabrous distally . . . . . . . . 12. *A. virginicus*
          7. Blades 13–109 cm long; sheaths usually scabrous; ligules 0.6–2.2 mm long; keels of the lower glumes sometimes scabrous below midlength . . . . . . . . . . . . . . . . . . . . . 14. *A. glomeratus* (in part)
    4. Rames sometimes exserted above their subtending sheaths at maturity; 1 or more peduncles more than 15 mm long at maturity.
      8. Anthers 3.
        9. Sessile spikelets 4.5–8.4 mm long; pedicellate spikelets 1.5–3.6 mm long, sterile; plants common and widespread in the southeastern United States . . . . . . . . . . . . . . . . . . . . . 4. *A. ternarius*
        9. Sessile spikelets 3–4 mm long; pedicellate spikelets mostly vestigial or absent, those of the terminal spikelet units well-developed and staminate; in the *Manual* region, known only from southern Florida . . . . . . . . . . . . . . . . . . . . . . . . . . . . . . . . . . . . . . . . . . . . 5. *A. bicornis*
      8. Anthers 1 (rarely 3).
      10. Peduncles with 2–13 rames . . . . . . . . . . . . . . . . . . . . . . . . . . . . . . . . . . . . . 8. *A. liebmannii*
      10. Peduncles usually with 2 (infrequently up to 4) rames or (in *A. gyrans* var. *gyrans* and *A. virginicus* var. *virginicus*), 2–5 (infrequently up to 7) rames.
        11. Culms 30–120(140) (usually less than 100) cm tall; blades 0.8–5 (usually less than 3) mm wide; inflorescence units 2–31 per culm.
          12. Peduncles with 2–5 rames; anthers 0.6–1.7 mm long; sessile spikelets (3)4.1–4.4(5.7) mm long . . . . . . . . . . . . . . . . . . . . . . . . . . . . . . . . . . . . . . . . . . . 9. *A. gyrans*
          12. Peduncles with 2 rames; anthers 1.2–2 mm long; sessile spikelets (4)4.8–5(5.5) mm long . . . . . . . . . . . . . . . . . . . . . . . . . . . . . . . . . . . . . . . . . . . . . . . 10. *A. tracyi*
        11. Culms (20)90–310 (usually more than 100) cm tall; blades 1.7–9.5 (usually more than 3) mm wide; inflorescence units 5–210 per culm.
          13. Upper portion of the plants open, the branches conspicuously arching . . . . . . . 11. *A. brachystachyus*
          13. Upper portion of the plants dense, the branches usually straight and erect to ascending.
            14. Rame internodes usually densely and uniformly pubescent over their entire length; anthers 1.3–3.5 mm long; sessile spikelets (3.8)4–6.1 mm long.
              15. Blades 15–35 cm long, often more or less pubescent; sheaths smooth, very rarely somewhat scabrous; anthers 2–3.5 mm long; inflorescence units 5–45 per culm . . . . . . . . . . . . . . . . . . . . . . . . . . . . . . . . . 6. *A. arctatus*
              15. Blades 32–61 cm long, usually glabrous; sheaths often scabrous; anthers 1.3–2 mm long; inflorescence units usually at least 50 (9–210) per culm . . . . . . . . . . . . . . . . . . . . . . . . . . . . . . . . . . . . . . . . . . . . . 7. *A. floridanus*
            14. Rame internodes sparsely pubescent basally, more densely pubescent distally; anthers 0.5–1.5 mm long; sessile spikelets 2.6–4(5).
              16. Blades 11–52 cm long; sheaths smooth, rarely somewhat scabrous; ligules 0.2–1 mm long; keels of the lower glumes usually smooth below midlength, scabrous distally . . . . . . . . . . . . . . . . . . . . . . . . . . . 12. *A. virginicus*
              16. Blades 13–109 cm long; sheaths usually scabrous, sometimes smooth; ligules 0.6–2.2 mm long; keels of the lower glumes sometimes scabrous below midlength . . . . . . . . . . . . . . . . . . . . . 14. *A. glomeratus* (in part)

# Andropogon L. sect. Andropogon

## 1. Andropogon gerardii Vitman BIG BLUESTEM, BARBON DE GERARD [p. 498, 556]

*Andropogon gerardii* grows in prairies, meadows, generally in dry soils. It is a widespread species, extending from southern Canada to Mexico, and was once dominant over much of its range. It is frequently planted for erosion control and restoration. It is also grown as an ornamental; the records from Washington and central Montana reflect such plantings. It hybridizes with *A. hallii*, the two sometimes being treated as conspecific subspecies.

## 2. Andropogon hallii Hack. SAND BLUESTEM [p. 498, 556]

*Andropogon hallii* grows on sandhills and in sandy soil. Its range extends through the central plains into northern Mexico. It differs from *A. gerardii* primarily in its rhizomatous habit, more densely pubescent rames and pedicels, and greater drought tolerance. *Andropogon hallii* and *A. gerardii* are sympatric in some locations. The two species can hybridize and are sometimes treated as conspecific subspecies.

# Andropogon sect. Leptopogon Stapf

## 3. Andropogon gracilis Spreng. WIRE BLUESTEM [p. 498, 556]

*Andropogon gracilis* grows on oölite in openings and rocky margins of pine woodlands of southern Florida and the West Indies. Although not uncommon, it is frequently overlooked. It has sometimes been placed in *Schizachyrium* because of its solitary rames.

## 4. Andropogon ternarius Michx. SPLIT BLUESTEM [p. 499, 556]

*Andropogon ternarius* grows in the southeastern United States and northern Mexico. It is planted as an ornamental and for erosion control on slopes in poor and sandy soils, and is tolerant of coastal conditions. It differs from *A. arctatus* in its possession of three anthers and usually in its longer spikelets, both sessile and pedicellate.

1. Rames densely villous, with hairs about twice as long as the sessile spikelets and more or less obscuring them; lower glumes of the sessile spikelets sometimes scabrous, without conspicuous veins between the keels . . . . . . . . . . . . . . . . . . . . . . . var. *cabanisii*
1. Rames sparsely villous, with hairs about as long as the sessile spikelets, but not obscuring them; lower glumes of the sessile spikelets scabrous, often conspicuously 2-veined between the keels . . . . . . . . var. *ternarius*

### Andropogon ternarius var. cabanisii (Hack.) Fernald & Griscom [p. 199]

*Andropogon ternarius* var. *cabanisii* grows in dry pine woods and scrublands of peninsular Florida.

### Andropogon ternarius Michx. var. ternarius [p. 499]

*Andropogon ternarius* var. *ternarius* grows in dry, sandy woods, fields, openings, and roadsides of the southeastern United States and Mexico.

## 5. Andropogon bicornis L. BARBAS DE INDIO [p. 499, 556]

*Andropogon bicornis* is a widespread species of the Western Hemisphere tropics. It was collected in the early 1960s in Dade County, Florida, near the track of a major hurricane, but may not be established in the *Manual* region.

## 6. Andropogon arctatus Chapm. PINEWOODS BLUESTEM [p. 499, 557]

*Andropogon arctatus* grows in flatwoods, bogs, and scrublands of southern Alabama and Florida. Its flowering appears to be stimulated by fire but, unlike other members of sect. *Leptopogon* in the *Manual* region, the effect lasts only one or two years, the plants then remaining vegetative until the next fire occurs. It differs from *A. ternarius* in its long, usually solitary anther and shorter spikelets.

## 7. Andropogon floridanus Scribn. FLORIDA BLUESTEM [p. 499, 557]

*Andropogon floridanus* grows on sandy soils in southeastern Georgia and Florida, being most abundant in *Pinus clausa* scrublands. It usually occurs in small stands, but stands of about a hundred individuals have been observed.

## 8. Andropogon liebmannii Hack. LIEBMANN'S BLUESTEM, MOHR'S BLUESTEM [p. 499, 557]

*Andropogon liebmannii* has two varieties. **Andropogon liebmannii var. pungensis** (Ashe) C.S. Campb., the variety found in the *Manual* region, differs from **A. liebmannii Hack. var. liebmannii**, which grows in Mexico, in having culms that are usually more than 80 cm tall, leaves that are more than 15 cm long, and sessile spikelets that are more than 4.2 mm long; in var. *pungensis* the culms are usually less than 90 cm tall, the leaves less than 15 cm long, and the sessile spikelets less than 4.2 mm long. *Andropogon liebmannii* var. *pungensis* grows along the coastal plain of the southeastern United States in bogs, swamps, savannahs, and flatwoods.

## 9. Andropogon gyrans Ashe [p. 499, 557]

*Andropogon gyrans* extends from the southeastern United States to the Caribbean and Central America.

1. Ligules 0.3–1.1 mm long; rames usually hidden within the more or less overlapping and inflated upper sheaths at maturity; plants usually of well-drained soils . . . . . . . . . . . . . . . . . . . . . . . . . . . . . . . var. *gyrans*
1. Ligules 0.8–1.5 mm long; rames usually exposed at maturity; plants of wet habitats . . . . . . . . . . . . . . var. *stenophyllus*

### Andropogon gyrans Ashe var. gyrans ELLIOTT'S BEARDGRASS [p. 499]

*Andropogon gyrans* var. *gyrans* generally grows in dry, sandy soil of roadsides, embankments, fields, and pine or oak woods, occasionally in moister soil. Its range extends south from the United States to the Caribbean and Central America. Plants from Florida and Mississippi do not have inflated sheaths.

### Andropogon gyrans var. stenophyllus (Hack.) C.S. Campb. [p. 499]

*Andropogon gyrans* var. *stenophyllus* grows in ditches, bogs, savannahs, and pond margins of the coastal plain, from eastern Texas to North Carolina.

## 10. Andropogon tracyi Nash TRACY'S BLUESTEM [p. 499, 557]

*Andropogon tracyi* grows on sandhills, sandy pinelands, and scrublands of the southeastern United States. It usually differs from *A. longiberbis* in having sparsely pubescent blades and a more slender appearance.

## 11. Andropogon brachystachyus Chapm. SHORTSPIKE BLUESTEM [p. 499, 557]

*Andropogon brachystachyus* grows in sandy, often seasonally wet soils of flatwoods, savannahs, pond margins, and scrublands of the southeastern United States. It sometimes forms large populations, but does not invade disturbed sites as do some morphologically similar forms of *A. virginicus* var. *virginicus*.

## 12. Andropogon virginicus L. Broomsedge Bluestem [p. 499, 557]

*Andropogon virginicus* is native from the southeastern United States to northern South America, but has become established outside its native range in California, Hawaii, Japan, and Australia. Three varieties are recognized, two of which contain morphologically distinct variants. *Andropogon virginicus* hybridizes with *A. glomeratus* and *A. longiberbis*.

1. Leaves bluish-green, more or less strongly glaucous
   . . . . . . . . . . . . . . . . . . . . . . . . . . . . . . . . . var. *glaucus*
1. Leaves green, sometimes somewhat glaucous.
   2. Sheaths subtending the inflorescence units (1.7)2.4–3.1(4) mm wide; inflorescences units usually with 2 rames; rames (1.3)1.5–2.3(3) cm long; peduncles (1) 4–9 (30) mm long . . . . . . . var. *decipiens*
   2. Sheaths subtending the inflorescences units (2.2)3.3–4.4(5.6) mm wide; inflorescence units with 2–5(7) rames; rames (0.5)1.9–3.3(4.4) cm long; peduncles (2)3–6(12) mm long . . . . . . . . var. *virginicus*

### Andropogon virginicus var. decipiens C.S. Campb. [p. 499]

*Andropogon virginicus* var. *decipiens* grows in flatwoods, scrublands, and disturbed sites, such as roadsides and cleared timberlands, of the southeastern coastal plain.

### Andropogon virginicus var. glaucus Hack. [p. 499]

*Andropogon virginicus* var. *glaucus* grows on moist or dry soils of the coastal plain, from southern New Jersey to eastern Texas. Plants growing on sandy, well-drained soils differ from those on poorly drained slopes in being glabrous (rather than pubescent) beneath the subtending sheaths of the inflorescence units, and in tending to have shorter rames.

### Andropogon virginicus L. var. virginicus [p. 499]

*Andropogon virginicus* var. *virginicus* is the widespread and weedy variety of *A. virginicus* that grows as a native from the central plains through Mexico and Central America to Colombia and, as a naturalized species, in California, Hawaii, Japan, and Australia. Plants colonizing openings in mature vegetation created by disturbance have green culms and green, pubescent leaves. Those growing in poorly drained soils of pond margins, swales, and cutover flatwoods have glaucous culms and glabrous, green to somewhat glaucous leaves. Glaucous plants of *A. virginicus* var. *virginicus* differ from those of var. *decipiens* in having no exposed rames and, often, wider sheaths subtending the inflorescence units.

## 13. Andropogon longiberbis Hack. Hairy Bluestem [p. 500, 557]

*Andropogon longiberbis* grows in sandy or rocky soils of roadsides, dunes, sandhills, pinelands, and fields, from the southeastern United States to the Bahamas. It usually differs from *A. tracyi* in having more densely pubescent blades and a less slender appearance. It appears to hybridize with both *A. virginicus* var. *virginicus* and *A. glomeratus* var. *pumilus*.

## 14. Andropogon glomeratus (Walter) Britton, Sterns & Poggenb. Bushy Bluestem, Bushy Beardgrass [p. 500, 557]

*Andropogon glomeratus* hybridizes with both *A. longiberbis* and *A. virginicus*. Some of its varieties are morphologically similar to the latter species.

1. Blades glaucous, glabrous, and smooth . . . . . . . . var. *glaucopsis*
1. Blades green, often pubescent or scabrous.
   2. Sheaths subtending the inflorescence units 1.5–3 mm wide; leaf sheaths usually smooth; ligules ciliate, the cilia 0.2–0.9 mm long . . . . . . . . . . . . var. *pumilus*
   2. Sheaths subtending the inflorescence units (1.5)2.3–3.4(4.4) mm wide; leaf sheaths often scabrous; ligules, when ciliate, with the cilia no more than 0.5 mm long.
      3. Keels of the lower glumes scabrous below and beyond midlength . . . . . . . . . . . . . . . . . . var. *scabriglumis*
      3. Keels of the lower glumes usually smooth below midlength, scabrous distally.
         4. Upper portion of the plants oblong to obpyramidal; mature peduncles (4)11–35 (60) mm long; anthers eventually falling . . . . . . . . . . . . . . . . . . . . . . . . . . . . var. *glomeratus*
         4. Upper portion of the plants cylindrical to oblong; mature peduncles 2–5 (8) mm long; withered remnants of anthers retained within the spikelets . . . . . . . . . . var. *hirsutior*

### Andropogon glomeratus var. glaucopsis (Elliott) C. Mohr [p. 500]

*Andropogon glomeratus* var. *glaucopsis* grows in flatwoods, bogs, ditches, swamps, pond margins, and swales of the southeastern coastal plain.

### Andropogon glomeratus (Walter) Britton, Sterns & Poggenb. var. glomeratus [p. 500]

*Andropogon glomeratus* var. *glomeratus* grows in bogs, swamps, savannahs, flatwoods, and ditches of the southeastern United States.

### Andropogon glomeratus var. hirsutior (Hack.) C. Mohr [p. 500]

*Andropogon glomeratus* var. *hirsutior* grows in ditches, swales, bogs, flatwoods, and savannahs of the southeastern coastal plain, often forming very large populations in cleared, low ground.

### Andropogon glomeratus var. pumilus (Vasey) L.H. Dewey [p. 500]

*Andropogon glomeratus* var. *pumilus* is weedy and grows in disturbed, wet or moist sites. It is abundant and widespread, extending from the southern United States through Central America to northern South America.

### Andropogon glomeratus var. scabriglumis C.S. Campb. [p. 500]

*Andropogon glomeratus* var. *scabriglumis* grows in moist soils of seepage slopes and the edges of springs, from California to New Mexico and southward into Mexico.

## 25.16 CYMBOPOGON Spreng.[1]

Pl usu per; csp. **Clm** 15–300 cm. **Lvs** aromatic, smelling of lemon oil or citronella; **shth** open, not strongly keeled except near the summit; **lig** memb; **bld** usu glab or mostly so, with long filiform apc. **Infl** tml and ax, false pan; **peduncles** often enclosed in the subtending lf shth at maturity, with 2 rames; **rames** with 4–7 heterogamous

[1]Mary E. Barkworth

spklt pairs, axes slender, without a median groove, lo rame of each pair with 1 homogamous spklt pair at the base, its ped swollen and more or less fused to the adjacent intnd, up rames with short, strl, flattened bases that are usu deflexed at maturity, without homogamous spklt units. **Heterogamous spklt units:** sessile spklt dorsally compressed, with 2 flt; **lo glm** chartaceous, concave or flat, 2-keeled, with or without intercostal veins, often streaked with oil glands; **up flt** with a short, glab awn (rarely unawned); **ped** linear, free from the rame axes; **pedlt spklt** well-developed.

*Cymbopogon* comprises 55 species, and is native to the tropics and subtropics of the Eastern Hemisphere. It is cultivated in southern Florida and California, sometimes persisting for a considerable period. Plants grown outdoors in the *Manual* region generally remain vegetative, but can usually be identified to genus by their lemony aroma. *Heteropogon melanocarpus* also smells lemony when fresh but, unlike the species of *Cymbopogon* that have been grown in the United States, it has a row of glandular depressions over the well-developed keels of the lower leaf sheaths.

Identification of any grass to species in the absence of reproductive parts is difficult. The key for use on vegetative plants of the three species of *Cymbopogon* reported from the *Manual* region should be used only as a last resort.

## REPRODUCTIVE KEY

1. Pedicels pilose on the margins, glabrous dorsally . . . . . . . . . . . . . . . . . . . . . . . . . . . . . . . . . . . . . . . . . . . . . . . . . . . . 2. *C. nardus*
1. Pedicels pilose on the margins and the dorsal surface.
    2. Lower glumes of the sessile spikelets shallowly concave distally; the keels not winged; ligules 2–6 mm long; blades whitish . . . . . . . . . . . . . . . . . . . . . . . . . . . . . . . . . . . . . . . . . . . . . . . . 1. *C. jwarancusa*
    2. Lower glumes of the sessile spikelets flat above, the keels narrowly winged; ligules 0.5–2 mm long; blades green . . . . . . . . . . . . . . . . . . . . . . . . . . . . . . . . . . . . . . . . . . . . . . . . . 3. *C. citratus*

## VEGETATIVE KEY

1. Ligules 0.5–2 mm long, truncate, the nodes not swollen . . . . . . . . . . . . . . . . . . . . . . . . . . . . . . . . . . . . . . . . 3. *C. citratus*
1. Ligules 2–6 mm long, truncate to acute, the nodes usually swollen.
    2. Basal sheaths purplish-red; blades 3–16 mm wide . . . . . . . . . . . . . . . . . . . . . . . . . . . . . . . . . . . . . . . . 2. *C. nardus*
    2. Basal sheaths whitish; blades 1.5–4 mm wide . . . . . . . . . . . . . . . . . . . . . . . . . . . . . . . . . . . . 1. *C. jwarancusa*

### 1. **Cymbopogon jwarancusa** (Jones) Schult.
IWARANCUSA GRASS [p. *500*]

*Cymbopogon jwarancusa* is native to Asia, where it is grown for perfume and as a medicine for fevers. It is grown as an ornamental in the United States, and may persist for a considerable time after planting in the warmest parts of the *Manual* region.

### 2. **Cymbopogon nardus** (L.) Rendle CITRONELLA GRASS, NARDGRASS [p. *500*]

*Cymbopogon nardus* has been cultivated in the United States, but the variety involved is not known.

### 3. **Cymbopogon citratus** (DC.) Stapf LEMON GRASS [p. *500*]

*Cymbopogon citratus* is now known only in cultivation, even in Asia. Young shoots are used as a spice, and the oils are extracted for lemon oil. It has been grown in Florida.

## 25.17 **SCHIZACHYRIUM** Nees[1]

**Pl** ann or per; csp or rhz, smt both csp and shortly rhz. **Clm** 7–210 cm, brchd above the bases, often purplish near the nd. **Lvs** not aromatic, **shth** open; **aur** usu absent; **lig** memb; **bld** flat, folded, or involute, those of the upmost lvs often greatly rdcd. **Infl** ax and tml, of 1, rarely 2, rames, peduncles subtended by a modified lf; **rames** not reflexed, with spklt in heterogamous sessile-pedlt spklt pairs, intnd more or less flattened, filiform to clavate, without a median groove, apc cupulate or fimbriate; **dis** in the rame axes, below the sessile spklt. **Spklt** somewhat dorsiventrally compressed. **Sessile spklt** with 2 flt; **glm** exceeding the flt, lanceolate to linear, memb; **lo glm** enclosing the up glm, convex, weakly 2-keeled, with several (smt inconspicuous) intercostal veins; **lo flt** rdcd to hyaline lm; **up flt** bisx, lm hyaline, bilobed or bifid to $^7/_8$ of their length (rarely entire), awned from the sinuses; **anth** 3. Ped free of the rame axes, usu pubescent. **Pedlt spklt** usu shorter than to as long as the sessile spklt, occ longer, strl or stmt, with 1 flt, often disarticulating as the rame matures; **lm** present in stmt spklt, hyaline, unawned or with a straight awn of less than 10 mm.

*Schizachyrium* is a genus of approximately 60 species that are native to tropical and subtropical regions of the world; nine are native to the *Manual* region. Most species of *Schizachyrium* differ from species of *Bothriochloa* and *Andropogon* in having only one rame per peduncle, but *S. spadiceum* has two. More reliable, but less conspicuous distinguishing features of *Schizachyrium* are the cupulate tips of the rame internodes, the convex lower glumes, and the presence of veins between the keels of the lower glumes. A few species of *Andropogon* have solitary rames, but they do not have these other features.

[1]J.K. Wipff

1. Peduncles with 2 rames . . . . . . . . . . . . . . . . . . . . . . . . . . . . . . . . . . . . . . . . . . . . . . . . . . . . . . . . . . . . . . . 1. *S. spadiceum*
1. Peduncles with only 1 rame.
  2. Leaf blades 0.5–2 mm wide, with a longitudinal stripe of white, spongy tissue (formed of bulliform
    cells) on their adaxial surfaces; plants cespitose; pedicellate spikelets about as long as the sessile
    spikelets . . . . . . . . . . . . . . . . . . . . . . . . . . . . . . . . . . . . . . . . . . . . . . . . . . . . . . . . . . . . . . . . . . . . . . . 6. *S. tenerum*
  2. Leaf blades (1)1.5–9 mm wide, without a longitudinal stripe of white, spongy tissue on their adaxial
    surfaces; plants cespitose or rhizomatous; pedicellate spikelets equal to or smaller than the sessile
    spikelets.
    3. Plants rooting and branching at the lower nodes and at aerial nodes in contact with the soil; leaf
      collars usually elongate and narrow; plants of sandy coastal habitats.
      4. Ligules 0.5–1 mm long, pedicellate spikelets 4.5–8.5 mm long . . . . . . . . . . . . . . . . . . . . . . . 4. *S. maritimum*
      4. Ligules 1.5–2 mm long, pedicellate spikelets 1.5–5 mm long . . . . . . . . . . . . . . . . . . . . . . . . . 5. *S. littorale*
    3. Plants not rooting or branching at the lower nodes; leaf collars neither elongate nor particularly
      narrow; plants of varied habitats.
      5. Pedicel bases 0.2–0.5 mm wide, gradually widening to 0.3–1 mm distally, straight, often
        somewhat stiff, not tending to curve outward; rames appearing linear.
        6. Pedicellate spikelets 6–8 mm long, about as long as the sessile spikelets, usually staminate,
          sometimes sterile, unawned . . . . . . . . . . . . . . . . . . . . . . . . . . . . . . . . . . . . . . . . 9. *S. cirratum*
        6. Pedicellate spikelets 0.7–10 mm long, usually shorter than the sessile spikelets, sterile,
          unawned or awned, the awns up to 6 mm long.
          7. Upper lemmas cleft for ²⁄₃–⁷⁄₈ of their length; lower glumes glabrous or pubescent . . . . . . . . 8. *S. sanguineum*
          7. Upper lemmas cleft for up to ½ of their length; lower glumes glabrous . . . . . . . . . . . 2. *S. scoparium* (in part)
      5. Pedicel bases 0.1–0.2 mm wide, flaring above midlength to about 0.5 mm wide, tending to
        curve outward; rames appearing somewhat open.
        8. Upper lemmas indurate at the base, cleft ³⁄₄–⁷⁄₈ of their length; leaf blades 2.5–10 cm long;
          pedicellate spikelets 0.5–2 mm long; plants cespitose; known only from peninsular Florida . . . . . . . . 7. *S. niveum*
        8. Upper lemmas membranous at the base, cleft for up to ½ of their length; leaf blades 7–105
          cm long; pedicellate spikelets 0.7–10 mm long; plants cespitose or not; widespread,
          including Florida.
          9. Plants cespitose, not or shortly rhizomatous . . . . . . . . . . . . . . . . . . . . . . . . . . . . . 2. *S. scoparium* (in part)
          9. Plants not cespitose, strongly rhizomatous.
            10. Pedicellate spikelets awned, awns to 4 mm; leaf blades usually 3.5–9 mm wide;
              culms usually 1–3 mm thick; plants of sandy soils . . . . . . . . . . . . . . . . . . . . . 2. *S. scoparium* (in part)
            10. Pedicellate spikelets unawned or the awns less than 1 mm; leaf blades 1–3 mm
              wide; culms usually less than 1 mm thick; plants of oölitic soil . . . . . . . . . . . . . . . . . 3. *S. rhizomatum*

## 1. Schizachyrium spadiceum (Swallen) Wipff HONEY BLUESTEM [p. *500*, *557*]

*Schizachyrium spadiceum* was once thought to be a Mexican endemic, but it is now known from limestone slopes in Brewster County, Texas.

## 2.  Schizachyrium scoparium (Michx.) Nash [p. *500*, *557*]

*Schizachyrium scoparium* is a widespread grassland species extending from Canada to Mexico. It exhibits considerable variation, much of it clinal. The following varieties are recognized because they are morphologically, ecologically, and geographically distinctive.

1. Plants not cespitose, strongly rhizomatous; pedicellate; spikelets sterile . . . . . . . . . . . . . . . . var. *stoloniferum*
1. Plants usually cespitose, not or shortly rhizomatous; pedicellate spikelets staminate or sterile.
  2. Pedicellate spikelets of the proximal spikelet units on each rame staminate, 5–10 mm long, with a lemma, pedicellate spikelets of the distal units usually smaller (1–4 mm) and sterile; sheaths and blades densely tomentose to glabrate . . . . . . . . . . . . . . . . . . . . . . . . . . . var. *divergens*
  2. Most pedicellate spikelets sterile, 1–6 mm long, without a lemma; sheaths and blades usually glabrous, occasionally pubescent . . . . . . . . var. *scoparium*

## Schizachyrium scoparium var. divergens (Hack.) Gould PINEHILL BLUESTEM [p. *500*]

*Schizachyrium scoparium* var. *divergens* is common in the south-central pinelands of the United States. The pubescence of the leaves varies across its range, western plants having longer and more villous leaves than those in the east and, towards Mississippi, the pubescence is confined to the sheaths. *Schizachyrium scoparium* var. *divergens* intergrades with var. *scoparium*.

## Schizachyrium scoparium (Michx.) Nash var. scoparium

LITTLE BLUESTEM, BROOM BEARDGRASS, BROOM BLUESTEM, SCHIZACHYRIUM À BALAIS [p. *500*]

*Schizachyrium scoparium* var. *scoparium* grows in a variety of soils and in open habitats. It is the most variable of the varieties recognized within *S. scoparium*, with morphological features that vary independently and continuously across its range, coming together in distinctive combinations in some regions.

## Schizachyrium scoparium var. stoloniferum (Nash) Wipff CREEPING BLUESTEM [p. *500*]

*Schizachyrium scoparium* var. *stoloniferum* grows in sandy soils of woodland openings and roadsides from southern Alabama and Georgia south to the Everglades. Northern populations consist of widely spaced, weak culms growing in rather bare sand; southern populations consist of dense, vigorous stands with taller, more robust culms growing primarily along roadsides, possibly spread by grading equipment. Some clones, particularly in the south, are largely sterile.

### 3. Schizachyrium rhizomatum (Swallen) Gould
FLORIDA LITTLE BLUESTEM [p. *501, 557*]

*Schizachyrium rhizomatum* grows in open glades and on the margins of pine woodlands and is endemic to Florida. It is restricted to thin, oölitic soils that are often saturated with water, and forms sparse stands, occasionally mixed with *Andropogon gracilis*, in the Florida Keys.

### 4. Schizachyrium maritimum (Chapm.) Nash GULF
BLUESTEM [p. *501, 557*]

*Schizachyrium maritimum* is endemic to the southeastern United States, growing in sandy areas, usually at the ocean waterline but also along roads in low, dune areas, from Louisiana to the Florida panhandle. The plants often appear rhizomatous because the lower, decumbent portions of the culms are frequently covered by sand.

### 5. Schizachyrium littorale (Nash) E.P. Bicknell SHORE
BLUESTEM, DUNE BLUESTEM [p. *501, 557*]

*Schizachyrium littorale* is restricted to shifting, coastal sand dunes of the Gulf, Atlantic, and Great Lakes coasts of the United States. It often appears rhizomatous because the lower nodes are frequently covered by sand.

### 6. Schizachyrium tenerum Nees SLENDER BLUESTEM [p.
*501, 557*]

*Schizachyrium tenerum* is an uncommon species in the southeastern United States, where it grows on sandy soils in pine forest openings and coastal prairies. Its range extends through Central America into South America.

### 7. Schizachyrium niveum (Swallen) Gould PINESCRUB
BLUESTEM [p. *501, 557*]

*Schizachyrium niveum* is an endangered, rare species known only from central peninsular Florida, where it occurs in openings and sandhills of *Ceratiola*-pine-oak woodlands.

### 8. Schizachyrium sanguineum (Retz.) Alston [p. *501, 557*]

*Schizachyrium sanguineum* extends from the southern United States to Chile, Paraguay, and Uruguay.

1. Lower glumes of the sessile spikelets glabrous or scabrous; pedicels ciliate on 1 edge . . . . . . . . . . var. *sanguineum*
1. Lower glumes of the sessile spikelets pubescent to hirsute; pedicels ciliate on both edges . . . . . . . . . var. *hirtiflorum*

#### Schizachyrium sanguineum var. hirtiflorum (Nees) S.L.
Hatch HAIRY CRIMSON BLUESTEM [p. *501*]

In the *Manual* region, *Schizachyrium sanguineum* var. *hirtiflorum* grows on rocky slopes and well-drained soils from Arizona to southwestern Texas and Florida. Its range extends through Central America to Chile, Paraguay, and Uruguay.

#### Schizachyrium sanguineum (Retz.) Alston var.
sanguineum CRIMSON BLUESTEM [p. *501*]

*Schizachyrium sanguineum* var. *sanguineum* grows in tropical and subtropical regions of America, Africa, and Asia. Within the *Manual* region, it is known only from the pine woods of Alabama and Florida.

### 9. Schizachyrium cirratum (Hack.) Wooton & Standl.
TEXAS SCHIZACHYRIUM, TEXAS BEARDGRASS [p. *501, 557*]

*Schizachyrium cirratum* grows on rocky slopes, mostly at elevations of 5000 feet or higher, from southern California to western Texas into Mexico, and is known from South America.

---

## 25.18 ARTHRAXON P. Beauv.[1]

**Pl** ann or per; scrambling. **Clm** 0.5–2 m, ascending to decumbent, often rooting at the nd, brchd. **Lvs** not aromatic; **shth** open, at least the outer mrg pubescent, usu with papillose-based hairs; **lig** memb, fimbriate or ciliate; **bld** ovate to ovate-lanceolate. **Infl** tml and ax, pan of subdigitate, often flabellate, clusters of rames; **rame intnd** not sulcate; **dis** in the rames, beneath the sessile spklt. **Spklt** in heteromorphic sessile-pedlt pairs or appearing solitary and sessile, pcd greatly rdcd and lacking spklt.

**Sessile spklt** bisx, with 2 flt; **cal** absent or blunt; **glm** equal or subequal; **lo flt** strl, rdcd to an unawned lm; **up flt** bisx, awned (rarely unawned); **anth** 2 or 3. **Ped** 0.2-3 mm, not thickened, not fused to the rame axes. **Pedlt spklt** absent or rdmt.

*Arthraxon* is a genus of seven species that are native to tropical and subtropical regions of the Eastern Hemisphere; one species is established in the *Manual* region.

### 1. Arthraxon hispidus (Thunb.) Makino JOINTHEAD,
SMALL CARPETGRASS [p. *501, 557*]

*Arthraxon hispidus* is native to Asia, but is naturalized and spreading along roadsides, shores, ditches and in low woods and fields of the eastern United States. It is also naturalized in Mexico, Central America, and the West Indies. Plants in the *Manual* region belong to **A. hispidus** (Thunb.) Makino var. **hispidus**, the most widespread and variable of the four varieties.

---

## 25.19 HYPARRHENIA Andersson *ex* E. Fourn.[2]

**Pl** ann or per; csp, often with short rhz. **Clm** 30–350(400) cm, usu erect, much brchd above the bases. **Lvs** not aromatic; **lig** memb, not ciliate; **bld** usu flat or folded. **Infl** false pan with numerous inflorescence units; **peduncles** with 2 rames in digitate clusters; **rames** with naked, often deflexed bases, axes without a translucent median groove; **dis** in the rames, beneath the bisx spklt. **Spklt** in sessile-pedlt pairs, bas 1–2 pairs on each rame homogamous, morphologically similar to the heterogamous pairs, stmt or strl, unawned, not forming an involucre, tardily deciduous, remaining pairs heterogamous. **Heterogamous spklt**

[1]John W. Thieret†  [2]Mary E. Barkworth

units: sessile spklt dorsally compressed or subterete; **cal** blunt to sharp, strigose; **glm** equal, pubescent; **lo glm** coriaceous, rounded, without keels, truncate to slightly bilobed; **up glm** narrower, shallowly keeled; **lo flt** strl, rdcd; **up flt** bisx, awned from between the teeth of the bifid lm; **awns** usu present, to 3.5(19) cm, pubescent on the lo portion. **Car** oblong, subterete. **Ped** slender, not adnate to the rame axes. **Pedlt spklt** usu slightly longer than the sessile spklt, stmt or strl, usu unawned, lo glm smt aristulate.

*Hyparrhenia* is a genus of approximately 55 mostly African species. Two have been introduced into the *Manual* region, but only one is known to be established.

1. Spikelets with whitish to dark yellow hairs . . . . . . . . . . . . . . . . . . . . . . . . . . . . . . . . . . . . . . . . . . . . . . . . . . . . 1. *H. hirta*
1. Spikelets with reddish hairs . . . . . . . . . . . . . . . . . . . . . . . . . . . . . . . . . . . . . . . . . . . . . . . . . . . . . . . . . . . . . . 2. *H. rufa*

### 1. Hyparrhenia hirta (L.) Stapf THATCHING GRASS [p. 502]

*Hyparrhenia hirta* is native to southern Africa, where it grows on stony soils and is sometimes used for thatching. It has been cultivated in Texas and Florida, but is not currently known to be established in the *Manual* region. A report of its occurrence in Los Angeles County, California, has not been verified.

### 2. Hyparrhenia rufa (Nees) Stapf JARAGUA GRASS [p. 502, 557]

*Hyparrhenia rufa* is native to the Eastern Hemisphere tropics, but is now established in tropical America. It grows in ditches, pastures, swamps, and pine flatwoods, and along roadsides, in the southeastern United States.

## 25.20 HETEROPOGON Pers.[1]

**Pl** ann or per; csp. **Clm** 20–200 cm, simple or brchd. **Lvs** smt aromatic and smelling of lemon oil or citronella; **shth** keeled, smt with a row of glandular depressions on the keel; **lig** memb, glab or ciliate. **Infl** tml and ax; **peduncles** usu with 1 rame, smt with several in a digitate cluster; **rames** with 3–10 homogamous, unawned, sessile-pedlt spklt pairs on the lo ¼–⅔ and heterogamous, awned, sessile-pedlt spklt pairs distally, axes slender, without a translucent median groove; **dis** in the rames, beneath the sessile spklt of the heterogamous spklt pairs, smt also below their pedlt spklt. **Homogamous spklt units** strl or stmt; **cal** poorly developed; **glm** memb, many-veined, keels winged above. **Heterogamous spklt units:** sessile spklt bisx, terete; **cal** 1.5–3 mm, sharp, antrorsely strigose, hairs golden brown; **glm** coriaceous, pubescent, concealing the flt; **lo glm** enclosing the up glm, obscurely 5–9-veined; **up glm** sulcate, 3-veined; **lo flt** strl, rdcd to a hyaline lm; **up flt** bisx, lm with conspicuous, geniculate awns; **awns** 5–15 cm, with hairs. **Car** lanceolate, sulcate on 1 side. **Ped** short, free of the rame axes, not grooved; **pedlt spklt** strl or stmt, larger than the sessile spklt; **cal** long, glab, functioning as ped; **glm** memb, many-veined, keels winged above.

*Heteropogon* is a pantropical genus of eight to ten species. Many grow well on poor soils. Two species grow in the *Manual* region; both are probably introduced.

1. Glumes of the pedicellate spikelets of the heterogamous spikelet units without glandular pits; plants
     perennial . . . . . . . . . . . . . . . . . . . . . . . . . . . . . . . . . . . . . . . . . . . . . . . . . . . . . . . . . . . . . . . . . . . . . . . . . 1. *H. contortus*
1. Glumes of the pedicellate spikelets of the heterogamous spikelet units with a row of glandular pits along
     the midvein; plants annual . . . . . . . . . . . . . . . . . . . . . . . . . . . . . . . . . . . . . . . . . . . . . . . . . . . . . . 2. *H. melanocarpus*

### 1. Heteropogon contortus (L.) P. Beauv. *ex* Roem. & Schult. TANGLEHEAD [p. 502, 557]

*Heteropogon contortus* grows on rocky hills and canyons in the southern United States into Mexico, and worldwide in subtropical and tropical areas, occupying a variety of different habitats, including disturbed habitats. It is probably native to the Eastern Hemisphere but is now found in tropical and subtropical areas throughout the world. It is considered a weed, being able to establish itself in newly disturbed and poor soils.

### 2. Heteropogon melanocarpus (Elliott) Benth. SWEET TANGLEHEAD [p. 502, 557]

*Heteropogon melanocarpus* is probably native to the Eastern Hemisphere, but is now found in tropical regions throughout the world. It grows in pine woods, fields, and disturbed areas of the southern United States. When fresh, plants of *H. melanocarpus* smell like citronella oil.

## 25.21 THEMEDA Forssk.[2]

**Pl** ann or per; usu csp, rarely stln. **Clm** 30–310 cm, erect. **Lvs** not aromatic; **shth** open; **lig** memb, smt ciliate. **Infl** numerous, tml and ax, false pan; **peduncles** shorter than the subtending shth, with 1–8 rames; **rames** spklt-bearing to the base, axes slender, without a longitudinal translucent groove, with 2 large, solitary, homogamous sessile-pedlt spklt pairs at the base and 1–4 smaller, heterogamous sessile-pedlt spklt pairs distally, tml or only unit smt a triplet with 1 sessile and 2 pedlt spklt; **dis** in the rames below the sessile spklt of the heterogamous spklt units, occ beneath the homogamous spklt. **Homogamous spklt pairs** distinctive, forming an involucre around the rame bases, separated by intnd less than ½ as long as the spklt; **spklt** subequal, strongly compressed dorsally, stmt or strl, unawned; **lo glm** memb, 2-keeled. **Heterogamous spklt pairs:** sessile spklt

[1,2]Mary E. Barkworth

subterete or dorsally compressed, awned; **cal** bearded, usu sharp; **glm** coriaceous; **lo glm** wrapped around and concealing the up glm, obscurely veined but not keeled, truncate; **up glm** sulcate, with thin mrg; **lo flt** highly rdcd, strl; **up flt** bisx, up lm usu terminating in a geniculate awn; **awns** usu present, to 9 cm, puberulent to pubescent, smt absent. **Ped** slender, not sulcate, not fused to the rame axes; **pedlt spklt** similar to the homogamous spklt except

narrower, stmt or strl, and unawned. **Car** narrowly ovate or linear, subterete or channeled on 1 side.

*Themeda* has approximately 18 species, all of which are native to tropical and subtropical regions of the Eastern Hemisphere, primarily southeast Asia. One species has been established in the *Manual* region, one is grown as an ornamental, and a third was introduced but is not known to be established.

1. Awns 7–9 cm long; calluses 3–4 mm long ................................................ 1. *T. arguens*
1. Awns 2.5–7 cm long; calluses 0.5–4 mm long.
   2. Plants annual; homogamous spikelets 4–7 mm long; sessile spikelets of heterogamous pairs 4–6 mm long ........................................................................ 2. *T. quadrivalvis*
   2. Plants perennial; homogamous spikelets 6–14 mm long; sessile spikelets of heterogamous pairs 6–14 mm long ........................................................................ 3. *T. triandra*

### 1. Themeda arguens (L.) Hack. CHRISTMAS GRASS [p. 502, 557]

*Themeda arguens* is native to northern Australia and southeastern Asia. It has been reported on ore piles in Maryland and Virginia.

### 2. Themeda quadrivalvis (L.) Kuntze KANGAROO GRASS [p. 502, 557]

A native of Malaysia, *Themeda quadrivalvis* has been found at scattered locations in the contiguous United States. It is established in

St. Landry and Iberia parishes, Louisiana, in addition to having escaped from cultivation in Florida.

### 3. Themeda triandra Forssk. ROOIGRAS [p. 502, 557]

*Themeda triandra* is native to India, Korea, China, and Japan. *Themeda triandra* subsp. *japonica* (Willd.) T. Koyama is sold as an ornamental in the *Manual* region.

---

## 25.22 ELIONURUS Humb. & Bonpl. *ex* Willd.[1]

**Pl** per, occ ann; csp, smt with short rhz. **Clm** 10–150 cm, erect, smt brchg above the base. **Lvs** smt aromatic; **shth** without glandular pits; **lig** shortly memb and densely ciliate or of hairs; **bld** involute, flat, or folded. **Infl** tml, smt also ax, composed of solitary, flexuous rames; **rame intnd** columnar to clavate, apc strongly oblique, not hollowed or rimmed; **dis** in the rames, below the sessile spklt. **Spklt** in sessile-pedlt pairs. **Sessile spklt** dorsally compressed; **cal** blunt, smt resembling a short ped; **lo glm** enclosing the up glm,

subcoriaceous, 2-keeled, keels prominently ciliate, intercarinal surface smooth, apc cuspidate to bilobed, rarely entire; **lo flt** rdcd, strl; **up flt** bisx, unawned. **Ped** stout, appressed but not fused to the rame axes, pubescent or ciliate on the angles. **Pedlt spklt** 3–8 mm, about equal to the sessile spklt, stmt, muticous to awntipped.

*Elionurus* has 15 species. Most are native to tropical Africa and America; one is Australian. Two species are native to the *Manual* region.

1. Lower glumes densely pilose; pedicels pilose dorsally; culms antrorsely hirsute below the nodes ........ 1. *E. barbiculmis*
1. Lower glumes glabrous or nearly so; pedicels ciliate on the angles, usually glabrous elsewhere; culms glabrous throughout ........................................................................ 2. *E. tripsacoides*

### 1. Elionurus barbiculmis Hack. WOOLYSPIKE BALSAMSCALE [p. 502, 558]

*Elionurus barbiculmis* grows on mesas, rocky slopes, hills, and in canyons, usually above 1200 m. Its range extends from southern Arizona and southwestern Texas into northern Mexico.

### 2. Elionurus tripsacoides Humb. & Bonpl. *ex* Willd. PAN-AMERICAN BALSAMSCALE [p. 502, 558]

*Elionurus tripsacoides* grows in moist pine woods and low prairies around southern Texas and the Gulf Coast to Georgia, and south through Mexico and Central America to Argentina.

---

## 25.23 HEMARTHRIA R. Br.[2]

**Pl** per. **Clm** to 150 cm, erect or decumbent, rooting at the nd, usu brchd above the bases. **Lvs** not aromatic; **shth** mostly glab, smt ciliate near the base; **lig** memb, ciliate; **bld** usu linear-lanceolate, smt linear. **Infl** tml and ax, with 1(2) flattened rames borne on a common peduncle, spklt partially embedded in the rame axes; **dis** in the rames, usu oblique and often tardy. **Spklt** in heterogamous sessile-pedlt pairs, dorsally compressed. **Sessile spklt** with 2 flt;

**cal** blunt; **lo glm** coriaceous, smooth; **up glm** equaling the lo glm, chartaceous to memb, smt partially adnate to the rame axes, smt awned; **lo flt** rdcd to hyaline lm; **up flt** bisx, lm unawned. **Ped** thick, fused to the rame axes. **Pedlt spklt** morphologically similar to the sessile spklt, stmt or strl.

*Hemarthria* is a genus of 12 species, native to the tropics and subtropics of the Eastern Hemisphere, and possibly to the

[1]Mary E. Barkworth   [2]Charles M. Allen

Western Hemisphere. All the species grow in or near water. One species has been introduced into the *Manual* region.

### 1. Hemarthria altissima (Poir.) Stapf & C.E. Hubb. [p. 502, <u>558</u>]

*Hemarthria altissima* grows in tropical and subtropical regions throughout the world, including southern Texas and Florida. It is considered native to the Mediterranean region.

## 25.24 COELORACHIS Brongn.[1]

Pl per; csp or rhz. Clm 60–400 cm, erect. Lvs not aromatic; bas and cauline; shth open, glab, mrg scarious; aur lacking; lig memb, ciliate; bld flat to conduplicate, glab or sparsely pubescent, mrg scarious, smt scabrous. Infl tml and ax, composed of a solitary, pedunculate rame; rames stout; dis in the rames, below the sessile spklt. Spklt dorsally compressed, in heterogamous sessile-pedlt pairs. Sessile spklt embedded in the rame axes, ovate, with 2 flt, unawned; lo glm indurate, smooth, rugose, or pitted, 7–11-veined, not keeled; up glm coriaceous, keeled, 1-veined; lo flt strl; up flt bisx, unawned; anth 3. Ped short, thick, appressed or partly fused to the side of the rame axes. Pedlt spklt 1–3 mm, usu rdcd. Car ellipsoid to broadly ellipsoid, yellow.

*Coelorachis* is a tropical genus of approximately 20 species; four are native to the southeastern United States. Most species tend to favor damp soils.

1. Culms and sheaths terete; lower glumes of the sessile spikelets with circular pits on the sides, the central region initially smooth, usually developing rectangular pits at maturity, occasionally remaining smooth . . . . . . . . . . . . . . . . . . . . . . . . . . . . . . . . . . . . . . . . . . . . . . . . . . . . . . . . . . . . . . . . . . . 1. *C. cylindrica*
1. Culms and sheaths compressed-keeled; lower glumes of the sessile spikelets transversely rugose, rectangular-pitted, or smooth.
    2. Lower glumes of the sessile spikelets rectangular-pitted . . . . . . . . . . . . . . . . . . . . . . . . . . . . . . . . . 2. *C. tessellata*
    2. Lower glumes of the sessile spikelets transversely rugose or smooth.
        3. Lower glumes of the sessile spikelets distinctly transversely rugose; rachises distinctly indented below the sessile spikelets . . . . . . . . . . . . . . . . . . . . . . . . . . . . . . . . . . . . . . . . . . . . . . . . . . 3. *C. rugosa*
        3. Lower glumes of the sessile spikelets smooth to slightly transversely rugose; rachises not, or only slightly, indented below the sessile spikelets . . . . . . . . . . . . . . . . . . . . . . . . . . . . . . . . . . . 4. *C. tuberculosa*

### 1. Coelorachis cylindrica (Michx.) Nash CAROLINA JOINTGRASS [p. 503, <u>558</u>]

*Coelorachis cylindrica* is native to the southeastern United States, where it grows in tallgrass prairies, the edges of forests, and roadsides. The specimen from Michigan was found in an old field, in association with many native species. Its source is unknown.

### 2. Coelorachis tessellata (Steud.) Nash PITTED JOINTGRASS [p. 503, <u>558</u>]

*Coelorachis tessellata* is endemic to the southern coastal plain of the United States, extending from Louisiana to northern Florida, although it is rare in Florida. It grows in bogs and moist pine woods, especially flatwoods.

### 3. Coelorachis rugosa (Nutt.) Nash WRINKLED JOINTGRASS [p. 503, <u>558</u>]

*Coelorachis rugosa* is endemic to the southeastern United States. It grows in moist to wet areas in prairies, bogs, and pine woods, especially flatwoods and savannahs.

### 4. Coelorachis tuberculosa (Nash) Nash SMOOTH JOINTGRASS [p. 503, <u>558</u>]

*Coelorachis tuberculosa* is an uncommon species, endemic to the southeastern United States. It grows in moist to wet areas such as bogs and pine woods, especially flatwoods and savannahs.

## 25.25 EREMOCHLOA Büse[2]

Pl per; csp, smt stln. Clm 10–70 cm, hrb, erect to decumbent, bas brchg exvag. Lvs not aromatic; shth open; aur absent; lig memb, truncate; bld flaccid, linear to lanceolate. Infl tml, of long-pedunculate, solitary, 1-sided rames (ax rames occ present) with more than 1 spklt unit; rame intnd clavate, glab; dis in the rames. Spklt in heteromorphic sessile-pedlt pairs, pedlt spklt absent or rdmt. Sessile spklt imbricate, not embedded in the rame axes, dorsally compressed, with 2 flt, unawned; cal truncate; glm exceeding the flt, differing in shape; lo glm 4–9-veined, 2-keeled, keels with spinelike projections and often winged distally, smooth between the keels, mrg folded inward; up glm often shorter than the lo glm, 3–5-veined, keels entire; lo flt stmt; lo pal present; up flt bisx or pist; lm and pal hyaline, unawned; anth 3; sty br 2, red, free to the base. Ped closely appressed but not fused to the rame axes, flattened, thick, widening above the bases, glab. Pedlt spklt usu absent or rdmt, occ well-developed.

*Eremochloa* has 11 species that are native to Asia and Australia. One species is naturalized in the southeastern United States; another was found once in California but is not established in the *Manual* region.

---

[1]Charles M. Allen  [2]John W. Thieret†

1. Keels of the lower glumes of the sessile spikelets winged distally, with 1–several 0.2–0.3 mm hooklike
   spines at the base . . . . . . . . . . . . . . . . . . . . . . . . . . . . . . . . . . . . . . . . . . . . . . . . . . . . . . . . . . . . . . . . 1. *E. ophiuroides*
1. Keels of the lower glumes of the sessile spikelets not winged, spine-bearing throughout, the basal spines
   1–3 mm . . . . . . . . . . . . . . . . . . . . . . . . . . . . . . . . . . . . . . . . . . . . . . . . . . . . . . . . . . . . . . . . . . . . . . 2. *E. ciliaris*

### 1. Eremochloa ophiuroides (Munro) Hack. Centipede Grass [p. *503*, *558*]

*Eremochloa ophiuroides*, an east Asian species, was introduced into the southeastern United States as a lawn grass about 1920. It is now established along roadsides and in woods, fallow fields, and dunes in the region.

### 2. Eremochloa ciliaris (L.) Merr. [p. *503*]

*Eremochloa ciliaris* is native to southeast Asia. It was collected in San Francisco in the nineteenth century, but has not been reported since from the *Manual* region.

## 25.26 ROTTBOELLIA L. f.[1]

Pl ann; csp. **Clm** 30–300+ cm, glab or sparsely pubescent below the nd, brchg above the bases. **Lvs** mostly cauline, not aromatic; **shth** smt with papillose-based hairs; **aur** absent; **lig** memb, ciliate; **bld** flat. **Infl** tml and ax, solitary rames with more than 1 spklt unit, spklt partially embedded in the rame axes; **dis** in the rame axes. **Spklt** heterogamous, in sessile-pedlt pairs, dorsally compressed, unawned. **Sessile spklt** with 2 flt; **lo glm** coriaceous, smooth or scabridulous, not pitted, 2-keeled, narrowly winged above; **up glm** coriaceous, 1-keeled, winged; **lo flt** stmt or strl; **up flt** bisx; **lm** and **pal** hyaline; **anth** 3; **ov** glab. **Ped** thick, fused to the rame axes. **Pedlt spklt** strl or stmt; **glm** hrb. **Car** with a hard endosperm.

*Rottboellia* is a genus of five species, all native to tropical Africa and Asia, represented in all tropical regions of the world by *R. cochinchinensis*, the species that has been introduced into the *Manual* region.

### 1. Rottboellia cochinchinensis (Lour.) Clayton Itchgrass [p. *503*, *558*]

*Rottboellia cochinchinensis* is a native of southeast Asia. The species is considered one of the world's worst weeds, and is classified as a noxious weed by the U.S. Department of Agriculture. It is established in the southeastern United States, and has been reported from scattered locations elsewhere in the contiguous United States. 'Itchgrass' aptly describes the effects of the hairs on the skin.

## 25.27 HACKELOCHLOA Kuntze[2]

Pl ann; csp. **Clm** 20–120 cm, erect to decumbent, often rooting at the lo nd, brchg above the bases. **Lvs** not aromatic; **shth** open; **aur** absent; **lig** memb, ciliate. **Infl** tml and ax, solitary, 2-sided rames, these smt fascicled and partially enclosed in subtending lf shth at maturity; **dis** in the rames, beneath the sessile spklt. **Spklt** in heterogamous sessile-pedlt pairs. **Sessile spklt** hemispherical, partly embedded in the rame axes; **lo glm** as long as the spklt, indurate, alveolate, indistinctly 7–11-veined, not keeled, mrg involute; **up glm** chartaceous, 3-veined, usu adherent to the rame axes; **lo flt** strl; **up flt** bisx; **anth** 3. **Ped** adnate to the rame axes, concealed by the sessile spklt. **Pedlt spklt** as long as or longer than the sessile spklt, ovate; **lo glm** dorsally compressed, 5–9-veined; **up glm** lat compressed, 5–7-veined; **lo flt** strl; **up flt** stmt; **anth** 3.

*Hackelochloa* is treated here as a unispecific genus that is widely distributed in warm regions of the world, often as a weed.

### 1. Hackelochloa granularis (L.) Kuntze Pitscale Grass [p. *503*, *558*]

*Hackelochloa granularis* is a native of the Eastern Hemisphere that has become established in cultivated land, roadsides, and weedy areas of the southern region of the United States. Its range extends south through Mexico and Central and South America.

## 25.28 TRIPSACUM L.[3]

Pl per; monoecious, stmt and pist spklt evidently distinct, located in the same infl, pist spklt below the stmt spklt. **Clm** 0.7–5 m. **Lvs** not aromatic; **shth** open; **lig** memb, erose to ciliate. **Infl** tml and ax, pan of 1–several subdigitate to rcm rames; **rames** with pist spklt proximally and stmt spklt distally; **dis** in the rames, beneath the pist spklt and at the base of the stmt portions. **Pist spklt** exposed, solitary, embedded in the indurate rame axes; **lo glm** coriaceous, closing the hollows in the rchs and concealing the flt; **up glm** similar but smaller; **lo flt** strl; **up flt** pist; **lm** and **pal** hyaline, unawned; **sty** 2, not fused. **Stmt spklt** paired, both sessile or both subsessile, or 1 sessile and the other pedlt; **glm** coriaceous, chartaceous, or memb; **lm** and **pal** hyaline, unawned. **Ped** (when present) not fused to the rame axes.

*Tripsacum* has 12 species, all of which are native to tropical and subtropical regions of the Western Hemisphere; three are native to the *Manual* region.

[1]J.K. Wipff  [2]John W. Thieret†  [3]Mary E. Barkworth

Measurements of the pistillate spikelets are based on measurements of the lower glumes of the sessile spikelets, the remainder of the spikelet being concealed between the rachis and the lower glume.

1. Staminate spikelets in sessile-pedicellate pairs, the pedicels almost flat to plano-convex in cross section, 2–5 mm long, less than 0.3 mm wide; glumes usually membranous (sect. *Fasciculata*) . . . . . . . . . . . . . . . . .1. *T. lanceolatum*
1. Staminate spikelets sessile, subsessile, or in sessile-pedicellate pairs, the pedicels triangular in cross section, to 2 mm long and about 0.5–0.8 mm wide; glumes somewhat coriaceous (sect. *Tripsacum*).
   2. Blades 9–35(45) mm wide, flat; culms 1–2(4) m tall . . . . . . . . . . . . . . . . . . . . . . . . . . . . . . . . .2. *T. dactyloides*
   2. Blades 1–7(15) mm wide, involute or folded; culms to 1 m tall . . . . . . . . . . . . . . . . . . . . . . . . . . .3. *T. floridanum*

## Tripsacum sect. Fasciculata Hitchc.

### 1. Tripsacum lanceolatum Rupr. *ex* E. Fourn.
MEXICAN GAMAGRASS [p. *504*, *558*]

*Tripsacum lanceolatum* grows in moist soil (often in canyon bottoms) of mountains from southeastern Arizona and southwestern New Mexico through Mexico to Guatemala. It has not been found in New Mexico since the 1800s.

## Tripsacum L. sect. Tripsacum

### 2. Tripsacum dactyloides (L.) L. EASTERN GAMAGRASS [p. *504*, *558*]

*Tripsacum dactyloides* grows in water courses and limestone outcrops from the central and eastern United States through Mexico to northern South America. Plants from the United States and northern Mexico belong to **Tripsacum dactyloides** var. **dactyloides**. They differ from those of the other two varieties in their erect stems and sessile staminate spikelets. Narrow-bladed plants of *T. dactyloides* from Texas resemble *T. floridanum*, but on transplanting to favorable conditions develop the wider blades characteristic of *T. dactyloides*. The two species can hybridize; the hybrids are partially sterile. *Tripsacum dactyloides* is also used as an ornamental grass, the chief attraction being its foliage.

### 3. Tripsacum floridanum Porter *ex* Vasey FLORIDA GAMAGRASS [p. *504*, *558*]

*Tripsacum floridanum* grows along roadsides and in pine woods, often in wet soils, of Florida and Cuba. It is grown as an ornamental. Reports of *T. floridanum* from Texas are based on narrow-bladed specimens of *T. dactyloides*.

# 25.29 ZEA L.[1]

**Pl** ann or per; monoecious, infl unisx or bisx with the pist spklt bas and the stmt spklt distal. **Clm** (0.2)0.5–6 m tall, 1–5 cm thick, solitary or several to many together, monopodial, often brchg (br frequently highly rdcd and hidden within the subtending lf shth), usu succulent when young, becoming wd with age; **lo nd** with prop roots; **intnd** pith-filled. **Lvs** not aromatic, cauline, distichous; **shth** open; **aur** smt present; **lig** memb, shortly ciliate; **bld** 2–12 cm wide, flat. **Pist or partially pist infl** tml on ax br; **stmt infl** (*tassels*) pan, of 1-many br or rames, smt with sec and tertiary brchg. WILD TAXA: **Pist infl** solitary, distichous rames (*ears*), these often tightly clustered in false pan, each usu wholly or partially enclosed by a thin prophyll and an equally thin bldless lf shth; **rames** composed of 5–15 spklt in 2 ranks; **dis** in the rame axes, dispersal units (*fruitcases*) consisting of an indurate, shiny rame segment and its embedded spklt. **Pist spklt** solitary, sessile, with 1 flt; **ped** and **pedlt spklt** suppressed; **lo glm** exceeding the flt, indurate on the cent, exposed portion, hyaline on the mrg, concealing the car at maturity. DOMESTICATED TAXON: **Pist infl** solitary, polystichous spikes (*ears*) terminating rdcd br, each spike surrounded by several to many, often bldless lf shth and a prophyll (*husks*), with 60–1000+ spklt in 8–24 rows, neither spikes nor spklt disarticulating at maturity. **Pist spklt** in subsessile pairs, each spklt with 1 fnctl flt; **glm** shorter than the spklt, indurate bas, hyaline distally; **lo flt** suppressed. ALL TAXA: **lm** and **pal** hyaline, unawned; **lod** absent; **ov** glab; **sty** (*silks*) 2, appearing solitary by being fused except at the very tip, filamentous, sides stigmatic. **Car** subspherical to dorsally compressed; **hila** round; **emb** about $2/3$ as long as the car. WILD TAXA: **Stmt pan** tml on the clm and pri br, smt also on the sec br and pist infl; **rames** distichous, similar in thickness and structure, axes disarticulating below the sessile spklt after pollination, abscission layers evident. DOMESTICATED TAXON: **Stmt pan** tml on the clm, cent axes always much thicker than the lat br and irregularly polystichous, lat br distichous to more or less polystichous, not disarticulating, without abscission layers below the sessile spklt. ALL TAXA: **Stmt spklt** in sessile-pedlt pairs, each with 2 stmt flt; **glm** memb to chartaceous, stiff to flexible, smt with a pair of winged keels, 5–14(28)-veined, acute; **lm** and **pal** hyaline; **lod** 2; **anth** 3.

*Zea* is an American genus of five species, four of which are native to montane Mexico and Central America. The fifth species, *Z. nicaraguensis* H.H. Iltis & B.F. Benz, is said to have been ubiquitous at one time in coastal Pacific Nicaragua, but is now known from only four or five small populations near sea level in seasonally flooded savannahs and riverine forests inland from the Bay of Fonseca, Nicaragua.

1. Pistillate inflorescences terete, with 2+ rows of paired spikelets, each spikelet with a functional floret, hence the spikelets in 4+ rank; staminate spikelets of wild taxa only slightly imbricate; lower glumes of the staminate spikelets flexible and translucent, loosely enclosing the upper glumes before anthesis, rounded on the back, the lateral veins not more prominent than those between, never forming winged keels; plants annual (sect. *Zea*) . . . . . . . . . . . . . . . . . . . . . . . . . . . . . . . . . . . . . . . . . . . .4. *Z. mays*

[1]Hugh H. Iltis

1. Pistillate inflorescences somewhat flattened, with 2 rows of solitary spikelets, hence the spikelets appearing 2-ranked; staminate inflorescences with densely imbricate spikelets; lower glumes of the staminate spikelets stiff, not translucent, strongly enclosing the upper glumes before anthesis, with more or less flat backs, the lateral veins evidently more prominent than those between, keeled and winged distally; plants annual or perennial (sect. *Luxuriantes*).

  2. Plants annual; lower glumes of the staminate spikelets narrowly winged; pistillate inflorescence units 1–many per node, usually enclosed by their subtending leaf sheaths, occasionally with 1–2 rames on naked peduncles that exceed the subtending leaf sheaths . . . . . . . . . . . . . . . . . . . . . . . . . 3. *Z. luxurians*

  2. Plants perennial, rhizomatous; lower glumes of the staminate spikelets strongly winged; pistillate inflorescence units 1–4(5) per node, almost always exceeding the subtending leaf sheaths.

    3. Rhizomes to 15 cm long, with internodes 0.2–0.6 cm long, often forming scaly, tuberous short shoots; culms to 3.5 m tall and 3 cm thick; leaf blades to 40 cm long and 4–5.5 cm wide . . . . . . . . 1. *Z. diploperennis*

    3. Rhizomes to 40 cm long or more, with internodes 1–6 cm long, lacking tuberlike shoots; culms to 2.5 m tall and 1.5–2 cm thick; leaf blades often to 65(80) cm long and 2–4.5 cm wide . . . . . . . . . . . 2. *Z. perennis*

## 1. Zea diploperennis H.H. Iltis, Doebley & R. Guzmán [p. *504*]

*Zea diploperennis*, although locally abundant, is rare in the wild, being known only from a few populations in the Sierra de Manantlán, Jalisco, Mexico. It grows at elevations of 1400–2400 m, sometimes forming large clones or extensive colonies in old maize fields and on the edges of oak-pine cloud forests. It is grown for genetic research and plant breeding in many countries and occasionally as an ornamental plant in warmer parts of the contiguous United States. It hybridizes infrequently with *Z. mays* subsp. *mays* in its native range.

## 2. Zea perennis (Hitchc.) Reeves & Mangelsd.
PERENNIAL TEOSINTE [p. *504*]

*Zea perennis* is parapatric to *Z. diploperennis*, being native to the northern base of the Volcán de Colima, Jalisco, Mexico, at elevations of 1520–2200 m. It is rare, although locally abundant, in and around maize fields and orchards in former open oak and pine forests and savannahs. *Zea perennis* crosses infrequently with *Z. mays* subsp. *mays*. The hybrids, being triploid, are sterile. It has also been cultivated at research stations in the United States for many years and it was reported to be established at James Island, South Carolina. It is not known if the population has persisted.

## 3. Zea luxurians (Durieu & Asch.) R.M. Bird
GUATEMALA TEOSINTE, FLORIDA TEOSINTE [p. *504*]

*Zea luxurians* is endemic to Central America, growing from Guatemala to Honduras, at elevations of 600–1200 m, and may extend into Oaxaca, Mexico. It was frequently grown for forage about a century ago, and is still sometimes grown for this purpose in the southern United States. Although it can hybridize with *Z. mays* subsp. *mays*, *Z. luxurians* rarely does so in the wild.

## 4. Zea mays L. [p. *504*]

Of the five subspecies of *Zea mays*, only the domesticated subspecies, *Z. mays* subsp. *mays*, is widely grown outside of research programs. Three wild subspecies are treated here, albeit briefly, because of their importance as genetic resources for *Z. mays* subsp. *mays*.

1. Pistillate inflorescences cylindrical spikes, 2–5(10) cm thick, with 8–24+ rows of spikelets pairs, each inflorescence tightly and permanently enclosed by several leaf sheaths and a large prophyll, not disarticulating at maturity; caryopses 60–1000+, not concealed by the glumes; staminate panicle branches not disarticulating below the sessile spikelets, lacking abscission layers; central axis of the staminate panicles polystichous, much thicker than the lateral branches; obligate domesticate . . . subsp. *mays*

1. Pistillate inflorescences cylindrical, distichous rames, less than 1 cm thick, with 2 rows of spikelet pairs, each rame usually enclosed by a single leaf sheath and a prophyll, disarticulating at maturity into fruitcases;

caryopses 4–15, each one concealed within a fruitcase; staminate panicles composed of rames that disarticulate below the sessile spikelets and have evident abscission layers; central axis of the staminate panicles similar in width to the rames; in the *Manual* region, wild taxa are known only from research plantings.

  2. Staminate spikelets (6.6)7.5–10.5 mm long; fruitcases 6–10 mm long, 4–6 mm wide; staminate panicles with 1–35+ ascending to divergent, rather stiff branches . . . . . . . . . . subsp. *mexicana*

  2. Staminate spikelets 4.6–7.2(7.9) mm long; fruitcases 5–8 mm long, 3–5 mm wide; staminate panicles usually with 10–100(235) divergent to nodding branches.

    3. Leaves pubescent; staminate panicles with (2)10–100(235) branches . . . . . . . . . . . . subsp. *parviglumis*

    3. Leaves glabrous or almost so; staminate panicles usually with fewer than 40 branches . . . . . . . . . . . . . . . . . . . . . . . subsp. *huehuetenangensis*

## Zea mays subsp. huehuetenangensis (H.H. Iltis & Doebley) Doebley HUEHUETENANGO TEOSINTE [p. *504*]

*Zea mays* subsp. *huehuetenangensis* is morphologically similar to subsp. *parviglumis*, but it often grows more than 5 m tall, and has essentially glabrous leaves, and smaller staminate panicles with fewer (less than 40), firmer branches, and different ecological, phenological, and molecular characteristics. It is endemic to the province of Huehuetenango, Guatemala, where it grows as a common weed on the edges of, and in, maize fields, and in seasonally moist oak cloud and tropical deciduous forests, at elevations from 900–1650 m. In its native range, it commonly hybridizes with *Z. mays* subsp. *mays*, both subspecies flowering from mid-December to mid-January, at the end of the wet season.

## Zea mays L. subsp. mays CORN, INDIAN CORN, MAIZE, MAÏS [p. *504*]

*Zea mays* subsp. *mays* is the familiar domesticated corn (or maize), from which around 400 indigenous races and many different kinds of cultivars have been developed. It is an obligate cultigen, unable to persist outside of cultivation because the caryopses are permanently attached to the rachis and enclosed by the subtending leaf sheaths.

## Zea mays subsp. mexicana (Schrad.) H.H. Iltis
CHALCO TEOSINTE, CENTRAL-PLATEAU TEOSINTE, NOBOGAME TEOSINTE [p. *504*]

*Zea mays* subsp. *mexicana* is a weedy taxon, native to upland Mexico. It is most abundant in the Meseta Central of the Mexican neo-volcanic plateau, at elevations of 1700–2500 m. It grows almost entirely in and around cornfields, and readily hybridizes with subsp. *mays*.

**Zea mays** subsp. **parviglumis** H.H. Iltis & Doebley
BALSAS TEOSINTE, GUERRERO TEOSINTE [p. *504*]

*Zea mays* subsp. *parviglumis*, which has the smallest fruitcases of all the wild taxa, is endemic to the Pacific slope of southern Mexico, from Oaxaca to Jalisco, being most abundant in the Balsas River drainage. It grows in highly seasonal, sunny thorn scrub, and open tropical deciduous forests and savannahs, at elevations of (450)600–1400(1950) m. One of its higher elevation populations appears to be the ancestor of subsp. *mays*. In the southern United States, *Z. mays* subsp. *parviglumis* is grown as part of breeding programs. In its native habitat, it tends to be seasonally isolated from subsp. *mays*, flowering a few weeks later, but the two sometimes form abundant hybrids in local areas.

## 25.30 COIX L.[1]

Pl ann or per; monoecious, pist and stmt spklt on separate rames in the same inflorescence. **Clm** to 3 m, erect, creeping, or floating, brchd; **intnd** solid. **Lvs** not aromatic; **lig** memb. **Infl** ax, of 2(3) rames, 1 pist, the other(s) stmt, pist rames completely enclosed in indurate, globose to cylindric, modified lf shth, termed *involucres*, from which the stmt rames protrude. **Pist rames** each with 3 spklt, 1 sessile and pist, the other 2 pedlt and rdmt; **sessile spklt** somewhat dorsally compressed; **glm** coriaceous, beaked; **stigmas** protruding from the involucres. **Car** more or less globose. **Stmt rames** flexible, exserted from the involucre; **spklt** in pairs or triplets, 1 sessile, the other(s) pedlt, rdcd, or absent; **lo glm** chartaceous, with 15 or more veins, 2-keeled, keels winged above; **up glm** similar, with 1 keel; **lo flt** smt strl; **up flt** stmt; **stamens** 0 or 3; **lod** 2. **Ped** not fused to the rame axes.

*Coix* is a genus of about five species, one of which has been introduced into the *Manual* region. All the species are native to tropical Asia.

1. **Coix lacryma-jobi** L. JOB'S-TEARS [p. *504*, *558*]

*Coix lacryma-jobi* is a tall, maize-like plant. In North America, it is usually grown as an ornamental, but it has become established at scattered locations in the *Manual* region. The involucres, which can be used as beads, may be white, blue, pink, straw, gray, brown, or black, with the color being distributed evenly, irregularly, or in stripes. Cultivars with easily removed involucres are grown for food and beverage, especially in Asia.

## LITERATURE CITED

Aliscioni, S.S., L.M. Guissani, F.O. Zuloaga, and E.A. Kellogg. 2003. A molecular phylogeny of *Panicum* (Poaceae:Paniceae): tests of monophyly and phylogenetic placement witin the Panicoideae. Amer. J. Bot. 90:796–821.

Bess, E.C., A.N. Doust, G. Davidse, and E.A. Kellogg. 2006. *Zuloagaea*, a new genus of neotropical grass within the "Bristle Clade" (Poaceae: Paniceae). Syst. Bot. 31:656–670.

Finot, V.L., P.M. Peterson, R.J.Soreng, and F.O. Zuloaga. 2004. A revision of *Trisetum, Peyritschia,* and *Sphenopholis* (Poaceae: Pooideae: Aveninae) in Mexico and Central America. Ann Missouri Bot. Gard. 91:1–30.

Finot, V.L., P.M. Peterson, F.O. Zuloaga, R.J. Soreng, and O. Matthei. 2005. A revision of *Trisetum* (Poaceae: Pooideae: Aveninae) in South America. Ann. Missouri Bot. Gard. In Press.

Finot, V.L., P.M. Peterson, R.J. Soreng, and F.O. Zuloaga. 2005. A revision of *Trisetum* and *Graphephorum* (Poaceae: Pooideae: Aveninae) in North America north of Mexico. SIDA 21(3):1419–1453.

Guo, Y.-L. and S. Ge. 2005. Molecular phylogeny of Oryzeae (Poaceae) based on DNA sequences from chloroplast, mitochondrial, and nuclear genomes. Amer. J. Bot. 92:1548–1558.

Londo, J.P., Y.-C. Chiang, K.-H. Hung, T.-Y. Chiang, and B.A. Schaal. 2006. Phylogeography of Asian wild rice, *Oryza rufipogon,* reveals multiple independent domestications of cultivated rice, *Oryza sativa.* PNAS 103:9578–9583.

Saltonstall, K. 2002. Cryptic invasion by a non-native genotype of the common reed, *Phragmites australis,* into North America. PNAS 99:2445–2449.

Saltonstall, K. and IAN Group. *Phragmites:* native or introduced. http://ian.umces.edu/pdfs/iannewsletter7.pdf

Saltonstall, K. 2003a. Genetic variation among North American Populations of *Phragmites australis*; implications for management. Estuaries 26:444–451.

Saltonstall, K. 2003b. Microsatellite variation within and among North American lineages of *Phragmites australis.* Mol. Ecol. 12:1689–1702.

Simon, B.K. and S.W.L. Jacobs. 2003. *Megathyrsus,* a new generic name for *Panicum* subgenus *Megathyrsus.* Austrobaileya 6:33, 571-574.

Tsvelev, N.N. 2006. Kratkii obzor roda mannik *Glyceria* (Poaceae) [Synopsis of the mannagrass genus *Glyceria* (Poaceae)]. Bot. Zhurn. (Moscow & Leningrad) 91:255–276. [Translation being posted to http://utc/usu/edu/Meliceae/Glyceria.Tsvelev.htm].

Wilson, B.L. 2007. A new variety of *Festuca roemeri* (Poaceae) of California Floristic Province of North America. J. Bot. Res. Inst. Texas 1:59–67.

Zuloaga, F.O., L.M. Giussani, and O. Morrone. 2007. *Hopia,* a new monotypic genus segregated from *Panicum* (*Poaceae*). Taxon 56: 145-156.

[1]John W. Thieret†

# Illustrations

The illustrators of the plates are indicated by the initials placed on them. In many instances, two sets of initials will be found on a single plate. The first set of initials belongs to the illustrator who designed the plate and prepared the rough illustrations; the second set belongs to the illustrator who converted the rough pencil illustration into a finished inked plate. Because of the number of illustrators involved, only those who served as the primary illustrator for one or more plates are listed on the cover page. The initials of all illustrators are listed below in alphabetical order.

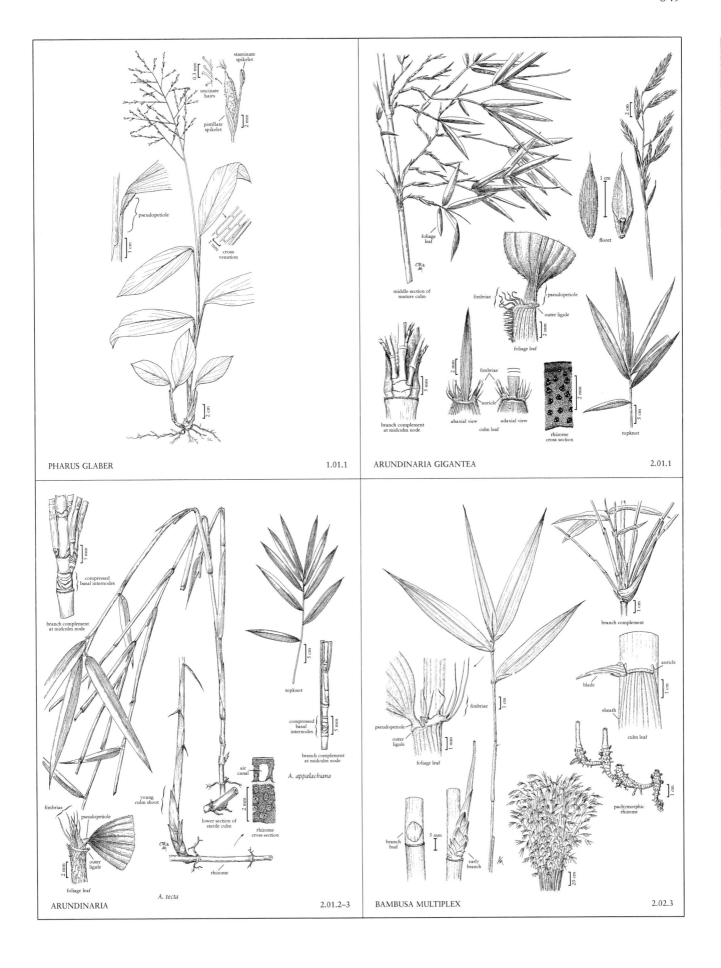

PHARUS GLABER                                                    1.01.1

ARUNDINARIA GIGANTEA                                            2.01.1

ARUNDINARIA                                                    2.01.2–3

BAMBUSA MULTIPLEX                                              2.02.3

350

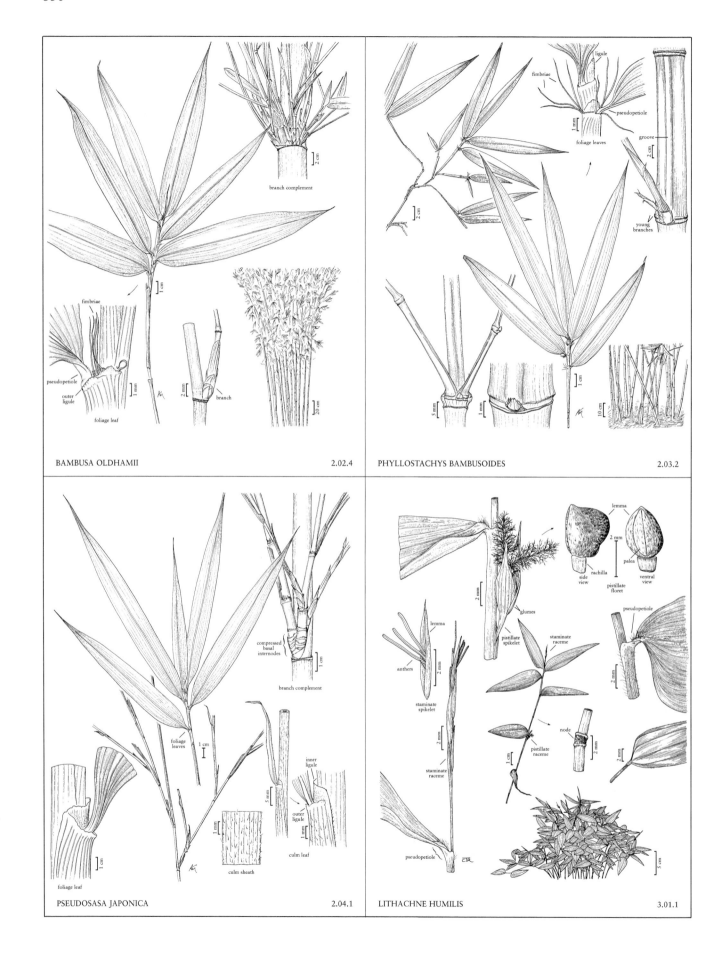

BAMBUSA OLDHAMII                                           2.02.4

PHYLLOSTACHYS BAMBUSOIDES                                  2.03.2

PSEUDOSASA JAPONICA                                        2.04.1

LITHACHNE HUMILIS                                          3.01.1

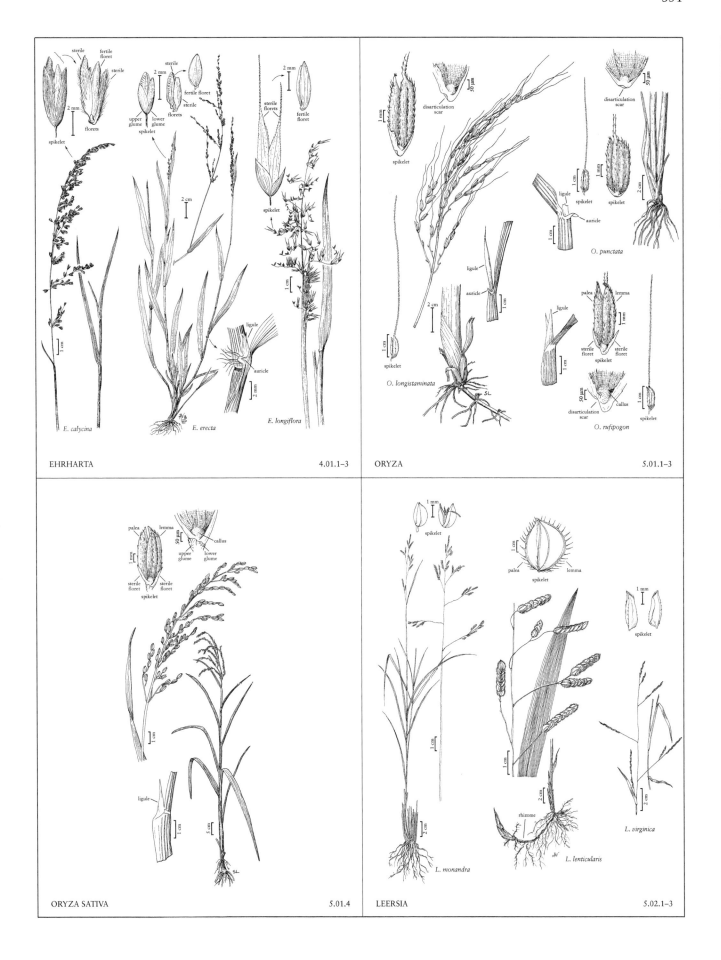

EHRHARTA 4.01.1–3

ORYZA 5.01.1–3

ORYZA SATIVA 5.01.4

LEERSIA 5.02.1–3

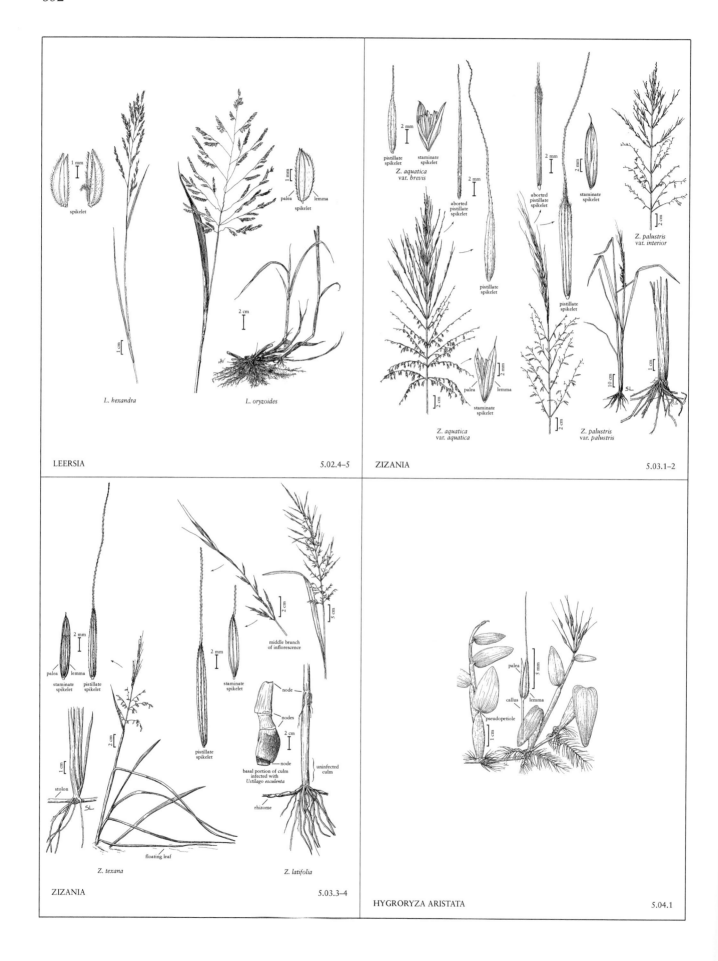

L. hexandra

L. oryzoides

palea   lemma
spikelet

spikelet

LEERSIA                                    5.02.4–5

pistillate   staminate
spikelet     spikelet
Z. aquatica
var. brevis

aborted
pistillate
spikelet

pistillate
spikelet

aborted
pistillate
spikelet

staminate
spikelet

Z. palustris
var. interior

pistillate
spikelet

palea   lemma
staminate
spikelet

Z. aquatica
var. aquatica

Z. palustris
var. palustris

ZIZANIA                                    5.03.1–2

palea   lemma
staminate    pistillate
spikelet     spikelet

middle branch
of inflorescence

staminate
spikelet

pistillate
spikelet

node

nodes

node

basal portion of culm
infected with
Ustilago esculenta

uninfected
culm

rhizome

stolon

floating leaf

Z. texana

Z. latifolia

ZIZANIA                                    5.03.3–4

palea

callus   lemma

pseudopetiole

HYGRORYZA ARISTATA                         5.04.1

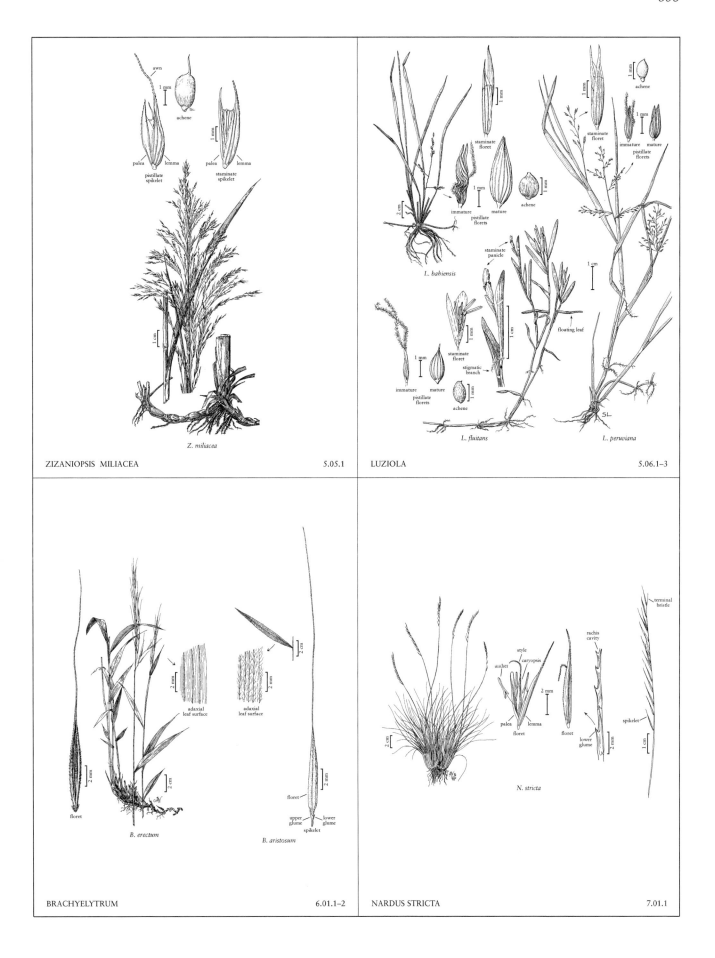

ZIZANIOPSIS MILIACEA 5.05.1

LUZIOLA 5.06.1–3

BRACHYELYTRUM 6.01.1–2

NARDUS STRICTA 7.01.1

354

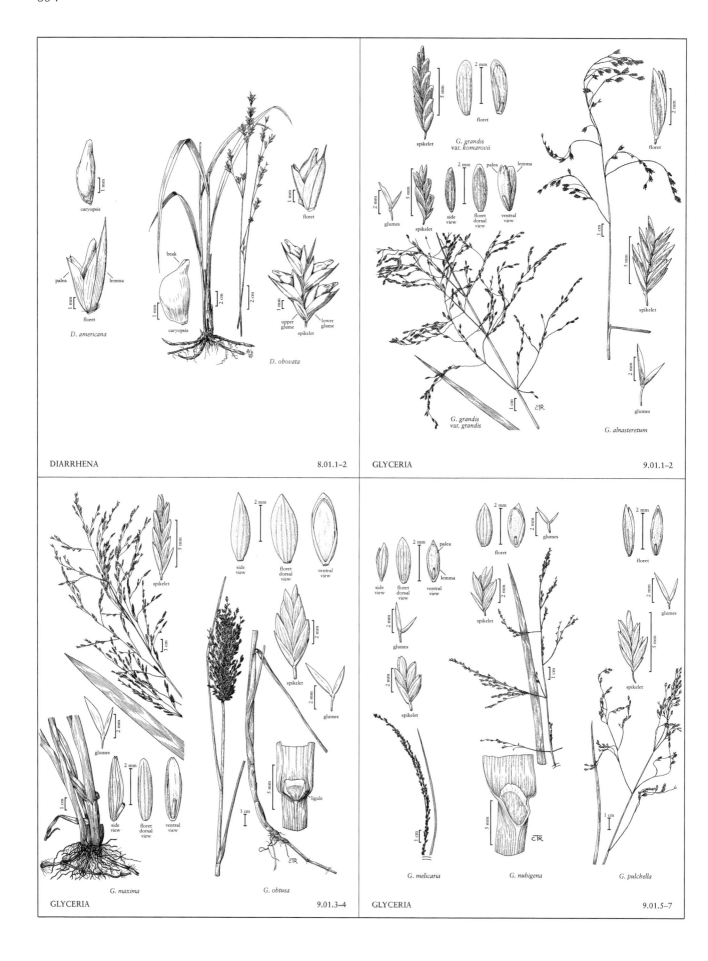

D. americana

beak

palea          lemma

floret

caryopsis

D. obovata

2 cm

2 cm

floret

upper          lower
glume          glume
spikelet

caryopsis

DIARRHENA                                      8.01.1–2

spikelet

2 mm

floret

G. grandis
var. komarovii

glumes          spikelet          side
view

palea          lemma

floret
dorsal
view          ventral
view

G. grandis
var. grandis

floret

spikelet

glumes

G. alnasteretum

GLYCERIA                                        9.01.1–2

spikelet

glumes

side
view          floret
dorsal
view          ventral
view

G. maxima

2 mm

spikelet

glumes

ligule

G. obtusa

GLYCERIA                                        9.01.3–4

side
view          floret
dorsal
view          ventral
view          palea

lemma

glumes

spikelet

G. melicaria

floret

spikelet

glumes

G. nubigena

floret

glumes

spikelet

G. pulchella

GLYCERIA                                        9.01.5–7

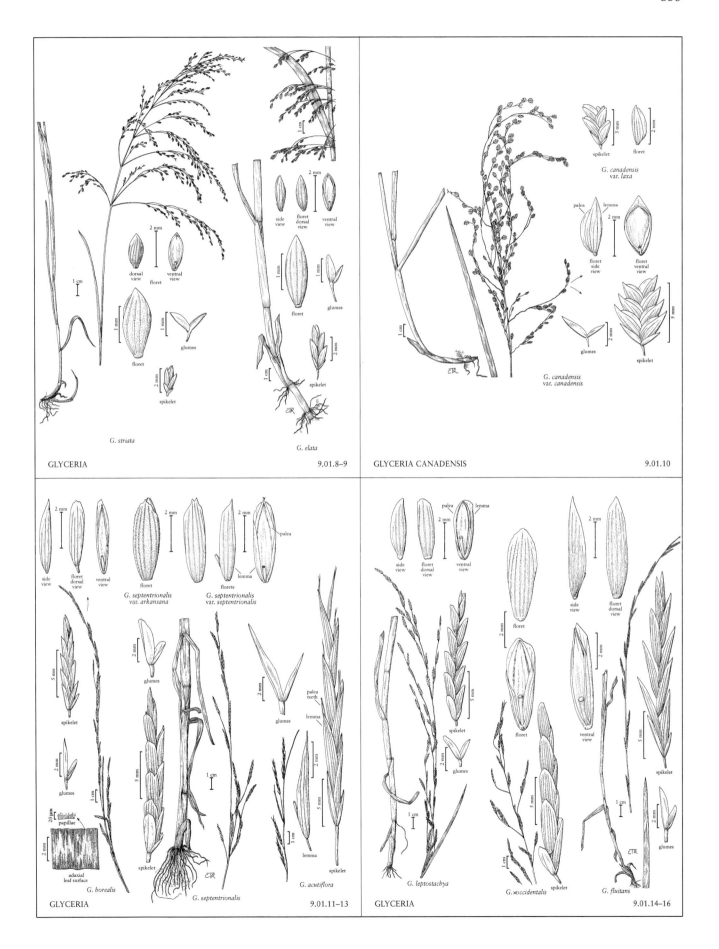

GLYCERIA

9.01.8–9

GLYCERIA CANADENSIS

9.01.10

*G. striata*

*G. elata*

*G. canadensis*
var. *laxa*

*G. canadensis*
var. *canadensis*

*G. septentrionalis*
var. *arkansana*

*G. septentrionalis*
var. *septentrionalis*

*G. borealis*

*G. septentrionalis*

*G. acutiflora*

GLYCERIA

9.01.11–13

*G. leptostachya*

*G. ×occidentalis*

*G. fluitans*

GLYCERIA

9.01.14–16

356

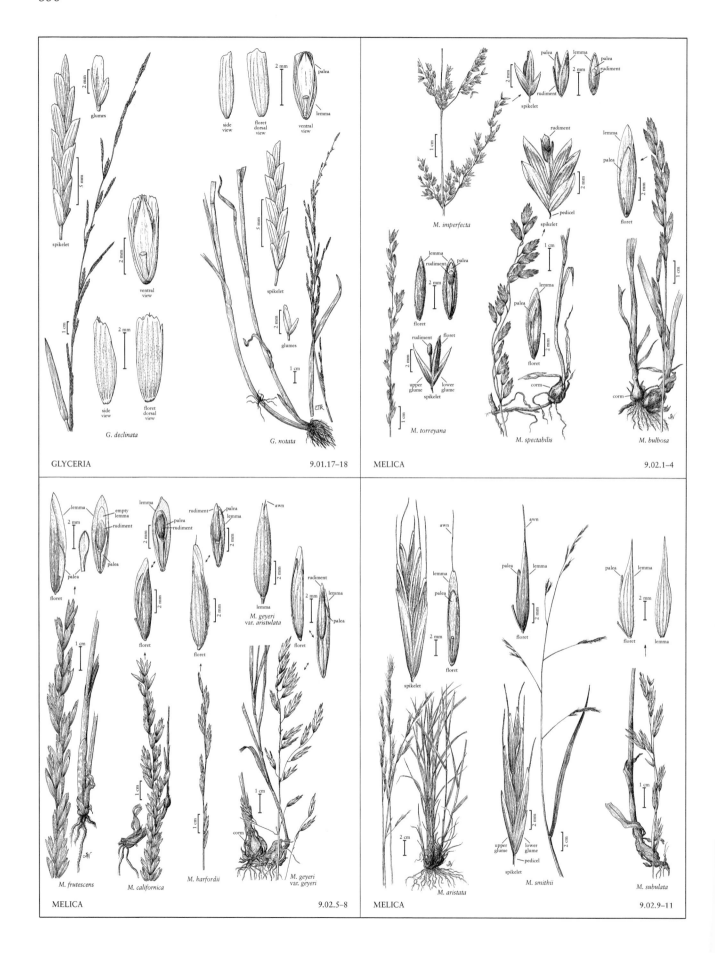

G. declinata

spikelet

glumes

2 mm

5 mm

side view

floret dorsal view

ventral view

1 cm

2 mm

side view

floret dorsal view

2 mm

2 mm

palea

lemma

ventral view

5 mm

spikelet

glumes

2 mm

1 cm

G. notata

GLYCERIA

9.01.17–18

M. imperfecta

palea

lemma

palea

rudiment

spikelet

2 mm

2 mm

1 cm

rudiment

spikelet

pedicel

2 mm

lemma

palea

floret

2 mm

lemma

rudiment

palea

floret

2 mm

rudiment

floret

upper glume

lower glume

spikelet

2 mm

1 cm

M. torreyana

palea

lemma

floret

2 mm

corm

1 cm

M. spectabilis

corm

1 cm

M. bulbosa

MELICA

9.02.1–4

lemma

empty lemma

rudiment

palea

floret

2 mm

lemma

palea

rudiment

floret

2 mm

rudiment

palea

lemma

floret

2 mm

floret

2 mm

awn

lemma

M. geyeri var. aristulata

2 mm

rudiment

lemma

palea

floret

2 mm

1 cm

1 cm

1 cm

1 cm

corm

M. frutescens

M. californica

M. harfordii

M. geyeri var. geyeri

MELICA

9.02.5–8

awn

lemma

palea

spikelet

floret

2 mm

awn

palea

lemma

floret

2 mm

palea

lemma

floret

lemma

2 mm

2 cm

upper glume

lower glume

pedicel

spikelet

2 mm

2 cm

1 cm

M. aristata

M. smithii

M. subulata

MELICA

9.02.9–11

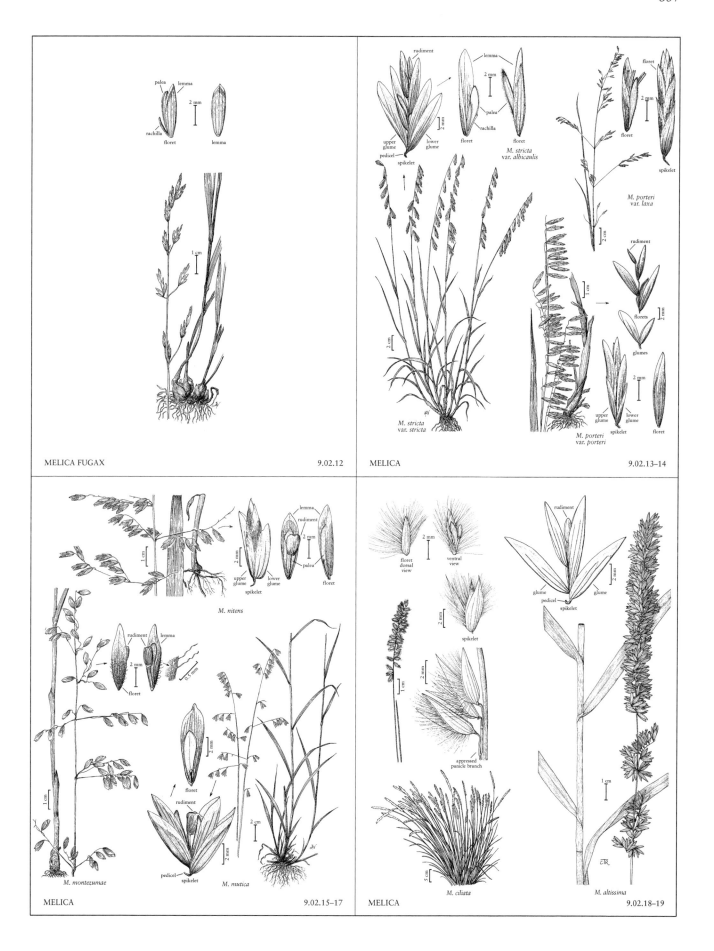

MELICA FUGAX

9.02.12

MELICA

9.02.13–14

MELICA

9.02.15–17

MELICA

9.02.18–19

358

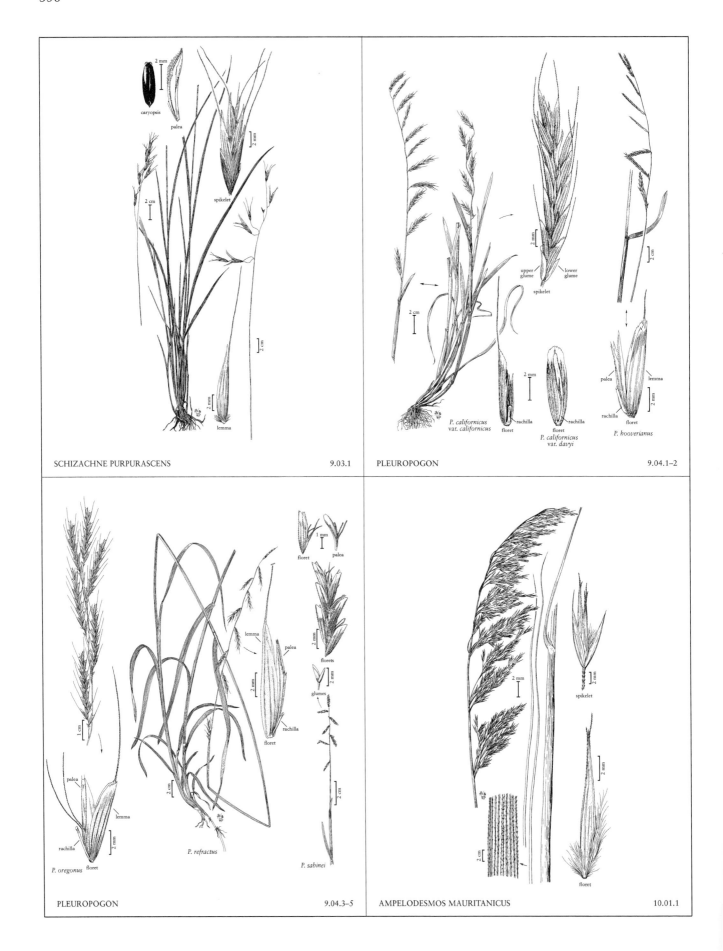

SCHIZACHNE PURPURASCENS                               9.03.1

PLEUROPOGON                                            9.04.1–2

PLEUROPOGON                                            9.04.3–5

AMPELODESMOS MAURITANICUS                              10.01.1

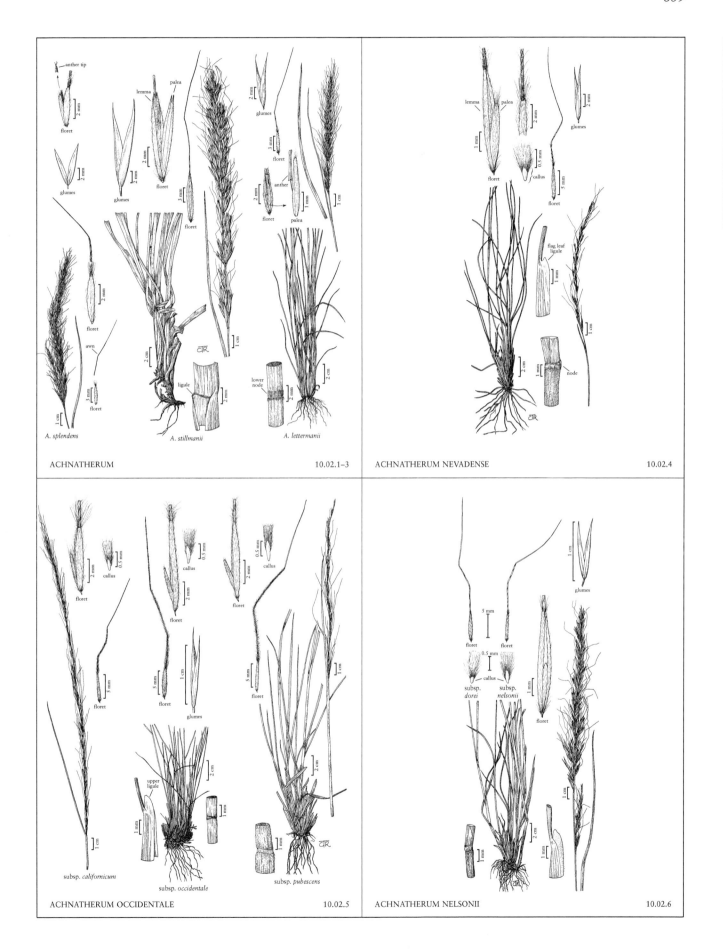

ACHNATHERUM                                              10.02.1–3

ACHNATHERUM NEVADENSE                                    10.02.4

ACHNATHERUM OCCIDENTALE                                  10.02.5

ACHNATHERUM NELSONII                                     10.02.6

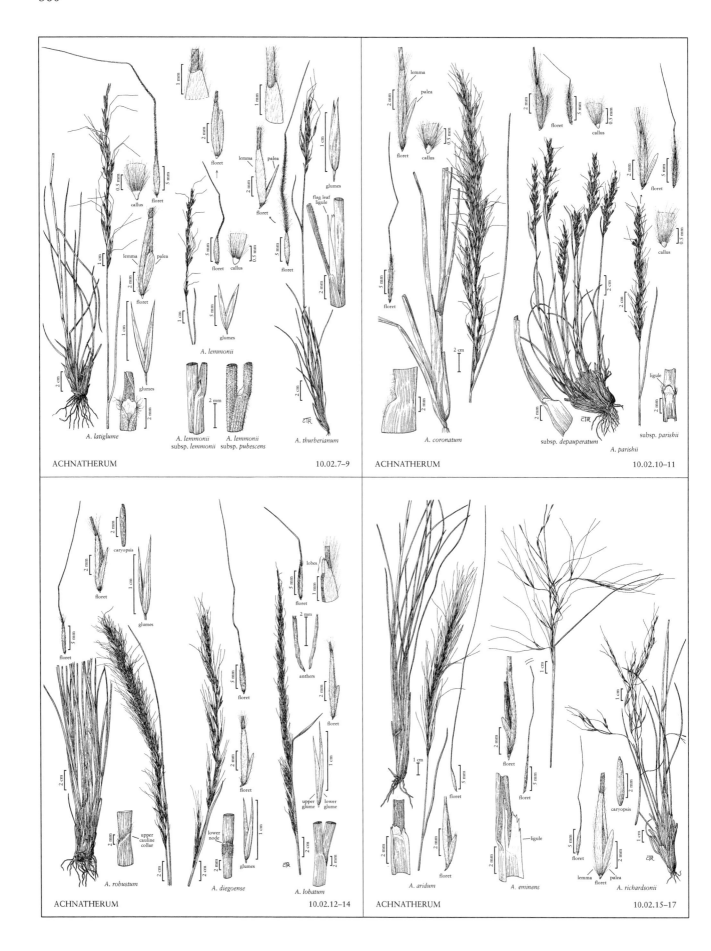

360

ACHNATHERUM

10.02.7–9

ACHNATHERUM

10.02.10–11

ACHNATHERUM

10.02.12–14

ACHNATHERUM

10.02.15–17

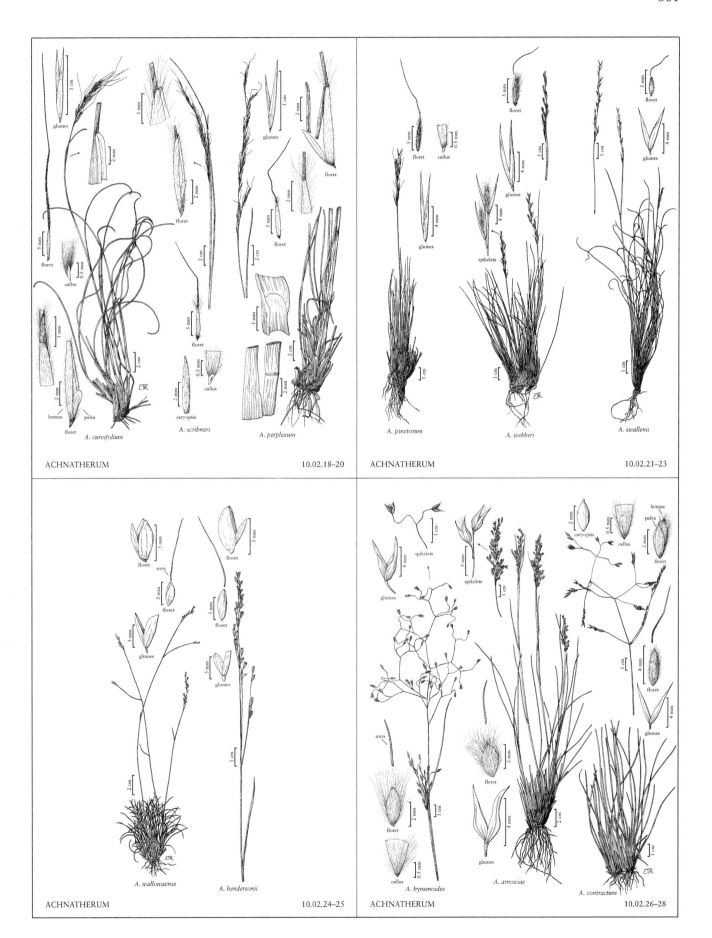

ACHNATHERUM 10.02.18–20

*A. curvifolium*

*A. scribneri*

*A. perplexum*

ACHNATHERUM 10.02.21–23

*A. pinetorum*

*A. webberi*

*A. swallenii*

ACHNATHERUM 10.02.24–25

*A. wallowaense*

*A. hendersonii*

ACHNATHERUM 10.02.26–28

*A. hymenoides*

*A. arnowiae*

*A. contractum*

361

362

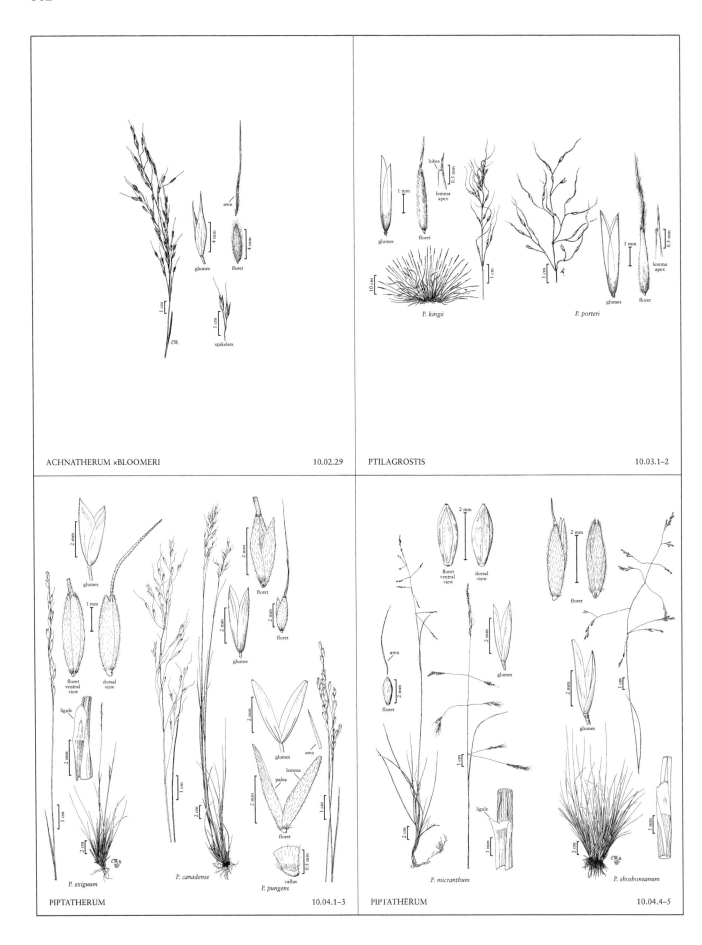

*P. kingii*

*P. porteri*

*P. exiguum*

*P. canadense*

*P. pungens*

*P. micranthum*

*P. shoshoneanum*

PIPTATHERUM                                    10.04.6–7

MACROCHLOA TENACISSIMA                          10.05.1

CELTICA GIGANTEA                               10.06.1

STIPA                                          10.07.1–2

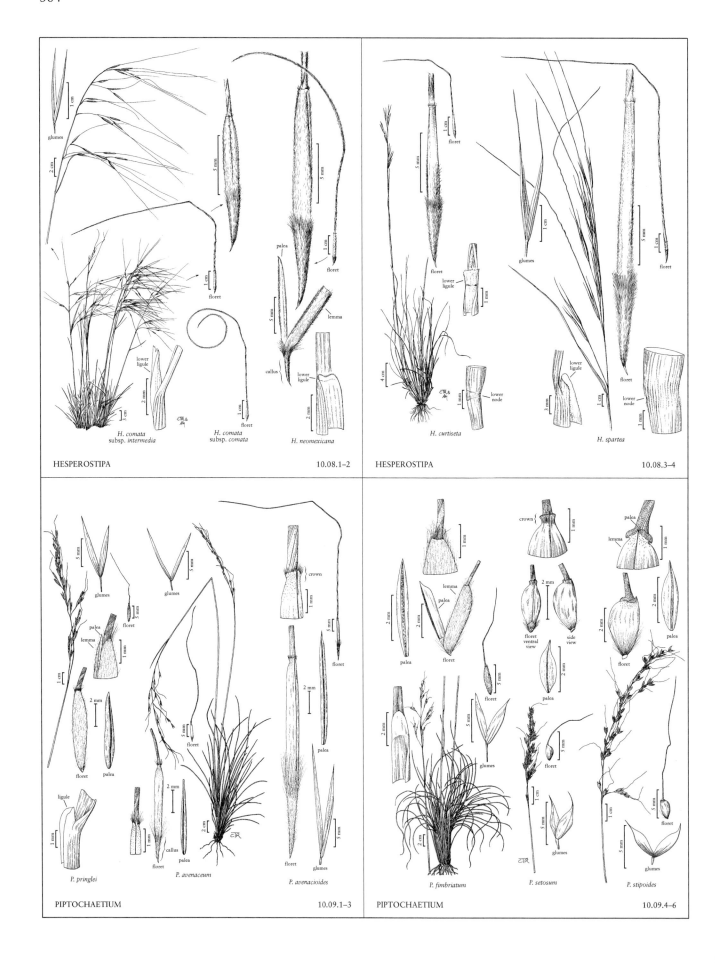

HESPEROSTIPA

10.08.1–2

HESPEROSTIPA

10.08.3–4

PIPTOCHAETIUM

10.09.1–3

PIPTOCHAETIUM

10.09.4–6

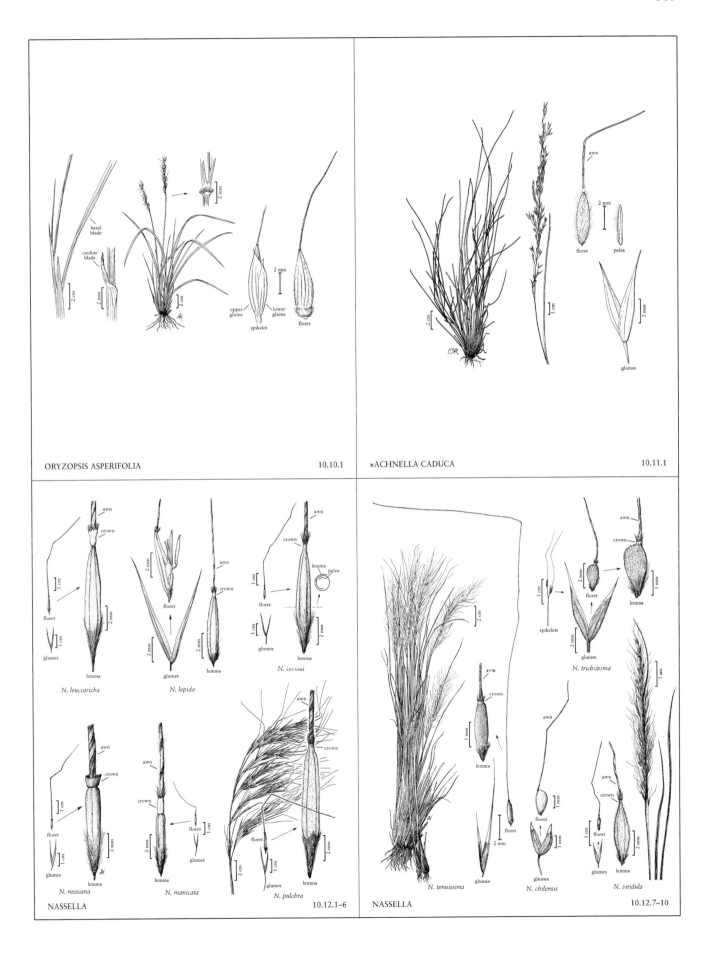

ORYZOPSIS ASPERIFOLIA                    10.10.1

×ACHNELLA CADUCA                    10.11.1

NASSELLA                    10.12.1–6

NASSELLA                    10.12.7–10

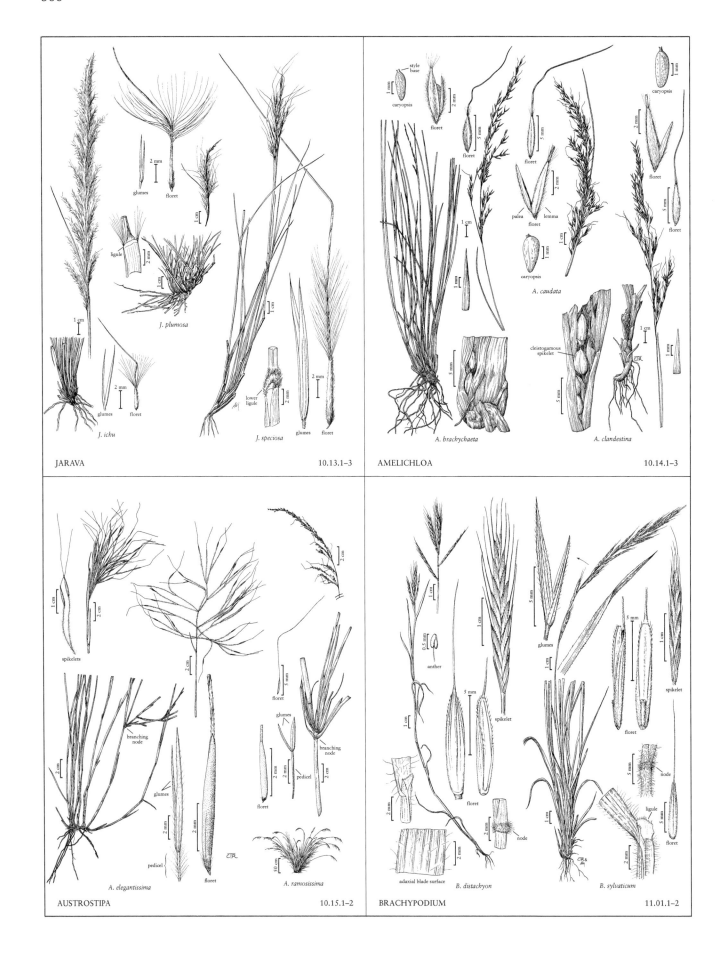

JARAVA                                                    10.13.1–3

AMELICHLOA                                                10.14.1–3

AUSTROSTIPA                                               10.15.1–2

BRACHYPODIUM                                             11.01.1–2

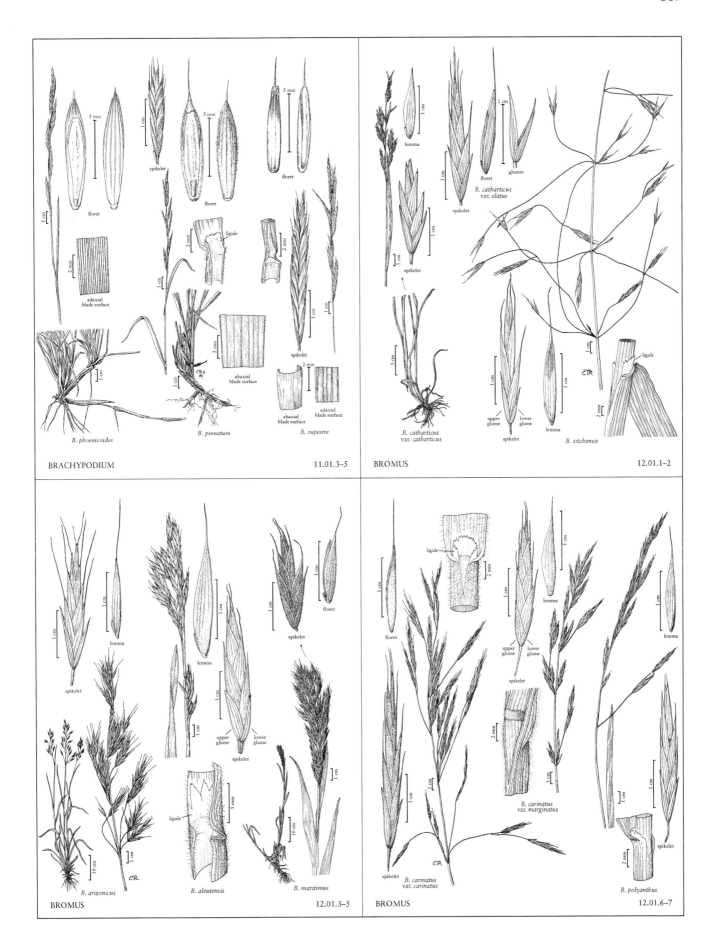

BRACHYPODIUM                    11.01.3–5

*B. phoenicoides*          *B. pinnatum*          *B. rupestre*

BROMUS                         12.01.1–2

*B. catharticus*
var. *catharticus*          *B. catharticus*
var. *elatus*          *B. sitchensis*

BROMUS                         12.01.3–5

*B. arizonicus*          *B. aleutensis*          *B. maritimus*

BROMUS                         12.01.6–7

*B. carinatus*
var. *carinatus*          *B. carinatus*
var. *marginatus*          *B. polyanthus*

BROMUS

12.01.8–10

B. riparius

B. inermis

B. pumpellianus subsp. dicksonii

B. pumpellianus subsp. pumpelliani

lemmas

spikelet

lemma

upper glume

lower glume

spikelet

ligule

auricle

node

CTR

BROMUS

12.01.11–13

B. kalmii

B. latiglumis

B. laevipes

glumes

spikelet

lemma

lemma

upper glume

lower glume

spikelet

auricle

flange

ligule

CTR

BROMUS

12.01.14–16

B. pseudolaevipes

B. hallii

B. orcuttianus

spikelet

lemma

lemma

lemma

spikelet

spikelet

auricle

ligule

lower leaf

upper leaf

CTR

BROMUS

12.01.17–18

B. grandis

B. frondosus

glumes

lemma

spikelet

lemma

upper glume

lower glume

spikelet

ligule

CTR

BROMUS                                                        12.01.19–21

BROMUS                                                        12.01.22–24

BROMUS                                                        12.01.25–27

BROMUS                                                        12.01.28–30

370

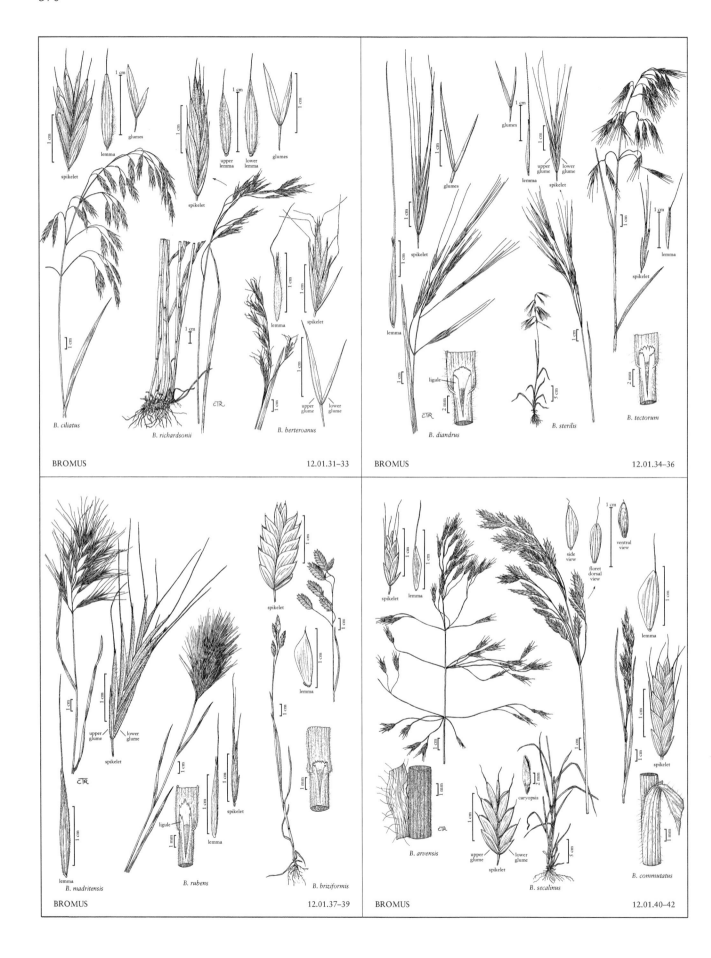

B. ciliatus

B. richardsonii

B. berteroanus

B. diandrus

B. sterilis

B. tectorum

B. madritensis

B. rubens

B. briziformis

B. arvensis

B. secalinus

B. commutatus

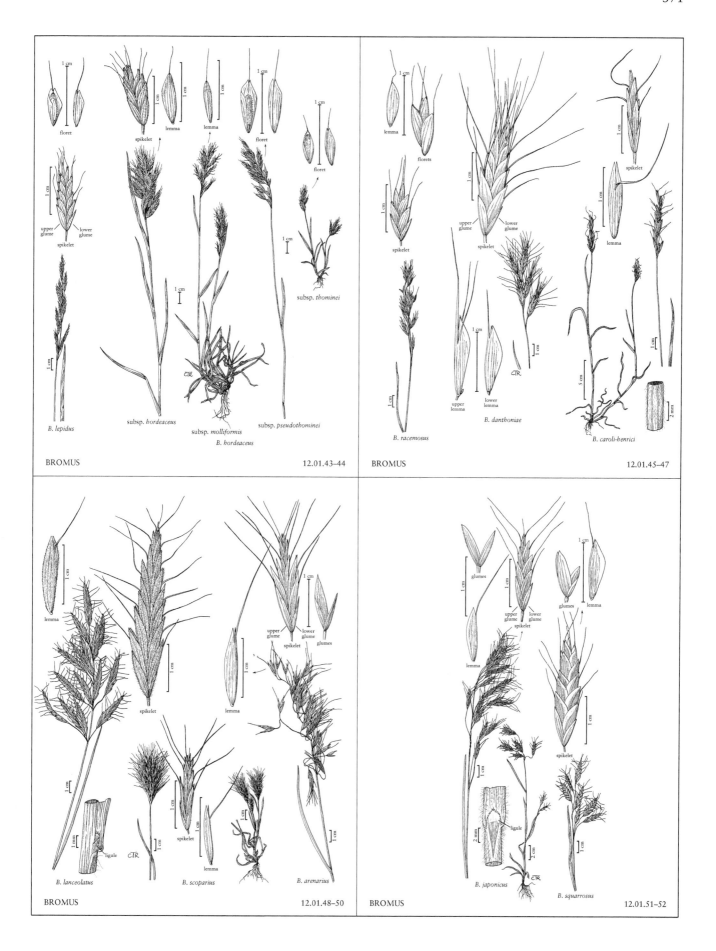

BROMUS                                    12.01.43–44

B. lepidus

subsp. hordeaceus

subsp. molliformis
B. hordeaceus

subsp. pseudothominei

subsp. thominei

floret
lemma
lemma
floret
floret
spikelet
upper glume
lower glume
spikelet

BROMUS                                    12.01.45–47

B. racemosus

B. danthoniae

B. caroli-henrici

lemma
florets
spikelet
upper glume
lower glume
spikelet
upper lemma
lower lemma
spikelet
lemma

BROMUS                                    12.01.48–50

B. lanceolatus

B. scoparius

B. arenarius

lemma
spikelet
ligule
upper glume
lower glume
spikelet
glumes
lemma
spikelet
lemma

BROMUS                                    12.01.51–52

B. japonicus

B. squarrosus

glumes
upper glume
lower glume
spikelet
glumes
lemma
lemma
ligule
spikelet

372

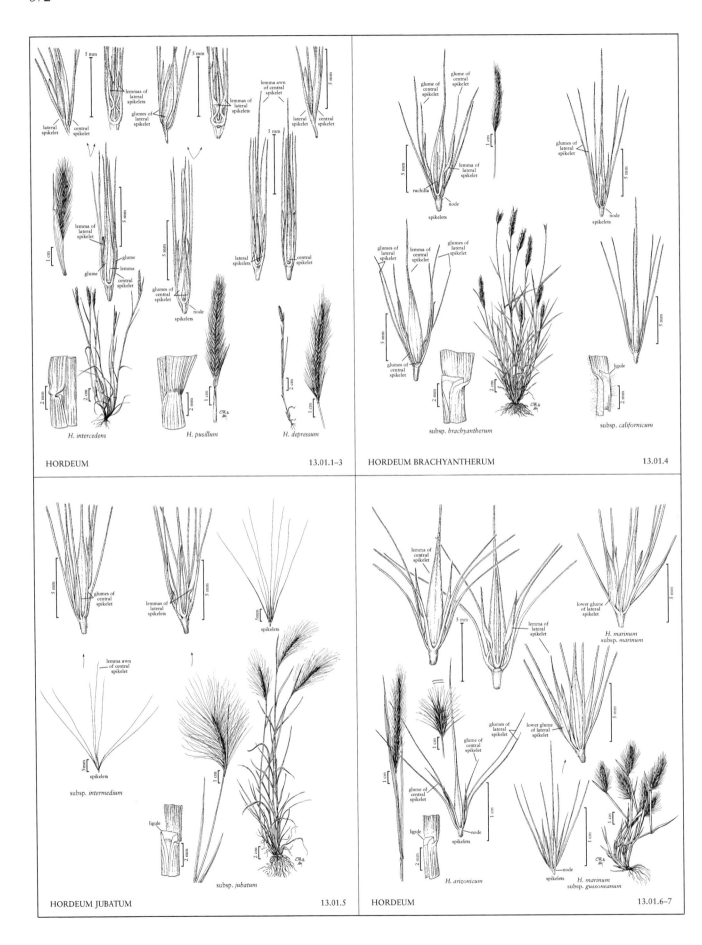

HORDEUM        13.01.1–3

HORDEUM BRACHYANTHERUM        13.01.4

HORDEUM JUBATUM        13.01.5

HORDEUM        13.01.6–7

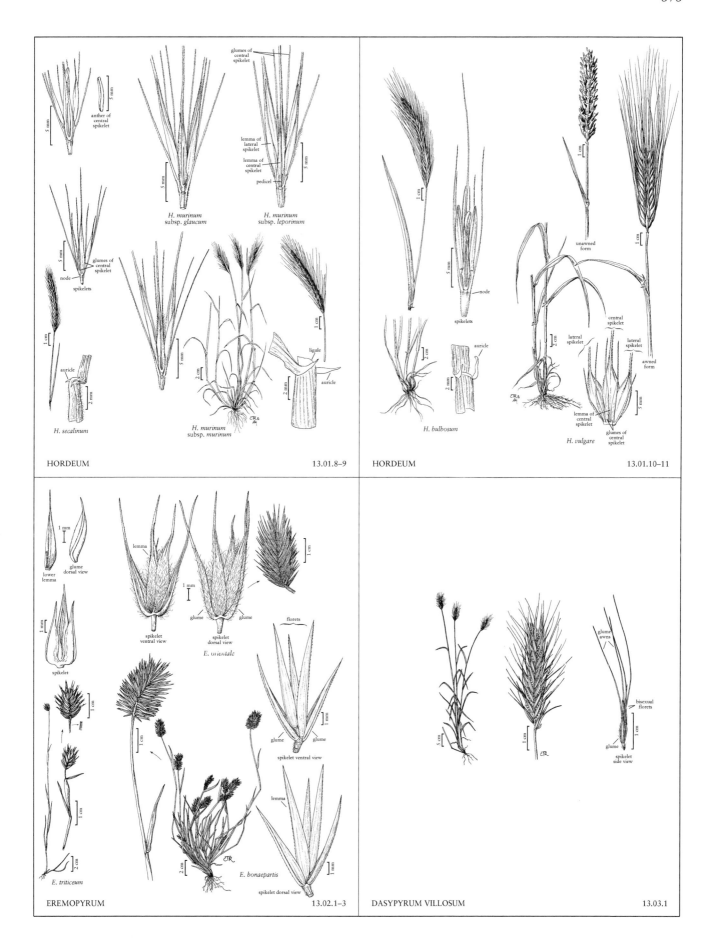

HORDEUM            13.01.8–9

HORDEUM            13.01.10–11

EREMOPYRUM            13.02.1–3

DASYPYRUM VILLOSUM            13.03.1

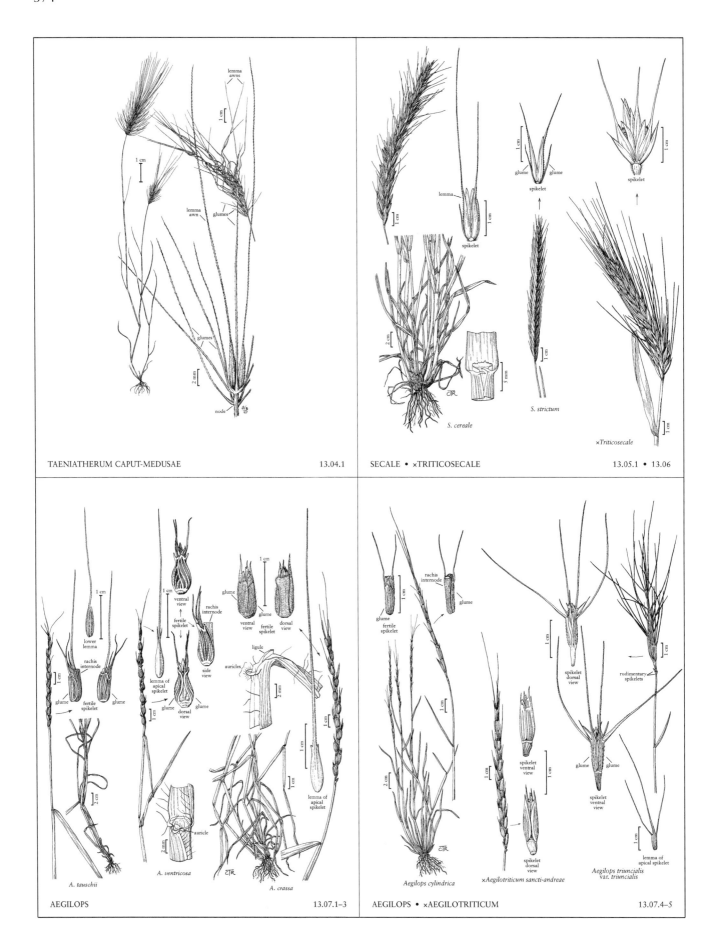

TAENIATHERUM CAPUT-MEDUSAE 13.04.1

SECALE • ×TRITICOSECALE 13.05.1 • 13.06

AEGILOPS 13.07.1–3

AEGILOPS • ×AEGILOTRITICUM 13.07.4–5

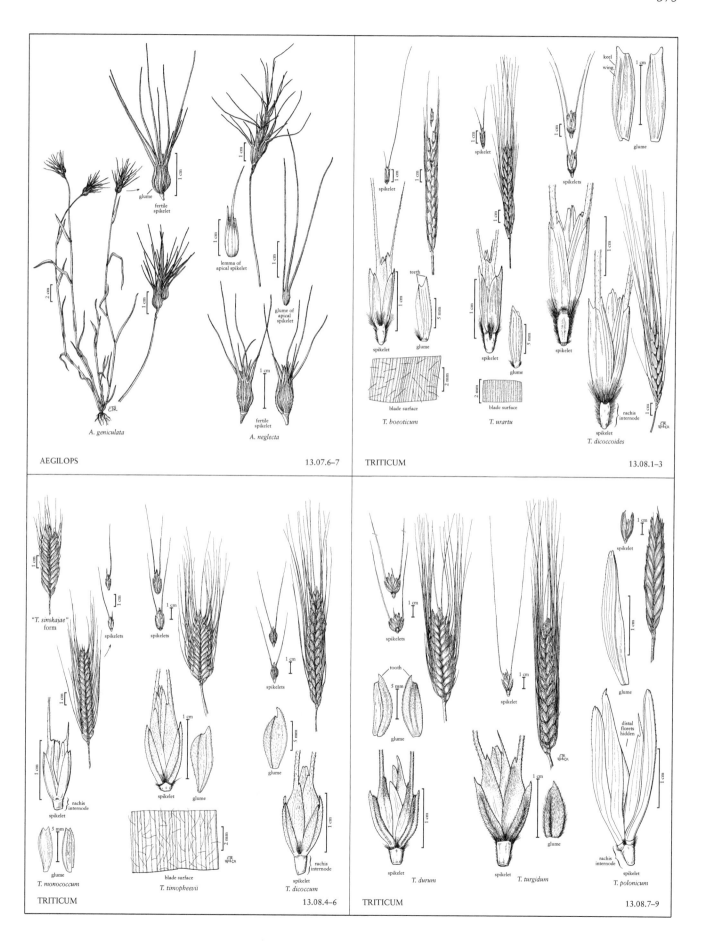

AEGILOPS

A. geniculata

glume

fertile
spikelet

lemma of
apical spikelet

glume of
apical spikelet

fertile
spikelet

A. neglecta

13.07.6–7

TRITICUM

spikelet

spikelet

teeth

glume

blade surface

T. boeoticum

spikelet

spikelets

spikelet

glume

blade surface

T. urartu

keel
wing

glume

spikelets

spikelet

rachis
internode

spikelet

T. dicoccoides

13.08.1–3

TRITICUM

"T. sinskajae"
form

spikelets

spikelets

spikelet

rachis
internode

glume

spikelet

T. monococcum

blade surface

T. timopheevii

spikelet

glume

spikelets

glume

spikelet

rachis
internode

T. dicoccum

13.08.4–6

TRITICUM

spikelets

tooth

glume

spikelet

T. durum

spikelet

spikelet

glume

spikelet

T. turgidum

spikelet

glume

distal
florets
hidden

glume

rachis
internode

spikelet

T. polonicum

13.08.7–9

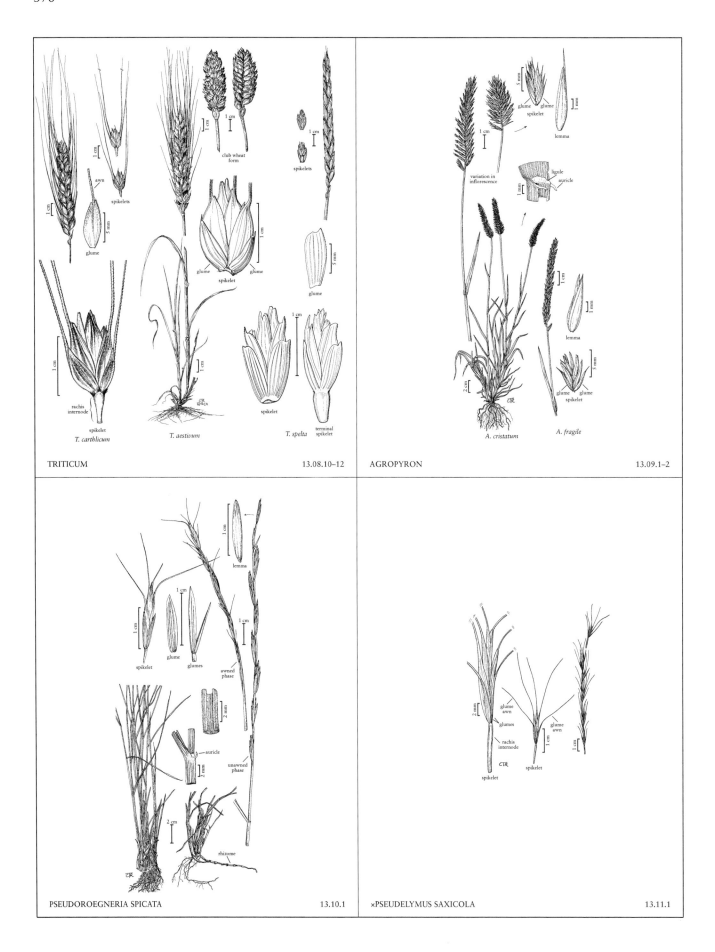

TRITICUM                                    13.08.10–12

AGROPYRON                                   13.09.1–2

PSEUDOROEGNERIA SPICATA                     13.10.1

×PSEUDELYMUS SAXICOLA                       13.11.1

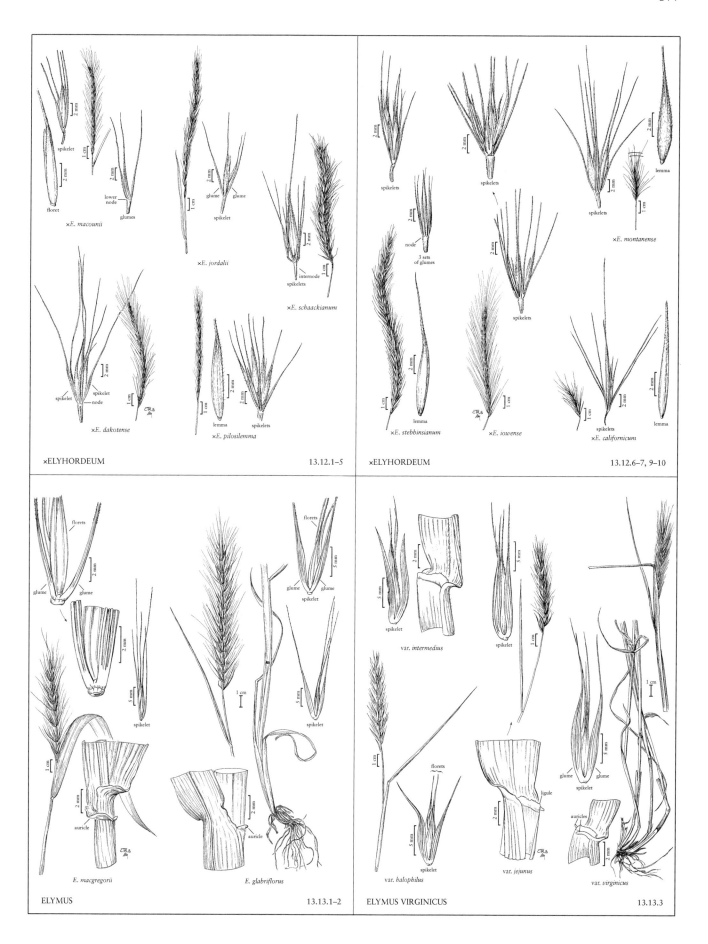

×E. macounii

lower node

spikelet

floret

glumes

×E. jordalii

glume glume

spikelet

×E. schaackianum

internode

spikelets

×E. dakotense

spikelet spikelet

node

×E. pilosilemma

lemma

spikelets

×ELYHORDEUM

13.12.1–5

spikelets

node

3 sets of glumes

spikelets

×E. stebbinsianum

lemma

×E. iowense

spikelets

×E. montanense

lemma

spikelets

×E. californicum

spikelets

lemma

×ELYHORDEUM

13.12.6–7, 9–10

florets

glume glume

spikelet

florets

glume glume

spikelet

spikelet

auricle

auricle

E. macgregorii

E. glabriflorus

ELYMUS

13.13.1–2

var. intermedius

spikelet

spikelet

florets

spikelet

var. halophilus

ligule

var. jejunus

spikelet

glume glume

spikelet

auricles

var. virginicus

ELYMUS VIRGINICUS

13.13.3

378

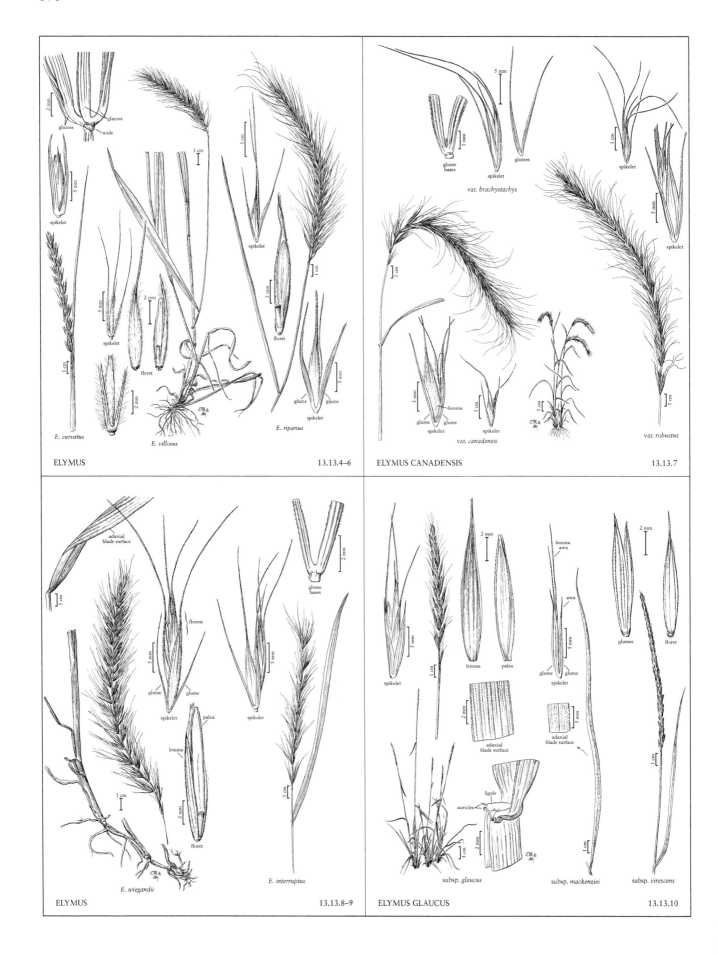

E. curvatus

E. villosus

E. riparius

ELYMUS                                                        13.13.4–6

var. brachystachys

var. canadensis                                    var. robustus

ELYMUS CANADENSIS                                             13.13.7

E. wiegandii

E. interruptus

ELYMUS                                                        13.13.8–9

subsp. glaucus                    subsp. mackenziei        subsp. virescens

ELYMUS GLAUCUS                                                13.13.10

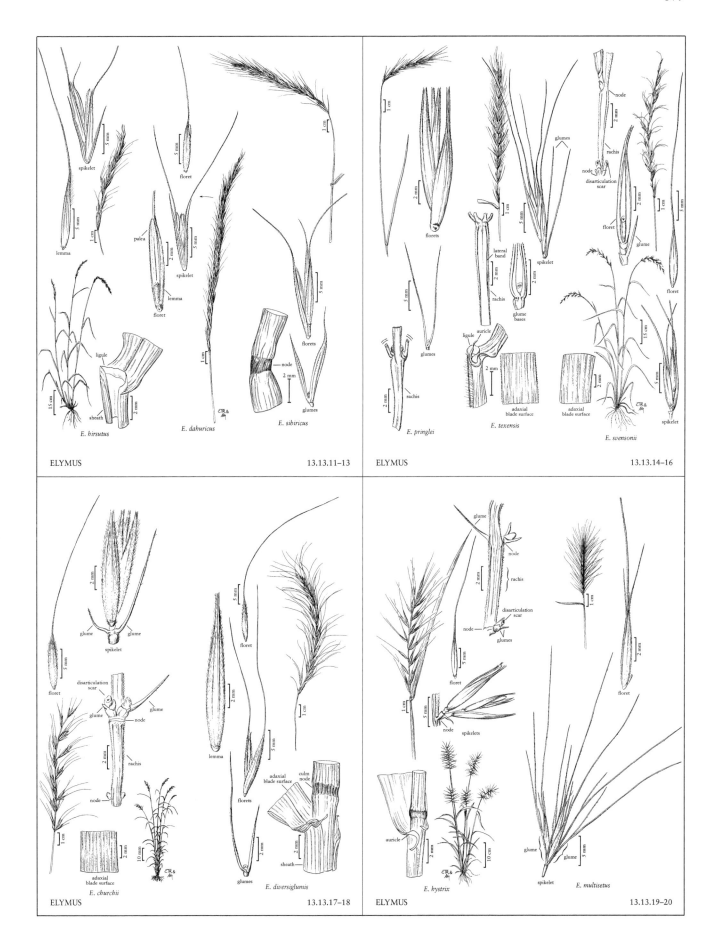

ELYMUS                                    13.13.11–13

ELYMUS                                    13.13.14–16

ELYMUS                                    13.13.17–18

ELYMUS                                    13.13.19–20

380

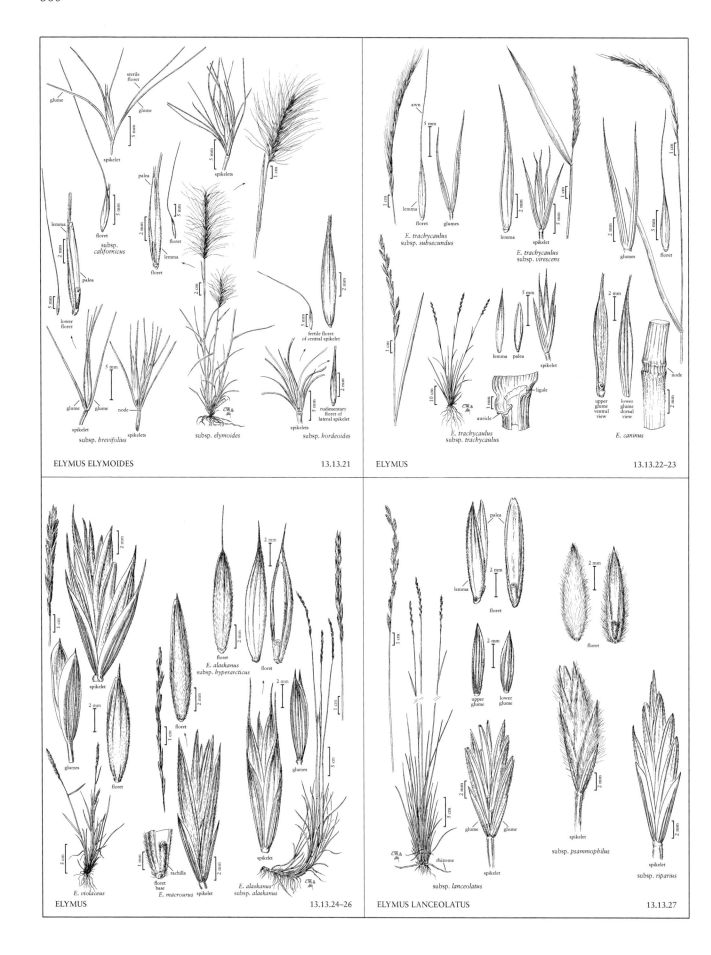

ELYMUS ELYMOIDES

13.13.21

ELYMUS

13.13.22–23

ELYMUS

13.13.24–26

ELYMUS LANCEOLATUS

13.13.27

381

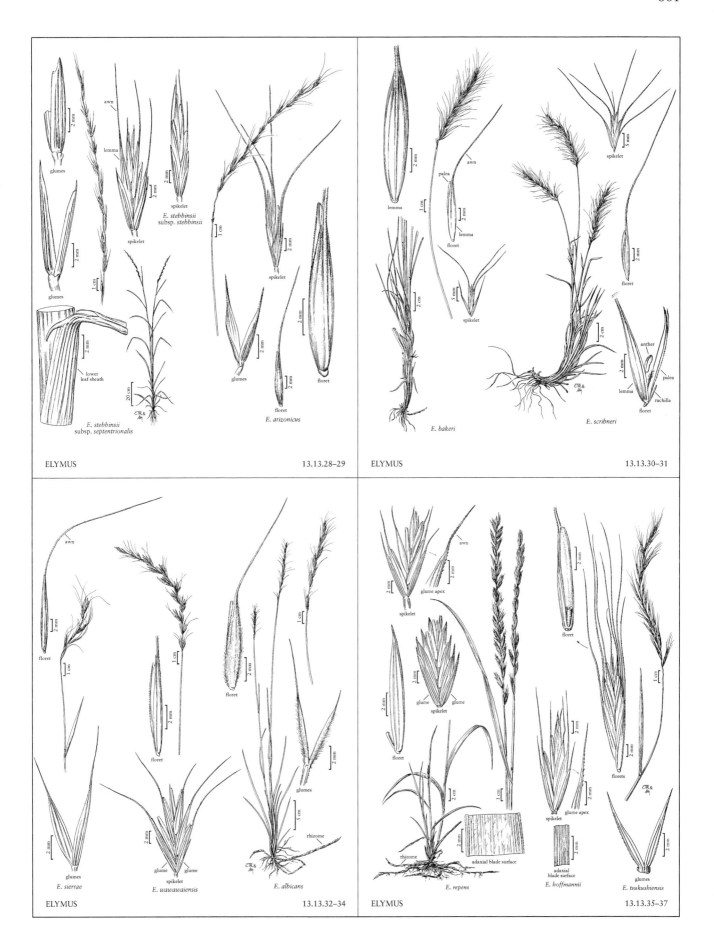

382

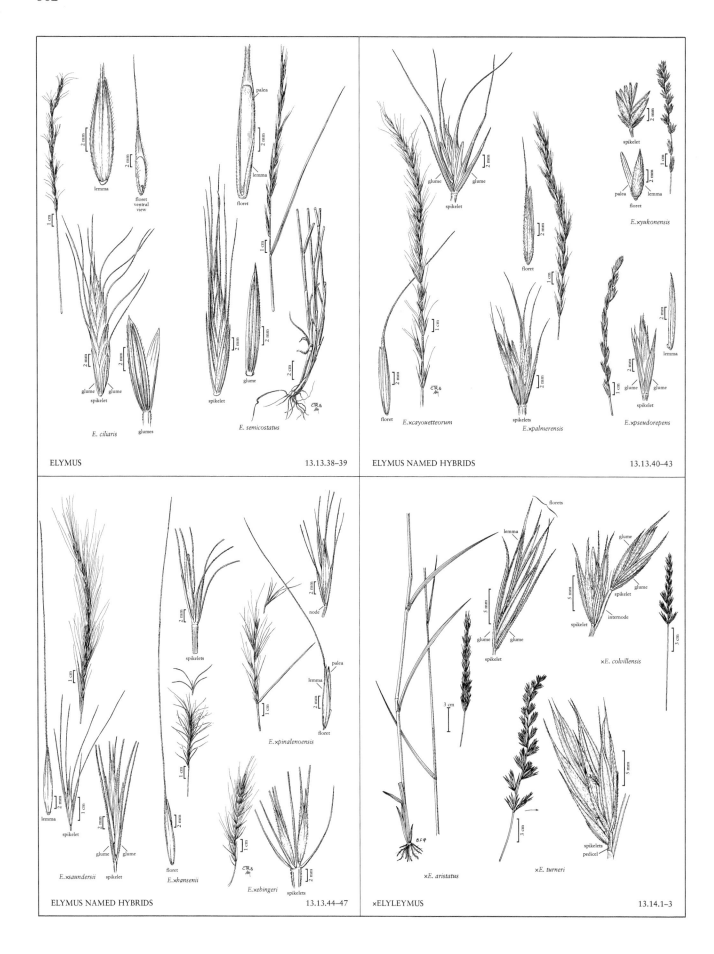

ELYMUS                                                          13.13.38–39

ELYMUS NAMED HYBRIDS                                            13.13.40–43

ELYMUS NAMED HYBRIDS                                            13.13.44–47

×ELYLEYMUS                                                      13.14.1–3

E. ciliaris

E. semicostatus

E.×cayouetteorum

E.×palmerensis

E.×pseudorepens

E.xyukonensis

E.×saundersii

E.×hansenii

E.×ebingeri

E.×pinalenoensis

×E. aristatus

×E. turneri

×E. colvillensis

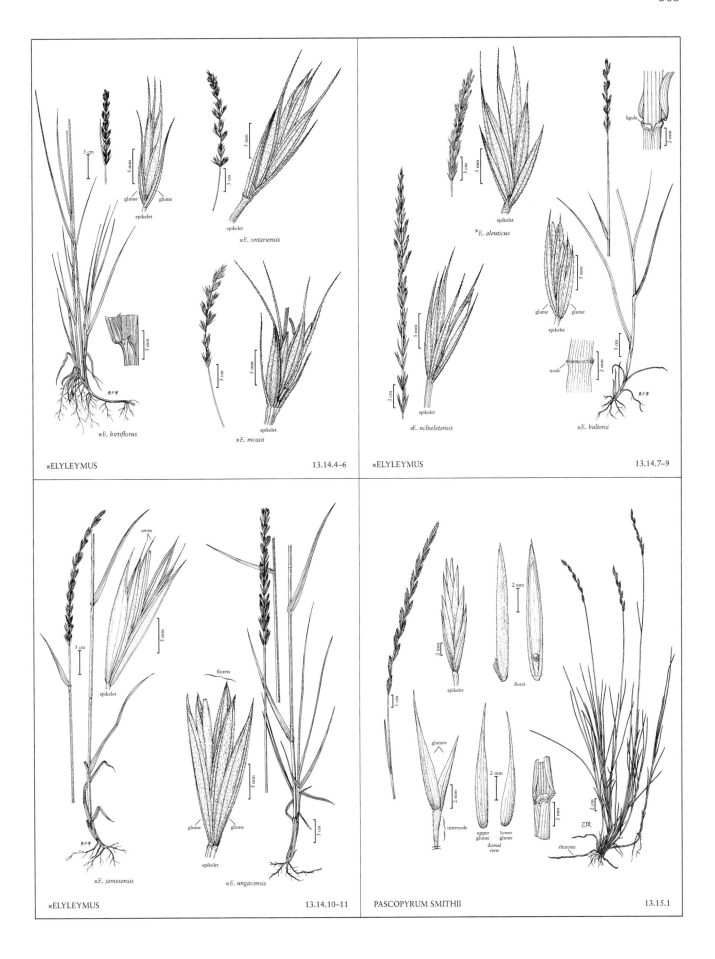

×ELYLEYMUS

13.14.4–6

×ELYLEYMUS

13.14.7–9

×ELYLEYMUS

13.14.10–11

PASCOPYRUM SMITHII

13.15.1

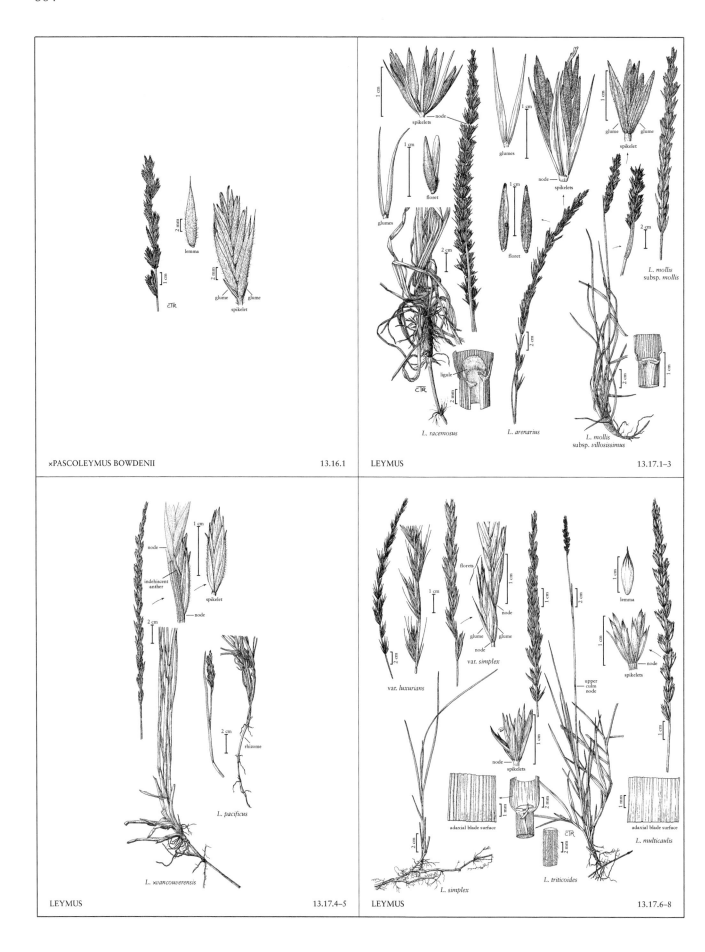

×PASCOLEYMUS BOWDENII                                    13.16.1

LEYMUS                                                   13.17.1–3

LEYMUS                                                   13.17.4–5

LEYMUS                                                   13.17.6–8

385

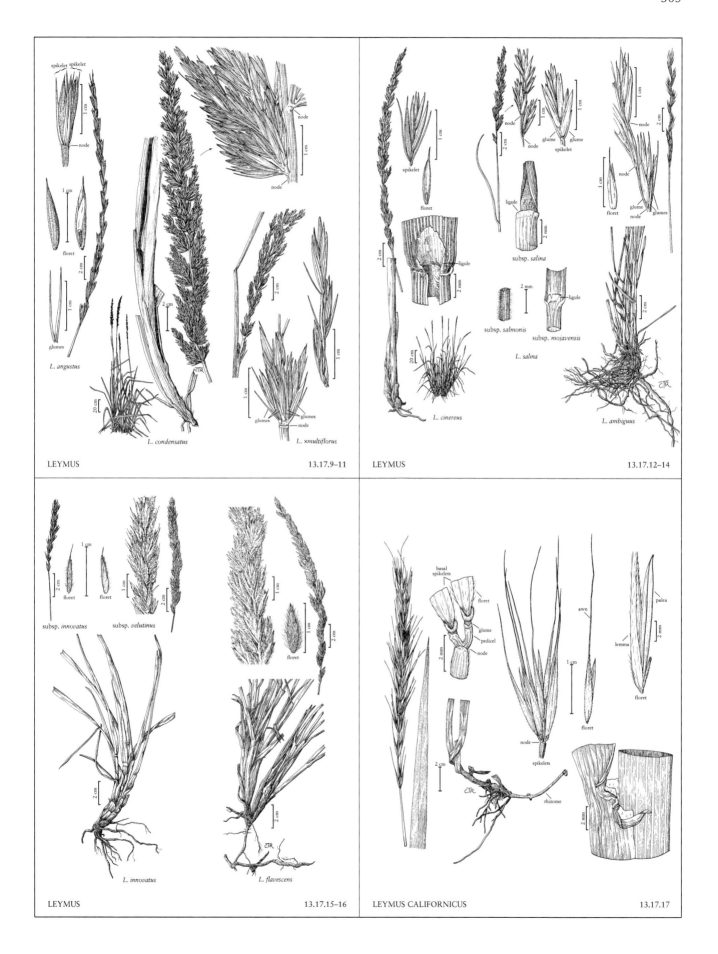

LEYMUS 13.17.9–11

LEYMUS 13.17.12–14

LEYMUS 13.17.15–16

LEYMUS CALIFORNICUS 13.17.17

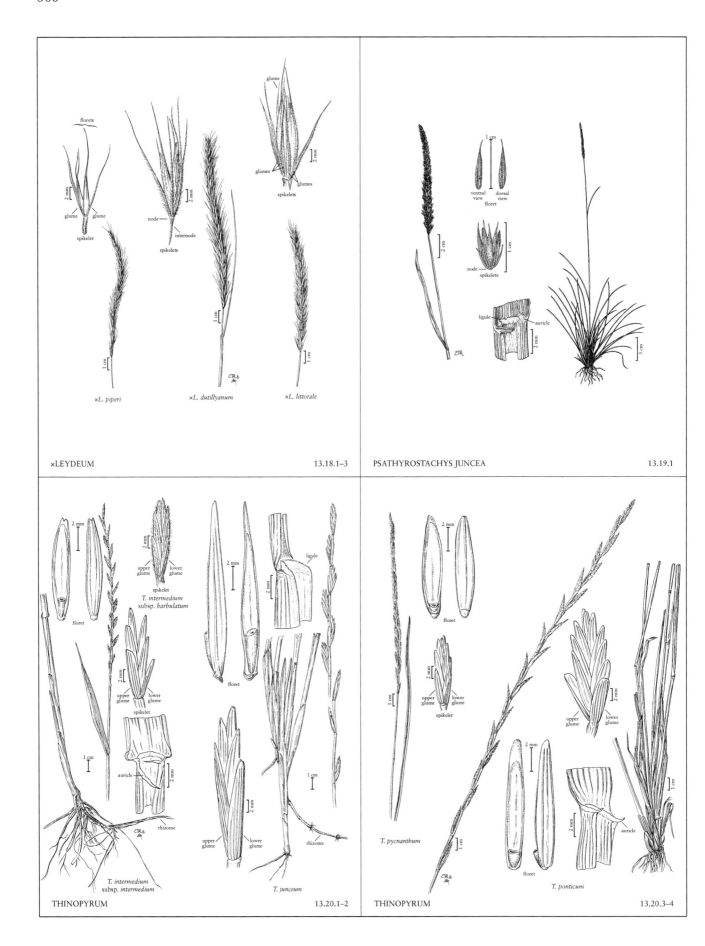

×LEYDEUM

13.18.1–3

PSATHYROSTACHYS JUNCEA

13.19.1

THINOPYRUM

13.20.1–2

THINOPYRUM

13.20.3–4

FESTUCA

14.01.1–3

FESTUCA

14.01.4–5

FESTUCA

14.01.6–8

FESTUCA

14.01.9–11

387

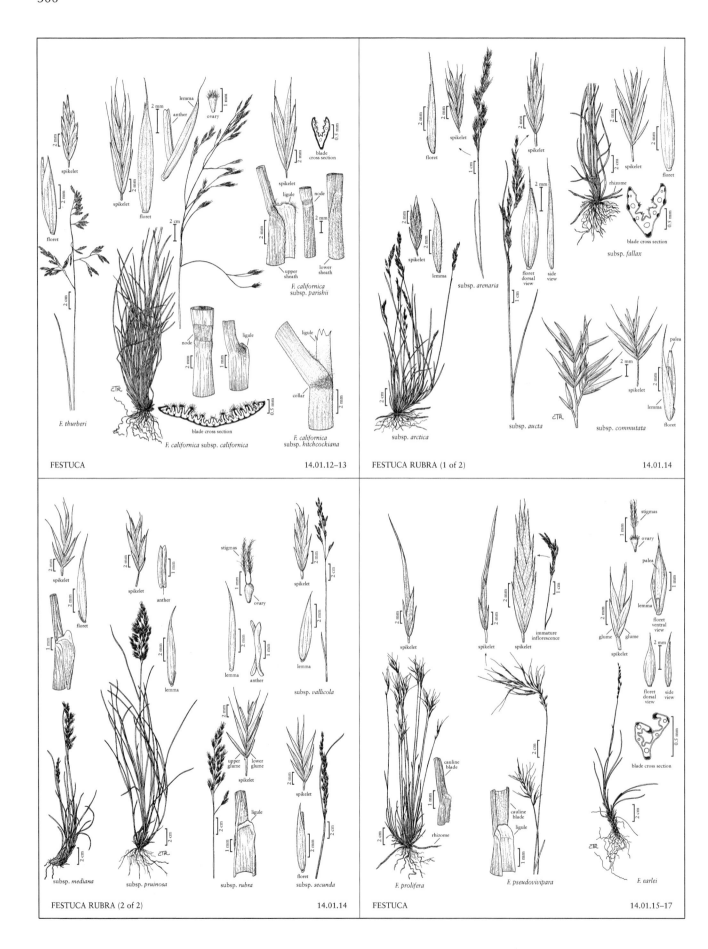

FESTUCA                                    14.01.12–13

FESTUCA RUBRA (1 of 2)                     14.01.14

FESTUCA RUBRA (2 of 2)                     14.01.14

FESTUCA                                    14.01.15–17

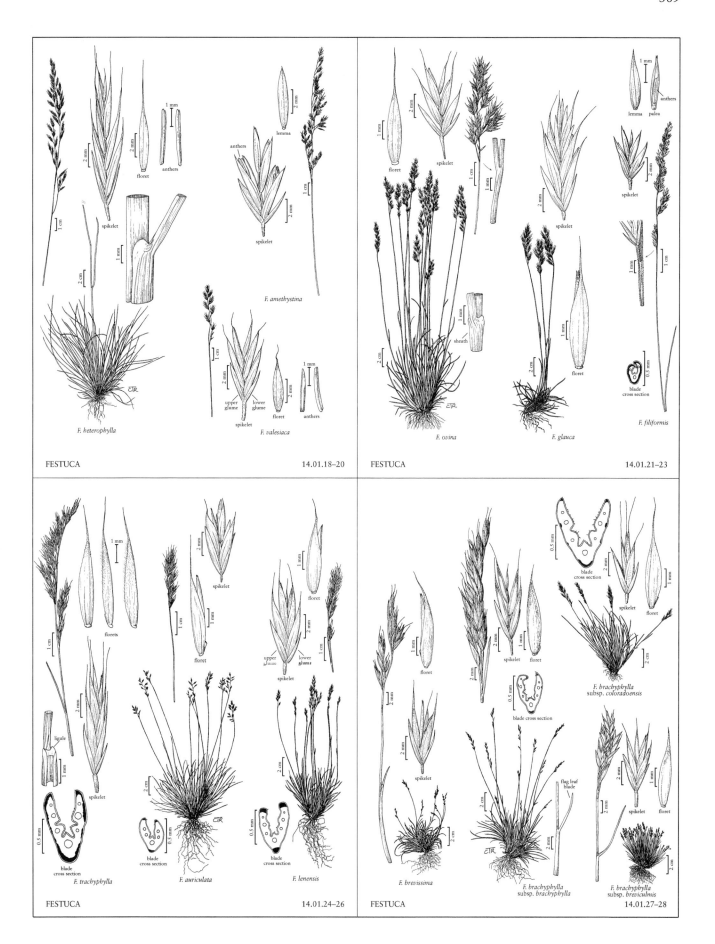

FESTUCA       14.01.18–20

FESTUCA       14.01.21–23

FESTUCA       14.01.24–26

FESTUCA       14.01.27–28

*F. heterophylla*

*F. amethystina*

*F. valesiaca*

*F. ovina*

*F. glauca*

*F. filiformis*

*F. trachyphylla*

*F. auriculata*

*F. lenensis*

*F. brevissima*

*F. brachyphylla* subsp. *brachyphylla*

*F. brachyphylla* subsp. *coloradoensis*

*F. brachyphylla* subsp. *breviculmis*

390

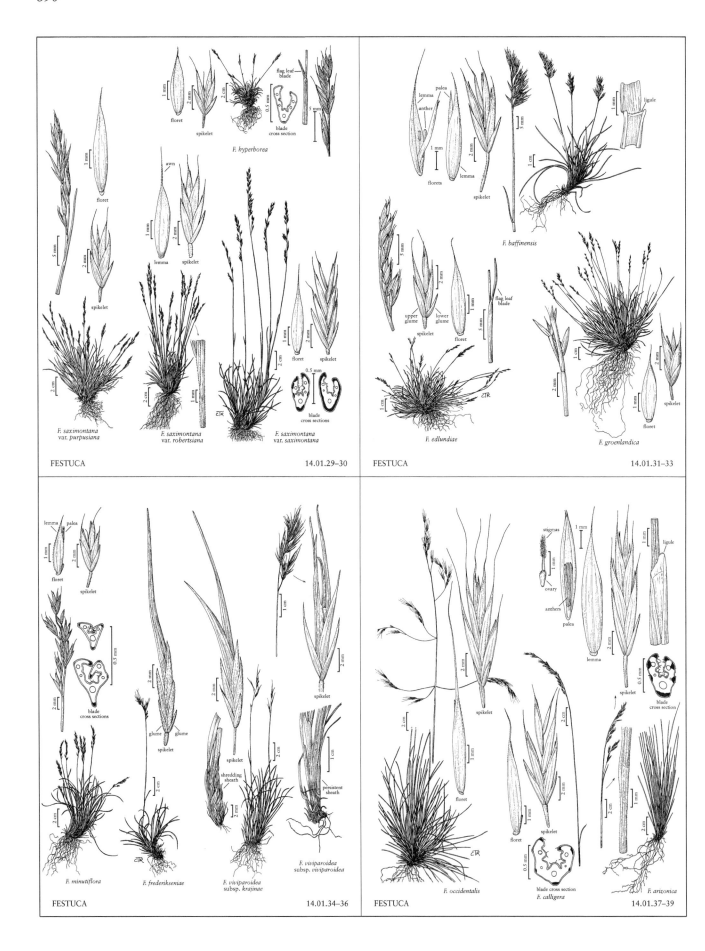

F. hyperborea

F. baffinensis

F. saximontana
var. purpusiana

F. saximontana
var. robertsiana

F. saximontana
var. saximontana

F. edlundiae

F. groenlandica

F. minutiflora

F. frederikseniae

F. viviparoidea
subsp. krajinae

F. viviparoidea
subsp. viviparoidea

F. occidentalis

F. calligera

F. arizonica

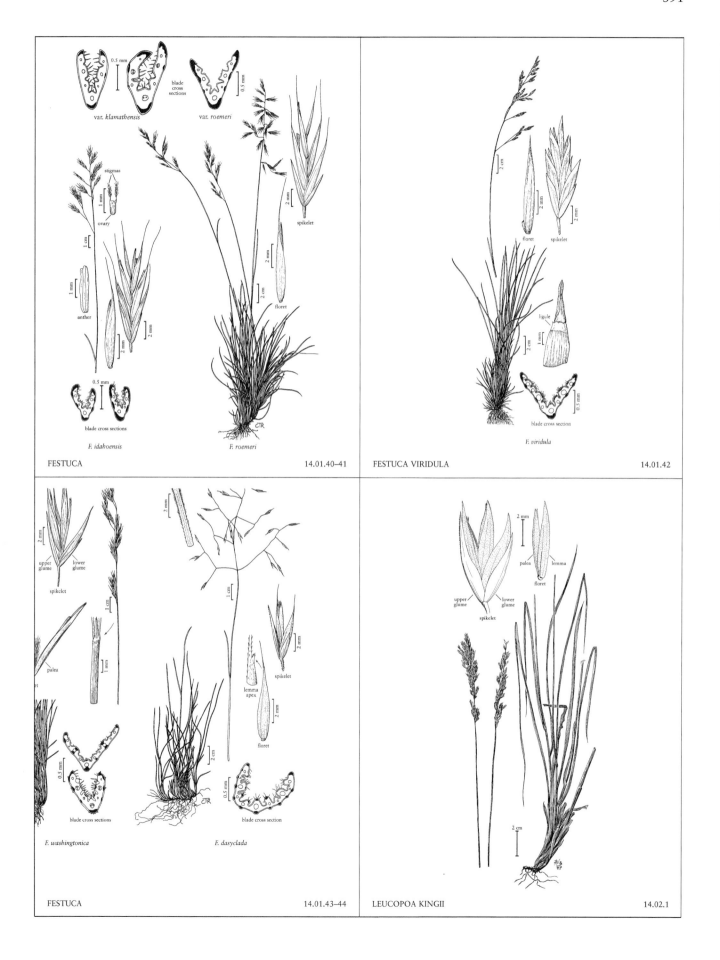

FESTUCA

14.01.40–41

FESTUCA VIRIDULA

14.01.42

FESTUCA

14.01.43–44

LEUCOPOA KINGII

14.02.1

392

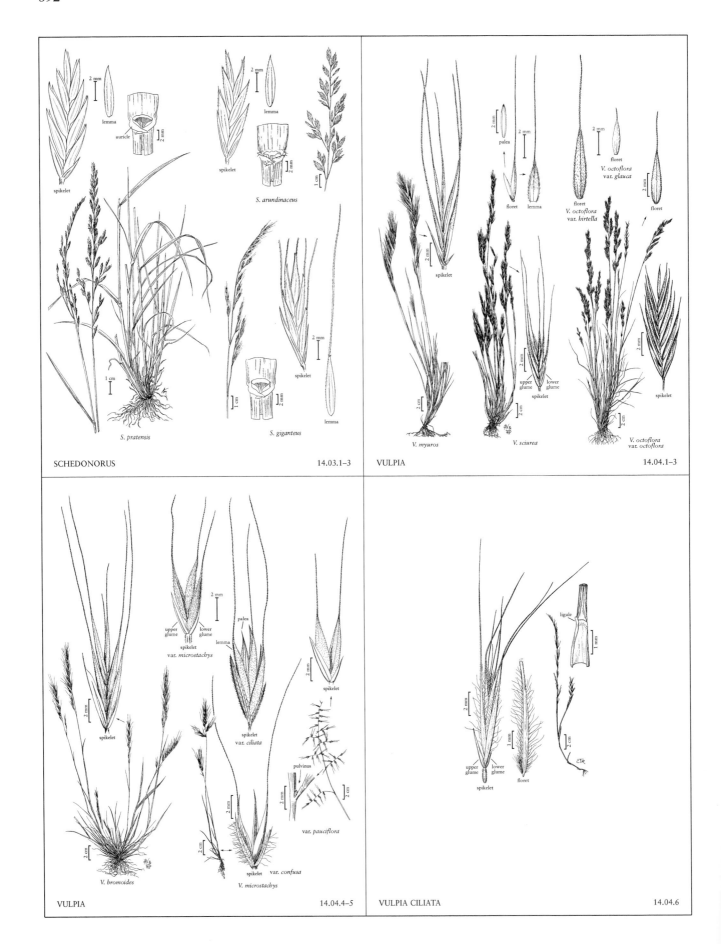

SCHEDONORUS                                    14.03.1–3

VULPIA                                         14.04.1–3

VULPIA                                         14.04.4–5

VULPIA CILIATA                                 14.04.6

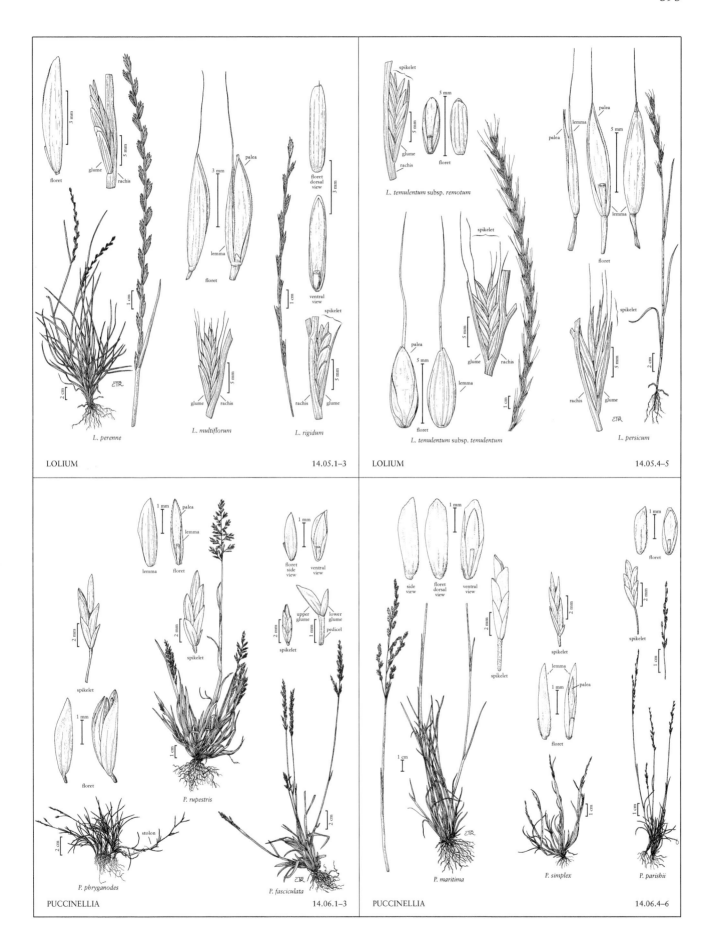

394

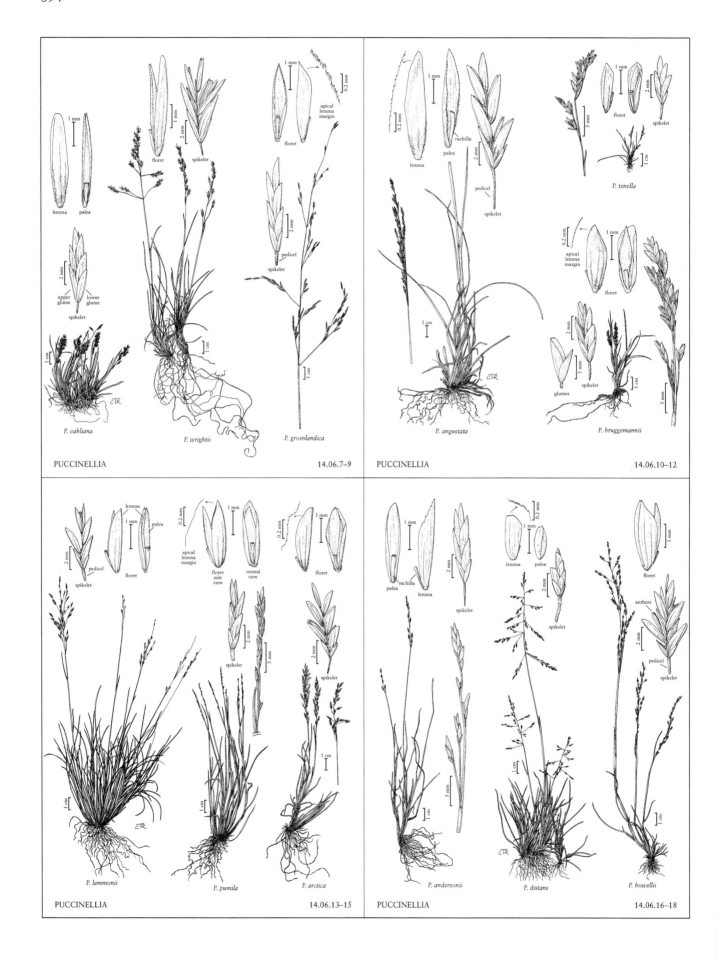

P. vahliana

P. wrightii

P. groenlandica

P. tenella

P. angustata

P. bruggemannii

P. lemmonii

P. pumila

P. arctica

P. andersonii

P. distans

P. howellii

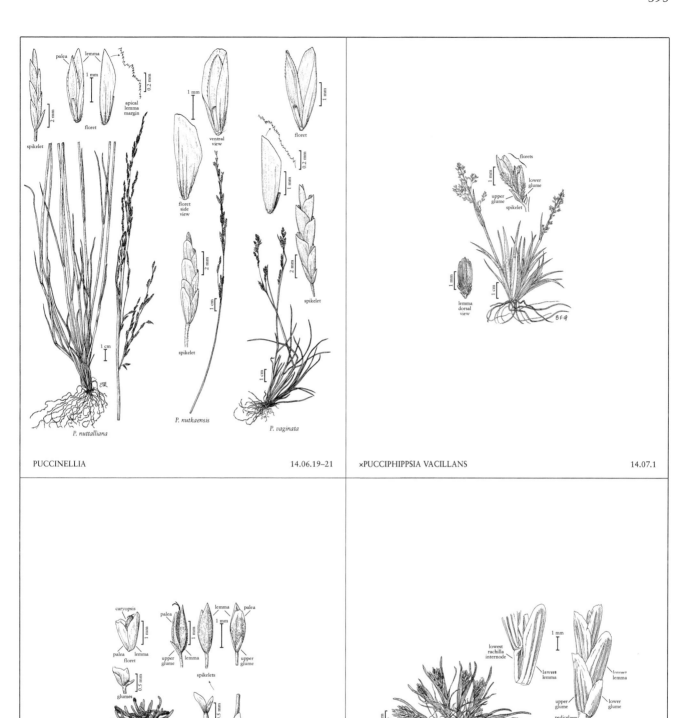

PUCCINELLIA                                          14.06.19–21

×PUCCIPHIPPSIA VACILLANS                             14.07.1

PHIPPSIA                              14.08.1–2        SCLEROCHLOA DURA                    14.09.1

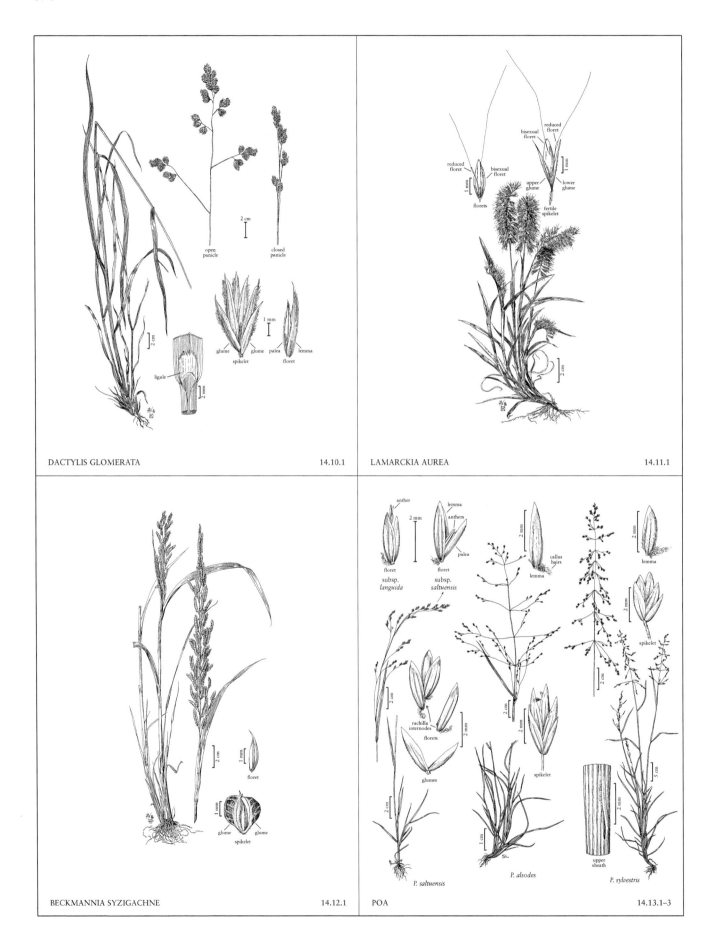

DACTYLIS GLOMERATA                    14.10.1

LAMARCKIA AUREA                       14.11.1

BECKMANNIA SYZIGACHNE                 14.12.1

POA                                   14.13.1–3

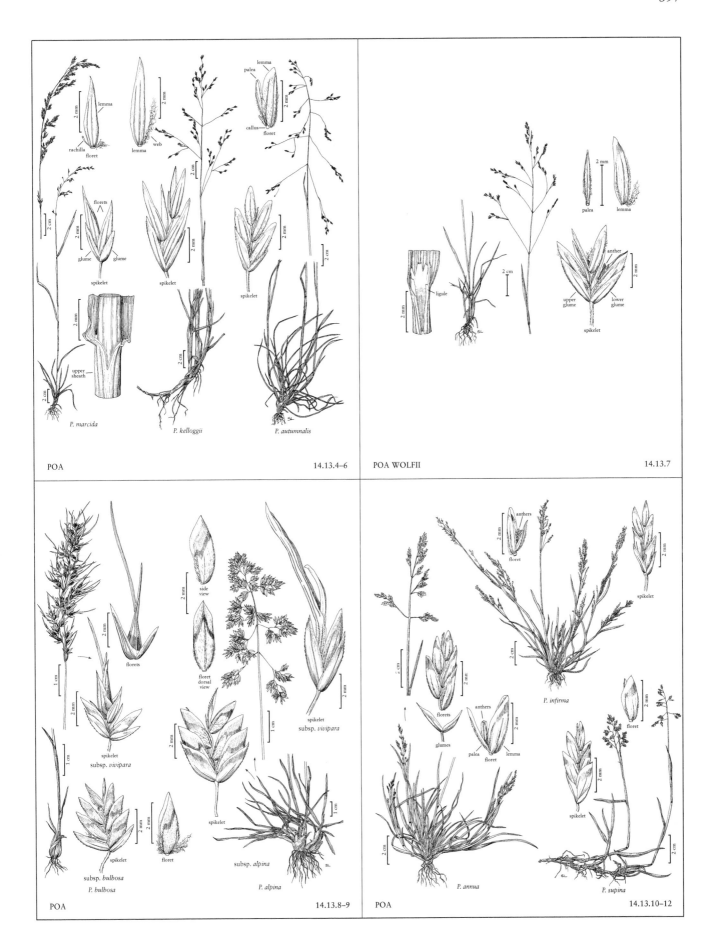

POA

14.13.4–6

POA WOLFII

14.13.7

POA

14.13.8–9

POA

14.13.10–12

398

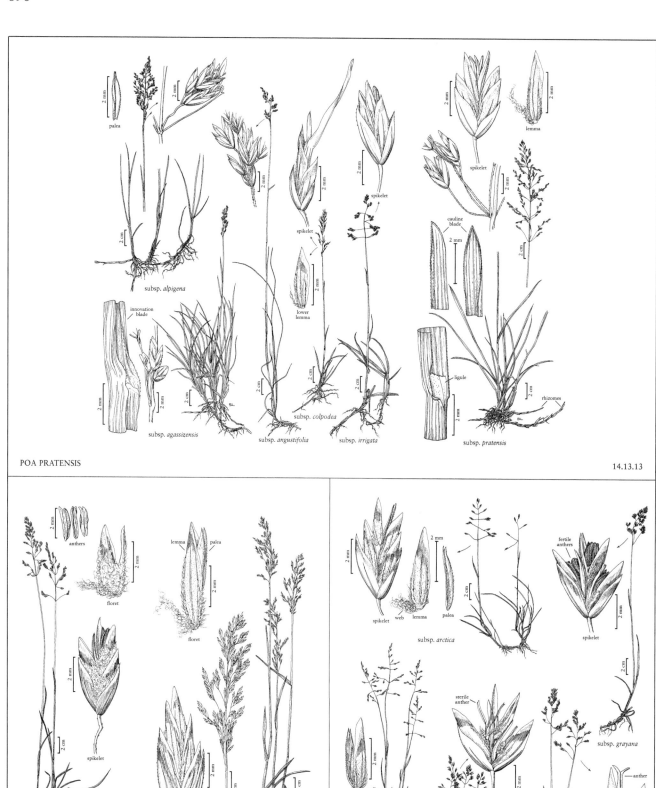

POA PRATENSIS

14.13.13

POA

14.13.14–15

POA ARCTICA

14.13.16

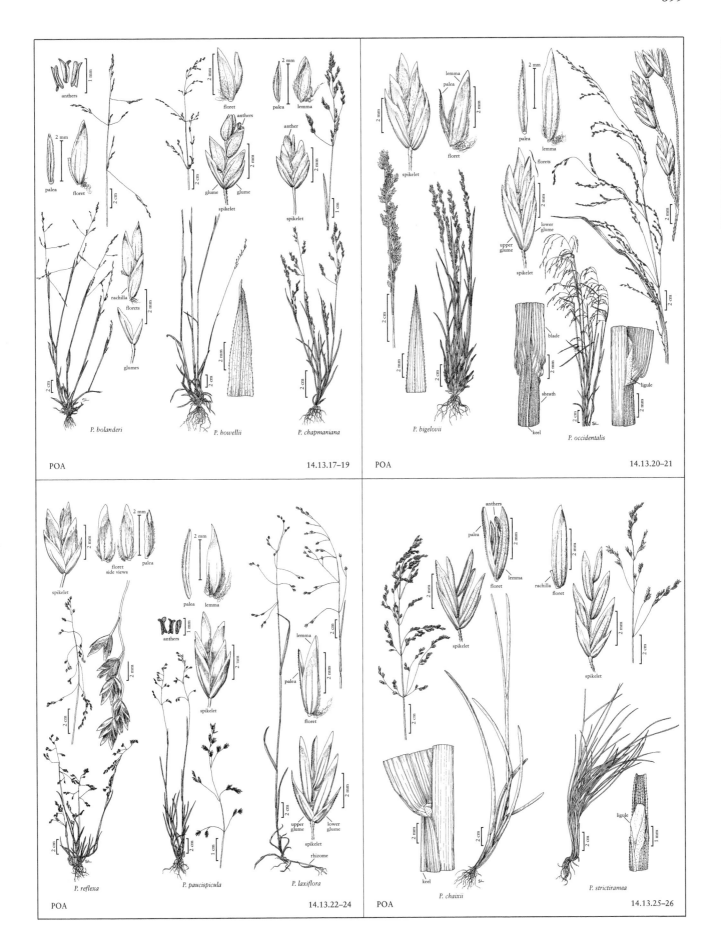

POA

14.13.17–19

POA

14.13.20–21

POA

14.13.22–24

POA

14.13.25–26

*P. bolanderi*

*P. howellii*

*P. chapmaniana*

*P. bigelovii*

*P. occidentalis*

*P. reflexa*

*P. paucispicula*

*P. laxiflora*

*P. chaixii*

*P. strictiramea*

400

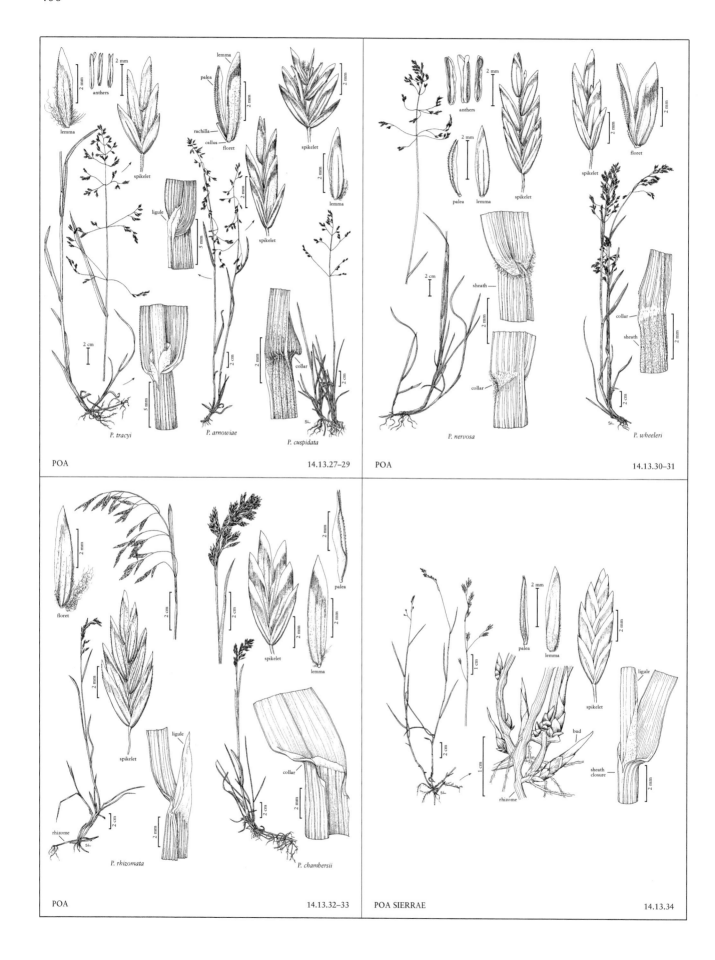

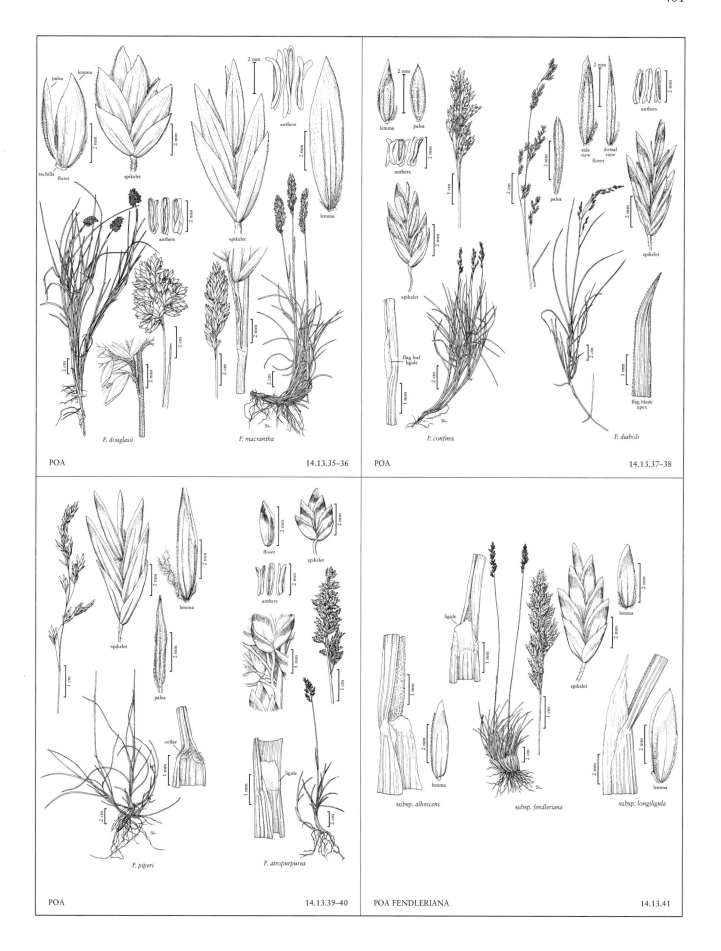

401

POA 14.13.35–36

*P. douglasii*

*P. macrantha*

palea
lemma
rachilla
floret
spikelet
anthers
spikelet
anthers
lemma
spikelet
2 mm

POA 14.13.37–38

*P. confinis*

*P. diaboli*

lemma
palea
anthers
spikelet
spikelet
flag leaf
ligule
side view
dorsal view
floret
anthers
palea
spikelet
flag blade apex
2 mm

POA 14.13.39–40

*P. piperi*

*P. atropurpurea*

spikelet
lemma
palea
collar
floret
spikelet
anthers
ligule

POA FENDLERIANA 14.13.41

subsp. *albescens*

subsp. *fendleriana*

subsp. *longiligula*

ligule
spikelet
lemma
lemma
lemma

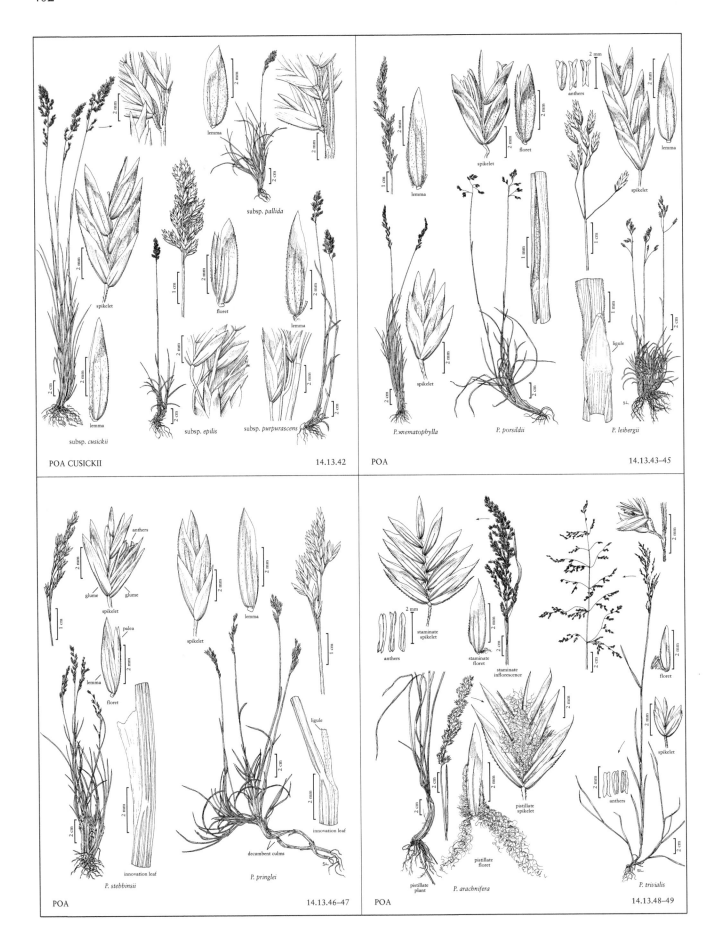

subsp. *pallida*

lemma

subsp. *cusickii*

spikelet

lemma

floret

spikelet

lemma

subsp. *epilis*

lemma

subsp. *purpurascens*

POA CUSICKII                                              14.13.42

lemma

spikelet

floret

spikelet

*P. ×nematophylla*

*P. porsildii*

anthers

lemma

spikelet

ligule

*P. leibergii*

POA                                                      14.13.43–45

anthers

glume          glume

spikelet

palea

lemma

floret

spikelet

lemma

innovation leaf

ligule

innovation leaf

decumbent culms

*P. stebbinsii*

*P. pringlei*

POA                                                      14.13.46–47

anthers

staminate
spikelet

staminate
floret

staminate
inflorescence

pistillate
spikelet

pistillate
floret

pistillate
plant

*P. arachnifera*

floret

spikelet

anthers

*P. trivialis*

POA                                                      14.13.48–49

403

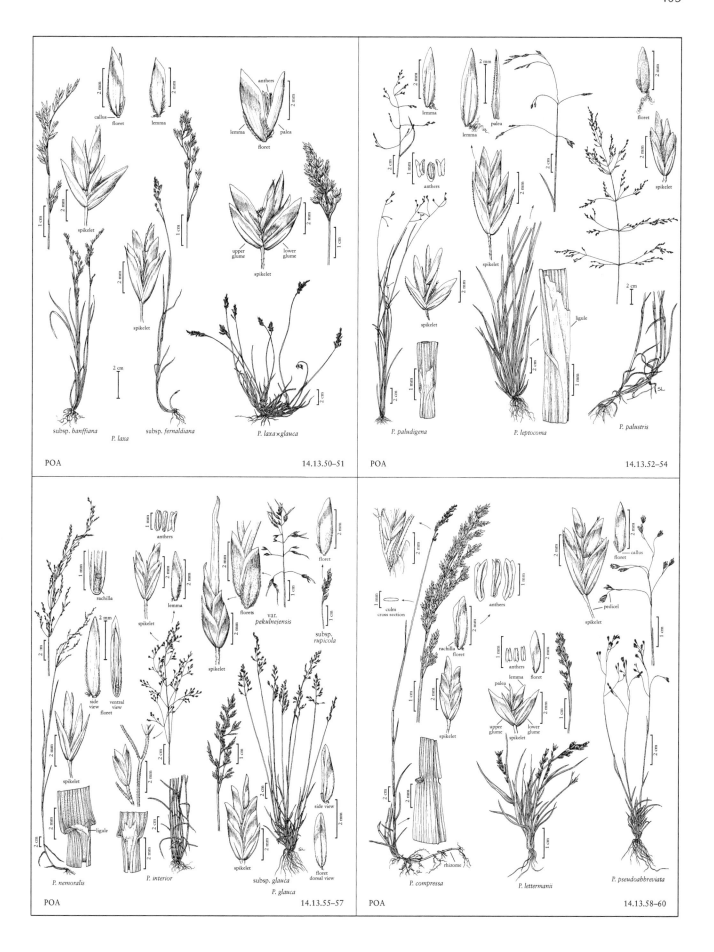

404

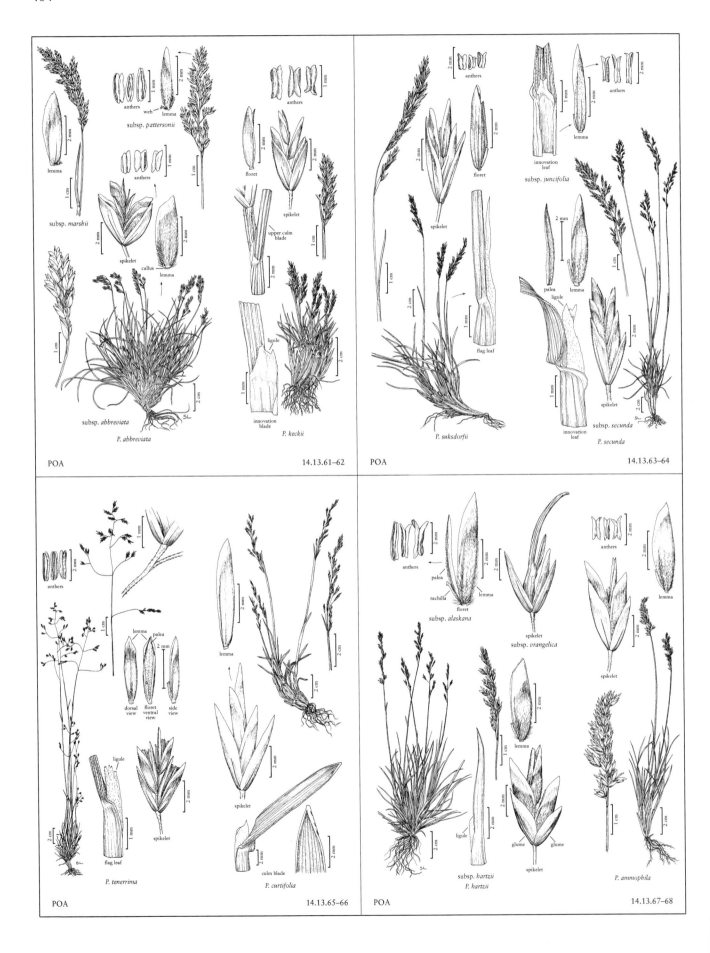

subsp. *pattersonii*

subsp. *marshii*

anthers

web

lemma

anthers

lemma

floret

spikelet

anthers

spikelet

upper culm blade

callus

lemma

subsp. *abbreviata*

spikelet

ligule

innovation blade

P. *abbreviata*

P. *keckii*

anthers

spikelet

floret

innovation leaf

subsp. *juncifolia*

lemma

anthers

palea

lemma

ligule

P. *suksdorfii*

flag leaf

innovation leaf

spikelet

subsp. *secunda*

P. *secunda*

anthers

lemma

palea

dorsal view

floret ventral view

side view

lemma

ligule

flag leaf

spikelet

P. *tenerrima*

spikelet

culm blade

P. *curtifolia*

anthers

palea

rachilla

lemma

floret

subsp. *alaskana*

spikelet

subsp. *vrangelica*

anthers

lemma

spikelet

lemma

ligule

glume

glume

spikelet

subsp. *hartzii*

P. *hartzii*

P. *ammophila*

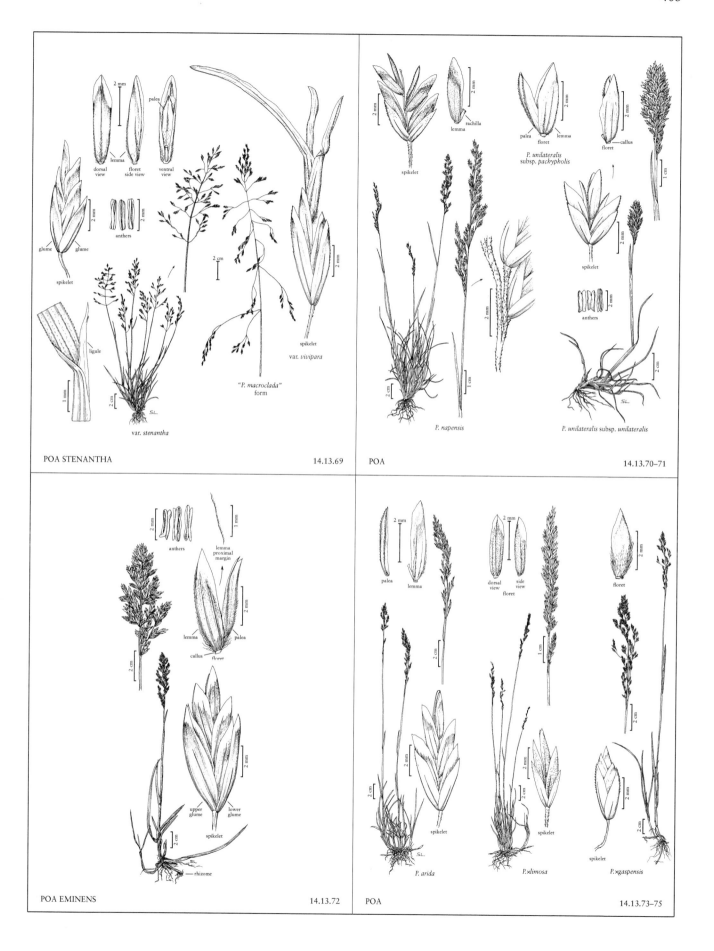

POA STENANTHA         14.13.69

POA         14.13.70–71

POA EMINENS         14.13.72

POA         14.13.73–75

406

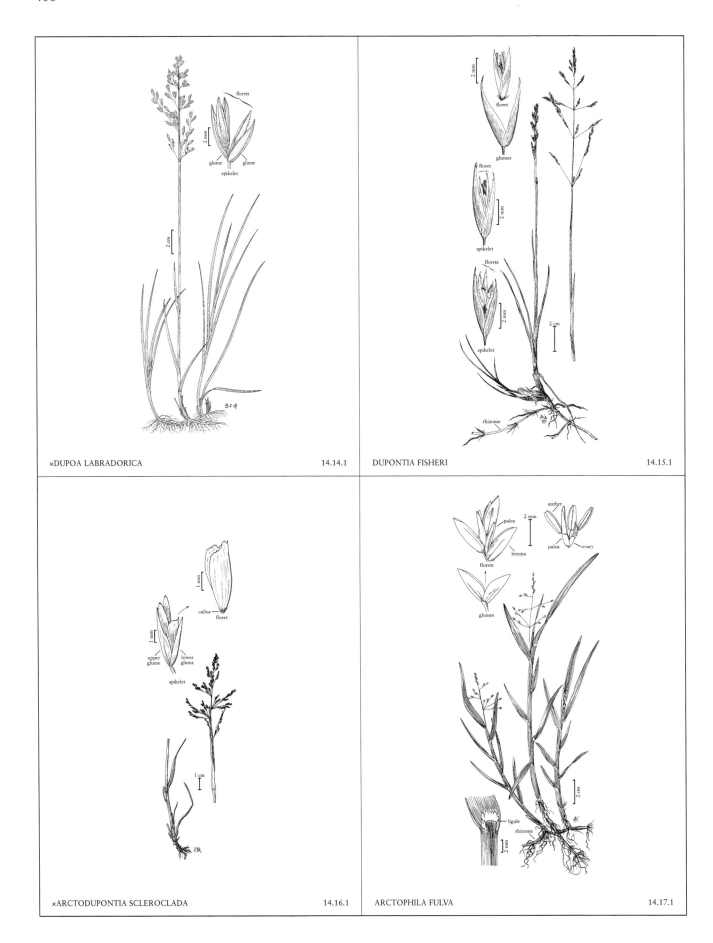

×DUPOA LABRADORICA                    14.14.1

DUPONTIA FISHERI                      14.15.1

×ARCTODUPONTIA SCLEROCLADA            14.16.1

ARCTOPHILA FULVA                      14.17.1

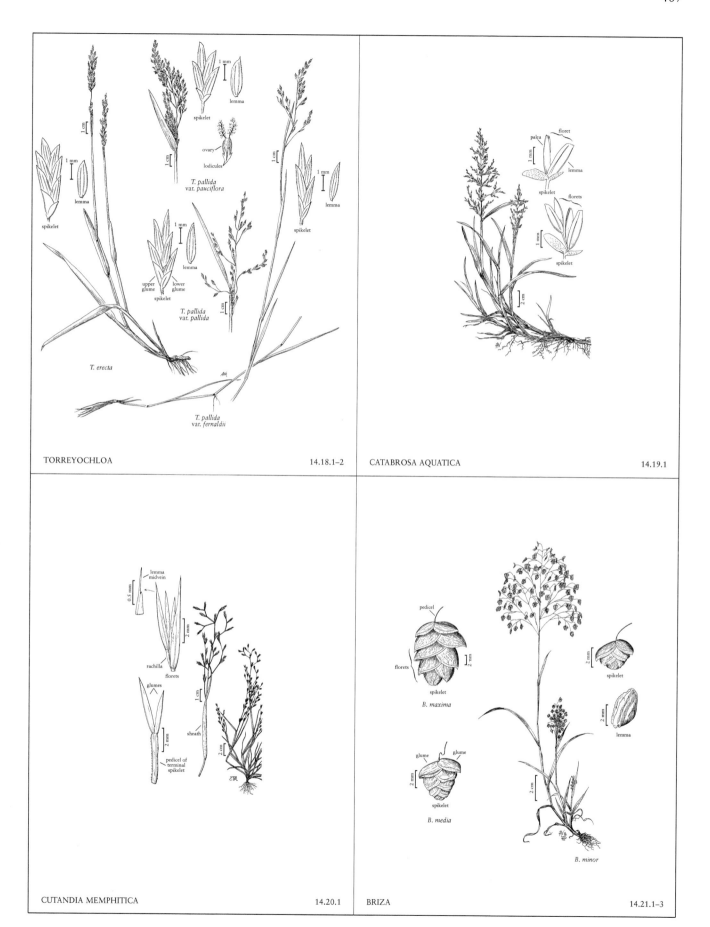

T. pallida
var. pauciflora

spikelet

lemma

ovary

lodicules

spikelet

lemma

spikelet

lemma

upper
glume

lower
glume

spikelet

T. pallida
var. pallida

spikelet

lemma

T. erecta

T. pallida
var. fernaldii

TORREYOCHLOA

14.18.1–2

palea

floret

lemma

spikelet

florets

spikelet

CATABROSA AQUATICA

14.19.1

lemma
midvein

rachilla

florets

glumes

sheath

pedicel of
terminal
spikelet

CUTANDIA MEMPHITICA

14.20.1

pedicel

florets

spikelet

B. maxima

spikelet

lemma

glume

glume

spikelet

B. media

B. minor

BRIZA

14.21.1–3

408

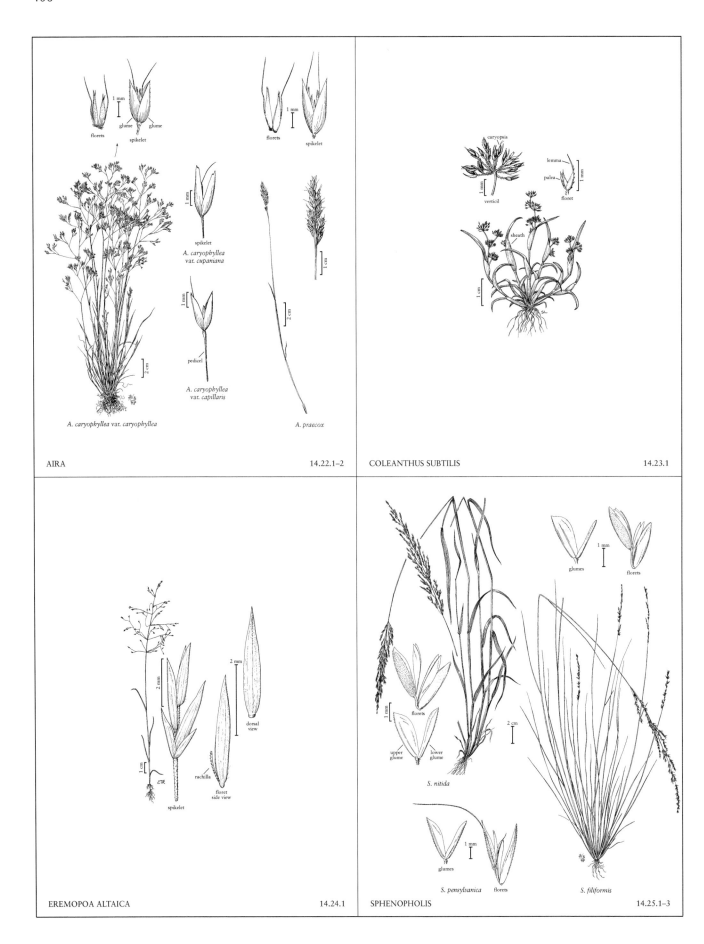

AIRA

14.22.1–2

COLEANTHUS SUBTILIS

14.23.1

EREMOPOA ALTAICA

14.24.1

SPHENOPHOLIS

14.25.1–3

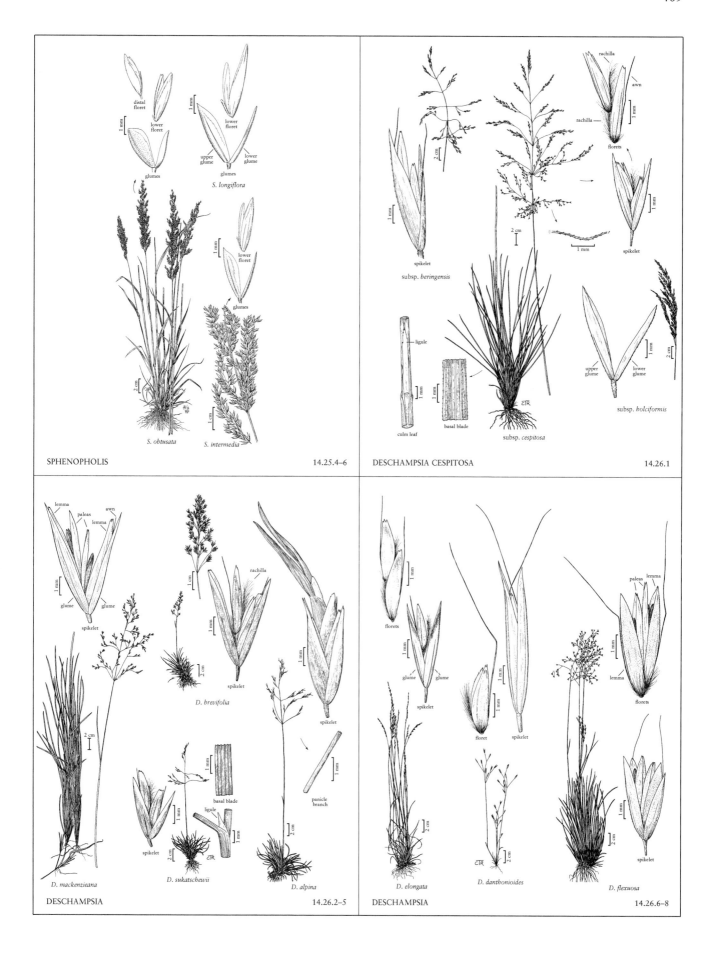

SPHENOPHOLIS           14.25.4–6

DESCHAMPSIA CESPITOSA       14.26.1

DESCHAMPSIA           14.26.2–5

DESCHAMPSIA           14.26.6–8

410

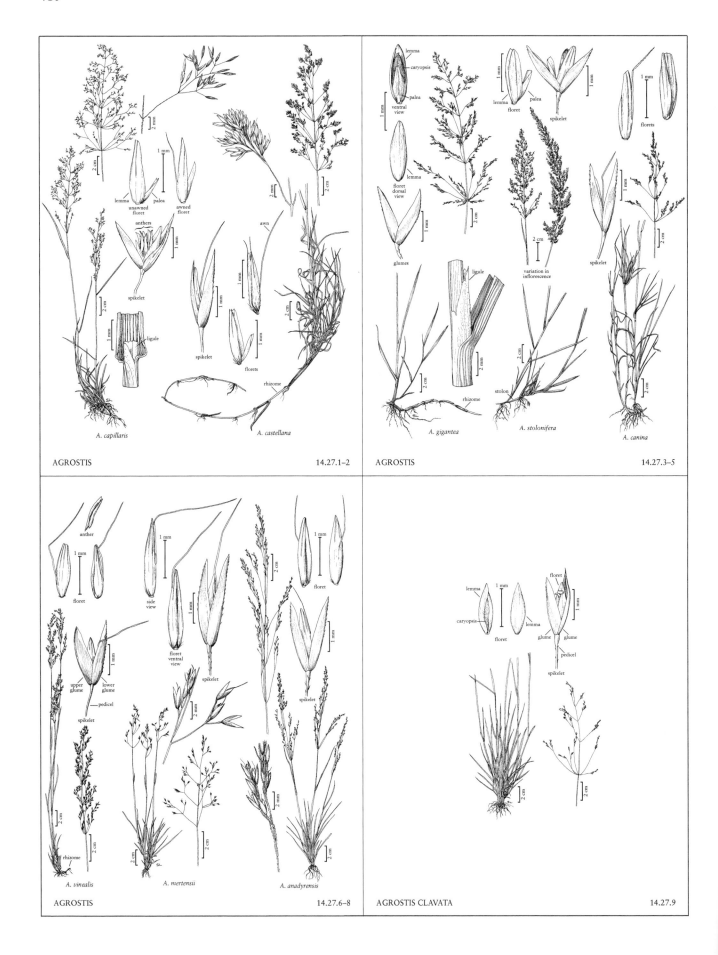

A. capillaris

A. castellana

AGROSTIS

14.27.1–2

A. gigantea

A. stolonifera

A. canina

AGROSTIS

14.27.3–5

A. vinealis

A. mertensii

A. anadyrensis

AGROSTIS

14.27.6–8

AGROSTIS CLAVATA

14.27.9

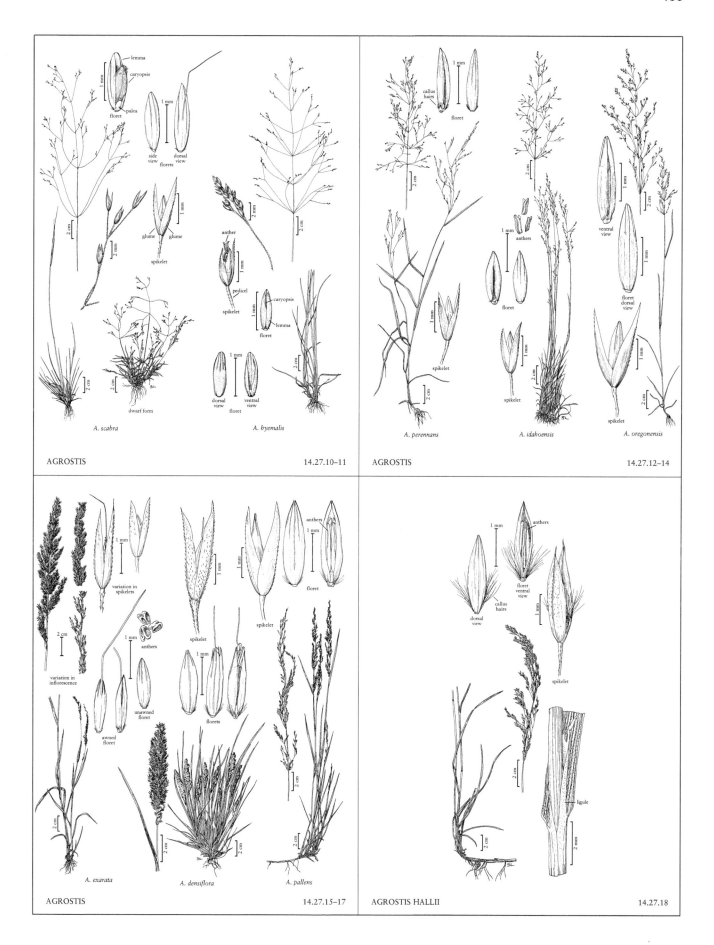

AGROSTIS

14.27.10–11

A. scabra

dwarf form

A. hyemalis

lemma

caryopsis

palea

floret

side view

dorsal view

florets

glume

glume

spikelet

2 cm

anther

pedicel

spikelet

caryopsis

lemma

floret

dorsal view

ventral view

floret

1 mm

AGROSTIS

14.27.12–14

A. perennans

A. idahoensis

A. oregonensis

callus hairs

floret

anthers

floret

spikelet

spikelet

ventral view

floret dorsal view

spikelet

2 cm

1 mm

AGROSTIS

14.27.15–17

A. exarata

A. densiflora

A. pallens

variation in spikelets

anthers

awned floret

unawned floret

variation in inflorescence

spikelet

spikelet

florets

floret

2 cm

1 mm

AGROSTIS HALLII

14.27.18

dorsal view

callus hairs

anthers

floret ventral view

spikelet

ligule

2 cm

1 mm

2 mm

412

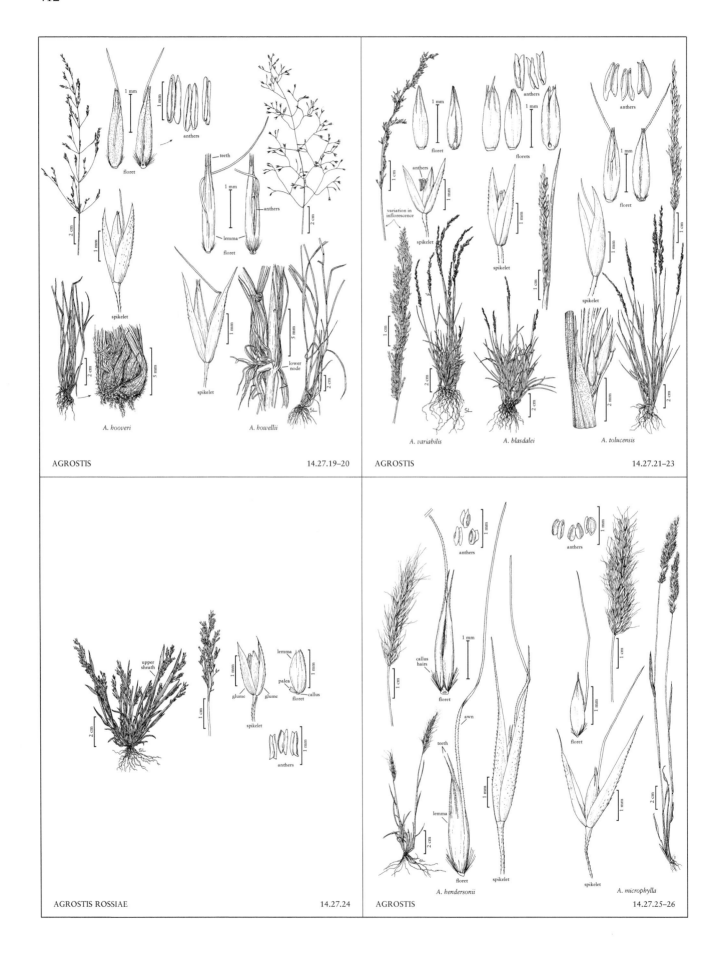

A. hooveri

A. howellii

AGROSTIS

A. variabilis

A. blasdalei

A. tolucensis

AGROSTIS ROSSIAE

AGROSTIS

A. hendersonii

A. microphylla

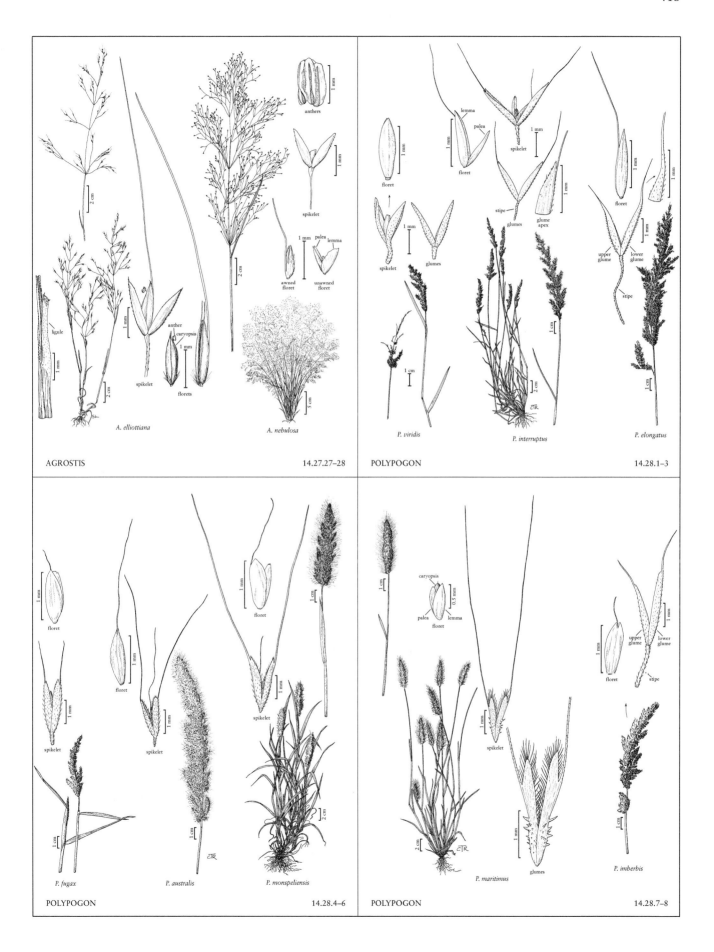

AGROSTIS 14.27.27–28

*A. elliottiana*

*A. nebulosa*

POLYPOGON 14.28.1–3

*P. viridis*

*P. interruptus*

*P. elongatus*

POLYPOGON 14.28.4–6

*P. fugax*

*P. australis*

*P. monspeliensis*

POLYPOGON 14.28.7–8

*P. maritimus*

*P. imberbis*

414

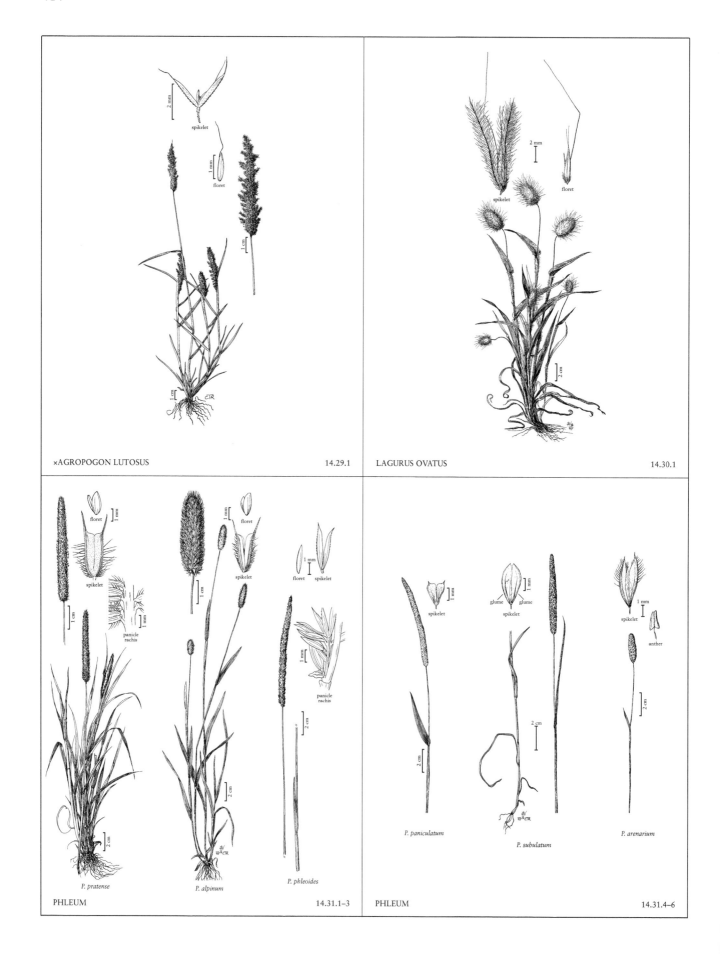

×AGROPOGON LUTOSUS                                        14.29.1

LAGURUS OVATUS                                           14.30.1

*P. pratense*                   *P. alpinum*                *P. phleoides*

PHLEUM                                                   14.31.1–3

*P. paniculatum*

*P. subulatum*

*P. arenarium*

PHLEUM                                                   14.31.4–6

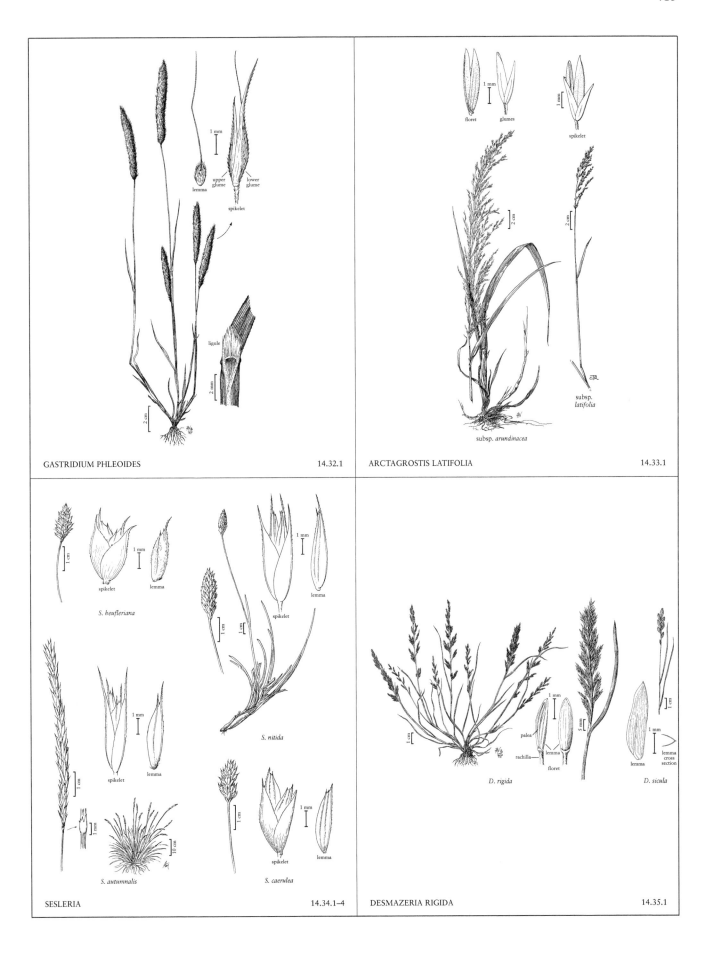

GASTRIDIUM PHLEOIDES                                                14.32.1

ARCTAGROSTIS LATIFOLIA                                              14.33.1

SESLERIA                                                         14.34.1–4

DESMAZERIA RIGIDA                                                  14.35.1

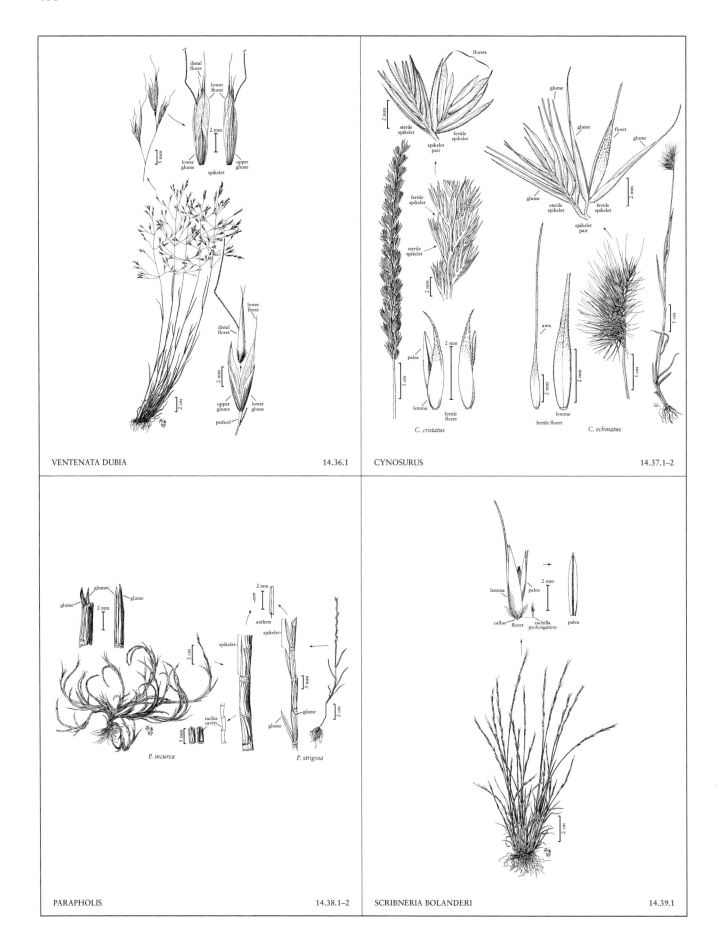

VENTENATA DUBIA                                    14.36.1

CYNOSURUS                                          14.37.1–2

C. cristatus

C. echinatus

PARAPHOLIS                                         14.38.1–2

P. incurva

P. strigosa

SCRIBNERIA BOLANDERI                              14.39.1

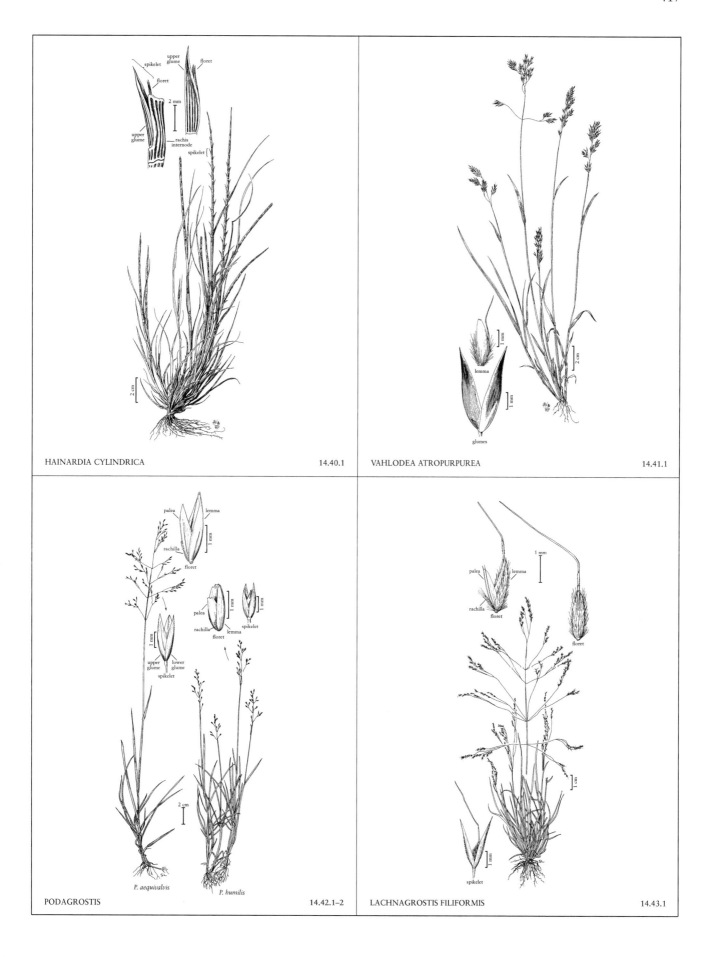

HAINARDIA CYLINDRICA

14.40.1

VAHLODEA ATROPURPUREA

14.41.1

PODAGROSTIS

14.42.1–2

LACHNAGROSTIS FILIFORMIS

14.43.1

418

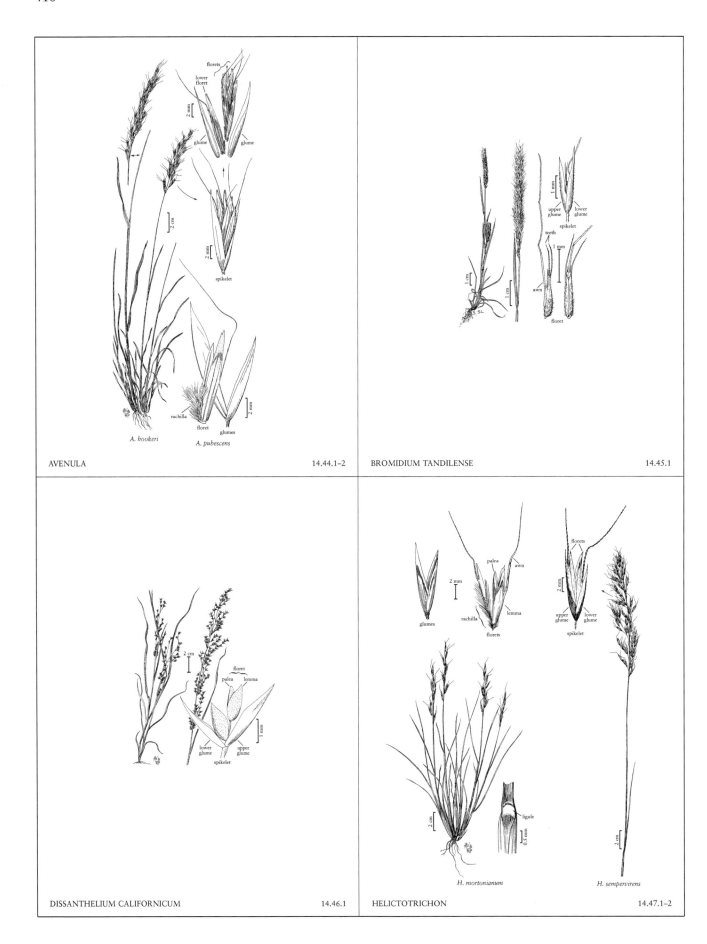

AVENULA                           14.44.1–2

BROMIDIUM TANDILENSE            14.45.1

DISSANTHELIUM CALIFORNICUM      14.46.1

HELICTOTRICHON                    14.47.1–2

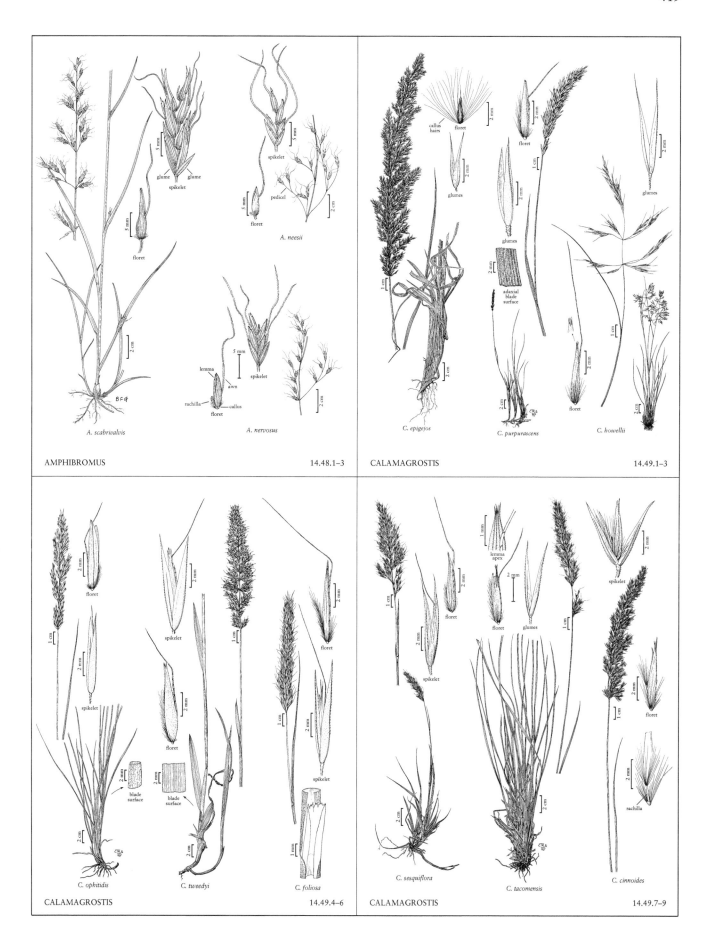

AMPHIBROMUS 14.48.1–3

CALAMAGROSTIS 14.49.1–3

CALAMAGROSTIS 14.49.4–6

CALAMAGROSTIS 14.49.7–9

420

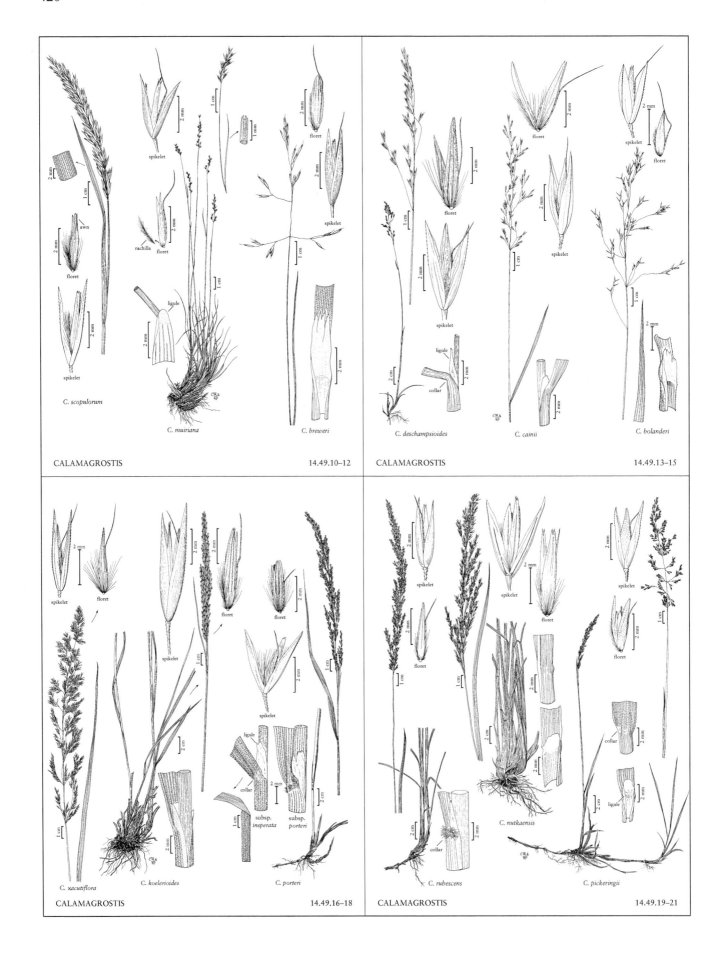

C. scopulorum

C. muiriana

C. breweri

CALAMAGROSTIS                                    14.49.10–12

C. deschampsioides

C. cainii

C. bolanderi

CALAMAGROSTIS                                    14.49.13–15

C. xacutiflora

C. koelerioides

subsp. insperata

subsp. porteri

C. porteri

CALAMAGROSTIS                                    14.49.16–18

C. rubescens

C. nutkaensis

C. pickeringii

CALAMAGROSTIS                                    14.49.19–21

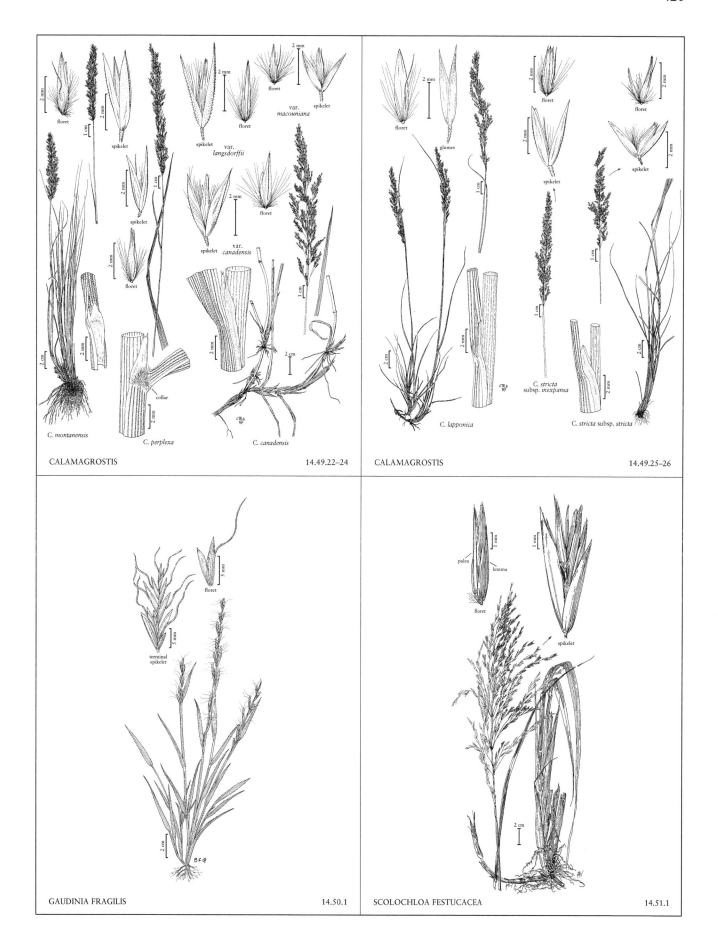

CALAMAGROSTIS                    14.49.22–24

CALAMAGROSTIS                    14.49.25–26

var. *macouniana*

var. *langsdorffii*

var. *canadensis*

*C. montanensis*

*C. perplexa*

*C. canadensis*

*C. lapponica*

*C. stricta* subsp. *inexpansa*

*C. stricta* subsp. *stricta*

GAUDINIA FRAGILIS                14.50.1

SCOLOCHLOA FESTUCACEA            14.51.1

422

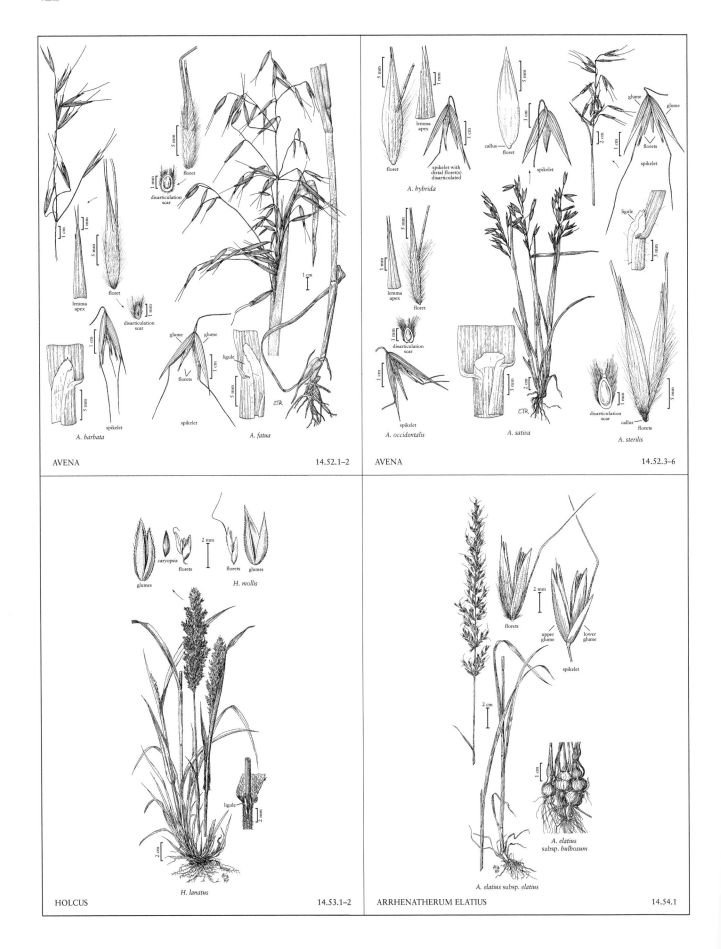

AVENA                           14.52.1–2

*A. barbata*

*A. fatua*

floret

disarticulation scar

lemma apex

floret

disarticulation scar

glume    glume

ligule

florets

spikelet

spikelet

AVENA                           14.52.3–6

*A. hybrida*

floret

lemma apex

spikelet with distal floret(s) disarticulated

callus — floret

spikelet

glume

glume

florets

spikelet

ligule

lemma apex

floret

disarticulation scar

*A. occidentalis*

spikelet

*A. sativa*

disarticulation scar

callus

florets

*A. sterilis*

HOLCUS                          14.53.1–2

glumes

caryopsis

florets

florets

glumes

*H. mollis*

ligule

*H. lanatus*

ARRHENATHERUM ELATIUS          14.54.1

florets

upper glume

lower glume

spikelet

*A. elatius* subsp. *bulbosum*

*A. elatius* subsp. *elatius*

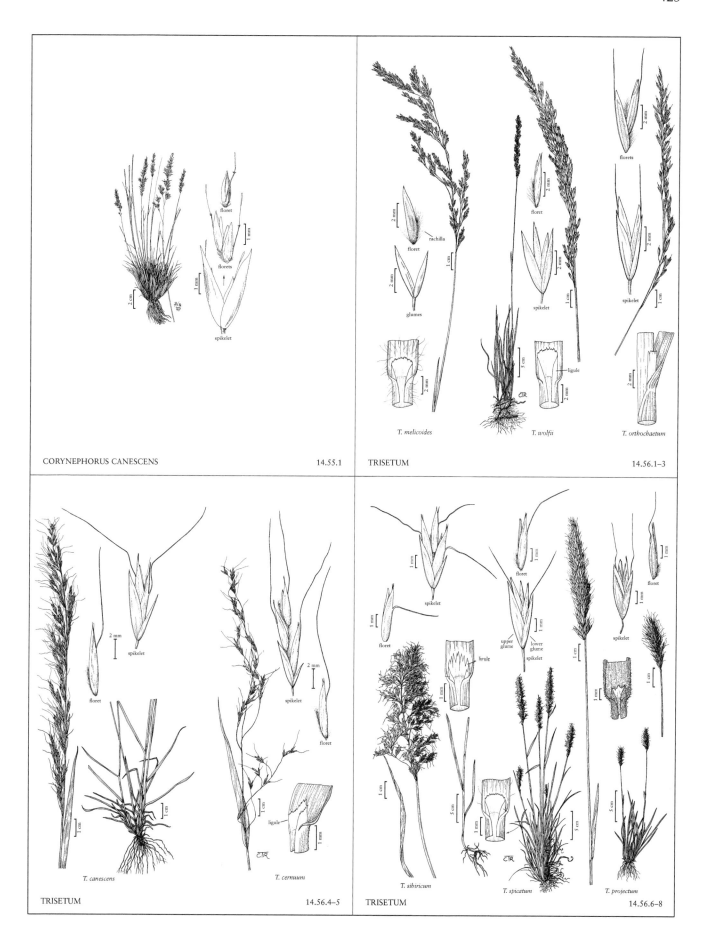

CORYNEPHORUS CANESCENS          14.55.1

TRISETUM          14.56.1–3

TRISETUM          14.56.4–5

TRISETUM          14.56.6–8

424

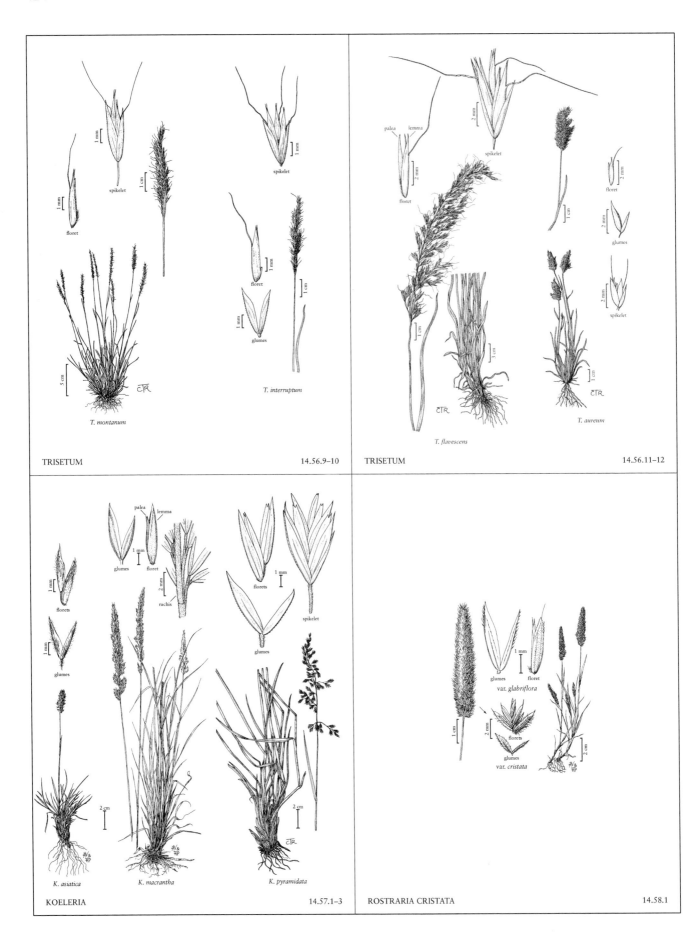

T. montanum

spikelet

floret

T. interruptum

spikelet

floret

glumes

palea   lemma

spikelet

floret

T. flavescens

floret

glumes

spikelet

T. aureum

palea   lemma

glumes   floret

rachis

florets

glumes

florets

glumes

spikelet

K. asiatica

K. macrantha

K. pyramidata

glumes   floret

var. glabriflora

florets

glumes

var. cristata

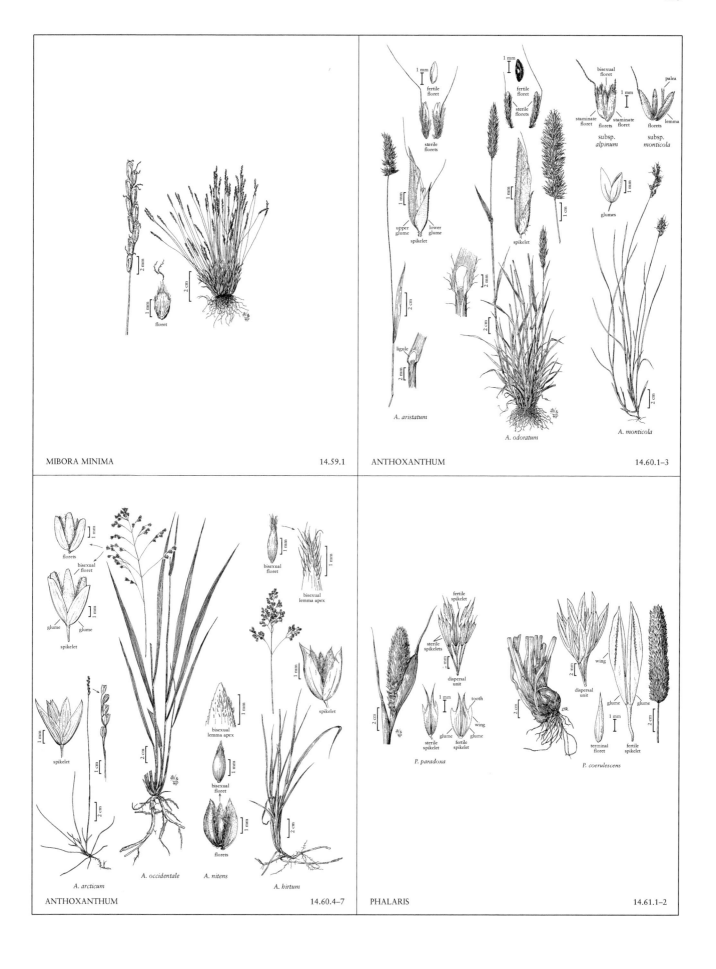

425

MIBORA MINIMA      14.59.1

ANTHOXANTHUM      14.60.1–3

ANTHOXANTHUM      14.60.4–7

PHALARIS      14.61.1–2

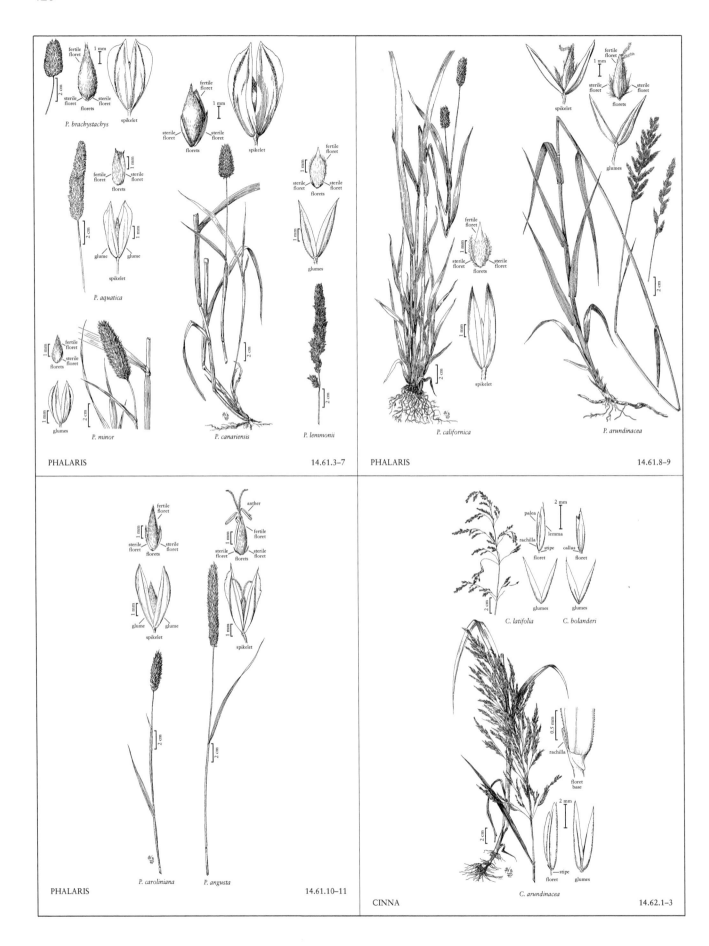

PHALARIS                                         14.61.3–7

PHALARIS                                         14.61.8–9

PHALARIS                                         14.61.10–11

CINNA                                            14.62.1–3

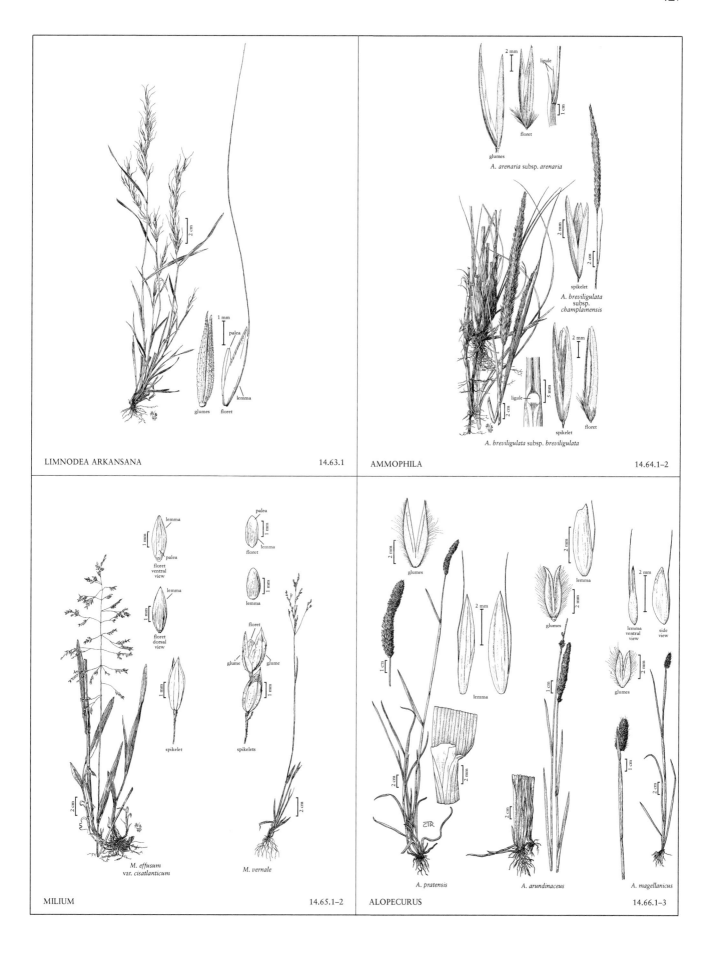

A. arenaria subsp. arenaria

glumes

floret

2 mm

ligule

1 cm

spikelet

A. breviligulata
subsp.
champlainensis

2 mm

2 mm

ligule

spikelet

5 mm

floret

2 mm

A. breviligulata subsp. breviligulata

2 cm

LIMNODEA ARKANSANA                    14.63.1

palea

lemma

glumes    floret

1 mm

2 cm

AMMOPHILA                              14.64.1–2

lemma

palea

floret
ventral
view

1 mm

lemma

floret
dorsal
view

1 mm

spikelet

1 mm

palea

lemma

floret

1 mm

lemma

1 mm

floret

glume    glume

spikelets

1 mm

2 cm

M. effusum
var. cisatlanticum

M. vernale

MILIUM                                 14.65.1–2

glumes

2 mm

lemma

2 mm

1 cm

2 mm

glumes

lemma

1 cm

lemma
ventral
view

2 mm

side
view

glumes

2 mm

1 cm

2 cm

2 cm

2 mm

2 mm

CTR

A. pratensis              A. arundinaceus              A. magellanicus

ALOPECURUS                             14.66.1–3

428

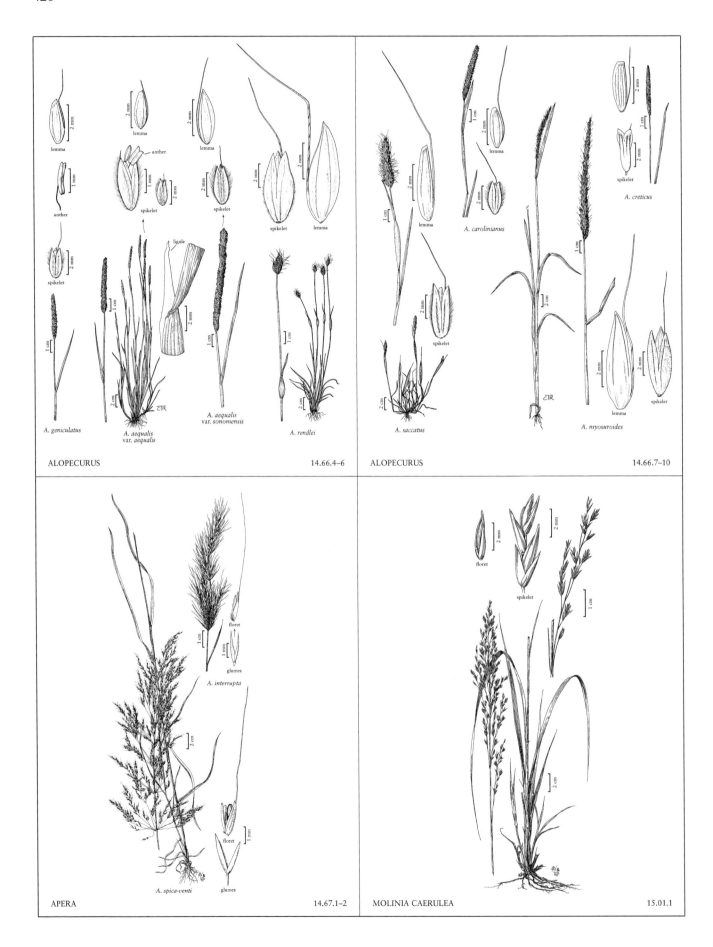

ALOPECURUS

14.66.4–6

ALOPECURUS

14.66.7–10

APERA

14.67.1–2

MOLINIA CAERULEA

15.01.1

*A. geniculatus*

*A. aequalis*
var. *aequalis*

*A. aequalis*
var. *sonomensis*

*A. rendlei*

*A. saccatus*

*A. carolinianus*

*A. creticus*

*A. myosuroides*

*A. interrupta*

*A. spica-venti*

429

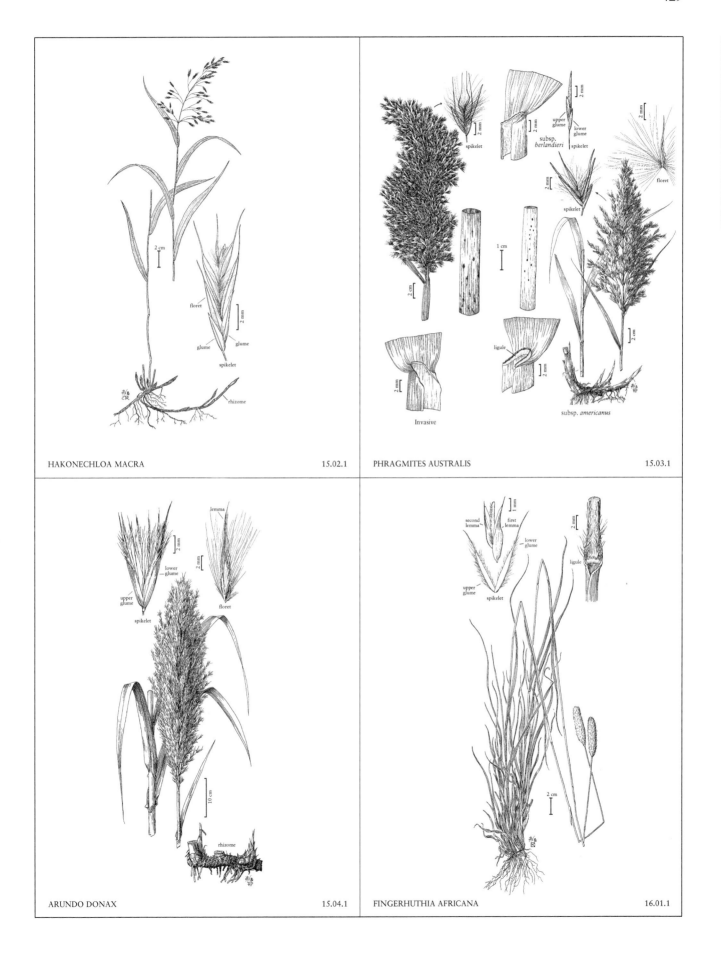

HAKONECHLOA MACRA                    15.02.1

PHRAGMITES AUSTRALIS                 15.03.1

ARUNDO DONAX                         15.04.1

FINGERHUTHIA AFRICANA                16.01.1

430

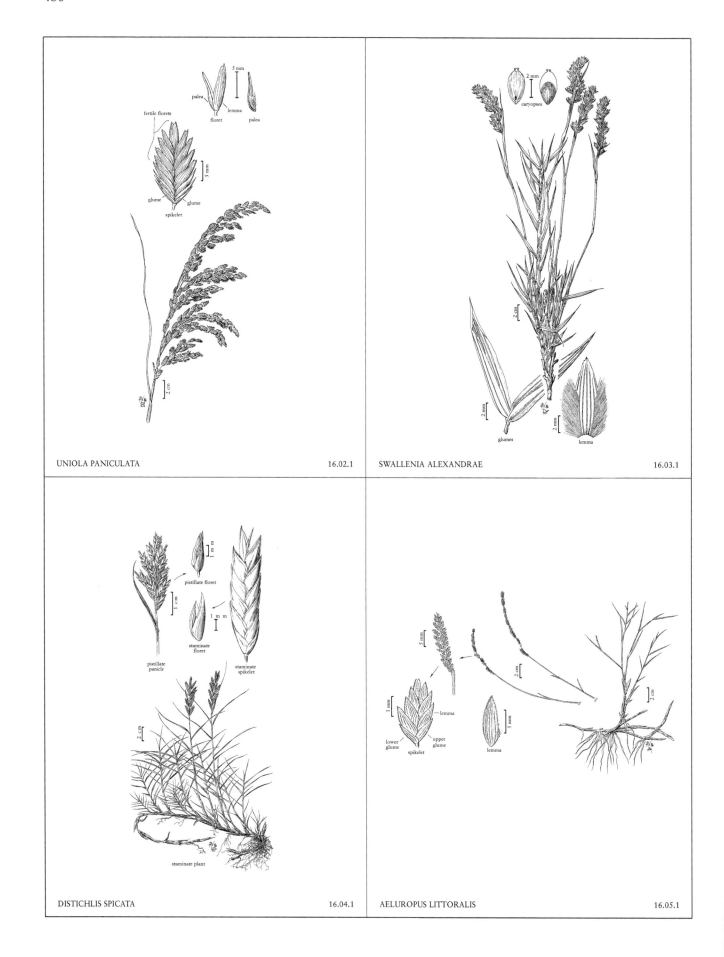

UNIOLA PANICULATA                                          16.02.1

SWALLENIA ALEXANDRAE                                       16.03.1

DISTICHLIS SPICATA                                         16.04.1

AELUROPUS LITTORALIS                                       16.05.1

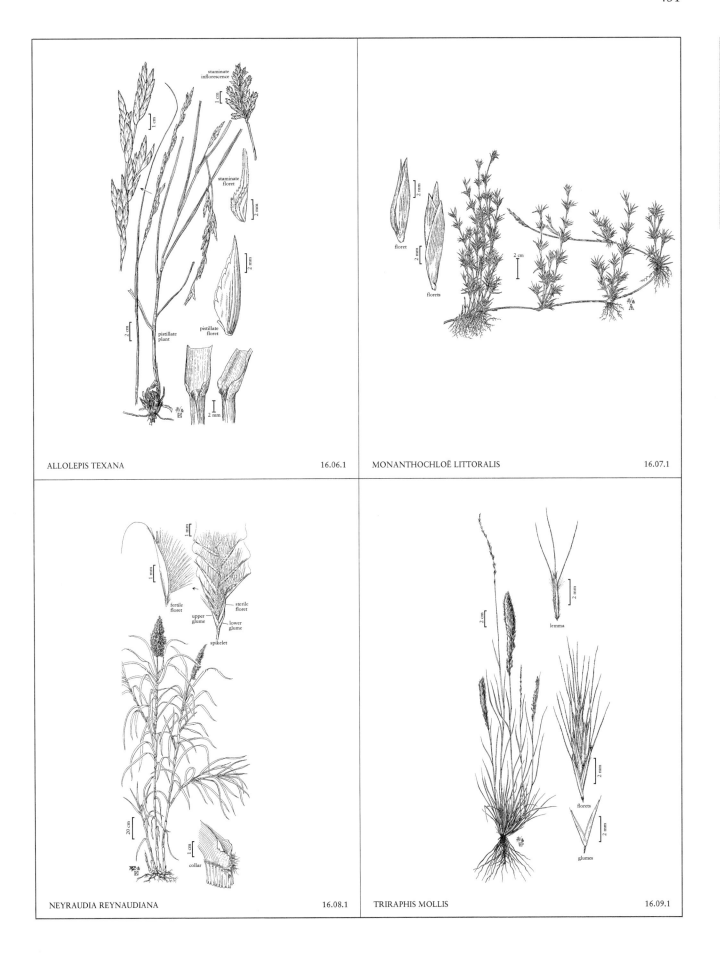

ALLOLEPIS TEXANA 16.06.1

MONANTHOCHLOË LITTORALIS 16.07.1

NEYRAUDIA REYNAUDIANA 16.08.1

TRIRAPHIS MOLLIS 16.09.1

432

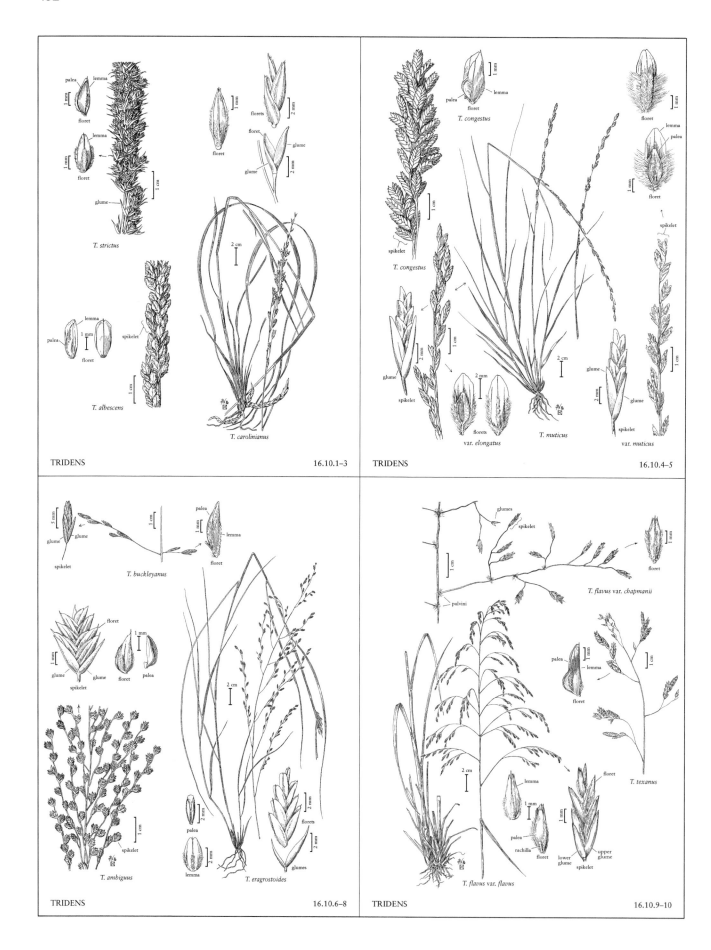

T. strictus

T. albescens

T. carolinianus

T. congestus

T. congestus

var. elongatus

T. muticus

var. muticus

T. buckleyanus

T. ambiguus

T. eragrostoides

T. flavus var. chapmanii

T. texanus

T. flavus var. flavus

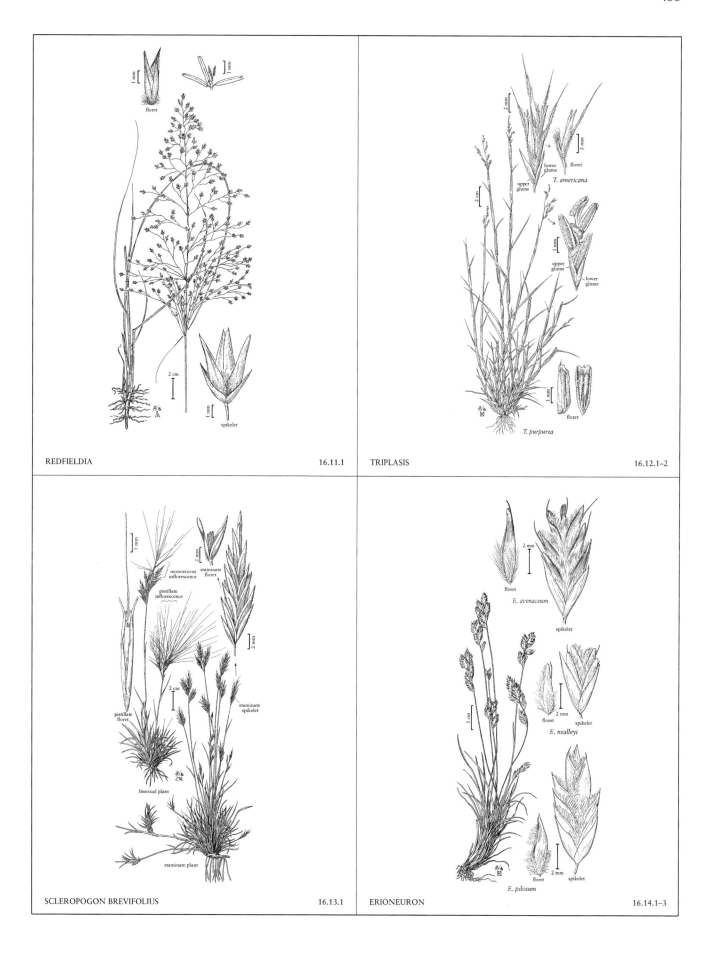

REDFIELDIA                                    16.11.1

TRIPLASIS                                     16.12.1–2

SCLEROPOGON BREVIFOLIUS                        16.13.1

ERIONEURON                                    16.14.1–3

434

DASYOCHLOA PULCHELLA                                    16.15.1

BLEPHARONEURON TRICHOLEPIS                              16.16.1

BLEPHARIDACHNE                                         16.17.1–2

MUNROA SQUARROSA                                        16.18.1

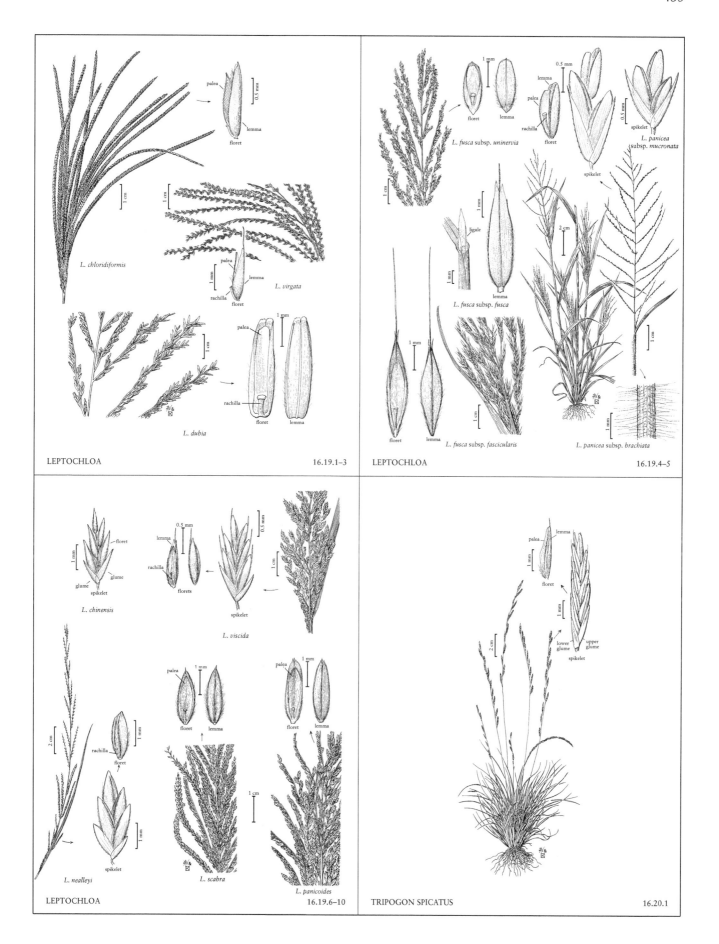

435

LEPTOCHLOA

L. chloridiformis

palea
0.5 mm
lemma
floret

1 cm
1 cm

L. virgata

palea
1 mm
lemma
rachilla
floret

L. dubia

palea
1 mm
rachilla
floret
lemma

LEPTOCHLOA                                    16.19.1–3

LEPTOCHLOA

L. fusca subsp. uninervia

1 mm
floret     lemma

0.5 mm
lemma
palea
rachilla
floret

L. panicea
subsp. mucronata
0.5 mm
spikelet

spikelet

1 mm
ligule
lemma

L. fusca subsp. fusca

2 cm

1 mm

1 cm

1 mm
floret     lemma

L. fusca subsp. fascicularis

1 cm

1 cm

1 mm
L. panicea subsp. brachiata

LEPTOCHLOA                                    16.19.4–5

floret
1 mm
lemma
glume
glume
spikelet

L. chinensis

0.5 mm
lemma
rachilla
florets

0.5 mm
1 cm
spikelet

L. viscida

2 cm
rachilla
floret

1 mm

spikelet
1 mm

L. nealleyi

palea
1 mm
floret     lemma

1 cm

L. scabra

palea
1 mm
floret     lemma

L. panicoides

LEPTOCHLOA                                    16.19.6–10

palea
lemma
1 mm
floret

1 mm

lower
glume
upper
glume
spikelet

2 cm

TRIPOGON SPICATUS                            16.20.1

436

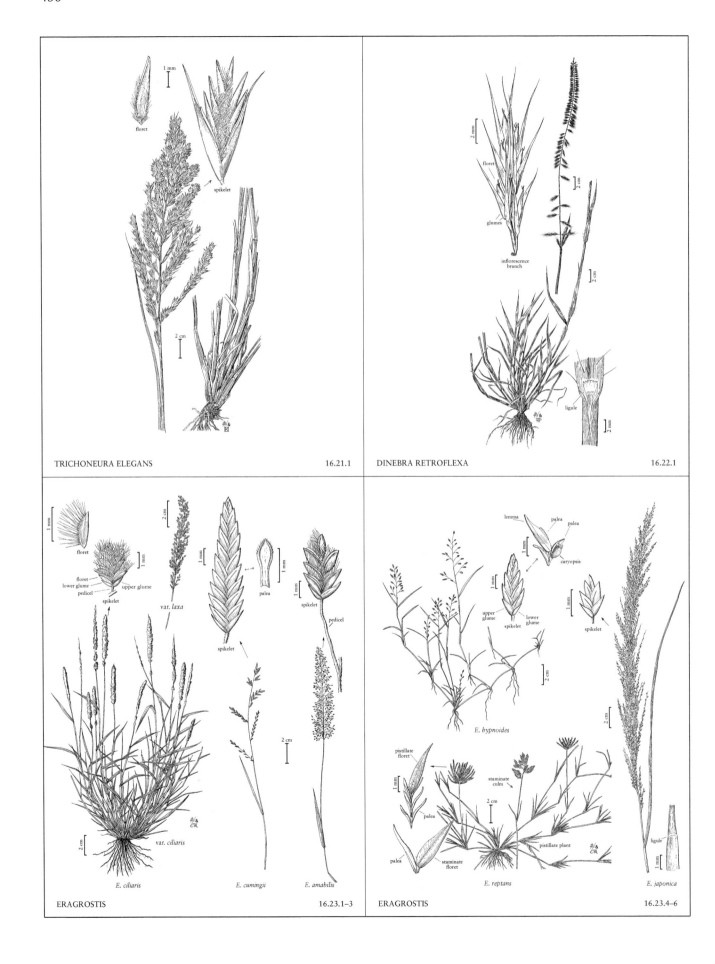

TRICHONEURA ELEGANS                16.21.1

DINEBRA RETROFLEXA                16.22.1

ERAGROSTIS                16.23.1–3

ERAGROSTIS                16.23.4–6

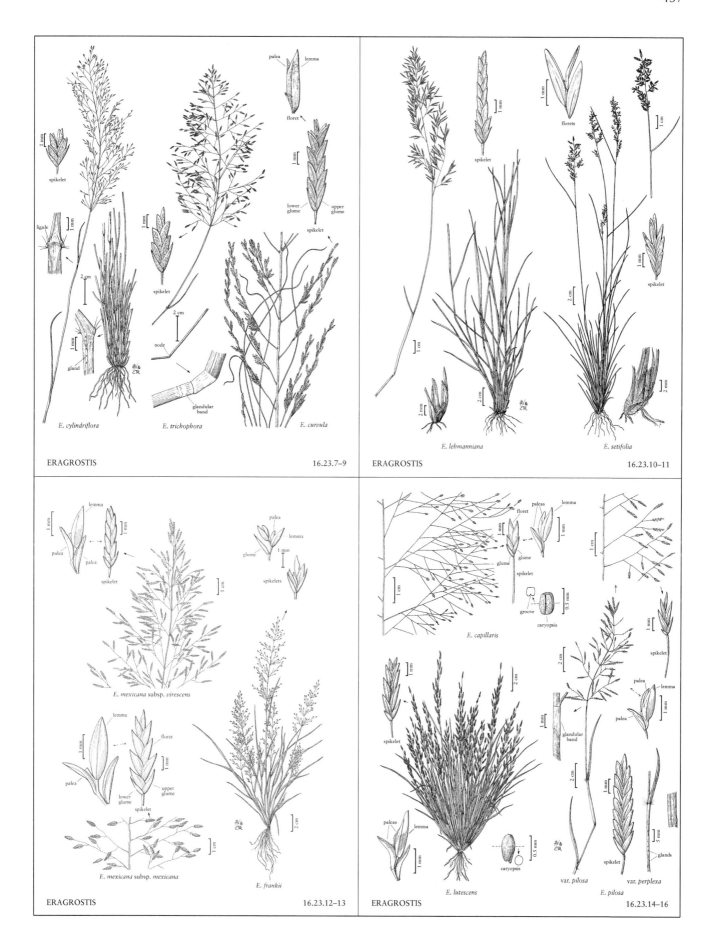

ERAGROSTIS 16.23.7–9

ERAGROSTIS 16.23.10–11

ERAGROSTIS 16.23.12–13

ERAGROSTIS 16.23.14–16

E. cylindriflora

E. trichophora

E. curvula

E. lehmanniana

E. setifolia

E. mexicana subsp. virescens

E. mexicana subsp. mexicana

E. frankii

E. capillaris

E. lutescens

var. pilosa

var. perplexa

E. pilosa

438

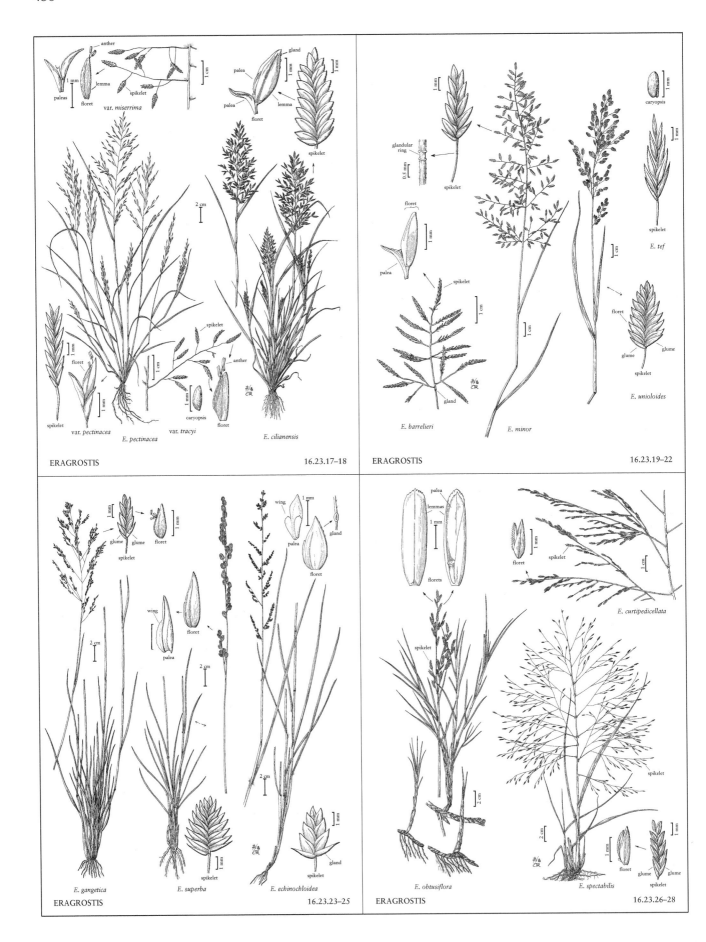

var. miserrima

spikelet

floret

lemma

anther

paleas

1 mm

lemma

palea

floret

gland

1 mm

palea

floret

lemma

spikelet

1 mm

2 cm

spikelet

1 mm

floret

1 mm

var. pectinacea

1 cm

spikelet

anther

caryopsis

1 mm

var. tracyi

floret

1 cm

E. pectinacea

E. cilianensis

1 mm

glandular ring

0.5 mm

spikelet

floret

palea

1 mm

spikelet

1 cm

gland

E. barrelieri

1 cm

E. minor

1 cm

caryopsis

1 mm

1 mm

spikelet

E. tef

floret

glume

glume

spikelet

E. unioloides

1 mm

glume  glume  floret

spikelet

2 cm

wing

floret

palea

E. gangetica

2 cm

E. superba

spikelet

1 mm

wing

palea

1 mm

gland

floret

2 cm

E. echinochloidea

spikelet

1 mm

gland

palea

lemmas

1 mm

florets

floret

1 mm

spikelet

spikelet

E. curtipedicellata

1 cm

spikelet

2 cm

E. obtusiflora

2 cm

spikelet

floret

1 mm

glume  glume

spikelet

1 mm

E. spectabilis

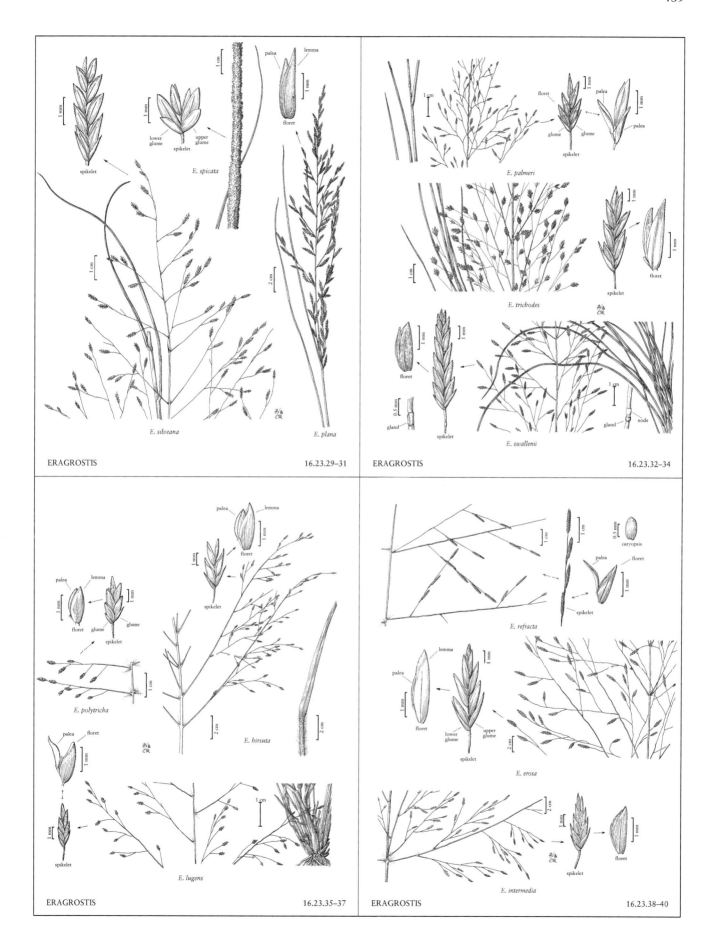

E. spicata

E. silveana

E. plana

E. palmeri

E. trichodes

E. swallenii

E. polytricha

E. hirsuta

E. lugens

E. refracta

E. erosa

E. intermedia

440

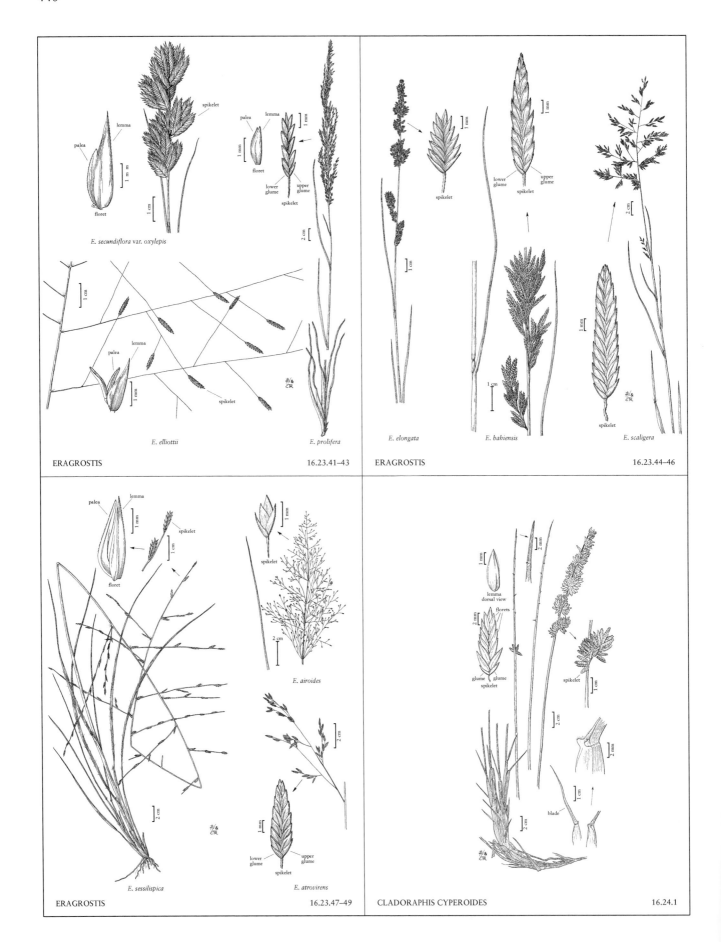

E. secundiflora var. oxylepis

E. elliottii

E. prolifera

ERAGROSTIS                                    16.23.41–43

E. elongata

E. bahiensis

E. scaligera

ERAGROSTIS                                    16.23.44–46

E. sessilispica

E. airoides

E. atrovirens

ERAGROSTIS                                    16.23.47–49

CLADORAPHIS CYPEROIDES            16.24.1

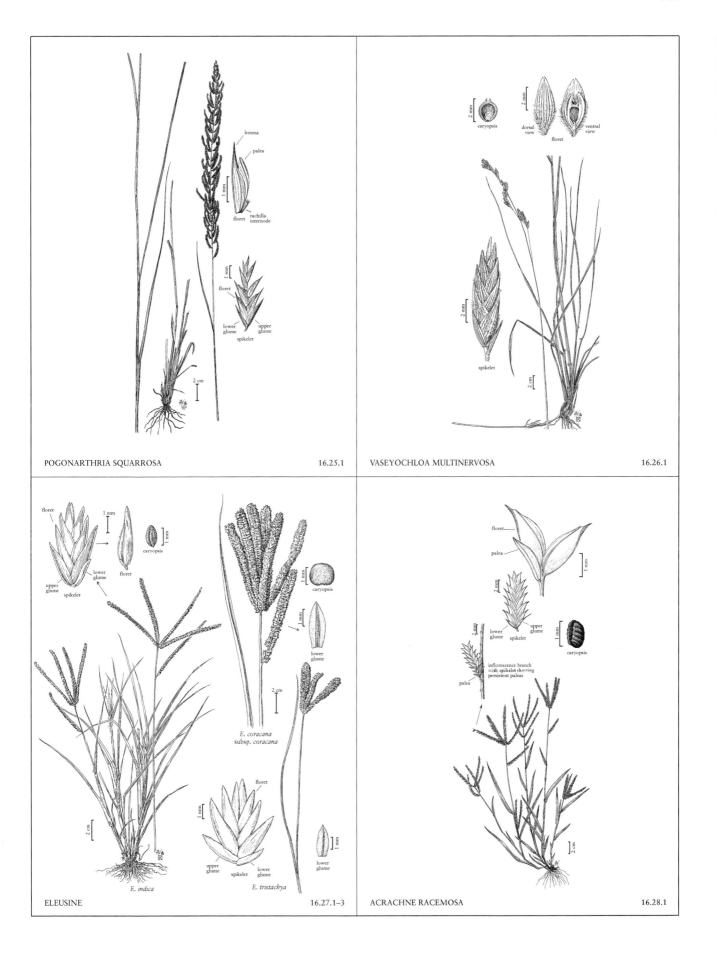

POGONARTHRIA SQUARROSA 16.25.1

VASEYOCHLOA MULTINERVOSA 16.26.1

ELEUSINE 16.27.1–3

ACRACHNE RACEMOSA 16.28.1

442

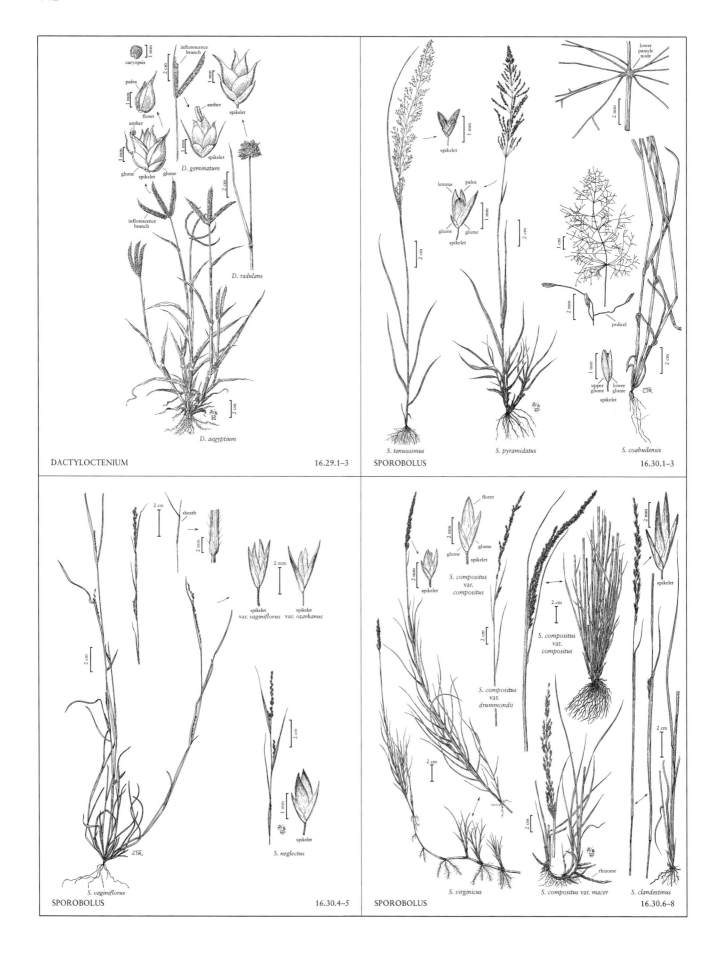

DACTYLOCTENIUM                                          16.29.1–3

*D. geminatum*
caryopsis
inflorescence branch
1 mm
2 mm
palea
anther
spikelet
floret
anther
1 mm
glume
spikelet
glume
*D. radulans*
2 cm
inflorescence branch
*D. aegyptium*
2 cm

SPOROBOLUS                                             16.30.1–3

lower panicle node
2 mm
spikelet
1 mm
lemma    palea
glume    glume
spikelet
1 mm
2 cm
2 cm
1 cm
pedicel
2 mm
upper    lower
glume    glume
spikelet
1 mm
2 cm
*S. tenuissimus*         *S. pyramidatus*         *S. coahuilensis*

SPOROBOLUS                                             16.30.4–5

2 cm
sheath
2 mm
2 mm
spikelet         spikelet
var. *vaginiflorus*   var. *ozarkanus*
2 cm
2 cm
1 mm
spikelet
*S. neglectus*
*S. vaginiflorus*

SPOROBOLUS                                             16.30.6–8

floret
2 mm
glume    glume
spikelet
2 mm
spikelet
*S. compositus*
var.
*compositus*
2 cm
*S. compositus*
var.
*drummondii*
2 cm
2 mm
spikelet
*S. compositus*
var.
*compositus*
2 cm
2 cm
rhizome
2 cm
*S. virginicus*     *S. compositus* var. *macer*   *S. clandestinus*

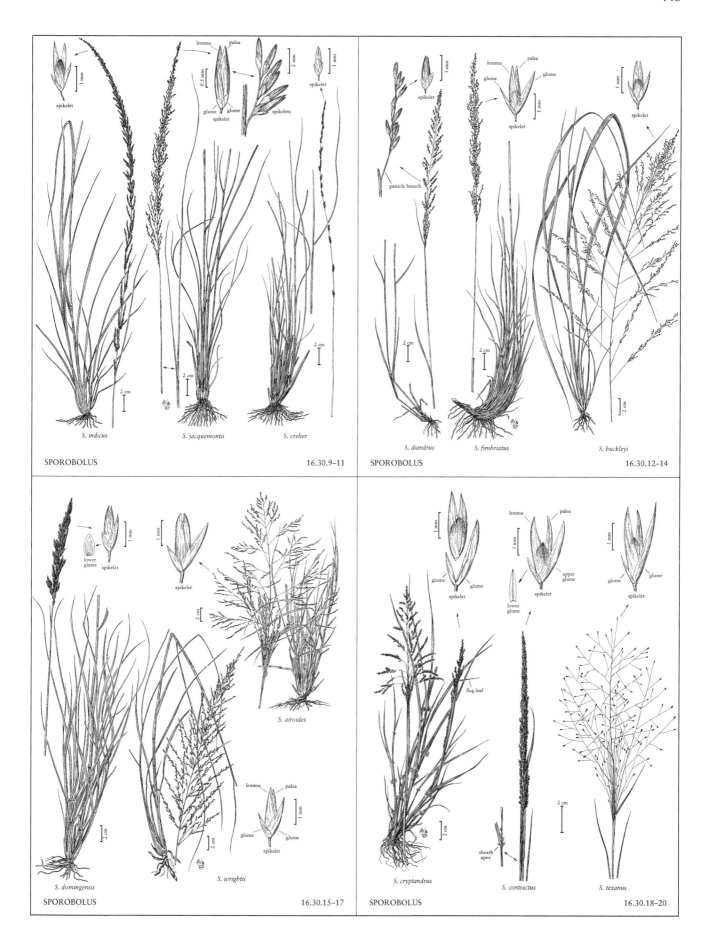

443

SPOROBOLUS                                           16.30.9–11

SPOROBOLUS                                           16.30.12–14

SPOROBOLUS                                           16.30.15–17

SPOROBOLUS                                           16.30.18–20

444

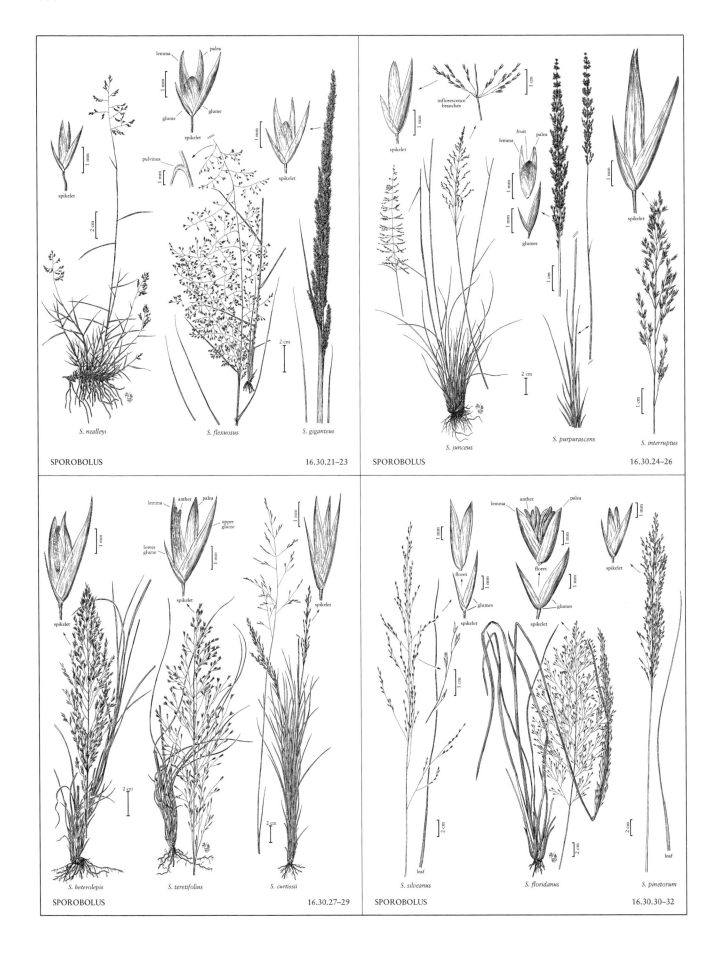

SPOROBOLUS

S. nealleyi

S. flexuosus

S. giganteus

16.30.21–23

SPOROBOLUS

S. junceus

S. purpurascens

S. interruptus

16.30.24–26

SPOROBOLUS

S. heterolepis

S. teretifolius

S. curtissii

16.30.27–29

SPOROBOLUS

S. silveanus

S. floridanus

S. pinetorum

16.30.30–32

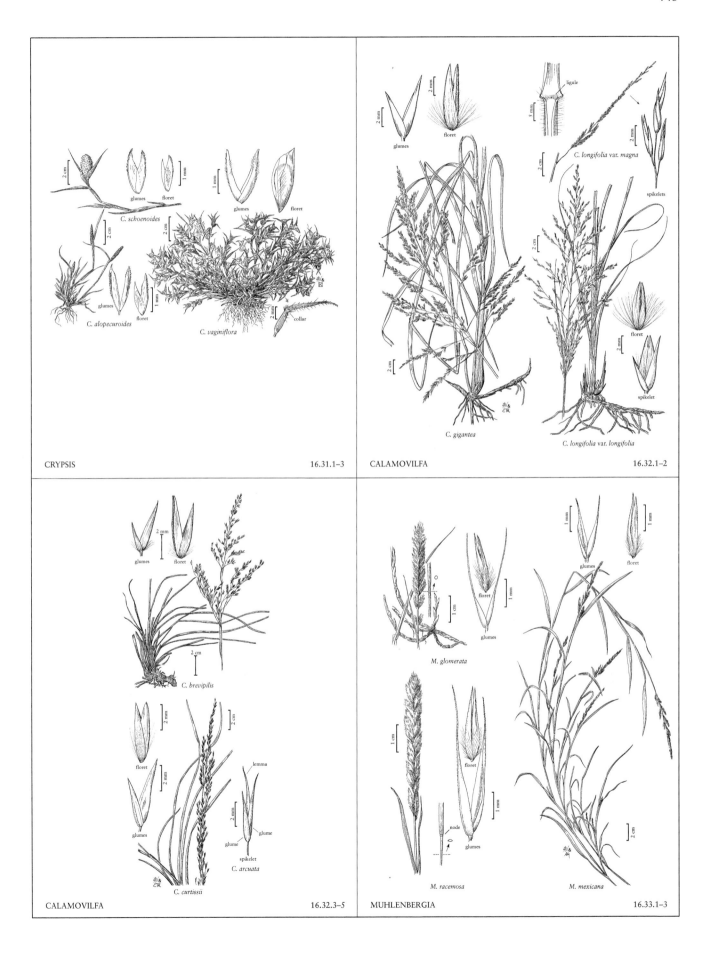

CRYPSIS 16.31.1–3

CALAMOVILFA 16.32.1–2

CALAMOVILFA 16.32.3–5

MUHLENBERGIA 16.33.1–3

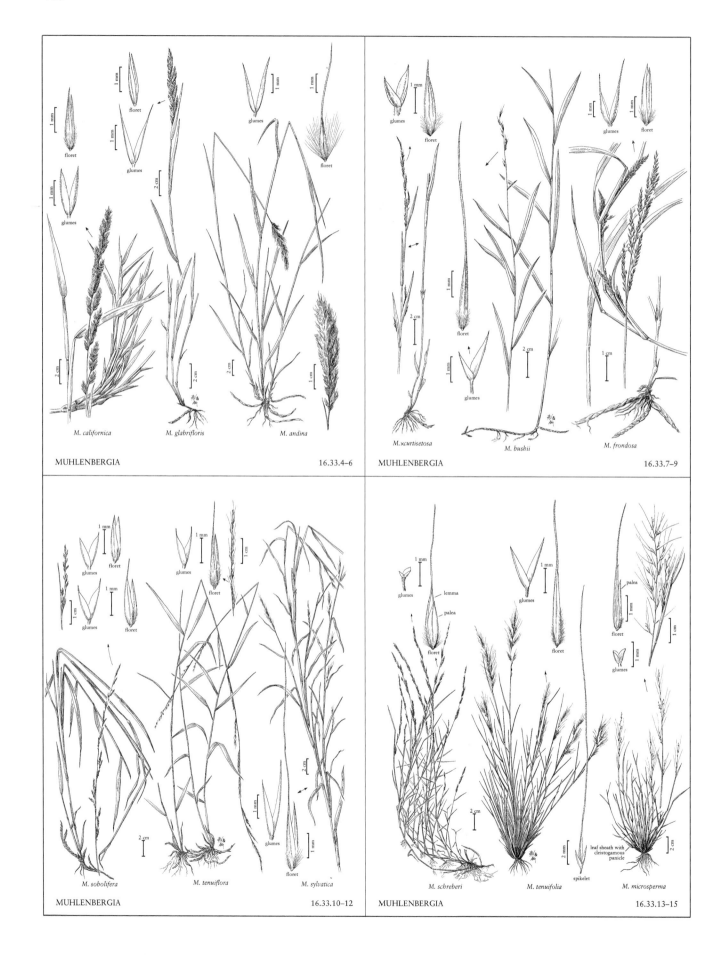

446

MUHLENBERGIA

*M. californica*    *M. glabrifloris*    *M. andina*

16.33.4–6

MUHLENBERGIA

*M. xcurtisetosa*    *M. bushii*    *M. frondosa*

16.33.7–9

MUHLENBERGIA

*M. sobolifera*    *M. tenuiflora*    *M. sylvatica*

16.33.10–12

MUHLENBERGIA

*M. schreberi*    *M. tenuifolia*    *M. microsperma*

16.33.13–15

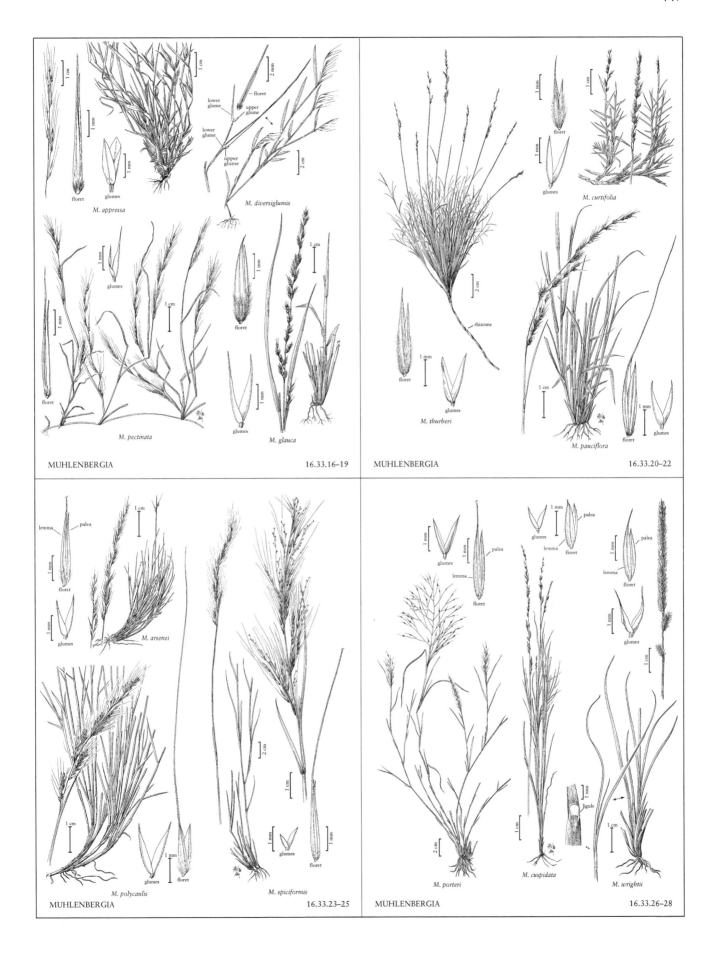

MUHLENBERGIA

16.33.16–19

MUHLENBERGIA

16.33.20–22

MUHLENBERGIA

16.33.23–25

MUHLENBERGIA

16.33.26–28

448

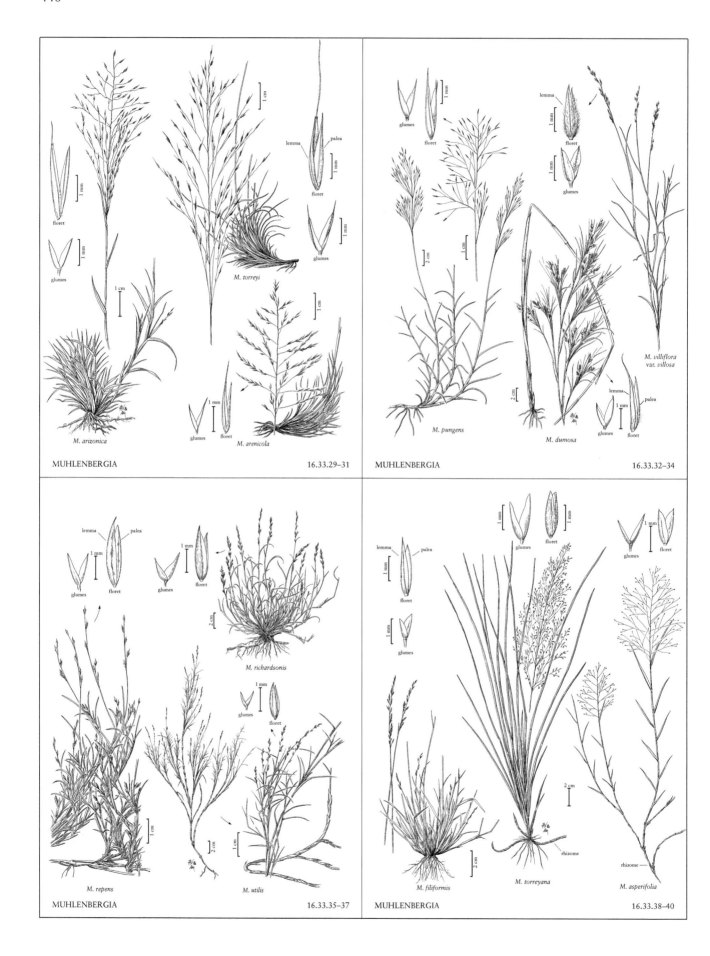

M. arizonica

floret

glumes

M. torreyi

lemma    palea

floret

glumes

glumes    floret

M. arenicola

M. pungens

glumes

floret

lemma

floret

glumes

M. villiflora
var. villosa

lemma    palea

glumes    floret

M. dumosa

lemma    palea

glumes    floret

M. richardsonis

glumes    floret

M. repens

glumes    floret

M. utilis

lemma    palea

floret

glumes

glumes    floret

glumes    floret

M. filiformis

M. torreyana

rhizome

rhizome

M. asperifolia

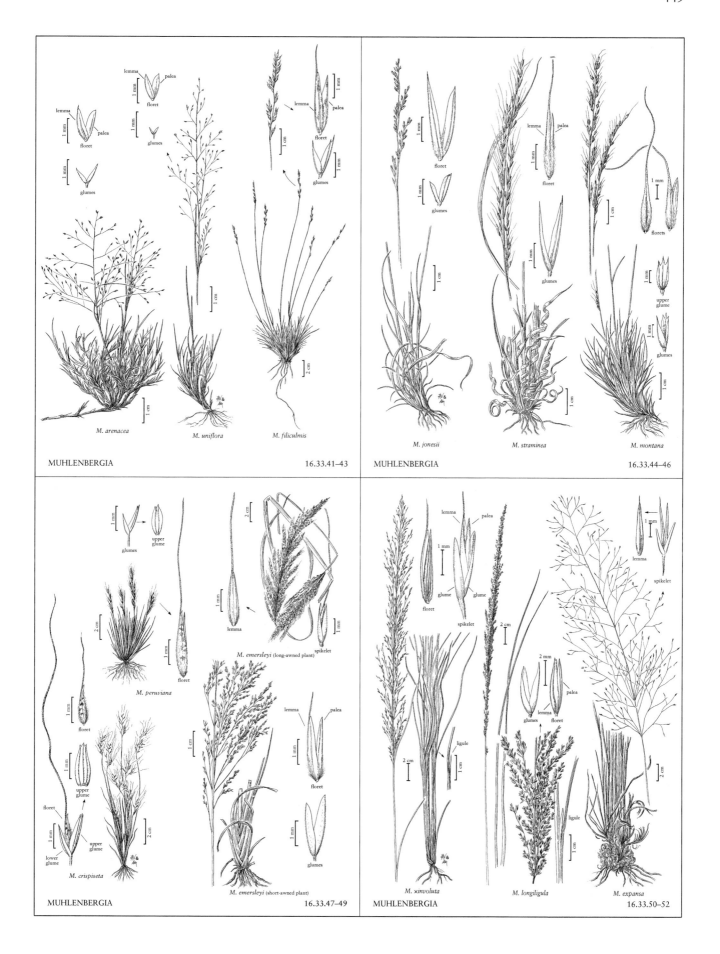

449

MUHLENBERGIA

*M. arenacea*

*M. uniflora*

*M. filiculmis*

MUHLENBERGIA                                    16.33.41–43

MUHLENBERGIA

*M. jonesii*

*M. straminea*

*M. montana*

MUHLENBERGIA                                    16.33.44–46

MUHLENBERGIA

*M. peruviana*

*M. crispiseta*

*M. emersleyi* (long-awned plant)

*M. emersleyi* (short-awned plant)

MUHLENBERGIA                                    16.33.47–49

MUHLENBERGIA

*M. ×involuta*

*M. longiligula*

*M. expansa*

MUHLENBERGIA                                    16.33.50–52

450

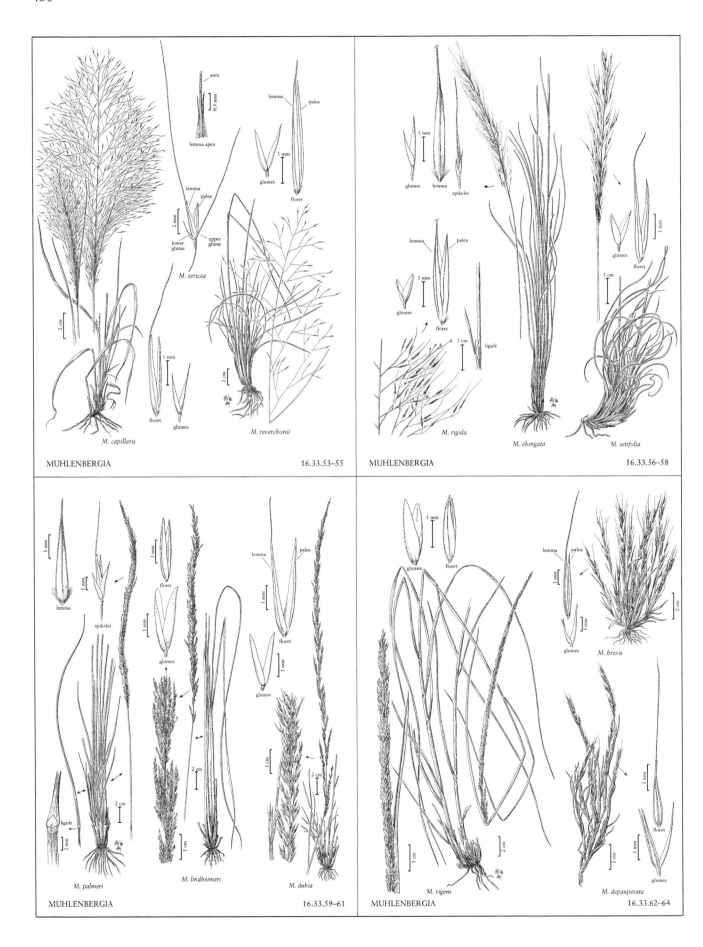

M. sericea

M. capillaris

M. reverchonii

MUHLENBERGIA                                           16.33.53–55

M. rigida

M. elongata

M. setifolia

MUHLENBERGIA                                           16.33.56–58

M. palmeri

M. lindheimeri

M. dubia

MUHLENBERGIA                                           16.33.59–61

M. rigens

M. brevis

M. depauperata

MUHLENBERGIA                                           16.33.62–64

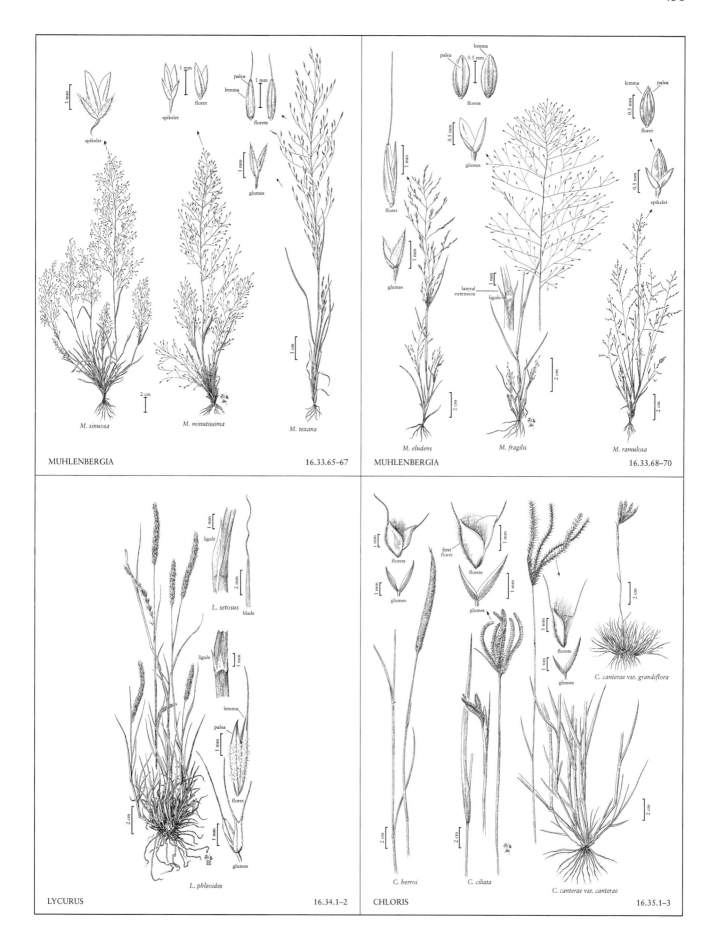

MUHLENBERGIA

*M. sinuosa*   *M. minutissima*   *M. texana*

16.33.65–67

MUHLENBERGIA

*M. eludens*   *M. fragilis*   *M. ramulosa*

16.33.68–70

LYCURUS

*L. setosus*

*L. phleoides*

16.34.1–2

CHLORIS

*C. berroi*   *C. ciliata*

*C. canterae* var. *grandiflora*

*C. canterae* var. *canterae*

16.35.1–3

452

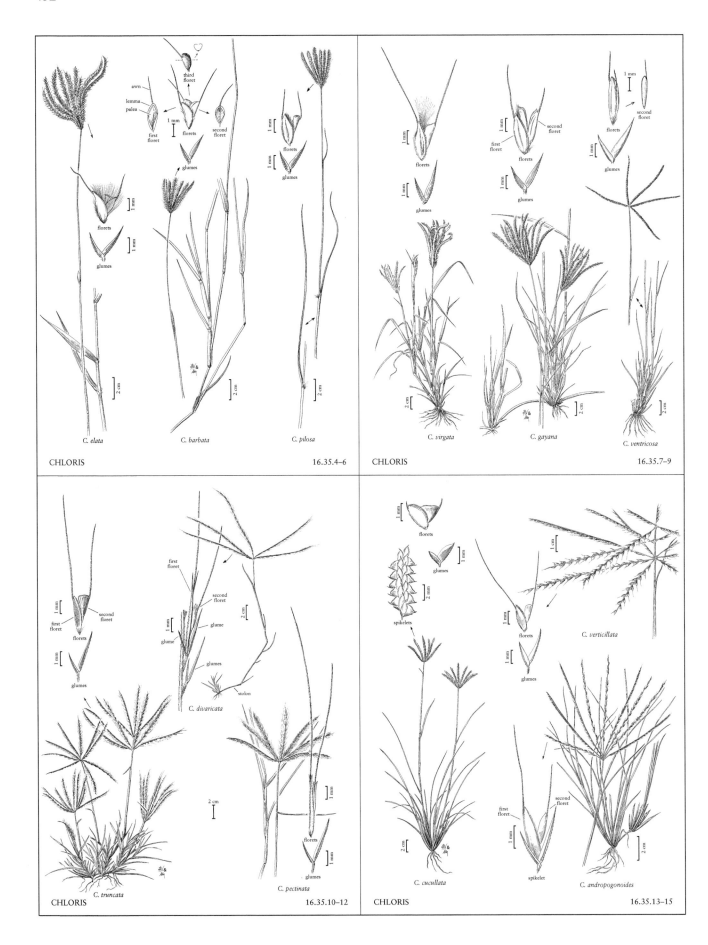

C. elata

C. barbata

C. pilosa

third
floret

awn

lemma
palea

first
floret

florets

second
floret

glumes

florets

glumes

florets

glumes

C. virgata

C. gayana

C. ventricosa

first
floret

florets

glumes

second
floret

florets

glumes

florets

second
floret

glumes

C. truncata

C. divaricata

C. pectinata

first
floret

second
floret

florets

glumes

first
floret

second
floret

glume

glume

glumes

stolon

florets

glumes

C. cucullata

C. verticillata

C. andropogonoides

florets

spikelets

florets

glumes

florets

glumes

florets

glumes

first
floret

second
floret

spikelet

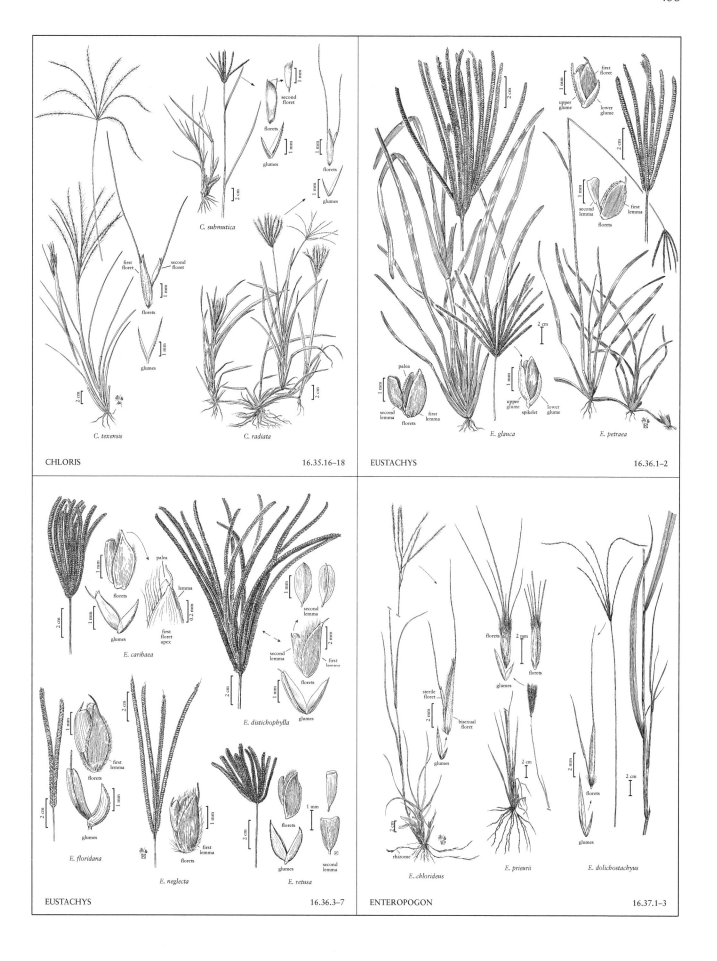

CHLORIS                                    16.35.16–18

EUSTACHYS                                  16.36.1–2

EUSTACHYS                                  16.36.3–7

ENTEROPOGON                                16.37.1–3

454

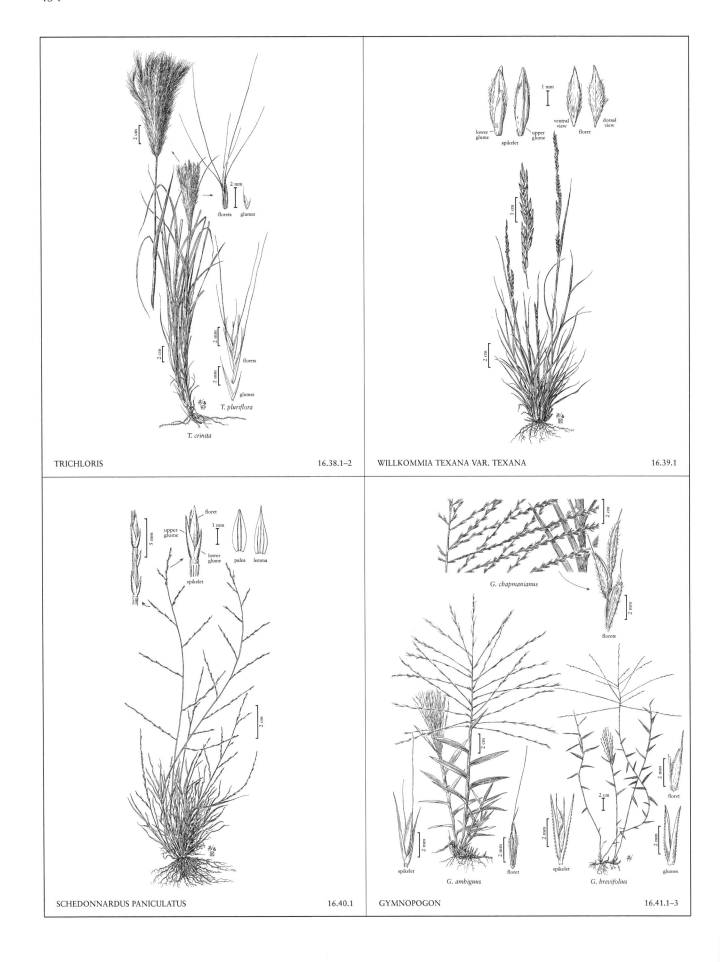

TRICHLORIS                                          16.38.1–2

WILLKOMMIA TEXANA VAR. TEXANA                       16.39.1

SCHEDONNARDUS PANICULATUS                           16.40.1

GYMNOPOGON                                          16.41.1–3

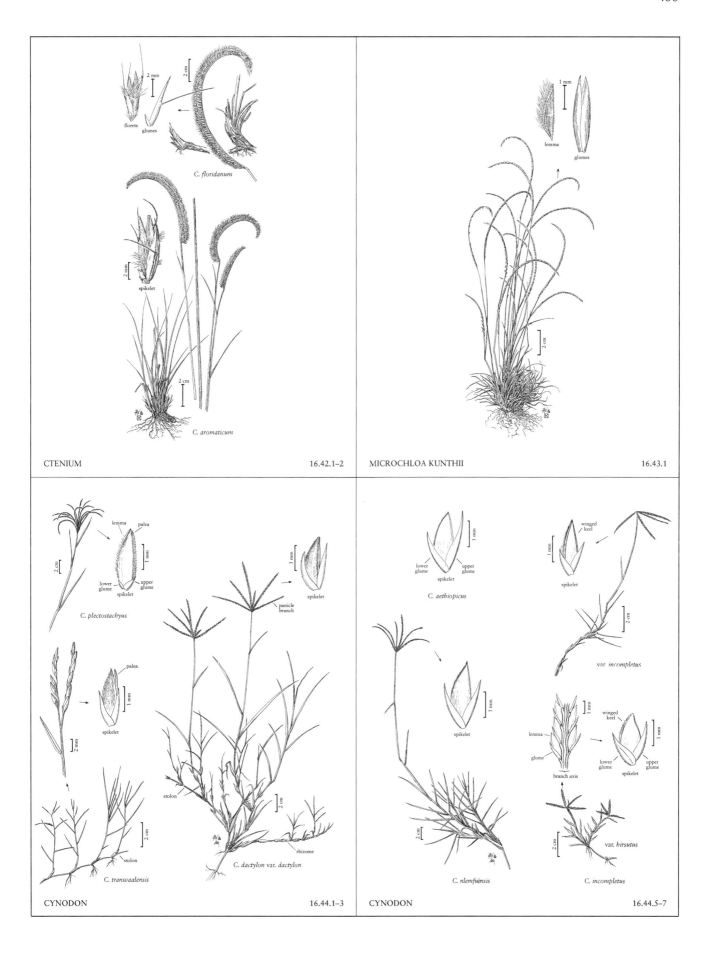

C. floridanum

florets  glumes

spikelet

C. aromaticum

lemma  glumes

MICROCHLOA KUNTHII

lemma  palea

lower  upper
glume  glume
spikelet

C. plectostachyus

palea

spikelet

C. transvaalensis

stolon

spikelet

panicle
branch

stolon

rhizome

C. dactylon var. dactylon

CYNODON

lower  upper
glume  glume
spikelet

C. aethiopicus

winged
keel

spikelet

var. incompletus

spikelet

C. nlemfuensis

lemma

glume

branch axis

winged
keel

lower  upper
glume  glume
spikelet

var. hirsutus

C. incompletus

CYNODON

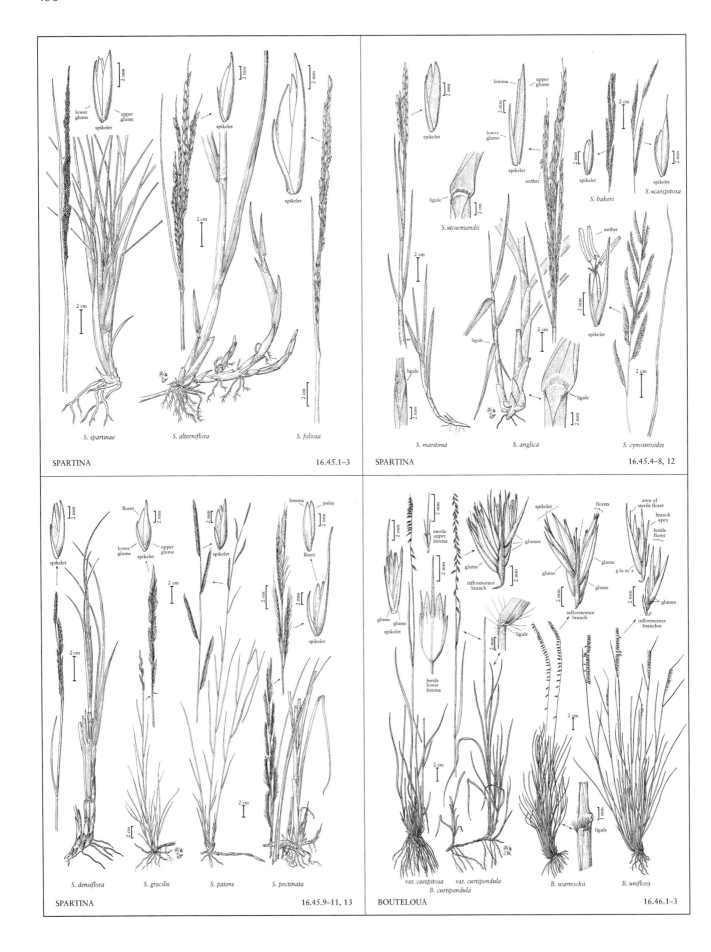

lower glume
upper glume
spikelet
2 mm
spikelet
2 mm
spikelet
2 mm
2 cm
2 cm
2 cm
S. spartinae
S. alterniflora
S. foliosa

SPARTINA 16.45.1–3

spikelet
2 mm
lemma
upper glume
lower glume
spikelet
anther
2 mm
ligule
2 cm
S.×townsendii
2 cm
spikelet
S. bakeri
spikelet
S.×caespitosa
2 mm
2 mm
anther
spikelet
2 mm
2 cm
ligule
2 cm
ligule
ligule
2 mm
2 mm
2 cm
S. maritima
S. anglica
S. cynosuroides

SPARTINA 16.45.4–8, 12

2 mm
spikelet
floret
2 mm
lower glume
spikelet
upper glume
2 cm
2 cm
2 mm
spikelet
lemma
palea
2 mm
floret
2 cm
2 mm
spikelet
2 cm
2 cm
S. densiflora
S. gracilis
S. patens
S. pectinata

SPARTINA 16.45.9–11, 13

2 mm
2 mm
sterile upper lemma
2 mm
glume
glume
spikelet
fertile lower lemma
glumes
glume
inflorescence branch
2 mm
ligule
2 mm
spikelet
florets
glume
glume
glume
2 mm
inflorescence branch
awn of sterile floret
branch apex
fertile floret
glume
2 mm
glumes
inflorescence branches
2 cm
2 cm
1 mm
ligule
var. caespitosa
var. curtipendula
B. curtipendula
B. warnockii
B. uniflora

BOUTELOUA 16.46.1–3

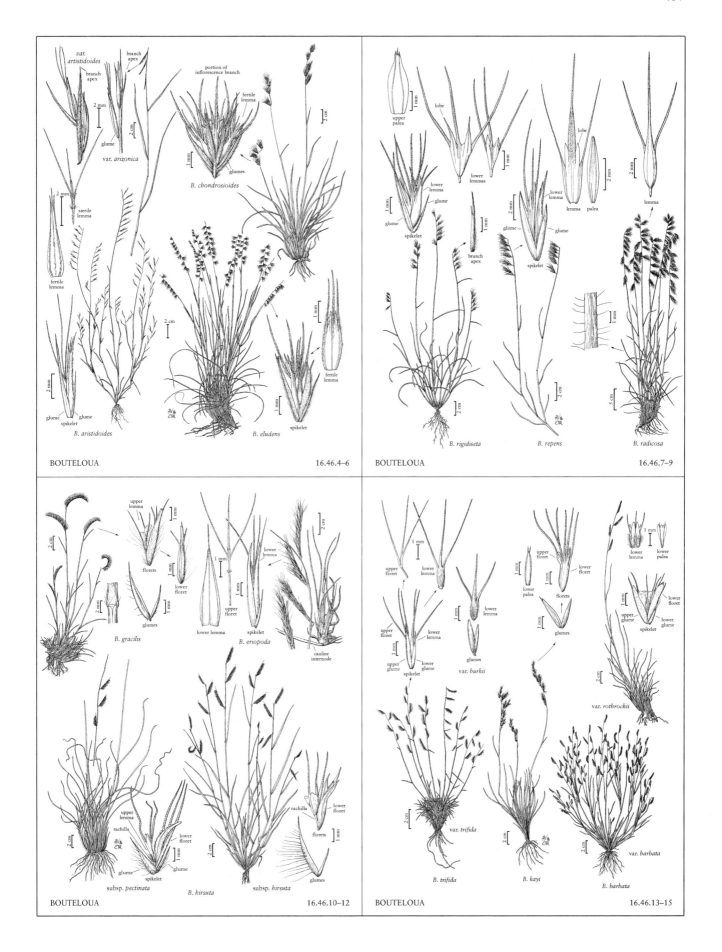

BOUTELOUA

16.46.4–6

var.
*artistidoides*
branch
apex
branch
apex
2 mm
glume
var. *arizonica*
2 cm

portion of
inflorescence branch
fertile
lemma
1 mm
glumes
*B. chondrosioides*
2 cm

2 mm
sterile
lemma
fertile
lemma

2 cm

2 mm
1 cm
fertile
lemma
glume
glume
spikelet
*B. aristidoides*
spikelet
*B. eludens*
1 mm

BOUTELOUA

16.46.7–9

upper
palea
1 mm
lobe
lower lemmas
1 mm
lobe
lemma palea
1 mm
2 mm
lemma

1 mm
lower
lemma
glume
glume
lower
lemma
spikelet
2 mm
glume glume
branch
apex
spikelet
1 mm

1 mm

2 cm
*B. rigidiseta*
2 cm
*B. repens*
5 cm
*B. radicosa*

BOUTELOUA

16.46.10–12

2 cm
upper
lemma
1 mm
florets
1 mm
glumes
lower
floret
1 mm
2 mm
*B. gracilis*

1 mm
lower lemma
upper
floret
spikelet
1 mm
lower
lemma
cauline
internode
*B. eriopoda*
2 cm

upper
lemma
rachilla
2 cm
glume glume
spikelet
subsp. *pectinata*
lower
floret
1 mm
*B. hirsuta*

rachilla
lower
floret
florets
1 mm
2 cm
glumes
subsp. *hirsuta*

BOUTELOUA

16.46.13–15

upper
floret
1 mm
lower
lemma
upper
floret
lower
lemma
1 mm
glumes
1 mm
upper
floret
lower
lemma
1 mm
upper
glume lower
spikelet glume
var. *burkii*
upper
floret
lower
palea
florets
1 mm
glumes
lower
lemma lower
palea
1 mm
lower
floret
1 mm
lower
floret
upper lower
glume glume
spikelet
var. *rothrockii*
2 cm

2 cm
var. *trifida*
*B. trifida*
2 cm
*B. kayi*
2 cm
var. *barbata*
*B. barbata*

458

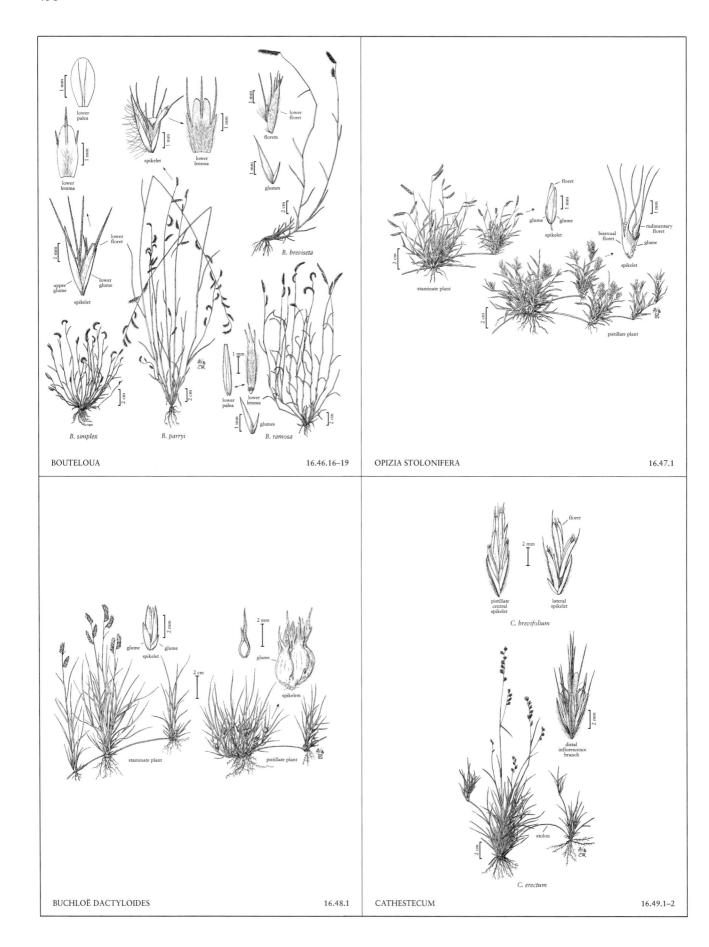

BOUTELOUA                                16.46.16–19

OPIZIA STOLONIFERA                       16.47.1

BUCHLOË DACTYLOIDES                      16.48.1

CATHESTECUM                              16.49.1–2

B. breviseta

B. simplex          B. parryi          B. ramosa

staminate plant                pistillate plant

staminate plant                pistillate plant

C. brevifolium

C. erectum

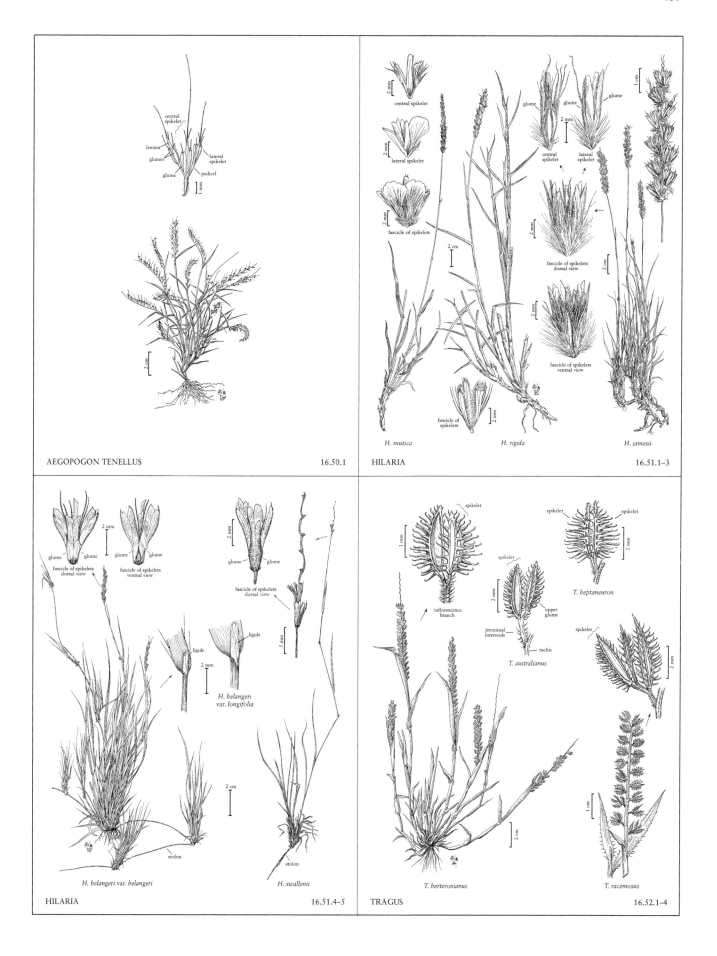

AEGOPOGON TENELLUS            16.50.1

central
spikelet

lemma

glumes      lateral
spikelet

glume     pedicel

1 mm

2 cm

HILARIA            16.51.1–3

central spikelet

2 mm

lateral spikelet

2 mm

fascicle of spikelets

2 mm

fascicle of
spikelets

2 mm

*H. mutica*

glume    glume

2 mm

central
spikelet

glume    glume

lateral
spikelet

fascicle of spikelets
dorsal view

2 mm

fascicle of spikelets
ventral view

2 mm

2 cm

1 cm

2 cm

*H. rigida*          *H. jamesii*

HILARIA            16.51.4–5

glume     glume

2 mm

fascicle of spikelets
dorsal view

glume     glume

fascicle of spikelets
ventral view

glume     glume

2 mm

fascicle of spikelets
dorsal view

5 mm

ligule

ligule

2 mm

*H. belangeri*
var. *longifolia*

2 cm

stolon

stolon

*H. belangeri* var. *belangeri*          *H. swallenii*

TRAGUS            16.52.1–4

spikelet

1 mm

inflorescence
branch

spikelet

spikelet

spikelet

2 mm

*T. heptaneuron*

spikelet

2 mm

proximal
internode

upper
glume

rachis

*T. australianus*

spikelet

2 mm

spikelet

2 cm

1 cm

*T. berteronianus*          *T. racemosus*

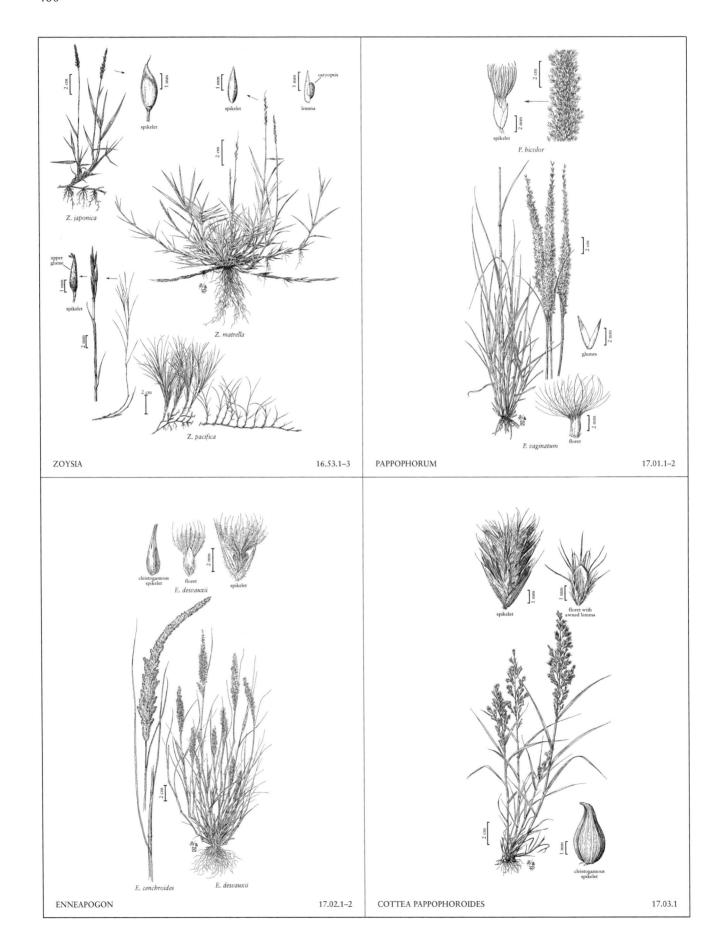

ZOYSIA                                                    16.53.1–3

PAPPOPHORUM                                               17.01.1–2

ENNEAPOGON                                                17.02.1–2

COTTEA PAPPOPHOROIDES                                     17.03.1

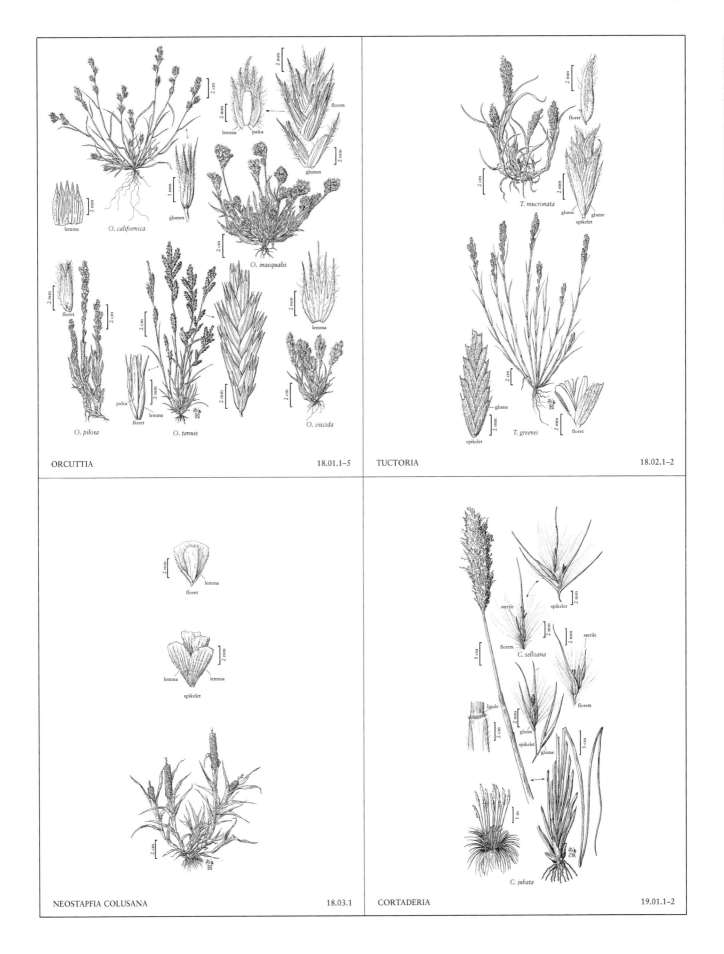

ORCUTTIA

18.01.1–5

TUCTORIA

18.02.1–2

NEOSTAPFIA COLUSANA

18.03.1

CORTADERIA

19.01.1–2

462

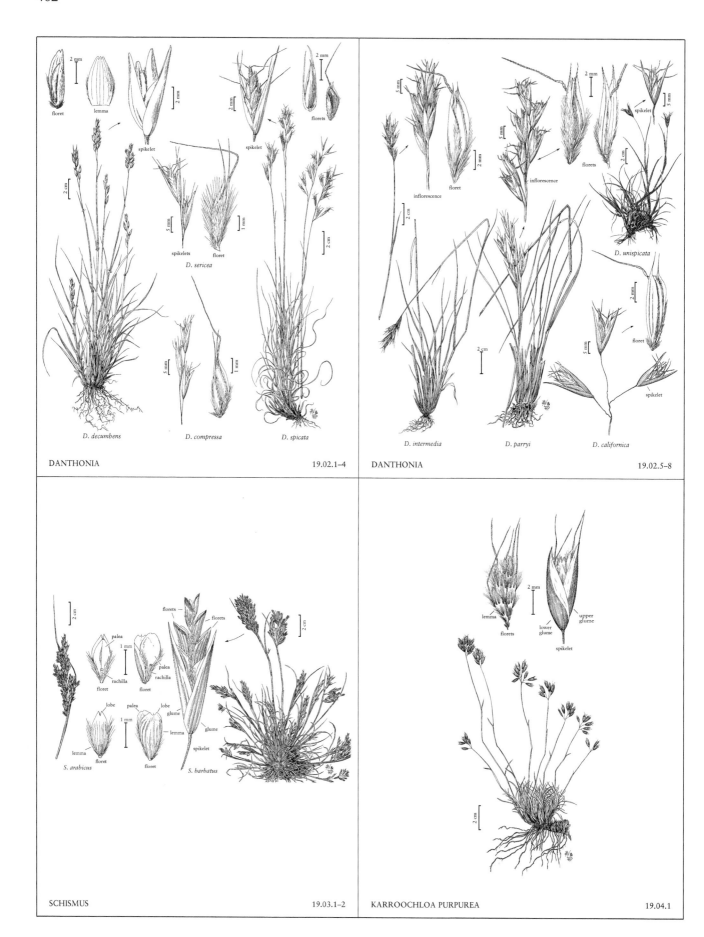

DANTHONIA                                        19.02.1–4

DANTHONIA                                        19.02.5–8

D. sericea

D. decumbens          D. compressa          D. spicata

D. intermedia          D. parryi          D. californica

D. unispicata

SCHISMUS                                         19.03.1–2

S. arabicus          S. barbatus

KARROOCHLOA PURPUREA                             19.04.1

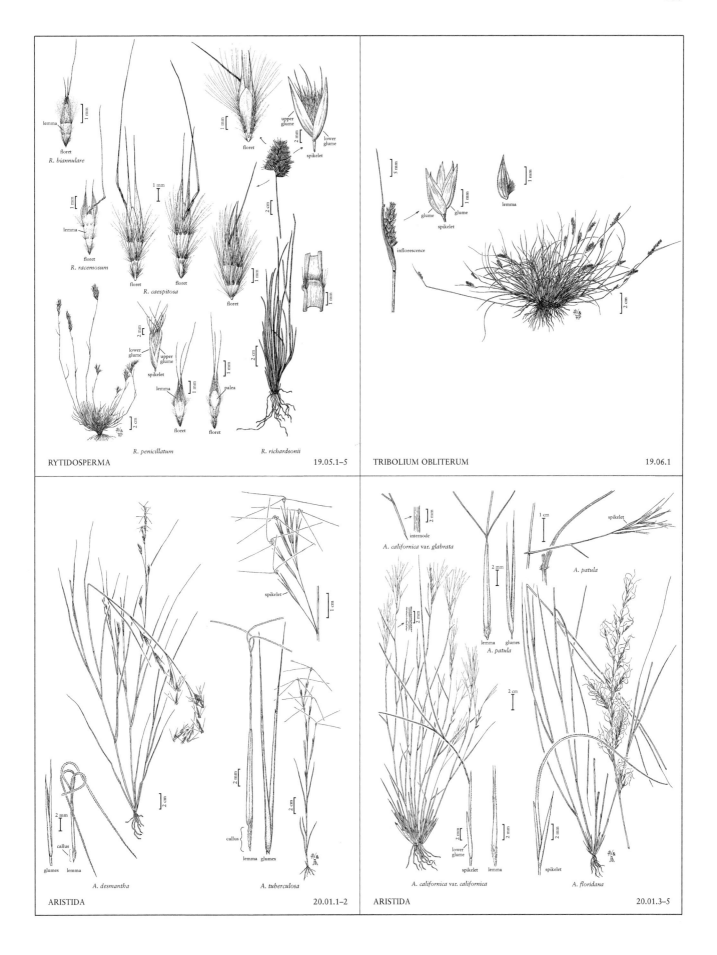

463

RYTIDOSPERMA

lemma
floret
*R. biannulare*

1 mm

lemma
floret
*R. racemosum*

1 mm

floret
floret
*R. caespitosa*

floret

floret

upper glume
lower glume
spikelet

floret

2 mm

1 mm

1 mm

2 cm

lower glume
upper glume
spikelet

lemma
floret

palea
floret

2 mm

1 mm

*R. penicillatum*

2 cm

*R. richardsonii*

RYTIDOSPERMA                    19.05.1–5

TRIBOLIUM OBLITERUM

5 mm

glume
glume
spikelet

lemma

1 mm

inflorescence

2 cm

TRIBOLIUM OBLITERUM              19.06.1

ARISTIDA

spikelet

1 cm

2 mm

2 cm

callus
glumes    lemma
*A. desmantha*

2 mm

callus
lemma  glumes
*A. tuberculosa*

2 cm

ARISTIDA                         20.01.1–2

internode
*A. californica* var. *glabrata*

2 mm

spikelet
*A. patula*

1 cm

lemma    glumes
*A. patula*

2 mm

2 mm

2 cm

lower glume
spikelet    lemma
*A. californica* var. *californica*

2 mm    2 mm

spikelet
*A. floridana*

2 mm

ARISTIDA                         20.01.3–5

464

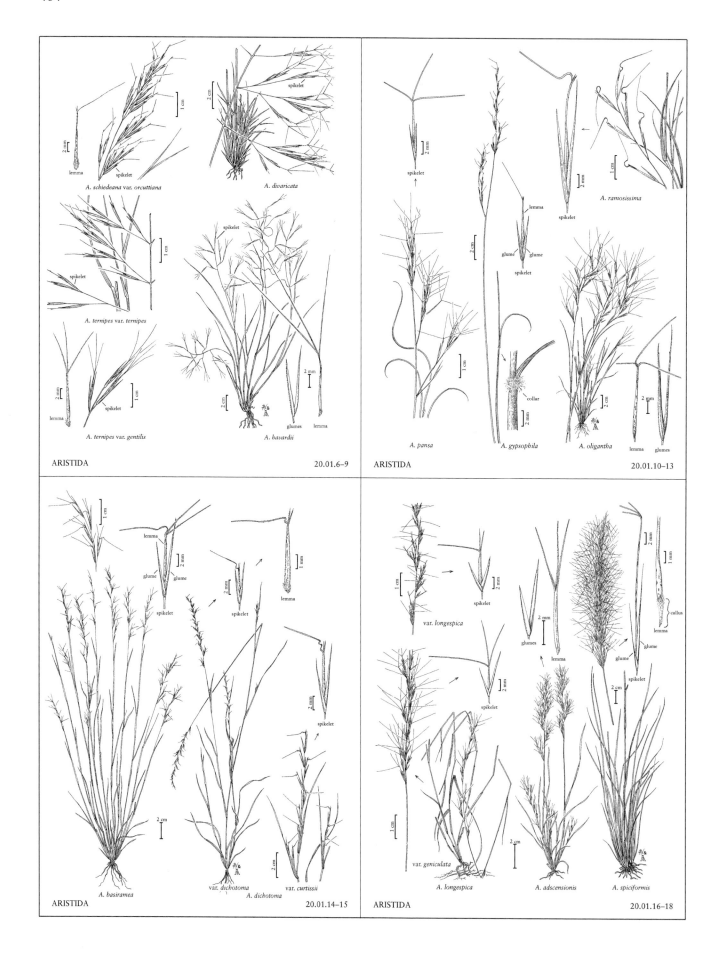

A. schiedeana var. orcuttiana

A. divaricata

A. ternipes var. ternipes

A. ternipes var. gentilis

A. havardii

A. ramosissima

A. pansa

A. gypsophila

A. oligantha

A. basiramea

var. dichotoma
A. dichotoma

var. curtissii

var. longespica

var. geniculata

A. longespica

A. adscensionis

A. spiciformis

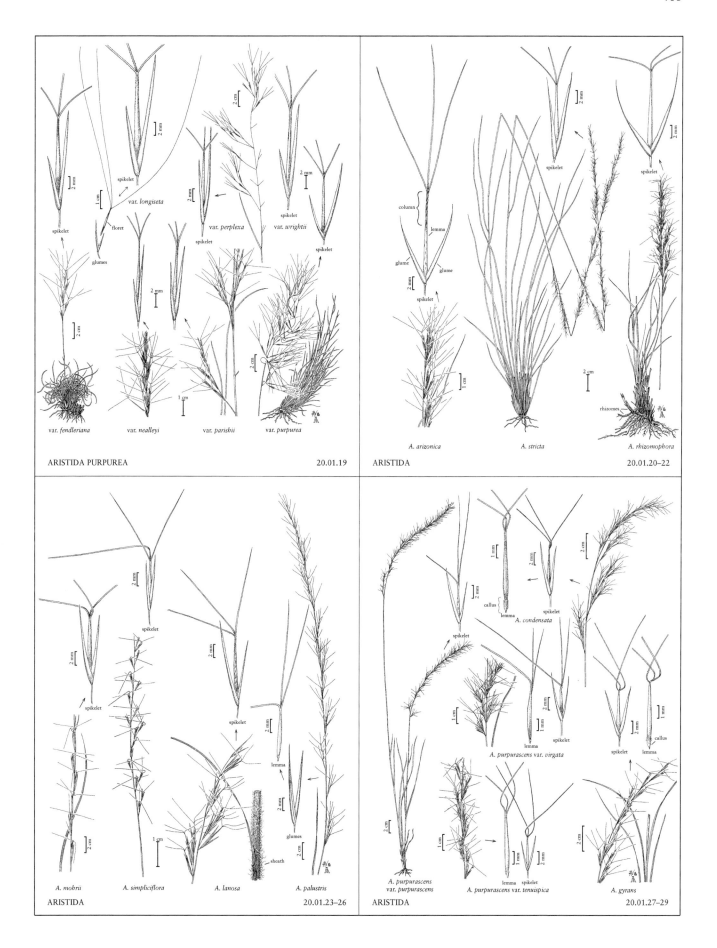

var. *longiseta*

spikelet

var. *perplexa*

var. *wrightii*

spikelet

spikelet

spikelet

floret

glumes

spikelet

var. *fendleriana*     var. *nealleyi*     var. *parishii*     var. *purpurea*

ARISTIDA PURPUREA                                    20.01.19

column

lemma

glume

glume

spikelet

spikelet

spikelet

rhizomes

*A. arizonica*          *A. stricta*          *A. rhizomophora*

ARISTIDA                                    20.01.20–22

spikelet

spikelet

*A. mohrii*          *A. simpliciflora*          *A. lanosa*          *A. palustris*

lemma

glumes

sheath

ARISTIDA                                    20.01.23–26

callus

lemma

spikelet

*A. condensata*

spikelet

lemma

spikelet

*A. purpurascens* var. *virgata*

spikelet

callus

lemma

*A. purpurascens*
var. *purpurascens*

lemma     spikelet

*A. purpurascens* var. *tenuispica*

*A. gyrans*

ARISTIDA                                    20.01.27–29

466

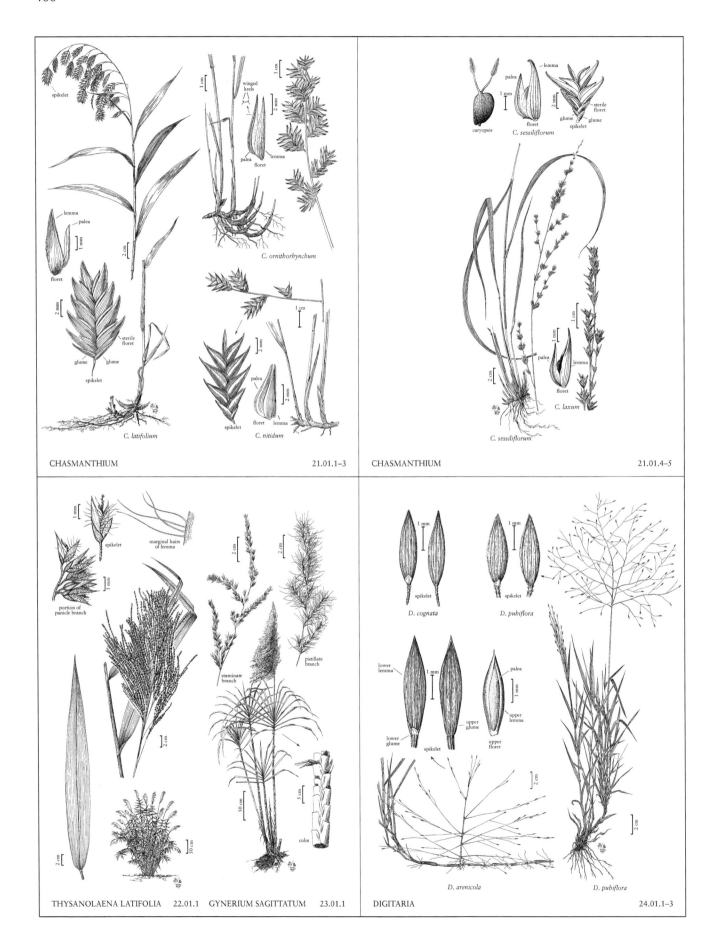

caryopsis  palea  lemma  *C. sessiliflorum*  floret  sterile floret  glume  glume  spikelet

spikelet

winged keels

palea  floret  lemma

*C. ornithorhynchum*

lemma  palea  floret

sterile floret  glume  glume  spikelet

palea  floret  lemma  spikelet

*C. nitidum*

*C. latifolium*

palea  lemma  floret  *C. laxum*

*C. sessiliflorum*

CHASMANTHIUM                    21.01.1–3

CHASMANTHIUM                    21.01.4–5

spikelet

marginal hairs of lemma

portion of panicle branch

staminate branch

pistillate branch

culm

THYSANOLAENA LATIFOLIA    22.01.1    GYNERIUM SAGITTATUM    23.01.1

spikelet  *D. cognata*

spikelet  *D. pubiflora*

lower lemma  upper glume  palea  upper lemma

lower glume  spikelet  upper floret

*D. arenicola*

*D. pubiflora*

DIGITARIA                    24.01.1–3

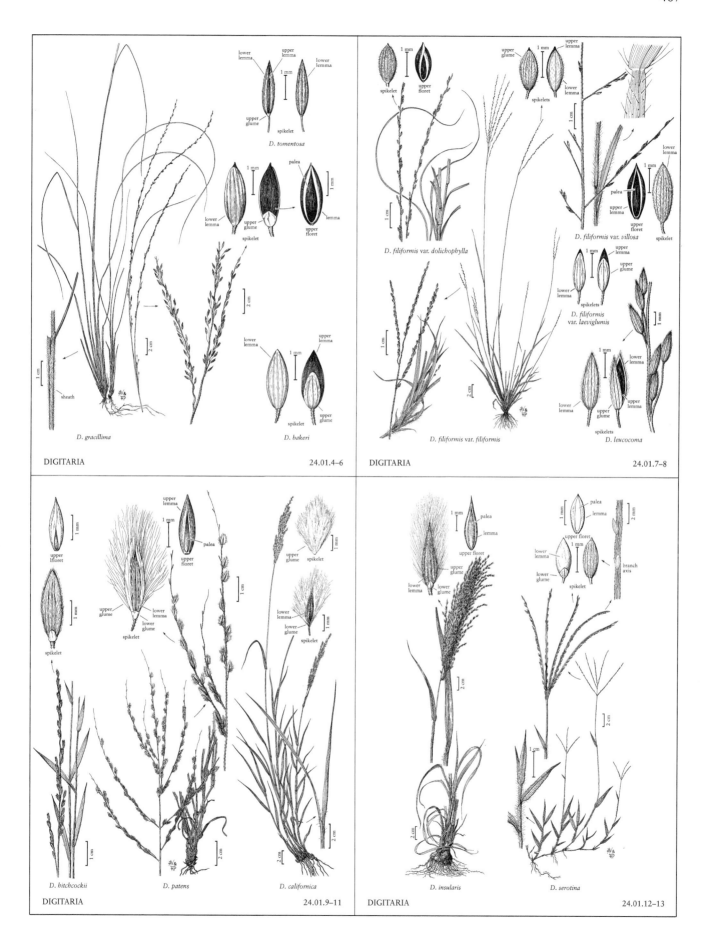

DIGITARIA

24.01.4–6

DIGITARIA

24.01.7–8

DIGITARIA

24.01.9–11

DIGITARIA

24.01.12–13

468

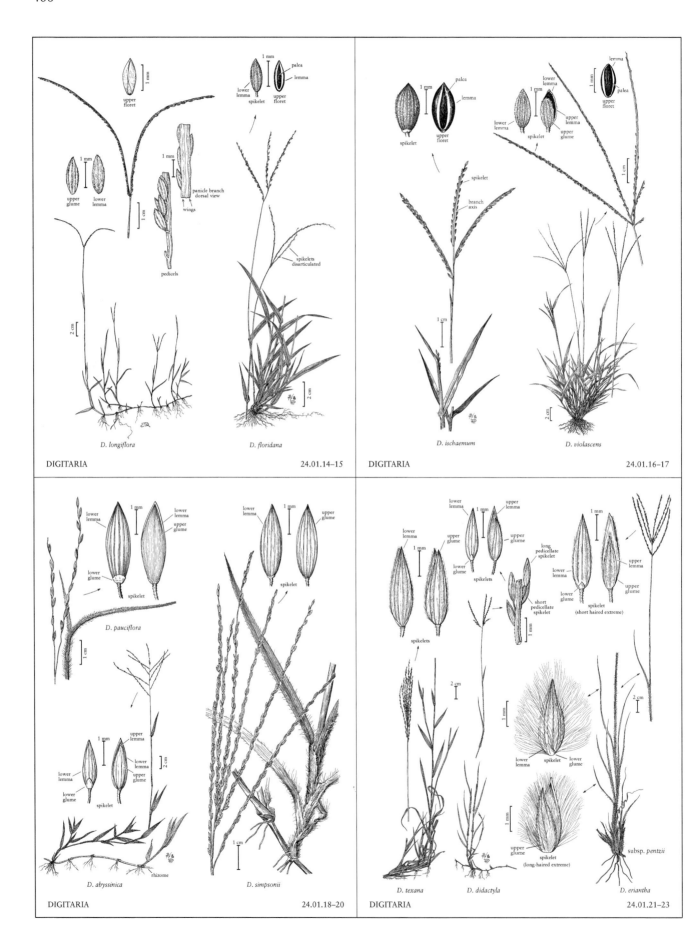

DIGITARIA

24.01.14–15

DIGITARIA

24.01.16–17

DIGITARIA

24.01.18–20

DIGITARIA

24.01.21–23

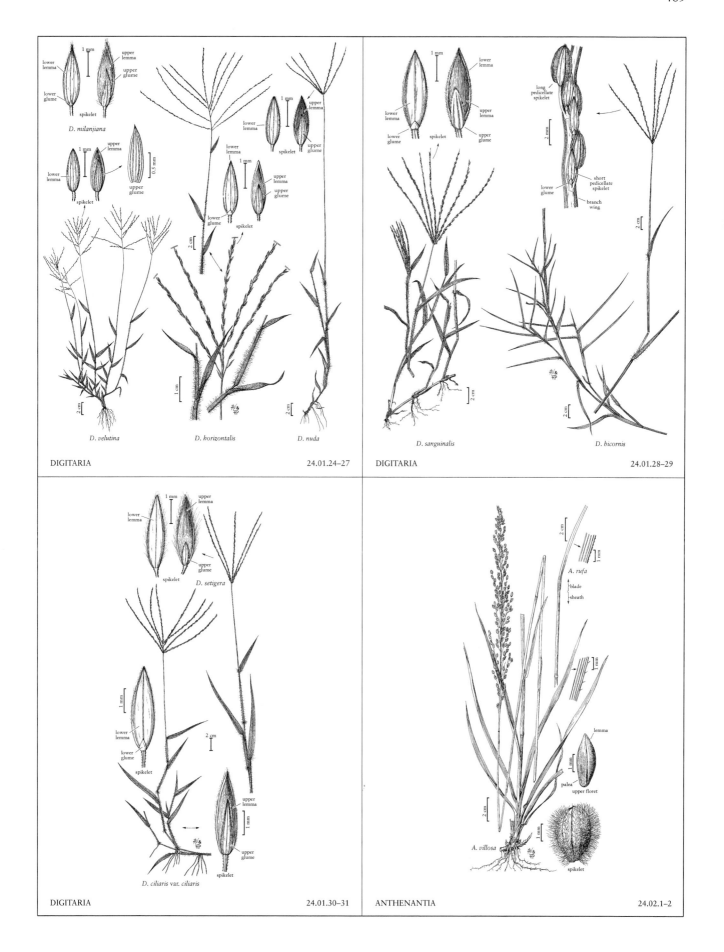

DIGITARIA                    24.01.24–27

DIGITARIA                    24.01.28–29

DIGITARIA                    24.01.30–31

ANTHENANTIA                  24.02.1–2

470

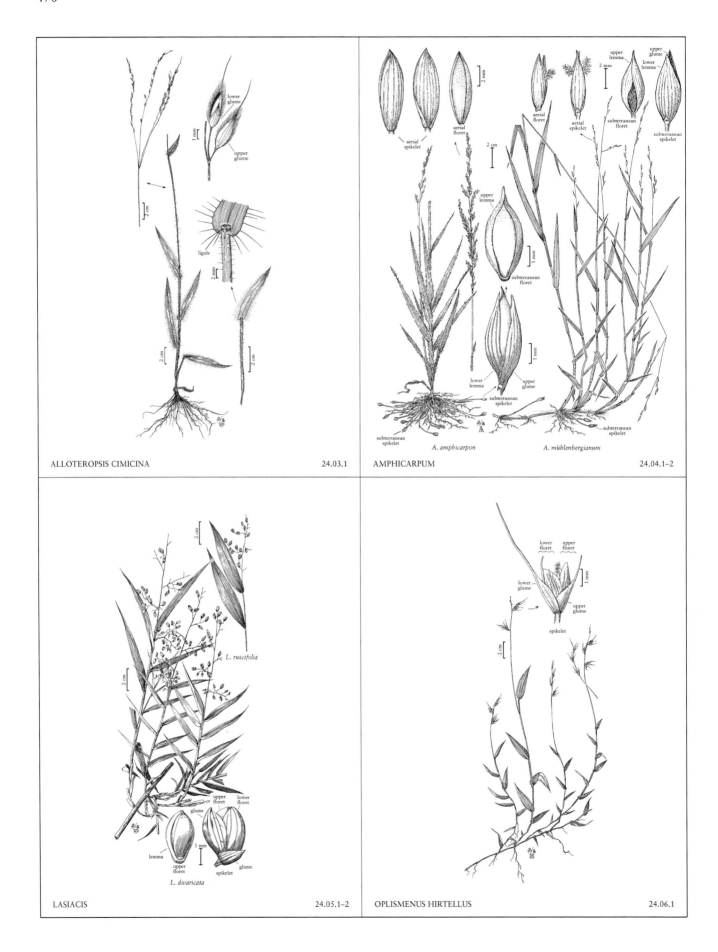

ALLOTEROPSIS CIMICINA                                    24.03.1

AMPHICARPUM                                              24.04.1–2

A. amphicarpon          A. mühlenbergianum

LASIACIS                                                 24.05.1–2

L. ruscifolia

L. divaricata

OPLISMENUS HIRTELLUS                                     24.06.1

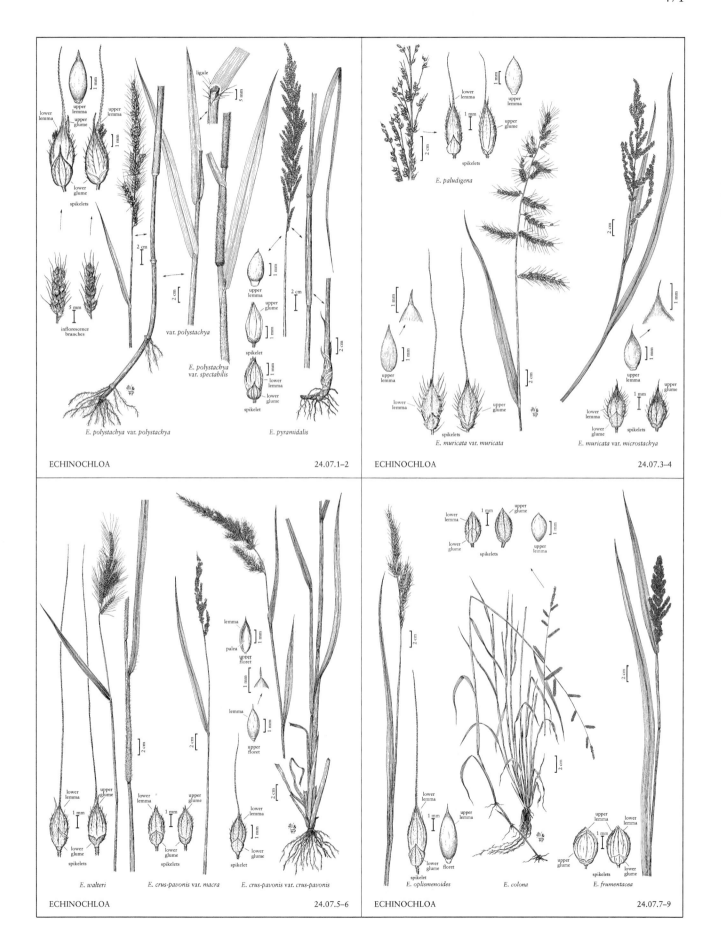

471

ECHINOCHLOA
24.07.1–2

ECHINOCHLOA
24.07.3–4

ECHINOCHLOA
24.07.5–6

ECHINOCHLOA
24.07.7–9

472

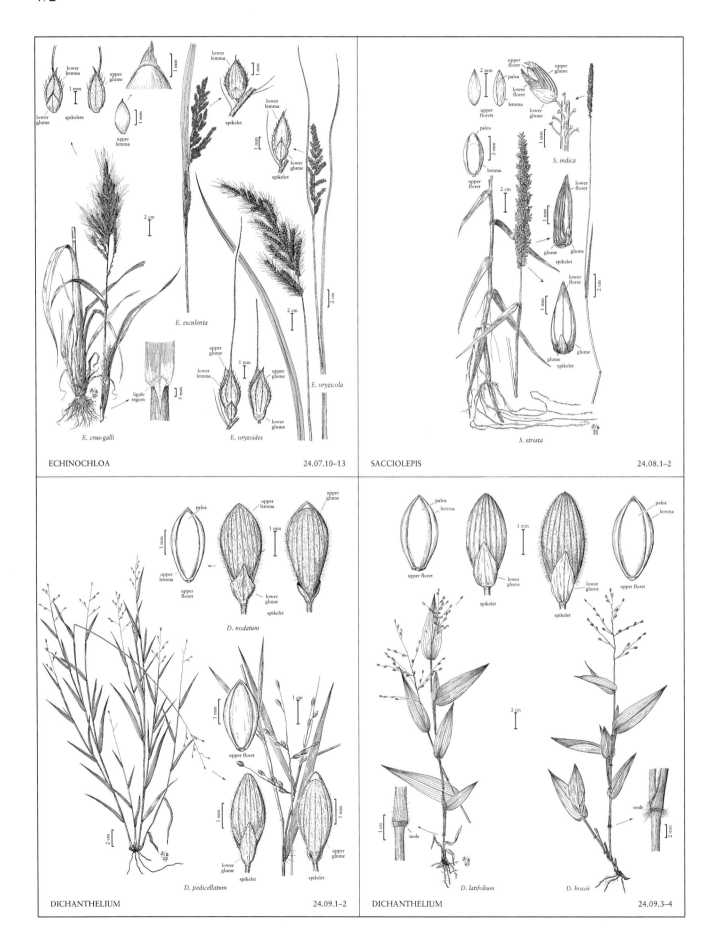

ECHINOCHLOA 24.07.10–13

SACCIOLEPIS 24.08.1–2

DICHANTHELIUM 24.09.1–2

DICHANTHELIUM 24.09.3–4

E. esculenta

E. crus-galli

E. oryzoides

E. oryzicola

S. indica

S. striata

D. nodatum

D. pedicellatum

D. latifolium

D. boscii

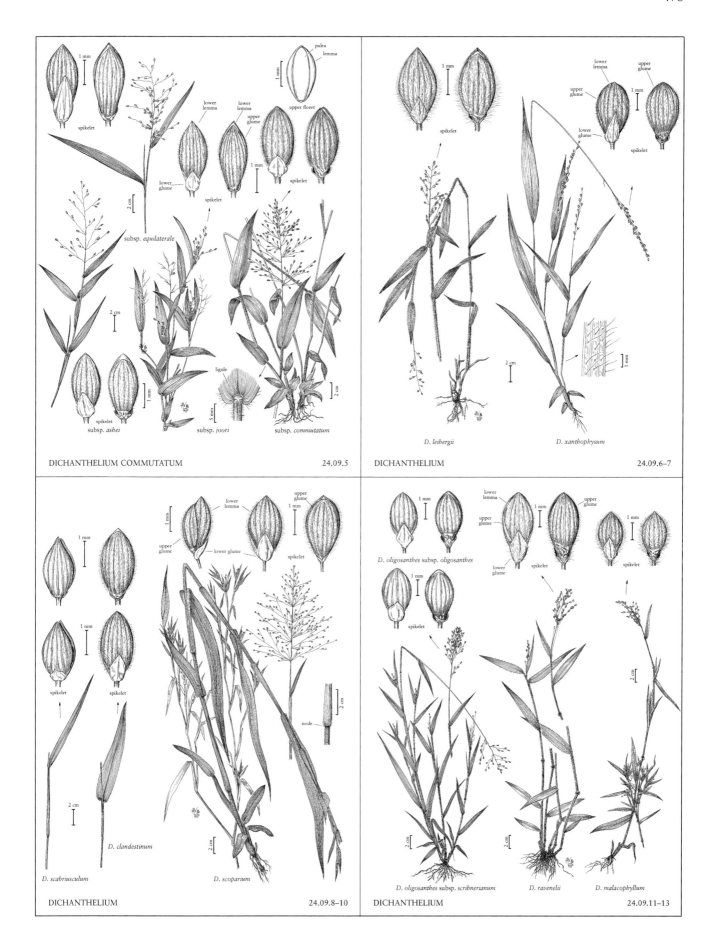

473

subsp. *equilaterale*

subsp. *ashei*

subsp. *joori*

subsp. *commutatum*

DICHANTHELIUM COMMUTATUM 24.09.5

*D. leibergii*

*D. xanthophysum*

DICHANTHELIUM 24.09.6–7

*D. scabriusculum*

*D. clandestinum*

*D. scoparium*

DICHANTHELIUM 24.09.8–10

*D. oligosanthes* subsp. *oligosanthes*

*D. oligosanthes* subsp. *scribnerianum*

*D. ravenelii*

*D. malacophyllum*

DICHANTHELIUM 24.09.11–13

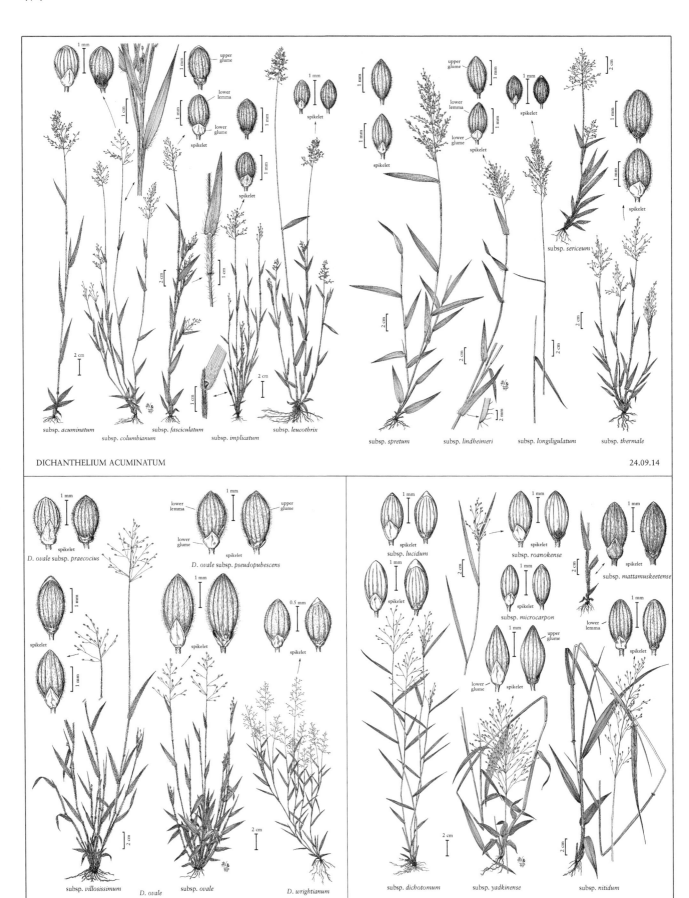

subsp. *acuminatum*

subsp. *columbianum*

subsp. *fasciculatum*

subsp. *implicatum*

subsp. *leucothrix*

upper glume

lower lemma

lower glume

spikelet

DICHANTHELIUM ACUMINATUM

24.09.14

subsp. *sericeum*

subsp. *spretum*

subsp. *lindheimeri*

subsp. *longiligulatum*

subsp. *thermale*

upper glume

lower lemma

lower glume

spikelet

*D. ovale* subsp. *praecocius*

spikelet

*D. ovale* subsp. *pseudopubescens*

lower lemma

lower glume

upper glume

spikelet

subsp. *villosissimum*

*D. ovale*

subsp. *ovale*

*D. wrightianum*

DICHANTHELIUM

24.09.15–16

subsp. *lucidum*

spikelet

subsp. *roanokense*

spikelet

subsp. *mattamuskeetense*

lower lemma

spikelet

subsp. *microcarpon*

spikelet

upper glume

lower glume

spikelet

subsp. *dichotomum*

subsp. *yadkinense*

subsp. *nitidum*

DICHANTHELIUM DICHOTOMUM

24.09.17

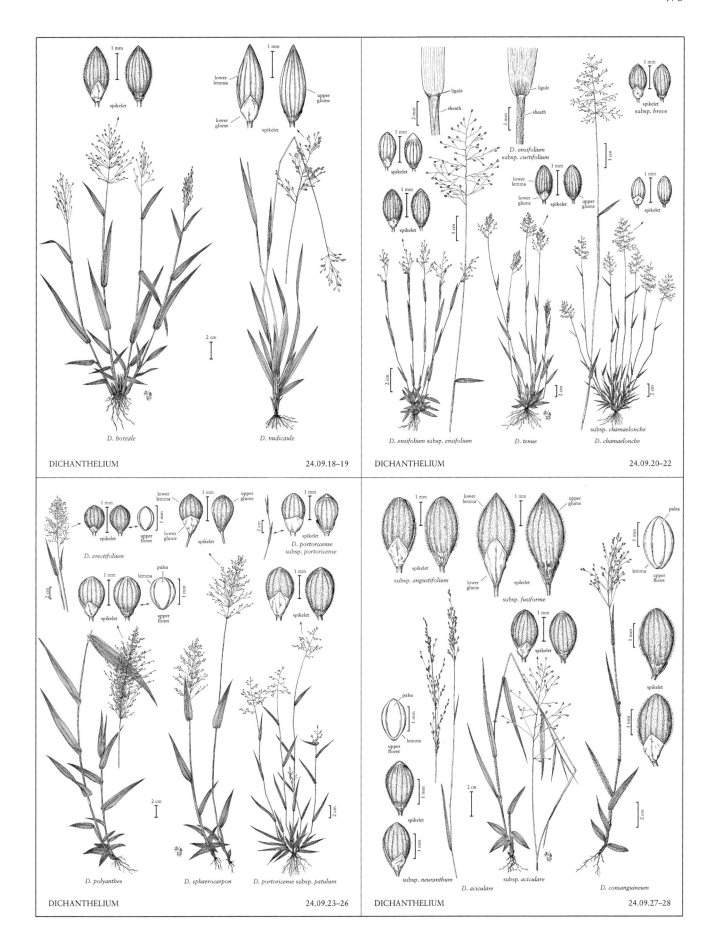

DICHANTHELIUM

24.09.18–19

*D. boreale*

*D. nudicaule*

DICHANTHELIUM

24.09.20–22

*D. ensifolium* subsp. *curtifolium*

subsp. *breve*

subsp. *chamaelonche*

*D. ensifolium* subsp. *ensifolium*

*D. tenue*

*D. chamaelonche*

DICHANTHELIUM

24.09.23–26

*D. erectifolium*

*D. portoricense* subsp. *portoricense*

*D. polyanthes*

*D. sphaerocarpon*

*D. portoricense* subsp. *patulum*

DICHANTHELIUM

24.09.27–28

subsp. *angustifolium*

subsp. *fusiforme*

subsp. *neuranthum*

*D. aciculare*

subsp. *aciculare*

*D. consanguineum*

476

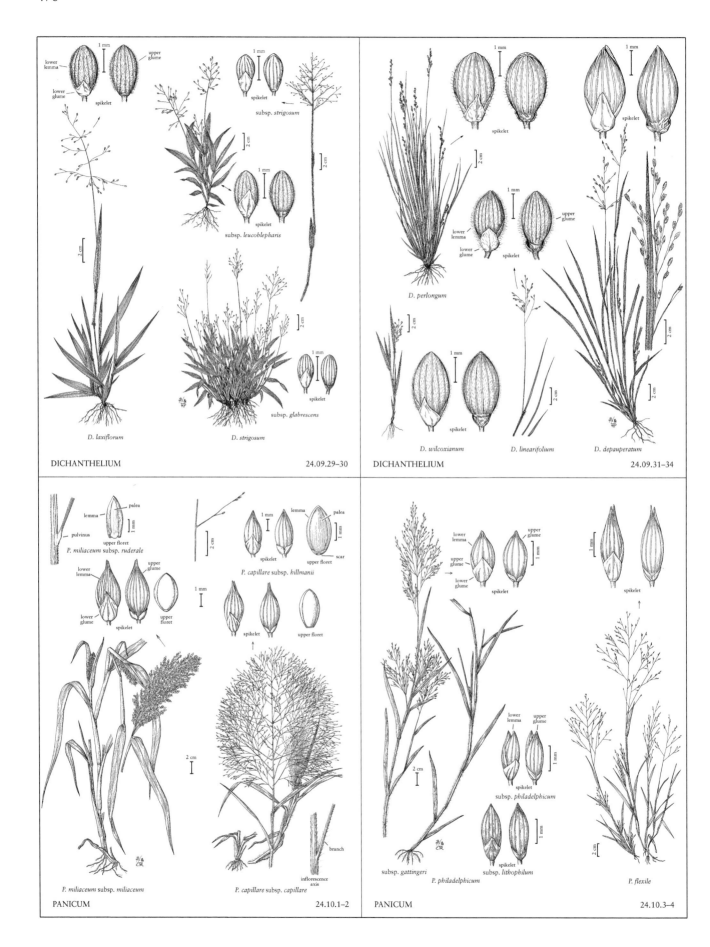

subsp. *strigosum*

lower lemma
upper glume
lower glume
spikelet

subsp. *leucoblepharis*

spikelet

*D. laxiflorum*

*D. strigosum*

subsp. *glabrescens*

spikelet

DICHANTHELIUM

24.09.29–30

*D. perlongum*

upper glume
lower lemma
lower glume
spikelet

*D. wilcoxianum*

*D. linearifolium*

*D. depauperatum*

spikelet

DICHANTHELIUM

24.09.31–34

lemma
palea
pulvinus
upper floret
*P. miliaceum* subsp. *ruderale*

lemma
palea
spikelet
upper floret
scar
*P. capillare* subsp. *hillmanii*

upper glume
lower lemma
lower glume
spikelet
upper floret

spikelet
upper floret

*P. miliaceum* subsp. *miliaceum*

*P. capillare* subsp. *capillare*

branch

inflorescence axis

PANICUM

24.10.1–2

lower lemma
upper glume
upper glume
lower glume
spikelet

spikelet

lower lemma
upper glume
spikelet
subsp. *philadelphicum*

spikelet
subsp. *lithophilum*

subsp. *gattingeri*

*P. philadelphicum*

*P. flexile*

PANICUM

24.10.3–4

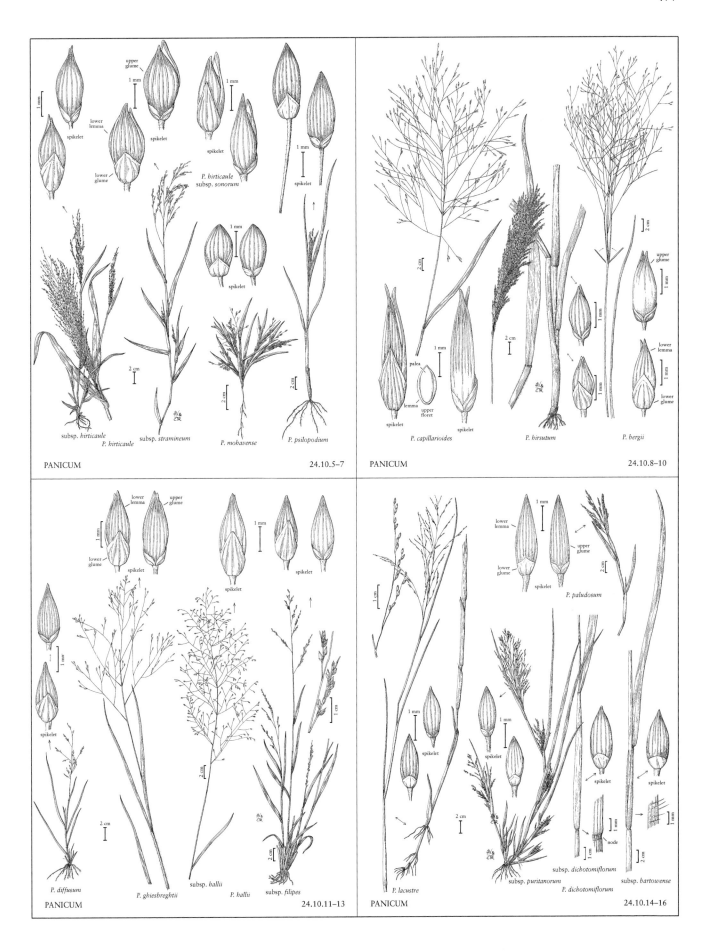

PANICUM                             24.10.5–7

PANICUM                             24.10.8–10

PANICUM                             24.10.11–13

PANICUM                             24.10.14–16

478

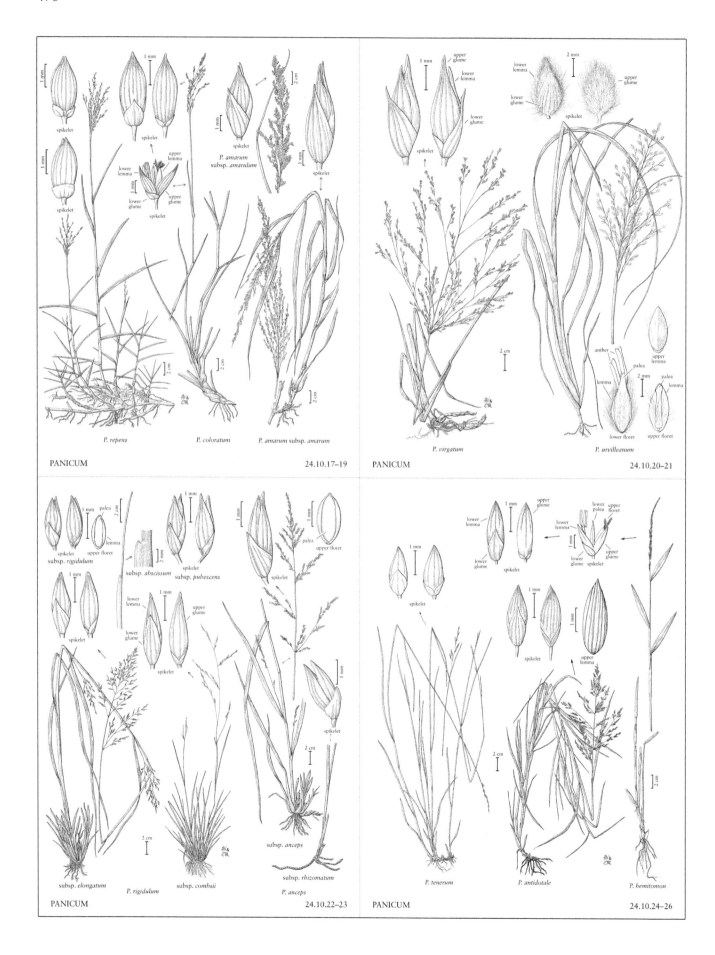

PANICUM                                    24.10.17–19

PANICUM                                    24.10.20–21

PANICUM                                    24.10.22–23

PANICUM                                    24.10.24–26

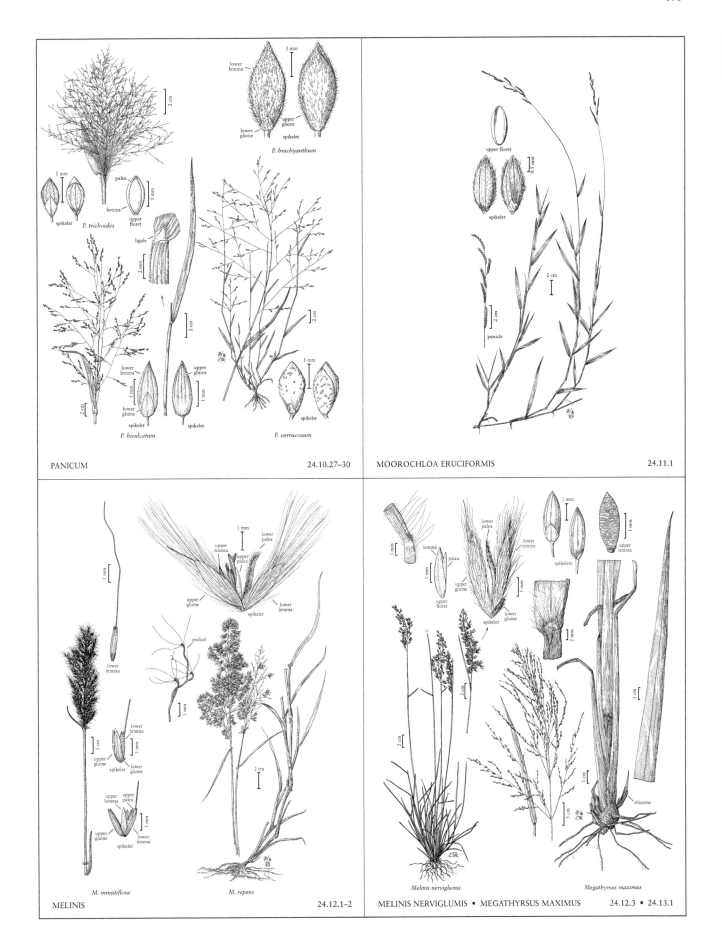

PANICUM 24.10.27–30

MOOROCHLOA ERUCIFORMIS 24.11.1

MELINIS 24.12.1–2

MELINIS NERVIGLUMIS • MEGATHYRSUS MAXIMUS 24.12.3 • 24.13.1

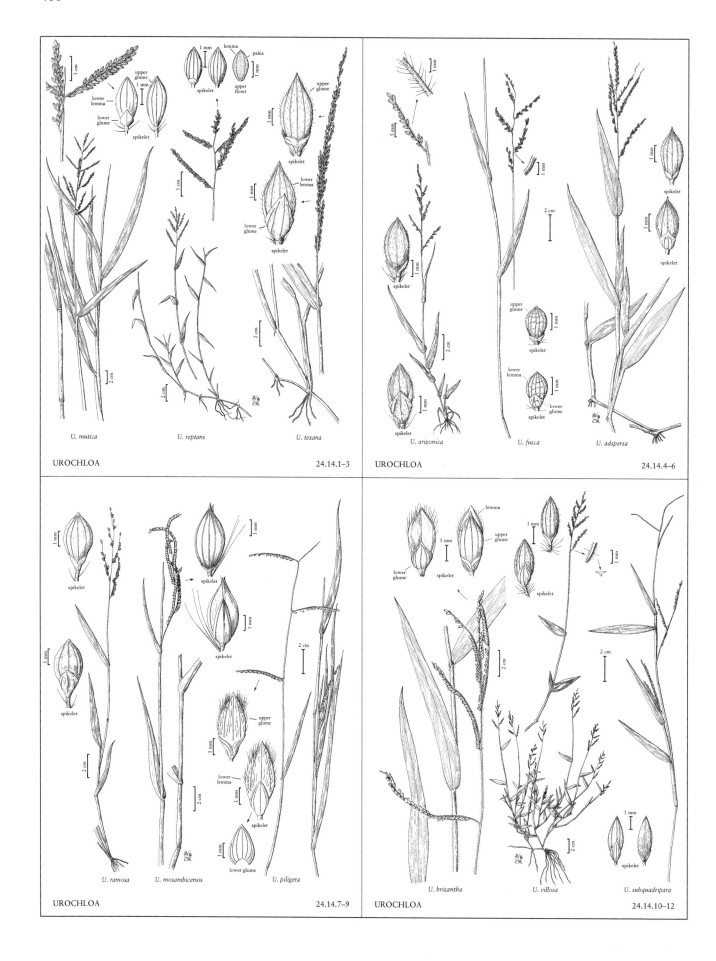

U. mutica

U. reptans

U. texana

UROCHLOA

24.14.1–3

U. arizonica

U. fusca

U. adspersa

UROCHLOA

24.14.4–6

U. ramosa

U. mosambicensis

U. piligera

UROCHLOA

24.14.7–9

U. brizantha

U. villosa

U. subquadripara

UROCHLOA

24.14.10–12

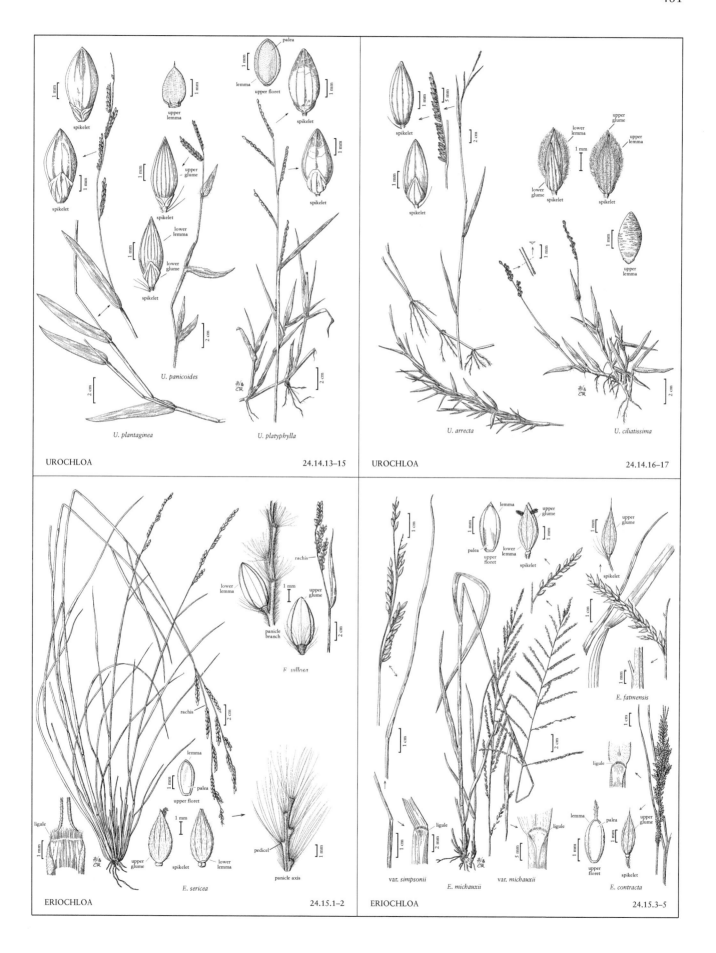

UROCHLOA

24.14.13–15

UROCHLOA

24.14.16–17

U. panicoides

U. plantaginea

U. platyphylla

U. arrecta

U. ciliatissima

E. villosa

E. sericea

ERIOCHLOA

24.15.1–2

var. simpsonii

E. michauxii

var. michauxii

E. fatmensis

E. contracta

ERIOCHLOA

24.15.3–5

482

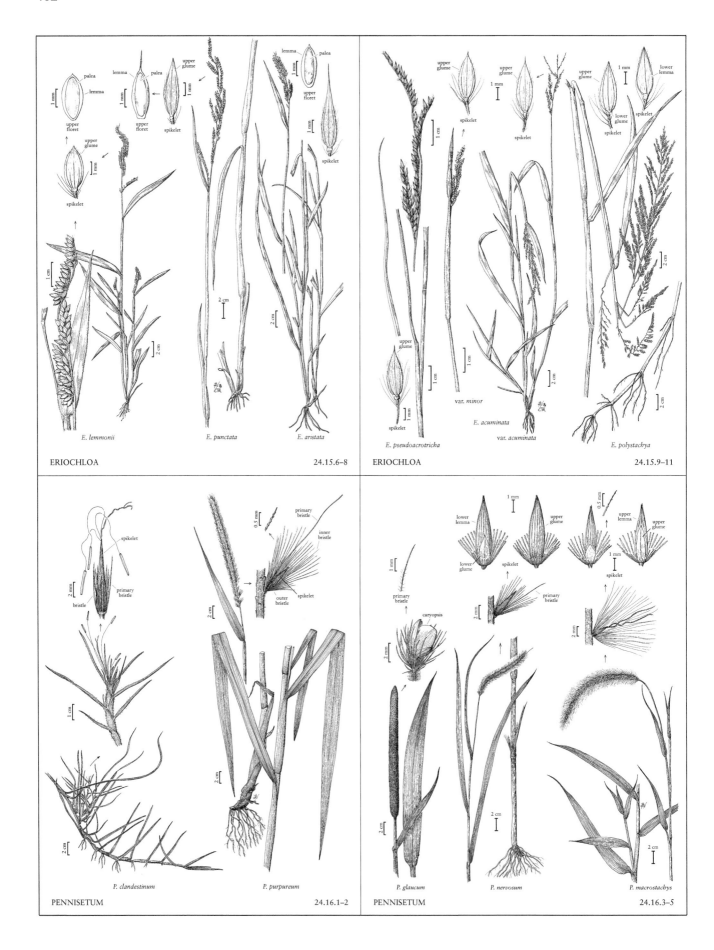

E. lemmonii

E. punctata

E. aristata

ERIOCHLOA

24.15.6–8

E. pseudoacrotricha

var. minor

E. acuminata

var. acuminata

E. polystachya

ERIOCHLOA

24.15.9–11

P. clandestinum

P. purpureum

PENNISETUM

24.16.1–2

P. glaucum

P. nervosum

P. macrostachys

PENNISETUM

24.16.3–5

PENNISETUM

24.16.6–8

PENNISETUM

24.16.9–11

PENNISETUM

24.16.12–15

PENNISETUM

24.16.16–18

484

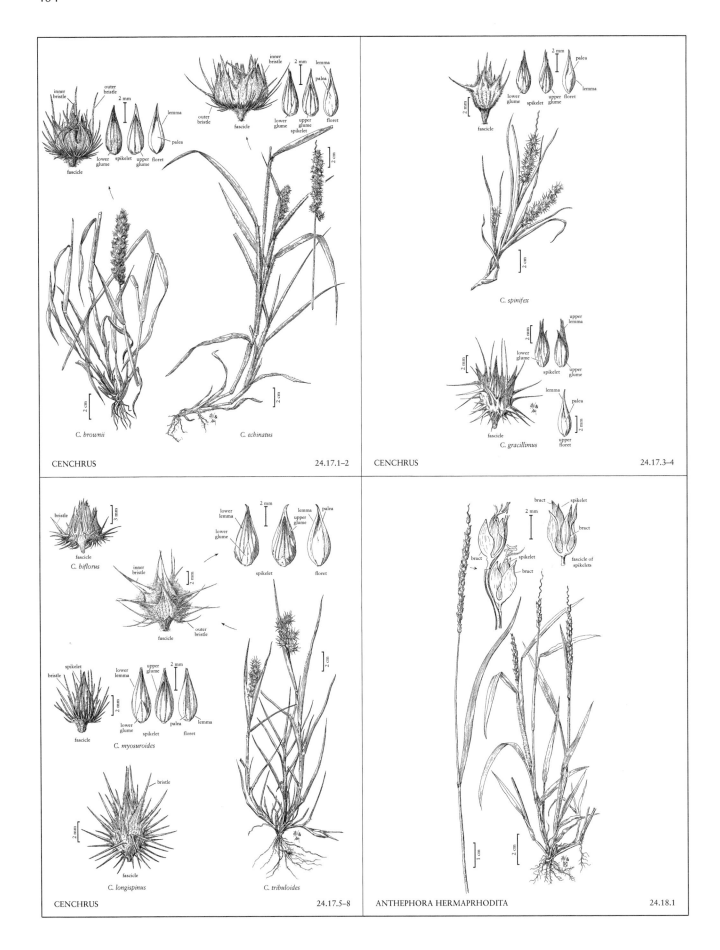

CENCHRUS                                    24.17.1–2

CENCHRUS                                    24.17.3–4

CENCHRUS                                    24.17.5–8

ANTHEPHORA HERMAPRHODITA                    24.18.1

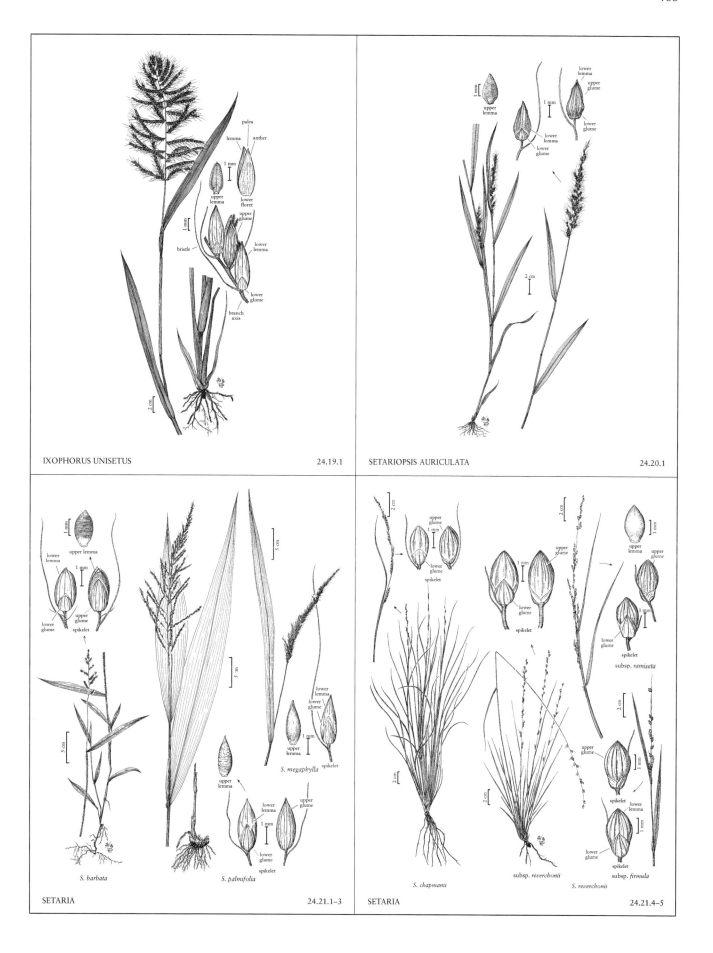

IXOPHORUS UNISETUS 24.19.1

SETARIOPSIS AURICULATA 24.20.1

SETARIA 24.21.1–3

SETARIA 24.21.4–5

*S. barbata*

*S. palmifolia*

*S. megaphylla*

*S. chapmanii*

subsp. *reverchonii*

*S. reverchonii*

subsp. *firmula*

subsp. *ramiseta*

486

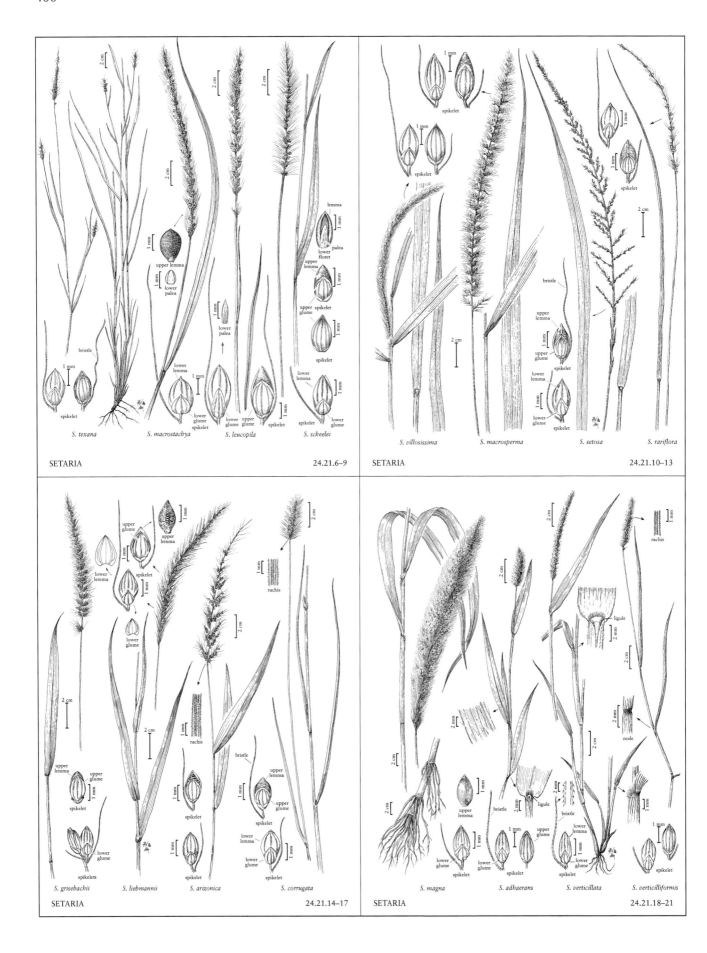

S. texana          S. macrostachya     S. leucopila        S. scheelei

SETARIA                                                    24.21.6–9

S. villosissima    S. macrosperma      S. setosa           S. rariflora

SETARIA                                                    24.21.10–13

S. grisebachii     S. liebmannii       S. arizonica        S. corrugata

SETARIA                                                    24.21.14–17

S. magna           S. adhaerans        S. verticillata     S. verticilliformis

SETARIA                                                    24.21.18–21

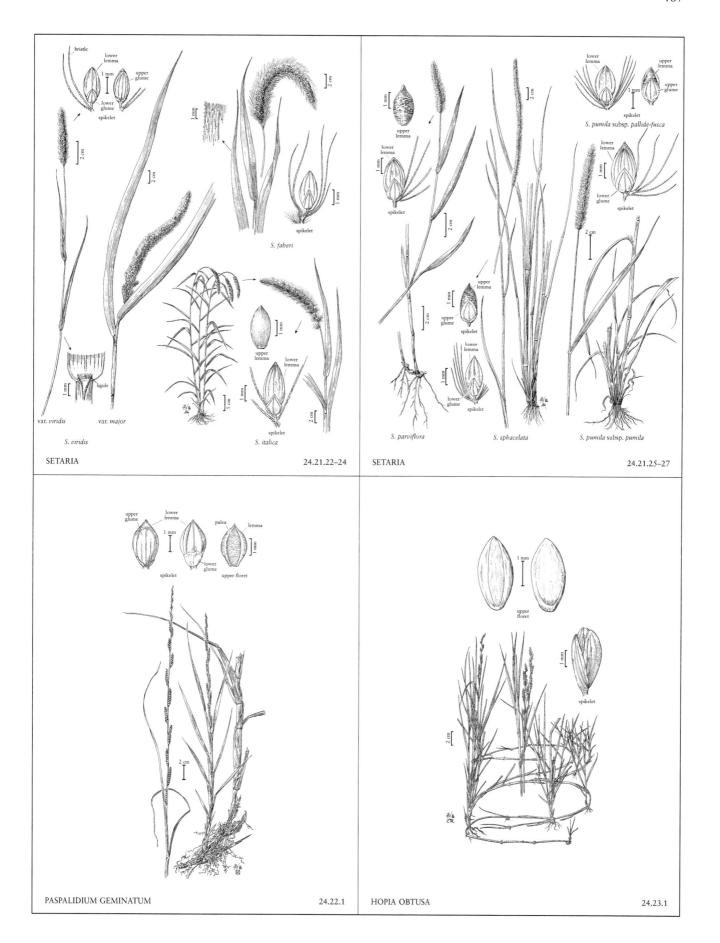

bristle
lower
lemma
upper
glume
lower
glume
spikelet

2 cm

1 mm

ligule

var. *viridis*          var. *major*

*S. viridis*

SETARIA                    24.21.22–24

1 mm

2 cm

2 cm

1 mm

spikelet

*S. faberi*

1 mm

upper
lemma

lower
lemma

spikelet

*S. italica*

5 cm

2 cm

1 mm

upper
lemma

lower
lemma

lower
lemma

spikelet

upper
lemma

upper
glume

spikelet

*S. parviflora*

2 cm

2 cm

lower
lemma

lower
glume

spikelet

*S. sphacelata*

2 cm

*S. pumila* subsp. *pumila*

SETARIA                    24.21.25–27

lower
lemma

upper
lemma
upper
glume

spikelet

*S. pumila* subsp. *pallide-fusca*

lower
lemma

lower
glume

spikelet

1 mm

2 cm

upper
glume

lower
lemma

palea

lemma

1 mm

1 mm

spikelet

lower
glume

upper floret

2 cm

PASPALIDIUM GEMINATUM              24.22.1

1 mm

upper
floret

1 mm

spikelet

2 cm

HOPIA OBTUSA                        24.23.1

488

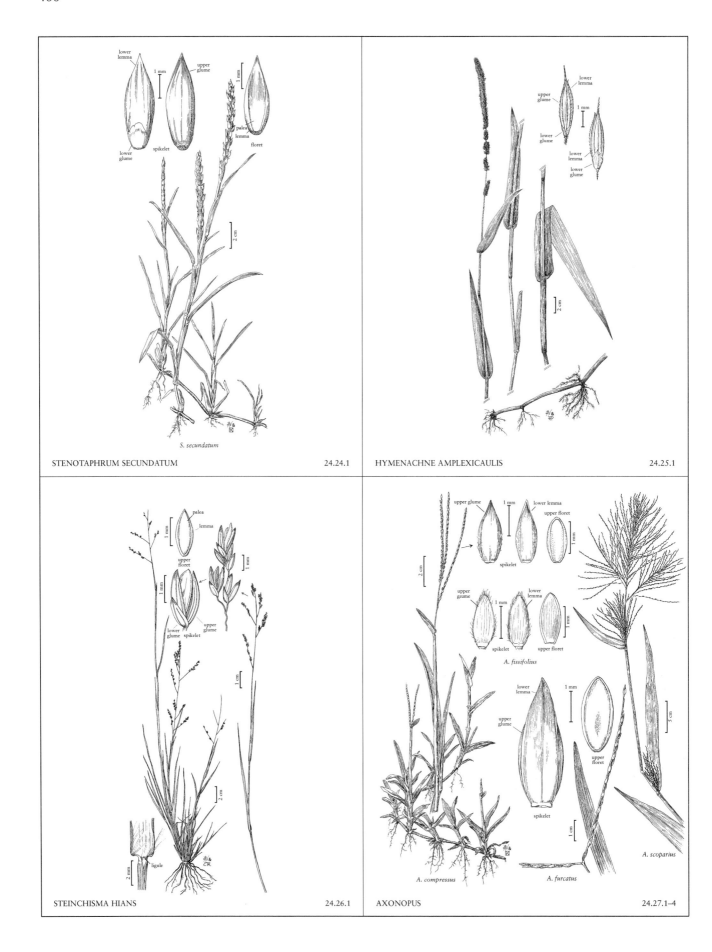

STENOTAPHRUM SECUNDATUM                                      24.24.1

HYMENACHNE AMPLEXICAULIS                                     24.25.1

STEINCHISMA HIANS                                            24.26.1

AXONOPUS                                                     24.27.1–4

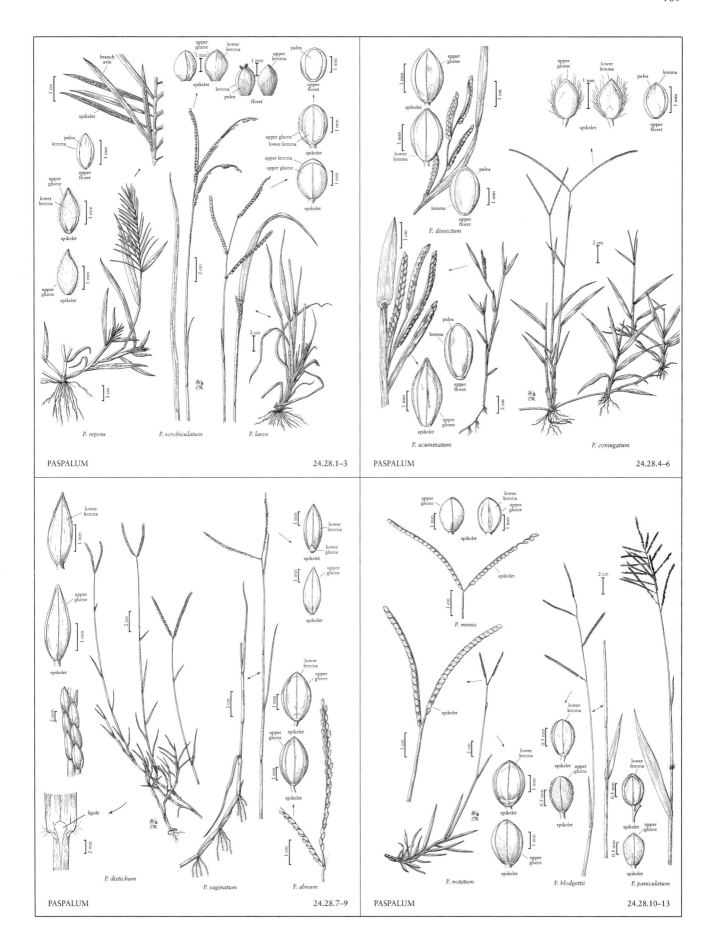

PASPALUM

24.28.1–3

PASPALUM

24.28.4–6

PASPALUM

24.28.7–9

PASPALUM

24.28.10–13

P. repens

P. scrobiculatum

P. laeve

P. acuminatum

P. conjugatum

P. dissectum

P. distichum

P. vaginatum

P. almum

P. minus

P. notatum

P. blodgettii

P. paniculatum

490

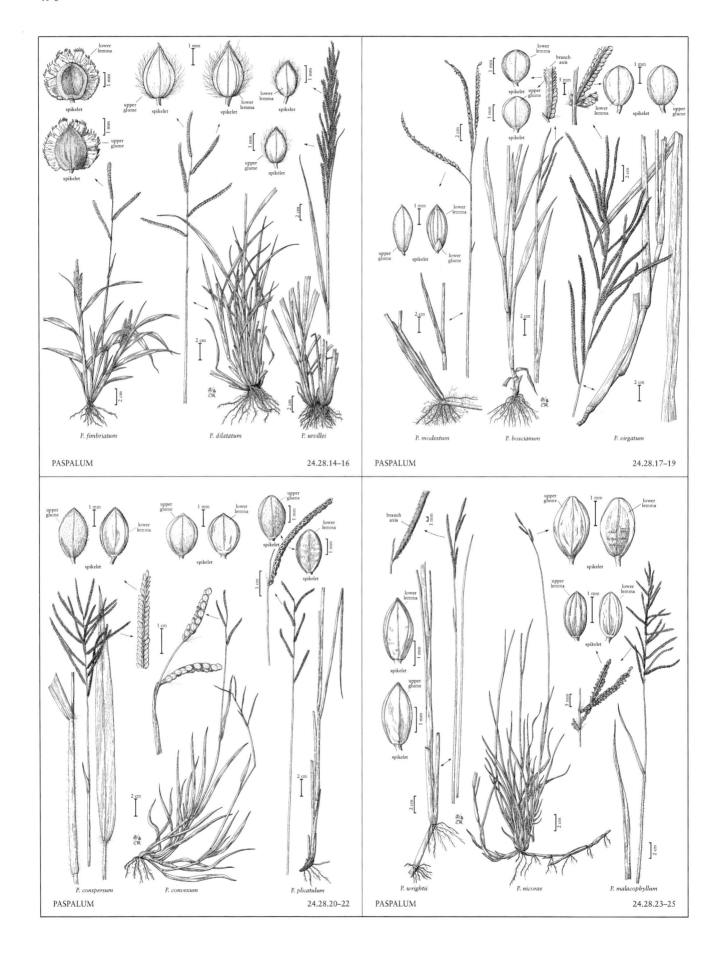

PASPALUM

24.28.14–16

PASPALUM

24.28.17–19

PASPALUM

24.28.20–22

PASPALUM

24.28.23–25

*P. fimbriatum*

*P. dilatatum*

*P. urvillei*

*P. modestum*

*P. boscianum*

*P. virgatum*

*P. conspersum*

*P. convexum*

*P. plicatulum*

*P. wrightii*

*P. nicorae*

*P. malacophyllum*

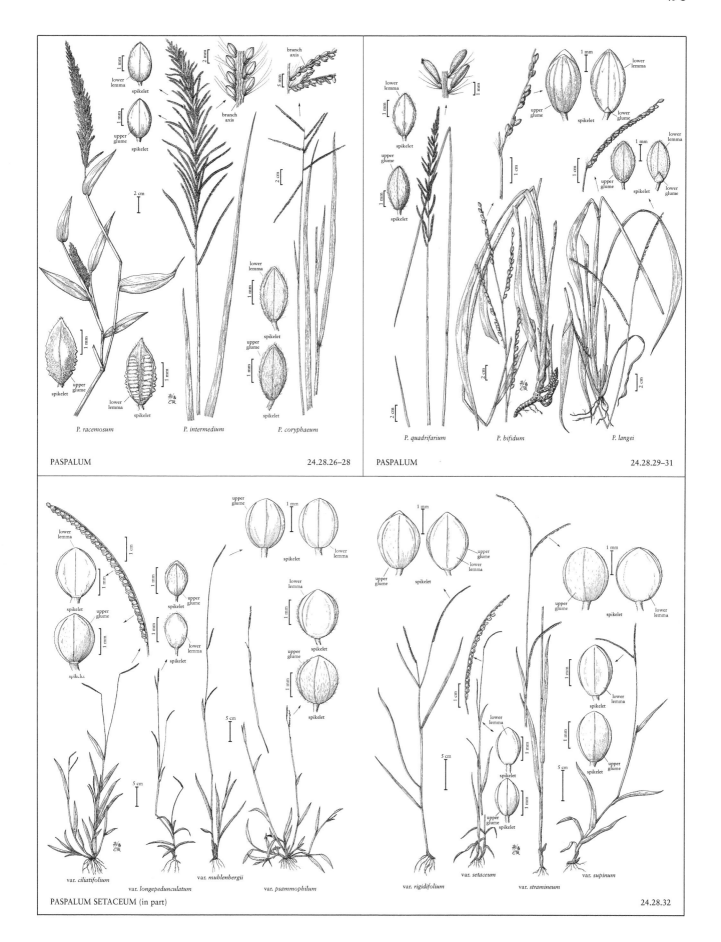

PASPALUM

P. racemosum    P. intermedium    P. coryphaeum

PASPALUM                                    24.28.26–28

P. quadrifarium    P. bifidum    P. langei

PASPALUM                                    24.28.29–31

var. ciliatifolium    var. muhlenbergii    var. psammophilum    var. rigidifolium    var. setaceum    var. stramineum    var. supinum

var. longepedunculatum

PASPALUM SETACEUM (in part)                                    24.28.32

492

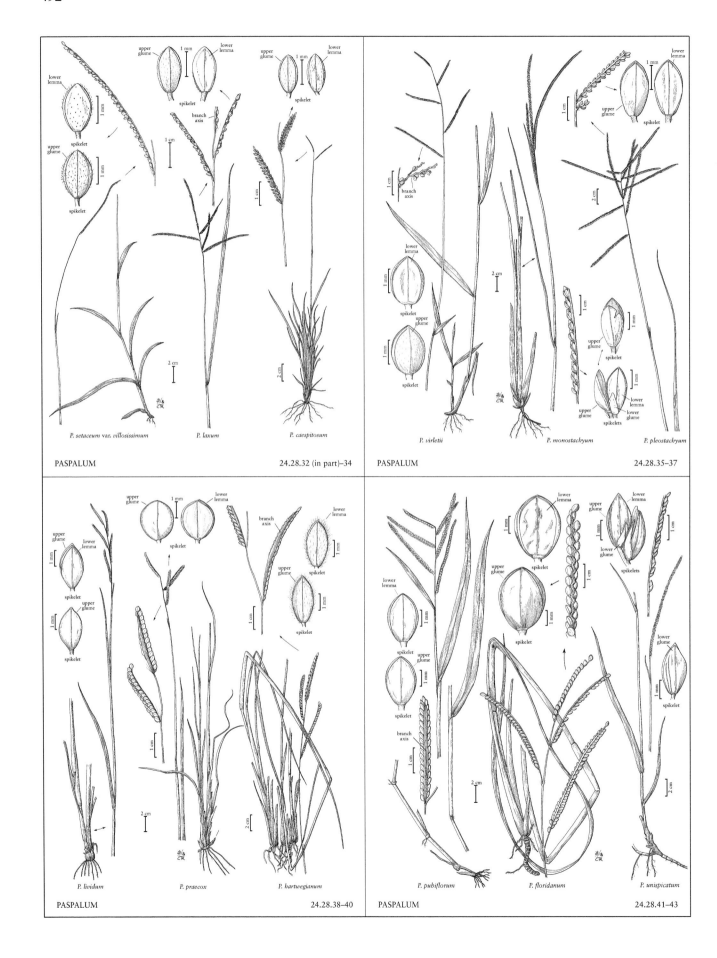

P. setaceum var. villosissimum

P. laxum

P. caespitosum

P. virletii

P. monostachyum

P. pleostachyum

P. lividum

P. praecox

P. hartwegianum

P. pubiflorum

P. floridanum

P. unispicatum

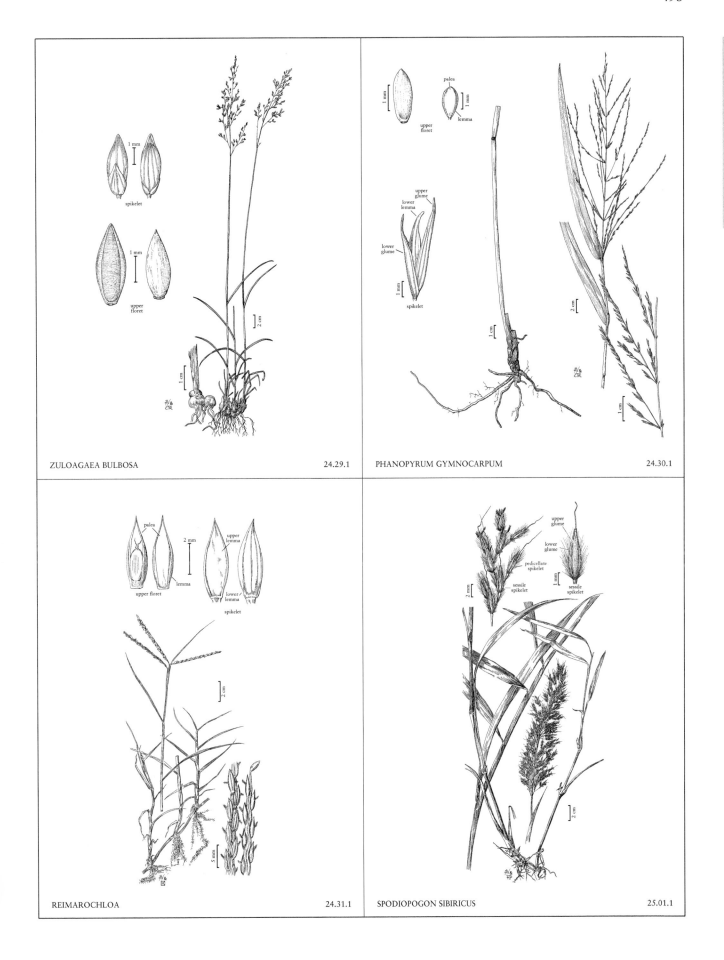

ZULOAGAEA BULBOSA                                      24.29.1

PHANOPYRUM GYMNOCARPUM                                 24.30.1

REIMAROCHLOA                                           24.31.1

SPODIOPOGON SIBIRICUS                                  25.01.1

494

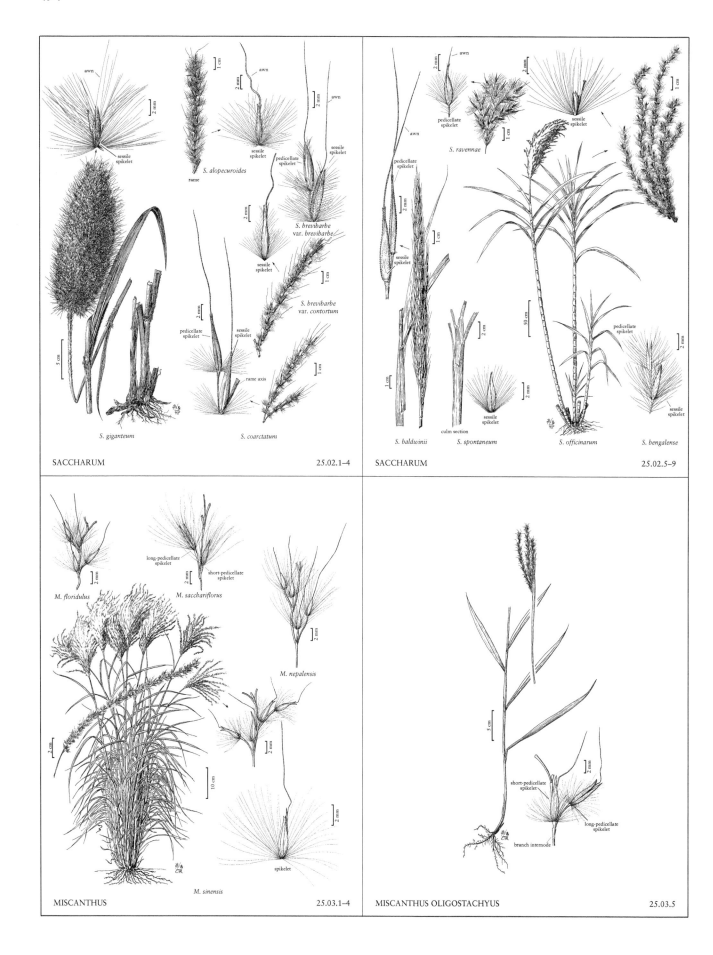

SACCHARUM                                    25.02.1–4

SACCHARUM                                    25.02.5–9

MISCANTHUS                                   25.03.1–4

MISCANTHUS OLIGOSTACHYUS                      25.03.5

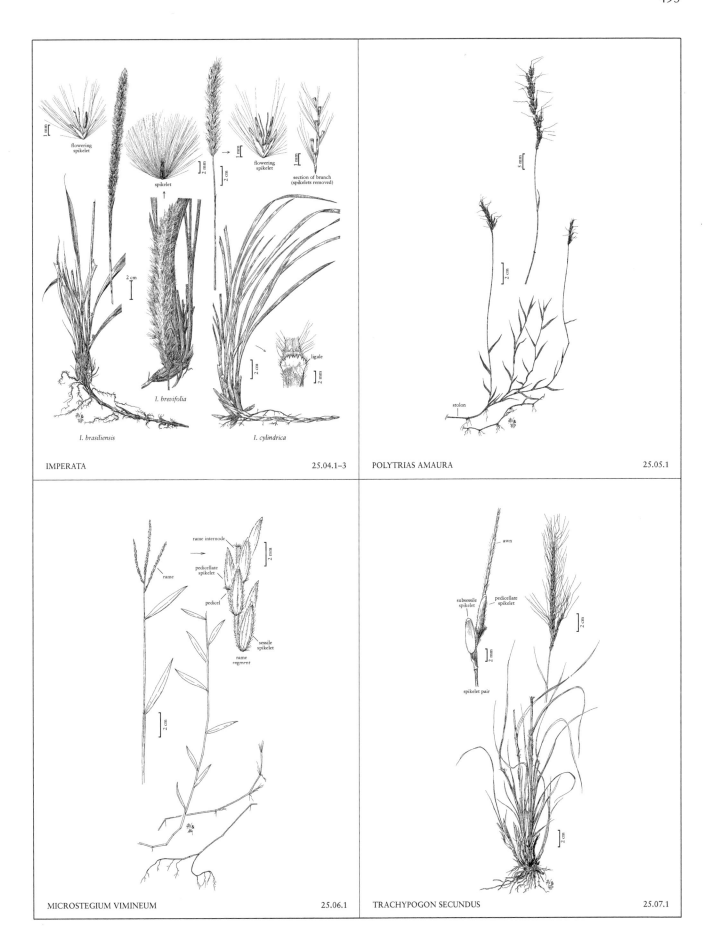

IMPERATA                                    25.04.1–3

POLYTRIAS AMAURA                            25.05.1

MICROSTEGIUM VIMINEUM                        25.06.1

TRACHYPOGON SECUNDUS                         25.07.1

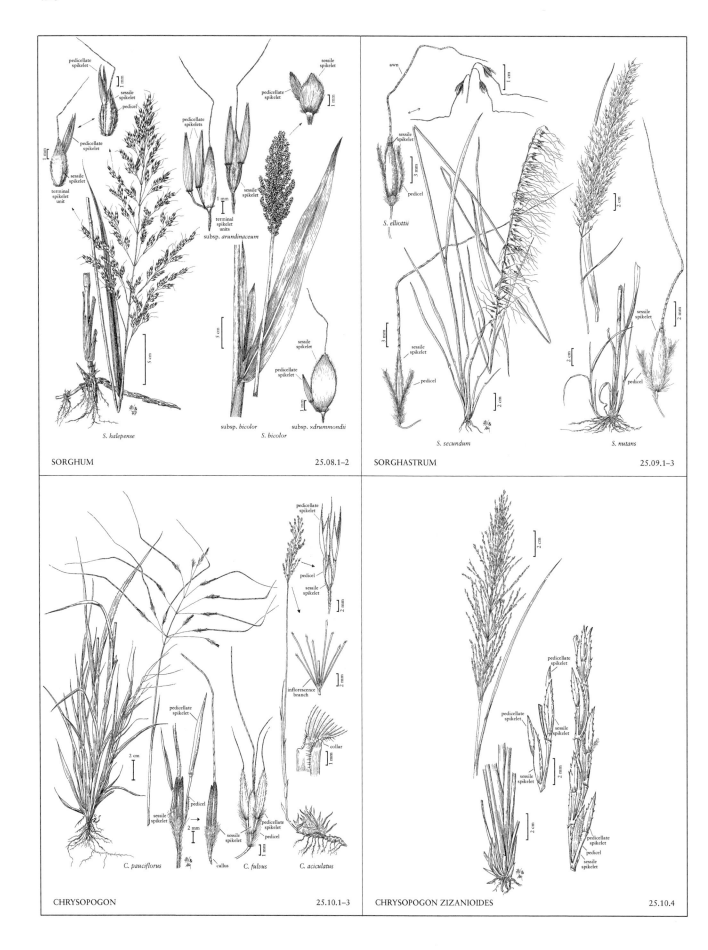

SORGHUM                                      25.08.1–2

SORGHASTRUM                                  25.09.1–3

CHRYSOPOGON                                  25.10.1–3

CHRYSOPOGON ZIZANIOIDES                       25.10.4

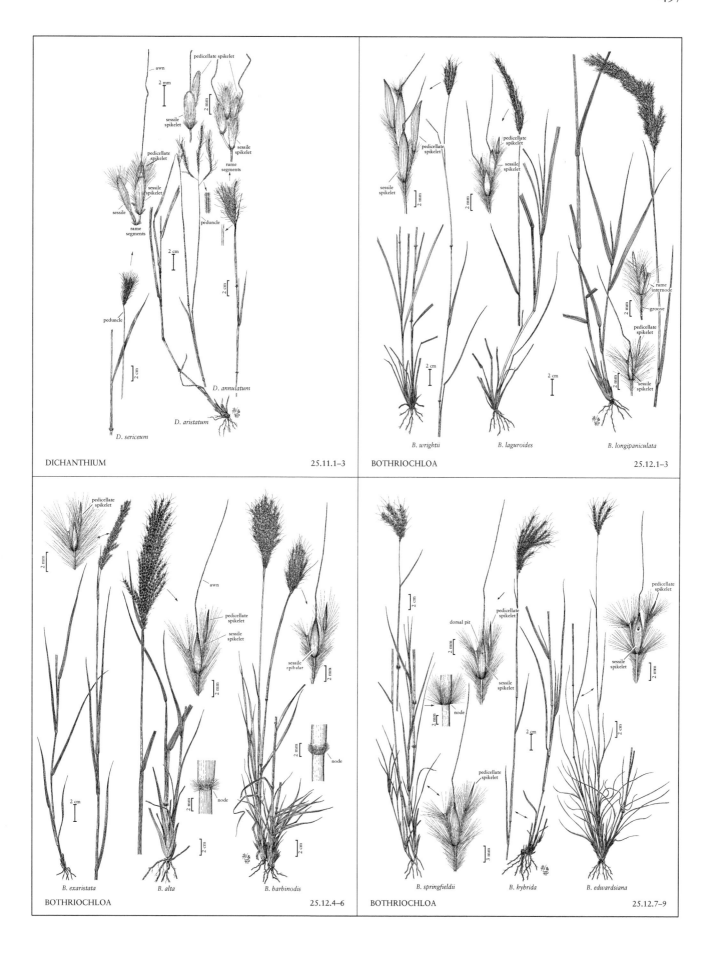

DICHANTHIUM 25.11.1–3

BOTHRIOCHLOA 25.12.1–3

BOTHRIOCHLOA 25.12.4–6

BOTHRIOCHLOA 25.12.7–9

498

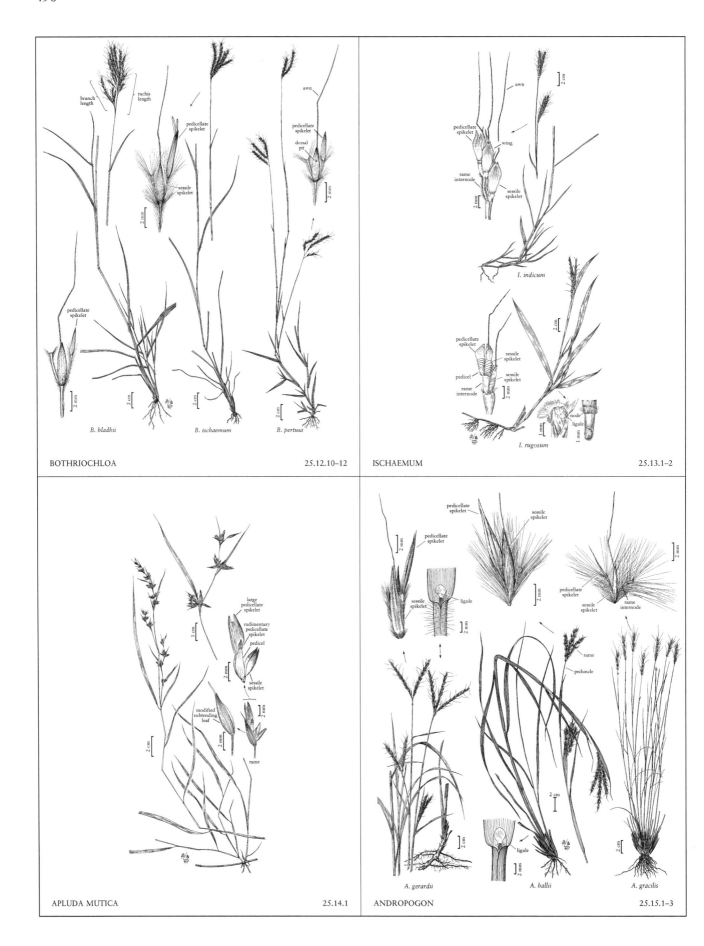

BOTHRIOCHLOA                                          25.12.10–12

ISCHAEMUM                                             25.13.1–2

APLUDA MUTICA                                         25.14.1

ANDROPOGON                                           25.15.1–3

B. bladhii        B. ischaemum        B. pertusa

I. indicum

I. rugosum

A. gerardii        A. hallii        A. gracilis

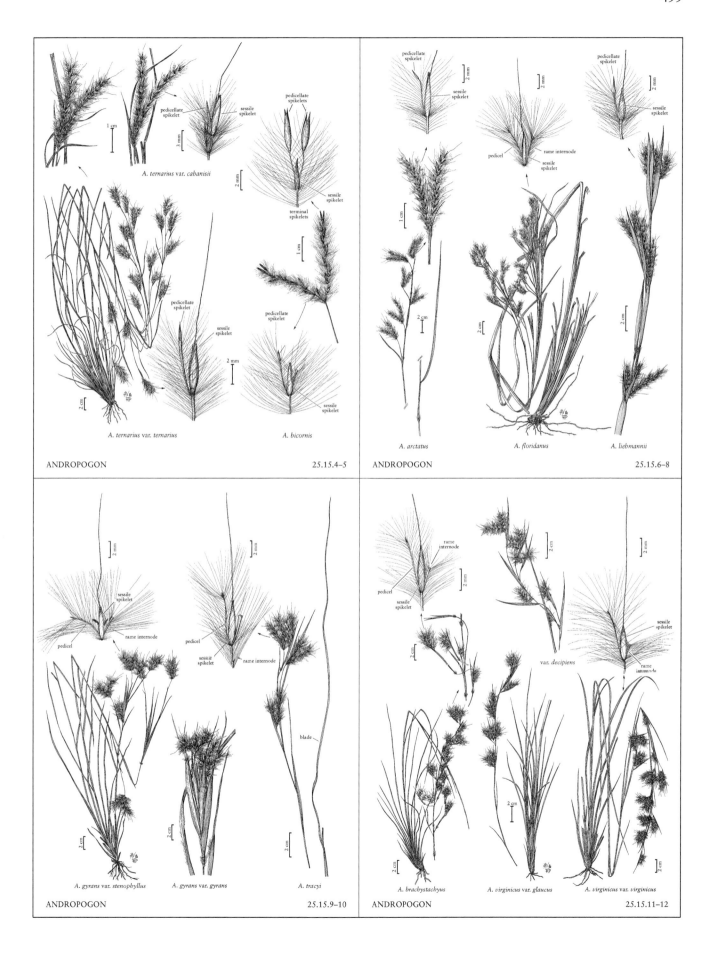

ANDROPOGON                                    25.15.4–5

*A. ternarius* var. *cabanisii*

pedicellate
spikelet

sessile
spikelet

terminal
spikelets

pedicellate
spikelet

sessile
spikelet

*A. ternarius* var. *ternarius*

pedicellate
spikelet

sessile
spikelet

*A. bicornis*

ANDROPOGON                                    25.15.6–8

pedicellate
spikelet

sessile
spikelet

pedicel    rame internode
sessile
spikelet

pedicellate
spikelet

sessile
spikelet

*A. arctatus*        *A. floridanus*        *A. liebmannii*

ANDROPOGON                                    25.15.9–10

sessile
spikelet

rame internode

pedicel

pedicel
sessile
spikelet

rame internode

blade

*A. gyrans* var. *stenophyllus*     *A. gyrans* var. *gyrans*     *A. tracyi*

ANDROPOGON                                    25.15.11–12

rame
internode

pedicel

sessile
spikelet

var. *decipiens*

sessile
spikelet

rame
internode

*A. brachystachyus*     *A. virginicus* var. *glaucus*     *A. virginicus* var. *virginicus*

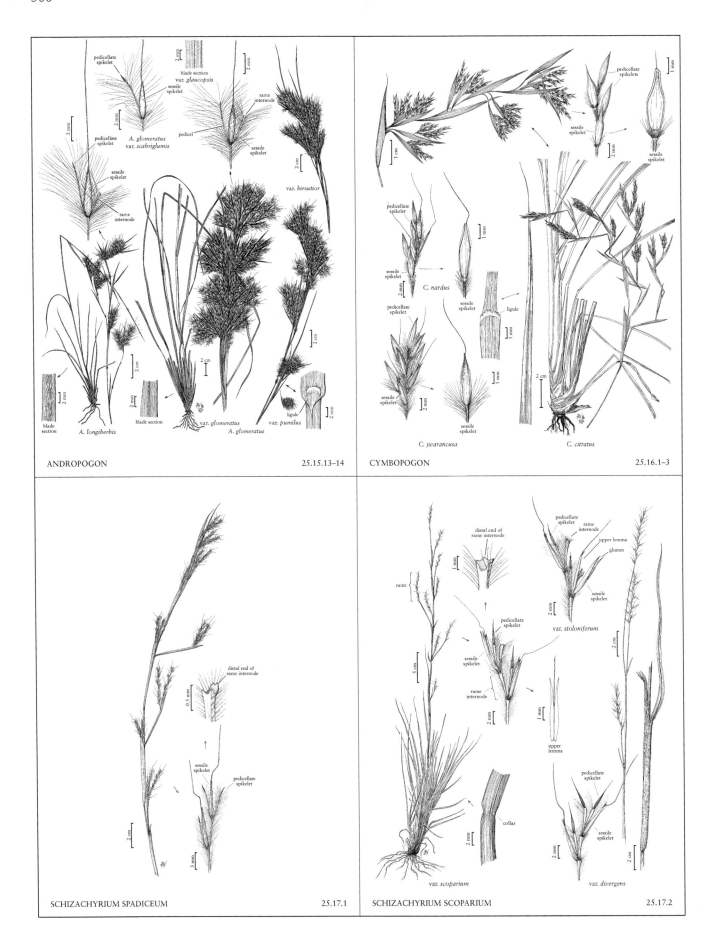

ANDROPOGON

25.15.13–14

CYMBOPOGON

25.16.1–3

SCHIZACHYRIUM SPADICEUM

25.17.1

SCHIZACHYRIUM SCOPARIUM

25.17.2

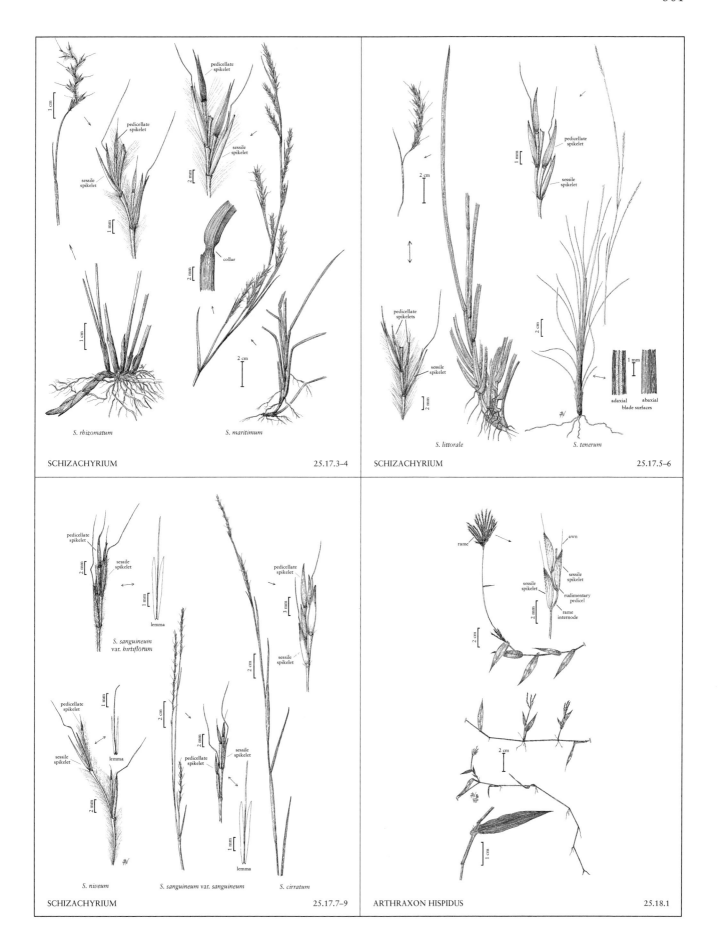

S. rhizomatum

S. maritimum

SCHIZACHYRIUM 25.17.3–4

S. littorale

S. tenerum

SCHIZACHYRIUM 25.17.5–6

S. sanguineum var. hirtiflorum

S. niveum

S. sanguineum var. sanguineum

S. cirratum

SCHIZACHYRIUM 25.17.7–9

ARTHRAXON HISPIDUS 25.18.1

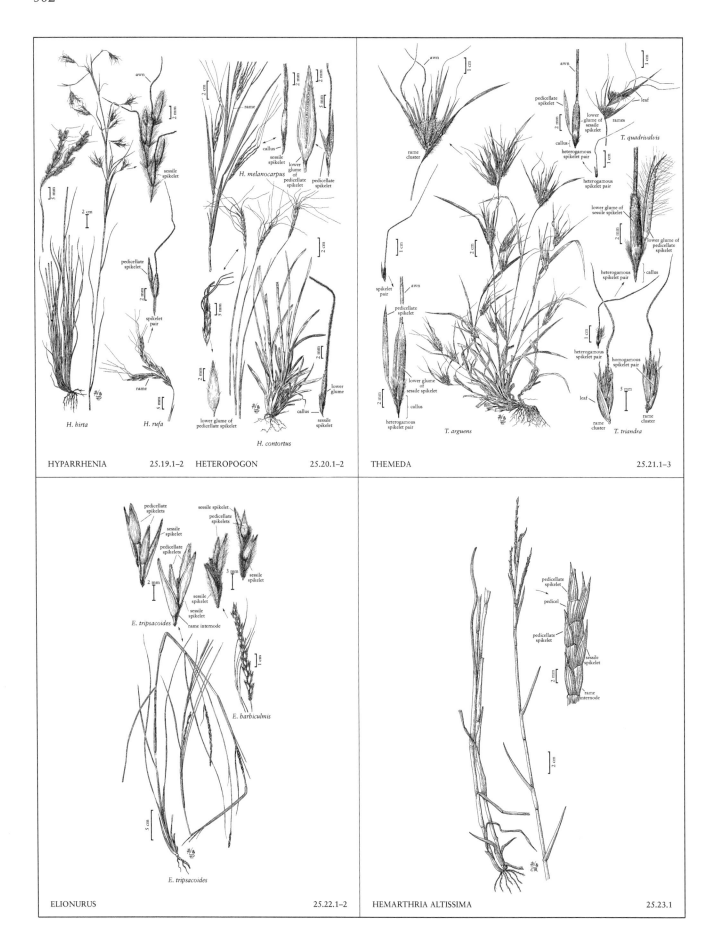

HYPARRHENIA  25.19.1–2  HETEROPOGON  25.20.1–2  THEMEDA  25.21.1–3

ELIONURUS  25.22.1–2  HEMARTHRIA ALTISSIMA  25.23.1

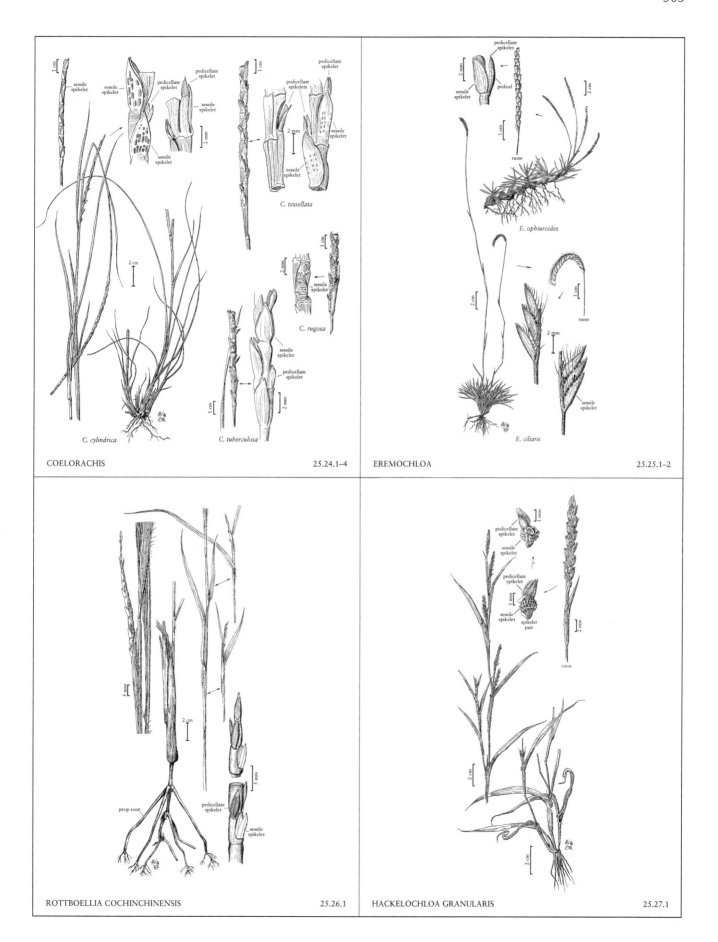

COELORACHIS 25.24.1–4

EREMOCHLOA 25.25.1–2

ROTTBOELLIA COCHINCHINENSIS 25.26.1

HACKELOCHLOA GRANULARIS 25.27.1

504

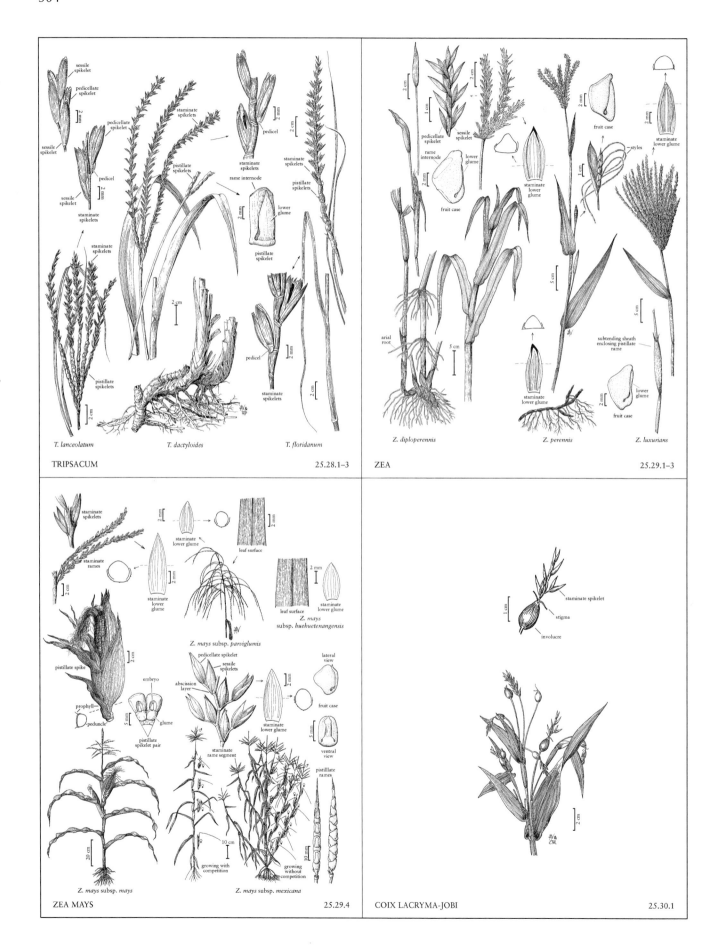

TRIPSACUM                                    25.28.1–3

T. lanceolatum        T. dactyloides        T. floridanum

ZEA                                          25.29.1–3

Z. diploperennis      Z. perennis           Z. luxurians

ZEA MAYS                                     25.29.4

Z. mays subsp. mays   Z. mays subsp. mexicana

Z. mays subsp. parviglumis

Z. mays subsp. huehuetenangensis

COIX LACRYMA-JOBI                            25.30.1

# Distribution Maps

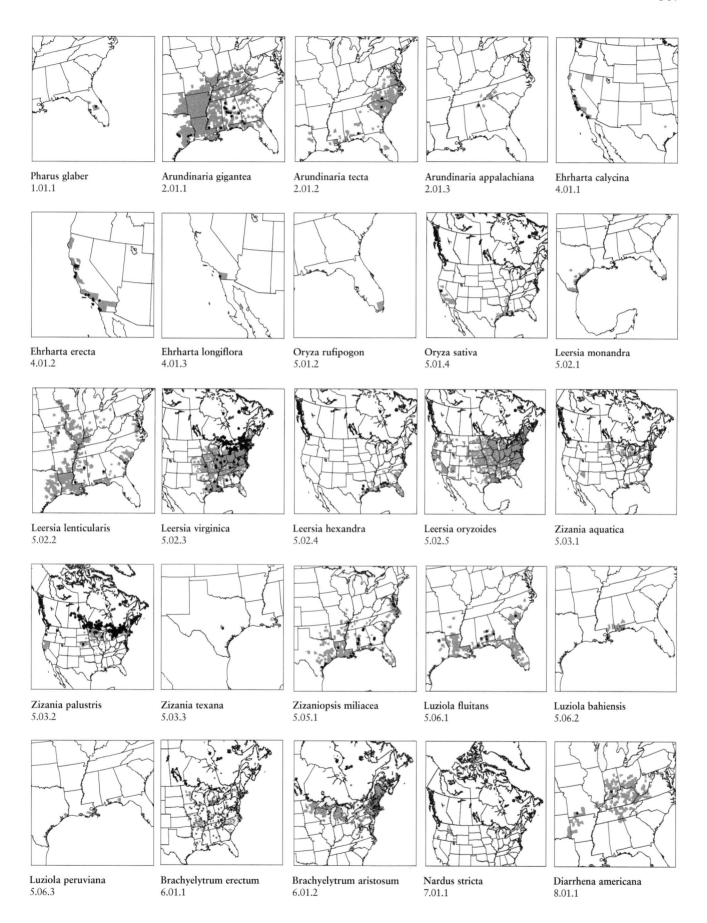

Pharus glaber
1.01.1

Arundinaria gigantea
2.01.1

Arundinaria tecta
2.01.2

Arundinaria appalachiana
2.01.3

Ehrharta calycina
4.01.1

Ehrharta erecta
4.01.2

Ehrharta longiflora
4.01.3

Oryza rufipogon
5.01.2

Oryza sativa
5.01.4

Leersia monandra
5.02.1

Leersia lenticularis
5.02.2

Leersia virginica
5.02.3

Leersia hexandra
5.02.4

Leersia oryzoides
5.02.5

Zizania aquatica
5.03.1

Zizania palustris
5.03.2

Zizania texana
5.03.3

Zizaniopsis miliacea
5.05.1

Luziola fluitans
5.06.1

Luziola bahiensis
5.06.2

Luziola peruviana
5.06.3

Brachyelytrum erectum
6.01.1

Brachyelytrum aristosum
6.01.2

Nardus stricta
7.01.1

Diarrhena americana
8.01.1

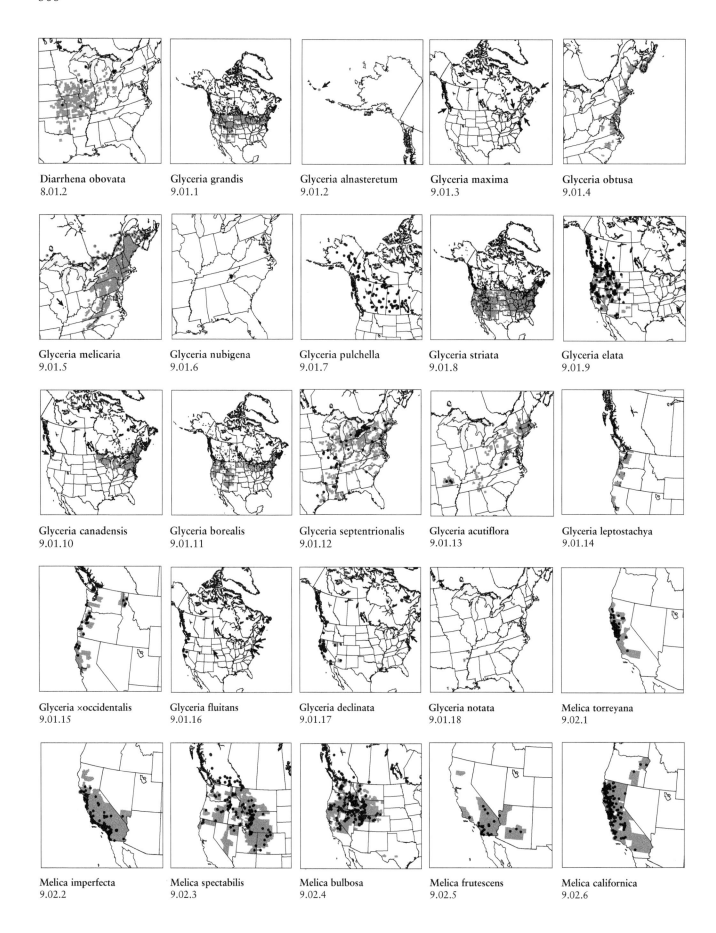

Diarrhena obovata
8.01.2

Glyceria grandis
9.01.1

Glyceria alnasteretum
9.01.2

Glyceria maxima
9.01.3

Glyceria obtusa
9.01.4

Glyceria melicaria
9.01.5

Glyceria nubigena
9.01.6

Glyceria pulchella
9.01.7

Glyceria striata
9.01.8

Glyceria elata
9.01.9

Glyceria canadensis
9.01.10

Glyceria borealis
9.01.11

Glyceria septentrionalis
9.01.12

Glyceria acutiflora
9.01.13

Glyceria leptostachya
9.01.14

Glyceria ×occidentalis
9.01.15

Glyceria fluitans
9.01.16

Glyceria declinata
9.01.17

Glyceria notata
9.01.18

Melica torreyana
9.02.1

Melica imperfecta
9.02.2

Melica spectabilis
9.02.3

Melica bulbosa
9.02.4

Melica frutescens
9.02.5

Melica californica
9.02.6

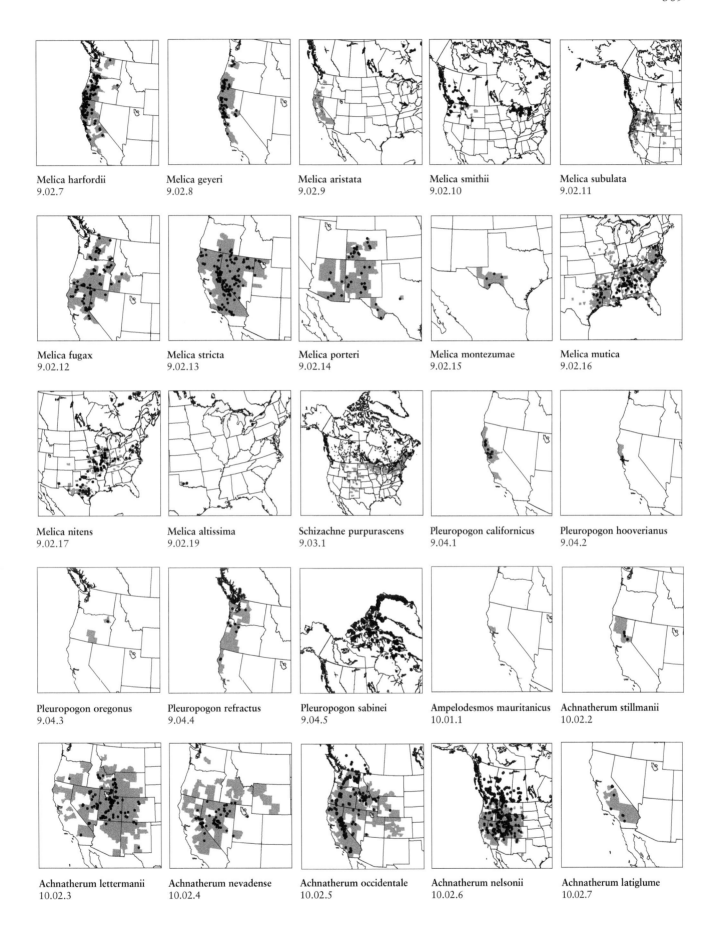

Melica harfordii
9.02.7

Melica geyeri
9.02.8

Melica aristata
9.02.9

Melica smithii
9.02.10

Melica subulata
9.02.11

Melica fugax
9.02.12

Melica stricta
9.02.13

Melica porteri
9.02.14

Melica montezumae
9.02.15

Melica mutica
9.02.16

Melica nitens
9.02.17

Melica altissima
9.02.19

Schizachne purpurascens
9.03.1

Pleuropogon californicus
9.04.1

Pleuropogon hooverianus
9.04.2

Pleuropogon oregonus
9.04.3

Pleuropogon refractus
9.04.4

Pleuropogon sabinei
9.04.5

Ampelodesmos mauritanicus
10.01.1

Achnatherum stillmanii
10.02.2

Achnatherum lettermanii
10.02.3

Achnatherum nevadense
10.02.4

Achnatherum occidentale
10.02.5

Achnatherum nelsonii
10.02.6

Achnatherum latiglume
10.02.7

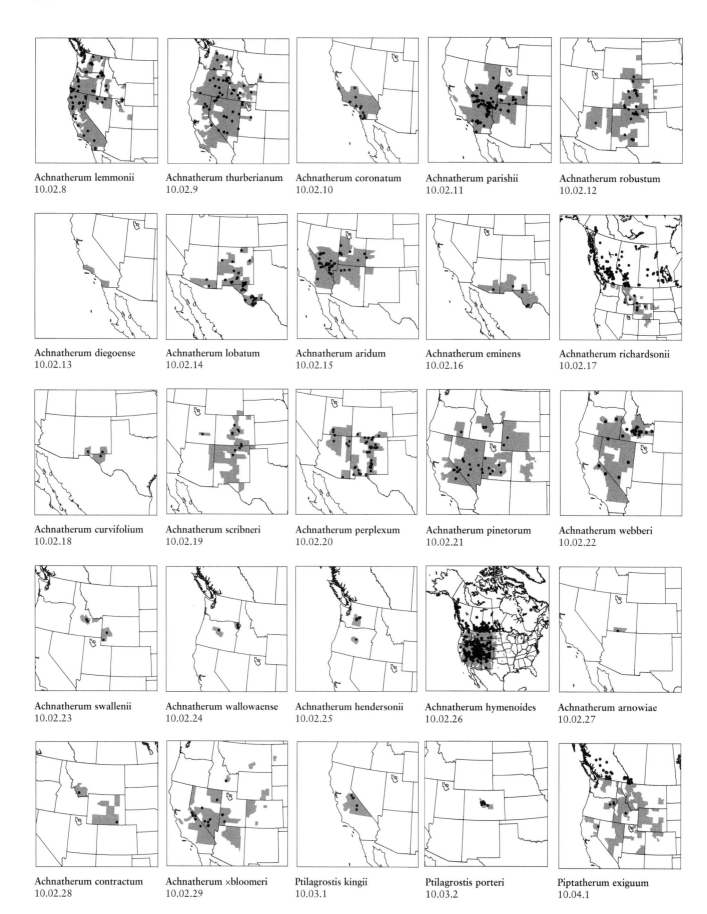

Achnatherum lemmonii
10.02.8

Achnatherum thurberianum
10.02.9

Achnatherum coronatum
10.02.10

Achnatherum parishii
10.02.11

Achnatherum robustum
10.02.12

Achnatherum diegoense
10.02.13

Achnatherum lobatum
10.02.14

Achnatherum aridum
10.02.15

Achnatherum eminens
10.02.16

Achnatherum richardsonii
10.02.17

Achnatherum curvifolium
10.02.18

Achnatherum scribneri
10.02.19

Achnatherum perplexum
10.02.20

Achnatherum pinetorum
10.02.21

Achnatherum webberi
10.02.22

Achnatherum swallenii
10.02.23

Achnatherum wallowaense
10.02.24

Achnatherum hendersonii
10.02.25

Achnatherum hymenoides
10.02.26

Achnatherum arnowiae
10.02.27

Achnatherum contractum
10.02.28

Achnatherum ×bloomeri
10.02.29

Ptilagrostis kingii
10.03.1

Ptilagrostis porteri
10.03.2

Piptatherum exiguum
10.04.1

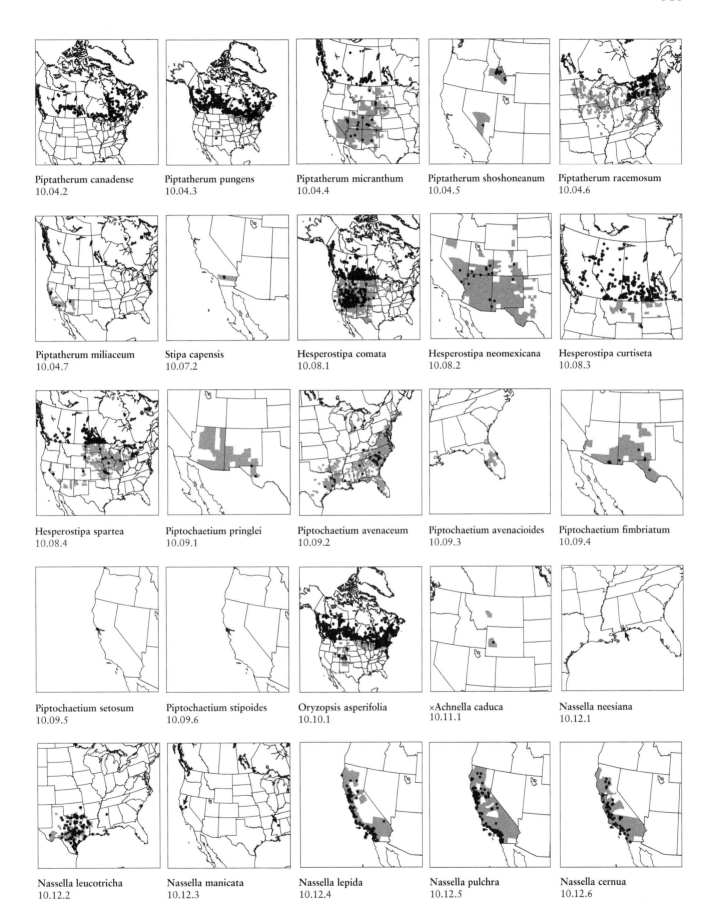

Piptatherum canadense
10.04.2

Piptatherum pungens
10.04.3

Piptatherum micranthum
10.04.4

Piptatherum shoshoneanum
10.04.5

Piptatherum racemosum
10.04.6

Piptatherum miliaceum
10.04.7

Stipa capensis
10.07.2

Hesperostipa comata
10.08.1

Hesperostipa neomexicana
10.08.2

Hesperostipa curtiseta
10.08.3

Hesperostipa spartea
10.08.4

Piptochaetium pringlei
10.09.1

Piptochaetium avenaceum
10.09.2

Piptochaetium avenacioides
10.09.3

Piptochaetium fimbriatum
10.09.4

Piptochaetium setosum
10.09.5

Piptochaetium stipoides
10.09.6

Oryzopsis asperifolia
10.10.1

×Achnella caduca
10.11.1

Nassella neesiana
10.12.1

Nassella leucotricha
10.12.2

Nassella manicata
10.12.3

Nassella lepida
10.12.4

Nassella pulchra
10.12.5

Nassella cernua
10.12.6

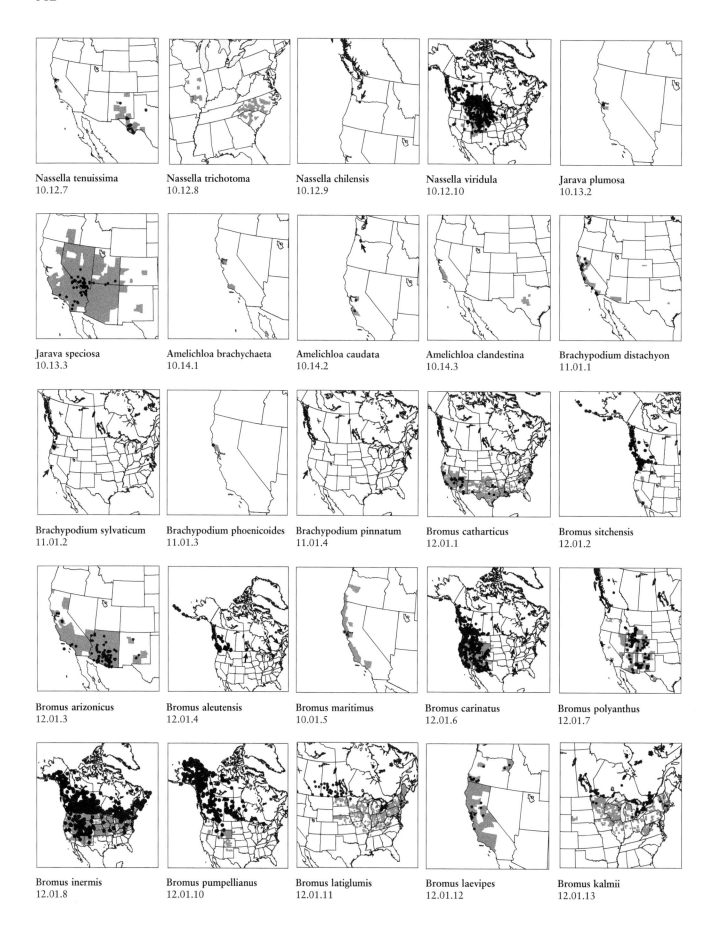

Nassella tenuissima
10.12.7

Nassella trichotoma
10.12.8

Nassella chilensis
10.12.9

Nassella viridula
10.12.10

Jarava plumosa
10.13.2

Jarava speciosa
10.13.3

Amelichloa brachychaeta
10.14.1

Amelichloa caudata
10.14.2

Amelichloa clandestina
10.14.3

Brachypodium distachyon
11.01.1

Brachypodium sylvaticum
11.01.2

Brachypodium phoenicoides
11.01.3

Brachypodium pinnatum
11.01.4

Bromus catharticus
12.01.1

Bromus sitchensis
12.01.2

Bromus arizonicus
12.01.3

Bromus aleutensis
12.01.4

Bromus maritimus
10.01.5

Bromus carinatus
12.01.6

Bromus polyanthus
12.01.7

Bromus inermis
12.01.8

Bromus pumpellianus
12.01.10

Bromus latiglumis
12.01.11

Bromus laevipes
12.01.12

Bromus kalmii
12.01.13

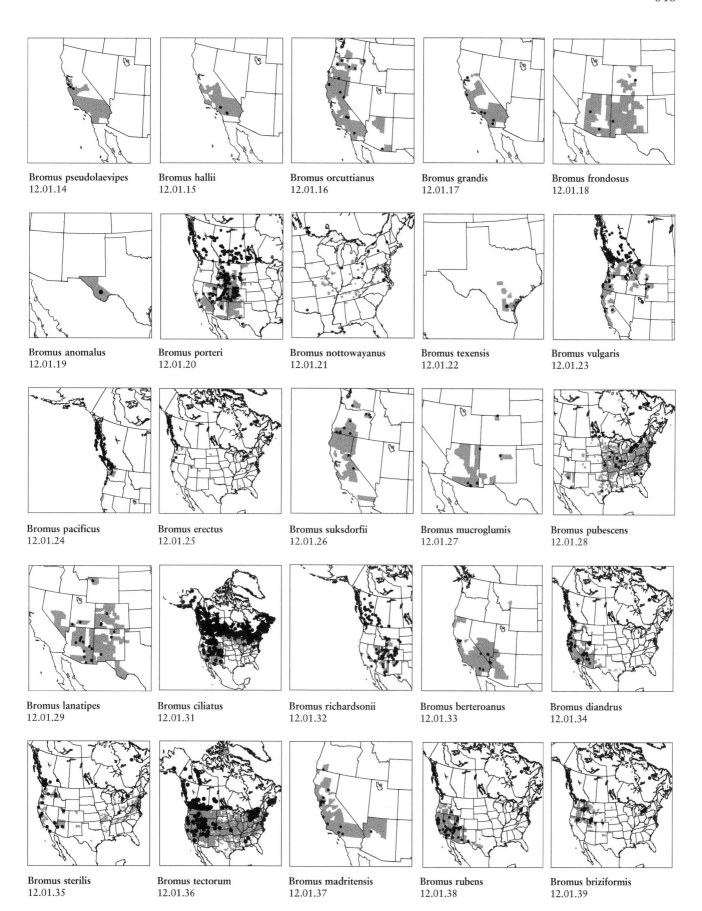

Bromus pseudolaevipes
12.01.14

Bromus hallii
12.01.15

Bromus orcuttianus
12.01.16

Bromus grandis
12.01.17

Bromus frondosus
12.01.18

Bromus anomalus
12.01.19

Bromus porteri
12.01.20

Bromus nottowayanus
12.01.21

Bromus texensis
12.01.22

Bromus vulgaris
12.01.23

Bromus pacificus
12.01.24

Bromus erectus
12.01.25

Bromus suksdorfii
12.01.26

Bromus mucroglumis
12.01.27

Bromus pubescens
12.01.28

Bromus lanatipes
12.01.29

Bromus ciliatus
12.01.31

Bromus richardsonii
12.01.32

Bromus berteroanus
12.01.33

Bromus diandrus
12.01.34

Bromus sterilis
12.01.35

Bromus tectorum
12.01.36

Bromus madritensis
12.01.37

Bromus rubens
12.01.38

Bromus briziformis
12.01.39

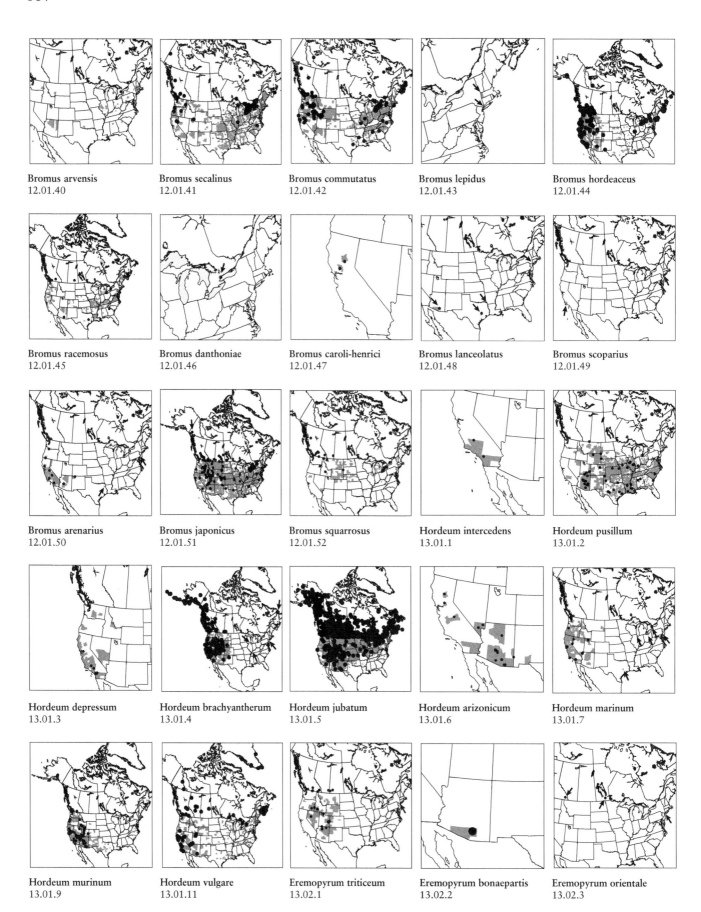

Bromus arvensis
12.01.40

Bromus secalinus
12.01.41

Bromus commutatus
12.01.42

Bromus lepidus
12.01.43

Bromus hordeaceus
12.01.44

Bromus racemosus
12.01.45

Bromus danthoniae
12.01.46

Bromus caroli-henrici
12.01.47

Bromus lanceolatus
12.01.48

Bromus scoparius
12.01.49

Bromus arenarius
12.01.50

Bromus japonicus
12.01.51

Bromus squarrosus
12.01.52

Hordeum intercedens
13.01.1

Hordeum pusillum
13.01.2

Hordeum depressum
13.01.3

Hordeum brachyantherum
13.01.4

Hordeum jubatum
13.01.5

Hordeum arizonicum
13.01.6

Hordeum marinum
13.01.7

Hordeum murinum
13.01.9

Hordeum vulgare
13.01.11

Eremopyrum triticeum
13.02.1

Eremopyrum bonaepartis
13.02.2

Eremopyrum orientale
13.02.3

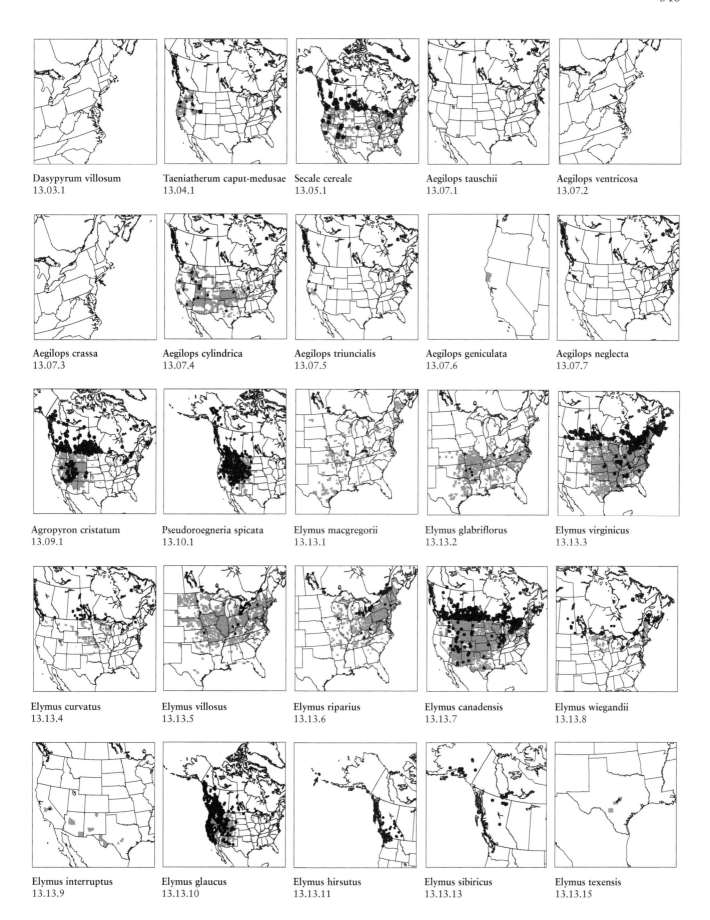

Dasypyrum villosum
13.03.1

Taeniatherum caput-medusae
13.04.1

Secale cereale
13.05.1

Aegilops tauschii
13.07.1

Aegilops ventricosa
13.07.2

Aegilops crassa
13.07.3

Aegilops cylindrica
13.07.4

Aegilops triuncialis
13.07.5

Aegilops geniculata
13.07.6

Aegilops neglecta
13.07.7

Agropyron cristatum
13.09.1

Pseudoroegneria spicata
13.10.1

Elymus macgregorii
13.13.1

Elymus glabriflorus
13.13.2

Elymus virginicus
13.13.3

Elymus curvatus
13.13.4

Elymus villosus
13.13.5

Elymus riparius
13.13.6

Elymus canadensis
13.13.7

Elymus wiegandii
13.13.8

Elymus interruptus
13.13.9

Elymus glaucus
13.13.10

Elymus hirsutus
13.13.11

Elymus sibiricus
13.13.13

Elymus texensis
13.13.15

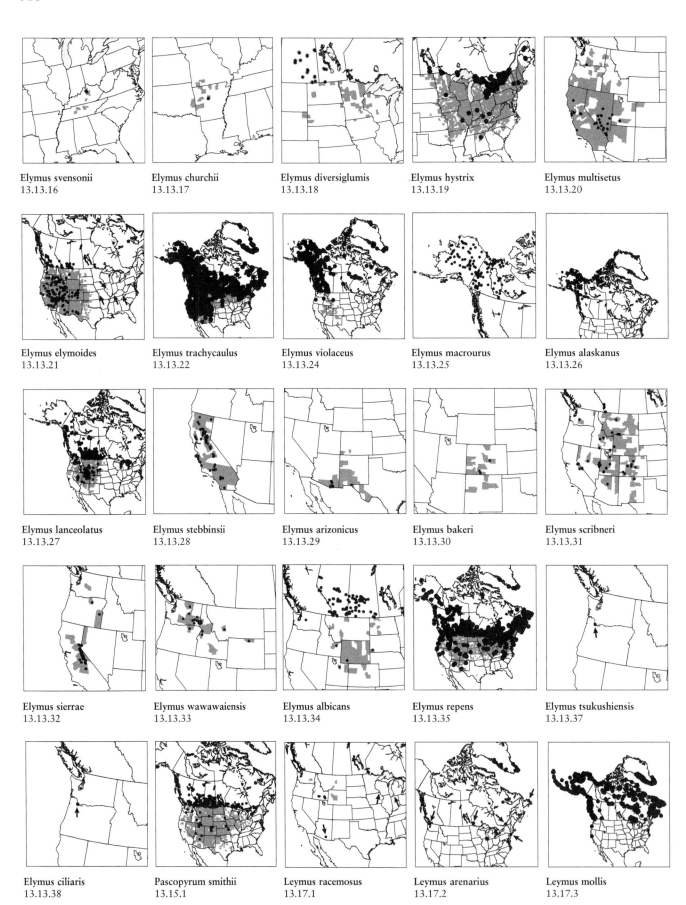

Elymus svensonii
13.13.16

Elymus churchii
13.13.17

Elymus diversiglumis
13.13.18

Elymus hystrix
13.13.19

Elymus multisetus
13.13.20

Elymus elymoides
13.13.21

Elymus trachycaulus
13.13.22

Elymus violaceus
13.13.24

Elymus macrourus
13.13.25

Elymus alaskanus
13.13.26

Elymus lanceolatus
13.13.27

Elymus stebbinsii
13.13.28

Elymus arizonicus
13.13.29

Elymus bakeri
13.13.30

Elymus scribneri
13.13.31

Elymus sierrae
13.13.32

Elymus wawawaiensis
13.13.33

Elymus albicans
13.13.34

Elymus repens
13.13.35

Elymus tsukushiensis
13.13.37

Elymus ciliaris
13.13.38

Pascopyrum smithii
13.15.1

Leymus racemosus
13.17.1

Leymus arenarius
13.17.2

Leymus mollis
13.17.3

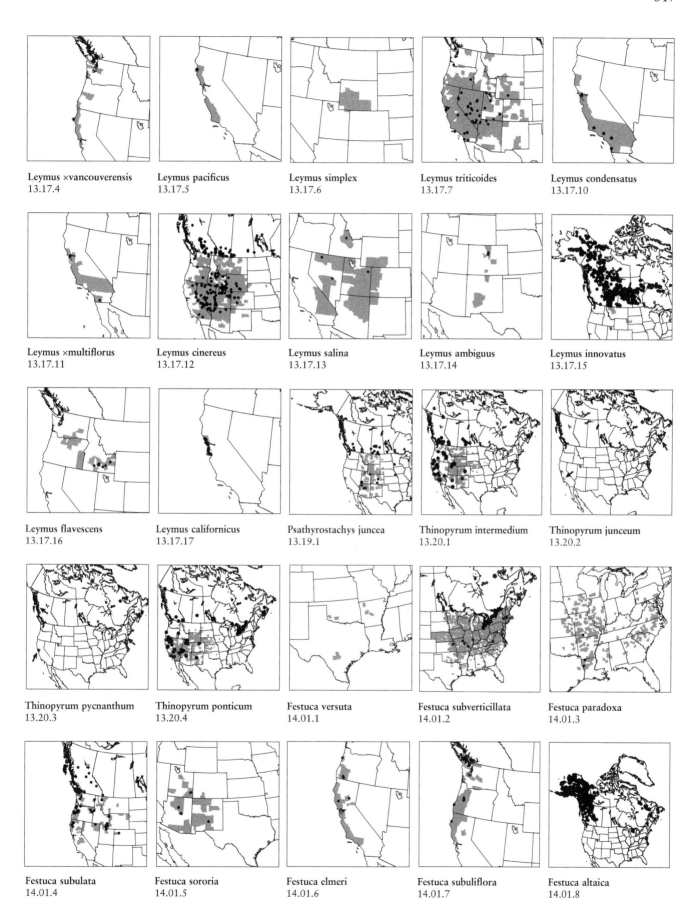

Leymus ×vancouverensis
13.17.4

Leymus pacificus
13.17.5

Leymus simplex
13.17.6

Leymus triticoides
13.17.7

Leymus condensatus
13.17.10

Leymus ×multiflorus
13.17.11

Leymus cinereus
13.17.12

Leymus salina
13.17.13

Leymus ambiguus
13.17.14

Leymus innovatus
13.17.15

Leymus flavescens
13.17.16

Leymus californicus
13.17.17

Psathyrostachys juncea
13.19.1

Thinopyrum intermedium
13.20.1

Thinopyrum junceum
13.20.2

Thinopyrum pycnanthum
13.20.3

Thinopyrum ponticum
13.20.4

Festuca versuta
14.01.1

Festuca subverticillata
14.01.2

Festuca paradoxa
14.01.3

Festuca subulata
14.01.4

Festuca sororia
14.01.5

Festuca elmeri
14.01.6

Festuca subuliflora
14.01.7

Festuca altaica
14.01.8

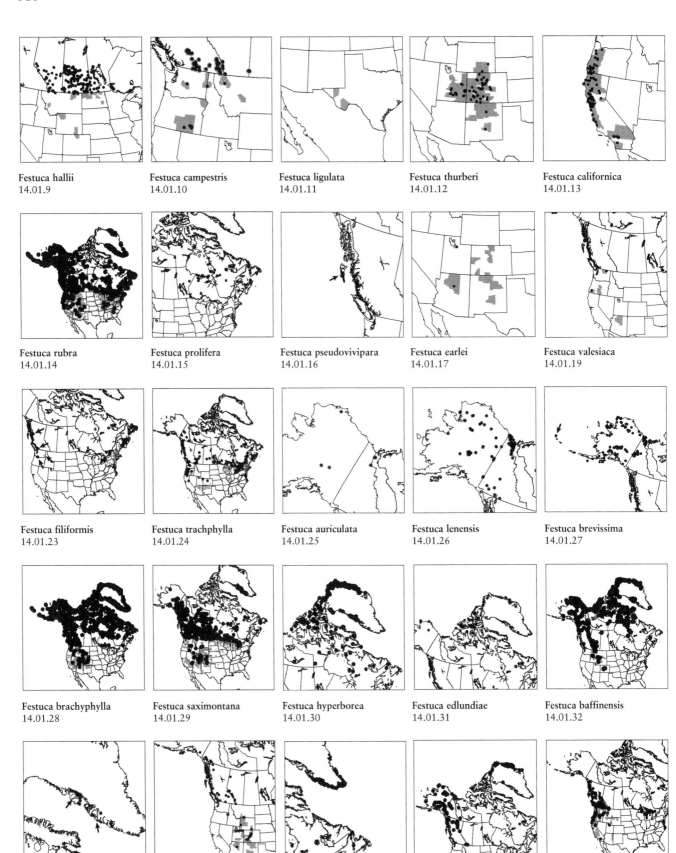

518

Festuca hallii
14.01.9

Festuca campestris
14.01.10

Festuca ligulata
14.01.11

Festuca thurberi
14.01.12

Festuca californica
14.01.13

Festuca rubra
14.01.14

Festuca prolifera
14.01.15

Festuca pseudovivipara
14.01.16

Festuca earlei
14.01.17

Festuca valesiaca
14.01.19

Festuca filiformis
14.01.23

Festuca trachphylla
14.01.24

Festuca auriculata
14.01.25

Festuca lenensis
14.01.26

Festuca brevissima
14.01.27

Festuca brachyphylla
14.01.28

Festuca saximontana
14.01.29

Festuca hyperborea
14.01.30

Festuca edlundiae
14.01.31

Festuca baffinensis
14.01.32

Festuca groenlandica
14.01.33

Festuca minutiflora
14.01.34

Festuca frederikseniae
14.01.35

Festuca viviparoidea
14.01.36

Festuca occidentalis
14.01.37

519

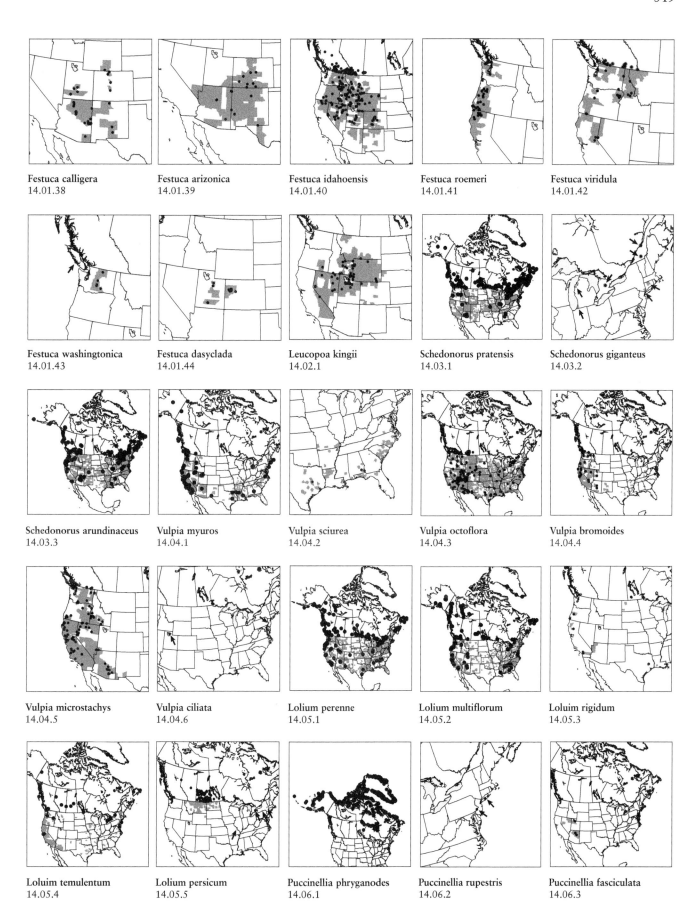

Festuca calligera
14.01.38

Festuca arizonica
14.01.39

Festuca idahoensis
14.01.40

Festuca roemeri
14.01.41

Festuca viridula
14.01.42

Festuca washingtonica
14.01.43

Festuca dasyclada
14.01.44

Leucopoa kingii
14.02.1

Schedonorus pratensis
14.03.1

Schedonorus giganteus
14.03.2

Schedonorus arundinaceus
14.03.3

Vulpia myuros
14.04.1

Vulpia sciurea
14.04.2

Vulpia octoflora
14.04.3

Vulpia bromoides
14.04.4

Vulpia microstachys
14.04.5

Vulpia ciliata
14.04.6

Lolium perenne
14.05.1

Lolium multiflorum
14.05.2

Loluim rigidum
14.05.3

Loluim temulentum
14.05.4

Lolium persicum
14.05.5

Puccinellia phryganodes
14.06.1

Puccinellia rupestris
14.06.2

Puccinellia fasciculata
14.06.3

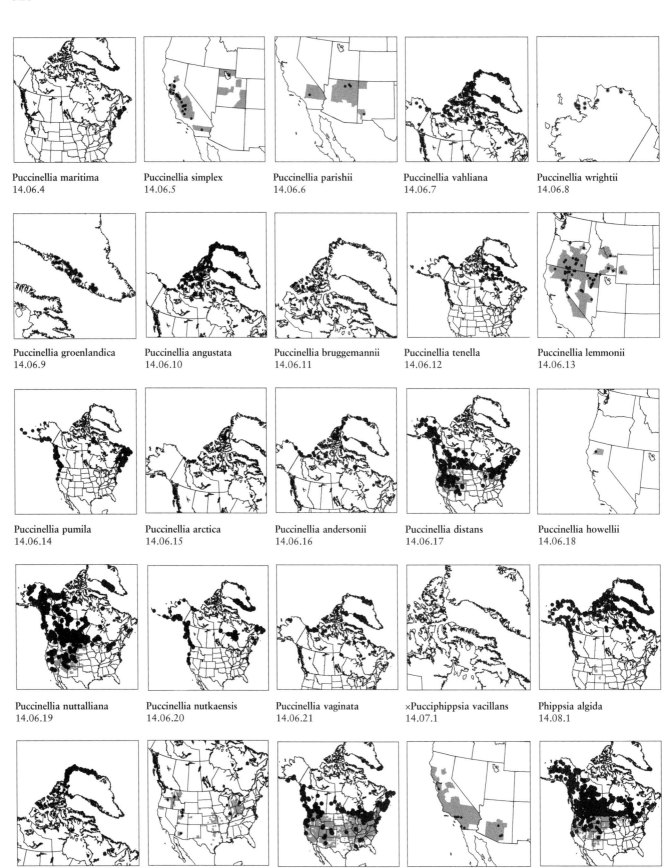

Puccinellia maritima
14.06.4

Puccinellia simplex
14.06.5

Puccinellia parishii
14.06.6

Puccinellia vahliana
14.06.7

Puccinellia wrightii
14.06.8

Puccinellia groenlandica
14.06.9

Puccinellia angustata
14.06.10

Puccinellia bruggemannii
14.06.11

Puccinellia tenella
14.06.12

Puccinellia lemmonii
14.06.13

Puccinellia pumila
14.06.14

Puccinellia arctica
14.06.15

Puccinellia andersonii
14.06.16

Puccinellia distans
14.06.17

Puccinellia howellii
14.06.18

Puccinellia nuttalliana
14.06.19

Puccinellia nutkaensis
14.06.20

Puccinellia vaginata
14.06.21

×Pucciphippsia vacillans
14.07.1

Phippsia algida
14.08.1

Phippsia concinna
14.08.2

Sclerochloa dura
14.09.1

Dactylis glomerata
14.10.1

Lamarckia aurea
14.11.1

Beckmannia syzigachne
14.12.1

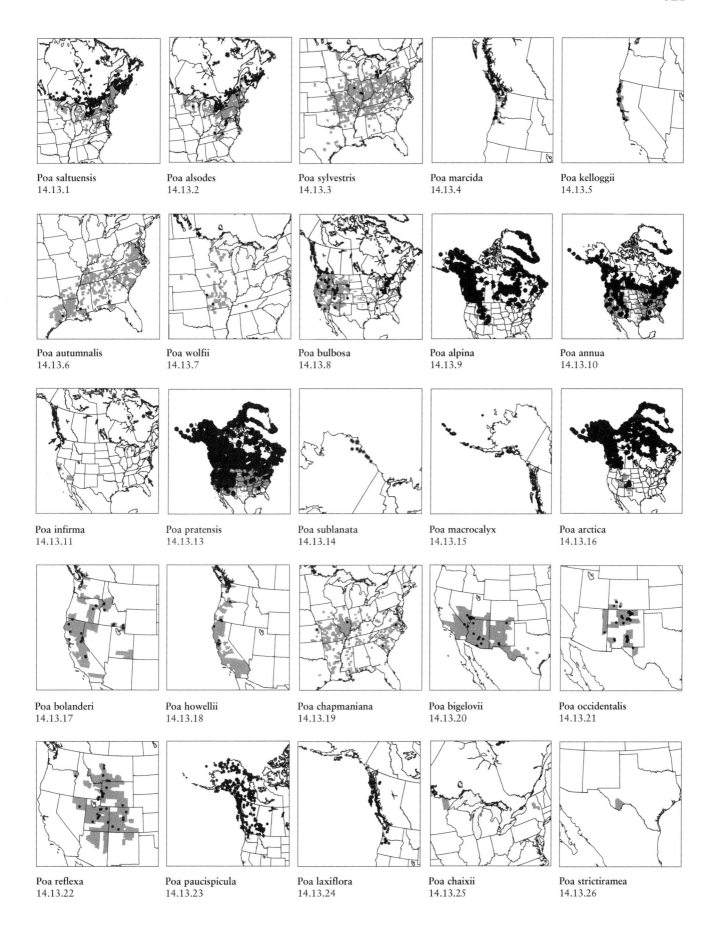

Poa saltuensis
14.13.1

Poa alsodes
14.13.2

Poa sylvestris
14.13.3

Poa marcida
14.13.4

Poa kelloggii
14.13.5

Poa autumnalis
14.13.6

Poa wolfii
14.13.7

Poa bulbosa
14.13.8

Poa alpina
14.13.9

Poa annua
14.13.10

Poa infirma
14.13.11

Poa pratensis
14.13.13

Poa sublanata
14.13.14

Poa macrocalyx
14.13.15

Poa arctica
14.13.16

Poa bolanderi
14.13.17

Poa howellii
14.13.18

Poa chapmaniana
14.13.19

Poa bigelovii
14.13.20

Poa occidentalis
14.13.21

Poa reflexa
14.13.22

Poa paucispicula
14.13.23

Poa laxiflora
14.13.24

Poa chaixii
14.13.25

Poa strictiramea
14.13.26

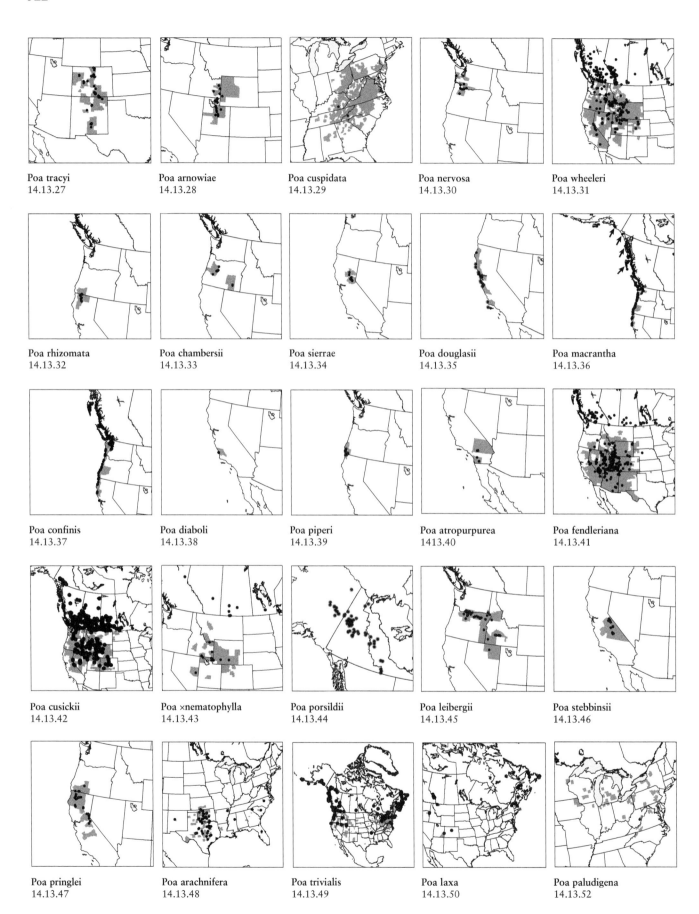

Poa tracyi
14.13.27

Poa arnowiae
14.13.28

Poa cuspidata
14.13.29

Poa nervosa
14.13.30

Poa wheeleri
14.13.31

Poa rhizomata
14.13.32

Poa chambersii
14.13.33

Poa sierrae
14.13.34

Poa douglasii
14.13.35

Poa macrantha
14.13.36

Poa confinis
14.13.37

Poa diaboli
14.13.38

Poa piperi
14.13.39

Poa atropurpurea
1413.40

Poa fendleriana
14.13.41

Poa cusickii
14.13.42

Poa ×nematophylla
14.13.43

Poa porsildii
14.13.44

Poa leibergii
14.13.45

Poa stebbinsii
14.13.46

Poa pringlei
14.13.47

Poa arachnifera
14.13.48

Poa trivialis
14.13.49

Poa laxa
14.13.50

Poa paludigena
14.13.52

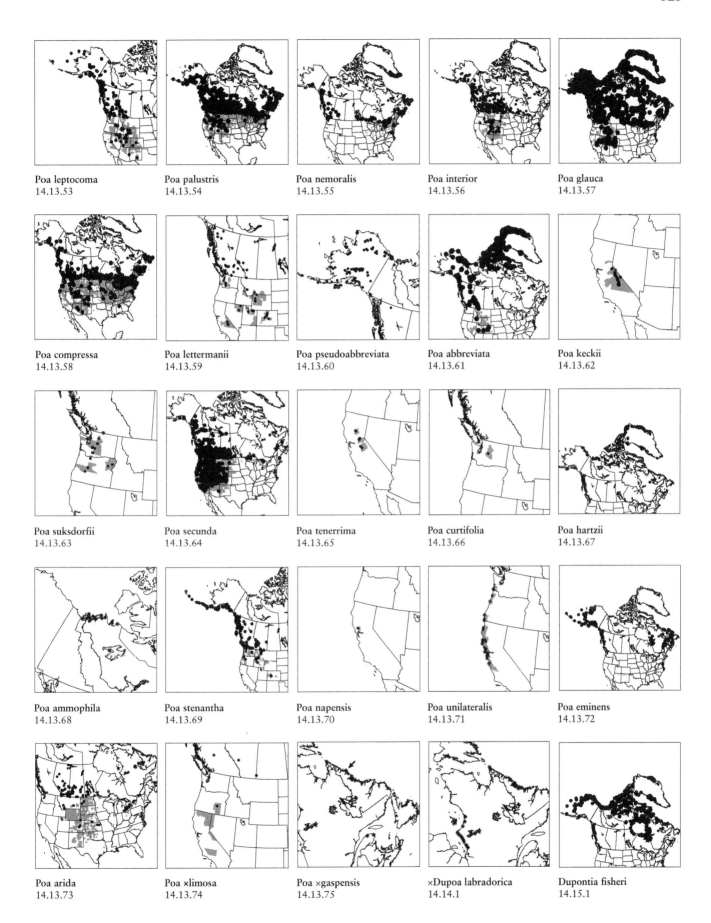

Poa leptocoma
14.13.53

Poa palustris
14.13.54

Poa nemoralis
14.13.55

Poa interior
14.13.56

Poa glauca
14.13.57

Poa compressa
14.13.58

Poa lettermanii
14.13.59

Poa pseudoabbreviata
14.13.60

Poa abbreviata
14.13.61

Poa keckii
14.13.62

Poa suksdorfii
14.13.63

Poa secunda
14.13.64

Poa tenerrima
14.13.65

Poa curtifolia
14.13.66

Poa hartzii
14.13.67

Poa ammophila
14.13.68

Poa stenantha
14.13.69

Poa napensis
14.13.70

Poa unilateralis
14.13.71

Poa eminens
14.13.72

Poa arida
14.13.73

Poa ×limosa
14.13.74

Poa ×gaspensis
14.13.75

×Dupoa labradorica
14.14.1

Dupontia fisheri
14.15.1

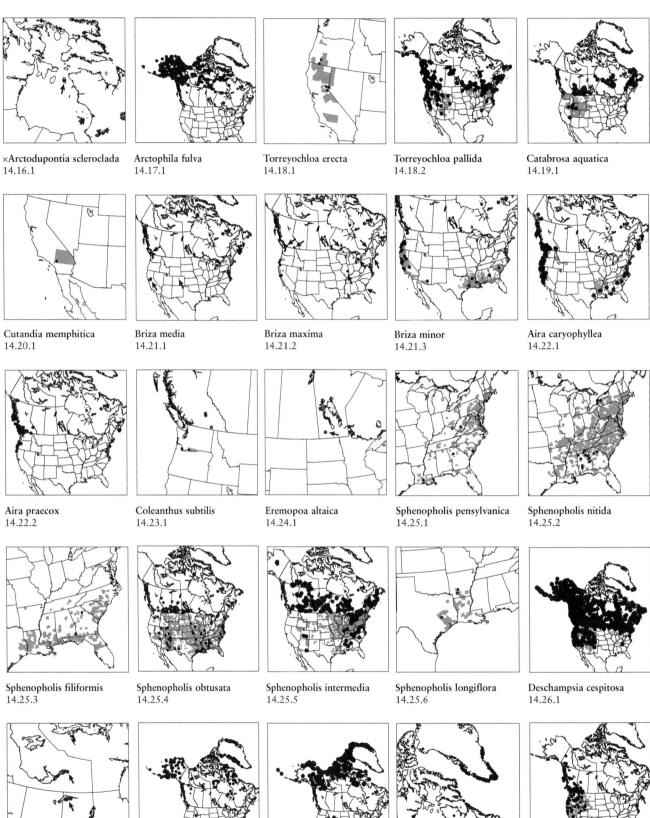

×Arctodupontia scleroclada
14.16.1

Arctophila fulva
14.17.1

Torreyochloa erecta
14.18.1

Torreyochloa pallida
14.18.2

Catabrosa aquatica
14.19.1

Cutandia memphitica
14.20.1

Briza media
14.21.1

Briza maxima
14.21.2

Briza minor
14.21.3

Aira caryophyllea
14.22.1

Aira praecox
14.22.2

Coleanthus subtilis
14.23.1

Eremopoa altaica
14.24.1

Sphenopholis pensylvanica
14.25.1

Sphenopholis nitida
14.25.2

Sphenopholis filiformis
14.25.3

Sphenopholis obtusata
14.25.4

Sphenopholis intermedia
14.25.5

Sphenopholis longiflora
14.25.6

Deschampsia cespitosa
14.26.1

Deschampsia mackenzieana
14.26.2

Deschampsia sukatschewii
14.26.3

Deschampsia brevifolia
14.26.4

Deschampsia alpina
14.26.5

Deschampsia elongata
14.26.6

525

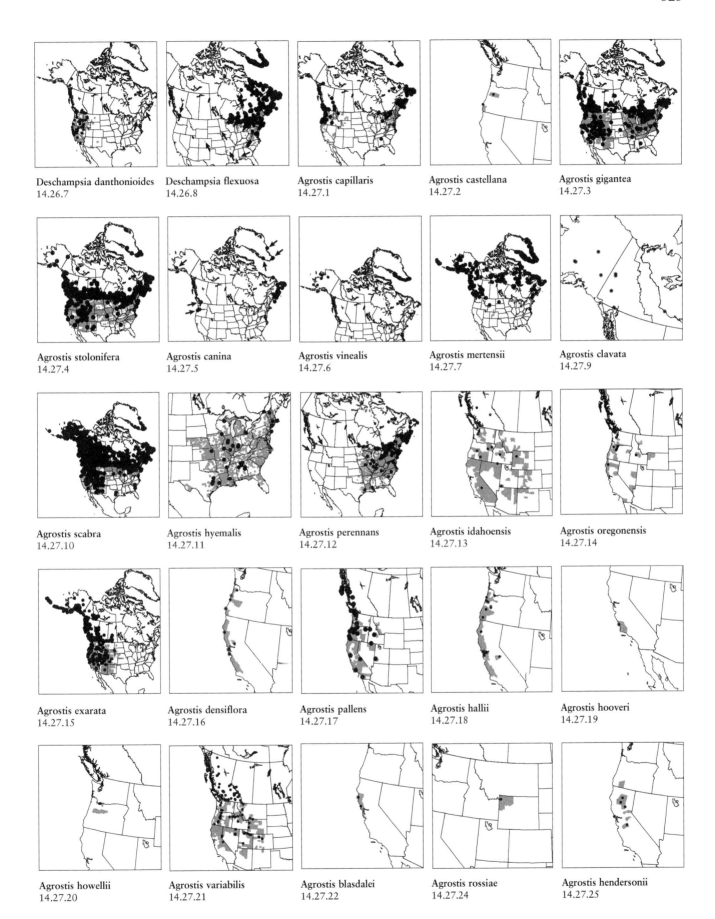

Deschampsia danthonioides
14.26.7

Deschampsia flexuosa
14.26.8

Agrostis capillaris
14.27.1

Agrostis castellana
14.27.2

Agrostis gigantea
14.27.3

Agrostis stolonifera
14.27.4

Agrostis canina
14.27.5

Agrostis vinealis
14.27.6

Agrostis mertensii
14.27.7

Agrostis clavata
14.27.9

Agrostis scabra
14.27.10

Agrostis hyemalis
14.27.11

Agrostis perennans
14.27.12

Agrostis idahoensis
14.27.13

Agrostis oregonensis
14.27.14

Agrostis exarata
14.27.15

Agrostis densiflora
14.27.16

Agrostis pallens
14.27.17

Agrostis hallii
14.27.18

Agrostis hooveri
14.27.19

Agrostis howellii
14.27.20

Agrostis variabilis
14.27.21

Agrostis blasdalei
14.27.22

Agrostis rossiae
14.27.24

Agrostis hendersonii
14.27.25

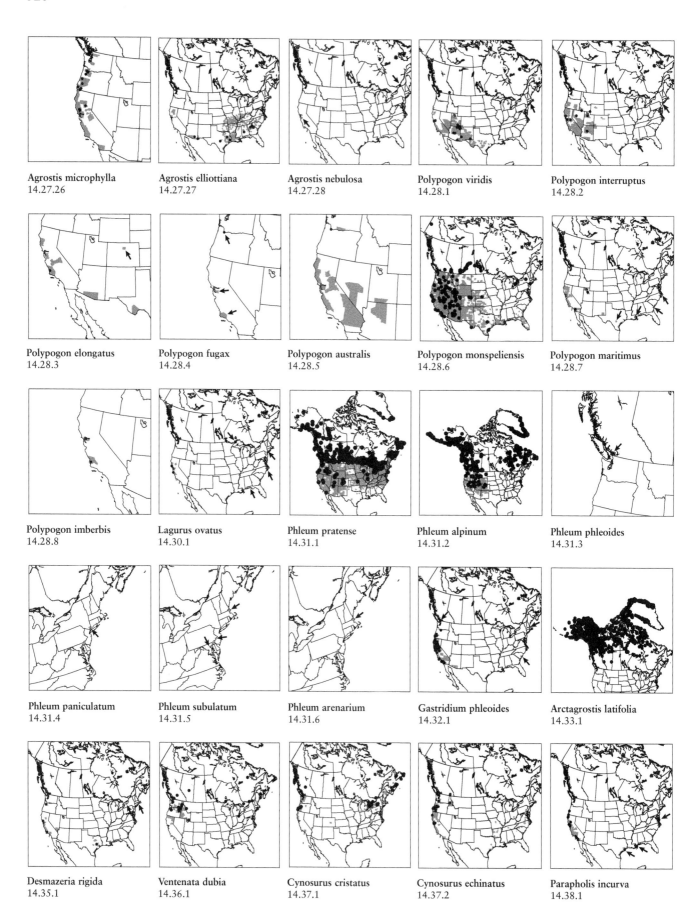

Agrostis microphylla
14.27.26

Agrostis elliottiana
14.27.27

Agrostis nebulosa
14.27.28

Polypogon viridis
14.28.1

Polypogon interruptus
14.28.2

Polypogon elongatus
14.28.3

Polypogon fugax
14.28.4

Polypogon australis
14.28.5

Polypogon monspeliensis
14.28.6

Polypogon maritimus
14.28.7

Polypogon imberbis
14.28.8

Lagurus ovatus
14.30.1

Phleum pratense
14.31.1

Phleum alpinum
14.31.2

Phleum phleoides
14.31.3

Phleum paniculatum
14.31.4

Phleum subulatum
14.31.5

Phleum arenarium
14.31.6

Gastridium phleoides
14.32.1

Arctagrostis latifolia
14.33.1

Desmazeria rigida
14.35.1

Ventenata dubia
14.36.1

Cynosurus cristatus
14.37.1

Cynosurus echinatus
14.37.2

Parapholis incurva
14.38.1

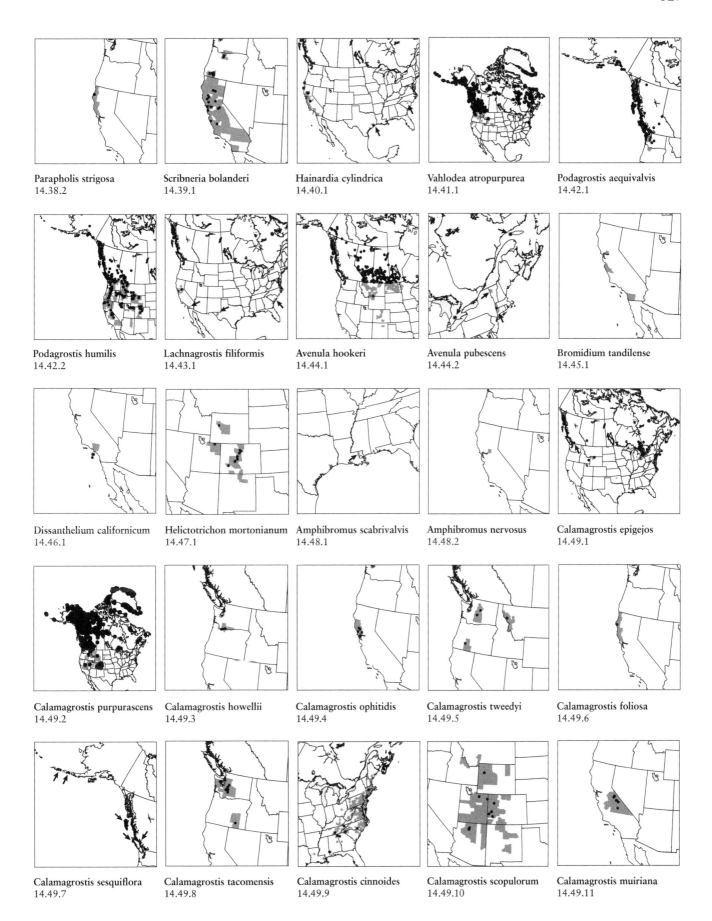

Parapholis strigosa
14.38.2

Scribneria bolanderi
14.39.1

Hainardia cylindrica
14.40.1

Vahlodea atropurpurea
14.41.1

Podagrostis aequivalis
14.42.1

Podagrostis humilis
14.42.2

Lachnagrostis filiformis
14.43.1

Avenula hookeri
14.44.1

Avenula pubescens
14.44.2

Bromidium tandilense
14.45.1

Dissanthelium californicum
14.46.1

Helictotrichon mortonianum
14.47.1

Amphibromus scabrivalvis
14.48.1

Amphibromus nervosus
14.48.2

Calamagrostis epigejos
14.49.1

Calamagrostis purpurascens
14.49.2

Calamagrostis howellii
14.49.3

Calamagrostis ophitidis
14.49.4

Calamagrostis tweedyi
14.49.5

Calamagrostis foliosa
14.49.6

Calamagrostis sesquiflora
14.49.7

Calamagrostis tacomensis
14.49.8

Calamagrostis cinnoides
14.49.9

Calamagrostis scopulorum
14.49.10

Calamagrostis muiriana
14.49.11

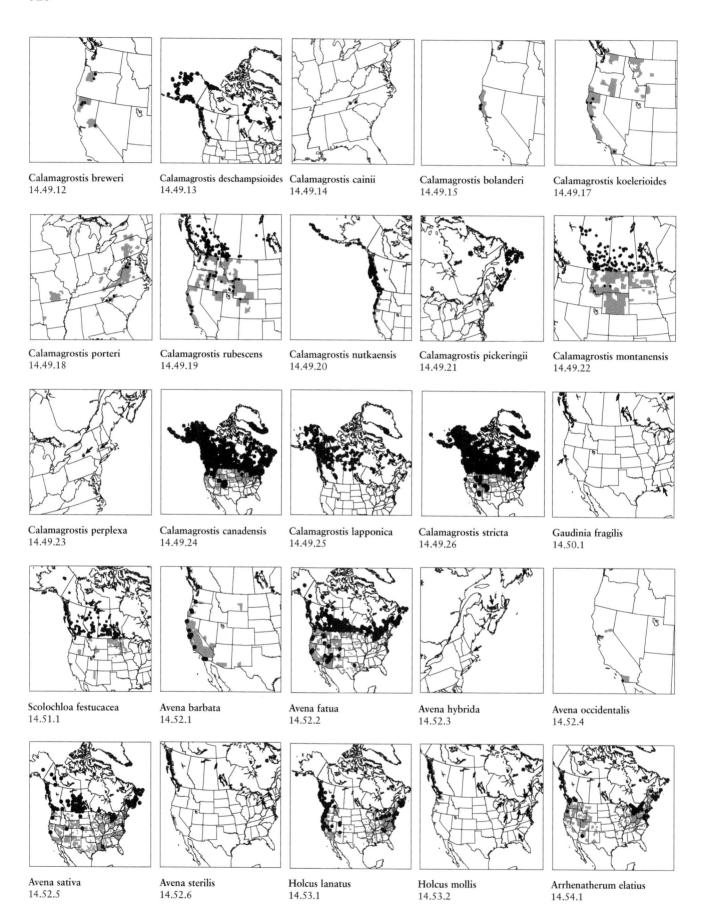

Calamagrostis breweri
14.49.12

Calamagrostis deschampsioides
14.49.13

Calamagrostis cainii
14.49.14

Calamagrostis bolanderi
14.49.15

Calamagrostis koelerioides
14.49.17

Calamagrostis porteri
14.49.18

Calamagrostis rubescens
14.49.19

Calamagrostis nutkaensis
14.49.20

Calamagrostis pickeringii
14.49.21

Calamagrostis montanensis
14.49.22

Calamagrostis perplexa
14.49.23

Calamagrostis canadensis
14.49.24

Calamagrostis lapponica
14.49.25

Calamagrostis stricta
14.49.26

Gaudinia fragilis
14.50.1

Scolochloa festucacea
14.51.1

Avena barbata
14.52.1

Avena fatua
14.52.2

Avena hybrida
14.52.3

Avena occidentalis
14.52.4

Avena sativa
14.52.5

Avena sterilis
14.52.6

Holcus lanatus
14.53.1

Holcus mollis
14.53.2

Arrhenatherum elatius
14.54.1

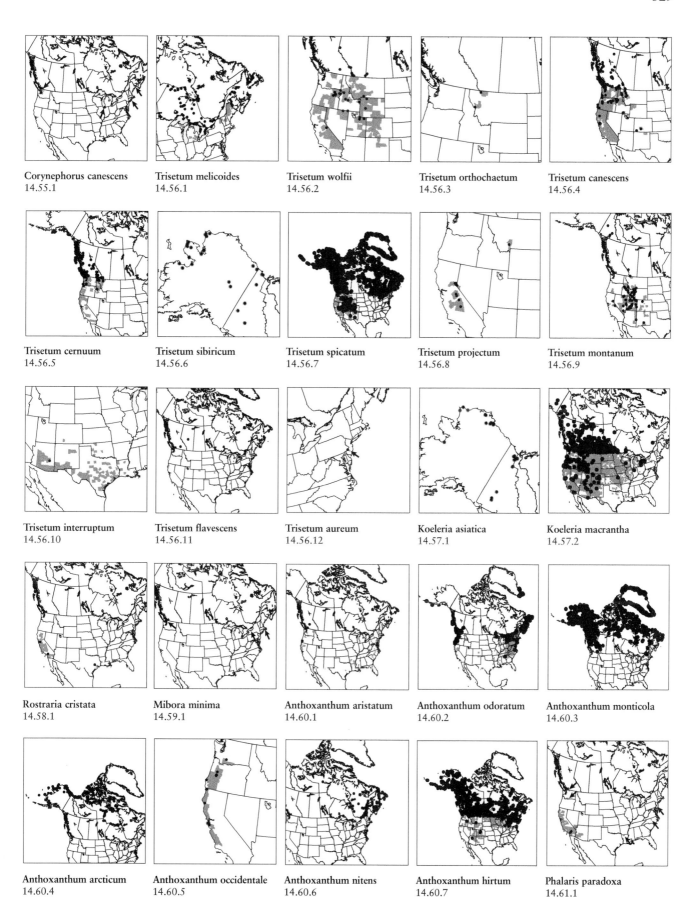

Corynephorus canescens
14.55.1

Trisetum melicoides
14.56.1

Trisetum wolfii
14.56.2

Trisetum orthochaetum
14.56.3

Trisetum canescens
14.56.4

Trisetum cernuum
14.56.5

Trisetum sibiricum
14.56.6

Trisetum spicatum
14.56.7

Trisetum projectum
14.56.8

Trisetum montanum
14.56.9

Trisetum interruptum
14.56.10

Trisetum flavescens
14.56.11

Trisetum aureum
14.56.12

Koeleria asiatica
14.57.1

Koeleria macrantha
14.57.2

Rostraria cristata
14.58.1

Mibora minima
14.59.1

Anthoxanthum aristatum
14.60.1

Anthoxanthum odoratum
14.60.2

Anthoxanthum monticola
14.60.3

Anthoxanthum arcticum
14.60.4

Anthoxanthum occidentale
14.60.5

Anthoxanthum nitens
14.60.6

Anthoxanthum hirtum
14.60.7

Phalaris paradoxa
14.61.1

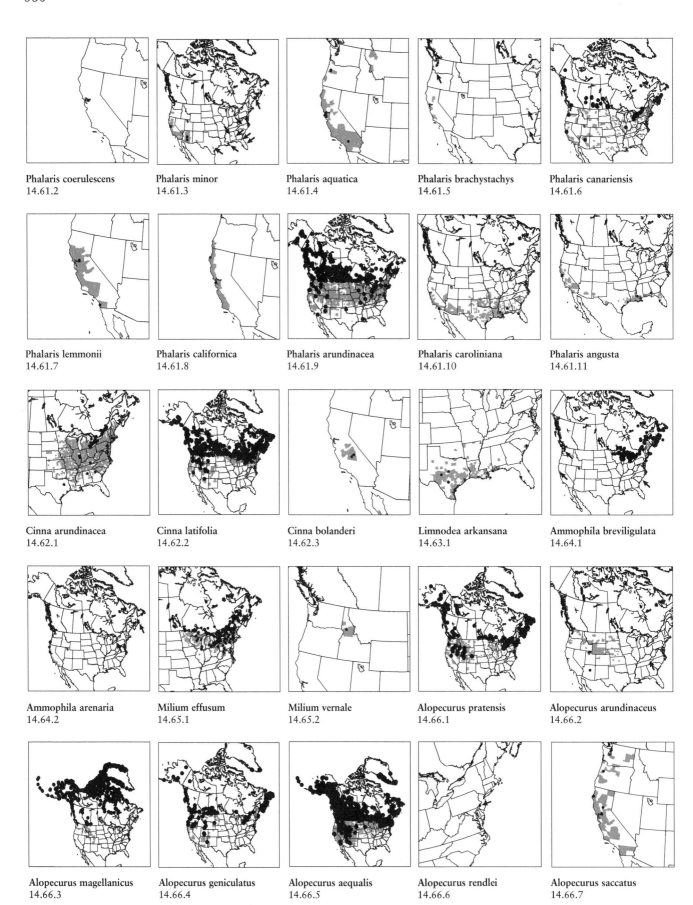

Phalaris coerulescens
14.61.2

Phalaris minor
14.61.3

Phalaris aquatica
14.61.4

Phalaris brachystachys
14.61.5

Phalaris canariensis
14.61.6

Phalaris lemmonii
14.61.7

Phalaris californica
14.61.8

Phalaris arundinacea
14.61.9

Phalaris caroliniana
14.61.10

Phalaris angusta
14.61.11

Cinna arundinacea
14.62.1

Cinna latifolia
14.62.2

Cinna bolanderi
14.62.3

Limnodea arkansana
14.63.1

Ammophila breviligulata
14.64.1

Ammophila arenaria
14.64.2

Milium effusum
14.65.1

Milium vernale
14.65.2

Alopecurus pratensis
14.66.1

Alopecurus arundinaceus
14.66.2

Alopecurus magellanicus
14.66.3

Alopecurus geniculatus
14.66.4

Alopecurus aequalis
14.66.5

Alopecurus rendlei
14.66.6

Alopecurus saccatus
14.66.7

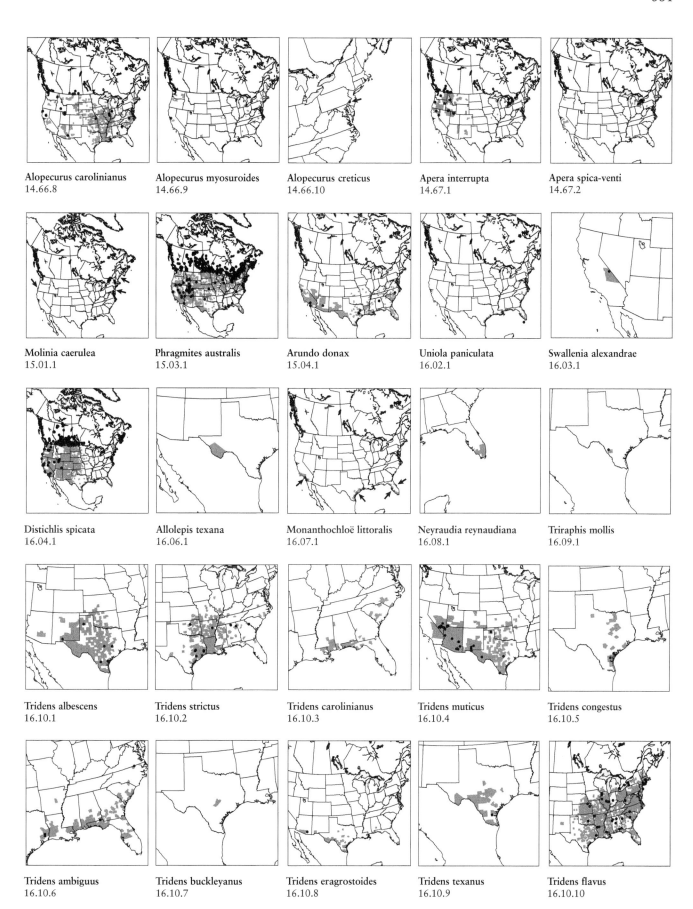

Alopecurus carolinianus
14.66.8

Alopecurus myosuroides
14.66.9

Alopecurus creticus
14.66.10

Apera interrupta
14.67.1

Apera spica-venti
14.67.2

Molinia caerulea
15.01.1

Phragmites australis
15.03.1

Arundo donax
15.04.1

Uniola paniculata
16.02.1

Swallenia alexandrae
16.03.1

Distichlis spicata
16.04.1

Allolepis texana
16.06.1

Monanthochloë littoralis
16.07.1

Neyraudia reynaudiana
16.08.1

Triraphis mollis
16.09.1

Tridens albescens
16.10.1

Tridens strictus
16.10.2

Tridens carolinianus
16.10.3

Tridens muticus
16.10.4

Tridens congestus
16.10.5

Tridens ambiguus
16.10.6

Tridens buckleyanus
16.10.7

Tridens eragrostoides
16.10.8

Tridens texanus
16.10.9

Tridens flavus
16.10.10

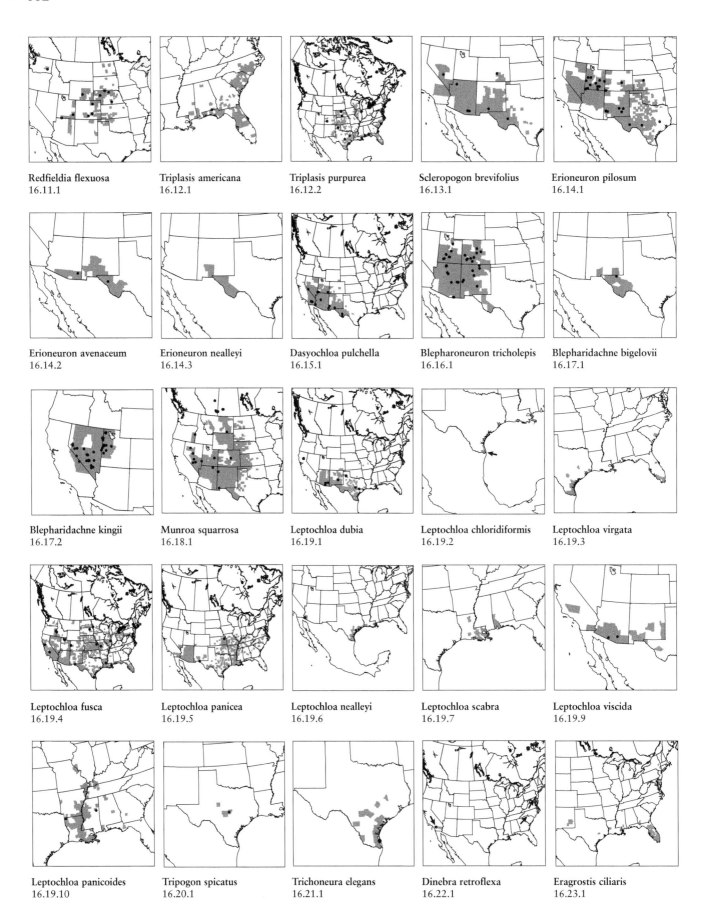

Redfieldia flexuosa
16.11.1

Triplasis americana
16.12.1

Triplasis purpurea
16.12.2

Scleropogon brevifolius
16.13.1

Erioneuron pilosum
16.14.1

Erioneuron avenaceum
16.14.2

Erioneuron nealleyi
16.14.3

Dasyochloa pulchella
16.15.1

Blepharoneuron tricholepis
16.16.1

Blepharidachne bigelovii
16.17.1

Blepharidachne kingii
16.17.2

Munroa squarrosa
16.18.1

Leptochloa dubia
16.19.1

Leptochloa chloridiformis
16.19.2

Leptochloa virgata
16.19.3

Leptochloa fusca
16.19.4

Leptochloa panicea
16.19.5

Leptochloa nealleyi
16.19.6

Leptochloa scabra
16.19.7

Leptochloa viscida
16.19.9

Leptochloa panicoides
16.19.10

Tripogon spicatus
16.20.1

Trichoneura elegans
16.21.1

Dinebra retroflexa
16.22.1

Eragrostis ciliaris
16.23.1

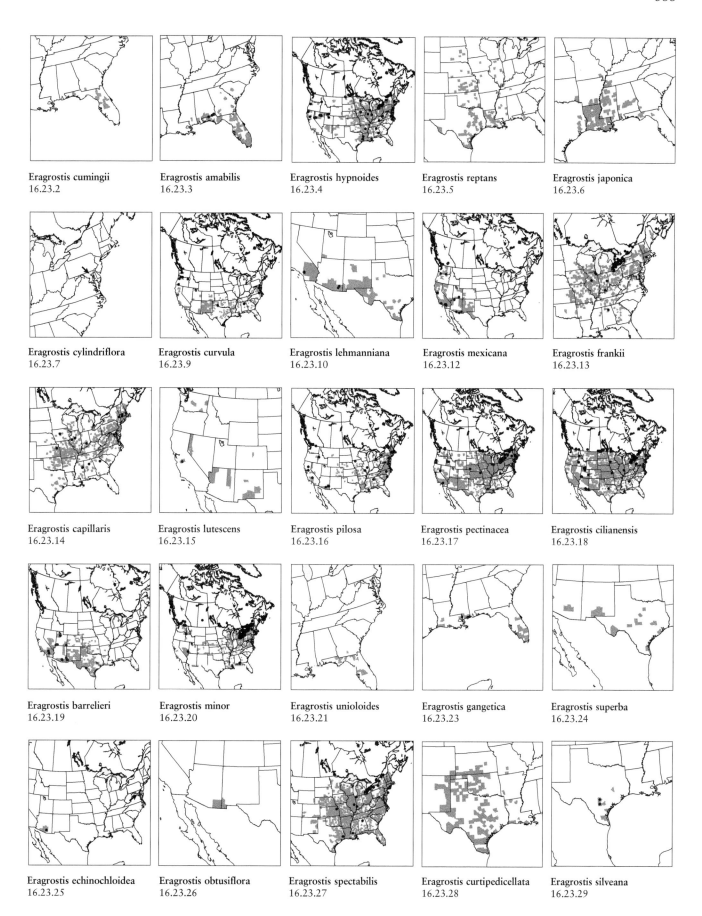

Eragrostis cumingii
16.23.2

Eragrostis amabilis
16.23.3

Eragrostis hypnoides
16.23.4

Eragrostis reptans
16.23.5

Eragrostis japonica
16.23.6

Eragrostis cylindriflora
16.23.7

Eragrostis curvula
16.23.9

Eragrostis lehmanniana
16.23.10

Eragrostis mexicana
16.23.12

Eragrostis frankii
16.23.13

Eragrostis capillaris
16.23.14

Eragrostis lutescens
16.23.15

Eragrostis pilosa
16.23.16

Eragrostis pectinacea
16.23.17

Eragrostis cilianensis
16.23.18

Eragrostis barrelieri
16.23.19

Eragrostis minor
16.23.20

Eragrostis unioloides
16.23.21

Eragrostis gangetica
16.23.23

Eragrostis superba
16.23.24

Eragrostis echinochloidea
16.23.25

Eragrostis obtusiflora
16.23.26

Eragrostis spectabilis
16.23.27

Eragrostis curtipedicellata
16.23.28

Eragrostis silveana
16.23.29

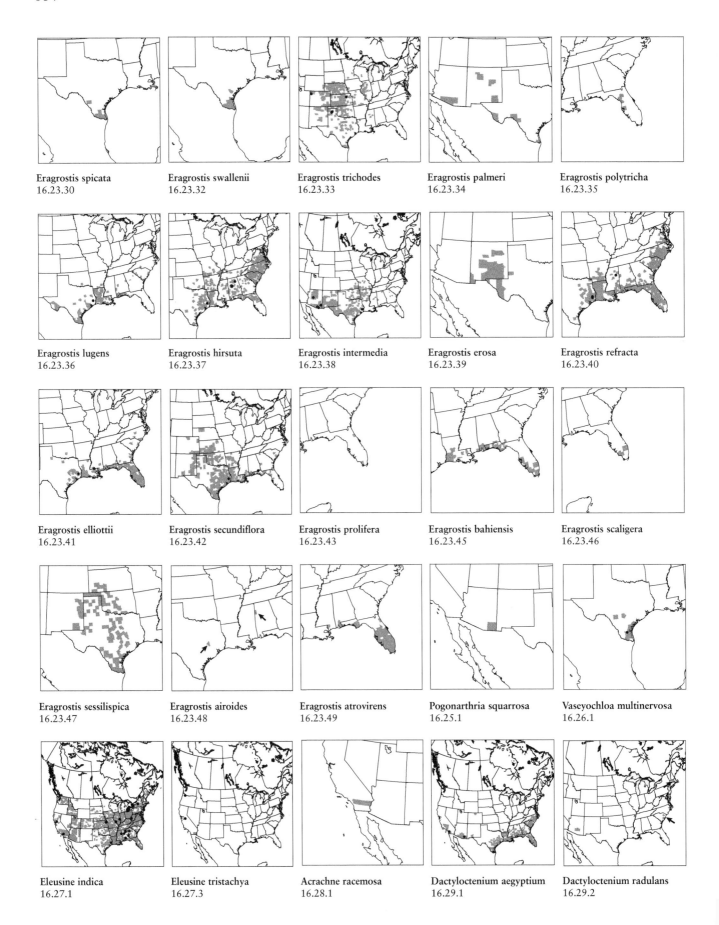

Eragrostis spicata
16.23.30

Eragrostis swallenii
16.23.32

Eragrostis trichodes
16.23.33

Eragrostis palmeri
16.23.34

Eragrostis polytricha
16.23.35

Eragrostis lugens
16.23.36

Eragrostis hirsuta
16.23.37

Eragrostis intermedia
16.23.38

Eragrostis erosa
16.23.39

Eragrostis refracta
16.23.40

Eragrostis elliottii
16.23.41

Eragrostis secundiflora
16.23.42

Eragrostis prolifera
16.23.43

Eragrostis bahiensis
16.23.45

Eragrostis scaligera
16.23.46

Eragrostis sessilispica
16.23.47

Eragrostis airoides
16.23.48

Eragrostis atrovirens
16.23.49

Pogonarthria squarrosa
16.25.1

Vaseyochloa multinervosa
16.26.1

Eleusine indica
16.27.1

Eleusine tristachya
16.27.3

Acrachne racemosa
16.28.1

Dactyloctenium aegyptium
16.29.1

Dactyloctenium radulans
16.29.2

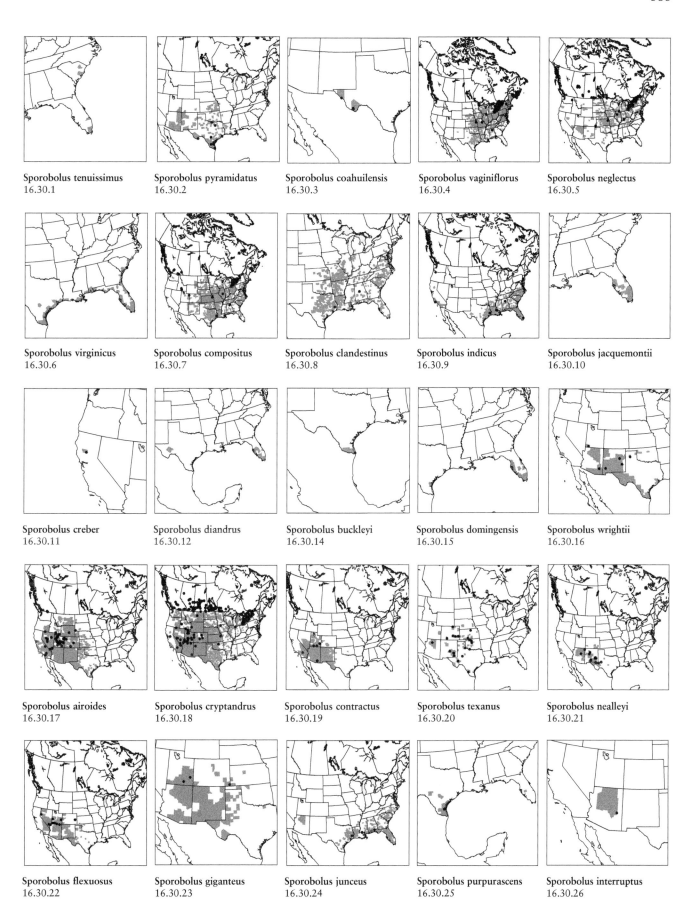

Sporobolus tenuissimus
16.30.1

Sporobolus pyramidatus
16.30.2

Sporobolus coahuilensis
16.30.3

Sporobolus vaginiflorus
16.30.4

Sporobolus neglectus
16.30.5

Sporobolus virginicus
16.30.6

Sporobolus compositus
16.30.7

Sporobolus clandestinus
16.30.8

Sporobolus indicus
16.30.9

Sporobolus jacquemontii
16.30.10

Sporobolus creber
16.30.11

Sporobolus diandrus
16.30.12

Sporobolus buckleyi
16.30.14

Sporobolus domingensis
16.30.15

Sporobolus wrightii
16.30.16

Sporobolus airoides
16.30.17

Sporobolus cryptandrus
16.30.18

Sporobolus contractus
16.30.19

Sporobolus texanus
16.30.20

Sporobolus nealleyi
16.30.21

Sporobolus flexuosus
16.30.22

Sporobolus giganteus
16.30.23

Sporobolus junceus
16.30.24

Sporobolus purpurascens
16.30.25

Sporobolus interruptus
16.30.26

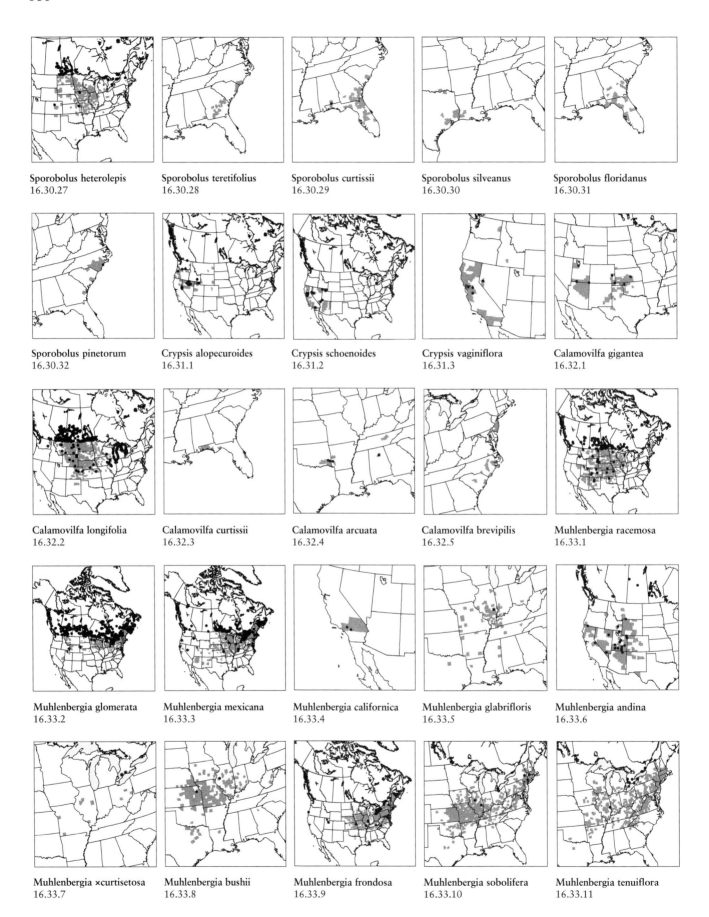

Sporobolus heterolepis
16.30.27

Sporobolus teretifolius
16.30.28

Sporobolus curtissii
16.30.29

Sporobolus silveanus
16.30.30

Sporobolus floridanus
16.30.31

Sporobolus pinetorum
16.30.32

Crypsis alopecuroides
16.31.1

Crypsis schoenoides
16.31.2

Crypsis vaginiflora
16.31.3

Calamovilfa gigantea
16.32.1

Calamovilfa longifolia
16.32.2

Calamovilfa curtissii
16.32.3

Calamovilfa arcuata
16.32.4

Calamovilfa brevipilis
16.32.5

Muhlenbergia racemosa
16.33.1

Muhlenbergia glomerata
16.33.2

Muhlenbergia mexicana
16.33.3

Muhlenbergia californica
16.33.4

Muhlenbergia glabrifloris
16.33.5

Muhlenbergia andina
16.33.6

Muhlenbergia ×curtisetosa
16.33.7

Muhlenbergia bushii
16.33.8

Muhlenbergia frondosa
16.33.9

Muhlenbergia sobolifera
16.33.10

Muhlenbergia tenuiflora
16.33.11

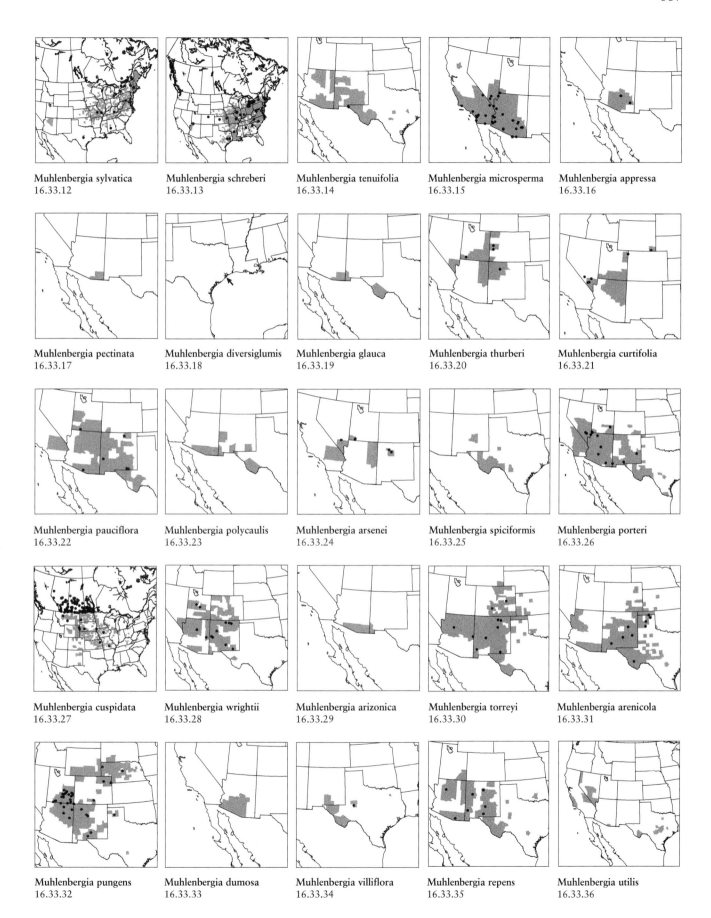

Muhlenbergia sylvatica
16.33.12

Muhlenbergia schreberi
16.33.13

Muhlenbergia tenuifolia
16.33.14

Muhlenbergia microsperma
16.33.15

Muhlenbergia appressa
16.33.16

Muhlenbergia pectinata
16.33.17

Muhlenbergia diversiglumis
16.33.18

Muhlenbergia glauca
16.33.19

Muhlenbergia thurberi
16.33.20

Muhlenbergia curtifolia
16.33.21

Muhlenbergia pauciflora
16.33.22

Muhlenbergia polycaulis
16.33.23

Muhlenbergia arsenei
16.33.24

Muhlenbergia spiciformis
16.33.25

Muhlenbergia porteri
16.33.26

Muhlenbergia cuspidata
16.33.27

Muhlenbergia wrightii
16.33.28

Muhlenbergia arizonica
16.33.29

Muhlenbergia torreyi
16.33.30

Muhlenbergia arenicola
16.33.31

Muhlenbergia pungens
16.33.32

Muhlenbergia dumosa
16.33.33

Muhlenbergia villiflora
16.33.34

Muhlenbergia repens
16.33.35

Muhlenbergia utilis
16.33.36

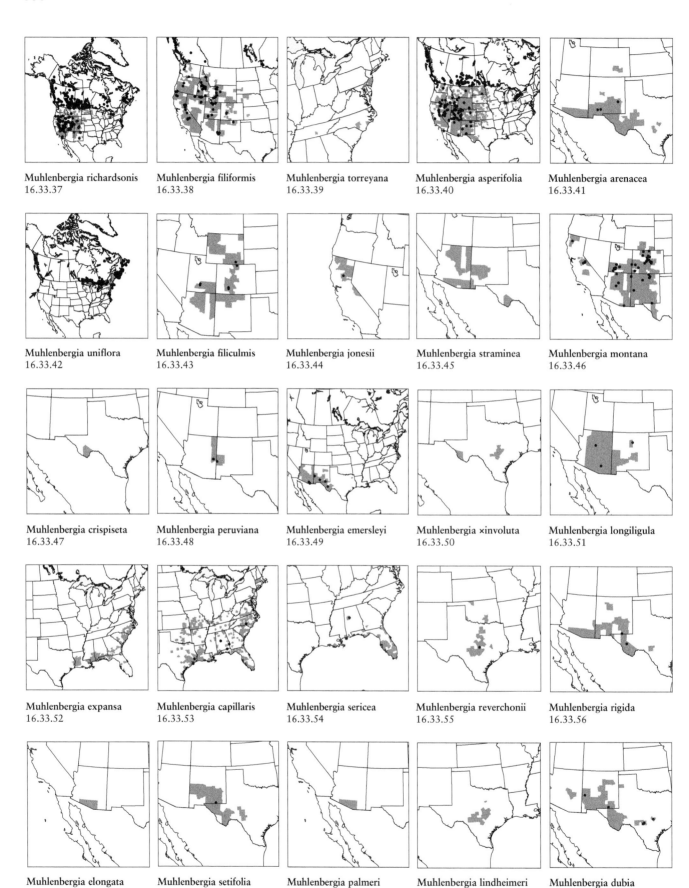

Muhlenbergia richardsonis
16.33.37

Muhlenbergia filiformis
16.33.38

Muhlenbergia torreyana
16.33.39

Muhlenbergia asperifolia
16.33.40

Muhlenbergia arenacea
16.33.41

Muhlenbergia uniflora
16.33.42

Muhlenbergia filiculmis
16.33.43

Muhlenbergia jonesii
16.33.44

Muhlenbergia straminea
16.33.45

Muhlenbergia montana
16.33.46

Muhlenbergia crispiseta
16.33.47

Muhlenbergia peruviana
16.33.48

Muhlenbergia emersleyi
16.33.49

Muhlenbergia ×involuta
16.33.50

Muhlenbergia longiligula
16.33.51

Muhlenbergia expansa
16.33.52

Muhlenbergia capillaris
16.33.53

Muhlenbergia sericea
16.33.54

Muhlenbergia reverchonii
16.33.55

Muhlenbergia rigida
16.33.56

Muhlenbergia elongata
16.33.57

Muhlenbergia setifolia
16.33.58

Muhlenbergia palmeri
16.33.59

Muhlenbergia lindheimeri
16.33.60

Muhlenbergia dubia
16.33.61

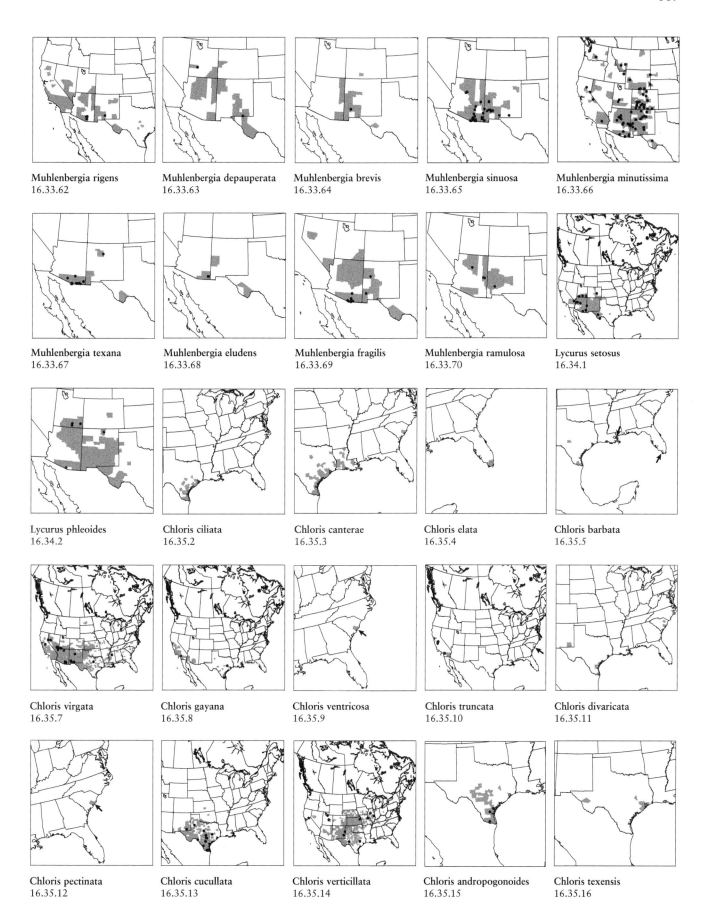

539

Muhlenbergia rigens
16.33.62

Muhlenbergia depauperata
16.33.63

Muhlenbergia brevis
16.33.64

Muhlenbergia sinuosa
16.33.65

Muhlenbergia minutissima
16.33.66

Muhlenbergia texana
16.33.67

Muhlenbergia eludens
16.33.68

Muhlenbergia fragilis
16.33.69

Muhlenbergia ramulosa
16.33.70

Lycurus setosus
16.34.1

Lycurus phleoides
16.34.2

Chloris ciliata
16.35.2

Chloris canterae
16.35.3

Chloris elata
16.35.4

Chloris barbata
16.35.5

Chloris virgata
16.35.7

Chloris gayana
16.35.8

Chloris ventricosa
16.35.9

Chloris truncata
16.35.10

Chloris divaricata
16.35.11

Chloris pectinata
16.35.12

Chloris cucullata
16.35.13

Chloris verticillata
16.35.14

Chloris andropogonoides
16.35.15

Chloris texensis
16.35.16

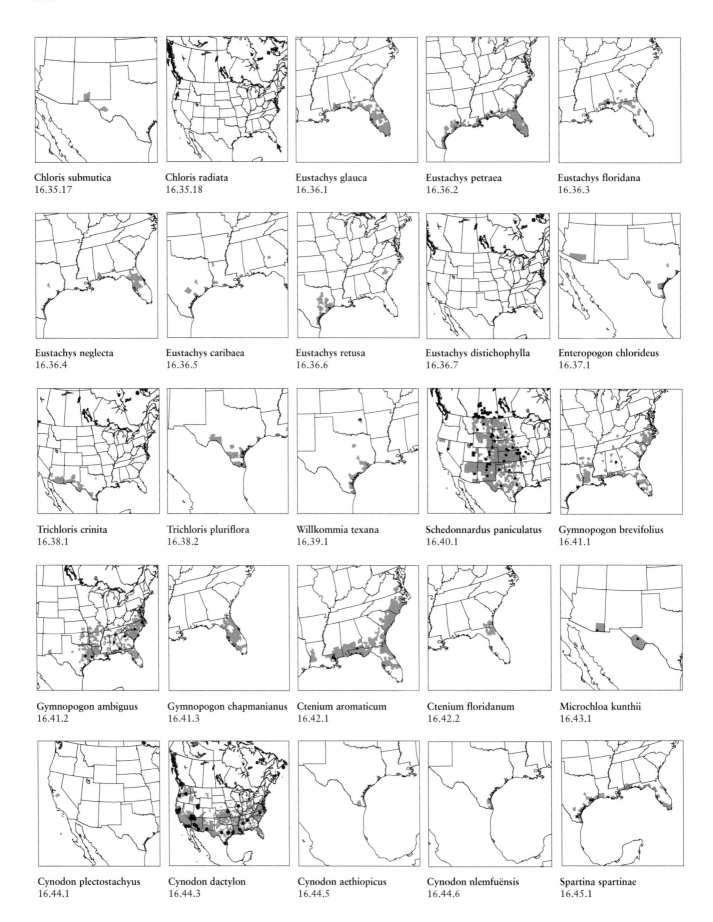

Chloris submutica
16.35.17

Chloris radiata
16.35.18

Eustachys glauca
16.36.1

Eustachys petraea
16.36.2

Eustachys floridana
16.36.3

Eustachys neglecta
16.36.4

Eustachys caribaea
16.36.5

Eustachys retusa
16.36.6

Eustachys distichophylla
16.36.7

Enteropogon chlorideus
16.37.1

Trichloris crinita
16.38.1

Trichloris pluriflora
16.38.2

Willkommia texana
16.39.1

Schedonnardus paniculatus
16.40.1

Gymnopogon brevifolius
16.41.1

Gymnopogon ambiguus
16.41.2

Gymnopogon chapmanianus
16.41.3

Ctenium aromaticum
16.42.1

Ctenium floridanum
16.42.2

Microchloa kunthii
16.43.1

Cynodon plectostachyus
16.44.1

Cynodon dactylon
16.44.3

Cynodon aethiopicus
16.44.5

Cynodon nlemfuënsis
16.44.6

Spartina spartinae
16.45.1

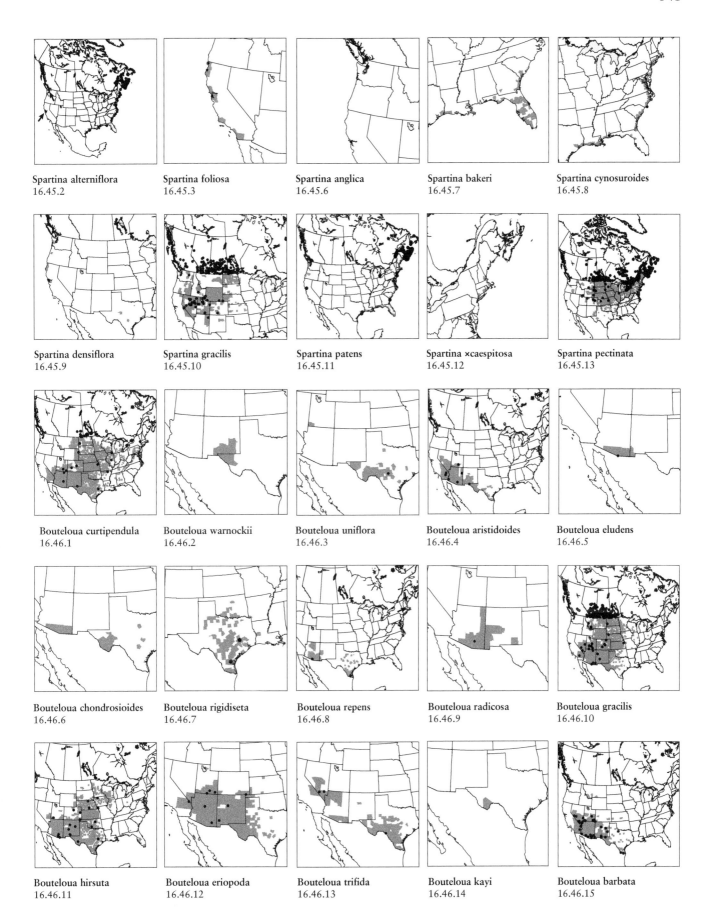

Spartina alterniflora
16.45.2

Spartina foliosa
16.45.3

Spartina anglica
16.45.6

Spartina bakeri
16.45.7

Spartina cynosuroides
16.45.8

Spartina densiflora
16.45.9

Spartina gracilis
16.45.10

Spartina patens
16.45.11

Spartina ×caespitosa
16.45.12

Spartina pectinata
16.45.13

Bouteloua curtipendula
16.46.1

Bouteloua warnockii
16.46.2

Bouteloua uniflora
16.46.3

Bouteloua aristidoides
16.46.4

Bouteloua eludens
16.46.5

Bouteloua chondrosioides
16.46.6

Bouteloua rigidiseta
16.46.7

Bouteloua repens
16.46.8

Bouteloua radicosa
16.46.9

Bouteloua gracilis
16.46.10

Bouteloua hirsuta
16.46.11

Bouteloua eriopoda
16.46.12

Bouteloua trifida
16.46.13

Bouteloua kayi
16.46.14

Bouteloua barbata
16.46.15

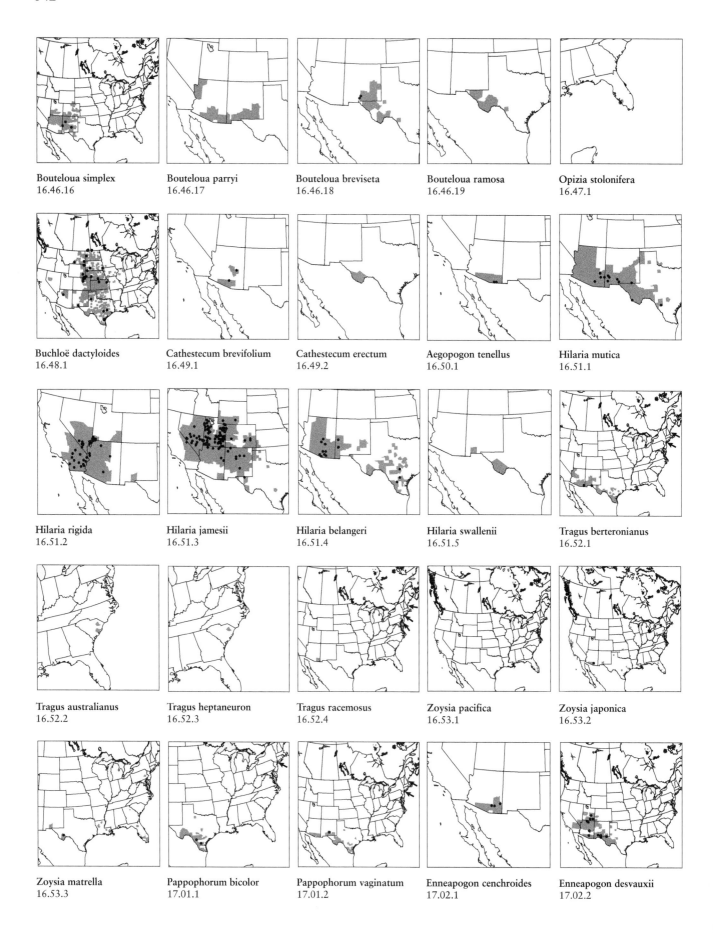

Bouteloua simplex
16.46.16

Bouteloua parryi
16.46.17

Bouteloua breviseta
16.46.18

Bouteloua ramosa
16.46.19

Opizia stolonifera
16.47.1

Buchloë dactyloides
16.48.1

Cathestecum brevifolium
16.49.1

Cathestecum erectum
16.49.2

Aegopogon tenellus
16.50.1

Hilaria mutica
16.51.1

Hilaria rigida
16.51.2

Hilaria jamesii
16.51.3

Hilaria belangeri
16.51.4

Hilaria swallenii
16.51.5

Tragus berteronianus
16.52.1

Tragus australianus
16.52.2

Tragus heptaneuron
16.52.3

Tragus racemosus
16.52.4

Zoysia pacifica
16.53.1

Zoysia japonica
16.53.2

Zoysia matrella
16.53.3

Pappophorum bicolor
17.01.1

Pappophorum vaginatum
17.01.2

Enneapogon cenchroides
17.02.1

Enneapogon desvauxii
17.02.2

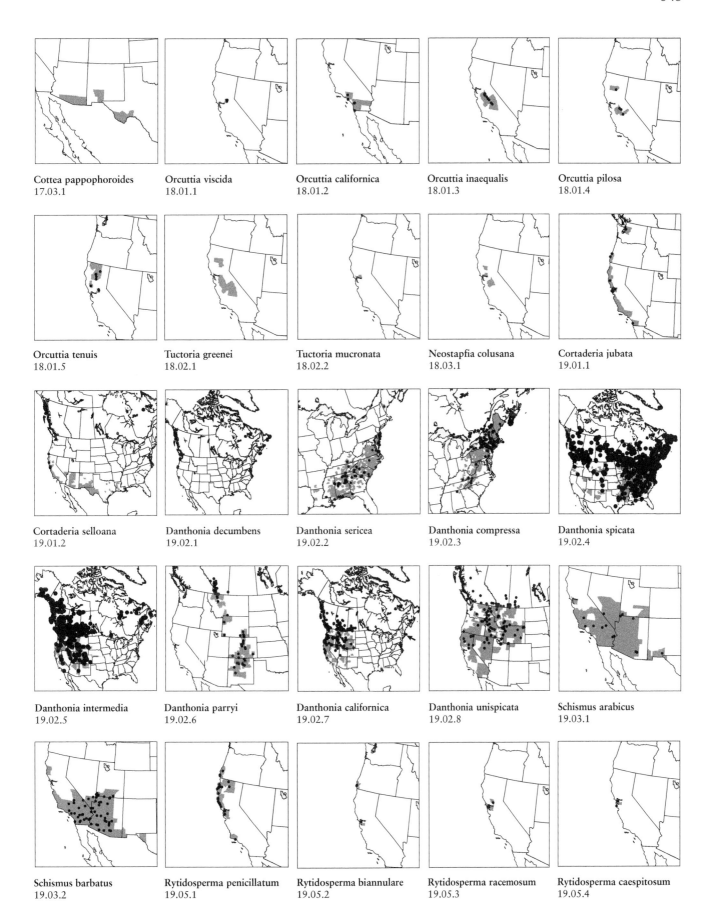

Cottea pappophoroides
17.03.1

Orcuttia viscida
18.01.1

Orcuttia californica
18.01.2

Orcuttia inaequalis
18.01.3

Orcuttia pilosa
18.01.4

Orcuttia tenuis
18.01.5

Tuctoria greenei
18.02.1

Tuctoria mucronata
18.02.2

Neostapfia colusana
18.03.1

Cortaderia jubata
19.01.1

Cortaderia selloana
19.01.2

Danthonia decumbens
19.02.1

Danthonia sericea
19.02.2

Danthonia compressa
19.02.3

Danthonia spicata
19.02.4

Danthonia intermedia
19.02.5

Danthonia parryi
19.02.6

Danthonia californica
19.02.7

Danthonia unispicata
19.02.8

Schismus arabicus
19.03.1

Schismus barbatus
19.03.2

Rytidosperma penicillatum
19.05.1

Rytidosperma biannulare
19.05.2

Rytidosperma racemosum
19.05.3

Rytidosperma caespitosum
19.05.4

544

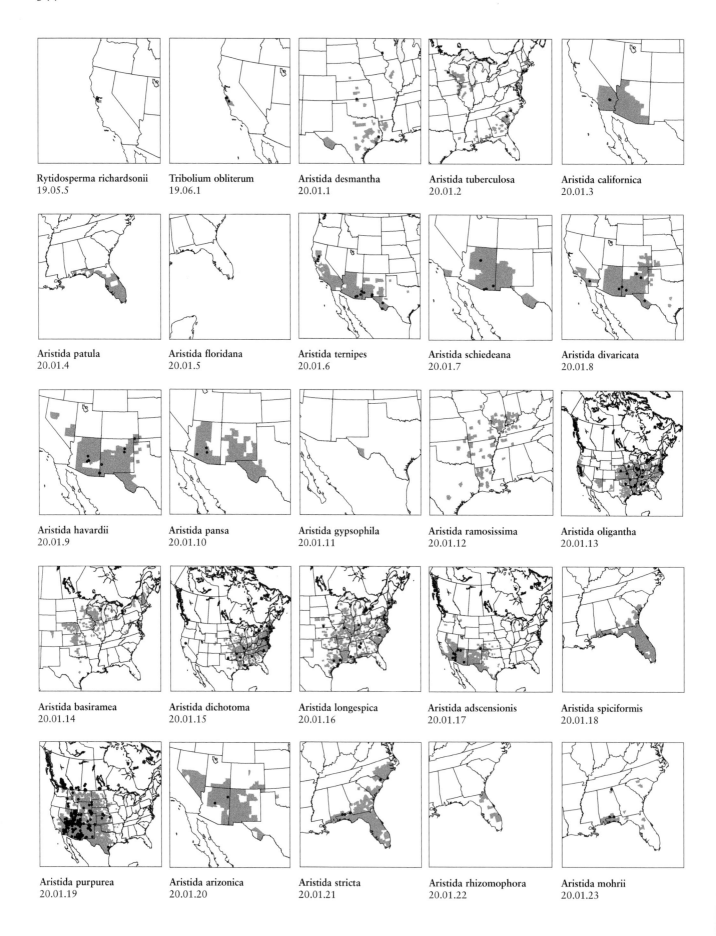

Rytidosperma richardsonii
19.05.5

Tribolium obliterum
19.06.1

Aristida desmantha
20.01.1

Aristida tuberculosa
20.01.2

Aristida californica
20.01.3

Aristida patula
20.01.4

Aristida floridana
20.01.5

Aristida ternipes
20.01.6

Aristida schiedeana
20.01.7

Aristida divaricata
20.01.8

Aristida havardii
20.01.9

Aristida pansa
20.01.10

Aristida gypsophila
20.01.11

Aristida ramosissima
20.01.12

Aristida oligantha
20.01.13

Aristida basiramea
20.01.14

Aristida dichotoma
20.01.15

Aristida longespica
20.01.16

Aristida adscensionis
20.01.17

Aristida spiciformis
20.01.18

Aristida purpurea
20.01.19

Aristida arizonica
20.01.20

Aristida stricta
20.01.21

Aristida rhizomophora
20.01.22

Aristida mohrii
20.01.23

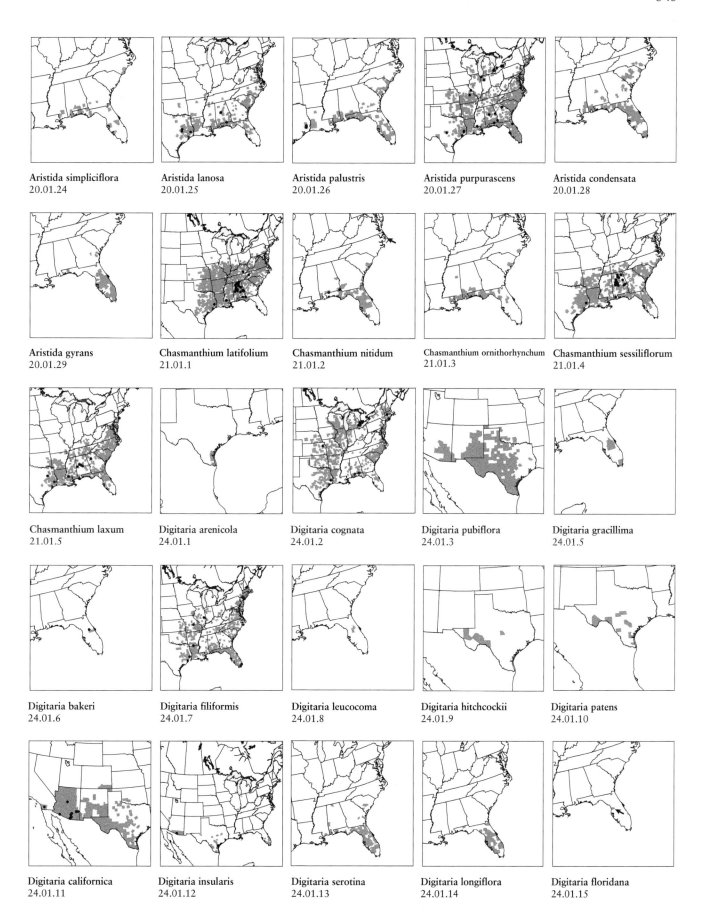

Aristida simpliciflora
20.01.24

Aristida lanosa
20.01.25

Aristida palustris
20.01.26

Aristida purpurascens
20.01.27

Aristida condensata
20.01.28

Aristida gyrans
20.01.29

Chasmanthium latifolium
21.01.1

Chasmanthium nitidum
21.01.2

Chasmanthium ornithorhynchum
21.01.3

Chasmanthium sessiliflorum
21.01.4

Chasmanthium laxum
21.01.5

Digitaria arenicola
24.01.1

Digitaria cognata
24.01.2

Digitaria pubiflora
24.01.3

Digitaria gracillima
24.01.5

Digitaria bakeri
24.01.6

Digitaria filiformis
24.01.7

Digitaria leucocoma
24.01.8

Digitaria hitchcockii
24.01.9

Digitaria patens
24.01.10

Digitaria californica
24.01.11

Digitaria insularis
24.01.12

Digitaria serotina
24.01.13

Digitaria longiflora
24.01.14

Digitaria floridana
24.01.15

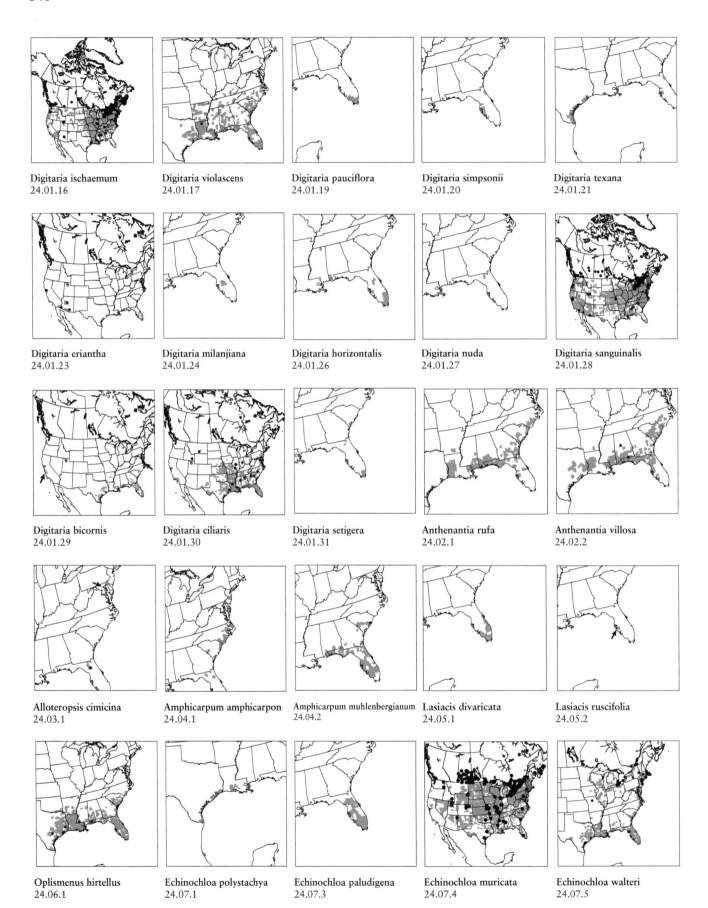

Digitaria ischaemum
24.01.16

Digitaria violascens
24.01.17

Digitaria pauciflora
24.01.19

Digitaria simpsonii
24.01.20

Digitaria texana
24.01.21

Digitaria eriantha
24.01.23

Digitaria milanjiana
24.01.24

Digitaria horizontalis
24.01.26

Digitaria nuda
24.01.27

Digitaria sanguinalis
24.01.28

Digitaria bicornis
24.01.29

Digitaria ciliaris
24.01.30

Digitaria setigera
24.01.31

Anthenantia rufa
24.02.1

Anthenantia villosa
24.02.2

Alloteropsis cimicina
24.03.1

Amphicarpum amphicarpon
24.04.1

Amphicarpum muhlenbergianum
24.04.2

Lasiacis divaricata
24.05.1

Lasiacis ruscifolia
24.05.2

Oplismenus hirtellus
24.06.1

Echinochloa polystachya
24.07.1

Echinochloa paludigena
24.07.3

Echinochloa muricata
24.07.4

Echinochloa walteri
24.07.5

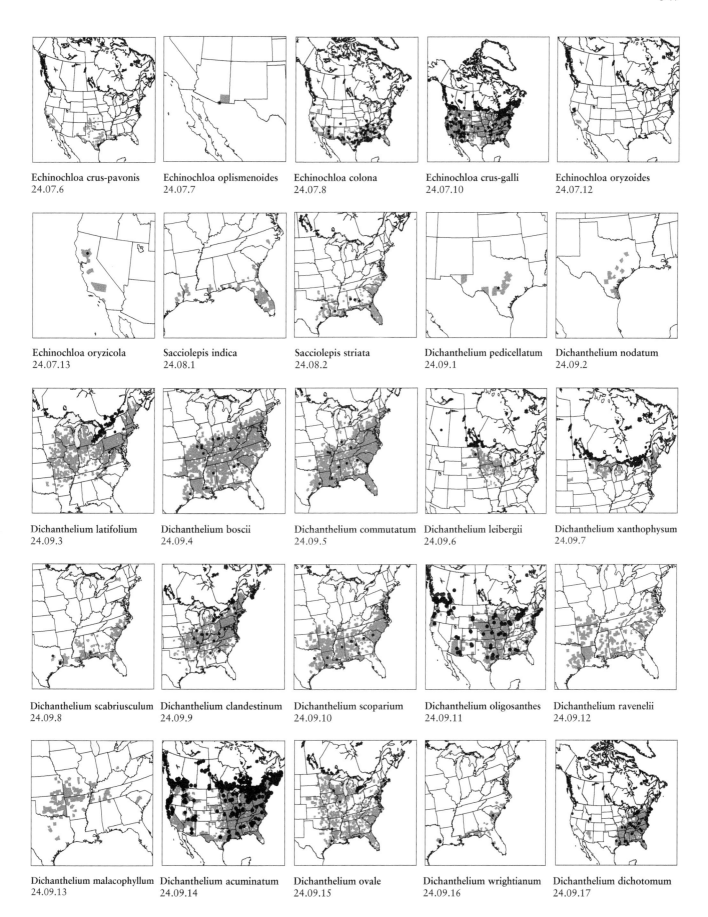

Echinochloa crus-pavonis
24.07.6

Echinochloa oplismenoides
24.07.7

Echinochloa colona
24.07.8

Echinochloa crus-galli
24.07.10

Echinochloa oryzoides
24.07.12

Echinochloa oryzicola
24.07.13

Sacciolepis indica
24.08.1

Sacciolepis striata
24.08.2

Dichanthelium pedicellatum
24.09.1

Dichanthelium nodatum
24.09.2

Dichanthelium latifolium
24.09.3

Dichanthelium boscii
24.09.4

Dichanthelium commutatum
24.09.5

Dichanthelium leibergii
24.09.6

Dichanthelium xanthophysum
24.09.7

Dichanthelium scabriusculum
24.09.8

Dichanthelium clandestinum
24.09.9

Dichanthelium scoparium
24.09.10

Dichanthelium oligosanthes
24.09.11

Dichanthelium ravenelii
24.09.12

Dichanthelium malacophyllum
24.09.13

Dichanthelium acuminatum
24.09.14

Dichanthelium ovale
24.09.15

Dichanthelium wrightianum
24.09.16

Dichanthelium dichotomum
24.09.17

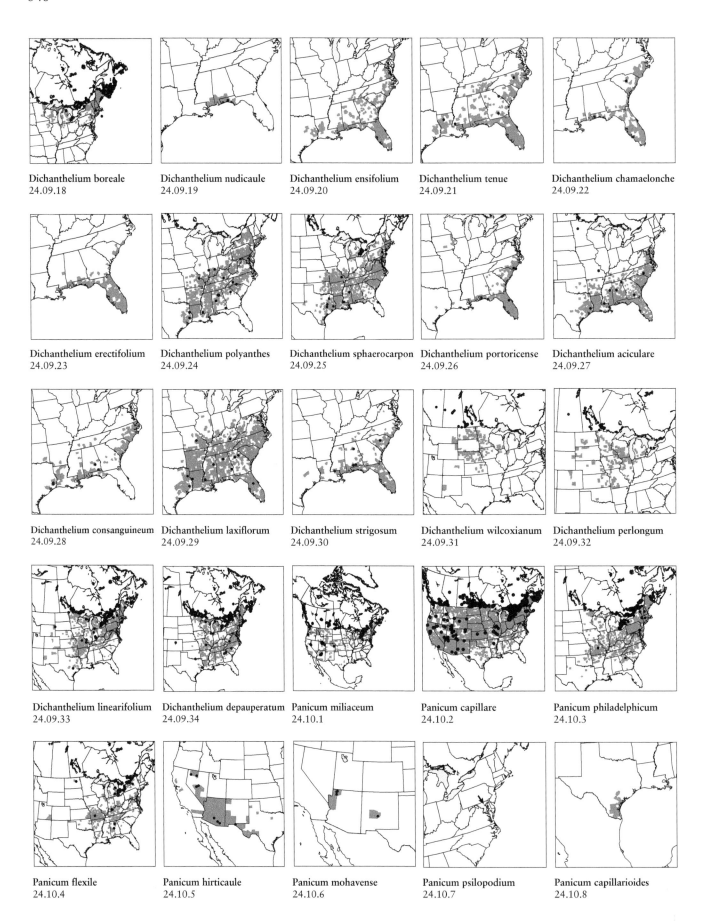

Dichanthelium boreale
24.09.18

Dichanthelium nudicaule
24.09.19

Dichanthelium ensifolium
24.09.20

Dichanthelium tenue
24.09.21

Dichanthelium chamaelonche
24.09.22

Dichanthelium erectifolium
24.09.23

Dichanthelium polyanthes
24.09.24

Dichanthelium sphaerocarpon
24.09.25

Dichanthelium portoricense
24.09.26

Dichanthelium aciculare
24.09.27

Dichanthelium consanguineum
24.09.28

Dichanthelium laxiflorum
24.09.29

Dichanthelium strigosum
24.09.30

Dichanthelium wilcoxianum
24.09.31

Dichanthelium perlongum
24.09.32

Dichanthelium linearifolium
24.09.33

Dichanthelium depauperatum
24.09.34

Panicum miliaceum
24.10.1

Panicum capillare
24.10.2

Panicum philadelphicum
24.10.3

Panicum flexile
24.10.4

Panicum hirticaule
24.10.5

Panicum mohavense
24.10.6

Panicum psilopodium
24.10.7

Panicum capillarioides
24.10.8

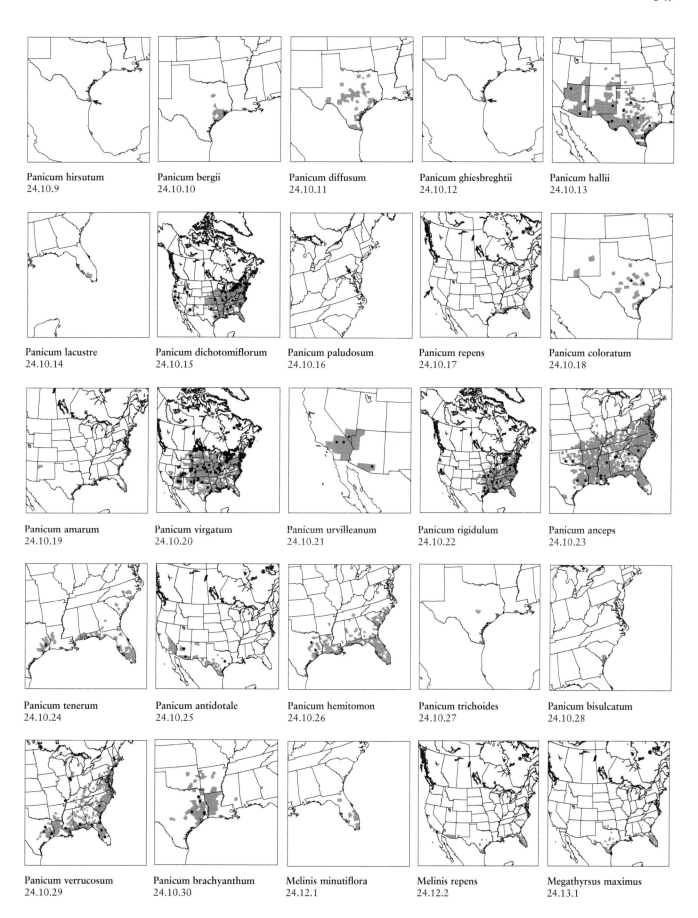

Panicum hirsutum
24.10.9

Panicum bergii
24.10.10

Panicum diffusum
24.10.11

Panicum ghiesbreghtii
24.10.12

Panicum hallii
24.10.13

Panicum lacustre
24.10.14

Panicum dichotomiflorum
24.10.15

Panicum paludosum
24.10.16

Panicum repens
24.10.17

Panicum coloratum
24.10.18

Panicum amarum
24.10.19

Panicum virgatum
24.10.20

Panicum urvilleanum
24.10.21

Panicum rigidulum
24.10.22

Panicum anceps
24.10.23

Panicum tenerum
24.10.24

Panicum antidotale
24.10.25

Panicum hemitomon
24.10.26

Panicum trichoides
24.10.27

Panicum bisulcatum
24.10.28

Panicum verrucosum
24.10.29

Panicum brachyanthum
24.10.30

Melinis minutiflora
24.12.1

Melinis repens
24.12.2

Megathyrsus maximus
24.13.1

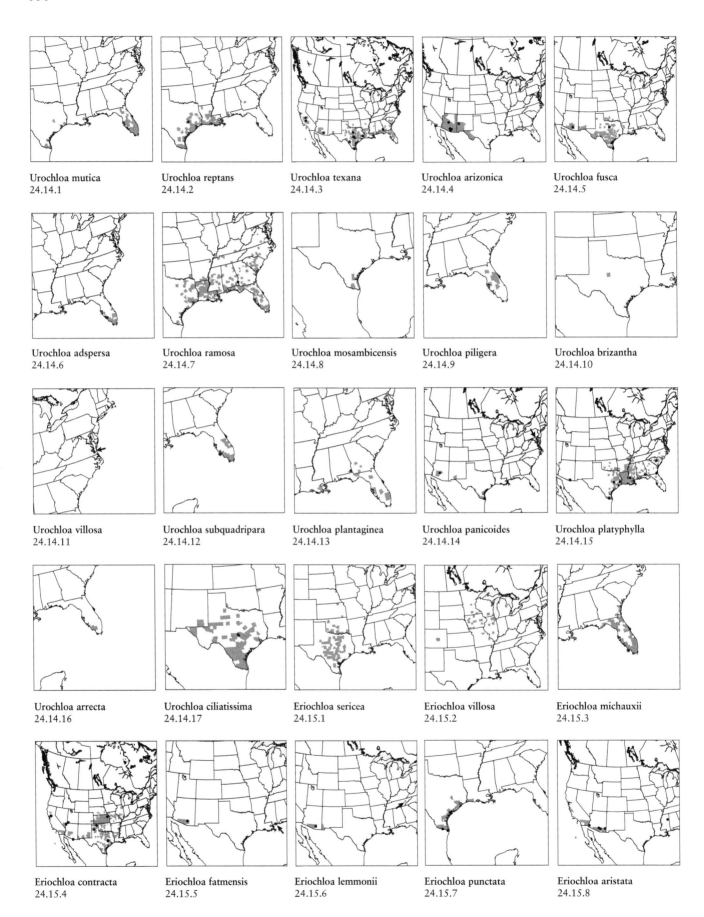

Urochloa mutica
24.14.1

Urochloa reptans
24.14.2

Urochloa texana
24.14.3

Urochloa arizonica
24.14.4

Urochloa fusca
24.14.5

Urochloa adspersa
24.14.6

Urochloa ramosa
24.14.7

Urochloa mosambicensis
24.14.8

Urochloa piligera
24.14.9

Urochloa brizantha
24.14.10

Urochloa villosa
24.14.11

Urochloa subquadripara
24.14.12

Urochloa plantaginea
24.14.13

Urochloa panicoides
24.14.14

Urochloa platyphylla
24.14.15

Urochloa arrecta
24.14.16

Urochloa ciliatissima
24.14.17

Eriochloa sericea
24.15.1

Eriochloa villosa
24.15.2

Eriochloa michauxii
24.15.3

Eriochloa contracta
24.15.4

Eriochloa fatmensis
24.15.5

Eriochloa lemmonii
24.15.6

Eriochloa punctata
24.15.7

Eriochloa aristata
24.15.8

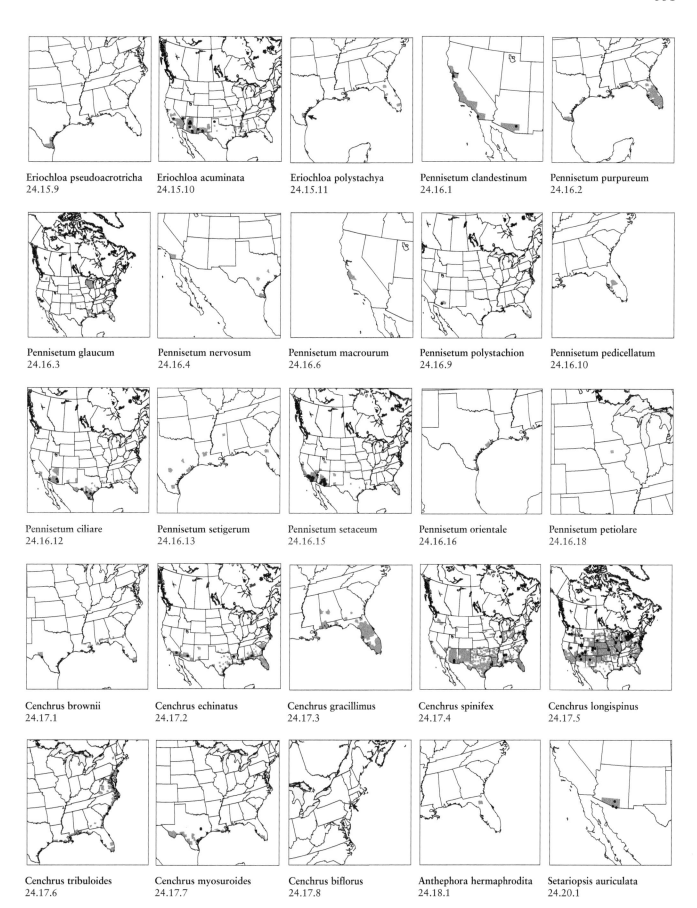

Eriochloa pseudoacrotricha
24.15.9

Eriochloa acuminata
24.15.10

Eriochloa polystachya
24.15.11

Pennisetum clandestinum
24.16.1

Pennisetum purpureum
24.16.2

Pennisetum glaucum
24.16.3

Pennisetum nervosum
24.16.4

Pennisetum macrourum
24.16.6

Pennisetum polystachion
24.16.9

Pennisetum pedicellatum
24.16.10

Pennisetum ciliare
24.16.12

Pennisetum setigerum
24.16.13

Pennisetum setaceum
24.16.15

Pennisetum orientale
24.16.16

Pennisetum petiolare
24.16.18

Cenchrus brownii
24.17.1

Cenchrus echinatus
24.17.2

Cenchrus gracillimus
24.17.3

Cenchrus spinifex
24.17.4

Cenchrus longispinus
24.17.5

Cenchrus tribuloides
24.17.6

Cenchrus myosuroides
24.17.7

Cenchrus biflorus
24.17.8

Anthephora hermaphrodita
24.18.1

Setariopsis auriculata
24.20.1

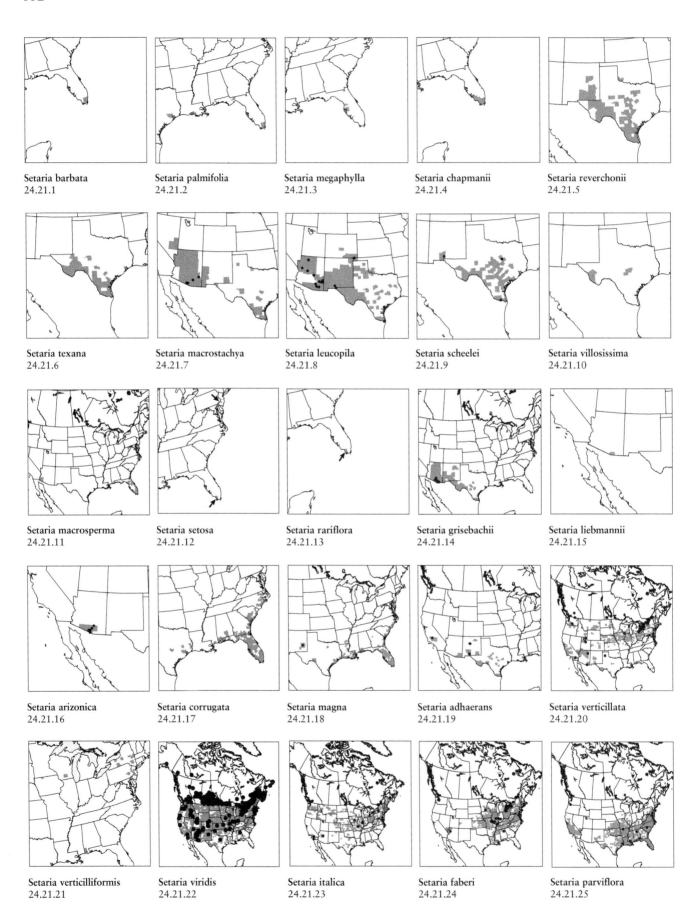

Setaria barbata
24.21.1

Setaria palmifolia
24.21.2

Setaria megaphylla
24.21.3

Setaria chapmanii
24.21.4

Setaria reverchonii
24.21.5

Setaria texana
24.21.6

Setaria macrostachya
24.21.7

Setaria leucopila
24.21.8

Setaria scheelei
24.21.9

Setaria villosissima
24.21.10

Setaria macrosperma
24.21.11

Setaria setosa
24.21.12

Setaria rariflora
24.21.13

Setaria grisebachii
24.21.14

Setaria liebmannii
24.21.15

Setaria arizonica
24.21.16

Setaria corrugata
24.21.17

Setaria magna
24.21.18

Setaria adhaerans
24.21.19

Setaria verticillata
24.21.20

Setaria verticilliformis
24.21.21

Setaria viridis
24.21.22

Setaria italica
24.21.23

Setaria faberi
24.21.24

Setaria parviflora
24.21.25

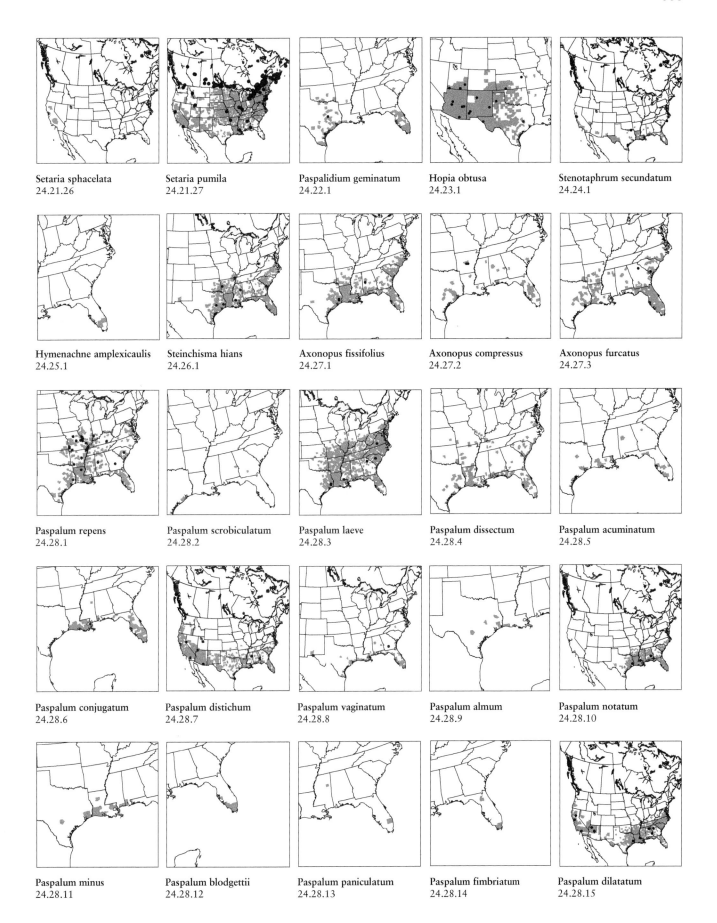

Setaria sphacelata
24.21.26

Setaria pumila
24.21.27

Paspalidium geminatum
24.22.1

Hopia obtusa
24.23.1

Stenotaphrum secundatum
24.24.1

Hymenachne amplexicaulis
24.25.1

Steinchisma hians
24.26.1

Axonopus fissifolius
24.27.1

Axonopus compressus
24.27.2

Axonopus furcatus
24.27.3

Paspalum repens
24.28.1

Paspalum scrobiculatum
24.28.2

Paspalum laeve
24.28.3

Paspalum dissectum
24.28.4

Paspalum acuminatum
24.28.5

Paspalum conjugatum
24.28.6

Paspalum distichum
24.28.7

Paspalum vaginatum
24.28.8

Paspalum almum
24.28.9

Paspalum notatum
24.28.10

Paspalum minus
24.28.11

Paspalum blodgettii
24.28.12

Paspalum paniculatum
24.28.13

Paspalum fimbriatum
24.28.14

Paspalum dilatatum
24.28.15

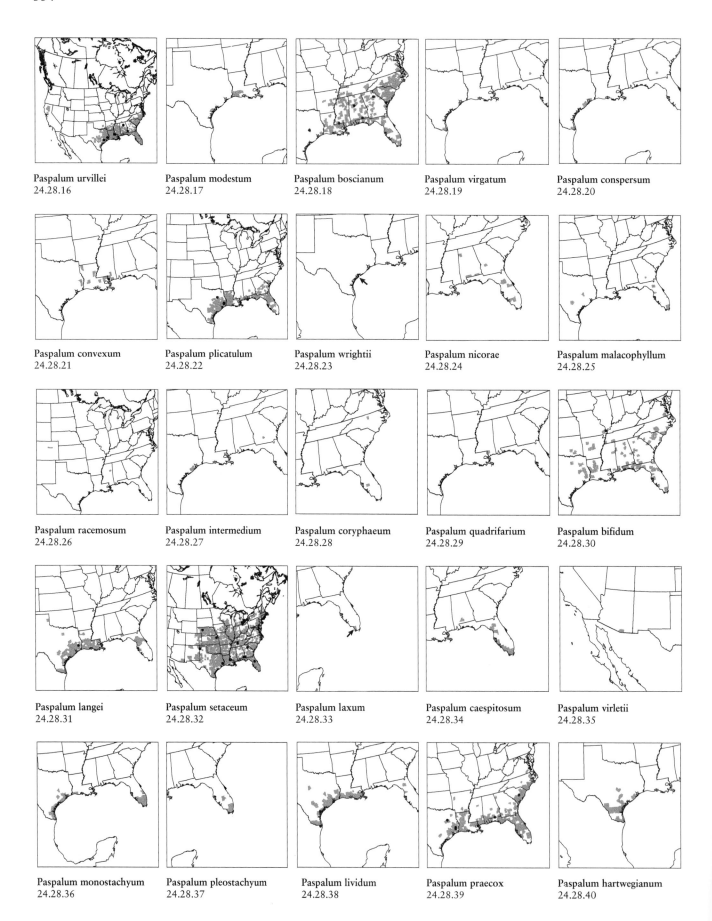

Paspalum urvillei
24.28.16

Paspalum modestum
24.28.17

Paspalum boscianum
24.28.18

Paspalum virgatum
24.28.19

Paspalum conspersum
24.28.20

Paspalum convexum
24.28.21

Paspalum plicatulum
24.28.22

Paspalum wrightii
24.28.23

Paspalum nicorae
24.28.24

Paspalum malacophyllum
24.28.25

Paspalum racemosum
24.28.26

Paspalum intermedium
24.28.27

Paspalum coryphaeum
24.28.28

Paspalum quadrifarium
24.28.29

Paspalum bifidum
24.28.30

Paspalum langei
24.28.31

Paspalum setaceum
24.28.32

Paspalum laxum
24.28.33

Paspalum caespitosum
24.28.34

Paspalum virletii
24.28.35

Paspalum monostachyum
24.28.36

Paspalum pleostachyum
24.28.37

Paspalum lividum
24.28.38

Paspalum praecox
24.28.39

Paspalum hartwegianum
24.28.40

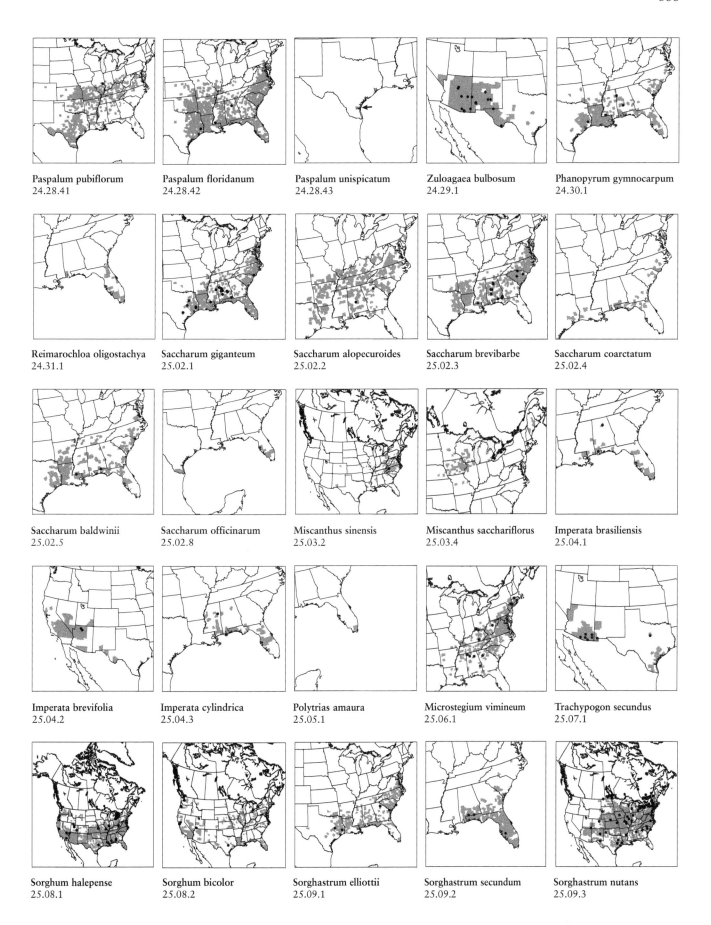

Paspalum pubiflorum
24.28.41

Paspalum floridanum
24.28.42

Paspalum unispicatum
24.28.43

Zuloagaea bulbosum
24.29.1

Phanopyrum gymnocarpum
24.30.1

Reimarochloa oligostachya
24.31.1

Saccharum giganteum
25.02.1

Saccharum alopecuroides
25.02.2

Saccharum brevibarbe
25.02.3

Saccharum coarctatum
25.02.4

Saccharum baldwinii
25.02.5

Saccharum officinarum
25.02.8

Miscanthus sinensis
25.03.2

Miscanthus sacchariflorus
25.03.4

Imperata brasiliensis
25.04.1

Imperata brevifolia
25.04.2

Imperata cylindrica
25.04.3

Polytrias amaura
25.05.1

Microstegium vimineum
25.06.1

Trachypogon secundus
25.07.1

Sorghum halepense
25.08.1

Sorghum bicolor
25.08.2

Sorghastrum elliottii
25.09.1

Sorghastrum secundum
25.09.2

Sorghastrum nutans
25.09.3

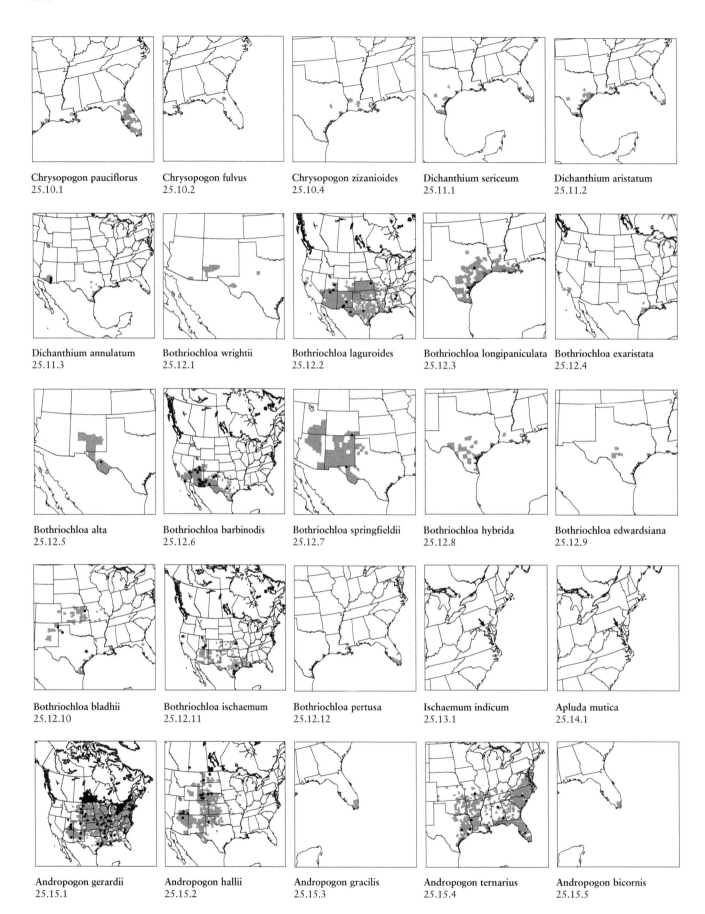

Chrysopogon pauciflorus
25.10.1

Chrysopogon fulvus
25.10.2

Chrysopogon zizanioides
25.10.4

Dichanthium sericeum
25.11.1

Dichanthium aristatum
25.11.2

Dichanthium annulatum
25.11.3

Bothriochloa wrightii
25.12.1

Bothriochloa laguroides
25.12.2

Bothriochloa longipaniculata
25.12.3

Bothriochloa exaristata
25.12.4

Bothriochloa alta
25.12.5

Bothriochloa barbinodis
25.12.6

Bothriochloa springfieldii
25.12.7

Bothriochloa hybrida
25.12.8

Bothriochloa edwardsiana
25.12.9

Bothriochloa bladhii
25.12.10

Bothriochloa ischaemum
25.12.11

Bothriochloa pertusa
25.12.12

Ischaemum indicum
25.13.1

Apluda mutica
25.14.1

Andropogon gerardii
25.15.1

Andropogon hallii
25.15.2

Andropogon gracilis
25.15.3

Andropogon ternarius
25.15.4

Andropogon bicornis
25.15.5

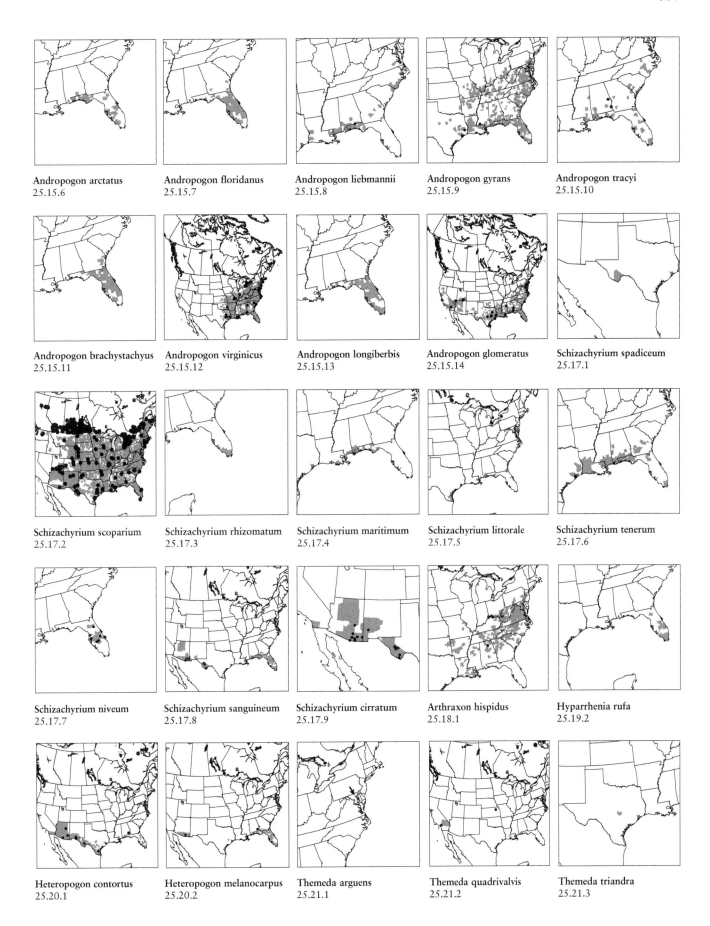

Andropogon arctatus
25.15.6

Andropogon floridanus
25.15.7

Andropogon liebmannii
25.15.8

Andropogon gyrans
25.15.9

Andropogon tracyi
25.15.10

Andropogon brachystachyus
25.15.11

Andropogon virginicus
25.15.12

Andropogon longiberbis
25.15.13

Andropogon glomeratus
25.15.14

Schizachyrium spadiceum
25.17.1

Schizachyrium scoparium
25.17.2

Schizachyrium rhizomatum
25.17.3

Schizachyrium maritimum
25.17.4

Schizachyrium littorale
25.17.5

Schizachyrium tenerum
25.17.6

Schizachyrium niveum
25.17.7

Schizachyrium sanguineum
25.17.8

Schizachyrium cirratum
25.17.9

Arthraxon hispidus
25.18.1

Hyparrhenia rufa
25.19.2

Heteropogon contortus
25.20.1

Heteropogon melanocarpus
25.20.2

Themeda arguens
25.21.1

Themeda quadrivalvis
25.21.2

Themeda triandra
25.21.3

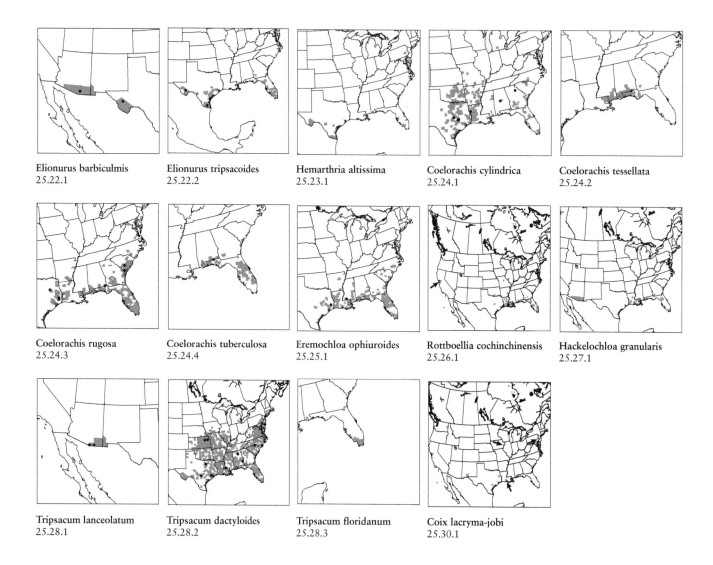

Elionurus barbiculmis
25.22.1

Elionurus tripsacoides
25.22.2

Hemarthria altissima
25.23.1

Coelorachis cylindrica
25.24.1

Coelorachis tessellata
25.24.2

Coelorachis rugosa
25.24.3

Coelorachis tuberculosa
25.24.4

Eremochloa ophiuroides
25.25.1

Rottboellia cochinchinensis
25.26.1

Hackelochloa granularis
25.27.1

Tripsacum lanceolatum
25.28.1

Tripsacum dactyloides
25.28.2

Tripsacum floridanum
25.28.3

Coix lacryma-jobi
25.30.1

*Index*

# Abbreviations

abx . . . . . . . . . . . . . abaxial
adx . . . . . . . . . . . . . adaxial
adxly . . . . . . . . . . adaxially
ann . . . . . . . . . . . . . . annual
anth . . . . . . . . . . . anthers
apc . . . . . . . . . . . . . apices
aur . . . . . . . . . . . . . auricles
ax . . . . . . . . . . . . . axillary
bas . . . . . . . . basal, basally
bisx . . . . . . . . . . . bisexual
bld . . . . . . . . . . . . . blades
bldless . . . . . . . . bladeless
br . . . . . . . . . . . . . branches
brchd . . . . . . . . . branched
brchg . . . . . . . . . branching
brlets . . . . . . . . branchlets
bu sc . . . . . . . . . bud scales
cal . . . . . . . . . . . . calluses
car . . . . . . . . . . . caryopses
cent . . . . . . . . . . . . central
ces . . . . . . . . . . . cespitose
clm . . . . . . . . . . . . . culms
clstgn . . . . . . . cleistogenes
col . . . . . . . . . . . . . collars
cmp . . . . . . . . complements
crwn . . . . . . . . . . . crowns
dis . . . . . . . . disarticulation
emb . . . . . . . . . . embryos
emgt . . . . . . . . . emarginate
epdm . . . . . . . . . epidermes
exvag . . . . . . . extravaginal
flt . . . . . . . . . . . . . . florets
fmb . . . . . . . . . . . fimbriae
fnctl . . . . . . . . . functional
fol . . . . . . . . . . . . . foliage
ftl . . . . . . . . . . . . . . fertile

glab . . . . . . . . . . . glabrous
glm . . . . . . . . . . . . glumes
hrb . . . . . . . . . herbaceous
infl . . . . . . . inflorescences
infvag . . . . . . . infravaginal
intnd . . . . . . . . . internodes
invag . . . . . . . . intravaginal
jnct . . . . . . . . . . . junction
lat . . . . . . . lateral, laterally
lf . . . . . . . . . . . . . . . leaf
lig . . . . . . . . . . . . . ligules
lm . . . . . . . . . . . . . lemmas
lo . . . . . . . . . . . . . lower
lod . . . . . . . . . . . lodicules
lvs . . . . . . . . . . . . leaves
memb . . . . . . . membranous
mid . . . . . . . . . . . middle
mrg . . . . . . . . . . margins
mrgl . . . . . . . . . marginal
nd . . . . . . . . . . . . . nodes
occ . . . . . . . . . occasionally
ov . . . . . . . . . . . . ovaries
pal . . . . . . . . . . . . paleas
pan . . . . panicles, paniculate
ped . . . . . . . . . . . pedicels
pedlt . . . . . . . . . pedicellate
per . . . . . . . . . . . perennial
pist . . . . . . . . . . . pistillate
pl . . . . . . . . . . . . . plants
plcsp . . . . . . . pluricespitose
pri . . . . . . . . . . . primary
prcp . . . . . . . . . . pericarp
prphl . . . . . . . . . prophylls
psdlig . . . . . . . pseudoligules
psdpet . . . . . psuedopetioles
psdspk . . . . . pseudospikelets

psinva . . pseudointravaginal
rchl . . . . . . . . . . rachillas
rchs . . . . . . . . . . rachises
rcm . . . . . racemes, racemose
rcmly . . . . . . . . racemosely
rdcd . . . . . . . . . reduced
rdg . . . . . . . . . . . . ridges
rdgd . . . . . . . . . . ridged
rdmt . . . . . . . . . rudiments,
                     rudimentary
rhz . . rhizomes, rhizomatous
sd . . . . . . . . . . . seeds
sdlg . . . . . . . . . . . seedling
sec . . . . . . . . . . secondary
sht . . . . . . . . . . . shoots
shth . . . . . . . . . . sheaths
smt . . . . . . . . . sometimes
spklt . . . . . . . . . spikelets
spnd . . . . . . . . . supranodal
sta . . . . . . . . . . . stamens
st g . . . . . . . . . starch grains
stln . . . stolons, stoloniferous
stmd . . . . . . . . staminodes
stmt . . . . . . . . . staminate
strl . . . . . . . . . . . sterile
sty . . . . . . . . . . . styles
sx . . . . . . . . . . . sexual
tml . . . . terminal, terminally
ult . . . . . . . . . . ultimate
unilat . . . . . . . . unilateral
unisx . . . . . . . . . unisexual
up . . . . . . . . . . . upper
usu . . . . . . . . . . usually
wd . . . . . . . . . . . woody

# Alphabetical Listing of the Family, Subfamilies, Tribes, and Genera

The list below shows the family, subfamilies, tribes, and genera treated in this manual in alphabetical order, followed by the page number on which they are treated.